Holocene Climate Change and Environment

Holocene Climate Change and Environment

Edited by

Navnith Kumaran
Formerly of Agharkar Research Institute, India

Damodaran Padmalal
National Centre for Earth Science Studies, India

Elsevier
Radarweg 29, PO Box 211, 1000 AE Amsterdam, Netherlands
The Boulevard, Langford Lane, Kidlington, Oxford OX5 1GB, United Kingdom
50 Hampshire Street, 5th Floor, Cambridge, MA 02139, United States

Notices
Knowledge and best practice in this field are constantly changing. As new research and experience broaden our understanding, changes in research methods, professional practices, or medical treatment may become necessary.

Practitioners and researchers must always rely on their own experience and knowledge in evaluating and using any information, methods, compounds, or experiments described herein. In using such information or methods they should be mindful of their own safety and the safety of others, including parties for whom they have a professional responsibility.

To the fullest extent of the law, neither the Publisher nor the authors, contributors, or editors, assume any liability for any injury and/or damage to persons or property as a matter of products liability, negligence or otherwise, or from any use or operation of any methods, products, instructions, or ideas contained in the material herein.

British Library Cataloguing-in-Publication Data
A catalogue record for this book is available from the British Library

Library of Congress Cataloging-in-Publication Data
A catalog record for this book is available from the Library of Congress

ISBN: 978-0-323-90085-0

For Information on all Elsevier publications visit our website at
https://www.elsevier.com/books-and-journals

Publisher: Candice Janco
Acquisitions Editor: Marisa LaFleur
Editorial Project Manager: Ruby Gammell
Production Project Manager: Kumar Anbazhagan
Cover Designer: Greg Harris

Typeset by Aptara, New Delhi, India

Contents

Contributors

Amol Abhale
School of Earth Sciences, Savitribai Phule Pune University, Pune, India

Hema Achyuthan
Institute for Ocean Management (IOM), Anna University, India

M.S. Aneesh
National Centre for Earth Science Studies, Ministry of Earth Sciences, Thiruvananthapuram, Kerala, India

Ambili Anoop
Indian Institute of Science Education and Research Mohali, Manauli, Punjab, India

Aqib J. Ansari
Department of Earth Sciences, Indian Institute of Technology Kanpur, Kanpur, India

Erwin Appel
Department of Geosciences, University of Tübingen, Tübingen, Germany

Upasana S. Banerji
National Centre for Earth Science Studies, Ministry of Earth Sciences, Thiruvananthpuram, Kerala, India

N. Basavaiah
Indian Institute of Geomagnetism, New Panvel, Navi Mumbai, India; Department of Physics, Krishnaveni Degree & P.G. College, Narasaraopet, Guntur, India

Purnima Bejugam
Marine Sciences, School of Earth, Ocean and Atmospheric Sciences, Goa University, Goa, India

A. Brauer
German Research Centre for Geosciences, Potsdam, Germany

Elora Chakraborty
Centre for the Study of Regional Development, JNU, New Delhi, India

Supriyo Chakraborty
Indian Institute of Tropical Meteorology, Ministry of Earth Sciences, Pune, India

Gaurav Chauhan
Department of Earth Sciences, K.S.K.V. Kachchh University, Bhuj, Gujarat, India

Amey Datye
Indian Institute of Tropical Meteorology, Ministry of Earth Sciences, Pune, India

Hritika Deopa
Department of Earth Sciences, Banasthali Vidyapith, Rajasthan, India

B.C. Deotare
Formerly of Deccan College Post-Graduate and Research Institute (now Deemed to be University), Pune, India

Narayan P. Gaire
Key Laboratory of Tropical Forest Ecology, Xishuangbanna Tropical Botanical Garden, Chinese Academy of Sciences Menglun, Mengla, Yunnan, China; Central Department of Environmental Science, Tribhuvan University, Kirtipur, Kathmandu, Nepal

Naveen Gandhi
Indian Institute of Tropical Meteorology, Pune, India

S.S. Gudadhe
Yashwantrao Chavan Arts, Commerce and Science College, Lakhandur, India

Subrota Halder
Indian Institute of Tropical Meteorology, Ministry of Earth Sciences, Pune, India

S.M. Hussain
Department of Geology, School of Earth and Atmospheric Sciences, University of Madras, Chennai, India

Priyanka Joshi
Birbal Sahni Institute of Palaeosciences, Lucknow, UP, India

Navin Juyal
Physical Research Laboratory, Navrangpura, Ahmedabad, India

Vishwas S. Kale
Formerly at Department of Geography, SP Pune University, Pune, India

A.M. Kandekar
School of Earth Sciences, Savitribai Phule Pune University, Pune, India

Janhavi Kangane
Marine Sciences, School of Earth, Ocean and Atmospheric Sciences, Goa University, Goa, India

Om Katel
College of Natural Resources, Royal University of Bhutan, Lobesa, Punakha, Bhutan

A.S. Khadkikar
Råcksta, Stockholm stad, Sweden

Ravi Korisettar
ICHR Senior Academic Fellow, Sri Saranakripa, Sivagiri, Dharwad, Karnataka , India

K.P.N. Kumaran
PalynoVision, Mon Amour, Erandaane, Pune, Maharashtra, India; Shreenath Hermitage, Pune, Maharashtra, India

Ruta B. Limaye
PalynoVision, Mon Amour, Erandaane, Pune, Maharashtra, India; Formerly at Biodiversity & Palaeobiology Group, Agharkar Research Institute, India

Ishita Manna
Centre for the Study of Regional Development, JNU, New Delhi, India

K. Maya
National Centre for Earth Science Studies, Ministry of Earth Sciences, Thiruvananthapuram, India

Nivedita Mehrotra
Birbal Sahni Institute of Palaeosciences, Lucknow, India

Bulbul Mehta
Indian Institute of Science Education and Research Mohali, Manauli, Punjab, India

D.C. Meshram
School of Earth Sciences, Savitribai Phule Pune University, Pune, India

Praveen K. Mishra
Wadia Institute of Himalayan Geology, Dehradun, Uttarakhand, India

Sheila Mishra
Formerly of Deccan College Post-Graduate and Research Institute (now Deemed to be University), Pune, India

Siba Prasad Mishra
Department of Civil Engineering, Centurion University of Technology and Management, Jatni, Bhubaneswar, Odisha, India

Krishna G. Misra
Birbal Sahni Institute of Palaeosciences, Lucknow, India

Pavani Misra
Department of Earth Sciences, Indian Institute of Technology Kanpur, Kanpur, India

P.M. Mohan
Pondicherry University, Port Blair, India

S. Vishnu Mohan
Department of Geology, Sree Narayana College, Chempazhanthy, Kerala, India; Department of Geology, University of Kerala, Thiruvananthapuram, India

Harsanti P. Morley
Palynova Ltd, Littleport, United Kingdom

Robert J. Morley
Palynova Ltd, Littleport, United Kingdom; Earth Sciences Department, Royal Holloway, University of London, Egham, United Kingdom

Charuta Murkute
Indian Institute of Tropical Meteorology, Ministry of Earth Sciences, Pune, India

Debarati Nag
Birbal Sahni Institute of Palaeosciences, Lucknow, UP, India

Ganapati N. Nayak
Marine Sciences, School of Earth, Ocean and Atmospheric Sciences, Goa University, Goa, India

Mohammed Noohu Nazeer
Department of Geology, School of Earth and Atmospheric Sciences, University of Madras, Chennai, India

N. Nowaczyk
German Research Centre for Geosciences, Potsdam, Germany

D. Padmalal
National Centre for Earth Science Studies, Ministry of Earth Sciences, Thiruvananthpuram, Kerala, India

Uttam Pandey
Birbal Sahni Institute of Palaeosciences, Lucknow, India

Anant Parekh
Indian Institute of Tropical Meteorology, Ministry of Earth Sciences, Pune, India

Binita Phartiyal
Birbal Sahni Institute of Palaeosciences, Lucknow, UP, India

Sushma Prasad
ERA Scientific Editing, Potsdam, Germany; Institute for Earth Science, University of Potsdam, Potsdam, Germany

Md. Firoze Quamar
Birbal Sahni Institute of Palaeosciences, Lucknow, Uttar Pradesh, India

K. Radhakrishnan
Department of Geology, School of Earth and Atmospheric Sciences, University of Madras, Chennai, India

S.N. Rajaguru
KamalSudha Apartment, Shaniwar Peth, Pune, Maharashtra, India; Formerly of Deccan College Post-Graduate and Research Institute (now Deemed to be University), Pune, India

A. Rajkumar
Department of Geology, School of Earth and Atmospheric Sciences, University of Madras, Chennai, India

K. Nageswara Rao
Department of Geo-Engineering, Andhra University, Visakhapatnam, Andhra Pradesh, India

Suman Rawat
Wadia Institute of Himalayan Geology, Dehradun, India

Phanindra Reddy A
Indian Institute of Tropical Meteorology, Pune, India; Savitribai Phule Pune University, Pune, India

M.R. Resmi
Department of Earth Sciences, Banasthali Vidyapith, Rajasthan, India

S.J. Sangode
School of Earth Sciences, Savitribai Phule Pune University, Pune, India

J. Seetharamaiah
Department of Geology, The University of Dodoma, Dodoma, Tanzania

Kumar Chandra Sethi
Department of Civil Engineering, Centurion University of Technology and Management, Jatni, Bhubaneswar, Odisha, India

Santosh K. Shah
Birbal Sahni Institute of Palaeosciences, Lucknow, India

Bimal Sharma
Key Laboratory of Tropical Forest Ecology, Xishuangbanna Tropical Botanical Garden, Chinese Academy of Sciences Menglun, Mengla, Yunnan, China

Milap Chand Sharma
Centre for the Study of Regional Development, JNU, New Delhi, India

Vikram Singh
Birbal Sahni Institute of Palaeosciences, Lucknow, India

Nitesh Sinha
Center for Climate Physics, Institute for Basic Science, Busan, Republic of Korea; Pusan National University, Busan, Republic of Korea

V. Sivapriya
Department of Geology, School of Earth and Atmospheric Sciences, University of Madras, Chennai, India

Lamginsang Thomte
Birbal Sahni Institute of Palaeosciences, Lucknow, India; Department of Geography, Gauhati University, Guwahati, India

Anjali Trivedi
Birbal Sahni Institute of Palaeosciences, Lucknow , Uttar Pradesh, India

M. Vandana
National Centre for Earth Science Studies, Ministry of Earth Sciences, Thiruvananthapuram, India

V.R. Vivek
National Centre for Earth Science Studies, Ministry of Earth Sciences, Thiruvananthapuram, India

Ankit Yadav
Indian Institute of Science Education and Research Mohali, Manauli, Punjab, India

Ram R. Yadav
Wadia Institute of Himalayan Geology, Dehradun, India

Akhilesh K. Yadava
Birbal Sahni Institute of Palaeosciences, Lucknow, India

Preface

The Holocene is the current geological epoch and it began approximately 11,700 cal years BP after the last glacial period. Though Holocene is a relatively short duration in terms of the planet's total age, this period is vital in many aspects, particularly in man's development. From the hunter-gatherer status, it was during the beginning of Holocene that man evolved to begin a settled form of life. Since then, rapid proliferation, growth, and impacts of the human species worldwide, development of major civilizations, technological revolutions, and overall significant transition toward urban living in the present have taken place. This was possible since man adopted the avocations of agriculture and livestock rearing. Small-scale settlements evolved into villages and centers of development of major civilizations, including the Indus Valley (Harappa) and the Maya.

Along with human development, the monsoon vicissitudes also significantly impacted climate patterns that evolved to the current state. It has become progressively clear that the civilizations developed and perished due to gradual or abrupt climatic changes. This prompted scientists to look at the climatic changes of Holocene as recorded in the marine and terrestrial sediments.

Holocene climate change study has attained great importance in recent years because the present-day landscape and environment are the products of the biosphere–geosphere interactions. The Holocene period has witnessed the most dramatic changes in climate and sea level. These, coupled with local and regional tectonics, had a pronounced impact on plant and animal life, landforms, water, other nonliving resources, and human evolution and civilization. The findings of these vast and varied investigations are brought about by several international journals exclusively devoted to Quaternary studies. Besides the international committees such as International Union for Quaternary Research (INQUA) and International Geosphere-Biosphere Programme (IGBP), numerous universities worldwide have set up departments solely dedicated to Quaternary research. Compared to these, and notwithstanding the large volume of the literature created during the last few decades, the study of Quaternary science, especially the Holocene research in the tropics of South Asia, lags behind and not at all par with the rest of the world. This prompted us to closely look into the imprints of climate change left in the diverse archives for decoding the signatures using multiple proxies, new analytical tools, and dating methods.

Developments in dating techniques, isotope geology, and other modern analytical tools have aided the study of Holocene sediments as a repository of climate proxies. Efforts to decode signatures of climate dynamics during Holocene have been made in the past decade and a lot of information, including data set and models, has been available within the tropical regions, such as the Indian subcontinent. Further, the development of various physical, chemical, and biological proxies provided finely tuned information on the environmental conditions in recent times. Considering the Holocene being dynamic, an appraisal is being attempted in this compendium "Holocene Climate Change and Environment" to overview the recent developments in Holocene studies in the Indian subcontinent and identify the significant gaps and issues for strengthening our efforts to tackle climate change and the environment. The volume is a compendium of 25 research papers, beginning with the chapter "The Prelude to the Holocene: Tropical Asia during the Pleistocene," which, feasibly provides a befitting introduction to this volume. In fact, all the contributions are classified under four major themes—a) Climate change, b) Indian summer monsoon and teleconnections, c) Ecosystems and Environment, and d) Geoarchaeology and Human culture.

a) Climate Change

The biophysical environment of tropical Asia is unique in many aspects, and any climate change could adversely affect the region's ecological and socioenvironmental fabric. A better understanding of the past, climate change and vegetation dynamics of the region will be of immense use for addressing future climate change scenarios through reliable predictions and developing appropriate climate adaptation strategies for human survival. This theme embodies a total of ten chapters (1–10). The introductory chapter by Robert Morley and Harsanti Morely gives an overview of climate and vegetation history of tropical Asia for the last three million years. The Pleistocene, according to them, had witnessed the disappearance of humid tropical forests. In contrast, the appearance of extensive open grasslands hosting a stock of diverse fauna, a few of them, have migrated to Java during the early Pleistocene. The widespread occurrences and abundances of megafauna and the climate perturbations might have considerably modified the vegetal landscape across the region fed essentially by monsoon precipitation.

The Indian subcontinent has experienced significant climate and environmental changes in the last glacial and interglacial periods. To understand these changes, one of the most reliable tools is to examine proxy records in various continental and marine archives in the light of instrumental climate records and scientific/technological advancements.

Sharma et al. in their paper "Contemporary dynamics and Holocene extent of glaciers in the Himalayas" describe the reduction in the thickness and aerial extent of glaciers in the "Third Pole"—the Himalaya. The study revealed that most glaciers had expanded 4–6 km concerning the present position, between 9 and 8 ka and almost stabilized till 6.0 ka. However, later on, they retreated close to the new boundaries by 6–4 ka. The analysis of satellite data reiterates that the retreating trend continues because of climate change and other human interventions. The study from Ladakh in the trans-Himalaya by Phartiyal et al. showed five major arid phases during Holocene that had been intervened by comparatively warmer periods. Sangode et al., in their study from Chandra valley basin in the northwestern Himalaya, showed sustainability of valley environments during the last glacial/interglacial transition and the Holocene. Sediment production in a sizeable denudational catchment area is locked up due to fewer tributary and hanging valley passages. The trunk system characterizes the glacial to interglacial change to the tributary system. Excessive sediment productivity favored by the current warming has enhanced the Chandra valley's susceptibility to damming and the other effects of extreme weather conditions.

Apart from the glacial-influenced rivers, the other rivers in the country also responded significantly to the Holocene climate change and palaeomonsoon variability. In the paper "Holocene regional-scale behavior of the rivers of Indian Peninsula," Vishwas S. Kale accounted for evidence of a regional phase of aggradation coinciding with the monsoon's mid-Holocene weakening. The deposits occur as inset terraces in the river basin environment. In the last 3–4 millennia, intermittent deposition, superimposed on a slow, steady incision, has dominated the fluvial regimes. Paleoflood records show clustering of large floods during medieval climate anomaly around the onset and end of the little ice age and the current warm period. Quamar conducted a study on "Holocene vegetation and climate change from central India: An updated and a detailed pollen-based review" and disclosed the role of vegetation in maintaining the water and energy balance in the global climate system.

Like the continental archives, the coastal and marine sediments also record the Holocene climate change variables in its different proxies. Padmalal et al. give a detailed account of "Holocene climate and sea-level changes and their impact on ecology, vegetation and landforms" taking the South Kerala Sedimentary Basin (SKSB), southwest India as a case study site. Each climate episode in the Holocene epoch significantly affected the floral diversity and richness (as represented by its palynofacies) in the SKSB and its hinterlands. The evergreen forest was converted into wetlands, aquifers/groundwater resources were severely affected, and the sensitive ecosystems such as mangroves and freshwater swamps of *Myristica* become relicts resulting from the changing climate scenarios brought in the hydrological processes and climate variability of the Holocene epoch. Nayak et al. conducted a study of surface and core sediments in the continental shelf and slope of Krishna, Godavari, and Hooghly rivers. High illite content in the sediment core indicates deposition of sediments from the cold condition by altering muscovite mica in the Himalayan region, suggesting northeast monsoon enhancement. Simultaneously, the presence of smectite, Fe, and Ti in the sediments shows sediment contribution from mafic source rocks and red beds supplied by the peninsular rivers, especially Krishna and Godavari rivers during the increased spells of southwest monsoon.

Recently many advanced and reliable proxies such as environmental magnetism, tree rings, etc., are used to decode the climate change signals in a region. A few papers on these lines are also included in the section to get acquainted with the latest tools and their applications in climate change science, taking a few case studies from selected regions of the Indian subcontinent. It is now widely accepted that environmental magnetism in sedimentary deposits can be used for the study of Paleomonsoon records. However, a study on environmental magnetic records of the Holocene History of Indian Monsoon by Basavaiah et al. cautioned against the use of the magnetic susceptibility (χ) for palaeomonsoon reconstruction. Instead, they argue that the S-ratio parameter is a more efficient proxy for environmental, paleomonsoon, and climate change studies. Tree rings offer another excellent proxy for understanding long-term variability in river discharge characteristics. Highlighting a case study in the treed vegetation in the Sutlej basin of the Himalaya, Singh et al. emphasized the use of this powerful tool for the hydrological studies of the Himalayan trove of biological and geological diversities that host home to many major perennial rivers in the Indian subcontinent.

b) Indian Summer Monsoon (ISM) and Teleconnections

India is located at the center of the monsoon domain. The ISM's progression over India manifests the migration of the Inter-Tropical Convergence Zone (ITCZ) over the country and the northern Indian Ocean. The recent incidences of anomalous spells of climate events over the Indian subcontinent warrant the need to understand ISM's spatiotemporal changes in the Holocene epoch. There is an imminent need to fully understand the climate variables and their linkages/teleconnections with the global forcing factors. Therefore, the theme "Indian Summer Monsoon (ISM) and Teleconnections" of this volume is devoted exclusively to exploring the recent advances and thoughts in ISM phenomena and its teleconnections with the global forcing/driving.

A total of six papers are included under this theme (chapters 11–16). The article of Shah et al. on "Potential utility of the Himalayan tree-ring $\delta 18O$" explains how the isotope measurements can track the spatial distribution of drought reconstruction in historical drought concerning monsoon failures in the region. Their study could reveal the decreasing strength of ISM in the western and the high-altitude, northwestern parts of the Indian subcontinent. The established teleconnections with the sea surface temperature variation in the Pacific Ocean and the Indian Ocean–related forcing phenomena demarcated the present spatial reconstruction's strength. During Holocene, the ISM showed marked variations in the forcing factors and the oceanographic processes. The signatures of Bond events registered in the north Atlantic and the Mediterranean sedimentary archives have also been deciphered in the northern Indian Ocean and Indian subcontinent proxies, revealing a teleconnection of ISM with the Atlantic climate system. Moreover, El Nino Southern Oscillation (ENSO) of the Pacific Ocean with ISM has been established previously. The present review of Upasana and Padmalal on the "Bond events and monsoon variability during Holocene- Evidence from marine and continental archives" reiterates a close connection and interplay of North Atlantic Oscillations (NAO) and ENSO in modulating ISM intensities.

Reddy and Gandhi's study on "Indian Summer Monsoon, variability on different timescales and deciphering its oscillations from irregularly spaced paleoclimate data using different spectral techniques" shows the presence of interannual to millennial-scale ISM variability. Interannual variations in ISM seem to be associated with the slowly evolving external boundary conditions—the Pacific Decadal Oscillations, ENSO, and the Northern hemisphere temperature. Yadav et al., in their paper "Monsoon variability in Indian Subcontinent—A review based on proxy and observational datasets" deal essentially with the changes in Indian monsoon rainfall on spatial and temporal scales, which in turn is closely linked to meridional shifts of the ITCZ and irregular, periodic events such as ENSO, NAO, and Indian Ocean Dipole (IOD). In their paper, "Holocene hydroclimatic shifts across the Indian subcontinent: A review based on interarchival coherences," Mishra et al. present a comprehensive analysis of paleoclimate data available from the Indian subcontinent, especially the speleothems and lake and marine sediment records across the Indian subcontinent. The assessment gives significant insights into the interarchival and interproxy coherences observed in various climatic and geomorphic zones of India and links with the local and regional forcing factors as explained earlier. Nowadays, stable isotopes are being used effectively to unfold the climate change signals of any region. In their research, Chakraborthy et al. used precipitation isotopic records from the Andaman Islands in the Bay of Bengal to investigate how the isotopes respond to monsoon rainfall over a varying spatial domain. The relation between precipitation isotopes and moisture dynamics remains robust in encompassing a large portion of the Bay of Bengal but weakens beyond its geographical territory. The study predicts that point-scale observations such as speleothem isotopic records, though respond to extensive scale processes, the rainfall contribution, however, for a vast region may be underestimated.

c) Ecosystems and Environment

Under the theme "Ecosystems and Environment" a total of five papers (chapters 17–21) are included dealing with ostracods in mangrove environments, geomorphic markers of the major rivers of the Peninsular India and some case studies on coastal wetlands and their evolution. The paper "Mangrove Ostracoda species fluctuations, habitual adaptation, and its environmental implications—A review" by Hussain et al. discussed how the mangrove ostracod species respond to the diverse environmental fluctuations. The dominance of highly ornamental forms with intricate hinge patterns in the ostracod species explains the tidal influence and wave turbidity; precipitation and runoff from the land aids in less distribution of ostracod species. Among the different geological agents at work, rivers are more responsive to environmental and tectonic changes. In a study "Late Quaternary landscape evolution of Peninsular India: A review using fluvial archives," Resmi et al. evaluated the tectonic controls on major peninsular rivers based on investigations of geomorphic markers such as terrace formation, gorges, longitudinal profiles, morphotectonic indices, paleochannels, and delta, and provided a geologic and geomorphic history of the alluvial reaches of the west- and east-flowing rivers of the Peninsular India. The Holocene period witnessed dramatic changes in climate and sea levels in the coastal lowlands of India. In their study "The Imprints of Holocene climate and environmental changes in the South Mahanadi Delta and Chilika lagoon in Odisha, India - An overview" Mishra and Sethi discussed the Holocene geomorphic changes along Mahanadi delta, northerly shifts in tidal inlets, and gradual drying up of the lagoon along with anthropogenic interferences in the delta. In a multiproxy study from southwest India, Maya et al. described the fluvial geomorphic changes and evolution of the Kuttanad Kole Wetland (KKW)—the expansive Southern part of the Vembanad lake, the largest Ramsar wetland in India. The KKW acts as a room for floodwaters during episodic events and serves as a rice bowl of Southern Kerala, where paddy cultivation is being carried out below the mean sea level. In another study, Vishnu et al. revealed the Holocene evolution of the coastal wetlands in the Trivandrum block located south of the Achankovil Shear Zone, using sedimentological and geochemical proxies.

d) Geoarchaeology and Human culture

The necessity to understand the ancient settlement history and human interactions with their environment is not merely an academic exercise, but such information would help us evaluate even the micro-level changes in climate and background of the region in which they survived. On many occasions, there is a need to discriminate the influence of the environment and human activity on vegetation to identify the more likely causes for environmental degradation. Therefore, a special theme "Geoarchaeology and Human culture" is included in this volume (chapters 22–25) which comprises four papers—two papers are from geoarchaeology and the remaining two discuss mainly vegetation, climate change, and cultural records.

In the paper "Late Quaternary geoarchaeology and palynological studies of some saline lakes of the Thar Desert, Rajasthan, India" Deotare et al. revealed the man-land relationship existed between hunter-gatherers and shallow lakes surviving more than 15,000 years. The arid Thar Desert remained in the shadow semiarid/dry climatic zone almost throughout the Quaternary period. The hunter-gatherers continued their existence and activity for the last 600 years, with a likely decline during the Last Glacial Maximum (LGM). Availability of suitable rocks and minerals for making stone artifacts, availability of surface freshwater resources, and favorable geomorphic features for nourishing deep-rooted vegetation growth were the critical environmental parameters useful for human activity/survival even in the harsh arid climate. In another paper, Ravi Korisettar revealed that the Indian Peninsula preserves a continuous record of human prehistory as compared with that of the Indo-Gangetic and extra peninsular regions. The study also stresses the scope for geoarchaeological investigations for reconstructing the man-land relationship in a multidisciplinary and absolute chronological framework showing the refugium for Pleistocene and Holocene hunter-gatherer and early agro-pastoral communities.

Using pollen proxy records, Anjali Trivedi, came with a review paper on Holocene vegetation, climate, and culture in northeast India: a pollen data-based review. As the pollen proxy is a useful tool for detecting vegetation fluctuations, the information could effectively be used to reconstruct forest expansion and monsoon strength. According to her, the impact of hominoid interaction on forest cover and agricultural expansion in Holocene is a great concern in northeast India, which needs to be addressed adequately. Another study from the Konkan coast, by Limaye et al., revealed that the natural forcing has a considerable impact on the landscape evolution and vegetation cover along India's coast. On the Konkan coast, the mangrove cover has been drastically declined due to the hydrodynamic changes and land modifications. Evidence of anthropogenic impacts has been deciphered at many locations in changing forest cover into agricultural lands and human occupation, and maritime trade.

In short, many pieces of evidence exist indicating the existence of a culturally evolved human settlement in many parts of India. They are confined to mountains, valleys, plains, and even unfavorable areas such as deserts and coastal lowlands.

The main appeal of the book is that it focuses on one of the most interesting regions of the world in terms of climate variability, ecosystem, and human evolution as well as their underlying dynamic processes during the Holocene. These processes are acting today, and their understanding can make the greatest contribution to future-oriented concerns. This is especially crucial for the countries, including India, falling within the monsoon domain that is expected to face substantial impacts during the 21st century in the light of a changing climate. The audience from various fields such as geomorphologists, hydro-climatologists, coastal scientists, climate scientists, environmental scientists, and physical geographers would certainly be benefited from this volume.

Acknowledgments

This book is a contribution of a team of dedicated researchers actively engaged in the field of Quaternary research with much focus on the Holocene. Our contact and exchange of information with the national and international experts together with the critical comments of the renowned reviewers of Elsevier helped immensely in improving our idea and coming with the contents of chapter presented in this volume.

We are greatly indebted to Late Dr. K.M. Nair, Former Director, Centre for Earth Science Studies (CESS), and our Mentor for guiding us in the field of Quaternary Research. We are also grateful to Professor S.N. Rajaguru, Formerly of Deccan College Post-graduate and Research Institute (now Deemed to be University), Pune, Maharashtra, India, for his constant encouragement and support. The academic and technical inputs extended by Dr. Mrs. Ruta Limaye and Dr. B. Karthick are also very much appreciated.

Dr. Damodaran Padmalal is thankful to Dr. M. Rajeevan, Secretary, Ministry of Earth Sciences (MoES), Government of India and the Director, National Centre for Earth Science Studies (NCESS), Thiruvananthapuram for encouragement and support. Many experts of the field helped us in reviewing the research contributions in this volume, their names are mentioned in a list given at the end of this book (Annexure I).

We are also indebted to Elsevier Publishing Ltd for making our dream of this book project a reality. Special thanks are also due to Ms. Candice Janco, Ms. Marisa LaFeur, Ms. Ruby Gammell, Ms. Indhumathi Mani, Mr. Kumar Anbazhagan, and Mr. Greg Harris from Elsevier for their continuous support and understanding while executing the book project. Typeset by Aptara, New Delhi is also acknowledged.

Finally, we thank all those who are directly or indirectly helped us in the successful completion of this book project with Elsevier.

Navnith Kumaran
Damodaran Padmalal

CHAPTER

The prelude to the Holocene: tropical Asia during the Pleistocene

1

Robert J. Morley[a,b], **Harsanti P. Morley**[a]

[a] *Palynova Ltd, Littleport, United Kingdom.* [b] *Earth Sciences Department, Royal Holloway, University of London, Egham, United Kingdom*

1.1 Introduction

India and Southeast Asia are closely related biologically, but in terms of geology and tectonics contrast dramatically. The Southeast Asian rainforest flora was largely derived from India (Morley, 1998, 2000, 2018) as it collided with Asia during the Eocene, and many elements of Southeast Asia's fauna dispersed from India during the late Neogene. In addition, two species of *Homo, H. erectus* and *H. sapiens*, found their way into Southeast Asia from Africa via India during the Early and Middle Pleistocene respectively. From a geological perspective India is an ancient craton, formerly part of Gondwana, mostly comprising a large and stable elevated plateau composed of ancient rocks, but with the Himalayan range, the world's highest and most extensive mountain system, along its northern margin, formed as a result of India's collision with Asia. This mountain range effectively controls the regional climate across the whole of tropical Asia by driving the Indian Monsoon. Southeast Asia, on the other hand is a region with island arcs characterized by intense geological activity, initially molded by the Eocene extrusion of Indochina following the India–Asia collision, and later squeezed by the Neogene northward drift of Australasia. However, in contrast with India much of the maritime continent of Southeast Asia is at a very low altitude, with large areas forming shallow seas which may become exposed when global sea levels fell. These two areas have thus experienced contrasting histories in terms of regional geography and climate in the period leading up to the Holocene.

1.2 Scope, methodology, and data sources

This chapter discusses how the flora and fauna interacted with the differing tectonic, eustatic, geomorphological, and climatic scenarios for the period immediately preceding the Holocene through the evaluation of palynological data, supported where appropriate by macrofloral and faunal records. The main issues raised are: (1) How did vegetation change in India and Southeast Asia during the Pliocene as the Himalaya were reaching their highest elevation and ice caps expanded in the northern hemisphere? (2) What was the impact of falling sea levels on vegetation and climate in the two areas over this period? (3) Was there a savanna corridor across Sunda during Pleistocene periods of low sea level, and at what stage was *Homo erectus* able to migrate across Sunda? and (4) What were the effects of global climate

Holocene Climate Change and Environment. DOI: https://doi.org/10.1016/B978-0-323-90085-0.00022-X

change on vegetation in India and Southeast Asia during the last full glacial period, and at what stage did *Homo sapiens* migrate across the region? Also, did the Toba eruption have any lasting effects on either the vegetation, climate, or the establishment of humans across the region? In discussion, issues are raised concerning the interpretation of palynomorph assemblages in terms of changing *terra firma* vegetation in these regions, as most localities analyzed palynologically are not the traditional small pollen catchments which are efficient in capturing regional pollen signals, but are either marine cores, where there are many additional factors that might influence recovered pollen assemblages specific to marine environments, or are fluvial deposits, where there may be difficulty in separating pollen signals of the local edaphic (swamp) vegetation from the regional signals which reflect the surrounding dry-land vegetation.

The review has been achieved through the evaluation of published palynological datasets from across the region and is based on the detailed examination and comparison of the actual published pollen diagrams from each locality.

1.3 Pliocene demise of Indian wet tropical forests

Prior to the Pliocene, the Indian subcontinent was clothed with extensive perhumid and seasonal tropical forests (Guleria, 1992). These forests were characterized by members of the great tropical tree family Dipterocarpaceae, which today dominate the rainforests of the Sunda region and Indochina. Today, apart from the sal (*Shorea robusta*) which is widespread across the deciduous forests of northern India west of the Indo-Burmese Range, members of the Dipterocarpaceae are confined to a few wet refugia in the Western Ghats and Sri Lanka. During the Pliocene, however, species such as *Shorea tumbuggaia*, now confined to deciduous forests of the Western Ghats, were widespread across much of the Indian subcontinent. This is shown by the presence of diverse and well-preserved fossil leaf and fruit assemblages which include *S. tumbuggaia* as well as many dipterocarps belonging to the genera *Anisoptera, Dipterocarpus, Hopea*, and *Vatica,* found in the foreland basin deposits of the Siwaliks (Khan et al., 2016; Ashton et al., 2021).

The change from expansive C3-dominated forests to the strongly monsoonal, dominantly C4 vegetation that characterizes India today can be seen in Siwalik stable isotope records (Quade et al., 1989). The change is best illustrated however by the Siwalik faunal succession which is summarized by Patnaik (2015). Forest-dwelling frugivores and browsers dominated mammalian faunas prior to 8 Ma followed by mainly mixed feeders until 2–3 Ma, and with grazers dominant in the Pleistocene (Fig. 1.1). These changes relate to successively drier climates caused by the strengthening of the Indian monsoon following Himalayan uplift, which accelerated during the Quaternary (Govin et al., 2020) at the time of greatest expansion of savanna-dwelling taxa.

The Southeast Asian region on the other hand was characterized by widespread tropical forests throughout this period, with extensive evergreen forests at lower latitudes and seasonally dry forests in Indochina. Across Indochina areas of more open vegetation with grasslands expanded in distribution during the later late Miocene and Pliocene, especially in the lowlands and rain shadow areas, whereas evergreen dipterocarp and montane rainforests were present in upland areas (Morley, 2018).

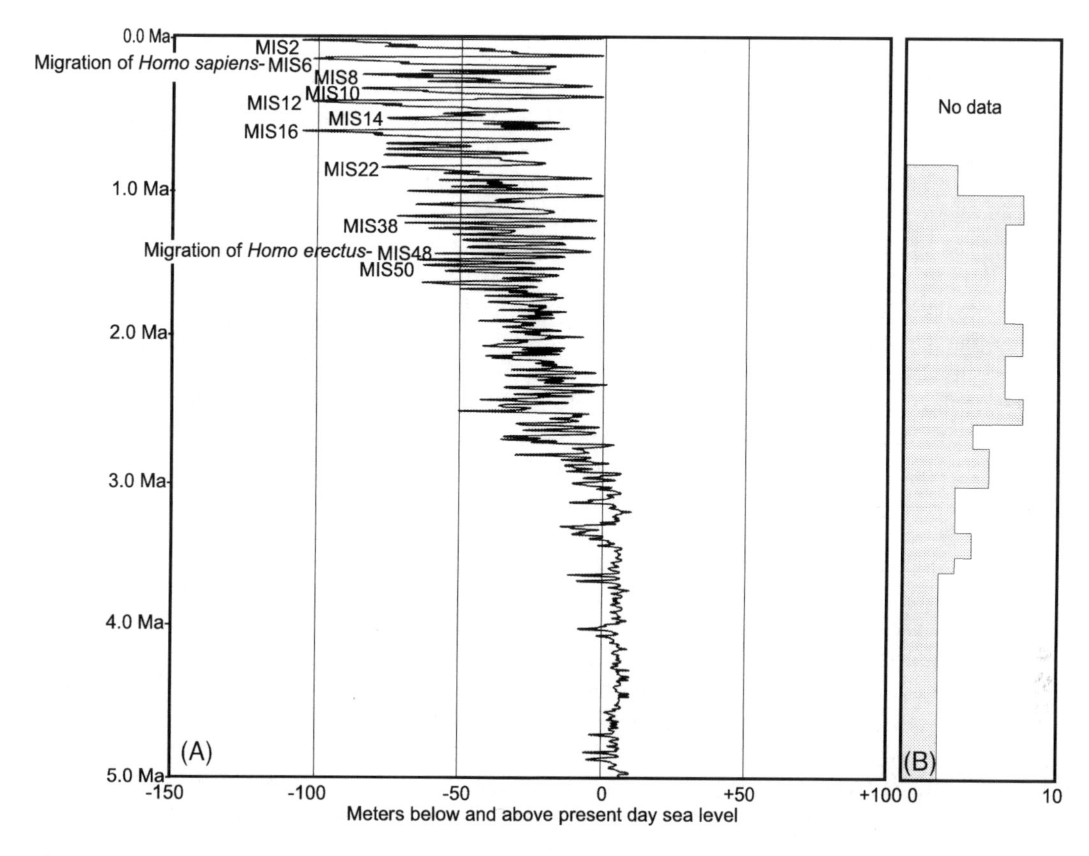

FIGURE 1.1

(A) Global sea level change (reflected by ice-volume calculations from the ^{18}O curve) during the Plio–Pleistocene (from De Boer et al., 2010) in relation to the timing of immigration of *Homo erectus* and *H. sapiens.* (B) Number of grazing taxa in Siwalik fauna over time according to Patnaik (2015), extracted from Fig. 1.8.

1.4 Pleistocene vegetation and climate change across India and Southeast Asia

The Pleistocene differs from other geological periods in that sea levels were intermittently lower, and more land was exposed than at any time during the earlier Cenozoic. This created opportunities for dispersal of taxa, including hominins, in a manner not possible in the Neogene, and the increased land area resulted in more continental climates than during the Pliocene.

1.4.1 Falling sea levels and topographic change across tropical Asia

With the onset of northern Hemisphere glaciation after 2.6 Ma, at the beginning of the Quaternary, global sea levels began to fall in relation to obliquity-, and later eccentricity-driven climate cycles. Amplitudes of the sea level oscillations were initially somewhat irregular, with maxima of the order of 40 m (De Boer et al., 2010), but after about 1.7 Ma the amplitude of sea level oscillations increased to about 70 m in relation to regular 41 ka obliquity cycles, sufficient to result in a land connection between Indochina and Java. There was a further increase in amplitude to about 120 m from 900 ka onward, and a shift in cyclicity to 106 ka eccentricity-driven cycles which continues to the present day, when we reside in the phase of high sea levels which followed Glacial Termination 1 at 14 ka at the end of the Weichselian glaciation (Fig. 1.1). The current short time interval of high sea levels, corresponding to Marine Isotope Stage (MIS) 1, is the period that comprises the Holocene.

During the Quaternary tectonic changes resulted in the modification of topography in some areas. The Himalaya underwent further uplift, and there was also uplift associated with the Indo-Burmese Ranges. Sunda began to show its present form and many islands of Wallacea became established, or elevated. Mount Kinabalu in Borneo underwent its final stage of uplift (Merckx et al., 2015), Sumatra expanded in size and much of the Barisan Range became elevated. The island of Java became established, first in the west, and later in the east (Morley et al., 2016), initially with low lying islands characterized by insular faunas which bore dwarf elephants, hippos and giant tortoises (Van den Bergh, 2001). It was also during this time that many of the islands of Wallacea became established, or greatly enlarged or uplifted, including Sulawesi (Nugraha and Hall, 2017) and Timor (Nguyen et al., 2013). However, the overriding factor which determined the fate of both floras and faunas over this period was the sudden change to intermittently falling sea levels, dramatically affecting some parts of the region.

By the Middle Quaternary, when sea levels intermittently fell to 120 m below current levels, the impact on the general outline of the Indian subcontinent was relatively minor, due to the limited representation of continental shelves around the subcontinent, although Sri Lanka would have had a direct land connection with mainland India, and the large deltas of the Indus, Ganges and Brahmaputra Rivers extended seaward. The effect of these sea level falls across Southeast Asia, on the other hand, was dramatic, with vast swathes of the Sundanian continental shelf becoming dry-land and the islands of Java, Borneo and Sumatra being connected to mainland Malay Peninsula and Vietnam. The Andaman Islands became a major island arc and together with the Nicobars resulted in the Andaman Sea being almost isolated from the Indian Ocean, and the archipelago of the Philippines was transformed into just six major islands. The islands of Wallacea, on the other hand changed relatively little, due to the absence of surrounding shelves, and remained isolated from Sunda by deep troughs, from the Lombok Straits to the east of Bali, the Makassar Straits separating Sulawesi from Borneo, and the Mindoro Straits south of the Philippines. However, all these straits were much narrower at the time of the LGM, with the Straits of Makassar being just 45 km wide at 18,000 ka (Fig. 1.2).

1.4.2 Sundanian vegetation during periods of low sea level

The presence of a substantial vertebrate savanna fauna from the Pleistocene of Java which bears close affinities with the Siwalik fauna of northern India, and which occurs together with *Pithecanthropus erectus* fossils, has long generated discussion as to its origin. The suggestion that the fauna found its

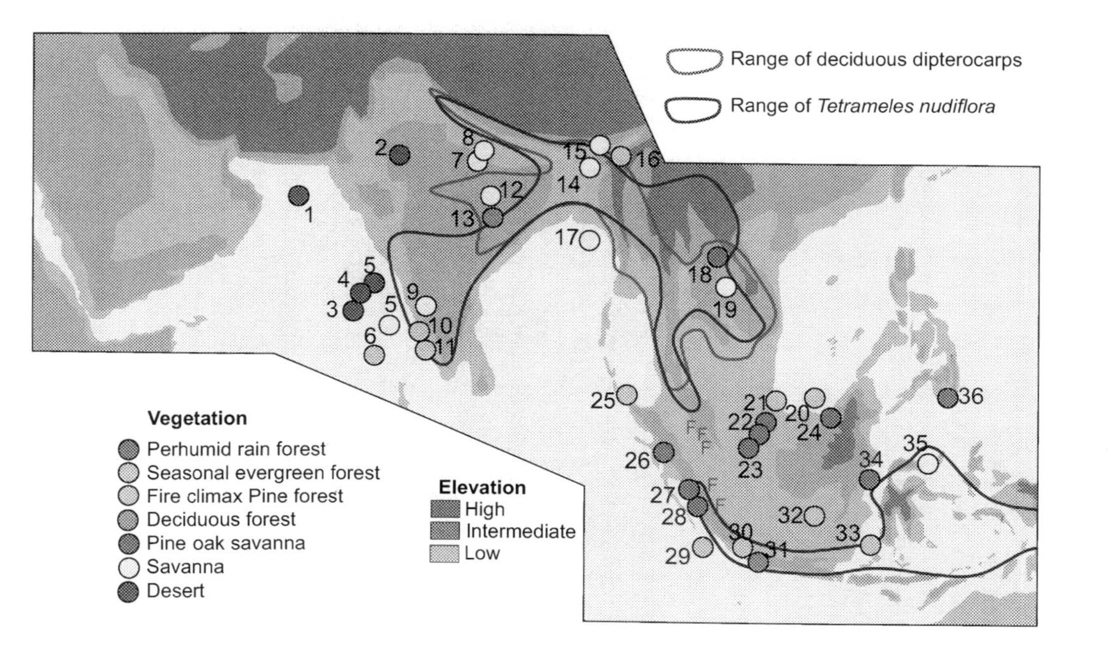

FIGURE 1.2

South and Southeast Asia land distribution at the peak of the last glaciation showing interpreted vegetation for MIS2, and the distribution of deciduous dipterocarps and the semi-evergreen forest species *Tetrameles nudiflora* (Datiscaceae). "F" shows the position of ferricretes (which predate the LGM). Colored circles show the vegetation cover for pollen catchment areas for localities mentioned in the text. 1, Core 119 KA, Ansari and Vink (2007); 2, Thar Desert, Singh et al. (1972); 3, Core SK128A-30, Prabhu et al. (2004); 4, Core SK128A31, Prabhu et al. (2004); 5, Core MD76131/77194, Van Campo (1986); 6, Core SK129 CR05, Farooqui et al. (2014); 7, Lashoda Tal, Trivedi et al. (2019); 8, Karela Jheel, Chauhan et al. (2015); 9, Nilgiris, Vishnu-Mittre and Gupta, 1971, Blasco and Thanikaimoni (1974); 10, Nilgiris, Caner et al. (2007); 11, Chahganachery, Farooqui et al. (2010); 12, Lakadandh Swamp, Quamar and Bera (2017); 13, Nakta Lake, Quamar and Kar, 2020; 14, Deepor Swamp, Tripathi et al. (2019); 15, Subankhata Swamp, Basumatary et al., 2015; 16, Ziro Valley, Battacharyya et al. (2014); 17, SO188-342KL, Williams et al. (2009); 18, Nong Pa Kho, Penny (2001); 19, Khorat Plateau, Yang and Grote (2017); 20, Core NS07-25, Luo et al. (2019); 21, Core CG-2 ; 22, Core 18300, 18302, Wang et al. (2009); 23, Core 18323, Wang et al. (2009); 24, Niah, Hunt et al. (2012) ; 25, Core BAR94-25, Van der Kaars (2012); 26, Pea Bullock, (Maloney and McCormac, 1995); 27, Danau Di Atas, (Stuijts et al., 1988); 28, Danau Padang, Morley (1982); 29, BAR94.42, Van der Kaars (2010); 30, Rawa Danau, Van der Kaars et al. (2001); 31, Banding Lake, Dam and Van der Kaars (1995); 32, Sebangau, Morley (1981); 33, Sangkarang-16, Morley et al. (2004); 34, Papalang-10, Morley et al. (2004); 35, Lake Tonasa, Dam et al. (2001); 36, Bian et al. (2011). Map outline based on maps downloaded from pinterest.com.

way to Java via a "savanna corridor" was mooted by the Earl of Cranbrook (Medway, 1972) and its presence subsequently discussed by Morley and Flenley (1987) and developed by Heaney (1991), Bird et al. (2005), Wuster et al. (2014), and Louys and Roberts (2020).

For the last and penultimate glacial periods, there is now increasing evidence for seasonally dry conditions in different parts of Sundaland. In Indochina there was open vegetation with abundant grasses, open oak-pine woodland and gallery forests (Penny, 2001; Yang and Grote, 2017) and also large numbers of browsing and grazing vertebrates (Tougard and Montieure, 2006; Louys et al., 2007; Louys and Meijaard et al., 2010) which had diets consistent with open vegetation (Louys and Roberts, 2020). Correspondingly from south of the equator in areas surrounding the Java Sea there is evidence for climate seasonality for the LGM over the same period, from pollen from a low altitude crater lake in West Java (Van der Kaars and Dam, 2001), from Bandung Lake (van der Kaars et al., 1995), from seasonal swamps in southern Kalimantan (Morley, 2013) and from a marine core from the East Java Sea (Morley et al., 2004; Morley and Morley, 2010). There is little faunal evidence from later Pleistocene glacial maxima, but the Punung fauna, referred to the last interglacial by Westerway et al. (2007) has a rainforest-dependent fauna which would be in keeping with evidence for perhumid conditions from Bandung Lake during the last interglacial (van der Kaars et al., 1995) and also from the East Java Sea (Morley and Morley, 2010).

However, there are few vertebrate, or palynological localities from equatorial latitudes which would resolve the nature of the former vegetation in the central Sunda Shelf region. Climate modeling (Cannon et al., 2009) and modeling former distributions of Dipterocarpaceae (Raes et al., 2014) suggest continuous perhumid rainforests across the equatorial zone for the LGM, although molecular studies including rats suggest that there was indeed a dispersal barrier of some sort from west to east (Gorog et al., 2004). Palynological studies which penetrate the LGM from south of Natuna (Wang et al., 2009) and offshore Sumatra (Van der Kaars et al., 2010) suggest rainforest during the LGM and provide no support for strongly seasonal climates.

From the faunal perspective, Medway (1972), referring to the work of Hoojier (1949) noted that the Javanese fauna was much less diverse than the Siwalik Early Pleistocene fauna, where there are many additional genera. These included *Equus*, giraffids, camels, and several antelopes and bovids, which Medway considered to be adapted to arid or open seasonal vegetation or savanna. The large Early Pleistocene Javanese herbivores mainly included forms that would have typically frequented riverine, forest edge and forest habitats. From the floristic perspective, it is noteworthy that although there are elements of the Indochinese semi-evergreen forest flora in Java (Ashton, 2014), such as *Tetrameles nudiflora* (Fig. 1.2) and *Reevesia*, deciduous-forest elements, especially of Indochinese deciduous dipterocarps, such as *Dipterocarpus intricatus, D. obtusifolius, D. tuberculatus*, and *Shorea obtusa* (Fig. 1.2), also *Pinus*, which occupied savanna as far south as Kuala Lumpur during one Early Pleistocene dry phase (Morley, 1998), are conspicuously absent (Ashton, 2014; Ashton et al., 2021; Morley, 2018) suggesting a corresponding north-south barrier for deciduous forest plants. The conclusion to be drawn from these observations is that it is unlikely that there was an actual savanna corridor across Sundaland, but a belt of semi-evergreen forests is a possibility, but there is no evidence to suggest that this corridor was present during the LGM.

The Javanese megafauna could have found their way to Java along riverine floodplains, which would have been widespread across Sundaland, but it should also be considered that Pleistocene forests across the Southeast Asian region may have been significantly different to those present during the Holocene (Johnson, 2009; Corlett, 2013), because the forests, whether perhumid or seasonal, were likely to have been much more open and patchy than present day primary perhumid forests since they maintained an extensive megafauna, and so lowland forests would have been much more accessible to diverse vertebrates, kept open by trails and clearings created by elephants and rhinoceros. The seeds of many cauliflorous rainforest trees are thought to have evolved to permit dispersal by these large mammals and such mechanisms would not be in place unless the vectors were sufficiently common to facilitate regular dispersal.

A factor that has been given little consideration in the discussion regarding the former vegetation of central Sunda is the likelihood that the geography of the broad region may have been different during the earlier Quaternary due to tectonic activity and that this may have had an effect on paleoclimate and biotic dispersal. The likelihood of subsidence of the Malay Peninsula has been previously discussed by Haile (1975), Batchelor (1979), and Parham (2016), but recent discussions by Husson et al. (2019) and Sarr et al. (2019) have also considered the biogeographical significance of such subsidence. They suggest that central Sundaland may have been subsiding since MIS11 at 400 ka, and that prior to that time, the area would have been dry-land irrespective of sea level change. Periods of low sea level would not have acted as dispersal barriers prior to 400 ka and to support this idea they demonstrate from molecular evidence that rates of vicariance has increased dramatically after this time in many taxa.

The Early Pleistocene, corresponding to this period of more widespread emergence is likely to have been drier during periods of low sea level as suggested from the presence of palynomorph assemblages suggesting pine savanna from the Malay Peninsula (Morley, 1998), which is associated with the Old Alluvium, dated by Batchelor (2015) as Early Pleistocene, and roughly time-equivalent to Early Pleistocene savanna from East Java, dated to MIS47–51 (Morley et al., 2020), consistent with evidence for megafaunal diets based on δ^{13}C data (Louys and Roberts (2020). However, bearing in mind the increasing volume of palynological data from across the region, and the contrasting distributions of deciduous and semi-evergreen taxa discussed above, a savanna corridor across Sundaland at this time was also unlikely, a conclusion recently reached from studies of dry tropical forest dynamics by Hamilton et al. (2020). The central Sunda region was more likely to have borne semi-evergreen forests, its position possibly marked by the occurrence of ferricretes from southern Malay Peninsula through Bintan to Central Sumatra, as suggested by Morley (2018).

Seasonally dry conditions in Java were less likely during the Pliocene, during which time diverse dipterocarp forests with strictly perhumid taxa such as *Dryobalanops* were abundant in West Java and South Sumatra, buried in volcanic ash (Mandang and Kagemori, 2004; Van Gorsel et al., 2014). The seasonally dry climate of East Java therefore developed following the Early Pleistocene uplift of east Java.

1.4.3 Early Pleistocene and the dispersal of *Homo erectus* to India and Sunda

Homo erectus evolved in Africa about 2 Ma, and within a relatively short time-span expanded its range outside Africa and found its way to Java via the Middle East and India. There is only a single skull tentatively attributed to *H. erectus* from the Middle Pleistocene of India, (Patnaik and Nanda, 2010), but there are many discoveries from Java and these have attracted international attention since Eugene Dubois discovered a hominin skull and femur at Trinil, which he named *Pithecanthropus erectus* (Dubois, 1896). The skull of the Modjokerto child from Perning (Von Koenigswald, 1936) has long been considered the oldest, as it is from the Pucangan Formation that underlies the hominin-bearing Kabuh Formation to the west. The Perning locality, was dated using potassium-argon at 1.81 Ma by Swisher et al. (1994). This date conflicted with a magnetostratigraphic date of 0.97 Ma by Hyodo et al. (1992) and was deemed too old by Morwood et al. (2003) who obtained an age of 1.4 Ma based on fission track dating. These three different age scenarios were tested using a sequence biostratigraphic approach by Morley et al. (2020), who suggested an age of 1.43 Ma within MIS47 by reference to the Quaternary isotope curve of Gibbard et al. (2005). *Homo erectus* specimens of similar age have also been found at Sangiran, dated to 1.51 ± 0.8 Ma (Larik et al., 2001). It is thus possible that these two fossils record the same immigration event in MIS47, which would indicate that the Kabuh/Pucangan formation boundary is diachronous. The appearance of *Homo erectus* in Java very closely followed a phase of successive global sea level falls after about 1.6 Ma from MIS54 onward (Fig. 1.1) suggesting that *H. erectus* was very much an opportunist and was able to colonize appropriate areas more or less as soon as they became available, since prior to this time, East Java was still in the process of formation, and formed a series of low-lying islands, as discussed above.

The environment during the Early Pleistocene provided many habitats that would have formed rich foraging grounds for the hominins who were living on the floodplain of a small sand-dominated delta (Fig. 1.3). The seasonally dry climate maintained open lowland savanna grasslands, which supported abundant fauna including deer, bovids and proboscids, with broad-leaf montane forests present on nearby volcanoes, and widespread *Nypa* (Arecaceae) swamps on muddy deltas along the coastline.

Van den Bergh et al. (2001) reviewed the Javanese faunal succession and indicated that the Ci Saat and Trinil faunas (until about 1.4 Ma) were unbalanced, and that it was not until the time of Kedung Brubus at about 1.3 Ma (Larick et al., 2001) that more diverse and better structured faunas were present. Van den Bergh et al. (2001) proposed that the Ci Saat and Trinil faunas reflect filter dispersal, whereas the Kudung Brubus, continuing to the Ngandong Fauna at about 110 ka (Rizal et al., 2019), reflects a period of intermittent corridor dispersal between Java and the Asian mainland. It is thus difficult to see how the possible exposure of Central Sundaland as suggested by Sarr et al. (2019) significantly impacted on the dispersal of *Homo erectus* and the Siva fauna to Java.

1.4.4 Southeast Asia and India during the last glacial/interglacial cycle

1.4.4.1 Southeast Asia

Late Quaternary vegetational history in Southeast Asia was initially studied using cores from lakes in altitudinal sequence to determine the extent of temperature change since the LGM by estimating the altitudinal movement of montane vegetation zones (Flenley, 1979; Morley, 1982; Stuijts et al., 1988; Morley and Flenley, 1987; Maloney and McCormac, 1995). The sites selected were mostly from upland areas, which attract orographic precipitation, and so only a few, such as the Bandung Lake succession from West Java (Van der Kaars and Dam, 1995) provide information about moisture change. Some lowland sites, such as the extinct volcanic crater Rawa Danau in West Java provide an excellent record of the history of seasonal swamp vegetation which formed around the crater lake, but little information on the former nature of surrounding forests. A 50 ka record from cave deposits from Niah in Sarawak by Hunt et al. (2012) yields tantalizing pollen assemblages but in addition to transportation by wind and water, pollen may have been brought into caves by additional vectors such as nectivorous bats and humans, adding further complexity to the process of interpretation in terms of vegetation.

The analysis of marine cores, pioneered by the Monash group headed by Peter Kershaw opened up a new dimension to the interpretation of vegetation and climate history since the publication of a pollen diagram from Lombok Ridge, from 2000 m water depth from south of Flores (Van der Kaars, 1991). This study provided a record of vegetation change for northern Australia and southern Indonesia

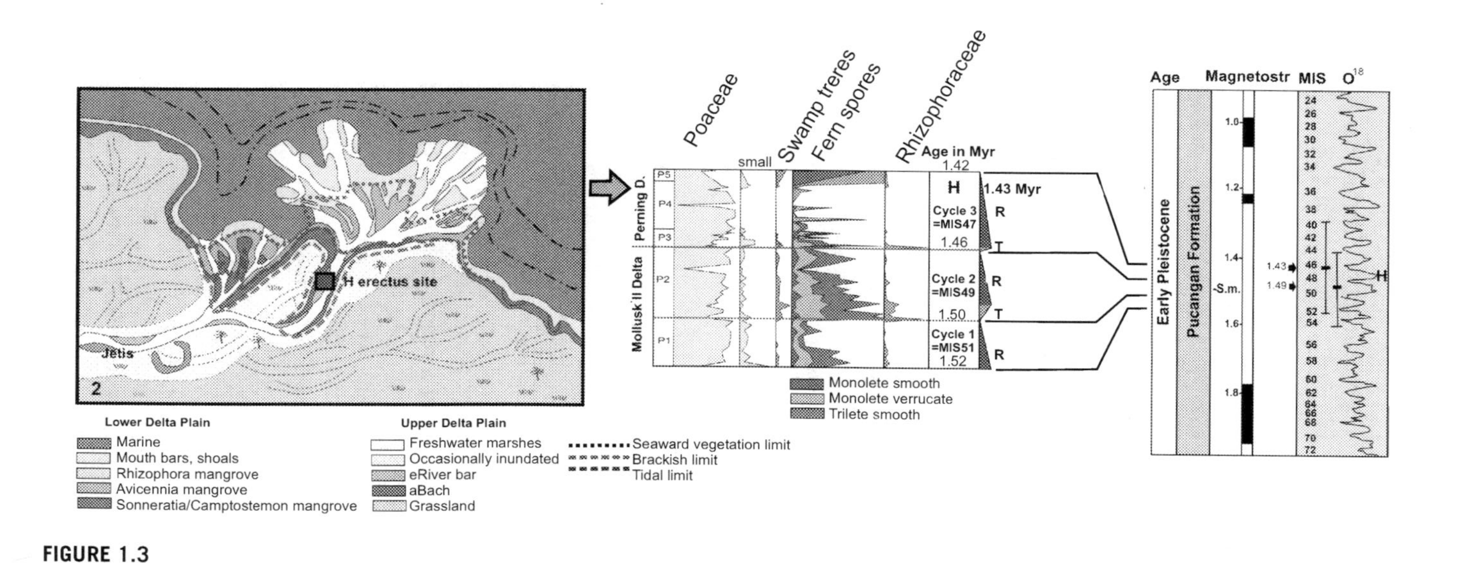

FIGURE 1.3

(A) The Perning Delta from near Mojokerto, site of the Modjokerto child skull of *Homo erectus*. (B) 41 ka transgressive-regressive depositional cycles from Perning, showing the time of occurrence of the *H. erectus* skull at 1.43 Ma (modified from Morley et al., 2020). H, hominin bed; R, regressive; T, transgressive.

stretching back to MIS6 within the Middle Pleistocene. Marine cores have the advantage that stable isotope analyses, such as ^{18}O, can provide an accurate chronostratigraphy, and other proxy analyses can be undertaken on the mineralogy of the core. Dupont and Wyputta (2003) emphasized that when interpreting marine cores it is important to take account of the distribution of palynomorphs in modern marine sediments and their mode of transport from adjacent land areas, but there are additional often unforeseen factors that also need to be taken into account. Transportation patterns will be different between times of marine transgression and sea level fall (e.g., Morley, 1996), and bottom currents can considerably modify palynomorph assemblages by sorting grains of different sizes (Traverse, 1988). Also, in deep water settings much of the pollen present may be brought into the marine environment by a variety of mass transport processes (Morley et al., 2021). Perhaps most important is core location; the pollen "catchment" for a marine core could involve multiple river catchments, or the catchment may change between phases of transgression and regression, as is the case for the modern Amazon fan (Hoorn, 1997), or middle Miocene deltas from the Northern Malay Basin (Morley et al., 2021). A good example of a well-located core is the Papalang-10 core from the Mahakam Fan (Morley et al., 2004; Morley and Morley, 2010), which precisely captures the pollen record of the Mahakam river catchment through the later part of the last glacial and the Holocene.

In this review we have selected three marine cores from contrasting parts of Sundaland (Fig. 1.4) each of which spans the entire last glacial, and allow a broad judgment to be made regarding changing vegetation and climate in relation to fluctuating sea level during each of the marine isotope stages of the last glacial period. The chosen cores are: Core BAR94-42 from offshore South Sumatra (Van der Kaars et al., 2010), BAR94-25 from offshore North Sumatra (Van der Kaars et al., 2012) and Sangkarang-16, from offshore Sulawesi in the East Java Sea, by Morley et al. (2004).

The three profiles show some clear parallels, but several major differences in terms of palynomorph assemblages, the nature of vegetation that is suggested, and the pattern of climate change. The similarities are at the scale of eccentricity-driven glacio-eustatic cyclicity, whereas the differences are at the scale of the shorter MIS 2–4 stages. In all three cores Stages MIS 1 and 5 are characterized by the dominance of pollen from lowland rain forests, with low percentages of pollen from lower montane forests, abundant pteridophyte spores and the limited representation of herbaceous pollen, in keeping with the present day warm and wet climate of the Sunda region. Common mangrove pollen reflects the development of mangrove swamps prograding along coastlines over the gently sloping continental shelf. The intervening Stages 2–4 contain increased pollen from lower montane forests, and the reduced representation of pollen from lowland forests, suggesting generally cooler climates across the region, and very low values for mangrove pollen, in keeping with limited opportunities for mangrove swamps to form when the shoreline is close to the shelf edge. The pattern of assemblage changes at the scale of eccentricity driven climate cycles thus follows the predicted succession based on sequence biostratigraphic reasoning (Morley, 1996, Morley et al., 2021).

The differences between the three cores, particularly relating to MIS2–4, need discussion as follows.

1.4.4.1.1 North Sumatra, core BAR94-25

Core BAR94-25 from offshore North Sumatra is characterized by abundant *Pinus* (Pinaceae) pollen and moderate numbers of spores and herbaceous pollen. The occurrence of *Pinus merkusii* in North Sumatra has long attracted attention, growing on mountains between 800 and 2000 m elevation; it includes the only natural stand of Laurasian conifers in the southern hemisphere, at 2°S in Kerinci (Cooling, 1968). *Pinus merkusii* is more typically a species of strongly seasonally dry climates in more

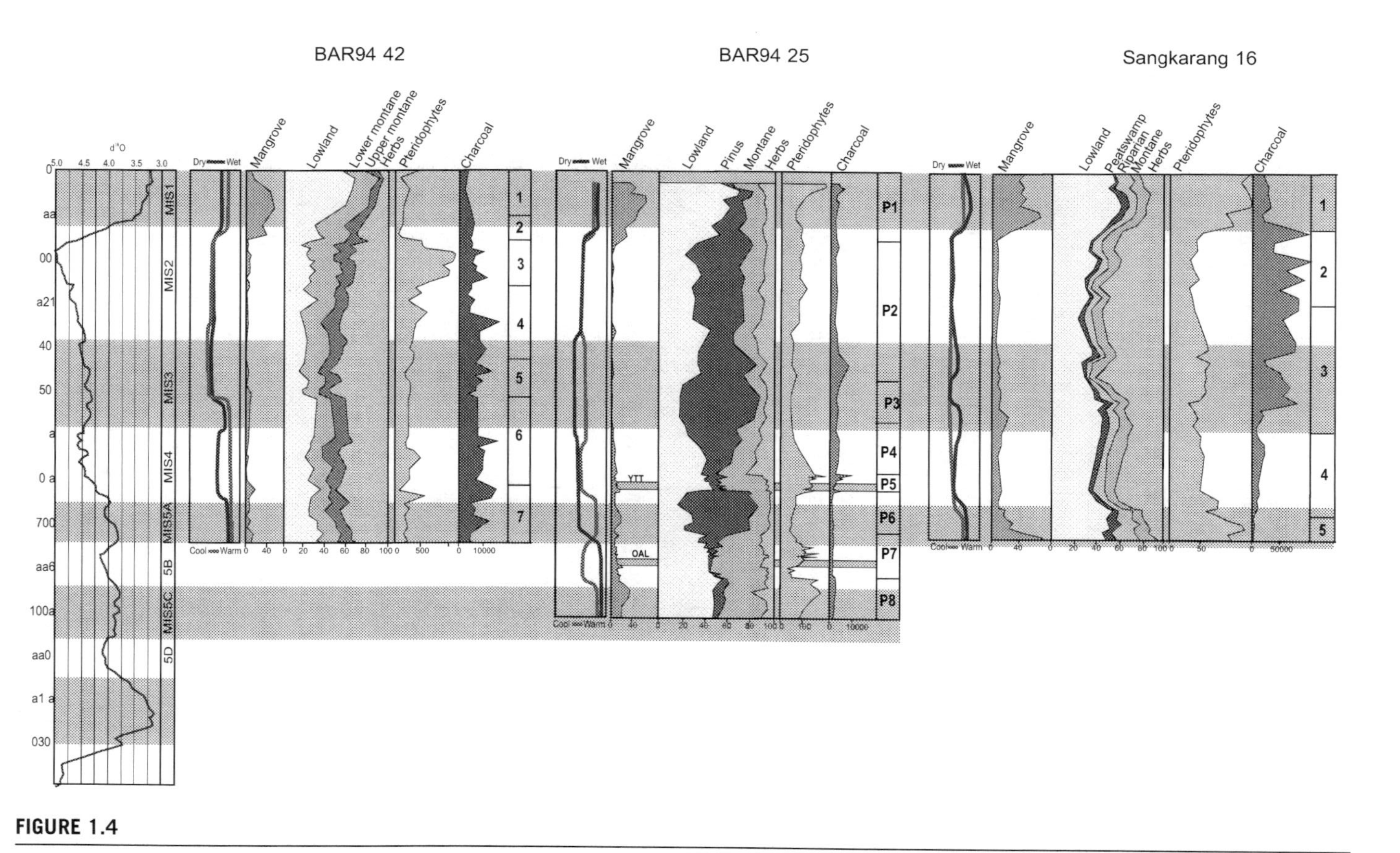

FIGURE 1.4

Summaries of three palynological profiles from Sunda: S Sumatra core BAR94-42 (Van der Kaars, 2010), N Sumatra core BAR94-25 (Van der Kaars et al., 2012) and East Java Sea/S Sulawesi core Sangkarang 16 (Morley et al., 2004; Morley and Morley, 2010). OAL, Older Ash Layer; YTT, Younger Toba Tuff.

lowland settings (below 600 m) in Indochina and the Philippines (Whitmore, 1975), preferring nutrient-poor, well-drained soils and occurs widely in savanna in central Thailand and formerly was much more widespread (Werner, 1997; Ashton, 2014; Ratnam et al., 2016). *Pinus merkusii* seedlings cannot tolerate shade, and stands are maintained by fire (Goldhammer and Penfiel, 1990). The natural Sumatran stands occur in rain shadow pockets on dry sites and lahars and occurrences of mature trees in primary forest are rare, but its range has extended dramatically in North Sumatra and elsewhere by felling and burning. Abundant *Pinus* and Poaceae pollen recorded from the Early Pleistocene Old Alluvium near Kuala Lumpur (Morley, 1998) shows that *Pinus* formerly grew close to this locality at low altitudes during the Early Pleistocene. To explain the abundant occurrence of *Pinus* pollen in Core BAR94-25 one needs to invoke seasonally dry climates, widespread burning, and growth mainly within the lowlands rather than in the mountains, as *Pinus* displays a more or less inverse distribution to pollen of montane trees such as *Quercus*.

Pinus shows its first prominence within MIS5A, indicating widespread fire-climax seasonally dry forest, but following deposition of the Toba Tuff, at the beginning of MIS4, *Pinus* reduced in abundance, whereas *Quercus* (Fagaceae) and charcoal fragments, followed by Poaceae, show an increase suggesting a cooler and drier climate. The reduction of *Pinus* above the tuff thus is more likely to be driven by global climate change rather than disturbance of vegetation caused by volcanic activity. Within MIS3, *Pinus* is again abundant, charcoal fragments become more common, and montane elements and pteridophyte spores decrease in abundance, suggesting a warmer and drier climate and an expansion of seasonally dry fire-climax pine forests. There is no mangrove pollen maximum associated with the sea level rise at the beginning of MIS3, but it may be that there was only minor exposure of the shelf at this time along the North Sumatra coast. Stage MIS2 is characterized by a slight reduction of *Pinus* pollen relative to MIS3, increased Leguminosae and montane pollen, and a slight increase in pteridophyte spores, suggesting a cooler, and slightly more humid climate, but still with widespread fire-climax forests with *Pinus merkusii*.

1.4.4.1.2 South Sumatra, core BAR95-42

The occurrence of abundant pteridophyte spores together with common Poaceae pollen within MIS2–4 in BAR95-42 was interpreted by Van der Kaars et al. (2010) as reflecting open herbaceous swamps lining river courses or surrounding lakes, and that the superabundant pteridophyte spores reflect species-rich and fern-rich closed canopy rain forest. The superabundance of fern spores in association with Poaceae pollen is more suggestive of the fern-dominated swamps that were widespread in coastal areas during the early Miocene of the Cuu Long and Thao Chu basins offshore southern Vietnam (Morley and Morley, 2013; Morley et al., 2019). Also, since pollen of peatswamp taxa, such as *Durio* (Bombacaceae), *Campnosperma* (Anacardiaceae), *Cephalomappa* (Euphorbiaceae), and *Gonystylus* (Gonystylaceae) is surprisingly rare in this section, the rain forests during MIS 2–4 may be better envisaged as seasonal evergreen rainforests (because peat swamps were not forming (Morley, 2018)), with grass and fern swamps along rivers and coastlines. This suggestion is further supported since with sea level rise at the beginning of the Holocene, the fern-dominated communities were entirely replaced by mangroves, whereas all dry-land pollen groups show only gradual change, suggesting that the fern communities occurred in the areas subsequently occupied by mangroves. The spore assemblage with abundant smooth monolete spores, *Nephrolepis* type, several *Lycopodium* spp. (Lycopodiaceae), and *Selaginella* (Selaginellaceae) is very reminiscent or the assemblage of fern spores present through the Holocene at Danau Padang from the Kerinci area of Sumatra (Morley, 1982).

1.4.4.1.3 East Java Sea, core Sangkarang 16

The Sangkarang-16 core was taken from the East Java Sea, close to the southwestern arm of Sulawesi. The pollen catchment included part of South Sulawesi, and the currently submerged catchment of the Java Sea River and its delta, which is thought to be the main source. Poaceae pollen is abundant and within Stages MIS2–4 may form over 50% of the dry-land pollen. The remaining dry-land assemblage consists mainly of pollen from lowland rain forests, with very low numbers of pollen such as *Celtis*, and scattered *Pterospermum* (Morley and Morley, 2010), which may be derived in part from deciduous forests. Pollen from montane forests is present in small percentages reflecting the overall low-lying nature of the exposed Java Sea and surrounding catchment. Peat swamp elements, such as *Campnosperma, Cephalomappa, Combretocarpus* (Rhizophoraceae), and *Gluta* (Anacardiaceae) are present through most of the section in low numbers, together with typically riparian taxa such as *Ilex* (Aquifoliaceae) and *Pandanus* (Pandanaceae). When considered together with the common herb pollen it is suggested that alluvial and widespread grass swamps may have formed part of the vegetation of the exposed Java Sea. The Sebangau PR-8 core by Morley (1981) shows the same pattern; the basal interval with abundant Poaceae pollen and laevigate spores also contains a full complement of taxa that occur in peat swamps, and thus an alluvial grass-dominated swamp is likely. The dominance of rain forest elements but without evidence for peat swamp formation and minimal indications of deciduous taxa suggest that the dominant lowland *terra firma* vegetation was seasonal evergreen forest, with extensive grass swamps in flood plains, rather than open savanna, as would have been present in the Early Pleistocene of East Java at Perning (Morley et al., 2020).

Comparison of assemblage changes between BAR94-25 and Sangkarang-16 shows that the highest Poaceae values occur in the latter part of MIS3, and that in both sections follow a wetter period which included a secondary maximum of mangrove pollen. The MIS3 assemblages in these cores suggest that sea level change may be the primary driver of the assemblage changes, with rising sea levels enabling a return of mangrove swamps and with more common grass pollen during the MIS3 "highstand." MIS2 was the period of lowest global sea levels, and for this period, the Poaceae records from the two profiles are very similar, initially being high, and then reducing at the time of lowest global sea levels. Examination of shallow seismic from areas such as the Java Sea shows that the Late Pleistocene river system was incised and that incision took place mid MIS2 (Posamentier, 2001). Prior to incision, the meandering river system would have provided extensive terrain for alluvial swamps, but following incision as sea levels fell, the sudden change in base levels due to river incision would have reduced the opportunities for swamp habitats to the incised floodplain, although with more land exposed at the LGM the climate may have been drier (Fig. 1.5).

In all three sections reviewed, there is consistency of climate change, with coolest conditions in MIS4 and MIS2, and the driest conditions in late MIS3.

This review indicates that lowland seasonally dry fire-climax pine forest occurred in North Sumatra during the last glacial, but the evidence for "savanna" across Sunda during the Late Pleistocene glacial maximum is lacking, without any support from palynology or plant biogeography. In a recent paper, Luo et al. (2019) for a Sunda Shelf core, which also yielded evidence for pine forests, mentioned that there is still controversy as to whether the LGM was characterized by tropical forests or herbs. To further address this problem, four basic issues with respect to the interpretation of marine cores needs to be addressed which are rarely considered. Firstly, marine cores need to be located in positions which are likely to capture the pollen production of clearly defined river catchments. Second, it is noted that several studies of marine cores do not find significant numbers of mangrove pollen which are the most important

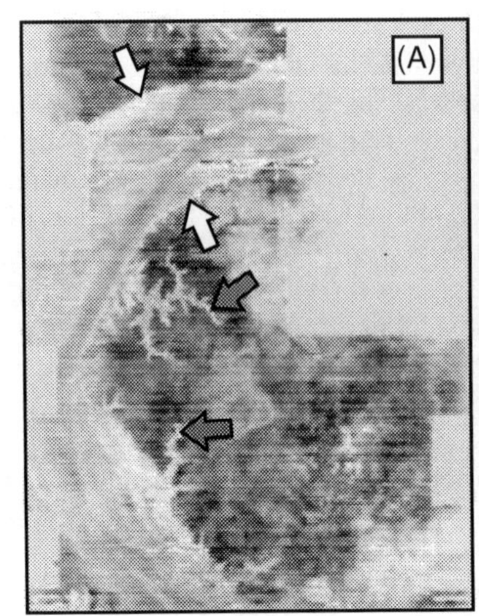

FIGURE 1.5

Incised valleys in the Java Sea, from shallow seismic 3D volumes (Posamentier, 2001), (A) 54 milliseconds (ms) interval transit time, (B) 60 ms subsea (location shown in Fig. 2). Slice A shows the lateral extent of the incised valley with a width of 5 km (white arrows) and numerous incised tributary valleys (gray arrows), showing a dendritic drainage pattern.

group of palynomorphs for interpreting transgressive–regressive marine successions in Southeast Asia (Morley, 1996; Morley, et al. 2021). Several of those studies involved processing using a 10 micron sieve, which is very effective in removing most pollen of the family Rhizophoraceae (Morley et al., 2007). Third, there seem to be issues with pollen and spore identification, for instance, a component of most palynological studies that involve mangroves in Southeast Asia show common *Acrostichum* (Pteridaceae) spores as part of the mangrove sporomorph component (e.g., Anderson and Muller, 1975; Morley, 1996; Yulianto, 2004; Morley et al., 2019), but many deep marine profiles, including Luo et al. (2019) give no indication of the presence of this spore type, but show many other trilete spore genera that tend not to be seen in coastal profiles from Sunda. Also, although the pollen present in the lower part of this core is interpreted to have been sourced from Borneo, no peat swamp indicators were determined, which elsewhere are ubiquitous in circum-Bornean sediments (Muller, 1972; Anderson and Muller, 1975). Fourthly, palynomorph assemblages can be modified by marine processes (e.g., Traverse, 1988), for instance pollen and spores of different size may be separated by bottom currents. Sedimentological issues also need to be carefully considered in deep marine profiles. In core NS07-25, analyzed by Luo et al. (2019), the pollen influx maximum at 300 cm coinciding with a sedimentation rate change strongly suggests transportation in a turbidite. Such processes introduce dramatic changes to deep marine pollen assemblages (Morley et al., 2021) and unless these processes are understood, misinterpretations will follow.

1.4.4.2 India

A surprising number of palynological studies have been undertaken on marine cores from offshore India (Fig. 1.2), and although giving often incomplete data, provide valuable perspectives of the Quaternary history of Indian vegetation. Along the western seaboard, Core SK129 CR05 from 9°21′N from 2300 m water depth yielded rich pollen assemblages from the latter part of the last interglacial, when sea levels were high, during MIS5A (Farooqui et al., 2014). The section yielded assemblages consisting of mangrove pollen, together with pollen indicative of the lowland and Shola montane forests of the Western Ghats, suggesting a vegetational setting similar to that of the Holocene. The more common pollen types in this flora suggest the presence of members of many tropical families including Anacardiaceae, Burseraceae, Clusiaceae, and Dipterocarpaceae. The presence of common *Myristica malabarica* type (Myristicaceae) pollen is indicative of the seasonal swamps that occur at the base of Western Ghat valleys (Pascal, 1986). Farooqui et al. (2010) found a similar assemblage from an estuarine deposit from a shallow well from Chahganachery, to the south of Cochin, which can be dated to the time of the Toba Tuff at 75 ka based on the presence of glass shards from the Toba eruption. Here they were able to identify many of the evergreen Western Ghat taxa, such as *Cullenia* (Bombacaceae) and *Garcinia* (Clusiaceae), but also pollen of deciduous taxa such as *Bombax* and *Lagerstroemia*, suggesting that during MIS5A, the local pollen catchment would have included both moist deciduous and semi-evergreen lowland forests, as suggested for the present day prior to deforestation by Ashton (2014). Well sections from further north, from north of Mangalore, considered to relate to the last interglacial, analyzed by Caratini et al. (1990) yielded similar, but less diverse assemblages with abundant mangrove and rain forest pollen, suggesting that during MIS5 the Western Ghat forests extended northward to at least 13°50′N, similar to the present day.

Notably, Farooqui et al. (2010, 2014, 2020) identified pollen of two taxa currently extinct in India, *Ongokia gore* (Aptandraceae), a tree of evergreen and semi-evergreen forests in tropical West Africa, and an extinct *Basella* (Basellaceae), named *B. keralensis,* from MIS5 sediments. *Basella* is a small genus of perennial twining herbs or vines with one widespread species, and four species restricted to Madagascar and East Africa. This emphasizes that the refugial Western Ghat rain forest flora has continued to lose its endemic diversity due to Pleistocene climate change and northward drift. The higher diversity of evergreen rainforests in Sri Lanka, compared to the Western Ghats is probably due to its more equatorial position, with more influence by rainfall associated with the intertropical convergence.

What was the fate of these forests during stages MIS 2–4? Several palynological studies have been undertaken from the Nilgiri Hills, a plateau typically reaching 2000 m asl (Vishnu-Mittre and Gupta, 1971; Blasco and Thanikaimoni, 1974) show that the present day Shola forests replaced grasslands after the LGM and this conclusion is supported by carbon isotope studies on peats (Sukumar et al.,1993). However, studies of carbon isotopes from soil profiles by Caner et al. (2007) show that whereas some LGM soils yielded C4 signatures, indicating grasslands, others formed under C3 vegetation, reflecting former forests. They suggest that the LGM grasslands were maintained by low temperatures rather than moisture availability, and the occurrence of frosts (which occur in this area today) and that forests were restricted to sheltered valleys along the western Nilgiri Hills. The presence of forest cover during the LGM at low altitudes is suggested by the occurrence of laterites between the organic-rich sediments yielding Toba ash (discussed above) and the Holocene. Laterites form under strongly seasonal tropical climates beneath forest cover (Ghosh and Guchhait, 2015). A similar history probably applies to the perhumid forests of Sri Lanka. A pollen study from the Horton Plains (Premithalake and Risberg, 2003) suggests that the LGM climate was "semiarid," but this is based on just a few pollen grains from a probable paleosol and it is likely that the original pollen content has been lost through oxidation.

A palynological study by Singh et al. (1972) from the Thar Desert in northwestern India suggested that desert climates were much more extensive during the LGM. This was also indicated by Prabhu et al. (2004) and Ansari and Vink (2007) who studied deep marine cores from offshore Mumbai and the Indus fan. Prahbu et al. (2004) recorded assemblages which were dominated by Poaceae with common Chenopodiaceae type and *Artemisia* (Asteraceae) pollen, whereas Ansari and Vink (2007) found similar assemblages but with the addition of *Ephedra* (Ephedraceae) pollen, also suggesting increased aridity. Van Campo (1995) examined two cores from 1200 m from offshore Mumbai and Cochin. The LGM of the northern core gave similar results to the Prabhu et al. (2004) and Ansari and Vink (2007) studies, whereas the southern core was dominated by Poaceae pollen during the LGM, suggesting extensive grasslands or savanna within the pollen catchment area. The southern boundary of desert conditions thus probably occurred between the latitude of these two localities

A very broad perspective of the vegetation of the Ganges plain for the period immediately following the Toba eruption is suggested from a marine core offshore the Ganges Delta by Williams et al. (2009) which suggests that woodland to open grassland was widespread across central India at that time. A more detailed picture of vegetation for the latter part of the last glacial in the vicinity of Lucknow is provided from the analysis of a core from Lashoda Tal, a lake which formed in an abandoned river channel in the floodplain of the Ganges River (Trivedi et al., 2019). The record extends back to 25.5 ka. Open savanna is suggested until 22 ka, after which time trees became better represented, especially *Holoptelea integrifolia* (Ulmaceae), *Madhuca indica* (Sapotaceae), and *Acacia* (Mimosaceae). After 14.3 ka trees expanded significantly, and forest groves interspersed with open grassland is suggested (Fig. 1.6). During this period there were major hydrological changes in the development of the area, and to differentiate the pollen signal from hinterland vegetation with that of the ephemeral topography of the floodplain is open to different interpretations from a single section. Chauhan et al. (2015) studied the nearby Karela Jheel which suggested a similar picture, with open grasslands with sparse *Holoptea integrifolia* prior to 12.5 ka, and forest groves with *H. integrifolia, Acacia*, and *Bombax* (Bombacaceae) after that date.

Two profiles from central India also suggest more open vegetation during the LGM. Quamar and Bera (2017) evaluated a core from Lakadandh Swamp from Koriya District, Chhattisgarh, and suggest savanna woodland vegetation prior to 9 ka, which was replaced by deciduous forest during the Holocene. The savanna included the trees *Holoptelea* and Sapotaceae with sparse *Acacia, Emblica officinalis* (Phyllanthaceae), *Lagerstroemia* (Lythraceae), *Madhuca indica*, and *Syzygium* (Myrtaceae). Quamar and Kar, 2020, however, looking at Nakta Lake to the southeast in Mahasamund District found a similar trend over the same time period from open grass-dominated vegetation suggesting savanna, with scattered trees, similar to those at Lakadandh Swamp, prior to 11.7 ka, followed by deciduous forest with the same tree flora until 8.5 ka. These studies show that deciduous forests also retracted their range across India during the last glacial.

From northeast India, montane localities mainly suggest ameliorating climates following the LGM, whereas lowland localities either show little change, or suggest drier LGM climates in the manner of localities in Peninsula India. A study by Battacharyya et al. (2014) from a valley site at 1580 m at the northern tropical margin presents a long record, spanning the period from 66 ka to present, but with the LGM probably missing. Stage MIS4, from 66 to 36 ka, was characterized by widespread *Pinus* and *Tsuga* (Pinaceae), suggesting a cold climate, whereas during MIS3 the region bore oak forests

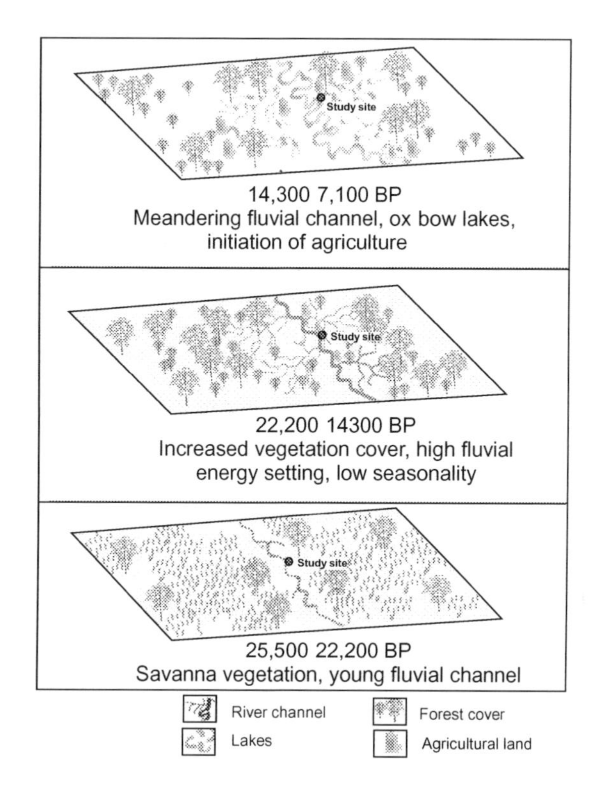

FIGURE 1.6

Interpreted vegetation for MIS2 and early MIS1, from Trivedi et al. (2019).

during a period of climatic amelioration. There is no record for MIS2 which was probably a period of nondeposition or nonpreservation, as it coincides with a succession of sandy lithologies suggesting a fluvial channel. Battacharyya et al. (2014) suggest that savanna was widespread in the area surrounding the site during MIS4, and that this reduced during MIS3 when the climate was wetter. However, a perusal of the pollen diagram shows that there is a close relationship between the abundance of Poaceae and some aquatics, such as *Potomogeton* (Potomogetonaceae) and *Impatiens* (Balsaminaceae), and this suggests that a significant proportion of the grass pollen was probably derived from swamp vegetation rather than true savanna. Tripathi et al. (2019) studied the Deepor swamp located in the Brahmaputra floodplain of Assam. They suggest a cooler and drier climate for the LGM than at present, followed by a warmer and wetter setting with deciduous forest including *Shorea robusta* and *Lagerstroemia* during the early Holocene. This site is characterized today by vast grass-dominated wetlands, and again a major issue is to differentiate local pollen sources from the hydrosere, and that from surrounding *terra firma* forests. A further study, from the Assam foothills by (Basumatary et al., 2015), from the Subankhata swamp, north of the Brahmaputra floodplain, suggests a similar pattern, with widespread lowland savanna forest during MIS2, and a dry climate regime, followed by the expansion of deciduous and semi-evergreen forests in the Holocene. Again, a major issue is to separate local changes relating to the hydrosere with regional vegetation changes driven by climate change.

1.4.5 Immigration of *Homo sapiens*

There is a very poor Pleistocene record of fossil *Homo sapiens* from the Indian subcontinent, with the earliest definitive specimens from Sri Lanka dated to 28 ka (Kennedy and Deraniyagala, 1989), in contrast with Southeast Asia, where *H sapiens* fossils are claimed in the early MIS5 Punung fauna from Java (Storm et al., 2005), dated to 128–118 ka (Westerway et al., 2007) and there are well-substantiated discoveries from the Philippines, dated to 68 ka, and Niah Cave in Sarawak, dated by Barker et al. (2007) to 46 ka. However, India is rich in well-dated Paleolithic remains, reviewed by Blinkhorn and Petraglia (2017). Late Acheulian sites (bifaces and limited flake tools) are dated to 140 ± 11 ka and 137 ± 10 ka, and suggest that *H. sapiens* migrated to India before glacial Termination II during the period of low sea levels of stage MIS6 (Fig. 1.1), and thus *H. sapiens* would have been well placed for a further eastward migration to Java. There are also many Middle Paleolithic sites (more diverse flake tools), the oldest artifacts being dated to 95.6 ± 13.1 ka. *Homo sapiens* was thus well established in India by the end of the Eemian interglacial, before the eruption of Mount Toba in Sumatra.

1.4.6 The Toba eruption

After 400 ka of quiescence, Gunung Toba in Sumatra began to erupt in earnest, dated at 75.0 ± 0.9 ka (Mark et al., 2014) leading to the largest single volcanic eruption of the Quaternary, at least two orders of magnitude bigger than the Gunung Tambora eruption of 1816, which caused a global "volcanic winter" with crop failures across Europe and North America (Rampino and Self, 1982). The Toba eruption resulted in the ejection of 2.800 km^3 of rock equivalent and left a 6-year spike of volcanic sulfate in the Greenland GISP ice core 12,000 km distant (Zeilinski et al., 1996). The sulfate spike tied closely to the beginning of a 1000-year stadial event. It has been suggested that the eruption was the cause of this period of cooler global climates (Zeilinski et al., 1996) although a direct connection with this event is disputed by Haslam and Petraglia (2010). The eruption resulted in lava flows extending from coast to coast in Sumatra with a typical thickness of 50–150 m, and with 400 m thick ignimbrites resulting from pyroclastic flows. Sufficient ash was produced to form a 5–15 cm thick layer across much of tropical Asia, and this can be found in India and the Malay Peninsula, and in cores from the Indian Ocean, Arabian and East China Seas, and as a thin layer as far west as Lake Malawi in Africa (Lane et al., 2013). Erosion of the ash immediately following deposition would have led to choking of stream channels, by which process it reaches 6 m in some Indian localities.

There has been much discussion regarding the impact of the eruption on flora, fauna and early humans across South Asia. The lava flows and ignimbrites would have caused destruction of any biota in areas where they made direct contact, but it is unlikely that the event significantly impacted the faunal diversity of North Sumatra (Louys, 2012). The palynological study of the BARS94-25 marine core from offshore Aceh (Van der Kaars et al., 2012) that includes the Toba Tuff shows a dramatic reduction in *Pinus* pollen immediately following the tuff which Van der Kaars et al. (2012) have difficulty in explaining, but as noted above, would be more in fitting with climate deterioration at the beginning of MIS4, rather than relating to vegetation destruction caused by the Toba ashfall. There is a small maximum of *Dacrycarpus imbricatus* (Podocarpaceae) and *Dacrydium* (Podocarpaceae) pollen immediately following the tuff, which may relate to a volcanosere (e.g., Polhaupessy, 1990) and this small acme may reflect the impact of the Toba eruption in the pollen rain offshore North Sumatra. Clearly, deposition of the tuff inflicted no serious damage on the regional vegetation on either the short- or long term.

The Toba Tuff has been studied palynologically in two further localities, Core BAR-42 from South Sumatra (van der Kaars, 2010) shows no change across the tuff, whereas Core S0188-342KL in the Bay of Bengal (Williams et al., 2009) shows persistently increased percentages of Chenopodiaceae-type pollen and reduced *Stenochlaena palustris* (Blechnaceae) spores following the tuff, suggesting that the climate for the pollen catchment, which would have been the Ganges River system, may have been drier following deposition of the tuff. Williams et al. (2009) also examined carbon isotope data from above and below the Toba Tuff from three sites in India which show that C3 forest was replaced by C4 grassland immediately above the tuff and suggest that the Toba ashfall caused climatic cooling and prolonged deforestation across India. However, correlation need not reflect causation and this suggestion is disputed by Haslam and Petraglia (2010) who consider that the drying suggested from pollen data is more likely due to changing climate rather than a long-term effect of the Toba Tuff eruption.

It has been suggested that the eruption and its climatic and environmental consequences resulted in a global ecological disaster and led to a bottleneck in human evolution which occurred at about 70 ka (Ambrose, 1998). Although there is clear molecular evidence for such a bottleneck (Hawks et al., 2000), Gathorne-Hardy and Harcourt-Smith (2003) found no evidence to support the suggestion that the bottleneck could be linked to the Toba eruption. A similar conclusion was reached by Clarkson et al. (2011) as human occupation sites from southern India which span the Toba Tuff horizon show minimal change in lithic artifacts below and above the tuff.

1.4.7 *Homo sapiens* after Toba

Records of human occupation employing Late Paleolithic technologies increase dramatically across India after 45 ka (Blinkhorn and Petraglia, 2017) suggesting that the density of *Homo sapiens* would have shown a similar increase, modeled by Jones (2012). Such a population increase of hunter gatherers, would have imposed a major burden on the indigenous fauna, and many taxa may have suffered population decline as a result, with their populations eventually becoming restricted to the refugial areas in which they occur today (Fig. 1.7). Stuart (2015) reviews global Late Quaternary megafaunal extinctions by continent and indicate fewer extinctions across the Indo-Malay ecoregion than the Americas, northern Eurasia and Australasia, but more than in Africa, whereas Roberts et al. (2016) evaluated a cave fauna from Billasurgam in India and found that 20 of 21 identified mammalian taxa from 100 ka remain extant. They suggest that although local extinctions occurred, most taxa adapted to new ecological pressures in fragmented habitats in areas that offered protection from humans.

Johnson (2009) and Bakker et al. (2015) suggest that the human induced (Sandom, 2014) decline of large herbivores is likely to have triggered large changes in plant communities and induce major alterations in landscape structure and ecosystem functions. For India, was the possible increased representation of savanna landscapes of the last glacial period a reflection of the much more extensive representation of megafauna prior to the expansion of *Homo sapiens*? For Sunda, were the rain forests more open, with clearings and open areas?

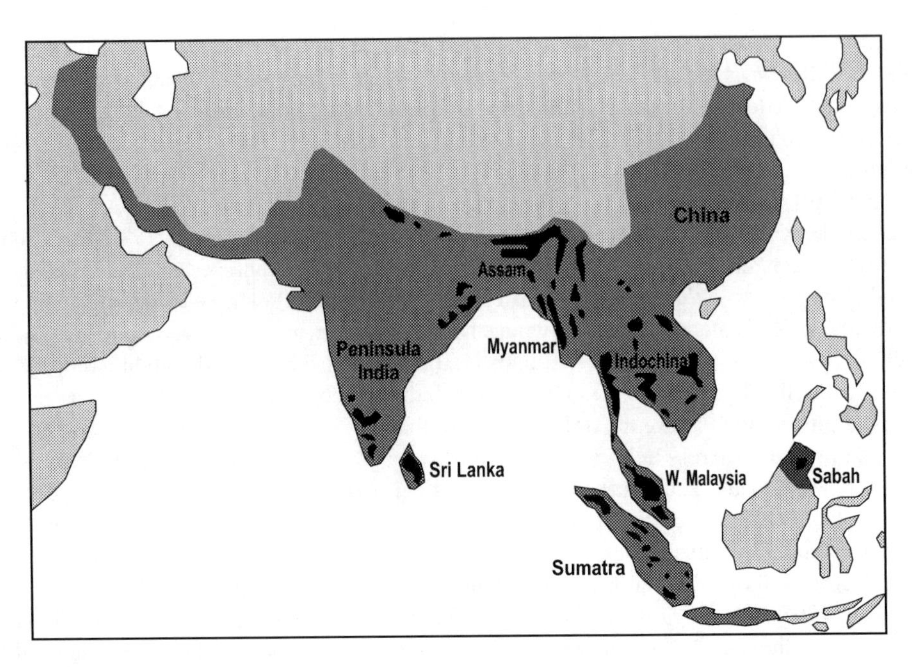

FIGURE 1.7

Former (gray) and current (black) range of Asian elephant (from various sources). The "current" range probably overestimates its actual present day distribution.

1.5 Learning and knowledge outcomes

An evaluation of 36 palynological profiles from across Southeast Asia and India greatly clarifies the nature of vegetation across tropical Asia during the Pleistocene following the demise of wet tropical forests across India after the Himalaya reached their highest elevations. The general pattern in both regions is of expansion and contraction of the geographical range of perhumid rain forests with glacio-eustatic sea level oscillations, with perhumid forests showing their greatest expansion during periods of high sea levels during MIS5 and 1 and presumably older periods of high sea level.

Examination of the plant biogeography of Indochina and Java suggests that it is unlikely that there was ever a savanna corridor connecting the two areas, but since elements of Indochinese semi-evergreen forests occur in Java, but deciduous forest elements are conspicuously missing, it is possible that at some stage there was a corridor of seasonally dry semi-evergreen forests linking the two areas. The timing of this was more likely to be during the Early Pleistocene, with widespread open savanna in East Java during MIS51-47 (Morley et al., 2020), and there was a permanent land exposure in central Sundaland (Sarr et al., 2019) possibly extending into the Middle Pleistocene. The corridor may have occurred at the time of formation of ferricretes that are widespread in the southern part of the Malay Peninsula and West Sumatra (Fig. 1.2), which formed during the Early Pleistocene or possibly earlier (Batchelor, 2015).The time window for this corridor is likely to be the time of the "intermittent corridor" from 1.3 Ma to 110 ka, suggested by Van den Bergh et al. (2001).

For the last glacial period, and the LGM in particular (Fig. 1.2), the pattern of climate and vegetation change is becoming increasingly clear. At the time of the LGM, there is evidence for cooler climates at virtually all studied localities. Based on the few profiles that extend beyond the LGM, MIS4 may have been the coldest, and MIS3 highstand the driest, at least in the Sunda region. In Sunda the LGM witnessed the retraction of the range of perhumid forests to the equatorial belt. To the north, fire-climax pine forests became widespread, extending from North Sumatra to the Sunda Shelf north of Natuna, mirroring the expansion of pine savanna from the Early Pleistocene deposit near to Kuala Lumpur (Morley, 1998). To the north, in Indochina, there was a major expansion of savanna grassland and pine-oak savanna. To the south of the equator, there was a parallel belt of seasonal evergreen forests, extending from South Sumatra to the East Java Sea, with extensive alluvial swamps, dominated by grasses in poorly drained floodplains. Deciduous forest may have been present to the east of Bandung Lake based on similarities of the *terra firma* vegetation surrounding the lake with East Java (van der Kaars and Dam, 1995) and some suggestions of subordinate deciduous forest in the catchment of the Java Sea River. There is no evidence for a savanna or seasonally dry corridor across central Sundaland. In the Indian subcontinent, seasonal rain forests persisted in sheltered sites along the southern Western Ghats and on the adjacent floodplains, and in western Sri Lanka, but to the north, savanna grasslands and desert expanded to the east and south at the expense of deciduous and evergreen forests.

To better understand the vegetational, and climate history of South and Southeast Asia, additional studies are always useful, but equally critical is to ensure that existing datasets are appropriately interpreted. The method of vegetational history interpretation and climate inferences based on palynology began from the examination of cores from peat bogs and small lakes in temperate areas where vegetation patterns and pollen transfer mechanisms are relatively simple (Von Post, 1916). Major advances in interpretation were possible with the understanding of how pollen productivity varies in different plant taxa (Janssen, 1966) how it is transported in air and water (Tauber, 1967), the concept of a pollen budget (Peck, 1973) and calculation of pollen concentration and influx (Davis, 1964). These issues were reviewed for the tropics by Flenley (1973) who emphasized that differences between tropical and temperate assemblage interpretation was simply a matter of degree and that the dominance of entomophily in the tropics did not present an insurmountable issue. To extend the method into marine environments, additional factors need to be considered that are only relevant in the largest of lakes. The dominance of water over wind transport (Morley, 1991), the significance of the Neves effect (Chaloner and Muir, 1968; Riding et al., 2017), the sorting and winnowing effect of marine currents (Traverse, 1988), different transportation methods during transgressive periods as opposed to the present time when most coastlines are essentially regressive (Morley, 1996) and the issues concerned with pollen transportation, onto the shelf and into deep water settings, especially by mass flow processes. In addition, it is clear that for tropical palynomorph assemblages, there are pollen identification issues, and also anomalies with sample processing, since although sieving with a small mesh sieve cleans the slides of annoying fine debris, using too coarse a sieve size, or even too rigorous sieving with a fine mesh sieve, can result in the removal of small pollen (Morley et al., 2007). Our own observations show that sieving tropical marine residues with a 10 micron sieve will remove virtually all pollen of the family Rhizophoraceae, which is by far the most useful indicator of sea level change across the tropics (Morley, 1996; Grindrod et al., 2002; Morley et al., 2020; Morley et al., 2021). The authors' experience is that large amounts of small mangrove pollen may be lost by vigorous processing with a 7 micron and also a 5 micron sieve (e.g., by speeding up the sieving process using a pump or even ultrasonic treatment).

Gently sieving with a 5 micron sieve, or without a small mesh sieve, is preferred to maximize recovery of this important pollen type.

In South China Sea studies, there seem to be major issues with pollen identification, as pollen of several key taxa are missing from many profiles, especially pollen of peat swamp trees, which according to Muller (1972) are ubiquitous in circum-Borneo marine sediments. No effort has been made in any profiles to differentiate pollen of species of the mangrove tree *Sonneratia*, despite the fact that they are easy to identify (Muller, 1969) and have different ecologies in mangrove swamps making them very useful for evaluating patterns of sea level change (Morley et al., 2021). Also, spores of the mangrove fern *Acrostichum* are missing from most profiles. Without the mangrove component properly recorded and evaluated, it is difficult to properly interpret patterns of sea level change, and to interpret the pollen from *terra firma* settings.

Evaluation of marine cores from offshore South and North Sumatra suggests that the original interpretations placed far too much emphasis on the search for patterns of climate change, rather than geographical changes in the pollen catchment landscape and its spatial vegetation composition. Changes in the land area exposed across Sundaland will change the position of the shoreline relative to the core locality and this may have a significant effect on pollen composition without any need to invoke climate change (Morley, 1991; Chaloner and Muir, 1968). Also, much more emphasis needs to be placed on the differentiation of pollen from edaphic communities and *terra firma* sites, as edaphic communities provide an independent indicator of changing climates (Morley and Morley, 2013), and also help to unravel the character of the pollen catchment landscape.

With respect to Pleistocene pollen records from India, the main issue is concerned with the interpretation of maxima of grass pollen, on which basis several profiles from mainland India suggest the former wider representation of savanna at the expense of forests during a cooler and drier MIS2 climate. Bush (2002) warned of the inherent dangers in the interpretation of the Poaceae pollen record, since it may reflect either the increased representation of open-land grasslands or relate to the presence of grass-dominated swamps which may be purely edaphic. Most Indian LGM profiles are from fluvial riverine successions, where grass-dominated swamps are widespread, and differentiating the regional from the local edaphic grass pollen signal may be particularly problematic. Two approaches are suggested to clarify this. Firstly, palynological analyses should be sufficiently closely spaced, with emphasis being placed on both obligate and facultative wetland components, to allow the hydrosere to be accurately reconstructed over time. From this, elements of the Poaceae record which cannot be supported as hydrosere elements may be sourced from possible *terra firma* vegetation. Combining pollen with phytolith analysis may help to make this differentiation (Morley et al., 2020). The second approach is to build "resolved" diagrams (sensu Faegri and Iversen, 1964) by analyzing multiple profiles through a wetland area, from a medial to a peripheral setting, and evaluating the representation of edaphic and regional elements with distance from the *terra firma* source (cf. Morley, 1981).

If the regional landscape may have borne extensive savanna, was this savanna a result of a drier and cooler climate, or was it also maintained by megafaunal grazing and browsing, or a combination of the two? Studies from Africa of herbivory by mega-grazers, such as Rhinoceros (Waldram et al., 2008) and elephant (Asner et al., 2006) show that large mammals are important in maintaining the savanna landscape, and the same may have been the case prior to overhunting by humans in India. Studies need to be conceived that could clarify this, perhaps by integrating $\delta^{13}C$, palynological studies of the identification of changes in the record of the dung fungus *Sporormiella* (e.g., Burney et al., 2003) and other multiproxy approaches in lacustrine and marine sediments.

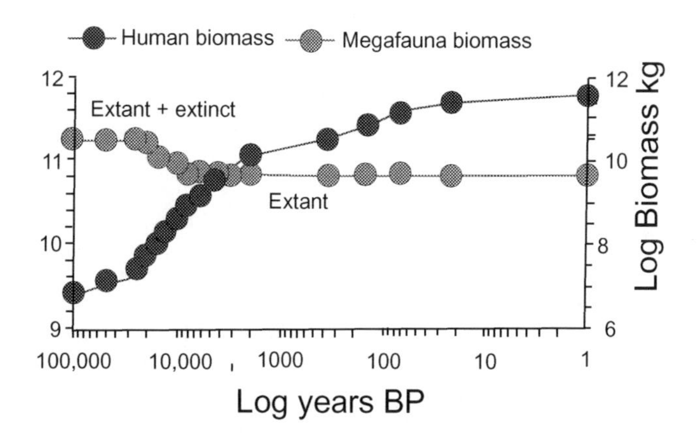

FIGURE 1.8

Megafauna biomass trade-off as a driver of Quaternary and future extinctions (Barnosky, A.D., 2008). Proceedings of the National Academy of Sciences USA 105, 11543–11548. Copyright (2008) National Academy of Sciences U.S.A.

Barnosky (2008) compared global estimates of the biomass of extinct and extant megafauna and humans over time (Fig. 1.8). The point at which megafaunal biomass reduces dramatically and human biomass shows a similar increase corresponds with the beginning of the Holocene, when hunter-gathering was replaced by agriculture. The removal of megafauna by overhunting, and their restriction to poorly accessible river floodplains, needs to be considered when interpreting Late Pleistocene palynological datasets.

From this review palynology is playing a major role in determining the Pleistocene environmental and climatic history of tropical Asia. With more rigorous consideration of the methods of processing and palynological analysis, and consideration of the different issues that affect pollen deposition, the precision of our understanding will undoubtedly improve, and such knowledge will be helpful in understanding and dealing with future changes relating to human pressure on natural ecosystems due to the unabated expansion of agriculture and urbanization, leading to excessive fragmentation or complete destruction of natural ecosystems.

Acknowledgments

The authors thank Navnith Kumaran for inviting us to write this review. Peter Ashton kindly read the manuscript and improved the discussion on seasonally dry vegetation. Henry Posamentier helped with discussion of the evolution of the Java Sea by comparing shallow seismic 3D data with palynological successions from marine cores. R.C. Mehrota and an anonymous reviewer helped with finalization of the manuscript.

References

Ambrose, S.H., Late Pleistocene human population bottlenecks, volcanic winter, and differentiation of modern humans, J. Hum. Evol. 34 (1998) 623–651.

Anderson, J.A.R., Muller, J., Palynological analysis of a Holocene peat and a Miocene coal from NW Borneo, Rev. Palaeobot. Palynol. 19 (1975) 291–351.

Ansari, M.H., Vink, A., Vegetation history and palaeoclimate of the past 30 kyr in Pakistan as inferred from the palynology of continental margin sediments off the Indus Delta, Rev. Palaeobot. Palynol. 145 (2007) 201–216.

Ashton, P.S., On the forests of Tropical Asia. Lest the memory fade, Royal Botanic Gardens Kew and Harvard University, London and Boston, 2014, pp. 1–670.

Ashton, P.S., Morley, R.J., Heckenhauer, J., Prasad, V., The magnificent Dipterocarps: an epitaph? Kew Bull. (2021) in press.

Asner, G.P., Vaughn, N., Smit, I.P.J., Levick, S., Elephant ecosystem-scale effects of megafauna in African savannas, Ecography 39 (2006) 240–252.

Bakker, E., Gill, J.L., Johnson, C.N., Vera, F.W.M., Sandom, C.J., Asner, G.P., Svenning, J-C., Combining paleo-data and modern exclosure experiments to assess the impact of megafauna extinctions on woody vegetation, Proc. Natl. Acad. Sci. USA 113 (2015) 847–855.

Barker, G., Barton, H., Bird, M., Daly, P., Datan, I., Dykes, A., Farr, L., Gilbertson, D., Garrisson, B., Hunt, C., Higham, T., Kealhofer, L., Krigbaum, J., Lewis, H., McLaren, S., Paz, V., Pike, A., Piper, P., Pyatt, B., Rabett, R., Reynolds, T., Rose, J., Rushworth, G., Stephens, M., Stringer, C., Thompson, J., Turney, C., The "human revolution" in lowland tropical Southeast Asia: the antiquity and behavior of anatomically modern humans at Niah Cave (Sarawak, Borneo), J. Hum. Evol. 52 (2007) 243–261.

Barnosky, A.D., Megafauna biomass trade-off as a driver of Quaternary and future extinctions, Proc. Natl. Acad. Sci. USA 105 (2008) 11543–11548.

Basumatary, S.K., Tripathi, C.M., Bera, S., Nautiyal, S.K., Devi, C.M., Sarma, N., G.C., Late Pleistocene palaeo-climate based on vegetation of the Eastern Himalayan foothills in the Indo-Burma Range, India, Palynology 39 (2015) 220–233.

Batchelor, B.C., Discontinuously rising Late Cainozoic eustatic sea levels, with special reference to Sundaland, Southeast Asia, Geol. Mijnbouw 58 (1979) 1–20.

Batchelor, D.A.F., Clarification of stratigraphic correlation and dating of Late Cainozoic alluvial units in Peninsular Malaysia, Bull. Geol. Soc. Malaysia 61 (2015) 75–84.

Bhattacharyya, A., Mehrotra, N., Shah, S.K., Basavaiah, N., Chaudhary, V., Singhet, I.B., Analysis of vegetation and climate change during Late Pleistocene from Ziro Valley, Arunachal Pradesh, Eastern Himalaya region, Quat. Sci. Rev. 101 (2014) 111–123.

Bian, Y-P., Jian, Z-M., Weng, C-Y., Khunt, W., Bolliet, T., Holbourn, A., A palynological and palaeoclimatological record from the southern Philippines since the Last Glacial Maximum, Chin. Sci. Bull. 56 (2011) 2359–2365.

Bird, M.I., Taylor, D., Hunt, C., Palaeoenvironments of insular Southeast Asia during the Last Glacial Period: a savanna corridor in Sundaland? Quat. Sci. Rev. 24 (2005) 2228–2242.

Blasco, F., Thanikaimoni, G., Late Quaternary vegetational history of southern region, in: Surange, K.R. (Ed.), Aspects and Appraisal of Indian Palaeobotany, Birbal Sahni Institute of Palaeobotany, Lucknow, 1974, pp. 632–643.

Blinkhorn, J., Petraglia, M.D., Environments and cultural change in the Indian subcontinent, Curr. Anthropol. 58 (Suppl. 17) (2017) S463–S479.

Burney, B.A., Robinson, G.S., Pigott Burney, L., *Sporormiella* and the late Holocene extinctions in Madagascar, Proc. Natl. Acad. Sci. USA 100 (2003) 10800–10805.

Bush, M.B., On the interpretation of fossil Poaceae pollen in the lowland humid neotropics, Palaeogeogr. Palaeo-clim. Palaeoecol. 177 (2002) 5–17.

Caner, L., Lo Seen, D., Gunnell, Y., Ramesh, B.R., Bourgeon, G., Spatial heterogeneity of land cover response to climatic change in the Nilgiri highlands (Southern India) since the last glacial maximum, Holocene 17 (2007) 195–205.

Cannon, C.H, Morley, R.J., Bush, A.B.G., The current refugial rainforests of Sundaland are unrepresentative of their biogeographic past and highly vulnerable to disturbance, Proc. Natl. Acad. Sci. USA 106 (2009) 11188–11193.

Caratini, C., Delibrias, G., Ragagopalan, G., Palaeomangroves of the Karnataka Coast, India, and their implications on Late Pleistocene sea level changes, in: Lain, K.P., Tiwari, R.S. (Eds.), Proceedings of a Symposium, 'Vistas in Indian Palaeobotany'. Palaeobotanist 38, 1990, pp. 370–378.

Chaloner, W.G., Muir, M., Spores and floras, in: Murchinson, D.G., Westall, T.S. (Eds.), Coal and Coal-Bearing Strata, Oliver and Boyd, Edinburgh, 1968, pp. 127–146.

Chauhan, M.S., Pokharia, A.K., Srivastava, R.K., Late Quaternary vegetation history, climatic variability and human activity in the Central Ganga Plain, deduced by pollen proxy records from Karela Jheel, India, Quat. Int. 371 (2015) 144–156.

Clarkson, C., Jones, S., Harris, C., Continuity and change in the lithic industries of the Jurreru Valley, India, before and after the Toba eruption, Quat. Int. 258 (2011) 165–179.

Cooling., E.N.G., *Pinus merkusii*, Fast-Growing Timber Tree of the Lowland Tropics 4, Department of Forestry, Oxford, 1968.

Corlett, R.T., The shifted baseline: prehistoric defaunation in the tropics and its consequences for biodiversity conservation, Biol. Conserv. 163 (2013) 13–21.

Dam, R.A.C., Fluin, J., Suparan, P., Van der Kaars, S., Palaeoenvironmental developments in the Lake Tonado area (N. Sulawesi, Indonesia) since 33,000 yr BP, Palaeogeogr. Palaeoclim. Palaeoecol. 171 (2001) 147–183.

Davis, M.B., Determination of absolute pollen frequency, Ecology 47 (1964) 310–311.

De Boer, B., Van de Wal, R.S.W., Bintanja, R., Lourens, L.J., Tuenter, E., Cenozoic global ice-volume and temperature simulations with 1-D ice-sheet models forced by benthic δ18O records, Ann. Glaciol. 51 (2010) 23–33.

Dubois, E., On *Pithecanthropus erectus*: A Transitional form Between Man and the Apes, J. Anthropol. Inst. Great Britain Ireland 25 (1896) 240–255.

Dupont, L.M., Wyputta, U., Reconstructing pathways of aeolian pollen transport to the marine sediments along the coastline of SW Africa, Quat. Sci. Rev. 22 (2003) 57–174.

Faegri, K., Iversen, J., Textbook of Pollen Analysis, second ed., Blackwell, Oxford, 1964.

Farooqui, A, Pattan, J.N., Parthiban, G., Srivastava, J., Palynological record of tropical rain forest vegetation and sea level fluctuations since 140 ka from sediment core, south-eastern Arabian Sea, Palaeogeogr. Paleoclim. Palaeoecol. 411 (2014) 95–109, doi:10.1016/j.palaeo.2014.06.020.

Farooqui, A., Ray, J.G., Farooqui, S.A., Tiwari, R.K., Khan, Z.A., Palyno-diversity, sea level and climate prior to 40 ka in Indian Peninsula, Kerala, Quat. Int. 213 (2010) 2–11.

Farooqui, A., Ray, J.G., Garg, A., An extinct species of *Basella* : pollen evidence from sediments (~80 ka) in Kerala, India, Grana 58 (2020) 399–407.

Flenley, J.R., The use of modern pollen rain samples in the study of the vegetational history of tropical regions, in: Birks, H.J.B., West, R.G. (Eds.), Quaternary Plant Ecology, Blackwell Scientific Publications, London, 1973, p. 13141.

Flenley, J.R., The Equatorial Rain Forest – A Geological History, Butterworths, London, 1979, p. 162.

Gathorne-Hardy, F.J., Harcourt-Smith, W.E.H., The super-eruption of Toba, did it cause a human bottleneck? J. Hum. Evol. 45 (2003) 227–230.

Ghosh, S., Guchhait, S.K., Characterization and evolution of laterites in West Bengal: implication on the geology of Northwest Bengal basin, Trans. Inst. Indian Geogr. 37 (2015) 93–119.

Gibbard, P.L., Boreham, S., Cohen, K.M., Moscariello, A., Global chronostratigraphical correlation table for the last 2.7 million years, Boreas 34 (2005) modified/updated 2007.

Goldhammer, J.G., Penfiel, S.R., Fire in the pine-grassland biomes of tropical and subtropical Asia, in: Goldammer, J.G. (Ed.), Fire in the Tropical Biota©, Springer-Verlag, Berlin, Heidelberg, 1990, pp. 45–62. 1990.

Gorog, A.J., Sinaga, M.H., Engstrom, M.D., Vicariance or dispersal? Historical biogeography of three Sunda shelf murine rodents (*Maxomys surifer, Leopoldamys sabanus* and *Maxomys whiteheadi*), Biol. J. Linnean Soc. Lond. 81 (2004) 91–109.

Govin, G., van der Beek, P., Najman, Y., Millar, I., Gemignani, L., Huyghe, P., Dupont-Nivet, G., Bernet, M., Mark, C., Wijbrans, J., Early onset and late acceleration of rapid exhumation in the Namche Barwa syntaxis, eastern Himalaya, Geology 48 (2020) https://doi.org/10.1130/G47720.1.

Grindrod, J.F., Late Quaternary mangrove pollen records from continental shelf and ocean cores in the north Australian-Indonesian region, in: Kershaw, P., David, B., Tapper, N., Penny, D, Brown, J. (Eds.), Bridging Wallace's Line: The Environmental and Cultural History and Dynamics of the Australian-Southeast Asian Region. Advances in Geoecology, 34, Catena Verlag, Reiskirchen, 2002, pp. 119–146.

Guleria, J.S., Neogene vegetation of peninsular India, Palaeobotanist 40 (1992) 285–331.

Haile, N.S., Postulated Late Cainozoic high sea levels in the Malay Peninsula, J. Malaysian Branch Royal Asiatic Soc. 48 (1975) 78–88.

Hamilton, R., Penny, D., Hall, T.L., Forest, fire & monsoon: investigating the long-term threshold dynamics of south-east Asia's seasonally dry tropical forests, Quat. Sci. Rev. 238 (1–11) (2020) 106334.

Haslam, M., Petraglia, M., Comment on "environmental impact of the 73 ka Toba super-eruption in South Asia" by M.A.J. Williams, S.H. Ambrose, S. van der Kaars, C. Ruehlemann, U. Chattopadhyaya, J. Pal, P.R. Chauhan Palaeogeogr. Palaeoclim. Palaeoecol. 284 (2009) 295–314, Palaeogeogr. Palaeoclim. Palaeoecol. 296 (2010) 199–203.

Hawks, J., Hunley, K., Lee, S.-H., Wolpoff, M., Population bottle necks and Pleistocene human evolution, Mol. Biol. Evol. 17 (2000) 2–22.

Heaney, L.R., A synopsis of climatic and vegetational change in Southeast Asia, Clim. Change 19 (1991) 53–61.

Hoojier, D.A., Mammalian evolution in the Quaternary of southern and eastern Asia, Evolution 3 (1949) 125–128.

Hoorn, C., Palynology of the Pleistocene glacial/interglacial cycles of the Amazon Fan (Holes 944A and 946A, in: Flood, R.D., Piper, D.J.W., Klaus, A., Peterson, L.C. (Eds.), Proceedings of the Ocean Drilling Program, Scientific Results 155, Texas A&M University, Houston, 1997, pp. 397–409.

Hunt, C.O., Gilbertson, D.D., Rushworth, G., A 50,000-year record of Late Pleistocene tropical vegetation and human impact in lowland Borneo, Quat. Sci. Rev. 37 (2012) 61–80.

Husson, L., Boucher, F.C., Sarr, A.-C., Sepulchre, P., Sri Yudawati, C., Evidence of Sundaland's subsidence requires revisiting its biogeography, J. Biogeogr. 47 (2019) 843–853.

Hyodo, M., Sunata, W., Susanto, E.E., A long-term geomagnetic excursion from Plio-Pleistocene sediments in Java, J. Geophys. Res. 97 (1992) 9323–9335.

Janssen, C.R., Recent pollen spectra from the deciduous and coniferous-deciduous forests of North-eastern Minnesota: a study in pollen dispersal, Ecology 47 (1966) 804–825.

Johnson, C.N., Ecological consequences of Late Quaternary extinctions of megafauna, Proc. R. Soc. B 276 (2009) 2509–2519.

Jones, S.C., Local- and regional-scale impacts of the 74 ka Toba supervolcanic eruption on hominin populations and habitats in India, Quat. Int. 258 (2012) 100–118.

Kennedy, K.A.R., Deraniyagala, S.U., Fossil remains of 28,000-year-old hominids from Sri Lanka, Curr. Anthropol. 30 (1989) 394–399.

Khan, A.M., Spicer, R.A., Spicer, T.E., Bera, S., Occurrence of *Shorea roxburghi* ex C. F. Gaertner (Dipterocarpaceae) in the Neogene Siwalik forests of eastern Himalaya and its biogeography during the Cenozoic of Southeast Asia, Rev. Palaeobot. Palynol. 233 (2016) 236–254.

Lane, C.S., Chorn, B.T., Johnson, T.C., Ash from the Toba supereruption in Lake Malawi shows no volcanic winter in East Africa at 75 ka, Proc. Natl. Acad. Sci. (2013) 1–5, doi:10.1073/pnas.1301474110.

Larick, R., Ciochon, R.L., Zaim, Y., Sudijono, S., Riza, Y., Aziz, F., Reagan, M., Heizler, M., Early Pleistocene 40Ar/39Ar ages for Bapang Formation hominins, Central Java, Indonesia, Proc. Natl. Acad. Sci. USA 98 (2001) 4866–4871.

Louys, J., Limited effect of the Quaternary's largest super-eruption (Toba) on land mammals from Southeast Asia, Quat. Sci. Rev. 26 (2012) 3108–3117.

Louys, J., Curnoe, D., Tong, H., Characteristics of Pleistocene megafauna extinctions in Southeast Asia, Palaeogeogr. Palaeoclim. Palaeoecol. 243 (2007) 152–173.

Louys, J., Meijaard, E., Palaeoecology of Southeast Asian megafauna-bearing sites from the Pleistocene and a review of environmental changes in the region, J. Biogeogr. 37 (2010) 1432–1449.

Louys, J., Roberts, P., Environmental drivers of megafauna and hominin extinction in Southeast Asia, Nature (2020) 1–5 doi.org/, doi:10.1038/s41586-020-2810-y.

Luo, C., Haberle, S., Zheng, Z., Xiang, R., Chen, C., Lin, G., Sazal, K., Environmental changes in the north-east Sunda region over the last 40 000 years, J. Quat. Sci. 34 (2019) 245–257.

Maloney, B.K., McCormac, F.G., A 30,000 year pollen and radiocarbon record from highland Sumatra as evidence for climatic change, Radiocarbon 37 (1995) 181–190.

Mandang, Y.I., Kagemori, N., A fossil wood of Dipterocarpaceae from Pliocene deposit in the West Region of Java Island, Indonesia, Biodiversitas 5 (2004) 28–35.

Mark, D.F., Petraglia, M., Smith, V.C., Morgan, L.E., Barfod, D.N., Ellis, B.S., Pearce, N.J., Pal, J.N., Korisettar, R., A high-precision 40Ar/39Ar age for the Young Toba Tuff and dating of ultra-distal tephra: forcing of Quaternary climate and implications for hominin occupation of India, Quat. Geochronol. 21 (2014) 90e103.

Medway, Lord, The Quaternary mammals of Malaysia: a review, in: Ashton, P.S., Ashton, M. (Eds.), The Quaternary Era in Malaysia, Geography Department University of Hull, Hull, 1972, pp. 63–98. Miscellaneous Series 13.

Merckx, V.S.F.T., Hendriks, K.P., Beentjes, K.K., et al., Evolution of endemism on a young tropical mountain, Nature 524 (2015) 347–350.

Morley, R.J., Development and vegetation dynamics of a lowland ombrogenous peat swamp in Kalimantan Tengah, Indonesia, J. Biogeogr. 8 (1981) 383–404.

Morley, R.J., A palaeoecological interpretation of a 10,000 year pollen record from Danau Padang, Central Sumatra, Indonesia, J. Biogeogr. 9 (1982) 151–190.

Morley, R.J., Tertiary stratigraphic palynology in Southeast Asia: current status and new directions, Geol. Soc. Malaysia Bull. 28 (1991) 1–36.

Morley, R.J., 1996. Biostratigraphic characterisation of systems tracts in Tertiary sedimentary basins. In: Proceedings of the International Symposium on Sequence Stratigraphy in SE Asia, IPA Jakarta, pp. 49–71.

Morley, R.J., Palynological evidence for Tertiary plant dispersals in the Southeast Asian region in relation to plate tectonics and climate, in: Hall, R., Holloway, J.D. (Eds.), Biogeography and Geological Evolution of SE Asia, Backhuys, Leiden, 1998, pp. 211–234.

Morley, R.J., Origin and Evolution of Tropical Rain Forests, Wiley, Chichester, 2000, p. 362.

Morley, R.J., Cenozoic ecological history of South East Asian peat mires based on the comparison of coals with present day and Late Quaternary peats, J. Limnol. 72 (2013) 36–59.

Morley, R.J., Assembly and division of the South and South-East Asian flora in relation to tectonics and climate change, J. Trop. Ecol. (2018) 209–234, doi:10.1017/S0266467418000202.

Morley, R.J., Dung, B.V., Tung, N.T., Kullman, A.J., Bird, R.T., Van Kieu, N., Chung, N.H., High-resolution Palaeogene sequence stratigraphic framework for the Cuu Long Basin, offshore Vietnam, driven by climate change and tectonics, established from sequence biostratigraphy, Palaeogeogr. Palaeoclim. Palaeoecol. 530 (2019) 113–135.

Morley, R.J., Flenley, J.R., Late Cainozoic vegetational and environmental changes in the Malay Archipelago, in: Whitmore, T.C. (Ed.), Biogeographical Evolution of the Malay Archipelago, Oxford Monographs on Biogeography 4, Oxford Scientific Publications, Oxford, vol. 4, 1987, pp. 50–59.

Morley, R.J., Hasan, S.S., Morley, H.P., Jais, J.H.M., Mansor, A., Raziken Aripin, M., Hafiz Nordin, M., Helmi Rohaizar, M., Sequence biostratigraphic framework for the Oligocene to Pliocene of Malaysia: High-frequency depositional cycles driven by polar glaciation, Palaeogeogr. Palaeoclim. Palaeoecol. 561 (2021) 110058.

Morley, R.J., Morley, H.P., 2010. Neogene climate history of the Makassar Straits with emphasis on the Attaka Field. In: Proceedings of the 34th Indonesian Petroleum Association IPA10-G-208.

Morley, R.J., Morley, H.P., Mid Cenozoic freshwater wetlands of the Sunda region, J. Limnol. 7 (2013) 218–235.

Morley, R.J., Morley, H.P., Swiecicki, T., 2016. Mio-Pliocene palaeogeography, uplands and river systems of the Sunda region based on mapping within a framework of VIM depositional cycles. In: IPA Fortieth Annual Convention & Exhibition. Jakarta IPA16-506-G.

Morley, R.J., Morley, H.P., Wonders, A.A.H.W., Sukarno, Van Der Kaars, S., 2004. Biostratigraphy of Modern (Holocene and Late Pleistocene) Sediment Cores from Makassar Straits, Deepwater and Frontier Exploration in Asia & Australasia Proceeding. Jakarta December 2004.

Morley, R.J., Morley, H.P., Zaim, Y., Huffman, O.F., Palaeoenvironmental setting of Mojokerto *Homo erectus*, the palynological expressions of Pleistocene marine deltas, open grasslands and volcanic mountains in East Java, J. Biogeogr. (2020) 1–18.

Morley, R.J, Salvador, P., Challis, M.L., Morris, W.R., Adyaksawan, I.R., Sequence biostratigraphic evaluation of the North Belut Field, West Natuna Basin. In: Proceedings of the Indonesian Petroleum Association 31st Annual convention, IPA07-G-120, 2007.

Morwood, M.J., O'Sullivan, P., Susanto, E.E., Aziz, F., Revised age for Mojokerto 1, an early *Homo erectus* cranium from East Java, Indonesia, Aust. Archaeol. 57 (2003) 1–4.

Muller, J., A palynological study of the genus *Sonneratia* (Sonneratiaceae), Pollen Spores 11 (1969) 223–298.

Muller, J., Palynological evidence for change in geomorphology, climate and vegetation in the Mio-Pliocene of Malesia, in: Ashton, P.S., Ashton, M. (Eds.), The Quaternary Era in Malaysia, Geography Department, University of Hull, Misc. Ser 13, Hull, 1972, pp. 6–34.

Nguyen, N., Duffy, B., Shulmeister, J., Quigley, M., Rapid Pliocene uplift of Timor, Geology 41 (2013) 179–182.

Nugraha, A.M.S., Hall, R., Late Cenozoic palaeogeography of Sulawesi, Indonesia, Palaeogeogr. Palaeoclim. Palaeoecol. 490 (2017) 20–217.

Parham, P.R., Late Cenozoic relative sea-level highstand record from Peninsular Malaysia and Malaysian Borneo: implications for vertical crustal movements, Bull. Geol. Soc. Malaysia 62 (2016) 91–115 https://doi.org/10.7186/bgsm62201612.

Pascal, J.P., Explanatory Booklet on the Forest Map of South India. Forest Department and the French Institute of Pondicherry, Karnataka, Inst Fr Pondichery, Trav Sec Sct Tech, Series 18, 1986.

Patnaik, R., Diet and habitat changes among Siwalik herbivorous mammals in response to Neogene and Quaternary climate changes: an appraisal in the light of new data, Quat. Int. 371 (2015) 232–243.

Patnaik, R., Nanda, A.C., et al., Early Pleistocene mammalian faunas of India and evidence of connections with other parts of the world, in: Fleagle, J.G., et al. (Eds.), Out of Africa I: The First Hominin Colonization of Eurasia, Vertebrate Paleobiology and Paleoanthropology, Springer Science+Business Media B.V, Dordrecht, 2010, pp. 129–143.

Peck, R.M., Pollen budget studies in a small Yorkshire catchment, in: Birks, H.J.B., West, R.G. (Eds.), Quaternary Plant Ecology, Blackwell Scientific Publications, London, 1973, pp. 43–60.

Penny, D., A 40,000 year palynological record from north-east Thailand; implications for biogeography and palaeo-environmental reconstruction, Palaeogeogr. Palaeoclim. Palaeoecol. 171 (2001) 97–128.

Polhaupessy, N., 1990. Late Cenozoic palynological studies on Java. Unpublished PhD thesis, University of Hull. 339 pp.

Posamentier, H., Lowstand alluvial bypass systems: incised vs. unincised, AAPG Bull. 85 (2001) 1771–1793.

Prabhu, C.N., Shankar, R., Anupam, K., Taieb, M., Bonnefille, R., Vidal, L., Prasad, S., A 200-ka pollen and oxygen-isotopic record from two sediment cores from the eastern Arabian Sea, Palaeogeogr. Palaeoclim. Palaeoecol. 214 (2004) 309–321.

Premathilake, R., Risberg, J., Late Quaternary climate history of the Horton Plains, central Sri Lanka, Quat. Sci. Rev. 22 (2003) 1525–1541.

Quade, J.J., Cerling, T.E., Bowman, J.R., Development of Asian monsoon revealed by marked ecological shift during the latest Miocene in northern Pakistan, Nature 342 (1989) 163–166.

Quamar, M.F., Bera, S.K., Pollen records related to vegetation and climate change from northern Chhattisgarh, central India during the late Quaternary, Palynology 41 (2017) 17–30.

Quamar, M.F., Kar, R., Prolonged warming over the last ca. 11,700 cal years from the central Indian Core Monsoon Zone: pollen evidence and a synoptic overview, Rev. Palaeobot. Palynol. 276 (2020) 104159.

Raes, N, Cannon, C.H., Hijmans, J., Piessense, T., Saw, L.G., van Welzen, P.C., Slik, J.W.F..

Rampino, M.R., Self, S., Historic eruptions of Tambora (1815), Krakatau (1883), and Agung (1963), their stratospheric aerosols, and climatic impact, Quat. Res. 18 (1982) 127–143.

Ratnam, J, Tomlinson, K.W., Rasquinha, D.N., Sankaran, M., Savannahs of Asia: antiquity, biogeography, and an uncertain future, Philos. Trans. R. Soc. B 371 (2016) 20150305.

Riding, J.B., Scott, A.C., Collinson, M., A biography and obituary of William G. Chaloner FRS (1928–2016), Palynology 44 (2017) 127–166.

Rizal, Y., Westaway, K.E., Zaim, Y., van den Bergh, G.D., Bettis, E.A., Morwood, M.J., Huffman, O.F., Grün, R., Joannes-Boyau, R., Bailey, R.M., Sidarto, Westaway, Kurniawan, M.C., Moore, I., Storey, M.W., Aziz, M., Suminto, F., Jian-xin Zhao, Aswan, Sipola, M.E., Larick, R., Zonneveld, J.-P., Scott, R., Putt, S., Ciochon, R.L., Last appearance of *Homo erectus* at Ngandong, Java, 117,000–108,000 years ago, Nature (2019) 1–5 2019, doi:10.1038/s41586-019-1863-2.

Roberts, P., Delson, E., Miracle, P., Ditchfield, P., Roberts, R.G., Jacobs, Z., Blinkhorn, J., Ciochon, R.L., Fleagle, J.G., Frost, S.R., Gilbert, C.C., Gunnell, G.F., Harrison, T., Korisettar, R., Petraglia, M.D., Continuity of mammalian fauna over the last 200,000 y in the Indian subcontinent, Proc. Natl. Acad. Sci. 111 (2016), doi:10.1073/pnas.1323465111.

Sandom, C., Faurby, S., Sandel, B., Svenning, J.C., Global late Quaternary megafauna extinctions linked to humans, not climate change, Proc. R. Soc. Lond. B Biol. Sci. 28 (1787) (2014) 20133254.

Sarr, A.-C., Husson, L., Sepulchre, P., Pastier, A.-M., Pedoja, K., Elliot, M., Arias-Ruiz, C., Solihuddin, T., Aribowo, S., Susilohadi, Subsiding Sundaland, Geology 47 (2019) 119–122 https://doi.org/10.1130/G45629.1.

Singh, G., Wasson, R.J., Agrawal, D.P., Vegetational and seasonal climatic changes since the last full glacial in the Thar Desert, northwestern India, Rev. Palaeobot. Palynol. 64 (1972) 351–358.

Storm, P., Aziz, F., de Vos, J., Kosasih, D., Baskoro, S., van den Hoek Ostende, L.W., Ngaliman, Late Pleistocene *Homo sapiens* in a tropical rainforest fauna in East Java, J. Hum. Evol. 49 (2005) 536–545.

Stuart, A.J., Late Quaternary megafaunal extinctions on the continents: a short review, Geol. J. 50 (2015) 338–363.

Stuijts, I., Newsome, J.C., Flenley, J.R., Evidence for late Quaternary vegetational change in the Sumatran and Javan highlands, Rev. Palaeobot. Palynol. 55 (1988) 207–216.

Sukumar, R., Ramesh, R., Pant, R.K., Ragagopalan, G., A $\delta^{13}C$ record of late Quaternary climate change from tropical peats in Southern India, Nature 364 (1993) 703–706.

Swisher III, C.C., Curtis, J.H., Jacob, T., Getty, A.G., Age of the earliest known hominids in Java, Indonesia, Science 263 (1994) 1118–1121.

Tauber, H., Investigations of the mode of pollen transfer in forested areas, Rev. Palaeobot. Palynol. 3 (1967) 277–286.

Tougard, C., Montuire, S., Pleistocene paleoenvironmental reconstructions and mammalian evolution in South-East Asia: focus on fossil faunas from Thailand, Quat. Sci. Rev. 25 (2006) 126–141.

Traverse, A., Production, dispersal and sedimentation of spores/pollen., Paleopalynology., Unwin Hyman, Boston, 1988, pp. 375–427.

Tripathi, S., Thakur, B., Nautiyal, C.M., Bera, S.K., Floristic and climatic reconstruction in the Indo-Burma region for the last 13,000 cal. yr: a palynological interpretation from the endangered wetlands of Assam, northeast India, Holocene 30 (2019) 315–331.

Trivedi, A., Saxena, A., Chauhan, M.S., Sharma, A., Farooqui, A., Nautiyal, C.M., Yao, Yi-Feng, Wang, Yu-Fei, Li, Cheng-Sen, Tiwari, D.P., Vegetation, climate and culture in Central Ganga plain, India: a multi-proxy record for Last Glacial Maximum, Quat. Int. 507 (2019) 134–147.

Van Campo, E., Monsoon fluctuations in two 20,000 yr B.P. oxygen isotope/pollen records off southwest India, Quat. Res. 26 (1986) 376–388.

van den Bergh, G., de Vos, J., Sondaar, P., Late Quaternary palaeogeography of mammal evolution in the Indonesian Archipelago, Palaeogeogr. Palaeoclim. Palaeoecol. 171 (2001) 385–408.

van der Kaars, S., Penny, D., Tibby, J., Fluin, J., Dam, R., Suparan, P., Late Quaternary palaeoecology, palynology and palaeolimnology of a tropical lowland swamp: Rawa Danau, West Java, Indonesia, Palaeogeogr. Palaeoclim. Palaeoecol. 171 (2001) 185–212.

van der Kaars, W.A., Palynology of eastern Indonesian marine piston-cores: a Late Quaternary vegetational and climatic record for Australasia, Palaeogeogr. Palaeoclim. Palaeoecol. 85 (1991) 239–302.

van der Kaars, W.A, Bassinot, F., De Deckker, P., Guichard, F., Changes in monsoon and ocean circulation and the vegetation cover of southwest Sumatra through the last 83,000 years: the record from marine core BAR94-42, Palaeogeogr. Palaeoclim. Palaeoecol. 296 (2010) 52–78.

van der Kaars, W.A., Dam, M.A.C., A 135,000-year old record of vegetation and climatic change from the Bandung area, West Java, Indonesia, Palaeogeogr. Palaeoclim. Palaeoecol. 117 (1995) 55–71.

Van der Kaars, S., Williams, M.A.J., Bassinot, F., Guichard, F., Moreno, E., Dewilde, F., Cook, E.J., The influence of the 73 ka Toba super-eruption on the ecosystems of northern Sumatra as recorded in marine core BAR94-25, Quat. Int. 258 (2012) 45–53.

Van Gorsel, J.T.(Han), Lunt, P., Morley, R.J., Introduction to Cenozoic biostratigraphy of Indonesia-SE Asia, Berita Sedimentol. 29 (2014) 6–40.

Vishnu-Mittre, Gupta, H.P., Origin of Shola forest in the Nilgiris, South India, Palaeobotanist 19 (1971) 110–114.

Von Koenigswald, G.H.R., Ein fossiler Hominide aus dem Alpleistocän Ostjavas. De Ingenieur in Nederlandsch-Indië, Mijnbouw Geol. De Mijningenieur 4 (1936) 149–157.

Von Post, L., Einiges 'idschwedischen Quellmoore, Bull. Geol. Inst. Univ. Upsala 15 (1916) 219–278.

Waldram, M.S., Bond, W.J., Stock, W.D., Ecological engineering by a Mega-Grazer: white rhino impacts on a South African Savanna, Ecosystems 11 (2008) 101–112.

Wang, X.-M., Sun, X.-J., Wang, P.-X., Stattegger, K., Vegetation on the Sunda Shelf, South China Sea, during the Last Glacial Maximum, Palaeogeogr. Palaeoclim. Palaeoecol. 278 (2009) 88–97.

Werner, W.L., Pines and other conifers in Thailand – a Quaternary relic? J. Quat. Sci. 12 (1997) 451–454.

Westaway, K.E., Morwood, M.J., Roberts, R.G., Rokus, A.D., Zhao, J.-x., Storm, P., Aziz, F., van den Bergh, G., Hadi, P., Jatmiko de Vos, J., Age and biostratigraphic significance of the Punung Rainforest Fauna, East Java, Indonesia, and implications for *Pongo* and *Homo*, J. Hum. Evol. 53 (2007) 709–717.

Whitmore, T.C., Tropical Rain Forests of the Far East, Clarendon Press, Oxford, 1975, p. 282.

Williams, M.A.J., Ambrose, S.H., van der Kaars, S., Ruehlemann, C., Chattopadhyaya, U., Pal, J., Chauhan, P.R., Environmental impact of the 73 ka Toba super-eruption in South Asia, Palaeogeogr. Palaeoclim. Palaeoecol. 284 (2009) 295–314.

Wuster, C.M., Bird, M.I., Bull, I.D., Creed, F., Bryant, C, Dungait, J.A.J., Paz, V., Forest contraction in north equatorial Southeast Asia during the Last Glacial Period, Proc. Natl. Acad. Sci. USA 107 (2014) 155008-155011.

Yang, F.-C., Grote, P.J., Riverine vegetation and environments of a Late Pleistocene river terrace, Khorat Plateau, Southeast Asia, Palynology 42 (2017) 158–167 https://doi.org/10.1080/01916122.2017.1296044.

Yulianto, E., Sukapti, W.S., Rahardjo, A.T., Noeradi, D., Siregar, D.A., Suparman, P., Hirakawa, K., Mangrove shoreline responses to Holocene environmental change, Makassar Strait, Indonesia, Rev. Palaeobot. Palynol. 131 (2004) 251–268.

Zeilinski, G.A., Mayewski, P.A., Meeker, L.D., Whitlow, S., Twickler, M.S., Potential atmospheric impact of the Toba mega-eruption ~71,000 years ago, Geophys. Res. Lett. 23 (1996) 837–840.

CHAPTER

2 Contemporary dynamics and Holocene extent of glaciers in the Himalayas

Milap Chand Sharma, Ishita Manna, Elora Chakraborty
Centre for the Study of Regional Development, JNU, New Delhi, India

2.1 Introduction

Glacier and glacier studies in India span over 200 years (Hodgson, 1822). Ironically, Hodgson in 1817 had considered Gaumukh to be a snowfield, covered with rubbish and rocks. It was not until the late 19th century that scientific expeditions were led to these locations (Greisbach, 1891). The staggering Himalayan altitudes not only provide an awe-inspiring aesthetic scenery, but are also endowed with thousands of glaciers, sources of perennial rivers within this mountain system and beyond in the foreland basins. Compounded by the regional tectonics and climate, processions of many cycles of glacial and interglacial events have shaped the landscape that exists today, be it in or outside the mountains. Glacier inventories for the Himalayas suggest a total of 32,392 glaciers with 9409 permanent snow fields and glacierets (Sharma et al., 2013) which reserve a vast pool of information pertaining not only to the contemporary dynamics but also of the geological past.

2.2 Scope

Not only the successive glacial episodes have helped in evolving the jagged terrain but also produced millions of tons of mineral-rich sediments over millions of years that make the foreland as some of the most fertile regions in the world. Millions of years of ongoing earth's surface forming processes in the mountain landscape have evolved precipitous slopes, leading to immediate down-slope mobilization of unconsolidated earth materials upon changes in the local base-level or the regional climate pattern. Relief is immense, and so is the overland flow of the snow-melt and downpour. Such relief and steep gradient in turn have obliterated much of the past records that could have been of immense significance in reconstructing the past climate-glacier events. Earnestly, studies on the past glacial reconstruction started by de Terra and Peterson (1939), categorizing glacial phases in Kashmir in line with the four-fold European model. It was not until the new techniques were invented and used for a chronological reconstruction to determine a near-precise timing and magnitude, that the regional picture of glaciation started emerging (Owen et al., 1992, 2001; Sharma and Owen, 1996; Singh and Agrawal, 1976). Multiple dating methods such as OSL and CRN are used in the glacier studies, having a limited scope of ^{14}C because of the absence of organic content in the sedimentary deposits in the Higher Himalaya.

Holocene Climate Change and Environment. DOI: https://doi.org/10.1016/B978-0-323-90085-0.00011-5

Although the phases have been dated to many 10s of kilo years over the past three decades, we would only discuss the dated glacial episodes of the Holocene in the present chapter.

2.3 Methodology and processes

Similar to the other mountain regions of the world, the Indian Himalaya has also undergone a substantial loss in ice-cover over the past century and a half. In accordance with the global standards in cryosphere research, we have built our database from the satellite products of similar spatial and spectral scale in order to keep the margin of error the least for the present century. An inventory of glaciers within the states of Himachal Pradesh and Uttarakhand with regard to multiple geomorphologic and geographic parameters has been created with repeated observations using optical satellite imageries within the years 2000–2001, 2016, and 2019, respectively. It may be noted that the changes in cryosphere would occur on account of changes in the climatic parameters. These changes bring in positive and negative impact on the Equilibrium-Line Altitude (ELA), thus, in effect, an expansion or recession. ELA, demarcating the boundary between the accumulation and ablation zones is squarely related to the Mass Balance in glaciers. Using geomatic methods, ELAs demarcated for glaciers exceeding 10 sq km in size in the catchments of Beas, Chenab, Ravi and Sutlej in Himachal Pradesh; Bhagirathi, Yamuna, Alaknanda, and Kaliganga, respectively in Uttarakhand. The considered ELA calculation methods include, Median Elevation in Glaciers (MEG), Accumulation Area Ratio Method (AAR). We have also taken help of the repeat photography method (Schmidt and Nuesser, 2009) and illustrations for an easy comprehension of the loss that have been witnessed over a century in the study region. Our analysis reveals that not all the glaciers are retreating, as flagged elsewhere. Such variability in change detection analysis still remains to be scientifically ascertained.

Climatologic data has been sourced from TerraMODIS/ AquaMODIS Land Surface Temperature Products (2000–2019) as well as Tropical Rainfall Measurement Mission (1998–2014) and Global Precipitation Measurement (2000–2019) for precipitation estimates for each basin. The assessments of long term climatologic estimates have been accomplished from gridded CRU data until 1990. Monthly mean values have been ascertained to study trends in the datasets with respect to each of the valleys concerned.

Miyar valley of the Chandrabhaga catchment and Upper Bhagirathi Basin of the Ganga catchment were studied for their past glacial stages since the initiation of Holocene. Field mapping, sedimentological differentiation, satellite remote-sensing, relative and absolute age determination techniques have been used in order to arrive to meaningful conclusions regarding the local climatic past.

2.4 Fieldwork and sampling techniques

Since early 2000, repeated field trips were carried out in Miyar (H.P.) and Bhagirathi (Uttarakhand) subwatersheds to assess the frontal retreat, radar based discharge dynamics and debris-ice melt relationship of the glaciers. Since 2016, a steady stream of data has been acquired for Miyar and Bhagirathi basins pertaining to glacier thickness using Ground Penetrating Radar, snout mapping using Total station and Differential Global Positioning System. Regular monitoring of the trunk and tributary glaciers have been carried out with differing orientations to understand the effect of micro regional forcings within the complicated topography of these two basins. Black Carbon concentration measurement from Menthosa glacier and village downstream in the Miyar Valley helped ascertain black carbon and ice melt relationships. Enquiries into the Holocene past have been conducted using the Thermo Luminescence-Optically Simulated Luminescence (TL-OSL) and Cosmogenic Radionuclide (CRN) from Lacustrine and Morainic deposits from the Miyar valley and Upper Bhagirathi Basin.

2.5 Observations and results

2.5.1 Contemporary dynamics in glaciers of the Western and Eastern Himalayas

Tectonics has played the greatest role in the manifestation of varying climatic regimes across the HKH Mountains. Until about 120 Ma BP, most of the presently known 'Tibetan Plateau' was below the mean sea level; validated by the presence of marine fossil containing limestone in the western and central Tibet (Searle, 2013). The collision of the Indian Plate with the Eurasian Plate was evidenced with the remarkable decrease in the rates of movement of the Indian plate from 100–112 mm/yr to 45–65 mm/yr at approximately 55–35 Ma BP (Copley et al., 2010; Molnar and Stock, 2009; Molnar and Tapponier, 1975; Patriat and Achache, 1984). The differing magnitude of convergence exerted by the Indian *terrane* caused folding of crust along the range to the heights that we witness today. Thus, the environmental condition for the formation of glaciers was set in the otherwise unlikely climate of the tropics and subtropics for its existence or for that matter, the very sustenance.

The Himalayas, marking the northernmost boundary of the Indian terrane, stretch over an extent of approximately 2400 km from West to East, having 10 of the world's 14 grandest peaks exceeding 8000 m above mean sea level. Formed due to the collision and subduction of the Indian terrane underneath the Eurasian Plate, the Himalayan landmass crumpled to form complicated sets of ridges and mountain systems with each event of crust upheaval due to folding. The magnificent and complex topographical setting of the Himalayas renders various micro climatic regimes. Each of these micro climatic regimes reflects characteristic glacier behavior, considering the enormous west to east extent of the Himalayas. The multiple watersheds and sub watersheds thus created bear a source of perennial supply of freshwater, vital to the lives of millions of North Indians, across the breadth of the country.

The Himalaya-Karakoram region has continued to play a determining role in the regional climate (Benn and Owen, 1998). The Himalaya-Tibet Orogen because of the high altitude, undulating relief, vast expansion and strategic location between two dominant regional climatic systems (the South West monsoon and the Mid-latitude Westerlies), play a significant role in modulating the local, regional and global climate gradients over different time scales. Local climate in the Western Himalayan region is either influenced by the Indian Summer Monsoon or the Mid latitude westerlies, while, the Eastern Himalayas reflect a tri junction of differing climatic regimes of the Mid Latitude Westerlies, Indian Summer Monsoons, and the Easterlies, respectively (Fig. 2.1). The role of the Polar vortex ingression has never been addressed although a huge glacier cover in the Karakoram may hold the key for such an assessment.

2.5.1.1 *Glaciers of the North Western Himalayan region*

The Polar and the Mountain glaciers are genetically different from each other. Due to the large variation in topography in the Himalayas; a large array of glacier types exists. The states of Himachal

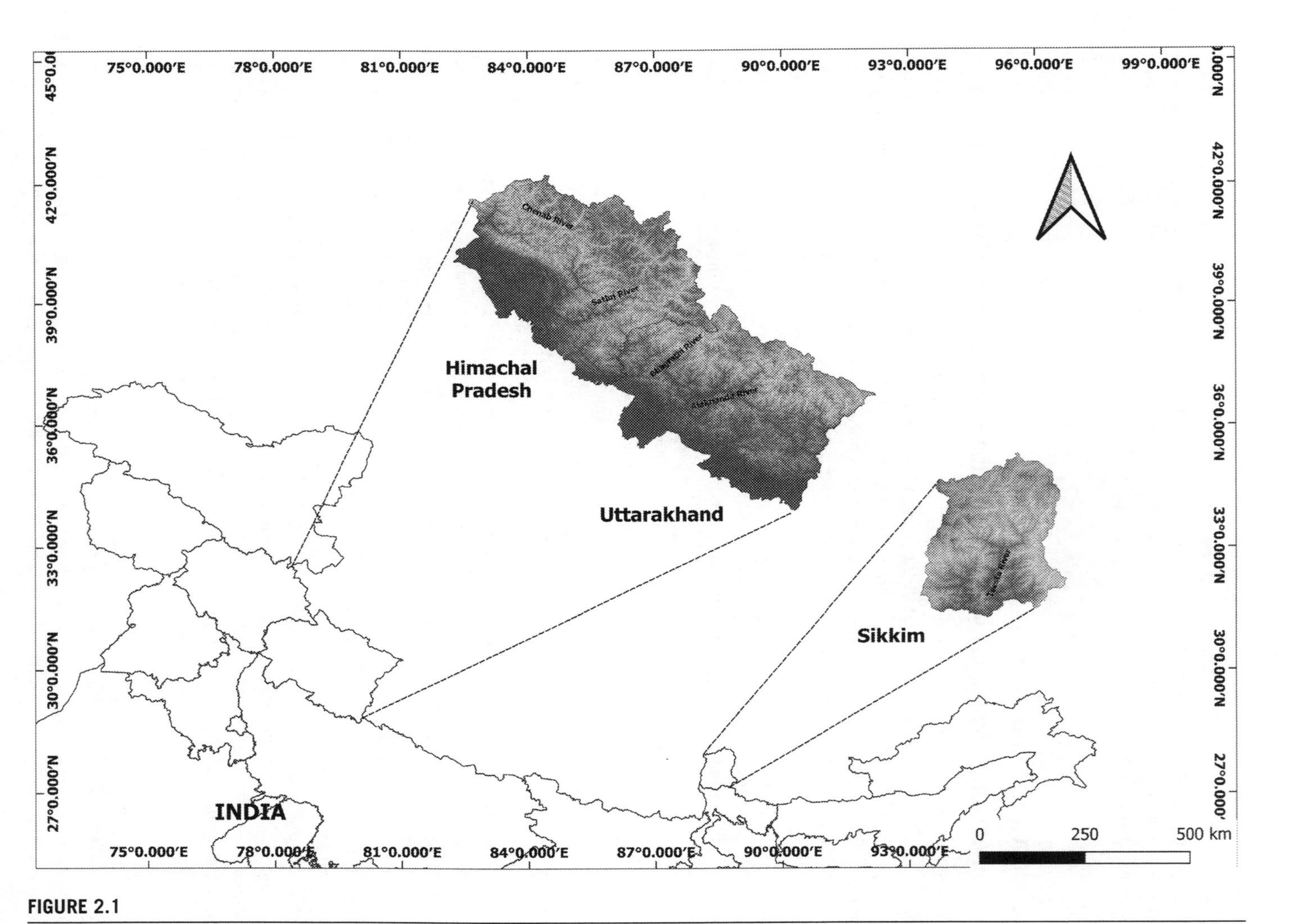

FIGURE 2.1
Location of the enquiry into glacial dynamics: Himachal Pradesh, Uttarakhand in the North Western Himalaya, and Sikkim in Eastern Himalaya.

Pradesh and Uttarakhand contain 2189 and 807, respectively, identifiable glaciers of varying dimensions (Fig. 2.2). The Chenab or Chandrabhaga basin contains the maximum number of glaciers among the basins evaluated, that is, 813, in the Himalayan states of Himachal Pradesh and Uttarakhand. Ruggedness of the topography, orientation of ranges and high altitudes render the Chenab basin haven for glacier formation. The health of these glaciers is sustained by the influx of solid precipitation during the long frigid winter months and even during the summers in the higher reaches.

Modern geo-informatics techniques have added teeth to glaciological studies in identifying change in cover time and space with much ease (Ajai 2018). The availability of crisp remotely sensed optical satellite imageries and Digital Elevation Models have furthered the possibility of delineating glacier boundaries notwithstanding errors in digital and manual demarcation of permanent ice beneath debris. The importance of such studies has emerged in response to the need for regional assessment of change in glaciers with varying climate conditions. A systematic enquiry into individual glacier dynamics reveals its relationship with climatic parameters vis a vis geomorphology. The glacier inventory in this study consists of the vectorization of all perennial ice masses irrespective of size, debris cover and other factors; generally for the end of ablation season to gauge the true magnitudes (Table 2.1 & 2.2).

The glaciers in Himachal Himalaya have been classified under 5 standard categories; namely (a) Cirque glaciers, (b) Hanging glaciers, (c) Mountain glaciers, (d) Mountain Rock type, and (e) Valley glaciers, respectively (Raina & Srivastava, 2009). The Mountain and Valley Glaciers are the most commonly observed glacier types, with 1663 and 381 glaciers in respective categories. Mountain Rock and Cirque type glaciers individually occupy the least area as well as number within the state with 12 and 24 glaciers, respectively. The Garhwal Himalaya in Uttarakhand also contain a large number of Mountain Niche and Valley type glaciers with 300 and 255 glaciers, respectively. The state contributes to considerably large amount of glacier cover in the Western Himalaya, with a break up to 197 Hanging glaciers, 08 Mountain Basin glaciers, 06 Mountain glacierets, and 05 Rock glaciers, respectively.

Among the select watershed of the North Western Himalaya, Chenab contained the largest area under glacier cover at approximately 1204 km^2 in the year 2000. The least area under glacier cover was exhibited the tiny watershed of Tsarap Chhu, a tributary to the Indus within the cold desert conditions, with a cover of approximately 58 km^2 for the year 2000. Among the Indian Summer Monsoon (ISM) dominated basins, Yamuna had a total glacier cover of approximately 98 km^2 in the year 2000. Such differences in glacier cover of Chenab and Yamuna can be accredited to the different climatic regimes, and the differences in altitudes in association with other micro regional controls such as aspect, slopes and percentage of debris cover.

2.5.1.1.1 Glaciers of the Yamuna basin (Ajai 2018)

There is an indisputable dependence of glacier dynamics on its individual climatic regime and any temporary perturbations may not impact a glacier. However, sustained changes like consecutive positive summer temperature anomalies may affect mass balance adversely, causing rapid reduction of glacier mass, if not compensated well with solid precipitation in accumulation zones in months when accumulation prevails. The Yamuna basin echoes such tragedy with a percentage reduction of 2.13% in glacier cover between 2000 and 2019; and therefore, this basin demonstrates vulnerability of its glacial cover. With only two glaciers exceeding 10 km^2 in size, the percentage reduction of these two is among the least (1.05%) in 2000–2019. Smaller glaciers are considered to have shorter lag in glacier response to climatic perturbations; as evidenced by a higher percentage loss of 4.06% in the Yamuna basin in

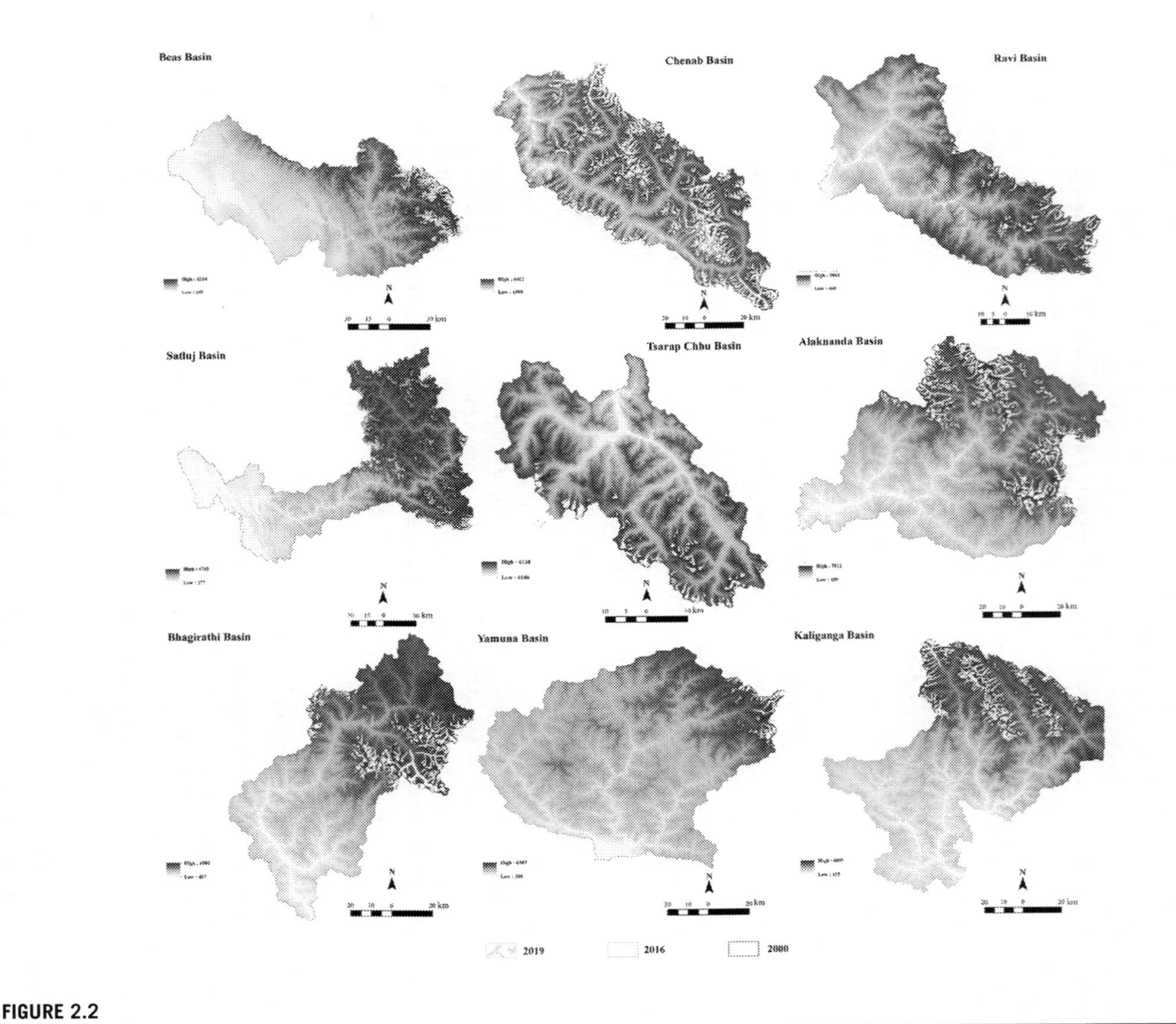

FIGURE 2.2
Basin-wise glacier distribution in the Himachal & Garhwal Himalaya (2000–2019).

2000–2019. On a micro analysis, most of the glacier cover reduction is traced with the glaciers bearing East, South East, and South West orientations, highlighting the control of aspect in glaciers' sustenance.

Most of the glacierized area in the Yamuna watershed is situated within a slope angle category of 15°-20°; also registering one of the lowest values of percentage changes (1.29%) in glacier cover between 2000 and 2019. Highest changes in glacierized area has been found with glaciers bearing a mean angle of 30°-35°, a change of 5.92%; mostly identified as Glacierets and Mountain Niche Glaciers with faster response times to climatic variabilities notwithstanding the role of aspect.

The larger valley glaciers are mostly covered with debris over more than 50% of the surface area. Supra glacial debris has a dubious role to play in controlling glacier retreat. In many cases, expansive debris cover insulates the glacier from stark temperature changes while also absorbing heat. Therefore, the role of debris in a glacier needs to be assessed alongside the number of hours that it receives direct insolation for. 45.74% of the contemporary glaciers in the Yamuna basin are clean, and reflect the highest rates of reduction in glacierized area of 5.25% over 2000–2019. Lowest percentage reduction in glacier cover has been observed in glaciers (0.80% in 2000–2019) with 76–100% debris covered surface area. There has not been a pronounced change in glacier ELA from 2000 to 2016.

2.5.1.1.2 Glaciers of the Beas basin

Bordered by the Pir Panjal to the north and the Dhauladhar to the south, the Beas basin bears the second highest rate of glacier retreat in Himachal and Garhwal Himalayas. The monsoon dominated catchment of basin contains a reserve of 346 glaciers, most of which are small in size ($<0.5\ km^2$). The catchment nests 8 medium sized ($5.01–10.00\ km^2$) glaciers and 8 big glaciers ($>10\ km^2$), with little changes (0.24 and 1.25%) in 2000–2019 in glacierized area compared to smaller glaciers ($< 5.00\ km^2$). Large glaciers typically occupy troughs with gentler slopes, and are less readily responsive to short-term climatic perturbations. The most rapid change occurred in the glaciers, with size category of 1.01 to 5.00 km^2, a reduction of 3.21 km^2 over 2000–2019; which also registered 0.02% area reduction over 2016–2019 itself. The year 2015–2016 has been deemed to be an El Nino event, characterized by higher than mean temperatures and feeble precipitation and snow cover in the higher altitudes. The impact of sudden changes in ambient conditions manifests most strongly in glaciers with lower masses on mountainous depressions bearing high mean slope angles.

Micro regional factors like the orientation of a glacier determines the duration of sunshine and intensity of precipitation received by individual glaciers. Within the Beas basin maximum reduction occurred in the glaciers with a south western aspect, recording a 2.37% decrease between 2000 and 2019. Most of the glaciers in the Beas catchment are situated at average angles ranging from 10° to 25°. Glaciers situated on gentler slopes of 15° - 20° reflect increase in glacier area by small margin (1.98%) over a span of two decades (2000–2019). Glaciers with mean slopes ranging between 20° - 25° have been assessed to be the most vulnerable with a maximum loss of 3.33% of its glacierized area in 2000–2019. Out of the 123 glaciers categorized under this angle category, a total loss of 6.78 km^2 has been recorded in the past two decades (2000–2019).

Extensive studies of the Himalayan glaciers reveal a complicated role of debris with the glacial dimensions. It is understood that glaciers bearing thick mantle of debris behave differently than those without. Studies on glacier modeling reveal that debris free glaciers are more likely to retain their thickness profile better than those with debris cove. The waning away of glacial mass depends on the magnitude of energy fluxes and thermal resistance. Along with debris cover, a multitude of glacier surface patterns, presence of supra-glacial lakes, as well as Geomorphology (including complexity of mountain ridges hence orientations). Clean glaciers reflect the approximate true health of the glacier through the measurement of their area; while in the case of debris covered glaciers, the potential of vertical reduction in glacier profile with respect to the depth of debris cover on the glacier remains unaccounted for in two dimensional measurements. However, snouts of larger glaciers especially in the Himalayas may reflect debris thickness of nearly up to 1 m; increasing debris pressure load alongside decreased albedo, destabilized energy balance and basal melt-water conditions. Thin debris cover insulates the glacier from immediate temperature variations, while thicker debris may assist in ebbing away of glacier mass following thermodynamic forcing. However, to arrive at any plausible conditions regarding glacier health necessitates the assessment of the effects of local meteorology and geomorphology along with debris characters. Most of the glaciers in the Beas basin are clean, reflecting a steady gain in glacierized area in 2000–2019 by 2.13%. However, glaciers with 1–25% of surface covered by debris show an astonishing reduction of 9.31% of the glacier area in the same span. However, larger glaciers with large debris cover (76–100%) show no perceptible change for the said period.

Upon enquiry into the ELA of eight glaciers exceeding the size of 10 sq km, not much difference has been observed in 2000–2016. Most of the larger glaciers show no change in the ELA, placed within the elevations of 5030–5367 m asl. The highest magnitude of change in ELA of about 16 m has been noticed with one glacier of 19.5 sq km area.

2.5.1.1.3 Glaciers of the Chenab basin

The upper catchment of Chenab or Chandrabhaga basin largely covers the Lahaul and Spiti district of Himachal Pradesh. The Chenab emerges out of the confluence of two major streams namely, Chandra and Bhaga, originating from the proximity of Baralacha Pass (4800 m asl), with confluence at Tandi. Countless rivers and rivulets of glacial origin don the vast basin. Oriented from the east to west, major portions of the river basin shy away from the direct influence of the Indian Summer Monsoon; rather, the climate is determined by the interaction of the landscape with the Westerlies, as well as the weaker ISM. The Land Surface Temperature pattern here is notably different from other basins in the region due to a peculiar orographic location. The watershed lying at the juncture of two dissimilar climatic types namely; the Mid latitude Westerly and associated winter precipitation; and ISM and associated summer precipitation. After the initial downpour on the southern slopes of Dhauladhar and Pir Panjal, the moisture laden Indian Summer Monsoon winds weaken crossing over to the Great Himalayan Range in this region. The ISM orographic barrier thus renders most of the Chenab Basin a cold desert type climate (July-August), with precipitation of approximately 150–200 mm.[1] Heavy premonsoon (Jan–May) precipitation (100–300 mm[1]) especially in the form of snow sustains the discharge of rivers through the short summer season, and is crucial to agriculture of a primarily agro based economy of the region.

Containing 813 glaciers, the basin exhibits glacial systems with trunk and tributary glaciers of colossal magnitudes. Contrastingly, the valley also abuts 432 small glacial bodies below 0.5 km^2 area, covering 96.38 km^2 area currently. The small glaciers have increased in total glacial cover by 6.06% during 2000–2019. Approximately 29 glaciers classified in grouped area category of 5.01-10.00 km^2 occupy an area of 191.86 km^2 in 2019; show the highest reduction of 3.06% area in 2000–2019. While 416.18 km^2 of the total glacierized area in the Chenab basin falls in size class of 1.01–5.00 km^2, and

[1] Estimated from Tropical Rainfall Measurement Mission (TRMM) monthly rainfall product (1998-2014), data downloadable at https://search.earthdata.nasa.gov/

reveals a reduction of 1.88% in areal coverage, yet 33% belongs to 17 glaciers of size-class of 10 km^2 but had a limited reduction of only 0.13% in area in two decades (2000–2019). Undeniably, the larger glaciers of Chenab basin show apparent stability compared to the medium sized glaciers may be due to a larger glacier mass, and therefore a larger response time.

Glaciers bearing a southern aspect largely underwent loss, and have registered reduction to the tune of 1.71%, 2.06%, and 2.57% of glacier area in each category of glacier aspect, that is, South East, South, and South West, respectively. Glaciers oriented toward the East covered a total of 207.86 km^2 in the year 2019 but registered 0% loss since 2000.

In this basin, a rapid rate of reduction of glacier area has been noticed on the steep slopes; decadal monitoring of the glaciers reveal 14.61%, 11.62%, and 68.77% area reduction from the very few glaciers bearing mean slope angles of 35°–40° (15 glaciers with combined glacierized area of 0.97 km^2 in 2019), 40°–45° (6 glaciers with combined glacierized area of 0.29 km^2 in 2019) and 45°–50° (1 glacier with glacierized area of 0.01 km^2 in 2019). The higher rates of reduction coincide with the higher values of mean slope. Glaciers situated on average slope class of 15°–20°, contributing to the largest share of the glacier masses in the basin (623.36 km^2 in 2019) reveal an insignificant increase in glacierized area by 0.04% in 2000–2019. Progressively higher values of loss are registered to glaciers belonging to higher mean slope angle groups.

As a matter of fact the larger valley glaciers which cover larger distance exhibit debris cover. Debris cover on glaciers can occur due to a multitude of factors; such as mass movement, lithology; feeding mechanism of avalanching etc. Debris cover influences rates of ablation in specific season in accordance with other micro regional parameters and exerts combined impact. Most of the glacierized area (428.90 km^2) in the basin reflects no significant increase in debris cover over a period of 20 years (2000–2019) yet show a marginal increase in area by 0.11%. Contrary to this, glaciers with surface debris cover of 75–100% contributing 60.22 km^2 in 2019 show a 1.75% loss over since 2000, the largest percentage reduction in the watershed. The ELAs of 17 glaciers in the Chenab valley are placed on elevations ranging from 4641 to 5576 m asl, which amounts to insignificant changes in the annual mean ELA in 2000–2016.

2.5.1.1.4 Glaciers of the Ravi basin

The Ravi basin, with 225 glaciers of varying sizes is spread over Chamba district and north eastern part of Kangra district, covering three subwatersheds of Siul, Budhil, and Ravi. The valley is bordered by the Pir Panjal and Dhauladhar to the south, with elevation ranging from 468 to 5861 m asl, and an area of ~4900 km^2. The climate reflects a transition between Indian Summer Monsoon type, with a prominent dry winter and Mid Latitude Westerly with dry summer season (Chand & Sharma, 2015). In the basin, TRMM estimates reflect a bimodal distribution with a prominent peak in the premonsoon months of March–May and subsequently in June to September. Most of the glaciers in the upper Ravi catchment are located towards the South east whereby the elevation reaches above ~4000 m asl.

Most of these glaciers (146 out of 225) in the valley are small, with sizes below 0.5 km^2, with a combined glacierized area of only 25.04 km^2. A cumulative change of 0.12 km^2 (0.48%) reduction is observed in this category over the period from 2000 to 2019. The maximum reduction in glacierized area however has been observed in the size category of 0.51–1.00 km^2; reflecting a decrease of 2.81% by 2019, over a combined area of 28.82 km^2 that existed in 2000. Three glaciers of size ranging from 5.01-10.00 km^2 reflected a marginal increase in cumulative glacierized area (by 0.05%) between 2000 and

2019 when compared through a time series; while the largest glacier containing an area of 13.01 km^2 show no change over the same period of time. Therefore, our analyses reveal that the smaller glaciers are susceptible to change faster due to variations in climate, given its specific regional controlling parameters.

As noticed earlier, the role of aspect becomes very important in the assessment of glacier dynamics in the alpine systems because of the complexity of its terrain, and resultant micro regional characteristics. Glaciers oriented toward the north and south have indicated marginal increase in areas by 0.13% and 0.08%, respectively, over these two decades (combined area of 11.15 km^2 of North oriented glaciers, and 16.29 km^2 of South oriented glaciers). Glaciers bearing the Eastern aspect have exhibited no change in total glacierized area; while those oriented toward the South East, South West and West show percentage reductions of area ~0.81%, 1.29%, and 1.34% respectively in 2019 over the combined areas of 8.13 km^2, 13.03 km^2, and 54.16 km^2, respectively, since 2000. Notably the glaciers oriented to the West, also showed large amount of areal coverage of glaciers at 53.43 km^2 in the year 2019, found to undergo a percentage reduction of 1.34% over its area of year 2000, amounting to a total loss of 0.72 km^2 in a period of two decades.

Ten glaciers that belong to the lowest mean slope angle class of 10°–15° occupied a total area of 43.33 km^2 in the year 2000, but have undergone a reduction of 0.85% by 2019. However, the largest reduction in area was noticed in the 29 glaciers bearing mean slope class of 30°–35° (1.89% reduction over 5.49 km^2 in year 2000). Twenty-five small glaciers with mean slope angles above 35° showed no change in area during the same period.

Glacier statistics with respect to debris in the Ravi basin reveal that glaciers with debris cover ranging between 26% and 50% of individual glacier area reflected phenomenal increase of 7.88% of glacier area cover by 2019 over the combined total of 19.98 km^2 in 2000. Of the 26 clean glaciers, and 20 debris covered glaciers (76–100%) covered by debris, have undergone losses of 0.87% and 0.84% in area since 2000. Thirty six glaciers with 26–50% of area debris cover necessitate for a closer look, as it emboldens the overarching role of regional factors in causing an increase of 7.88% in glacierized area. There has been no noticeable change in the ELA placed at 4968 m asl, of the single glacier of 13.01 sq km area.

2.5.1.1.5 Glaciers of the Satluj basin

The Satluj basin has an extensive catchment, covering an area of 20566.76 km^2, influenced equally by the differing climatic regimes of the Mid Latitude Westerlies and the ISM. While the eastern half of the upper Satluj catchment hosts the largest glacier of the catchment, Baspa, the North Western part of Upper Satluj catchment practically experiences a cold desert type climate as the rugged Pir Panjal restricts the access of the moisture laden ISM winds, resulting in only small glaciers. By virtue of the enormity of the basin, it accommodates 657 glaciers, half of which are smaller than 0.5 km^2 in area. The combined glacial area of 85.62 km^2 in the year 2000, subsequently lost 2.26% of area by 2019, highest among all the size classes. Of 131 glaciers of size between 0.51 and 1.00 km^2 showed the second highest loss of 1.49% to combined glacier area in two decades. Assessment of monthly land surface temperature using MODIS LST[2] product revealed an insignificant negative trend of temperature both in the Accumulation (October- March) and Ablation Seasons (April–September) in Satluj basin within 2000–2019. The bigger glaciers in the valley with size greater than 5.01 km^2 surprisingly registered no reduction in the combined area; in fact registered small but notable increase in glacier size classes by

[2] MOD11B3v2 monthly mean LST composites downloadable from https://earthexplorer.usgs.gov/

0.10% by 2019 (size class between 5.01–10.00 km^2) or a 0.12 km^2 over its area of year 2000. Seven of the biggest glaciers in the watershed having a total of 118.16 km^2 area registered an increase by 0.30% over 20 years, by 2019; that is, addition of 0.43 km^2 to the total glacierized area in the category in two decades.

Large numbers of the glaciers in the catchment are oriented toward the North East (216) and North West (170) directions with combined glacierized areas of 276.69 km^2 and 200.58 km^2 respectively for the year 2000. These glaciers registered losses of 0.44% and 0.76%, respectively, until 2019. The highest losses in area, however, have been registered with the glaciers bearing a Northern aspect with a percentage loss of 1.14% or 0.25 km^2 in absolute terms since, 2000. Glaciers with higher mean slopes in Satluj basin registered higher rates of area loss. Glaciers located on gentler mean angles (282 glaciers in the category 15°-20° of mean slope angle) reportedly showed a decrease of 0.52% during the period of analysis. Twenty-eight glaciers, with higher mean angle of slope of 30°-35° on the other hand, showed a combined increase of 0.16% over the total glacierized area of 37.74 km^2 that existed in 2000, to 37.81 km^2 by 2019.

Surprisingly, the glaciers with surface debris cover of 26–50% area reflected a gain of 0.26% in 2000–2019 to the total glacial cover of 118.28 km^2 of year 2000. However, the glaciers with least debris cover showed marginal gains in the area 2016–2019; while clean glaciers showed a gain of 1.05 km^2, glaciers with 1–25% debris cover showed a gain of 0.05 km^2 over the period 2016–2019. In the glaciers of the Sutlej watershed, the ELAs are placed at altitudes exceeding 5221 m asl and showed no change since 2000 till 2016.

2.5.1.1.6 Glaciers of the Bhagirathi basin

Located in the North Western Garhwal Himalayas, the Bhagirathi with its tributary rivers are essential to the supply of water to the expansive Great North Indian Plains for sustaining a rich ecology of flora and fauna; more contemporarily, to some of the most populated states of the Union of India. Climatologically, the watershed is largely controlled by the South West Indian Summer Monsoon and annually receives approximately 2000 mm of rainfall.[3] With the history of 18 million years of Monsoon in the Indian subcontinent, the Bhagirathi catchment provides splendid opportunities to study the role of Palaeo-Monsoon on glaciations with regards astronomical cycles as well as events of uplift. Glaciers in the Bhagirathi basin thrive in climate characterized by a bimodal annual temperature curve truncated by the annual monsoon. While the mean accumulation period Land Surface Temperature (LST) has been observed to reflect a negative trend, the mean ablation period LST has been noticed to be rising; notwithstanding general cyclic climatic patterns. Year 2016 recorded the highest mean accumulation period LST, but registered a sharp decline in the successive years.

Containing a total of 185 glaciers of varying dimensions, the valley has undergone a loss of 0.48% to total glacierized area in 2000, by 2019; and in absolute terms, 2.35 km^2. In the Bhagirathi basin, 67 glaciers belonged to the size category of 1.01–5.00 km^2, covering a total area of 130.70 km^2 as in year 2019, after having lost 1.01 km^2 since 2000. However, the maximum percentage of loss of 2.36% in area was noticed in the smaller glacier size category of 0.51–1.00 km^2; with a combined area loss of 0.54 km^2. The least amount of loss in total glacierized area (0.01 km^2) was recorded in the 55 glaciers belonging to size category of less than 0.50 km^2. Total glacierized area in the catchment showed a loss

[3] Estimated from Tropical Rainfall Measurement Mission (TRMM) monthly rainfall product (1998-2014), data downloadable at https://search.earthdata.nasa.gov/

of 0.39 km^2 between 2000 and 2016, and subsequently, 0.03 km^2 between 2016 and 2019, accounting for a total loss of 0.42 km^2, that is, 0.18% of the total area of 2000.

The varying rates of reduction is somewhat related to the aspect in these glaciers. Of the 56 glaciers with North-Eastern aspect, absolute loss of 0.77 km^2 was recorded in the area (0.55% of the total of 2000). The highest percentage loss of 1.84% or 0.11 km^2 glacier area was registered on 18 glaciers bearing a South-Eastern aspect. No loss was noticed in the 06 glaciers with Northern aspect, while 03 oriented to the North-West, yielded a loss of 0.53 km^2 over 2000 by 2019. Ironically, the highest percentage of reduction in ice cover area is that on the glaciers bearing South and South-Eastern aspect. The change in glaciers with regard to mean angles of slopes revealed the maximum loss in glacierized area (0.98 km^2) in glaciers pertaining to the mean slope angle of 20°–25° between 2000 and 2019. Glaciers with lowest mean slope angle belonging to the category of 10°–15° constituteed the largest accounted glacierized area among all slope classes in the watershed (240.75 km^2 in the year 2000); these glaciers showed a total reduction of 0.61 km^2 till 2019, while the reduction was limited to 0.02 km^2 in 2016–2019.

This basin houses a total of 59 clean glaciers, highest among any category within the basin. As a matter of fact, only the larger valley glaciers which cover larger distance, exhibit considerable debris cover. The debris free glaciers have been observed to have the maximum percentage change of 1.74% amounting to the total area change of 1.74 km^2 between 2000 and 2019. The glaciers bearing high debris cover of 76–100% showed an overall positive change in the cumulative glacierized area with an addition of 0.06 km^2 between 2000 and 2016, which subsequently showed an increase of 0.16 km^2 till 2019; proving the dynamicity of the small bodies of glacier ice to the local climatic fluctuations. However, glaciers with debris cover of 51–75% of glacier area in 2016–2019 registered an increase of 0.02 km^2 in total area. In the Bhagirathi basin seven glaciers exceeding the size of 10 sq km showed pronounced changes in their ELA in 2000–2016. Placed at altitudes greater than 5129 m asl, there was a maximum contraction of 21 m in the ELA; while most glaciers reflected lowering of the ELA with a maximum magnitude of 13 m.

2.5.1.1.7 Glaciers of the Kaliganga basin

The Kaliganga basin is the easternmost major water-line in the Garhwal Himalaya, bordering Nepal. The Kaliganga hosts 262 glaciers of varying sizes, covering 477.02 km^2 (2019), and show a reduction of 1.58 km^2 since 2000. The largest glacier in the watershed covers an whopping 28.5 km^2, and surprisingly, showed no change in its area since 2000; while the second largest glacier bearing an area of 25.03 km^2 in the year 2000, added 0.04 km^2 in its area by 2016, and an additional 0.14 km^2 in the year 2019. Similar trends of increase in glacier area was noticed in two more glaciers among a total of seven glaciers above 10 km^2 in the basin; having added 0.21 and 0.07 km^2 to glacierized area between 2000 and 2019. Cumulatively, in the bigger glaciers of the valley (140.31 km^2 in the year 2019) an increase of 0.76% in area was recorded in the last two decades (2000–2019). Sixteen glaciers, categorized in the range of 5.01–10.00 km^2 in area, cumulatively registered a gain of 0.45 km^2 between 2000 and 2019 (cumulative total glacierized areas in the category in 2000, 2016, and 2019 were 109.67 km^2, 109.45 km^2, and 110.13 km^2, respectively). The maximum loss of 0.78 km^2 in 2000–2019 was registered on 116 glaciers in the watershed which have size areas of less than 0.05 km^2. However, the total glacierized area in this size category was the least among all, with only 28.49 km^2 as in 2019.

The largest numbers of glaciers (70) in the Kaliganga basin are oriented toward the north-east, cumulatively accounts for a total of 160.33 km^2 in 2019; registering a loss of 0.34 km^2 of glacierized area since 2000. However, a maximum loss of 0.39 km^2 of area was noticed in 33 glaciers bearing South-Eastern aspect; accounting for a total of 35.01 km^2 in the year 2019. There are relatively lesser number of glaciers (24) oriented toward South-West in Kaliganga basin; and cumulatively registered 0.20 km^2 of glacier area reduction between 2000 and 2019.

The largest glacier in the Kaliganga basin has a total area of 48.5 km^2 (2000), and has shown only a marginal reduction of 0.04 km^2 between 2000 and 2016. Similar value of reduction has been calculated for the second largest glacier in the valley covering an area of 32.0 km^2 in 2000. Both these glaciers are situated on a mean slope angle between 15° and 20°, more precisely, 18.87° and 17.54°, respectively. This slope angle category of 15°–20° has a total of 70 glaciers, with the largest cumulative area of 247.47 km^2 in the year 2019. Noticeably, these glaciers cumulatively have lost 0.49 km^2 of glacierized area in 2000–2016; while registering a gain of 0.27 km^2 in 2016–2019. The percentage of glacierized area lost in 2000–2019 also increases with the increase in mean angle of slope; the highest of a 4.18% decrease is registered with 09 glaciers in slope category 35°–40° occupying 2.16 km^2 in the year 2019. The second highest total glacierized area (153.88 km^2), belonged to the 94 individual glaciers categorized in the mean slope angle category of 20°–25°; show a total loss of 0.40 km^2 between 2000 and 2019, amounting to a 0.26% loss in cumulative area since 2000.

Clean glaciers in Kaliganga basin covered the largest area (117.92 km^2 in 2000); showed a cumulative loss in glacial area by 1.27% in two decades. Thirty seven glaciers belonging to the debris cover category of 1–25% of area, covering an area of 97.63 km^2 in 2019, lost 0.81% of cumulative glacier area since 2000. Glaciers with debris cover over 26–75% registered a combined growth of 0.80 km^2 areas in 2000–2019. However, glaciers with large debris cover, that is, > 76% of area, have showed an insignificant loss of 0.06 km^2 for the same period. Changes in the ELA ranged from a lowering of 16 m to a retreat of 19 m in 2000–2016.

2.5.1.1.8 Glaciers of the Alaknanda basin

The Alaknanda River forms the second important headstream to the mighty Ganges which serves millions of people downstream. Issuing from the melt-waters of the Satopanth and the Bhagirath Kharak glaciers, the stream is joined by the Mandakini at Rudraprayag and the combined waters meet the Bhagirathi at Devprayag. From the crests of the mountains till its mouth, the Alaknanda basin displays a wide range of elevation from 459 to 7512 m asl. Influenced by the Indian Summer Monsoon, Alaknanda basin hosts an assemblage of 316 glaciers of varying sizes, accounting for a total glacierized area of 774.47 km^2 in 2000; has undergone a loss of 0.91 km^2 till 2016, and further by 1.02 km^2 in 2016–2019, thus registering a total loss of 0.25% of the glacier cover in 20 years (2000–2019). Among the 316 glaciers of the watershed, 110 glaciers have showed no changes in area in the last two decades (2000–2019), 92 glaciers individually have vacated areas up to 0.01 km^2 and, 62 glaciers underwent reduction of individual glacier areas of 0.01 and 0.10 km^2; while only two glaciers have showed area reduction of 0.105 km^2 and 0.265 km^2, respectively. The largest glacier in the watershed with 42.4 km^2 (in the year 2000) yielded an increment of 0.22 km^2 in glacierized area between 2000 and 2019. The second largest glacier of the basin with 35.25 km^2 of area initially reduced by of 0.0002 km^2 between 2000 and 2016, but did not change in area in 2016–2019. Cumulatively, 20 glaciers above the size of 10 km^2, covering a total of 366.23 km^2 in 2000 witnessed a reduction of 0.02% or an absolute area of 0.07 km^2 by 2019.

The largest percentage of area (0.48%) however was lost by 122 glaciers in the size category of < 0.50 km^2 in 2000–2019, with an absolute area reduction of 0.15 km^2, out of 32.45 km^2 of glacierized area of 2000.

With respect to mean angle of slope, twenty of the biggest glaciers of the Alaknanda are situated with mean slopes of 14°-30°; registered individual losses in area ranging from 0.0002 to 0.0767 km^2, but later additions of 0.002 to 0.22 km^2. Eleven glaciers belonging to mean slope category of 35°-40° showed minimum reduction of 0.11% in area amounting up to a vacation of 0.001 km^2 in 2000–2019. Since 2000, the percentage of reduction has generally been high in glaciers with higher mean angles of slope (30°-35°) with the maximum of 0.45% in 33 glaciers with an equivalent area of 0.19 km^2 by 2019. The glaciers situated at higher mean slope angles are generally smaller in size, and have showed quicker response time to immediate changes in any of the drivers of glaciations. Eleven glaciers in the slope category of 35°-40° reflected a reduction of 0.001 km^2 in glacierized area amounting up to a loss of 0.11% in 20 years, since 2000, while in the same period, 3 glaciers in the highest mean slope angle category of 40°–45° reflected a gain of 0.01 km^2 of glacierized area or a 0.31% increase.

In the Alaknanda basin, glaciers with Eastern and South-Western aspect have registered losses of 0.49 km^2 and 0.63%, respectively, with 0.45 km^2 and 0.23 km^2 of area lost over 2000–2019. The glaciers oriented toward the North, North-East, and North-West have registered the least percentages of area loss, that is, 19%, 0.06%, and 0.20% on areas of 8.33, 228.29, and 210.77 km^2 between 2000 and 2019. This statistics probably point toward to the fact that those glaciers which fall directly under the influence of the ISM winds w.r.t. aspects, reflect a greater loss in glacier area than others at least in this century.

Clean glaciers accounted for the maximum number of glaciers in the valley (126) and for a total area of 126.86 km^2 in 2019, registered a loss of a total of 0.45 km^2 of glacierized area in 2000–2019. The highest loss in cumulative area was noticed in the debris cover category of 1–25% (representing individual glaciers which have a debris cover of 1–25%) with a percentage reduction of 0.40% or an equivalent of 0.68 km^2 over 2000–2019. The least glacierized area lost in 2000–2019 (0.01%) was recorded with 46 glaciers belonging to the debris cover category of 51–75% in 2000–2019. The same glacier category cumulatively reflected an increase of 0.07 km^2 in 2016–2019. However, in the same time period, the glaciers with most debris cover have showed a total loss of 0.04 km^2 or 0.14% of the cumulative glacierized area in the category. Twenty glaciers with areas above 10 sq km in area have been studied for their ELA in 2000–2019; and found to be placed at elevations exceeding 4814 m asl. While only two glaciers showed no change in ELA, nine showed retreat with a maximum value of 50 m. The largest magnitude of lowering in ELA (found in a total of nine glaciers of the catchment) was assessed at 21 m from the former.

2.5.1.1.9 Glaciers of the Tsarap Chhu basin

The smallest among all watersheds within the political limits of our study area, is occupied by the Tsarap Chhu to the extreme north of the State of Himachal Pradesh. The 182 km Tsarap River flows past Sarchu in Spiti subdivision, providing waters to small scale subsistence agriculture in the Zanskar valley. The elevation in the watershed ranges from 4146 to 6138 m asl. Eighty glaciers of the watershed, mostly clean, had a cumulative total glacierized area of 57.75 km^2 in 2000 that reduced by 0.43 km^2 by 2016, but again gained an area of 0.22 km^2 by 2019. While 29 glaciers in the watershed reflected growth in their individual glacier areas up to 0.09 km^2, 06 have practically showed no change between 2000 and 2019. Thirty three glaciers in the watershed have lost glacierized surface area of less than 0.01 km^2, while 10 glaciers have registered reduction of 0.01–0.14 km^2 in 2000–2019, respectively.

Most of the glaciers (45) in the watershed have been categorized under size category of 0.50 km^2, and have registered a cumulative increase of 0.10 km^2, accounting for a 0.93% gain in glacier area between 2000 and 2019. Covering a total of 31.98 km^2 of area in the year 2019, the 14 glaciers with sizes ranging from 0.51 – 1.00 km^2 showed an increase in the total glacierized area since 2000 by 0.38%. Eighteen glaciers with size category of 1.01–5.00 km^2 covering 31.98 km^2 in 2019, reduced by 0.41 km^2 till 2016, followed by an increase of 0.06 km^2 between 2016 and 2019. This glacier size category accounted for the maximum percentage loss of 1.12% among all categories. The biggest glacier in the watershed encompassed an area of 5.04 km^2; and showed no changes in the last two decades (2000–2019).

It is important to note that in the Tsarap Chhu watershed, none of the glaciers bear South, South-Western or Western aspect. Only one glacier oriented to the South-East, registered a reduction on 1.15%, or 0.62 km^2 since 2000, that is in the last 20 years. However, the largest reduction was noted with the six glaciers in the watershed oriented toward the East, with a percentage reduction of 3.93% over these two decades; in absolute terms the loss was 0.15 km^2. Glaciers oriented to the North and North-West (12 and 20 glaciers, respectively) have recorded an increase in glacierized area by 4.45% and 0.74% in 2000–2019, highlighting the role of the climatic regime, different from rest of the study region.

Most of these glaciers (56) of the basin have mean angle of slope of $< 25°$, and show variety of contrastingly changes, 22 glaciers with mean angles $> 25°$ reflect gains to total glacierized area; thirteen glaciers with mean angles of slope ranging between 10°-15° glacierized area of 21.62 km^2 in 2000 registered the highest area loss with 1.56% or 0.34 km^2 during 2000–2019. The subsequent category of glacier individually possessing mean slope angles of 15° -20° reflected a loss of 0.11 km^2 to total area of 20.22 km^2 (2000) in these two decades.

2.5.1.2 *Glaciers in the Sikkim Himalayas*

The ICIMOD demarcates the Eastern Himalayas from the Kaligandaki river of Central Nepal to North West Yunnan in China and, encompassing Bhutan and the Indian states of Sikkim, Northern West Bengal and Arunachal Pradesh covering a total area of 525,000 km^2 (ICIMOD, 2009). Unlike the Western Himalayan region, glaciers of the Eastern Himalayas are the Summer Accumulation type. Glaciers of the region are fed by the solid precipitation upon the arrival of the Indian Summer Monsoon in the higher reaches. The Mid Latitude Westerlies and the North Eastern Monsoon contribute little to the glaciations in the region (Kumar et al., 2020; Debnath et al., 2019). The region contains more than a hundred glaciers with elevations ranging between 280–8586 m asl. Compared to the Western Himalayas, the climatic characteristics in the complexity of interaction of differing climatic systems remain ill studied. Temperatures in the Eastern Himalayan region is on a steady rise at the rate of 0.01°C a year for elevations $<$1000 m asl; with alarming rates of 0.02°C and 0.04°C/year in elevation categories of 1000–4000 m asl and $>$4000 m asl, respectively. However in the summer months of June-August there is lowering of annual temperature by 0.01°C in elevation classes $<$1000 m asl and 1000–4000 m asl.

Observations from Tadong and Gangtok suggest of the existence of positive trend in minimum temperature in 1961–2017 in both Tadong and Gangtok; decrease in summer temperature in Gangtok while it remained approximately the same in Tadong (Kumar et al., 2020). Gangtok has reported an insignificant positive trend in mean annual temperatures, while the positive trend is more pronounced at Tadong. The mean monthly temperatures in a year in the district of North Sikkim ranges from 5°C to 20°C (Debnath et al., 2018); months of June and July being the warmest. The region receives an average of 550 mm in the month of July as an obvious impact of the Indian Summer Monsoon, while in the winter months there is little precipitation. Accumulation in the glaciers of the region is due to the summer precipitation that arrives at higher elevations in the form of snow. A positive change in summer temperature is understandably detrimental to glacier health in the region.

2.5.1.2.1 Glaciers of the Upper Teesta watershed

Ajai (2018) classified the glaciers according to the watersheds of Indus, Ganga, and Brahmaputra, in nonconsonance with the region wise approach; with the help of optical satellite resources, Ajai (2018) marks 34 glaciers within the Teesta watershed in the Eastern Himalayas among which 23 have been reported to have lost glacierized area at the rate of mean annual loss of 0.07% in 1989–1990 to 2001–2004. The above 34 glaciers within the Teesta watershed were seen to have cumulatively lost 1% of glacierized area since 1990 till 2004; numerically it came down from 305 km^2 to 304 km^2. Raina and Srivastava (2008) have delineated a total of 449 glaciers with a total glacierized area of 706 km^2 and ice volume of 39.61 cubic km. in the Upper Teesta watershed, a major contributor to the Brahmaputra river system. Due to the proximity of the Eastern Himalayas to the Bay of Bengal, ample summer precipitation adds to the accumulation of glaciers (Debnath et al., 2019).

There are a total of 81 glaciers of the Changme Khangpu watershed, a subwatershed of the Upper Teesta catchment (Debnath et al., 2019), ranging in sizes from 0.02 to 13.8 km^2,while the number of glaciers belonging to size category of lesser than 0.5 km^2 have been found to be the highest (more than sixty), fewer glaciers belonging to areas of more than 5 km^2 possess the largest area in the watershed. The aspect of the glaciers help somewhat elucidate effect of the Indian Summer Monsoon; most of the glaciers (43%) in the Changme Khangpu watershed have been reported to be oriented toward the South-East, South, and South-West, respectively, occupying almost 73% of the total glacierized area. Glaciers oriented toward North, North-Eastern and North-Western aspects are markedly smaller in area; point out to the weak effect of the ISM (Tables 2.1 and 2.2).

The larger glaciers with southern aspect are situated at the lowest elevations of >5000 a msl, over gentler mean angles of slopes of less than 20°. However, the larger glaciers impact on the local geomorphology more due to large mass and erosive capacity. Understandably, the glaciers located on rain shadow of incumbent precipitation are much smaller in sizes, lacking the ability to carve larger and gentler valleys. Studies thus reveal the prime significance of local geomorphology in determining glacier size as well as dynamics. Clean glaciers in the ISM dominated Eastern Himalayan region are much more vulnerable to area loss than partially and mostly debris covered glaciers. Since 1975, the annual rate of reduction in cumulative glacierized area in the Upper Teesta subcatchment stands at 0.453 km^2 per annum (Debnath et al., 2019). The rate of retreat of cumulative glacier area in the region was studied to be lowest in 1988–2001, possibly due to cooler late summer temperatures along with ample rainfall to arrest the rate of deglaciation. The rate of change in glacierized area has been the highest in 2001–2016, with a loss rate of 0.66 km^2 per annum in the Changme Khangpu Basin; although in the North-Western Himalayan counterparts have reportedly shown a decline in the rates of retreat (Debnath et al., 2019; Sharma et al., 2009; Brahmbhatt et al., 2017).

Table 2.1 Contemporary glacier statistics in Himachal and Garhwal Himalaya.

Basin	No. of glaciers	Glacierized area in square kilometers (2000)	Glacierized area in square kilometers (2016)	Change in glacierized area (Δ) in square kilometer (2000–2016)	Percentage change (Δ) in glacierized area (2000–2016)	Glacierized area in square kilometers (2019)	Change in glacierized area (Δ) in square kilometer (2016–2019)	Percentage change (Δ) in glacierized area (2016–2019)	Change in glacierized area (Δ) in square kilometer (2000–2019)	Percentage change (Δ) in glacierized area (2000–2019)
Beas	346	440.15 ± 0.06	435.61 ± 0.07	4.54 ± 0.07	1.03	432.34 ± 0.06	3.27 ± 0.06	0.75	7.81 ± 0.10	1.77
Chenab	813	1204.91 ± 0.08	1192.71 ± 0.08	12.20 ± 0.08	1.01	1194.28 ± 0.08	-1.58 ± 0.07	-1.58	10.63 ± 0.10	0.88
Ravi	225	158.84 ± 0.03	157.56 ± 0.03	1.28 ± 0.03	0.81	157.57 ± 0.03	-0.03 ± 0.03	-0.02	1.26 ± 0.03	0.79
Satluj	657	748.85 ± 0.05	743.96 ±0.05	3.89 ± 0.05	0.52	744.6 ± 0.05	0.37 ± 0.05	0.05	4.26 ± 0.05	0.56
Tsarap Chhu	80	57.75 ± 0.03	57.38 ± 0.03	0.43 ± 0.03	0.74	57.6 ± 0.03	-0.22 ± 0.03	-0.38	0.15 ± 0.03	0.25
Bhagirathi	185	481.82 ± 0.14	479.75 ±0.14	2.074 ± 0.14	0.43	479.47 ± 0.14	0.28 ± 0.01	0.05	2.35 ± 0.14	0.48
Yamuna	94	98.66 ± 0.05	96.71 ± 0.05	1.94 ± 0.05	1.97	96.57 ± 0.05	0.15 ± 0.01	0.16	2.10 ± 0.11	2.13
Alaknanda	316	774.47 ± 0.05	773.71 ± 0.05	0.91 ± 0.02	0.12	772.71 ± 0.05	1.02 ± 0.04	0.13	1.93 ± 0.07	0.25
Kaliganga	262	478.60 ± 0.29	476.44 ±0.29	2.16 ± 0.02	0.45	477.02 ± 0.29	-0.58 ± 0.02	-0.12	1.58 ± 0.04	0.33

Table 2.2 Basin-wise glacier change statistics for the Himachal and Garhwal Himalaya (2000–2019).

Basin	No. of glaciers	Glacierized area in square kilometers 2000	Glacierized area in square kilometers 2016	Change in glacierized area (Δ) in square kilometer 2000–2016	Percentage change (Δ) in glacierized area 2000–2016	Glacierized area in square kilometers 2019	Change in glacierized area (Δ) in square kilometer 2016–2019	Percentage change (Δ) in glacierized area (2016–2019)	Change in glacierized area (Δ) in square kilometer (2000–2019)	Percentage change (Δ) in glacierized area (2000–2019)	Mean annual change (Δ) in glacierized area in Sq km (2000–2019)
Beas	346	440.15 ± 0.06	435.61 ± 0.07	4.54 ± 0.07	1.03	432.34 ± 0.06	3.27 ± 0.06	0.75	7.81 ± 0.10	1.77	0.39
Chenab	813	1204.91 ± 0.08	1192.71 ± 0.08	12.20 ± 0.08	1.01	1194.28 ± 0.08	-1.58 ± 0.07	-1.58	10.63 ± 0.10	0.88	0.53
Ravi	225	158.84 ± 0.03	157.56 ± 0.03	1.28 ± 0.03	0.81	157.57 ± 0.03	-0.03 ± 0.03	-0.02	1.26 ± 0.03	0.79	0.06
Satluj	657	748.85 ± 0.05	743.96 ±0.05	3.89 ± 0.05	0.52	744.6 ± 0.05	0.37 ± 0.05	0.05	4.26 ± 0.05	0.56	0.21
Tsarap Chhu	80	57.75 ± 0.03	57.38 ± 0.03	0.43 ± 0.03	0.74	57.6 ± 0.03	-0.22 ± 0.03	-0.38	0.15 ± 0.03	0.25	0.01
Bhagirathi	185	481.82 ± 0.14	479.75 ±0.14	2.074 ± 0.14	0.43	479.47 ± 0.14	0.28 ± 0.01	0.05	2.35 ± 0.14	0.48	0.12
Yamuna	94	98.66 ± 0.05	96.71 ± 0.05	1.94 ± 0.05	1.97	96.57 ± 0.05	0.15 ± 0.01	0.16	2.10 ± 0.11	2.13	0.11
Alaknanda	316	774.47 ± 0.05	773.71 ± 0.05	0.91 ± 0.02	0.12	772.71 ± 0.05	1.02 ± 0.04	0.13	1.93 ± 0.07	0.25	0.10
Kaliganga	262	478.60 ± 0.29	476.44 ±0.29	2.16 ± 0.02	0.45	477.02 ± 0.29	-0.58 ± 0.02	-0.12	1.58 ± 0.04	0.33	0.08

Table 2.3 Optically stimulated luminescence ages in the Chandrabhaga (MSM and MST) and Bhagirathi (MSG) River basins.

Sample id	U (ppm)	Th (ppm)	K (%)	Dose rate (gy/ka)	Age (ka)
MSM 1	5.9	21.6	2.9	5.0 ± 0.5	**992 ± 120**[a]
MSM 2	4.7	38.8	2.8	5.7 ± 0.5	***8 ± 2 ka***
MSM 3	6.73	19.1	2.40	4.6 ± 0.6	**4.3 ± 0.6**
MSM 4	5.4	19.9	2.9	4.9 ± 0.6	**1.0 ± 0.1**
MSM 5	9.63	25.6	3.4	6.5 ± 1.0	***6.6 ± 1.0***
MSM 6	6.31	2.97	2.97	5.8 ± 0.4	***8 ± 1***
MSM 7	6.4	2.89	2.89	5.7 ± 0.4	***10 ± 1***
MSM 8	11.0	12.7	2.4	5.4 ± 0.4	***6.2 ± 0.5***
MSM 9	10.1	17.3	2.6	5.4 ± 0.5	***6.4 ± 0.6***
MST 10	3.72 ± 0.13	2.567 ± 0.76	2.583	3.557 ± 0.171	88.82 ± 2.59
MST 11	4.242 ± 0.11	7.598 ± 0.85	2.917	4.495 ± 0.144	74.272 ± 5.529
MSG 12	10.2	18.4	2.6	5.4 ± 0.5	5 ± 1
MSG 13	7.4	20.7	2.5	5.5 ± 0.5	**16 ± 2**
MSG 14	4.4	19.8	2.9	4.7 ± 0.3	***34.5 ± 10.8***
MSG 15	3.8	14.4	2.6	3.9 ± 0.3	***40.6 ± 6.4***
MSG 16	4.5	11.5	2.9	4.5 ± 0.3	***45.7 ± 4.3***
MSG 17	2.9	16.9	3.1	4.3 ± 0.3	***66.3 ± 11.5***

[a] *annum*

2.5.2 The Holocene glacial extent and timing

Glaciers and ice bodies, being sensitive indicators to variabilities in climate parameters, are in a phase of general recession for past one and a half century world over. For the past records, we have used the state-of-art techniques of the Optical and Exposure dating methods for determining the timing of extent of many glaciers in the Central and Western Himalaya (Tables 2.3 and 2.4). We have extended our present day understandings, supported by field evidences and chronologically constrained landforms to reconstruct the past extent of glaciers in the different climatic zones along the Himalaya range. Since the present day landforms are conspicuous and staggering in all the contemporary glacier margins, large assemblages of landforms of previous episodes of expansion abound the landscape in most of these locations, and have been used to recreate and reconstruct the extents, and thus the magnitude of each glacial episodes in either of the basins. We have adopted the multiple techniques (*field mapping, sedimentological differentiation, satellite remote-sensing, relative and absolute age determination*). These previous, yet younger, advances throughout the Himalaya are well represented by extensive depositional landforms such as lateral and terminal moraines, hummocks, erratics, drumlins, polished bedrock and lacustrine fills, typically constrained within 9–6 ka; bracketing it to the early Holocene glacial event; and have erosional U-shaped valleys with defined trim-lines (Table 2.3). This event seems to have been followed by a major phase of recession between ~6–4 ka, resulting in evolving a new local base-level to create the staggering paraglacial fan complexes in the basins (Fig. 2.3). This phase of large waning ~6–4 ka had probably caused glaciers to recede to higher altitudes as compared to the contemporary levels.

Table 2.4 Exposure ages in the Gangotri basin.

Sample No	Latitude (DD)	Longitude (DD)	Elevation (m)	Shielding correction	[Be-10] (atoms/g)	Age (ka)
MCS 1	30.940	79.063	3908	0.946973	1.025 + 04	**0.21 ± 0.038**
MCS 2	30.763	79.078	4343	0.980066	1.018 + 05	**1.62 ± 0.15**
MCS 3	30.913	79.078	4335	0.980066	8.305 + 04	**1.32 ± 0.13**
MCS 4	30.996	78.931	3029	0.918425	1.129 + 05	**3.81 ± 0.36**
MCS 5	30.997	78.927	3012	0.907476	1.057 + 05	**3.65 ± 0.36**
MCS 6	30.952	79.058	4008	0.947383	2.298 + 03	**0.04 ± 0.03**
MCS 7	30.945	79.060	3935	0.949056	2.230 + 04	**0.44 ± 0.05**
MCS 8	30.997	78.931	3021	0.899156	1.103 + 05	**3.73 ± 0.35**
MCS 9	30.997	78.929	3017	0.910892	1.111 + 05	**3.71 ± 0.33**
MCS 10	30.997	78.928	3022	0.906537	1.082 + 05	**3.62 ± 0.32**
MCS 11	30.997	78.928	3017	0.905309	1.001 + 05	**3.37 ± 0.36**
MCS 12	30.997	78.927	3039	0.907085	9.869 + 04	**3.27 ± 0.32**

Succeeding advance ~1.5 ka (Table 2.4) had descended to altitudes of 3785 m asl as dated erratics on lateral margins at Tapovan near Gaumukh indicate. The terminus appears to have remained stand-still for almost a millennium before the contemporary recession which started some 440 ± 40 years ago. The precision of the CRN technique, tested on the erratic dated 1889 (Greisbach, 1891) yields an age of 210 ± 0.038[a], robust and near-accurate. These landforms and extents were earlier thought to have been contemporaneous the Little Ice Age expansions elsewhere (Sharma and Owen, 1996).

Prior to the Holocene advance, glacier cover grew to a whopping ~680, and ~500 km^2, respectively, in the Chandrabhaga and Bhagirathi basins, that is, almost 2.5 times compared to the contemporary cover (Table 2.3). This largest of the advances that is recorded in the glacial landform extent and chronostratigraphy is constrained beyond >60 ka and 85 ka; that is, the largest late Quaternary local glacial maximum during the MIS 3. The ice thickness at places at this event remained over 250 m from the present valley floor, expanding the Gangotri glacier terminus almost by 40 km downstream to Jhala (2400m asl) as compared to the present Gaumukh at 4000 m asl. Glaciers of the Chandrabhaga probably terminated at 3000 m at ~85 ka, some 70 km of length as a major trunk cover. Such was the magnitude of this climate event that almost all the tributaries glaciers in the basin coalesced to form a gigantic glacier system. It is obvious from the dated sediments that this event of expansion predated the Global Last Glacial Maximum of other regions by almost three-times, making it asynchronous with the rest of high latitude glacial expansions at and around 22–18 ka years ago. Similarly, the glaciers in Miyar basin (3980 m asl) covered a large area after coalescing to reach down valley and terminating at Karpat (3020 m asl).

This largest episode has yet remained undated but is considered to be contemporaneous to the largest expansion of Gangotri glacier in the late Quaternary (~60–40 ka). But what remains intriguing for such large expansions is the source of precipitation and cause of cooling that might have turned these regions into ice-lands. In addition, it is puzzling to find large scale landforms of this episode largely preserved in the Bhagirathi basin, whereas almost a completely obliterating similar landforms in the Miyar basin. Does this fact indicate toward causing latitudinal fluctuations of the ISM over last 20 ka by some unknown forcing mechanism, thus destroying the landform records in one basin but leaving it intact in the other? The issue still remains speculative. We suggest that the period of inter-glacial in these regions lasted for over 25 ka which would have given sufficient time to slope and other processes to modify glacial landscape substantially, and in return, mobilize huge sediment load to the frontal regions (Fig. 2.3).

The Holocene glaciation in these basins is largely constrained within ~9–6 ka, with an expansion of 5–7 km from that of the contemporary terminuses. The high terrace and bench topography of Bhujbas-Gaumukh and lateral moraines of the tributary glaciers reaching the main trunk valley floor, are the atypical feature of such an expansion (Fig. 2.4). However, these small scale expansions around mid-Holocene resulted in expanding tributary glaciers substantially which choked the trunk river at many places to evolve large lakes that exist as thick lacustrine deposits. These lacustrine sediments have yielded very promising results with the use of OSL method of age determination. Unique assemblage of till materials indicate that the deglaciation had begun just after ~6.0 ka. These drumlins are dated within ~6.2 to ~4.0 ka in the region, that is, formed at this time; while lacustrine sediments are dated at ~8 ka (which infers that the filling was in progress, therefore, ice-blockade was still on in the trunk river). Therefore, we conclude that the streamlined drumlin assemblage in the broad glacial valley relates more to the gradual expansion than retreat in the Himalaya (Fig. 2.5).

Glaciers have since been in a phase of retreat in either of the basins, except for a small expansion (Fig. 2.6). As the conventional wisdom tells us, had there been any advance succeeding this episode, it would have wiped out the landform deposits of the mid-Holocene. We conclude that the glaciers in both the basins have been in general retreat post this expansion (mid-Holocene, Fig. 2.4). The ages constrained within ~990[a] to 1.0 ka in the basin within glacial reach also validate our assumption that not much have changed in terms of the glacial positions at least in the past one millennium. It is interesting to note that a settled agriculture was practiced at this glacial reach at 3700 m asl in the Miyar basin (~800–400[a] ^{14}C age; Saini et al., 2019).

2.6 Discussion

The varying states of the glaciers in the North Western and Eastern Himalayas are results of the influence of different climatic regimes as well as the ambient climatic condition in the foreland regions. Glaciers of the Yamuna Basin show the most profound negative changes in their dimensions over two decades (2000–2019). The basin marks the southernmost boundary of the MLW and ISM, which may be a cause for the different behavioral pattern of these glaciers. The MLW in its South Eastward journey loses most of its moisture load before entering into the Yamuna watershed. While, ISM has been observed to have intensified over the years, an increase in mean ablation period ambient atmospheric temperature may cause many sq kms of area in the ablation areas to experience liquid precipitation leading to more melt than accumulation. The orientation of the basin allows for the summer monsoon to possibly travel directly to the glacier areas without much hindrance. Comparatively, the seasonality of the MLW causes moisture influx in the form of solid precipitation which sustains through the ablation months in North Western Himalayas restricting excessive glacial melt in certain catchments.

The characteristic differences in contemporary glacier dynamics of the North-Western to those in the Eastern Himalayas arise out of the differing climatic regimes. While the catchments of Bhagirathi, Alaknanda, Kaliganga, Beas, and Yamuna largely fall under the influence of ISM; Chenab, Satluj, Ravi,

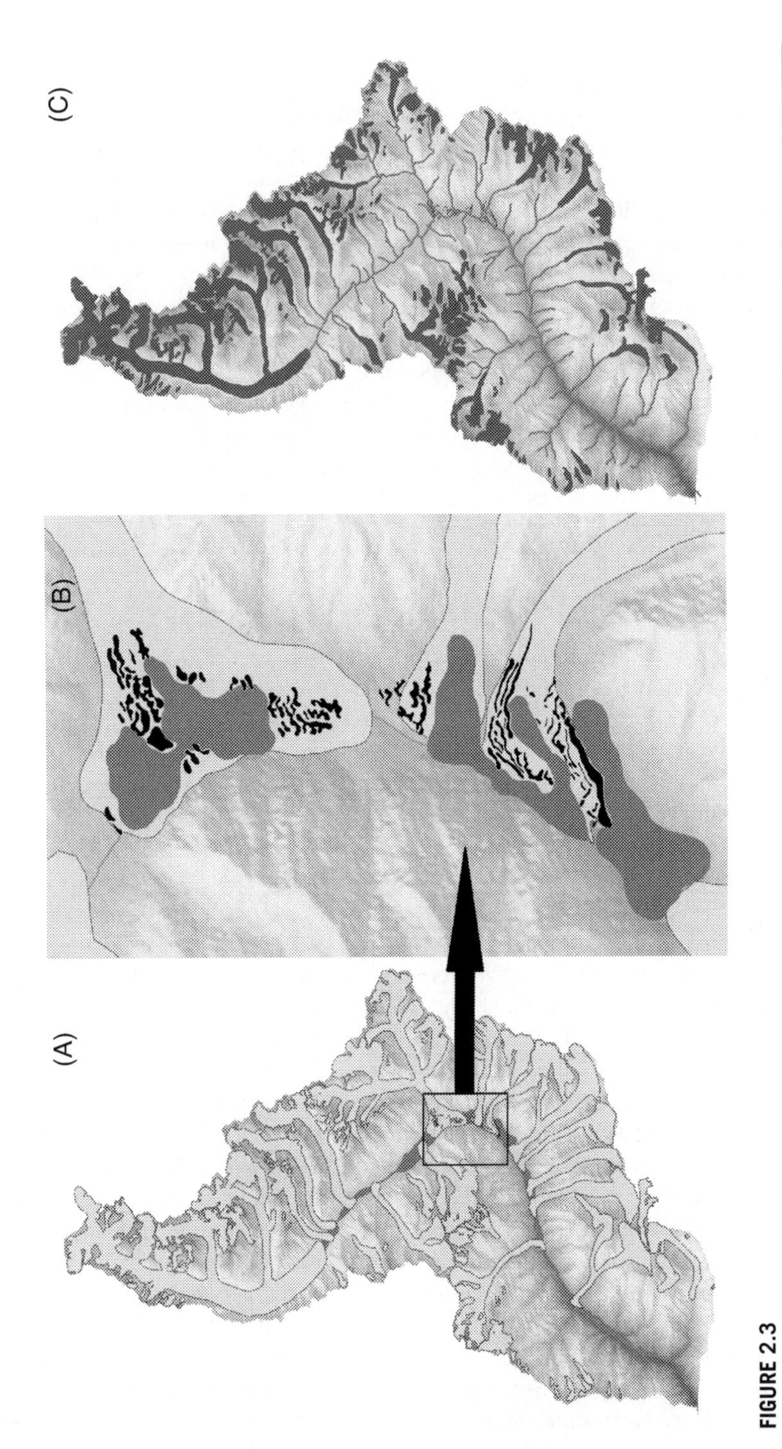

FIGURE 2.3

Style of landform evolution in the Miyar basin since the Holocene (adapted from Deswal et al., 2017).

FIGURE 2.4
The Gangotri glacier frontal landforms of the Holocene.

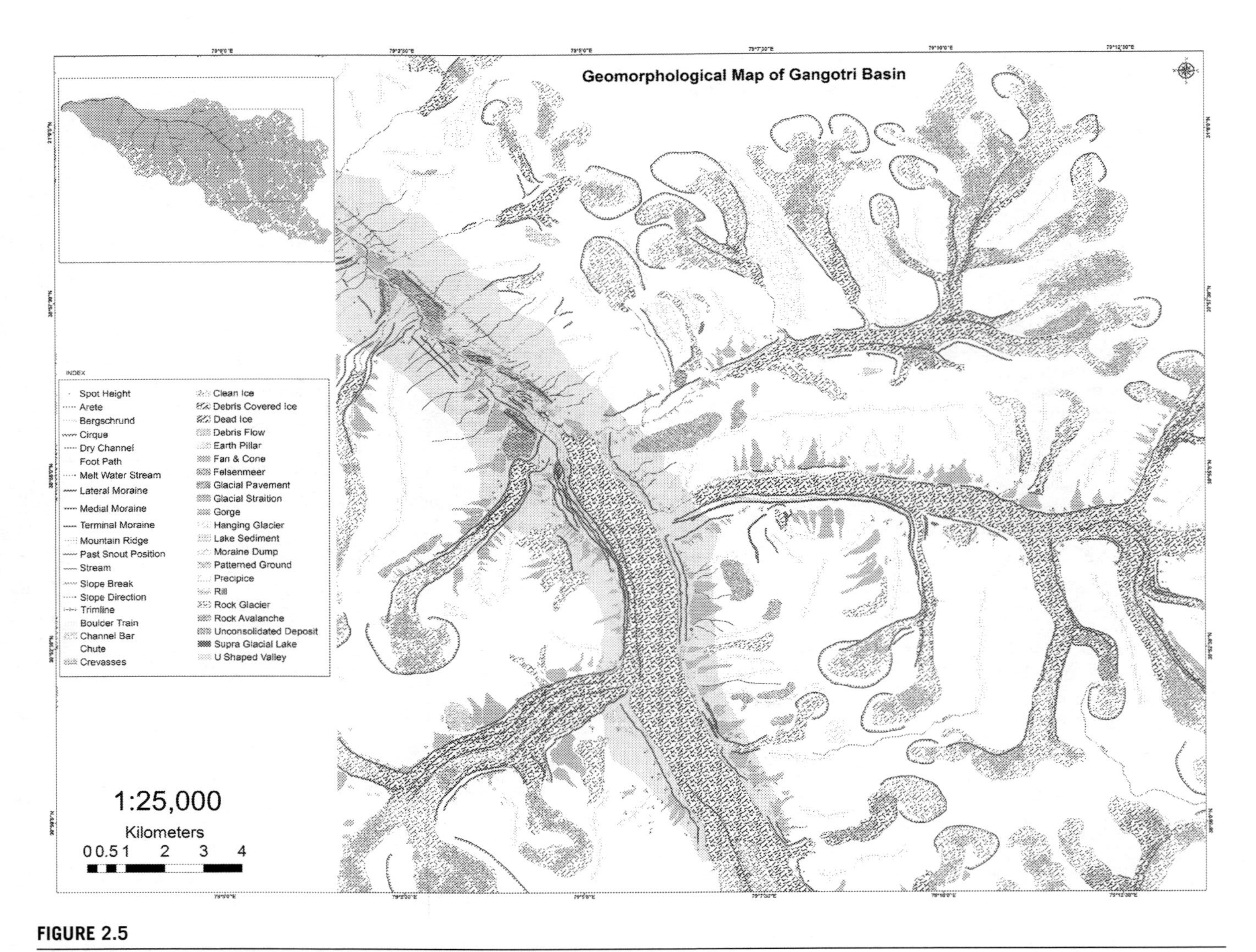

FIGURE 2.5
Geomorphology of Gangotri Glacier environs within the Holocene limits.

FIGURE 2.6

Holocene ice limit of Gangotri (upper image) and Neelkanth (lower image) Glaciers.

Tsarap Chhu come under added influence of MLW dominated winter accumulation, for a large part of the accumulation period. Contrastingly, the Eastern Himalayas are mainly summer accumulation type. In the North-Western context, the glaciers oriented toward the direction of the Indian summer monsoon have reflected faster rates of retreat between the times of analysis. Larger glaciers are found to be more stable in general than the smaller glaciers.

In the North-West, there are 34 glaciers in the State of Himachal Pradesh and 36 glaciers in Uttarakhand which are larger than 10 km^2 in area. While 725 are found to be receding in Himachal

Himalaya, 1578 were found to be stable and 52 advancing; 266 were found receding, 558 stable and 24 glaciers advancing in the State of Uttarakhand. Both the North-Western Himalayan States of Himachal and Uttarakhand have shown percentage deglaciation of 0.92% and 0.43%, respectively, in the last two decades. In the Eastern Himalayan Changme Khangpu basin on the other hand, was found to have a much higher rate of area change at 20.72% in nearly double the time frame (between 1975 to 2016). The seasonal difference in accumulation also shapes contemporary glacial dynamics along with individual micro regional characters between the North-Western from the Eastern Himalayas. Greater inputs pertaining to individual glacier on a time scale may help model a better and robust future dynamics.

2.7 Learning and knowledge outcomes

The location and orientation of glaciers on the landscape and local geomorphology play a major role in ascertaining whether they experience mass loss, gain or remain stable over a considerable period of time. Reconstruction of local Holocene extents requires a thorough understanding of the geomorphology of the glacier valley and foreland for etched characteristic features of the glacial past in relation to their geographic locations. In the recent decades, however, the rates of retreat in the large glaciers (>25 km) in Upper Bhagirathi and Miyar basin though reveal accelerated values, are dissimilar to each other and may be related to the response time which varies across regions, locations, size and micro-climate. Over a centennial scale, beginning of the glacial retreat is witnessed now, has been documented since 1850s in the Himalaya. But smaller glaciers in the recent years show complex trend of both retreat and stability. A similar style and trend is observed in the glaciers in the Ravi basin and the Sikkim Himalaya. The LST assessed for some of these glaciers for recent decades show that temperatures have dropped fractionally. It is interesting to observe that a few glaciers are showing recovery. The impact of human action on increased glacial recession, as proposed elsewhere, is extremely difficult to establish on these high altitude glaciers, as some of these glaciers show sheer disregard by standing still and tall to increased activity along the national highways and the Higher Himalaya. Glacier dynamics to a large degree is determined by the Micro regional parameters in rugged and lofty high altitudes they rest in. The differences in glacial dynamics are often driven by long term climatic parameters rather than short term along with individual glacier mass. Contemporary or long term glacial dynamics is best understood acknowledging the dominant climatic regime at glacier locations. Some of the resultant characteristic glacier behaviors with regard to glacier size, mean angle of slope, aspect and debris cover have been discussed in this chapter.

Acknowledgments

We would like to thank the Department of Science and Technology and the Space Application Centre ISRO, Govt. of India for financially supporting our studies on the Himalayan Cryosphere over the years.

References

Ajai, Inventory and monitoring of snow and glaciers of the Himalaya using space data, in: Goel, P., Ravindra, R., Chattopadhyay, S. (Eds.), Science and Geopolitics of the White World, Springer International Publishing AG 2018, Ahmedabad, 2018, pp. 5–113 https://doi.org/10.1007/978-3-319-57765-4_8.

Benn, D.I., Owen, L.A., The role of the Indian summer monsoon and the mid-latitude westerlies in Himalayan glaciation review and speculative discussion, J. Geol. Soc. Lond. 155 (1998) 353–364 https://doi.org/10.1144/gsjgs.155.2.0353.

Brahmbhatt, R., Bahuguna, I., Rathore, B., Significance of glacio-morphological factors in glacier retreat: a case study of part of Chenab basin, Himalayan J. Mt. Sci. 14 (2017) 128–141 https://doi.org/10.1007/s11629-015-3548-0.

Chand, P., Sharma, M., Glacier changes in the Ravi Basin, North Western Himalayas (India) during the last four decades (1971-2010/13), Glob. Planet. Changes 135 (2015) 133–147 https://doi.org/10.1016/j.gloplacha.2015.10.013.

Copley, A., Avouac, J., Royer, J., The India-Asia collision and the Cenozoic slowdown of the indian plate; implications for the forces driving plate motions, J. Geophys. Res. B 115 (3) (2010) https://doi.org/10.1029/2009JB006634.

Debnath, M., Sharma, M., Syiemlieh, H., Glacier dynamics in Changme Khangpu basin, Sikkim Himalaya, India, between 1975 and 2016, Geoscience 9 (6) (2019) 259 https://doi.org/10.3390/geosciences9060259.

Debnath, M., Syiemlieh, H., Sharma, M., Kumar, R., Chowdhury, A., Lal, U., Glacial lake dynamics and lake surface temperature assessment along the Kangchengayo-Pauhunri Massif, Sikkim Himalaya, 1988-2014, Remote Sens. Appl. Soc. Environ. 9 (2018) 26–41 https://doi.org/10.1016/j.rsase.2017.11.002.

Fu, X., Wang, B., Annamalai, H., Kikuchi, K., Interaction of MJO with Asian Summer Monsoon, 2007 https://usclivar.org/working-groups/mjo/science/interaction-mjo-with-asian-summer-monsoon accessed 10 January 2020.

Griesbach, C.L., Geology of the Central Himalayas, Geol. Surv. India Mem. 170 (1891) 208–209.

Hodgson, J.A., Journal of a survey to the heads of the rivers Ganges and Jumna, Asiatic Res. 14 (1822) 60–152.

ICIMOD, The Changing Himalayas; Impact of Climate Change on Water Resources and Livelihoods in the Greater Himalayas. ICIMOD, Kathmandu 2009.

Kumar, P., Sharma, M., Saini, R., Singh, G., Climatic variability at Gangtok and Tadong weather observaatories in Sikkim, India, during 1961-2017, Sci. Rep. 10 (1) (2020) 15177 10.1038/s41598-020-71163-y.

Molnar, P., Stock, J., Slowing of India's convergence with Eurasia since 20 Ma and its implications for Tibetan mantle dynamics, Tectonics TC3001 (2009) 1–11 https://doi.org/10.1029/2008TC002271.

Molnar, P., Tapponier, P., Cenozoic tectonics of Asia: effects of a continental collision, Science 189 (4201) (1975) pp. 419–426 doi:10.1126/science.189.4201.419.

Owen, L., Gualtieri, L., Finkel, R., Finkel, R., Caffee, M., Benn, D., Sharma, M., Cosmogenic radionuclide dating of glacial landforms in the Lahul Himalaya, Northern India: defining the timing of Late Quaternary glaciation, J. Quat. Sci. 17 (3) (2001) 279–281 https://doi.org/10.1002/jqs.621.

Owen, L., White, B., Derbyshire, E., Rendell, H., Loessic silt deposits in the Western Himalayas: their sedimentology, genesis and age, Catena 19 (6) (1992) 493–509 https://doi.org/10.1016/0341-8162(92)90049-H.

Patriat, P., Achache, J., India-Eurasia collision chronology has implications for crustal shorterning and driving mechanism of plates, Nature 311 (1984) 615–621 https://doi.org/10.1038/311615a0.

Raina, V., Srivastava, D., Glacier Atlas of India, Geological Society of India, Bangalore, 2009.

Saini, R., Sharma, M.C., Deswal, S., Barr, D.I., Kumar, P., Kumar, P., Kumar, P., Chopra, S., Glacio-archaeological evidence of permanent settlements within a glacier end moraine complex during 980-1840 AD: The Miyar Basin, Lahaul Himalaya, India. Anthropocene 26 (2019), DOI:10.1016/j.ancene.2019.100197.

Schmidt, S., Nusser, M., Fluctuations of Raikot glacier during the past 70 years: a case study from the Nanga Parbat Massif, Northern Pakistan, J. Glaciol. 55 (194) (2009) 949–959 10.3189/002214309790794878.

Searle, M., Colliding Continents, first ed., Oxford University Press, Oxford, 2013.

Sharma, A., Singh, S., Kulkarni, A., Ajai, Glacier inventory in Indus, Ganga and Brahmaputra basins of the Himalayas, Natl. Acad. Sci. Lett. 36 (5) (2013) 497–505.

Sharma, E., Chettri, N., Tse-ring, K., Shrestha, A., Jing, F., Mool, P., 2009. Climate change impacts and vulnerabilities in the Eastern Himalayas. In: ICIMOD. Kathmandu.

Sharma, M., Owen, L., Quaternary glacial history of NW Garhwal, Central Himalayas, Quat. Sci. Rev. 15 (1996) 335–365 https://doi.org/10.1016/0277-3791(95)00061-5.

Singh, J.S., Agrawal, D.P., Radiocarbon evidence for deglaciation in north-western Himalaya, Nature 260 (1976) 232 https://doi.org/10.1038/260232a0.

Terra, H., de Paterson, T.T., *Studies on the Ice Age in India and Associated Human Cultures* (Carnegie Institution of Washington Publication No. 493. ed.), Carnegie Institution of Washington, Wasington, DC, 1939.

CHAPTER

Holocene climatic record of Ladakh, Trans-Himalaya

3

Binita Phartiyal, Debarati Nag, Priyanka Joshi
Birbal Sahni Institute of Palaeosciences, Lucknow, UP, India

3.1 Introduction

The Trans-Himalaya, with the Indus Suture Zone (ISZ) and the Karakorum Fault (KF) trailing the high mountain ranges (4000–7000 m a.s.l.) is tectonically active (Brown et al., 2003; Phartiyal and Sharma, 2009) as well as falls in the rain shadow region of the Indian summer monsoon (ISM). This region has a typical high-altitude desertic topography with approximately less than 100 mm of annual precipitation (Fig. 3.1). It is characterized by glaciated mountain ranges, valleys occupied by mighty rivers (Indus, Hanle, Zanskar, Shyok, Tangtse, etc.) preserving voluminous sediment of glacial, fluvial, lacustrine and eolian origin and also hosts several massive valley lakes viz., Pangong Tso, Tso Moriri, and Tsokar. These lakes (paleo and glacial) are good archives for studying the paleoclimate of the region. The lake expansion and/or shrinkage here result from changing precipitation and evaporation controlled by the interplay of the westerlies strength and Asian monsoon (Bookhagen et al., 2005; Chen et al., 2008; Long et al., 2010; Hudson and Quade, 2013; Dimri et al., 2016). These lakes have experienced significant changes in water level during the Holocene, that is, inferred mainly from lacustrine sediment records and the chronostratigraphy of lake terraces/paleo-shorelines.

Several lakes were formed during the late Pleistocene, in under different time spans and stretches (Fort et al., 1989; Kotlia et al., 1997; Phartiyal et al., 2005 2013 2015, 2020a, 2020b; Sangode et al.,2013; Nag and Phartiyal, 2015; Clift et al., 2014; and references cited therein). Most of them are formed just after the glacial stages suggesting climate is the major driving force. Apart from these several paleolakes, were formed by a coupled effect of neotectonic and climatic activity along the ISZ (Phartiyal et al., 2005 2013; Phartiyal and Sharma, 2009; Sant et al., 2011; Sangode et al., 2011, 2013; Nag and Phartiyal, 2015; Nag et al., 2016, 2021). Damming of the Indus River by massive landslides (Abbott 1849; Becher, 1859) and catastrophic flooding in higher precipitation regimes (Dortch et al., 2009) and breaching of these landslides generated dams/lakes (Burbank, 1983), suggest that these events are not geologically uncommon in this cold high-altitude desert. Even the Pangong Tso (Bangong Co) extended 40 km eastward along the KF was once draining into the Shyok River (Phartiyal et al., 2015), the Khalsar paleolake existed along the KF (Phartiyal et al., 2005, Phartiyal and Sharma, 2009) and the Khalsi-Saspol paleolake along River Indus (Nag and Phartiyal, 2015; Nag et al., 2016, 2021).

A complete picture of the climatic variation during Holocene is recorded in the paleolake sequences and the glacial lakes of the Indus valley and the northward Ladakh range. We present here multiproxy records of the paleolake sections at Spituk; Khalsi and published glacial lake records from North Pulu (NP) and South Pulu (SP; Phartiyal et al., 2020a, 2020b); Tsoltak and YayaTso (Joshi et al., 2020) of the

Holocene Climate Change and Environment. DOI: https://doi.org/10.1016/B978-0-323-90085-0.00023-1

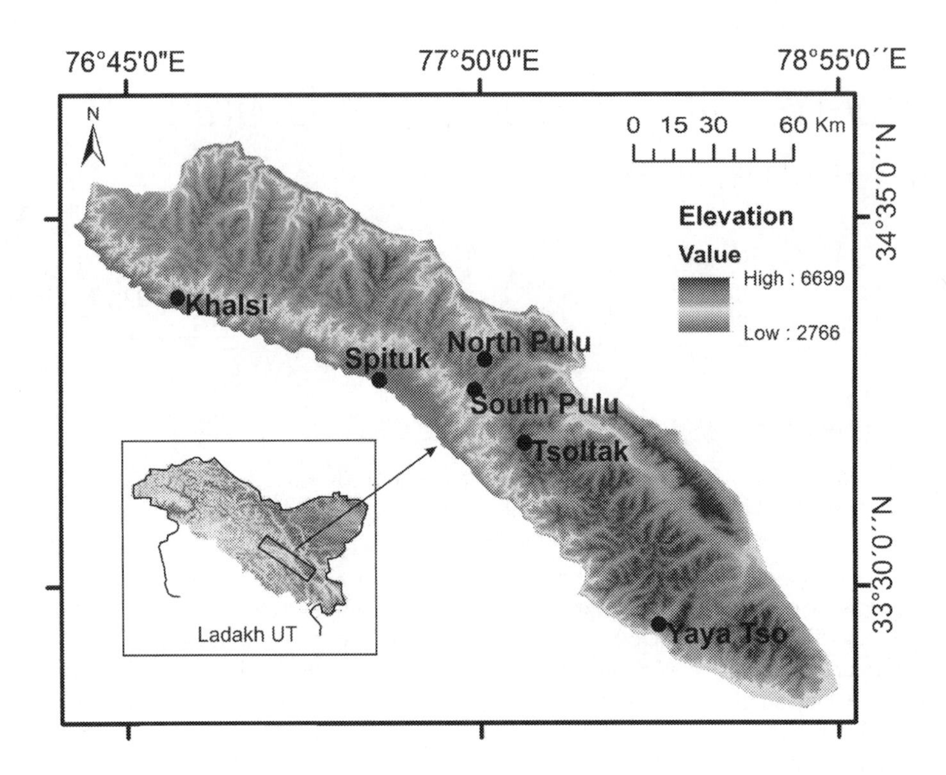

FIGURE 3.1

Map of India showing the transact selected to generate the swath profiles for elevation and precipitation from the foothills to the Karakoram Range in the Northwestern Himalaya along the line AA' and study area marked in black rectangle.

Ladakh Range (Fig. 3.2). These records cover the entire Holocene period and a spatio-temporal picture of the Ladakh sector as far as Holocene climatic history is concerned.

3.1.1 Study area

Ladakh sector lies in the extreme north-western part of India in the Trans-Himalaya and is the western fringe of the Tibetan Plateau. It falls under the rain-shadow zone for monsoon clouds and therefore experiences cold desertic climatic conditions. The study area lies along the River Indus that river originates in the Tibetan plateau (from the northern slope of Mount Kailas (6714 m) in the Gangdise range) near Lake Mansarovar (Allen, 1984). Indus River is one of the most important rivers of the Ladakh region and flows for a length of ~500 km in India (mostly in Ladakh region) (Fig. 3.3).It is geographically the backbone of Ladakh, and flows in the NE direction along the ISZ. The main right bank tributary is the River Shyok which joins Indus at Skardu (in Pakistan). The left bank tributaries are Hanley, Zanskar and Suru and in the study area Zanskar river forms the main left bank tributary. Apart from the major tributaries, the Indus River is fed by numerous tributaries throughout its length

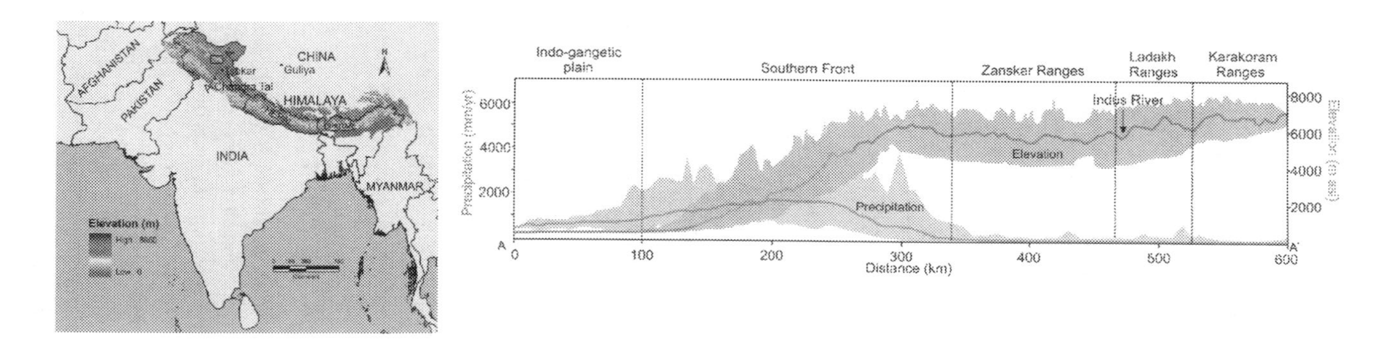

FIGURE 3.2

DEM of Himalaya with the studied sites and Elevation and Precipitation swath profile along AA'.

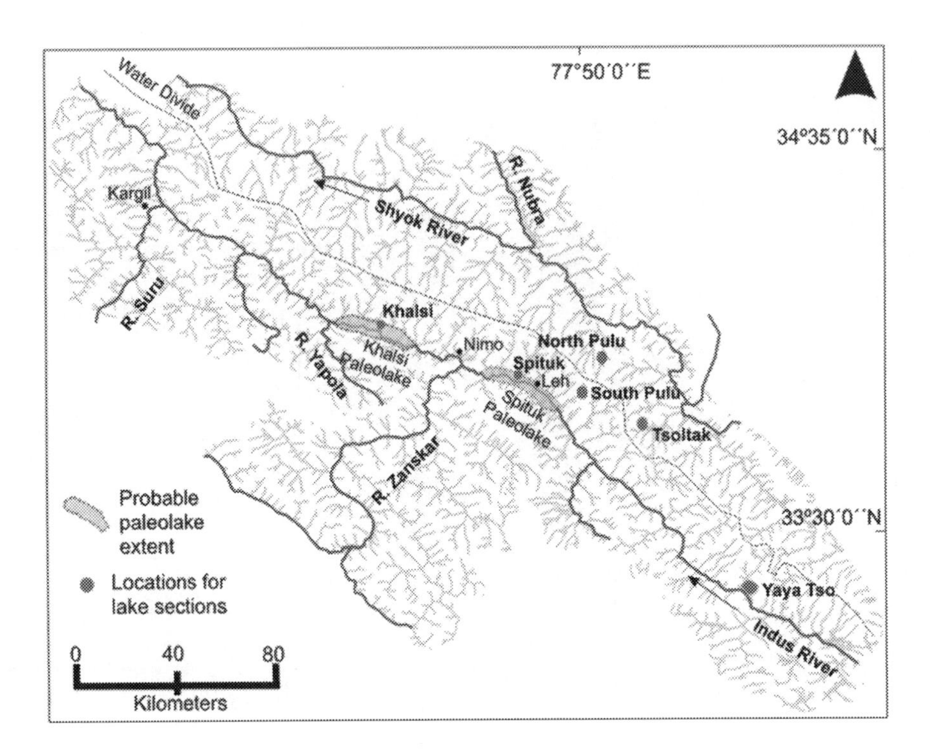

FIGURE 3.3

Drainage map of the study area with location of the study sites marked.

(Fig. 3.3). The river is narrow, constricted and meandering in most of the places except in Leh valley region where the valley becomes wide (2–3 km) changing the nature of the river to a braided stream. At times it is seen flowing through narrow gorges cutting deep into the country rock. The Indus seems not only to have eroded through the overlying sediments to superimpose itself upon the rocks of the Ladakh terrain beneath but also have maintained erosion as an antecedent stream across the newly rising structure further downstream (Shroder Jr., 1993).

3.1.2 Geomorphology, climate, and vegetation

Ladakh sector lies in the vicinity of the ISZ, a seismo-tectonically active fault trending in a NW–SE direction with Ladakh batholith in the north and Zanskar group of rocks toward the south (Searle, 1986). It has a rugged topography constituting barren mountains with altitudes more than 3000 m. Severe winters are experienced, with snow cover nearly for 3–6 months in the valley areas and peaks above 5000 m permanently capped with snow throughout the year. The prominent Ladakh Range trends NE–SW is ~350 km in length, making the north wall/bank of the Indus river in the study area. This range variesin width from ~50 km in the west and pinching toward the eastern side to 20 km. It is a tectonically active zone between the ISZ in the south and the Karakoram Thrust (KT) in the north, hosting several proglacial lakes.

Exhibiting a typical mountain geomorphology dominated by glacial and fluvial landforms, it has enormous Quaternary deposits of glacial, lacustrine, paleolacustrine, fluvial, and eolian origin are scattered in the numerous basins of this range. This region is an exceptional area studying the glacial and lacustrine deposits for their depositional environment and climatic variations. The terrain is rugged and the glacial cover is very less as nearly all the snow/ice covering the peaks of the Ladakh range melt in the summer months. The proglacial lakes occupy only the higher elevations (as of today) and the ones below are seen as remnant/paleolakes/dry lakes now scattered over the abandoned moraines. It is well established that the reconstruction of former glacial lake outlines based on geomorphologic evidence has long played a central role in reconstructing the past climate and hence four sections were studied (SP, NP, Tsoltak, and YayaTso) located at Khardung La pass, ChangLa pass, and the HorLa pass, respectively (Figs. 3.2 and 3.3).

The region lies in the rain shadow of NW Himalaya having an arid to hyper arid climate and dry steppe vegetation (Fig. 3.1). With two sources of precipitations (i.e., westerlies and ISM) and it is affected intensely by the ISM during abnormal monsoon years (Mayewski et al., 1980; Goudie et al., 1984). With 6 months of very cold weather conditions the area is almost a barren land that is exposed to physical weathering by frost action and shattering the rocks into rubble piles. This also contributes to a huge sediment budget in the area which lies loose in the high angled slopes. At time it can be triggered down slope either by climate or tectonic induced activity and slide down the slopes blocking the rivers and channels at constricted places, pounding them and forming lakes (Kotlia et al., 1997; Phartiyal and Sharma, 2009; Phartiyal et al., 2011, 2013, 2015; Nag and Phartiyal, 2015). The average summer temperature in the Ladakh region goes up to 25°C in July while the average winter temperature dips down to -15°C by January (Hussain et al., 2015).

Most of the region lies above the tree line with little or no vegetation. Semideserts and steppes type of vegetation prevail in this landscape (Dvorsky et al., 2010). The vegetation present on hill slopes is dwarfed and prickly (xerophytic shrubs Artemisia, Caragana, and herbaceous plants) and small trees (Alnus, Betula) grow on the valley floors (Phartiyal et al., 2005; Joshi et al., 2020). The scant precipitation makes Ladakh a cold-arid desert, harboring short-term scarce herbaceous vegetation, bryophytes, lichens, algae and microscopic heterotrophs etc., that are observed mainly along the margins of water courses and slightly raised grassy patches/cushions accreting finer sediments or rocky crevices and mounds. At present the human settlements in the lower reaches of valleys presently cultivate 15 species of gramineae for straw and seeds including *Triticum aestivum* (wheat), 3 of cyperaceae, 7 of salicaceae, 10 species of leguminosae for pods and leaves, 2 each of Elaeagnaceae and Saxifragaceae, Rosaceae for fruits, berries and leaves besides *Morus alba* for leaves (Ahmad and Singh 2020). Willows, apples, apricots, walnut trees in this region grow due to irrigation through various streams, springs and rivers in the valley. Several other plants used as fodder or for fruits are in cultivation. Juniper trees grow wild in some locations. *Tanecetum, Artemisia, Echinops*, and *Circium* of Asteraceae family are quite common in the valleys. However, only the seasonal herbs are present around the lakes under study as they are situated in altitudes above 5000 m.

3.2 Scope

Outside the polar regions, this mid latitude, high altitude glaciated region holds the key for understanding several climate uncertainties amidst concurrent planet warming scenario. As global surface temperatures over land and oceans congruently show a discernible warming trend that might be impacting glacier mass balance globally, it is understandable that low latitude, high altitude glaciated terrains are much more susceptible and prone, to respond adversely. Recent uprising trend in flash floods induced by abrupt and intense precipitation events in the high-altitude Himalaya are becoming worrisome. Taken together, hydroclimate variability of north-western Himalaya is undergoing plausibly unprecedented changes amidst global warming era. To understand concurrent variability, it is very important to have good knowledge of past variability especially Holocene and this study is trying to give a complete overview of the paleoclimate during the Holocene.

We present here the multiproxy, high-resolution records of the paleolake sections at Spituk (~13,500–1500 cal yr BP); Khalsi (~14,000–6500 cal yr BP) and published glacial lake records from NP (~4500–360 cal yr BP) and SP (~4100–266 cal yr BP) (Phartiyal et al., 2020a, 2021); YayaTso (~6300–220 cal yr BP) (Joshi et al., 2020) and Chang La (Tsoltak) (~7200–1000 cal yr BP) of the Ladakh Range (Fig. 3.2). These records cover the entire Holocene period and a spatio-temporal picture of the Ladakh sector as far as the Holocene climatic history of Ladakh is concerned. Moreover, the importance of the study area can be gauged from the fact that this region of Leh Ladakh had been utilized for international trade via famous Silk route during early historic period (between 206 BC to 220 AD) (Mischke et al., 2002). The climate and the environmental factors may have played an important role in fostering economic and socio-cultural changes along this route. In addition, this cold desert area is also recently chosen as analogue site for extra-terrestrial research (spacewardbound.astrobiologyindia.in) (Pandey et al., 2019). Hence a clear picture of the climate through Holocene will help not only the climatologists but also the historians and the archeologist.

3.3 Methodology and processes

Paleolake sediments, as well as the glacial lakes were identified and mapped along the Indus river (Ladakh) in the field with the help of Survey of India topographic maps and satellite data. Sampling was conducted sequentially by either channel or pit; core raising and from exposed sections and textural analysis, mineral magnetism, total organic content (TOC) and loss on ignition (LOI) analysis were carried out. Based on the frequency and variations of the above multiproxy dataset, TILIA 1.7 software for grouping/cluster analysis using constrained incremental sums of squares cluster analysis (CONISS) was run, and accordingly climatic zones were defined for different sections that allowed defining a particular environmental condition. TILIA and its graph software were used for the preparation of a range chart and CONISS was applied for cluster analysis (Grimm, 1987, 1990). Textural analysis, mineral magnetism, and LOI/TOC proxies were used.

For *textural analysis* standard technique for sedimentological parameters was used. In nature, the particles are generally separated by their sizes depending on the movements of air and water. The distribution of sizes is related to the availability of different sizes of the parent material, the processes operating, particularly the competency of flow and concentrations of particles in suspension. Hence, the textural analysis provides fundamental information in identifying the environment of deposition (Folk and Ward, 1957; Friedman, 1969; Inman, 1949; Krumbein, 1934; Spencer, 1963; Tanner, 1958; Visher, 1969). For river dammed lakes, current and wave motion form the prime controlling factor for variation in the textural characteristics. Grain size distribution study of the long, deep and narrow open valley lakes provide insight into the sediment transport and distribution of energy condition with respect to the position within the basin. However, for closed lakes, sediments are deposited via settling and is controlled by the lake level and size. Grain size data for closed lake reveal history of low and high stands. Higher energy conditions of sediment production or transport as well as low-level lake stands corresponds to increased grain size and vice versa. For *mineral magnetic studies* samples magnetic susceptibility (χ) at low (0.47 kHz) frequency was determined (Walden, 1999). Anhysteric remnant magnetization (ARM) was induced and the susceptibility of ARM (χ_{ARM}) was calculated. Isothermal remnant magnetization (IRM) was induced and interparametric ratios were calculated—the SIRM/χ_{lf}, χ_{ARM}/SIRM and S-ratio. The SIRM/χ_{lf} ratio gives a determination of grain size with higher ratio suggesting a coarse grain size and vice versa for χ_{ARM}/SIRM. The S-ratio is the negative of the ratio of the IRM induced at -300 mT to the SIRM induced at 700 mT (-IRM-300mT/SIRM700mT). For magnetic mineralogy measurements, the IRM acquisition was performed on all samples. The IRM acquisitions also show the contribution of magnetite which is the primary mineral contributing to the remanence in these sediments.

The formation, transport, deposition, and transformation of magnetic content in lake sediments indicate environmental conditions and geomorphic processes in the catchment (Oldfield, 1991; Versoub and Roberts, 1995). Higher values of χ_{lf} imply effectiveness of the erosional processes and increasing concentration of detrital input from the catchment (Williamson et al., 1998). During the high-intensity erosion, more magnetic particles come as clastic input to the basin, thus increasing the magnetic properties of the sediments. Glacial abrasion is the most influential and powerful mechanical force for the land surface erosion in glacial areas, for example, more the glacial expansion (enhanced glacial erosion), more the generation of magnetic clast, and vice versa. Therefore, glacier advance or retreat can be revealed well by the studies of proglacial lakes (Dahl et al., 2003; Matthews et al., 2000; Nesje et al., 2000). Similarly, S-ratio, SIRM are also selectively sensitive for the detrital influx which is the result of either glacial advancement or the glacial melting. χ_{ARM} is the reflection of authigenic contribution not the direct result of catchment processes.

LOI is an indirect proxy of organic matter in the sediment and where TOC when direct TOC was not available LOI was used for paleoenvironmental investigations (Korsman et al., 1999; Dodson and Ramrath, 2001; Bendell-Young et al., 2002). In lake and bog systems, the TOC concentration can serve as a direct indicator of the biomass of local vegetation (Chai, 1990; Zhou et al., 1997, 2004) and thereby potentially reflects changes in the regional precipitation. This method has been widely used for paleoenvironmental investigations (Bendell-Young et al., 2002; Dodson and Ramrath 2001; Korsman et al., 1999; Phartiyal et al., 2011).

Chronologies of the sections were obtained from accelerator mass spectrometer (AMS) radiocarbon dates from Silesian University of Technology, Gliwice, Poland and Direct AMS USA for all the sections (paleolake as well as glacial lakes; chronologies published in Phartiyal et al., 2013, 2020b, 2021; Nag and Phartiyal 2015; Joshi et al., 2020 from few of these paleolakes were used). The calibrated age depth graph for all the studied sections is shown in Fig. 3.4A and B. Table 3.ST1 shows the details of chronology which were obtained from accelerator mass spectrometer (AMS) radiocarbon dates from Silesian University of Technology, Gliwice, Poland and Direct AMS USA for all the sections (paleolake as well as glacial lakes). For all measured ^{14}C ages (of sediment organic matter), the calibration was performed using standard methods OxCal v4.2 (Piotrowska, 2013; Ramsey, 2009) and CALIB 7.1 Radiocarbon Calibration Software (Stuiver et al., 2017). The calibrated age ranges at 95% confidence interval (2σ) and their weighted means are given in Table 3.ST1 and Fig. 3.4A and B. A supplementary age-depth model with is provided which is made using Bchron which runs the Compound

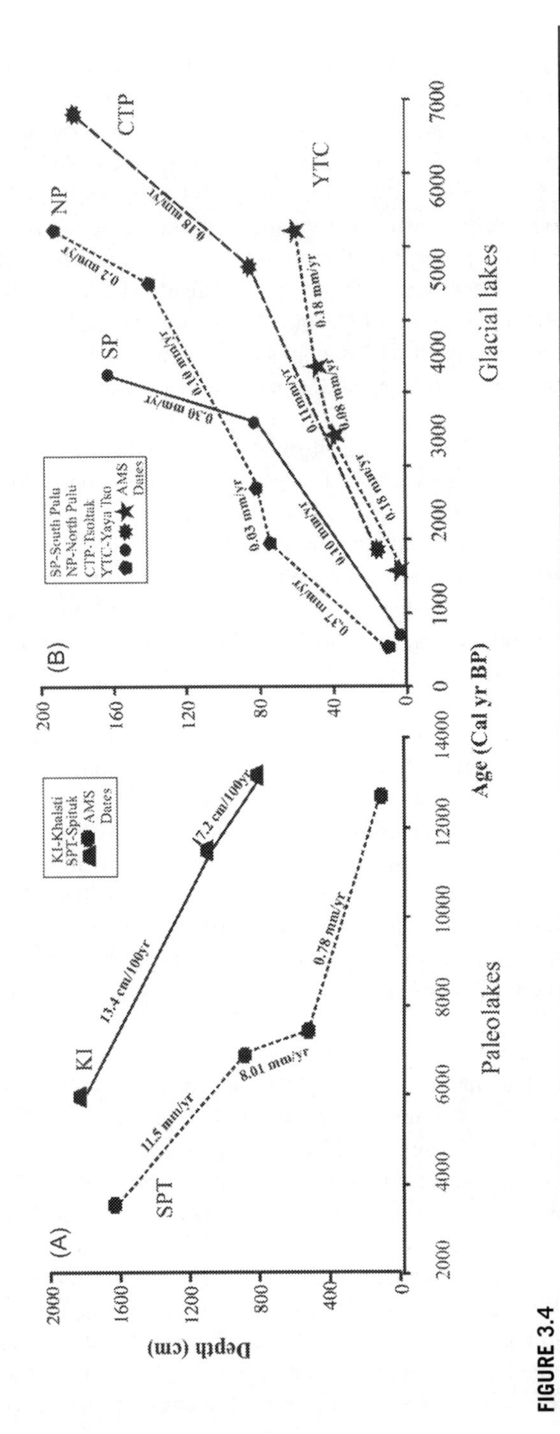

FIGURE 3.4

Age depth model of the studied sites with their calibrated age ranges and probable sedimentation rate (A) of the paleolakes and (B) glacial lakes.

Poisson-Gamma chronology model of Haslett and Parnell (2008) using IntCal13. The Bchron function fits a compound Poisson-Gamma distribution to the increments between the dated levels. This involves a stochastic linear interpolation step where the age gaps are gamma distributed, and the position gaps are exponential. The plot of the Bchron ages plot of all the sections is given in the supplementary material (Fig. 3.SF1).

3.4 Field work and sampling techniques

The field work was carried out between Upshi and Khalsi, a stretch of ~140 km that hosts several sedimentary facies types. Paleolake sediments, as well as the glacial lakes were identified and sampled. For the exposed sections (Spituk and Khalsi), the surface was cleaned and the samples were collected at every 5 cm level. For NP, Tsoltak dry lake bed, a jaw crushing bulldozer was employed to dig a trench and then channel and pit sampling was done. Pit was also dug at the SP site and sampling was carried out in a continuous way, of 2–2.5 cm as one sample layer. In YayaTso and NP a core was raised in by a PVC pipe and the samples were cut at 1.5–2.0 cm interval in the laboratory. Fig. 3.5A–K represents the field photographs of the studied sites and the different types of techniques involved during the sample collection procedure. Lithological variations of the studied sections are given in Fig. 3.5. The sections depth from NP (130 cm); SP (168 cm); Tsoltak (190 cm); Yaya Tso (100 cm); and for Spituk Section (24 m) and Khalsi (18 m) was studied/obtained.

3.5 Observations and results

3.5.1 Lithology and chronology

3.5.1.1 Khalsi section (34°20.03°N, 76°52.54°E; altitude 3250 m)

This 18 m thick section is a part of Saspol-Khalsi paleolake reported by Nag and Phartiyal (2015). The section is exposed above Khalsi village at approximately 100 m aprl. Based on grain size, lithofacies and color variation the entire section is divided into 13 units (Unit K1 to Unit K13) (Fig. 3.6) (Nag and Phartiyal 2015; Nag et al., 2016). The details of the facies description are given in Table 3.1.

The chronology for Khalsi section is established by dating three samples using ^{14}C AMS method from depth 8.5 m, 11.5 m and 18.5 m. Chronological details are given in Table 3.ST1, SF1, and Fig. 3.3A. The average sedimentation rate calculated for the period 11,154–5938 cal yr BP is and for the period 12,887–11,154 cal yr BP is 1.72 mm/yr. The ages for the whole profile are extrapolated using the mean sedimentation rate calculated as 1.34 mm/yr.

3.5.1.2 Spituk section (34.13246°N; 77.52563°E)

The Spituk section which was studied previously (Phartiyal et al., 2005; Sangode et al., 2011, 2013; Phartiyal and Sharma, 2009; Phartiyal et al., 2013) overlies the Ladakh batholith and is exposed on the right bank of the Indus River 4 km downstream Leh. The 24 m thick sequence (Phartiyal et al., 2013) of mainly fluviolacustrine facies (buff colored) has been taken up for these studies, however other workers have described ~38 m (Sant et al., 2011) and ~50 m (Sangode et al., 2013; Kumar and Srivastava, 2017) of this section. The details of the lithology are given in Table 3.2.

FIGURE 3.5

Field photographs of the studied sections; (A) Overview of the Khalsi paleolake; (B) Sampling in the Khalsi paleolake, the section is cleaned/scraped off thoroughly before samples collection; (C) Overview of Spituk paleolake; (D) Pit sampling being done manually at the South Pulu; (E) Jaw crusher bulldozer (JCB) used for digging pit at Tsoltak; (F) Core raising at Yaya Tso lake with a PVC pipe; (G and H) Channel and pit sampling at the Tsoltak glacial lake; (I and J) The studied sections from North Pulu and South Pulu glacial lake sites; (K) Core cutting done at the BSIP laboratory using the core cutter.

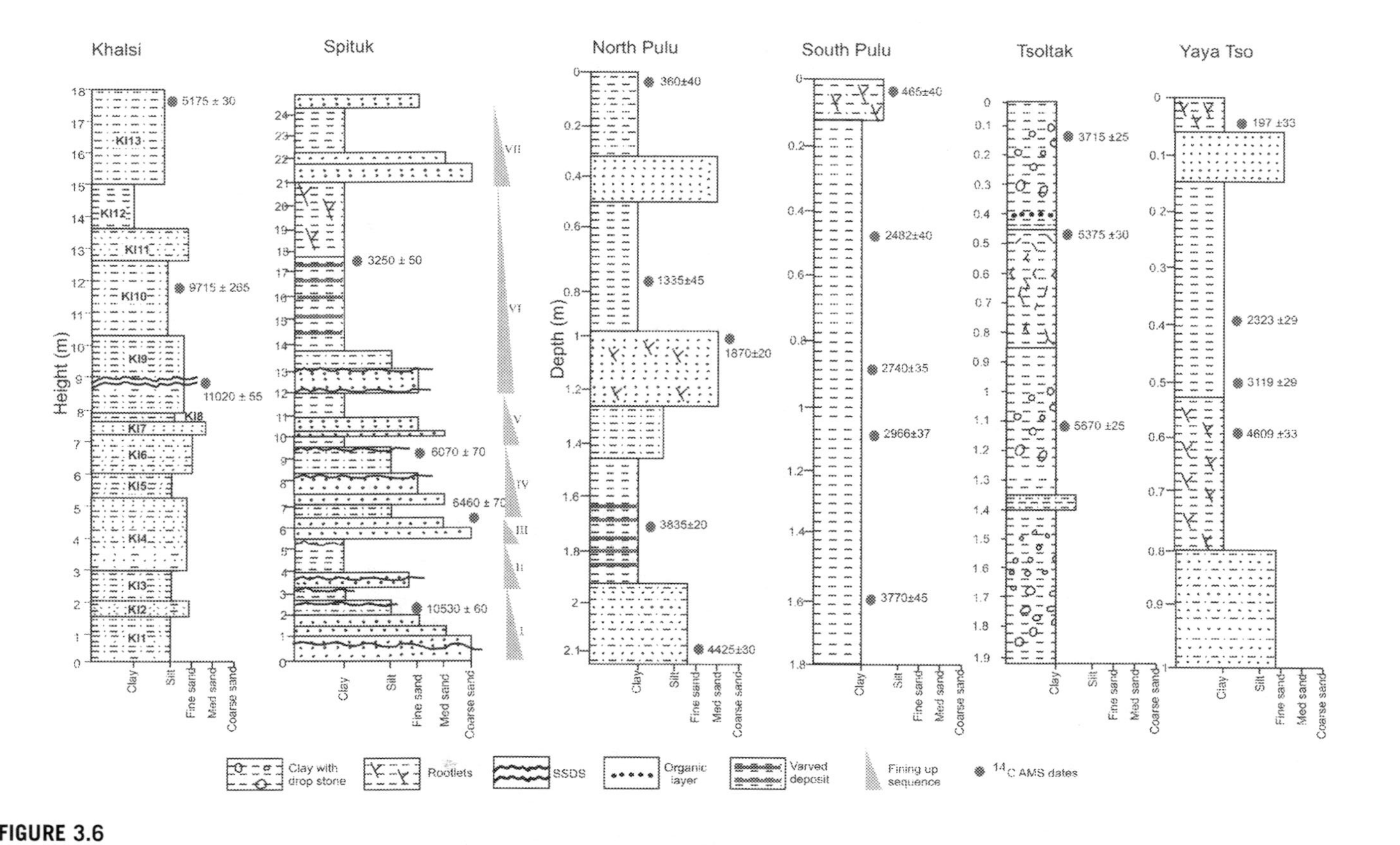

FIGURE 3.6

Lithosections of the paleolakes (height measured from ground surface) and glacial lakes (depth from ground surface) with marked ^{14}C AMS dated positions.

Table 3.1 Lithological description of the Khalsi section (after Nag et al., 2016).

Unit	Broad unit/facies	Detailed description and sedimentary characteristics
Unit K1	Coarse to fine silt with very coarse silty partings	~1.4 m thick; grayish medium to fine silty bed, punctuated by mm scale thick very coarse silty partings at two places; polymodal, poorly sorted very coarse to fine silt with very fine to fine skewed and mesokurtic to leptokurtic grain size distribution
Unit K2	Very fine sand bed	~1.6 m thick consisting of massive, grayish very fine sand deposit having a gradational contact with the underlying unit. Polymodal, moderate to poorly sorted very fine sand with very fine skewness value and very leptokurtic grain size distribution
Unit K3	Laminated medium to fine silt bed	~1 m thick laminated (few cm) fine silt sediments. Fine silt is interlayered with medium silt at few levels. Samples are polymodal, poorly sorted medium to fine silt with fine skewed and leptokurtic grain size distribution
Unit K4	Very coarse silt bed	~2.2 m thick, massive grayish very coarse silt, polymodal, poorly sorted very coarse and coarse silt with very fine skewed and mesokurtic to very leptokurtic grain size distribution
Unit K5	Fine silt bed	~0.7 m thick with polymodal, poorly sorted fine silt showing fine skewed and mesokurtic to leptokurtic grain size distribution
Unit K6	Coarse silt bed	~1.2 m in thickness and consists of grayish massive coarse silt deposit. Samples are polymodal, very poorly to poorly sorted coarse to medium silt with very fine to fine skewed, and platykurtic and leptokurtic grain size distribution
Unit K7	Small scale cross-bedded very fine sand bed	~0.5 m thick and made up of very fine sand showing small scale low angle cross-bedding. Samples are polymodal, poorly sorted very fine sand with very fine skewed, and very leptokurtic grain size distribution
Unit K8	Very fine silt bed	10 cm thick very fine silt and samples are polymodal, very poorly to poorly sorted fine to very fine silt with very fine skewness value and mesokurtic to leptokurtic grain size distribution
Unit K9	Coarse to medium silt bed	~2 m thick massive silty bed, shows deformation level (convolute and pseudonodules structures) and samples are polymodal, very poorly to poorly sorted coarse to medium silt with very fine to fine skewed and mesokurtic to leptokurtic grain size distribution
Unit K10	Fine silt bed	The unit consists of fine silt facies of ~3 m thickness, showing polymodal, poorly sorted fine to very fine silt with fine to very fine skewed and leptokurtic to very leptokurtic grain size distribution
Unit K11	Coarse silt bed	0.6 m of coarse silt and samples are trimodal, poorly to very poorly sorted with very fine skewed and leptokurtic grain size distribution
Unit K12	Very fine silt and clay facies	~1.2 m thick grayish yellow massive silty clay. Samples of the facies analyzed showed a bimodal nature of sediment, moderately to poorly sorted very fine silt and clay with coarse skewed and very leptokurtic grain size distribution
Unit K13	Coarse to fine silt bed	~3m grayish silty bed and consists of polymodal, very poorly to poorly sorted coarse to fine silt with very fine to fine skewness and mesokurtic to leptokurtic grain size distribution

Table 3.2 Lithology of the Spituk section.

Units	Height (m) from bottom to top	Description	Deformation levels (Phartiyal and Sharma, 2009)	Unit (as given in Sangode et al., 2013)
I	~3.2	Lower part pebbly grading to medium to coarse grained sand layer varying in thickness showing seismites at the top. Overlying is well sorted; gray, but intensely liquefied silt. This fine silt has thin lenticular laminations of mud followed by another liquefied mud layer. This upper part constitutes of a mudstone package (thinly laminated mud layers as well as light yellowish gray, massive mudstone) showing ball and pillow structures at the top. The thickness and lateral continuity of this mud dominated unit indicates probably the first extensive lacustrine phase in this basin.	Three	Units 3–8
II	~2	Massive sandstone, overlain by thinly laminated, interlayered sand and silt showing ripple marks. Overlying it is light cream colored mudstone, the upper part becomes grayish, showing thin laminations. Top of this unit shows liquefaction injecting into upper layer. The symmetrical ripples indicate oscillatory nature of deposition under shallow water conditions.	Two	Units 9–12
III	~1.2	Fine grained and thinly laminated sandstone layer containing coarse grained, laterally pinching-out sand layers in the middle part overlain by thinly laminated mud and silt.	–	Units 13, 14
IV	~2.8	Coarse and medium grained sands, well imbricated, rounded to subrounded pebbles. Overlain by dark gray silty sands of massive nature followed by interlayered mud showing liquefaction.	Two	Units 15–17
V	~2	Fining upward cycle of course to medium grained sand with trough cross stratifications overlain by meter thick massive mudstone.	–	Units 18–20
VI	~11	Thick coarse grained, stratified sandstone showing angular clasts at the bottom overlain by a course to medium grained sandstone showing parallel laminations at the base and ripple laminations at the top with convolute bedding. It grades into thick mudstone of yellow color followed by olive green mudstones. Four bands/layers of dark mudstone are distinct. This is overlain by a thick sequence of variegated mudstones having varves of gray, green, pink, and light brown color.	Two	Units 21–31
VII	>2.5	Coarse to fine grained, fining up cycle of sand, gradationally passing to mudstone.	–	Units 36–38

Chronology for five samples has been used after Phartiyal et al. (2013). Details of the chronologies are given in Table 3.ST1. This section has been studied by several workers and has had a discrepancy in the chronology which was resolved after re-dating the samples (Phartiyal et al., 2013). The age bracket of this section ranges from 12,000 to 1500 cal yr BP. The sedimentation rate between 12,512 and 7385 cal yr BP is 0.78 mm/yr, from 7385 to 6992 cal yr BP is 8.01 mm/yr and from 6992 to 3446 is 11.5 mm/yr (Fig. 3.4A, SF1, and ST1).

3.5.1.3 North Pulu (NP) (34°17.321′N, 77°35.714′E; altitude: 5098 m)

The two streams originating from the water divide are—Khardung Range (28 km) of the Ladakh Range, to the north joins the Shyok River and Sangto Tokpo (26 km) toward the south joining the Indus River. Both these streams host small intermorainic glacial lakes (<0.5 sq km area) mostly frozen except for few summer months at NP and SP. Samples were collected from the NP by channel and pit sampling (130 cm) and core raising (80 cm). The lake is bracketed from ~5400 to 350 cal yr BP. The sedimentation rate between 360 and 1335 cal yr BP is 0.37 mm/yr, from 1335 to 1870 cal yr BP is 0.03 mm/yr, from 1870 to 3835 cal yr BP is 0.11 mm/yr and between 3835 and 4425 cal yr BP is 0.2 mm/yr (Fig. 3.4B).

3.5.1.4 South Pulu (SP) (34°14′55.45″N; 77°37′03.46″E; altitude: 4900 m)

The SP lake is a small lake of an area of 0.00325 sq. km placed over the right lateral moraine of the Khardung Glacier. A 168 cm section of the SP Lake was sampled by pit and channel sampling, continuously with each sample thickness being 2.4–2.5 cm. The lake is bracketed from ~4100 to 266 cal yr BP. From ~4130 cal yr BP to ~2820 cal yr BP the sedimentation rate is around 0.10 mm/yr, between 2820 and 2546 cal yr BP. it is ~0.30 mm/yr and in the upper part till 266 cal yr BP it decreased to ~0.02 mm/year (Fig. 3.4B and SF1).

3.5.1.5 Tsoltak paleolake (CTP) (34°66′06.89″N and 78°00′14.89″E; 4719 m)

In the outwash slope (descending from ChangLa pass toward Durbuk; west slope) in the left bank a lateral undulating moraine is seen on which thick patches buff colored clay with mud cracks on the surface. A 190 cm section consists of clay, silty-clay beds with some dropstones till 40 cm level from the top, and then a blackish layer after 40 cm level. Further beneath in this dark layer, several rootlets were seen with the sediment till 85 cm level underlain by clays till 190 cm with a layer of silty clay at 135–140 cm level. Dropstones ranging in size from 1 cm to 2–3 cm are seen scattered throughout. At the base the section becomes coarser. Three accelerator mass spectrometer radiocarbon dates were obtained by dating the bulk sediment which bracket this section from ~7000 to 1000 cal yr BP. Sedimentation rates between 1535 and 5015 cal yr BP is 0.09 mm/yr, whereas from 5015 to 6822 it is 1.80 mm/yr for Tsoltak glacial lake (Fig. 3.4B and SF1).

3.5.1.6 Yaya Tso (YTC) (33°16.874″N and 78°29.172″E; 4701 m)

The Yaya Tso lake with an area of 6396 sq m is another pristine lake, a part of the Hor La-Mahe basin which is situated on the eastern side of the Ladakh Range. Situated enroute to Hor La, ~8 km from Kocha village this is a bigger lake system compared to the Tsoltak lake. The Yaya Tso is fed by several small channels running down from Hor La and on the northern side has lush green herbaceous meadows with a dense network of drainage and paleo strand lines on which a nomadic village exists. Even today the lake is not much disturbed by anthropogenic activities. A 1 m core was raised from the margin of the lake in a PVC pipe. The ^{14}C AMS chronology brackets the Yaya Tso between ~6400 and 220 cal yr BP. The sedimentation rate for Yaya Tso lake between 223 and 2337 cal yr BP is 0.18 mm/yr, from 2337 to 3356 cal yr BP it is 0.08 mm/yr while from 3356 to 5418 cal yr BP it is 0.18 mm/year (Fig. 3.4B and SF1).

3.5.2 Paleoclimate reconstruction

Based on the frequency and variations of the multiproxy dataset, TILIA 1.7 software for grouping/cluster analysis using CONISS was done, and accordingly climatic zones were defined for different sections that allowed defining a particular environmental condition (Figs. 3.7A–F).

3.5.2.1 *Paleolake sections*

3.5.2.1.1 Khalsi (KI)

The base of the Khalsi section is ~14,000 cal yr BP. Broadly nine (KI 1–9) climatic zones can be delineated at Khalsi (Fig. 3.7A). Nag and Phartiyal (2015) have placed the base of this paleolake at 17,800 cal yr BP, to have been formed after the LGM deglaciation. From14,000 to 13,200cal yr BP (KI 1–3) fluctuation between short pulses of cold-arid and warm wet climatic phases characterized by low mineral magnetic concentration and increased χ_{ARM}/SIRM denotes abundance of finer magnetic grain size. Low values of S-ratio indicate high proportion of antiferromagnetic minerals. Increasing sand size sediments, lowering of TOC is seen for these zones. This zone suggests fluctuations in climatic condition that alternate between cold-arid and warm-humid phases. The period between 13,200 and 12,000 yr BP (Zone KI 4 and 5) is distinguished by low χ_{lf}, low χ_{ARM}. Decreasing mineral magnetic concentration accompanied with decreasing coarser magnetic grain size. S-ratio reveals presence of both ferrimagnetic and antiferromagnetic. The zone is marked by silty sediment followed by sand. Sand content increases toward the top of the zone. Organic carbon content is low to moderately high. This zone is characterized by moderately cold and arid climate. From 12,000 to 10,800 cal yr BP (KI 6) low χ_{lf}, χ_{ARM}, and SIRM with increased S-ratio denoting mixed antiferro and ferromagnetic minerals and increase in χ_{ARM}/SIRM indicates increasing finer magnetic grain size. High silt and clay content is observed. Organic carbon content is moderately high. This zone denotes moderately warm and wet climate. The preceding zone 10,800–10,000 (K7) with rise in the SIRM concentration. Reducing trend if χ_{ARM}/SIRM shows coarser magnetic grain size and S-ratio indicates presence of both ferri and antiferro magnetic minerals. Increase in sand level and drop in clay content is seen, however silt is the dominant sediment size with decreasing organic content. This zone indicates a cold and arid climate phase. The zone KI 8 falling between 10,000 and 8000 cal yr BP is marked by low χ_{lf} and SIRM. χ_{ARM} is slightly increasing to the top of the zone. Finer magnetic grain size increases as seen from increased χ_{ARM}/SIRM values. S-ratio values depict a shift toward antiferromagnetic minerals. High proportion of clay and silt and high organic content is seen. This zone is warm and wet. The period between 9000 and 8500 cal yr BP (KI 9) is characterized by increasing mineral magnetic concentration accompanied with further decrease in finer magnetic grain size. S-ratio reveals increase toward more ferromagnetic content. Dominant silt sediment followed by sand and negligible clay. Organic carbon content is low for this zone. This zone is characterized by a cold and arid climatic condition. From 8500 to 6500 cal yr BP, the concentration-dependent parameters show low values. Increasing values of SIRM/χ_{lf} and S-ratio show shift toward ferrimagnetic minerals. Low χ_{ARM}/SIRM also supports the presence of coarser magnetic grains. Mineral magnetic parameters suggest dominance of multi domain ferrimagnetic minerals which is characteristics of detrital grains. The climate shows a warm wet phase with increased melt water supply to the basin.

3.5.2.1.2 Spituk (SPT)

Broadly six climatic zones are seen (SPT1-SPT6) (Fig. 3.7B) SP. The cold arid zone from 13,500 to 12,500 (Zone SPT1A) shows low values all the concentration-dependent parameters (χ_{lf}, χ_{ARM},

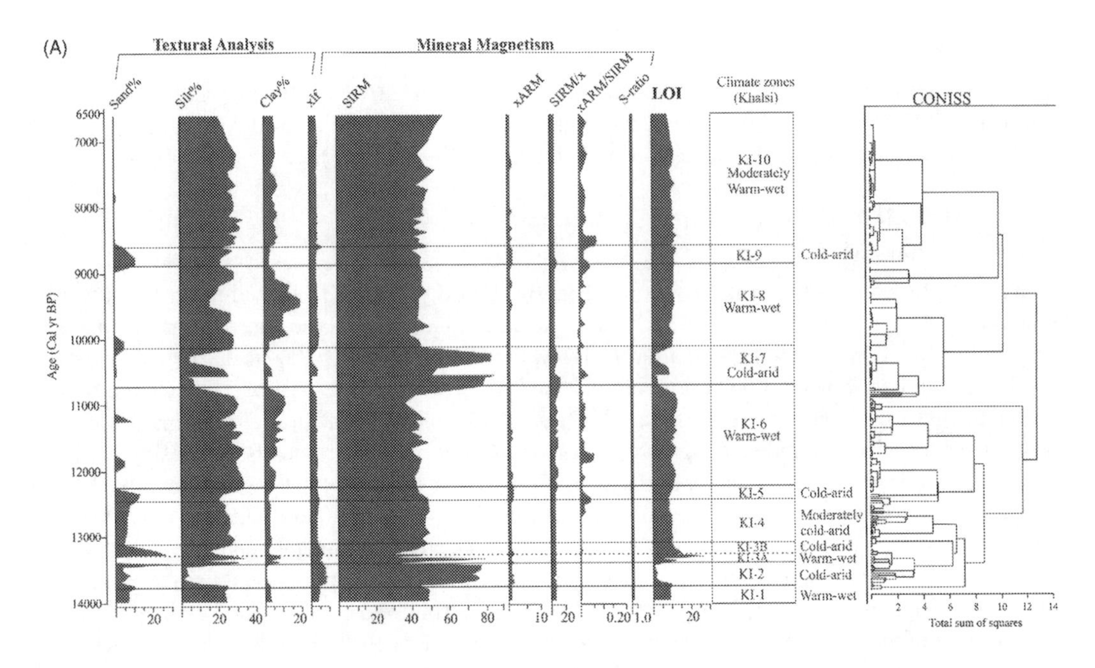

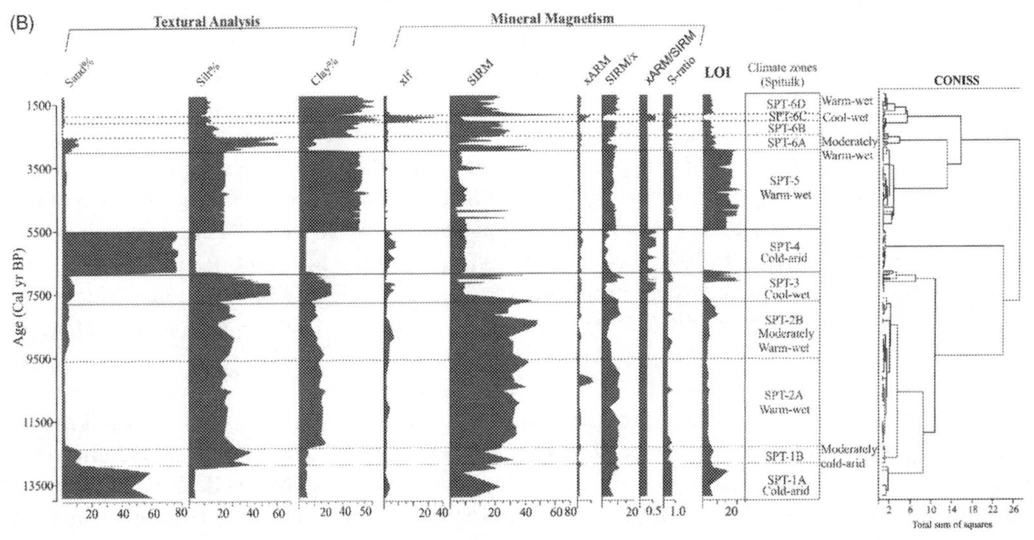

FIGURE 3.7

The TELIA CONISS plots of the studied proxy records (textural analysis, mineral magnetism, and LOI/TOC) and their respective climatic divisions from (A) Khalsi paleolake; (B) Spituk paleolake; (C) North Pulu glacial lake; (D) South Pulu glacial lake; (E) Tsoltak glacial lake, and (F) Yaya Tso paleolake.

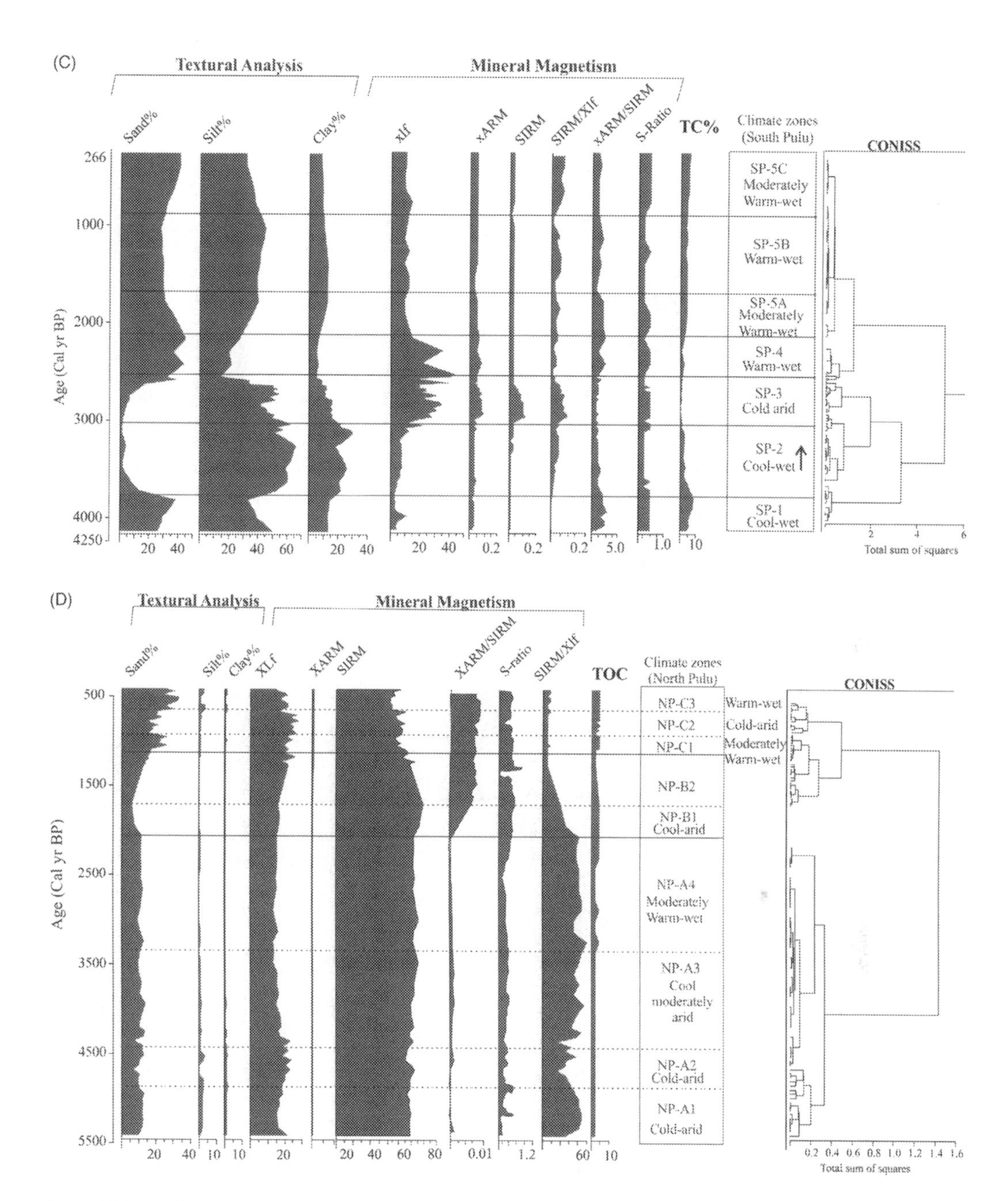

FIGURE 3.7 Continued

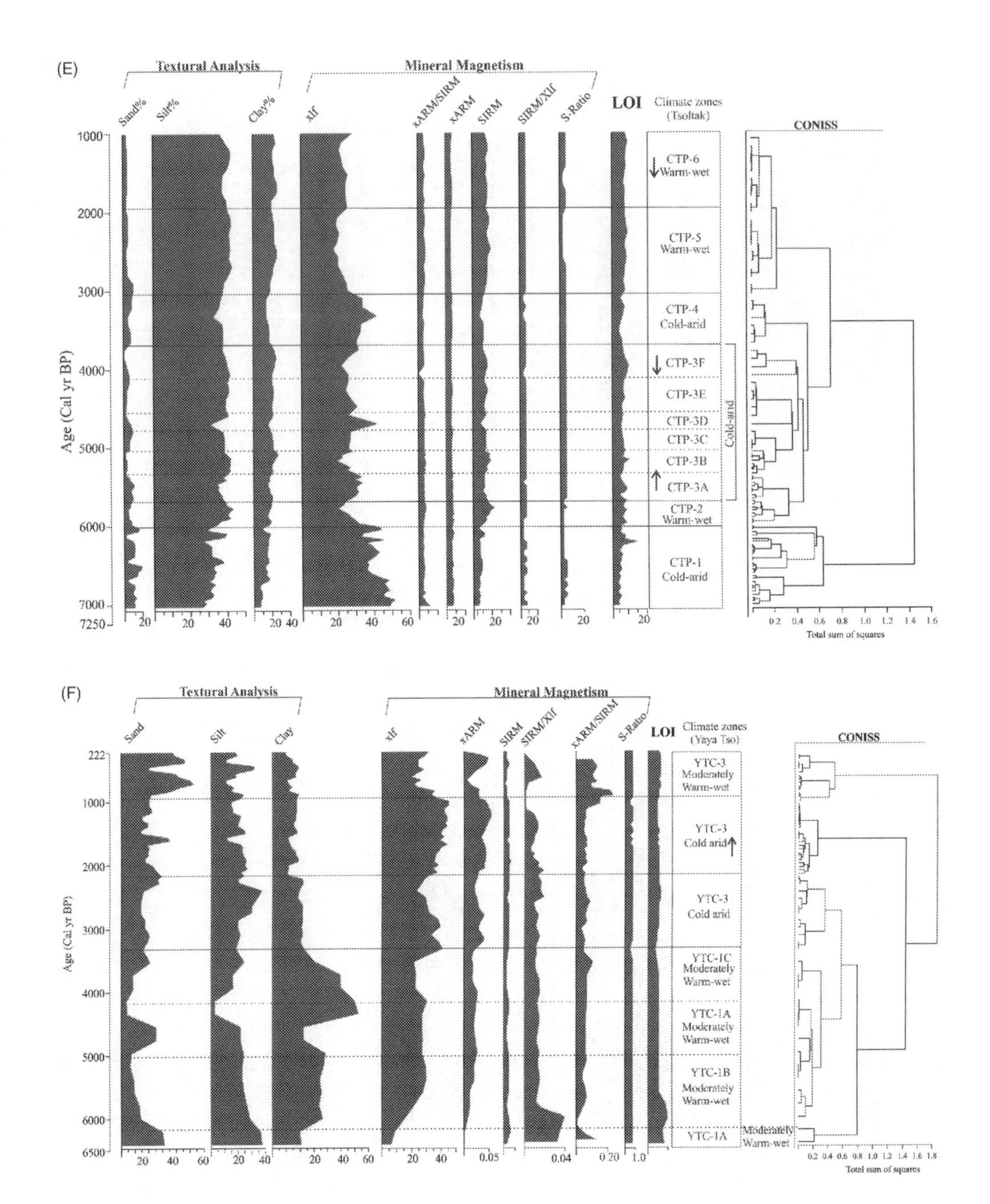

FIGURE 3.7 Continued

SIRM) and higher values of SIRM/χ_{lf} and S-ratio shows dominance of ferrimagnetic minerals. Low χ_{ARM}/SIRM also support the presence of coarser magnetic grains. Mineral magnetic parameters suggest dominance of multi domain ferrimagnetic minerals. Sand values remain very low and largely compensated by the silt with little increment in clay also. From 12,500 to ~9500 cal yr BP, the climate shows a warming trend with moderately cold arid phase from 12,500 to 11,800 cal yr BP (Zone SPT1B) to moderately warm wet from 11,800 to ~ 9500 cal yr BP (Zone SPT2), increasing melt water supply to the basin. Return of cold arid condition is noted from 9500 to 7000 cal yr BP (Zone SPT3) with increased χ_{lf} and SIRM from preceding zone. SIRM/χ_{lf} and S-ratio shows dominance of ferrimagnetic minerals. Low χ_{ARM}/SIRM and increased sand content indicate coarser magnetic grains indicating an increased detrital input from catchment to the lake basin. Top of this zone shows amelioration in climate and turning to warm wet from 7000 to 5500 cal yr BP (Zone SPT4). Proxy records indicate that from 7000 to 5500 cal yr BP time bracket, the sediment became much coarser as a result of tectonic activity recorded by four levels of seismites (Phartiyal and Sharma, 2009), and increased rainfall triggering sediment supply to the basin resulting in extremely high sedimentation rates. The bio-turbated surfaces and clay partings having coarser fraction removed by wind suggest that the lake was much shallower at this time turning into fluvial regime. Due to this the absolute values of organic carbon are reduced, however, a little increase in carbonate abundance is a result of contribution from detrital source. During 5500 to 3000 cal yr BP (Zone SPT 5), the lake environment became calmer allowing finer sediment to accumulate forming varves, initially rich in clay and became silty afterward and reflected in form of increased organic in the lower part and carbonate in the upper part. During this phase lowest value of χ_{lf}, χ_{ARM}, SIRM is seen. SIRM/χ_{lf} and S-ratio shows dominance ferromagnetic minerals. However antiferromagnetic components are recorded toward the base of this zone. Lowering of χ_{ARM} /SIRM points to coarser multi domain magnetic particles indicating increased detrital influx. This phase started with warm wet condition and turning to moderately cold with increased glacial melt water supply. From 3000 to 2500 cal yr BP (Zone SPT 6A), sand input has increased and organic carbon percentage is lowered in the sediment. Overall, the χ_{lf}, χ_{ARM}, SIRM values have increased than preceding zone. Increased χ_{ARM} and lowering of χ_{ARM}/SIRM indicates mixture of both single domain and multidomain grains; which points toward reduced detrital influx as climate returned to comparatively arid colder conditions. From 2500 to 1500 cal yr BP (Zone SPT 6B-D) amelioration of climate is seen with a short cold pulse from ~1800 to 1500 cal yr BP (Zone SPT6C).

3.5.2.2 *Glacial lakes*

3.5.2.2.1 North Pulu (NP)

This section has been earlier reported by Phartiyal et al. (2020b) and the same climatic zones are mentioned (Fig. 3.7C). The climate started with a cold arid phase from 5400 to 3400 cal yr BP (NP A1) as indicated by the magnetic parameters, high sand percent, increase in silt and clay and very low TOC. The freeze–thaw action and enhanced weathering during this cold phase generated a higher amount of sediment. Climate turned moderately warm and wet from 3400 to ~1200 cal yr BP (Zones NPA4 and NP B) with decreasing trends in χlf, SIRM, sand, silt indicates warmer conditions. A cold pulse is seen from ~2000 to 1700 cal yr BP (NPB1 and 2) and conditions improve during 1200–900 cal yr BP (NP C1 and 2). Thereafter from 900 (NPC) the climate is mostly oscillating between warm and cold phases. From 900 to 6000 (NP C3) the return of cold climate is seen followed by a warm phase from 600 to ~400 cal yr BP.

3.5.2.2.2 South Pulu (SP)

From ~4100 to 3700 cal yr BP(SP1) colder climate with wetter condition is seen (Fig. 3.7D). This zone has very low χ_{lf}, χ_{ARM}, and SIRM, varying S-ratios, increasing sand content, varying silt, constantly increasing clay and TC. The period from 33,700 to 3000 cal yr BP (SP2) is similar to the one preceding but with continuous increase in magnetic concentration and decreased S-ratio, sand and TOC. From 3000 to ~2600 cal yr BP (SP 3) there is a sharp change in the proxy records. This phase is characterized by increase in values of χ, χ_{ARM}, SIRM indicate shift toward colder climatic conditions. S-ratios also increase showing a shift toward magnetite. Sand and silt have higher, clay and TOC shows consistently decreasing. From 2600 to 2100 cal yr BP (SP4) the climate turned warm and wet with sharp fluctuations in magnetic concentration is seen, S-ratios decreasing toward the top, sand increases while silt and clay percent show a decreasing trend and organic matter shows an increasing trend. From 2100 to ~ 260 cal yr BP (SP5) marks a transition phase with low χ_{ARM}, moderate S-ratios, sand percentage are high with an increasing trend ~770 cal yr BP (SP5A-B) onward while silt and clay decrease from ~770 cal yr BP (SP5C). TOC percentages increasing further.

3.5.2.2.3 Tsoltak (CTP)

Six broad zones are defined in Tsoltak record (CTP1–6). An overall cold-arid climate from ~7000 to ~2700 cal yr BP (CTP1–4) with short warm pulses in between (CTP2) (Fig. 3.7E). This zone shows increased magnetic concentration, increased input of detrital ferromagnetic grains, and higher sand content. However, a warm–wet pulse is seen at ~6000–5800 cal yr BP (CTP2). The wet zone is characterized by decreasing magnetic concentration flux in the basin and increased silt and clay content. The climate between 5800 to 3800 cal yr BP (CTP 3) is mostly fluctuating between moderately arid and wet phases. From 2700 cal yr BP (CTP 5–6), the proxy analysis shows shift of climate to warm–wet condition with very high clay percentage, reduced S-ratio indicating antiferromagnetic minerals and increased organic content.

3.5.2.2.4 Yaya Tso (YTC)

The record of Yaya Tso is from 6400 to 266 cal yr BP with four broad zones (YTC1–YTC4). The climate phase commences with warm–wet from ~6400 to 3300 cal yr BP (YTC1) (Fig. 3.7F). This zone is characterized by highest percentage of clay, relatively reduced magnetic concentration and good organic content. Following this the period from 3300 to ~1300 cal yr BP the climate turned cold and arid. This zone records high sand influx reduced organic content and increasing trend of χlf, ARM, and SIRM values. The increase in magnetic concentration points to enhanced clastic input. The climate from ~1300 to ~200 cal yr BP (YTC4) records a return of warm–wet condition with a sharp decrease in the magnetic parameters and increased organic matter. Increased sand content indicates increased melt water supply to the basin.

3.6 Discussion

On the basis of the proxy data (textural analysis, mineral magnetism, and LOI/TOC) from the exposed sections of two paleolakes (Khalsi and Spituk) and four glacial lakes (NP, SP, Tsoltak dry lake, and Yaya Tso) in the Indus valley and its northward Ladakh Range in a spatial area of ~5400 sq km, we have

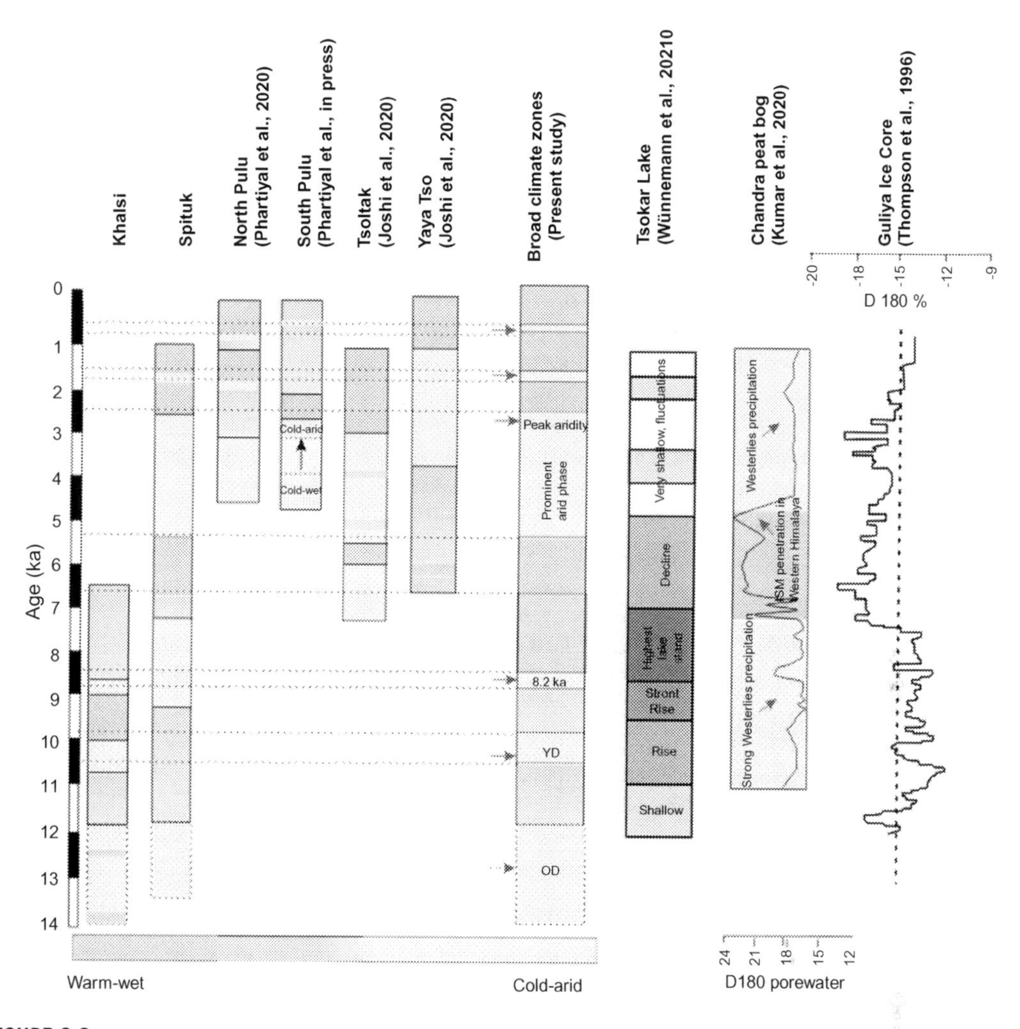

FIGURE 3.8

The overall climatic picture of the climatic variations during the Holocene from Ladakh. Sketch shows the studied records and their comparison with Tsokar lake record (Wünnemann et al., 2010); Chandra Tal peat bog (Kumar et al., 2020); and Guliyaicecore (Thompson et al., 1997) for a regional picture in the Tethyan, Trans-Himalaya, and the Tibetan plateau.

reconstructed the complete Holocene climatic record. Fig. 3.8 gives the comparative data along with the important records published from recently from the same laboratory and proxies following the same protocol. Prior to 11,700 cal yr BP (commencement of Holocene) both the Spituk paleolake and the Khlasi paleolake were existing occupying a major part of the Indus main river valley in the study area (Fig. 3.3). The Khalsi paleolake (named Khalsi-Saspol lake by Nag et al., 2016) was formed after the

LGM and the Spituk lake after the Older Dryas deglaciation. A short cold pulse is recorded at ~12,000–11,700 cal yr BP, which is the effect Younger Dryas of the northern regions. Five prominent arid phases mark the Holocene period in Ladakh (~10,800–10,000; ~8800–8600; a longer arid phase at ~5200–2600 with increasing aridity toward the top part; ~1700–1500 and ~400 cal yr BP) intervened by warmer conditions in between. Presently, the region lies in the rain shadow of NW Himalaya having an arid to hyper arid climate and dry steppe vegetation. Two sources of precipitations (westerlies and ISM) are responsible for governing the hydroclimate of the region. Hence, Fig. 3.8 shows the comparison of the data to the d18O pore water (Kumar et al., 2020), from Chandratal, south of the study area and to Guliya ice core data (Thompson et al., 1997), third pole region. The westerly dominates in the beginning of Holocene while the mid-Holocene sees the advent of ISM to the region. Westerlies further take over from mid-Holocene to ~3200 cal yr BP and from this period onward the ISM again seem to dominate the hydroclimate of the region. In the paleolake (Spituk and Khalsi), the older lake phase (12,600–10,800 cal yr BP) is indicative of warmer and humid conditions resulting from the ISM (summer precipitation) for the sustenance of this lake, whereas the younger lake phase (5500–3200 cal yr BP) shows a westerly influence as is evident from the presence of varve deposits in Spituk section. Our records show dry conditions prior to ~12,600 cal yr BP and this period and matches the records from Tsokar (southeast of study area) data indicating increased monsoon circulation between 12,900 and 12,500 cal yr BP (Demske et al., 2009). Decreased temperature and monsoon strength is also inferred starting at ~12.5–12.0 ka from Central Himalayas (Juyal et al., 2004; Beukema et al., 2011). Chandra peat bog (south of the study area) and Tsokar lake records from the region show a considerable summer monsoon moisture supply occurred between ca. 11,600 and 8800 cal yr BP (Wünnemann et al., 2010; Rawat et al., 2015), early Holocene high stand between 11,200 and 8600 yr BP (Hudson et al., 2015) and maximum monsoon activity between ca. 10,900 and 9200 yr BP (Demske et al., 2009). In Bangong Co (west of the study area), warm/wet conditions, attributed to the monsoon influence were established at ca. 9600 yr BP (Fontes et al., 1996) and a clear-cut warm pulse is observed between ca. 10,000 and 8000 yr BP (Gasse et al., 1991). Bangong Lake had its highest water level around ca. 9600–8700 yr BP (Hui et al., 1996) and moist conditions prevailed until ca. 6200 yr BP (Van Campo et al., 1996). The decline of the lake levels had started ~6800 ka and arid phase was experienced in the region between 6800 and 5200 cal years BP. The bigger lakes did exist but the lake level lowered. Aridity in the period has been continuously recorded till 2600 cal yr BP. A very distinct arid event at 4000 cal yr BP is recorded from South Asia (Madella and Fuller, 2006; Willis and MacDonald, 2011) and reduced precipitation and aridity from Himalaya (Kotlia et al., 1997, 2015; Phadtare, 2000). In the SP record the variability between ~4100 and 2600 cal yr BP (Phartiyal et al., 2021) is perfectly synchronous (mimics) with the GISP record (Grootes and Stuiver, 1997) indicative of the north Atlantic forcing (westerly) being dominant and controlling the climatic variability during this period of time in the Ladakh region. During ~2600–1700 cal yr BP a warmer climatic condition is recorded. Hence during this time span, this region appears to have a minimal effect of the north Atlantic forcing and was dominantly controlled by ISM variability. Between 2100 and ~500 cal yr BP, a moderately warm phase, which is consistent with the tropical south-western (ISM) forcing again, although a short spell of colder climate of minor amplitude is recorded at ~500 cal yr BP. It is for long believed that the moisture source is dominantly contributed by the westerlies against ISM contribution 2/3 versus 1/3, respectively, of the total precipitation received the Ladakh, Karakorum (Mayewski et al., 1980; Goudie et al., 1984), and the westerlies seems to have controlled the winter precipitation in the entire North-west Himalaya, Trans-Himalaya, and Tibet (Benn and Owen, 1998; Kumar et al., 2020) and also penetrated the Central Himalaya during mid-late

Holocene (Joshi et al., 2017). The Indus water isotopic composition (Pande et al., 2000; Sharma et al., 2017) in Ladakh presently shows ~26% contribution from the Mediterranean (North Atlantic Forcing; i.e., westerlies) and rest from the ISM source (Tropical South-western Forcings, i.e., ISM) in Indus waters (main channel flow). Our lake records and other published records from the region show that this change in the moisture source has occurred several times in the Holocene—7200, 5200, and 2600 cal yr BP although the westerly has been a dominant precipitation source in the Ladakh region through most of Holocene. This observation hints Trans-Himalayan region could be a hotspot for understanding mutually interacting northern (high latitudinal) and tropical forcing factors that might be responsible for the observed surge of extreme rainfall related hazards in last few decades (e.g., 2010, 2013) and the threat in the future to the evident changes in the ecology of this arid cold high-altitude desert.

3.7 Learning and knowledge outcomes

A composite and complete picture of the Holocene climatic variation and forcing is presented.

Five prominent arid phases mark the Holocene period in Ladakh (~10,500–10,000; ~8800–8400; a longer arid phase at ~5200–2600 with increasing aridity toward the top part; ~1700–1500 and ~500 cal yr BP) intervened by warmer conditions. The lake records and other published records from the region show that this change in the moisture source has occurred several times in the Holocene—7200, 5200, and 2600 cal yr BP although the westerly has been a dominant precipitation source in the Ladakh region through most of Holocene. Apart from the changes in the circulation pattern, the snow/ice-albedo feedback, changes in cloud cover, surface water vapor, and anthropogenic land use changes cannot be negated to contribute to the observed climatic variations.

Acknowledgments

Our sincere thanks to Director, BSIP, Lucknow for encouragement (BSIP/RDCC/Publication No. 63/2020-21). Dr. Prashant Das, BSIP is thanked for helping in laboratory analysis. Laboratory facilities were provided by SAIF, BSIP; Gliwice Radiocarbon Laboratory, Poland, Direct AMS, USA. DC, Leh-Ladakh and Department of Wild Life Protection, Jammu (J&K) is thanked for permission to carry out studies in the area.

References

Abbott, J., Report on the cataclysm of the Indus taken from the lips of an eyewitness, J. Asiatic Soc. Bengal 17 (1849) 231–233.

Ahmad, S., Singh, J.P., Verma, D.K., 2020. Inventory of important fodder plants of Ladakh Himalaya. In: International Grassland Congress Proceedings, 3 (2020)

Allen, C., A Mountain in Tibet, Futura, London, 1984.

Becher, J., The flooding of the Indus. Letter addressed to R. H. Davis, Secretary of the Government of the Punjab and its dependencies, J. Asiatic Soc. Bengal 28 (1859) 219–228.

Bendell-Young, L.I., Thomas, C.A., Stecko, J.R.P., Contrasting the geochemistry of toxic sediments across ecosystems: a synthesis, Appl. Geochem. 17 (2002) 1563–1582.

Benn, D.I., Owen, L.A., The role of the Indian summer monsoon and the midlatitude westerlies in Himalayan glaciation: review and speculative discussion, J. Geol. Soc. Lond. 155 (1998) 353–363.

Beukema, S.P., Krishnamurthy, R.V., Juyal, N., Basavaiah, N., Singhvi, A.K., Monsoon variability and chemical weathering during the late Pleistocene in the Goriganga basin, higher central Himalaya, India, Quat. Res. 75 (3) (2011) 597–604.

Bookhagen, B., Thiede, R.C., Strecker, M.R., Late Quaternary intensified monsoon phases control landscape evolution in the northwest Himalaya, Geology 33 (2) (2005) 149–152.

Brown, E.T., Bendick, R., Bourles, D.L., Gaur, V., Molnar, P., Raisbeck, G.M., Yiou, F., Early Holocene climate recorded in geomorphological features in Western Tibet, Palaeogeogr. Palaeoclim. Palaeoecol. 199 (2003) 141–151.

Burbank, D.W., Multiple episodes of catartropic flooding in the Peshawar basin during the past 700,000 years, Geol. Bull. Univ. Peshawar 16 (1983) 43–49.

Chai, X., Peat-Geology, Geological Publishing House, Beijing, 1990.

Chen, F., Yu, Z., Yang, M., Ito, E., Wang, S., Madsen, D.B., Huang, X., Zhao, Y., Sato, T., Birks, H.J.B., Boomer, I., Chen, J., An, C., Wünnemann, B., Holocene moisture evolution in arid central Asia and its out-of-phase relationship with Asian monsoon history, Quat. Sci. Rev. 27 (2008) 351–364.

Clift, P.D., Wan, S., Blusztajn, J., Reconstructing chemical weathering, physical erosion and monsoon intensity since 25 Ma in the northern South China Sea: a review of competing proxies, Earth Sci. Rev. 130 (2014) 86–102.

Dahl, S.O., Bakke, J., Lie, Ø., Nesje, A., Reconstruction of former glacier equilibrium-line altitudes based on proglacial sites: an evaluation of approaches and selection of sites, Quat. Sci. Rev. 22 (2003) 275–287.

Demske, D., Tarasov, P.E., Wunnemann, B., Riedel, F., Late glacial and Holocene vegetation, Indian monsoon and westerly circulation dynamics in the Trans-Himalaya recorded in the pollen profile from high altitude Tso Kar Lake, Ladakh, NW India, Palaeogeogr. Palaeoclim. Palaeoecol. 279 (2009) 172–185.

Dimri, A.P., Yasunari, T., Kotlia, B., Mohanty, U.C., Sikka, D.R., Indian winter monsoon: present and past, Earth Sci. Rev. 163 (2016) 297–322.

Dodson, J.R., Ramrath, A., An Upper Pliocene lacustrine environmental record from south-Western Australia—preliminary results, Palaeogeogr. Palaeoclim. Palaeoecol. 167 (2001) 309–320.

Dortch, J.M., Owen, L.A., Haneberg, W.C., Caffee, M.W., Dietsch, C., Kamp, U., Nature and timing of large landslides in the Himalaya and Trans-Himalaya of northern India, Quat. Sci. Rev. 28 (2009) 1037–1054.

Dvorsky´, M., Dole, J., de Bello, F., KlimesÇova´, J., Klimes, L., Vegetation types of East Ladakh: species and growth form composition along main environmental gradients, Appl. Veg. Sci. 14 (2010) 132–147.

Folk, R.L., Ward, W.C., Brazos River Bar: a study in the significance of grain size parameters, J. Sediment. Petrol. 27 (1957) 3–26.

Fontes, J.C., Gasse, F., Gibert, E., Holocene environmental changes in Lake Bangong basin (Western Tibet) part 1: chronology and stable isotopes of carbonates of a Holocene lacustrine core, Palaeogeogr. Palaeoclim. Palaeoecol. 120 (1996) 25–47.

Fort, M., Burbank, D.W., Freytet, P., Lacustrine sedimentation in a semi-arid alpine setting: an example from Ladakh, northwestern Himalaya, Quat. Res. 31 (1989) 332–352.

Friedman, G.M., Trace elements as possible environmental indicators in carbonate sediments, in: Friedman, G.M. (Ed.), Depositional Environments in Carbonate Rocks, The Society Economic Paleontologists and Mineralogists, Oklahoma, USA, 1969, pp. 193–200. Depositional Environments in Carbonate Special Publication No. 14.

Gasse, F., Arnold, M., Fontes, J.C., Fort, M., Gilbert, E., Huc, A., Li, B., Li, Y., Qing, L., Mélières, F., Van Campo, E., Fubao, W., Qingsong, Z..

Goudie, A.S., Brunsden, D., Collins, D.N., Derbyshire, E., Ferguson, R.I., Hashnet, Z., Jones, D.K.C., Perrott, F.A., Said, M., Waters, R.S., Whalley, W.B., The geomorphology of the Hunza valley, Karakoram Mountains,

Pakistan, in: Miller, K. (Ed.), International Karakoram Project, Cambridge University Press, Cambridge, 1984, pp. 339–411.

Grimm, E.C., CONISS: a FORTRAN 77 program for stratigraphically constrained cluster analysis by the method of incremental sum of squares, Comput. Geosci. 13 (1987) 13–35.

Grimm, E.C., TILIA and TILIA.GRAPH, PC spreadsheet and graphics software for pollen data, INQUA Working Group Data Handling Meth. Newslett. 4 (1990) 5–7.

Grootes, P.M., Stuiver, M., $^{18}O/^{16}O$ variability in Greenland snow and ice with 103 to 105-year time resolution, J. Geophys. Res. 26 (1997) 455–470.

Haslett, J., Parnell, A., A simple monotone process with application to radiocarbon-dated depth chronologies, J. R. Stat. Soc.: Ser. C (Appl. Stat.) 57 (4) (2008) 399–418.

Hudson, A.M., Quade, J., Long-term east-west asymmetry in monsoon rainfall on the Tibetan Plateau, Geology 41 (2013) 351–354.

Hudson, A.M., Quade, J., Huth, T.E., Lei, G., Cheng, H., Edwards, E.R., Olsen, J.W., Zhang, H., Lake level reconstruction for 12.8-2.3 ka of the Ngangla Ring Tso closed-basin lake system, southwest Tibetan Plateau, Quat. Res. 83 (2015) 66–79.

Hui, F., Gasse, F., Huc, A., Li, Y., Siffedine, A., Soulié-Märsche, I., Holocene environmental changes in Bangong Co basin (western Tibet). Part 3: biogenic remains, Palaeogeogr. Palaeoclim. Palaeoecol. 120 (1996) 65–78.

Hussain, G., Singh, Y., Bhat, G.M., Geotechnical investigation of slopes along the national highway (NH-1D) from Kargil to Leh, Jammu and Kashmir (India), Geomaterials 5 (2015) 56–67.

Inman, D.L., Sorting of sediments in the light of fluid mechanics, J. Sediment. Petrol. 19 (1949) 51–70.

Joshi, L.M., Kotlia, B.S., Ahmad, S.M., Wu, C.C., Sanwal, J., Raza, W., Singh, A.K., Shen, C.C., Long, T., Sharma, A.K., Reconstruction of Indian monsoon precipitation variability between 4.0 and 1.6 ka BP using speleothem δ 18 O records from the Central Lesser Himalaya, India, Arab. J. Geosci. 10 (16) (2017) 356.

Joshi, P., Phartiyal, B., Joshi, M., Hydro-climatic variability during last five thousand years and its impact on human colonization and cultural transition in Ladakh sector, India, Quat. Int. (2020) In press.https://doi.org/10.1016/j.quaint.2020.09.053.

Juyal, N., Pant, R.K., Basavaiah, N., Yadava, M.G., Saini, N.K., Singhvi, A.K., Climate and seismicity in the higher Central Himalaya during 20-10 ka: evidence from the Garbyang basin, Uttaranchal, India, Palaeogeogr. Palaeoclim. Palaeoecol. 213 (2004) 315–330.

Korsman, T., Nilsson, M.B., Landgren, K., Renberg, I., Spatial variability in surface sedimentcompositioncharacterised by near-infrared (NIR) reflectance spectroscopy, J. Paleolimnol. 21 (1999) 61–71.

Kotlia, B.S., Shukla, U.K., Bhalla, M.S., Mathur, P.D., Pant, C.C., Quaternary fluvio-lacustrine deposits of Lamayuru basin, Ladakh Himalaya: preliminary multi-diciplinary investigations, Geol. Mag. 134 (1997) 807–812.

Kotlia, B.S., Singh, A.K., Joshi, L.M., Dhaila, B.S., Precipitation variability in the Indian Central Himalaya during last ca. 4,000 years inferred from a speleothem record: impact of Indian Summer Monsoon (ISM) and Westerlies, Quat. Int. 371 (2015) 244–253.

Krumbein, W.C., Size frequency distributions of sediments, J. Sediment. Petrol. 4 (1934) 65–77.

Kumar, A., Srivastava, P., The role of climate and tectonics in aggradation and incision of the Indus River in the Ladakh Himalaya during the late Quaternary, Quat. Res. 87 (2017) 363–385.

Kumar, O., Ramanathan, A.L., Bakke, J., Kotlia, B.S., Shrivastava, J.P., Disentangling source of moisture driving glacier dynamics and identification of 8.2 ka event: evidence from pore water isotopes, Western Himalaya, Sci. Rep. (2020) 15324. https://doi.org/10.1038/s41598-020-71686-4.

Long, H., Lai, Z.P., Wang, N.A., Li, Y., Holocene climate variations from Zhuyeze terminal lake records in East Asian monsoon margin in arid Northern China, Quat. Res. 74 (2010) 46–56.

Madella, M., Fuller, D.Q., Palaeoecology and the Harappan civilisation of South Asia: a reconsideration, Quat. Sci. Rev. 25 (11-12) (2006) 1283–1301.

Matthews, J.A., Dahl, S.O., Nesje, A., Berrisford, M.S., Andersson, C., Holocene glacier variations in central Jotunheimen, Southern Norway based on distal glaciolacustrine sediment cores, Quat. Sci. Rev. 19 (2000) 1625–1647.

Mayewski, P.A., Pregent, G.P., Jeschke, P.A., Ahmad, N., Himalayan and Trans–Himalayan glacier fluctuations and the South Asian monsoon record, Arctic Alpine Res. 12 (1980) 171–182.

Mischke, S., Fuchs, D., Riedel, F., Schudack, M.E., Mid to Late Holocene palaeoenvironment of Lake Eastern Juyanze (north-western China) based on ostracods and stable isotopes, Geobios 35 (2002) 99–110.

Nag, D., Phartiyal, B., Climatic variations and geomorphology of the Indus River Valley, between Spituk and Batalik, Ladakh (NW Trans Himalayas) in Late Quaternary, Quat. Int. 371 (2015) 87–101.

Nag, D., Phartiyal, B., Joshi, M., Late Quaternary tectono-geomorphic forcing vis-a-vis topographic evolution of Indus catchment, Ladakh, India, Catena 199 (2021) 105103. https://doi.org/10.1016/j.catena.2020.105103.

Nag, D., Phartiyal, B., Sen, D.S., Sedimentary characteristics of palaeolake deposits along Indus river valley, Ladakh, Trans-Himalaya: implications to depositional environment, Sedimentology 63 (2016) 1765–1785.

Nesje, A., Dahl, S.O., Andersson, C., Matthews, J.A., The lacustrine sedimentary sequence in Sygneskardvatnet, Western Norway: a continuous, high-resolution record of the Jostedalsbreen ice cap during the Holocene, Quat. Sci. Rev. 19 (2000) 1047–1065.

Oldfield, F., Environmental magnetism: a personal perspective, Quat. Sci. Rev. 10 (1991) 73–85.

Pande, K., Padia, J.T., Ramesh, R., Sharma, K.K., Stable isotope systematics of surface water bodies in the Himalayans and Trans-Himalayan (Kashmir) region, J. Earth Syst. Sci. 109 (2000) 109–115.

Pandey, S., Clarke, J., Nema, P., Bonaccorsi, R., Som, S., Sharma, M., Phartiyal, B., Rajamani, S., Mogul, R., Martin-Torres, J., Vaishampayan, P., Blank, J., Steller, L., Srivastava, A., Singh, R., McGuirk, S., Zorzano, M., Güttler, J.M., Mendaza, T., Soria-Salinas, A., Ahmed, S., Ansari, A., Singh, V.K., Mungi, C., Bapat, N., Ladakh: diverse, high-altitude extreme environments for off earth analogue and astrobiology research, Int. J. Astrobiol. 19 (2019) 78–98.

Phadtare, N.R., Sharp decrease in summer monsoon strength 4000–3500 cal yr B.P. in the Central Higher Himalaya of India based on pollen evidence from alpine peat, Quat. Res. 53 (2000) 122–129.

Phartiyal, B., Kapur, V.V., Nag, N., Sharma, A., Spatio-temporal climatic variations during the last five millennia in Ladakh Himalaya (India) and its links to archaeological finding(s) (including co-prolites) in a palaeoecological and palaeoenvironmental context: a reappraisal, Quat. Int. (2020b) In press.https://doi.org/10.1016/j.quaint.2020.11.025.

Phartiyal, B., Sharma, A., Soft-sediment deformation structures in the Late Quaternary sediments of Ladakh: evidence for multiple phases of seismic tremors in the North western Himalayan Region, J. Asian Earth Sci. 34 (2009) 761–770 .

Phartiyal, B., Sharma, A., Kothyari, G.C., Existence of late Quaternary and Holocene lakes along the River Indus in Ladakh Region of Trans Himalaya, NW India: implications to climate and tectonics, Chin. Sci. Bull. 58 (2013) 142–155.

Phartiyal, B., Sharma, A., Nautiyal, C.M., Interpretation of the apparent ages in the Ladakh and Lahaul-Spiti Quaternary lacustrine sediments, in: Singh, D.S., Chabra, N.L. (Eds.), Geological Processes and Climate Change, Macmillan Publisher India Ltd., India, 2011, pp. 105–116.

Phartiyal, B., Sharma, A., Upadhyay, R., Ram-Awatar, Sinha, A.K., Quaternary geology, tectonics and distribution of palaeo- and present fluvio/glacio lacustrine deposits in Ladakh, NW Indian Himalaya – a study based on field observations, Geomorphology 65 (2005) 241–256.

Phartiyal, B., Singh, R., Joshi, P., Nag, D., Late Holocene climatic record in glacial lake of Ladakh Range, Trans-Himalaya, India, Holocene 30 (2020a) 1029–1042.

Phartiyal, B., Singh, R., Kothyari, G.C., Late-Quaternary geomorphic scenario due to changing depositional regimes in the Tangtse Valley, Trans-Himalaya, NW India, Palaeogeogr. Palaeoclim. Palaeoecol. 422 (2015) 11–24.

Phartiyal, B., Singh, R., Nag, D., Sharma, A., Agnihotri, R., Prasad, V., Yao, T., Yao, P., Balasubramanian, K., Joshi, P., Gahlaud, S.K.S., Thakur, B., Reconstructing climate variability during last four millennia from Trans–Himalaya (Ladakh–Karakorum, India) using multiple proxies, Palaeogeogr. Palaeoclim. Palaeoecol. 562 (2021) 110142. https://doi.org/10.1016/j.palaeo.2020.110142.

Piotrowska, N., Status report of AMS sample preparation laboratory at GADAM Centre, Gliwice, Poland, Nucl. Instrum. Methods Phys. Res. 294 (2013) 176–181.

Ramsey, C.B., Bayesian analysis of radiocarbon dates, Radiocarbon 51 (2009) 337–360.

Rawat, S., Gupta, A.K., Sangode, S.J., Srivastava, P., Nainwal, H.C., Late Pleistocene Holocene vegetation and Indian summer monsoon record from the Lahaul, Northwest Himalaya, India, Quat. Sci. Rev. 114 (2015) 167–181.

Sangode, S.J., Phadtare, N.R., Meshram, D.C., Rawat, S., Suresh, N., A record of Lake Outburst in the Indus valley of Ladakh Himalaya, India, Curr. Sci. 100 (2011) 1712–1718.

Sangode, S.J., Rawat, S., Meshram, D.C., Phadtare, N.R., Suresh, N., Integrated mineral magnetic and lithologic studies to delineate dynamic modes of depositional conditions in the Leh valley Basin, Ladakh Himalaya, India, J. Geol. Soc. India 82 (2013) 107–120.

Sant, D.A., Wadhwan, S.K., Ganjoo, R.K., Basavaiah, N., Sukumaran, P., Bhattacharya, A.S., Morphostratigraphy and palaeoclimate appraisal of the Leh Valley, Ladakh Himalayas, J. Geol. Soc. India **77** (2011) 499–510.

Searle, M.P., Structural evolution and sequence of thrusting in the High Himalayan, Tibetan-Tethys and Indus suture zones of Zanskar and Ladakh, Western Himalaya, J. Struct. Geol. 8 (1986) 923–936.

Sharma, A., Kumar, K., Laskar, A., Singh, S.K., Mehta, P., Oxygen, deuterium, and strontium isotope characteristics of the Indus River water system, Geomorphology 284 (2017) 5–16.

Shroder Jr., J.F., Himalaya to the sea: geomorphology and the Quaternary of Pakistan in the regional context, in: Shroder, J.F. (Ed.), Himalaya to the Sea: Geology, Geomorphology and the Quaternary, Routledge, London, 1993, pp. 1–27.

Spencer, D.W., The interpretation of grain size distribution curves of clastic sediments, J. Sediment. Res. 33 (1963) 180–190.

Stuiver, M., Reimer, P.J. and Reimer, R.W., 2017. CALIB 7.1. [WWW program] http://calib.org.

Tanner, W.F., The zig-zag nature of type 1 and type IV curves, J. Sediment. Petrol. 28 (1958) 372–375.

Thompson, L.G., Yao, T., Davis, M.E., Henderson, K.A., Mosley-Thompson, E., Lin, P.N., Beer, J., Synal, H.A., Cole-Dai, J., Bolzan, J.F., Tropical climate instability: the last glacial cycle from a Qinghai-Tibetan ice core, Science 276 (1997) 1821–1825.

Van Campo, E., Cour, P., Sixuan, H., Holocene environmental changes in Bangong Co basin (Western Tibet). Part 2: the pollen record, Palaeogeogr. Palaeoclim. Palaeoecol. 120 (1996) 49–63.

Verosub, K.L., Roberts, A.P., Environmental magnetism: past, present and future, J. Geophys. Res. 100 (1995) 2175–2192.

Visher, G.S., Grain size distributions and depositional processes, J. Sediment. Petrol. 39 (1969) 1074–1106.

Walden, J., Sample collection and preparation, in: Walden, J., Oldfield, F., Smith, J. (Eds.), Environmental Magnetism: A Practical Guide, Quaternary Research Association, Great Britain, 1999, pp. 63–88.

Williamson, D., Jelinowska, A., Kissel, C., Tucholka, P., Gibert, E., Gasse, F., Massault, M., Taieb, M., Van Campo, E., Wieckowski, E., Mineral magnetic proxies of erosion/oxidation cycles in tropical marr-lake sediments (Lake Tritivakely, Madagascar): palaeoenvironmental implications, Earth Planet. Sci. Lett. 155 (1998) 205–219.

Willis, K.J., MacDonald, G.M., Long-term ecological records and their relevance to climate change predictions for a warmer world, Annu. Rev. Ecol. Evol. Syst. 42 (2011) 267–287.

Wünnemann, B., Demske, D., Tarasov, P., Kotlia, B.S., Reinhardt, C., Bloemendal, J., Diekmann, B., Hartmann, K., Krois, J., Riedel, F., Arya, N., Hydrological evolution during the last 15 ky in the Tso Kar lake basin (Ladakh,

India), derived from geomorphological, sedimentological and palynological records, Quat. Sci. Rev. 29 (2010) 1138–1155.

Zhou, W.J., Donahue, D., Jull, A.J., Radiocarbon AMS dating of pollen concentrated from eolian sediments, Radiocarbon 39 (1997) 19–26.

Zhou, W.J., Yu, X.F., Jull, A.J., Burr, G., Xiao, J.Y., Lu, X., Xian, F., High-resolution evidence from southern China of an early Holocene optimum and a mid-Holocene dry event during the past 18,000 years, Quat. Res. 62 (2004) 39–48.

CHAPTER

4 Late Holocene advancements of denudational and depositional fronts in the Higher Himalaya: A case study from Chandra valley, Himachal Pradesh, India

S.J. Sangode[a], D.C. Meshram[a], A.M. Kandekar[a], Amol Abhale[a], S.S. Gudadhe[b], Suman Rawat[c]

[a]*School of Earth Sciences, Savitribai Phule Pune University, Pune, India.* [b]*Yashwantrao Chavan Arts, Commerce and Science College, Lakhandur, India.* [c]*Wadia Institute of Himalayan Geology, Dehradun, India*

4.1 Introduction

A significant amount of area in the Higher and Trans Himalaya presently manifests deglaciated valleys connected to glacierized catchments. These valleys and their transitional environments preserve a collage of glacial, fluviatile, eolian, and colluvial deposits reflecting various processes including the recycling and incision of postglacial valley fills during most of the Late Holocene. In an emerging climate change scenario, these valley basins play an important role in evaluating their response to extreme conditions. This demands a better understanding of the sediment–geomorphic environments of the valley-catchment setups in terms of documentation of ongoing processes and their implications to future changes. Detailed studies on the records of such change, within a given geomorphic setup would allow better assessment of their vulnerability to episodic and spasmodic controls of climate and tectonics.

Sedimentation and morphological changes in the Himalaya are primarily governed by repeated glacial/interglacial cycles (Shroder and Higgins, 1989; Owen et al., 1998; Benn and Owen, 2002; Bookhagen et al., 2005). A combination of varied orographic and climatic factors resulted in contrasting glacial patterns across the Himalayan and Karakorum mountain belts (Owen et al., 2005, 2009). Mapping, documentation and correlation of the sediment records across these deglaciated valleys are therefore fundamental aspects to identify their connection with climatic and/or tectonic changes, particularly during the Late Quaternary times (e.g., Mitchell et al., 1999; Owen et al., 2009; Sangode and Gupta, 2010).

The Late Pleistocene–Holocene time interval in the Higher Himalayas has recorded major change in the valley system from glacial to glaciofluvial with typical misfit-fluvial valley morphology. This time has also witnessed major change in the rates of erosion, sediment production, sediment transport, recycling, and deposition; demanding their interrelations to be achieved in modern and predictive context. Continued warming may result into various morphological changes such as creation and merger of smaller-deglaciated valleys. The altered hydrometeorology encourages conditions like GLOF,

Holocene Climate Change and Environment. DOI: https://doi.org/10.1016/B978-0-323-90085-0.00021-8

river damming, slope instability, landslides, and mass flow through flash floods. New insights are therefore necessary to observe and document the valley basins in relation to their catchments. The Late Holocene period having direct connection to Recent, such valley conditions are significant for developing parametric inputs to predictive models.

The Chandra river segment in the Higher Himalaya (Fig. 4.1) represents an ideally interconnected glacial/glaciofluvial conditions. Fringing on the climatic domains of mid-latitude westerlies (MLW) and ISM, the basin also falls in the distinct lithologic boundary between Tethyan sedimentary sequence (TSS) and the Greater Himalayan crystalline complex (GHC). The Chandra basin preserves some of the most pronounced imprints of dynamic changes occurred during the Late Quaternary glacial–interglacial stages and are very well documented as accounted below.

The recent work in this region is highlighted with abundant cosmogenic radionuclide ages that provided data on regional variability of Late Quaternary glaciation which show marked contrasts in the extent and timing of glaciation between various mountain ranges (Owen et al., 1997, 2001, 2006; Dortch et al., 2010; Hedrick et al., 2011; Eugster et al., 2016). Owen et al. (2005) suggested this as a reflection of temporal and spatial variability in the moisture regimes of South Asian Monsoon (SAM) and the MLW, in addition to the variation in regional precipitation gradients. These authors observe that the older glacial landforms are well preserved as the successive glacial extents were increasingly restricted throughout the Late Quaternary. Whereas, in the regions that are intensely influenced by monsoon, the preservation potential of pre-Late Glacial moraine successions is generally poor. The intense denudation by fluvial and gravitational movements in wetter climates would result in rapid erosion and recycling of the glacial sediments (op. cit.). Very recently, Chand et al. (2017) documented the post-Last Ice Age (LIA) fluctuations in the Bara Shigri glacier, which also dammed the Chandra river during a glacier surge in early 19th century causing floods downstream. This indicates the sensitivity of Chandra river to impulsive nature of its glacial catchment.

4.2 Scope

The rates of denudation, deposition, and sediment mobility can be parameterized through various direct and indirect methods available in standard texts. Their interrelations however demand closer examination of given case study such as a valley basin. In mountainous regions of the Himalaya, these rates are laterally varied due to differential tectonic, structural, lithologic, geomorphic, and climatic factors; and such parameterization and their interrelations are to be established based on field observations among individual basin and catchments. The rates of uplift and to some extent the incision are independent of basins in the Himalayas. However the first order rates of denudation and deposition are to be actuated for independent valley basins (/subbasins) before their interbasinal mixing. The emergent climate warming and developmental work have increased the frequency of extreme conditions in the Himalayan valley basins demanding a closer look on the above aspects. Whereas the long-term effects of denudation are to be examined to understand the sediment transfer into marginal basins in the Himalaya. The parameterization of field observations helps in developing robust database for modeling and predictions of the extreme conditions. In the time scales such as Late Holocene, the climate factors become more prevalent to operate these weathering/denudational processes. This paper makes a comprehensive assessment of the weathering and denudational surfaces to interrelate with

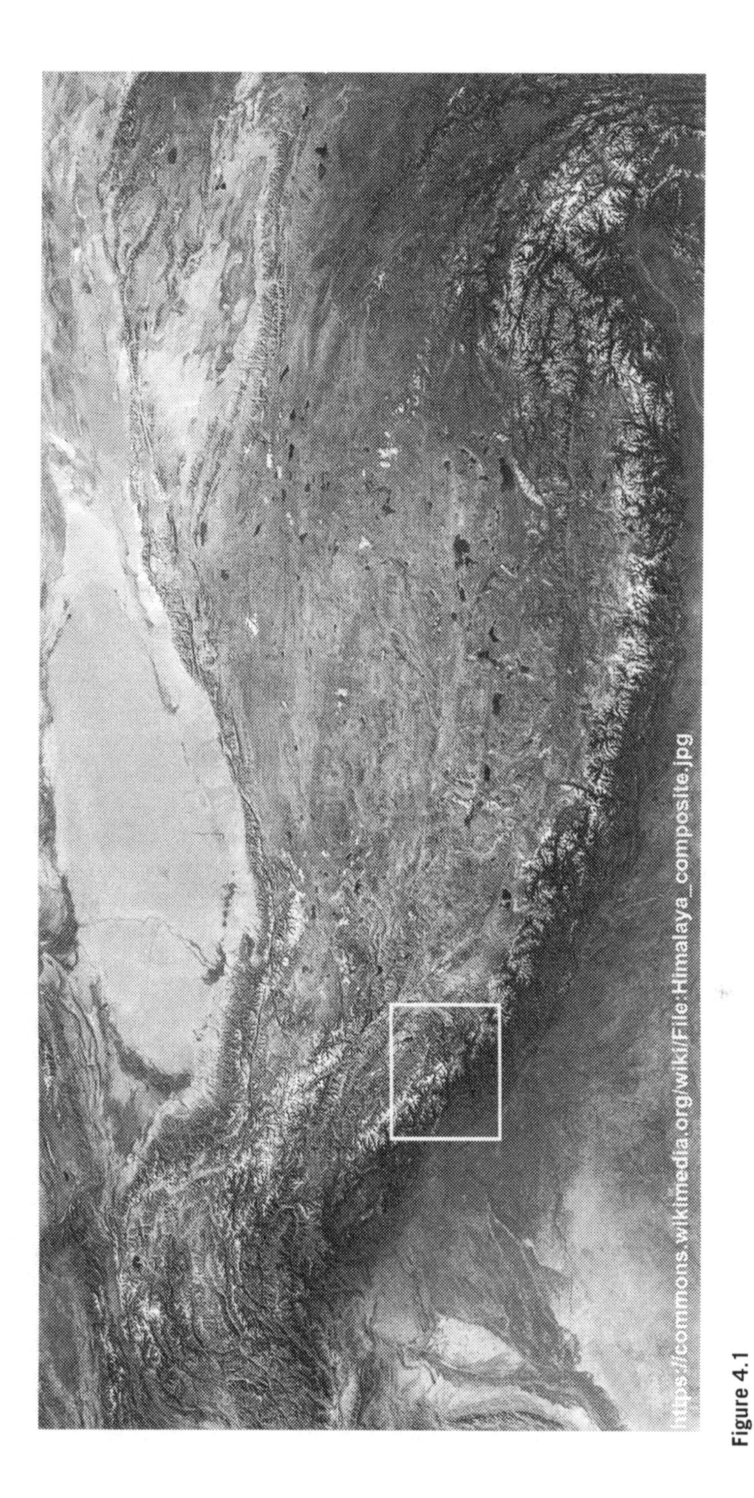

Figure 4.1
The study area (rectangle) represents similar glacial and glaciofluvial valleys for large part of the Higher and Trans-Himalaya (source printed on photo).

sedimentation and sediment mobility conditions under the given geomorphic setting of Chandra valley in Lahaul region of northwest Himalaya.

4.3 Methods

The field documentation was completed in two field seasons, documenting the lithologic characters of remnant, dissected and recycled moraines, drumlins, colluvial fans, erratic boulders, and obstacle dunes; followed by morphological characteristics such as roche moutonee, glacial trimlines, horns, incised fluvial terraces and pot-holes. This was followed by the next field season of mapping and systematic profiling across the valley for the segment from Batal to Tandi which was further subdivided into two segments (˜30 km each) as segment I from Batal to Khoksar and segment II from Khoksar to Tandi as shown in Fig. 4.2. This segmentation is based on change in geomorphic and sediment attributes (described later). We find that the Upper reaches of Chandra valley is marked by the Tethyan Sedimentary Sequence (TSS) clasts while rest of the valley shows dynamic input of the granitic–gneissic domains both from the trunk valley mainly during glacial cycles, and the tributary/hanging valleys chiefly during the interglacial.

Cross profiles in field were prepared using tape (/staff) and compass mapping aided with GPS measurements. We mapped the lithofacies associations, location and orientation of boulders with their modal distribution. These field observations allowed us to divide the Chandra river valley into the two segments that are used for classification on imagery. In the segment I, the river is of unconfined nature at places with shallow incision. Afterward the river is mostly confined with increasing incision toward the lower reaches. In segment I, the river prefers to align the left bank developing an asymmetric thalweg-valley base relationship.

The field studies were followed by satellite data produced for the river long profile at 300 m interval and more than 60 cross profiles, 10 km each (at 200 m interval) across the river valley and covering the ridge/catchment area. The cloud-free ALOS (Advance Land Observing Satellite) PALSAR (Phased Array Type L-band Synthetic Aperture Radar) Digital Elevation Model (ALOS DEM) data were used to identify variations in the river long profile and cross profiles river valley. The spatial resolution of 12.5 m ALOS PALSAR Radiometric Terrain Corrected DEM data with wide-swath images 250–350 km^2 of August 2010 (historical datasets of 2006–2011) was downloaded from the ALOS Research and Application Project of EORC, Japan Aerospace Exploration Agency (https://search.asf.alaska.edu/#/). The database in excel was then used for various statistical attributes, plotting, and interpretations in this work.

4.4 Field observations and documentation

The Lahaul Himalaya is located at the junction of the monsoon-influenced Pir Panjal–Higher Himalayan–Lesser Himalayan ranges and the semiarid mountains of the Trans-Himalaya influenced by the MLW. The region presently receives most of its precipitation as winter snow, due to the MLW that brings moisture from the Mediterranean, Black-, and Caspian Seas. However, during stronger summer monsoon, presently, heavy precipitation occurs throughout the Chandra valley, even in the elevated regions as far north of Chandra Tal and the Bhaga valley. The Chandra valley shows massive sediment production and recycling of the products of older glacial stages.

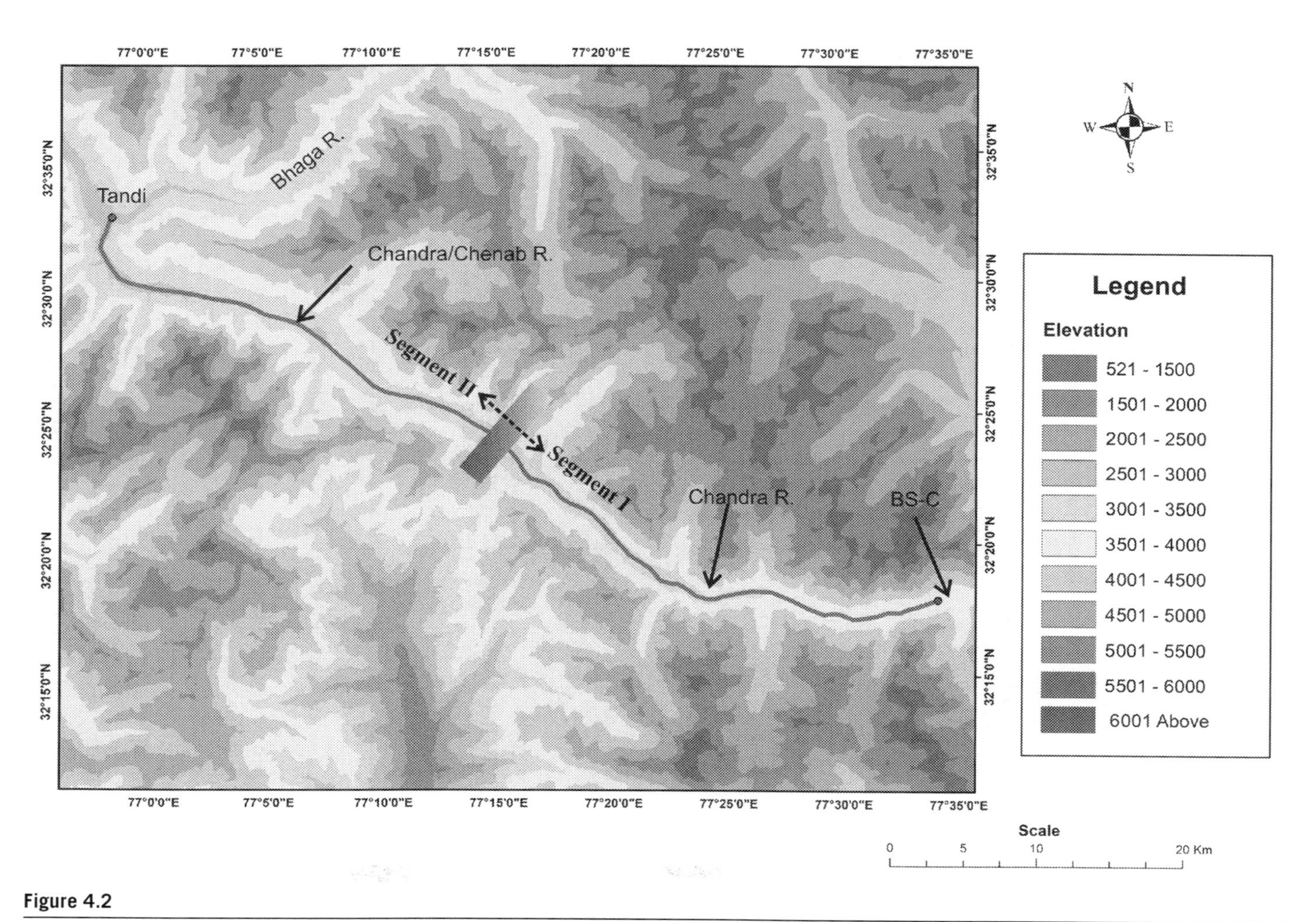

Figure 4.2

The segment of long profile for the Chandra river depicting the broad topographic association of the study area (source and method given in the text).

On the basis of morphostratigraphy, Owen et al. (1995, 1996, 1997) assigned five glacial stages: Chandra (oldest), Batal, Kulti, and Sonapani I and II (youngest). Owen et al. (1997) dated deltaic sediments using OSL method to determine that they were formed during the Batal glacial stage at 40 ka. Reinterpretation of the morphostratigraphic relationships of these deltaic deposits and a comprehensive TCN dating by Owen et al. (2001) assigned a younger age for the Batal glacial stage (i.e., 12–15.5 ka: Late Glacial Interstadial). Owen et al. (2001) also showed that Kulti glacial stage occurred in the early Holocene (10–11.4 ka). Owen et al. (1996) documented Sonapani-II glacial stage during the latter part of the 19th century, by comparing photographs of glaciers taken by Walker and Pascoe (1907) in 1905. Eugster et al. (2016) produced new cosmogenic ^{10}Be exposure ages with improved production rate scaling model to show the LGM ice extent and subsequent deglaciation in the Chandra Valley. They (Eugster et al., 2016) inferred that ">1000 m thick valley glacier retreated >150 km within a few thousand years after the onset of LGM deglaciation (quoted)."

In the upper reaches, the Chandra river shows majority of clasts of TSS dominated by the pink colored Shan quartzites, white colored Muth quartzites, dark gray limestone, and gray carbonaceous shales derived from the ridges of Spiti valley (Batal formation, Kunzum la formation, and the Haimanta series). The ridges show pink-, gray-, green splintery shale, siltstone, and limestones as scree deposits encroaching the wide flood plains of Chandra.

The upper reaches can be observed for the advancing denudational fronts both from top topographic/denudational surface and the sloping lateral erosional surface (Fig. 4.3A). The sedimentation governed by the colluvial/scree fans appears to be advancing toward the topographic denudational front. This observation has been maintained further in our documentation for the entire segment of Chandra valley. This is based on the premises that the rate of weathering/erosion, deposition, and incision are interrelated.

4.4.1 Segment I

Near the locality Batal, the floodplain shows significant mass input from the right as well the left bank (e.g., Fig. 4.3B). The confluence of the Bara Shigri valley with Chandra has been described in detail as the glacial surge during LIA by (Chand et al., 2017). In this area the valley becomes wider with fluvially modified sediment- filled planes. The large, wide-, open channel also accommodates outsized clasts and rock units indicating high energy and low transportation. The fluvial channel-cut profile in the river shows clast composition dominated by TSS. However, it also shows largest (outsized) clast of disk-shaped granite (32×10 cm) along with some well-rounded granitic pebbles (2×3 cm). The largest TSS-pink sandstone clast comprises ~29×29 cm. The modal composition of a 1×5m grid thus indicate that the pebbles are dominated by pink and green sandstones, granites, and phyllites with a coarse sand of granular matrix. There is significant sediment input over the banks from both the ridges (left and right) as scree deposits and recycling of moraines. The floodplains are frequently encroached by shallow streams (crevasse) cutting the older floodplains. The bald ridge topography on the tributary valley depict three stages of glacial valley development. In the lower half of this segment I, there are prominent records of the lateral moraines (Fig. 4.3C) mainly comprising of the TSS boulders/clasts indicating the record of older glaciation. These lateral moraines experienced significant erosion during the subsequent glacial/interglacial stages and the erosion is undergoing through Late Holocene.

Figure 4.3

(A) This site at Chandra lake (upstream of the segment) shows a combination of denudational and depositional fronts. (B) At the confluence of Bara Shigri glacial valley with the Chandra valley, there is significant sediment dynamics as observed. (C) The active erosion of lateral moraines in the lower part of segment I of the river profile. (D) A hanging valley glacier depicting the mechanism of active sediment input to Chandra river. (E) Advanced denudation, excessive deposition, valley overfill, and development of terraces in the region of depocenter-1. (F) Prominent pothole occurrences in a restricted part of the segment I. (G) Typical convergence of the denudational and depositional fronts within the terminal part of segment I. (H) Upward migration of the stages of cirque glaciation along with older sedimentation surface being incised in the segment II.

Fig. 4.3D shows an active hanging valley glacier producing inputs to the Chandra river along with excessive sediment input in the form of scree into the valley. The scree formation makes one of the most significant contribution to the valley, at least during the Holocene and Recent. The screes are generated seasonally due to snowfall over the ridges. This scree material is free flowing and lighter to be transported under various modes within the high energy channel conditions of Chandra river. The tributary valleys too are seen to produce significant scree that is transported into the main valley. This lower part of segment I shows complex interaction of the tributary valley inputs with the active channel producing terraces. Overall the valley is overfilled with sedimentation producing a major

Figure 4.3

(continued)

depocenter (Fig. 4.3E). In the terminal region of this segment we find significant changes in morphology, sedimentation, and river gradient. The river gradient is dropped and coincides with occurrence of prominent potholes over the valley side ridges (Fig. 4.3F). In the absence of surface dating we propose this to be a long-term snout of an older glacial stage. This lower part of the segment also depicts higher denudational erosion, higher river incision, and thicker sedimentation with a relative convergence of the denudational and depositional fronts (Fig. 4.3G).

4.4.2 Segment II

The Chandra river channel characterizes significant entrenchment (incision) of the pre-existing moraines or glacial till in this segment. It is marked by higher valley fill with deep entrenchment/incision. The valley appears to be more V shaped relative to the segment I. The cirque glaciers appear to have been the predominant mechanism for the complex lateral as well as denudational erosion in different stages (Fig. 4.3H) in segment II. The topographic lowering of the denudational surfaces appear to be governed by complex association of the cirque glacier weathering. The upper part of this segment is dominated by

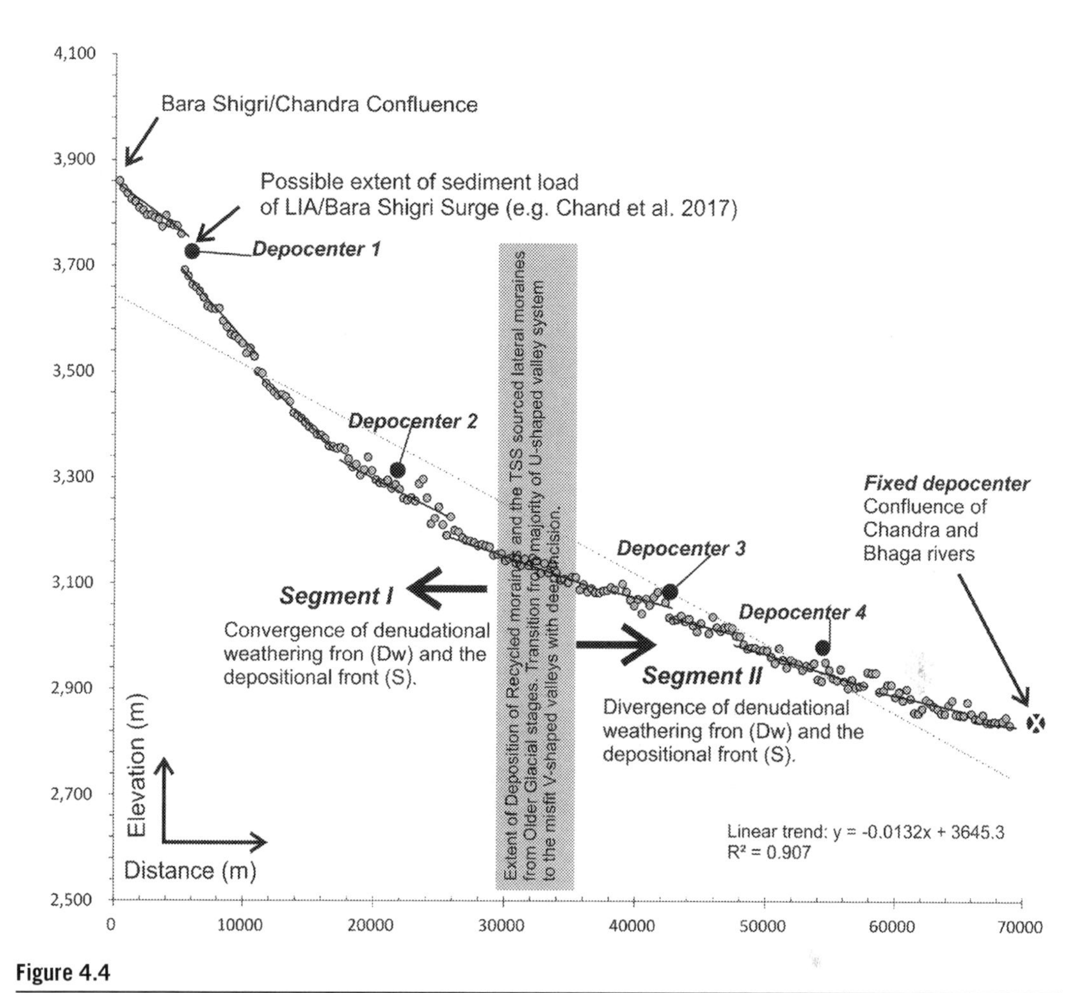

Figure 4.4

The long profile for the segment of Chandra river valley subdivided into two based on field information as described in the text.

incision over sedimentation, while the lower part shows significant sedimentation over the floodplains and banks before the confluence with the Bhaga river near Tandi.

4.5 Inferences from profiling

We further plot the long profile for the entire segment at an interval of 300 m (Fig. 4.4). The long profile shows a major nick point marked in the figure near depocenter 1. The record and extent of a Bara Shigri glacier surge has been documented by Chand et al. (2017), and hence this depocenter can be marked for the possible limit of LIA. A mix of set of granite dominant and TSS dominant facies, their recycling and the large boulders in this very wide valley base, before this nick point, substantiate the extent of sediment load related to LIA depicting the locale for the uppermost depocenter in this segment. Further the depocenter 2 is marked by significant inputs from the tributary valleys over the wide valley base along with recycling of the older lateral moraines coinciding with the occurrence of the potholes and sharp change in valley gradient. Based on field observation on the thickness of recycled moraine deposits in the valley, we further identify two more depocenters (3 and 4) in the segment II. In the absence of chronologic data, these depocenters can be inferred as the artifacts of combination of the spasmodic and episodic depositions during Late Holocene. These depocenters provide the locales for first order estimates on the stable, metastable, and mobile sediment mass and demands more detailed field documentation along with geophysical surveys for shallow subsurface mapping of valley base.

Finally we analyze the cross-profile data (Fig. 4.5) from both these segments to depict the scales and levels of denudational weathering surface (Dw), the sedimentation base (S), the lateral denudational extent (Lw) with reference to the base level (bn). It can be noted that the valley ridges in both the segments show asymmetric elevations, while the Dw is compensated to large extent. The accurate profiling of Dw and S can be inferred for convergence and divergence of these two surfaces as a result of variation in the rates of denudation and deposition. These observations along with the field documentation are further used to suggest the conceptual model for the valley processes (Fig. 4.6).

4.6 Discussion

The Late Pleistocene/Holocene transition in the Higher and Trans Himalaya is marked by the changeover from extensive glacial fields to dissected/dispersed multiple glacial valley fields. This glacial/interglacial changeover is morphologically reflected in change from majority of "*trunk glacial*" system to "*tributary, hanging valley, and cirque glaciers*" system. In Chandra valley, majority of the glacial trimlines and lateral moraines related to the trunk system are poorly preserved due to extensive lateral valley erosion along with the advancement of denudation. This has resulted in both the lowering of catchment topography and eroding the sidewalls of the trunk valley. The denudational surfaces are marked by abundant eroded horn and arete phases depicting the style of erosion after deglaciation. This denudational surface receives seasonal snowfall causing further physical weathering during Late Holocene. The longer and warmer summers of current climate change have accelerated the weathering and sediment transport over this deglaciated catchment/denudational surface. Considering the high catchment surface area to valley area ratio (visualized in Fig. 4.5), this denudational surface appears to make chief contribution to sediment production in the valley. This sediment mass is channelized from hanging valley, tributary valleys, and the scree faces. If this mobilization occurs episodically under extreme weather conditions, the current channel and valley widths are resulting into avalanches and flash flood like events. We therefore denote this surface as the major denudational front or topographic weathering front named as denudational surface front (Dw). The instantaneous rate of denudation may be calculated by exposure dating of the peaks and the base of this denudational topography.

The second important surface is the line of highest sediment deposition (*S1*) which can be marked by the scree deposits higher up in the valley; or the surface of valley floor (S2) merging with the floodplain (e.g., see Fig. 4.6). The rate of growth of the scree fans and the denudation needs to be established to examine their interrelations. The highest level of scree deposition appears to be an active process hand in hand with the lowering of denudational surface due to weathering. These sediment deposition lines

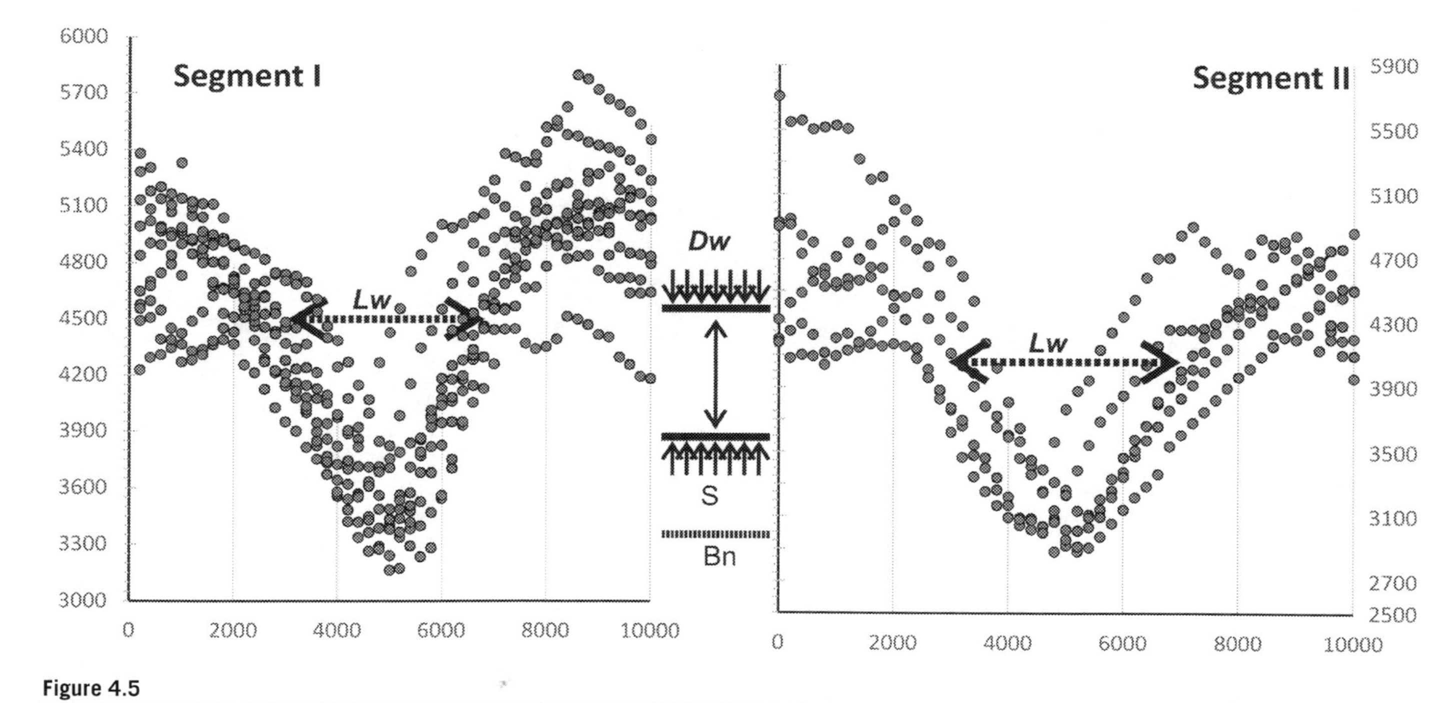

Figure 4.5
The elevations data for segments I and II to express the positions and magnitudes of denudational and depositional fronts as basis for the conceptual model given in Fig. 4.6.

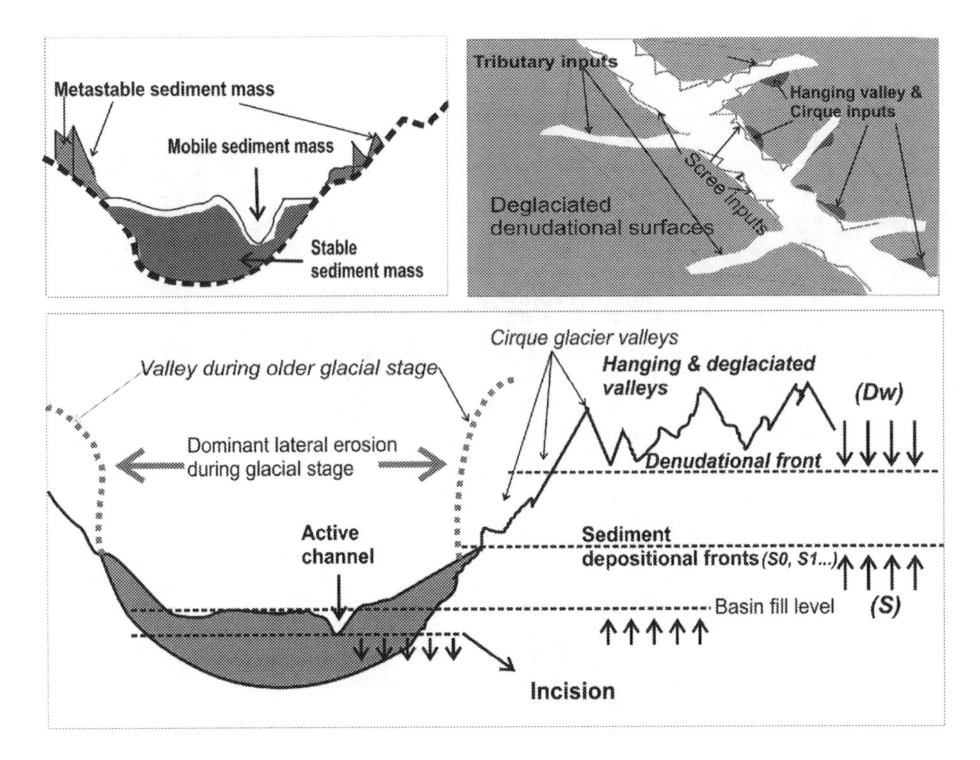

Figure 4.6

Top left diagram shows cartoon depicting the broad status of sediment mass in the Chandra valley. The top right diagram shows approximate proportion of catchment-channel distribution and the tributary, hanging valley, and scree inputs. The lower diagram depicts the model to visualize the concept of different denudational and depositional fronts and their interrelationship. Details discussed in the text.

(*S1 and S2*) can be anticipated to advance higher up whereas the denudational surface is lowered down producing the convergence of the two. On the contrary, the higher rate of incision, lowering the base level may appear to diverge these surfaces.

The rate of valley fill is cumulative and is governed by the rate of sediment transfer; and both are further altered by the episodic events such as during extreme conditions. The field observations indicate that the Chandra river is being obstructed and fed by the recycled and morainic sediment inputs from tributary valleys damming the channel at several places. This tendency of Chandra river to be blocked and breached episodically, mobilizes the sediments under higher energy regimes with redeposition creating multiple depocenters as mapped in field and shown in the long profile (Fig. 4.4).

Increase in precipitation as well as melting due to the emergent climate warming may enhance the chances of such events in the near future. This demands parameterization of these sediment–geomorphic components to enable the estimation of rates of denudation, deposition and sediment transfer, and their interaction throughout the long profile. On this premises we presented the conceptual model as shown in Fig. 4.6. This model demands detailed studies (database) on the available sediment mass in the form of mobile, stable, metastable states (Fig. 4.6) for projections under extreme conditions. In general, the glacial stages are predominant mass producer while the interglacial periods are the mobilizers. It is therefore necessary to understand the interglacial mass transfer (pathways) in the Himalayan region. With the experience of Chandra valley we propose this general and conceptual model to evaluate the Holocene interglacial sediment dynamics (Fig. 4.6).

4.7 Learning and knowledge outcomes

During the current warming trend in the Himalaya, the duration of summers has increased significantly relative to the winter duration. As a result, the melting as well as summer precipitation have increased producing higher inputs from the tributary valleys. In the studied segment of the Chandra river, the changes in valley gradient and occurrence of multiple depocenters depicting overfilled valley conditions are observed. This indicates inflexions in down valley gradient of the sediment mass transfer depicting the significance of episodic mass transfer through extreme events. Whereas in segment II, the river has attempted to encroach the flood plains and higher up due to the cumulative increase in mass flows. With insufficient width of the floodplains under competitive geomorphic accommodation space, the river may respond forced incision and the enhanced floodwater levels. The metastable sediment mass entrapped in the valley sides therefore can produce inputs at various levels in the profile during such episodes. Thus a correct estimation of these different masses of sediments (e.g., stable, metastable, and mobile as shown in Fig. 4.6) is necessary.

Higher incision causing deepening of the valley will result into forced divergence of the Dw and S surfaces creating landslides like conditions. The extended sediment inputs from the tributaries and hanging valleys into the trunk valley at several places in the Chandra valley indicates the blockage of river occurred several times during past. Since the lateral widening is relatively slow, there is no immediate scope for increasing the accommodation space, although the advancement of sedimentation front would increase the surface area for the floodplains by attaining new elevation. This will also result in further incision by the active channel.

Sediment budgeting by evaluation of mobile and immobile sediments along with the rates of weathering and transport for various segments of valley is fundamental to develop predictive models for extreme conditions. The thickness and mass of mobile and immobile sediment packages, susceptibility of both these classes to normal and extreme conditions, routine sediment/water input and output records, sediment accumulation, and erosion rates are the fundamental parameters to be evolved and integrated. Thus the present work may lead to insights into more detailed approaches in the Late Holocene assessment of the valley basins in the Himalaya and their vulnerability to extreme conditions.

Acknowledgments

We acknowledge Savitribai Phule Pune University (SPPU) for partial funding and the Head, Department of Geology and Geography, SPPU for the encouragement and necessary permissions. S.R. acknowledges Director, Wadia Institute of Himalayan Geology, Dehradun for guidance and encouragement. Critical reviews from two anonymous referees have greatly improved the manuscript.

References

Benn, D.I., Owen, L.A., Himalayan glacial sedimentary environments: a framework for reconstructing and dating the former extent of glaciers in high mountains, Quat. Int. 97–98 (2002) 3–25.

Bookhagen, B., Rasmus, C., Thiede, M., Strecker, R., Late Quaternary intensified monsoon phases control landscape evolution in the northwest Himalaya, Geology 33 (2) (2005) 149–152.

Chand, P., Sharma, M.C., Bhamdri, R., Sangewar, C.V., Juyal, N., Reconstructing the pattern of the Bara Shigri Glacier fluctuation since the end of the Little Ice Age, Chandra valley, north-western Himalaya, Progresses in Physical Geography 41 (2017) 643–675, https://doi.org/10.1177/0309133317728017. In this issue.

Dortch, J.M., Owen, L.A., Caffee, M.W., Quaternary glaciationinthe Nubra and Shyok confluence, northernmost Ladakh, India Quat. Res. 74 (2010) 132–144.

Eugster, P., Scherler, D., Thiede, R.C., Codilean, A.T., Strecker, M.R., Rapid last glacial maximum deglaciation in the Indian Himalaya coeval with midlatitude glaciers: new insights from ^{10}Be-dating of ice-polished bedrock surfaces in the Chandra Valley, NW Himalaya, Geophys. Res. Lett. 43 (2016) 1589–1597, doi:10.1002/2015GL066077.

Hedrick, K.A., Seong, Y.B., Owen, L.A., Caffee, M.C., Dietsch, C., Towards defining the transition in style and timing of Quaternary glaciation between the monsoon-influenced Greater Himalaya and the semi-arid Transhimalaya of Northern India, Quat. Int. 236 (2011) 21–33.

Owen, L.A., Finkel, R.C., Barnard, P.I., Ma, H., Asahi, K., Caffee, M.W. and Derbyshire, E., Climatic and topographic controls on the style and timing of Late Quaternary glaciation throughout Tibet and the Himalaya defined by ^{10}Be cosmogenic radionuclide surface exposure dating. Quat. Sci. Rev. 24 (2005) 1391–1411.

Owen, L.A., Thackray, G., Anderson, R.S., Briner, J., Kaufman, D., Roe, G., Pfeffer, W., Yi, C., Integrated research on mountain glaciers: current status, priorities and future prospects, Geomorphology 103 (2009) 158–171.

Owen, L.A., Benn, D.I., Derbyshire, E., Evans, D.J.A., Mitchell, W., Sharma, M., et al., The geomorphology and landscape evolutionof the Lahul Himalaya, Northern India, Z. Geomorphol. 39 (1995) 145–174.

Mitchell, W.A., Taylor, P.J., Osmaston, H., Quaternary geology in Zanskar, NW Indian Himalaya: evidence for restricted glaciation and preglacial topography, Journal of Asian Earth Sciences 17 (1999) 307–318.

Owen, L.A., Benn, D.I., Derbyshire, E., Evans, D.J.A., Mitchell, W.A., Richardson, S., The Quaternary glacial history of the LahulHimalaya, Northern India, J. Quatern. Sci. 11 (1996) 25–42.

Owen, L.A., Caffee, M., Bovard, K., Finkel, R.C., Sharma, M., Terrestrial cosmogenic surface exposure dating of the oldest glacial successions in the Himalayan orogen, Geol. Soc. Am. Bull. 118 (2006) 383–392.

Owen, L.A., Derbyshire, E., Fort, M., The Quaternary glacial historyof the Himalaya, Quat. Proc. 6 (1998) 91–120.

Owen, L.A., Gualtieri, L., Finkel, R.C., Caffee, M.W., Benn, D.I., Sharma, M.C., Cosmogenic radionuclide dating of glacial landforms in the Lahul Himalaya, Northern India: defining the timing of Late Quaternary glaciation, J. Quat. Sci. 16 (2001) 555–563.

Owen, L.A., Mitchell, W., Lehmkuhl, F., Bailey, R.M., Coxon, P., Rhodes, E., Style and timing of Glaciation in the Lahul Himalaya, Northern India, J. Quat. Res. 12 (1997) 83–110.

Sangode, S.J., Gupta, K.R., An Overview of two Decades of Quaternary Research in India: Some reflections based on Bibliographic Analysis, Episodes 33 (2010) 109–115.

Shroder, J.F., Higgins, S.M., Quaternary glacial chronology and neotectonics in the Himalaya of northern Pakistan, Special Pap. Geol. Soc. Am. 232 (1989) 275–294.

Walker, H., Pascoe, E.N., Notes on certain glaciers in Lahaul, Rec. Geol. Surv. India 35 (1907) 139–147.

CHAPTER

Holocene regional-scale behavior of the rivers of Indian Peninsula

5

Vishwas S. Kale

Formerly at Department of Geography, SP Pune University, Pune, India

5.1 Introduction

The triangular-shaped Indian Peninsula (Fig. 5.1), underlain by rocks of Archean to Cretaceous–Eocene age, displays an erosional landscape dominated by bedrock landforms and partially to deeply weathered rocks (Kale and Vaidyanadhan, 2014). Recent studies suggest that the Cenozoic geomorphic history of the Peninsular landscape was dominated by slow and steady-state denudation, primarily controlled by climate rather than by tectonic movements (Jean et al., 2020 and references therein). The fluvially sculpted landscape of the Peninsula shows poor preservation of the late Cenozoic fluvial archives. Exposed and well-dated Neogene fluvial deposits are conspicuously absent, and the exposed Quaternary fluvial deposits are confined to a narrow belt in certain stretches of the river valleys (Kale and Vaidyanadhan, 2014).

The principal focus of the Quaternary fluvial studies in India since 1970s has been on the Pleistocene alluvial fills yielding fossils and abundant stone tools (Corvinus et al., 1973; Williams and Clarke, 1984; Kale and Rajaguru, 1987; Chamyal et al., 2003; Jain and Tandon, 2003; Mishra et al., 2003; Juyal et al., 2006; Williams et al., 2006; Patnaik et al., 2009). Existing stratigraphical, geoarcheological, and geochronological studies indicate that the bulk of the exposed Quaternary deposits in the river valleys were laid down during mid to late Pleistocene. In comparison, the Holocene deposits are patchy and discontinuous even in the alluvial reaches. As a result, Holocene fluvial archives have received much less attention from the Quaternary geologists and geomorphologists. The reconstruction of the Holocene climatic and environmental conditions on different timescales over the Peninsula is primarily based on the interpretation of the lacustrine and speleothem records as well as the offshore sediments (Misra et al., 2019; Mishra et al., 2020; Gupta et al., 2020).

5.2 Scope of the study

Although limited in areal extent and thickness, the Holocene alluvial deposits are present in many valleys of the Peninsula as inset deposits (Fig. 5.2). In addition, overbank flood sequences and bedrock channel slackwater deposits of mid to late Holocene age occur in specific geomorphic settings and provide information about the flooding history on different timescales. The overall objective of this paper is to broadly reconstruct the Holocene fluvial history of the Peninsular river basins by synthesizing and summarizing the results of numerous site-specific and basin-specific fluvial reconstruction studies of

Holocene Climate Change and Environment. DOI: https://doi.org/10.1016/B978-0-323-90085-0.00017-6

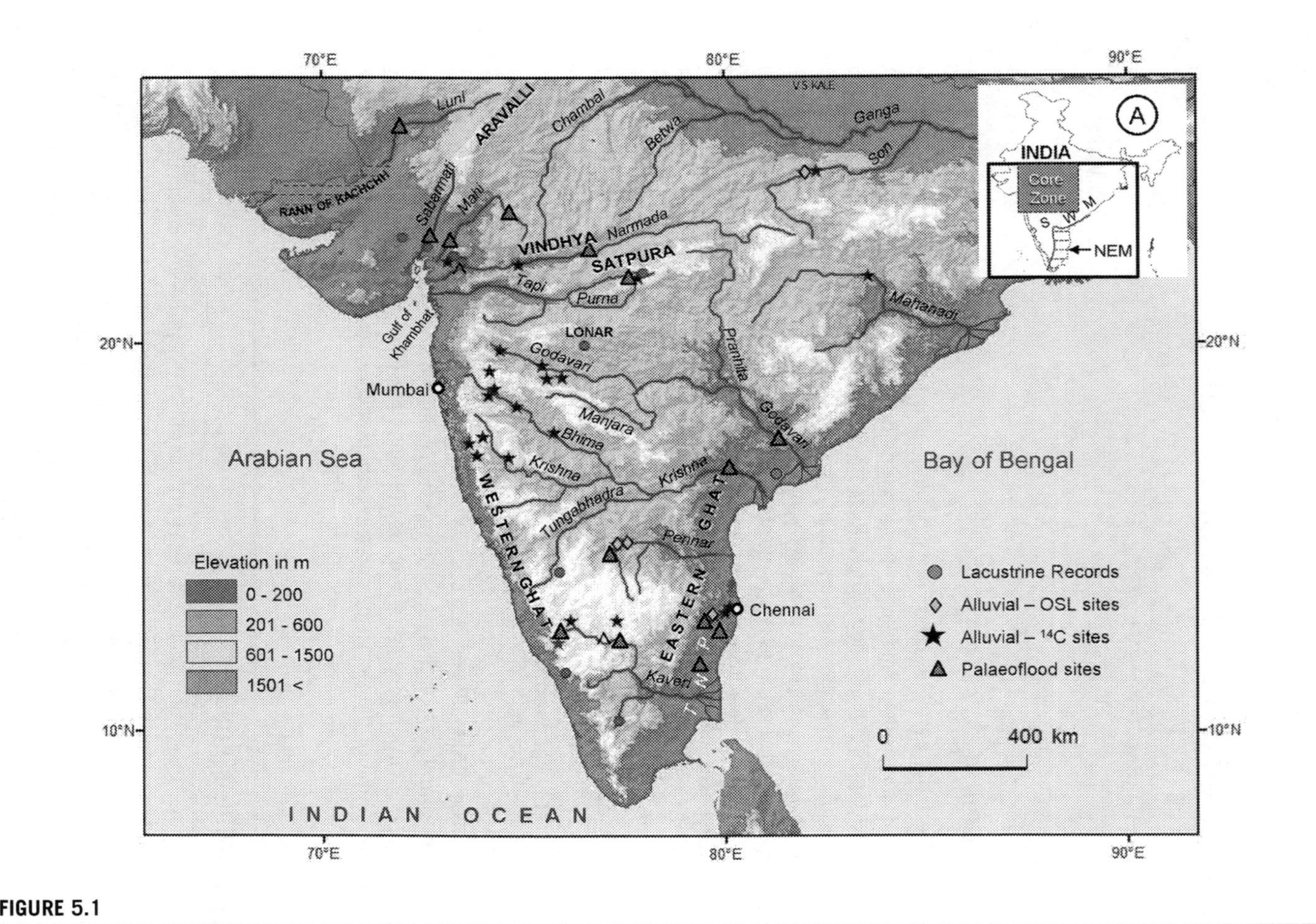

FIGURE 5.1

Map of the Indian Peninsula showing the locations of the dated Holocene alluvial, paleoflood, and lacustrine sites included in the present study. *TNP*, Tamil Nadu Coastal Plains. The inset map (A) shows the core monsoon zone or CMZ. The core area dominated by northeast monsoon (NEM) is shown by horizontal dashed blue lines. Unshaded areas within the box are dominated by southwest monsoon (SWM).

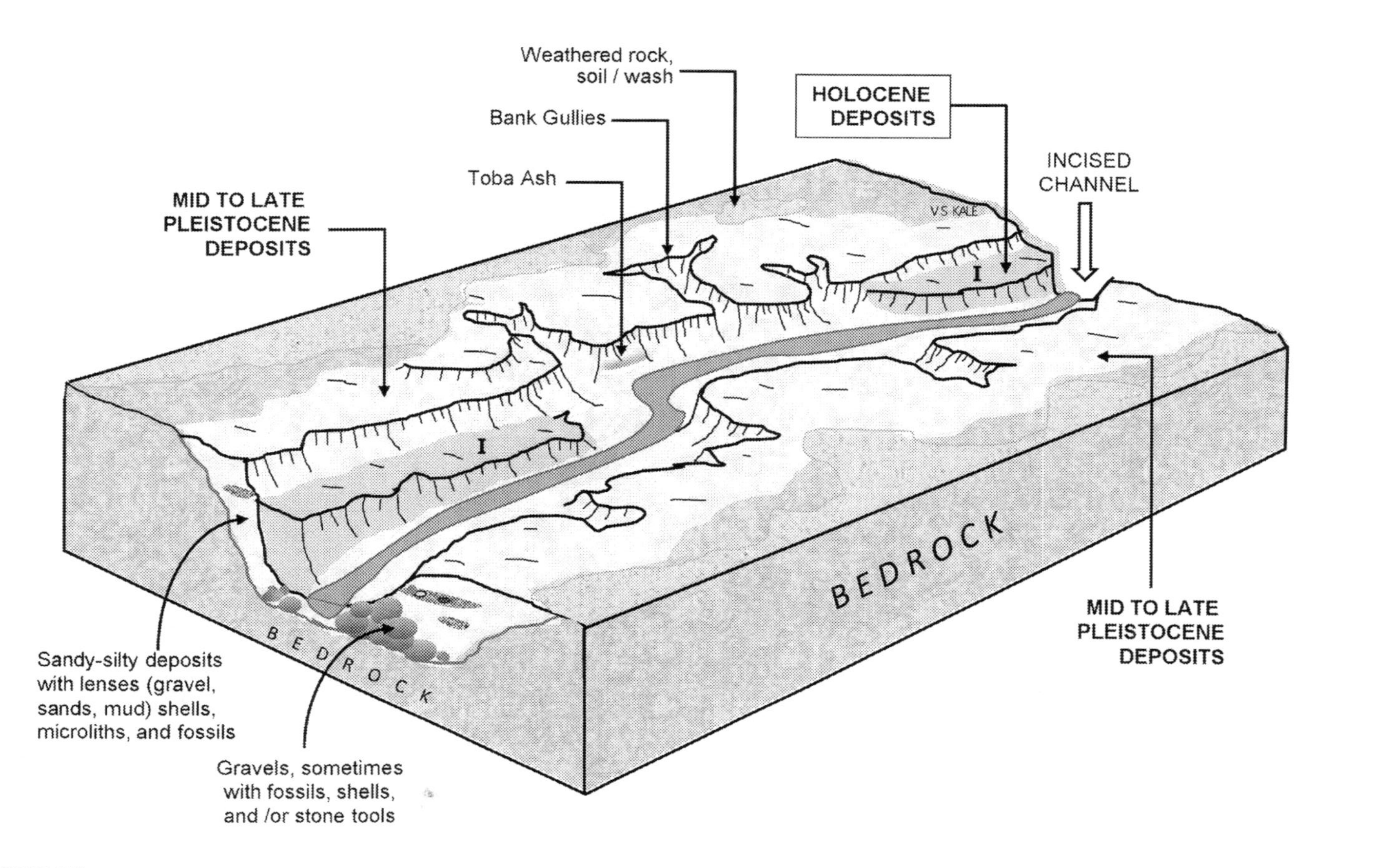

FIGURE 5.2

Schematic block diagram illustrating the geomorphic setting of the Holocene deposits as observed in some Peninsular rivers. The diagram shows up to three river terraces/surfaces, the incised river channel within the mid to late Pleistocene alluvial deposits and the location of younger Holocene inset deposits (I). Not to scale.

the alluvial and paleoflood deposits across the Indian Peninsula. Overall, the synthesis indicates minor to significant inter- and intrabasin differences in the fluvial activity (i.e., degradation, aggradation) and flooding regimes as well as variable patterns of synchronicity across the Peninsula.

In order to appreciate the differences in the responses of the different river basins, it is imperative to understand the differences in the topographic, climatic (monsoon rainfall pattern), geomorphic, and tectono-structural settings of the Peninsular rivers as well as the spatio-temporal variations in the monsoon strength during the late Quaternary. Therefore, first the geomorphic and hydrologic characteristics of the rivers are briefly described. This is followed by a brief summary of the pre-Holocene fluvial records and the inferred Holocene paleo-monsoon history. The fluvial activity during the Holocene is then discussed on the basis of the data and inferences from nearly two dozen dated alluvial sites and about a dozen paleoflood sites spread across the Indian Peninsula.

5.3 The Peninsular rivers: fluvial geomorphology and hydrology

The Indian Peninsula is the largest and the oldest geomorphic province of India (Kale and Vaidyanadhan, 2014). The monsoon-fed river systems draining the Peninsula (Fig. 5.1) are classified into three principal groups:

(a) The east-flowing large rivers, such as the Godavari, Krishna, and Kaveri that originate in the Western Ghat and drain into the Bay of Bengal. This group also includes Mahanadi and Pennar rivers and other smaller rivers that drain into the Bay of Bengal.

(b) The west and southwest flowing rivers, such as the Narmada, Tapi, Mahi, and Sabarmati that rise in the highlands or hill ranges of central or western India and debouch into the Arabian Sea via Gulf of Khambhat. This group also includes the short and swift-flowing west coast rivers that begin in the Western Ghat and drain into the Arabian Sea.

(c) The north or northeast-flowing rivers, such as the Son, Betwa, and Chambal, that form part of the Ganga–Yamuna drainage system.

Although sandstones, quartzites, and limestones are present in the catchments of some of the rivers, the two dominant types of geological terranes in the peninsula are—the granite-gneissic (including charnockites and khondalites) terrane and the basaltic (Deccan Traps) terrane (Kale and Vaidyanadhan, 2014). Existing alluvial deposits in the former terrane are sand-dominated, and the fluvial sediments in the Deccan Traps region generally show preponderance of finer fractions (silt and clay) and gravels. Sand-bed rivers are more common in the granite-gneissic terrane.

The west or south-west flowing rivers, such as the Narmada, Tapi, Mahi, and Sabarmati and the northeast flowing Son river occupy ancient structural depressions represented by the Cambay Graben and the Son–Narmada–Tapi (SONATA) rift zone (Chamyal et al., 2003; Kale and Vaidyanadhan, 2014, and references therein). These rivers have some of the thickest (up to 300–800 m) and laterally extensive Quaternary alluvial fills in Peninsular India (Maurya et al., 2000 and references therein).

Monsoon is the defining characteristic of the tropical climate over India, at least during the last 8–10 Ma (Kale and Vaidyanadhan, 2014; Gupta et al., 2020). Though the southwest summer monsoon overwhelmingly dominates the rainfall supply of the Peninsular basins, considerable spatial and temporal variability exists in the distribution of precipitation. Analyses of the instrumental data available for the last approximately two centuries indicate multidecade episodes of wetter (above-average) and drier (below-average) monsoon conditions at the all-India as well as the basin level (Kale, 1999, 2012). Decadal-scale variability in the Indian monsoon since the late 15th century is also reflected in the tree ring records with annual resolution from Kerala, the India's gateway to the southwest monsoon (Borgaonkar et al., 2010).

The southwest summer monsoon (June to September) contributes >80% of the annual rainfall over a major part of the Indian Peninsula. Only in the southeastern part of the Peninsula (Fig. 5.1A), the share of the northeast winter monsoon (October to December) in the annual rainfall increases. Rivers such as, Narmada, Tapi, Mahanadi, Mahi, Son, Chambal, and Betwa have their sources in the CMZ, that is, the core monsoon zone (Fig. 5.1A). Here the southwest monsoon contributes up to 90% of the annual rainfall. The ~1500-km long Western Ghat, with orographic rainfall, is the source of three principal Peninsular rivers, namely, the Godavari, Krishna, and Kaveri. Although these rivers and many of their tributaries (such as Pravara, Bhima, Tungabhadra, etc.) begin in the high rainfall zone (6000–3000 mm) of the Western Ghat, they flow through the drier, rainshadow zone (1000–500 mm) for a few tens to hundreds of kilometers. Such rivers have been classified as allochthonous rivers (Kale and Rajaguru, 1987), because the tributaries of the rainshadow zone do not contribute much discharge to the Western Ghat rivers downstream. The contribution of the northeast monsoon is significant in the lower reaches of the Kaveri, Pennar, and over the catchments of smaller east-flowing rivers of the Tamil Nadu Coastal Plains (Fig. 5.1), such as, the Palar, Ponnaiyar, Koratallaiyar, Cooum, Vaigai, etc.

Many of the rivers in the CMZ fall in the path of the Bay of Bengal low-pressure systems (depressions and cyclones) that move in the west, northwest, or north direction. Some of the extraordinary floods (peak discharges on record between 70,000 and 99,000 m^3/s) on these rivers in the last 100–150 years were produced by the Bay depressions and cyclones (Kale, 2012).

5.4 The Pleistocene fluvial records in the Peninsular river valleys

Although the focus of the present review is on the Holocene fluvial activity, it would be helpful to have a general idea about the Pleistocene fluvial records of the Peninsular rivers. This is because the Holocene formations often occur within deep channels, carved into the mid- to late-Pleistocene alluvium, as inset deposits or terraces (Fig. 5.2). The only exceptions are the coastal delta plain rivers, such as the Palar, Ponnaiyar, Koratallaiyar, Cooum, and lower Kaveri in the south and lower Sabarmati in the west. Only a brief description is presented below.

In comparison with the Indus–Ganga–Brahmaputra Alluvial Plains, the Quaternary alluvial formations in Peninsular India are remarkably limited in areal extent. Even a cursory glance at the geological map of India published by the Geological Survey of India (GSI) will confirm this. Further, there are noteworthy inter- and intrabasin differences in the thickness and lateral extent of the alluvial fills mainly due to differences in the topographic, geomorphic and tectonic settings.

As stated earlier, the structurally controlled rivers draining into the Gulf of Khambhat, such as the Narmada, Tapi-Purna, Mahi, and Sabarmati (Fig. 5.1) have the thickest (up to 300–800 m) and areally extensive alluvial deposits (Maurya et al., 2000; Deshpande and Pitale, 2014). These rivers, particularly in the middle and lower reaches, flow over alluvium and have well-defined incised alluvial channels. The river channels are bounded by high alluvial cliffs (up to 30 m), which are presently dissected by bank gullies (Fig. 5.2). The northeast flowing Son river as well has thick alluvial deposits in the middle and lower reaches (Williams et al., 2006). This tributary of the Ganga river occupies the SONATA zone.

The upper reaches of the Godavari and Krishna and many of their major tributaries such as Bhima, Manjara, Penganga, and Wainganga also have relatively thicker (up to 40–50 m) alluvial fill deposits (Corvinus et al., 1973; Kale and Rajaguru 1987; Rajaguru et al., 1993; Deshpande and Pitale, 2014). Elsewhere the formations are patchy, shallow, or absent. As a result, the rivers flow through gorges or bedrock reaches or semicontrolled reaches (bedrock exposed on the bed or banks).

Pleistocene fluvial deposits are one of the most extensively explored archives in Peninsular India. From the existing ^{14}C, OSL/TL, ESR, and U/Th dates as well as fossil fauna and stone age tools (Acheulian to Mesolithic) occurring within these deposits, it is evident that the bulk of the exposed alluvial fills in the valleys belong to the mid to late Pleistocene Epoch (Rajaguru et al., 1993; Juyal et al., 2006; Williams et al., 2006; Patnaik et al., 2009). There are indications that even the bedrock gorges may have been completely filled and buried by the Pleistocene deposits (Gupta et al., 2007).

These pre-Holocene formations often contain nodular and tubular calcretes and sometimes carbonate cemented sands and gravels. Incipient to well-developed soil horizons (vertisols or red soils) are not uncommon. Furthermore, the 75-ka Young Toba Tuff (YTT), which is considered as a distinct stratigraphic marker horizon, occurs within the alluvial deposits of many Peninsular river basins (Acharyya and Basu, 1993), such as the Son, Narmada, Tapi, Mahanadi, Bhima (Kukadi and Karha), Pennar (Sagileru), etc.

The Toba ash bed and a good number of radiometric dates indicate that aggradation was the dominant fluvial process in the Peninsula valleys during Marine Isotope Stage or MIS 5/4 transition as well as LGM (Last Glacial Maximum centered on ~21 ka) of MIS 2 and even earlier (Rajaguru et al., 1993; Juyal et al., 2006; Williams et al., 2006). Both these global climatic events (LGM and MIS 5/4 glaciation) were associated with weaker monsoon or arid conditions over the Indian subcontinent (Kale et al., 2003a and references therein; Juyal et al., 2006; Zorzi et al., 2015). Presently, the river channels are entrenched into these mid to late Pleistocene alluvial formations, commonly referred to as "Old Alluvium" in many of the 20th century publications as well as reports of the GSI.

5.5 Variability in the monsoon regime during the Holocene

It is a very well established fact that the fluvial records on land generally lack the continuity and the time-resolution required to reconstruct submillennial to subcentenary scale changes in the monsoon rainfall (Kale et al., 2003a). Speleothems and lacustrine deposits provide a near continuous terrestrial record of regional variations in the monsoon strength. Although there are a few studies on speleothems of Peninsular India, the existing high-resolution records do not cover the entire Holocene and there are long temporal gaps (Band and Yadava, 2020 and references therein). The available tree ring records from the peninsular region are even shorter (Borgaonkar et al., 2010). In comparison, the lacustrine records are relatively better resolved in terms of the chronology and cover the entire Holocene or a better part of the Holocene Epoch (Chauhan et al., 2013; Prasad et al., 2014a, 2014b; Veena et al., 2014; Raj et al., 2015; Basavaiah et al., 2015; Sandeep et al., 2017; Rajmanickam et al., 2017; Sridhar et al., 2020, and others). However, due to differences in chronological resolution and due to minor to significant inconsistencies in the timing and duration of the wetter and drier periods inferred by different workers (Misra et al., 2019; Mishra et al., 2020; Phartiyal et al., 2020) only very broad characteristics of the Holocene monsoon climate can be deduced at this stage. In general, the existing lacustrine records indicate moist conditions during the first half, and drier conditions during the second half of the Holocene.

It is logical to consider the Lonar Meteor Lake (Fig. 5.1) record as the representative record of the Indian Peninsular region because the lake is a large natural waterbody located within the CMZ and covers the entire Holocene. The high resolution, well-dated, multiproxy reconstruction of Holocene paleoclimate reveals that the monsoon was much wetter than the present between ~10 and 6 ka BP and drier conditions prevailed over the region during the second half of the Holocene in general, and severe arid conditions occurred between ~4.8 and 4.0 ka BP in particular (Prasad et al., 2014a; Sarkar et al., 2015). The plot of NADM or normalized accumulated departure from the mean $\delta^{13}C_{org}$ for Lonar (Fig. 5.3E) clearly shows three prominent climatic shifts—around 10.1, 6.1, and 4.6 ka BP, and three minor shifts at 1.7, 1.2, and 0.2 ka BP. The interval from 10.1 to 6.1 ka BP was the wettest period, and the interval from 6.1 to 4.6 ka BP was the driest period of the Holocene. In comparison, the time interval between ~4.5 and 1.7 ka BP appears to be a period of remarkably moderate and low-variability conditions, that is, neither too wet nor too dry. A short, but prominent event around 4.5 ka BP is also evident (Fig. 5.3E).

The evidence of a stronger monsoon during the first half of the Holocene (~10–6 ka BP) than the second half (<6 ka BP) is also provided by other proxy records (Fig. 5.3). Marine records from the Arabian Sea and the Bay of Bengal (Schulz et al., 1998; Gupta et al., 2003; Rashid et al., 2011; Gill et al., 2017; Gupta et al., 2020; Gautam et al., 2021, and references therein) generally indicate wetter conditions in the early to mid-Holocene times (Fig. 5.3). U/Th dates of thick calc-tufas in the rainshadow zone of the Western Ghat (Pawar et al., 1988) and radiocarbon dates of peat from Nilgiri Hills (Rajagopalan et al., 1997) also suggest stronger monsoon regime between ~10 and 8 ka.

A number of offshore records further indicate consistent weakening of the monsoon after about 6 ka (Sarkar et al., 2000; Gupta et al., 2020; Gautam et al., 2021, and references therein; Fig. 5.3). Peat records from the Nilgiri Hills also show weakening of the monsoon during mid to late Holocene (Rajagopalan et al., 1997). Similar to the Lonar record, the Andaman Sea and the Bay of Bengal sediment records show a sharp change and weakening of the monsoon from ~4.8 to 4.2 ka BP (Rashid et al., 2011). Here it is pertinent to mention that recently Griffiths et al. (2020) have presented stalagmite-based evidence for driest periods of the Holocene from 5.11 to 3.25 ka in the monsoon-dominated SE Asia, coinciding with the drying of the Sahara Desert.

It is generally presumed that the southwest monsoon (summer) and the northeast monsoon (winter) co-vary on different timescales (Gunnell et al., 2007). There are only a few studies that have been able to determine the Holocene temporal variations in the northeast monsoon (Misra et al., 2019). A pollen-based study of a sediment core from the Kaveri Delta indicates partial climatic amelioration and humid conditions between ~11.5 and 7.0 ka BP, followed by weaker northeast monsoon from ~7.0 to 3.5 BP (Mohapatra et al., 2019). However, Gautam et al. (2021), based on the isotopic data of a sediment core off the Kaveri Delta, have observed evidence of strengthening of the southwest monsoon but subdued nature of the northeast monsoon during the onset of Holocene.

From the archeological perspective, the mid–late Holocene is an extremely important time interval. Archeological evidence from a large number of sites suggests that while the early to mid-Holocene was dominated by Mesolithic culture, the Deccan Chalcolithic cultures flourished approximately between 2500 and 700 BCE (~4.5 and 2.6 ka BP) in central and western India (Fig. 5.3) and the Deccan Neolithic culture thrived in southern India between 2800 and 1200 BCE (~4.8–3.1 ka BP; Mishra, 2001; Fuller et al., 2004). The summed probability plots of calibrated ^{14}C dates from Chalcolithic and Neolithic sites in the Peninsula show that both the cultures flourished in the respective areas between ~4.5 and 2.8 ka BP (Ponton et al., 2012, supplementary material). The Lonar NADM graph (Fig. 5.3E) shows that this is the phase of unusually low-variability and moderate monsoon conditions in the CMZ.

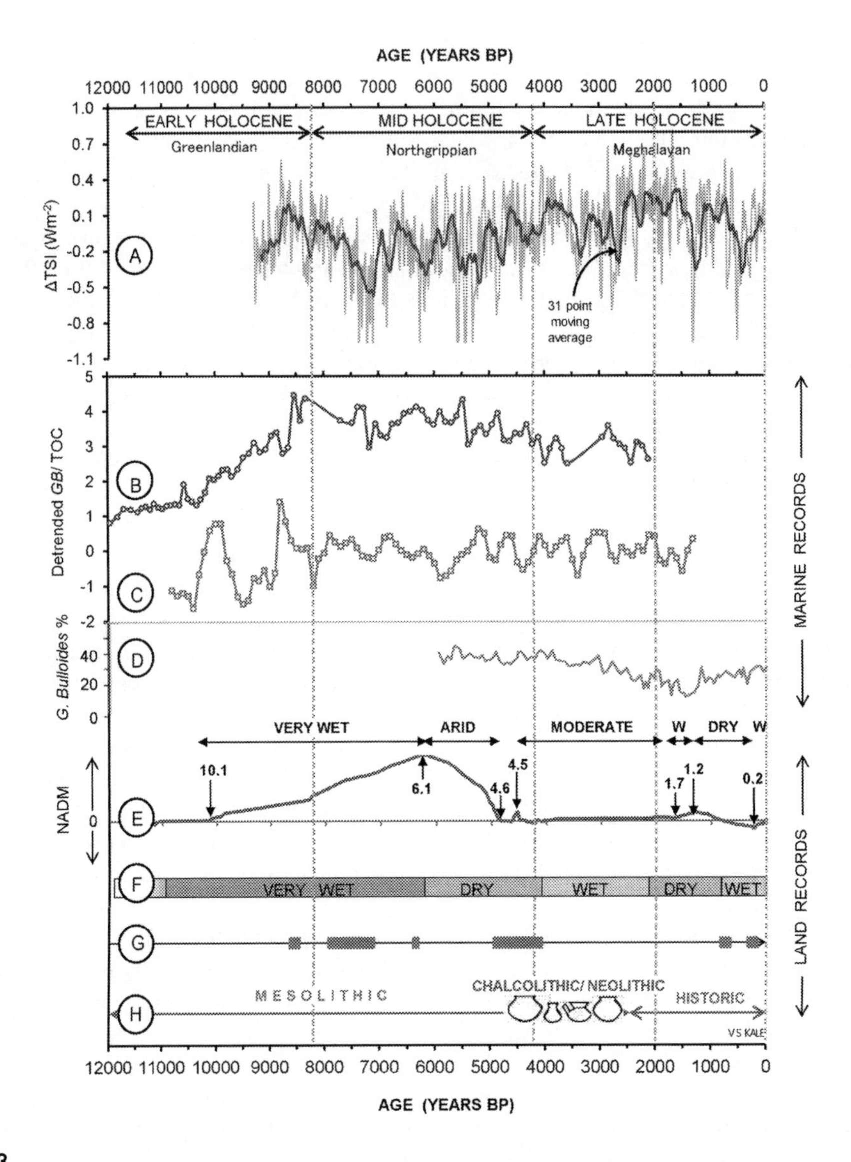

FIGURE 5.3

Plot of various multiproxy records from the Arabian Sea and the Indian Peninsula. (A) ΔTSI in watts per square meter, difference of total solar irradiance from the value of the PMOD composite during the solar cycle minimum of the year 1986 (data source: Steinhilber et al., 2009). (B) Percent total organic carbon (TOC) in core 136 KL (Schulz et al., 1998). (C) Detrended *G. bulloides* for core 723A (Gupta et al., 2003). (D) *G. bulloides* abundance in RC2735 (Anderson et al., 2010). (E) Normalized accumulated departure from mean (NADM) $\delta^{13}C_{org}$ (‰) for Lonar. The numbers at the end of black arrows are inferred radiocarbon ages (in ka) associated with major changes in the monsoon conditions. Basic data from Prasad et al. (2014a), supplementary material. (F) Generalized monsoon regime conditions inferred from Lonar multiproxy record by Prasad et al. (2014a) and Sarkar et al. (2015), $W=$ wetter. (G) Major periods of alluviation (thick lines) in the Indian Peninsula based on radiocarbon clusters with high probability (see Fig. 5.5). Basic data from Kale (2007). (H) Human cultures after Mishra (2001) and Ponton et al. (2012), supplementary material.

The gradual decline of agriculture-based Deccan Chalcolithic settlements, approximately between ~1500 and 700 BCE (~3.5–2.6 ka BP), has been attributed to the onset of drier conditions over central and western India (Mishra, 2001; Shinde and Deshpande, 2015). The summed probability distributions of calibrated ^{14}C dates from Chalcolithic and Neolithic sites show a marked decline after ~3.5 ka BP (Ponton et al., 2012, supplementary material). However, this shift in the monsoon conditions is not reflected prominently in any of the existing lake records, including the Lonar Lake record (Fig. 5.3E). Only proxy records with annual resolution can determine whether this decline in the agriculture-based Chalcolithic settlements was due to increase in inter- or intra-annual variability or some other type of hydroclimatic change. A recent summary of cave studies by Band and Yadava (2020) reveals that such high-resolution speleothem records for the interval 3.5 ± 0.5 ka or more are presently not available for any cave in the Indian Peninsula. Bentaleb et al. (1997), Sarkar et al. (2000), and Mohapatra et al. (2019), however, have detected a change around this time in the marine and delta records.

Approximately a couple of centuries after the last Deccan Chalcolithic settlement was abandoned, amelioration of monsoon in the Early Historic period (beginning at ~2.5 ka BP) is indicated by geoarcheological studies and pollen data. In the middle Narmada domain, for example, the excavated Early Historic settlements are located away from the high alluvial banks of this perennial, but flood-prone river to avoid frequent inundation by floods (Mishra et al., 1999). Further, pollen data from marine sediment cores provide evidence of a wetter phase between ~2.0 and 2.4 ka BP in the Godavari and Mahanadi basins (Zorzi et al., 2015).

It appears from the available proxy records that as compared to the prominent climatic shifts in the early to mid-Holocene, the changes in the monsoon regime conditions during the Common Era were not very dramatic, but very modest in nature (Bentaleb et al., 1997; Misra et al., 2019; Mishra et al., 2020). On a coarse temporal resolution, the monsoon behavior over the Peninsula broadly corresponds with the globally recognized distinct climate intervals, such as the Medieval Climate Anomaly (MCA; ~950–1250 CE or ~1.0–0.75 ka BP) and the Little Ice Age (LIA; ~1450–1850 CE or ~0.75–0.2 ka BP; Dixit and Tandon, 2016).

The Lonar NADM graph (Fig. 5.3E) shows low-magnitude shifts in the precipitation regime at 1.7, 1.2, and 0.2 ka BP, roughly coinciding with the beginning of the Dark Ages Cold Period (DAC), the end of MCA and LIA, respectively. A similar plot of NADM of the composite $\delta^{18}O$ record of the Jhumar and Dandak speleothems (Sinha et al., 2011) indicates noteworthy shifts in the monsoon conditions at ~930, ~1270, and ~1670 CE (Fig. 5.6). The first 2 years approximately represent the beginning and end of the MCA, and the latter year (1672 CE) is close to the mid-point of LIA.

Vegetation records also reflect an alternation of moist and dry phases in central India that broadly correlate with the MCA and LIA, respectively (Zorzi et al., 2015). Relatively wetter and warmer conditions during MCA are also suggested by the Arabian Sea records (Anderson et al., 2010) and model simulations (Tejavath et al., 2019). In comparison, relatively cooler and drier conditions prevailed during the LIA. Multiple proxy records and model simulations indicate that the LIA was characterized by—(a) lower monsoon rainfall, (b) increased frequency of droughts and stronger El Niños, (c) decrease in the frequency of Bay of Bengal cyclones, and (d) lower sea level (Anderson et al., 2002; Kale and Baker, 2006; Sinha et al., 2011; Dixit and Tandon, 2016; Tejavath et al., 2019; Kench et al., 2019).

In addition to these climatic intervals, there is evidence of a modest shift in the monsoon toward drier conditions, approximately from 300 to 1000 CE (~1.6–0.95 ka BP), roughly coinciding with the DCA. This time interval is archeologically documented as—a period of noteworthy decline in the urban settlements in northern and western India (Kale et al., 1997 and references therein), a phase of multiple severe droughts in central India (Sinha et al., 2011), roughly the period of increase in the tank irrigation in the dry zone of south India (Gunnell et al., 2007; Ponton et al., 2012), and lower sea level in the Indian Ocean (Kench et al., 2019).

To sum up, five main inferences, which are relevant from the standpoint of Holocene fluvial activity, could be drawn from the synthesis of the available lacustrine, marine and speleothem records as well as the archeological evidence.

- The Holocene climate over the Peninsula was characterized by two climatic states of the Indian monsoon—the wetter first half (~10–6 ka BP) and the relatively drier second half (<6 ka BP).
- The time interval from 6.1 to 4.6 ka BP was the driest period of the Holocene and a distinct multicentury episode of aridity occurred between ~4.8 and 4.0 ka BP, straddling the 4.2 Meghalayan-event.
- The time interval between ~4.5 and 1.7 ka BP was marked by remarkably low-variability and moderate monsoon conditions in the CMZ (as compared to the wetter early and drier mid-Holocene).
- A probable shift in the hydroclimatic conditions after ~3.5 ka BP that perhaps lasted almost until the beginning of Early Historic Period (~2.5 ka BP) as indicated by the puzzling decline in the agriculture-based Deccan Chalcolithic settlements.
- A phase beginning around the Early Historic period (~2.5 ka BP), characterized by modest variations in the summer monsoon regime conditions and broadly corresponding with the globally recognized two distinct climate intervals, namely the wetter MCA (~1.0–0.75 ka BP) and the drier LIA (~0.75–0.2 ka BP) as well as a regionally recognized weaker monsoon phase from ~1.6–0.95 ka BP, roughly coinciding with the DAC.

5.6 Observations and discussion

The Holocene regional-scale behavior of the Peninsular rivers has been reconstructed here by compiling stratigraphical, geochronological (^{14}C and OSL), and geomorphic data from several site-specific or basin-specific studies undertaken by various workers since the 1970s. Radiocarbon dates and luminescence (OSL) ages of the Holocene fluvial deposits from over two dozen sites, paleoflood records from about a dozen sites (Fig. 5.1), and archeological evidence from a large number of sites have allowed us to broadly reconstruct the Holocene regional fluvial history. The flooding history from the Early Historic to Current Period is briefly discussed separately in the later part of this section.

5.6.1 Holocene regional-scale river behavior of the Peninsular rivers

It will become evident from Fig. 5.1 that the density of the Holocene sites, with at least one radiocarbon date, is the highest in the upper Godavari and Krishna basins underlain by Deccan Traps. Kale and Rajaguru (1987) described the Holocene deposits (~4.5–3 ka BP) occurring within the catchment of these two rivers. These deposits, designated as Post-Black Soil Formation (PBF), are inset against the older, moderately-to-heavily calcretized late Pleistocene alluvial formations. The PBF sediments are up to 15 m thick, and are generally noncalcareous, dark brown sandy-silts, sometimes with gravelly units at the base (Fig. 5.4A–C). The PBF sediments appear to be reworked Pleistocene sediments (Kale and Rajaguru, 1987). Driftwoods, bones, hearths, potsherds, and stone artifacts of Neolithic–Chalcolithic age occur within these deposits. It is important to note here that some of the tributaries of

FIGURE 5.4

Holocene sedimentary deposits exposed on the banks of Godavari, Krishna, Kaveri, and Mahi rivers. (A) Godavari river at Kalavi, near Nasik, Maharashtra. (B) Krishna river upstream of Wai near Balkawdi, Maharashtra. (C) Krishna river at Manjri, near Shiraguppi, Karnataka. (D) Kaveri river at Siddapur, Karnataka. (E) Mahi river at Bhungda, Rajasthan. (F) Godavari river at Kodepudi, upstream of Polavaraum, Andhra Pradesh. Photo A and E courtesy of Pramodkumar S. Hire. All other photos by Vishwas S. Kale.

the Godavari and the Bhima river that have their sources in the rainshadow zone also have Holocene fill deposits, but the ^{14}C ages range between ~11 and 8 ka BP (Mishra et al., 2003).

The mid-Holocene alluvial deposits, occurring in the middle Son Valley, have been designated as Khetaunhi Formation. The formation occurs as an inset aggradational terrace (Williams et al., 2006). These deposits are up to 10 m thick and constitute interbedded silts, clays and fine sands. Radiocarbon ages of ~5.6–3.3 ka BP and IRSL ages of ~1.9–2.7 ka were reported by Williams and Clarke (1984) and Haslam et al. (2012), respectively. Microlithic and Neolithic artifacts are present in these deposits (Williams et al., 2006; Haslam et al., 2012).

In the lower Mahi and the middle reaches of the Sabarmati river the mid-Holocene deposits, represented by younger surface (S3), occur as unpaired terraces (3–6 m high) along the channel and terminate abruptly against the older alluvial deposits (Maurya et al., 2000). The deposits consist of horizontally laminated very fine silty sands overlain by thick clay-rich sediments.

In the lower Narmada Valley, the tops of the Holocene sediments are represented by S4 surface, the lowest surface level (Chamyal et al., 2002). The deposits are inset against mid to late Pleistocene formations. The S4 surface is underlain by two distinct lithofacies—the tidal estuarine facies close to the mouth and the fluvial sandy facies upstream. These mid to late Holocene riverine deposits consist mainly of horizontally stratified fluvial silty sands (Chamyal et al., 2002). Upstream, in the middle domain (the Nimar stretch), alluvial fill deposits with Chalcolithic pottery (~4.4–4.0 ka BP) have been reported by Mishra et al. (2003). These sediments are sandy-silty in texture, less calcareous, and browner in color and occur within the incised alluvial channel of the Narmada river (Mishra et al., 1999). Further upstream (the Hoshangabad stretch), Holocene deposits occur along the Narmada river and have been labeled as Ramnagar and Bauras Formations (Patnaik et al., 2009).

Holocene deposits within the entrenched bedrock channel of the upper Tapi river were reported by Kale et al. (2003b). The sand-dominated sedimentary sequence is about 4 m thick and is moderately lithified by calcium carbonate. Shells (~9.5–9.4 ka BP) and a few Mesolithic tools occur within the sediments. Although there are no radiometric ages from the lower Tapi basin, archeological studies indicate that the Holocene deposits are present in the form of an inset terrace (Pappu and Shinde, 1990).

Mahanadi is one of the major rivers in eastern India. However, there is very little information about the late Quaternary records along this large river. The alluvial deposits occurring on the bank of middle reaches of the Mahanadi river were investigated by Tripathi et al. (2014). The sandy to silty deposits, covering the older alluvial deposits, were dated between ~1.9 and 4.5 ka BP.

The Pennar river in southern India originates in the semiarid zone and drains into the Bay of Bengal (Fig. 5.1). The sedimentary fluvial archives, dominated by sands and gravels, are present only at a few sheltered localities in the upper reaches of this highly seasonal river. The phase of deposition on this river was bracketed between 3 and 2 ka on the basis of OSL dates (Thomas et al., 2007).

Further south, evidence of aggradation around 8.4–7.9 ka and after 1.7 ka is provided by thick structureless sandy silts in the upper reaches of the Kaveri river (Kale et al., 2010; Fig. 5.4D). Further downstream, in the Kollegal area, Valdiya and Rajagopalan (2000) have reported Kaveri deposits dominated by clays of mid-Holocene age (~5.2–4.8 ka BP).

Sandy-silty alluvial deposits, about 4–6 m thick, are exposed along the Koratallaiyar and the Cooum rivers, draining the coastal plains of Tamil Nadu (Fig. 5.1). Unlike the rivers of central and western India, these rivers of the core of the northeast monsoon are not incised into the older alluvial deposits, but directly overlie the Upper Gondwana sandstones and shales. Based on radiocarbon dates, Nagalakshmi and Achyuthan (2004) have inferred three aggradational phases in these coastal rivers: 9.7–8.4,

5.9–5.5, and ~0.9 ka BP. Similarly, phases of aggradation after ~3.59 and around ~ 2.42 ka in the lower Palar basin have been deduced by Resmi et al. (2017) on the basis of OSL dates. It is important to note here that the studies undertaken till date have not been able to establish conclusively whether the Holocene fluvial response in northeast monsoon-dominated rivers was in anyway different from southwest monsoon-dominated rivers.

Ideally, radiometric dates are required from the top as well as from the bottom alluvial units in the same section to constrain the timing and duration of the aggradational episode on the basin or regional scale. However, this is not the case with almost all the dated Holocene sites in the Indian Peninsula. This is mainly due to the absence of material suitable for radiocarbon dating, besides other reasons. In the case of luminescence dating, it is due to other constraints, such as availability of resources for OSL dating, objective of the sampling, poorly bleached sediments, etc.

One way to acquire a better understanding of the Holocene regional-scale behavior of the rivers is to derive the summed probability distribution plots of the calibrated radiocarbon dates (Kale, 2007). In a summed probability distribution plot, distinct periods of alluviation are represented by clusters of calibrated radiocarbon dates with higher probabilities, and erosion-dominant (i.e., nondeposition) phases are indirectly indicated by low probabilities (Kale, 2007; Macklin et al., 2012). The plot given in Fig. 5.5 shows two significant and four smaller clusters of the radiocarbon dates. The two major intervals of region-wide alluviation occurred during mid-Holocene at ~8.0–7.2 ka BP and ~4.9–4.1 ka BP. These two clusters either follow or straddle the two well-established global climatic events of the Holocene at 8.2 and 4.2 ka (Walker et al., 2018). The gap between these two clusters does not necessary implies break in aggradation, but could be simply due to absence of ^{14}C dates of this interval. The highest probability is observed at 0.7 ka BP during the Common Era and a cluster of radiocarbon dates occurs between ~8.8 and 8.4 ka BP within the early Holocene (Fig. 5.5).

A review of the Holocene radiocarbon dates, their geographical distribution and the probability peaks of alluviation (Fig. 5.5) leads to the following observations—(a) the number of radiocarbon dates of the early Holocene period from the principal rivers is small, (b) most of the radiocarbon dates from the main rivers are of mid-late Holocene age, (c) a high proportion of the ^{14}C dates of the early Holocene period are from the tributaries in the rainshadow zone of the Western Ghat or the small coastal rivers of the Tamil Nadu Coastal Plains, and (d) some of the late Holocene ^{14}C dates are on buried materials that were collected from the channel bed deposits, and thus do not necessarily represent aggradation. As a result, multiple smaller and a few major clusters of ^{14}C dates are observed in Fig. 5.5.

Fig. 5.5 also shows long intervals of low probability (below mean probability) during Holocene. The intervals of low probability may be either due to nondeposition or due to nonsampling of alluvial units of this time interval (Kale, 2007 and references therein). However, after taking into consideration the inferences based on other climate proxies (Fig. 5.3), it appears that these low-probability intervals were very likely dominated by fluvial erosion and incision, rather than aggradation.

5.6.2 Regional-scale river behavior: Early Historic Period to Current Period (last ~2.5 ka)

The fluvial sedimentary archives of the last two and half millennia in the Indian Peninsula dominantly include overbank or slackwater flood sequences (Fig. 5.4E and F). There are no indications of a distinct aggradational phase or phases on the basin or regional scale during the Common Era, except a short-duration peak around 0.7 ka BP (Fig. 5.5). There are over a dozen sites of this period in the Peninsula

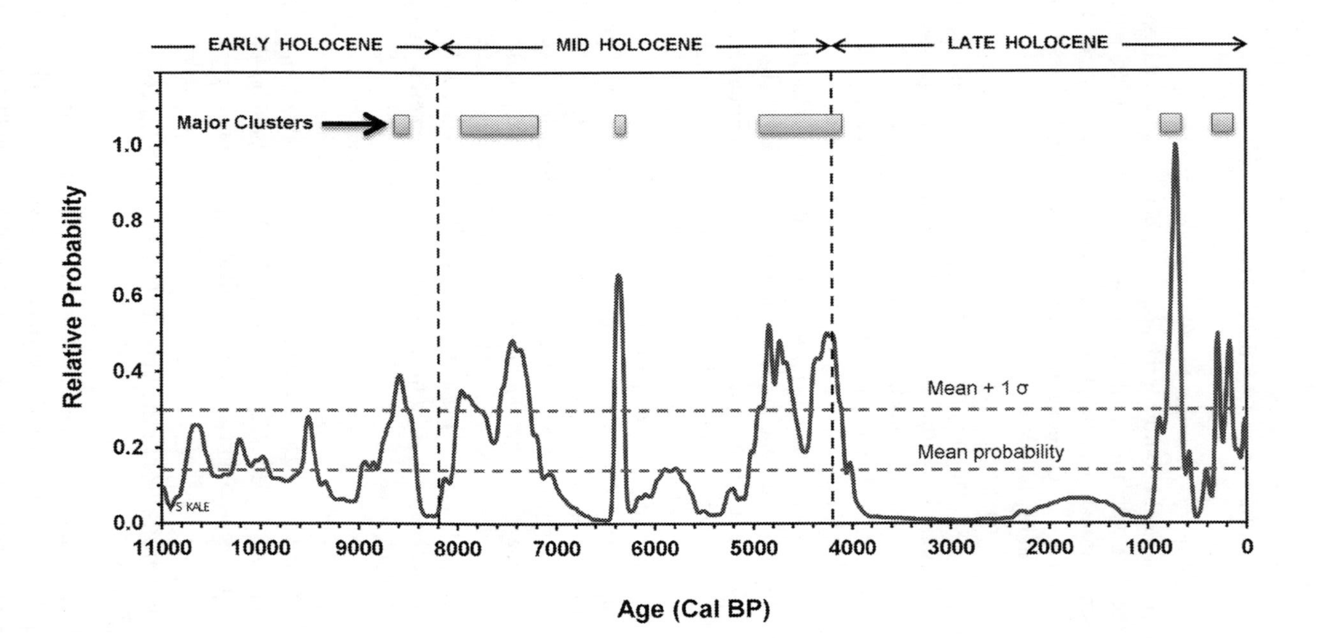

FIGURE 5.5

Summed probability distribution plot for calibrated radiocarbon dates from alluvial deposits from different Peninsular rivers. The horizontal bars at the top represent clusters with higher probability (above mean probability + 1 standard deviation (σ)). Probability data from Kale (2007).

(Fig. 5.1) from where radiocarbon or luminescence ages are available. The flood deposited sediments occur both in bedrock and alluvial reaches (Fig. 5.6).

Investigations of the slackwater flood deposits in the bedrock gorges of the Narmada, Tapi, Godavari, Krishna, Pennar, and Luni rivers (Kale and Baker, 2006 and references therein) reveal clustering of successively large floods during the Common Era, and a noteworthy absence of large-magnitude floods during the LIA (Fig. 5.6).

Slackwater deposits within the confined alluvial channels of the lower Narmada, Mahi, and Sabarmati rivers have also been investigated (Sridhar and Chamyal, 2018 and references therein). Fine-grained flood sediments accumulate in backwater areas in the bank gullies developed within the mid to late Pleistocene calcareous alluvium. Synthesis of the paleoflood data from the lower reaches of these three rivers indicates clustering of floods at 3.0–2.8, 2.2–1.6, 1.3–1.1, and 0.65–0.2 ka BP, and occurrence of widespread flooding events at 1.7 and 0.5 ka (Sridhar and Chamyal, 2018; Fig. 5.6). The last two events coincide with the weak monsoon phases and/or climatic transitions in this part of the Peninsula (Sridhar and Chamyal, 2018).

Stratigraphical studies and OSL ages of flood deposits and slackwater sediments occurring in the upper Kaveri river reveal flood rich clusters centered on ~0.7 and ~0.22 ka (Goswami et al., 2019). The study has also identified another cluster of floods coinciding with the Early Historic period (~2.1–2.4 ka). Another study of flood units along the Palar, Gingee, Then-Pennai, and Vellar rivers, draining the Tamil Nadu Plains, also suggests flood rich periods at the climatic transitions (Mahadev et al., 2019). Results of this study provide evidence of multiple decade-long flood rich episodes during the last three centuries and a distinct period of large floods about 750–800 years ago (Fig. 5.6).

In addition, there are references in several studies and reports to flood deposits ranging in age from Early Historic Period to the 19th century. For example, Rajaguru and Badam (1984) have reported flood silts over an Early Historical site near Paithan along the bank of the upper Godavari river. Mishra et al. (1999) described the effects of floods on Early Historical settlements in the middle Narmada domain. Thick accumulations of flood-deposited sands occur along a 50 km stretch of the Hagari river, a tributary of the Tungabhadra river. These sands were deposited by a storm-induced flash flood in the mid-19th century (Kale et al., 2020). Flood deposits of the same event have also been found in the Pennar river (Thomas et al., 2007). Similarly, historical reports indicate that thick sediments were deposited by an extraordinary flood on the Tapi river in 1829. These flood deposits are present on the southern bank of the Tapi river near Prakasha (Kale et al., 2020).

5.7 Holocene regional-scale river behavior vis-à-vis the monsoon variability

It is now a well-established fact that the behavior of the Indian Peninsula rivers is closely linked to changes in the monsoon strength on different timescales (Kale et al., 2003a). Fluctuations in the intensity of Asian monsoon, connected with global cooling and warming during the Quaternary (Corrick et al., 2020), have had a striking impact on the hydrological characteristics and fluvial responses of the Peninsular rivers (Kale et al., 2003a).

Two relevant points should be noted here that have important implications for the reconstruction of the fluvial history. First, there is sufficient evidence from the marine records that the monsoon was

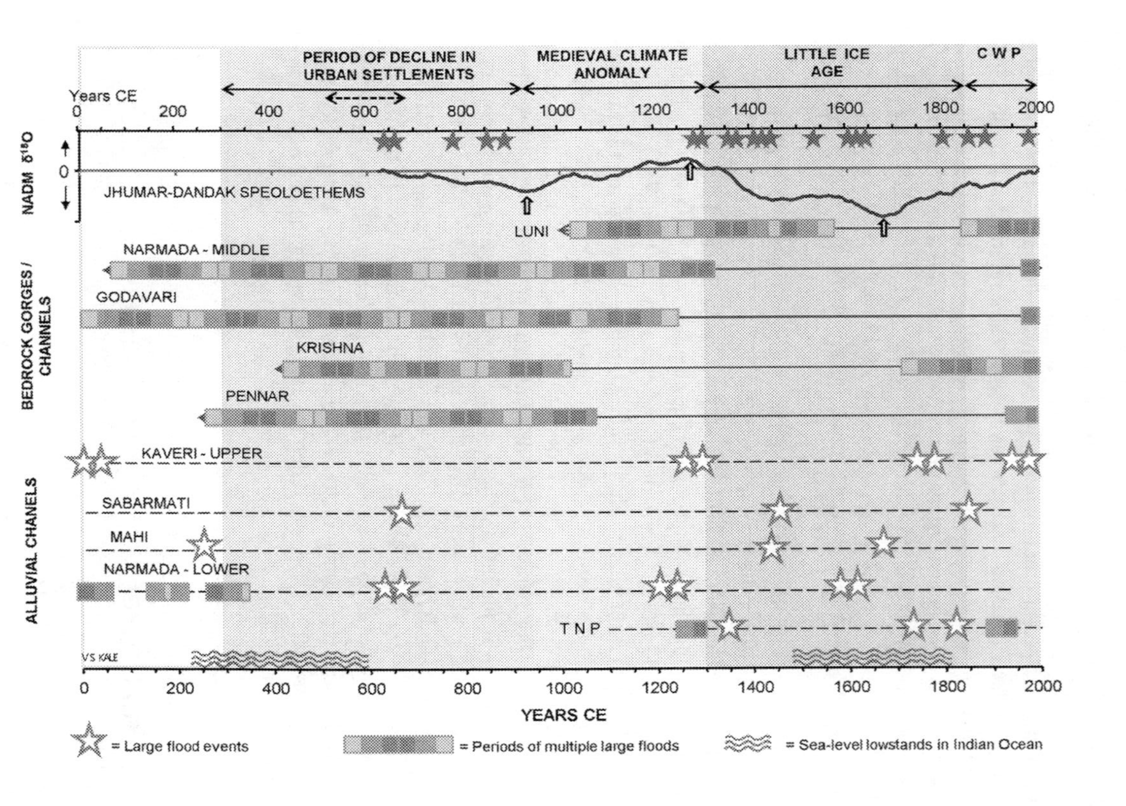

FIGURE 5.6

Summary of the paleoflood records from the Indian Peninsula rivers during the past two millennia. Records from bedrock gorges and channels from Kale and Baker (2006). Records from alluvial and/or semicontrolled channels from Sridhar and Chamyal (2018), Goswami et al. (2019), and Mahadev et al. (2019). Dark red lines represent the approximate length of the paleoflood record in bedrock gorges/channels. In case of ^{14}C dated events, the base year is 1950. The dark blue line on the top is the normalized accumulated departure from the mean (NADM) of the composite $\delta^{18}O$ record of the Jhumar and Dandak speleothems and the red stars represent years/periods of inferred droughts. Basic data from Sinha et al. (2011). Arrows represent major shifts in the monsoon conditions over CMZ. Sea level lowstands in the Indian Ocean and late Antique Little Ice Age (dashed double arrow) after Kench et al. (2019). Period of decline in urban settlements after Kale et al. (1997) and references therein. *CWP*, Current Warm Period; *TNP*, rivers of the Tamil Nadu Coastal Plains.

much stronger during the early–mid-Holocene (between ~ 10 and 6 ka BP) as compared to the second half of the Holocene (<6 ka BP; Fig. 5.3). However, even a cursory examination of the continental records, particularly the lake records, indicates that the evidence of wetter or drier intervals on land is spatially variable. Even considering the fact that to some extent this may be due to chronological uncertainties, some differences are expected. Instrumental data for the homogeneous regions of India clearly show large spatial variability in monsoon rainfall on different timescales (Kale, 2012; Praveen et al., 2020, and references therein). Analyses of the annual peak discharge data for the large Peninsular rivers show noteworthy interbasin differences in the flood-rich and flood-poor periods (Kale, 1999; Kale 2012). Therefore, some lags and leads as well as inter- and intrabasin differences in the duration of the river responses and fluvial activity during the Holocene should not be considered as anomalous. Second, unlike the phases of aggradation, the erosion and incision events are not usually well constrained chronologically. This is because the erosion events in alluvial terrains cannot be dated directly. The radiometric ages of the alluvial deposits are often used to bracket phases of river incision and degradation (Macklin et al., 2012).

The early Holocene enhanced monsoon, primarily driven by maximum obliquity and minimum precession conditions (Zorzi et al., 2015) and concomitant increase in the summer insolation (Fig. 5.6), was responsible for increased runoff in the catchments and higher discharges in the rivers (Kale et al., 2003a). Consequently, the rejuvenated rivers responded by excavating and incising their channels into the Pleistocene alluvial formations. This is also reflected by fewer early Holocene radiocarbon dates from the principal Peninsular rivers. This period of valley-scale incision and excavation, although not radiometrically well constrained, is distinctly observed in many large rivers of the Indian Peninsula, such as the Son (Williams et al., 2006), the Mahi, Sabarmati, and Narmada (Chamyal et al., 2003; Jain and Tandon, 2003; Sridhar et al., 2013), as well as the upper Godavari, Bhima, and Krishna basins (Kale and Rajaguru, 1987; Mishra et al., 2003). The rivers incised and enlarged their channels in alluvial reaches to accommodate large discharges (Kale et al., 2003a). In case of the Narmada river, it has been hypothesized that the alluvium-filled bedrock gorge near Punasa was re-exposed by stripping of the late Pleistocene alluvial cover during the Holocene incision (Gupta et al., 2007).

However, the summed probability distribution plot (Fig. 5.5) provides some hints of aggradation also during this wetter phase. Fig. 5.5 shows a cluster of radiocarbon dates between ~8.8 and 8.4 ka BP and three other smaller clusters (above mean) beginning around 11 ka BP. As stated earlier, this is the case with the tributaries of Godavari and Krishna located in the rainshadow zone of the Western Ghat (Mishra et al., 2003). Similar river response could be extrapolated to other tributaries of the rainshadow zone, although there are no radiometrically dated sites to confirm this. Further, radiocarbon ages of the deposits occurring along the Koratallaiyar and the Cooum rivers indicate a phase of aggradations from 9.7 to 8.4 ka BP (Nagalakshmi and Achyuthan, 2004). It is likely that this aggradation in these small, coastal rivers was in response to the rising erosional base-level (i.e., sea level) during the early Holocene (Loveson and Nigam, 2019) rather than climate, especially if the northeast monsoon was subdued during the early Holocene (Gautam et al., 2021).

Aggradational phases generally appear coincident with drier phases. Previous studies have shown a reasonably good linkage between aggradation and weaker monsoon conditions in the Peninsula rivers (particularly during the LGM and MIS 5/4 transition). Therefore, the occurrence of middle Holocene deposits in many valleys provides indications of a phase of weaker monsoon and drier conditions. The probability-based clusters of radiocarbon dates suggest two phases of alluviation at ~8.0–7.2 ka BP and ~4.9–4.1 ka BP (Fig. 5.5). Although paleoclimatic evidence from the Lonar

Lake record for the first phase (~8.0–7.2 ka BP) is lacking, high-resolution oxygen isotopic data of the Kotumsar Cave indicates an overall weaker monsoon between ~8.5 and 7.0 ka in the CMZ (Band et al., 2018). The second phase of aggradation coincides with the prolonged aridity between 4.8 and 4.0 ka BP, inferred from Lonar Lake and other records (Sarkar et al., 2015). This was also the time of widespread, mega-drought conditions experienced from Africa to Southeast Asia (Griffiths et al., 2020). Furthermore, the lower rate of progradation of the Godavari Delta during ~6–4 ka BP (Nageswara Rao et al., 2015) reflects reduced sediment flux due to upstream sediment sequestration within the channels.

There are no indications that the river channels, which were deeply incised and carved during early Holocene rejuvenation, were completely filled with the mid-Holocene sediments, or the Pleistocene sediments were completely covered by younger deposits. Even in the structurally controlled valleys of the Narmada, Son, Tapi, Mahi, and Sabarmati, very thick sequences of deposits of this subepoch are surprisingly absent. The sediments were accreted mostly within the entrenched channels in the alluvial reaches. The mid-Holocene sediments appear to be reworked and redeposited Pleistocene sediments (Kale and Rajaguru, 1987; Maurya et al., 2000). Today these deposits occur as low, discontinuous inset terraces or benches due to subsequent river rejuvenation, incision, and fluvial excavation. The high proportion of fine sediments implies low-energy fluvial conditions. Considering all the available information from different river valleys, it is reasonable to state that the middle Holocene aggradational phase was not a major geomorphic episode, in comparison with the prolonged phase of aggradation during the late Pleistocene in general and LGM in particular (Kale and Rajaguru, 1987; Williams et al., 2006).

Synthesis of all the available data further indicates that in comparison with the early and middle Holocene, the late Holocene stage (4.2 ka to present) was not a very eventful phase in terms of fluvial activity. The NADM plot (Fig. 5.3E) clearly shows a period of low-variability and moderate monsoon conditions between ~4.5 and 1.7 ka BP. This is approximately the time when Deccan Chalcolithic cultures flourished (and also perished) in Peninsular India, particularly over the Malwa Plateau (the Kayatha and Malwa traditions) and in the lower Tapi and Upper Krishna and Godavari basins (the Savalda and Jorwe traditions; Mishra, 2001; Shinde and Deshpande, 2015).

Fig. 5.5 shows a long period of low probability during the late Holocene. This, as stated earlier, could be either due to nondeposition or due to nonsampling of alluvial units of this time interval (Kale, 2007). However, the fact that the middle Holocene deposits occur at majority of the sites today as low terraces or benches, there is a strong possibility of river rejuvenation leading to incision and excavation of the younger inset deposits in this last stage of Holocene. Offshore data show that the Godavari river, the largest river of Peninsular India, was transferring higher amounts of sediment to the Bay of Bengal during the late Holocene (Giosan et al., 2017). The rapid increase in the rate of progradation of the Godavari river delta after ~4 ka BP (Nageswara Rao et al., 2015) is yet another evidence in support of the argument that degradation was the dominant fluvial activity in the Godavari's catchment in the late Holocene. Apart from the erosion and excavation of the mid-Holocene deposits, the remarkable increase in the sediment flux could also be attributed to the enhanced erosion of the Pleistocene sediments via bank gullies.

It is apparent from the available geomorphic data from various rivers that the gradual, but remarkable increase in the anthropogenic activities (agriculture, deforestation, etc.) in the catchment areas during the past 3–4 millennia have not resulted in a dramatic shift in the fluvial processes from degradation to aggradation (Giosan et al., 2017). The channel morphology of the principal rivers and their main tributaries do not provide any evidence of transport-limited conditions and decline in the sediment fluxes. This may be due to lower sediment to discharge ratio and frequent occurrence of geomorphically effective monsoon floods.

The fluvial activity in the Indian Peninsula from the Early Historic Period to the Current Period (the last ~2.5 ka BP) is dominantly recorded in the form of overbank and/or slackwater flood deposits. The flooding history very broadly corresponds with the globally recognized distinct climate intervals, such as the MCA, the LIA, and the Current Warm Period (CWP). Relatively warmer and wetter climatic intervals (Early Historical, MCA, and CWP) in general and times of climatic transitions in particular show clustering of large-magnitude floods (Fig. 5.6), with some lags and leads. The summed probability distribution plot given in Fig. 5.5 shows a short peak of possible alluviation at 0.7 ka BP, roughly coinciding with MCA–LIA transition. At most paleoflood sites on the Indian Peninsular rivers, the modern floods have been recorded to be higher than the late historical floods indicating increase in the extreme flood-generating rainfall events during the CWP (Kale and Baker, 2006).

Paleoflood records from the bedrock gorges, particularly those falling within the path of the Bay of Bengal low-pressure systems, show a significant decline in the high-magnitude floods during LIA (Fig. 5.6). In addition, the slackwater flood archives from central Narmada provide indications of a period of infrequent but extreme floods between ~400 and ~1000 CE (~1.5–0.95 ka BP). This time interval, as stated earlier, is a period of multiple droughts and overall decline in the human settlements.

Considering all the geomorphic, multiproxy paleoclimatic, archeological and chronological data, it is logical to infer that the late Holocene stage was characterized by intermittent minor deposition, superimposed on a slow, steady incision (Kale and Rajaguru, 1987).

5.8 Learning and knowledge outcomes

Similar to other regions of the monsoon Asia, the Indian Peninsula also experienced fluctuations in the monsoon strength on different timescales within the Holocene. This is amply reflected in the variations in the fluvial response and activity of the Peninsular rivers. While the predominant response to monsoon intensification was degradation, aggradation was induced by weaker monsoon. However, there are spatial differences in the fluvial responses due to location-related and catchment-specific factors. The following important conclusions emerge from the brief review of the Holocene records presented in the preceding sections (Fig. 5.7).

- The river rejuvenation, incision, and excavation during the early Holocene (~11–8 ka BP) was broadly synchronous with a strong summer monsoon regime. Significant increase in the discharges and increased flood activity promoted excavation and vertical incision into the mid to late Pleistocene formations, especially in the alluvial reaches of the principal rivers (Fig. 5.7B). Fewer radiocarbon dates of this interval from the principal rivers denote predominance of fluvial erosion over deposition. Aggradation, particularly in the catchments of tributaries in the rainshadow zone of the Western Ghat and over the coastal plains during this wetter phase, on the other hand, implies out-of-phase response and transport-limited conditions. This early Holocene wetter phase was dominated by Mesolithic culture in Peninsular India.

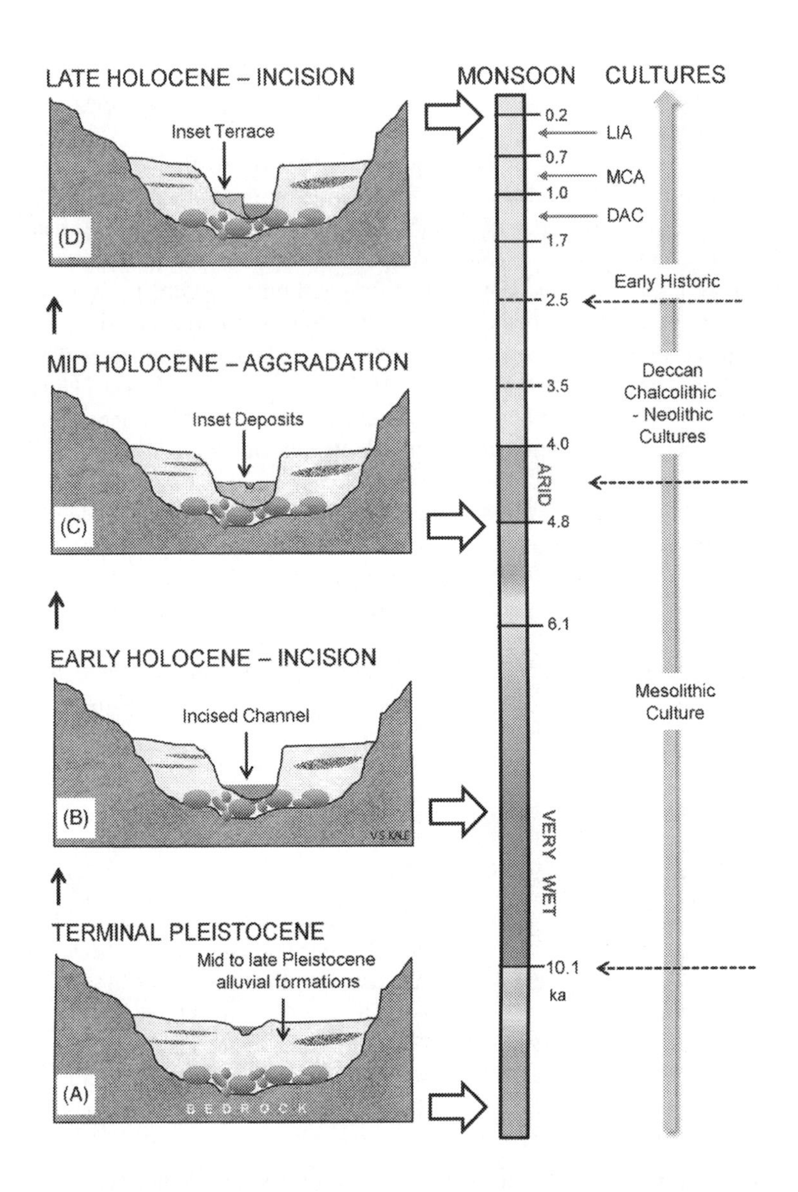

FIGURE 5.7

A schematic representation of the major fluvial episodes in the Indian Peninsula during the Holocene. Inferred Holocene monsoon regime conditions (refer the text for explanation) and the major cultural periods are also given. The time intervals, 2.5 and 3.5 ka are based on archeological evidence. *DAC*, Dark Ages Cold Period; *LIA*, Little Ice Age; *MCA*, Medieval Climate Anomaly.

- With the onset of weaker monsoon and increased frequency of mega-droughts in the middle Holocene (particularly between 6.1 and 4.6 ka BP), the principal rivers responded to the reduced discharge by accreting sediments within the incised and excavated channels (Fig. 5.7C). The centuries between 4.8 and 4.0 ka BP experienced severe aridity and dramatic reduction in the streamflows. Occurrence of inset deposits of the middle Holocene age suggests higher sediment to discharge ratio and relative dominance of aggradation over fluvial incision. However, the high proportion of fine sediments implies low-energy fluvial conditions. Furthermore, the limited extent and inset nature of these deposits suggests that this was a low-magnitude aggradational event. The transition from Mesolithic to Neolithic–Chalcolithic cultures occurred during this regionally widespread weaker monsoon phase, particularly in the northern and northwestern parts of the Peninsula.
- The late Holocene stage (after the end of the arid phase) was not a very eventful time interval in terms of fluvial activity. The interval between ~4.5 and 1.7 ka BP was marked by remarkably low-variability and moderate monsoon conditions. All the available evidence indicates that this latest interval was characterized by intermittent deposition, superimposed on a slow, steady incision. The incision, induced by low sediment to discharge ratio and frequent occurrence of geomorphically effective monsoon floods, was responsible for channel entrenchment within the mid-Holocene deposits. This low-magnitude, second phase of rejuvenation and incision was responsible for the formation of the inset terraces (Fig. 5.7D). Agriculture-based Chalcolithic cultures flourished during the first half of this subepoch, but the settlements were gradually abandoned, perhaps with the onset of relatively drier conditions or some change in the hydroclimatic conditions. The beginning of the Early Historic Period roughly coincides with the amelioration of monsoon.
- The fluvial history since the beginning of the Early Historic Period (~2.5 ka BP) to the current period is primarily recorded in the flood and slackwater sedimentary sequences in the bedrock and confined alluvial reaches. Synthesis of the slackwater data provides evidence of clustering of large-magnitude floods in the Early Historic Period, the MCA, around the onset and end of the LIA as well as in the CWP. By and large, fluvial erosion and degradation has dominated the fluvial activity because of geomorphically effective floods, in spite of gradual increase in the anthropogenic activities in the catchments.

It is apparent from the above discussion that in spite of the remarkable increase in the number of multiproxy based studies of the continental and oceanic records in the last two decades, there are large geographical as well as temporal gaps in our understanding of the Holocene monsoon variability as well as the fluvial activity on the regional scale. The reconstruction of the nature of variations in the monsoon strength during Holocene over land on the century or submillennial scale is still awaited. Further, the role of climate (monsoon variability) in the growth and decay of Bronze, Chalcolithic, Neolithic, and later cultures in the Indian subcontinent in general and the Peninsular India in particular is also yet to be unequivocally established.

Several other issues have emerged from the present review that remain less well understood or completely unanswered. These deal with—(a) the small to significant differences in the timing and duration of the monsoon phases inferred from different proxies, (b) the variable correlation between marine and continental records, (c) the inter- and intrabasin differences in the monsoon-river response relationships, (d) the nature of synchronicity between southwest and northeast monsoon, (e) the role of Holocene sea level changes in the lower reaches of the principal Peninsula rivers, (f) the poor or near-absence of unequivocal signals of globally recognized major climatic events (such as the 8.2 and 4.2 ka events, the mega-drought of mid-Holocene experienced from Africa to SE Asia, etc.) in the lacustrine as well as fluvial records, and (g) the impact of profound and abrupt shifts in the monsoon conditions on the prehistoric and historic settlements. These issues merit attention from Quaternary geologists and geomorphologists as well as climate modelers for the better understanding of the consequences of the human-induced climate changes in the near and distant future.

In the last few years, the focus of the Quaternary terrestrial studies has gradually shifted away from alluvial archives toward speleothem and lake records. However, many of the above-mentioned issues still remain unresolved. A better and more comprehensive understanding of the Holocene climate-river response and reliable reconstruction of the monsoon variability must await more high quality, well-dated, multiproxy studies involving higher resolution (submillennial to century scale) lacustrine, speleothem and fluvial geochronologies from multiple sites in different geomorphic settings of this ancient landmass of India.

Acknowledgments

Special thanks are due to N. Kumaran for the invitation to contribute to this monograph, to Alpa Sridhar, Sheila Mishra, and S.N. Rajaguru for discussions, and to L.S. Chamyal and A.C. Narayana for comments and suggestions.

References

Acharyya, S.K., Basu, P.K., Toba ash on the Indian Subcontinent and its implications for co-relation of late Pleistocene alluvium, Quat. Res. 40 (1993) 10–19.

Anderson, D.M., Overpeck, J.T., Gupta, A.K., Increase in the Asian SW Monsoon during the past four centuries, Science 297 (2002) 596–599.

Anderson, D.M., Baulcomb, C.K., Duvivier, A.K., Gupta, A.K., Indian summer monsoon during the last two millennia, J. Quat. Sci. 25 (2010) 911–917.

Band, S., Yadava, M.G., Addressing past monsoon variability from speleothems, Curr. Sci. 119 (2020) 244–254.

Band, S., Yadava, M.G., Lone, M.A., Shen, C.C., Sree, K., Ramesh, R., High-resolution mid-Holocene Indian Summer Monsoon recorded in a stalagmite from the Kotumsar Cave, Central India, Quat. Int. 479 (2018) 19–24.

Basavaiah, N., Mahesh Babu, J.L.V., Gawali, P.B., Naga Kumar, K.C.V., Demudu, G., Prizomwala, S.P., Hanamgond, P.T., Nageswara Rao, K., Late Quaternary environmental and sea level changes from Kolleru Lake, SE India: inferences from mineral magnetic, geochemical and textural analyses, Quat. Int. 371 (2015) 197–208.

Bentaleb, I., Claude Caratini, C., Fontugne, M., Morzadec-Kerfourn, M.T., Pascal, J.P., Tissot, C., Monsoon regime variations during the Late Holocene in the Southwestern India, in: Dalfes, H.N., Kukla, G., Weiss, H. (Eds.), Third Millennium BC Climate Change and Old World Collapse. NATO ASI Series I49, Springer-Verlag, Berlin, Heidelberg, 1997, pp. 475–488.

Borgaonkar, H.P., Sikder, A.B., Ram, S., Pant, G.B., El Niño and related monsoon drought signals in 523-year-long ring width records of teak (*Tectona grandis* L.F.) trees from South India, Palaeogeogr. Palaeoclim. Palaeoecol. 285 (2010) 74–84.

Chamyal, L.S., Maurya, D.M., Bhandari, S., Rai, R., Late Quaternary geomorphic evolution of the Lower Narmada Valley, Western India: implications for neotectonic activity along the Narmada-Son Fault, Geomorphology 46 (2002) 177–202.

Chamyal, L.S., Maurya, D.M., Raj, R., Fluvial systems of the drylands of Western India: a synthesis of Late Quaternary environmental and tectonic changes, Quat. Int. 104 (2003) 69–86.

Chauhan, M.S., Sharma, A, Phartiyal, B., Kumar, K., Holocene vegetation and climatic variations in Central India: a study based on multiproxy evidences, J. Asian Earth Sci. 77 (2013) 45–58.

Corrick, E.C., Drysdale, R.N., Hellstrom, J.C., Capron, E., Rasmussen, S.O., Synchronous timing of abrupt climate changes during the last glacial period, Science 369 (2020) 963–969.

Corvinus, G., Rajaguru, S.N., Mujumdar, G.G., Some observations on the Quaternary formations of western Maharashtra (India), Quarter 23-24 (1973) 53–69.

Deshpande, G.G., Pitale, U.L., Geology of Maharashtra, Geological Society of India, Bangalore, 2014, pp. 184–187.

Dixit, Y., Tandon, S.K., Hydroclimatic variability on the Indian subcontinent in the past millennium: review and assessment, Earth Sci. Rev. 161 (2016) 1–15.

Fuller, D.Q., Korisettar, R., Venkatasubbaiah, P.C., Jones, M.K., Early plant domestications in southern India: some preliminary archaeobotanical results, Veg. Hist. Archaeobot. 13 (2004) 115–129.

Gautam, P.K., Narayana, A.C., Kiran Kumar, P., Bhavani, P.G., Yadava, M.G., Jull, A.J.T., Indian monsoon variability during the last 46 kyr: isotopic records of planktic foraminifera from southwestern Bay of Bengal, J. Quat. Sci. 36 (2021) 138–151.

Gill, E.C., Rajagopalan, B., Molnar, P.H., Kushnir, Y., Marchitto, T.M., Reconstruction of Indian summer monsoon winds and precipitation over the past 10,000 years using equatorial Pacific SST proxy records, Paleoceanography 32 (2017) 195–216.

Giosan, L, Ponton, C., Usman, M., Blusztajn, J., Fuller, D.Q., Galy, V., Haghipour, N., Johnson, J.E., McIntyre, C., Wacker, L., Eglinton, T.I., Massive erosion in monsoonal central India linked to late Holocene land cover degradation, Earth Surf. Dyn. 5 (2017) 781–789.

Goswami, K., Rawat, M., Jaiswal, M.K., Kale, V.S., Luminescence chronology of late Holocene palaeofloods in the upper Kaveri basin, India: an insight into the climate–flood relationship, Holocene 29 (2019) 1094–1104.

Griffiths, M.L., Johnson, K.R., Pausata, F.S.R., White, J.C., Henderson, G.M., Wood, C.T., Yang, H., Ersek, V., Conrad, C., Sekhon, N., End of Green Sahara amplified mid- to late Holocene mega droughts in mainland Southeast Asia, Nat. Commun. (2020), doi:10.1038/s41467-020-17927-6.

Gunnell, Y., Anupama, K., Sultan, B., Response of the South Indian runoff-harvesting civilization to northeast monsoon rainfall variability during the last 2000 years: instrumental records and indirect evidence, Holocene 17 (2007) 207–215.

Gupta, A., Kale, V.S., Owen, L.A., Singhvi, A.K., Late Quaternary bedrock incision in the Narmada River at Dardi Falls, Curr. Sci. 93 (2007) 564–567.

Gupta, A.K., Anderson, D.M., Overpeck, J.T., Abrupt changes in the Asian southwest monsoon during the Holocene and their links to the North Atlantic Ocean, Nature 421 (2003) 354–357.

Gupta, A.K., Prakasam, M., Dutt, S., Clift, P.D., Yadav, R.R., Evolution and development of the Indian monsoon, in: Gupta, N., Tandon, S.K. (Eds.), Geodynamics of the Indian Plate, Springer Nature, Switzerland, 2020, pp. 499–535.

Haslam, M., Harris, C., Clarkson, C., Pal, J.N., Shipton, C., Crowther, A., Koshy, J., Bora, J., Ditchfield, P., Ram, H.P., Price, K., Dubey, A.K., Petraglia, M., Dhaba: an initial report on an Acheulian, Middle Palaeolithic and microlithic locality in the Middle Son Valley, North-Central India, Quat. Int. 258 (2012) 191–199.

Jain, M., Tandon, S.K., Fluvial response to Late Quaternary climate changes, Western India, Quat. Sci. Rev. 22 (2003) 2223–2235.

Jean, A., Beauvais, A., Chardon, D., Arnaud, N., Jayananda, M., Mathe, P.E., Weathering history and landscape evolution of Western Ghats (India) from $^{40}Ar/^{39}Ar$ dating of supergene K–Mn oxides, J. Geol. Soc. Lond. 177 (2020) 523–536.

Juyal, N., Chamyal, L.S., Bhandari, S., Bhusnan, R., Singhvi, A.K., Continental record of the southwest monsoon during the last 130 ka evidence from the southern margin of the Thar Desert, India, Quat. Sci. Rev. 25 (2006) 2632–2650.

Kale, V.S., Long-period fluctuations in monsoon floods in the Deccan Peninsula, India, J. Geol. Soc. India 53 (1999) 5–15.

Kale, V.S., Fluvio–sedimentary response of the monsoon-fed Indian rivers to late Pleistocene–Holocene changes in monsoon strength: reconstruction based on existing ^{14}C dates, Quat. Sci. Rev. 26 (2007) 1610–1620.

Kale, V.S., On the link between extreme floods and excess monsoon epochs in South Asia, Clim. Dyn. 39 (2012) 1107–1122.

Kale, V.S., Achyuthan, H., Jaiswal, M.K., Sengupta, S., Palaeoflood records from upper Kaveri River, Southern India: evidence for discrete floods during Holocene, Geochronometria 37 (2010) 49–55.

Kale, V.S., Baker, V.R., An extraordinary period of low-magnitude floods coinciding with the Little Ice Age: palaeoflood evidence from central and western India, J. Geol. Soc. India 68 (2006) 477–483.

Kale, V.S., Gupta, A., Singhvi, A.K., Late Pleistocene-Holocene palaeohydrology of monsoon Asia, in: Gregory, K.J., Benito, G. (Eds.), Palaeohydrology: Understanding Global Change, Wiley, Chichester, UK, 2003a, pp. 213–232.

Kale, V.S., Mishra, S., Baker, V.R., A 2000-year palaeofloods record from Sakarghat on Narmada, Central India, J. Geol. Soc. India 50 (1997) 283–288.

Kale, V.S., Mishra, S., Baker, V.R., Sedimentary records of palaeofloods in the bedrock gorges of the Tapi and Narmada Rivers, Central India, Curr. Sci. 84 (2003b) 1072–1079.

Kale, V.S., Narayana, A.C., Jaiswal, M.K., Wind-blown, flash flood-deposited sands of Hagari River, Anantapur district, Andhra Pradesh, India, Curr. Sci. 119 (2020) 556–558.

Kale, V.S., Rajaguru, S.N., Late Quaternary alluvial history of the northwestern Deccan Upland region, Nature 325 (1987) 612614.

Kale, V.S., Vaidyanadhan, R., The Indian Peninsula: geomorphic landscapes, in: Kale, V.S. (Ed.), Landscapes and Landforms of India, Springer, Dordrecht, 2014, pp. 65–78.

Kench, P.S., McLean, R.F., Owen, S.D., Ryan, E., Morgan, K.M., Ke, L., Wang, X., Roy, K., Climate-forced sea-level lowstands in the Indian Ocean during the last two millennia, Nat. Geosci. (2019), doi:10.1038/s41561-019-0503-7.

Loveson, V.J., Nigam, R., Reconstruction of Late Pleistocene and Holocene sea level curve for the east coast of India, J. Geol. Soc. India 93 (2019) 507–514.

Macklin, M.G., Lewin, J., Woodward, J.C., The fluvial record of climate change, Philos. Trans. R. Soc. 370 (2012) 2143–2172.

Mahadev, Singh, A.K., Jaiswal, M.K., Application of luminescence age models to heterogeneously bleached quartz grains from flood deposits in Tamil Nadu, southern India: reconstruction of past flooding, Quat. Int. 513 (2019) 95–106.

Maurya, D.M., Raj, R., Chamyal, L.S., History of tectonic evolution of Gujarat alluvial Plains, Western India during Quaternary: a review, J. Geol. Soc. India 55 (2000) 343–366.

Misra, P., Tandon, S.K., Sinha, R., Holocene climate records from lake sediments in India: assessment of coherence across climate zones, Earth-Sci. Rev. 190 (2019) 370–397.

Mishra, P.K., Chauhan, P.R., Diwate, P., Parth, S., Anoop, A., Holocene climate variability and cultural dynamics in the Indian subcontinent, Episodes 43 (2020) 552–562.

Mishra, S., Ota, S.B., Shete, G., Naik, S., Deotare, B.C., Late Quaternary alluvial history and archaeological sites in the Nimar Region of Western Madhya Pradesh, India, Man Environ. 24 (1999) 149–157.

Mishra, S., Naik, S., Rajaguru, S.N., Deo, S., Ghate, S., Fluvial response to late Quaternary climatic change: case studies from upland Western India, Proc. Indian Natl. Sci. Acad. 69 (2003) 185–200.

Mishra, V.N., Prehistoric human colonization of India, J. Biosci. 26 (2001) 491–531.

Mohapatra, P.P, Stephen A., S. Prasad, S., Singh, P, Anupama, K, Late Pleistocene and Holocene vegetation changes and anthropogenic impacts in the Cauvery delta plains, southern India, Quat. Int 507 (2019) 249–261.

Nagalakshmi, T., Achyuthan, H..

Nageswara Rao, K., Saito, Y., Nagakumar, K.C.V., Demudu, G., Rajawat, A.S., Kubo, S., Li, Z., Palaeogeography and evolution of the Godavari delta, east coast of India during the Holocene – an example of wave-dominated and fan-delta settings, Palaeogeogr. Palaeoclim. Palaeoecol. 440 (2015) 213–233.

Pappu, R., Shinde, V., Site catchment analysis of the Deccan Chalcolithic in the central Tapi basin, Bull. Deccan Coll. Res. Inst. 49 (1990) 317–338.

Patnaik, R., Chauhan, P.R, Rao, M.R., Blackwell, B.A.B., Skinner, A.R., Sahni, A., Chauhan, M.S., Khan, H.S., New geochronological, paleoclimatological, and archaeological data from the Narmada Valley hominin locality, Central India, J. Hum. Evol. 56 (2009) 114–133.

Pawar, N.J., Kale, V.S., Atkinson, T.C., Rowe, P.J., Early Holocene waterfall tufa from semiarid Maharashtra Plateau, India, J. Geol. Soc. India 32 (1988) 513515.

Phartiyal, B., Farooqui, A., Bose, T., Climate change variability through lacustrine records published during 2016-2019. Implications, new approaches and future directions, Proc. Indian Natl. Sci. Acad. 86 (2020) 389–403.

Ponton, C., Giosan, L., Eglinton, T.I., Fuller, D.Q., Johnson, J.E., Kumar, P., Collett, T.S., Holocene aridification of India, Geophys. Res. Lett. 39 (2012) L03704, doi:10.1029/2011GL050722.

Prasad, S., Anoop, A., Riedel, N., Sarkar, S., Menzel, P., Basavaiah, N., Krishnan, R., Fuller, D., Plessen, B., Gaye, B., Rohl, U., Wilkes, H., Sachse, D., Sawant, R., Wiesner, M.G., Stebich, M., Prolonged monsoon droughts and links to Indo-Pacific warm pool: a Holocene record from Lonar Lake, central India, Earth Planet. Sci. Lett. 391 (2014a) 171–182.

Prasad, V., Farooqui, A., Sharma, A., Phartiyal, B., Chakraborty, S., Bhandari, S., Raj, R., Singh, A., Mid–late Holocene monsoonal variations from mainland Gujarat, India: a multiproxy study for evaluating climate culture relationship, Palaeogeogr. Palaeoclim. Palaeoecol. 397 (2014b) 38–51.

Praveen, B., Talukdar, S., Shahfahad, Mahato, S., Mondal, J., Sharma, P., Md, A.R., Islam, T., Rahman, A., Analyzing trend and forecasting of rainfall changes in India using non-parametrical and machine learning approaches, Sci. Rep. 10 (2020) 10342, https://doi.org/10.1038/s41598-020-67228-7.

Raj, R., Chamyal, L.S., Prasad, V., Sharma, A., Thakur, B., Verma, P., Holocene climatic fluctuations in the Gujarat Alluvial Plains based on a multiproxy study of the Pariyaj Lake archive, Western India, Palaeogeogr. Palaeoclim. Palaeoecol. 42 (2015) 60–74.

Rajagopalan, G., Sukumar, R., Ramesh, R., Pant, R.K., Rajagopalan, G., Late Quaternary vegetational and climatic changes from tropical peats in southern India – an extended record up to 40,000 years BP, Curr. Sci. 73 (1997) 60–63.

Rajaguru, S.N., Badam, G.L., Litho and biostratigraphy of Quaternary formations, Central Godavari Valley, Geological Survey of India, Maharashtra, 1984, pp. 89–96. Special Publication No. 14.

Rajaguru, S.N., Kale, V.S., Badam, G.L., Quaternary fluvial systems in upland Maharashtra, Curr. Sci. 64 (1993) 817822.

Rajmanickam, V., Achyuthan, H., Eastoe, C., Farooqui, A., Early Holocene to present palaeoenvironmental shifts and short climate events from the tropical wetland and lake sediments, Kukkal Lake, Southern India. Geochemistry and palynology, Holocene 27 (2017) 404–417.

Rashid, H., England, E., Thompson, L., Polyak, L., Late glacial to Holocene Indian summer monsoon variability based upon sediment records taken from the Bay of Bengal, Terres. Atmos. Oceanic Sci. 22 (2011) 215–228.

Resmi, M.R., Achyuthan, H., Jaiswal, M.K., Middle to late Holocene paleochannels and migration of the Palar River, Tamil Nadu: implications of neotectonic activity, Quat. Int. 443 (2017) 211–222.

Sandeep, K., Shankar, R., Warrier, A.K., Yadava, M.G., Ramesh, R., Jani, R.A., Weijian, Z., Xuefeng, L., A multi-proxy lake sediment record of Indian summer monsoon variability during the Holocene in southern India, Palaeogeogr. Palaeoclim. Palaeoecol. 476 (2017) 1–14.

Sarkar, A., Ramesh, R., Somayajulu, B.L.K., Agnihotri, R., Jull, A.J.T., Burr, G.S., High-resolution Holocene monsoon record from the eastern Arabian Sea, Earth Planet. Sci. Lett. 177 (2000) 209–218.

Sarkar, S., Prasad, S., Wilkes, H., Riedel, N., Stebich, M., Basavaiah, N., Sachse, D., Monsoon source shifts during the drying mid-Holocene: biomarker isotope based evidence from the core monsoon zone of India, Quat. Sci. Rev. 123 (2015) 144–157.

Schulz, H., von Rad, U., Erlenkeuser, H., Correlation between Arabian Sea and Greenland climate oscillations of the past 110,000 years, Nature 393 (1998) 54–57.

Shinde, V., Deshpande, S.S., Crafts and technologies of the Chalcolithic people of South Asia: an overview, Indian J. Hist. Sci. 50 (2015) 42–54.

Sinha, A., Berkelhammer, M., Stott, L., Mudelsee, M., Cheng, H., Biswas, J., The leading mode of Indian summer monsoon precipitation variability during the last millennium, Geophys. Res. Lett. 38 (2011) L15703, doi:10.1029/2011GL047713.

Sridhar, A., Chamyal, L.S., Implications of palaeohydrological proxies on late Holocene Indian summer monsoon variability, Western India, Quat. Int. 479 (2018) 25–33.

Sridhar, A., Chamyal, L.S., Bhattacharjee, F., Singhvi, A.K., Early Holocene fluvial activity from the sedimentology and palaeohydrology of gravel terrace in the semi-arid Mahi River Basin, India, J. Asian Earth Sci. 66 (2013) 240–248.

Sridhar, A., Thakur, B., Basavaiah, N., Seth, P., Tiwari, P., Chamyal, L.S., Lacustrine record of high magnitude flood events and climate variability during mid to late Holocene in the semiarid alluvial plains, Western India, Palaeogeogr. Palaeoclim. Palaeoecol. 542 (2020) 109581.

Steinhilber, F., Beer, J., Fröhlich, C., Total solar irradiance during the Holocene, Geophys. Res. Lett. 36 (2009) L19704, doi:10.1029/2009GL040142.

Tejavath, C.T., Ashok, K., Chakraborty, S., Ramesh, R., A PMIP3 narrative of modulation of ENSO teleconnections to the Indian summer monsoon by background changes in the last millennium, Clim. Dyn. 53 (2019) 3445–3461.

Thomas, P.J., Juyal, N., Kale, V.S., Singhvi, A.K., Luminescence chronology of late Holocene extreme hydrological events in the Upper Penner River basin, South India, J. Quat. Sci. 22 (2007) 747–753.

Tripathi, S., Basumatary, S.K., Singh, V.K., Bera, S.K., Nautiyal, C.M., Thakur, B., Palaeovegetation and climate oscillation of western Odisha, India: a pollen data-based synthesis for the mid-late Holocene, Quat. Int. 325 (2014) 83–92.

Valdiya, K.S., Rajagopalan, G., Large palaeolakes in Kaveri basin in Mysore Plateau: late Quaternary fault reactivation, Curr. Sci. 78 (2000) 101–105.

Veena, M.P., Achyuthan, H., Eastoe, C., Farooqui, A., A multi-proxy reconstruction of monsoon variability in the late Holocene, South India, Quat. Int. 325 (2014) 63–73.

Walker, M., Head, M.J., Berkelhammer, M., Bjorck, S., Cheng, H., Cwynar, L., Fisher, D., Gkinis, V., Long, A., Lowe, J., Newnham, R., Rasmussen, S.O., Weiss, H., Formal ratification of the sub-division of the Holocene Series/Epoch (Quaternary System/Period): two new Global Boundary Stratotype Sections and Points (GSSPs) and three new stages/subseries, Episodes 41 (2018) 213–223.

Williams, M.A.J., Clarke, M.F., Late Quaternary environments in north-central India, Nature 308 (1984) 633–635.

Williams, M.A.J., Pal, J.N., Jaiswal, M., Singhvi, A.K., River response to Quaternary climatic fluctuations: evidence from the Son and Belan valleys, north-central India, Quat. Sci. Rev. 25 (2006) 2619–2631.

Zorzi, C., Goñi, M.F.S., Anupama, K., Prasad, S., Hanquiez, V., Johnson, J., Giosan, L., Indian monsoon variations during three contrasting climatic periods: the Holocene, Heinrich Stadial 2 and the last interglacial–glacial transition, Quat. Sci. Rev. 125 (2015) 50–60.

CHAPTER

6 Holocene vegetation and climate change from central India: An updated and a detailed pollen-based review

Md. Firoze Quamar

Birbal Sahni Institute of Palaeosciences, Lucknow, Uttar Pradesh, India

6.1 Introduction

Vegetation dynamics and climate are strappingly related in a way that regional climate affects land surface processes over a range of scales with unprecedented speed (IPCC, 2007) and vegetation, in turn, affects climate through feedbacks via photosynthesis and evapotranspiration, changes in albedo and biogenic volatile organic compound emissions (Henderson-Sellers, 1993; Fang et al., 2003; Meng et al., 2011; Faubert et al., 2012; Wang and Dickinson, 2012; Henden et al., 2013; Li et al., 2014). Pollen grains and spores, produced by the vegetation itself, constitute a pollen assemblage after transport and mixing by wind and/or water, which represent vegetation characteristics, and contemporary climatic conditions or sedimentary environment at a specific time or area (Erdtman, 1952; Birks and Birks, 1980; Faegri et al., 1989). Pollen-derived vegetation records from an area will reveal variations in monsoon (here the Indian Summer Monsoon; ISM/Southwest Monsoon; SWM) rainfall (Kar and Quamar, 2019, 2020; Quamar, 2019, 2021; Quamar and Kar, 2020a; Quamar and Bera, 2020; Quamar et al., 2021, and references cited therein). India, being an agricultural country, is mostly dependent on the monsoon rain for its agricultural productivity, economy, and societal well-being (Webster et al., 1998; Gadgil, 2003; Gadgil and Gadgil, 2006).

The present communication reviews the studies on vegetation dynamics, associated climate change and the ISM variability during the Holocene from central India (the core of the monsoon zone-CMZ), comprising mainly the States of Madhya Pradesh (*Central Province/Central Territory*; also known as "*the heart of India*") and Chhattisgarh (*Thirty-Six Forts*; Fig. 6.1). The Holocene (~11.7 kyr BP to the Present), recently classified by the International Commission on Stratigraphy (ICS, 2018) into three ages—the Greelandian (11.7–8.2 kyr BP; Early Holocene); the Northgrippian (8.2–4.2 kyr BP; Middle Holocene); and the Meghalayan (4.2 kr BP to Present; Late Holocene), is the most recent Geological Epoch and the present Interglacial stage (Walker et al., 2019). The summer monsoon strengthened during the Early Holocene and subsequently weakened during the Mid- and Late-Holocene, especially at 8.2 ka BP and 4.2 ka BP (sudden rainfall reduction; Berkelhammer et al., 2012; Dixit et al., 2014a, 2018). The major forcing factors, which control the variability of the ISM during the Holocene are solar insolation and migration of the Inter Tropical Convergence Zone (ITCZ), North Atlantic Oscillation (NAO), and El-Niño and Southern Oscillation (ENSO; Fleitmann et al., 2003; Gupta et al., 2003;

Holocene Climate Change and Environment. DOI: https://doi.org/10.1016/B978-0-323-90085-0.00013-9

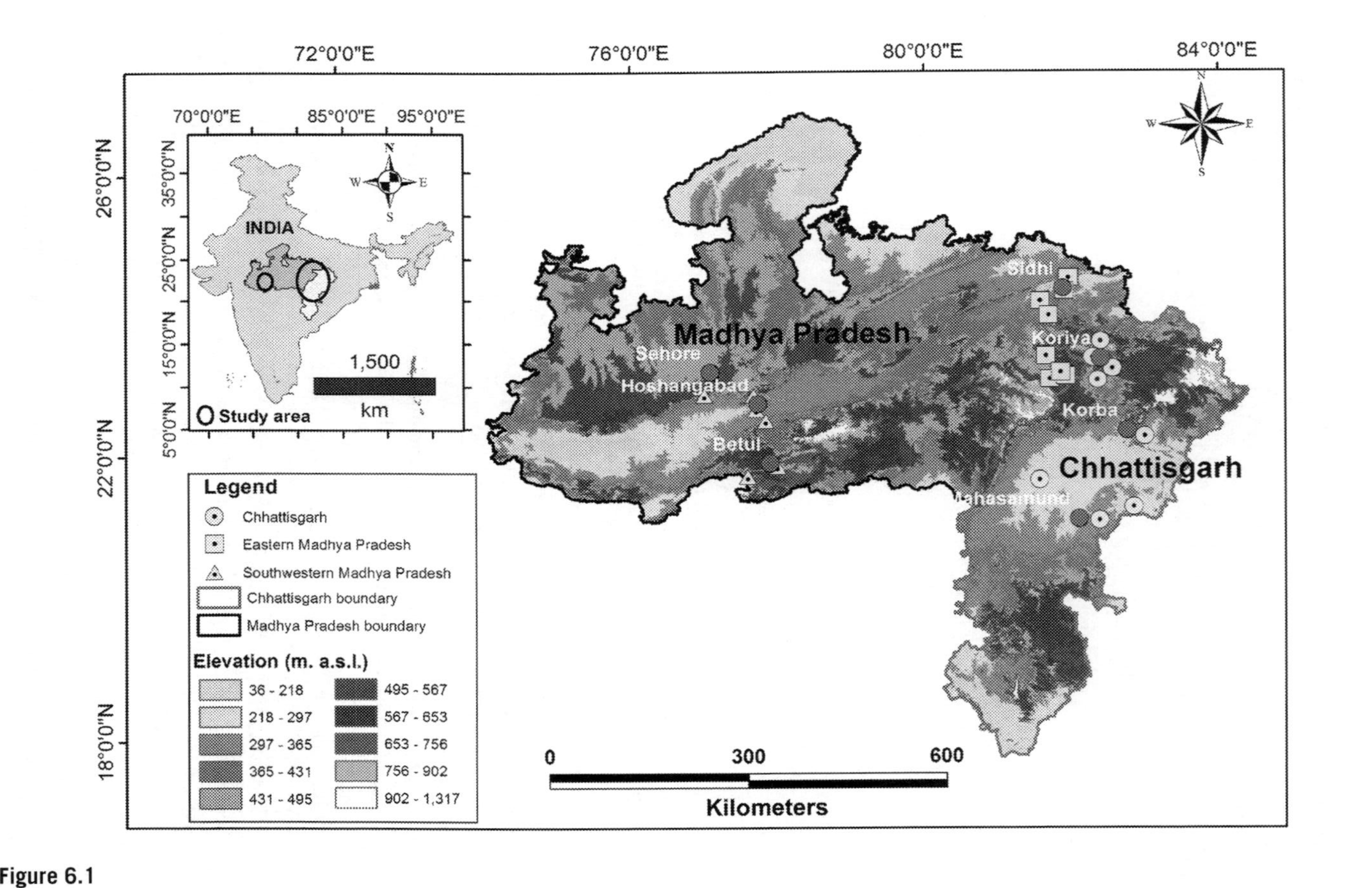

Figure 6.1

Shuttle radar topographic mission (SRTM) digital elevation map (DEM) of the States of Madhya Pradesh and Chhattisgarh, central India, showing the location of the study areas ("yellow triangle" shows the sampling site in Southwestern Madhya Pradesh; "yellow rectangle" shows the sampling site in Southeastern Madhya Pradesh, and "yellow circle" shows the sampling site in Chhattisgarh State; "Magenta-colored circle" shows some of the districts in the States of Madhya Pradesh and Chhattisgarh); geographic map of India showing the States of Madhya Pradesh and Chhattisgarh (inset left), as well as the study site ("nonsquare rectangle"). (For interpretation of the references to colour in this figure legend, the reader is advised to see the web version of this review article). Source of Fig. 6.1: The figure has been made using ArcGIS 10.3. The details of the studies are given in Table 6.1.

Dutt et al., 2015). Further, the strong ISM during the Early Holocene is ascribed to increased solar insolation and associated enhanced evaporation in the oceans and northward movement of the ITCZ (Fleitmann et al., 2003). The ISM during the Mid-Holocene (8.2–4.2 kyr BP) is asynchronous (Gupta et al., 2020), which could be due to the different responses of the two branches of the ISM (the Bay of Bengal: BoB and the Arabian Sea: AS branches) and to changes in the moisture source and position of the ITCZ (Fleitmann et al., 2003; Prasad et al., 2014). Meanwhile, the abrupt weakening of the ISM (abrupt drought event) at 8.2 kyr BP (known as the 8.2 kyr event) could be due to the large influx of the freshwater into the Labrador Sea from the melting of the Laurentide Ice Sheet (LIS), which resulted into a decrease in the Atlantic Meridional Overturning Circulation (AMOC; Bond et al., 2001; Wang et al., 2013). The ISM weakening and the associated 4.2 kr drought event could be linked to the stronger phase of El-Niño and a shift of the Indian Ocean Dipole to a strong negative state (Morrill et al., 2003; MacDonald, 2011).

The Holocene is significant in many respects, particularly in terms of human development and establishment of centers of civilizations. The Holocene, in fact, allowed civilizations to develop and evolve, as well as to perish owing to the gradual and/or abrupt climatic changes during the epoch (Petit et al., 1999; Kumaran and Limaye, 2014; Misra et al., 2020). The abrupt monsoon reduction at 4.2 ka BP had significant societal impact, resulting into the migration of population of Indus-Harappan Valley Civilization to new areas to the east for domestic and agricultural use, as reliable seasonal flooding of the river system around the edge of Thar desert disappeared (Giosan et al., 2012) and also the agricultural productivity of the Harappan settlements decreased in response to the weakened summer monsoon (Berkelhammer et al., 2012; Dutt et al., 2018; Gupta et al., 2020). Thus, knowledge and understanding of the climate change and the ISM variability during this epoch could be of immense interest in order to strengthen our understanding of the present ISM-influenced climatic conditions, as well as of possible future climatic trends and projections (Royer, 2008; Cai et al., 2010; Singhvi et al., 2010).

6.2 Scope

Pollen-based paleoclimatic studies from central India provided insights into the spatio-temporal discrepancy in vegetation dynamics, associated climate change and the ISM rainfall variability during the Holocene. Most of the studies have coarser resolution due to the paucity of good number of dates/ages, as well as large sampling intervals. The poor sampling resolution affects the stratigraphic resolution of the records and the varying time lengths of the records (poor chronological control) limit the comparative assessment of the studied archives completely. Further, these limitations create hindrance in ascertaining the notion whether the vegetation response to climatic change has been insignificant or whether the pollen proxy responded differentially (various proxies have different responses). Also, due to the absence of long-term high-resolution pollen-based records/poor time control, different proxy response time, spatial inhomogeneity/variability of the monsoon precipitation, and the complex forcing mechanisms (solar insolation, internal teleconnections: ENSO, tropical-midlatitude interactions), a comprehensive picture of the continental records of monsoon variability, compared to the marine records, are sparse from the wetlands (lakes, swamps and/or various other archives) in India and simultaneously has rather proved elusive. This limits our understanding of the ISM rainfall variability and its impact on cultures at community levels. So, multiproxy studies with large number of dates (conventional ^{14}C, accelator mass spectrometer [AMS ^{14}C] and/or optical simulated luminescence

[OSL]), as well as with good sampling resolution (1 or 2 cm interval) are further required to be conducted for high-resolution paleoclimatic studies from central India, which could be helpful in future climatic predictions.

6.3 Study area, vegetation, and climate

Madhya Pradesh (21°N–27°N: 74°E–82°E) is one of the richest botanical provinces of the country, which possesses about 24–26% of the total forest the country. On the other hand, Chhattisgarh (18°N–24°N: 79°E–85°E) is also one of the greenest states of India with over 44% of its total area under lush forests and about 8–12% of the total forest cover of the country (Chauhan and Quamar, 2012; Quamar and Bera, 2014; Kar and Quamar, 2019, and references cited therein; Fig. 6.1). Both the states of central India (Madhya Pradesh and Chhattisgarh) have tropical deciduous forests (both dry and moist types; Champion and Seth, 1968; Quamar and Kar, 2020b; Fig. 6.2) and enjoys a Tropical savannah-type climate (Aw), as well as Mesothermal climate–Gangetic Plain-type climate (Cwg; Köppen, 1936; Quamar and Kar, 2020b; Fig 6.3).

Nearest Climate Research Unit Timeseries (CRU TS) 4.01, 0.5 × 0.5 gridded climate data points, 1901–2019, showing mean monthly precipitation and temperature around Madhya Pradesh and Chhattisgarh, India (Harris et al., 2014) has been represented in Fig. 6.4. In Madhya Pradesh, the mean annual temperature is 25.66°C and the mean annual precipitation is 972.44 mm, whereas the mean annual temperature and mean annual precipitation in Chhattisgarh State are 25.69°C and 1242.56 mm, respectively (Supplementary Files 1a, b; 2a, b). Most of the precipitation (~90%) occurs through the SWM/ISM during the months of June to September (JJAS); whereas, some precipitation also takes place in these two States during the months of October, November, and December (OND) due to the North East Monsoon (NEM) or Winter Monsoon.

6.4 Results and discussion: regional and global contextualization

Significant contributions with reference to the vegetation dynamics and associated climate change, as well as the ISM rainfall variability are discussed below (see Table 6.1 for details). For ease of presentation, the subject in question has been discussed with the subheads: Early Holocene to Mid Holocene, Mid Holocene to Late Holocene, and Late Holocene to the Present. The pollen-based review work on the aforesaid topic provide insights into the effect of the global climatic events, such as the Younger Dryas (YD), Holocene Climatic Optimum (HCO), Roman Warm Period (RWP), Dark Ages Cold Period (DACP), Medieval Warm Period (MWP)/Medieval Climatic Anomaly (MCA), Little Ice Age (LIA), and Current Warm Period (CWP) around the areas in question. The inclusion of YD in the present review lies with the fact that the signatures of this cold event have been recorded around the studied areas, although roughly, during ~12.7–7.1 ka BP at the Hoshangabad District in southwestern Madhya Pradesh (Quamar and Chauhan, 2012), and during ~12.785- 9 ka BP at the Koriya District of Chhattisgarh State (Quamar and Bera, 2017), central India, broadly encompassing the termination of the YD at ~11.7 kyr BP and the global onset of the present interglacial stage (the Holocene epoch) when the ISM strengthened (Rasmussen et al., 2006; Pearce et al., 2013; Misra et al., 2019). Furthermore, the signatures of the global 8.2 ka and 4.2 ka events have not been captured around the studied areas in

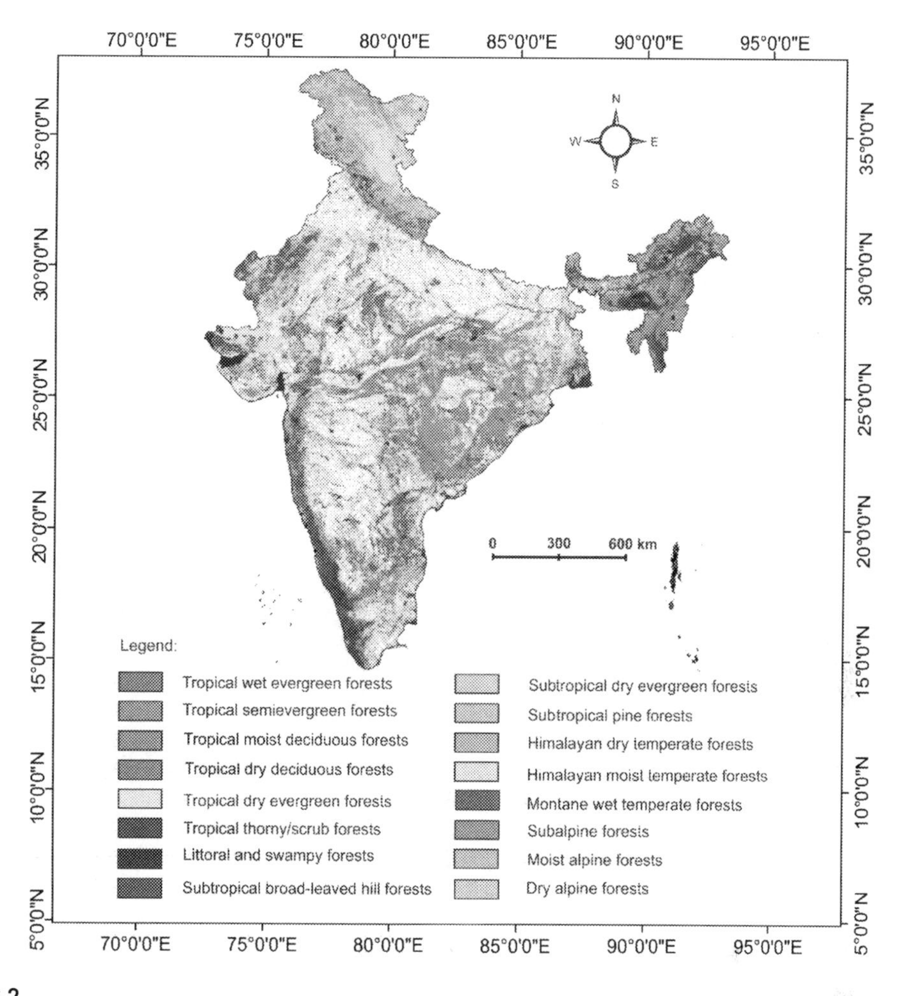

Figure 6.2

Map showing the major vegetation types of India (modified after Reddy et al., 2015; Quamar and Kar, 2020b).

central India till date, which could be probably ascribed to poor sampling resolution and chronological control.

6.4.1 Early-Holocene to Mid-Holocene (~11,700–8200 cal yr BP)

The pollen-based studies conducted from the States of Chhattisgarh and Madhya Pradesh, central India during this phase suggested that tree-savannah vegetation occupied the landscape in the Koriya District (Quamar and Bera, 2017) of Chhattisgarh State under a cool and dry climate with reduced monsoon rainfall between ~12,785 and 9035 cal yr BP. Earlier, tree-savannah vegetation was also suggested around the Hoshangabad District of Madhya Pradesh between ~12,700 and 7150 cal yr BP under a

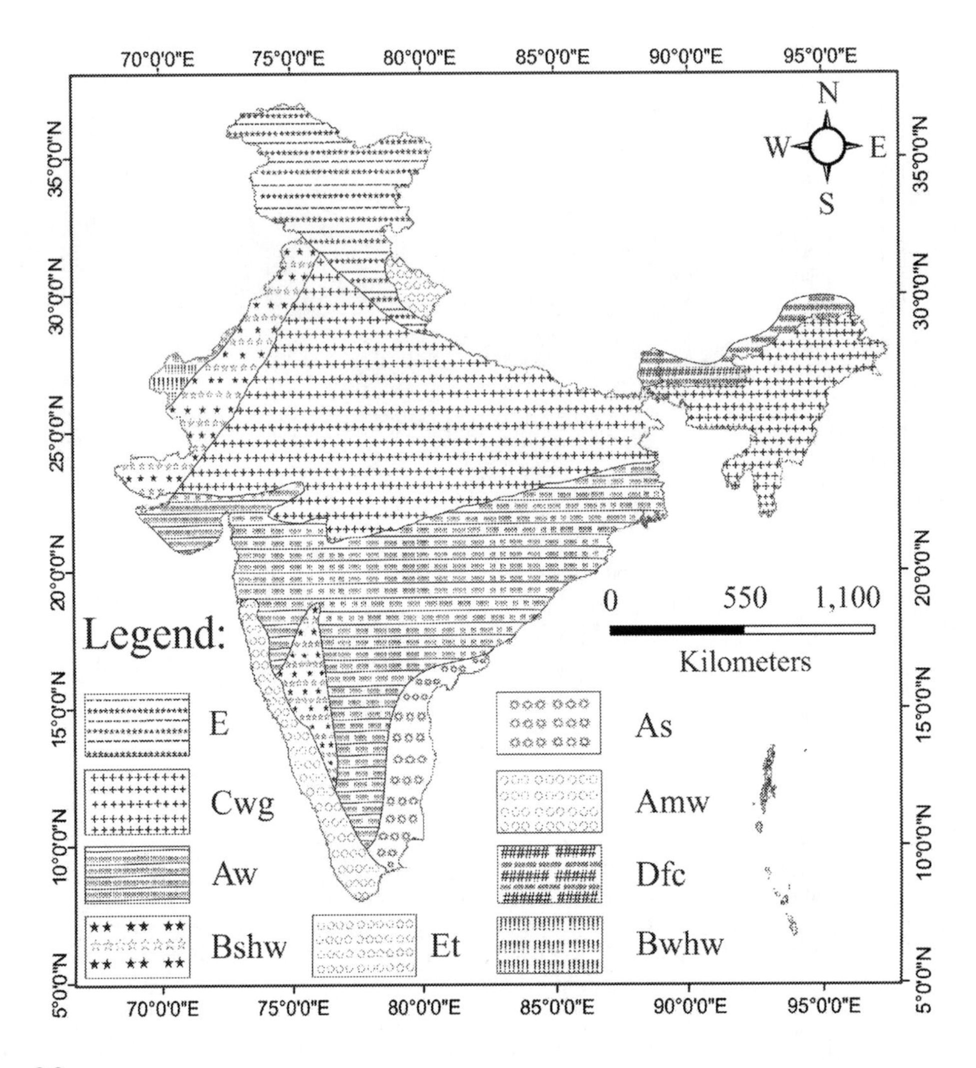

Figure 6.3

Map of India showing the climatic regions (proposed by Köppen 1936; Quamar and Kar, 2020b). Abbreviations used: *Amw*, tropical monsoon type climate; *Aw*, tropical savannah-type climate; *As: BShw*: tropical moist climate; semiarid steppe-type climate; *BWhw*, hot desert-type climate; *Cwg*, mesothermal climate—Gangetic Plain-type climate; *Dfc*, cold, humid winter-type climate; *E*, Polar-type climate; *Et*, Tundra-type climate.

cool and dry climate with reduced monsoon precipitation (Quamar and Chauhan, 2012). The monsoonal climatic inferences drawn (reduced ISM) in the early part of these two studies are correlatable with the Younger Dryas (YD) cold event, which is known globally between 12,900 and 11,600 yr BP (Cheng et al., 2020) and could be attributed to the weakening of the AMOC (McManus et al., 2004). A cool and dry climate, corresponding to that of global YD event, was reported between 13,040 and 11,700 cal

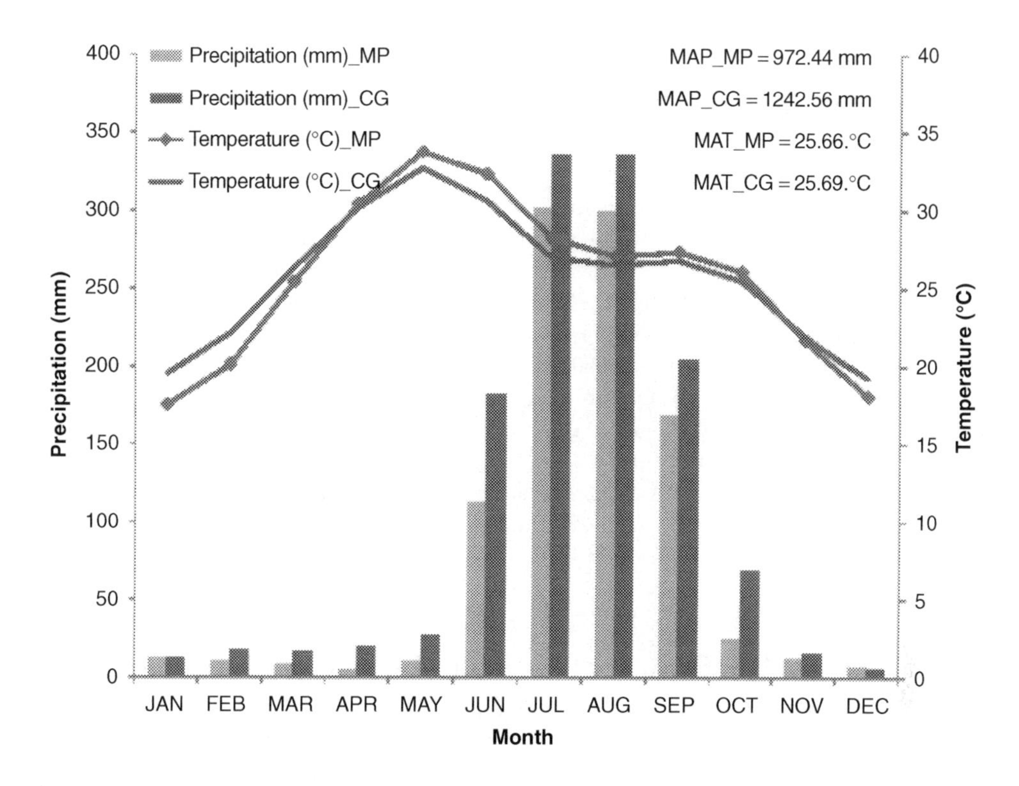

Figure 6.4

Nearest CRU TS 4.03, 0.5 × 0.5 gridded climate data point, 1901–2018, showing mean monthly precipitation and temperature around Madhya Pradesh and Chhattisgarh States, central India. These data are 119-year climate averages for the period 1901–2018. MAP, mean annual precipitation; MAT, mean annual temperature.

yr BP from Deepor wetland, Kamrup District of Assam State, Northeast India (Tripathi et al., 2020), however, the YD event was suggested between 12,450 and 10,810 cal BP from the Chhayagaon swamp, Kamrup District (Assam), Northeast India (Dixit and Bera, 2013), between 12,700 and 11,600 cal years BP from Dabaka swamp, Nagaon District of Assam, Northeast India (Dixit and Bera, 2012). The YD event (a prominent dry spell) has been identified between 11,500 and 10,500 ^{14}C yr BP from the Ganga Plain (Sharma et al., 2006). This event has been recognized from several Himalayan sites also, based on various proxies. Juyal et al. (2004) recorded a cooling phase between 12,000 and 11,000 cal yr BP at the Garbayang basin, Kumaon Himalaya, based on magnetic susceptibility values and elemental concentration in the sediments. Sinha et al. (2005) reported this YD event between 12,700 and 11,600 cal yr BP using oxygen isotope records from stalagmites analyzed from Timta cave, Kumaon Himalaya. Pant et al. (2005) reported this event, based on loess study, from Alakhnanda and Pindar basins between 12,000 and 9000 cal yr BP. Bhattacharayya et al. (2011a) suggested the YD event between 12,406 and 10,633 cal yr BP from the Garhwal Himalaya. However, Demske et al. (2009) suggested the time period 12,200–11,800 yr BP for the YD event from the Tso-Kar Lake,

Ladakh and Rawat et al. (2015) as 12,880–11,640 cal yr BP from Chandra peat, Lahaul, Himachal Pradesh, India. YD event has also been reported from many parts of the globe (Lamb, 1977; Bradley, 1999; Alley, 2000; Morrill et al., 2003).

Quamar and Kar (2020a) suggested mixed tropical deciduous forest around the Mahasamund District of Chhattisgarh State between ~11,700 and 8500 cal yr BP under a warm and humid climate with increased monsoon rainfall. Subsequently, the forest was transformed into dense mixed tropical deciduous forest since ~8500 cal yr BP onward under a warm and relatively more humid climate with further increase in monsoonal rainfall. A prolonged warming has been suggested around the study area since the last 11,700 years, which is well correlated with the HCO/Holocene Thermal Maximum (HTM). Similarly, Quamar and Bera (2017) suggested open mixed tropical deciduous forest around the study area in the Koriya District (Chhattisgarh State) between 9035 and 4535 cal yr BP under a warm and moderately humid climate with increased monsoon rainfall, corresponding to the HCO. Chauhan (1995) recorded open grassland vegetation between 10,000 and 8700 yr BP from the Sidhi District of eastern Madhya Pradesh under a cool and dry climate with reduced monsoon rainfall. However, tree-savannah vegetation was suggested around the study area in the Shahdol District, eastern Madhya Pradesh (Chauhan, 2002) between 9500 and 6800 yr BP under a cool and dry climate with an ameliorating trend in monsoon.

6.4.2 Mid-Holocene to Late-Holocene (~8200–4200 cal yr BP)

Pollen-based study conducted during this phase from the Kartala Forest Range of the Korba District in Chhattisgarh State, central India (Quamar and Bera, 2021) suggested open mixed tropical deciduous forest between ~8378 (=8400) and 1198 (=1200) cal yr BP around the landscape under a warm and moderately humid climate with moderate monsoon precipitation. However, Quamar and Bera (2020) suggested some improvement in the open mixed tropical deciduous forest especially some increase in a few tree taxa around the Baikunthpur Forest Range of the Koriya District in Chhattisgarh State between ~7340 and 1961 cal yr BP under a warm and relatively more humid climate with increased monsoon precipitation. In global perspective, this warming event and a simultaneous amelioration in climatic condition corresponds with the HCO or HTM, which falls broadly within the time interval of 7000–4000 BP (Benarde, 1992). The HCO is asynchronous, that is, the timings and magnitudes of maximum warming vary substantially among the different regions across the globe (Renssen et al., 2012) and is commonly associated with the orbitally forced summer insolation maximum (Wanner et al., 2008; Bartlein et al., 2011). Further, the HCO was affected by other forcings and feedbacks, as well as the remnant Laurentide Ice Sheet (LIS) (Jansen et al., 2007; Bartlein et al., 2011).

Premathilake and Risberg (2003) reported that ~8700 cal yr BP, the maximum intensity of the SWM rains is reached in central Sri Lanka due to the orbitally induced increment of summer insolation, which can be correlated with the HCO between 10,200 and 7800 cal yr BP (Petit-Maire et al., 2000). Band et al. (2018) on the basis of $\delta^{18}O$ records in a stalagmite from the Kotumsar Cave in Chhattisgarh State (central India) observed a steady increase in the ISM intensity between 6.3 and 5.6 ka. Kar et al. (2018) also observed a warming phase between the Mid- to Late-Holocene, as evidenced by low magnetic susceptibility values from Ny-Alesund area, Svalbard, Arctic. This decreased susceptibility could be due to increased weathering and redeposition of the existing sediments by melt-water streams. The better preserved biological activities in the upper part of the trench have also somewhat overwhelmed

Table 6.1 Details of the studies carried out on vegetation dynamics, associated climate change, and the Indian Summer Monsoon (ISM) rainfall variability from various regions of Madhya Pradesh and Chhattisgarh States, central India.

Area/locality with coordinates and references	Period (yr BP, unless otherwise stated)	Vegetation/forest dynamics	Climate change and ISM variability
Madhya Pradesh			
Eastern Madhya Pradesh			
1. Bastua swamp (24°33′3.600″N: 81°58′8.400″E), Chhui stream (24°32 ″38.400″N: 81°56′31.200″E), Amgaon swamp (24°33′N: 81°57′28.800″E), Sidhi District (Chauhan, 1995)			
	10,000–8700	Open grassland	Cool and dry
	~8000 (at Bastua)	Shrub-savannah	Climate amelioration
	6720–5010 (at Chhui)	Tree savannah	
	4500–3800 (at Bastua, Chhui and Amgaon)	Tropical deciduous forest	
	Prior to 1200 (at Amgaon)	First appearance of sal pollen	
	1200–Present (at Bastua and Amgaon)	Sal forest	
2. Jagmotha swamp (24°21′28.8″N: 81°53′ 49.2″E), Sidhi District (Chauhan, 2000)			
	6500–4250	Tree savannah	Cool and dry with amelioration of climate
	4250–2950	Open mixed tropical deciduous forest	Warm and moist
	2950–1050	Dense mixed tropical deciduous forest	Warm and moist with increased monsoon precipitation
	1050–Present	Tropical deciduous sal forest	Warm and more moist with further increase in monsoon precipitation
3. Dongar-Sarbar (23°6′N:81°42′E), Shahdol District (Chauhan, 2002)			
	9500–6800	Tree savannah	Cool and dry with amelioration in climate
	6800–4600	Open mixed tropical deciduous forest	Warm and moist with increased monsoon precipitation
	4600–1800	Dense mixed tropical deciduous forest	Warm and moist with the onset of SW monsoon
	1800–Present	Tropical deciduous sal forest	Warm and more moist with more active SW monsoon
4. Jarbokhoh swamp (24°12′28.8″N: 81°34′ 1.2″E), Sidhi District, (Chauhan, 2004)			
	1500–1160	Dense tropical deciduous forest	Warm and moist with increased summer monsoon
	1160–800	Open tropical deciduous sal forest	Warm and less moist with weak summer monsoon
	800–Present	Dense tropical deciduous sal forest	Warm and more moist with strong summer monsoon

(*continued on next page*)

Table 6.1 (*continued*)

Area/locality with coordinates and references	Period (yr BP, unless otherwise stated)	Vegetation/forest dynamics	Climate change and ISM variability
5. Kerha swamp (24°N: 81°41′E), Sidhi District, Chauhan, 2005)			
	1600–850	Open tropical deciduous sal forest	Warm and moist with moderate monsoon precipitation
	850–400	More open and less diversified tropical deciduous forest	Warm and dry climate attributable to late and weak summer monsoon
	400–Present	Tropical deciduous sal forest	Warm and moist (more) with timely arrival of strong summer monsoon
6. Kiktiha swamp (23°25′30′ ′ N: 81°38′38.4″E), Shahdol District, (Chauhan and Quamar, 2010)			
	1600–700	Tropical deciduous sal forest	Warm and moist climate
	700–300	Mixed tropical deciduous forest	Warm and less moist with weak SW (sparse and less diversified) monsoon
	300–Present	Tropical deciduous sal forest	Warm and moist with active SW monsoon
7. Padauna swamp (23°8′38.4″N: 81°54′54″E), Anuppur District (Chauhan et al., 2013)			
	8600–7500 cal yr BP	Open tree savannah	Warm and relatively less humid
	7500–6250 cal yr BP	Open mixed tropical deciduous forest	Warm and humid climate
	6250–2800 cal yr BP	Dense mixed tropical deciduous forest	Warm and relatively more humid climate with invigoration of SW monsoon
	2800 cal yrs BP–Present	Modern sal forest	Warm and more humid with timely advent of SW monsoon
8. Tula Jalda Lake (23°11′42″N: 81°50′56.4″E), Amarkantak, Anuppur District (Chauhan, 2015)			
	4500–3600 cal yr BP	Open mixed tropical deciduous forest	Warm and relatively less humid
	3600–2761 cal yr BP	Dense mixed tropical deciduous forest	Warm and moderately humid with increased monsoon precipitation
	2761–2200 cal yr BP	Much expansion of forest	Warm and more humid with intensification of the SW monsoon
	2200 cal yr BP-Present	Modern sal forest	Warm and relatively more humid with timely arrival of more active southwest monsoon
Southwestern Madhya Pradesh			
1. Kachhar lake (23°12′2.66″ N: 77°4′28.56″E; 550 m a.s.l.; KL), Sehore District (Quamar and Chauhan, 2011)			
	2050–1610	Acacia-scrub forest	Relatively warm and humid with moderate monsoon precipitation
	1610–600	Dense Acacia scrub forest	Warm and more humid with increased monsoon precipitation
	600 yr BP–Present	Open Acacia scrub forest	Warm and less humid

(*continued on next page*)

Table 6.1 (*continued*)

Area/locality with coordinates and references	Period (yr BP, unless otherwise stated)	Vegetation/forest dynamics	Climate change and ISM variability
2. Sapna Lake (21°51′35.75″N: 77°59′58.06″E; 275 m a.sl.; SL), Betul District (Chauhan and Quamar, 2012a)			
	3800–2700	Open Acacia-dominated scrub forest	Dry climate
	2700–1260	Open mixed tropical deciduous forest	Warm and humid with increased monsoon precipitation
	1260 yr BP-Present	Open tropical deciduous forest	Warm and less humid
3. Amjhera Swamp (22°30′ 25″N: 77°50′E; 548 m a.s.l.; AS), Hoshangabad District (Chauhan and Quamar, 2012b)			
	6000–5409	Open mixed tropical deciduous forest	Warm and relatively less humid
	5409–4011	Dense mixed tropical deciduous forest	Warm and relatively more humid
	4011–2178	Open mixed tropical deciduous forest	Warm and relatively less humid with weak SW monsoon
	2178 yr BP–Present	Dense mixed tropical deciduous forest	Warm and relatively more humid
4. Nitaya Lake (22°40′N: 77°42′E; 400 m a.s.l.; NL), Hoshangabad District (Quamar and Chauhan, 2012)			
	12,700–7150	Tree savannah	Cool and dry with reduced monsoon precipitation
	7150–4657	Open mixed tropical deciduous forest	Warm and moderately humid climate with increased monsoon precipitation
	4657–2807	Open mixed tropical deciduous forest	Warm and relatively less humid climate with weak monsoon precipitation
	2807–1125	Mixed tropical deciduous forest	Warm and more humid with increased monsoon precipitation
	1125 yr BP–Present	Open vegetation	Warm and comparative less humid climate with decreased monsoon precipitation
5. Khedla Quila Lake (21°43′25.585″N: 77°35′15.23″E; ~400 m a.s.l.; KQL), Betul District (Quamar and Chauhan, 2014)			
	1416–506	Open mixed tropical deciduous forest	Warm and moderately humid climate
	506–120	Dense mixed tropical deciduous forest	Warm and humid climate
	120 cal yr BP–Present	Open mixed tropical deciduous forest	Warm and less humid climate
6. Kachia-Jhora (Lake: KJ; 22°52′32″N:77°40′E; 600 m a.s.l.; K-J), Sehore District, MP (Quamar and Chauhan, 2015)			
	3350–2250	Open mixed tropical deciduous forest	Warm and less humid with reduced monsoon precipitation

(*continued on next page*)

Table 6.1 (***continued***)

Area/locality with coordinates and references	Period (yr BP, unless otherwise stated)	Vegetation/forest dynamics	Climate change and ISM variability
	2250–800	Dense mixed tropical deciduous forest	Warm and relatively more humid with increased monsoon precipitation
	800 yr BP–Present	Open mixed tropical deciduous forest	Warm and moderately humid with reduced monsoon precipitation
7. Manjarkui Lake (22°52′25″N: 77°E; 600 m a.s.l.; MKL), Sehore District (Quamar and Nautiyal, 2017)			
	5679–4939 cal yr BP	Open vegetation	Dry climate with reduced monsoon precipitation
	4939–3762 cal yr BP	Mixed tropical deciduous forest	Warm and humid climate with increased monsoon precipitation
	3762 cal yr BP–Present	Dense mixed tropical deciduous forest	Warm and relatively more humid climate with further increase in monsoon precipitation
Chhattisgarh			
1. Matijharia Lake (23°16′40.8″N: 82°33′36″E; 250 m a.s.l.; MJL), Koriya District (Quamar and Bera, 2014)			
	6410–4250 cal yr BP	Tree savannah	Cool and dry with reduced monsoon precipitation
	4250–1000 cal yr BP	Mixed tropical deciduous forest	Warm and humid with increased monsoon precipitation
	1000 cal yr BP–Present	Dense mixed tropical deciduous forest	Warm and relatively more humid climate with further increase in monsoon
2. Lakadandh swamp (23°14′59.98″N: 82°32′ 58.66″E; 529 m a.s.l.; LS), Koriya District (Quamar and Bera, 2017)			
	12,785–9035 cal yr BP	Tree savannah	Cool and dry climate with reduced monsoon precipitation
	9035–4535 cal yr BP	Open mixed tropical deciduous forest	Warm and moderately humid climate with increased monsoon precipitation
	4535 cal yr BP-Present	Mixed tropical deciduous forest	Warm and relatively more humid climate with further increase in monsoon precipitation
3. Chhuhi Lake (23°24.196′N: 82°16.965′E; 700 m a.s.l.; CL), Koriya District (Quamar et al., 2017)			
	3796–2428 cal yr BP	Open mixed tropical deciduous forest	Warm and moderately humid climate with moderate monsoon precipitation
	2428–1431 cal yrs BP	Dense mixed tropical deciduous forest	Warm and relatively more humid climate with increased monsoon precipitation
	1431 cal yr BP-Present	Open mixed tropical deciduous forest	Warm and relatively less humid climate with decreased monsoon precipitation
4. Nakta Lake (21°16.680′N:82°50.196′E; 273 m a.s.l.; NL), Mahasamund District (Quamar and Kar, 2020)			
	~14,100–11,700 cal yr BP	Open vegetation	Dry climate with reduced monsoon precipitation

(*continued on next page*)

Table 6.1 (*continued*)

Area/locality with coordinates and references	Period (yr BP, unless otherwise stated)	Vegetation/forest dynamics	Climate change and ISM variability
(first phase has not been discussed as this time interval is beyond the focus of the present review article)			
	~11,700–8500 cal yr BP	Mixed tropical deciduous forest	Warm and humid climate with increased monsoon precipitation
	~8500 cal yr BP–Present dense mixed tropical deciduous forest		Warm and relatively more humid climate with further increase in monsoon precipitation
5. Anandpur Nursery Lake (23°14′ 59.98″ N: 82°32′ 58.66″ E; 529 m a.sl.; ANL), Koriya District (Quamar and Bera, 2020)			
	22,200–18,658 cal yr BP	Open vegetation	Dry climate with reduced monsoon Precipitation
	18,658–7340 cal yr BP	Open mixed tropical deciduous forest	Warm and moderately humid climate with increase in monsoon precipitation
(first two phases have not been discussed as these time intervals are beyond the focus of the present review article)			
	7340–1961 cal yr BP	Open mixed tropical deciduous forest	Warm and less humid climate with weak monsoon
	1961 cal yr BP–Present	Dense mixed tropical deciduous forest	Warm and more humid climate with increased monsoon precipitation
6. Panguiha Lake (21°39.849′N: 81°33.249′E; 273 m a.s.l.; PL), Bemetra District (Quamar et al., 2021)			
	~3000–2600 cal yr BP	Tree savannah	Cool and dry with reduced monsoon rainfall
	~2600–2200 cal yr BP	Open mixed tropical deciduous forest	Warm and moderately humid climate with moderate monsoon rainfall
	~2200–2000 cal yr BP	Mixed tropical deciduous forest	Warm and humid climate with increased monsoon rainfall
	~2000–1800 cal yr BP	Dense mixed tropical deciduous forest	Warm and relatively more humid climate with further increase in monsoon rainfall
7. Kodamsar Lake (22°16.681′N: 82°59.389′E; ~300 m a.s.l.; KL), Korba District (Quamar and Bera, 2021)			
	8378–1198 cal yr BP	Open mixed tropical deciduous forest	Warm and moderately humid climate with moderate monsoon precipitation
	1198–478 cal yr BP	Mixed tropical deciduous forest	Warm and humid climate with increased monsoon precipitation
	478 cal yr BP–Present	Dense mixed tropical deciduous forest	Warm and relatively more humid climate with further increase in monsoon precipitation
8. Govinda Lake (21°05.225′N: 82°22.415′E; 389 m a.s.l.; GL), Mahasamund District (Quamar, 2021)			
	3600–2500 cal yr BP	Open forest vegetation	Dry climate with educed monsoon
	2500–1650 cal yr BP	Mixed tropical deciduous forest	Warm and humid; increased monsoon
	1650–950 cal yr BP	Open mixed tropical deciduous forest	Warm and relatively less humid; reduced monsoon
	950 cal yr BP–Present	Mixed tropical deciduous forest	Warm and relatively more humid; increased monsoon

the terriginous sediments leading to low susceptibility. An active SW monsoon during 10,000–5800 ^{14}C yr BP from the Ganga Plain was also inferred (Sharma et al., 2004), which could correspond to the early Mid-HCO. Analogous findings about the HCO was also reported from different regions in India, such as Son valley in north central India (Williams and Clarke, 1984) and Rajasthan in western India (Singh et al., 1972, 1974; Bryson and Swain, 1981; Swain et al., 1983). Tripathi et al. (2020) suggested HCO from Deepor wetland, Kamrup District (Assam), Northeast India between 8340 and 4640 cal yr BP, whereas between 6780 and 1980 cal yr BP from Chhayagaon swamp, Kamrup District (Assam) in Northeast India (Dixit and Bera, 2013). Dixit and Bera (2012) indicated the time period of HCO as 7100 and 1550 cal yr BP from Dabaka swamp, Nagaon District of Assam State, Northeast India. Sharma et al. (2004) reported HCO from the Ganga Plain between ~10,000 and 6000 yr BP. Further, Chauhan et al. (2009, 2015) suggested HCO from Lahuradeva Lake, Sant Kabir Nagar District of Uttar Pradesh (U.P.), Central Ganga Plain between 5000 and 2000 yr BP, from Karela Lake, Lucknow (U.P.) between 4800 and 2000 cal yr BP, respectively. The HCO has also been reported from Jalesar Lake, Unnao District, U.P., Central Ganga Plain between 4760 and 3200 cal yr BP (Trivedi et al., 2012). Saxena et al. (2015) suggested that the time period of HCO as 6422 and 3150 cal yr BP from Chaudhary-Ka-Tal, Raebareli District, U.P., Central Ganga Plain, however, Trivedi and Saxena (2017) recorded the HCO from Kikar Lake, Raebareli District of U.P., Central Ganga Plain between 7200 and 4200 cal yr BP. HCO has also been recorded from the Gharana wetland, Jammu (between ~5296 and 2776 cal yr BP; Quamar, 2019), Chandra Peat bog, Lahaul, Northwestern Himalaya (between 6732 and 3337 cal yr BP; Rawat et al., 2015), Tso Kar lake in Ladakh (~6.9–4.8 ka BP; Demske et al., 2009), the bogs in the temperate belt of Kashmir (Dodia et al., 1985), Dhakuri peat bog (~6000–4500 cal yr BP; Phadtare, 2000), the alpine belt of Marhi in Himachal Pradesh (~8000–3500 yr BP; Bhattacharyya, 1988), and Ziro Valley, Arunachal Pradesh (between 10,200 and 3800 cal yr BP; Bhattacharayya et al., 2014).

Open forest vegetation occupied the landscape around the study area (Manjarkui Lake) in the Sehore District of southwestern Madhya Pradesh between 5679 and 4939 cal yr BP under a dry climate with reduced monsoon precipitation. However, the forest became comparatively denser and mixed tropical deciduous forest came into being around the study area between 4939 and 3762 cal yr BP under a warm and relatively more humid climate with intensified monsoon precipitation (Quamar and Nautiyal, 2017). Quamar and Chauhan (2012) suggested open mixed tropical deciduous forest around the Nitaya Lake in the Hoshangabad District of southwestern Madhya Pradesh between 7150 and 4657 cal yr BP under a warm and moderately humid climate with increased monsoon precipitation. This amelioration in climatic condition is partly correlatable with the HCO. However, a bit thinning of the existing forest took place around the study area between 4657 and 2807 cal yr BP under a warm and relatively less humid climate with weak monsoon precipitation (Quamar and Chauhan, 2012). Chauhan and Quamar (2012a) suggested open mixed tropical deciduous forest around the Amjhera Swamp in the Hoshangabad District of southwestern Madhya Pradesh under a warm and relatively less humid climate with reduced monsoon precipitation, which culminated into dense mixed tropical deciduous forest between 5409 and 4011 cal yr BP under a warm and relatively more humid climate, (probably) indicating increased monsoon precipitation. The latter phase with an ameliorating monsoonal trend falls partially within the time-bracket of the Period of HCO. Chauhan (2000) indicated that tree-savannah vegetation occupied the landscape around the Jagmotha Swamp in the Sidhi District (southeastern Madhya Pradesh) between 6500 and 4250 yr BP under a cool and dry climate, but with some amelioration (in monsoonal climate). Chauhan (2002) also suggested tree savannah vegetation around Dongar-Sarbar in the Shahdol District of southeastern Madhya Pradesh between 9500 and 6800 yr BP under a cool and dry climate with amelioration (in monsoonal climate). Subsequently, between 6800 and 4600 yr BP, the tree savannah was transformed into open mixed tropical deciduous forest around the study area in the Shahdol District under a warm and moist climate with intensified monsoon precipitation. Chauhan et al. (2013), however, suggested open tree savannah vegetation around the Padauna Swamp in the Anuppur District, southeastern Madhya Pradesh between 8600 and 7500 cal yr BP under a warm and relatively less humid climate with reduced monsoon precipitation. Subsequently, open mixed tropical deciduous forest succeeds the open tree savannah vegetation around the study area between 7500 and 6250 cal yr BP under a warm and humid climate with a relatively intensified monsoon. Further, the forest became denser around the study area and dense mixed tropical deciduous forest came into existence between 6250 and 2800 cal yr BP under a warm and relatively more humid climate with further increase in the monsoon precipitation. Shrub-savannah vegetation was suggested ~8000 yr BP from Bastua in the Sidhi District (southeastern Madhya Pradesh) under ameliorating climatic trend (Chauhan, 1995). Between 6720 and 5010 cal yr BP, tree savannah vegetation was suggested and tropical deciduous forest was suggested between 4500 and 3800 yr BP from Bastua, Chhui, and Amgaon in the Sidhi District (Chauhan, 1995). Further, between 4500 and 3600 yr BP open mixed tropical deciduous forest was suggested from Amarkantak area in the Anuppur District (southeastern Madhya Pradesh) under a warm and relatively less humid climate with reduced monsoon precipitation (Chauhan, 2015).

6.4.3 Late-Holocene onward (~4200 cal yr BP–Present)

Pollen analytical studies conducted during the Late-Holocene from central India also show the spatio-temporal variability in vegetation composition and associated climate change, influenced by the ISM variability. The pollen evidence suggested that between 4250 and 1000 cal yr BP, mixed tropical deciduous forest vegetation occurred around the landscape of the study area (Majijharia Lake, Baikunthpur Forest Range) in the Koriya District of Chhattisgarh State under a cool and dry climate with increased monsoon precipitation. However, since 1000 cal yr BP onward the forest became denser and dense mixed tropical deciduous forest came into existence under a warm and relatively more humid climate with further increase in the monsoon precipitation (Quamar and Bera, 2014). Quamar and Bera (2017) suggested mixed tropical deciduous forest around the study area (Lakadandh Swamp, Baikunthpur Forest Range) in the Koriya District since 4535 cal yr BP onward under a warm and relatively more humid climate with intensified monsoonal precipitation. However, open mixed tropical deciduous forest was suggested around the Chhui Lake at the Manendragarh Forest Range, Koriya District between 3796 and 2428 cal yr BP under a warm and moderately humid climate with moderate monsoon precipitation. Subsequently, between 2428 and 1431 cal yr BP dense mixed tropical deciduous forest came into being around the study area under a warm and relatively more humid climate with increased monsoon precipitation. Gayantha et al. (2017) also indicated the climate to be warm and humid with intense precipitation during ~2941–2390 cal yr BP in Sri Lanka. Since 1431 cal yr BP onward, the existing dense mixed tropical deciduous forest was replaced by open mixed tropical deciduous forest under a warm and relatively less humid climate with decreased monsoon precipitation (Quamar et al., 2017). Dense mixed tropical deciduous forest was suggested around the Anandpur Nursery Lake at the Baikunthpur Forest Range of the Koriya District since ~1961 cal yr BP onward under a warm and relatively more humid climate with intensified monsoon precipitation (Quamar and Bera, 2020). Quamar and Bera (2021), however, suggested mixed tropical deciduous forest under a warm and humid climate with increased monsoon precipitation between 1198 and 478 cal yr BP from the Kodamsar Lake, Korba District, which culminated into a dense mixed tropical deciduous forest under a warm and relatively more humid climate with further intensification of the monsoon precipitation since 478 cal yr BP (AD 1472) onward (Quamar and Bera, 2021). This latest phase of warming falls within the time frame of the CWP (AD 1800 to the Present; Wu et al., 2012; Fleury et al., 2015; Gupta et al., 2020). Singh et al. (2020), based on grain size, $\delta^{13}C$ and $\delta^{15}N$, as well as TOC and TN studies from Himachal Pradesh (NW Himalaya), India, also recorded CWP during ~1600–2000 AD. Li et al. (2014), based on the AMS ^{14}C dating and the analysis of LOI, TOC, TN, grain size, and MS in the sediments from Basomtso Lake, southeastern Tibetan Plateau, suggested that the higher values of LOI, TOC, TN, coarse silt, sand, and the MS in the sediment indicated higher sediment input probably owing to warmer climatic conditions and higher glacial melt water input during AD 1790–2012, corresponding well with the CWP. The strengthening of the monsoon during the CWP is due to an increase in global temperature, coinciding with the industrial revolution and other anthropogenic factors (Masson-Delmotte et al., 2013), as well as the associated increase in surface evaporation and convection in the Indian Ocean (Anderson et al., 2002; Wang et al., 2005a, b; Sinha et al., 2011). Solar insolation and El Niño intensity have been the major forcing factors influencing the ISM strength during the last 1300 years (Cook et al., 2010; Sinha et al., 2011).

Tree savannah vegetation was suggested from a lacustrine site in the Bemetara District of Chhattisgarh State between ~3000 and 2600 cal yr BP under a cool and dry climate with reduced monsoon precipitation. Between~2600 and 2200 cal yr BP, open mixed tropical deciduous forest came into being under a warm and moderately humid climate with moderate monsoon precipitation. Subsequently, between ~2200 and 2000 cal yr BP, mixed tropical deciduous forest occupied the landscape under a warm and humid climate with increased monsoon precipitation. Finally, between~2000 and 1800 cal yr BP dense mixed tropical deciduous forest occupied the landscape around the study area under a warm and relatively more humid climate with further intensification in the monsoon precipitation (Quamar et al., 2021). The gradual warming and the climatic amelioration during ca. 2600–1800 cal yr BP (~750 BC-AD150) corresponds to the RWP, which is recorded globally between 2500–1600 cal yr BP (~550 BC to AD ~350; Wang et al., 2012). This study also provides insights into the gradual intensification of the monsoon since the last ca. 2600 cal yr BP (between ca. 2600–1800 cal yr BP), and an increase in the ISM strength, against the generally weakening trend during the Late Holocene. Quamar (2021) indicated that between ~3600 and 2500 cal yr BP, open forest vegetation occurred around the Govinda Lake in the Mahasamund District of Chhattisgarh State under a cool and dry climate with reduced monsoon precipitation. Between ~2500 and 1650 cal yr BP (~550 BC to AD ~300), mixed tropical deciduous forest occurred around the landscape under a warm and humid climate with increased monsoon precipitation and is correlatable with the RWP, recorded globally between 2500 and 1600 cal. yr BP (~550 BC to AD ~350; Wang et al., 2012). The Roman Empire in Europe risen during this phase of warm and humid climate, which also paved the way for the growth of an agricultural-based economy and societal welfare (McCormick et al., 2012). In India also, the RWP coincides with the rise of Maurya Empire (Singh, 2008). Increased ISM strength boosts the agricultural production, which brought socioeconomic prosperity in this densely populated region of the globe (Gupta et al., 2006; Singh, 2008). As the economy of the region was on peak as a result of agricultural expansion, and trade surplus during the RWP, the time interval is, therefore, also known as the "Golden Age of India" (McCrindle, 1877; Singh et al., 2020). The RWP has been described as an interval when the circum-North Atlantic region was warm, triggered either by solar variability or internal North Atlantic variability (Martin-Puertas et al., 2009; Büntgen et al., 2011). The high ISM rainfall during the RWP is contemporaneous with an increased solar insolation and associated northward shift of the ITCZ (Fig. 6.5; Haug et al., 2001; Solanki et al., 2004) and were also recorded as high wind intensity in the Arabian Sea (Gupta et al., 2003), more negative oxygen isotopes in speleothems from Uttarakhand

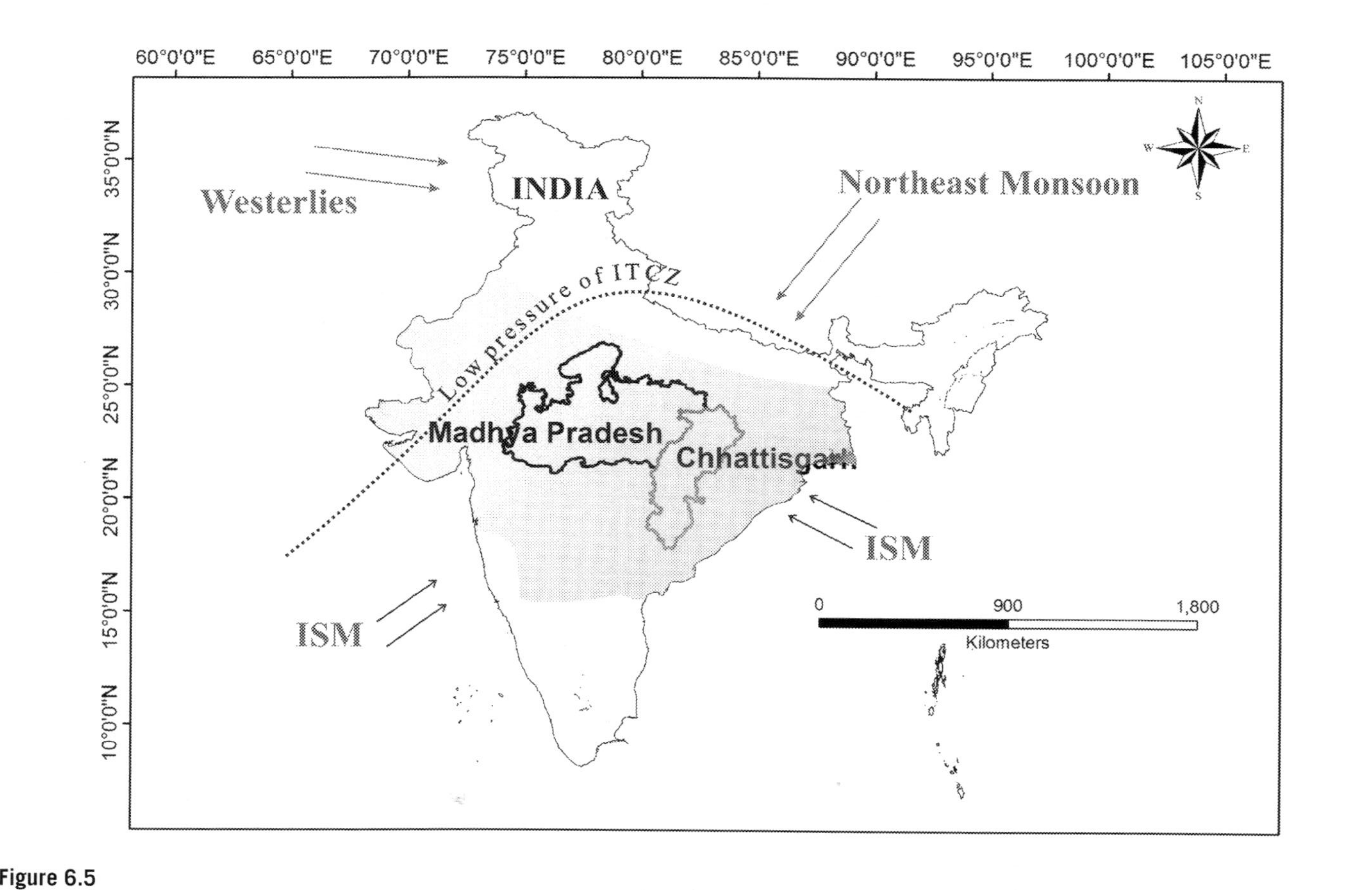

Figure 6.5

Map of India showing the light gray-shaded area as the core monsoon zone (CMZ); also, showing Madhya Pradesh and Chhattisgarh States, central India; the Indian Summer Monsoon (ISM), Northeast Monsoon, Westerlies, and the position of ITCZ (after Zorzi et al., 2015; Kotlia et al., 2015).

(Kotlia et al., 2015; Kathayat et al., 2017), high stream discharge in the Tso Moriri Lake, Ladakh (Dutt et al., 2018) and high leguminous plants in Banni grassland (Pillai et al., 2018). Singh et al. (2020), based on grain size, $\delta^{13}C$ and $\delta^{15}N$, as well as TOC and TN studies from the Himachal Pradesh (NW Himalaya), India also recorded RWP between ~550 BC and 450 AD. Naidu et al. (2020), based on their higher $\delta^{18}Ow$ values from the BoB, India, also suggested wetter ISM rainfall between ca. 1950 and 1550 cal yr BP and correlated the event with the RWP. Subsequently, between ~1650 and 950 cal yr BP (AD 300–1000), open mixed tropical deciduous forest came into existence under a warm and relatively less humid climate with reduced monsoon precipitation, and is correlatable with the DACP. Singh et al. (2020), based on their grain size, stable isotope ratio ($\delta^{13}C$ and $\delta^{15}N$), and organic geochemistry (TOC and TN) studies from Himachal Pradesh (NW Himalaya, India), suggested weak ISM conditions between AD 450 and 950 and termed (this phase of weak ISM) as Dark Age or Migration Period Cooling. This interval of weak ISM corresponds with several dynastic upheavals and decline of the Gupta dynasty in India (McCrindle, 1877; Singh et al., 2020). Also, the transition from Late Antiquity to the Early Middle Ages with large scale human migrations was observed in the Europe (Wanner et al., 2011). The period is also considered as the manifestation of the North Atlantic ice-rafting event at about 1400 years ago (Bond et al., 2001; Helama et al., 2017). Interestingly, solar proxy data also suggest low sun activity during this phase which is further linked to the negative phase of the NAO and/or ENSO and colder conditions in the Northern Hemisphere (Gray et al., 2010; Halema et al., 2017). Helama et al. (2017) suggested that the DCAP ranges from AD 400–765 (1185 and 1550 cal yr BP). They further suggested that a widespread cooling, the 'Late Antique Little Ice Age' (LALIA), overlaps with the DACP and has been tentatively linked with volcanic aerosol and solar irradiance variations reinforcing the climatic downturn since AD 536. Naidu et al. (2020), based on their higher $\delta^{18}Ow$ values from the BoB, India, also suggested reduced ISM rainfall between 1550 and 1250 years BP and correlated the phenomenon with the DACP. Since 950 cal yr BP (AD 1000) onward, mixed tropical deciduous forest transformed the open mixed tropical deciduous forest around the study area under a warm and relatively more humid climate with intensified monsoon, and falls within the time interval of the MWP, which is known to occur between 750 and 1200 AD worldwide (Lamb, 1997; Anderson et al., 2002) and coincides with the northward shift of the ITCZ (Fig. 6.5; Haug et al., 2001). MWP, also known as Medieval Climatic Anomaly (MCA) and Medieval Climate Optimum (MCO), is globally known between AD 740 and 1150 (Lamb, 1977; Anderson et al., 2002). Europe experienced warmer conditions during the MWP (Mann et al., 2008; Büntgen et al., 2011). The MWP has also been observed in other paleoclimate records from central India (Sinha et al., 2011), NW Himalaya (Dutt et al., 2018; Pillai et al., 2018; Singh et al., 2020), western Arabian Sea (Gupta et al., 2005), the Ganga basin (Singh et al., 2015), the BoB (Naidu et al., 2020).

Acacia-scrub forest was suggested around the Kachchhar Lake area in the Sehore District, southwestern Madhya Pradesh between 2050 and 1610 yr BP under a relatively warm and humid climate with moderate monsoon precipitation. Subsequently, dense *Acacia*-scrub forest occurred around the landscape between 1610 and 600 yr BP under a warm and more humid climate with increased monsoon precipitation. Since 600 yr BP onward, open *Acacia*-scrub forest came into being under a warm and less humid climate with moderate monsoon precipitation (Quamar and Chauhan, 2011). Chauhan and Quamar (2012b) suggested open *Acacia*-dominated scrub forest around the Sapna Lake, Betul District between 3800 and 2700 yr BP under a dry climate with reduced monsoon precipitation. Subsequently, open mixed tropical deciduous forest occurred around the landscape between 2700 and 1260 yr BP under a warm and humid climate with increased monsoon precipitation. Since 1260 yr BP onward, open tropical deciduous forest occurred around the study area under a warm and less humid climate with reduced monsoon precipitation. Moreover, open mixed tropical deciduous forest occurred around the landscape (around the Amjhera Swamp) in the Hoshangabad District between 4011 and 2178 yr BP under a warm and relatively less humid climate with weak monsoon, which, under the influence of an active SW monsoon, culminated into dense mixed tropical deciduous forest around the study area since 2178 yr BP onward with a warm and relatively more humid climate with increased monsoon precipitation (Chauhan and Quamar, 2012a). Furthermore, between 2807 and 1125 cal yr BP, mixed tropical deciduous forest occurred around the study area (Nitaya Lake) in the Hoshangabad District under a warm and more humid climate with increased monsoon precipitation, which, under the influence of a warm and relatively less humid climate and reduced monsoon precipitation, changed into open vegetation since 1125 yr BP onward (Quamar and Chauhan, 2012). Open mixed tropical deciduous forest was suggested around the landscape in the study area (Kachia-Jhora/Lake) in the Sehore District between 3350 and 2250 yr BP under a warm and less humid climate with reduced monsoon precipitation. Subsequently, dense mixed tropical deciduous forest occurred around the study area between 2250 and 800 yr BP under a warm and relatively more humid climate with increased monsoon precipitation. Since 800 yr BP onward, open mixed tropical deciduous forest again occupied the landscape around the region under a warm and moderately humid climate with decreased monsoon precipitation (Quamar and Chauhan, 2015). Quamar and Nautiyal (2017) indicated dense mixed tropical deciduous forest around the Manjarkui Lake in the Sehore District since 3762 cal yr BP onward under a warm and relatively more humid climate with intensified monsoon precipitation. Quamar and Chauhan (2014) suggested open mixed tropical deciduous forest around the landscape (at the Khedla Quila Lake) in the Betul District between 1416 and 506 yr BP (AD 534–1444) under a warm and moderately humid climate with moderate monsoon precipitation, partly coinciding with the MWP. Climate forcings during this period have been described as entirely natural and associated with volcanic aerosols and solar variability (Bradley et al., 2016). This short warm phase of the MCA has been linked to the changes in the strength of the thermohaline circulation in the North Atlantic region (Cronin et al., 2003), while the ENSO-modulated solar insolation has been postulated to positively impact the ISM strength during this period (Emile-Geay et al., 2007). Subsequently, between 506 and 120 yr BP dense mixed tropical deciduous forest came into existence around the study area under a warm and humid climate with increased monsoon precipitation. Since 120 yr BP onward, open mixed tropical deciduous forest again occurred in the region under a warm and moderately humid climate with reduced monsoon precipitation.

Chauhan (1995) reported the first appearance of sal (*Shorea robusta* Gaertn. f.) pollen prior to 1200 yr at Amgaon in the Sidhi District, southeastern Madhya Pradesh, which could be due to the timely arrival of active SW monsoon. However, since 1200 yr BP onward sal (*Shorea robusta* Gaertn. f.) forest occupied the landscape around the study area (at Bastua and Amgaon) under a warm and moist climate with intensified monsoon precipitation. However, open mixed tropical deciduous forest was suggested between 4250 and 2950 yr BP around the study area (at Jagmotha Swamp) in the Sidhi District under a warm and moist climate with moderate monsoon precipitation. Subsequently, between 2950 and 1050 yr BP, dense mixed tropical deciduous forest came into being around the study area under a warm and moist climate with increased monsoon precipitation. Since 1050 yr BP onward, tropical deciduous sal (*Shorea robusta* Gaertn. f.) forest occupied the landscape around the study area under a warm and more moist climate with further intensification in monsoon precipitation (Chauhan, 2000). Chauhan (2002) also indicated the existence of the tropical deciduous sal (*Shorea robusta* Gaertn. f.) forest around the Dongar-Sarbar area in the Shahdol District since 1800 yr BP onward under a warm and more moist climate with more active SW monsoon. Chauhan (2004) further indicated that dense tropical deciduous forest occurred around the study area (Jarbokhoh Swamp) in the Sidhi District between 1500 and 1160 yr BP under a warm and moist climate with increased summer monsoon precipitation. Subsequently, between 1160 and 800 yr BP, open mixed tropical deciduous forest occupied the study area under a warm and less moist climate with weak monsoon precipitation, which, under the influence of stronger summer monsoon since 800 yr BP (AD 1150) onward, culminated into dense tropical deciduous sal (*Shorea robusta* Gaertn. f.) forest under a warm and more moist climate, falling within the time-frame of the CWP (AD 1800 to the Present). Open tropical deciduous forest was suggested around the study area (Kerha Swamp) in the Sidhi District between 1600 and 850 yr BP under a warm and moist climate with moderate monsoon precipitation. Subsequently, the open tropical deciduous forest became more open and less diversified between 850 and 400 yr BP under a warm and dry climate in response to late and weak summer monsoon precipitation. Since 400 yr BP (AD 1650) onward, tropical deciduous sal (*Shorea robusta* Gaertn. f.) forest came into being around the landscape under a warm and more moist climate with timely arrival of strong summer monsoon, corresponding with the CWP (AD 1800 to the Present; Chauhan, 2005). Chauhan et al. (2013) suggested that modern sal (*Shorea robusta* Gaertn. f.) forest occupied the landscape around the study area (at Padauna Swamp) in the Anuppur District since 2800 cal yr BP onward under a warm and more humid climate with the timely arrival of active SW monsoon. Chauhan and Quamar (2010) indicated that between 1600 and 700 yr BP (AD 350 and 1250), tropical deciduous sal (*Shorea robusta* Gaertn. f.) forest occupied the landscape around the Kiktiha Swamp in the Shahdol District under a warm and moist climate with increased monsoon precipitation, corresponding with the global MWP (AD 750 and 1200). Subsequently, between 700 and 300 yr BP (AD 1250 and 1650), the forest became sparse and less-diversified (i.e., mixed tropical deciduous forest) around the study area under a warm and less moist climate with weakened SW monsoon, falling within the time interval of the LIA (AD 1440–1850; Bradley, 1985). This cold and dry spell of the LIA could be linked to the intensified Westerlies over NW India due to the intensification of the Asian westerly jet stream over Middle East during a positive phase of the NAO and migration of the Asian jet to the lower latitudes during the warm phase of ENSO (Dixit and Tandon, 2016), as well as the simultaneous weakened ISM. The General Circulation Model Simulations, however, show that an overall declining trend in the ISM strength toward the present day could be attributed to the increased aerosols in the atmosphere due to anthropogenic causes and to the reduced tropical Meridional Overturning Circulation (Ramanathan et al., 2005; Bollasina et al., 2011). The weakest phase of the ISM across the Indian subcontinent during the past three millennia and also during the LIA can be related to the southward shift of the ITCZ location owing to increased northward energy flux across the equator during a cold northern hemisphere (Singh et al., 2020; Dutt et al., 2018; Bischoff and Schneider, 2014; Kathayat et al., 2017; Prasad et al., 1997; Laskar et al., 2013; Rawat et al., 2015; Das et al., 2017). In fact, ENSO restricts the northward movement of the ITCZ that delays the onset of the summer monsoon causing a reduction in the rainfall (Sinha et al., 2011). Dutt et al. (2021) suggested that the low rainfall largely impacted the agricultural production and, in turn, the economy of the Indian subcontinent. Subsequent to the MWP, low rainfall led to the collapse of the economy. Foreign invasions, plunders and internal strife for resources caused socioeconomic upheavals and led to establishment of foreign rules in India. The Pala Empire and the Sena Empire collapsed in Bengal during the 11th century AD and 12th century AD onward, respectively (Majumdar et al., 1978). Thereafter, widespread droughts occurred during the LIA (AD 1550–1850), which was a period of aridity with low moisture levels in the atmosphere. The weakest phase of LIA lasts for about 100 years and coincides with the Maunder Minimum (AD 1645–1710), a period of very low temperature when the sun-spot activity was almost absent (Gupta et al., 2019; Eddy, 1976; Mann and Jones, 2003). During the era of British rule in India (1765–1947), 12 major famines occurred (in 1769–1770, 1783–1784, 1791–1792, 1837–1838, 1860–1861, 1865–1867, 1868–1870, 1873–1874, 1876–1878, 1896–1897, 1899–1900, and 1943–1944) which caused the deaths of millions people (Maharatna, 1996). These famines were a result of the failure of the summer monsoon, which led to widespread droughts and crop failures (Cook et al., 2010; Mishra et al., 2019). Since 300 yr BP (AD 1650) to the present, tropical deciduous sal (*Shorea robusta* Gaertn. f.) forest came into existence around the study area under a warm and moist climate with the advent of active SW monsoon, correlating with the CWP (AD 1800 to the Present). Chauhan (2015) indicated that between 3600 and 2761 cal yr BP, dense mixed tropical deciduous forest occurred around the study area (Tula Jalda lake) at Amarkantak in the Anuppur District under a warm and moderately humid climate with increased monsoon precipitation. Subsequently, between 2761 and 2200 cal yr BP, the existing forest expanded comparatively under a warm and more humid climate with further intensification of the SW monsoon precipitation. Since 2200 cal BP onward, modern sal (*Shorea robusta* Gaertn. f.) forest came into existence around the study area under a warm and relatively more humid climate with the timely arrival of more active SW monsoon.

MWP has also been reported from Deepor wetland, Kamrup District of Assam, Northeast India between 1500 and 710 cal yr BP (Tripathi et al., 2020), however, between 1980 and 989 cal yr BP from Chhayagaon swamp, Kamrup District (Assam; Dixit and Bera, 2013). The time period suggested for the MWP from Dabaka swamp, Nagaon District (Assam) as between 1550 and 768 cal yr BP (Dixit and Bera, 2012) and between 1510 and 540 cal yr BP from Deosila swamp, Goalpara District, Assam, Northeast India (Dixit and Bera, 2011). LIA has also been recorded from Deepor wetland, Kamrup District of Assam, Northeast India since 710 cal yr BP to the Present (Tripathi et al., 2020), however, since 989 cal yr BP to the Present from Chhayagaon swamp, Kamrup District (Assam; Dixit and Bera, 2013). The time period suggested for the LIA from Dabaka swamp, Nagaon District (Assam) as since 768 cal yr BP to the Present (Dixit and Bera, 2012) and since 540 cal yr BP to the Present from Deosila swamp, Goalpara District, Assam, Northeast India (Dixit and Bera, 2011). From the Himalaya also, the MWP and LIA was recorded. Sharma and Chauhan (2001) recorded the MWP from Kupup Lake, Sikkim Himalaya between ~1450 and 450 yr BP and LIA between ~450 and 200 yr BP. The MWP and LIA were recorded from the Gangotri Glacier, Uttarakhand between ~1750 and 850 yr BP and ~850 yr BP to the Present, respectively (Kar et al., 2002). Bali et al. (2015) recorded the MWP and LIA from the Pindari Glacier, Kumaun between 1750 and 900 yr BP and between 900 and 200 yr BP, respectively. Rawat et al. (2015) suggested the time period for MWP and LIA as between 1158 and 647 cal yr BP and 647 and 341 cal yr BP (~ CE1303 and 1609), respectively. Bali et al. (2017) indicated the signatures of MWP and LIA from Triloknath Glacier, Lahaul between ~962 and 300 cal yr BP and ~300 cal yr BP (CE ~1650) to the Present, respectively. Ghosh et al. (2018) suggested the time period for MWP and LIA from the Darjeeling Himalaya as 364 BCE to 131 CE and CE 1367 and 1802. From Jammu also, Quamar (2020a) suggested the time period for the MWP as since 865 cal yr BP to the Present.

The timings and durations of the major global climatic events from central India, as well as their correlations in the Indian and global (to some extent) scenario are shown in Figs. 6.6, 6.7, 6.8A, and 6.8B.

6.5 Learning and knowledge outcomes

Pollen analysis ranks very high in a wide range of methods and/or tools, which are used for reconstructing the paleovegetation dynamics and associated climate change, influenced by the ISM (especially the SWM) rainfall variability. A significant spatio-temporal shift in vegetation composition

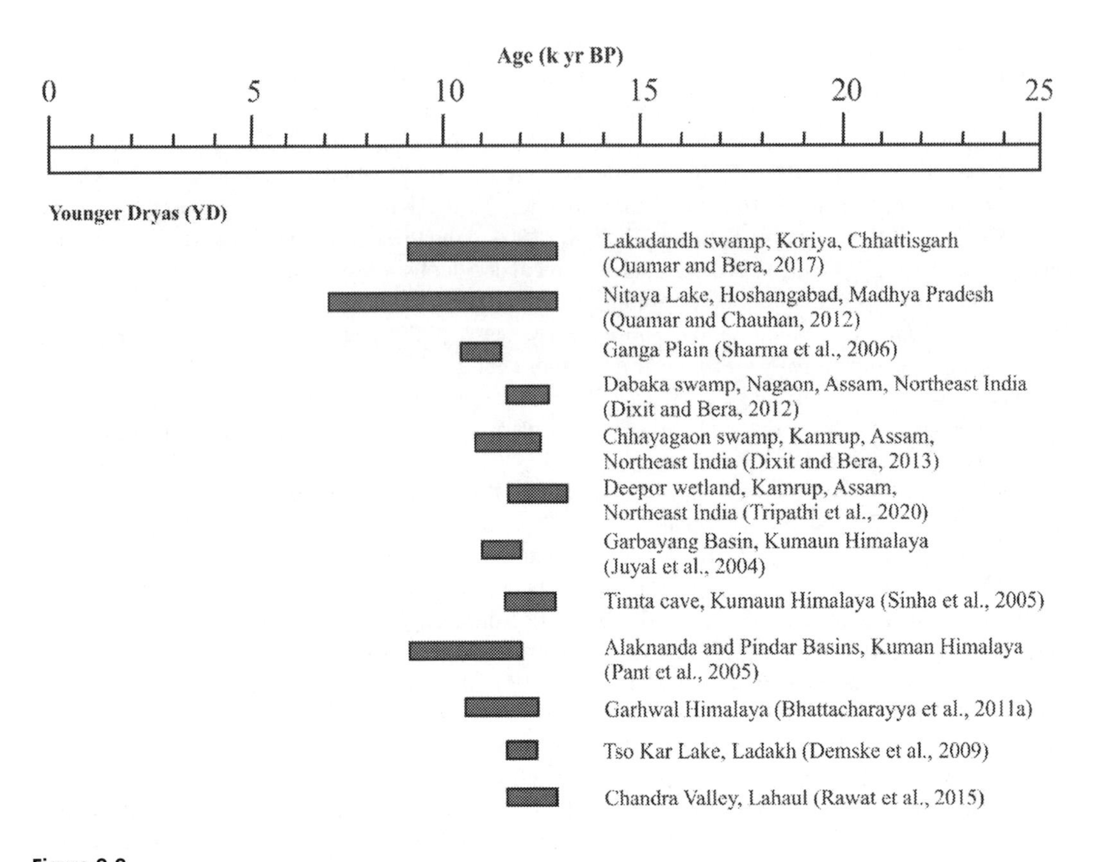

Figure 6.6

Diagram showing the regional correlation of the YD cold event.

around the study areas was observed with changing monsoonal (ISM/SWM) climatic behavior during the Holocene epoch. From the perusal of the aforementioned discussion on pollen-based vegetation dynamics, associated climate change and the ISM rainfall variability from the States of Madhya Pradesh and Chhattisgarh (central India), following conclusions can be drawn:

1. The impact of YD cold event (weak ISM) was observed and was manifested in the vegetation composition (tree savannah vegetation) around the study areas, broadly encompassing its termination at 11.7 kyr BP and its simultaneous coincidence with the global onset of the Holocene Epoch when the ISM strengthened.
2. HCO was manifested by the warming period and a simultaneous intensification of the ISM around the study areas.
3. Strong monsoon (ISM/SWM) and higher precipitation was suggested around the study areas during the RWP between 2500 and 1650 cal yr BP (~550 BC to AD ~300), MWP/MCA between AD 950 and 1250 (AD 750 and 1200).

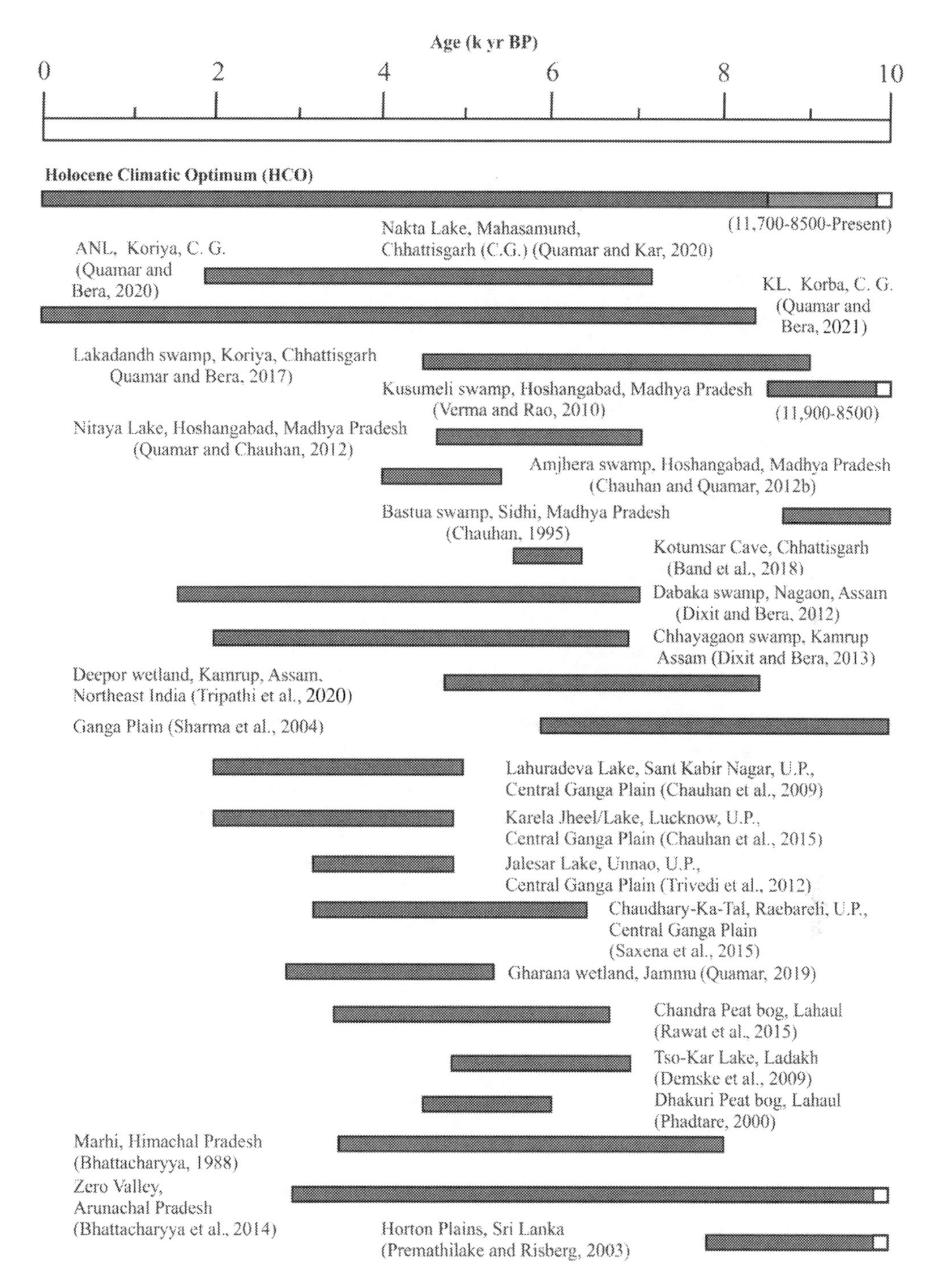

Figure 6.7

Diagram showing the regional correlation of the asynchronous HCO warming event.

Figure 6.8A

Diagram showing the regional correlation of the RWP, MWP, and LIA.

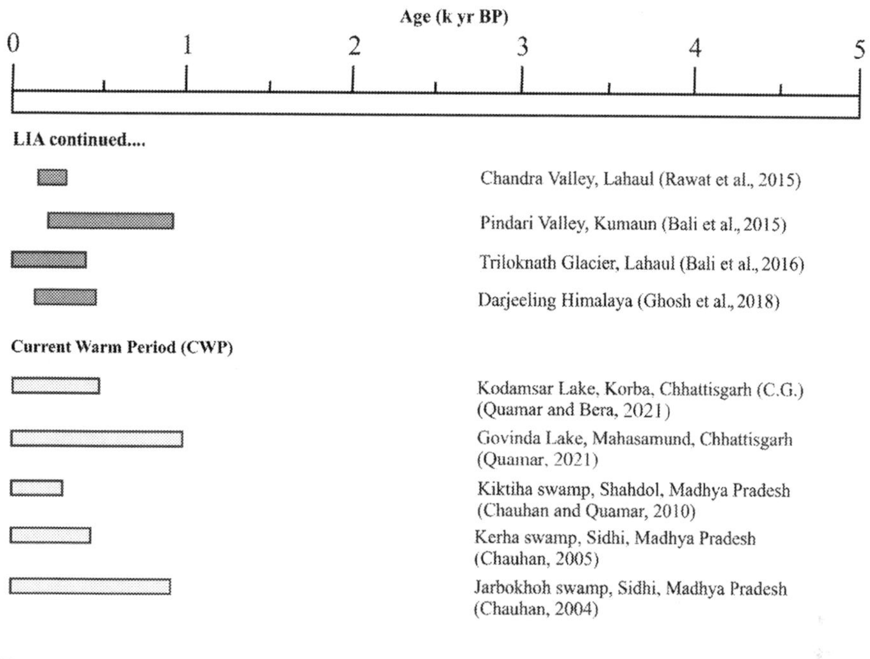

Figure 6.8B

Diagram showing the regional correlation of the LIA and CWP.

4. Weak monsoon (ISM/SWM) was observed during the DACP between ~1650 and 950 cal yr BP (AD 300–1000) and the LIA cold period between ~AD 1400 and 1700 (AD 1440–1850). Furthermore, a strong ISM since AD 1800 to the present was observed during the CWP.
5. 8.2 ka and 4.2 ka dry events were not discernible around the study areas, which could be owing to the poor sampling resolution, and chronological control.

Multiproxy studies are further required to be conducted in order to generate database for fine-resolution paleoclimatic studies from central India, which could be helpful in developing paleoclimatic models for future climatic predictions and also for a scientifically sound policy planning with a key aspect of societal relevance.

Acknowledgments

I am thankful to Dr. (Mrs.) Vandana Prasad, Director, Birbal Sahni Institute of Palaeosciences (BSIP), Lucknow, India for providing the infrastructure facilities needed to complete the review work, and also for permission to publish. Dr. K.P.N. Kumaran, Emeritus Scientist (CSIR), is sincerely thanked for the invitation to contribute this review article in the Book entitled "*Holocene Climate Change and Environment*," published by the Elsevier. All the researchers are also acknowledged for their contribution to the palynological studies in the States of Madhya Pradesh and Chhattisgarh, central India and if the work(s) of somebody is missed then that are inadvertent. Thanks are also due to the Editor and the learned reviewers for the helpful suggestions on the earlier version of the manuscript.

References

Alley, R., The Younger Dryas cold interval as viewed from central Greenland, Quat. Sci. Rev. 19 (2000) 485–492.

Anderson, D.M., Overpeck, J.T., Gupta, A.K., Increase in the Asian southwest monsoon during the past four centuries, Science 297 (2002) 596–599.

Bali, R., Ali, S.N., Bera, S.K., Patil, S.K., Agarwal, K.K. and Nautiyal, C.M., Impact of anthropocene vis-à-vis Holocene climatic changes on Central Indian Himalayan glaciers Lollino, G. et al. (Eds.), Engineering Geology for Society and Territory, Vol. 1, Springer International Publishing, Switzerland, 2015, 1, 467–471; doi:10.1007/978-3-319-09300-0-89.

Bali, R., Chauhan, M.S., Mishra, A.K., Ali, S.N., Tomar, A., Khan, I., Singh, D.S., Srivastava, P., Vegetation and climate change in the temperate-subalpine belt of Himachal Pradesh since 6300 cal. yrs. BP, inferred from pollen evidence of Triloknath palaeolake, Quat. Int. 444 (2017) 11–23, doi:10.1016/j.quaint.2016.07.057.

Band, S., et al., High-resolution mid-Holocene Indian Summer Monsoon recorded in a stalagmite from the Kotumsar Cave, Central India, Quat. Int. 479 (2018) 19–24 https://doi.org/10.1016/j.quaint.2018.01.026.

Bartlein, P.J., Harrison, S.P., Brewer, S., Connor, S., Davis, B.A.S., Gajewski, K., Guiot, J., Harrison-Prentice, T.I., Henderson, A., Peyron, O., Prentice, I.C., Scholze, M., Seppä, H., Shuman, B., Sugita, S., Thompson, R.S., Viau, A.E., Williams, J., Wu, H., Pollen-based continental climate reconstructions at 6 and 21 ka: a global synthesis, Clim. Dynam. 37 (2011) 755–802.

Benarde, M.A., Global Warming. John Wiley and Sons, New York, USA (1992).

Berkelhammer, M., Sinha, A., Stott, L., et al., An abrupt shift in the Indian monsoon 4000 years ago, Am. Geophys. Union Geophys. Monogr. Ser. 198 (2012) 75–87.

Bhattacharyya, A., Vegetation and climate during postglacial period in the vicinity of Rohtang Pass, Great Himalayan Range, Pollen Spores 30 (3–4) (1988) 417–427.

Bhattacharyya, A., Mehrotra, N., Shah, S.K., Basavaiah, N., Chaudhary, V., Singh, I.B., Analysis of vegetation and climate change during Late Pleistocene from Ziro Valley, Arunachal Pradesh, Eastern Himalaya region, Quat. Sci. Rev. 101 (2014) 111–123.

Bhattacharayya, A., Ranhotra, P.S., Gergan, J.T., Vegetation vis-a-vis climate and glacier history during 12,400 to 5,400 BP from Dokriani valley, Garhwal Himalaya, India, J. Geol. Soc. India 77 (2011a) 401–408.

Birks, H.J.B., Birks, H.H., Quaternary Palaeoecology, Edward Arnold, London, 1980 (reprinted 2004 by Blackburn Press, New Jersey).

Bischoff, T., Schneider, T., Energetic constraints on the position of the intertropical convergence zone, J. Clim. 27 (13) (2014) 4937–4951.

Bollasina, M.A., Ming, Y., Ramaswamy, V., Schwarzkopf, M.D., Naik, V., Contribution of local and remote anthropogenic aerosols to the twentieth century weakening of the south Asian monsoon, Geophys. Res. Lett. 41 (2) (2014) 680–687, doi:10.1002/2013GL058183.

Bond, G., Kromer, B., Beer, J., Muscheler, R., Evans, M.N., Showers, W., Hoffmann, S., Lotti-Bond, R., Hajdas, I., Bonani, G., Persistent solar influence on North Atlantic climate during the Holocene, Science 294 (5549) (2001) 2130–2136.

Bradely, R.S., Quaternary Palaeoclimatology: Methods of Palaeoclimatic Reconstruction, Boston, Allen & Unwin, London, 1985.

Bradley, R.S., Palaeoclimatology: Reconstructing Climates of the Quaternary, Harcourt Academic Press, USA, 1999, p. 2.

Bradley, R.S., Hughes, M.K., Diaz, H.F., Climate in medieval time, Science 302 (2003) 404–405, doi:10.1126/science.1090372.

Broecker, W.S., Peteet, D.M., Rind, D., Does the ocean–atmosphere system have more than one stable mode of operation? Nature 315 (1985) 21–26.

Bryson, R.A., Swain, A.M., Holocene variations of monsoon rainfall in Rajasthan, Quat. Res. 16 (1981) 135–145.

Büntgen, U., Tegel, W., Nicolussi, K., McCormick, M., Frank, D., Trouet, V., Kaplan, J.O., Herzig, F., Heussner, K.U., Wanner, H., Luterbacher, J., 2500 years of European climate variability and human susceptibility, Science 331 (6017) (2011) 578–582.

Cai, Y., Tan, L., Cheng, H., An, Z., Edward, R.L., Kelly Megan, J., Kong, X., Wang, X., The variation of summer monsoon precipitation in central China since the last deglaciation. Earth Planet. Sci. Lett. 291 (2010) 21–31.

Champion, H.G., Seth, S.K., A Revised Survey of Forest Types of India, Manager of Publications, Government of India Press, New Delhi, 1968.

Chauhan, M.S., Origin and history of tropical deciduous sal (*Shorea robusta* Gaertn.) forests in Madhya Pradesh, India, Palaeobotanist 43 (1995) 89–101.

Chauhan, M.S., Pollen evidence of late-quaternary vegetation and climate change in Northeastern Madhya Pradesh, India, Palaeobotanist 49 (2000) 491–500.

Chauhan, M.S., Holocene vegetation and climate changes in southeastern Madhya Pradesh, India, Curr. Sci. 83 (12) (2002) 1444–1445.

Chauhan, M.S., Late-Holocene vegetation and climatic changes in Eastern Madhya Pradesh, Gondwana Geol. Mag. 19 (2) (2004) 165–175.

Chauhan, M.S., Pollen record of vegetation and climatic changes in northeastern Madhya Pradesh during last 1,600 years, Trop. Ecol. 46 (2) (2005) 265.

Chauhan, M.S., Vegetation and climatic variability in southeastern Madhya Pradesh, India since Mid-Holocene, based on pollen records, Curr. Sci. 109 (5) (2015) 956–965, doi:10.18520/cs/v109/i5/956-965.

Chauhan, M.S., Pokharia, A.K., Singh, I.B., Pollen record of vegetation, climate change and human habitation from Lahuradewa lake, Sant Kabir Nagar District, Uttar Pradesh, India, Man Environ. 45 (2009) 125–129.

Chauhan, M.S., Pokharia, A.K., Singh, I.B., Late Quaternary vegetation history, climatic variability and human activity in the central Ganga Plain, deduced by pollen proxy records from Karela Jheel, India, Quat. Int. 371 (2015) 144–156.

Chauhan, M.S., Quamar, M.F., Vegetation and climate change in southeastern Madhya Pradesh during Late Holocene, based on pollen evidence, J. Geol. Soc. India 76 (2) (2010) 143–150, doi:10.1007/s12594-010-0084-y.

Chauhan, M.S., Quamar, M.F., Pollen records of vegetation and inferred climate change in southwestern Madhya Pradesh during the last ca. 3800 years, J. Geol. Soc. India 80 (4) (2012a) 470–480, doi:10.1007/s12594-012-0166-0.

Chauhan, M.S., Quamar, M.F., Mid-Holocene vegetation vis-a-vis climate change in southwestern Madhya Pradesh, Curr. Sci. 103 (2012b) 1455–1461.

Chauhan, M.S., Sharma, A., Phartiyal, B., Kumar, K., Holocene vegetation and climatic variations in central India: a study based on multiproxy evidences, J. Asian Earth Sci. 77 (2013) 45–58, doi:10.1016/j.jseaes.2013.08.005.

Cheng, H., Zhang, H., Christoph Spötld, C., Baker, J., Sinha, A., Li, H., Bartolomé, M., Morenog, A., Kathayat, G., Zhao, J., Dong, X., Li, Y., Ninga, Y., Jia, X., Zong, B., Brahim, YA., Pérez-Mejiaś, C., Cai, Y., Novelloh, VF., Cruzh, FW., Severinghausi, JP., An, Z., Edwards L., Timing and structure of the younger dryas event and its underlying climate dynamics. 117 (38) 2020, 23408–23417 www.pnas.org/cgi/doi/10.1073/pnas.2007869117

Cook, E.R., Anchukaitis, K.J., Buckley, B.M., D'Arrigo, R.D., Jacoby, G.C., Wright, W.E., Asian monsoon failure and mega drought during the last millennium, Science 328 (2010) 486–489.

Cronin, T.M., Dwyer, G.S., Kamiya, T., Schwede, S., Willard, D.A., Medieval Warm Period, Little Ice Age and 20th century temperature variability from Chesapeake Bay, Global Planet. Change 36 (2003) 17–29.

Das, M., Singh, R.K., Gupta, A.K., Bhaumik, A.K., Holocene strengthening of the Oxygen Minimum Zone in the northwestern Arabian Sea linked to changes in intermediate water circulation or Indian monsoon intensity? Palaeogeogr. Palaeoclim. Palaeoecol. 483 (2017) 125–135.

Demske, D., Tarasov, P.E., Wünnemann, B., et al., Late glacial and Holocene vegetation, Indian monsoon and westerly circulation dynamics in the Trans-Himalaya recorded in the pollen profile from high-altitude Tso Kar Lake, Ladakh, NW India, Palaeogeogr. Palaeoclim. Palaeoecol. 279 (3) (2009) 172–185.

Dixit, S., Bera, S.K., Mid-Holocene vegetation and climate variability in tropical deciduous sal (*Shorea robusta*) forest of Lower Brahmaputra Valley, Assam, J. Geol. Soc. India 77 (2011) 419–432.

Dixit, S., Bera, S.K., Holocene climatic fluctuations from Lower Brahmaputra flood plain of Assam, northeast India, J. Earth Syst. Sci. 121 (1) (2012) 135–147.

Dixit, S., Bera, S.K., Pollen-inferred vegetation vis-à-vis climate dynamics since Late Quaternary from Western Assam, Northeast India: signal of global climatic events, Quat. Int. 286 (2013) 56–68.

Dixit, Y., Hodell, D.A., Giesche, A., et al., Intensified summer monsoon and the urbanization of Indus Civilization in Northwest India, Sci. Rep. 8 (1) (2018) 4225.

Dixit, Y., Hodell, D.A., Petrie, C.A., Abrupt weakening of the summer monsoon in Northwest India ~4100 yr ago, Geology 42 (4) (2014a) 339–342.

Dixit, Y., Hodell, D.A., Sinha, R., et al., Abrupt weakening of the Indian summer monsoon at 8.2 kyr BP, Earth Planet. Sci. Lett. 391 (2014b) 16–23.

Dixit, Y., Tandon, S.K., Hydroclimatic variability on the Indian subcontinent in the past millennium: review and assessment, Earth Sci. Rev. 161 (2016) 1–15.

Dodia, R., Agrawal, D.P., Vora, A.B., New pollen data from the Kashmir bogs: a summary (Eds. D.P. Agarwal, S. Kusumgar and Krishnamurthy, R.V.), Current Trends in Geology: Climate and Geology of Kashmir, 6, Today and Tomorrow's Printers and Publishers, New Delhi, 1985, pp. 101–108.

Dutt, S., Gupta, A.K., Cheng, H., Clemens, S.C., Singh, R.K., Tewari, V.C., Indian summer monsoon variability in ortheastern India during the last two millennia, Quat. Int. 571 (2021) 73–80 https://doi.org/10.1016/j.quaint.2020.10.021.

Dutt, S., Gupta, A.K., Clemens, S.C., et al., Abrupt changes in Indian summer monsoon strength during 33,800 to 5500 years BP, Geophys. Res. Lett. 42 (13) (2015) 5526–5532.

Dutt, S., Gupta, A.K., Wünnemann, B., Yan, D., A long arid interlude in the Indian summer monsoon during ~ 4,350 to 3,450 cal. yr BP contemporaneous to displacement of the Indus valley civilization, Quat. Int. 482 (2018) 83–92.

Eddy, J.A., The maunder minimum, Science 192 (1976) 1189–1202.

Emile-Geay, J., Cane, M., Seager, R., Kaplan, A., Almasi, P., El Niño as a mediator of the solar influence on climate, Paleoceanography 22 (2007) PA3210, doi:10.1029/2006PA001304.

Erdtman, G., Pollen morphology and Plant Taxonomy; Angiosperms. An Introduction to Palynology, I, Almqist & Wiksell, Stockholm, 1952.

Faegri, K., Kaland, P.E., Krzywinski, K., et al., Textbook of Pollen Analysis, Chichester, Wiley, 1989.

Fang, J.Y., Piao, S.L., Wang, Q., Ma, W.H., Vegetation activity is strengthened during the past 20 years in China, Sci. China C 33 (2003) 554–565 (in Chinese).

Faubert, P., Tiiva, P., Michelsen, A., Rinnan, Å., Ro-Poulsen, H., Rinnan, R., The shift in plant species composition in a subarctic mountain birch forest floor due to climate change would modify the biogenic volatile organic compound emission profile, Plant Soil 352 (2012) 199–215.

Fleitmann, D., Burns, S.J., Mudelsee, M., et al., Holocene forcing of the Indian monsoon recorded in a stalagmite from southern Oman, Science 300 (2003) 1737–1739.

Fleury, S., Martinez, P., Crosta, X., Charlier, K., Billy, I., Hanquiez, V., Blanz, T., Schneider, R.R., Pervasive multidecadal variations in productivity within the Peruvian Upwelling System over the last millennium, Quat. Sci. Rev. 125 (2015) 78–90.

Gadgil, S., The Indian monsoon and its variability, Am. Rev. Earth Planet. Sci. 31 (2003) 429–467.

Gadgil, S., Gadgil, S., The Indian monsoon, GDP and agriculture, Econ. Polit. Weekly. 41 (47) (2006) 4887–4895.

Gayantha, K., Routh, J., Chandrajith, R., A mult-proxy reconstruction of the late Holocene climate evolution in Lake Bolgoda, Sri Lanka, Palaeogeogr. Palaeoclim. Palaeoecol. 473 (2017) 16–25.

Ghosh, R., Biswas, O., Paruya, D.K., Agrawal, S., Sharma, A., Hydroclimatic variability and corresponding vegetation response in the Darjeeling Himalaya, India over the past ~2400 years, Catena 170 (2018) 84–99.

Ghosh, S.K., Pant, M.C., Dewan, B.N., Influence of the Arabian Sea on the Indian summer monsoon, Tellus 30 (1978) 117–125, doi:10.1111/j.2153-3490.1978.tb00825.x.

Giosan, L., Clift, P.D., Macklin, M.G., et al., Fluvial landscapes of the Harappan civilization, Proc. Natl. Acad. Sci. U S A 109 (2012) E1688–E1694.

Gray, L.J., Beer, J., Geller, M., Haigh, J.D., Lockwood, M., Matthes, K., Cubasch, U., Fleitmann, D., Harrison, G., Hood, L., Luterbacher, J., Solar influences on climate, Rev. Geophys. 48 (4) (2010) RG4001, doi:10.1029/2009RG000282.

Gupta, A.K., Anderson, D.M., Overpeck, J.T., Abrupt changes in the Asian southwest monsoon during the Holocene and their links to the north Atlantic Ocean, Nature 421 (2003) 354–357.

Gupta, A.K., Anderson, D.M., Pandey, D.N., et al., Adaptation and human migration, and evidence of agriculture coincident with changes in the Indian summer monsoon during the Holocene, Curr. Sci. 90 (8) (2006) 1082–1090, doi:10.1029/2005GL022685.

Gupta, A.K., Das, M., Anderson, D.M., Solar influence on the Indian summer monsoon during the holocene, Geophys. Res. 32 (17) (2005) L17703 doi:10.1029/2005GL022685.

Gupta, A.K., Dutt, S., Cheng, H., Singh, R.K., Abrupt changes in Indian summer monsoon strength during the last~ 900 years and their linkages to socio-economic conditions in the Indian subcontinent, Palaeogeogr. Palaeoclim. Palaeoecol. 536 (2019) 109347.

Gupta, AK., Prakasam, M., Dutt, S., Clift, P.D., Yadav, R.R., Evolution and development of the Indian Monsoon, in: Gupta, N., Tandon, S.K. (Eds.), Geodynamics of the Indian Plate, Springer Geology, Springer International Publishing, (2020) 499–535. https://doi.org/10.1007/978-3-030-15989-4_14.

Helama, S., Jones, P.D., Briffa, K.R., Dark Ages cold period: a literature review and directions for future research, Holocene 27 (10) (2017) 1600–1606.

Harris, I., Jones, P.D., Osborn, T.J., Lister, D.H., Updated high-resolution grids of monthly climatic observations-the CRU TS 3.10, Int. J. Climatol. 34 (2014) 623–642.

Haug, G.H., Hughen, K.A., Sigman, D.M., Peterson, L.C., Rohl, U., Southward migration of the intertropical convergence zone through the Holocene, Science 293 (5533) (2001) 1304–1308.

Henden, J.A., Yoccoz, N.G., Ims, R.A., Langeland, K., How spatial variation in areal extent and configuration of labile vegetation states affect the riparian bird community in arctic tundra, PLoS One 8 (2013) e63312.

Henderson-Sellers, A., Continental vegetation as a dynamic component of a global climate model: a preliminary assessment, Clim. Change 4 (1993) 337–377.

ICS, 2018. International Commission on Stratigraphy. International Chronostratigraphic Chart, v.2018/07. http://www.stratigraphy.org/ICSchart/ChronostratChart2018-07.Pdf.

IPCC, Summary for policymakers, in: Solomon, S., Qin, D., Manning, M., Chen, Z., Marquis, M., et al. (Eds.), Climate Change: The Physical Science Basis Contribution of Working Group I to the Fourth Assessment Report of the Intergovernmental Panel on Climate Change, Cambridge University Press, Cambridge, 2007.

Jansen, E., Overpeck, J.T., Briffa, K.R., Duplessy, J.C., Joos, F., Masson-Delmotte, V., Olago, D., Otto-Bliesner, B., Peltier, W.R., Rahmstorf, S., Rengaswamy, R., Raynaud, D., Rind, D., Solomina, O., Villalba, R., Zhang, D., Palaeoclimate, in: Solomon, S., Qin, D., Manning, M., Chen, Z., Marquis, M., Averyt, K.B., Tignor, M., Miller, H.L. (Eds.), Climate Change 2007: The Physical Science Basis. 4th Assessment Report IPCC, Cambridge University Press, Cambridge, UK, 2007, pp. 433–498.

Juyal, N., Pant, R.K., Basavaiah, N., Yadava, M.G., Saini, N.K., Singhvi, A.K., Climate and seismicity in the higher central Himalaya during 20-1- ka: evidence from Garbayang basin, Uttaranchal, India, Palaeogeogr. Palaeoclim. Palaeoecol. 213 (2004) 315–330.

Kar, R., Mazumder, A., Mishra, K., Patil, S.K., Ravindra, R., Ranhotra, P.S., Govil, P., Bajpai, R., Singh, K., Climatic history of Ny-Alesund region, Svalbard, over the last 19,000 yr: insights from quartz grain microtexture and magnetic susceptibility, Polar Sci. 18 (2018) 189–196, doi:10.1016/j.polar.2018.04.004.

Kar, R., Ranhotra, P.S., Bhattacharayya, A., Sekar, B., Vegetation vis-à-vis climate and glacial fluctuations of the Gangotri glacier since last 2000 years, Curr. Sci. 82 (2002) 347–351.

Kar, R., Quamar, M.F., Pollen-based Quaternary palaeoclimatic studies in India: an overview of recent advances, Palynology 43 (1) (2019) 76–93.

Kar, R., Quamar, M.F., Late Pleistocene-Holocene vegetation and climate change from the Western and Eastern Himalaya (India): palynological perspective, Current Science 119 (2) (2000) 195–218.

Kathayat, G., Cheng, H., Sinha, A., Yi, L., Li, X., Zhang, H., Li, H., Ning, Y., Edwards, R.L., The Indian monsoon variability and civilization changes in the Indian subcontinent, Sci. Adv. 3 (12) (2017) 1701296.

Köppen, W., Das geographische System der Klimate. In: Köppen W., Geiger R. (Eds.), Handbuch der Klimatologie. Gebrüder Borntraeger, Berlin, 1936, p. 1−44.

Kotlia, B.S., Singh, A.K., Joshi, L.M., Dhaila, B.S., Precipitation variability in the Indian central Himalaya during last ca. 4,000 years inferred from a speleothem record: impact of Indian Summer Monsoon (ISM) and Westerlies, Quat. Int. 371 (2015) 244–253.

Kumaran, K.P.N., Limaye, R., Holocene palynology and tropical paleoecology (Editorial), Quat. Int. 325 (2014) 2–4.

Lamb, H.H., 1977. Climate: Present, Past and Future. Methuen, London.

Laskar, A.H., Yadava, M.G., Ramesh, R., Polyak, V.J., Asmerom, Y., A 4 kyr stalagmite oxygen isotopic record of the past Indian Summer Monsoon in the Andaman Islands, Geochem. Geophys. 14 (9) (2013) 3555–3566.

Li, C., Qi, J., Yang, L., Wang, S., Yang, W., Zhu, G., Zon, S., Zhang, F., Regional vegetation dynamics and its response to climate change—a case study in the Tao River Basin in Northwestern China, Environ. Res. Lett. 9 (2014) 125003.

Li, K., Liu, X., Herzschuh, U., Wang, Y., Rapid climate fluctuations over the past millennium: evidence from a lacustrine record of Basomtso Lake, southeastern Tibetan Plateau, Sci. Rep. 6 (2014) 24806.

Maharatna, A., The Demography of Famines: An Indian Historical Perspective. Doctoral Dissertation, London School of Economics and Political Science, United Kingdom (1996).

Mann, M.E., Jones, P.D., Global surface temperatures over the past two millennia, Geophys. Res. Lett. 30 (2003) 1820 https://doi.org/10.1029/2003GL017814.

Mann, M.E., Zhang, Z., Hughes, M.K., Bradley, R.S., Miller, S.K., Rutherford, S., Ni, F., Proxy-based reconstructions of hemispheric and global surface temperature variations over the past two millennia, Proc. Natl. Acad. Sci. 105 (36) (2008) 13252–13257.

Masson-Delmotte, V. et al. Information from palaeoclimate archives, in climate change 2013. The Physical Science Basis. In: T.F. Stocker et al. (Eds.), Contribution of Working Group I to the Fifth Assessment Report of the Intergovernmental Panel on Climate Change, Cambridge Univ. Press, U.K., 2013, pp. 383–464.

Martín-Puertas, C., Valero-Garcés, B.L., Brauer, A., Mata, M.P., Delgado-Huertas, A., Dulski, P., The Iberian–Roman humid period (2600–1600 cal yr BP) in the Zoñar Lake varve record (Andalucía, southern Spain), Quat. Res. 71 (2009) 108–120.

McCormick, M., Büntgen, U., Cane, M.A., et al., Climate change during and after the Roman Empire: reconstructing the past from scientific and historical evidence, J. Interdiscip. Hist. 43 (2) (2012) 169–220.

McCrindle, J.W., 1877. Ancient India as Described by Megasthenes and Arrian. Thacker, Calcutta, Spink.

McDonald, R.E., Understanding the impact of climate change on Northern Hemisphere extra-tropical cyclones, Clim. Dyn. 37 (2011) 1399–1425.

McManus, J.F., Francois, R., Gherardi, J.M., Keigwin, L.D., Brown Leger, S., Collapse and rapid resumption of Atlantic meridional circulation linked to deglacial climate changes, Nature 428 (6985) (2004) 834, doi:10.1038/nature02437.

McMichael, A.J., Woodruff, R.E., Hales, S., Climate change and human health: present and future risks, Lancet 367 (9513) (2006) 859–869.

Meehl, G.A., Coupled land-ocean-atmosphere processes and South Asian monsoon variability, Science 266 (5183) (1994) 263–267.

Meng, M., Ni, J., Zong, M.J., Impacts of changes in climate variability on regional vegetation in China: NDVI-based analysis from 1982 to 2000, Ecol. Res. 26 (2011) 421–428.

Misra, P., Tandon, S.K., Sinha, R., Holocene climate records from lake sediments in India: Assessment of coherence across climate zones. Earth Science Reviews. 190 (2019) 370–397. doi:10.1016/j.earscirev.2018.12.017.

Misra, P., Farooqui, A., Sinha, R., Khanolkar, S., Tandon, S.K., Millennial-scale vegetation and climatic changes from an Early to Mid-Holocene lacustrine archive in Central Ganga Plains using multiple biotic proxies, Quat. Sci. Rev. 243 (2020) 106474.

Morrill, C., Overpeck, J.Y., Cole, J.E., A synthesis of abrupt changes in the Asian summer monsoon since the last deglaciation, Holocene 13 (4) (2003) 465–476.

Naidu, P.D., Ganeshram, R., Bollasina, M.A., Panmei, C., Nürnberg, D., Donges, J.F., Coherent response of the Indian monsoonal rainfall to Atlantic multi-decadal variability over the last 2000 years, Sci. Rep. 10 (2020) 1302 https://doi.org/10.1038/s41598-020-58265-3.

Pant, R.K., Basavaiah, H., Juyal, N., Saini, N.K., Yadava, M.G., Appel, E., Singhvi, A.K., A 20-ka climate record from central Himalayan loess deposits, J. Quat. Sci. 20 (2005) 1–8.

Pearce, C., Seidenkrantz, M.S., Kuijpers, A., Massé, G., Reynisson, N.F., Kristiansen, S.M., Ocean lead at the termination of the Younger Dryas cold spell. Nat. Commun., 4 (1) (2013) 1–6.

Petit, J.R., Jouzel, J., Raynaud, D., Barkov, N.I., Barnola, J.M., Basile, I., Bender, M., Chappellaz, J., Davis, M., Delaygue, G., Delmotte, M., Climate and atmospheric history of the past 420,000 years from the Vostok ice core, Antarctica, Nature 399 (6735) (1999) 429.

Petit-Maire, N., Bouysse, P., Beaulieu de, J.L., Boulton, G., Guo, Z., Iriondo, M., Kershaw, P., Lisitsyna, O., Partridge, T., Pflaumann, U., Schulz, H., Soons, J., Viiet-Lanoe, B.V., Geological records of the recent past, a key to the near future world environments, Episodes 23 (4) (2000) 230–246.

Phadtare, N.R., Sharp decrease in summer monsoon strength 4000-3500 cal yr BP in the Central Higher Himalaya of India based on pollen evidence from alpine peat, Quat. Res. 53 (2000) 122–129.

Pillai, A.A., Anoop, A., Prasad, V., Manoj, M.C., Varghese, S., Sankaran, M., Ratnam, J., Multi-proxy evidence for an arid shift in the climate and vegetation of the Banni grasslands of western India during the mid-to late-Holocene, Holocene 28 (7) (2018) 1057–1070.

Prasad, S., Anoop, A., Riedel, N., Sarkar, S., Menzel, P., Basavaiah, N., Krishnan, R., Fuller, D., Plessen, B., Gaye, B., Rohl, U., Prolonged monsoon droughts and links to Indo-Pacific warm pool: a Holocene record from Lonar Lake, central India, Earth Planet Sci. Lett. 391 (2014) 171–182.

Premathilake, R., Risberg, J., Late Quaternary climate history from the Horton Plains, central Sri Lanka, Quat. Sci. Rev. 22 (2003) 1525–1541.

Quamar, M.F., Vegetation dynamics in response to climate change from the wetlands of Western Himalaya, India: Holocene Indian Summer Monsoon variability, Holocene 29 (2) (2019) 345–362.

Quamar, M.F., Late Holocene vegetation vis-á-vis climate change influenced by the ISM variability from the Western Himalaya, India, J. Palaeontol. Soc. India (2020) Under review.

Quamar, M.F., Monsoonal climatic reconstruction from central India during the last ca. 3600 cal yr: Signatures of global climatic events, based on lacustrine sediment pollen records, Palynology (Accepted) (2021) Published online; DOI:10.1177/09596836211003191.

Quamar, M.F., Ali, S.N., Nautiyal, C.M., Bera, S.K., Vegetation and climate reconstruction based on a ~4 ka pollen record from north Chhattisgarh, central India, Palynology 41 (4) (2017) 504–515.

Quamar, M.F., Bera, S.K., Vegetation and climate change during the mid and late Holocene in northern Chhattisgarh, central India inferred from pollen records, Quat. Int. 349 (2014) 357–366.

Quamar, M.F., Bera, S.K., Pollen records related to vegetation and climate change from northern Chhattisgarh, central India during the Late Quaternary, Palynology 41 (1) (2017) 17–23.

Quamar, M.F., Bera, S.K., Pollen records of vegetation dynamics, climate change and ISM variability since the LGM from Chhattisgarh State, central India, Rev. Palaeobot. Palynol. (2020) 104159.

Quamar, M.F., Bera, S.K., A 8400-year pollen record of vegetation dynamics and Indian Summer Monsoon climate from central Indian Core Monsoon Zone: signatures of global climatic events, J. Palaeontol. Soc. India (2021) (in press), (revised).

Quamar, M.F., Chauhan, M.S., Late Holocene vegetation, climate change and human impact in southwestern Madhya Pradesh, India, Palaeobotanist 60 (2) (2011) 281–289.

Quamar, M.F., Chauhan, M.S., Late Quaternary vegetation, climate as well as lakelevel changes and human occupation from Nitaya area in Hoshangabad District, southwestern Madhya Pradesh (India), based on pollen evidence, Quat. Int. 263 (2012) 104–113.

Quamar, M.F., Chauhan, M.S., Signals of Medieval Warm Period and Little Ice Age from southwestern Madhya Pradesh (India): a pollen-inferred Late-Holocene vegetation and climate change, Quat. Int. 325 (2014) 74–82.

Quamar, M.F., Chauhan, M.S., Pollen-based vegetation and climate change in southwestern Madhya Pradesh, central India during the last 3300 years, J. Palaeontol. Soc. India 60 (2) (2015) 47–55.

Quamar, M.F., Kar, R., Prolonged warming over the last ca. 11,700 cal years from the central Indian Core Monsoon Zone: pollen evidence and a synoptic overview, Rev. Palaeobot. Palynol. (2020a) 104159. https://doi.org/10.1016/j.revpalbo.2020.104159.

Quamar, M.F., Kar, R., Modern pollen dispersal studies in India: a detailed synthesis and review, Palynology 44 (2) (2020b) 270–279.

Quamar, M.F., Kar, T., Thakur, B., Vegetation response to the Indian Summer Monsoon (ISM) rainfall variability during the Late Holocene from the central Indian Core Monsoon Zone, Holocene 31 (7) (2021) 1197–1211.

Quamar, M.F., Nautiyal, C.M., Mid-Holocene pollen records from southwestern Madhya Pradesh, central India and their palaeoclimatic significance, Palynology 41 (3) (2017) 401–411, doi:10.1080/01916122.2016.1219973.

Ramanathan, V., Chung, C., Kim, D., et al., Atmospheric brown clouds: impacts on South Asian climate and hydrological cycle, Proc. Natl. Acad. Sci. U S A 102 (2005) 5326–5333.

Rasmussen, S.O., Andersen, K.K., Svensson, A.M., Steffensen, J.P., Vinther, B.M., Clausen, H.B., Siggaard-Andersen, M.L., Johnsen, S.J., Larsen, L.B., Dahl-Jensen, D., Bigler, M., A new Greenland ice core chronology for the last glacial terminationJ. Geophys. Res.: Atmosphere, 111 (D6) (2006).

Rawat, S., Gupta, A.K., Sangode, S.J., et al., Late Pleistocene–Holocene vegetation and Indian summer monsoon record from the Lahaul, Northwest Himalaya, India, Quat. Sci. Rev. 114 (2015) 167–181.

Reddy, C.S., Jha, C.S., Diwakar, P.G., Dadhwal, V.K., Nationwide classification of forest types of India using remote sensing and GIS. Environmental Monitoring and Assessment. 187 (12) (2015) 777.

Renssen, H., Seppä, H., Crosta, X., Goosse, H., Roche, D.M., Global characterization of the Holocene Thermal Maximum, Quat. Sci. Rev. 48 (2012) 7–19.

Royer, D.L., Linkages between CO_2, climate, and evolution in deep timeProc. Natl. Acad. Sci. Unit. States Am., 105 (2) (2008) 407–408.

Saxena, A., Trivedi, A., Chauhan, M.S., Sharma, A., Holocene vegetation and climate change in central Ganga Plain: a study based on multi-proxy records from Chaodhary-ka-Tal, Raebareli District, Uttar Pradesh, India, Quat. Int. 371 (2015) 164–174. Available at: https://doi.org/10.1016/j.quaint.2015.01.041.

Sharma, C., Chauhan, M.S., Late Holocene vegetation and climate from Kupup Lake, Sikkim Himalaya, India, J. Palaeontol.l Soc. India 46 (2001) 51–58.

Sharma, S., Joachimski, M., Sharma, M., Tobschall, H.J., Singh, I.B., Sharma, C., Chauhan, M.S., Morgeroth, G., Late Glacial and Holocene environmental changes in Ganga Plain, Northern India, Quat. Sci. Rev. 23 (2004) 45–159.

Sharma, S., Joachimski, M., Tobschali, H.J., Singh, I.B., Sharma, C., Chauhan, M.S., Correlative evidence of monsoon variability, vegetation change and human inhabitation in Sanai lake deposit: Ganga Plain, India, Curr. Sci. 90 (7) (2006) 973–978.

Singh, D.S., Gupta, A.K., Sangode, S.J., Clemens, S.C., Prakasam, M., Srivastava, P., Prajapati, S.K., Multiproxy record of monsoon variability from the Ganga Plain during 400–1200 AD, Quat. Int. 371 (2015) 157–163.

Singh, G., Joshi, R.D., Chopra, S.K., Singh, A.B., Late Quaternary history of vegeation and climate of the Rajasthan Desert, India, Phil. Trans. R. Soc. Lond. 267 (1974) 467–501.

Singh, G., Joshi, R.D., Singh, A.B., Stratigraphic and radiocarbon evidence for the age and development of three salt lake deposits in Rajasthan, India, Quat. Res. 2 (4) (1972) 496–505.

Singh, S., Gupta, A., Dutt, S., Bhaumik, A.K., Anderson, D.M., Abrupt shifts in the Indian summer monsoon during the last three millennia, Quat. Int. 558 (2020) 59–65 https://doi.org/10.1016/j.quaint.2020.08.033.

Singh, U., A History of Ancient and Early Medieval India: From the Stone Age to the 12th Century (PB), Pearson Education India, Noida, 2008.

Singhvi, A.K., Bhattacharyya, A., Kale, V.S., Quadir, D.A., Gupta, A.K., Phadtare, N.R., Shrestha, A.B., Chauhan, O.S., Kolli, R.K., Sheikh, M.M., Manzoor, N., Adnan, M., Ashraf, J., Khan, A.M., Chauhan, M.S., Thamban, M., Yadav, R.R., Chakraborty, S., Roy, P.D., Devkota, L.P., Instrumental, terrestrial and marine records of the climate of south Asia during the Holocene: present status, unresolved problems and societal aspects, (Eds. Mitra, A.P., and Sharma, C.) (National Physical Laboratory, New Delhi, India) Global Environmental Changes in South Asia, Springer, The Netherlands, 2010, pp. 54–124.

Sinha, A., Berkelhammer, M., Stott, L., Mudelsee, M., Cheng, H., Biswas, J., The leading mode of Indian Summer Monsoon precipitation variability during the last millennium, Geophys. Res. 38 (15) (2011) L15703.

Sinha, A., Cannariato, K.G., Stott, L.D., Li, H.C., You, C.F., Cheng, H., Edwards, R.L., Singh, I.B., Variability of southwest Indian summer monsoon precipitation during the Bolling-Allerod, Geology 33 (10) (2005) 813–816.

Solanki, S.K., Usoskin, I.G., Kromer, B., Schüssler, M., Beer, J., Unusual activity of the Sun during recent decades compared to the previous 11,000 years, Nature 431 (7012) (2004) 1084–1087.

Swain, A.M., Kutzbach, J.E., Hastenrath, S., Estimates of Holocene precipitation for Rajasthan, India, based on pollen and lake-level data, Quat. Res. 19 (1983) 1–17.

Tripathi, S., Thakur, B., Nautiyal, CM., Bera, S.K., Floristic and climatic reconstruction in the Indo-Burma region for the last 13,000 cal. yr: a palynological interpretation from the endangered wetlands of Assam, northeast India, Holocene 30 (2) (2020) 315–331.

Trivedi, A., Chauhan, M.S., Sharma, A., Nautiyal, C.M., Tiwari, D.P., Late Pleistocene-Holocene vegetation and climate change in the central Ganga Plain: a multiproxy study from Jalesar Tal, Unnao District, Uttar Pradesh, Curr. Sci. 103 (5) (2012) 555–562.

Trivedi, A., Saxena, A., Pollen based vegetation and climate change records deduced from the lacustrine sediments of Kikar Tal (Lake), Central Ganga Plain, India, Palaeobotanist 66 (2017) 37–46.

Walker, M., Gubbard, P., Head, M.J., Berkelhammer, M., Björck, S., Cheng, H., Cwynar, L.C., Fisher, David, Vasilios Gkinis, D., Antony Long, V., Lowe, A., Newnham, J., Rasmussen, R., Harvey, S.O., Weiss, H., Formal subdivision of the Holocene Series/Epoch: a summary, J. Geol. Soc. India 93 (2019) 135–141.

Wang, B., Ding, Q., Global monsoon: dominant mode of annual variation in the tropics, Dyn. Atmos. Oceans 44 (3) (2008) 165–183.

Wang, B., Liu, J., Kim, H.J., et al., Northern Hemisphere summer monsoon intensified by mega-El Niño/southern oscillation and Atlantic multidecadal oscillation, Proc. Natl. Acad. Sci. USA 110 (2013) 5347–5352.

Wang, K., Dickinson, R.E., A review of global terrestrial evapotranspiration: observation, modeling, climatology, and climatic variability, Rev. Geophys. 50 (2012) 1–54.

Wang, P., Clemens, S., Beaufort, L., et al., Evolution and variability of the Asian monsoon system: state of the art and outstanding issues, Quat. Sci. Rev. 24 (5) (2005b) 595–629.

Wang, T., Surge, D., Mithen, S., Seasonal temperature variability of the Neoglacial (3300–2500 BP) and Roman warm period (2500–1600 BP) reconstructed from oxygen isotope ratios of limpet shells (*Patella vulgata*), Northwest Scotland, Palaeogeogr. Palaeoclim. Palaeoecol. 317 (2012) 104–113.

Wang, Y., Cheng, H., Edwards, R.L., et al., The Holocene Asian monsoon: links to solar changes and North Atlantic climate, Science 308 (2005a) 854–857.

Wanner, H., Beer, J., Butikofer, J., Crowley, T.J., Cubasch, U., Fluckiger, J., Goosse, H., Grosjean, M., Joos, F., Kaplan, J.O., Kuttel, M., Muller, S.A., Prentice, I.C., Solomina, O., Stocker, T.F., Tarasov, P., Wagner, M., Widmann, M., Mid-to Late Holocene climate change: an overview, Quat. Sci. Rev. 27 (2008) 1791–1828.

Wanner, H., Solomina, O., Grosjean, M., Ritz, S.P., Jetel, M., Structure and origin of Holocene cold events, Quat. Sci. Rev. 30 (21–22) (2011) 3109–3123.

Webster, P.J., Magana, V.O., Palmer, T.N., Shukla, J., Tomas, R.A., Yanai, M.U., Yasunari, T., Monsoons: processes, predictability, and the prospects for prediction, J. Geophys. Res. Oceans 103 (C7) (1998) 14451–14510.

Williams, M.A.J., Clarke, M.F., Late Quaternary environments in North-Central India, Nature 308 (1984) 633–635.

Wright, R.P., Bryson, R., Schuldenrein, J., Water supply and history: Harappa and the Beas regional survey, Antiquity 82 (2008) 37–48.

Wu, W., Tan, W., Zhou, L., Yang, H., Xu, Y., Sea surface temperature variability in southern Okinawa Trough during last 2700 years, Geophys. Res. Lett. 39 (2012), L14705, doi:10.11029/2012GL052749.

Yadava, M.G., Ramesh, R., Monsoon reconstruction from radiocarbon dated tropical Indian speleothems, Holocene 15 (1) (2005) 48–59.

Zhao, D.S., Wu, S.H., Yin, Y., Yin, Z.Y., Vegetation distribution on Tibetan Plateau under climate change scenario, Reg. Environ. Change 11 (2011) 905–915.

Zorzi, C., Sanchez Goni, M.F., Anupama, K., Prasad, S., Hanquiez, V., Johnson, J., Giosan, L., Indian monsoon variations during three contrasting climatic periods: the Holocene, Heinrich stadials and the last interglacial-glacial transition, Quat. Sci. Rev. 215 (2015) 50–60.

CHAPTER

7 Holocene climate and sea-level changes and their impact on ecology, vegetation and landforms in South Kerala Sedimentary Basin, India

D. Padmalal[a], **Ruta B. Limaye**[b], **K.P.N. Kumaran**[b,c]

[a]*National Centre for Earth Science Studies, Thiruvananthapuram, Kerala, India.* [b]*PalynoVision, Mon Amour, Erandaane, Pune, Maharashtra, India.* [c]*Shreenath Hermitage, Pune, Maharashtra, India*

7.1 Introduction

Holocene comprises the last ~11000 years and is the shortest epoch in the history of our planet. This time span is important in many aspects, particularly in terms of man's development and subsequent interventions and modifications of the land and the natural resources. The Holocene epoch witnessed the most dramatic changes based on climate and sea level. These, coupled with local and regional tectonics, had a pronounced impact on plant and animal life, landforms, water, other nonliving resources, and human evolution and civilization. In fact, the present day landscape and environment are the products of interactions of the biosphere and geosphere. As such, the study of the Holocene climate has attained great importance in recent years. Developments in dating techniques, fossil remains, isotope geology, and other modern analytical tools aided the study of Holocene sediments as a repository of climate proxies. Through the study of Holocene events, one gets an insight into the unmasked and unobliterated results of the changes in climate and hydrology, sea-level oscillations and tectonics. In this sense, the dictum "present is the key to the past" becomes most meaningful through the study of Holocene represented by the sedimentary records. The subsurface sediments of South Kerala Sedimentary Basin (SKSB) are the focus of the present study while addressing the climate and sea-level changes and their impact on ecology, landforms, and vegetation. Signatures in the form of pollen and spores, subfossil woods, peat, nonpollen palynomorphs (NPP) as well as records of sedimentology, geochemistry, and geochronology were decoded to address the Holocene dynamics and its impact on the ecosystem and environment in the SKSB.

The SKSB is located in the coastal plain between Kollam and Kodungallur (8°45[1] and 10°15[1] N latitude) and it is unique due to the presence of a curvilinear landward extension of the offshore sedimentary basin with a sediment fill of ~700 m thickness (Figs. 7.1 and 7.2). About 600 m of these sediments are of Early to Middle Miocene and the remaining of Quaternary age. This stretch of the coastal plain includes the major wetlands, viz., Vembanad lagoon, Kayamkulam lagoon, the perennial to seasonal wetlands of Kuttanad, alluvial fans and mounds, paleoestuaries, and the characteristic ridge–runnel systems (Fig. 7.3). In the absence of outcrops, stratigraphic information in respect of Quaternary sequences has to be obtained from the boreholes only. Although an outline of Quaternary sedimentary

Holocene Climate Change and Environment. DOI: https://doi.org/10.1016/B978-0-323-90085-0.00010-3

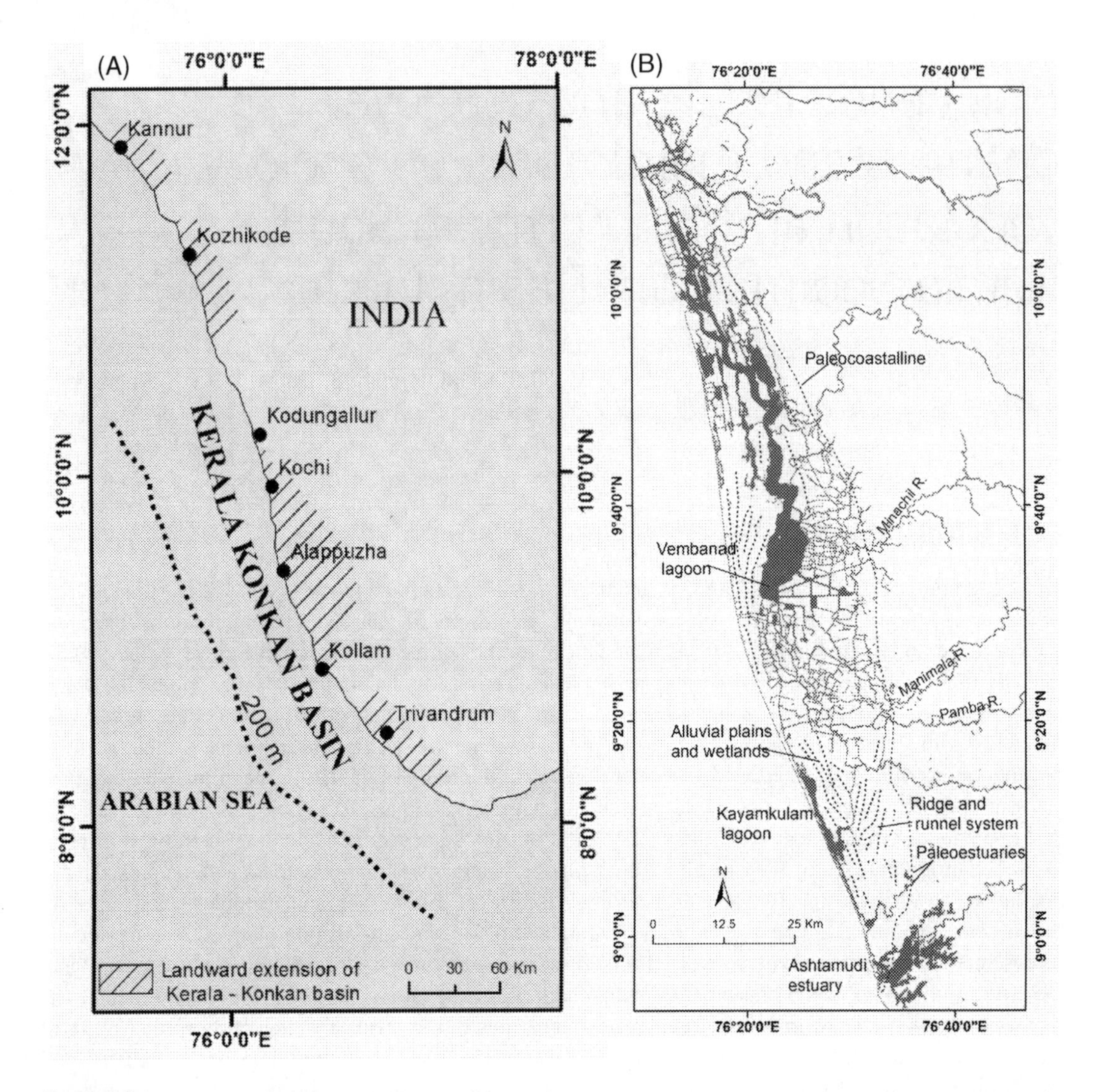

Figure 7.1

(A) Kerala Konkan Basin and South Kerala Sedimentary Basin—a landward extension of Kerala Konkan Basin. (B) South Kerala Sedimentary Basin and associated landforms showing present geomorphology (modified and adopted from Chattopadhyay, 2002)

stratigraphy and facies of SKSB became available, there exist many lacunae for understanding fully the sea level and climate changes and their impact on ecology and landforms. Subsurface sediments of two 25.0 m deep boreholes and five boreholes of 6.0–7.0 m depth were used to decode the paleoclimate and sea-level signatures preserved in the sediments while addressing the impact on the coastal area of southern Kerala. Table 7.1 gives the stratigraphic succession in the SKSB.

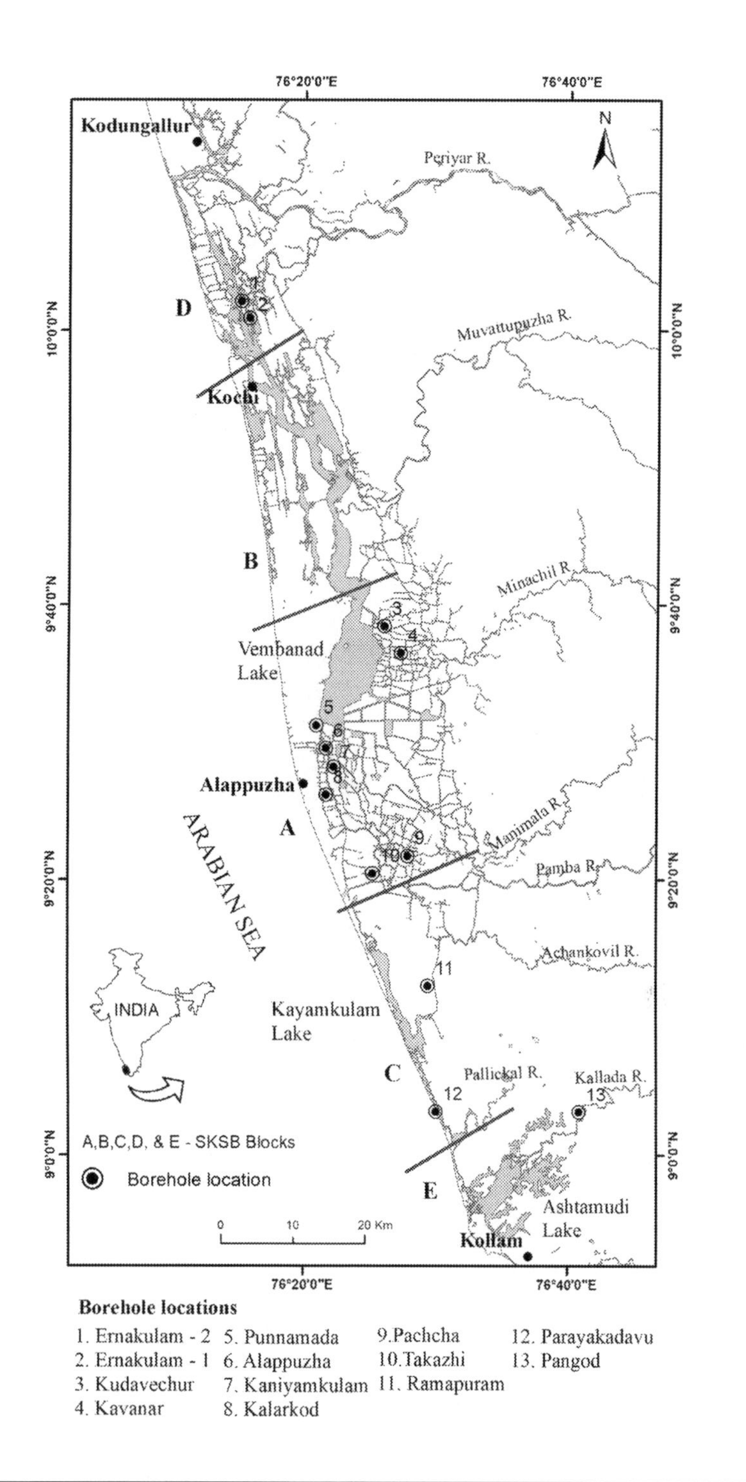

Figure 7.2

Location map showing borehole sites studied in South Kerala Sedimentary Basin.

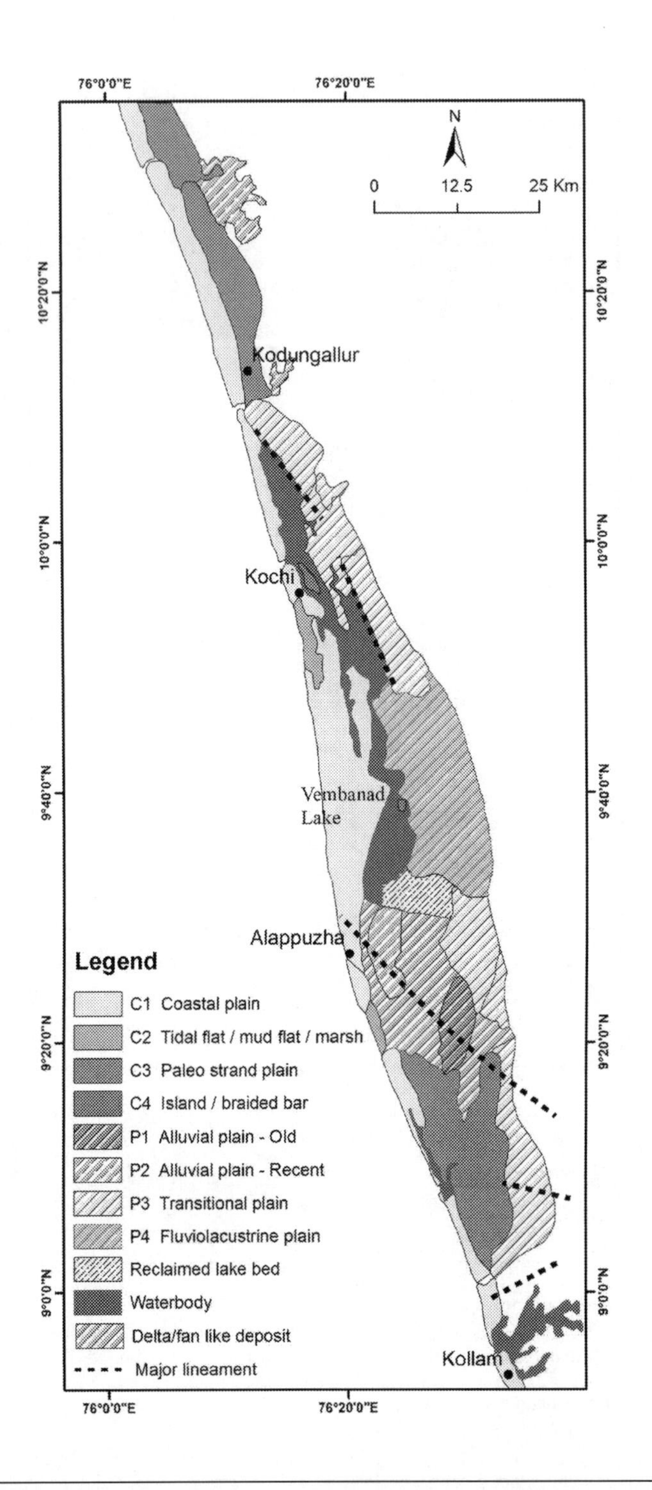

Figure 7.3

Geomorphology of coast of South Kerala Sedimentary Basin (Modified after Chattopadhyay 2002).

Table 7.1 Stratigraphic succession in South Kerala Sedimentary Basin.

Rock unit	Lithology	Age	Thickness in m
Vembanad Formation	Sand, clay, and peat/lignite	Quaternary	80
Laterite	Residual and reworked laterites	Late Miocene–Early Quaternary	30
Warkalli Formation	Sand, clay, and lignite	Middle Miocene	150
Quilon Formation	Calcareous clay, limestone	Early Miocene	130
Vaikom Formation	Pebbly sand, clay, and lignite	Early Miocene	300
Basement	Mainly khondalites and charnockites		

Modified and adopted from Najeeb (1999).
The clastics below the Quilon Formation have been divided into Alleppey beds and Vaikom beds in certain cases. This division is not justified and hence not adopted here. The Quaternary sand–clay sequence has been given an informal name "Vembanad Formation" by Raha et al. (1983) and is adopted herein.

7.2 Scope

The area covered in the present paper falls within the SKSB belonging to the coastal stretch between Kollam and Kodungallur. While addressing the theme of the contribution emphasis is made to establish the history of rainfall during Holocene Epoch and examine whether it was similar to, higher or lower than the present rate. The present study aims to decipher ecological and vegetational changes in the provenance and depositional environments due to climatic and sea-level changes inferred from the sediment archives by looking at the spores, pollen, NPP, and subfossil logs. The study synthesizes the data from all sources including those generated in the study for establishing the depositional environments and their changes in time in the SKSB has also been set forth while focusing on the overall theme of this chapter. Aspects of biogeoresources gifted by Holocene events and their relevance in the present environment are also being focused on.

7.3 Material and methods

Since the entire study is based on subsurface samples, boreholes were drilled from the following locations to obtain continuously undisturbed samples. 1. Ernakulam (25.0 m), 2. Alappuzha (25.0 m), 3. Kaniyamkulam (7.0 m), 4. Punnamada (7.0 m), 5. Kudavechur (7.0 m), 6. Thiruvarpu (7.0 m), 7. Pangod near Bharanikkavu (6.0 m; Figs. 7.2, 7.4, and 7.5). Boreholes (1) and (2) above were drilled mechanically using a hired rig. The others were collected manually using a specially fabricated push corer assembly. Sand–silt–clay fractions in the samples were estimated following the procedure given by Lewis (1994). The results are shown in Fig.7.6 – 7.8 (Tables 7.2 – 7.4). Samples of Ernakulam boreholes were subjected to estimation of total organic carbon, and $CaCO_3$ following El-Wakeel and Riley (1957) and Jackson (1967), respectively, and the down core variations are depicted in Figs. 7.7 and 7.9 (Tables 7.5 – 7.7). A ternary diagram showing different hydrodynamic fields of deposition in respect of samples from Alappuzha, Kaniyamkulam, and Ernakulam boreholes was prepared following Pejrup (1988; Fig. 7.10). Representative samples from the deep cores were subjected

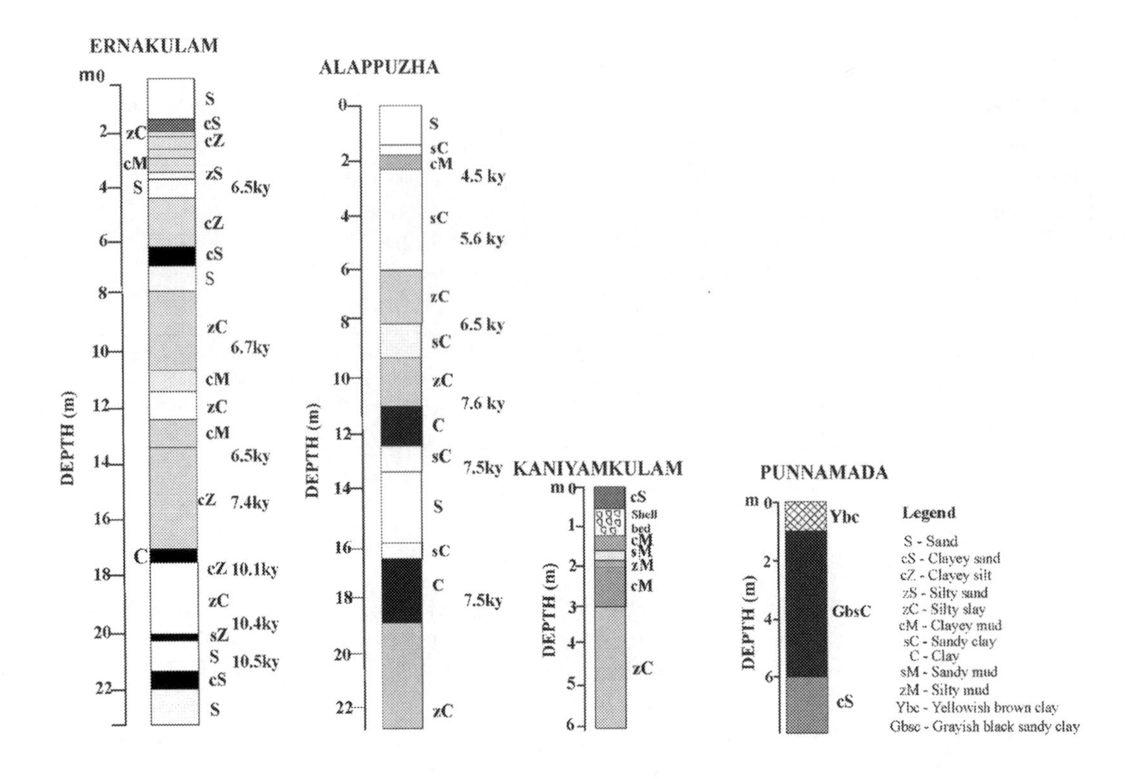

Figure 7.4

Lithosections of Ernakulam, Alappuzha, Kaniyamkulam, and Punnamada.

to palynological studies (Fig. 7.11). While making palynological observations, attention was also given to understand the microfauna and diagenetic aspects of organic matter.

The occurrence of fossil wood and subfossil logs in the seasonal–perennial wetlands in the SKSB and the adjacent laterite covered midland was known for generations. Except for sporadic ^{14}C dates (Rajendran et al., 1989; Pawar et al., 1983; Nair and Padmalal, 2003) the wetlands of Kerala have not been subjected to any studies. Subfossil logs from two locations—Karippuzha South in SKSB and a clay quarry in the midland at Pangod near Bharanikkavu, Kollam District—were collected for vegetation analysis as they constituted part of the evergreen forests. Another wood sample from the laterite covered midland was also retrieved from Vettiyar, 6.0 km east of Mavelikara. All the fossil wood samples were carbon dated for chronological analysis. In addition to the wood samples above, 16 samples of sediments, peat, and shells were also provided conventional ^{14}C dates. The results of this and some of the dates from earlier boreholes used in the present study are presented in Table 7.8 (Maya et al., 2021). In order to get a satisfactory picture of the Holocene facies and stratigraphy, the data from the following boreholes drilled earlier have also been used. (i) Ernakulam lagoon, ~4.0 km West of South West of Ernakulam borehole drilled; (ii) Kavanar, 10.0 km West of Kottayam; (iii) Kalarkod, ~12.0 km South of South East of Alappuzha; (iv) Pachcha, 10.0 km East of Ambalappuzha; (v) Ramapuram,

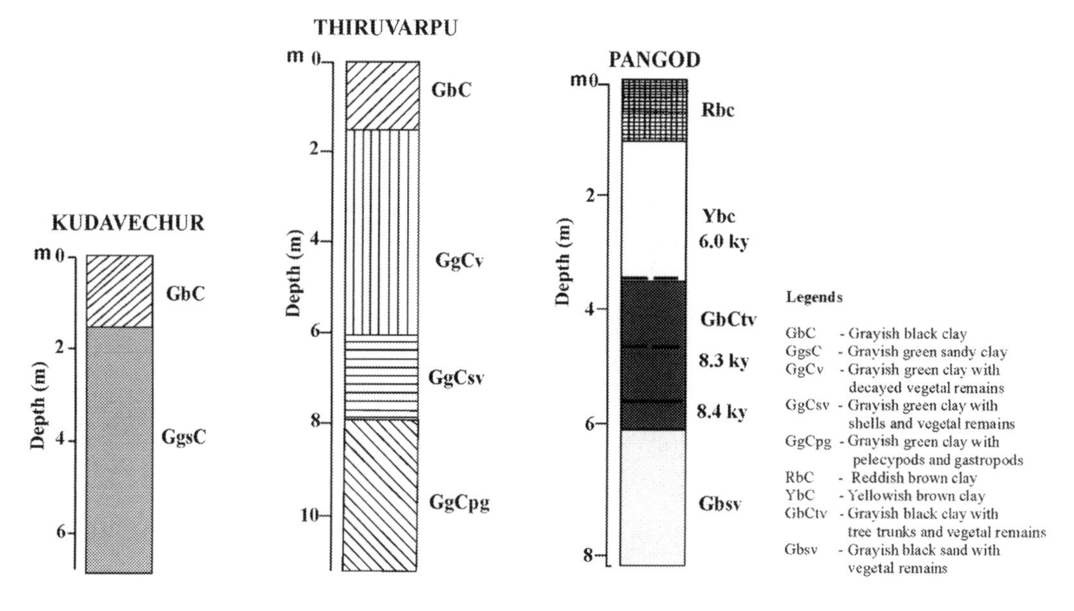

Figure 7.5

Lithosections of Kudavechur, Thiruvarpu, and Pangod.

~ 5.0 km North of Kayamkulam; (vi) Parayakadavu, ~ 7.0 km North of North West of Karunagapally. Locations of these boreholes and other boreholes that have used for getting a holistic picture of the Holocene sediments are shown in Fig. 7.2. All the relevant data such as organic carbon and calcium carbonate contents from other sources were compiled and integrated while addressing the Holocene climate and sea-level changes and their impact on ecology and landforms of the SKSB.

7.3.1 Borehole data

Since data collected from boreholes are to be used for the interpretation of the sedimentary facies of a particular area, it is appropriate that these data are presented on an area-wise basis. Accordingly, the borehole data are shown in the following order from north to south (Fig. 7.2):

(i) Ernakulam Area, (ii) Thiruvarpu–Kavanar–Kudavechur area to the east of the Vembanad lagoon, (iii) Alappuzha–Kalarkod area to the southwest of the Vembanad lagoon, (iv) Pachcha–Thakazhi area to the Southeast of Vembanad lagoon, and (v) The southern, the sand-dominated area represented by the Ramapuram and Parayakadavu borehole cores.

7.3.1.1 *Ernakulam area*

Two borehole cores were used for representing the Ernakulam area, one was drilled on land, and the other was drilled in the lagoon earlier.

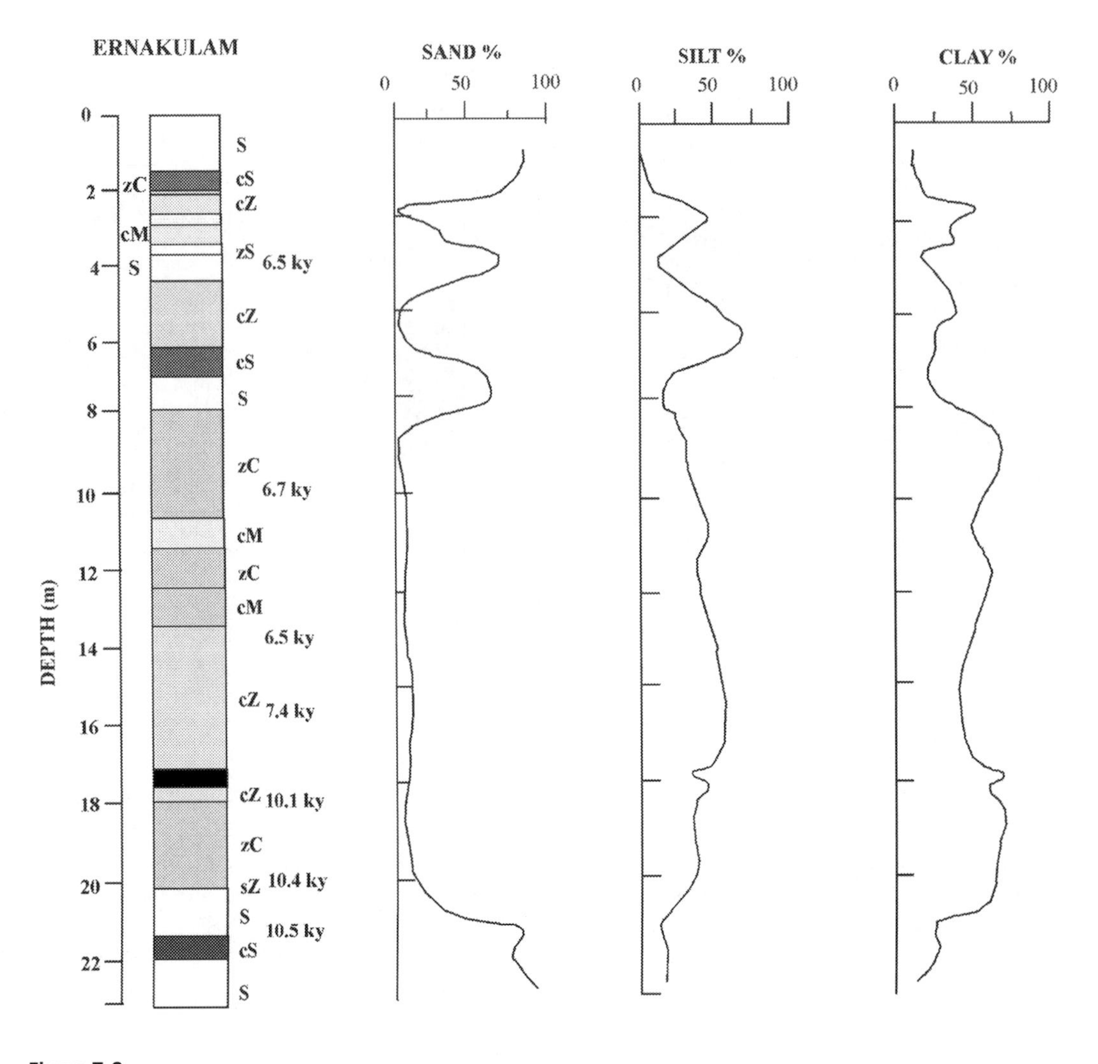

Figure 7.6

Downcore variation of sand, silt, and clay contents of Ernakulam borehole.

i. *Ernakulam borehole-1:* It was located in the heart of Ernakulam city and was ~7.0 km east of the Kochi bar mouth. The core was drilled down to 25.0 m using a rig fitted with a split sediment sampler. It is bottomed on laterite resting over Vaikom Formation. Since this borehole has provided a set of high-resolution data, the section is described in detail.
ii. *Ernakulam borehole-2:* This borehole was drilled in connection with soil investigation for the Ernakulam–Vypin bridge under the Goshri Island Development Project. It has revealed a 44.0 m thick section. The Holocene section is 22.0 m thick and rests on a lateritic clay underlain by the Vaikom Formation (Neogene). The Holocene sediments, in general, are mud-dominated with

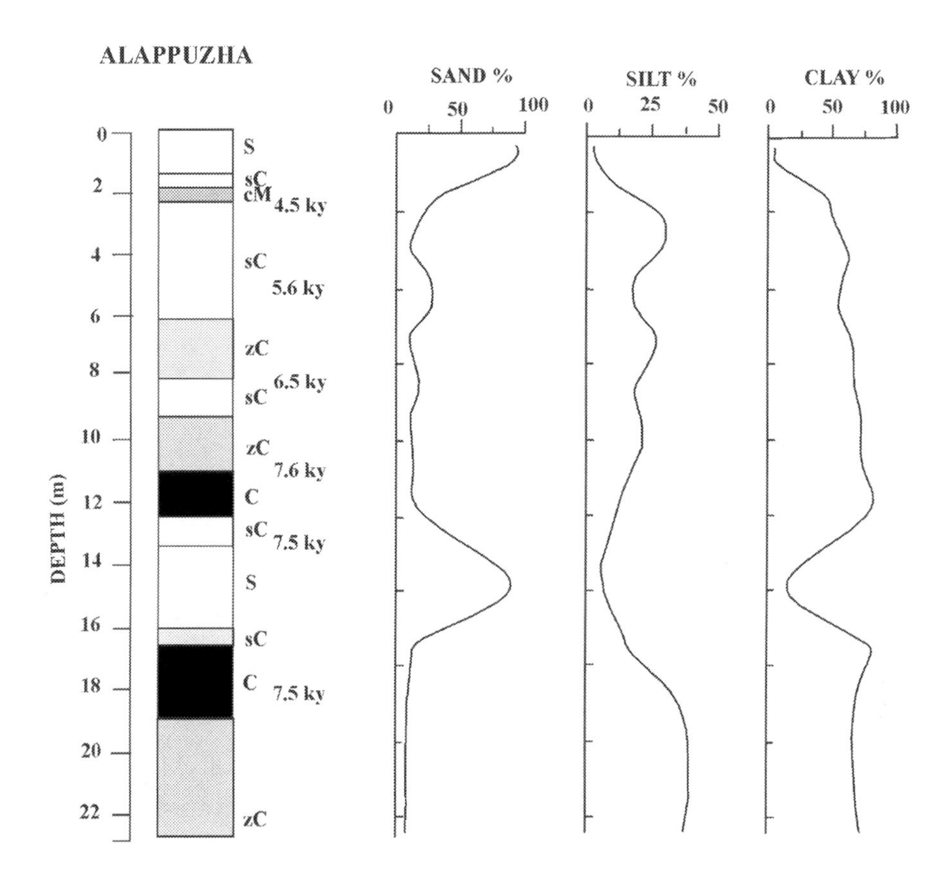

Figure 7.7

Downcore variation of sand, silt, and clay contents of Alappuzha borehole.

three clayey layers of sand at ~ 17.0–18.0 m, 7.0–9.5 m, and 3.5–1.5 m (top of sediment column). The sediments are generally dark gray to black with 1.0–6.0% organic carbon. The entire column contains varying contents of molluscan shells, and the $CaCO_3$ component ranges between 1.0% and 21.0%.

7.3.1.2 *Thiruvarpu–Kavanar–Kudavechur area*

All the boreholes along the east of Vembanad lagoon have revealed a mud-dominated Holocene sequence. Thiruvarpu, Kavanar, and Kudavechur boreholes have been continuously drilled, and therefore undisturbed samples were available. They are described below:

i. *Thiruvarpu borehole:* This 11.0 m deep borehole, drilled manually, has penetrated clay with various shades of gray. Broken and unbroken shells of pelecypods and gastropods are occasionally present throughout the clay. The clay is rich in the decayed vegetal matter between 7.0 and 2.0 m. The topmost 1.5 m is composed of grayish black clay.

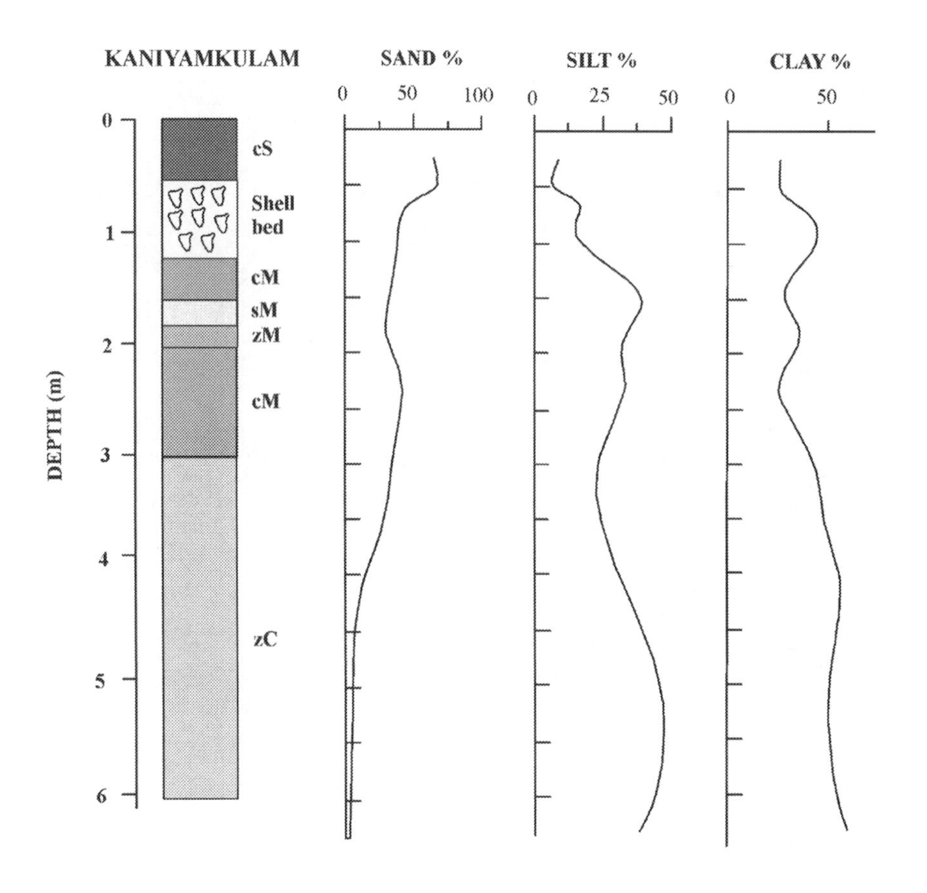

Figure 7.8

Downcore variation of sand, silt, and clay contents of Kaniyamkulam borehole.

ii. *Kavanar borehole:* The Kavanar borehole is drilled in a freshwater–swamp located ~3.0 km west of Kumarakom, the famous tourist village, west of Kottayam. The 10.5 km thick Holocene section comprises dark gray stiff clay between 2.0 and 10.5 m, with several thin inter layers of peat. The clay in the 8.5–9.5 m level embeds molluscan shells similar to those found in the present lagoon. In this borehole, the topmost 2.0 m contains semidecomposed remains of present day vegetation in an unconsolidated clay matrix. The samples of this borehole have been studied in detail, and the ecological and environmental significance of the section will be discussed in the subsequent section.

iii. *Kudavechur borehole:* It is located 10.0 km north of the Kavanar borehole. It has provided 6.0 m of undisturbed samples. Barring the uppermost 1.5 m, the section comprises grayish green sandy clay with small quantities of micaceous minerals. Very fine-grained sand and silt are found as small lenticels.

Table 7.2 Sand, silt, and clay contents in the Ernakulam-1 borehole.

Depth (cm)	Sand (%)	Silt (%)	Clay (%)	Sediment type (Picard, M.D., 1971)
65–75	85.32	3.23	11.45	Sand
130–140	85.02	5.44	9.55	Sand
180–190	68.84	7.06	24.10	Clayey sand
190–200	64.11	24.34	11.55	Silty sand
220–230	2.66	34.94	62.40	Silty clay
240–250	2.19	53.06	44.75	Clayey silt
280–290	34.74	33.62	31.65	Sandy mud
320–330	30.85	23.70	45.45	Clayey mud
330–340	68.05	20.65	11.30	Silty sand
370–380	75.18	2.08	22.75	Sand
480–490	2.67	55.08	42.25	Clayey silt
510–520	7.18	51.12	41.70	Clayey silt
530–540	2.04	74.41	23.55	Clayey silt
610–620	7.34	66.56	26.10	Clayey silt
630–640	50.48	29.07	20.45	Clayey sand
730–740	75.04	5.41	19.55	Sand
800–810	1.00	32.85	66.15	Silty clay
880–890	1.46	27.84	70.70	Silty clay
1010–1020	7.21	36.89	55.90	Silty clay
1060–1070	5.83	47.27	46.90	Silty mud
1110–1120	8.22	40.63	51.15	Clayey mud
1200–1210	4.44	29.71	65.85	Silty clay
1310–1320	4.77	45.38	49.85	Clayey mud
1410–1420	6.14	50.36	43.50	Clayey silt
1510–1520	11.42	52.93	35.65	Clayey silt
1620–1630	6.57	54.18	39.25	Clayey silt
1720–1730	6.14	47.82	46.05	Clayey silt
1740–1750	4.23	19.87	75.90	Clay
1760–1770	5.34	50.66	44.00	Clayey silt
1820–1830	1.04	26.82	72.15	Silty clay
1920–1930	7.21	32.04	60.75	Silty clay
2000–2010	6.80	34.15	59.05	Silty clay
2110–2115	31.60	13.25	55.15	Silty clay
2115–2125	90.10	2.95	6.95	Sand
2180–2190	61.90	10.10	28.00	Clayey sand
2280–2290	86.60	8.60	4.80	Sand

Table 7.3 Sand, silt, and clay contents in the Alappuzha borehole.

Depth (cm)	Sand (%)	Silt (%)	Clay (%)	Sediment type (Picard, M.D., 1971)
30–40	92.90	2.85	4.25	Sand
90–100	95.90	3.20	0.90	Sand
180–190	36.50	13.10	50.40	Sandy clay
240–250	21.10	31.55	47.35	Clayey mud
390–400	3.47	29.48	67.05	Silty clay
430–440	23.30	17.35	59.35	Sandy clay
580–590	30.60	16.30	53.10	Sandy clay
640–650	5.40	28.25	66.35	Silty clay
740–750	10.14	23.31	66.55	Silty clay
840–850	17.17	15.13	67.70	Sandy clay
920–930	5.56	20.29	74.15	Silty clay
1040–1050	10.06	20.59	69.35	Silty clay
1140–1150	9.64	12.96	77.40	Clay
1240–1250	3.20	10.30	86.50	Clay
1420–1430	77.70	2.95	19.35	Sand
1460–1480	84.75	3.95	11.30	Sand
1540–1550	84.18	5.22	10.60	Sand
1670–1680	3.05	13.95	83.00	Clay
1715–1730	10.22	13.43	76.35	Clay
1880–1890	0.48	42.28	57.25	Silty clay
2500	1.19	29.21	69.60	Silty clay

Table 7.4 Sand, silt, and clay contents in the Kaniyamkulam borehole.

Depth (cm)	Sand (%)	Silt (%)	Clay (%)	Sediment type (Picard, M.D., 1971)
20–30	64.85	9.45	25.70	Clayey sand
50–60	72.03	3.47	24.50	Clayey sand
60–65	42.01	22.84	35.15	Sandy mud
90–100	37.78	7.87	54.35	Sandy clay
140–150	33.77	45.98	20.25	Silty mud
170–180	24.28	34.27	41.45	Clayey mud
200–210	36.12	30.73	33.15	Sandy mud
230–240	43.33	36.10	20.58	Silty mud
290–300	31.19	21.32	47.49	Clayey mud
350–360	28.49	23.66	47.85	Clayey mud
410–420	3.32	33.28	63.39	Silty clay
500–510	2.69	49.52	47.79	Silty clay
600–610	0.26	44.45	55.29	Silty clay
630–635	0.80	38.25	60.95	Silty clay

Table 7.5 Organic carbon and calcium carbonate contents in the Ernakulam borehole.

Sr. no.	Depth (cm)	C-Org (%)	$CaCO_3$ (%)
1	65–75	0.126	0.5
2	130–140	0.097	0.5
3	280–290	1.105	14.5
4	480–490	2.560	10.0
5	618–620	0.390	11.5
6	800–810	3.310	11.0
7	1010–1020	2.500	8.0
8	1310–1320	3.920	10.0
9	1740–1750	1.590	0.5
10	2000–2010	5.290	3.0
11	2280–2290	0.292	1.0

Table 7.6 Organic carbon and calcium carbonate content in the Alappuzha borehole.

Sr. no.	Depth (cm)	C-org (%)	$CaCO_3$ (%)
1	30–40	2.750	3.50
2	180–190	1.240	0.75
3	430–440	2.410	4.50
4	740–750	2.780	7.50
5	1040–1050	2.600	10.00
6	1460–1470	0.495	4.50
7	1715–1730	0.526	1.00
8	2500	2.690	0.50

7.3.1.3 Alappuzha–Kalarkod area

In this area, there are four boreholes with undisturbed samples for the entire column investigated. These boreholes are Punnamada, Alappuzha, Kaniyamkulam, and Kalarkod and are described below.

i. *Punnamada borehole:* This 8.0 m deep borehole is located ~ 6.0 km northeast of Alappuzha along the southwest bank of Vembanad lagoon. The section can be divided into three intervals.
   1. 8.0–6.0 m: Light gray, clayey sand constitutes this interval. The sand grains are made up mostly of quartz and low quantities of heavy minerals.
   2. 6.0–1.0 m: This interval is made up of grayish black, organic-rich sandy clay. The decayed vegetal matter is noticed at certain levels. Shells of gastropods and pelecypods have been observed throughout the section.
   3. 1.0 m–ground level (GL): This section is composed of yellowish brown clay with the decayed vegetal matter.

ii. *Alappuzha borehole:* This 23.0 m deep borehole is located at ~ 250 m south of the Alappuzha Boat Jetty. The sedimentary column is divided into four intervals.

Table 7.7 Organic carbon and calcium carbonate contents in the Kaniyamkulam borehole.

Sr. no.	Depth (cm)	C-org (%)	$CaCO_3$ (%)
1	20–30	0.258	1.5
2	50–60	0.924	-
3	150–160	0.639	1.5
4	230–240	0.927	1
5	350–360	2.63	4.25
6	460–470	2.801	1.5
7	600–610	2.467	0.5

Table 7.8 Radiocarbon dates from South Kerala Sedimentary Basin.

Lab no.	Location	Depth in m	Material	^{14}C age	Calibrated age	Calibrated age range
2373	Pangode H1	6.0	Sediment	7550 ± 160	8370	8450–8170
2374	Pangode H2A	5.0	Sediment	7480 ± 80	8260	8350–8170
2386	Pangode H2B	5.0	Wood	7490 ± 90	8340	8390–8180
2384	Pangode H4	3.0	Wood	5260 ± 120	5990	6200–5910
2371	Vettiyar H6B	2.0	Wood	13880 ± 200	16650	16980–16330
2392	Alappuzha	2.35	Sediment	3990 ± 150	4520	4810–4240
2437	Alappuzha	5.25	Sediment	4840 ± 170	5590	5740–5330
2439	Alappuzha	8.25	Sediment	5670 ± 140	6470	6640–6300
2440	Alappuzha	10.95	Sediment	6780 ± 160	7610	7790–7510
2438	Alappuzha	12.95	Sediment	6630 ± 230	7500	7680–7320
2436	Alappuzha	15.45	Sediment	6620 ± 200	7500	7670–7320
2347	Ernakulam	3.8–3.9	Sediment	5680 ± 170	6450	6890–6000
2451	Ernakulam	10.0–10.1	Sediment	5870 ± 160	6680	6870–6490
2467	Ernakulam	13.9–14.0	Sediment	5700 ± 180	6510	6720–6300
2378	Ernakulam	15.7–15.8	Sediment	6530 ± 130	7430	7570–7320
2449	Ernakulam	17.9–18.0	Sediment	8940 ± 150	10150	10240–9780
2445	Ernakulam	19.4–19.5	Sediment	9170 ± 200	10410	10640–10180
2346	Ernakulam	20.2–20.3	Sediment	9250 ± 170	10450	10670–10220
1824	Ramapuram	7.6	Shell	24450 ± 710	Not applicable	Not applicable
1808	Ramapuram	12.0	Shell	39370 ± 1000	Not applicable	Not applicable
-	Ramapuram	3.0	Wood	2460 ± 120	Not applicable	Not applicable
1929	Kavanar	6.5	Sediment	2180 ± 70	Not applicable	Not applicable
	Kavanar	9.5	Shell	7090 ± 100	Not applicable	Not applicable
	Kalarkod	8.45	Sediment	6740 ± 120	Not applicable	Not applicable
	Kalarkod	40.0	Sediment	20380 ± 490	Not applicable	Not applicable
	Parayakadavu	6.5	Shell	4610 ± 100	Not applicable	Not applicable

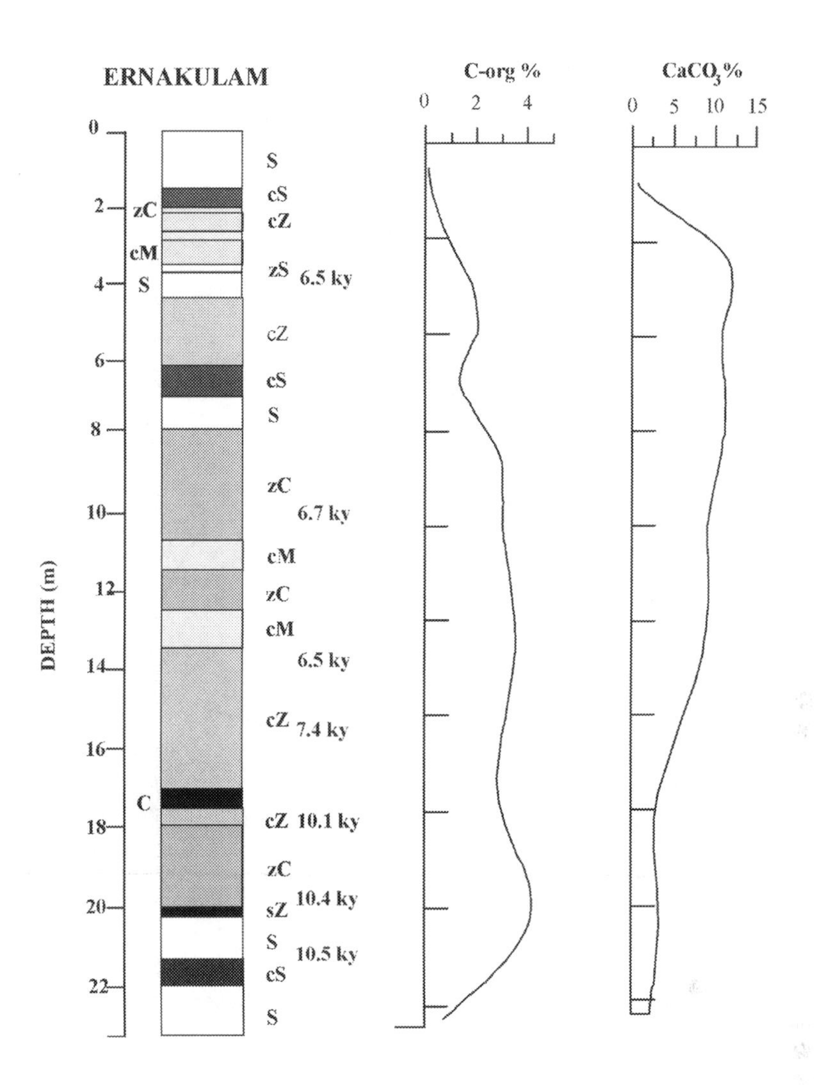

Figure 7.9

Downcore variation of organic carbon and calcium carbonate contents Ernakulam.

1. 23.0–16.0 m: It is composed of grayish black, silty sands, and very stiff clay.
2. 16.0–13.0 m: This layer is formed of brownish red to yellowish brown clayey sand.
3. 13.0–1.2 m: The layer is made up of clay, silt, and sandy clay with shades of gray.
4. 1.2–GL: Possibly earth fill.

iii. *Kaniyamkulam borehole:* Kaniyamkulam is a southern suburban village of Alappuzha Town. A 0.75–1.5 m thick shell bed, almost exclusively made up of the pelecypods (*Villorita*), occurs at ~ 1.0 m depth in this area. The shell does not show any preferred orientation. Intershell space is filled with clayey sand. The shell deposit is mined from a privately owned land and is a thriving

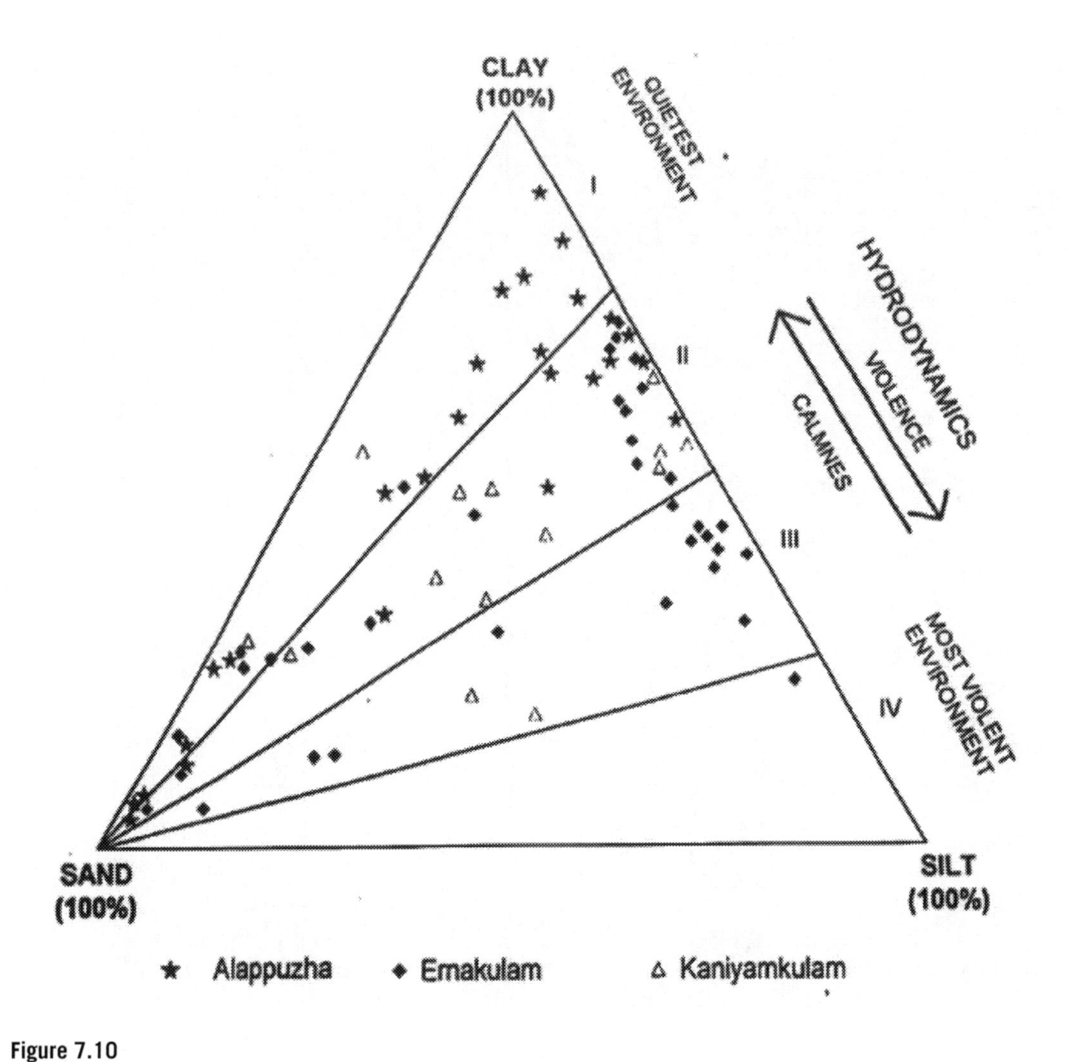

Figure 7.10

Ternary diagram of Alappuzha, Ernakulam, and Kaniyamkulam area.

small-scale industry. The 6.0 m deep borehole has been drilled manually. The interval 6.0–3.0 m is made up of dark gray to black clay containing molluscan shells' fragments. The section becomes progressively more sandy upward from 3.0 m onward.

iv. *Kalarkod borehole:* Kalarkod is situated to the southwest of the Vembanad lagoon and is dominated by freshwater swamps and marshes. A borehole was drilled in 2001 and the subsurface samples were collected. However, due to unconsolidated sugary sand in a major part of the borehole, the sample recovery was poor. This borehole, located in the deepest part of the SKSB, has been drilled down to 40.0 m. At 22.0 m, the sand is distinctly ferruginous and tends to be laterite. This might represent the exposure and aridity associated with the LGM. The sand becomes less ferruginous upward. There is

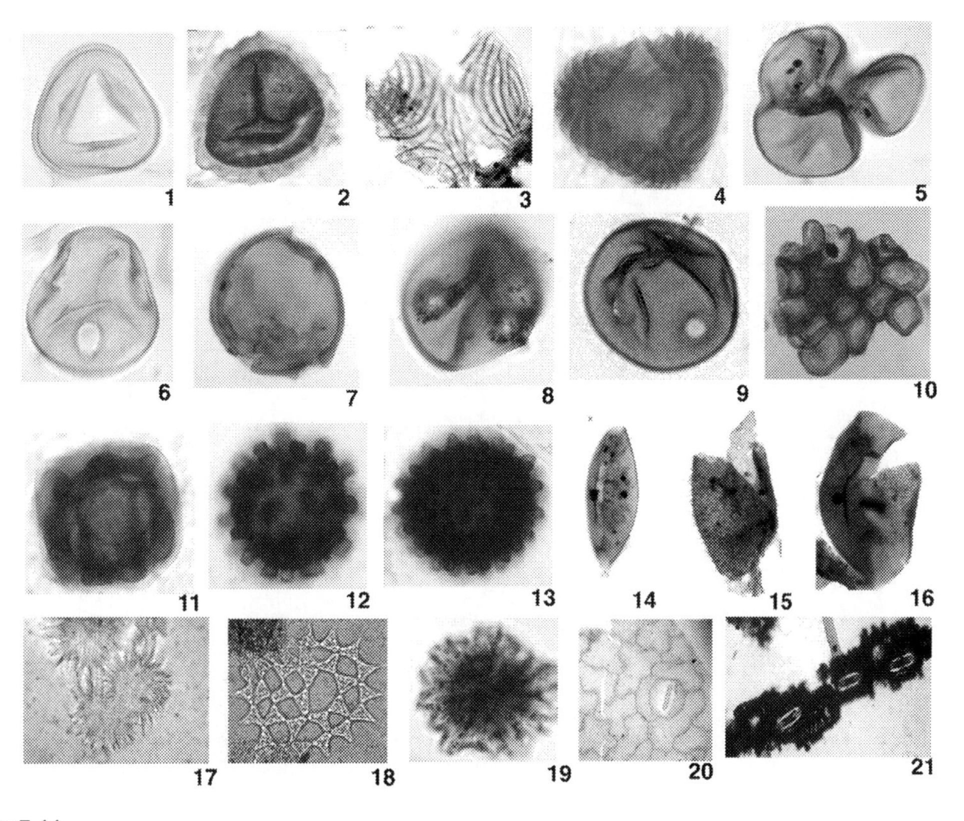

Figure 7.11

Selected representatives of palynomorphs and NPP assemblage of studied localities along the coastal stretch between Kollam and Kodungallur in South Kerala Sedimentary Basin (all photomicrographs are enlarged ca. ×500 and fungal remains enlarged ca. ×1000). 1. *Gleichenia* sp. (high humidity). 2. *Pteris* sp. (high humidity). 3, 4. *Ceratopteris thalictroides* (freshwater flood plains). 5. Sporangial mass of pteridophytic spores (high humidity). 6, 7, 8, 9. *Cullenia exarillata* (Bombacaceae, high rainfall). 10. *Rhizophora* sp. (Core Mangrove, Estuarine facies). 11. *Ctenolophonidites* sp. (reworked pollen from Warkalli Formation; high rainfall indicator, excessive humid conditions). 12. Euphorbiaceae pollen (wet evergreen element). 13. Asteraceae pollen (lowland element). 14, 16. Liliaceae pollen (moist marsh environment). 15. Palm pollen (coastal environment). 17. *Botryococcus* sp. (lacustrine environment). 18. *Pediastrum* sp. (fresh to brackish-water conditions). 19. *Heliospermopsis* sp. (salt glands; affinity with mangrove swamps, brackish to shallow marine). 20. Cuticle of fern leaf (Rubiaceae type of stomata, high humidity indicator). 21. Poaceae cuticle preserved in the form of charcoal (dumb-bell shaped stomata, dry conditions/fire activities). 22, 23. Morphotypes of *Rivularia* sp. (Cyanobacteria; biomarkers of hydrological changes—heavy rainfall and wet climate). 24, 25. Morphotypes of *Gloeotrichia* sp. (Cyanobacteria; biomarkers of hydrological changes—heavy rainfall and wet climate). 26. *Cirrenalia tropicalis* (Mangrove-associated fungus, high humidity). 27. *Lirasporites* sp. (fungal spore, high humidity). 28. *Kutchiathyrites eccentricus* (fungal fruiting body, high humidity). 29. *Notothyrites setiferus* (Microthyriaceous fungal fruiting body, high humidity). 30. *Tuberculodinium vancampoae* (Dinoflagellate, marine conditions). 31. Scolecodont (near-shelf or beach environment). 32, 33. *Spiniferites* sp. (Dinoflagellate, marine conditions). 34. *Lejeunecysta hyalina* (Dinoflagellate, probably reworked Miocene form, marine conditions). 35. Foraminiferal lining (shallow marine facies). 36. *Centropyxis aculeata* "discoides" (Thecamoebian cyst, dry and stressful environmental conditions).

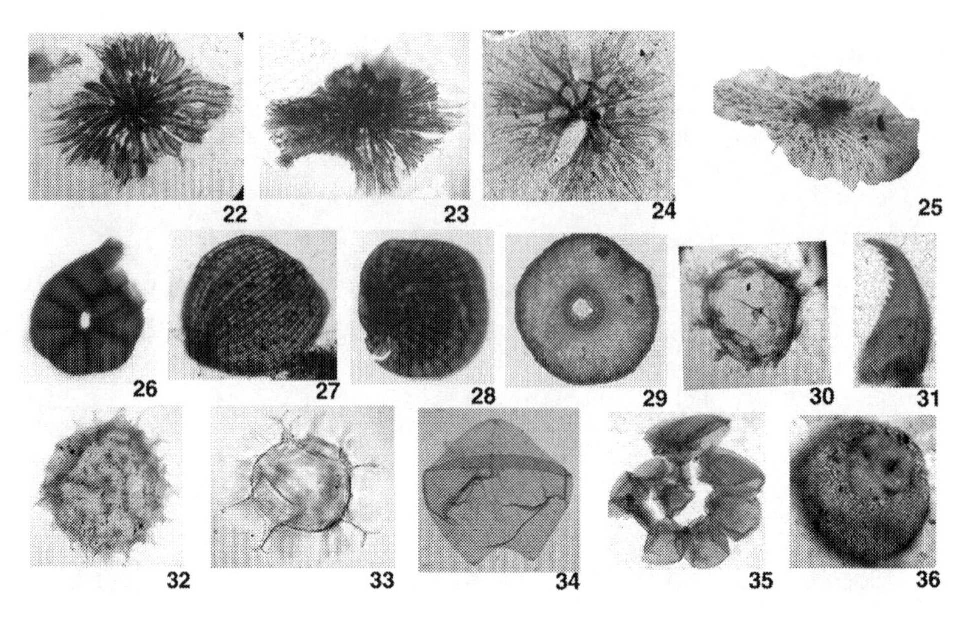

Figure 7.11 Continued

a clay bed at 7.0–10.0 m in this borehole. The clay is typically similar to those found in Ernakulam and Alappuzha boreholes and has molluscan shells dispersed throughout. The interval 7.0 m–GL is characterized by clayey unconsolidated sand and sandy clay in an approximate ratio of 4:1 typically found in modern lagoonal deltaic and fan complexes.

7.3.1.4 Pachcha–Thakazhi area

This area is situated along the southeast of Vembanad lagoon. Pachcha is along the eastern flank of the Central Depression, whereas Thakazhi is located closer to the deepest part of the SKSB.

i. *Pachcha borehole:* The borehole was drilled to a depth of 14.0 m, out of which the top 9.0 m section represents Holocene. The bottommost thin layer of silty clay is overlain by 0.6 m thick shell bearing clayey mud. A sediment sample from the depth of 8.5 m is dated 7.0 ky BP. Stiff soil with various percentages of sand overlies this. The uppermost layer, 4.0 m in thickness, is sand-rich. This layer is yellow to red in color and the sand grains have a coating of Fe–Mn oxide.
ii. *Thakazhi borehole:* Data from a fully cored borehole are available from the Thakazhi ferry site. By correlation, the Holocene bottom is inferred to be at 10.0 m. The clayey sand at the bottom is overlain by a coarse to very coarse sand with molluscan shells. This changes upward into a brown–yellow sandy mud, which extends to the surface. The section is largely similar to that of the Pachcha borehole.

7.3.1.5 Boreholes in the Southern sand-dominated area

The Holocene sediments in the SKSB south of the "U" line are sand-dominated with occasional layers of laterally discontinuous muddy sediments. Two boreholes with dates in this area are described below:

i. *Ramapuram borehole:* This borehole is located on one of the sand ridges of the ridge–runnel system ~5.0 km north of Kayamkulam. The ~7.6 m thick Holocene section is entirely made up of sand.
ii. *Parayakadavu borehole:* This borehole is located on the barrier separating the Kayamkulam lagoon from the sea. It has been drilled for soil investigation for building construction at the Mata Amritanandamayi Math, situated ~7.0 km North of Northwest of Karunagapally. The Holocene section is estimated to be ~9.0 m thick. Molluscan shells are dispersed in the sand, particularly in the bottom half of the unit. The beach sand with abundant heavy minerals is found from 2.0 to 6.0 m and the rest being clayey sand similar to the sediments in the adjacent lagoon.

7.3.2 Fossil wood/subfossil logs

Occurrence of fossil wood/subfossil wood in the seasonal perennial wetlands of Kerala has been known for generations. Besides, carbonized wood reportedly occurs in sediments constituting the terraces along most rivers. A preliminary reconnaissance was undertaken to ascertain their significance as sources of proxies for paleoclimate, paleohydrology, and sea-level changes. Tree trunks are found buried under 1.0–3.0 m clay in vast areas all along the coast from Karunagapally to Ernakulam. Along the banks of the Kayamkulam and the Vembanad lagoons, including the Mundakan, Kari, Kuttanad, and Kol lands, tree trunks are encountered while digging canals for navigation/irrigation. Most of the paddy fields, lakes, swamps, and marshes, in the midland region contiguous to SKSB and elsewhere, are reported to contain tree trunks. Some of these wetlands have 6.0–10.0 m sediment fills. The top clay in some of such wetlands is mined for tile-brick manufacture. These quarries expose tree trunks at several levels. Many of the river terraces, particularly those in the terrain, which has undergone uplift in Quaternary times, contain carbonized wood. On the erosion of the riverbank, this wood floats down the river and is collected and used as fuel. The "*in situ.*" presence of such wood has to be located. Occurrences of wood can be found virtually along the entire length of the state. Wood samples from three locations have been collected and studied.

i. *Karippuzha South Puncha:* The wetland of Kuttanad extends southward up to about $9^{\circ}10^{1}$ latitude. The southern extremity of this wetland system is at Pathiyoor. The Karippuzha South Puncha is about 5.0 km north of Pathiyoor. All over this wetland system, occurrence of fossil wood is noticed. While digging a short canal for navigation, several stems of fossil wood were encountered and most of them are with branches and roots.
ii. *Pangod near Bharanikkavu:* A 6.0 m deep clay quarry in the Pangod wetland near Bharanikavu exposes 3.0 m thick black clay which embeds subfossil logs of varied dimensions. The clay is underlain by coarse to medium-grained sand of at least 2.0 m thickness. The exposed section of the quarry can be clearly divided into two intervals: (1) 3.0 m thick black, too stiff, slightly silty clay with three horizons containing wood. A horizon of wood pieces marks the top of the 3.0 m thick stiff black clay. The presence of wood pieces at the same level has been noticed all over the quarry wall. This wood is dated 5.3 ky BP. (2) 3.0 m thick yellow sandy–silty clay overlying the black stiff clay is found all over the quarry wall. Similar clay is noticed in pits and trenches in the area. (3) Vettiyar 6.0 km east of Mavelikara.

There are tree trunks with roots buried under black clay in an abandoned tributary channel (now used for paddy cultivation). In the process of mining the black clay, the tree trunks and branches were encountered. One sample collected from a depth of 2.0 m below ground level in a paddy field gave ^{14}C age 13.9 ky BP. This shows that there may be cases of pre-Holocene fossil wood, the distribution of which has to be established. The wood appears to be more carbonized than the wood (i) and (ii) described above.

7.4 Fieldwork and sampling techniques

The source of samples was restricted to locations within the SKSB (Figs. 7.1 and 7.2). In the absence of outcrops and surface sediments in the present study, Holocene sediments can be studied only through subsurface samples. Besides, subfossil logs and carbonized wood remain retrieved from trenches, river channels, and wetlands have been taken into account for paleoclimate appraisal. The subsurface samples were recovered through boreholes by mechanical drilling for long cores and improvised coring techniques using push coring devises for shorter cores. The details of locations of the borehole sites and relevant information are dealt with in the succeeding paragraphs. Holocene sediments are found virtually all over the SKSB. The numerous river valleys abandoned river channels, and other wetlands in midland contain Holocene sediments. A rising sea level coupled with or without subsidence could have aided sedimentation in the lagoons and their peripheries. The riverine and lacustrine wetlands seem to have witnessed considerable flooding due to climatic reasons, which led to the accumulation of sizeable thickness of Holocene sediments. Since there are many boreholes, many of which have provided undisturbed samples and some geochronological data for correlation, a general idea about the geometry of the Holocene sediments in the SKSB could be obtained. The map of the Holocene units in Fig. 7.3 and the numerous borehole sections help in conveying a reasonably clear picture of the subsurface sediment architecture. Inferences on paleocurrents could be made from the boreholes lithological patterns (Figs. 7.4 and 7.5). Major parts of the Holocene sediments contained palynoflora, other plant remains, and microfauna and megafauna. A study of them provided ample data for the interpretation of sedimentary facies. While studying sedimentary facies, following the above criteria would automatically provide enough background to understand ecology. Since drilling has tested in all types of landforms, the study of the sedimentary facies has provided ample information about the impact of the above criteria in the evolution of these landforms (Figs. 7.1A and 7.1B). The hydrology of any region depends on the rainfall, and inferences could explain the river channels peculiarities and the riverbed sedimentation.

7.5 Observations and results

As mentioned earlier, the SKSB has been formed as an embayment during Early Miocene. Except the Quilon Formation of shallow marine origin, the Cenozoic sediments were deposited in fluviolacustrine and lagoonal environments (Nair and Rao, 1980). The next phase of sedimentation in SKSB took place during Late Pleistocene–Holocene. Our studies so far have indicated that the Late Pleistocene sediments are predominantly of littoral lagoonal origin, with deposits of continental environments underlying and overlying the former. With the substantial modification of the basin architecture by tectonics of Pleistocene–Holocene transitional times, the Holocene residues also appear to have been deposited in littoral lagoonal and other transitional environments. As is well known, a lagoon is sustained by a sea with an intimate connection. So, it can be considered a faithful recorder of sea-level changes.

A lagoon is a coastal water body into which the riverine drainage from the adjacent land empties its fresh water and the sediment load. Primarily a lagoon is a mix of saline water of marine origin and freshwater of land origin. This by itself creates a large and complex spectrum of sedimentary environments and facies. The sediments deposited in the lagoon reflect the weathering and erosion transportation settings of the land area. The land-derived residues include remnants of life forms of the land, significantly the vegetal organic debris. The lagoon and adjoining seasonal–perennial wetlands are the habitats for a broad spectrum of animals and plants. The preservation potential of these life forms in a lagoonal setting is very high. Thus, the study of lagoonal sediments can lead to inferences on sea level, climate, ecology, and landforms in the region. In the following paragraphs, an attempt is made to understand these from the sedimentary records.

7.5.1 Ernakulam area

The two boreholes sampled in this area together give a complete section of Holocene sediments.

7.5.1.1 *Ernakulam-1*

The oldest layer is laterite and lateritic, sandy paleosol. This surface appears to have been submerged to form a swamp marsh under the freshwater influence. This lowest sand dated 9.3 ky BP, grades upward into stiff, black clay, the top of which indicates an age of 8.9 ky BP. The section penetrated by this borehole can be divided into three, which are described from bottom to upward.

1. 25.0–21.0 m level: This interval is made up of light gray and reddish brown sand and clayey sand, which grade downward into laterite as revealed by several boreholes drilled in this area. This appears to be a residual paleosol devoid of any palynoflora and fauna.
2. 21.0–8.0 m level: This interval constitutes the most important part of the borehole. It is composed of stiff grayish green and dark gray–black clay.
3. This can be subdivided into three levels:
   i. 21.0–17.5 m: The section is composed of grayish green stiff clay. Dispersed in the clay are broken molluscan shells commonly found in the present day lagoons with active tidal action. The clay contains pellets of organic matter-rich, hardened clay, possibly formed by burrowing crabs or other life forms similar to those found in the tidal flats and mangrove swamps. Reworked Miocene sediments seem to have contributed to the clay, as can be inferred from the pollen of *Ctenolophonodites*.
   ii. 17.5–8.0 m: This interval is characterized by greenish black, stiff clay with littoral marine–lagoonal shells. Silty laminations are distinct in the lower part. The clay has 1.5 – >2.0% organic carbon. Palynoflora suggests fluctuating marine–freshwater influence with high terrestrial input intervals and indicating heavy rainfall and high humidity. The topmost 1.5 m contains typical littoral gastropods and lamellibranchs and an abundance of pollen of mangrove vegetation. The three carbon dates in this interval suggest that it includes reworked sediments, which can help infer the paleohydrology and paleoclimate to be discussed in the subsequent section.
   iii. 8.0 m–GL: This section is made up of sand–clay alternations. Two silty clay layers separate three sand layers. A part of the 2.0 m thick topmost sand layer could be attributed to landfill. The sand

is white, sugary, medium to fine grained with abundant mica flakes (Table 7.2). The bottommost and middle sand layers are grayish black with about 2% organic carbon. The lower clay layer has yielded an abundance of marine, broken molluscan shells, and foraminiferal tests. As compared to this, the upper clay layer is conspicuous by the absence of flora and fauna of any marine affinity. The clay at 3.8 m is dated 5680 ± 170 yrs BP.

7.5.1.2 Ernakulam-2 borehole

The corresponding layer is 1.5–2.0 m thick and gave a date of 8.3 ky BP. In both cases, the fossil content is dominated by freshwater plant remnants, such as *Botryococcus* and inland montane plant complex represented by *Cullenia* sp. and Euphorbiaceae. It is interesting to note that both cases include broken lagoonal–littoral molluscan shells. The sparse occurrence of *Spiniferites, Selenopemphia*, and *Legeunecysta* (dinoflagellates) and foraminiferal tests of marine–littoral origin has been observed in this interval. During 10–8 ky BP, the sea level was ~25–20 m lower, which means that this area was more or less at the same elevation as the ambient sea level. The coastal swamp marsh was possibly exposed to periodic storm surges. The organic carbon and the abundance of terrestrial and freshwater flora indicate profuse vegetation in the depocenter and the provenance. A comparison of this clay with the clays being deposited in the present day lakes, lagoons, and estuaries suggests that the Early Holocene clay was deposited much higher during a period of rainfall than the present rate. This also suggests that chemical weathering of the provenance was dominant.

The clay overlying this is lithologically similar; but with increasing littoral–lagoonal fauna and flora as indicated by the abundance of foraminiferal tests, dinoflagellates (*Lejeunecysta* and *Spiniferites*). However, there seems to be an abundance of pollen of *Cullenia exarillata*, fungal fruit bodies, pollen of Malvaceae and *Pandanus* suggesting the continuation of the heavy rainfall in the Middle Holocene. The lagoonal alga *Rivularia* is prominent. All these reiterate that the depositional environment is a lagoon with highly variable contributions from marine and freshwater. There is a clear indication that the marine fauna increases above Middle Holocene. The clayey sand–silty mud sequence with the base at 9.5 m in the Ernakulam-2 borehole is possibly Late Holocene as indicated by the date of 3.8 ky BP for the sample at 7.5 m. The common occurrence of foraminiferal tests, ostracod carapaces, dinoflagellates *Multispinula quanta*, and littoral blue–green alga *Rivularia* sp. proves the dominance of marine influence. An observation complimentary is that the freshwater input is minimal, distinctly due to a reduced monsoon activity and even spells of aridity. This episode is followed by a period of slightly higher input from freshwater. Besides foraminifera, diatoms like *Thalessioria oestrupii* and *Triceratium favus*, and *Polysphaeridium* sp., spores of *Ceratopteris thalictroides*, and *Lygodium* sp., occur in the shallowest part. Liliaceae/Arecaceae (pollen) and beach-sand forest plants like *Pandanus* sp. are well represented at the most superficial level. The typical montane and terrestrial floral components like *Cullenia exarillata* and Euphorbiaceae are present but far less than in the Middle Holocene part. The detailed palynological analysis shows the mixing of spores and pollens of terrestrial plants and littoral mangrove vegetation (Fig. 7.11).

There are two carbon-dated layers in this borehole. The lower one is from a depth of 21.5 m and is dated 8.3 ky BP. The second one is at 7.5 m giving an age of 3.8 ky BP. The samples above 10.0 m, possibly corresponding to approximately 5 ky BP, seems to contain more marine and lagoonal fauna and flora than terrestrial sources. Samples from 17.0 to 10.0 m interval show lagoonal life forms, but the terrestrial flora, including montane flora, is abundant. The Holocene section's bottom part shows the dominance of freshwater flora and fauna with the sporadic occurrence of flora and fauna of lagoonal and littoral origin. Downcore variation of sand, silt, and clay contents of Ernakulam is given in Fig. 7.6 and Table 7.2. The description of the depositional environments at Ernakulam will not be complete without mentioning reworked sediments. The Middle Holocene section with a ^{14}C age range of 6.5–5.7 ky BP contains a substantial thickness of reworked sediments in the Ernakulam-1 borehole. It may be there in the second borehole core too. The reworking could be due to the sediments already deposited in offshore sediment banks that are transported landward and deposited in the lagoon during storm surges and also due to the increased rainfall and resultant wider river channels, on being incised in response to lowered rainfall. This could lead to erosion of the river terraces and banks and such older sediments could be deposited along with younger sediments.

7.5.2 The area along the east of Vembanad lagoon

The three drilled cores that are correlatable with the cores of Kudavechur, Kavanar, Thiruvarpu, and several other boreholes prove that the Holocene section is mostly composed of clay. The Kavanar borehole section has been studied in considerable detail. Thiruvarpu and Kudavechur boreholes are closely similar to that at Kavanar in terms of gross lithology.

The Kavanar borehole reveals a 10.5 m thick Holocene section composed of clay with varying quantities of carbonaceous matter/peat. The clay in the 8.5–9.5 m interval contains molluscan shells. The palynological assemblage at 9.0 m indicates marine influence. It is dominated essentially by dinoflagellates (*Spiniferites* and *Multispinula quanta*). The blue–green alga *Rivularia* sp., spores and pollen of mangrove vegetation have also been observed. All these indicate mixing of lagoonal–mangrove flora and fauna with those of marine origin. The depositional environment is inferred to be mangrove–lagoon susceptible to tidal action. Molluscan shell at 9.5 m is carbon dated 7.1 ky BP. An abundance of organic matter characterizes the palynological assemblage at 7.0 m. Although there is no occurrence of lagoonal molluscan shells, there seems to be a marine influence in the form of dinoflagellate cysts of *Protoperidinium* sp. and foraminiferal linings. The sediment at 6.5 m is carbon dated 2.2 ky BP. It appears that the lagoon present during 7.0 ky BP got silted up to a mangrove system with occasional tidal flushing. The topmost 2.0 m of the borehole contain semidecomposed remains of present day vegetation in a matrix of unconsolidated clay.

The most significant point to be noted is lagoonal shells at 9.5 m, corresponding to an age of 7.1 ky BP. This roughly corresponds to when the sea level reached the present position according to the sea-level curve (Hashimi et al., 1995). It appears that within a short while, the lagoonal/marine influence was significantly reduced, leaving the area to become a mangrove swamp with occasional tidal flushing.

7.5.3 The area along the southwest of Vembanad lagoon

7.5.3.1 Alappuzha area

The Alappuzha area has been adequately sampled and supplemented by sufficient geochronological data. The 23.0–16.0 m interval in Alappuzha borehole is made up of stiff black silty clay with no micro or megafauna of littoral–marine or lagoonal affinity. This clay is obviously of lacustrine origin derived from a provenance undergoing chemical weathering. Besides, the lacustrine environment seems to have been the habitat for profuse vegetation. This clay is rich in organic matter and shows well-preserved cellular structure and preservation under an oxygen-deficient sediment–water interface. This clay is overlain by brownish red sand at 13.5–16.0 m interval, representing the oxidizing environment. A few very thin laminae of clay are present. One sample from 15.5 m interval gave a ^{14}C date of 6.6 ky BP. This sand appears to be related to marine influence. The foraminiferal tests and a major part of the sand could be attributable to the littoral activity. The sand is the first indication of the Holocene transgression and is seen to be similar to that of the Ernakulam area. The coastal terrain was likely tilted toward the north during the Early Holocene allowing earlier marine connection in the Ernakulam area. The deposition probably kept pace with sea-level rise, as discussed later.

The clay overlying this has been observed to contain at its lower part highly biodegraded vegetal matter with no identifiable cellular differentiation. Coiled foraminifera, ostracods, and blue–green algae are common. This is ^{14}C dated in ascending order 6.6, 6.8, and 5.7 ky BP, respectively. As noticed in the Ernakulam borehole, there is a clear indication of reworked clays in the section. This shows that a shallow lagoon was formed due to the formation of sandy and clayey barrier banks, which got eroded to provide part of the sediment deposit.

The grayish green clay overlying this from the interval 1.0–6.0 m is much less compacted than the clay below. This clay is characterized by the occasional presence of broken shells of pelecypods and biodegraded organic matter. Spores and pollen are scarce. Juvenile foraminiferal tests of *Bolivina* and *Pararotalina* are widespread. Micro planktons like *Polyspheridium* sp. and *Veryhachium* sp. also are common. All these prove the dominance of marine influence in the depocenter and a corresponding reduction of freshwater influx. This clay is ^{14}C dated 4.8 and 4.0 ky BP. It clearly indicates that the rainfall has gradually reduced from about 6.0–5.5 ky BP, allowing proportionately more substantial marine influence in the lagoon. The topmost 2.0 m of this borehole, comprising mostly sand, could be partly earth fill material. The muddy sediment at 2.4 m below this sandy interval is dated as 4 ky BP and is rich in molluscan shells. This suggests that the lagoon was completely silted up by ~4 ky BP. The sedimentary column can be divided into four intervals.

- **i.** 23.0–16.0 m: Molluscan shells are absent. Palynoflora suggests an environment more dominated by freshwater than brackish water.
- **ii.** 16.0–13.0 m: A few tests of foraminifera constitute the main fauna. This layer suggests subaerial exposure and oxidation. A sediment sample from 15.5 m level shows a carbon date of 6620 ± 200 yrs BP. In the absence of any dates below, the subaerial exposure/diastem duration cannot be inferred.
- **iii.** 13.0–1.2 m: The presence of broken lagoonal molluscan shells has been noticed throughout the section. However, the interval 2.0–4.0 m is a bed characterized by an abundance of pelecypods. The topmost part of this bed at ~2.0 m is virtually made up of pelecypod shells, and correlatable to the shell layer found Kaniyamkulam borehole and also in the Vembanad lagoon.
- **iv.** 1.2–GL: It is interesting to see that the interval 16.0–10.0 m includes reworked sediments as observed in Ernakulam-1 borehole.

Downcore variation of sand, silt, and clay content of Alappuzha is given in Fig. 7.7 and Table 7.3.

7.5.3.2 Kaniyamkulam borehole

A 6.0 m deep borehole drilled at Kaniyamkulam is about 4.0 km southeast of the Alappuzha borehole. However, the significant point is the presence of a shell bed. The shell is dated 3.3 ky BP. The sand overlying the shell bed contains weathered shells. It appears that the faunal diversity in the Vembanad lagoon gradually got reduced as a result of silting up with more influx of sand. It may be noticed that the pelecypod that makes up the shell bed is the genus *Villoritta* that thrives in sandy or sandy clay substrates and in less salinity than that of the lagoon (3–10 g/l). The shell bed is found as laterally discontinuous pockets buried under a yellowish brown sandy clay (GSI, 1976) in several parts of the Vembanad lagoon. This would mean that the lagoon has undergone through (a) stages of silting up, creating several semicontinuous water bodies; (b) deposition of sand from the possibly ephemeral streams along the axis of the lagoon; (c) developing the ecology for *Villoritta* to flourish; (d) drying up exposing the sediments to subaerial exposure and oxidation; (e) re-establishment of the lagoon to the present stage after ~3 ky BP. Process (a) to (d) would have been aided by a relatively continuous barrier system separating the sea from the lagoon. Down core variation of sand, silt, and clay content of Kaniyamkulam is shown in Figs. 7.8 and 7.4.

7.5.4 Thiruvarpu–Kavanar–Kudavechur area

7.5.4.1 Thiruvarpu

From the decayed vegetal matter, palynological assemblage, and megafauna, the clay is inferred to have been deposited in a mangrove swamp peripheral to a lagoon. It has been observed that this borehole at its topmost 1.0 m interval contains carbonaceous matter distinctly traceable to *Oryzasativa* (rice) and further investigation of this is likely to throw light on the history of human settlement and agriculture.

7.5.5 Kalarkod area

7.5.5.1 Kalarkod

The 40.0 m deep borehole has brought out a sand-dominated Holocene section overlying a clay-dominated Pleistocene section, the latter being inferred to be of lagoonal origin. The Holocene sediments are ~22.0 m thick. The clay at 7.0–10.0 m interval has yielded organic carbon up to 2.0% and an abundance of flora and fauna. Besides the commonly occurring lagoonal molluscan shells, the following microfossils have been observed in this clay:

Marine microplankton is represented by a few species of diatoms like *Coscinodicus* sp., *Triceratium favus, T. cinnamomeum, Bacteriastrum* sp., *Biddulphia aurita, Cyclostella stiata, Thalassiosira oestrupii, Navicula spectabilis, Actinocyclus ehrenbergii*, and *Campylodiscus clevei* and dinoflagellates like *Peridinium oblongum* (*Lejeunecysta* sp.) and are common. Foraminiferal linings are common, but tests are infrequent. Littoral/lagoonal blue–green algae-like *Rivularia* sp. are also found. The freshwater plant complex represented by the pollen of Liliaceae is widespread. The beach forest plant represented by pollen of *Pandanus* is frequent. Freshwater flood plain fern *Ceratopteris thalictroides* spores are also common. The fungal spore complex is well represented. This assemblage suggests mixing of marine facies with high quantities of terrestrial organic input. The clay sample at 8.45 m is dated 7.0 ky BP. This means that the Holocene marine transgression took place at this time, similar to that in the Alappuzha area.

7.5.6 The area along the east of the Vembanad lagoon

7.5.6.1 Pachcha

The ~9.0 m thick Holocene section in the Pachcha borehole starts with a ferruginous clay that tends to become lateritic, suggesting that sedimentation is formed by subsidence or inundation of a land surface. The earliest Holocene sediment is dark gray clay with abundant lagoonal molluscan fauna. The $CaCO_3$ in this is 26.0%, while TOC is 2.37%. The shells from this layer have been dated 7.0 ky BP. This clay is inferred to be deposited in the lagoon that formed during the main Holocene transgression. Microplanktons like *Spiniferites* sp., *Polysphaeridium zoharyi*, and *Lejeunecysta* sp. as well as foraminiferal linings are very common. Land-derived pteridophytic spores are more common than other spores and pollen. Pollen grains of *Cullenia exarillata* are prevalent. It is interesting to note that marine–lagoonal origin's fauna and flora are more dominant than those of terrestrial origin.

The silty and sandy clays (8.0–3.25 m) overlying this sediment are black, very stiff with an organic carbon content of 2.67–3.31% and $CaCO_3$ content of 2.5–4.0%. The flora of montane and swamp–marsh habitats coupled with calcareous shells rarity suggest that the lagoon got silted up fast and became a freshwater wetland with rare and short duration inundation by lagoonal waters. Further, the spores and pollen offer very thick tropical wet forest cover. These sediments characterize the wetlands south and southeast of Vembanad lagoon and the abandoned river channels in the midland, which presently are used for growing paddy. It should be pointed out that this clay that is rich in TOC is a product of chemical weathering of a provenance that is thickly forested and drained by rivers that are bereft of the potential energy to transport any coarser clastic if available. This loss of the rivers possible life is higher than the present sea level during 7.0–6.0 ky BP.

The topmost 3.0 m of the Pachcha borehole section comprising sandy mud contain rare pteridophytic spores and fungal hyphae suggesting much less humidity and precipitation in the provenance. More studies on these near-surface sediments are needed for confirming this. Being a wetland area, supporting abundant swamp–marsh vegetation before the arrival of man on the scene, the surface sediment is carbonaceous to various extents and covers a vast area of the wetland systems south of Vembanad lagoon.

7.5.6.2 Thakazhi

The Thakazhi borehole section, by correlation, is found to have ~ 10.0 m of Holocene sedimentary section, which is essentially similar to that at Pachcha. The section starts with a 2.0 m thick silty sand of possible fluvial origin. This is overlain by another 2.0 m thick layer of sand, which clearly indicates deposition in a lagoon where fluvial sediments were deposited. This is probably the indicator of extension of the Holocene transgression-induced lagoon. Like in the Pachcha area, the lagoonal condition did not last long. The lagoon became silted up with sandy and clayey mud of possible swamp–marsh origin. It appears that this borehole is located toward the southern end of the Holocene lagoon. The lagoonal conditions south of this were probably confined to the present day Kayamkulam lagoon and its immediate vicinity to the east. The formation and sustenance of Kaymkulam lagoon seem to be largely related to the construction of the discontinuous ridge–runnel system along the coast and, to some extent, to the tectonics.

7.5.7 Southern sand-dominated area

There is a vast area south of the "U" curve is shown in Fig. 7.3 in the SKSB, where Holocene sediments are composed of sands of various categories. While a major part of such area falls in the ridge–runnel landforms, there seem to be areas along with the Vembanad lagoon trend, but south of the latitude of

~9°5'–9°10', where clayey sand-dominated sections seem to be present. The sands in the numerous boreholes available in this area have not been investigated well.

7.5.7.1 *Ramapuram borehole*

In the Ramapuram borehole, sand overlies the eroded top of a Late Pleistocene clay and shell bed sequence. The only date available is 2.5 ky BP from a tree nut embedded in the sand column. Besides, the sand contains heavy mineral percentages similar to those found in the present day beaches. This unconformably overlies pebbly sand with abundant molluscan fauna, which has been dated 24.5 ky BP. The sand is unconsolidated, well rounded, and quartz-dominated with heavy mineral fractions ranging from 5.0% to 25.0%. The heavy minerals are composed of ilmenite, sillimanite, and zircon and are similar to those found in the modern beach in the nearby area. The sand at ~3.5 m depth is ferruginous and suggests a period of long-duration exposure to subaerial weathering and oxidation. Scanning electron micrographs of sand grains from this horizon suggest eolian activity. Another date available is from a molluscan shell in clayey sand in the Parayakadavu borehole. As indicated already, this date of 4.6 ky BP indicates the time of forming the present day beach ridge. It is worth reiterating that the vast Punchaland, extending from Pathiyoor in the south to Veeyapuram and Thottappally in the north fossil wood/subfossil logs are buried under 1.0–3.0 m of clayey sand or sandy clay. The fossil wood almost invariably is older than the main Holocene transgression. There are no indications for the presence of a Holocene lagoon in such areas. In other words, the two principal sedimentary environments are the swamps and marshes and the ridge–runnel system. Regressive phase of the sea is responsible for the formation of the NNE trending ridge–runnel system over the uplifted and eroded top of Pleistocene terrain.

7.6 Discussion

The sedimentological–geochemical and floral–faunal data collected from primary and secondary sources have been analyzed. A glimpse of the climate, sea level, and tectonics has also been taken into account. These data in the light of observations and inferences made on the Holocene sea-level changes and climatic variations in the study area have been discussed. The effects of these and the tectonics on the sedimentary environments in general and ecology and landforms in particular in SKSB and the adjacent hinterland are examined.

7.6.1 The relevance of tectonics in the study of sedimentary facies

The formation and sustenance of a sedimentary environment are intimately connected to tectonics. The study area forms a part of the passive, rifted continental margin. In the history of rifting, the NNW trending West Coast Fault had played a significant role. The Sahyadri escarpment was initially along this fault. Due to polycyclic denudational events, the cliff retreated to the present position (Balakrishnan, 1997, 2001). Nair (1990) and Varadarajan and Balakrishnan (1980) studied lineaments of Kerala and emphasized the importance of the NS and NNW trending lineaments in the formation of the coastline. Varadarajan and Balakrishnan (1980) believe that the coast has undergone possible upwarping by reactivation of faults or otherwise giving rise to an eastward tilt to the beach. They explain that the landward dip of Cenozoic sediments at several places along the coast is attributable to this upwarping event. The structural elements responsible for the formation of SKSB are superimposed on by this upwarping or tilting. Coupled with this is the northward tilt of the SKSB during the Pleistocene–Holocene transition. This is the primary reason for the formation and sustenance of the Vembanad lagoon. A perusal of the geomorphology map of the Holocene unit in Fig. 7.3 and the publihed reports of Nair and Padmalal (2003) reveals that the thickness of Holocene sediments is less along the coast than along the Vembanad lagoon trend. This is suggestive of the periodic upwarping along the coast. Nair (1999), while studying the coastal geomorphology, has opined that the ENE trending lineaments are the youngest or most recently reactivated ones since they truncate all the other lineaments. Bilham et al. (1998) observed that following India's collision–Eurasia plates, intracratonic cymatogenic arches, and depressions have been formed across the Indian Peninsula with an average wavelength of approximately 220 km. The Mulki–Pulicat Arch identified by Subrahmanya (1996) is probably among these arches. It cannot be a coincidence that the deepest part of the SKSB shows an approximate ENE trend, which might coincide with a cymatogenic low. Extended toward WSW, this low corresponds with the structurally most deficient part of the shelf edge. Thus, the formation and sustenance of the SKSB can be seen as a product of a regional tectonic event, which started during the earliest Miocene when the plate collision really took place (Aitchison and Davis, 2001). From these observations, it is clear that the sedimentary environment's interpretation is not practicable and appropriate without a proper understanding of tectonics.

7.6.2 Climate variability and sea-level changes and their impacts

There is a virtual consensus on the observation of Hashimi et al. (1995) on the post-LGM sea-level history along India's west coast. The sea level, according to this, has reached the present position of ~7.0 ky BP. A further rise in the sea level to over 4.0–6.0 m is suggested by Merh (1992). Pandarinath et al. (2001) inferred that the sea level rose by 6.0–10.0 m after 6 ky BP. After this transgression, the sea receded in stages and by ~4 ky BP the sea level has been oscillating about the present position.

On Holocene climatic changes, a sizeable literature is available. Kale et al. (2004) have reviewed this literature on Asian Monsoon and presented a synoptic picture while discussing the data retrieved from the SKSB. There is a convergence of ideas that the beginning of the Holocene witnessed the Monsoon's intensification lasting over a few centuries. From 8.5 to 5.5 ky BP, Monsoon Asia saw much higher precipitation, universally known as Holocene Climatic Optimum. The rainfall of this period is often stated to be about three times more than the present rate (Rajagopalan et al., 1997; Bryson and Swain, 1981). It is interesting to see that this excessive precipitation has taken place in Asia and all along the path of our summer monsoon. During this period, the present deserts of the Sahara witnessed the development of large lakes and prolific vegetal cover (Petit-Maire, 1995). According to Ghienne et al. (2002), the Chad Lake in Central Africa seems to have witnessed expansion to over greater than three times its present area.

The areas covered by the SKSB and its river catchments experience an average annual rainfall of 2500–4000 mm. The Early Holocene excessive precipitation, by itself, would have resulted in a substantial modification in the landforms. This is to be considered in conjunction with sea levels.

7.6.2.1 *Excess rainfall and higher sea level*

The sea level after reaching the present position is seen to have risen further, causing depositional environments as depicted conceptually in Fig. 7.3. Excessive sediments brought in by the Achankovil,

Pamba, Manimala, Meenachil, Moovattupuzha, and Periyar rivers appear to have caused a series of coastal sandbanks or ridges along the northern half of the SKSB, creating a lagoon that became the Vembanad lagoon. The areas to the east and southeast of this lagoon became mangrove swamps and related environments with infrequent inundation by tidal waters.

The area to the south of these came under a littoral sea during the Holocene transgression. This area had a paleoslope toward the northwest. This paleoslope has been created during Pleistocene–Holocene transition as a result of uplift and tilt, which is amply proved by the erosional loss of Pleistocene sections in the area. The NE and NNE trending ridge–runnel system developed during this regression from the site with a northwestward paleoslope. It has to be indicated here that the Achankovil and Kallada rivers flowing parallel WNW trending courses controlled by Achankovil and Thenmala shear zones. On entering the midland, diverge from each other, the former took a northwestward course and the latter a southwestward course, leaving a triangular area without the old drainage. The streams that developed to drain this area are responsible for forming several small estuaries south of 9°20[1] latitude. These have been called paleoestuaries by Nair and Padmalal (2003). The Achankovil river, on entering the lowland, takes a west–northwestward course due to the uplift indicated above and joins the Vembanad lagoon after joining the northward-flowing Pamba a few kilometers south of its confluence with the lagoon. The scenario of the uplift and the river course diversion and the formation of the NNE trending ridge–runnel system and paleoestuaries is a working hypothesis. It is subject to confirmation by detailed multiproxi studies.

After ~6.0 ky BP receded initially, the sea level created the NNE trending ridge–runnel system in the uplifted area. It appears very likely that after this, the entire coast witnessed identical tectonic–depositional processes resulting in the formation of the coast–parallel ridge–runnel system. From texture and mineralogy, it appears that most of the sand deposit that constitutes the ridges are of beach origin with some modification by wind action. There is virtual unanimity in the view that the ridge–runnel system developed due to Holocene marine regression (Nair and Padmalal, 2003, 2004). If regression is the process responsible for the ridges, those that are farthest from the coast should be the oldest. In the SKSB, the ridge system is found along a long stretch (Fig. 7.3). Since carbon datable material is seldom found in the sandy sediment, we could not date them. In this context, it is worth mentioning that in the Parayakadavu borehole, 6.0 m thick beach sand with heavy minerals is underlain by clayey sand with molluscan fauna of obvious lagoonal origin. A shell sample from this has given a carbon date of 4.6 ky BP. Therefore, it is reasonable to assume that the beach ridge is younger than this date. In the development of the coast, parallel ridge–runnel system, the coastal uplift referred to above seems to be of some importance. As seen along the coast between 9°45[1] and 10°12[1] latitudes, the gaps in this system are probably attributable to minor subsidence of the uplifted coastal segment influenced by the ENE trending faults. Such subsided areas host coastal swamps and marshes mostly with mangrove vegetation.

7.6.2.2 *Deficient rainfall/aridity and sea level as at present*

One of the most significant climatic events of Holocene is the reduction of rainfall starting from ~6.0 ky BP. There is no consensus on the exact time of initiation of this climatic spell, but it is within the range of ~5.0–4.0 ky BP. In the case of Southern Kerala, there is unmistakable evidence to state that the dry spell started at 5260 ± 120 yrs BP. This has been possible by dating the wood at the abrupt contact between a carbon-rich black clay and a ferruginous clayey sand in Pangod quarry in the Kallada river basin. The sections younger than this in the boreholes in SKSB clearly show that as compared to the old sections a progressive reduction of terrestrial inputs and a corresponding increase in the marine/lagoonal fauna and flora has taken place.

It appears almost certain that a major part of the Vembanad lagoon had started to be silted up by ~4.0 ky BP. As mentioned earlier, the lagoon is separated from the sea by landforms such as barrier beach ridges, coastal swamps, marshes, and mangroves. These were adjusted to the higher sea level until about 5.0 or 4.5 ky BP. These probably had many gaps allowing more free communication between the sea and the lagoon. With the fall in sea level, this barrier system became more continuous. The most direct result of this is the change in the ecology of the lagoon. With the reduced communication with the sea, several parts of the lagoon probably witnessed more freshwater influenced ecology. This aspect would have become further accentuated by the undulating bottom topography of the lagoon. Numerous depressions with a wide variety of salinity and other ecological conditions developed (Fig. 7.3). These depressions received mainly sandy sediments essentially eroded from the old wider river channels by the incising active narrow channels that formed due to reduced rainfall. The sandy substrates with near-fresh water conditions resulted in ecology, which suited mollusks proliferation like *Villorita*. The shell deposits such as those encountered in the Kaniyamkulam borehole dated 3.4 ky BP owe their origin to this specialized ecology. Rajendran et al. (1989) have reported date of 3.1 ky BP for the shell bed in the lagoon at Mohamma ~4.0 km north of Alappuzha. Another date of 3.7 ky BP reported by these authors pertains to a sample of shells near Vechoor, along the lagoon's eastern bank (Maya et al., 2021).

There is evidence for the fact that even these specialized and aerially restricted environments did not last long. In all probability, these localized depressions got completely dried up, exposing the thin sediment layer overlying the shell beds to subaerial exposure and oxidation. The yellow clay/sand acting as a tracer bed for the occurrence of shell beds below, as reported by Murthy et al. (1976), is significant in this context.

After ~2.5 ky BP, the climate started to become wetter to return to the present state. The monsoonal influx of water gradually increased, raising the water level in the lagoons higher than that of the sea level. This would have led to the breaching of the barrier system along several stretches. The barrier possibly would have looked like a chain of islands. With the stabilization of climate and deposition of more sands along with the barrier system, the present geomorphology of the coast emerged (Fig. 7.3).

7.6.3 Holocene deposits and the society

Holocene events are responsible for the generation of mineral resources like the beach placers, tile-brick clays, and glass sand. The aggregate grade river sand found in the active channels and flood plains are another example. These present active channels together with the river terraces constituted the channel width of rivers during the Holocene Climatic Optimum. Now that mining has exhausted the entire river sand, and all the river terraces are under intense cultivation, this resource's replenishments would remain a mirage. Most of the abandoned river channels in the midland are the locales for mining tile-brick clays. It is common to find that there is a variable thickness of river sand underneath the clay. This sand is facing the threat of illegal mining. The mining of clay and the sand underneath has the serious potential of causing severe environmental problems. The sands in the ridge–runnel system constitute excellent aquifers, which are tapped by lakhs of households. With the reclamation of the runnels, the principal source of recharging of this aquifer system will be nonexistent in a major part of the area covered by this landform.

7.7 Learning and knowledge outcomes

The SKSB is a curvilinear landward extension of the offshore sedimentary basin in the coastal plain between Kollam and Kodungallur (8°45' and 10°15'N latitude) in the southwestern part of India. In the absence of outcrops, stratigraphic information has been derived from the subsurface sediments. Holocene sedimentary sequences are clay-dominated in the Vembanad lagoon and its adjoining areas. The Holocene Epoch witnessed a rise in sea level from approximately 80.0 m below the present position to 4.0–6.0 m above. The rainfall varied from highly excessive (~ 2–3 times the current rate) to deficient. There have been noticeable tectonic movements, which have influenced the sedimentary environments to a considerable extent. These three factors acting in tandem have produced a complex combination of situations affecting the ecology and shaping the landforms. Most of the coastal landforms have been developed or substantially modified due to Holocene climate and sea-level changes. Similarly, the drainage system has undergone drastic modifications due to excessive rainfall coupled with rising and fall of sea level. There are pieces of evidence to prove that tectonic movements too have influenced drainage systems and landforms. The evergreen forests were converted into wetlands, water bodies got shrunken, aquifers and groundwater resources were severely affected, and sensitive ecosystems like the mangroves and the freshwater swamps of *Myristica* are becoming relics as a result of the changing scenario brought in by the hydrological processes and climate variability in the past 11 ka years.

Holocene events are also responsible for generating mineral resources like beach placers, tile-brick clays, and glass sand. The role of sandy aquifers is worth mentioning as it is a significant aspect of the environmental process. The sand of the ridge–runnel system and various other landforms supply water to lakhs of households all over Kerala's coastal areas. These aquifers are recharged from infiltration and the runnels, which are underwater for 9–12 months in a year. Given the fast-declining areas under paddy cultivation, a significant recharge source of the life-sustaining aquifer is likely to face problems sooner than later. A major part of the land, water, and mineral resources are the gifts of Holocene events. It is essential that ordinary people in general and the younger generation, in particular, are given at least a glimpse of the change in sea level, climate, and tectonic of the Holocene Epoch and their role in molding the present landscape and ecology.

Acknowledgments

The authors are indebted to Late Dr. K.M. Nair for getting them involved in the subject of Quaternary Geology of Kerala and therefore dedicate this contribution to his memory. Sincere thanks are expressed to Kerala State Council for Science, Technology, and Environment, Thiruvananthapuram, for granting a project through which we could do fieldwork and sample collection made for present contribution. The comments and suggestions of the reviewers helped improving the original manuscript considerably. D.P. thanks the Director, National Centre for Earth Science Studies, Thiruvananthapuram, for logistic support. R.B.L. and K.P.N.K. thank Director, Agharkar Research Institute, Pune, for help and encouragement.

References

Aitchison, J.C., Davis, A.M., When did the India-Asia collision really happen? Gondwana Res. 4 (2001) 560–561.

Balakrishnan, T.S., Major tectonic elements of the Indian subcontinent and contiguous areas, Geol. Soc. India Mem. 38 (1997) 155.

Balakrishnan, T.S., Tectonics of western India inferred from gravity patterns and geophysical exploration, Geol. Soc. India Sahyadri Mem. 47 (1) (2001) 67–68.

Bilham, R., Blume, F., Gaur, V.K., Geodeltic constraints on the translation and deformation of India: implications for future Himalayan earthquakes, Curr. Sci. 74 (1998) 213–224.

Bryson, R.A., Swain, A.M., Holocene variations of monsoon rainfall in Rajasthan, Quat. Res. 16 (2) (1981) 135–145.

Chattopadhyay, S., Emergence of central Kerala coastal plains: a geomorphic analysis, in: Tandon, S.K., Thakur, B. (Eds.), Recent Advances in Geomorphology, Quaternary Geology and Environmental Geoscience: Indian Case Studies, Manisha Publications, New Delhi, 2002, pp. 287–298.

El Wakeel, S.K., Riley, J.P., The determination of organic carbon in marine muds, J. conseil int. pour l'explor. 22 (1957) 181–183.

Ghienne, J.F., Schuster, M., Bernard, A., Duringer, P., Brunet, M., The Holocene giant lake Chad revealed by digital elevation models, Quat. Int. 87 (2002) 81–85.

GSI, Summary of discussions, conclusions and recommendations of the workshop on Tertiary and Quaternary sedimentary formations of Kerala. Rec. Geol. Surv. India, 112 (5) (1976) 8–14. Geological Survey of India.

Hashimi, N.H., Nigam, R., Nair, R.R., Rajagopalan, G., Holocene sea level fluctuations on western Indian continental margin: an update, J. Geol. Soc. India 46 (1995) 145–152.

Jackson, M.L., Soil Chemical Analysis, Prentice Hall of India Pvt. Ltd., New Delhi, 1967, p. 498.

Kale, V.S., Gupta, A., Singhvi, A.K., Late Pleistocene–Holocene palaeohydrology of monsoon Asia, J. Geol. Soc. India 64 (2004) 403.

Lewis, W.S., Practical Sedimentology, Hutchinson Publishing Co., Pennsylvania, 1994, p. 227.

Maya, K., Padmalal, D., Vandana, M., Vishnu Mohan, S., Vivek, V.R., Limaye, R.B., Kumaran, K.P.N., Holocene changes in fluvial geomorphology, depositional environments and evolution of Coastal Wetlands—a multiproxy study from Southwest India, in: Kumaran, K.P.N., Padmalal, D. (Eds.), Holocene Climate Change and Environment, Elsevier, 2021. In this issue.

Merh, S.S., Quaternary sea-level changes along Indian coast, Indian Natl. Sci. Acad. 58a (5) (1992) 461–472.

Murthy, M.V.N., Mazumder, S.K., Bhaumik, N., Significance of tectonic trends. In the geological evolution of the Meghalaya uplands since Precambrian, Geol. Surv. India Misc. Publ. 23 (1976) 471–484.

Nair, K.M., Padmalal, D., Quaternary sea-level oscillations, geological and geomorphological evolution of South Kerala Sedimentary Basin. In: PCR submitted to DST, Govt. of India, Thiruvananthapuram, 2003, pp. 1–60 (unpublished report).

Nair, K.M., Padmalal, D., Quaternary geology and geomorphic evolution of South Kerala Sedimentary Basin, west coast of India, in: Ravindrakumar, G.R., Subhash, N. (Eds.), Earth System Science and Natural Resources Management, CESS, 2004, pp. 69–82.

Nair, K.M., Rao, M.R., 1980. Stratigraphic analysis of Kerala Basin. In: Proceedings of Symposium of Geology—A Geomorphology of Kerala. Geological Survey of India, pp. 1–8 Special Publication No. 5.

Nair, M.M., Quaternary coastal geomorphology of Kerala, Indian J. Geomorphol. 4 (1999) 51–80.

Nair, M.M., Structural trend line patterns and lineaments of the Western Ghats south of 13′ latitude, J. Geol. Soc. India 35 (1990) 99–105.

Najeeb, K.Md., Ground Water Exploration in Kerala Region as on 31-3-1999, Central Ground Water Board, Kerala Region, Trivandrum, 1999 (Unpublished Report).

Pandarinath, K., Shankar, R., Yadava, M.G., Late Quaternary changes in sea-level and sedimentation rate along the sea coast of India: evidence from radiocarbon dates, Curr. Sci. 81 (2001) 594–600.

Pawar, S.D., Venkataraman, B., Mathai, T., Mallikarjuna, C., Systematic Geological Mapping Around Cherthalai-Vaikom-Alleppey -Kottayam and Panthalamareas in Parts of Alleppeyand Kottayam Districts, Kerala State, Geological Survey of India, 1983 Unpublished Report.

Pejrup, M., et al., Triangular diagram used for classification of estuarine sediments: a new approach, in: Boer, P.L., et al. (Eds.), Tide Influenced Sedimentary Environments and Facies, Reidal Publishing Co., New York, 1988, pp. 289–300.

Petit-Maire, N., Past global climatic changes and the tropical arid/semi-arid belt in the North Africa, J. Coastal Res. Special Issue 17 (1995) 87–92.

Picard, M.D., Classification of fine-grained sedimentary rocks, J. Sedimentol. Petrol. 41 (1971) 179–195.

Raha, P.K., Sinharoy, S., Rajendran, C.P..

Rajagopalan, G., Sukumar, R., Ramesh, R., Rajagopalan, G., Late Quaternary vegetal and climatic changes from Tropical peats in southern India, an extended record upto 40,000 yr BP, Curr. Sci. 73 (1997) 60–63.

Rajendran, C.P., Rajagopalan, G., Narayanaswamy, B., Quaternary geology of Kerala: evidence from radiocarbon dates, J. Geol. Soc. India 33 (1989) 318–322.

Subrahmanya, K.R., Active intra plate deformations in South India, Tectonophysics 262 (1996) 231–242.

Varadarajan, K., Balakrishnan, M.K., Kerala coast: landsat's view, Proceedings of Symposium of Geology and Geomorphology of Kerala, Geological Survey of India, Special Publication 5, 1980, pp. 67–68.

CHAPTER

8 Reading source, Holocene climate change and monsoon signatures in surface and core sediments from western Bay of Bengal

Ganapati N. Nayak, Purnima Bejugam, Janhavi Kangane

Marine Sciences, School of Earth, Ocean and Atmospheric Sciences, Goa University, Goa, India

8.1 Introduction

Understanding climate change on Earth is gaining importance as humans are worried about what will happen in the future. Before predicting future climate changes, it is essential to know (a) whether Earth had experienced such changes in the past, and if yes (b) what factors control these changes? The scientific results published in recent years on climate change based on isotope measurements, microfossils and others have largely answered the first question and proved that Earth has experienced climate change in the past (Ahmad et al., 2008; Rashid et al., 2011; Panchang and Nigam, 2012; Raza et al., 2014; Bejugam and Nayak, 2019a). The beginning of civilizations, settlement of people and societies, the growth of urban settlements and agriculture have all been related to changes in climatic conditions (Singhvi and Kale, 2010), directly or indirectly. In recent times largely on account of the impact of global warming and climate change on food production and lifestyle of the people, understanding climate change has become more important. The main factor is the solar energy that drives Earth's climate. The Earth on an average receives 430 quintillion joules (18 zeros after 430) of energy from the Sun. Solar radiation includes visible light, ultraviolet light, infrared, radio waves, X-rays, and gamma rays. This energy is redistributed within the atmosphere, hydrosphere, lithosphere, and biosphere. The three elements of the Milankovitch (Orbital) cycles, namely eccentricity, obliquity, and precession, contribute to the variation in energy received from the Sun to the Earth and therefore contribute to climate change. The heat or the energy is also added to Earth's surface by volcanic eruptions and heat transfer through conduction and convection within the Earth. Further, the climate of a location is affected by the land topography, elevation, distance from the sea, distance from the equator/polar, the direction of winds, ocean currents, water bodies, vegetation, and precipitation. The Earth's temperature is a balancing act of the radiation budget and higher temperature means that heat waves are likely to occur more often and possibly last longer. Global warming refers to the long-term rise of the planet's temperatures and is an aspect of climate change. It is caused by increased concentrations of greenhouse gases in the atmosphere, mainly from burning fossil fuels, deforestation and the type of farming that is from human activities. Carbon dioxide, a by-product of fossil fuel combustion, is the principal greenhouse gas contributing to global warming along with other gases namely methane, nitrous oxide, and several industrial gases. All these are important contributors to climate change.

Holocene Climate Change and Environment. DOI: https://doi.org/10.1016/B978-0-323-90085-0.00020-6

The paleoclimatology is a multidisciplinary subject, includes geology, physics, chemistry, biology, archaeology, and related fields that study earth, ocean, and atmosphere. The Earth has experienced many cold and warm climates in the past and the most recent warm period is referred to as Holocene. The Holocene started after a prolonged Ice Age called Last Glacial Maxima 22–18 kya. The Holocene was a glacial retreat/warming with many short cooler phases and was as follows: (1) 10,000–8500 BC—Younger-Dryas—cold period; (2) 5000–3000 BC—Holocene Optimum/Climatic Optimum—warm period—the average global temperature was maximum, that is, 1–2°C higher than that of today. The great ancient civilizations of the Earth flourished during climate optimum; (3) 3000–2000 BC—cold period—caused a large decrease in sea level; (4) 2000–1500 BC—warm period; (5) 1500–750 BC—cold period; (6) 750–150 BC—warm period; (7) 150 BC–900 AD—cold period—cooling began during the Roman Empire (150 BC–300 AD) but 600–900 AD was referred to as "The Dark Ages" as average global temperatures were extremely low; (8) 1100–1300 AD—the Medieval Warm Period also called Little Climatic Optimum; (9) Up to 1400 AD—a cooler and more extreme weather; (10) 1550–1850 AD—cold period—Little Ice Age (LIA); and (11) From 1900 AD warm period. It is important to note that global average temperatures were less than 1 degree warmer during the Medieval Warm Period and less than 1 degree cooler during the LIA than the temperature of 1900 AD.

The Indian Summer Monsoon (ISM; June–September) involving large exchanges of mass and energy from the ocean, atmosphere, and continents is the strongest climate expression of Earth and the monsoon will change in the face of global warming (Turner and Annamalai, 2012). Saha et al. (1979) stated that the summer monsoon affects the lives of millions of people who are dependent on monsoon rains for agriculture, hydroelectric generation, industrial development, and other basic human needs. Singhvi and Kale (2010) stated that agriculture in India is largely monsoon rainfall dependent and hence it is important to understand the future prediction of monsoon. The pressure gradient created by the solar heating causes cross-equatorial flow and affected by Earth's Coriolis force is responsible for the winds of the southwest (SW) summer monsoon. While blowing from ocean to land these winds pick up moisture from the warm Indian Ocean and bring the ISM or SW monsoon. Around September pressure gradient reversal occurs, the cooler high-pressure air from land starts moving toward the ocean during this period with lower pressure giving rise to the northeast (NE) monsoon. Therefore, the monsoon in the Indian Ocean is due to the shifting of winds seasonally from the SW in summer and the NE in winter. Most parts of India receive rains of the SW monsoon while Tamil Nadu state receives NE monsoon rains. Prell and Kutzbach (1992) and Higginson et al. (2004) have reported that the Indian SW and NE monsoons have changed their relative strengths with time and exhibited a relationship with cold (stadial) and warm (interstadial) events of the North Atlantic. Sirocko et al. (1996), however, related Indian monsoon to ENSO of the Pacific at centennial time scale. The ISM was related to the major global climatic events such as the Roman Warm Period, Medieval Climate Period, and the LIA (Naidu et al., 2020). The weaker ISM was reported during the LIA (Naidu et al., 2020) and the strengthening of the Monsoon during the retreating phase of LIA (Panchang and Nigam, 2012). However, the ISM response to forcing factors and climate variables has not yet been fully explored and understood.

The indirect forms of evidence, for climate-dependent natural processes/parameters, that can be used to infer climate are called proxy. Proxies are stored in different archives namely tree-rings, ice-cores, corals, ocean, or lake sediment cores. The sediment formed by the physical and chemical weathering of rocks present on the Earth's surface is transported through the glacier, air, or water medium and finally deposited. The rivers are carriers of material that include suspended and bed sediments, organic matter and nutrients from land to ocean. The lithogenic and biogenic elements are transferred from the

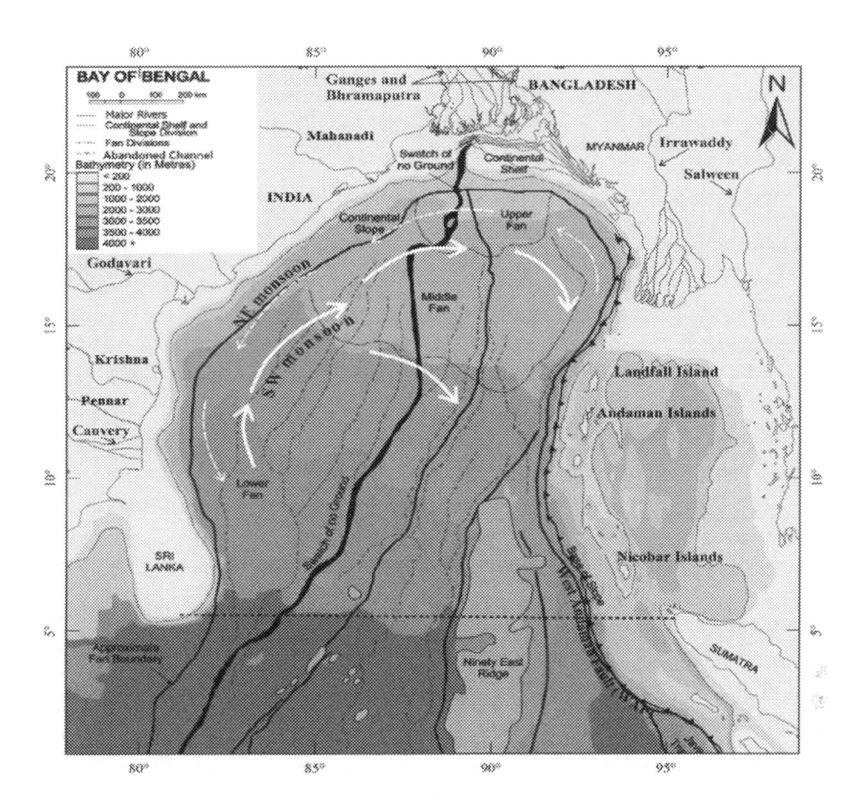

FIGURE 8.1

Seafloor morphology, Bay of Bengal (Bejugam, 2018).

terrestrial environment to the ocean which is the primary connection in their global cycle (Billen, 1993). The rates of sediment deposition depend on parent rock type, weathering activity, topography, the rate of sediment flux, tidal activity, pH conditions, biological activity, and environmental conditions (Xu et al., 2011; Tripathi, 2017). The sediment texture, diatoms, foraminifera, pollen, ice cores, tree rings, isotope geochemistry, the concentration of trace elements, and magnetic susceptibility are the commonly used proxies.

The continental shelf and continental margin preserve sediments and paleoclimate records and therefore form an important area to study the source, processes, and depositional environments. The Bay of Bengal, in the northeastern Indian Ocean a large and relatively shallow embayment, lies roughly between 5° and 22°N and 80° to 90°E and occupies an area of about 2,173,000 sq km (839,000 square miles) is one of the known locations for paleoclimate study. The Bay of Bengal seafloor morphology is presented in Fig. 8.1. The Bay annually receives a large amount of freshwater and sediment (2000 m tons) during the NE and the SW monsoons from the Himalayan and the Indian peninsular rivers, respectively. The arid dry conditions prevail in the Himalayan regions and nearly, temperate humid conditions prevail across the Indian subcontinent. The area is sensitive to changes in temperature and monsoon intensity due to this difference. These changes are recorded in the sediments of the Bay of

Bengal over the years. Further, the large sediment influx to the bay makes it an ideal location for deciphering the signature of the paleoclimate in the high sedimentation area.

Several researchers have studied different proxies in the sediments of the east coast of India and the western Bay of Bengal (Kolla and Rao, 1990; Chauhan et al., 2005; Pattan et al., 2008; Panchang and Nigam, 2012; Bejugam and Nayak, 2019a). Different paleoclimatic proxies such as detrital and organic material, clay minerals, carbonates, metals, pollens have been studied in the Bay of Bengal sediments (Wei et al., 2003). Based on the study carried out on suspended sediment transport and dynamics Barua et al. (1994) revealed that sediment transport is a function of tidal energy within the Ganges–Brahmaputra river system. Chauhan et al. (2005) reported different dispersal patterns for the terrigenous flux of the Ganges–Brahmaputra rivers during SW and NE monsoons. Goldberg and Griffin (1970) reported that river-borne solids control the clay mineral distributions in the Bay of Bengal and Rao et al. (1988) stated that clay minerals represent their sources but their preservation is controlled by the energy conditions. Kolla and Rao (1990) have related the presence of high smectite, sillimanite, garnet in the sediments to supply from peninsular Indian rivers and the presence of high illite, hornblende, epidote to supply from the Himalayan rivers. Raman et al. (1995) have stated that the clay mineral of the distal Bengal Fan is controlled by the relative rates of sediment supply from the Himalayan and Deccan sources. Reddy and Rao (2001) reported an increase in montmorillonite proportion during Holocene as the Ganges derived sediments are diluted by supply from the peninsular rivers. Chakrapani and Subramanian (1990) using the geochemical proxies of the Mahanadi basin observed that reworking caused the finer sediments to be deposited downstream. Raj and Jayaprakash (2008) revealed an association of higher content of trace metals with higher finer grain size and organic carbon. Pattan et al. (2008) based on geochemistry of surface sediments confirmed the minor contribution of Ganges–Brahmaputra rivers to the Krishna–Godavari basin compared to peninsular rivers.

Crowley et al. (1998) observed that contribution of Himalayan and Indian subcontinent sources to distal fan sedimentation varied with time and uplift, weathering and erosion rates, eustatic sea-level changes and switching of fan channels controls sediment supply. Kuehl et al. (1989) employing sedimentological and geochronological proxies reported that the highest sediment accumulation from the Ganges–Brahmaputra river takes place near the head of the Swatch of No Ground and suggested that the Swatch of No Ground is a major conduit for the seaward transport of sediments from the Bengal shelf. The variations in paleoclimate and monsoon variability in the Bay of Bengal from the Last Glacial Maximum (LGM) to the late Holocene period was explained by Govil and Naidu (2011), Rashid et al. (2011), Ponton et al. (2012), and Phillips et al. (2014). Govil and Naidu (2011) reconstructed SSS and SST of Bay of Bengal to understand the rainfall variability associated with SW monsoon over the past 32 ka. Their results showed that during LGM the western Bay of Bengal was ~3°C cooler than today. Rashid et al. (2011) studied the Late Glacial to Holocene ISM variability based on sediment records from the Bay of Bengal and using SST and $\delta^{18}O$ suggested that seawater was colder during the LGM compared to the early Holocene. Ponton et al. (2012) reconstructed the Holocene paleoclimate during Core Monsoon Zone off Godavari river and reported that Core Monsoon Zone aridification intensified in the late Holocene through a series of sub-Millenial dry episodes. Phillips et al. (2014) used multiproxy records to reconstruct monsoon induced variations in productivity and lithogenic sediment flux since past 110 ka from a northern Bay of Bengal core. Many researchers using various proxies of sediment cores have carried out the study in the Bay of Bengal to reconstruct the paleomonsoon intensity (Krishnamurthy and Goswami, 2000; Pothuri et al., 2014), paleocirculation (Gourlan et al., 2010), paleoproductivity (Prajih et al., 2018; Bejugam and Nayak, 2019a), and changing weathering conditions (Symphonia and Nathan, 2018; Bejugam and Nayak, 2019b; Kangane et al., 2019). Attempts also have been made to understand the monsoon precipitation by using oxygen isotope values of waters in the Bay of Bengal since the last glaciations (Ahmad et al., 2008; Rashid et al., 2011; Raza et al., 2014).

8.1.1 Scope of the study

Understanding the response of proxies off major rivers delivering huge amounts of freshwater and sediment to the ocean is important and is the scope of this article. This chapter describes the results of the analysis of sediment multicores and gravity core sampled and analyzed for sediment components namely sand, silt clay; clay minerals, and metals from the western continental shelf region in the Bay of Bengal to understand the source, processes, depositional environments and changing climatic conditions including monsoons.

8.2 Methodology

8.2.1 Sampling area

The western continental shelf and slope sediments of the Bay of Bengal off Hooghly, Mahanadi, Vamsadhara, Godavari, Krishna, and Pennar river mouth regions constitute the present study area. The continental shelf region here is with variable width, flat and broad near-shore, and steeper in deeper water depths. The western Bay of Bengal is prone to cyclonic storms in the postmonsoon season during prevalent NE monsoon winds resulting in high rainfall in coastal regions (Lin et al., 2009). However, the bulk of the precipitation is received during the SW monsoon (June–September) intensifying sediment influx into the coastal Bay of Bengal (Nair and Ittekkot, 1991). The surface hydrographic circulation in the Bay of Bengal gets complicated due to the seasonal reversal in wind directions and high freshwater influx (Varkey et al., 1996). The high freshwater influx reduces the salinity of the water off the river mouths (Suokhrie et al., 2018) and forms the stratified freshwater and saline water layers making a barrier for vertical mixing.

The Hoogly river along its course consists of Archean gneisses, sandstones, feldspathic quartzites, metamorphosed Archean–Proterozoic sediments, and recent alluvium drained from Ganges (Singh et al., 2007). The Mahanadi river emerges from the Eastern Ghats and the bulk of the geology along its course is composed of felsic rocks of khondalites, charnockites, granites, gneisses and the limestones, sandstones, and shales of the Gondwanas (Chakrapani and Subramanian, 1990). Krishna and Godavari rivers flow across the Indian peninsular shield constituting Precambrian, Deccan basalts, and Dharwar formations. The detailed geology of the river basins is presented in Fig. 8.2. The Peninsular rivers drain enormous sediment load onto the western continental shelf of the Bay of Bengal under the influence of NE and SW monsoons. The sediment flux data from 1986 to 2006 (CWC—Central Water Commission) for Mahanadi river is $30.6*10^6$ tons/yr, Godavari river $170*10^6$ tons/yr, Krishna river $9*10^6$ tons/yr, and Pennar river $3*10^6$ tons/yr (Panda et al., 2011) while for Hooghly river a tributary of the Ganges it is about $65*10^6$ tons/yr (Mukhopadhyay et al., 2006).

The high sediment accumulation rate during the late Holocene in the Bay of Bengal was reported. The rates measured using the ^{210}Pb method, Off Hooghly, was ~2.0 cm/year (Kuehl et al., 1989; Suckow et al., 2001). Off Godavari it was 0.25 cm/yr for cores between 30 and 100 m water depths (Kalesha et al., 1980; Kiran et al., 2015). In deeper water depths of 600–1400 m, rates measured by the

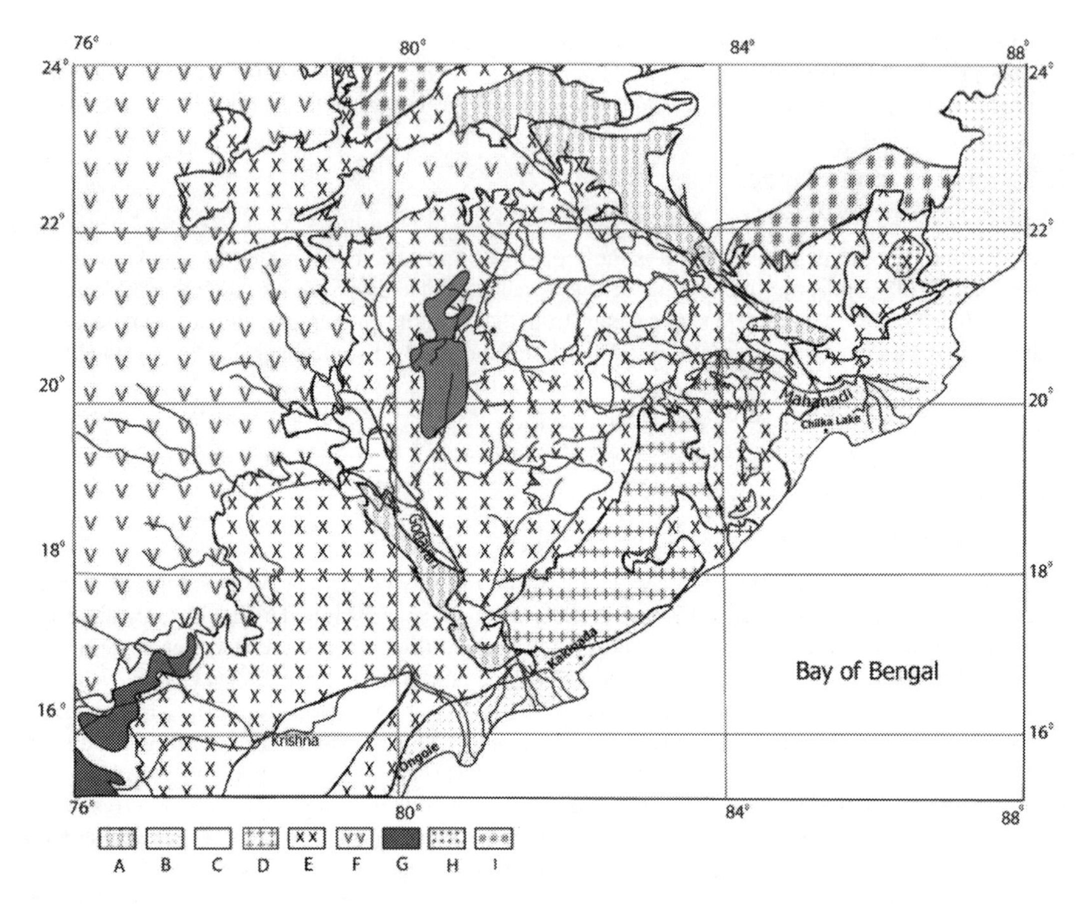

FIGURE 8.2

Geological map of Mahanadi, Godavari, and Krishna drainage basins. A, Gondwana sediments; B, Quaternary sediments; C, Mid–Late Proterozoic sediments; D, Archean charnockite and khondalite; E, Archean–Proterozoic gneissic complex; F, Deccan basalt; G, Early Proterozoic ELSST073-08-granites; H, Archean–Proterozoic Singhbhum metamorphic complex (after Mazumdar et al., 2015).

^{14}C method (Mazumdar et al., 2012, 2014; Usapkar et al., 2016) were 0.24 cm/yr off Mahanadi river and 0.34 cm/yr off Krishna river. In these studies, the dating was carried out on the foraminiferal species *Globigerina ruber* and *Globigerina sacculifer*.

8.2.2 Sampling methods

The sediment cores used for the present study were collected onboard RV Sindhu Sankalp (Cruise No. 35) in June 2012 off Hooghly using spade corer and onboard RV Sagar Kanya (Cruise No. 308) in January 2014 using multicorer off Mahanadi, Vamsadhara, Godavari, Krishna, and Pennar rivers and gravity corer off Mahanadi river. The cores off Hooghly river were collected from the western side of

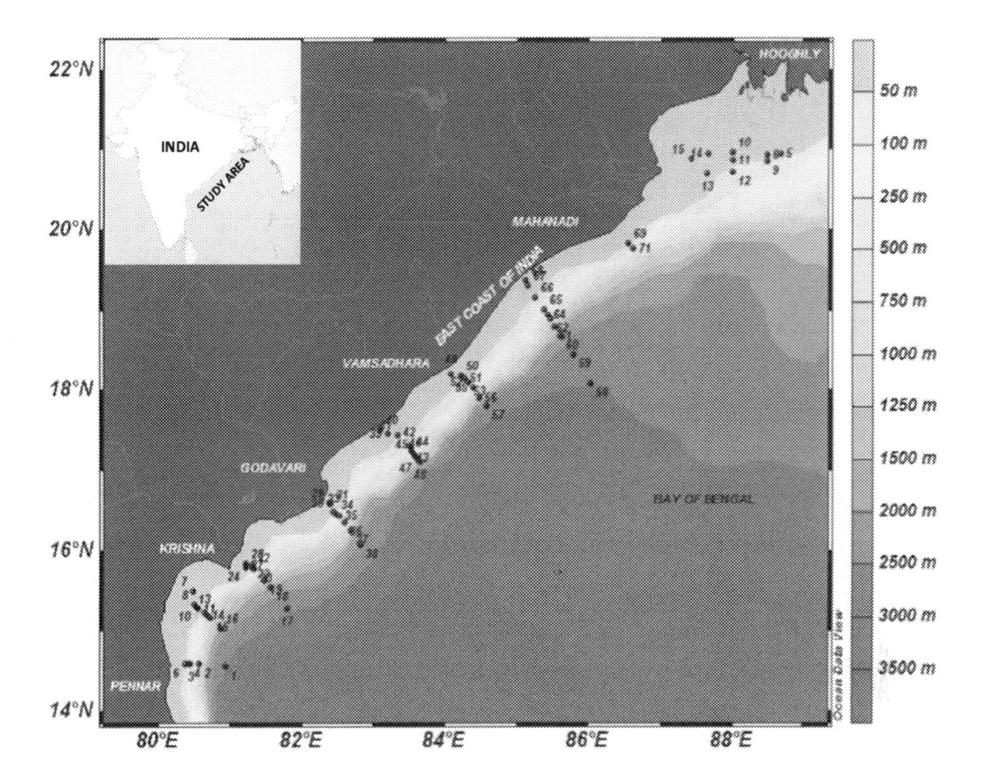

FIGURE 8.3

Sampling locations of sediment cores collected from western Bay of Bengal off Hooghly, Mahanadi, Godavari, Vamsadhara, Krishna, and Pennar River mouth regions (Bejugam, 2018).

the "Swatch of no ground" submarine canyon. From the several cores collected from 25 to 2500 m water depths, 70 surface samples, 25 short cores (Fig. 8.3), and one gravity core collected at 19°82[1] N and 86°56[1] E at 101.9 m water depth were utilized for the present study. The samples were collected along nearly latitudinal transects off major rivers and the water depth at the sampling location was recorded with the help of an echo-sounder. The multicores (Fig. 8.4) and gravity core were subsampled at 1cm interval onboard, preserved in cold storage and alternate samples were used for analysis.

8.2.3 Laboratory analysis

In the laboratory, the sediment sample was oven-dried and a part of the sample was used for grain size and clay mineral analysis. Another part of the sediment was ground into a fine powder using an agate mortar and pestle and used for the analysis of inorganic elements. For sediment grain size (sand: silt: clay) analysis, samples were processed to remove salinity and then treated with 10 ml of 10% sodium hexametaphosphate to dissociate clay particles and then to oxidize the organic matter using 5 ml hydrogen peroxide. The pipette method based on Stoke's settling velocity principle (Folk, 1968) was

FIGURE 8.4

Retrieving multicorer with sample onboard.

employed to determine the grain size of sediment samples. For the clay mineral analysis, the procedure given by Rao and Rao (1995) was adopted. On prenumbered slides, 1 ml of clay was evenly spread and these slides were gylcolated using ethyl glycol vapors for 1 hour at 100°C. The slides were later scanned on Rigaku Altima IV X-Ray diffractometer using nickel-filtered CuKα radiation from 3° to 30° $\Delta 2\Theta$ at 1.2° $\Delta 2\Theta$/min. To identify and quantify the clay minerals the procedure given by Biscaye (1965) was used.

Employing the procedure given by Jarvis and Jarvis (1985) ground sediment samples were digested to total decomposition for metal analysis. The acid mixture HF, HNO_3, and $HClO_4$ in a ratio of 7:3:1 was used to digest the sediment sample in a Teflon vessel. Along with the samples, a certified reference standard JLK-1 from the Geological Survey of Japan was also digested to test the analytical accuracy of the method for gravity core samples. Later the samples and standards were analyzed on atomic absorption spectrophotometer (Thermo Scientific- SOLAAR M6 AAS model). The average recovery for Mn was 95.3%, for Fe 96.2%, and for both Zn and Pb, it was 97.1%. For the multicore and spade core samples, the concentrations of major and trace elements (Al, Fe, Ti, Mn, Zn, Pb) were determined using Varian AA 240 FS flame atomic absorption spectrometry with an air/acetylene for Fe, Mn, Zn, and Pb; and for Al and Ti nitrous oxide/acetylene was employed at specific wavelengths. To test the analytical accuracy of the method, together with the samples, certified reference standard MAG-1 was analyzed for the same metals. The average recoveries for Al, Fe, Ti, Mn, Zn, and Pb were 88%, 94%, 95%, 95%, 96%, and 96%, respectively. The Merck chemical standards were used to calibrate the instruments and at regular intervals, the recalibration checks were performed.

Gravity core was processed for obtaining age dates. Mixed planktic foraminifera was picked at >63 μm size fraction from few sediment subsamples for measurement of AMS ^{14}C age. The dating procedure was carried out at the Centre for Applied Isotopic Studies, University of Georgia, USA.

Table 8.1 Range of sand, silt, and clay in surface sediments along transects from Hoogly to Pennar rivers.

Transect	Off river mouth	Sand (%)	Silt (%)	Clay (%)
1	Hooghly river mouth—North	3.97–94.43	1.39–56.59	3.20–42.08
2	Hooghly river mouth—South	1.43–48.94	22.58–59.67	28.48–62.64
3	Mahanadi river mouth—North	Highest at 102 m water depth	Highest at 102 m water depth	Highest at 507 m water depth
4	Mahanadi river mouth—South	0.30–42.43	6.00–25.74	44.64–64.20
5	Vamsadhara river mouth	0.46–61.14	10.34–32.93	27.76–89.20
6	Between Godavari and Vamsadhara river mouths	0.33–78.07	7.58–26.80	10.72–90.56
7	Godavari river mouth	0.27–2.70	29.05–59.57	40.16–83.30
8	Krishna river mouth	0.22–9.01	14.82–57.39	33.60–84.72
9	Between Krishna and Pennar river	0.41–4.54	18.68–24.90	34.88–80.08
10	Pennar river mouth	0.17–3.83	11.77–28.50	68.52–87.50

Calibration of ^{14}C ages was carried out using the Marine13 dataset (Reimer et al., 2013) drawn from Calib 7.1 (Stuiver et al., 2017) online calibration program; under the assumption that the top of the sediment core was undisturbed, it was assigned the year of core collection (2014). Further, considering the constant rate of sedimentation between the obtained dates, the ages of the remaining samples were calculated.

8.2.4 Data processing

The software's Grapher-8.0, Adobe Illustrator, and ODV-4.0 (Ocean Data View) were used for preparing the illustrations.

8.3 Results and discussion

8.3.1 Surface and spade/multicore sediments—source and processes

8.3.1.1 Sediment components

The range of sand, silt, and clay for each transect is provided in Table 8.1 and the data were plotted on Ocean Data View with the color bar on the right depicting the concentration (Fig. 8.5). The sediment components showed a decrease in grain size from north to south, that is, from off Mahanadi, through off Godavari and off Krishna to off Pennar river mouths. Sand content exhibited a decrease from off Hooghly river mouth to off Pennar river mouth in the continental shelf samples up to a depth of 200 m. Sand content (Fig. 8.5) was highest (94.43%) in transect 1 off Hooghly river mouth at 42 m water depth and least (0.63%) off Godavari river mouth at a water depth of 107 m. In the deeper regions in the slope and abyssal plains between water depths of 200 and 2500 m the sand content was low and

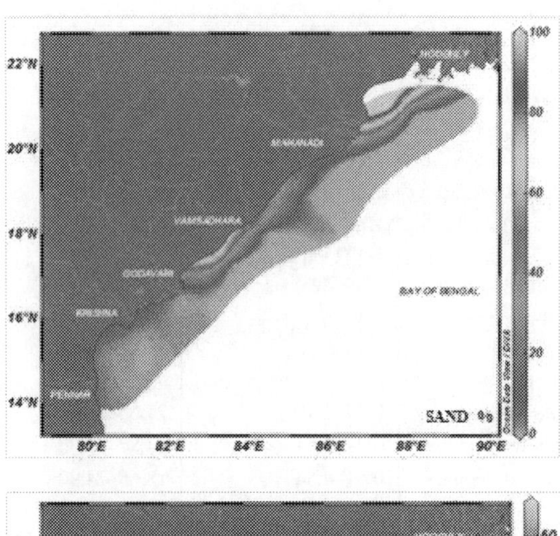

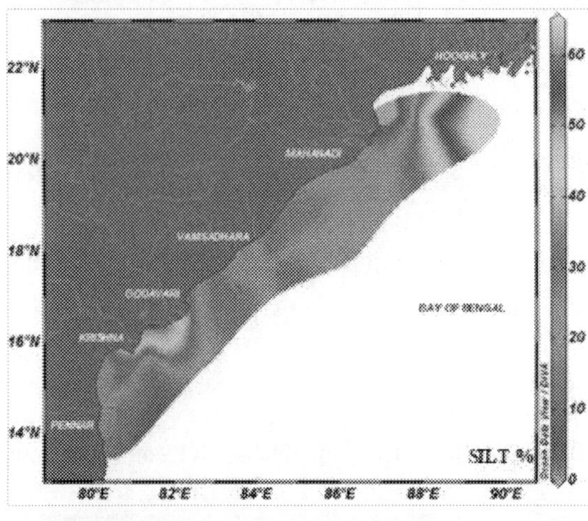

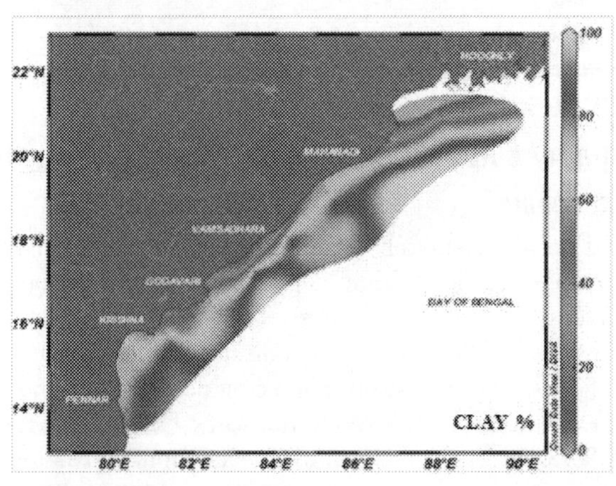

FIGURE 8.5

Distribution of sand, silt, and clay contents in the surface sediments of the study area.

varied in a small range of 0.17% and 2.98% except at 280 m water depth where the sand was slightly higher at 7.36%. Silt content (Fig. 8.5) also displayed a decreasing trend from north to south with an exception at few stations at shallower water depths up to a depth of 500 m especially off Godavari and off Krishna river mouth regions where its concentration was higher. Silt content was highest (59.67%) in transect 2 off Hooghly river mouth. Clay content (Fig. 8.5) increased from north to south with the lowest clay content (3.20%) off Hooghly river mouth and highest clay concentration (93.70%) at 1073 m water depth off Mahanadi river mouth. In the deeper water depths, clay content was consistently high varying between 65.04% and 93.70% in the majority of the samples. Along all transects from north to south at shallower water depths closer to the coast the clay content was low.

Further, when the average of down core values of sediment components are compared based on the water depths along the transects in the range of 31–83 m, 104–107 m, 202–260 m, 495–499 m, and 645–1005 m sand content exhibited an overall decrease from shallow to deeper water depths in the study area (Bejugam, 2018). The silt content showed an overall large range of variation in shallow regions compared to deeper regions and the clay distribution showed higher values toward deeper water depths compared to shallow depths in the study area.

The sediments released from a river basin provide useful information on their source and transportation history, and hydrodynamic conditions prevailed during the time of sediment deposition. Topography and climate in the source area are the main factors controlling the processes like weathering, erosion, and transportation, which determine the dispersal system connecting source and basin (Jackson and Nordstorm, 2011; Weltje and von Eynatten, 2004).

The grain size of sediment reveals the sediment transportation pathways, energy conditions, and the intensity of monsoon precipitation (Prizomwala et al., 2014; Basavaiah et al., 2015). Also, the grain size is related to the bathymetry of the depositing site, sediment input, and processes (Lopez, 2017). Sediment size analysis of surface and spade/multicore sediments along the western Bay of Bengal revealed that overall, sand content was low while silt and clay was the dominant sediment fraction. However, it was noted that along all transects the sand content was relatively high up to a water depth of 100 m and clay was dominant in sediments deeper than 1000 m in the continental slope region. The high silt content was noted in samples near the river mouth except off Hooghly river mouth. The sediment size, in general, decreased away from the coast and with increasing water depth in the study area. The occurrence of high sand content in the shallow sediments signifies higher hydrodynamic energy, strong wind activity, and closer source area (Babeesh et al., 2017). Furthermore, a loss of energy offshore may be unable to carry the sand toward deeper water depths explaining higher clay deposition in deeper regions. Finer sediments from the rivers may have been transported directly toward the continental slope and abyssal plain regions and deposited there. Further, grain size decreased from off Hooghly river mouth, through off Mahanadi, off Vamsadhara, off Godavari, and off Krishna to off Pennar river mouth regions indicating a north to south transport of sediments and relatively high energy environment prevailing off Hooghly and off Mahanadi river mouths which enabled the deposition of coarser material and prevented the accumulation of finer sediments. A probable loss of energy southward inhibited the transport of coarse grain sediments toward the south and facilitated the deposition of finer sediments. Kumar et al. (2010) suggested that the dominance of finer clay particles indicated low energy conditions and facilitated the deposition of finer sediments. In transects off Hooghly river mouth, overall sediment size decreased from west to east, that is, toward the "Swatch of No Ground"—submarine canyon, indicating that probable current direction was toward the east which transported the sediment. Fine-grained sediments were possibly carried by the currents formed at high energy conditions near the river mouth toward east parallel to the coast and also toward the south and southeast direction. Later these fine-grained sediments upon redistribution may have been deposited in the eastern and southeastern parts of this area. The distribution of sediment components in southern transects off peninsular rivers indicated a possible direction of transport from coastal to deeper water regions indicating sediment supply from these rivers to the bay.

The single most important climate factor in the south-east Asian regions is the monsoon and therefore, paleoclimatology in the Indian context is largely the reconstruction of the monsoon through time. During the LGM, the snow cover over the Tibetan plateau and central Asia decreased the land–ocean temperature contrast during summer and increased during winter leading to weaker SW monsoon and strengthened NE monsoon (Tiwari et al., 2009). Monsoon records from the Arabian Sea displayed similar patterns to the adjacent Asian and African regions (Duplessy, 1982; Sirocko et al., 1991; Overpeck et al., 1996) inferred from strong winter monsoon during LGM as observed by enhanced NE monsoon current during that period. This was probably because the Indian and Chinese monsoons have similar sub-Milankovitch periodicities (Sarkar et al., 2000). Unlike the Arabian Sea and the South China Sea, paleomonsoon interpretations from Bay of Bengal sediment cores are limited except few recent reports (Rashid et al., 2011; Govil and Naidu, 2011; Ponton et al., 2012; Tripathy et al., 2014). Further, the majority of the paleomonsoon records are based chiefly on foraminiferal species (*Globigerina ruber, Globigerina bulloides*), magnetic susceptibility studies, sea surface temperature and salinity but very seldom based upon lithogenous proxies such as grain size, clay minerals, and inorganic elements. These proxies are typical to terrestrial sources and are directly linked to weathering and the intensity of monsoon precipitation. The down-core sediment component variation in cores off Hooghly river mouth indicate that a change in the trend of the grain size distribution toward coarser size above 10 cm (Bejugam and Nayak, 2016a) indicating a probable change in hydrodynamic conditions due to floods, a storm, an increase in tidal current or a change in hydrodynamic conditions which might have taken place with a possible direction of the storm toward the east. In one of the earlier studies, McLaren and Bowles (1985) created a sediment transfer model and inferred that sediments tend to become finer in the direction of transport with a decreasing energy regime. A higher concentration of sand in the upper section of cores off Mahanadi and off Pennar river mouths is evidence of abundant terrestrial input in recent years probably due to higher rainfall. Also, higher sand from the bottom up to a depth of 10 cm and in some cores from the bottom up to a depth of 20 cm may be due to a flood event during higher rainfall which brought with it abundant terrestrial material. Basu (1976) emphasized that grain size is an indicator of climate change where an increase in sand fraction suggested intensification of the SW summer monsoon. The relative changes in concentration of grain size are a direct representation of monsoonal variation where an increase in coarser fraction indicates that the hinterland region received good monsoonal precipitation leading to enhanced weathering (Basavaiah et al., 2015). The velocity of the transporting medium increases during high rainfall periods resulting in enhanced entry of coarse sediments. Conversely, due to the reduction in stream velocity during low rainfall periods finer sediment influx is enhanced (Conroy et al., 2008).

8.3.1.2 Clay minerals

Off Hooghly and off Mahanadi river mouths, in all the surface samples illite was the predominant mineral followed by other clay minerals namely kaolinite, smectite and chlorite while off Godavari and off Krishna river mouths, smectite was the dominant clay mineral which was followed by kaolinite, illite, and chlorite. Off Pennar river mouth smectite was the dominant clay mineral followed by illite,

Table 8.2 Range of illite, smectite, kaolinite, and chlorite in surface sediments along transects from Hoogly to Pennar rivers.

Transect	Off river mouth	Illite (%)	Smectite (%)	Kaolinite (%)	Chlorite (%)
1	Hooghly river mouth—North	78.50–81.86	3.26–7.69	9.85–12.59	2.29–5.61
2	Hooghly river mouth—South	75.14–79.01	4.50–11.60	9.47–13.51	1.76–4.50
3	Mahanadi river mouth—North	Highest (63.03) noted at 507 m water depth			
4	Mahanadi river mouth—South	43.77–65.84	10.29–22.26	17.61–30.87	3.09–6.53
5	Vamsadhara river mouth	43.84–66.67	10.61–26.67	18.18–34.09	3.43–4.64
6	Between Godavari and Vamsadhara river mouths	30.77–72.34	5.32–40.57	17.55–35.16	3.12–6.50
7	Godavari river mouth	9.52–53.33	25.33–69.84	13.53–39.51	2.29–8.57
8	Krishna river mouth	7.41–28.92	37.35–79.84	9.61–26.24	0.87–7.50
9	Between Krishna and Pennar river	3.33–15.72	67.03–82.53	7.21–16.72	1.96–3.52
10	Pennar river mouth	14.48–28.76	51.63–70.14	11.61–18.04	2.25–5.60

kaolinite, and chlorite (Table 8.2). The variation in concentration of clay minerals when observed from north to south (Fig. 8.6) it was noted that smectite content was very low off Hooghly river mouth in the north and increased slightly off Mahanadi and off Vamsadhara river mouths, while off Godavari, off Krishna, and off Pennar river mouths, smectite increased significantly except for sample MC-06 off Pennar river mouth where smectite content was relatively low (Fig. 8.6). The lowest smectite content (3.26%) was noted at 31 m water depth off Hooghly river mouth while the highest smectite (82.53%) was observed at 76 m water depth south off Krishna river mouth. Illite concentration (Fig. 8.6) was high off Hooghly river mouth in the north and decreased southward but was the dominant clay mineral off Mahanadi and off Vamsadhara river mouths and further decreased significantly off Godavari, off Krishna and off Pennar river mouths except for sample MC-06 off Pennar river mouth where illite content was high. The highest illite content (81.86%) was observed at 31m water depth off Hooghly river mouth while the least illite content (3.33%) was observed at 275 m water depth south off Krishna river mouth. Kaolinite concentration (Fig. 8.6) was low off the Hooghly river mouth in the north and increased southward. It was also observed that in the shallow water regions off the river mouths kaolinite was enriched specifically off Godavari river mouth. Kaolinite content was highest (39.51%) at 107 m water depth off Godavari river mouth, while the least kaolinite content was observed at 500 m water depth off Krishna river mouth (south transect). Chlorite content was consistently low (Fig. 8.6) throughout the study area from north to south varying within a small range. However, off Hooghly, off Mahanadi, off Vamsadhara and up to off Godavari river mouth chlorite was slightly higher and decreased significantly off Krishna river mouth. The highest chlorite content (8.57%) was noted in sample MC-34 off Godavari

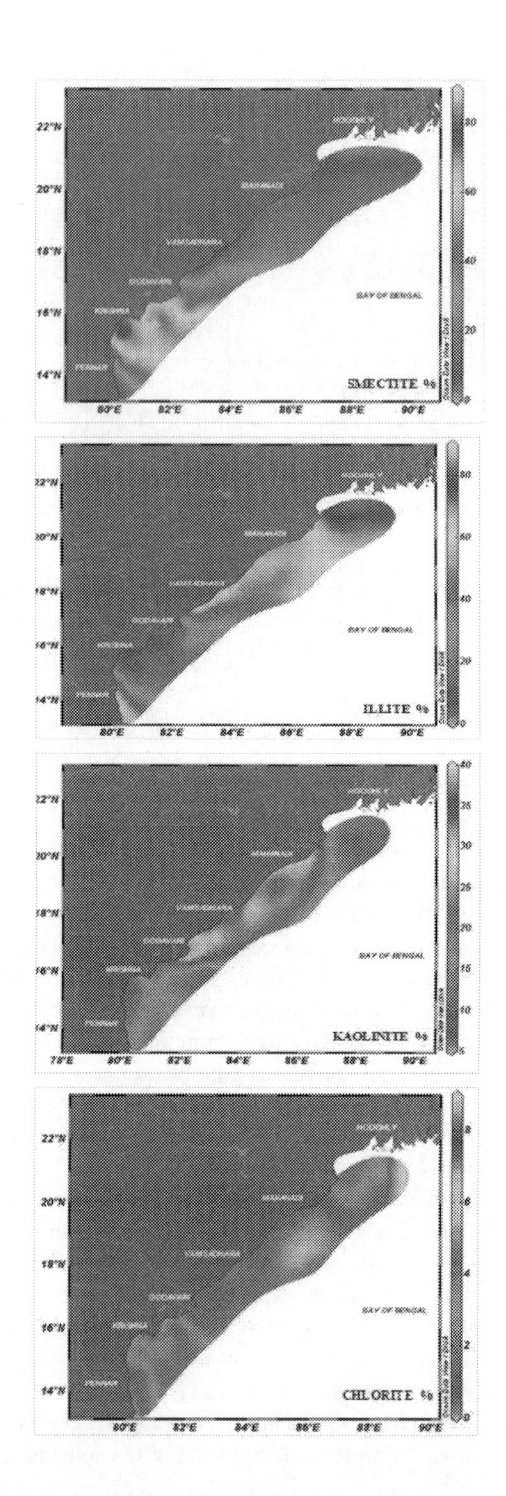

FIGURE 8.6

Distribution of clay minerals in the surface sediments of the study area.

river mouth at 744 m water depth while the least chlorite content (0.87%) was observed off Krishna river mouth at a deeper water depth of 2006 m.

When the average concentration of clay minerals was compared from north to south for a range of water depths 31–83 m, 104–107 m, 202–260 m, 495–499 m, 645–767 m, and 979–1005 m, the smectite content increased and illite content decreased from off Hooghly, through off Mahanadi, off Vamsadhara, off Godavari, off Krishna to off Pennar river mouths (Bejugam, 2018). Kaolinite content increased and chlorite decreased from off Hooghly river mouth in the north to off Pennar river mouth in the south at different depth ranges with few exceptions.

While illite was the predominant clay mineral in cores off Hooghly, off Mahanadi, and off Vamsadhara river mouths, abundant smectite characterized sediments off Godavari, off Krishna, and off Pennar river mouths. The distributions of the clay minerals in sediments off Hooghly river mouth represent large input from Himalayan sediments. Clay minerals are mainly formed by hydrolytic disintegration of primary aluminosilicates (Singer, 1984) and undergo weathering under particular climatic conditions and therefore have been successfully used as paleoclimate indicators (Wahsner et al., 1999; He et al., 2013). The Hooghly river is a tributary of the Ganges which originates in the Himalayas and drains under cold climatic conditions. On its course of transport, it erodes Precambrian formations in its upper channels and Rajmahal Traps and recent alluvium in its lower channels (Rao et al., 1988). Illite, a predominant mineral off Hoogly river, is a residual product of mechanical weathering under arid conditions with its source from the Himalayan region indicating the role of NE monsoon as a major factor in the transportation of sediments. Higher illite percentage in the sediments from transect 1 off Hooghly river mouth associated with more sand strongly supports the release of sediments from Himalaya during NE monsoon or through glacial melt due to the effect of climate change. Illite is an alteration product of muscovite which is associated with phyllites, schists, and shales; and muscovite mica which is a dominant mineral in the Himalayas must be the major source for illite.

Off Mahanadi river mouth as compared to off Krishna–Godavari river mouths high illite content obtained indicates its supply from felsic source rocks, from Archean–Proterozoic gneissic complex (Mazumdar et al., 2015) as K-feldspars are easily weathered to clay minerals (Chen et al., 2010). The concentration of Ganges transported illite decreased southward indicating its dilution by the input from peninsular rivers. Further, lower concentrations off Godavari, off Krishna, and off Pennar river mouth regions may be due to a loss of energy southward which restricts illite-rich sediments to reach toward the south. However, appreciable illite was also noted in few sections down the core off Krishna and off Godavari as the Krishna–Godavari river's flow over the charnockites and granites of Archean age. These rocks weather under warm and humid conditions can produce mixed clays (Vuba et al., 2013). Higher illite concentration was also noted in a surface sample off Pennar river mouth, however, possible conversion of smectite to illite could be ruled out as diagenetic conversion takes place at high temperatures under greater depths (Grim, 1968; Mazumdar et al., 2015).

Regional climatic conditions and variations in latitudes have a high influence on sources of marine clays (Naidu et al., 1995; Petschick et al., 1996). Smectite content increased abundantly in cores off Godavari, off Krishna, and off Pennar river mouths and supports the statement that it is considered as a low latitude clay mineral (Biscaye, 1965). High smectite off these southern rivers of the study area may be due to the release by chemical weathering of basic volcanic igneous rocks namely Deccan basalts (Raman et al., 1995; Kulkarni et al., 2015) under the prevalent humid tropical climatic conditions. Also, smectite must have been added from black cotton soils and crystalline Archean rocks. Kolla and Rao (1990), Somayajulu et al. (1993), and Pattan et al. (2008) also reported earlier high smectite in sediments off Godavari–Krishna river mouths. Minor smectite concentration observed in few sections in cores off Mahanadi river mouth may have been contributed from the weathering of eastern Deccan traps over which a tributary of Mahanadi passes through (Rickers et al., 2001). An appreciable concentration of kaolinite observed in few sections in cores off Mahanadi and off Vamsadhara river mouths may have been formed from alteration of aluminosilicate minerals like feldspar leaching from Archean granites, gneisses, charnockites, and khondalites which are the dominant rock types in the region.

Further, illite increased and kaolinite decreased at deeper than 1500 m water depth. High kaolinite concentration in shallow waters may be attributed to its higher settling velocity (Whitehouse et al., 1960; Prithviraj and Prakash, 1990). Less contribution of kaolinite from Ganges–Brahmaputra rivers was earlier attributed to possible equatorward dispersal (Chauhan and Vogeslang, 2006). Higher kaolinite content in cores located at shallower regions off Godavari river mouth may be contributed from the "Red beds" which is a major rock type near the Vishakapatnam coast as also reported by Malathi (2013) earlier. Among the clay minerals, chlorite concentration was lowest in the sediments and the values obtained were in range with the chlorite percentage reported earlier by Raman et al. (1995) and Philips et al. (2014). Chlorite content was low from cores in the region due to prevailing humid tropical conditions which makes it unstable (Thamban et al., 2002). It has therefore been observed from the clay distribution that the source of clays in sediments is largely from the weathering of rocks from the terrestrial region.

In cores off Godavari and off Pennar river mouths with increasing water depth average smectite increased which probably is due to its finer size that made smectite remain in suspension for a longer time and further even weak currents could transport it to deeper region. Salinity is low in waters near the river mouth. Illite is more stable than smectite under less saline conditions explaining the slightly higher illite content near the river mouth. Such inverse relation among clay minerals is observed in many studies worldwide due to their different settling rates (Aksu et al., 1998). Also, the flocculation of individual minerals varies for saline conditions (Rao et al., 1988; Patchineelam and De Figueiredo, 2000). Moore and Reynolds (1989) and Liu et al. (2010) reported that clay mineral distribution also depends on energy conditions. It is known that illite and kaolinite due to their coarser size, deposit under high energy conditions and smectite with relatively finer size gets deposited under low energy conditions. A higher concentration of smectite off Godavari, off Krishna, and off Pennar river mouths suggests that low energy prevailed in this region. Abundant illite off Hooghly and off Mahanadi river mouths indicates high energy conditions prevalent during the time of deposition. This hydrodynamic variation from north to south was observed in grain size variation as well.

Clay minerals in marine sediments provide information on overall climate impact (Singer, 1984). In cores off Hooghly river mouth, illite is the dominant clay mineral transported to the region probably during the NE monsoon period. However, an appreciable concentration of smectite was noted at 5 cm in all cores off Hooghly river mouth in the north which must have been brought in during intense SW monsoon winds blowing toward the region. In cores off Mahanadi and Vamsadhara river mouths as well illite was the dominant clay mineral. However, high smectite was noted in a few sections in these cores suggesting increased smectite discharge from the terrestrial region during increased SW monsoons. At 15 cm section in majority of the cores off Godavari river mouth, at 20 cm in cores off Krishna river mouth and; 25 and 5 cm in cores off Pennar river mouth smectite decreased and illite increased in concentration indicating the prevalence of NE monsoon during that period.

8.3.1.3 Metals

The bulk metals in marine sediment are useful indicators of terrigenous provenance and dispersal patterns (Kolla and Rao, 1990; Yuste et al., 2014), as well as paleoclimatic conditions in the source areas (Thiry, 2000; Thamban et al., 2002; Dou et al., 2010). Al, Ti, and Fe are the major lithogeneous contributors to the marine sediment and can be used to trace the geology, weathering history of source rocks, and diagenesis. The bulk metal concentration is also useful in understanding postdepositional processes. Further, the geochemistry of the individual elements (Rollinson, 2014) and their abundance varies conspicuously with the intensity of monsoonal precipitation (Pattan et al., 2012). The metal content in the surface sediments was analyzed for samples off Mahanadi river mouth to off Pennar river mouth (Table 8.3). In transect 3 element concentrations varied in a small range between stations.

From north to south Al content (Fig. 8.7) was largely consistent except at the shallow region off Vamsadhara river mouth where Al content was low. Al content ranged between 4.43% off Vamsadhara river mouth at 50 m water depth and 10.75% at 520 m water depth along transect 6. Fe concentration (Fig. 8.7) increased from north to south and varied between 2.95% at 55 m water depth in the transect south off Vamsadhara and 9.70% at 40 m water depth off Godavari river mouth. Ti was high in sediments off Godavari and off Krishna river mouths while off Mahanadi and off Vamsadhara river mouth in the north and off Pennar river mouth in the south Ti was low in the sediments (Fig. 8.7). Ti ranged from 0.28% off Mahanadi river mouth at 105 m water depth and 1.90% at 32 m water depth off Krishna river mouth. Mn varied in a small range from north to south (Fig. 8.7). In the intermediate water depths, Mn was however low as compared to deeper water depths. Mn ranged between 0.03% at 507 m water depth off Mahanadi river mouth and 1.20% at 2488 m water depth off Vamsadhara river mouth. Zn was largely constant from north to south (Fig. 8.7) except off Pennar river mouth and also at shallow water depths where Zn was slightly low. Pb decreased from north to south wherein off Mahanadi and off Vamsadhara river mouth transects Pb was high and decreased in the south off Godavari and off Krishna river mouth transects (Fig. 8.7).

The core average of Fe, Ti, and Mn concentration decreased from shallow to deeper water depths in all the transects except in core MC-66 off Mahanadi river mouth and also in shallow cores off Vamsadhara river mouth where average metal content was slightly low. Fe, Ti, and Mn content increased toward south from off Mahanadi to off Godavari and off Krishna river mouths in the shallow as well as deep-water sediment cores (Bejugam, 2018).

Elemental chemistry plays a significant role in the assessment of the sedimentary environment and provenance (Yiyang and Mingcai, 1994). Al is a major constituent of sediments attained from continents (Taylor and McLennan, 1985) and is commonly used to measure the extent of accumulation of the lithogenous component (Murray and Leinin, 1996). The average concentrations of Al in different rock types differ by about 10% from the average crustal value (Turekian and Wedepohl, 1961; Taylor and McLennan, 1985) and hence useful as a lithogenous source indicator. The Al concentration in the majority of the surface and core samples ranged from 7% to 9% indicating its uniform terrestrial source. This suggested that changes in Al accumulation along transects were limited. It is close to the crustal PAAS value of 9.94%. Fe concentration in sediments was higher off Godavari, off Krishna and off Pennar river mouths as compared to off Mahanadi and off Vamsadhara river mouths and also higher than crustal PAAS value of 5%. Taylor (1964) recorded an average of 5.63% Fe in continental crust sediments and Pattanayak and Shrivastava (1999) reported ~9–11% Fe content in Deccan basalts. The iron-rich clay minerals are produced from the weathering of iron-rich source rocks namely unaltered ferromagnesium minerals and Fe-rich oxyhydroxides (Das and Krishnaswami, 2007). The higher Fe

Table 8.3 Range of Al, Ti, Fe, Mn, Zn, and Pb in surface sediments along transects from Mahanadi to Pennar rivers.

Transect	Off river mouth	Al (%)	Ti (%)	Fe (%)	Mn (%)	Zn (ppm)	Pb (ppm)
1	Hooghly river mouth—North	Nd	Nd	Nd	Nd	Nd	Nd
2	Hooghly river mouth—South	Nd	Nd	Nd	Nd	Nd	Nd
3	Mahanadi river mouth—North						
4	Mahanadi river mouth—South	5.45–9.52	0.28–0.82	4.35–6.14	0.08–0.67	54.75–181.25	12.73–35.30
5	Vamsadhara river mouth	4.43–9.08	0.41–0.60	3.51–6.10	0.07–1.20	51.0–154.25	20.35–43.0
6	Between Godavari and Vamsadhara river mouths	4.79–10.75	0.33–0.85	2.95–5.91	0.05–0.58	60.25–154.0	12.45–45.05
7	Godavari river mouth	7.13- 8.57	0.63- 1.30	6.21- 9.70	0.10- 0.40	89.0–140.50	9.68–24.8
8	Krishna river mouth	7.18–9.03	0.76–1.90	5.74- 6.85	0.07- 0.28	99.0–141.50	8.65–23.43
9	Between Krishna and Pennar river	6.59–10.30	0.63–0.96	5.80–7.83	0.06–0.63	87.0–151.50	5.9–43.88
10	Pennar river mouth	6.86–8.85	0.63–0.84	5.32–8.80	0.15–0.27	72.75–129.25	12.58–31.05

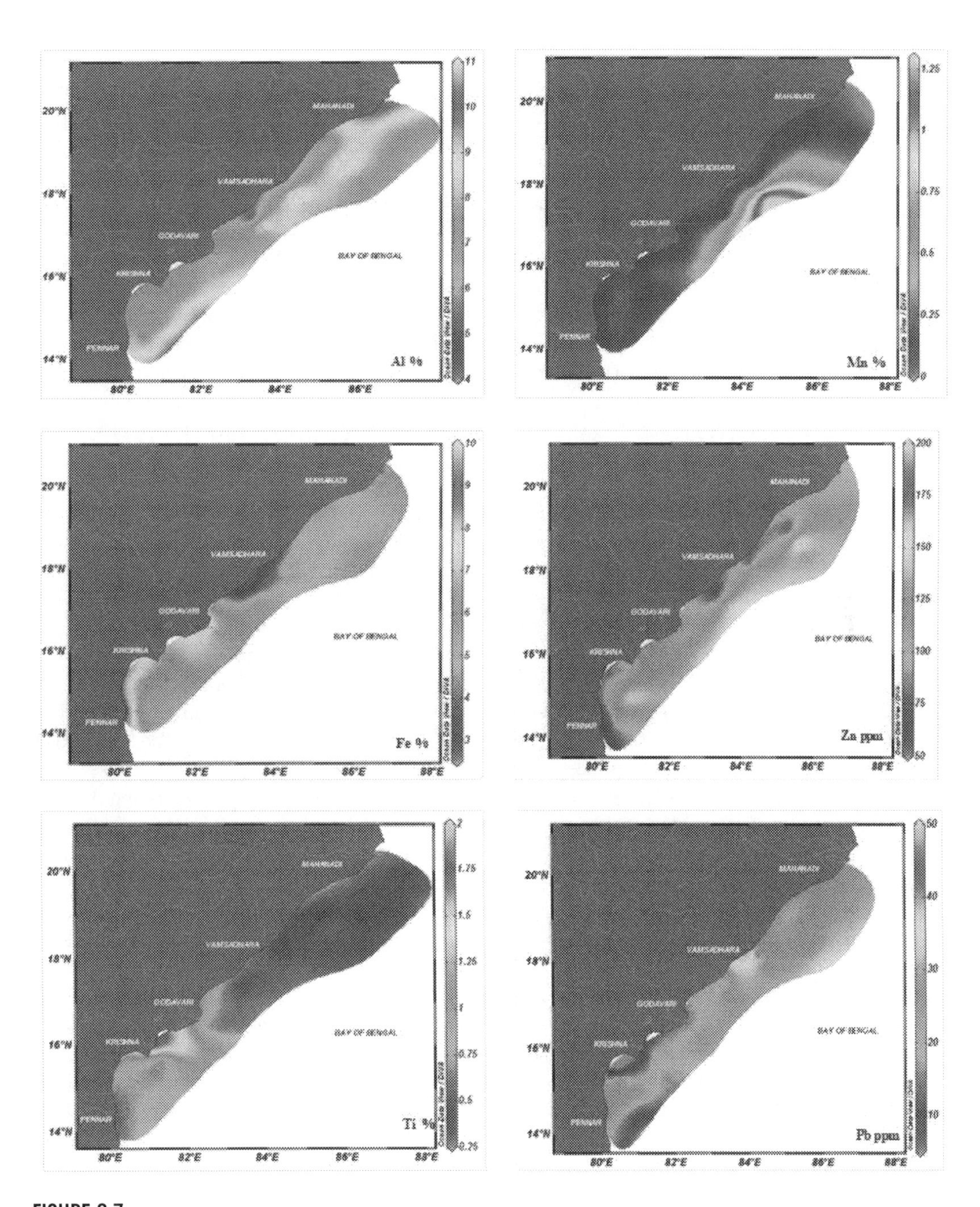

FIGURE 8.7

Distribution of a few major and trace elements in the surface sediments of the study area.

in cores off Godavari river mouth may have been contributed from the leaching of iron-rich sediments from "Red beds," dominant rocks near the Vishakapatnam coast. These red beds are formed under severe oxidizing conditions and are composed of ferric hydroxide. The red beds are usually associated with appreciable amounts of kaolinite explaining its higher concentration in the samples near the river mouth. Ti content in the samples was higher than the crustal PAAS value of 0.6%. However, Ti concentration in the sediments off Mahanadi and off Vamsadhara river mouths was lower indicating felsic source rocks. The higher Ti off Godavari, off Krishna and off Pennar river mouths as compared to off Mahanadi river mouth may have been contributed from the mafic source, that is, Deccan Trap basalt which is rich in titanium. The felsic rocks are the source of sediments off Mahanadi river while sediments off Krishna–Godavari river were mainly derived from Deccan basalts (Mazumdar et al., 2015). This is also supported by higher illite in the northern transects off Mahanadi and off Vamsadhara river mouths and high smectite content off Godavari, off Krishna, and off Pennar river mouths. The high Pb content also supported the source as felsic rocks in surface sediments off Mahanadi and off Vamsadhara river mouths and in the transect north off Godavari river mouth. Pb is usually hosted with rocks rich in feldspars (Prinz, 1967; Sensarma et al., 2016; Babeesh et al., 2017) such as granites. Pb is higher than crustal PAAS value of 20 ppm in most of the samples in the northern transects as well as few samples off Godavari, off Krishna, and off Pennar river mouths where illite is higher. Illite tends to adsorb Pb more readily into its structure explaining higher Pb enrichment in the sediments where illite was the dominant clay mineral. Several sorption experiments were carried out by researchers (Echeverria et al., 2005; Serrano et al., 2005) to understand the extent of Pb uptake on clay minerals. They revealed that due to cation exchange Pb tends to absorb on illite. However, smectite too is known to adsorb Pb into its lattice (Rybicka et al., 1995; Mhamdi et al., 2013) because of the negative charge on the surface and its shrink-swell property. This explains the appreciable Pb concentration in surface samples off Godavari, off Krishna, and off Pennar river mouths as well. In the majority of the samples, Mn and Zn concentrations were higher than crustal PAAS values of 0.09% and 85 ppm (Taylor and McLennan, 1985), respectively, indicating the presence of hydroxides.

The variations in average metal concentrations of the cores revealed that with increasing distance from the river mouth Fe, Ti, Mn, and Zn contents decreased indicating the abundant supply of these elements from lithogenous source rocks which after weathering are drained to the region through the peninsular river runoff. The chemical composition of marine sediments is the signature of materials derived from detrital, authigenic, and hydrothermal sources (Tripathy et al., 2014). Higher toxic waters probably lead to slightly higher metal enrichment in the samples close to the river mouth. However, average metal content is reduced off Mahanadi river mouth and at shallower water depth despite abundant terrestrial input as the dominant sand fraction in the core may not be able to preserve the metals within it. However, Al, Fe, Ti, Mn, Zn enrichment is noted in the upper few cm corresponding to higher clay value suggesting a metal association with the clay fraction. With a decrease in grain size trace element concentration increases reflecting an association with finer clay fraction due to its larger surface by volume ratio (Horowitz and Elrick, 1987; Burdige, 2006).

8.3.2 Gravity core sediments—sedimentary depositional environments and Holocene climate

The range and average values of sediment components, clay minerals, and metals are presented in Table 8.4A and zone wise average values in Table 8.4B for the gravity core GC-16 collected off

Table 8.4A Range and average values of sediment components, clay minerals, and metals in core GC-16.

Parameters	Range	Average
Sand (%)	0.43–30.11	3.49
Silt (%)	26.81–75.82	46.28
Clay (%)	22–66.16	50.23
Smectite (%)	0.0–21.89	5.98
Illite (%)	48.15–91.23	67.31
Kaolinite (%)	3.24–31.82	18.8
Chlorite (%)	0.0–25.1	7.91
Fe (%)	2.11–7.05	4.47
Mn (ppm)	229.5–698	412.14
Zn (ppm)	34.5–611.5	99.47
Pb (ppm)	0.02–9.59	2.61

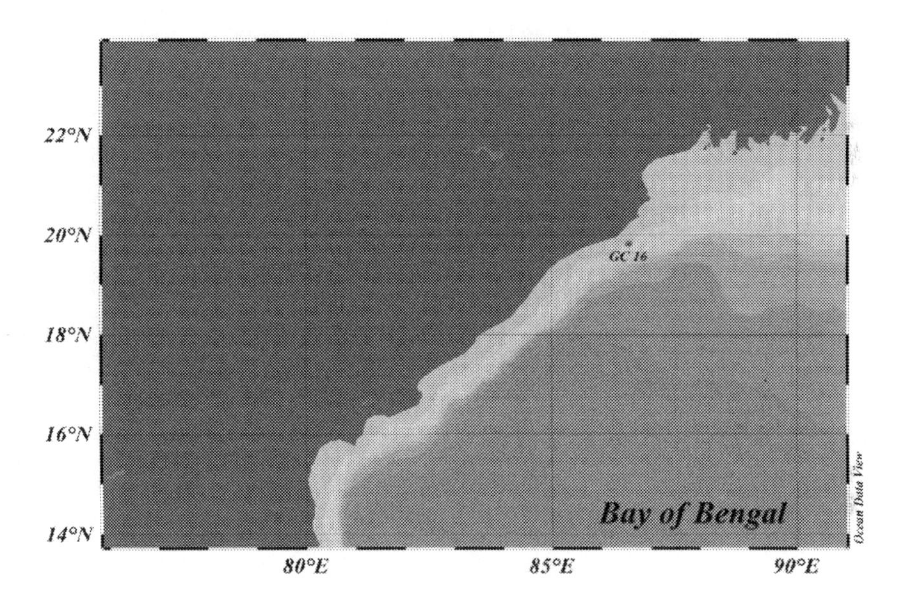

FIGURE 8.8

Map showing the core location.

Mahanadi river (Fig. 8.8). The rate of sedimentation (Fig. 8.9) computed using the ^{14}C dating method was 0.027 cm/yr from 200 to 100cm and 0.036 cm/yr from 100 cm to the surface of the sediment, indicated a relatively higher rate of sedimentation in recent years.

In the studied core sediment samples, the silt and clay together constituted over 96% with clay as the predominant sediment component. Based on the distribution of sediment components, clay minerals and metals the sediment core from bottom to the surface was divided into five zones they are, zone A to

Table 8.4B Average values of sediment components, clay minerals, and metals of zones in core GC-16.

Zone	Sand (%)	Silt (%)	Clay (%)	Smectite (%)	Illite (%)	Kaolinite (%)	Chlorite (%)	Fe (%)	Mn (ppm)	Pb (ppm)	Zn (ppm)
E	1.65	45.61	52.73	13.75	56.72	25.41	4.12	5.54	426.11	0.29	70.77
D	2.06	42.54	55.41	14.89	57.63	23.45	4.03	5.56	424.50	3.55	98.09
C	8.76	41.31	49.93	5.13	64.23	20.73	9.91	3.47	369.83	0.70	139.73
B	1.20	44.04	54.77	2.06	75.12	14.85	7.97	4.12	352.23	2.89	85.20
A	2.06	52.76	45.17	0.75	75.16	14.20	9.89	4.24	450.03	4.49	88.83

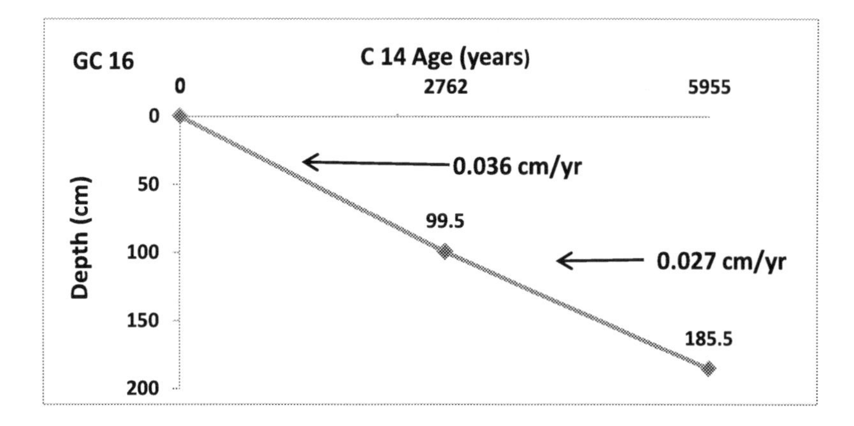

FIGURE 8.9

Age–depth curve of core GC-16.

zone E. The sand percentage was higher than the core average in zone C, silt was higher in zone A and clay was higher in zones B, D, E, and C very close to the average value (Table 8.4A–B). The variation observed in sediment components suggested the deposition of sediment in fluctuating depositional environments. The presence of coarser sediment in the sample indicates high energy environments and finer sediment suggests quiet hydrodynamic conditions (Fernandes and Nayak, 2009, 2020). The hydrodynamic conditions and sediment particle size control the rate of sedimentation at a given location. The types of rocks present in the catchment area, type of weathering, topography (Riebe et al., 2015), drainage network, climatic conditions, and monsoon intensity control the sediment size. The sediment is redistributed based on size upon entering the ocean which is controlled by the hydrodynamic conditions. The high clay content in the upper portion of the core reveals sediment deposition in quiet environments with a high rate of sedimentation.

Among the clay minerals, illite was the most abundant (avg. ~67%) in the core sediments being relatively resistant (Biscaye, 1965). As stated earlier high illite off Mahanadi river mouth indicates supply from felsic rocks, which belong to the Archean–Proterozoic gneissic complex (Mazumdar et al., 2015) and as K-feldspars are easily weathered to clay minerals (Chen et al., 2010). The smectite and chlorite were low in the sediments. The chemical weathering of mafic minerals present in the gneiss and charnockites (Bastia et al., 2020) must have contributed to smectite and chlorite was unstable in humid tropical conditions. So, the weathered materials from the catchment area and transported through the Mahanadi river was the major supplier of these minerals at the core location. Besides, illite was also contributed through Ganges and Brahmaputra as they are transported and redistributed by turbidity currents (Kolla and Rao, 1990; Bejugam and Nayak, 2016b) along the western Bay of Bengal. The distribution of clay minerals provides information on depositional environments as it is controlled by energy conditions (Rao et al., 1988; Moore and Reynolds, 1989; Liu et al., 2010). The smectite is finer in size and therefore can remain in suspension for a longer time and further even weak currents could transport it. Salinity is low in waters near the river mouth. Compared to smectite, illite has coarser size and more stable than smectite under less saline conditions explaining the higher illite content under

Table 8.5 Geochronology and Holocene climate periods.

Zone	Depth (cm)	Characteristic features	Rate of sedimentation (cm/yr)	Age (cal yr. BP)	Holocene climatic periods
E	0–26	High clay, smectite, kaolinite, Fe, Mn, Zn	0.036	0–708 (2014–1306 AD)	Recent global warming and retreating Little Ice Age
D	26–58	High clay, smectite, kaolinite, Fe, Mn, Zn, Pb		708–1597 (1306–417 AD)	Medieval warm period also called Little Climatic Optimum
C	58–100	High and fluctuating clay and smectite		1597–2762 (417 AD–748 BC)	Roman warm and cold period
B	100–122	High and fluctuating silt and illite; low Fe, Mn, Zn	0.027	2762–3097 (748–1083 BC)	Sub-boreal cold climatic period
A	122–184	High silt, illite, Mn, Pb		3097–5955 (1083–3941 BC)	Sub-boreal cold climatic period

high energy conditions near the river mouth. The presence of predominant illite at the core location off Mahanadi river mouths indicates high energy conditions prevalent during the time of deposition.

Further, illite and smectite contents present in sediments were related to supply of weathered material from Himalayan regions through Hoogly river and peninsular region by peninsular Indian rivers, respectively (Kolla and Rao, 1990; Raman et al., 1995; Reddy and Rao, 2001) concerning intensification of NE and SW monsoons. The clay mineral distribution in the core indicated high illite values in zones A and B, kaolinite in zones C, D, and E, smectite in zones D and E, and chlorite values in zones A, B, and C than the average. High illite and chlorite in zones A and B indicate their formation in cooler and dryer climatic periods and represent high latitude regions while high smectite and kaolinite in zones D and E indicate warmer and wetter climates (Griffin et al., 1968) with zone C representing intermediate.

In zone A higher than the average illite associated with silt and lower than the average smectite associated with clay suggested their deposition in relatively higher hydrodynamic conditions. Higher than average illite and chlorite (Table 8.4B) were noted in zone B and the kaolinite content was overall lower than average in both zones A and B. Illite associated with silt (Table 8.5) in these zones must have deposited in relatively high energy hydrodynamic conditions. The zones A and B varying from 5955 to 2762 cal yr B.P. largely represent the sub-boreal climatic period of northern Europe, between the Atlantic and sub-Atlantic stages about 5000–2800 years ago. Nagasundaram et al. (2020) have reported the intense weakening of the SW monsoon during the sub-boreal period in the Bay of Bengal. The reduced terrigenous material supply because of the weak SW monsoon to the study area during the period possibly was responsible for the reduced rate of sedimentation. The increased hydrodynamic regime in Galicia during this climatic period was observed earlier by Rocha et al. (2006). High sand content (8.76%) and clay (49.93%), higher than average kaolinite, chlorite (Table 8.4B), and smectite with lower than average silt and illite in zone C from 2762 to 1597 cal yr B.P. largely coincided with the Roman Warm and cold periods. High sand and clay possibly indicate change over from cold to warm climate and changed hydrodynamic conditions from high energy to low possibly facilitating the deposition of high clay and smectite in quiet hydrodynamic conditions. During the Roman Warm Period increased rainfall in the Indian Subcontinent was documented by Pothuri et al. (2014) and Pothuri (2017). The enhanced SW monsoon might have facilitated the chemical weathering of mafic rocks and helped in the formation of smectite. In zone D, higher than average clay, smectite, and kaolinite (Table 8.4B) and low silt, illite, and chlorite represent the Medieval warm period (Table 8.5). Easterbrook (2016) recorded 1°C warmer global temperature than recent temperature and Sridhar (2009) and Suokhrie et al. (2018) reported strengthening monsoon during the Medieval Warm Period in the Bay of Bengal. In zone E, higher than the average smectite, kaolinite (Table 8.4B) along with clay and, low illite, chlorite and silt suggested increased chemical weathering of mafic source rocks in the presence of warm and wet climate during the recent global warming period.

To identify the source, processes and conditions of deposition the metal content in sediments are an important proxy (Chakrapani and Subramanian, 1990; Raj and Jayaprakash, 2008; Pattan et al., 2008). Among the studied metals in the core, Fe values were high in zones D and E; Mn in A, D, and E; Pb in A, B, and D; and the value of Zn in zone C was higher than average. In zone A, high Mn and Pb are associated with high silt and Illite with low content of Fe and Zn associated with clay. Lower than average Fe, Mn, Zn, and fluctuating Pb were observed in zone B. In zone C, Zn content was high along with high sand percentage. Higher than average Fe, Mn, Zn, Pb was associated with high clay, smectite, kaolinite, and low silt, illite and chlorite in zone D. In zone E higher than average Fe and Mn were associated with high clay, smectite, and kaolinite. In zones A and B presence of low Fe, Mn, and Zn indicated the reduced release of material from the Peninsula rivers controlled by the reduced intensity of the SW monsoon. In zones D, E, and to some extent in zone C high Fe, Mn, and Zn support higher supply regulated by warm and wet climate and enhanced SW monsoon.

High silt, illite, and Mn with lower than average clay, kaolinite, smectite, Zn, and fluctuating Fe regulated by the cold and dry climate were responsible for lower sedimentation rate obtained from 186 to 100 cm. In the upper 100 cm of the core high clay, kaolinite, smectite, Fe, Mn, Zn with low silt, illite, chlorite coincided with the high rate of sedimentation regulated by warm and wet climate and intense SW monsoon. The signatures of the proxies used in the present study namely sediment components, clay minerals, and metals were effective in understanding change in sediment supply, depositional environments, and sedimentation rates which were regulated by the cold and dry, and warm and wet climate and intensity of SW and NE monsoons.

8.4 Learning and knowledge outcomes

The source, transportation history, climatic conditions at the time of sediment deposition and intensity of monsoons were studied using proxies namely grain size, clay minerals, and metals in the western Bay of Bengal which is fed by sediment flux from Ganges–Brahmaputra, Mahanadi, Godavari, Krishna, Cauvery, Irrawaddy, and Pennar rivers. The study on surface and multicore sediments reveals that among the sediment components clay is the dominant sediment fraction. However, higher sand was observed in sediments in shallow water regions due to the closer proximity to the river mouth. Higher sand content in shallow water regions and higher clay content in deeper water regions indicate transport direction from coast to offshore and material supply from the rivers to the bay. Also, grain size decreases from off Hooghly river mouth in the north, through off Mahanadi, off Vamsadhara, off Godavari, and off

Krishna to off Pennar river mouth regions in the south indicates high energy conditions prevalent off Hooghly and Mahanadi rivers which facilitates the deposition of relatively coarse grain sediments and prevents the accumulation of fine-grain sediments near the river mouth. Among the clay minerals, the smectite was dominant in samples off Godavari, Krishna, and Pennar rivers formed as a product of chemical weathering under humid tropical conditions and finds its source from the Deccan Trap basalts and associated black cotton soils. Illite was higher in sediments off Hooghly, Mahanadi, and Vamsadhara rivers formed as a product of mechanical weathering under arid cold climatic conditions by the alteration of muscovite mica which is a dominant mineral in the Himalayan region and illite may have drained through the Ganges–Brahmaputra rivers. Also, rocks from the Eastern Ghats rich in potash feldspar may have undergone alteration and provided illite to the region. A loss of energy southward could have prevented the downward transport of illite-rich sediments. Appreciable kaolinite content is probably from the Archean granites and gneisses and from red beds which is a dominant rock type along the coastal areas of Godavari river. Fe and Ti concentrations are higher in sediments off Godavari and Krishna derived from mafic-rich source rocks as well as from the red beds which are rich in ferric oxide.

The variation of proxies studied along the length of the gravity core represents changing Holocene climatic oscillations. The lower two zones of the core largely coincide with the sub-boreal climate periods during which dry conditions favored increased physical weathering of felsic source rocks. Zone C represents the intermediate climate conditions and coincides with the Roman cold–warm period, while, toward the surface of the core, the two zones represent warm periods of the Medieval Warm Period and recent global warming. During these periods a significant increase in warm–wet conditions must have facilitated the leaching of mafic rocks.

Acknowledgments

One of the authors (G.N.N.) thanks to the Ministry of Earth Sciences (MOES/SIBER/NIO(RN)/11) and Inter-University Accelerator Centre (IUAC) (Project No. 62314) New Delhi for the financial support to carry out this work under research projects. Prof. G.N. Nayak thanks CSIR New Delhi for awarding the CSIR Emeritus Scientist position. The authors thank Dr. S.W.A. Naqvi (Former Director NIO), Dr. R. Nigam (Former Deputy Director/Scientist, NIO), Dr. R. Saraswat (Scientist, NIO), Dr. Anoop Kumar Tiwari (Scientist, NCPOR), and Mr. Girish A. Prabhu for their support.

References

Ahmad, S.M., Anil Babu, G., Padmakumari, V.M., Waseem, R., Surface and deep water changes in the northeast Indian Ocean during the last 60 ka inferred from carbon and oxygen isotopes of planktonic and benthic foraminifera, Palaeogeogr. Palaeoclim. Palaeoecol. 262 (2008) 182–188.

Aksu, A.E., Yaşar, D., Orhan, U.S.L.U., Assessment of marine pollution in Izmir Bay: heavy metal and organic compound concentrations in surficial sediments, Turk. J. Eng. Environ. Sci. 22 (5) (1998) 387–416.

Babeesh, C., Achyuthan, H., Jaiswal, M.K., Lone, A., Late Quaternary loess-like paleosols and pedocomplexes, geochemistry, provenance and source area weathering, Manasbal, Kashmir Valley, India, Geomorphology 284 (2017) 191–205.

Barua, D.K., Kuehl, S.A., Miller, R.L., Moore, W.S., Suspended sediment distribution and residual transport in the coastal ocean off the Ganges-Brahmaputra river mouth, Mar. Geol. 120 (1-2) (1994) 41–61.

Basavaiah, N., Babu, J.M., Gawali, P.B., Kumar, K.C.V.N., Demudu, G., Prizomwala, S.P., Hanamgond, P.T., Rao, K.N., Late Quaternary environmental and sea level changes from Kolleru Lake, SE India: inferences from mineral magnetic, geochemical and textural analyses, Quat. Int. 371 (2015) 197–208.

Bastia, F., Equeenuddin, S.M., Roy, P.D., Hernández-Mendiola, E., Geochemical signatures of surface sediments from the Mahanadi river basin (India): chemical weathering, provenance, and tectonic settings, Geol. J. 55 (7) (2020) 5294–5307.

Basu, A., Petrology of Holocene fluvial sand derived from plutonic source rocks: implications to paleoclimatic interpretation, J. Sediment. Res. 46 (3) (1976) 694–709.

Bejugam, P., 2018. Paleoclimate signatures of recent past through sedimentological and geochemical investigations, western Bay of Bengal. Ph.D. Thesis, Goa University, Goa

Bejugam, P., Nayak, G.N., Changing depositional environment revealed from sediment components, west of "Swatch of No Ground", northern Bay of Bengal, Arab. J. Geosci. 9 (9) (2016a) 551–563, doi:10.1007/s12517-016-2563-y.

Bejugam, P., Nayak, G.N., Source and depositional processes of the surface sediments and their implications on productivity in recent past off Mahanadi to Pennar River mouths, western Bay of Bengal, Palaeogeogr. Palaeoclim. Palaeoecol. 483 (2016b) 58–69, doi:10.1016/j.palaeo.2016.12.006.

Bejugam, P., Nayak, G.N., Tracing source–sink processes and productivity from trace metals (Ba, Zn, Pb, Cd) of the surface sediments off Mahanadi to Pennar, western Bay of Bengal, Environ. Earth Sci. 78 (2019a) 107 https://doi.org/10.1007/s12665-019-8070-1.

Bejugam, P., Nayak, G.N., Source of sediments and response of clay minerals, organic matter and metals to fluctuating environmental conditions in western Bay of Bengal, Indian J. Geo-Mar. Sci. 48 (04) (2019b) 535–553.

Billen, G., The PHISON river system: a conceptual model of C, N and P transformations in the aquatic continuum from land to sea, Interactions of C, N, P and S Biogeochemical Cycles and Global Change, Springer, Berlin, Heidelberg, 1993, pp. 141–161.

Biscaye, P.E., Mineralogy and sedimentation of recent deep-sea clay in the Atlantic Ocean and adjacent seas and oceans, Geol. Soc. Am. Bull. 76 (7) (1965) 803–832.

Burdige, D.J., Geochemistry of Marine Sediments, Vol. 398, Princeton University Press, Princeton, 2006.

Chakrapani, G.J., Subramanian, V., Preliminary studies on the geochemistry of the Mahanadi river basin, India, Chem. Geol. 81 (3) (1990) 241–253.

Chauhan, O.S., Rajawat, A.S., Pradhan, Y., Suneethi, J., Nayak, S.R., Weekly observations on dispersal and sink pathways of the terrigenous flux of the Ganga–Brahmaputra in the Bay of Bengal during NE monsoon, Deep Sea Res. Part II: Top. Stud. Oceanogr. 52 (14) (2005) 2018–2030.

Chauhan, O.S., Vogelsang, E., Climate induced changes in the circulation and dispersal patterns of the fluvial sources during late Quaternary in the middle Bengal Fan, J. Earth Syst. Sci. 115 (3) (2006) 379–386.

Chen, H.F., Song, S.R., Lee, T.Q., Löwemark, L., Chi, Z., Wang, Y., Hong, E., A multiproxy lake record from Inner Mongolia displays a late Holocene teleconnection between Central Asian and North Atlantic climates, Quat. Int. 227 (2) (2010) 170–182.

Conroy, J.L., Overpeck, J.T., Cole, J.E., Shanahan, T.M., Steinitz-Kannan, M., Holocene changes in eastern tropical Pacific climate inferred from a Galápagos lake sediment record, Quat. Sci. Rev. 27 (11) (2008) 1166–1180.

Crowley, S.F., Stow, D.A., Croudace, I.W., Mineralogy and geochemistry of Bay of Bengal deep-sea fan sediments, ODP Leg 116: evidence for an Indian subcontinent contribution to distal fan sedimentation, Geol. Soc. Lond. Special Publ. 131 (1) (1998) 151–176.

Das, A., Krishnaswami, S., Elemental geochemistry of river sediments from the Deccan Traps, India: implications to sources of elements and their mobility during basalt–water interaction, Chem. Geol. 242 (1) (2007) 232–254.

Dou, Y., Yang, S., Liu, Z., Clift, P.D., Yu, H., Berne, S., Shi, X., Clay mineral evolution in the central Okinawa Trough since 28ka: implications for sediment provenance and paleoenvironmental change, Palaeogeogr. Palaeoclim. Palaeoecol. 288 (1) (2010) 108–117.

Duplessy, J.C., Glacial to interglacial contrasts in the northern Indian Ocean, Nature 295 (1982) 494–498.

Easterbrook, D.J., Cause of global climate changes: correlation of global temperature, sunspots, solar irradiance, cosmic rays, and radiocarbon and berylium production rates, Evidence-Based Climate Science, 2016, pp. 245–262. (Second Edition) Chapter 14, Editor: Don Easterbrook, Elsevier Inc., Radarweg 29, PO Box 211, 1000 AE Amsterdam, Netherlands..

Echeverría, J.C., Zarranz, I., Estella, J., Garrido, J.J., Simultaneous effect of pH, temperature, ionic strength, and initial concentration on the retention of lead on illite, Appl. Clay Sci. 30 (2) (2005) 103–115.

Fernandes, L., Nayak, G.N., Distribution of sediment parameters and depositional environment of mudflats of Mandovi Estuary, Goa, India, J. Coastal Res. 25 (2) (2009) 273–284 https://doi.org/10.2112/05-0614.1.

Fernandes, M.C., Nayak, G.N., Depositional environment and metal distribution in mangrove sediments within middle region of tropical estuaries, Karnataka, west coast of India, Reg. Stud. Mar. Sci. 39 (2020) 101473 https://doi.org/10.1016/j.rsma.2020.101473.

Folk, R.L., Petrology of Sedimentary Rocks, Hemphill Publishing Company, Texas, 1968, p. 117.

Goldberg, E.D., Griffin, J.J., The sediments of the northern Indian Ocean, Deep Sea Res. Oceanogr. Abstr. 17 (3) (1970) 513–537.

Gourlan, A.T., Meynadier, L., Allègre, C.J., Tapponnier, P., Birck, J.L., Joron, J.L., Northern Hemisphere climate control of the Bengali rivers discharge during the past 4 Ma, Quat. Sci. Rev. 29 (19-20) (2010) 2484–2498.

Govil, P., Naidu, P.D., Variations of Indian monsoon precipitation during the last 32 kyr reflected in the surface hydrography of the Western Bay of Bengal, Quat. Sci. Rev. 30 (27-28) (2011) 3871–3879, doi:10.1016/J.QUASCIREV.2011.10.004.

Griffin, J.J., Windom, H., Goldberg, E.D., The distribution of clay minerals in the world ocean, Deep Sea Res. Oceanogr. Abstr. 15 (4) (1968) 433–459.

Grim, R.E., Clay Mineralogy, McGraw-Hill, New York, 1968.

He, M., Zheng, H., Huang, X., Jia, J., Li, L., Yangtze River sediments from source to sink traced with clay mineralogy, J. Asian Earth Sci. 69 (2013) 60–69.

Higginson, M.J., Altabet, M.A., Wincze, L., Herbert, T.D., Murray, D.W., A solar (irradiance) trigger for millennial-scale abrupt changes in the southwest monsoon? Paleoceanogr. Paleoclim. 19 (3) (2004) PA3015, doi:10.1029/2004PA001031.

Horowitz, A.J., Elrick, K.A., The relation of stream sediment surface area, grain size and composition to trace element chemistry, Appl. Geochem. 2 (4) (1987) 437–451.

Jackson, N.L., Nordstrom, K.F., Aeolian sediment transport and landforms in managed coastal systems: a review, Aeolian Res. 3 (2) (2011) 181–196.

Jarvis, I., Jarvis, K.E., Rare-earth element geochemistry of standard sediments: a study using inductively coupled plasma spectrometry, Chem. Geol. 53 (1985) 335–344.

Kalesha, M., Rao, K.S., Somayajulu, B.L.K., Deposition rates in the Godavari delta, Mar. Geol. 34 (1-2) (1980) M57–M66.

Kangane, J., Nayak, G.N., Choudhary, S., Source and processes from core sediment samples off Mahanadi and Krishna rivers, western Bay of Bengal, J. Ind. Assoc. Sedimentol. 36 (1) (2019) 65–83.

Kiran, R., Krishna, V.G., Naik, B.G., Mahalakshmi, G., Rengarajan, R., Mazumdar, A., Sarma, N.S., Can hydrocarbons in coastal sediments be related to terrestrial flux? A case study of Godavari river discharge (Bay of Bengal), Curr. Sci. 108 (1) (2015) 96.

Kolla, V., Rao, N.M., Sedimentary sources in the surface and near-surface sediments of the Bay of Bengal, Geo-Mar. Lett. 10 (3) (1990) 129–135 .

Krishnamurthy, V., Goswami, B.N., Indian monsoon–ENSO relationship on interdecadal timescale, J. Clim. 13 (3) (2000) 579–595.

Kuehl, S.A., Hariu, T.M., Moore, W.S., Shelf sedimentation off the Ganges-Brahmaputra river system: evidence for sediment bypassing to the Bengal fan, Geology 17 (12) (1989) 1132–1135.

Kulkarni, Y.R., Sangode, S.J., Bloemendal, J., Meshram, D.C., Suresh, N., Mineral magnetic characterization of the Godavari river and western Bay of Bengal sediments: implications to source to sink relations, J. Geol. Soc. Ind. 85 (1) (2015) 71–78.

Kumar, G., Ramanathan, A.L., Rajkumar, K., Textural characteristics of the surface sediments of a Tropical mangrove ecosystem Gulf of Kachchh, Gujarat, India, Ind. J. Mar. Sci. 39 (3) (2010) 415–422.

Lin, I.I., Chen, C.H., Pun, I.F., Liu, W.T., Wu, C.C., Warm ocean anomaly, air sea fluxes, and the rapid intensification of tropical cyclone Nargis, Geophy. Res. Lett. 36 (3) (2009) L03817-21, doi:10.1029/2008GL035815.

Liu, Z., Colin, C., Li, X., Zhao, Y., Tuo, S., Chen, Z., Siringan, F.P., Liu, J.T., Huang, C.Y., You, C.F., Huang, K.F., Clay mineral distribution in surface sediments of the northeastern South China Sea and surrounding fluvial drainage basins: source and transport, Mar. Geol. 277 (1) (2010) 48–60.

López, G.I., Grain size analysis, Encyclopedia of Geoarchaeology, Springer, Netherlands, 2017, pp. 341–348.

Malathi, V., Geochemical studies of clays from coastal red sediments from Yerrapalem to Chepalakancheru, Vizianagaram District, Andhra Pradesh, Indian J. Mar. Sci. 42 (2013) 929–933.

Mazumdar, A., Joao, H.M., Peketi, A., Dewangan, P., Kocherla, M., Joshi, R.K., Ramprasad, T., Geochemical and geological constraints on the composition of marine sediment pore fluid: possible link to gas hydrate deposits, Mar. Petroleum Geol. 38 (1) (2012) 35–52.

Mazumdar, A., Kocherla, M., Carvalho, M.A., Peketi, A., Joshi, R.K., Mahalaxmi, P., Joao, H.M., Jisha, R., Geochemical characterization of the Krishna–Godavari and Mahanadi offshore basin (Bay of Bengal) sediments: a comparative study of provenance, Mar. Petroleum Geol. 60 (2015) 18–33.

Mazumdar, A., Peketi, A., Joao, H.M., Dewangan, P., Ramprasad, T., Pore-water chemistry of sediment cores off Mahanadi Basin, Bay of Bengal: possible link to deep seated methane hydrate deposit, Mar. Petroleum Geol. 49 (2014) 162–175.

McLaren, P., Bowles, D., The effects of sediment transport on grain-size distributions, J. Sediment. Res. 55 (4) (1985) 457–470, doi:10.1306/212F86FC-2B24-11D7-8648000102C1865D.

Mhamdi, M., Galai, H., Mnasri, N., Elaloui, E., Trabelsi-Ayadi, M., Adsorption of lead onto smectite from aqueous solution, Environ. Sci. Pollut. Res. 20 (3) (2013) 1686–1697.

Moore, D.M., Reynolds, R.C., X-ray Diffraction and the Identification and Analysis of Clay Minerals, Vol. 378, Oxford University Press, USA, 1989.

Mukhopadhyay, S.K., Biswas, H.D.T.K., De, T.K., Jana, T.K., Fluxes of nutrients from the tropical River Hooghly at the land–ocean boundary of Sundarbans, NE Coast of Bay of Bengal, India, J. Mar. Syst. 62 (1) (2006) 9–21.

Murray, R.W., Leinen, M., Scavenged excess aluminum and its relationship to bulk titanium in biogenic sediment from the central equatorial Pacific Ocean, Geochim. Cosmochim. Acta 60 (20) (1996) 3869–3878.

Nagasundaram, M., Achyuthan, H., Rai, J., Mid to Late Holocene reconstruction of the Southwest Monsoonal shifts based on a marine sediment Core, off the Landfall Island, Bay of Bengal, The Andaman Islands and Adjoining Offshore: Geology, Tectonics and Palaeoclimate, Springer, Jyotiranjan S. Ray and M. Radhakrishna; Springer Nature Switzerland AG., 2020, pp. 315–400.

Naidu, A.S., Han, M.W., Mowatt, T.C., Wajda, W., Clay minerals as indicators of sources of terrigenous sediments, their transportation and deposition: Bering Basin, Russian-Alaskan Arctic, Mar. Geol. 127 (1-4) (1995) 87–104.

Naidu, P.D., Ganeshram, R., Bollasina, M.A., Panmei, C., Nürnberg, D., Donges, J.F., Coherent response of the Indian monsoon rainfall to Atlantic Multi-decadal variability over the last 2000 years, Sci. Rep. 10 (2020) 1302.

Nair, R.R., Ittekkot, V., Enhanced particle fluxes in Bay of Bengal induced by injection of fresh water, Nature 351 (6325) (1991) 385.

Overpeck, J., Anderson, D., Trumbore, S., Prell, W., The southwest Indian Monsoon over the last 18 000 years, Clim. Dyn. 12 (3) (1996) 213–225.

Panchang, R., Nigam, R., High resolution climatic records of the past ~489 years from Central Asia as derived from benthic foraminiferal species, *Asterorotalia trispinosa*, Mar. Geol. 307-310 (2012) 88–104, doi:10.1016/j.margeo.2012.01.006.

Panda, D.K., Kumar, A., Mohanty, S., Recent trends in sediment load of the tropical (Peninsular) river basins of India, Glob. Planet. Change 75 (3-4) (2011) 108–118.

Patchineelam, S.M., de Figueiredo, A.G., Preferential settling of smectite on the Amazon continental shelf, Geo-Mar. Lett. 20 (1) (2000) 37–42.

Pattan, J.N., Parthiban, G., Gupta, S.M., Mir, I.A., Fe speciation and Fe/Al ratio in the sediments of southeastern Arabian Sea as an indicator of climate change, Quat. Int. 250 (2012) 19–26.

Pattan, J.N., Parthiban, G., PrakashBabu, C., Khadge, N.H., Paropkari, A.L., Kodagali, V.N., A note on geochemistry of surface sediments from Krishna-Godavari basin, East Coast of India, J. Geol. Soc. Ind. 71 (2008) 107–114.

Pattanayak, S.K., Shrivastava, J.P., Petrography and major-oxide geochemistry of basalts from the Eastern Deccan Volcanic Province, India, Mem.-Geol. Soc. India 2 (1999) 233–270.

Petschick, R., Kuhn, G., Gingele, F., Clay mineral distribution in surface sediments of the South Atlantic: sources, transport, and relation to oceanography, Mar. Geol. 130 (3-4) (1996) 203–229.

Phillips, S.C., Johnson, J.E., Giosan, L., Rose, K., Monsoon-influenced variation in productivity and lithogenic sediment flux since 110 ka in the offshore Mahanadi Basin, northern Bay of Bengal, Mar. Petroleum Geol. 58 (Part A) (2014) 502–525 https://doi.org/10.1016/j.marpetgeo.2014.05.007.

Ponton, C., Giosan, L., Eglinton, T.I., Fuller, D.Q., Johnson, J.E., Kumar, P., Collett, T.S., Holocene aridification of India, Geophy. Res. Lett. 39 (3) (2012) 1–6.

Pothuri, D., Indian monsoon rainfall variability during the common era: implications on the ancient civilization, AGU Fall Meeting Abstr. PP31A (2017) 1273.

Pothuri, D., Nürnberg, D., Mohtadi, M., Multi-decadal variation of the Indian monsoon rainfall: implications of ENSO, AGUFM PP43 (2014) A–1448.

Prajith, A., Tyagi, A., Kurian, P.J., Changing sediment sources in the Bay of Bengal: evidence of summer monsoon intensification and ice-melt over Himalaya during the Late Quaternary, Palaeogeogr. Palaeoclim. Palaeoecol. 511 (2018) 309–318.

Prell, W.L., Kutzbach, J.E., Sensitivity of the Indian monsoon to forcing parameters and implications for its evolution, Nature 360 (1992) 647–652.

Prinz, M., Geochemistry of basaltic rocks: trace elements, Basalts 1 (1967) 271–323.

Prithviraj, M., Prakash, T.N., Distribution and geochemical association of clay minerals on the inner shelf of central Kerala, India, Mar. Geol. 92 (3-4) (1990) 285–290.

Prizomwala, S.P., Bhatt, N., Basavaiah, N., Provenance discrimination and source-to-sink studies from a dryland fluvial regime: an example from Kachchh, western India, Int. J. Sediment. Res. 29 (1) (2014) 99–109.

Raj, S.M., Jayaprakash, M., Distribution and enrichment of trace metals in marine sediments of Bay of Bengal, off Ennore, south-east coast of India, Environ. Geol. 56 (1) (2008) 207–217.

Raman, C.V., Rao, G.K., Reddy, K.S.N., Ramesh, M.V., Clay mineral distributions in the continental shelf sediments between the Ganges mouths and Madras, east coast of India, Cont. Shelf Res. 15 (14) (1995) 1773–1793.

Rao, V.P., Rao, B.R., Provenance and distribution of clay minerals in the sediments of the western continental shelf and slope of India, Cont. Shelf Res. 15 (14) (1995) 1757–1771.

Rao, V.P., Reddy, N.P., Rao, C.M., Clay mineral distribution in the shelf sediments off the northern part of the east coast of India, Cont. Shelf Res. 8 (2) (1988) 145–151.

Rashid, H., England, E., Thompson, L., Polyak, L., Late Glacial to Holocene Indian summer monsoon variability based upon sediment records taken from the Bay of Bengal, Terr. Atmos. Ocean. Sci. 22 (2) (2011) 215–228.

Raza, T., Ahmad, S.M., Sahoo, M., Banerjee, B., Bal, I., Dash, S., Suseela, G., Mukherjee, I., Hydrographic changes in the southern Bay of Bengal during the last ~65,000 y inferred from carbon and oxygen isotopes of foraminiferal fossil shells, Quat. Int. 333 (2014) 77–85.

Reddy, N.P.C., Rao, K.M., Heavy sediment influx during early Holocene: inference from clay mineral studies in a core from the Western Bay of Bengal, Curr. Sci. 81 (10) (2001) 1361–1364.

Reimer, P.J., Bard, E., Bayliss, A., Beck, J.W., Blackwell, P.G., Ramsey, C.B., Buck, C.E., Cheng, H., Edwards, R.L., Friedrich, M., Grootes, P.M., IntCal13 and Marine13 radiocarbon age calibration curves 0–50,000 years cal BP, Radiocarbon 55 (4) (2013) 1869–1887.

Rickers, K., Mezger, K., Raith, M.M., Evolution of the continental crust in the Proterozoic Eastern Ghats Belt, India and new constraints for Rodinia reconstruction: implications from Sm–Nd, Rb–Sr and Pb–Pb isotopes, Precambr. Res 112 (3) (2001) 183–210.

Riebe, C.S., Sklar, L.S., Lukens, C.E., Shuster, D.L., Climate and topography control the size and flux of sediment produced on steep mountain slopes, Proc. Natl. Acad. Sci. USA 112 (51) (2015) 15574–15579, doi:10.1073/pnas.1503567112.

Rocha, F., Martins, V., Jouanneau, J.M., Weber, O., Gomes, C., A 5000-years history of climate change based on variations in the composition of the clay fraction from oceanic sediments off Galicia, Clay Sci. 12 (S2) (2006) 205–210.

Rollinson, H.R., Using Geochemical Data: Evaluation, Presentation, Interpretation, Routledge, Taylor & Francis Group, London., 2014.

Rybicka, E.H., Calmano, W., Breeger, A., Heavy metals sorption/desorption on competing clay minerals; an experimental study, Appl. Clay Sci. 9 (5) (1995) 369–381.

Saha, K.R., Mooley, D.A., Saha, S., The Indian monsoon and its economic impact. In the Asian monsoon and its economic consequences, Geol. J. 3 (2) (1979) 171–178 https://www.jstor.org/stable/41142210.

Sarkar, A., Ramesh, R., Somayajulu, B.L.K., Agnihotri, R., Jull, A.J.T., Burr, G.S., High resolution Holocene monsoon record from the eastern Arabian Sea, Earth Planet. Sci. Lett. 177 (3) (2000) 209–218.

Sensarma, S., Chakraborty, P., Banerjee, R., Mukhopadhyay, S., Geochemical fractionation of Ni, Cu and Pb in the deep sea sediments from the Central Indian Ocean Basin: an insight into the mechanism of metal enrichment in sediment, Chem. Erde-Geochem. 76 (1) (2016) 39–48.

Serrano, S., Garrido, F., Campbell, C.G., Garcıa-González, M.T., Competitive sorption of cadmium and lead in acid soils of Central Spain, Geoderma 124 (1) (2005) 91–104.

Singer, A., The paleoclimatic interpretation of clay minerals in sediments—a review, Earth-Sci. Rev. 21 (4) (1984) 251–293.

Singh, M., Singh, I.B., Müller, G., Sediment characteristics and transportation dynamics of the Ganga River, Geomorphology 86 (1) (2007) 144–175.

Singhvi, A.K., Kale, V.S., Paleoclimate Studies in India: Last Ice Age to the Present, Indian National Science Academy, Bahadur Shah Zafar Marg, Delhi-110002, 2010.

Sirocko, F., Garbe-Schonberg, D., McIntyre, A., Molfino, B., Teleconnections between the subtropical monsoons and high-latitude climates during the last deglaciation, Science 272 (5261) (1996) 526–529, doi:10.1126/science.272.5261.526.

Sirocko, F., Sarnthein, M., Lange, H., Erlenkeuser, H., Atmospheric summer circulation and coastal upwelling in the Arabian Sea during the Holocene and the last glaciation, Quat. Res. 36 (1) (1991) 72–93.

Somayajulu, B.L.K., Martin, J.M., Eisma, D., Thomas, A.J., Borole, D.V., Rao, K.S., Geochemical studies in the Godavari estuary, India, Mar. Chem. 43 (1-4) (1993) 83–93.

Sridhar, A., Evidence of a late-medieval mega flood event in the upper reaches of the Mahi River basin, Gujarat, Curr. Sci. 96 (11) (2009) 1517–1520.

Stuiver, M., Reimer, P.J., Reimer, R.W., Calib 7.1 [www program] Marine Reservoir Correction Database, 2017 http://calib.org.

Suckow, A., Morgenstern, U., Kudrass, H.R., Absolute dating of recent sediments in the cyclone-influenced shelf area off Bangladesh: comparison of gamma spectrometric (137 Cs, 210 Pb, 228 Ra), radiocarbon, and 32 Si ages, Radiocarbon 43 (2B) (2001) 917–927.

Suokhrie, T., Saalim, S.M., Saraswat, R., Nigam, R., Indian monsoon variability in the last 2000 years as inferred from benthic foraminifera, Quat. Int. 479 (2018) 128–140.

Symphonia, K.T., Nathan, D.S., Geochemistry and distribution of sediments in the East Indian shelf, SW Bay of Bengal: implications on weathering, transport and depositional environment, J. Earth Syst. Sci. 127 (7) (2018) 96.

Taylor, S.R., Abundance of chemical elements in the continental crust: a new table, Geochim. Cosmochim. Acta 28 (8) (1964) 1273–1285.

Taylor, S.R., McLennan, S.M., The Continental Crust: Its Composition and Evolution, Blackwell, Oxford, 1985, pp. 1–312.

Thamban, M., Rao, V.P., Schneider, R.R., Reconstruction of late Quaternary monsoon oscillations based on clay mineral proxies using sediment cores from the western margin of India, Mar. Geol. 186 (3) (2002) 527–539.

Thiry, M., Palaeoclimatic interpretation of clay minerals in marine deposits: an outlook from the continental origin, Earth-Sci. Rev. 49 (1) (2000) 201–221.

Tiwari, M., Managave, S., Yadava, M.G., Ramesh, R., Spatial and Temporal Coherence of Paleomonsoon Records From Marine and Land Proxies in the Indian Region During the Past 30 ka, Platinum Jubilee publication, Indian Academy of Sciences, Bangalore, India, 2009, pp. 1–19.

Tripathi, J.K., Recent contributions in the field of sediment geochemistry, Proc. Ind. Nat. Sci. Acad. 82 (3) (2017) 805–815.

Tripathy, G.R., Singh, S.K., Ramaswamy, V., Major and trace element geochemistry of Bay of Bengal sediments: implications to provenances and their controlling factors, Palaeogeogr. Palaeoclim. Palaeoecol. 397 (2014) 20–30.

Turekian, K.K., Wedepohl, K.H., Distribution of the elements in some major units of the earth's crust, Geol. Soc. Am. Bull. 72 (2) (1961) 175–192.

Turner, A.G., Annamalai, H., Climate change and the South Asian summer monsoon, Nat. Clim. Change 2 (2012) 587–595.

Usapkar, A., Dewangan, P., Badesab, F.K., Mazumdar, A., Ramprasad, T., Krishna, K.S., Basavaiah, N., High resolution Holocene paleomagnetic secular variation records from Bay of Bengal, Phys. Earth Planet. Interiors 252 (2016) 49–76.

Varkey, M., Murty, V., Suryanarayana, A., Physical oceanography of the Bay of Bengal and Andaman Sea, Oceanogr. Mar. Biol. 34 (1996) 1–70.

Vuba, S., Farnaaz, S., Sagar, N., Ahmad, S.M., Geochemical and mineralogical characteristics of recent clastic sediments from lower Godavari river: implications of source rock weathering, J. Geol. Soc. Ind. 82 (3) (2013) 217–226.

Wahsner, M., Müller, C., Stein, R., Ivanov, G., Levitan, M., Shelekhova, E., Tarasov, G., Clay-mineral distribution in surface sediments of the Eurasian Arctic Ocean and continental margin as indicator for source areas and transport pathways—a synthesis, Boreas 28 (1) (1999) 215–233.

Wei, G., Liu, Y., Li, X., Shao, L., Liang, X., Climatic impact on Al, K, Sc and Ti in marine sediments: evidence from ODP Site 1144, South China Sea, Geochem. J. 37 (5) (2003) 593–602.

Weltje, G.J., von Eynatten, H., Quantitative provenance analysis of sediments: review and outlook, Sediment. Geol. 171 (1) (2004) 1–11.

Whitehouse, U.G., Jeffrey, L.M., Debbrecht, J.D., Differential settling tendencies of clay minerals in saline waters, Clays and Clay minerals, Fifth National Conference on Clay and Clay Minerals, Vol. 81119, 1960.

Xu, Z., Lu, H., Zhao, C., Wang, X., Su, Z., Wang, Z., Liu, H., Wang, L., Lu, Q., Composition, origin and weathering process of surface sediment in Kumtagh Desert, Northwest China, J. Geogr. Sci. 21 (6) (2011) 1062–1076.

Yiyang, Z., Mingcai, Y., Geochemistry of Sediments of the China Shelf Sea, Science Press, Beijing, 1994.

Yuste, A., Luzón, A., Bauluz, B., Provenance of Oligocene–Miocene alluvial and fluvial fans of the northern Ebro Basin (NE Spain): an XRD, petrographic and SEM study, Sediment. Geol. 172 (3) (2014) 251–268.

CHAPTER

Holocene environmental magnetic records of Indian monsoon fluctuations

9

N. Basavaiah[a], J. Seetharamaiah[b], Erwin Appel[c], Navin Juyal[d], Sushma Prasad[e,f], K. Nageswara Rao[g], A.S. Khadkikar[h], N. Nowaczyk[i], A. Brauer[i]

[a]*Indian Institute of Geomagnetism, New Panvel, Navi Mumbai, India; Department of Physics, Krishnaveni Degree & P.G. College, Narasaraopet, Guntur, India.* [b]*Department of Geology, The University of Dodoma, Dodoma, Tanzania.* [c]*Department of Geosciences, University of Tübingen, Tübingen, Germany.* [d]*Physical Research Laboratory, Navrangpura, Ahmedabad, India.* [e]*ERA Scientific Editing, Potsdam, Germany.* [f]*Institute for Earth Science, University of Potsdam, Potsdam, Germany.* [g]*Department of Geo-Engineering, Andhra University, Visakhapatnam, Andhra Pradesh, India.* [h]*Råcksta, Stockholm stad, Sweden.* [i]*German Research Centre for Geosciences, Potsdam, Germany*

9.1 Introduction

The Indian Summer Monsoon (ISM) is a major component of the tropical climate system. The ISM is dominantly modified by differential land–sea thermal contrast, producing seasonal reversal in winds and intense rainfall during the summer (Colin et al., 1998; McGregor and Nieuwolt, 1998). The ISM plays an important role in the global hydrological cycle. Broadly, ~60% of the global population depends on the monsoon as it provides vital precipitation for agriculture. Therefore, understanding the nature of ISM variability during the Holocene is essential both to understand the present climatic conditions and to predict future climatic processes (Cai et al., 2010). The monsoon in India has two distinct phases (Saraswat et al., 2014): the southwest monsoon from June to September, also termed as the ISM which brings most of the precipitation over India, and the northeast monsoon (NEM) from November to February, also termed as the winter monsoon, that causes heavy rains over the equatorial Indian Ocean (Fig. 9.1). The evolution patterns of ISM and NEM in India led to controversial views on related driving mechanisms, including interplays of external solar insolation and the Northern Hemisphere (NH) ice volume.

A number of studies based on proxy data and modeling have documented that the Indian monsoonal maximum lagged behind the insolation maximum due to glacial boundary conditions in the NH (Fleitmann et al., 2007; Wang et al., 2010). After the initial phase of intensification, ISM strength had weakened, and its cause is being debated with two contrasting hypotheses. First, an abrupt weakening of ISM at ~5.0–4.5 cal ka BP, and second gradual weakening of the ISM following the decrease of the NH summer insolation (Fleitmann et al., 2007). Reconstruction of Holocene ISM is largely relied upon the lacustrine sediments investigated from the Core Monsoon Zone (CMZ) (Basavaiah et al., 2014; Prasad et al., 2014; Menzel et al., 2014), western India (Basavaiah and Khadkikar, 2004; Prasad and Enzel, 2006), and the Himalayan region (Juyal et al., 2004, 2009; Basavaiah et al., 2004, 2010; Beukema et al., 2011; Anoop et al., 2013; Kotlia et al., 2010; Kotlia and Joshi, 2013; Mishra et al., 2015, 2018;

Holocene Climate Change and Environment. DOI: https://doi.org/10.1016/B978-0-323-90085-0.00004-8

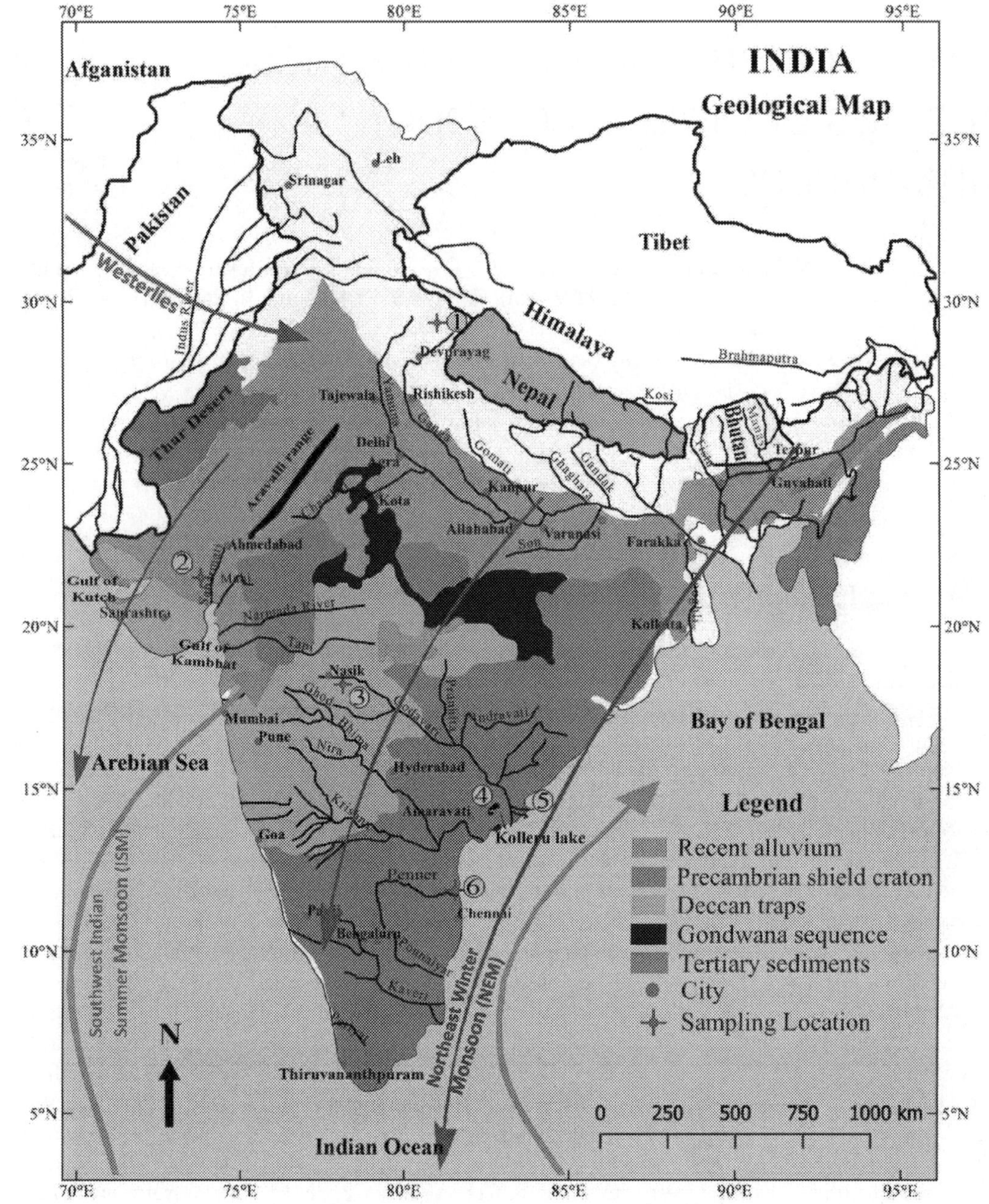

INDIA
Geological Map
Afganistan
Pakistan
Tibet
Himalaya
Nepal
Bhutan
Westerlies
Thar Desert
Aravalli range
Leh
Srinagar
Devprayag
Rishikesh
Tajewala
Delhi
Agra
Kota
Kanpur
Allahabad
Varanasi
Farakka
Kosi
Brahmaputra
Tezpur
Guwahati
Ahmedabad
Gulf of Kutch
Saurashtra
Gulf of Kambhat
Narmada River
Tapi
Nasik
Mumbai
Pune
Hyderabad
Amaravati
Kolleru lake
Goa
Penner
Chennai
Bengaluru
Kaveri
Thiruvananthpuram
Kolkata
Bay of Bengal
Arebian Sea
Indian Ocean
Southwest Indian Summer Monsoon (ISM)
Northeast Winter Monsoon (NEM)
N
Legend
Recent alluvium
Precambrian shield craton
Deccan traps
Gondwana sequence
Tertiary sediments
City
Sampling Location
0 250 500 750 1000 km
70°E 75°E 80°E 85°E 90°E 95°E
35°N 30°N 25°N 20°N 15°N 10°N 5°N

Figure 9.1

Locations of study areas 1–6. (1) Dhakuri Loess deposits from the Himalayan region, (2) Nal Sarovar Lake from western India, (3) Lonar Crater Lake from Central India, and (4) Kolleru Lake, (5) Godavari Delta, (6) Iskapalli lagoon from the central east coast of India (sources: Geological Survey of India, Google Earth Images and Google Maps).

Bohra et al., 2017; Ranhotra et al., 2017; Mehrotra et al., 2018). In contrast, the Ramsar Kolleru Lake is the only Holocene lake record that falls under the influence of the NEM zone (Basavaiah et al., 2015).

The accuracy of the climatic reconstructions greatly depends on the quality of the database and a discreet use of proxy indicators. For instance, carbon isotopes of sedimentary leaf waxes collected from sediment core recovered from the continental shelf off the Godavari river mouth provided an integrated and regionally extensive record of the flora in the CMZ by the persistence of aridity-adapted plants (Ponton et al., 2012; Ghosh et al., 2020). Several other studies have found intensified subseasonal extremes across parts of India and an increase in spatial variability of rainfall, despite an overall weakening of seasonal rainfall in the CMZ (Krishnamurthy et al., 2009; Singh et al., 2019). Segregation of the paleoclimate records from the sediment cores, however, takes inordinate time and efforts. Besides, the laboratory techniques are cumbersome, relatively large quantity of sample materials is needed for carrying out different nonmagnetic measurements.

Environmental magnetic methods have also been used towards understanding the pattern of global climate change reconstruction, paleoenvironment as well as modern environmental pollution as they are relatively simple, fast, efficient, repeatable, and nondestructive to samples whether in loess, lake, deltaic, and lagoonal sediments (Basavaiah and Khadkikar, 2004; Basavaiah et al., 2010, 2012, 2015, 2019; Sridhar et al., 2020). Variation in the magnetic mineralogy parameters reflects changes in climate, which has vital role in weathering along with other nonclimatic factors (Thompson and Oldfield, 1986; Basavaiah, 2011). The ubiquity of iron-bearing minerals being sensitive to lacustrine, fluvial, and deltaic processes such as runoff, sediment flux, accumulation, diagenesis of iron (hydr)oxides, and low temperature oxidation (LTO) carry important environmental information (Thompson and Oldfield, 1986; Evans and Heller, 2003; Basavaiah, 2011). By measuring magnetic parameters, such as magnetic susceptibility (χ), anhysteretic remanent magnetization (ARM), saturation isothermal remanent magnetization (SIRM), and magnetomineralogical S-ratio, we identify where sediments originated, and the intensity of weathering in the catchment area (Basavaiah and Khadkikar, 2004; Basavaiah, 2011).

The most important challenge for magnetism-based proxies is determining the proper interpretation of the magnetic sensitive parameters for paleoclimate signals revealed by sedimentary records. In addition, sedimentary environment contains a number of magnetic mineral subpopulations with different origin history that can reflect multiple environmental processes, and come from varying geological and depositional settings (Fig. 9.1). Hence, the analysis of these sediments with environmental magnetic methods is a challenging task because of the contribution from various material sources resulting in a multicomponent magnetic mineralogy. So, finding a way to assess the relationship between alterations of magnetic minerals and environmental change, driven by climate, is crucial. Here, we propose testing the reliability of magnetic S-ratio proxy by comparing environmental magnetic records from the six sediment archives in ISM, NEM, and westerly wind dominated regions of India (Fig. 9.1). Combined magnetic mineralogical records from the loess-paleosols (central Himalaya), lake (western India), delta, and lagoon sequences (eastern coast Godavari and Iskapalli lagoon) are used to construct a first-order Holocene climate model of the Indian subcontinent.

9.2 Scope of the study

Evaluation of climate models using paleoclimatic data suggested that abrupt changes of Indian monsoon were in phase with North Atlantic climate variations, Inter Tropical Convergence Zone, Indian Ocean

Sea Surface Temperature gradient, and Himalayan snow cover (Hahn and Shukla, 1976; Agnihotri et al., 2002; Gupta et al., 2003; Broccoli et al., 2006; Fleitmann et al., 2007; Sinha et al., 2007; Ponton et al., 2012; Mohtadi et al., 2014). Only a few high temporal resolution studies have been carried out on different past climate regimes (Pant et al., 2005; Ponton et al., 2012; Prasad et al., 2014; Cui et al., 2017). Given their potential for substantial societal impacts, there has been a flurry of recent research on subseasonal processes associated with extreme events (Robertson et al., 2020). However, understanding of these changes remains challenging because of uncertainties in observations and climate models.

Here, the magnetic data of different sedimentary records such as loess, lake, deltaic, and lagoonal sequences were combined to glean the paleoclimate information frozen in magnetic minerals in the form of magnetic parameters such as χ, ARM, SIRM, and S-ratio. Importantly, the S-ratio is highlighted as a new paleoprecipitation proxy to understanding the spatial and temporal variations of magnetic mineral assemblages, linked to ISM and NEM variations. The composite climate map of India is prepared by using S-ratio records in order to synthesize country-wide scheme of monsoonal rainfall events since Last Glacial Maximum (LGM). This environmental magnetic S-ratio proxy method is hoped to gain momentum in the paleoclimate research in near future, and that may also be expanded to incorporate additional paleoclimate proxies within the geologic record.

9.3 Methodology and processes

9.3.1 Study locations

We carried out paleoclimatic studies on six selected study areas across the Indian subcontinent. These are the Dhakuri loess from the Himalayan region, Nal Sarovar Lake sediments of the western region, Lonar Crater Lake sediments from central region of CMZ, the sediments from Godavari delta mangroves, Kolleru Lake, and Iskapalli lagoon from the central eastern coast of India (Fig. 9.2A–F). The details of fieldwork plan, analytical methods, and outcome of study areas were reported in earlier studies (Seetharamaiah et al., 2004, 2005; Pant et al., 2005; Prasad et al., 1997, 2014; Basavaiah et al., 2014, 2015).

The morphology of the six study areas was discussed using the satellite images (Fig. 9.2). Fig. 9.2A shows loess deposits (30°12′N, 79°50′E) at altitudes ranging from 1.8 to 2.5 km in the central Himalaya between Dhakuri in Bageshwar district in Pindar river basin and Chopta in Chamoli district in Alaknanda river basin of Uttarakhand (Pant et al., 2005). Nal Sarovar Bird Sanctuary, consisting primarily of a 120.82 km^2 lake and adjacent marshes, is situated ~64 km to the west of Ahmedabad in Gujarat (Fig. 9.2B). The Holocene stratigraphy of the Nal Sarovar Lake (22°48′N, 72°E) sediments from western India is reported by Prasad et al. (1997). Lonar Lake, the third largest saline lake in the world, is formed in a basaltic terrain in Maharashtra by a meteoritic impact (Fig. 9.2C). The paleoclimate databases are being generated from a 10 m long core, encompassing the Holocene that has been raised from this lake as well as investigations of the modern catchment and lake environment (Menzel et al., 2013; Basavaiah et al., 2014; Sarkar et al., 2014).

The Godavari river is India's second longest river after the Ganga having extensive wave-dominated delta covering an area of 5820 km^2, tidal flat and lagoonal environments along with the Kolleru Lake (Fig. 9.2D). Kolleru Lake is a shallow freshwater body that covers an area of 254 km^2 and is located between the two deltas of the Krishna and Godavari rivers, on the east coast of India

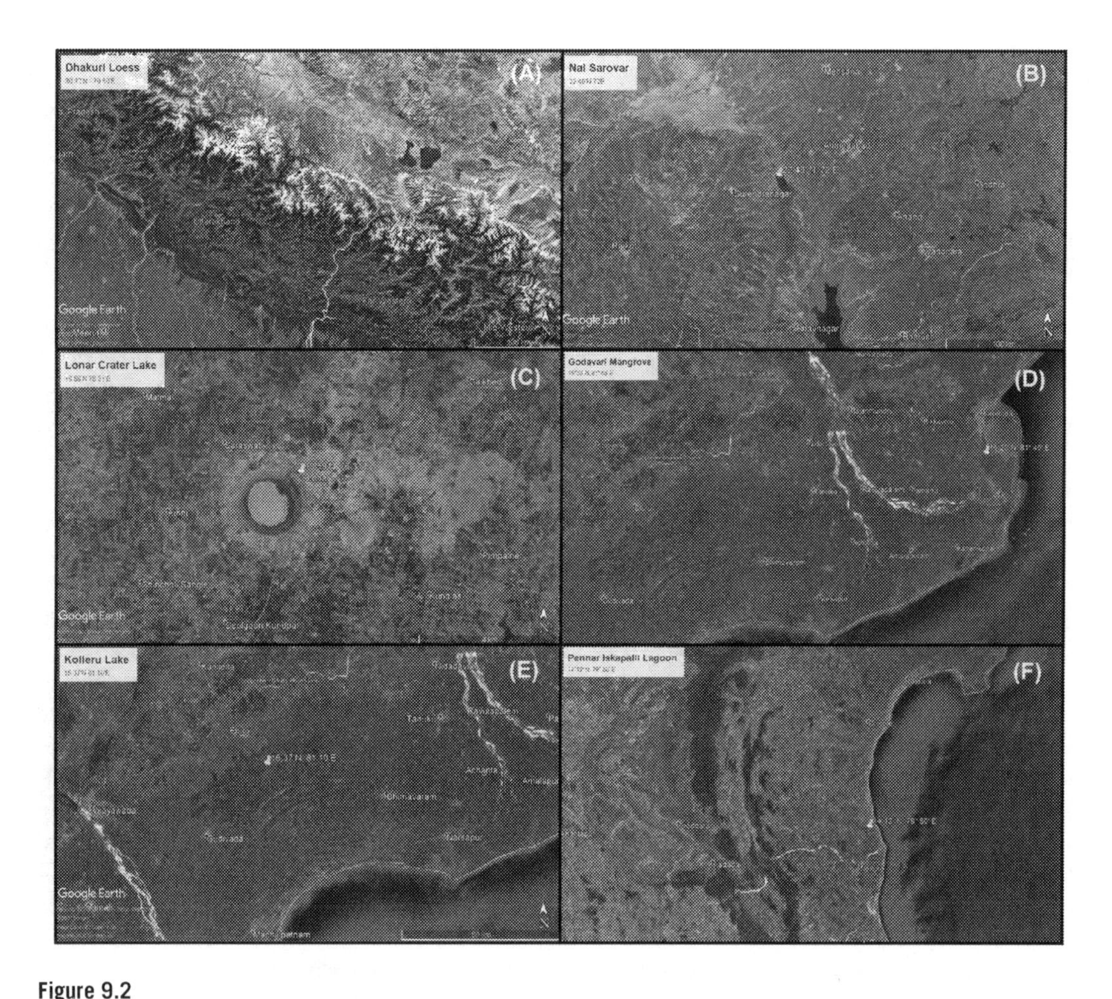

Figure 9.2

(A–F) Google Earth satellite images of study areas: (A) Dhakuri Loess, (B) Nal Sarovar Lake, (C) Lonar Crater Lake, (D) Godavari Mangroves, (E) Kolleru Lake, and (F) Iskapalli lagoon.

(Basavaiah et al., 2015; Nageswara Rao et al., 2020). The presence of beach ridges, tidal flat, and lagoon to lacustrine facies in the 30 km wide coastal belt between Kolleru Lake and the shoreline suggested that the Kolleru Lake was a coastal lagoon during Mid-Holocene and subsequently turned into a freshwater body since Late Holocene with the diminished marine influence and continued freshwater inputs from several rivulets (Nageswara Rao et al., 2010, 2020; Basavaiah et al., 2015; Fig. 9.2E).

Similarly, ~10 km^2 Iskapalli coastal lagoon, which is ~8 km long and ~2.5 km wide, developed during the Holocene evolution of the Penner delta, and is located in northern part of the Penner delta, east coast of India (Fig. 9.1). It has been suggested that the paleoriver courses of the Penner river (Seetharamaiah et al., 2005) are used to drain into this lagoon (Fig. 9.2F) implying that the sediments in Iskapalli lagoon were largely contributed by the Penner river drainage system having an area ~55,200 km^2 in the states of Karnataka and Andhra Pradesh (Fig. 9.2F).

9.3.2 Environmental magnetic measurements

To characterize the magnetic properties, intrinsic magnetic characteristics which arise due to magnetic crystal structure and magnetic grain sizes present in the sediments were measured. The magnetic parameters like χ, frequency dependent χ (χFD), ARM, SIRM, hysteresis loops, low and high temperature magnetic properties, and the Curie and Verwey transition temperatures are measured by magnetic susceptibility meters, spinner magnetometers, pulse magnetizers, demagnetizers, Vibrating Sample Magnetometer, and Kappabridge KLY-4 combined with CS-L cryostat and CS-4 furnace. All magnetic measurements (Table 9.1) were conducted at Environmental Magnetism Laboratory, Indian Institute of Geomagnetism (IIG), Navi Mumbai, India, following the procedures given in Basavaiah (2011).

Analyses of magnetic parameters, such as χ, ARM, SIRM, and their interparametric ratios of ARM/χ, SIRM/χ, and ARM/SIRM, reflect magnetic mineral concentration and/or grain size variations in sediments, respectively, although they are not necessarily well suited for identifying magnetic components within individual magnetic mineral assemblages. For example, SIRM and χ are useful indicators of magnetic mineral concentration. The χFD% ARM/χ, SIRM/χ, and ARM/SIRM represent the relative amount of magnetic grain size variations given as superparamagnetic, SP (<0.03 μm), single domain, SD (0.025–0.05 μm), pseudo single domain, PSD (0.05–10 μm), and multidomain, MD (>10 μm) for cubic and/or slightly elongated magnetite grains (Thompson and Oldfield, 1986; Dunlop and Ozdemir, 2001; Basavaiah, 2011).

S-ratio: This parameter is based on the strong magnetic fields needed to saturate hematite compared to magnetite, which saturates <0.3 T. Typically, maximum fields of 1 T (SIRM) are used, so hematite likely remains unsaturated, and any contribution >0.3 T is attributed to hematite and/or goethite, despite oxidized phases such as titanomagnetite and/or pyrrhotite, often having coercivities >0.3 T. After acquiring SIRM, specimens were subjected to reverse fields of 0.1 T and 0.3 T, and the resulting remanence (IRM) was measured using Molspin's Pulse Magnetizer and Spinner Magnetometer, and denoted as $IRM_{-0.1T}$ and $IRM_{-0.3T}$, respectively. The S-ratio = ($IRM_{-0.3T}$/SIRM), where IRM_{-300} denotes an IRM acquired in a reverse field of 300 mT, can be used to gain information about the magnetic minerals (Bloemendal et al., 1992; Kruiver and Passier, 2001; Basavaiah and Khadkikar, 2004). However, arithmetic means and standard deviations do not apply when using S-ratio data because the S-ratio provides relative rather than absolute information represented by the interdependent low- and high-coercivity magnetic mineral contributions (Heslop, 2009, 2015). In order to reduce nonuniqueness in the interpretation of S-ratio results, regional environmental magnetic records are combined and assessed for identifying mineral magnetic assemblages, and their alteration in diverse sediment archives.

9.4 Results

9.4.1 Himalayan loess-paleosol deposits

Dhakuri loess stratigraphy indicated several well-defined sedimentary deposits with ~30 cm thick pale yellow loessic silt at the base (200–170 cm), which is followed upward by a 30 cm thick moderately weathered reddish-yellow loess horizon at 170–140 cm (Pant et al., 2005). Above this, 50 cm thick unweathered loess (140–90 cm) occurred, which is overlain by a 40 cm thick paleosol (S1).

Table 9.1 The suite of rock magnetic parameters measured, and ratios between them yielded information about concentration, grain size, and composition of magnetic minerals and their environmental and climate significance (Thomson and Oldfield, 1986; Basavaiah and Khadkikar, 2004; Basavaiah, 2011; Basavaiah et al., 2018, 2019; Sridhar et al., 2020).

Parameter	Description	Reported environmental and climate Associations
Magnetic concentration parameters		
Magnetic susceptibility, χ (normalized to mass) (10^{-7} m^3/kg)	Measured as the ratio of volume susceptibility (k) to density $\chi = k/\rho$. Concentration of magnetic minerals, but is often roughly proportional to the concentration of strongly magnetic (e.g., magnetite-like) minerals. Weakly magnetic minerals, like hematite, have much lower χ values and water, organic matter have negative χ	χ is indicative of magnetizability of sediments differentiated by weathering and pedogenic processes. High values of χ represent catchment erosion inputs of primary Fe–Ti oxides from the Deccan Trap basalts of the catchment or unweathered inorganic material as a result of soil disturbance due to land-use
Anhysteretic remnant magnetization, ARM (10^{-5} Am^2/kg) or anhysteretic susceptibility, χ_{ARM} (m^3/kg)	Magnetic remanence imparted in a peak 100 mT demagnetizing alternating field. When normalized by the biasing field strength, ARM is termed as the χ_{ARM}	Allows estimation of the concentration of fine-grained stable single-domain (SSD 0.025 μm $<d<$0.05 μm) ferrimagnetic (magnetite-like) minerals and highest for grains at superparamagnetic (SP)/SSD boundary and lowest for coarser multidomain (MD) grains ($d >$ 10 μm in magnetite). The distribution of MD, SSD, and SP grains in sediment shows different patterns reflecting a change in the runoff processes and a possible weathered and unweathered processes
Saturation isothermal remanent magnetization SIRM (10^{-5} Am^2/kg)	Maximum obtainable IRM depends on concentration but also responds less sensitively than ARM to magnetic grain size	Concentration of all remanence holding minerals in a sample
Domain state		
Coefficient of frequency — dependency susceptibility χfd (%)	Variation in χ between low frequency (χlf 0.976 kHz) and high frequency (15.616 kHz) measurements. $\chi fd\% = 100 (\chi lf - \chi hf)/\chi lf$ Indicator for proportion of ultrafine SP grains	~10% or 5–10% would indicate large fine-grained, secondary SP magnetic mineral s (magnetite-like) that are the result of either weathering or pedogenesis
ARM/SIRM	Relative grain size of magnetite	High values suggest finer SSD and SP grains whereas low values indicate coarser MD minerals of detrital origin
ARM/χ	SSD and SP ferrimagnetic grains can affect the ratio	High values suggest significant SSD (magnetite-like) grains

(continued on next page)

Table 9.1 (*continued*)

Parameter	Description	Reported environmental and climate Associations
SIRM/χ	Could distinguish different types of magnetic behavior	Lower ratio suggests a greater proportion of diamagnetic and paramagnetic minerals. If χ and SIRM values are low, but SIRM/χ is high, there may be a large amount of hematite-like minerals
Mineralogical composition		
Isothermal remnant magnetization (IRM) (10^{-5} Am^2/kg)	Acquired in different DC forward and back fields	Abundance of minerals capable of remanence acquisition at the denoted field: (i) Low fields (<100 mT) magnetically "soft" material and (ii) High fields (>300 mT) magnetically "hard" material
Soft IRM (SIRM-IRM_{-20mT})	Remanent magnetization after a magnetization in the 20 mT reverse field	Mass specific indication of Fe–Ti magnetic minerals (magnetite-like) and used as a proxy to MD ferrimagnetic minerals of detrital origin
Hard IRM (HIRM) (SIRM-$IRM_{-0.3T}$)	Difference between SIRM and IRM measured after magnetization in the 300 mT reverse field	Mass specific indication of the concentration of high coercivity minerals (hematite-like), whereas "Hard" is a measure of the relative abundance of high coercivity (hematite-like) minerals in a mixture with ferrimagnetic (magnetite-like) minerals
S-ratio = ($IRM_{-0.3T}$/SIRM)	Ratio of saturated to nonsaturated minerals at the -300 mT reverse field. Indicator for coercivity, depends on mineralogy, to a lesser degree on grain size	Higher values close to 1 are dominated by soft Fe–Ti magnetic minerals (magnetite and titanomagnetite) with a low coercive force and lower values below 0.9 reflect increasing hard anti-Fe–Ti magnetic component (hematite) with a relatively high coercive force

Stratigraphically, S1 consisting of 20 cm thick layer of dark brown humus rich horizon (Ah) at the top and another 20 cm thick layer of clay illuviated horizon at bottom. The S1 layer is followed by S2 bed consisting of moderately weathered loess horizon (10 cm thick) covered weakly weathered soil (35 cm thick) at the top. Compared to S1 (Ah horizon), S2 layer has lower humus content (Fig. 9.3A).

Fig. 9.3A showing at bottom (200–140 cm) consistent lower values of χ, χFD%, and S-ratio indicate dry periods and higher values at the top denote wetter periods. Magnetic susceptibility changes show progressive increase from the LGM to a rapid increase in χ, approximately 9 ka in line with formation of soil layer S1 followed by a drop at approximately 4 ka, and an increase associated with younger soil layer S2 approximately 1 ka. The shift of χ value between episodes of loess-paleosols is highly correlated with similar trends of mineral magnetic results of χFD% and S-ratio (Fig. 9.3A).

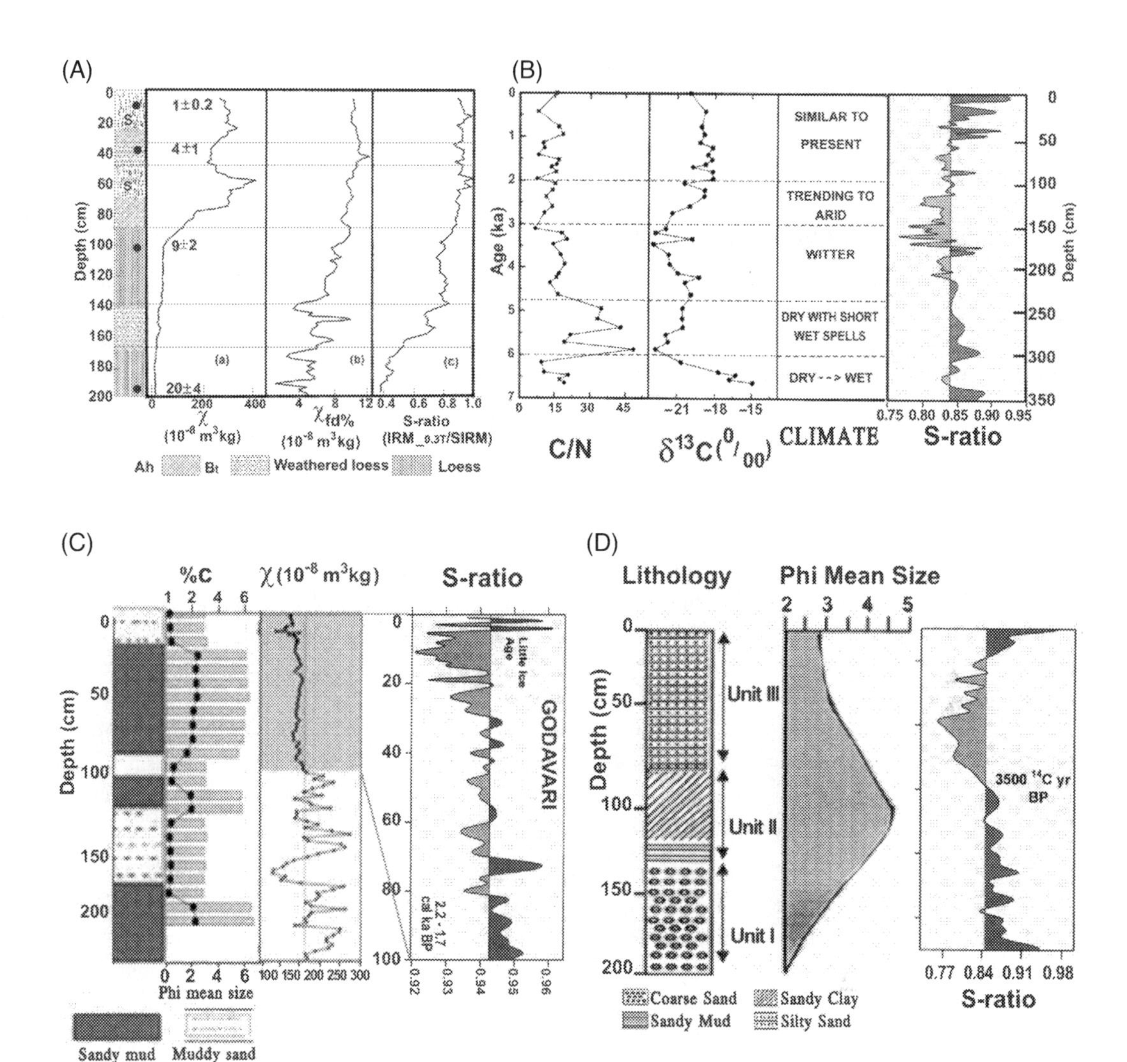

Figure 9.3

(A–D) Proxy climate records of study areas: (A) Dhakuri Loess (Pant et al., 2005), (B) Nal Sarovar (Prasad et al., 1997; Basavaiah and Khadkikar, 2004), (C) Godavari Mangroves (Seetharamaiah et al., 2004), and (D) Iskapalli lagoon (Seetharamaiah et al., 2005; Basavaiah and Khadkikar, 2004).

9.4.2 Nal Sarovar Lake sediments

Fig. 9.3B shows the results of a 3.5 m long sediment core, which was previously analyzed for climate proxies of $\delta^{13}C$, C/N, and S-ratios versus radiocarbon age (Prasad et al., 1997; Basavaiah and Khadkikar, 2004). At this location, the S-ratio ranged ~0.85–0.90 at the bottom of the core between 3.5 m and 2.1 m depths during 7.0–4.5 ka, except one deviation of 0.84 at 3.25 m depth ~6.5 ka (Fig. 9.3B). At the top of the core, between the depths of 2.1 m and 0.60 m (during 4.5–1.5 ka), S-ratio values were lower than 0.84, except at two locations at 1.75 m and 0.80 m depth levels corresponding to 3.5 ka and 2.0 ka, respectively. Subsequently, from 0.60 m depth to the surface, S-ratio values increased again above the mean value of 0.84 during the past two millennia (Fig. 9.3B).

9.4.3 Godavari delta mangrove sediments

The mangrove sediments that were recovered using a 2.2 m long core in the Godavari delta showed alternating sandy mud and muddy sand layers showing distinct sediment characteristics (Seetharamaiah et al., 2004). The sandy muds have a mean size of ~6 Φ, and that of the muddy sand is ~3 Φ. The organic carbon content was also variable in these two types of sediment. The sandy mud contained >2.0% of organic carbon, whereas the muddy sand contained only <0.5% of organic carbon.

The χ values in general were higher at 160–290 ($\times 10^{-8}$ m^3/kg) in the sandy mud layers and <150 ($\times 10^{-8}$ m^3/kg) in the muddy sand layers (Fig. 9.3C). From 100 cm to the top of the core, χ values were consistently <150 ($\times 10^{-8}$ m^3/kg) both in sandy mud and muddy sand layers, while S-ratio showed a low and narrow distribution varying between 0.92 and 0.96 (Fig. 9.3C). The radiocarbon dating of organic matter in this layer from a depth of 83 cm below the surface indicated an age of 2.2–1.7 cal ka BP (Seetharamaiah et al., 2004). Forester et al. (1994) argued that these fluctuations in Holocene χ records reflect climatic change in sediment provenance areas. Considering this possibility, the variations in χ values in the Godavari delta mangrove sediments are inferred to be due to a similar regional climatic and sedimentation pattern. For example, sudden decrease in χ examined by Seetharamaiah et al. (2004) reflects calm energy and slow sedimentation at ~2.0 cal ka BP along the Godavari, Krishna, and Cauvery deltas on the eastern coast of India.

9.4.4 Iskapalii lagoon sediments

The distribution pattern of phi mean grain size of the sediments in one of the cores (14°40′N, 80°09′E) collected from the shallow part of the Iskapalli lagoon indicated three-distinct sedimentary units, Units I–III (Fig. 9.3D). Unit I (200–125 cm), which constitutes the lower part of the core is composed of fine to medium sand with phi mean size varying from 2 to 4. Unit II (125–80 cm) in the middle part of the core is characterized by sandy mud to sandy clay with phi mean size varied from 3.5 at its bottom to 4.5 at its middle and then to 3.5 at its top, whereas Unit III (80–0) exhibits silty sand with phi mean size of 3.5–2.9 (Seetharamaiah et al., 2005).

The upward-fining sequence of sediments from the bottom of Unit I to the middle of Unit II is characterized by the abrupt drop in S-ratio at ~100 cm depth separating coarse sediments of lower and upper parts of the core (Basavaiah and Khadkikar, 2004). The coarser sediments in Unit I showed high S-ratio values (0.84–0.95), but relatively lower values were noticed at depths ~170 cm and 120 cm (Fig. 9.3D). These fluctuations in S-ratio are accompanied by a sudden drop in S-ratio (0.84–0.77) at 80 cm depth followed by a peak value of 0.92 at 10 cm depth (Fig. 9.3D).

9.5 Composite S-ratio map and paleomonsoon rainfall fluctuations

The multiple S-ratio records from the Dhakuri loess (Himalayan region), Nal Sarovar Lake (western India), and the sediments from the Godavari delta mangrove swamp and Iskapalli lagoon in the east coast

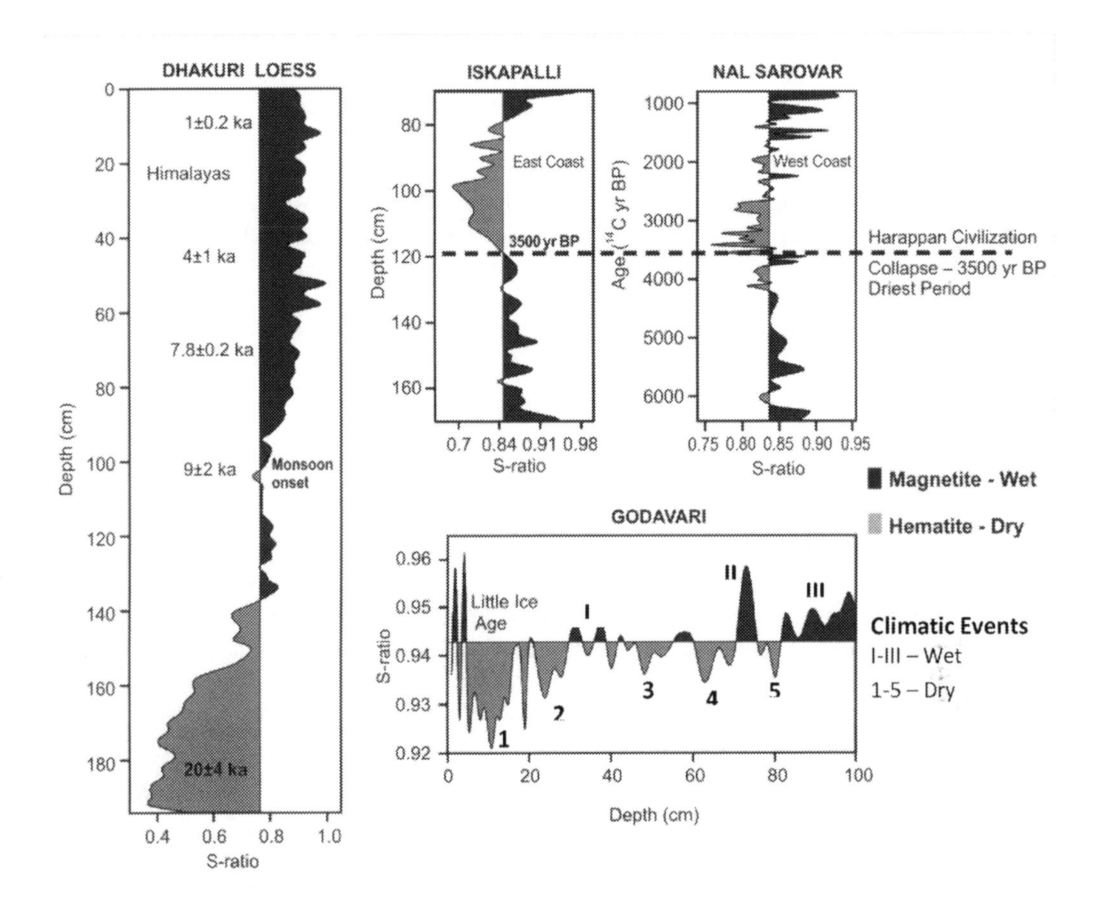

Figure 9.4

The composite S-ratio map of diverse sedimentary archives. Low S-ratios identify periods of reduced rainfall (shown in gray) on various time scales. This mineral magnetic parameter has enabled in identifying several key events hitherto unreported for the ISM (see the text).

of India are plotted in Fig. 9.4. The combined S-ratio results are expected to produce the regional scale paleoclimate changes rather than studying them in isolation of any localized context. The similarity of the signal in the magnetic minerals through S-ratio is assessed to reflect some regional forcing mechanism on the sediment archives as opposed to catchment-specific signals. Therefore, we prepared a composite paleoclimate map showing S-ratio variability of magnetic minerals against stratigraphic depth-age models at four locations as described in Section 9.3 (Fig. 9.3). This stacked S-ratio map was used to infer the regional paleoclimate changes in the Indian subcontinent (Fig. 9.4).

In Fig. 9.4, high S-ratio values of 0.92–0.97 indicated the abundance of soft ferrimagnetic minerals like magnetite/Ti-poor titanomagnetite, while values <0.92 suggested significant quantities of LTO weathered products of Ti-magnetite such as titanohematite, titanomaghemite, and hemoilmentites

(Basavaiah and Khadkikar, 2004; Basavaiah, 2011; Basavaiah et al., 2015, 2018, 2019). Generally, S-ratio values range from 0 to 1, which represent the contribution both from only high coercivity components to low coercivity components, respectively (Bloemendal et al., 1992; Basavaiah and Khadkikar, 2004). Nevertheless, variations in the relative concentration of low and high coercivity phases revealed by the S-ratio are nonlinear and the interpretation of the S-ratio is not unique (Heslop, 2009). As the mass concentration of the high-coercivity mineral increases, S-ratios do not decrease linearly (Frank and Nowaczyk, 2008). Kruiver and Passier (2001) demonstrated that S-ratios <1.0 may occur without the addition of a high-coercivity mineral like hematite, and can be due to coercivity hardening by surface oxidation of magnetite (Van Velzen and Zijderveld, 1995) or LTO of Ti-rich titanomagnetite (Basavaiah and Khadkikar, 2004; Basavaiah et al., 2010, 2015, 2018; Sridhar et al., 2020). The S-ratio value of 0.95 indicates the presence of substantial quantities of high-coercivity minerals (Bloemendal et al., 1992).

The S-ratio parameter, as described in Table 9.1, essentially documents changes in the relative abundances of two oxidation states of Fe in the upper layers of the soil environment leading to the formation of either the reduced or oxidized mineral phase (Basavaiah, 2011). Empirical evidence from paleorecords has demonstrated that the upper soil environment responds to changing redox environment, which in turn is governed by mean annual rainfall. We hypothesize that this could be due to the variable degrees of water saturation of the soil that eventually modulates the redox (oxidation) state leading to the formation of magnetite (possible through a microbial pathway) during periods of enhanced rainfall, whereas the formation of hematite and goethite is made possible through oxidative processes during periods of reduced rainfall. This may obviously be complicated if there is a detrital source of hematite/goethite. Hence, the Deccan Traps being a rich source of magnetite and titanomagnetite contribute to the success of the S-ratio as a proxy for paleomonsoon rainfall reconstruction (Fig. 9.4). With this observation, the S-ratio may be used in principle to reconstruct climate changes in diverse geological settings, rather than magnetic susceptibility.

9.5.1 A 20 ka climate record from Himalayan loess deposits

The continental records from the Himalayan loess-paleosol deposits reflected the efficiency of the S-ratio in accurately outlining major climatic events over the past 20 ka (Fig. 9.4). A synthesis of monsoonal variability revealed that lower S-ratio values during the LGM increase steadily indicting gradual improvement in ISM with the highest monsoon intensity between 6.0 and 4.5 ka (Fig. 9.4). Higher S-ratios from 0.85 to 0.91 in paleosols S1 and S2 ($\sim$9 to $>$4 ka) suggested a stronger monsoon caused by secondary formation of ultrafine-grained SP ferrimagnetic minerals (maghemite and possibly magnetite). Lower S-ratio corresponding to loess indicated a higher contribution of hematite (Fig. 9.4). The global LGM and Younger Dryas (YD) cold events on the basis of lower S-ratios have also been recognized from several Himalayan sites (Basavaiah et al., 2004; Pant et al., 2005; Juyal et al., 2004, 2009). The identified features of LGM and YD cooling were interpreted to indicate that the Himalayan monsoon climate was influenced by the NH glaciation.

Environmental magnetic studies of Pant et al. (2005) revealed that these fine sediments showed a higher magnetic content. The problem was that the anticipated source area of the glacial flour was a granite terrain and granites are poor in iron content. Pant et al. (2005) conjectured that the source of this flour was far away, and they quoted some researchers having obtained loess from the drill-cores taken from the sea floor (Oldfield et al., 1985; Harrison et al., 2001). The other reason Pant et al. (2005) indicated that the presence of magnetic content was due to oxidation of the rock flour after it was deposited as loess. Unless confirmed both explanations are valid. A detailed chronological history reported by Pant et al. (2005) indicated that a drier and a dustier climate from 20 ka to >15 ka, 12 ka to >9 ka, and 4 ka to >1 ka, suggesting these to be phases of weaker SW monsoon. Enhanced SW monsoon existed between 16 ka and 12 ka and even stronger monsoon conditions existed during 9 ka to >4 ka.

9.5.2 Holocene climate records in Western Lake and Eastern Coastal sediments

The period of wetter climatic conditions is characterized by high S-ratios above the mean value of 0.84, varying between 0.84 and 0.95, indicative of low-Ti titanomagnetite and magnetite, while drier climatic conditions showed decreased S-ratio values between 0.75 and 0.84, suggesting a greater proportion of high coercivity minerals such as Ti-rich titanomagnetite, titanomaghemite, titanohematite, hematite, and hemoilmenites (Fig. 9.4). Sediment S-ratio results from western Nal Sarovar Lake and eastern Iskapalli lagoon were provided evidence for the presence of a prolonged drought period between 4.2 and 2.0 ka marked by hematite and LTO weathered products (Fig. 9.4). This period of reduced rainfall is reflected well in the S-ratio that provided the relative proportions of reduced and oxidized state of iron oxides in the sediments. These oxidation states are primarily governed by weathering and water saturation of soils, which in turn is sensitive to mean annual rainfall.

The persistence of 4.2 ka dry event in Nal Sarovar Lake sediments was marked by lower S-ratio below the mean value of 0.84 implying the presence of hematite. The weakening and drying period of the ISM was confirmed between 3.5 ka and 2 ka from eastern (Iskapalli Lagoon) and western (Nal Sraovar Lake) coasts of India (Fig. 9.4). The Lonar Lake record from central India (Anoop et al., 2013; Menzel et al., 2014; Prasad et al., 2014; Mishra et al., 2018) indicated the driest conditions ~4.2 ka based on the maximum enrichment in $\delta^{18}O$ of separated gaylussite crystals. The corroborative evidence of reduction in monsoon rainfall is observed in pollen records of Himalayan Lake (Phadtare, 2000) and from Karwar (Caratini et al., 1994). The limited variability in the S-ratio indicates that the contribution from high-coercivity component, and therefore the relative concentration of low-coercivity remanence carriers vary by less than an order of magnitude (Frank and Nowaczyk, 2008; Heslop, 2009).

Nal Lake data showed an arid interval at approximately 3.5 ka (Prasad et al., 1997). The observed decline of the Bronze age Indus Valley civilization was ascribed to this arid phase, suggesting archeological significance of the data. From 3.0 ka onward, the aridity seems to increase until the present-day climate was established at 2 ka (Bryson and Swain, 1981; Swain et al., 1983; Prasad et al., 1997). Also, drier periods in Nal Sarovar data are correlated well with periods of glacier expansion in Eurasia indicating that the paleoclimatic variations recorded in Nal Sarovar are a regional feature (Prasad et al., 1997).

Similarly, luminescence chronology of Late Holocene extreme hydrological events in the Upper Penner river basin suggested the prevalence of a weak hydrological regime ~3.0 ka and indicated high magnitude flood deposits, overbank sedimentation, and pedogenesis during 1.0–0.6 ka implying intensification of the SW monsoon in the basin (Thomas et al., 2007). A regional flood plain aggradation corresponding to the Medieval Warm Period was observed in the southern (Thomas et al., 2007), central (Kale et al., 2003), and western India (Bhattacharya et al., 2013) implying hydrological response to strengthened ISM. Hence, a comparison of sedimentary archives from eastern and western coasts of

India suggested that ISM and NEM records might be correlated during the Holocene, which provided the first climate model derived from magnetic S-ratio proxy of paleomonsoon change related with magnetic mineralogy changes.

9.5.3 Late Holocene climate records from East Coast mangrove deposits

The S-ratio behavior of Godavari mangrove sediments is characterized by relatively lower values below the mean value of 0.945, marked by S-ratio Events (1–5), while higher S-ratio Events (I–III) are above the mean value (Fig. 9.4). The S-ratios derived for Godavari sediments ranging from 0.92 to 0.96 might be attributed to different stages of oxidation states of titanomagnetite having varying amounts of Ti oxide. The difference (0.04) between higher and lower S-ratio values is found to be a narrow distribution, thus provided possible links to changing rainfall pattern in magnetic mineral composition for the last 3.0 ka (Late Holocene). For example, lower S-ratio values <0.945–0.92 in Event 1 (Fig. 9.4) during the Little Ice Age (ca. 1400–1800 AD) was suggestive of reduced Indian monsoon rainfall, whereas higher S-ratios during Events I–III indicated a wetter monsoon conditions. Further, a low and narrow distribution in Godavari S-ratio record might reflect humidity related magnetic mineral alteration due to LTO process in the catchment area of Godavari river, which lies in the CMZ and western Bengal Fan. It is observed that S-ratio provided a better paleomonsoon rainfall proxy than χ which is found to be less sensitive to sediment provenance and/or climate change (Figs. 9.3C and 9.4).

9.6 Learning and knowledge outcomes

A first-order synthesis of environmental magnetic data from loess, lake, deltaic, and lagoonal sediments from the Indian subcontinent provided continental-scale paleoclimate/paleomonsoon variability related to ISM and NEM and provenance variation. High S-ratio values between 0.92 and 0.96 in the Godavari delta sediments indicated the presence of (titano)magnetite, and LTO weathered products derived from Deccan provenance variation. Whereas relatively low and narrow range of S-ratio values indicated reduced monsoon rainfall due to stronger LTO weathered (drier?) conditions in the Deccan Traps region.

Importantly, the magnetic properties of magnetic mineralogy variations through time revealed four distinct regional climatic phases. A general weakening of the monsoon rainfall was reflected by lower S-ratio values in the Himalayan loess deposits between 20 to >15 ka and 12–9 ka, when cold and dry conditions were prevailed in the region, which in fact favored the formation of loess deposits since 20 ka. This was in tune with the higher latitude climate events such as LGM and YD. In contrast, higher S-ratio values from 0.76 to 0.92 in loess-paleosols indicated a gradual improvement in the climate, and onset of ISM at 9.0 ka with its maximum intensity between 6.0 and 4.5 ka. Subsequently, the lower S-ratio than its mean value of 0.85 during 4.0–2.5 ka, indicated a weakened monsoon rainfall, and climate aridification leading to a continental scale oxidation of magnetic minerals along the western India (Nal Sarovar Lake) and the east coast (Iskapalli lagoon), which also led to the collapse Harappan civilization $\sim$3.5 ka BP. Further, the occurrence of lower S-ratio values in the Godavari mangrove sediments indicated a phase of reduced monsoon rainfall coinciding with the Little Ice Age in Late Holocene.

Thus, the assessment of response of the S-ratio signal of magnetic minerals in sedimentary records from different environments and climatic domains was linked to track climate change by following paleomonsoonal pattern of the subcontinent. Applications of novel composite S-ratio record proved to be an effective proxy to screen different climatic domains of India, and to predict climatic changes both at centennial and millennial time scales more accurately than existing methods including magnetic susceptibility (χ).

Acknowledgments

We are very grateful for the editorial comments and emphasis to contribute this chapter on magnetic proxies by Dr. Navnith K.P. Kumaran. We thank two anonymous reviewers for their helpful comments that improved the content and reorganization of the presentation given in the chapter.

References

Agnihotri, R., Dutta, K., Bhushan, R., Somayajulu, B.L.K, Evidence for solar forcing on the Indian monsoon during the last millennium, Earth Planet. Sci. Lett. 198 (2002) 521–527.

Anoop, A., Prasad, S., Plessen, B., Basavaiah, N., Gaye, B., Naumann, R., Menzel, P., Weise, S., Brauer, A., Palaeoenvironmental implications of evaporative Gaylussite crystals from Lonar lake, Central India, J. Quat. Sci. 28 (4) (2013) 349–359.

Basavaiah, N., Geomagnetism: Solid Earth and Upper Atmosphere Perspectives, Springer, Netherlands, 2011.

Basavaiah, N., Appel, E., Lakshmi, B.V., Deenadayalan, K., Satyanarayana, K.V.V., Misra, S., Juyal, N., Malik, M.A, Revised magnetostratigraphy and nature of the fluvio-lacustrine sedimentation of the Kashmir basin, India, during Pliocene-Pleistocene, J. Geophys. Res. 115 (2010) B08105.

Basavaiah, N., Blaha, U., Das, P.K., Deenadayalan, K., Schulz, H., Evaluation of environmental magnetic pollution screening in soils of basaltic origin: results from Nashik thermal power station, Maharashtra, India, Environ. Sci. Pollut. Res. 19 (2012) 3028–3038.

Basavaiah, N., Juyal, N., Pant, R.K., Yadava, M.G., Singhvi, A.K., Appel, A., Late Quaternary climatic changes reconstructed from mineral magnetic studies from proglacial lake deposits of Higher Central Himalaya, J. Indian Geophys. Union 8 (1) (2004) 27–37.

Basavaiah, N., Khadkikar, A., Environmental magnetism and its application towards palaeomonsoon reconstruction, J. Indian Geophys. Union 8 (1) (2004) 1–14.

Basavaiah, N., Mahesh Babu, J.L.V., Gawali, P.B., Naga Kumar, K.C., Demudu, G., Prizomwala, S.P., Hanamgond, P.T., NageswaraRao, K, Late Quaternary environmental and sea level changes from Kolleru Lake, SE India: inferences from mineral magnetic, geochemical and textural analyses, Quat. Int. 371 (2015) 197–208.

Basavaiah, N., Mahesh Babu, J.L.V., Prizomwala, S.P., Achyuthan, H., Siva, V.H.R., Boral, Pranab, Proxy mineral magnetic and elemental analyses for 2004 tsunami impact deposit along the Muttukadu backwater, East Coast of India–Scope of the palaeotsunami studies, Quat. Int. 507 (2019) 224–232.

Basavaiah, N., Satyanarayana, K.V.V., Deenadayalan, K., Prasad, J.N, Does Deccan Volcanic Sequence contain more reversals than the three-ChronN-R-N flow magnetostratigraphy? – a paleomagnetic evidence from the dyke-swarm near Mumbai, Geophys. J. Int. 213 (2018) 1503–1523.

Basavaiah, N., Wiesner, M.R., Anoop, A., Menzel, P., Nowaczyk, N.R., Deenadayalan, K., Brauer, A., Gaye, B., Naumann, R., Riedel, N., Stebich, M., Prasad, S., Physicochemical analyses of surface sediments from the Lonar Lake, central India – implications for palaeoenvironmental reconstruction, Fundam. Appl. Limnol. 184/1 (2014) 51–68.

Beukema, S.P., Krishnamurthy, R.V., Juyal, N., Basavaiah, N., Singhvi, A.K., Monsoon variability and chemical weathering during the late Pleistocene in the Goriganga basin, higher central Himalaya, India. Quat. Res. 75 (3) (2011) 597–604.

Bhattacharya, F., Rastogi, B.K., Ngangom, M., Thakkar, M.G., Patel, R.C., Late Quaternary climate and seismicity in the Katrol hill range, Kachchh, western India, J. Asian Earth Sci. 73 (2013) 114–120.

Bloemendal, J., King, J.W., Hall, F.R., Doh, S.J., Rock magnetism of Late Neogene and Pleistocene deep-sea sediments: relationship to sediment source, diagenetic processes and sediment lithology, J. Geophys. Res. 97 (1992) 4361–4375.

Bohra, A., Kotlia, B.S., Basavaiah, N., Palaeoclimatic reconstruction by using the varvite sediments of Bharatpur, Upper Lahaul Valley, NW Himalaya, India, Quat. Int. 443 (2017) 39–48.

Broccoli, A.J., Dahl, K.A., Stouffer, R.J., Response of the ITCZ to Northern Hemisphere cooling, Geophys. Res. Lett. 33 (2006) L01702, doi:10.1029/2005GL024546.

Bryson, R.A., Swain, A.K., Holocene variations in monsoonal rainfall in Rajasthan, Quat. Res. 16 (1981) 135–145.

Cai, Y., Tan, L., Cheng, H., An, Z., Edward, R.L., Kelly Megan, J., Kong, X., Wang, X., The variation of summer monsoon precipitation in central China since the last deglaciation, Earth Planet. Sci. Lett. 291 (2010) 21–31.

Caratini, C., Bentaleb, I., Fontugne, M., et al., A less humid climate since ca. 3500 yr B.P. from marine cores off Karwar, western India, Palaeogeogr. Palaeoclim. Palaeoecol. 109 (1994) 371–384.

Colin, C., Kissel, C., Blamart, D., Turpin, L., Magnetic properties of sediments in the Bay of Bengal and the Andaman Sea: impact of rapid North Atlantic Ocean climatic events on the strength of the Indian monsoon, Earth Planet. Sci. Lett. 160 (1998) 623–635.

Cui, M., Wang, Z., Nageswara Rao, K., Sangode, S.J., Saito, Y., Chen, T., Kulkarni, Y., Nagakumar, K.Ch.V., Demudu, G, A mid- to late-Holocene record of vegetation decline and erosion triggered by monsoon weakening and human adaptations in south-east Indian peninsula, Holocene 27 (12) (2017) 1976–1987.

Dunlop, D.J., Ozdemir, O., Rock Magnetism. Fundamentals and Frontiers, Cambridge University Press, Cambridge, UK, 2001.

Evans, M.E., Heller, F., Environmental Magnetism: Principles and Applications of Enviromagnetics, Academic Press, London, UK, 2003.

Fleitmann, D., Burns, S.J., Mangini, A., et al., Holocene ITCZ and Indian monsoon dynamics recorded in stalagmites from Oman and Yemen (Socotra), Quat. Sci. Rev. 26 (2007) 170–188.

Forster, T., Evans, M.E., Heller, F., The frequency dependence of low field susceptibility in loess sediments, Geophys. J. Int. 118 (1994) 636–642.

Frank, U., Nowaczyk, N.R., Mineral magnetic properties of artificial samples systematically mixed from haematite and magnetite, Geophys. J. Int. 175 (2008) 449–461.

Ghosh, S., Prasanta, S., Roy, S., et al., Early Holocene Indian summer monsoon and its impact on vegetation in the Central Himalaya: insight from δD and $\delta^{13}C$ values of leaf wax lipid, Holocene 30 (70) (2020) 1063–1074.

Gupta, A.K., Anderson, D.M., Overpeck, J.T., Abrupt changes in the Asian southwest monsoon during the Holocene and their links to the North Atlantic Ocean, Nature 421 (2003) 354–356.

Hahn, D.G., Shukla, J., An apparent relationship between Eurasian snow cover and Indian monsoon rainfall, J. Atmos. Sci. 33 (1976) 246–2462.

Harrison, S.P., Kohfeld, K.E., Roelandt, C., Claquin, T., The role of dust in climate changes today, at the last glacial maximum and in future, Earth Sci. Rev. 54 (2001) 43–80.

Heslop, D., On the statistical analysis of the rock magnetic S-ratio, Geophys. J. Int. 178 (1) (2009) 159–161.

Heslop, D., Numerical strategies for magnetic mineral unmixing, Earth Sci. Rev. 150 (2015) 256–284.

Juyal, N., Pant, R.K., Basavaiah, N., Bhushan, R., Jain, M., Saini, N.K., Yadava, M.G., Singhvi, A.K., Reconstruction of last glacial to early Holocene monsoon variability from relict lake sediments of the Higher Central Himalaya, Uttrakhand, India, J. Asian Earth Sci. 34 (3) (2009) 227–492.

Juyal, N., Pant, R.K., Basavaiah, N., Yadava, M.G., Saini, N.K., Singhvi, A.K., Climate and seismicity in the Higher Central Himalaya during the last 20 ka: evidences from Garbyang basin, Uttaranchal, Palaeogeogr. Palaeoclim. Palaeoecol. 213 (2004) 315–330.

Kale, V.S., Mishra, S., Baker, V.R., Sedimentary records of palaeofloods in the bedrock gorges of the Tapi and Narmada rivers, central India, Curr. Sci. 84 (2003) 1072–1079.

Kotlia, B.S., Joshi, L.M., Late Holocene climatic changes in Garhwal Himalaya, Curr. Sci. 104 (7) (2013) 911–919.

Kotlia, B.S., Sanwal, J., Phartiyal, B., Joshi, L.M., Trivedi, A., Sharma, C., Late Quaternary climatic changes in the eastern Kumaun Himalaya, India, as deduced from multi-proxy studies, Quat. Int. 213 (1–2) (2010) 44–55.

Krishnamurthy, C.K.B., Lall, U., Kwon, H.H, Changing frequency and intensity of rainfall extremes over India from 1951 to 2003, J. Clim. 22 (18) (2009) 4737–4746, doi:10.1175/2009JCLI2896.1.

Kruiver, P., Passier, H.F., Coercivity analysis of magnetic phases in sapropel S1 related to variations in redox conditions, including an investigation of the S-ratio, Geochem. Geophys. Geosyst. 2 (18) (2001) 1063, doi:10.1029/2001gc000181.

McGregor, G.R., Nieuwolt, S., Tropical Climatology, Wiley, Chichester, England, 1998, p. 339.

Mehrotra, N., Shah, S.K., Basavaiah, N., Laskar, A.H., Yadava, M.G., Resonance of the '4.2ka event' and terminations of global civilizations during the Holocene, in the palaeoclimate records around PT Tso Lake, Eastern Himalaya, Quat. Int. 507 (2018) 206–216.

Menzel, P., Gaye, B., Mishra, P.K., Anoop, A., Basavaiah, N., Marwan, N., Plessen, B., Prasad, S., Riedel, N., Stebich, M., Wiesner, M.G., Linking Holocene drying trends from Lonar Lake in monsoonal central India to North Atlantic cooling events, Palaeogeogr. Palaeoclim. Palaeoecol. 410 (2014) 164–178.

Menzel, P., Gaye, B., Wiesner, M.G., Prasad, S., Stebich, M., Das, B.K., Anoop, A., Riedel, N., Basavaiah, N., Influence of bottom water anoxia on nitrogen isotopic ratios and amino acid contributions of recent sediments from small eutrophic Lonar Lake, Central India, Limnol. Oceanogr. 58 (3) (2013) 1061–1074.

Mishra, P.K., Anoop, A., Schettler, G., Prasad, S., Jehangir, A., Menzel, P., Naumann, R., Yousuf, A.R., Basavaiah, N., Deenadayalan, K., Wiesner, M.G., Gaye, G., Reconstructed late Quaternary hydrological changes from Lake TsoMoriri, NW Himalaya, Quat. Int. 371 (2015) 76–86.

Mishra, P.K., Prasad, S., Marwan, N., Anoop, A., Krishnan, R., Gaye, B., Basavaiah, N., Stebich, M., Menzel, P., Nils, R., Contrasting pattern of hydrological changes during the past two millennia from central and northern India: regional climate differences or anthropogenic impact? Glob. Planet. Change 161 (2018) 97–107.

Mohtadi, M., Prange, M., Oppo, D.W., De Pol-Holz, R., Merkel, U., Zhang, X., Steinke, S., Luckge, A., North Atlantic forcing of tropical Indian Ocean climate, Nature 509 (2014) 76–80.

Nageswara Rao, K., Nagakumar, K.Ch.V., Subraelu, P., et al., Kolleru Lake revisited: the post 'Operation Kolleru' scenario, Curr. Sci. 98 (2010) 1289–1291.

Nageswara Rao, K., Pandey, S., Kubo, S., et al., Paleoclimate and Holocene relative sea-level history of the east coast of India, Paleolimnology 64 (2020) 71–89.

Oldfield, F., Hunt, A., Jones, M.D.H., Chester, R., Dearing, J.A., Olsson, L., Prospero, J.M., Magnetic differentiation of atmospheric dust, Nature 317 (1985) 516–518.

Pant, R.K., Basavaiah, N., Juyal, N., et al., A 20-ka climate record from Central Himalayan loess deposits, J. Quat. Sci. 20 (2005) 485–492.

Phadtare, N.R., Sharp decrease in summer monsoon strength 4000-3500 cal yr B.P. in the central higher Himalaya of India based on pollen evidence from Alpine peat, Quat. Res. 53 (2000) 122–129.

Ponton, C., Giosan, L., Eglinton, T.I., et al., Holocene aridification of India, Geophys. Res. Lett. 39 (3) (2012) L03704, doi:10.1029/2011GL050722.

Prasad, S., Anoop, A., Riedel, N., et al., Prolonged monsoon droughts and links to Indo-Pacific warm pool: a Holocene record from Lonar Lake, central India, Earth Planet. Sci. Lett. 391 (2014) 171–182.

Prasad, S., Enzel, Y., Holocene paleoclimates of India, Quat. Res. 66 (3) (2006) 442–453.

Prasad, S., Kusumgar, S., Gupta, S.K., A mid to late Holocene record of palaeoclimatic changes from Nal Sarovar: a palaeodesert margin lake in western India, J. Quat. Sci. 12 (2) (1997) 153–159.

Ranhotra, P.S., Sharma, J., Bhattacharyya, A., Basavaiah, N., Dutta, K., Late Pleistocene-Holocene vegetation and climate from the palaeolake sediments, Rukti valley, Kinnaur, Himachal Himalaya, Quat. Int. 479 (2018) 79–89.

Robertson, A., Frederic, W., Suzana, V., Camargo, J., Subseasonal to seasonal prediction of weather to climate with application to tropical cyclones, J. Geophys. Res. Atmos. 125 (6) (2020), doi:10.1029/2018JD029375.

Saraswat, R., Nigam, R., Correge, T., A glimpse of the Quaternary monsoon history from India and adjoining seas, Palaeogeogr. Palaeoclim. Palaeoecol. 397 (2014) 1–6.

Sarker, S., Wilkes, H., Prasad, S., Brauer, A., Riedel, N., Stebich, M., Basavaiah, N., Sachse, D, Spatial heterogeneity in lipid biomarker distributions in the catchment and sediments of a crater lake in central India, Org. Geochem. 6 (2014) 125–136.

Seetharamaiah, J., Basavaiah, N., Chakraborty, S., et al., Use of susceptibility of mangroves deposits in vibracores from deltaic environments, J. Indian Geophys. Res. 8 (1) (2004) 65–70.

Seetharamaiah, J., Farooqui, A., Suryabhagavan, K.V., NageswaraRao, K., Evolution of Iskapalli Lagoon in Penner delta region, Andhra Pradesh – a sedimentological and palynological approach, Indian J. Mar. Sci. 34 (2005) 25–30.

Singh, D., Ghosh, S., Roxy, M.K., Sonali, M.D., Indian summer monsoon: Extreme events, historical changes, and role of anthropogenic forcings, WIREs Clim. Change 10 (2019) e571, doi:10.1002/wcc.571.

Sinha, A., Cannariato, K.G., Stott, L.D., Cheng, H., Edwards, L.D., Yadav, M.G., Ramesh, R., Singh, I.B., A 900-year (600 to 1500 A.D.) record of the Indian summer monsoon precipitation from the core monsoon zone of India, Geophys. Res. Lett. 34 (16) (2007) L16707, doi:10.1029/2007GL030431.

Sridhar, A., Thakur, B., Basavaiah, N., Seth, P., Tiwari, P., Chamyal, L.S., Lacustrine record of high magnitude flood events and climate variability during mid to late Holocene in the semiarid alluvial plains, western India, Palaeogeogr. Palaeoclim. Palaeoecol. 542 (2020) 109581, doi:10.1016/j.palaeo.2019.109581.

Swain, A.M., Kutzbach, J.E., Hastenrath, S., Estimates of Holocene precipitation for Rajasthan, India, based on pollen and lake level data, Quat. Res. 19 (1983) 1–17.

Thomas, P.J., Juyal, N., Kale, V.S., Singhvi, A.K., Luminescence chronology of late Holocene extreme hydrological events in the upper Penner River basin, South India, J. Quat. Sci. 22 (2007) 747–753.

Thompson, R., Oldfield, F., Environmental Magnetism, Allen and Unwin, London, UK, 1986.

Van Velzen, A.J., Zijderveld, J.D.A., Effects of weathering on single-domain magnetite in Early Pliocene marine marls, Geophys. J. Int. 121 (1995) 267–278.

Wang, P., Tian, J., Lourens, L.J., Obscuring of long eccentricity cyclicity in Pleistocene oceanic carbon isotope records, Earth Planet. Sci. Lett. 290 (2010) 319–330.

CHAPTER

10 Application of tree rings in understanding long-term variability in river discharge of high Himalayas, India

Vikram Singh[a], **Krishna G. Misra**[a], **Akhilesh K. Yadava**[a], **Ram R. Yadav**[b]

[a] *Birbal Sahni Institute of Palaeosciences, Lucknow, India.* [b] *Wadia Institute of Himalayan Geology, Dehradun, India*

10.1 Introduction

Rivers providing the hydrological needs of mankind have been the backbone of human civilization. The societal relevance of rivers is well recognized through food production, hydropower generation and providing trade routes. However, excess/reduced river water supplies bring ecological catastrophe causing immense loss of life and property. Extreme hydrological events, the frequency of which is expected to increase in the future (IPCC, 2012; Visser et al., 2014) with global warming, may have severe socioeconomic impacts. Vulnerability to extreme hydrological events has further increased in recent times due to increased human settlements and developmental activities along the river courses. In view of this, better understanding of the recurrence behavior of extreme hydrological events in long-term perspective is important to adopt appropriate mitigation measures.

The Himalayan mountain system is unique due to its mid-latitude geographical location as well as dynamic geological characteristics. The mountain system having vast climatic, ecological and biodiversity ranges is vulnerable to climate change. High mountain ranges of the Himalaya, an abode of large number of glaciers, have the largest amount of snow/ice cover outside the polar regions. These glaciers are the source of several rivers, which are the lifeline of the downstream population. The high-altitude glaciers in the western Himalaya generally come under the monsoon shadow zone where Indian Summer Monsoon does not reach due to leeward monsoon shadow effects, whereas in the central and eastern Himalaya glaciers are largely fed by summer monsoon precipitation (Lang and Barros, 2004; Bookhagen and Burbank, 2010; Azam et al., 2016). The glaciers in the monsoon shadow zone are replenished by precipitation largely brought by western disturbances (WDs) during the winter and spring seasons (Yadav and Bhutiyani, 2013). In the western Himalaya snow and ice meltwater of glaciers largely contribute to the discharge of rivers. However, climatic warming and changes in precipitation pattern across the Himalaya is of growing concern to society due to its impact on river discharge and water availability on sustained basis. Observational discharge data of rivers originating from the Himalaya, though limited to the past few decades, show decreasing trend in recent decades (Bhutiyani et al., 2008; Romshoo et al., 2017). However, such short-term instrumental records restrict our understanding of the natural variability of river discharge on multidecadal-to-centennial

Holocene Climate Change and Environment. DOI: https://doi.org/10.1016/B978-0-323-90085-0.00018-8

scale perspective. Researches from different geographic regions of the Earth (Stockton and Jacoby, 1976; Woodhouse and Lukas, 2006; Gou et al., 2010; Wise, 2010; D'Arrigo et al., 2011; Margolis et al., 2011; Maxwell et al., 2011; Urrutia et al., 2011; Yang et al., 2012; Sun et al., 2013; Cook et al., 2013; Harley et al., 2016; Rao et al., 2018; MartínezSifuentes et al., 2020) have established that the annually resolved tree-rings from moisture stressed sites in river basins provide valuable proxy to supplement the existing river discharge records back to several centuries/millennia. Such studies, hitherto limited in the western (Singh and Yadav, 2013; Shah et al., 2013; Misra et al., 2015) and the eastern (Shah et al., 2014; Shekhar and Bhattacharyya, 2015) Himalayan region in India have also demonstrated the potential of developing long-term river discharge reconstructions using tree-ring chronologies developed from moisture stressed sites from the river basins.

10.2 Scope

The Himalayan mountain system has a large network of rivers that originate from glacier and snow-fed valleys. These rivers fulfill the socioeconomic needs of the millions of people in the adjacent and low lying areas in different ways. The changing climate of the globe is adversely affecting the natural discharge of rivers and threatening the ecosystem. In view of this a thorough understanding of the river discharge in long-term perspective is required to prepare for future challenges. Annually resolved tree-ring series from moisture stressed sites in the river basins provide a valuable proxy to supplement the existing records back to the past several centuries/millennia. Such long-term records should provide valuable baseline data to understand the hydrological variability of rivers on which the sustenance of human society largely depends.

10.3 Methodology

10.3.1 Field work and sampling techniques

Trees are the natural archives, consistently recording the ambient environmental changes affecting tree growth. The growth of a tree over a specific region is limited by various climatic factors and among them temperature and precipitation are largely the most common. These two climatic parameters in combination with others like solar radiation, wind, humidity, soil thickness and slope make the regional environmental settings of the growing trees. Trees growing over climatically stressed sites where at least one of the climatic variables acts as a growth limiting factor are ideal for dendroclimatic studies. Hence, as per the objective of the study sampling locations and sites are selected for the collection of tree-ring materials. Generally, in cold high-altitude regions of the Himalaya, close to the snowline tree growth is limited by temperature. However, in the semiarid–arid low lying areas precipitation, regulating the soil moisture availability, limits the tree-growth. Along with the elevation and aspect, soil thickness and slope are also important criteria for the site selection when reconstruction of hydrological parameters is aimed for. At gentle slopes or flat terrains soil cover is generally thick with good moisture holding capacity and therefore tree roots can get sufficient moisture. However, on steep slopes, where soil cover is generally thin with low moisture holding capacity, tree growth is very sensitive to fluctuations in precipitation. High moisture stress conditions over steep slopes make the trees a sensitive natural ombrometer, as the influence of precipitation is sensibly reflected on radial growth of trees compared to that on gentle slopes or flat terrains where water table is close to the tree's root zone. The annual growth ring series obtained from trees growing over climatically stressed sites show high variability in ring-width patterns (narrow and wide). However, the trees growing on sites with thick soil cover produce almost similar growth ring patterns without much variation in widths of annual growth ring sequences. For tree-ring-based hydroclimatic study moisture stressed sites are suitable where precipitation largely controls the tree-growth. Along with the climatic conditions tree-growth is also influenced by various other external and internal factors like competition among trees for nutrients, sunlight, diseases and physical injuries. Cutting and lopping are very common in forests close to human settlements disturbing natural tree growth patterns. Therefore, utmost care needs to be taken during the selection of sites and trees in sampling for hydroclimatic studies. Generally, healthy trees without any visible physical marks of injury are selected for the collection of tree-ring samples. Two increment cores from the opposite direction of the stem and perpendicular to the slope are usually taken to retrieve the pith of the target trees to get the beginning year of growth. Sometimes, in years with extreme climatic conditions trees do not produce growth rings uniformly in all the directions of the stem and hence the cores are taken from opposite directions to overcome the growth ring dating problems. Usually, the trees are cored at breast height (1.4 m) with the help of an increment borer. Due to the fragile nature of cores, utmost care is taken in the handling of cores in further analyses.

Himalayan cedar (*Cedrus deodara* (Roxb.) G. Don) is the most common constituent of the dry evergreen conifer forest in the Sutlej river basin in Kinnaur, Himachal Pradesh. This species has been found growing on moist to semiarid sites in the river basin at an altitude ranging from 2200 to 3500 m asl (Misra et al., 2015). Himalayan cedar has wide ecological range and usually prefers good amount of winter snowpack, well-drained soil and limited summer monsoon rainfall (Champion and Seth, 1968). Over moist sites trees have been found to grow faster but usually do not attain long age (Gamble, 1902) largely due to heartwood rot. However, trees growing over semiarid sites attain longer age due to slow annual growth with low moisture content resisting the heartwood rot. Over millennium old Himalayan cedar trees have been found still growing in healthy conditions at moisture stressed sites in the western Himalaya (Singh et al., 2004). In the Sutlej river basin, the other conifer species attaining over millennium age are of neoza pine (*Pinus gerardiana* Wall. ex Lamb.; Singh and Yadav, 2007; Yadava et al., 2016). For river discharge studies Misra et al. (2015) analyzed Himalayan cedar samples from a network of seven homogenous moisture stressed sites in the Sutlej river basin in Kinnaur, Himachal Pradesh (Fig. 10.1). Himalayan cedar trees are the important constituent of dry conifer forests in the river basin (Fig. 10.2A). Trees growing on steep rocky slopes with very thin soil cover (Fig. 10.2B), at many occasions attaining long age, provide the ideal material for hydrological studies as their radial growth is very sensitive to changes in precipitation due to low moisture holding capacity of the soil.

10.3.2 Crossdating of growth ring sequences

To assign exact calendar age to growth ring sequences crossdating, a pattern matching technique is used (Stokes and Smiley, 1968). Under crossdating procedure, sequences of ring-width patterns (narrow and wide) are matched among samples of individual trees and of the whole site. This procedure helps in identifying the missing/false rings if any in growth ring sequences of samples and assign exact year to each ring. After precise dating of tree-ring sequences ring widths are measured using linear

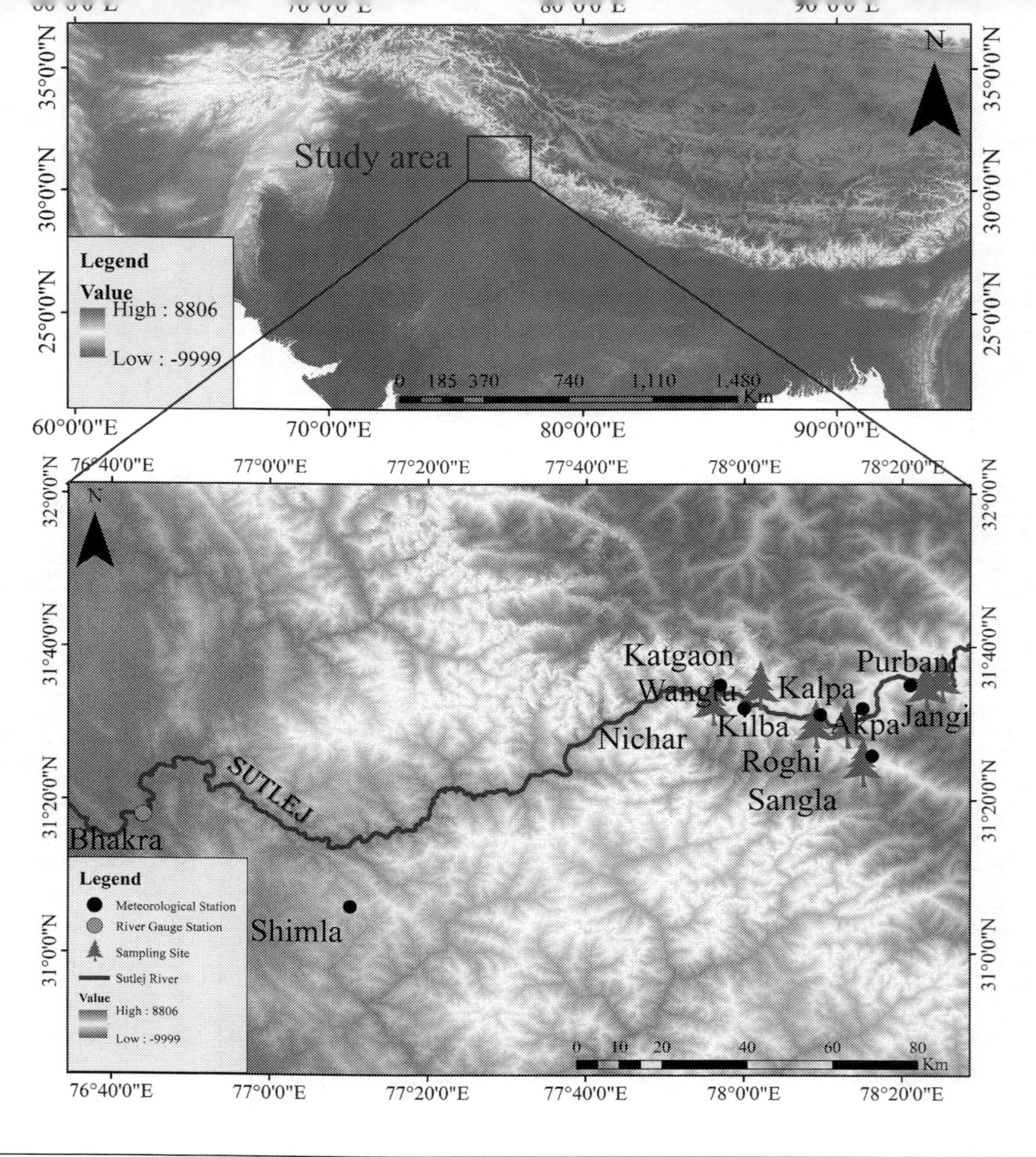

Figure 10.1

Location map of tree-ring sampling sites, meteorological and river gauge stations used in the Sutlej river discharge reconstruction (Misra et al., 2015).

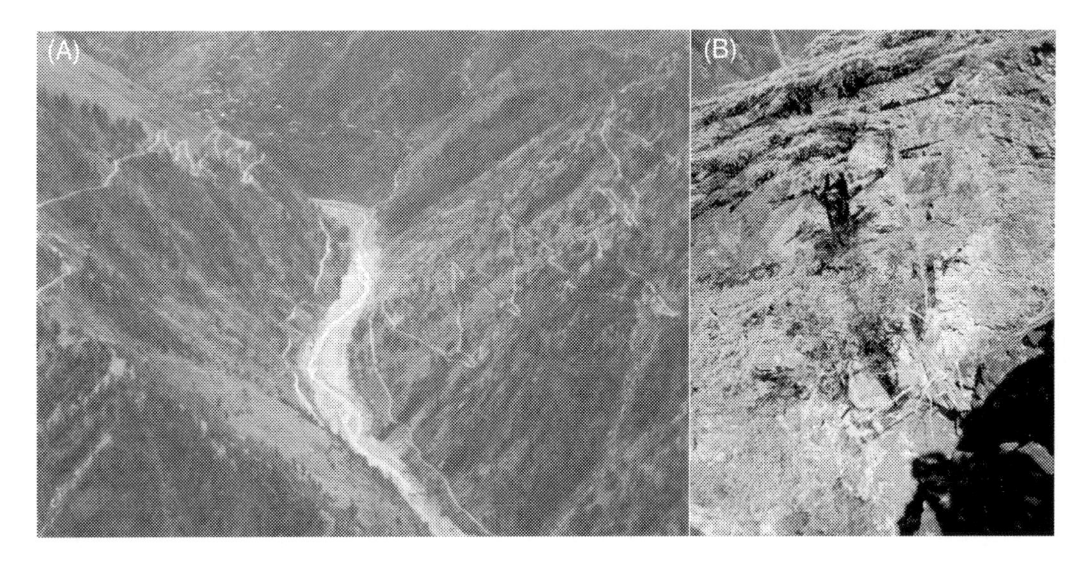

Figure 10.2

(A) General view of the Sutlej river basin with open dry conifer forest in Kinnaur District, Himachal Pradesh and (B) Himalayan cedar tree growing on moisture stressed rocky site with very thin soil cover.

encoder coupled with computer. The ring-width measurement series of samples are analyzed with program COFECHA (Holmes, 1983) for dating quality control check. In case of any errors indicated in COFECHA analyses tree-ring samples and ring-width measurements are carefully examined and errors, if any, corrected.

10.3.3 Chronology preparation

Ring-width measurement series of all the samples from a site are combined to develop mean tree-ring chronology. However, as the measurement series are affected by several environmental factors other than climate as well as internal factors (genetic constitution of trees, biological age) and external factors like competition among neighboring trees and diseases, data are standardized to minimize tree to tree variations and maximize the common signal before averaging to prepare the mean chronology (Fritts, 1976). For standardization of ring-width measurement series, various curve fitting options appropriate for the type of growth curve are used. Basic software packages commonly used in standardization of ring width measurement series are freely available on Laboratory of Tree-Ring Research, Arizona (https://ltrr.arizona.edu/research/software) and Tree-Ring Laboratory, Lamont-Doherty Earth Observatory, University of Columbia (https://www.ldeo.columbia.edu/tree-ring-laboratory/resources/software) websites. Mean tree-ring chronologies representing the population signal (Wigley et al., 1984) are further analyzed to understand environmental signal strength and climate variable for reconstruction.

Misra et al. (2015) analyzed over 302 increment core samples collected from 212 trees growing over seven moisture stressed sites in the Sutlej river basin (Fig. 10.1). To prepare ring-width chronologies

Table 10.1 Detail statistics of Himalayan cedar tree-ring width chronologies developed from seven sites in Kinnaur, Himachal Pradesh (Misra et al., 2015).

S. no.	Site	Location	Elevation (m)	Core/ trees	SY	Mean correlation with master dating series	Chronology with EPS >0.85	MS	SD	AR1
1.	Akpa	31°35′ N, 78°23′ E	3127–3266	23/17	1440	0.868	1500–2005	0.45	0.40	0.28
2.	Katgaon	31°35′ N, 78°02′ E	2639–2949	82/55	1480	0.835	1540–2005	0.42	0.40	0.33
3.	Kilba	31°30′ N, 78°09′ E	2260–2500	33/25	1432	0.830	1515–2005	0.42	0.43	0.36
4.	Nichar	31°33′ N, 77°56′ E	2182–2277	48/32	1580	0.800	1640–2005	0.35	0.32	0.22
5.	Roghi	31°30′ N, 78°13′ E	2851–3062	62/45	1388	0.797	1440–2005	0.39	0.35	0.28
6.	Sangla	31°25′ N, 78°15′ E	2345–2590	25/19	1607	0.821	1660–2005	0.37	0.36	0.37
7.	Jangi	31°36′ N, 78°25′ E	3318–3471	29/19	1353	0.813	1535–2005	0.35	0.35	0.39

AR1, *first-order autocorrelation;* EPS, *expressed population signal;* MS, *mean sensitivity;* SD, *standard deviation;* SY, *start year of the chronology.*

Table 10.2 Pearson correlation among Himalayan cedar ring-width chronologies AD 1660–2005 (Misra et al., 2015; two tailed $P \leq .000001$).

Site-chronologies	Akpa	Katgaon	Kilba	Nichar	Roghi	Sangla
Katgaon	.812					
Kilba	.805	.926				
Nichar	.696	.919	.862			
Roghi	.822	.879	.898	.838		
Sangla	.815	.803	.800	.739	.844	
Jangi	.904	.782	.785	.677	.826	.852

they standardized the measurement series to maximize the common climatic signal and minimize tree-to-tree differences, that is, noise. The chronology length varied in different sites (Table 10.1) and for that reason, they used the common chronology period AD 1660–2005 for climate studies. The ring-width chronologies from all the sites analyzed by Misra et al. (2015) revealed strong year-to-year similarity and significant correlations (Table 10.2) indicating common climatic forcing regulating tree-growth over the region.

10.3.4 Climatic settings along the course of Sutlej River

The Sutlej river is a tributary of the Indus river system and the largest in Himachal Pradesh. The river originates from the Mansarover and Rakshastal lake in Tibet at an elevation of about 4572 m. The river comes to the Indian Territory near Shipki La, Himachal Pradesh by the gorge and then flows through the arid to semiarid Kinnaur region of Himachal Pradesh. The river crosses three mountain ranges, Zanskar, Great Himalaya and Dhauladhar in its way across the Kinnaur. After flowing through Himachal Pradesh it enters the plains of Punjab at Bhakra. The major part of the river lies in the Greater Himalaya compared to the Lesser and Outer Himalayan ranges and as the river covers long distance from the Greater Himalaya to the Outer Himalaya, diverse climatic conditions are experienced throughout the river basin. Severe cold climate exists in the upper part of the river basin while the middle and lower part have cold temperate and tropical-warm climate, respectively. The river basin gets precipitation by different means viz., the upper part receives precipitation largely due to WDs during winter in the form of snow and the middle part receives both rain and snow while the lower part gets rain only. Kinnaur lies in the monsoon shadow zone where maximum annual precipitation occurs due to westerly disturbances during winter and therefore the role of winter precipitation is significant in controlling the Sutlej river discharge. According to Singh and Bengtsson (2004), during winter 65% of the catchment of Sutlej is covered by snow, which is reduced to about 11% after the ablation period. Snowmelt of winter and spring contribute about 59% of the Sutlej river discharge (Singh and Jain, 2002).

10.3.5 Identification of climate signal in ring-width chronologies

Once the chronology is developed from tree-ring-width measurement series its relationship with climate parameters is explored. For river discharge studies ring-width chronologies are first analyzed to understand the strength of monthly precipitation signal. If tree-ring chronologies have strong relationship with precipitation, available river discharge data are then analyzed to understand the relationship with tree-ring chronologies. We present here the Sutlej river discharge reconstruction as an example as it is the only study on river discharge reconstruction using the largest tree-ring data network prepared so far from any river basin in India (Misra et al., 2015). For this study, the recorded Sutlej river discharge data were obtained from Bhakra Beas Management Board and Punjab Irrigation Department, Chandigarh, India. The gauge of the recorded data is located before the Govind Sagar reservoir near Bhakra (Fig. 10.1). The data length ranges from 1922 to 2004 and the abnormal values in specific months (higher/lower than three standard deviations) were replaced by mean values of that month. Strong coherence was observed between total monthly precipitation and total monthly river discharge (Fig. 10.4). The coherence between precipitation and river discharge was broken during 1953–1968 with the correlation dropping to 0.27 as compared to the other periods (1922–1952 and 1969–2004, $r = 0.45$). The river discharge is significantly influenced by the monsoon rainfall as the gauge station is located in the monsoon dominated zone. The possible reason associated with the breakdown in correlation between tree-ring indices and river discharge data could be possibly due to the impact of above-average monsoon precipitation during 1953–1968 (Misra et al., 2015) on river discharge. In view of this, the river discharge records from 1953 to 1968 were excluded from calibration analyses. The mean river discharge data show that the maximum discharge occurred in the monsoon season (June, July and August; Fig. 10.3). It could be due to the location of the river gauge station in the monsoon zone. Response function and correlation analyses of Himalayan cedar chronologies with total monthly precipitation and

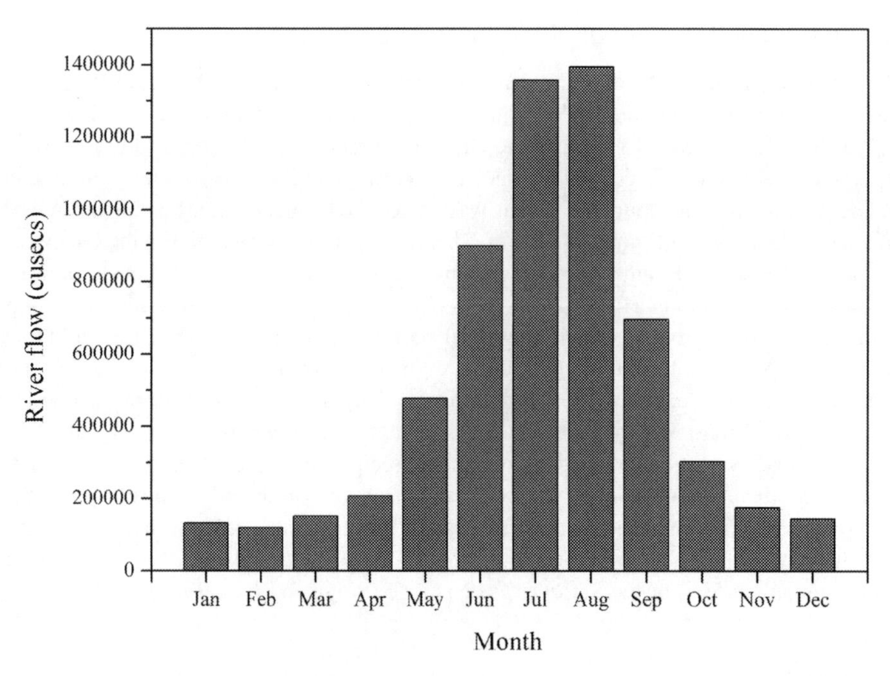

Figure 10.3

The monthly hydrograph of the Sutlej river based on gauge data of the Govind Sagar reservoir near Bhakra. The gauge data showed maximum discharge during summer monsoon months (June–July–August–September).

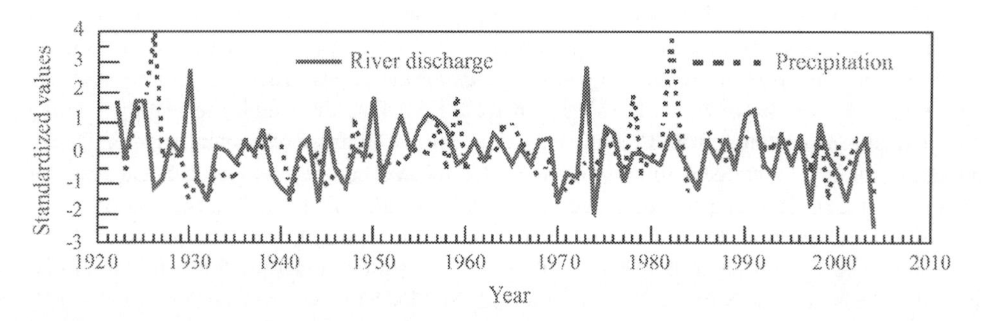

Figure 10.4

The Sutlej river discharge and regional mean precipitation of consecutive months (pOcJune) plotted together show good year-to-year consistency.

mean monthly temperature for the common period 1931–1998 showed that precipitation of the previous year October to current year June has positive relationship with tree-growth, whereas, temperature of the prior year October to current year May, except June, showed inverse relationship with tree growth. The negative impact of temperature on tree growth could be caused by increased moisture stress on tree growth due to temperature induced enhanced evapotranspiration. The correlations between ring-width indices and precipitation and temperature of the monsoon months are not significant, showing that neither precipitation nor temperature is limiting for tree-growth during monsoon months (Misra et al., 2015). The response function and correlation analyses revealed that wet and cool springs favor tree growth. As precipitation of previous year October to current year June, having direct and significant Pearson correlation with tree-growth ($r = 0.45$; two tailed $P = .003$, 1969–2004) also contributes to river discharge, this relationship was taken as a guide to reconstruct the river discharge variable. High temperature during April, May and June, which also promote higher river discharge due to added contribution of snow meltwater, shows an inverse relationship with tree-growth. The quantification of snow meltwater contributing to river discharge is not possible, as there is no long-term empirical study on the relationship between temperature and snowmelt in the upper Sutlej basin. As tree growth is influenced by precipitation and snow meltwater, tree-ring chronologies reliably reflect variations in river discharge contributed by liquid and solid precipitation largely occurring during the previous year October to the current year June (pOcJune).

10.3.6 River discharge reconstruction

Tree-ring chronologies from seven moisture stressed sites widely distributed in the Sutlej river basin were used by Misra et al. (2015) to develop river discharge reconstruction for pOcJune. Misra et al. (2015) applied principal component analysis to reduce multidimensional tree-ring data sets to lower dimensions for calibration and reconstruction (Peters et al., 1981). Under this procedure, a set of correlated chronology variables are transformed to a new set of uncorrelated variables, where the new variables (PCs) are linear combinations of the original variables. Varimax technique was performed for principal component analysis over the common chronology period of AD 1660–2005 (Misra et al., 2015). The first principal component (PC#1) with eigenvalue 5.9 explained 84.6% variance and was subsequently selected for further analyses.

The Sutlej river discharge (pOcJune) was reconstructed for 345 years (AD 1660–2004) using linear regression analysis. Split period calibration and verification analysis were applied to test the statistical veracity of the models. The analyses were performed for two subperiods (1923–1952 and 1969–2004) and calibration models were developed for each subperiod. The river discharge data of the period 1953–1968 showed inconsistency with the tree-ring data series and therefore excluded from the calibration and verification analyses. Although, the first calibration model (1923–1952) explained 40% and the second calibration model (1969–2004) 41% variance of the instrumental data, the second model has relatively lower t value compared to the first model and therefore first calibration model was taken for the river discharge reconstruction.

10.4 Observations and results

The Sutlej river discharge reconstructed from prior year October to current year June (pOcJune) developed using network of tree-ring-width chronologies from moisture stressed sites revealed high year-to-year and multidecadal variations (Fig. 10.5). In the whole 345-years record (AD 1660–2004), the highest river discharge was observed in 1957 when pOcJune discharge was 3,215,446 cusecs and lowest in 1941 with discharge of 1,804,058 cusecs. The other low and high river discharge years were 1782,

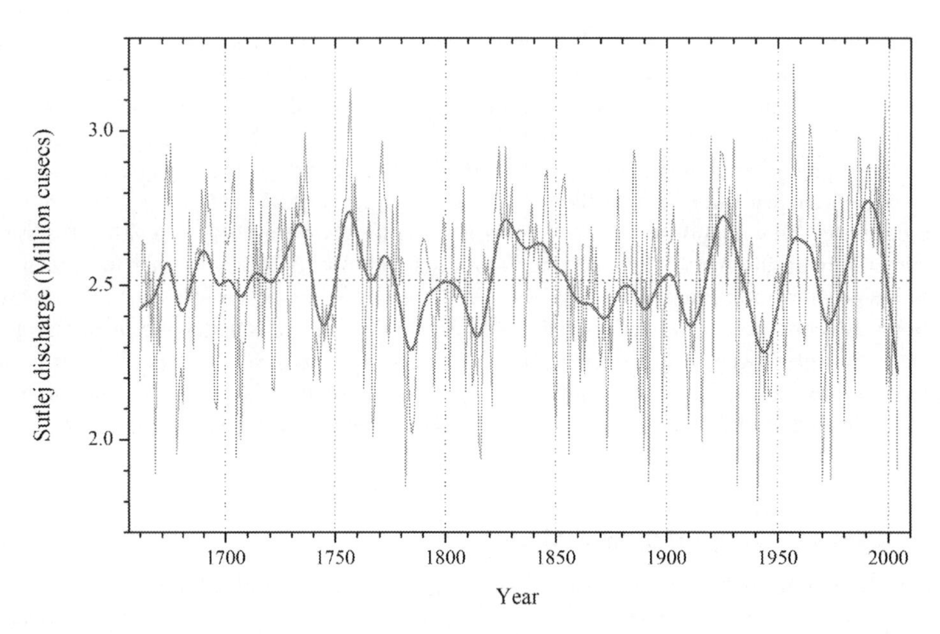

Figure 10.5

The Sutlej river discharge reconstruction (pOcJune; AD 1660–2004). The thick smooth line represents 20-year spline version of the reconstruction to show variations on scale of 20 years and above.

1932, 1892, 1970 and 1757, 1998, 1964, 1736, respectively. On decadal scale, the high river discharge periods were recorded in 1720s–1730s, 1750s, 1820s–1840s, 1920s, 1950s and 1980s. However, the low river discharge periods were noted in 1740s, 1780s-1810s, mid-1850s–late-1890s, 1900s–1910s, 1930s–1940s and 1970s. Reconstructed and observed Sutlej river discharge data revealed distinct decreasing trend since last decade of the 20th century and such a sharp decrease of the river discharge could be linked with decreasing trend in winter precipitation (Yadav and Bhutiyani, 2013) coupled with glacial thinning in the Sutlej basin (Bhutiyani et al., 2008). The 10-year running mean of reconstructed river discharge revealed low discharge in 1939–1948 followed by 1779–1788, 1812–1821, 1740–1749 and 1907–1916, while high discharge in 1987–1996 followed by 1751–1760, 1922–1931, 1730–1739 and 1822–1831. Similarly, the 20-year running mean of river discharge reconstruction showed low discharge in 1932–1951, 1779–1798, 1856–1875, 1802–1821 and high in 1979–1998, 1822–1841, 1754–1773, 1720–1739 in the whole record.

Low and high river discharge periods observed in the reconstruction showed strong resemblance with the other tree-ring-based hydrological records available from the western Himalayan region. Palmer Drought Severity Index reconstruction for the previous year October to the current year May (pOcMay) developed earlier for the western Himalayan region by Yadav (2013) revealed very close similarity with the variations in the Sutlej river discharge reconstruction (Misra et al., 2015). Dry/wet periods are closely associated with low/high river discharge values. The low/high Sutlej river discharge values have been also found to be synchronic with dry/wet periods indicated in tree-ring-based precipitation records from the western Himalayan region of Uttarakhand (Singh et al., 2006; Yadav et al., 2014). Further, on decadal scale, the low/high Sutlej river discharge values are also consistent with the tree-ring-based drought records in Kishtwar, Jammu & Kashmir (Singh et al., 2017).

To analyze the regional strength of the Sutlej river discharge, the reconstruction was compared with other river discharge records available from the western Himalaya. The Sutlej river discharge (pOcJune) developed by Misra et al. (2015) revealed strong year-to-year consistency and significant Pearson correlation ($r = 0.92$, $P < .0001$, 1660–2004) with prior year December to current year July (pDcJuly) discharge of the Sutlej river (Singh and Yadav, 2013; Fig. 10.6A). Though the Sutlej river discharge developed by Singh and Yadav (2013) involved only single predictor chronology, that is, a composite chronology of Himalayan cedar from three and neoza pine from one site in the Sutlej river basin. Tree-ring-based March–April discharge reconstruction of Beas river, a tributary of the Sutlej river by Shah et al. (2013) also revealed significant Pearson correlation with the pOcJune Sutlej river discharge reconstruction (Misra et al., 2015; $r = 0.55$, $P < .0001$, AD 1834–1984; Fig. 10.6B). Similarly, the Sutlej river discharge reconstruction (pOcJune; Misra et al., 2015) also revealed strong similarity

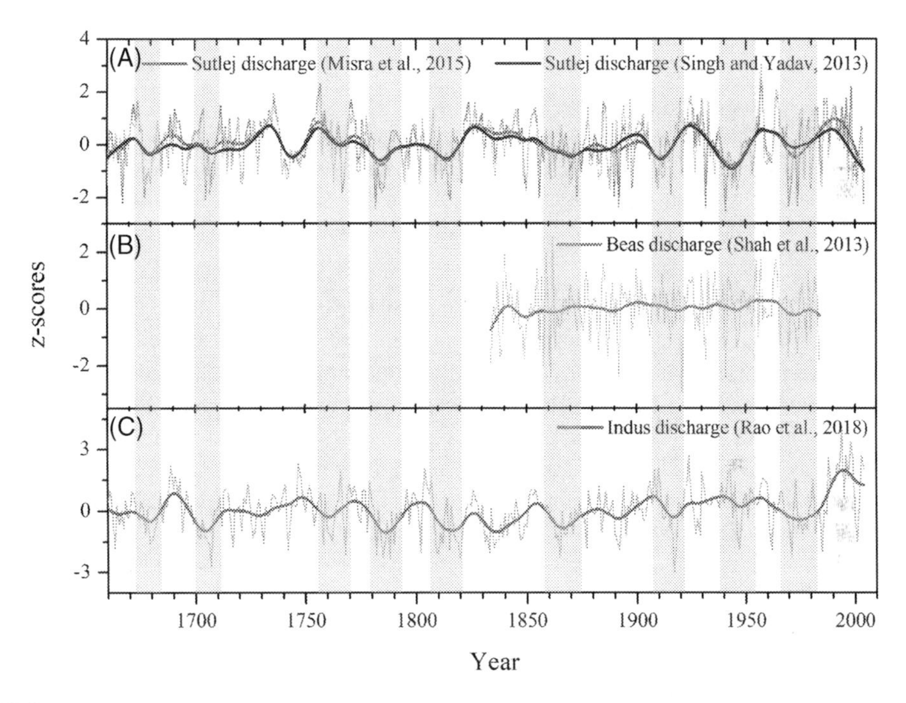

Figure 10.6

(A) Comparison of the Sutlej river discharge (pOcJune; Misra et al., 2015) with the Sutlej river discharge (pDcJuly, Singh and Yadav, 2013), (B) March–April Beas river discharge (Shah et al., 2013) and (C) May–September Indus river discharge (Rao et al., 2018). The thick smoothed line is 20-year spline version of the reconstruction and data were normalized with respect to the mean and standard deviation of the common period. The Beas river discharge data were digitized from published paper (Shah et al., 2013). Consistency in low river discharge periods is highlighted by vertical color bars.

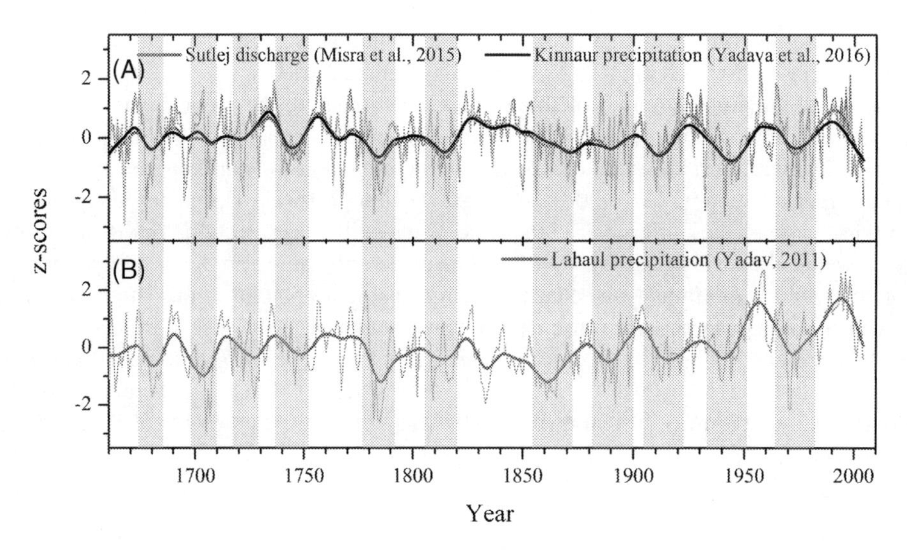

Figure 10.7

(A) The Sutlej river discharge compared with March–May precipitation of Kinnaur, Himalaya (Yadava et al., 2016), and (B) previous year August to current year July precipitation of cold arid Lahaul Himalaya (Yadav, 2011). The thick smoothed line is 20-year spline version of the reconstruction and data were normalized with respect to the mean and standard deviation of the common period. The consistent low river discharge and low precipitation periods are highlighted by vertical color bars.

with the tree-ring-based Indus river discharge developed by Rao et al. (2018; $r=0.29$, $P<.0001$; 1660–2004; Fig. 10.6C). Such close similarities in the above mentioned reconstructions indicate the influence of WD associated precipitation variations on river discharge in the western Himalayan region.

Further, to understand the role of precipitation on river discharge, we compared the Sutlej river discharge data with the tree-ring-based precipitation reconstructions available from the western Himalaya. Tree-ring-based boreal spring (March–April–May) precipitation record developed using a large tree-ring data network from the Kinnaur region (Yadava et al., 2016) showed strong and significant Pearson correlation ($r=0.94$, $P<.0001$, AD 1660–2004) with the pOcJune Sutlej river discharge (Misra et al., 2015; Fig. 10.7A). The Sutlej river discharge also showed close resemblance and significant Pearson correlation ($r=0.52$, $P<.0001$, AD 1660–2004) with the tree-ring-based prior year August to current year July (pAcJuly) precipitation developed from the cold arid Lahaul-Spiti region in the western Himalaya (Yadav, 2011; Fig. 10.7B). The close resemblance in the Sutlej river discharge and other hydrological records available from the western Himalayan region underpins the relevance of precipitation controlling the discharge of rivers in the western Himalaya.

Detailed analyses of the Sutlej river discharge reconstructed using the network of tree-ring-width chronologies from moisture stressed sites in the river basin revealed distinct association with the precipitation variability in the western Himalaya. Crossfield correlations between the Sutlej river discharge reconstruction and gridded precipitation data for the corresponding months (available in http://climexp.knmi.nl; Oldenborgh and Burgers, 2005) for the period 1970–2003 revealed

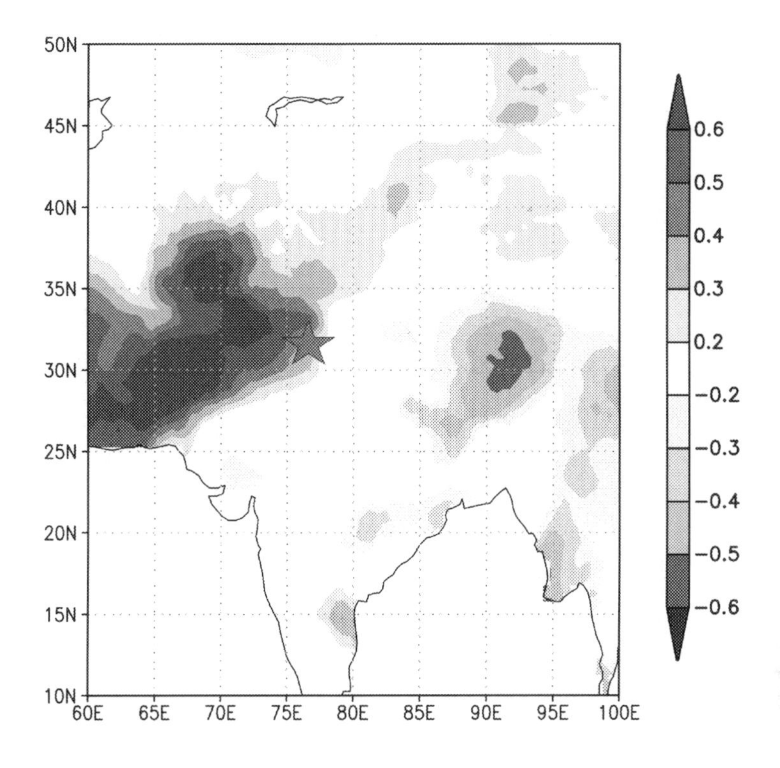

Figure 10.8

Spatial correlation of the Sutlej river discharge with prior year October to current year June gridded precipitation, calculated for 1970–2003. Tree-ring study location is indicated with star.

strong correlation for the westerly dominated western Himalaya and adjoining Central Asian regions (Fig. 10.8). The analyses revealed that winter and spring precipitations in the high altitudinal monsoon shadow zone of the western Himalaya have significant influence in modulating the discharge of rivers in the region.

Spectral properties of the reconstructed pOcJune Sutlej river discharge series were analyzed using Multi Taper Method (Mann and Lees, 1996) by Misra et al. (2015; Fig. 10.9). The high-frequency biennial periodicities observed in the spectral analysis are the dominant feature of the tropical Asia-Pacific climate system, which develop through the interaction of atmosphere–ocean monsoon system and extratropics (Meehl, 1997). In the spectral analyses other high- and low-frequency peaks were of 2.1–3.7, 9.0, 10.5, 29–33 years. The low-frequency peaks at 9.0 to 29–33 years observed in the spectral analysis fall in the range of North Atlantic Oscillation and Pacific Decadal Oscillation. These periodicities recorded in the Sutlej river discharge reconstruction also showed consistency with earlier developed river discharge records (Pederson et al., 2001; Singh and Yadav, 2013) and precipitation (Diaz and Pulwarty, 1994; Briffa et al., 2001; Diaz et al., 2001; Singh et al., 2006, 2009; Yadav et al., 2014).

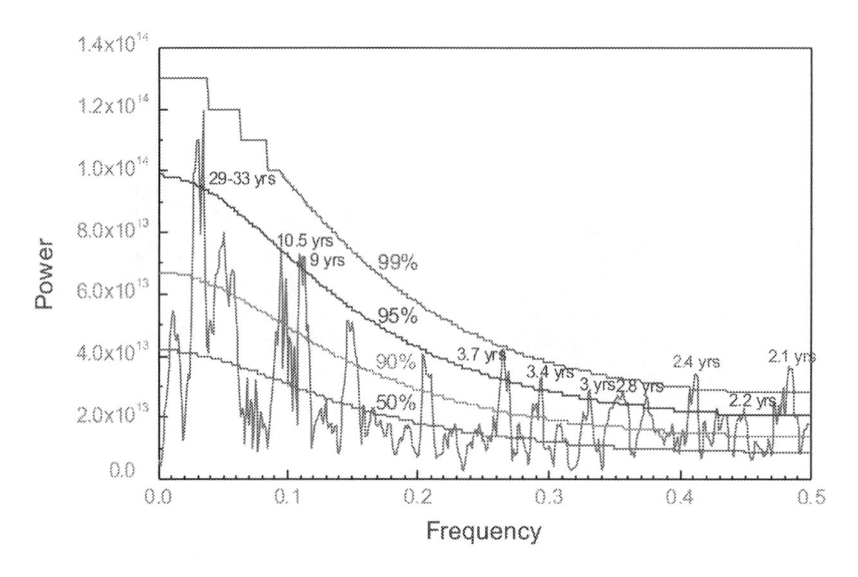

Figure 10.9

Multitaper power spectra of the reconstructed Sutlej river discharge (AD 1660–2004). Significance levels of powers are indicated in the inset with different curves.

10.5 Learning and knowledge outcomes

The rivers are the fundamental element of the climate system and provide the sole source to millions of lives on Earth. High Himalayan mountain ranges have the largest area of snow/ice caps outside the poles, hence the source of the large number of perennial rivers. These rivers transporting an immense quantity of water from the High Himalayas sustain the biodiversity downstream and provide livelihood to millions of people. However, with the accelerated pace of developmental activities and growing human needs sustained water supply is a major concern to society. In view of this it is very important to understand natural variability in river discharge in long-term perspective. As the observational records are too short and limited in context of the Himalayan region, the application of high-resolution proxies such as tree-rings should be explored to extend the river discharge records back to the past several centuries/millennia. The Sutlej river discharge reconstruction developed using tree-ring series from the moisture stressed sites revealed temporal consistency with the other hydrological records available from the western Himalayan region. The study has revealed high dependence of river discharge on winter and spring season precipitation in the westerly dominated western Himalayan region. Such long-term hydrological records should provide valuable data base to understand natural variability in river discharge in long-term perspective and the recurrence behavior of extreme events. It is further opined that the robust tree-ring-based river discharge data should help in the development of appropriate hydrological resource management plans in long-term perspective.

Acknowledgments

V.S., K.G.M. and A.K.Y. sincerely acknowledge Dr. Vandana Prasad, Director, Birbal Sahni Institute of Palaeosciences, Lucknow for providing all the necessary facilities and support. V.S. and K.G.M. are thankful to Jayendra Singh and Mukund P. Rao for sharing their river discharge data. V.S. is sincerely thankful to the Council of Scientific and Industrial research (CSIR), New Delhi for support through CSIR-Senior Research Fellowship (09/528(0021)/2018-EMR-I). A.K.Y. expresses sincere thanks to the Council of Scientific and Industrial Research (CSIR), New Delhi for the grant of Senior Research Associateship under scientists pool scheme (Pool No. 9015-A). R.R.Y. sincerely acknowledges support of the Council of Scientific and Industrial Research (CSIR), New Delhi under Emeritus Scientist scheme (No. 21(1010)/15/EMR-II).

References

Azam, M.F., Ramanathan, A.L., Wagnon, P., Vincent, C., Linda, A., Berthier, E., Sharma, P., Mandal, A., Angchuk, T., Singh, V.B., Pottakkal, J.G., Meteorological conditions, seasonal and annual mass balances of Chhota Shigri Glacier, western Himalaya, India, Ann. Glaciol. 57 (2016) 328–338.

Bhutiyani, M.R., Kale, V.S., Pawar, N.J., Changing streamflow patterns in the rivers of northwestern Himalaya: implications of global warming in the 20th century, Curr. Sci. 95 (5) (2008) 618–626.

Bookhagen, B., Burbank, D.W., Toward a complete Himalayan hydrological budget: spatiotemporal distribution of snowmelt and rainfall and their impact on river discharge, J. Geophys. Res.: Earth Surf. 115 (F3) (2010) 1–25.

Briffa, K.R., Osborn, T.J., Schweingruber, F.H., Harris, I.C., Jones, P.D., Shiyatov, S.G., Vaganov, E.A., Low-frequency temperature variations from a northern tree-ring density network, J. Geophys. Res. 106 (2001) 2929–2941.

Champion, H.G., Seth, S.K., A Revised Survey of the Forest Types of India, Manager of Publications, Delhi, 1968.

Cook, E.R., Palmer, J.G., Ahmed, M., Woodhouse, C.A., Fenwick, P., Zafar, M.U., Wahab, M., Khan, N., Five centuries of Upper Indus River flow from tree rings, J. Hydrol. 486 (2013) 365–375.

D'Arrigo, R., Abram, N., Ummenhofer, C., Palmer, J., Mudelsee, M., Reconstructed streamflow for Citarum River, Java, Indonesia: linkages to tropical climate dynamics, Clim. Dyn. 36 (2011) 451–462.

Diaz, H.F., Pulwarty, R.S., An analysis of the time scales of variability in centuries-long enso-sensitive records in the 1000 years, Clim. Change 26 (1994) 317–342.

Diaz, S.C., Touchan, R., Swetnam, T.W., A tree-ring reconstruction of past precipitation for Baja California Sur, Mexico, Int. J. Climatol. 21 (2001) 1007–1019.

Fritts, H.C., Tree Rings and Climate, Academic Press, London, 1976.

Gamble, J.S., A Manual of Indian Timbers, Sampson Low Marston and Co. Ltd., London, 1902.

Gou, X.H., Deng, Y., Chen, F.H., Yang, M.X., Fang, K.Y., Gao, L.L., Yang, T., Zhang, F., Tree ring based streamflow reconstruction for the upper Yellow river over the past 1234 years, Chin. Sci. Bull. 55 (2010) 4179–4186.

Harley, G., Maxwell, J., Larson, E., Grissino-Mayer, H.D., Henderson, J., Huffman, J., Suwannee River flow variability 1550-2005 CE reconstructed from a multispecies tree-ring network, J. Hydrol. 544 (2016) 438–451.

Holmes, R.L., A computer-assisted quality control program, Tree-Ring Bull. 43 (1983) 69–78.

IPCC, Managing the Risks of Extreme Events and Disasters to Advance Climate Change Adaptation. A Special Report of Working Groups I and II of the Intergovernmental Panel on Climate Change [Field, C.B., V. Barros, T.F. Stocker, D. Qin, D.J. Dokken, K.L. Ebi, M.D. Mastrandrea, K.J. Mach, G.-K. Plattner, S.K. Allen, M. Tignor, and P.M. Midgley (eds.)]. Cambridge University Press, Cambridge, UK, and New York, NY, USA, 2012, pp. 582.

Lang, T.J., Barros, A.P., Winter storms in the central Himalayas, J. Meteorol. Soc. Jpn. 82 (2004) 829–844.

Mann, M.E., Lees, J.M., Robust estimation of background noise and signal detection in climatic time series, Clim. Change 33 (1996) 409–445.

Margolis, E.Q., Meko, D.M., Touchan, R., A tree-ring reconstruction of streamflow in the Santa Fe River, New Mexico, J. Hydrol. 397 (2011) 118–127.

Martínez-Sifuentes, A.R., Villanueva-Díaz, J., Carlón-Allende, T., Estrada-Ávalos, J., 243 years of reconstructed streamflow volume and identification of extreme hydroclimatic events in the Conchos River Basin, Chihuahua, Mexico, Tress 34 (2020) 1347–1361.

Maxwell, R.S., Hessl, A.E., Cook, E.R., Pederson, N., A multispecies tree ring reconstruction of Potomac River stream flow (950-2001), Water Resour. Res. 47 (2011) W05512.

Meehl, G.A., The south Asian monsoon and the tropospheric biennial oscillation, J. Clim. 10 (1997) 1921–1943.

Misra, K.G., Yadav, R.R., Misra, S., Satluj river flow variations since AD 1660 based on tree-ring network of Himalayan cedar from western Himalaya, India, Quat. Int. 371 (2015) 135–143.

Oldenborgh, G.J., Burgers, G., Searching for decadal variations in ENSO precipitation teleconnections, Geophys. Res. Lett. 32 (2005) L15701.

Pederson, N., Jacoby, G.C., D'Arrigo, R.D., Cook, E.R., Buckley, B.M., Dugarjav, C., Mijiddorj, R., Hydrometeorological reconstructions for northeastern Mongolia derived from tree-rings: AD 1651-1995, J. Clim. 14 (2001) 872–881.

Peters, K., Jacoby, G.C., Cook, E.R., Principal components analysis of tree-ring sites, Tree-ring Bull. 41 (1981) 1–19.

Rao, M.P., Cook, E.R., Cook, B.I., Palmer, J.G., Uriarte, M., Devineni, N., Lall, U., D'Arrigo, R.D., Woodhouse, C.A., Ahmed, M., Zafar, M.U., Six centuries of Upper Indus Basin streamflow variability and its climatic drivers, Water Resour. Res. 54 (8) (2018) 5687–5701.

Romshoo, S.A., Zaz, S., Ali, N., Recent climate variability in Kashmir Valley, India and its impact on streamflows of the Jhelum River, J. Res. Dev. 17 (2017) 01–22.

Shah, S.K., Bhattacharyya, A., Chaudhary, V., Streamflow reconstruction of Eastern Himalaya River, Lachen 'Chhu', North Sikkim, based on tree-ring data of *Larix griffithiana* from Zemu Glacier basin, Dendrochronologia 32 (2) (2014) 97–106.

Shah, S.K., Bhattacharyya, A., Shekhar, M., Reconstructing discharge of Beas river basin, Kullu valley, western Himalaya, based on tree-ring data, Quat. Int. 286 (2013) 138–147.

Shekhar, M., Bhattacharyya, A., Reconstruction of January–April discharge of Zemu Chuu—a first stage of Teesta River North Sikkim Eastern Himalaya based on tree-ring data of fir, J. Hydrol.: Reg. Stud. 4 (2015) 776–786.

Singh, J., Park, W.-K., Yadav, R.R., Tree-ring-based hydrological records for western Himalaya, India, since AD 1560, Clim. Dyn. 26 (2006) 295–303.

Singh, J., Yadav, R.R., Dendroclimatic potential of millennium-long ring-width chronology of *Pinus gerardiana* from Himachal Pradesh, India, Curr. Sci. 93 (2007) 833–836.

Singh, J., Yadav, R.R., Tree-ring based seven century long flow records of Satluj river, western Himalaya, India, Quat. Int. 304 (2013) 156–162.

Singh, J., Yadav, R.R., Dubey, B., Chaturvedi, R., Millennium-long ring-width chronology of Himalayan cedar from Garhwal Himalaya and its potential in climate change studies, Curr. Sci. 86 (2004) 590–593.

Singh, J., Yadav, R.R., Wilmking, M., A 694-year tree-ring based rainfall reconstruction from Himachal Pradesh, India, Clim. Dyn. 33 (2009) 1149–1158.

Singh, P., Bengtsson, L., Effect of warmer climate on the depletion of snow covered area in the Satluj basin in the Western Himalayan region, Hydrol. Sci. J. 48 (2004) 413–425.

Singh, P., Jain, S.K., Snow and glacier contribution in the Satluj river at Bhakra Dam in the Western Himalayan region, Hydrol. Sci. J. 47 (2002) 93–106.

Singh, V., Yadav, R.R., Gupta, A.K., Kotlia, B.S., Singh, J., Yadava, A.K., Singh, A.K., Misra, K.G., Tree ring drought records from Kishtwar, Jammu and Kashmir, northwest Himalaya, India, Quat. Int. 444 (2017) 53–64.

Stockton, C.W., Jacoby, G.C., Long-Term Surface-Water Supply and Streamflow Trends in the Upper Colorado River Basin, Lake Powell Research Project Bulletin, National Science Foundation, Arlington, VA, 18 (1976) 1–70.

Stokes, M.A., Smiley, T.L., An Introduction to Tree-Ring Dating, The University of Chicago Press, Chicago, 1968.

Sun, J., Liu, Y., Wang, Y., Bao, G., Sun, B., Tree-ring based runoff reconstruction of the upper Fenhe river basin, north China, since 1799 AD, Quat. Int. 283 (2013) 117–124.

Urrutia, R.B., Lara, A., Villalba, R., Christie, D.A., Le Quesne, C., Cuq, A., Multicentury tree ring reconstruction of annual streamflow for the Maule River watershed in south central Chile, Water Resour. Res. 47 (2011) W06527.

Visser, H., Petersen, A.C., Ligtvoet, W., On the relation between weather-related disaster impacts, vulnerability and climate change, Clim. Change 125 (3-4) (2014) 461–477.

Wigley, T.M.L., Briffa, K.R., Jones, P.D., On the average value of correlated time series, with applications in dendroclimatology and hydrometeorology, J. Clim. Appl. Meteorol. 23 (1984) 201–213.

Wise, E.K., Tree ring record of streamflow and drought in the upper Snake River, Water Resour. Res. 46 (2010) W11529.

Woodhouse, C.A., Lukas, J.J., Multi-century tree-ring reconstructions of Colorado streamflow for water resource planning, Clim. Change 78 (2006) 293–315.

Yadav, R.R., Tree-ring evidence of 20th century precipitation surge in monsoon shadow zone of western Himalaya, India, J. Geophys. Res. 116 (2011) 1–10.

Yadav, R.R., Tree ring-based seven-century drought records for the western Himalaya, India, J. Geophys. Res. 118 (2013) 4318–4325.

Yadav, R.R., Bhutiyani, M.R., Tree-ring-based snowfall record for cold arid western Himalaya, India since A.D. 1460, J. Geophys. Res. Atmos. 118 (2013) 7516–7522.

Yadav, R.R., Misra, K.G., Kotlia, B.S., Upreti, N., Premonsoon precipitation variability in Kumaon Himalaya, India over a perspective of ~300 years, Quat. Int. 325 (2014) 213–219.

Yadava, A.K., Braeuning, A., Singh, J., Yadav, R.R., Boreal spring precipitation variability in the cold arid western Himalaya during the last millennium, regional linkages, and socio-economic implications, Quat. Sci. Rev. 144 (2016) 28–43.

Yang, B., Qin, C., Shi, F., Sonechkin, D.M., Tree ring-based annual streamflow reconstruction for the Hiehe river in arid northwestern China from AD 575 and its implications for water resource management, Holocene 22 (2012) 773–784.

CHAPTER

11 Potential utility of Himalayan tree-ring $\delta^{18}O$ to reveal spatial patterns of past drought variability—Its assessments and implications

Santosh K. Shah[a], **Nivedita Mehrotra**[a], **Narayan P. Gaire**[b,c], **Lamginsang Thomte**[a,d], **Bimal Sharma**[b], **Uttam Pandey**[a], **Om Katel**[e]

[a] *Birbal Sahni Institute of Palaeosciences, Lucknow, India.* [b] *Key Laboratory of Tropical Forest Ecology, Xishuangbanna Tropical Botanical Garden, Chinese Academy of Sciences Menglun, Mengla, Yunnan, China.* [c] *Central Department of Environmental Science, Tribhuvan University, Kirtipur, Kathmandu, Nepal.* [d] *Department of Geography, Gauhati University, Guwahati, India.* [e] *College of Natural Resources, Royal University of Bhutan, Lobesa, Punakha, Bhutan*

11.1 Introduction

The Indian Summer Monsoon (ISM) rainfall contributes about 80% of the annual rainfall of South Asia. A change in the timing and amount of rainfall consequently impact the society, water security, and economy of the region (Gadgil and Gadgil, 2006; Asoka et al., 2017; Mishra et al., 2020). The ISM is known as a large-scale atmospheric pattern controlled by its variability on a spatial and temporal scale (Goswami, 2012; Sinha et al., 2018). This ISM system exists through two main branches, namely the Bay of Bengal branch and the Arabian Sea branch. The former has been considered as isotopically depleted, whereas the latter is observed as isotopically enriched branch in the ISM system (Sinha et al., 2015). Within the ISM system the origins of the relative contribution of source water from the Bay of Bengal and the Arabian Sea to the Himalaya is variable from west to east (Medina et al., 2010). The sparse spatio-temporal coverage of instrumental observations has limited our understanding of such monsoon variability patterns. The future trends of monsoon variability can be interpreted through conclusive evidences from longer past records.

Annually formed tree-rings are a suitable proxy archive for past hydroclimate variability due to its relative abundance and precise dating (Fritts, 1976). Trees from the Himalaya have been used to reconstruct past climatic conditions beyond existing meteorological records in the region (Cook et al., 2003; Bhattacharyya and Shah, 2009; Shah et al., 2014; Pandey et al., 2016; Gaire et al., 2017, 2019, 2020; Shah et al., 2017, 2019). The Monsoon Asia Drought Atlas (MADA) developed by Cook et al. (2010), based on seasonally resolved gridded spatial reconstruction of Asian monsoon droughts and pluvials during the last Millennium was based on a network of tree-ring width records in Asia. The Palmer Drought Severity Index (PDSI) for the summer season (June–August) was reconstructed spatially using

Holocene Climate Change and Environment. DOI: https://doi.org/10.1016/B978-0-323-90085-0.00003-6

gridded measure of relative wetness for land area of the globe. The MADA reconstructions revealed the monsoon failure and concurrent historical droughts documented across Asian civilizations such as the Ming Dynasty drought (1638–1641 C.E.), the Strange Parallels drought (1756–1768 C.E.), the East India drought (1790–1796 C.E.), and the late Victorian Great Drought (1876–1878 C.E.). The impact of El Niño Southern Oscillation (ENSO) on monsoonal variability was not the only attribute but influenced by various other forcing factors contributing toward mega drought occurrences and monsoon failures in the region (Cook et al., 2010). In different parts of the globe tree-ring records have also been used to reconstruct spatial drought and termed as drought atlas, for example, North American Drought Atlas (Cook et al., 2004); Living Blended Drought Atlas (Cook et al., 2010), Old World Drought Atlas (Cook et al., 2015); New Zealand Summer Drought Atlas (Palmer et al., 2015); Mexican Drought Atlas (Stahle et al., 2016); European Russia Drought Atlas (Cook et al., 2020). Similar spatial drought reconstruction has been carried out from continental United States (Cook et al., 1999) and NW Africa (Touchan et al., 2011). All these spatial drought reconstruction (Cook et al., 2004, 2010, 2015; Touchan et al., 2011; Palmer et al., 2015; Stahle et al., 2016) are based on drought index known as PDSI (Palmer, 1965) and self-calibrated PDSI (scPDSI). In addition, spatial hydroclimatic variability based on Standardized Precipitation Evapotranspiration Index is available from Fennoscandian, Europe (Seftigen et al., 2015a, 2015b). There are few spatial reconstruction available for temperature such as spatial temperature reconstruction for summer month computed for East Asia (Cook et al., 2013), which incorporates few tree-ring chronologies from Himalaya region. In addition, Rydval et al. (2017) prepared spatial reconstruction of Scottish summer temperature across Scotland in Europe. All these spatial drought and temperature reconstructions are carried out using the technique of climate field reconstruction (CFR). More specifically it was done using standard and revised form of the CFR known as point-by-point regression (PPR) methodology of CFR (Cook et al., 1999). All these reconstructions were verified using rigorous calibration-verification statistics within PPR method.

Most of the tree-ring-based climate reconstructions from the various parts of the Indian region are available for a temporal scale (Bhattacharyya and Shah, 2009; Shah et al., 2014; Pandey et al., 2016). In these reconstructions, the climate data are taken either from the nearest single meteorological station or from nearest climate grid point. In spite of the available tree-ring-based climate reconstructions from the Himalaya region in a temporal scale, these tree-ring data have not been utilized to reconstruct climate in a spatial scale to assess the multidecadal climatic variability in the region. However, experimental large-scale summer monsoon drought variability has been investigated for India along with Tibetan Plateau using tree-ring chronologies from the "High Asia" region of the Himalaya, Karakoram, Tien Shan, and Tibetan Plateau (Cook and Krusic, 2008). This reconstruction has been carried out using reduced-space multivariate regression method, and both calibration and verification tests indicate that the reconstructions contain significant hindcast skill in substantial subregions of the grid, especially over India.

The tree-ring proxies are known for their widespread application and valuable high-resolution climate archiving. But tree-ring stable isotopes measurements of oxygen, hydrogen and carbon are now also known to contribute toward dendroclimatic studies (McCarroll and Loader, 2004; Gagen et al., 2011). The isotopic composition of water absorbed by the tree's roots overprinted by evapotranspiration in the leaf can be recorded in the stable oxygen and hydrogen isotope measurements from tree-rings. These measurements are a proxy for the isotopic water absorbed, where the evapotranspiration in leaf signal is dominated by vapor pressure deficit (relative humidity; Gagen et al., 2011). The $\delta^{18}O$ composition in the wood can record the hydrological or meteorological origin of the water absorption by the tree, evapotranspiration, variation in relative humidity, or the periodic quantity of rainfall (Roden et al., 2000; McCarroll and Loader, 2004; Gagen et al., 2011; Anchukaitis and Evans, 2010; Anchukaitis, 2017). Additionally, in lower latitudes, there is a negative correlation between the amount of rainfall and $\delta^{18}O$ of precipitation, known as the "amount effect" (Dansgaard et al., 1964) and therefore, ISM rainfall also influences the $\delta^{18}O$ in trees (Vuille et al., 2005). Studies on $\delta^{18}O$ in tree-rings have been carried out in the Indian subcontinent and in the entire Himalaya (Ramesh et al., 1985, 1986; Brunello et al., 2019; Singh et al., 2019; Sano et al., 2020; Managave et al., 2020).

On the basis of the forgoing discussion, we make an attempt to reconstruct and study the patterns of large-scale past drought variability over the India, Nepal, and Bhutan region from the available Himalayan tree-ring $\delta^{18}O$ chronologies (Fig. 11.1). The details of the tree-ring $\delta^{18}O$ records are discussed below and also provided in Table 11.1.

11.2 The scope of the present study

To understand our perspective of the present study an overview of the existing tree-ring $\delta^{18}O$ records covering Himalaya region of India, Nepal, and Bhutan was compiled. The representation of these studies to create an understanding of the climatic inferences and variations in the precipitation regime influenced by the ENSO phenomenon was described in detail in the present study. This is done below to direct toward the scope of the present study.

In the Himalayan transect the isotopic study was initiated on tree-rings of *Abies pindrow* from Kashmir Himalaya, which showed the sensitivity of hydrogen isotopes (δD) toward precipitation and mean maximum temperature, whereas carbon isotope ($\delta^{13}C$) and oxygen isotope ($\delta^{18}O$) are sensitive to temperature and cloud amount and humidity, respectively (Ramesh et al., 1985, 1986). The isotopic study was also conducted on tree-ring of *Cedrus deodara* and *Pinus wallichiana* from the same region to assess the species effect on the coherence of the isotope signal in different species of the region (Ramesh et al., 1985). The most recent tree-ring $\delta^{18}O$ from Keylong, Lahaul-Spiti, Western Himalaya over a period of 1146–2006 C.E. reveals the climatological changes in the region (Managave et al., 2020). The record reveals primarily scPDSI signal from previous year October to current year September in the tree-ring $\delta^{18}O$ of *Juniperus polycarpos*. Another 242-year tree-ring $\delta^{18}O$ record of *A. pindrow* from Manali, western Himalaya, was studied and reconstruction of scPDSI for the months of June to September during 1767–2008 C.E. was made (Sano et al., 2012). This record indicated the strength of the monsoon system in the study region and the influence of the Arabian Sea branch of the moisture source during the weaker monsoon years induced by the increased sea surface temperature (SST) of the Indian Ocean (Sano et al., 2017). The tree-ring $\delta^{18}O$ record based on multiple taxa, that is, *A. pindrow, Picea smithiana*, and *Aesculus india* from Dingad valley, Uttarakhand, western Himalaya spanning from 1743–2015 C.E., was used for reconstruction of the rainfall during monsoon months of June–July (Singh et al., 2019). The tree-ring $\delta^{18}O$ studies from Nepal Himalaya were carried out in Humla and Ganesh. The record of *A. spectabilis* from Humla, Nepal Himalaya revealed the increasing aridity during 223 years (1778–2000 C.E.) for the monsoon months (June–September) based on PDSI variations (Sano et al., 2012). The eastern Himalaya has tree-ring $\delta^{18}O$ records from Wache, Bhutan, based on multiple taxa such as *Larix griffithiana, J. indica*, and *P. indica* (Sano et al., 2013). The *L. griffithiana* tree-ring $\delta^{18}O$ based precipitation reconstruction for 269 years (1743–2011 C.E.) for the months of May–September (Sano et al., 2013), revealed the stable state of the ISM in the 20th century which is not

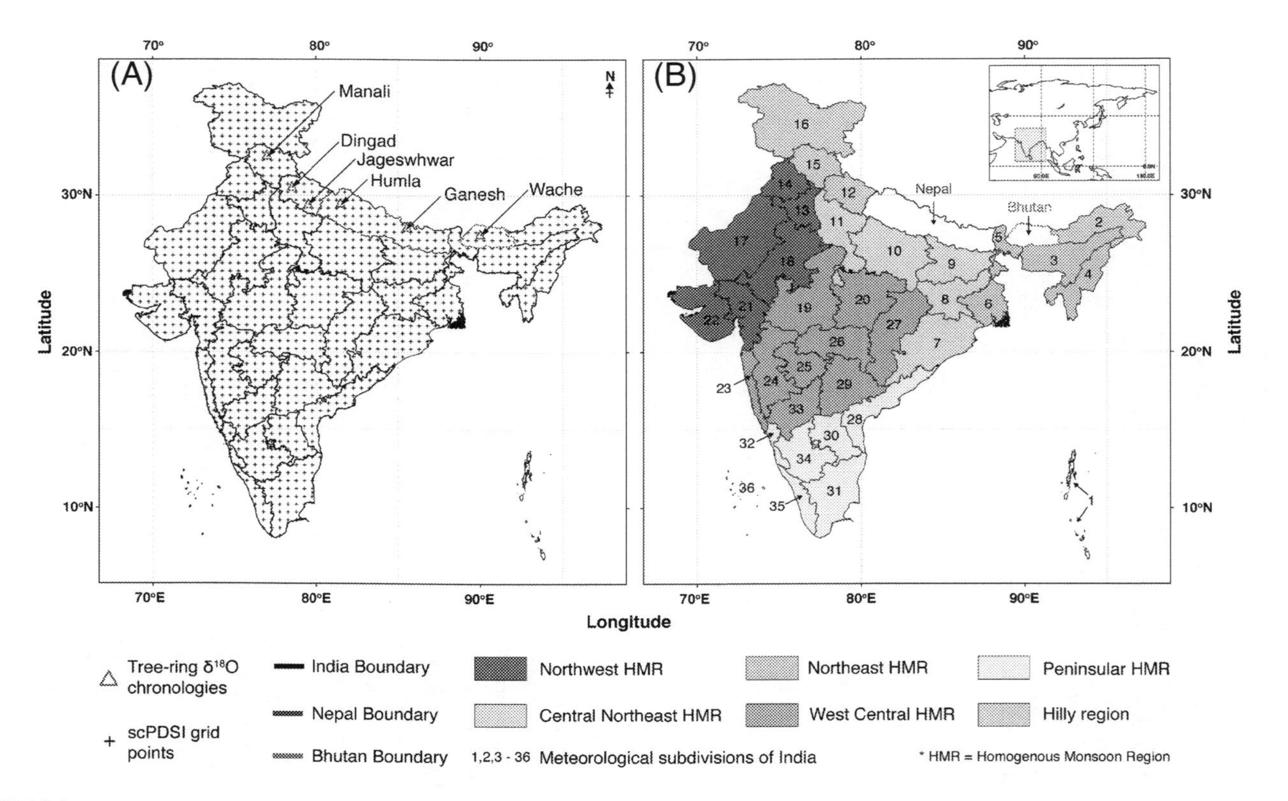

FIGURE 11.1

(A) Map showing locations of tree-ring sites as open triangles (the details of each site are given in Table 11.1) and JJAS-scPDSI target field (1234 grid points as black + sign) as half degree latitude-by-longitude. (B) Map showing homogenous monsoon region of India and categorization of meteorological subdivisions, location of Nepal and Bhutan (modified from http://www.tropmet.res.in/IITM/india-subdiv-rev1.png). The details of meteorological subdivisions numbered in map are as (1. Andaman and Nicobar Islands, 2. Arunachal Pradesh, 3. Assam and Meghalaya, 4. Nagaland, Manipur, Mizoram, and Tripura, 5. Sub-Himalayan West Bengal and Sikkim, 6. Gangetic West Bengal, 7. Orissa, 8. Jharkhand, 9. Bihar, 10. East Uttar Pradesh, 11. West Uttar Pradesh, 12. Uttaranchal, 13. Haryana, Chandigarh and Delhi, 14. Punjab, 15. Himachal Pradesh, 16. Jammu & Kashmir, 17. West Rajasthan, 18. East Rajasthan, 19. West Madhya Pradesh, 20. East Madhya Pradesh, 21. Gujarat, 22. Saurashtra, Kutch, and Diu, 23. Konkan and Goa, 24. Madhya Maharashtra, 25. Marathwada, 26. Vidarbha, 27. Chattisgarh, 28. Coastal Andhra Pradesh, 29. Telangana, 30. Rayalaseema, 31. Tamil Nadu and Pondicherry, 32. Coastal Karnataka, 33. North Interior Karnataka, 34. South Interior Karnataka, 35. Kerala, and 36. Lakshadweep).

Table 11.1 Details of tree-ring $\delta^{18}O$ chronologies used in the present study.

SN, CNT	SPEC	LAT	LON	ELEV	FYR	LYR	References	Data source
Manali, India	ABPI	32°13′	77°13′	2700	1768	2008	Sano et al., 2017	https://doi.org/10.25921/xxcd-4236
Dingad, India	ABPI, PISM, ABIN	30°50′	78°46′	1600	1743	2015	Singh et al., 2019	Digitized
Jageshwar, India	CEDE	29°38′	79°51′	1870	1621	2008	Xu et al., 2018	https://doi.org/10.25921/jdmh-bm92
Humla, Nepal	ABSP	29°51′	81°56′	3850	1778	2000	Sano et al., 2012	https://doi.org/10.25921/9w68-zf20
Ganesh, Nepal	ABSP	28°10′	85°11′	3550	1801	2000	Xu et al., 2018	https://doi.org/10.25921/heyh-ky04
Wache, Bhutan	LAGR	27°59′	90°00′	3500	1743	2011	Sano et al., 2013	https://doi.org/10.25921/hqcg-gy96

The species (SPEC) are Abies pindrow *(ABPI),* Abies spectabilis *(ABSP),* Aesculus indica *(ABIN),* Cedrus deodara *(CEDE),* Larix griffithiana *(LAGR), and* Picea smithiana *(PISM).*

ELEV, *site elevation in meters;* FYR, *first year;* LAT, *site latitude in degree north;* LON, *site longitude in degree east;* LYR, *last year;* NYR, *chronology length;* SN, CNT, *site name, country.*

resonating with the $\delta^{18}O$ records from Nepal (Sano et al., 2012) and Tibet (Xu et al., 2012; Grießinger et al., 2017), that have weaker ISM signals influenced by ENSO (Sano et al., 2013). A comparative record of tree taxa, *C. deodara* and *A. spectabilis* from Jageshwar, western Himalaya and Ganesh, Nepal Himalaya, respectively, whose tree-ring $\delta^{18}O$ series in combination with other $\delta^{18}O$ records from Humla, Nepal Himalaya (Sano et al., 2012), Manali, western Himalaya (Sano et al., 2017), Wache, Bhutan Himalaya (Sano et al., 2013) were studied to assess the decreasing ISM strength in the last 180 years in northern Indian subcontinent (Xu et al., 2018).

So most of these records captured the variation in the ISM (June–September) season and the change in the strength and duration causing wide spread aridity in the Indian subcontinent. This caused large-scale droughts, which were revealed in the tree-ring $\delta^{18}O$ record and its comparison to drought matrices, which had teleconnection with ENSO and SST changes in the Indian Ocean and remote regions of the Eastern Pacific Ocean. These reconstructions were made on a regional scale of the study sites distributed and tree-ring $\delta^{18}O$ records, which were analyzed to understand climatic forcing and teleconnections influencing the moisture availability in the regions. The moisture source was subjected to variation depending upon the strength of the ISM, which was impacting the origins of the source indirectly influenced by distant oceanic phenomenon such as ENSO and SST variability. So there was a need to summarize the variations simultaneously and on a spatial scale. Nevertheless these reconstructions were definitively indicating toward a drought signal and fluctuations in the relative humidity at each site. But all these were not spatially represented in terms of drought matrices. In the present study we have made the first attempt to spatially reconstruct the scPSDI based on the tree-ring $\delta^{18}O$ records available in the Himalaya and its spatial extents in the Indian subcontinent.

11.3 Fieldwork and sampling technique

11.3.1 Tree-ring $\delta^{18}O$ data sets

For the present study, we used six tree-ring $\delta^{18}O$ chronologies (Fig. 11.2) covering west to east transect of the Himalaya region viz., Manali (Sano et al., 2017), Dingad (Singh et al., 2019), and Jageshwar (Xu et al., 2018) from western Himalayan region of India, Humla (Sano et al., 2012) and Ganesh (Xu et al., 2018) from Nepal Himalaya and Wache from Bhutan Himalaya (Sano et al., 2013). For $\delta^{18}O$ measurement in trees all the six sites (Sano et al., 2012, 2013, 2017; Xu et al., 2018; Singh et al., 2019), tree cores has been collected by authors of the paper with the help of Increment borer. Except Dingad, all other five tree-ring $\delta^{18}O$ records are accessed from NOAA/WDC-Paleoclimatology database (https://www.ncdc.noaa.gov/data-access/paleoclimatology-data; Table 11.1). The Dingad tree-ring $\delta^{18}O$ record has been digitized using program ring-width curve converter in software tools CDendro (http://www.cybis.se/forfun/dendro/). In addition, other three additional tree-ring $\delta^{18}O$ records are also available in the same Himalayan transact viz., Kashmir valley (Ramesh et al., 1985), upper Kali Gandaki valley in the central Nepal Himalayas (Brunello et al., 2019), and Keylong, western Himalaya region of India (Managave et al., 2020). However, these three records are not used in the present study since the record of Ramesh et al. (1985) and Brunello et al. (2019) are available for shorter time span from 1903 to 1932 and from 1980 to 2015, respectively. The tree-ring $\delta^{18}O$ record of Managave et al. (2020) is although available for longer time interval but captured mixed annual drought signal of westerly and

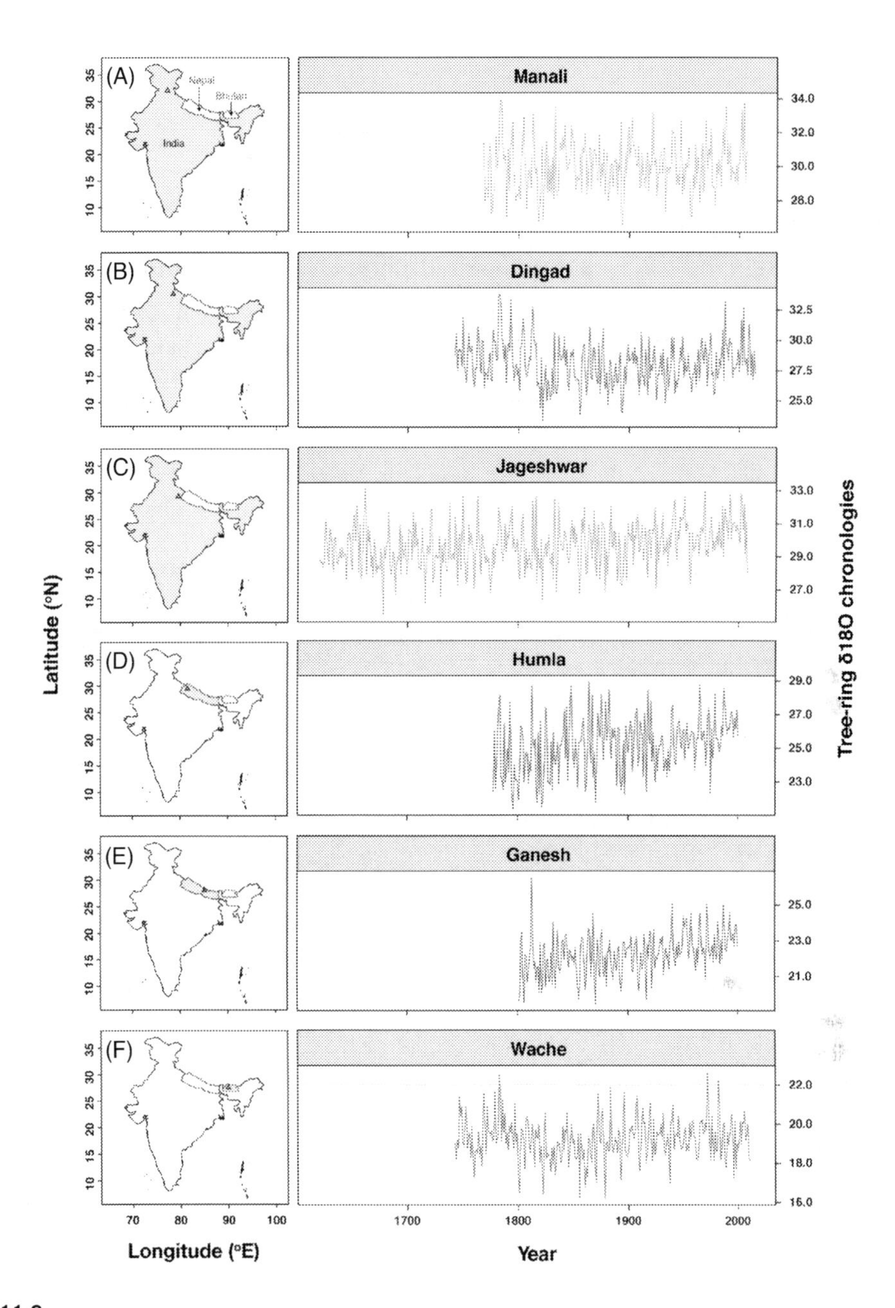

FIGURE 11.2

Tree-ring $\delta^{18}O$ isotope chronologies from six sites in right panel and their corresponding location in the left panel, and arranged in west to east Himalayan transect. These chronologies are for (A) Manali (Sano et al., 2017), (B) Dingad (Singh et al., 2019), (C) Jageshwar (Xu et al., 2018), (D) Humla (Sano et al., 2012), Ganesh (Xu et al., 2018), and (E) Wache (Sano et al., 2013). Records (A–C) are from Indian Himalaya region, (D and E) are from Nepal Himalaya, and (F) is from Bhutan Himalaya.

ISM. The details such as sampling sites, locations, species, and data source of each tree-ring $\delta^{18}O$ records incorporated in the present study are given in Table 11.1.

11.4 Methodology and processes

11.4.1 Cellulose extraction, isotope measurement, and chronology development

The collected samples for $\delta^{18}O$ measurement in trees of all the six sites (Sano et al., 2012, 2013, 2017; Xu et al., 2018; Singh et al., 2019), in each study are cross-dated (Stokes and Smiley, 1968) in the laboratory in order to find out the absolute dates of each ring by the authors of respective studies. The tree-ring widths were then measured and the quality of cross-dated series was checked using the program COFECHA (Holmes, 1983). The general criteria for sample selection toward isotopic measurement are primarily based on the width of the tree-rings to obtain sufficient amount of α-cellulose. Thus, for isotopic analysis, minimum four to five tree cores with relatively wide rings were selected. From the samples collected at all the six sampling sites, each annual ring was separated with the help of a scalpel or sharp blade by the said investigators. The samples are then adequately grinded, and cellulose is extracted based on the protocols established in Brendel et al. (2000) and Anchukaitis et al. (2008) as followed in the studies of Sano et al. (2010, 2012). Singh et al. (2019) adopted the methods for cellulose extraction as described in Wieloch et al. (2011), which involves homogenization/grinding of the wood shavings by way of ultrasonic treatment (Laumer et al., 2009). In case of Wache (Sano et al., 2013), Manali (Sano et al., 2017), Jageshwar, and Ganesh (Xu et al., 2018), for α-cellulose extraction, they employed the modified plate method (Xu et al., 2011; Kagawa et al., 2015), where a 1 mm thick wood lath of the samples can be used to directly extract cellulose from hundreds of rings at once rather than from each individual ring, making it more efficient. A modified chemical treatment protocol based on the Jayme–Wise method (Loader et al., 1997) was used to extract the cellulose components from the wood lath. A noteworthy feature of the modified plate method is that the cellulose from each annual rings is separated after cellulose extraction of the whole wood lath, as opposed to other conventional methods, in which case each annual ring is essentially separated before cellulose extraction. In addition, there is also no requirement for grinding in this method. Further, it has also been shown that the differences in $\delta^{18}O$ values obtained by both the plate method as well as standard method are statistically not significant (Kagawa et al., 2015). The extraction of cellulose from the various methods discussed earlier is then followed by measurement of the oxygen isotope ratios ($^{18}O/^{16}O$) with an isotope ratio mass spectrometer. The $^{18}O/^{16}O$ ratios were expressed in δ notation in parts per mil (‰), deviation relative to the Vienna Standard Mean Ocean Water: $\delta^{18}O = ([R_{sample}/R_{standard}] - 1) \times 1000‰$, where R_{sample} and $R_{standard}$, respectively, stand for the $^{18}O/^{16}O$ ratios of the sample and standard. The tree-ring $\delta^{18}O$ site chronology was developed, by averaging all the individual series for a given site.

11.4.2 Paleoclimate reconstruction methodology

We assessed the spatial field correlation between six tree-ring $\delta^{18}O$ chronologies (Sano et al., 2012, 2013, 2017; Xu et al., 2018; Singh et al., 2019) and a drought metric, scPDSI (Wells et al., 2004; van der Schrier et al., 2013; Osborn et al., 2017) covering entire India, Nepal, and Bhutan using a total of 1234 grids. The scPDSI is considered as more consistent range of relative variability in wetness and dryness across diverse climatic regions (Cook et al., 2015) compared to the original PDSI (Palmer, 1965). In the present study, gridded monthly scPDSI data (version 4.03) has been used which is developed based on CRU-TS 4.03 (Harris et al., 2020) monthly temperature and precipitation data sets using more realistic Penman–Monteith method of estimating potential evapotranspiration. The scPDSI data sets is available for global land area in spatial resolution of 0.5° longitude by 0.5° latitude and covers the time period 1901–2018 C.E., which can be accessed at https://crudata.uea.ac.uk/cru/data/drought/. The previous studies on the Himalayan tree-ring $\delta^{18}O$ chronologies for paleoclimatic study (Sano et al., 2012, 2013, 2017; Xu et al., 2018; Singh et al., 2019) have shown that the drought records of the ISM season have a negative response. Thus we chose June–September scPDSI (henceforth, JJAS-scPDSI) as the climatic predictor to establish correlation with tree-ring $\delta^{18}O$ chronologies. The correlation was carried out as one-tailed test using negative correlations. Although in the record from Dingad, reconstruction was carried for the rainfall during monsoon months of June–July (Singh et al., 2019), but was considered in the present study for JJAS-scPDSI. The meteorological subdivisions based on homogenous monsoon distribution regions in India (Fig. 11.1) and Indian subcontinent was demarcated to assess the spatial relationships.

A range of technique has been developed to investigate large-scale spatial patterns of recent and preinstrumental climatic variability. Researchers in the past utilized various statistical techniques for the development of spatial climate reconstructions such as canonical regression, CR/canonical correlation analysis, CCA (Fritts, 1976; Barnett and Preisendorfer, 1987), orthogonal spatial regression, OSR (Briffa et al., 1986; Cook et al., 1994), and regularized expectation maximization (Schneider, 2001; Zhang et al., 2004). The OSR method is essentially a principal components regression procedure (Draper and Smith, 1981), whereas CR is based on canonical correlation analysis (Cooley and Lohnes, 1971; Gittens, 1985). However, both methods are based on least-squares theory, which forces OSR and CR to be identical for the limiting case of complete multiple regression (Cook et al., 1994). The detail review and comparative description of OSR and CR approaches for paleoclimate reconstruction are given in Cook et al. (1994). The method of regularized expectation maximization offers some theoretical advantages over CR and OSR. This method calibrates the proxy data set against the instrumental climate record by treating the reconstruction as initially missing data in the combined matrix of proxy/instrumental data. The mean and covariance of the combined data matrix is then optimally calculated through an iterative technique, which yields a reconstruction of the climate field having minimal error variance (Shcneider, 2001; Rutherford et al., 2003; Mann and Rutherford, 2002). In the present study, the modified PPR method (Cook et al., 1999, 2004, 2010, 2013, 2015, 2020; Palmer et al., 2015; Stahle et al., 2016) was used to reconstruct the JJAS-scPDSI sequentially at each of its 1234 grid points using network of six tree-ring $\delta^{18}O$ chronologies of the Himalayan region of India, Nepal, and Bhutan. The PPR approach has been widely used to reconstruct climate from tree-rings (Cook et al., 1999, 2004, 2010, 2013, 2015, 2020; Palmer et al., 2015; Stahle et al., 2016). The PPR methodology is based on principal component regression (PCR) procedure and basically designed for CFRs (Cook et al., 1999). The PCR methodology is also designed to produce a "nested" form of reconstructions (Meko, 1997; Cook et al., 2002; Wilson et al., 2007) in which shorter predictor (tree-ring series) is sequentially removed from each PCR steps. The full "nested" reconstruction is then created by appending each subset-reconstruction extension back in time to the beginning of pre-existing shorter reconstruction after appropriate scaling to recover lost variance due to regression in each reconstruction, thus producing the longest possible reconstruction from the available tree-ring data. The scaling is done to insure that no artificial variability due to differences in regression R^2 from nest to nest is present in the full nested reconstruction. Each 1234 grid point nested-PCR model used to produce the spatial JJAS-scPDSI reconstruction was estimated from an overlapping subset of tree-ring $\delta^{18}O$ chronologies that fell within a search radius of 500 km. A single tree-ring $\delta^{18}O$ chronology was set to produce each grid point based on the field correlation. However, the overall geographical coverage area of India, Nepal, and Bhutan are considered here is much larger than the example search area implied in PPR model, the search radius is dynamically expanded from 500 km to achieve the entire grid point.

The skill of tree-ring-based climate reconstruction model for each nested subset was assessed using calibration and verification statistics. The calibration statistics calculated are, coefficient of determination or R^2 (CRSQ) and cross validation reduction of error (CVRE). Among these CVRE is based on the "leave-one-out" procedure analogous to R^2 based on Allen's *PRESS* statistic (Allen, 1974) and is a more conservative measure of explained variance than CRSQ (Cook et al., 2015). The validation period statistics calculated are, Pearson correlation coefficient squared (VRSQ), the reduction of error statistic (VRE), and the coefficient of efficiency (VCE). The validation statistics, VRSQ should be positive for significant reconstruction skill at the 95% level. However, this may go negative which indicates some lack of skill in the reconstructed values. For VRE and VCE there are no theoretical significance tests available, but if VRE > 0 or VCE > 0 then the reconstruction has some skill in excess of the calibration or verification period means, respectively, of the instrumental data (Cook et al., 1999, 2015). The detail description of each calibration and verification statistics based on PPR methodology is given in Cook et al. (1999).

The calibration time period was selected to estimate each PCR model to reconstruct spatial JJAS-scPDSI. In principal component analysis a common time period for estimating the covariance matrix is required and that can be achieved by considering the earliest last year of the time series used. In our case, based on the network of six tree-ring $\delta^{18}O$ chronologies, the earliest last year is 2000 C.E. Thus we considered year 2000 C.E. as the last year of the calibration period, which is common for all chronologies. The calibration period of 1931–2000 C.E. (70 years) was considered, which was 70% of the instrumental data selected for calibration analysis for the present study. The scPDSI available prior to 1931 was withheld from the calibration period to test the tree-ring $\delta^{18}O$ chronologies estimates of scPDSI for validation skill. Thus, the validation period considered for PPR model is 1901–1930 (30 years) considered as 30% of the instrumental data. In addition, the modified PPR methodology to produce nested grid point reconstructions helps to develop paleoclimatic records extremely similar to those produced by the "composite-plus-scale" (CPS) method (Smerdon et al., 2015). Thus the present JJAS-scPDSI reconstruction lead to the reconstructions to be updated from 2000 to 2018 with instrumental scPDSI data.

The reconstruction carried out in the present study was compared spatio-temporarily with the historic drought years documented in the records from the Indian subcontinent. The spatial reconstruction evidently showed a wide spread distribution of these droughts known to occur in particular regions or spatially distributed across the region. The tree-ring $\delta^{18}O$ chronologies record signal of the spatially distributed drought, which was captured in the scPDSI reconstruction. These wide spread droughts impacted a large population and resulted in massive famine that were documented and observed by government authorities. They were mainly the Strange Parallels drought, the Chalisa famine, the East India drought, and the late Victorian Great Drought, and other worse droughts recorded in India, which are further described. The scPSDI reconstruction was analyzed to perceive the teleconnection with remote moisture sources and/or phenomenon controlling the same such as ENSO, SST in the Oceanic realm. The spatial correlations analysis was performed using the KNMI Climate Explorer (Trouet and Oldenborgh, 2013; http://climexp.knmi.nl/).

11.5 Results

The data of tree-ring $\delta^{18}O$ chronologies (Fig. 11.2) used in the present study are based on the different conifer taxa, that is, *A. pindrow* in Manali, *C. deodara* in Jageshwar, *A. spectabilis* in Humla and Ganesh, and *L. griffithiana* in Wache. In case of site Dingad the $\delta^{18}O$ isotope data is based on both conifer (*A. pindrow* and *P. smithiana)* and broad-leaved taxa (*A. indica*). Despite the difference in taxa, the $\delta^{18}O$ records have similar response with JJAS-scPDSI data sets in spatial scale (Fig. 11.3). The tendency of each site $\delta^{18}O$ record in these tests showed higher negative correlations with some common subdivisions apart from the rest of the Indian subcontinent. These were mostly concentrated in the Northwest (NW), Hilly region (Meteorological subdivisions of India, Uttaranchal and Himachal Pradesh), central Northeast (CNE) except Orissa meteorological subdivision, and Nepal (NEP) for most of the records except Bhutan record, which observed a tendency to show higher negative correlation with Eastern Himalayan region (EHR) and Northeast (NE) subdivision. It was observed that for most of the tree-ring $\delta^{18}O$ sites, the negative correlation was commonly higher in the WH subdivision but at the site Wache and Ganesh lower negative correlation was observed. The JJAS-scPDSI spatial reconstruction based on the Himalayan tree-ring $\delta^{18}O$ records from six sites demarcated as year-by-year maps on a 1234 grid point half degree longitude-by-latitude covering Indian meteorological subdivision, Nepal and Bhutan. This spatial drought reconstruction for ISM monsoon month's reflected soil moisture condition and provides spatio-temporal hydroclimatic variability over the wide homogenous monsoon division.

We have developed the spatio-temporal drought reconstruction to assess the history of droughts and pluvials based on existing Himalayan tree-ring $\delta^{18}O$ chronologies for the past 379 years (1621–2000 C.E.), which was further, extended till 2018 (397 years) based on CPS method. This spatial JJAS-scPDSI reconstruction has been assessed through calibration and verification statistics calculated in modified PPR methodology and are shown in Fig. 11.4. For the validation statistics, only those values that passed (VRSQ with $P < .10$, one tailed, VRE, and VCE > 0) are plotted. The total number of 1234 grid point reconstructed for JJAS-scPDSI plotted in Fig. 11.4 has their mean and median values recorded as: CRSQ (mean = 0.166, median = 0.159), CVRE (mean = 0.162, median = 0.150), VRSQ (mean = 0.228, median = 0.210), VRE (mean = 0.185, median = 0.191), VCE (mean = 0.191 median = 0.196). The maximum values observed for CRSQ, CVRE, VRSQ, VRE, and VCE are 0.456, 0.426, 0.563, 0.492, and 0.436, respectively. The strong statistical verifications are seen in homogenous monsoon region of India, that is, NW, CNE excluding Orissa meteorological subdivision, western Himalaya region (Uttaranchal and Himachal meteorological subdivision), and Nepal. Based on this observation we selected regions having strong verification statistics as a best representation of the spatial JJAS-scPDSI reconstruction (Fig. 11.4). For this we averaged the JJAS-scPDSI reconstruction considering all the scPDSI grids, making the average JJAS-scPDSI reconstruction that extends from 1621 to 2018 C.E. (Fig. 11.5). In this reconstruction, the time period for the reconstruction during 1801–2018 C.E., 1778–1800 C.E., 1767–1777 C.E., 1743–1776 C.E., and 1621–1742 C.E. are based on average grids points of 622, 478, 463, 441, and 288, respectively. The total number of 622 grid point considered to make the average JJAS-scPDSI as a best reconstruction, have their mean and median values recorded as: CRSQ (mean = 0.241, median = 0.230), CVRE (mean = 0.200, median = 0.188), VRSQ (mean = 0.264,

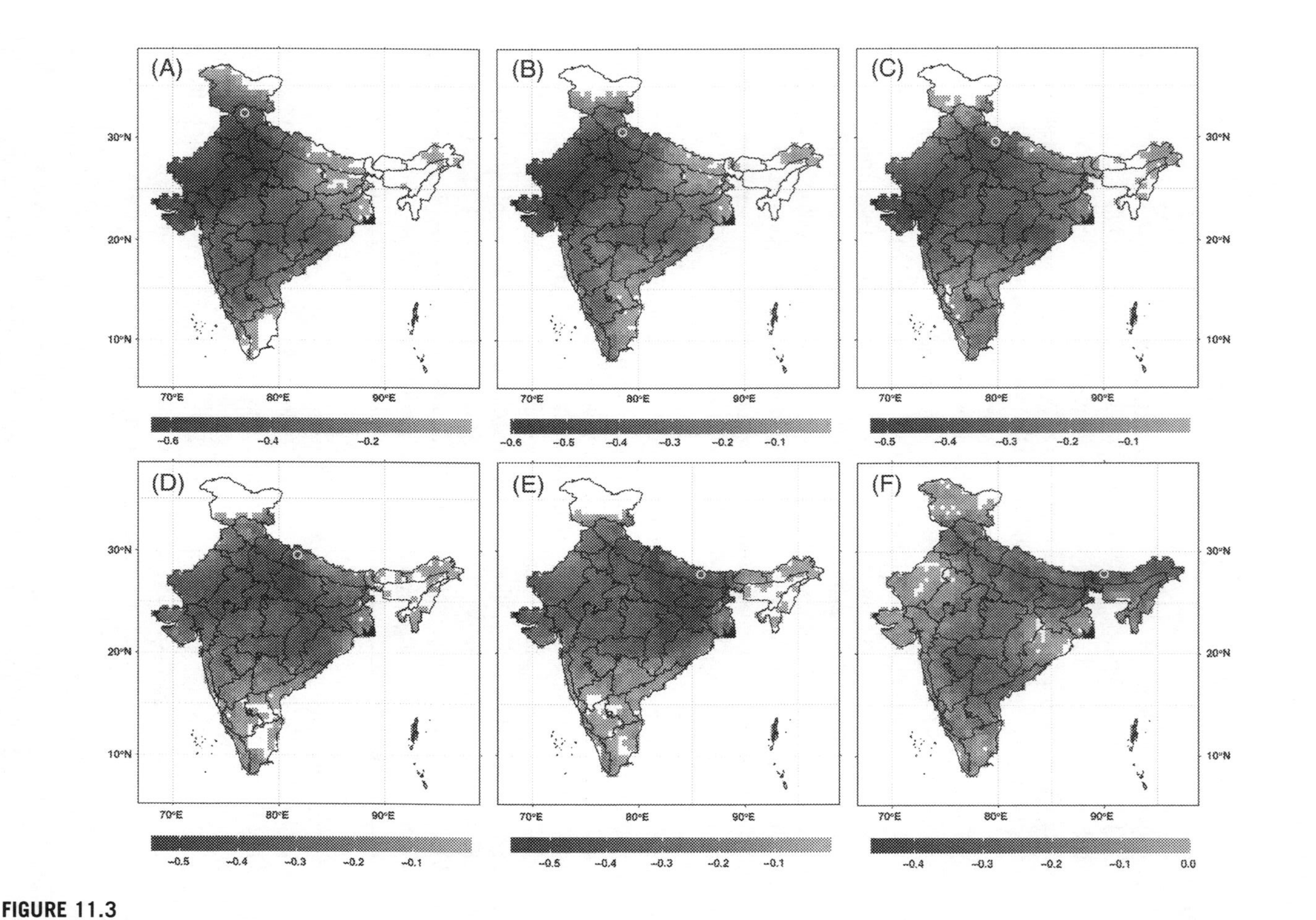

FIGURE 11.3

Field correlation of tree-ring $\delta^{18}O$ isotope chronologies of (A) Manali (Sano et al., 2017), (B) Dingad (Singh et al., 2019), (C) Jageshwar (Xu et al., 2018), (D) Humla (Sano et al., 2012), Ganesh (Xu et al., 2018), and (E) Wache (Sano et al., 2013) with 1234 scPDSI target grid points. The correlations are calculated and represented only for one-tailed test. Yellow circle in (A–F) marks the sampling site for the particular study.

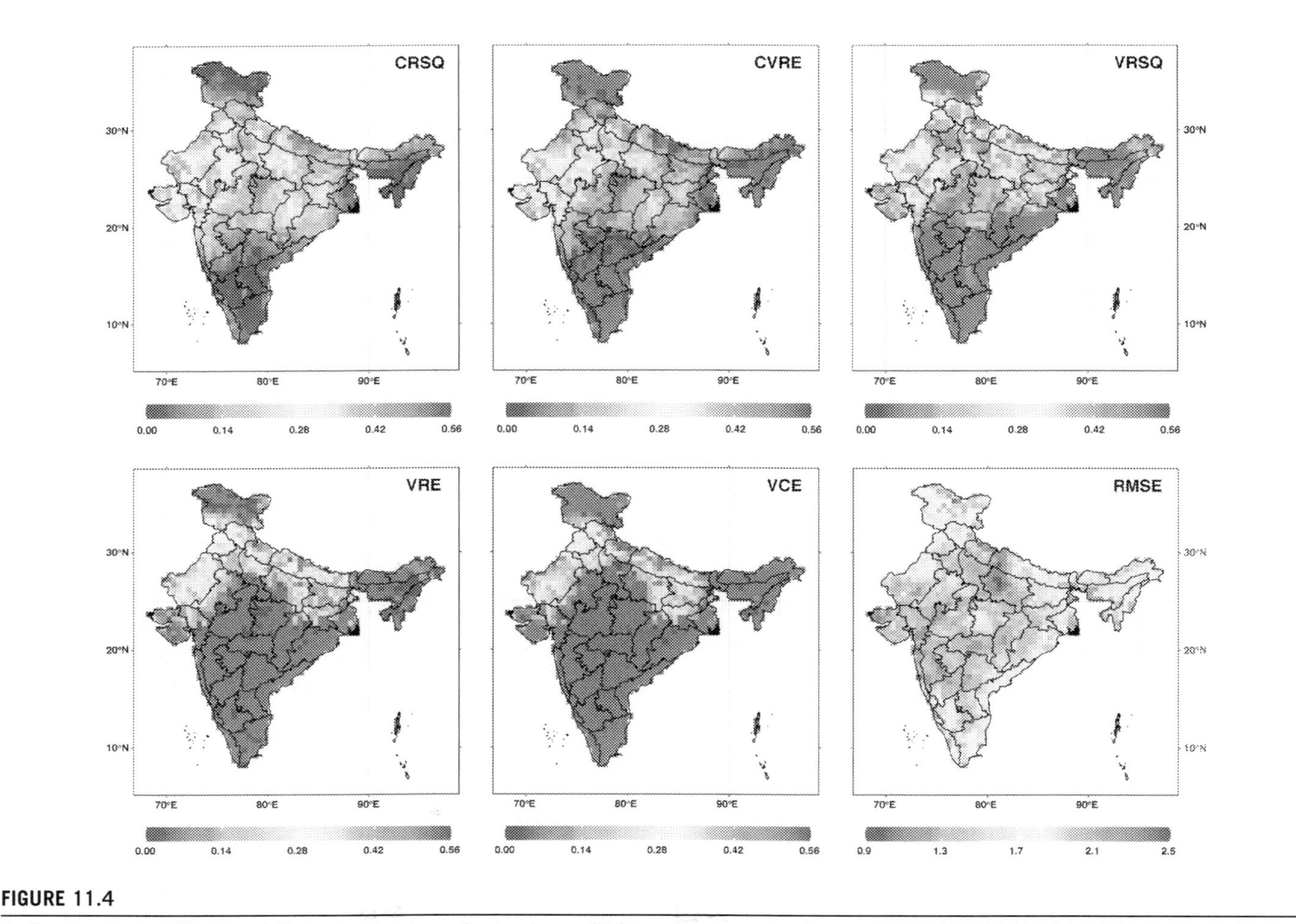

FIGURE 11.4

Calibration and verification statistics for the June–September scPDSI spatial reconstruction carried out for 1234 grid points. Calibrated statistics (prefixed with C) are *CRSQ*, coefficient of determination; *CVRE*, cross validation reduction of error, and verification statistic (prefixed with V) are *VRSQ*, Pearson correlation coefficient squared; *VRE*, reduction of error statistic; *VCE*, coefficient of efficiency. RMSE is root mean square error. All statistics are represented in units of fractional variance. The verification statistic, VRSQ is plotted only for those grid points which passed with $P < .10$ and VRE and VCE values with >0.

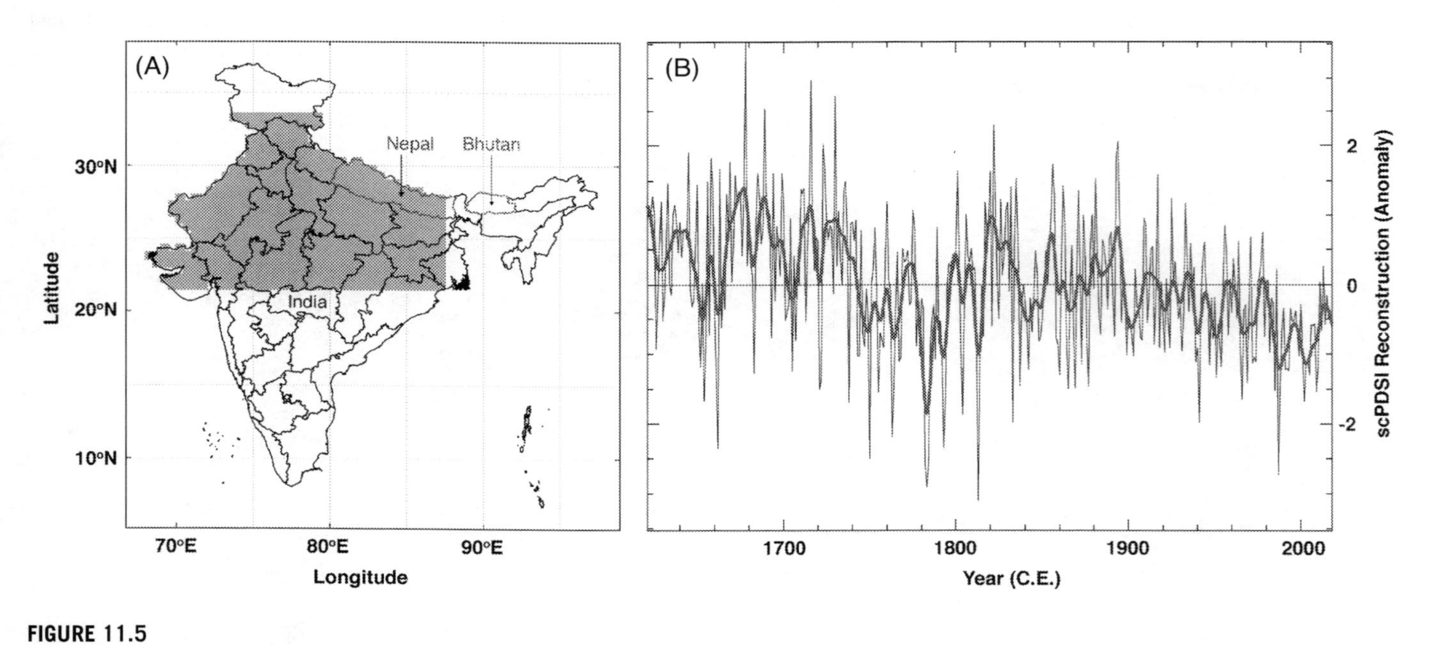

FIGURE 11.5

(A) Area marked in gray is best region, which validated well for verification statistics and considered to average the June–September (JJAS) scPDSI reconstruction. (B) Average JJAS-scPDSI reconstructions extending from 1621 to 2018 C.E. for marked gray area of (A).

median = 0.264), VRE (mean = 0.222, median = 0.239), VCE (mean = 0.204, median = 0.212). The maximum values observed for CRSQ, CVRE, VRSQ, VRE, and VCE that are based on all grid points (n = 1234) considered for present study falls within this region. The region of poorest validation has been observed in the region of southern India.

The spatial JJAS-scPDSI reconstructed from the tree-ring $\delta^{18}O$ chronologies when compared with the documented droughts years in the Indian subcontinent revealed the spatio-temporal distribution of the drought signals recorded. The severe droughts recorded in MADA (Cook et al., 2010) namely The Strange Parallels drought of 1756–1768 C.E. (Lieberman et al., 2003), the East India drought of 1790–1796 C.E. (Grove, 2007), and the late Victorian Great Drought of 1876–1877 C.E. (Davis et al., 2001) and also the Chalisa famine of 1783–1784 C.E. (Imperial Gazetteer of India, 1907) were observed in the present reconstruction (Fig. 11.6). The prominent drought signal of the Strange Parallels drought, the East India drought, and the Chalisa famine were captured in the NWI and the late Victorian Great Drought was wide spread across the subcontinent. The late Victorian Great Drought had a weaker signal in the reconstruction in the Nepal region from were two $\delta^{18}O$ chronologies were taken for the present study. Apart from these severe events other extreme drought years that impacted large parts of India (1877, 1899, 1905, 1911, 1918, 1920, 1951, 1965, 1966, 1972, 1974, and 1979) were well documented and compared in their spatio-temporal extent (Parthasarathy et al., 1987) across India covering an average of 45% area of the country. The present JJAS-scPDSI reconstruction records the spatial distribution of the drought signals in these particular years with variable strength across the region (Fig. 11.7). The study by Adamson (2014) on western India drought years documented to cause large-scale famine and social unrest, which were during 1790–1792, 1803, 1812, 1823–1824, 1833, 1838, 1845, 1847–1848, 1850, and 1855. The present JJAS-scPDSI reconstruction also recorded these drought years and their spatial distributions across the region (Fig. 11.8) whose signals are present in the $\delta^{18}O$ chronologies. But the drought years 1823–1824, 1845, 1855 recorded in western India (Adamson, 2014) was not prominent in the present reconstruction. The comparison of the reconstructed JJAS-scPDSI with historical and well-documented droughts database also serve as validation tests for the accuracy of the spatial drought reconstruction in the present study. These past drought records used for comparison are completely independent of present spatial drought reconstructions and most of them are prior to the statistical calibration period (1931–2000 C.E.) used for reconstructing scPDSI from tree-ring $\delta^{18}O$ chronologies.

The reconstructed JJAS-scPDSI was spatially compared with global SST (Hadley Centre Sea Ice and Sea Surface Temperature, HADISST1 data sets (Rayner et al., 2003) to understand the teleconnections with remotely occurring oceanic forcing phenomenon. The reconstruction correlated negatively with the SST over the tropical Pacific Ocean and Indian Ocean (Fig. 11.9). The correlation is particularly strong in the central Pacific belt-covering region of Nino 4, Nino 3.4, and Nino 3, which suggest that the Nino 3.4 region play a role in modulating the hydroclimate of the studied region, and enhance the drought events.

11.6 Discussion

The Indian monsoon season (June–September) scPDSI had the most significant correlations with the Himalayan tree-ring $\delta^{18}O$ records. Previous studies based on these tree-rings $\delta^{18}O$ records (Sano et al., 2012, 2013, 2017; Xu et al., 2018; Singh et al., 2019) used in present study also showed that tree-ring

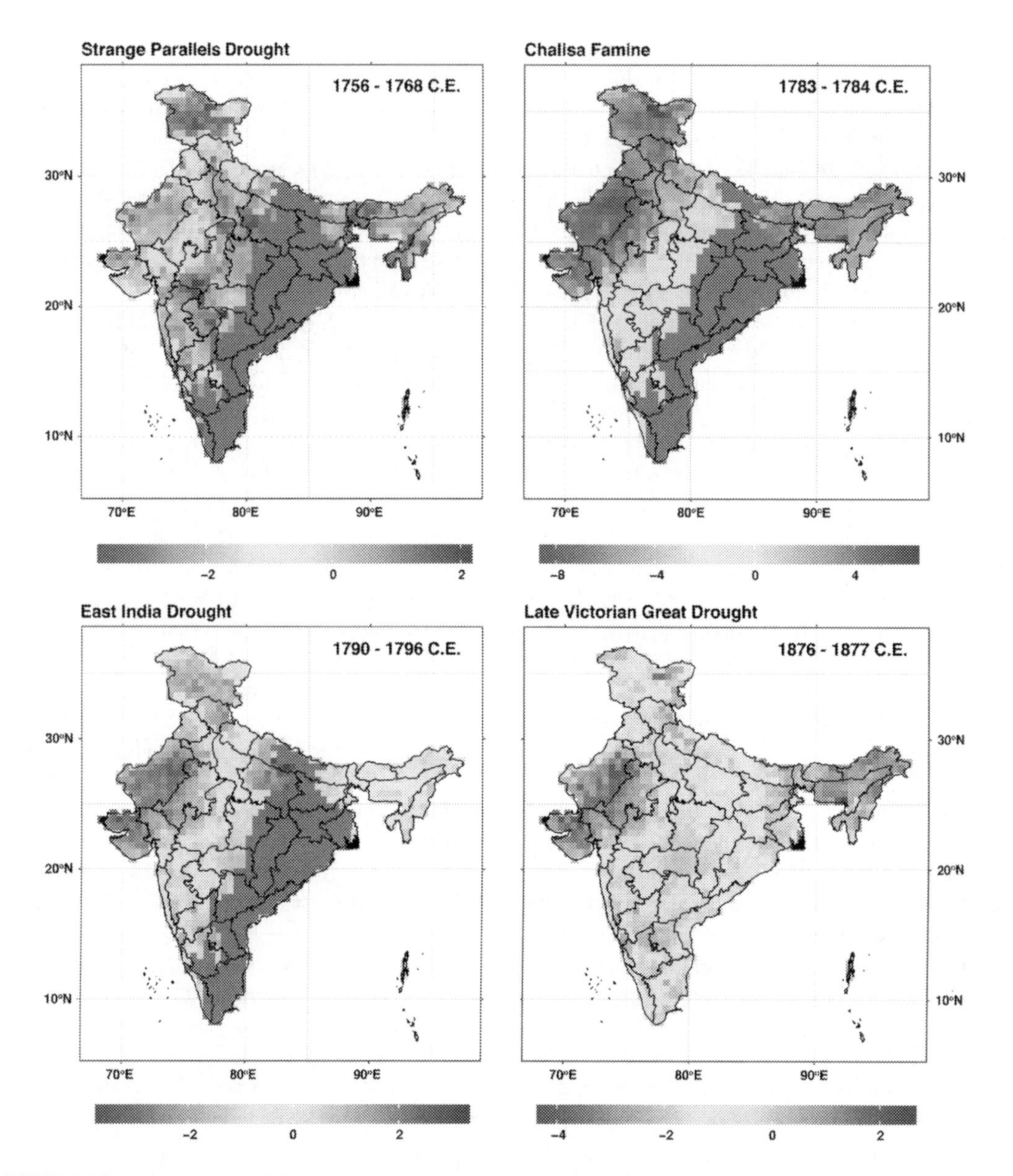

FIGURE 11.6

The reconstructed spatial drought (JJAS-scPDSI) patterns during well-documented historical droughts of Indian subcontinent. (A) The Strange Parallels drought (1756–1768 C.E.; Lieberman et al., 2003; Cook et al., 2010), (B) Chalisa famine (1783–1784 C.E.; Imperial Gazetteer of India, 1907; Grove, 2007), (C) The East India drought (1790–1796 C.E.; Grove, 2007; Cook et al., 2010), and (D) The late Victorian Great Drought (1876–1877; Davis et al., 2001).

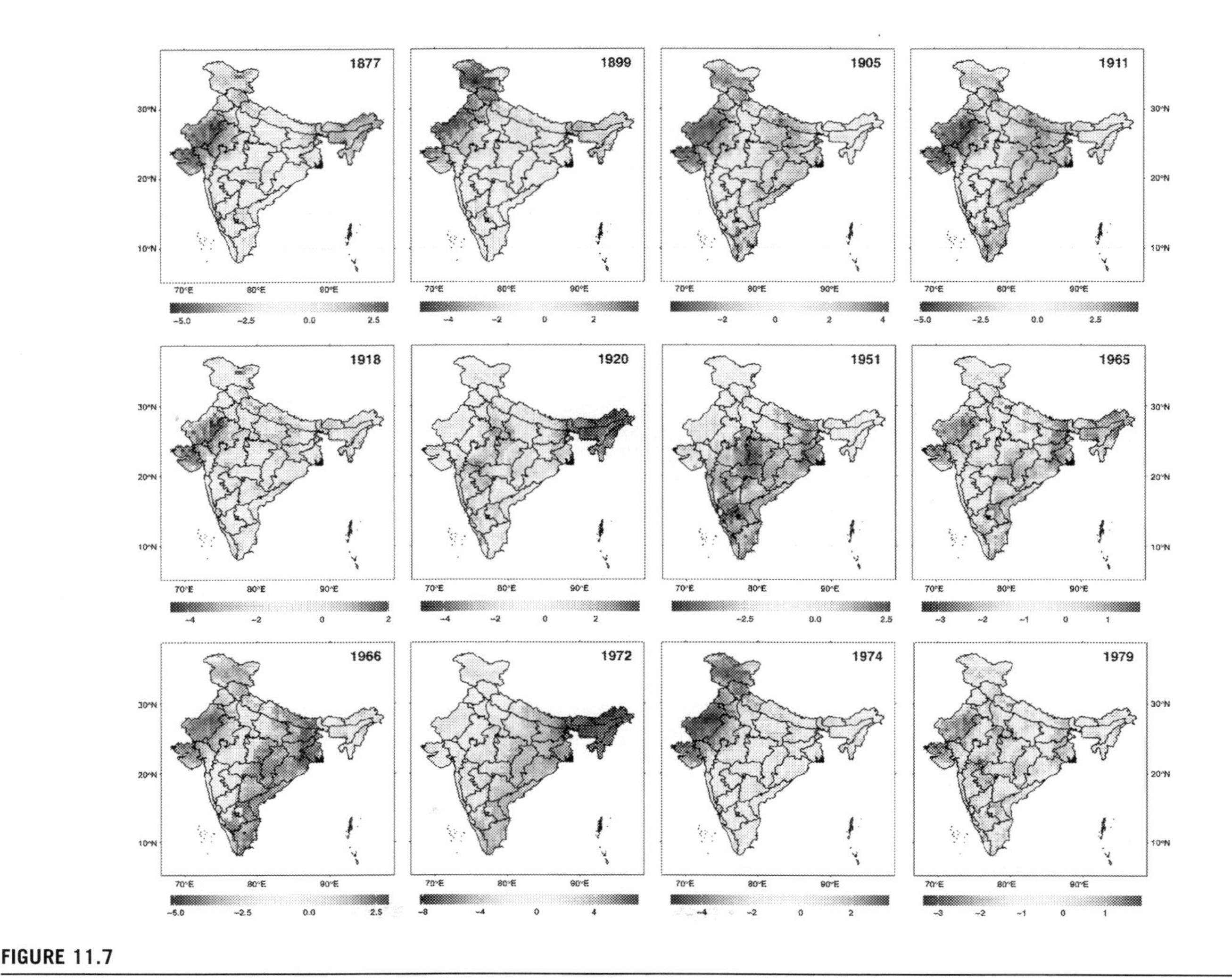

FIGURE 11.7

The reconstructed spatial drought (JJAS-scPDSI) patterns during worst drought years (Parthasarathy et al., 1987) in Indian meteorological subdivision occurred during summer monsoon season.

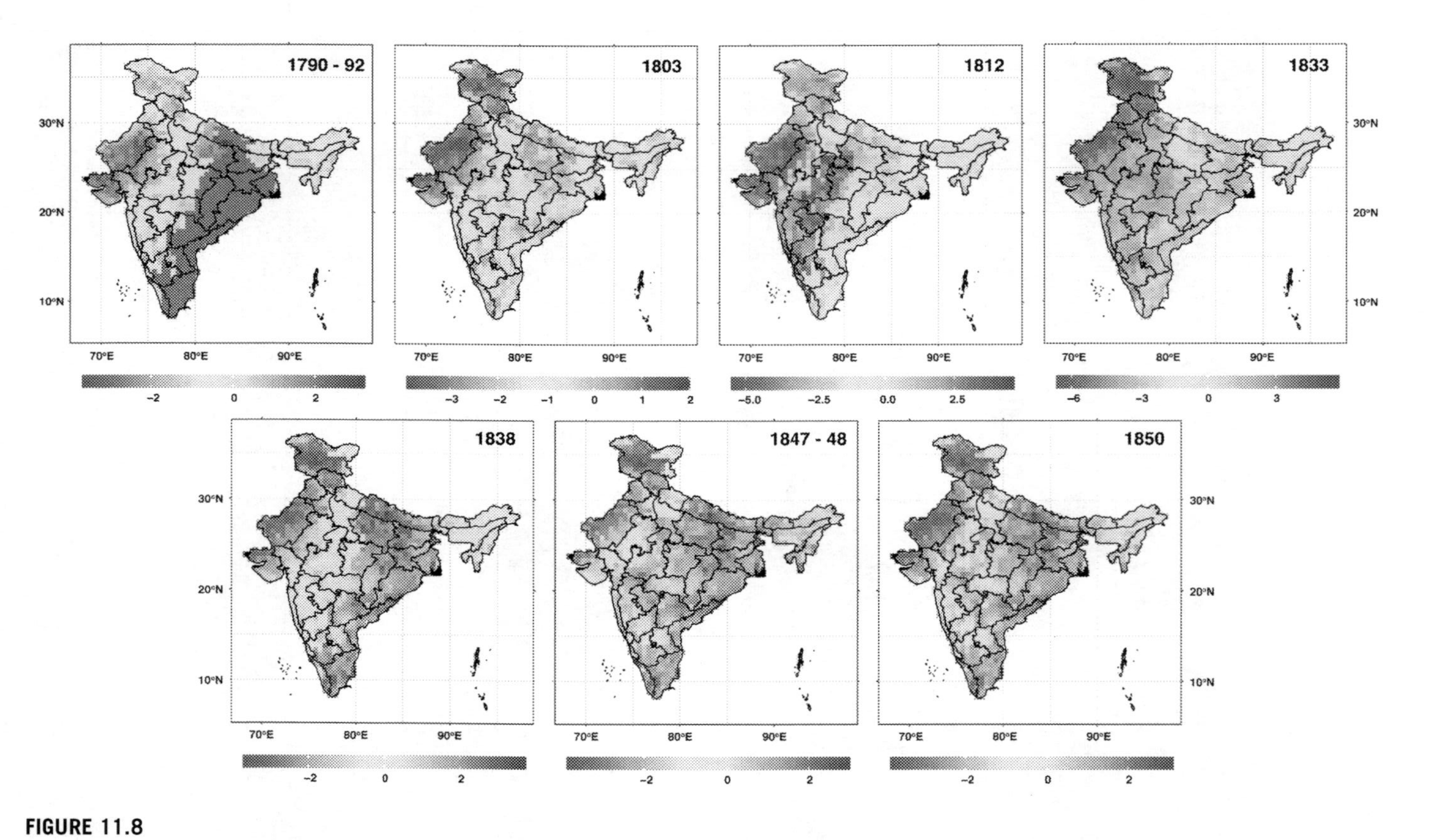

FIGURE 11.8

The reconstructed spatial drought (JJAS-scPDSI) patterns during widespread meteorological drought years recorded in western India (Adamson, 2014).

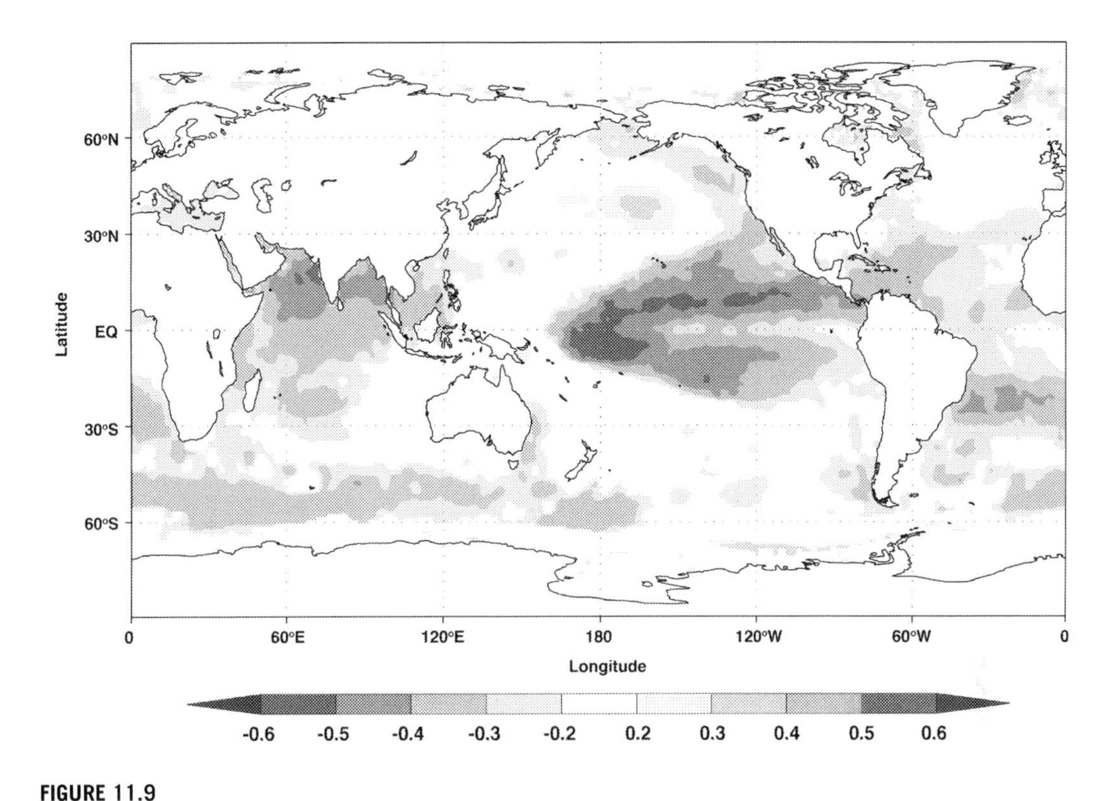

FIGURE 11.9

Spatial correlations between average reconstructed JJAS-scPDSI and Had1SST1 (Rayner et al., 2003) calculated for time period of 1870–2018 C.E.

$\delta^{18}O$ contain moisture-related signals (generally rainfall and drought indices) predominantly during the ISM from June to September. The tree-ring $\delta^{18}O$ chronologies, Manali (Sano et al., 2017), Dingad (Singh et al., 2019), and Jageshwar (Xu et al., 2018) from the Indian western Himalaya region and Humla (Sano et al., 2012) from western Nepal Himalaya shows stronger negative correlation with scPDSI for NWI homogenous monsoon regions and western Himalayan region of India. This suggested that, these sites $\delta^{18}O$ measurements are mostly controlled by the source water originating in the Arabian Sea ISM branch. The tree-ring $\delta^{18}O$ chronologies of Ganesh from Nepal (Xu et al., 2018), and Wache from Bhutan (Sano et al., 2013) had stronger correlation signal in eastern India and northeastern Himalaya region, respectively, which might be due to the influence of moisture source prevailing from Bay of Bengal ISM branch. The influence of the Arabian Sea ISM branch is more prevalent in the extreme drought years when there are higher negative correlation within the Arabian Sea region and further teleconnection with Eastern Pacific Ocean SSTs. Thus the strength of this branch also known as the isotopically enriched branch of the ISM shows a very strong impact in the NWI homogenous monsoon regions and western Himalayan region of India in the scPDSI reconstruction.

In the present study, the $\delta^{18}O$ signal measured, have been well translated in the spatial scale drought metrics, scPDSI covering entire validated grid points of the Indian subcontinent. The spatial JJAS-scPDSI reconstruction allows us to identify and analyze the regional pattern of drought and pluvial over the reconstructed JJAS-scPDSI time period (Figs. 11.6–11.8). The reconstruction captured regional, historical, and well-recorded droughts of the past in spatial scale and thus additionally validated its reliability apart from traditional calibration verification statistics (Fritts, 1976).

The Strange Parallels drought of mid-18th-century covering the time period of 1756–1768 C.E. is well represented in the present spatial JJAS-scPDSI reconstruction and particularly well observed in western India, NWI homogenous monsoon region and NW Himalaya region (Fig. 11.6). This observation is similar to that of tree-ring width-based MADA and this multidecadal wide spread and persistent "megadrought" from India to Southeast Asia is considered as one of the most critical periods of monsoon failure recorded in MADA (Cook et al., 2010). During this period a strong sociopolitical impact was observed across Southeast Asia and across the Siberian plains (Lieberman et al., 2003; Cook et al., 2010).

The Chalisa famine occurred during 1783–1784 C.E. in the Indian subcontinent followed unusual El Niño events that began in 1780. This event caused droughts throughout the region (Imperial Gazetteer of India, 1907; Grove, 2007) and ~11 million people probably died and large areas were deserted (Pattanayak and Puttaiah, 2014). This famine affected numerous parts of North India, particularly the Delhi territories, present-day Uttar Pradesh, Eastern Punjab, Rajputana, and Kashmir and was preceded by a famine in the previous year, 1782–1783, in some parts of South India. This Chalisa famine (1783–1784 C.E.) is well represented in the present spatial JJAS-scPDSI reconstruction and with the same regions of India where it actually affected during this period (Fig. 11.6). This famine was recorded in the tree-ring $\delta^{18}O$ record from the Dingad site by Singh et al. (2019), which was also considered in the present study.

The drought of 1790–1796 C.E., which is popularly known as the East India drought (Grove, 2007) occurred during the great El Niño of the late 18th century. This drought had a worldwide occurrence and resulted in socioeconomic disorder across the globe (Grove, 2007; Cook et al., 2010). In the present JJAS-scPDSI reconstruction, this drought period was captured in wide areas of western India, western and northwestern Himalayan region, northeastern Indian region, and some parts of Peninsular India (Fig. 11.6). However, the MADA record does not record its severity over India in comparison to other droughts records possibly due to the limited tree-ring coverage in India analyzed in the development of MADA (Cook et al., 2010). But in the present JJAS-scPDSI reconstruction is well recording, the East India drought, which effected India, known through various studies on severe famine (Grove, 2007) across India. The spatio-temporal distribution and strength of the present tree-ring $\delta^{18}O$ are higher in comparison to MADA records toward captured moisture-related signals from a larger region. The Dasuopo ice core record (Thomson et al., 2000) based on dust and geochemical analysis evidenced severe drought in the late 18th century which is also observed in our present JJAS-scPDSI reconstruction with severe droughts in the region to the west and north of the Himalayas.

The late Victorian Great Drought of 1876–1878 C.E. is one of the severe droughts, which were felt across much of the tropics, and were particularly acute in India. This extreme drought also occurred during one of the most severe El Niño events of the past 150 years (Davis et al., 2001). In the present JJAS-scPDSI reconstruction, this severe drought has been captured during 1876–1877 C.E. (Fig. 11.6). This drought has been captured across the entire region considered for present study as compared to the other three historical droughts shown in Fig. 11.6. The late Victorian Great Drought is also considered as the worst drought among all four historical drought captured in MADA records (Cook et al., 2010).

The worst drought years of the ISM season over different meteorological subdivision of India observed during the time period of 1871–1984 C.E. (Parthasarathy et al., 1987) are also well recorded in spatially reconstructed JJAS-scPDSI (Fig. 11.7). In addition, the widespread meteorological drought (i.e., deficient rainfall across the region) occurred in western India during 1790–1860 C.E. (Adamson, 2014) are also well observed in the spatial JJAS-scPDSI reconstruction particularly in the NWI homogenous monsoon region which include the part of western India (Fig. 11.8). During these drought periods of western India, the region has historic reports of food price rise, widespread migration, and death and debility due to starvation. This further provides additional validation of the present reconstruction carried out in spatial scale.

The inverse relationships of reconstructed JJAS-scPDSI from the present study with SST of the tropical Pacific Ocean (Fig. 11.9) may modulate the drought of the region due to ENSO events. The inverse relationship between the ENSO and the ISM is a well-known, where El Niño events characterized by anomalous SST warming over the equatorial Pacific Ocean cause a significant reduction in monsoon rainfall (Kumar et al., 1995, 1999, 2006; Webster et al., 1998) often contemporaneous with hot and dry weather extremes (Mishra et al., 2020). However, it should also be noted that the temporal stability of the correlation between ENSO and the Himalaya $\delta^{18}O$ chronologies are not consistent across sites (Sano et al., 2013; 2017). The decoupling of the ENSO and monsoon teleconnection post-1980 reported earlier by Kumar et al. (1999) was also reflected in the Humla $\delta^{18}O$ chronology from Nepal (Sano et al., 2010) as indicated by the lack of any significant correlations with June–September Nino 3.4 SST during 1981–2000 C.E. Interestingly, while the Wache $\delta^{18}O$ chronology from Bhutan (Sano et al., 2013) showed that ENSO–monsoon relationship was stable during 1971–2011 C.E., there was little or no association during the period 1951–1970 C.E. The period of the decoupled ENSO–monsoon relationship as seen in the Wache $\delta^{18}O$ chronology was noted to have coincided with the negative phase of the Pacific Decadal Oscillation, implying regionally variable periods of ENSO impact. Sano et al. (2017) associated the unstable ENSO–monsoon relationship to the increasing SST of the Indian Ocean during periods of weak correlations, which was contemporaneous with an increase in the $\delta^{18}O$ values. Irrespective of the timespan and study sites, nearly all the $\delta^{18}O$-based reconstructions thus far have revealed an overall drying trend in summer monsoon precipitation during the recent century or so (Sano et al., 2012, 2017; Xu et al., 2018; Singh et al., 2019; Managave et al., 2020) with the exception of the Wache chronology from Bhutan, that showed a normal monsoon during the 20th century (Sano et al., 2013). This contradictory trend was attributed to the fact that the $\delta^{18}O$ chronologies from the Indian and Nepal sites were much further inland and therefore, much more sensitive to the weakening monsoon circulation as opposed to the Wache site in Bhutan proximal to the Bay of Bengal. Decreasing trends in monsoon intensity have also been observed in the records across different proxies and sites such as tree-ring $\delta^{18}O$ since the 19th century from eastern Tibet and Southeastern Tibet (Xu et al., 2012; Grießinger et al., 2017), North China since the 1930s (Wang et al., 2020) and Southeast Asia since the 19th century which is more pronounced from the mid-20th century (Xu et al., 2013), Dasuopo glacier snow accumulation records from an ice-core in Southern Tibet since the 1840s (Zhao and Moore, 2006), Speleothem $\delta^{18}O$ from North India since the 1950s (Sinha et al., 2015), since the 1840s in varve-thickness records from lake sediments in Tibet (Chu et al., 2011) and Myanmar (Sun et al., 2016). Studies based on instrumental records also shows decreasing trend in global monsoon precipitation during 1949–2002 C.E. (Zhou et al., 2008), an overall decreasing trend in ISM rainfall during the recent three decades (Naidu et al., 2015) and in the Northwest Himalaya during 1866–2006 C.E. (Bhutiyani et al., 2010), though spatial variability in the trends are also observed. Empirically the JJAS-scPDSI reconstruction strength was proven by the spatial coherence with extreme events regionally and remotely occurring environmental phenomenon in the Pacific Ocean and Indian Ocean. The recurrence of the drought events in the JJAS-scPDSI reconstruction establishes the diminishing strength of the ISM, predominantly in the NWI homogenous monsoon regions and northwest Himalaya region.

11.7 Learning and knowledge outcomes

A $\delta^{18}O$ measurements based on six tree-ring chronologies distributed across the Himalaya in the Indian subcontinent successful analyzed toward drought field reconstruction. The available tree-ring $\delta^{18}O$ chronology in the region has been previously analyzed to reconstruct moisture signal in terms of rainfall and drought. But for the first time scPDSI for June–September has been reconstructed and compared through its spatial distribution based on the network of Himalayan tree-ring $\delta^{18}O$ chronologies. The primary outcome of the present study was to identify the spatial distribution of the present drought reconstruction in terms of historical drought in relation to ISM failures in the study region in a spatial scale. These historical modes of spatio-temporal variability of drought can provide more in-depth dynamics of the monsoon variability for the region. These widespread drought events have had exuberant socioeconomic impacts on the agriculture-dependent economy of the Indian subcontinent. This impact of droughts throughout the 18th century and 19th century has been well documented by the governments mostly due to the subsequent famines and monsoon-dependent agriculture in those times. The present study reveals the decreasing strength of ISM in the western and the high altitude northwestern parts of the Indian subcontinent. The established teleconnections with SST variations in Pacific Ocean and Indian Ocean related forcing phenomenon demarcated the strength of the present spatial reconstruction.

Acknowledgments

Authors would like to express their gratitude to Dr. Vandana Prasad, Director, Birbal Sahni Institute of Palaeosciences, Lucknow for encouragement and permission to publish this work (BSIP contribution No. 64/2020-21). The authors are deeply indebted to the various workers who dedicated a lifetime to measure stable oxygen isotope data in tree-rings from various localities in the Himalaya region. We are also indebted to Editors of the book Dr. Navnith K.P. Kumaran and Dr. D. Padmalal for the invitation to contribute this chapter. The author (N.M.) is highly grateful to DST WOS-A for the support given to her. The authors are indebted to the two anonymous reviewers and the editors for their valuable suggestions to improve our contribution in this book.

References

Adamson, G.C.D., Institutional and community adaptation from the archives: a study of drought in western India, 1790–1860, Geoforum 55 (2014) 110–119, https://doi.org/10.1016/j.geoforum.2014.05.010.

Allen, D.M., The relationship between variable selection and data augmentation and a method for prediction, Technometrics 16 (1) (1974) 125–127, https://doi.org/10.1080/00401706.1974.10489157.

Anchukaitis, K.J., Tree rings reveal climate change past, present, and future 1, Proc. Am. Philos. Soc. 161 (3) (2017) 244–263.

Anchukaitis, K.J., Evans, M.N., Tropical cloud forest climate variability and the demise of the Monteverde golden toad, Proc. Natl. Acad. Sci. 107 (11) (2010) 5036–5040, https://doi.org/10.1073/pnas.0908572107.

Anchukaitis, K.J., Evans, M.N., Lange, T., Smith, D.R., Leavitt, S.W., Schrag, D.P., Consequences of a rapid cellulose extraction technique for oxygen isotope and radiocarbon analyses, Anal. Chem. 80 (6) (2008) 2035–2041, https://doi.org/10.1021/ac7020272.

Asoka, A., Gleeson, T., Wada, Y., Mishra, V., Relative contribution of monsoon precipitation and pumping to changes in groundwater storage in India, Nat. Geosci. 10 (2) (2017) 109–117, https://doi.org/10.1038/ngeo2869.

Barnett, T.P., Preisendorfer, R., Origins and levels of monthly and seasonal forecast skill for United States surface air temperatures determined by canonical correlation analysis, Monthly Weather Rev. 115 (1987) 1825–1850, https://doi.org/10.1175/1520-0493(1987)115<1825:OALOMA>2.0.CO;2.

Bhattacharyya, A., Shah, S.K., Tree-ring studies in India past appraisal, present status and future prospects, IAWA J. 30 (4) (2009) 361–370, https://doi.org/10.1163/22941932-90000224.

Bhutiyani, M.R., Kale, V.S., Pawar, N.J., Climate change and the precipitation variations in the northwestern Himalaya: 1866–2006, Int. J. Climatol. 30 (4) (2010) 535–548, https://doi.org/10.1002/joc.1920.

Brendel, O., Iannetta, P.P.M., Stewart, D., A rapid and simple method to isolate pure alpha-cellulose, Phytochem. Anal. 11 (1) (2000) 7–10, https://doi.org/10.1002/(SICI)1099-1565(200001/02)11:1<7::AID-PCA488>3.0.CO;2-U.

Briffa, K.R., Jones, P.D., Wigley, T.M.L., Pilcher, J.R., Baillie, M.G.L., Climate reconstruction from tree rings: part 2, spatial reconstruction of summer mean sea-level pressure patterns over Great Britain, J. Climatol. 6 (1) (1986) 1–15, https://doi.org/10.1002/joc.3370060102.

Brunello, C.F., Andermann, C., Helle, G., Comiti, F., Tonon, G., Tiwari, A., Hovius, N., Hydroclimatic seasonality recorded by tree ring δ^{18}O signature across a Himalayan altitudinal transect, Earth Planet. Sci. Lett. 518 (2019) 148–159, https://doi.org/10.1016/j.epsl.2019.04.030.

Chu, G., Sun, Q., Yang, K., Li, A., Yu, X., Xu, T., Yan, F., Wang, H., Liu, M., Wang, X., Evidence for decreasing South Asian summer monsoon in the past 160 years from varved sediment in Lake Xinluhai, Tibetan Plateau, J. Geophys. Res. 116 (2011) D02116, https://doi.org/10.1029/2010JD014454.

Cook, E.R., Anchukaitis, K.J., Buckley, B.M., D'Arrigo, R.D., Jacoby, G.C., Wright, W.E., Asian monsoon failure and megadrought during the last millennium, Science 328 (5977) (2010) 486–489, https://doi.org/10.1126/science.1185188.

Cook, E.R., Briffa, K.R., Jones, P.D., Spatial regression methods in dendroclimatology: a review and comparison of two techniques, Int. J. Climatol. 14 (4) (1994) 379–402, https://doi.org/10.1002/joc.3370140404.

Cook, E.R., D'Arrigo, R.D., Mann, M.E., A well-verified, multiproxy reconstruction of the winter North Atlantic Oscillation Index since A.D. 1400, J. Clim. 15 (13) (2002) 1754–1764, https://doi.org/10.1175/1520-0442(2002)015<1754:AWVMRO>2.0.CO;2.

Cook, E.R., Krusic, P.J., Experimental reconstruction of large-scale summer monsoon drought over India and the Tibetan Plateau using tree rings from 'High Asia', Palaeobotany 57 (3) (2008) 515–528.

Cook, E.R., Krusic, P.J., Anchukaitis, K.J., Buckley, B.M., Nakatsuka, T., Sano, M., Tree-ring reconstructed summer temperature anomalies for temperate East Asia since 800 C.E., Clim. Dyn. 41 (2013) 2957–2972, https://doi.org/10.1007/s00382-012-1611-x.

Cook, E.R., Krusic, P.J., Jones, P.D., Dendroclimatic signals in long tree-ring chronologies from the Himalayas of Nepal, Int. J. Climatol. 23 (7) (2003) 707–732, https://doi.org/10.1002/joc.911.

Cook, E.R., Meko, D.M., Stahle, D.W., Cleaveland, M.K., Drought reconstructions for the continental United States, J. Clim. 12 (1999) 1145–1162, https://doi.org/10.1175/1520-0442(1999)012<1145:DRFTCU>2.0.CO;2.

Cook, E.R., Seager, R., Kushnir, Y., et al., Old world megadroughts and pluvials during the Common Era, Sci. Adv. 1 (10) (2015) e1500561, https://doi.org/10.1126/sciadv.1500561.

Cook, E.R., Solomina, O., Matskovsky, V., et al., The European Russia Drought Atlas (1400-2016 CE), Clim. Dyn. 54 (2020) 2317–2335, https://doi.org/10.1007/s00382-019-05115-2.

Cook, E.R., Woodhouse, C.A., Eakin, C.M., Meko, D.M., Stahle, D.W., Long-term aridity changes in the western United States, Science 306 (5698) (2004) 1015–1018, https://doi.org/10.1126/science.1102586.

Cooley, W.W., Lohnes, P.R., Multivariate Data Analysis, Wiley, New York, 1971.

Dansgaard, W., Stable isotopes in precipitation, Tellus 16 (4) (1964) 436–468, https://doi.org/10.1111/j.2153-3490.1964.tb00181.x.

Davis, M., Late Victorian Holocausts: El Niño Famines and the Making of the Third World, Verso, London, 2001.

Draper, N.R., Smith, H., Applied Regression Analysis, Wiley, New York, 1981.

Fritts, H.C., Tree Rings and Climate, Academic Press Inc., New York, 1976.

Gadgil, S., Gadgil, S., The Indian monsoon, GDP and agriculture, Econ. Polit. Wkly. 41 (47) (2006) 4887–4895.

Gagen, M., McCarroll, D., Loader, N.J., Robertson, I., Stable isotopes in dendroclimatology: moving beyond 'potential', in: Hughes, M.K., Swetnam, T., Diaz, H.F. (Eds.), Dendroclimatology: Progress and Prospects, 11, Springer, Dordrecht, 2011, pp. 147–172, https://doi.org/10.1007/978-1-4020-5725-0_6.

Gaire, N.P., Bhuju, D.R., Koirala, M., Shah, S.K., Carrer, M., Timilsena, R., Tree-ring based spring precipitation reconstruction in western Nepal Himalaya since AD 1840, Dendrochronologia 42 (2017) 21–30, https://doi.org/10.1016/j.dendro.2016.12.004.

Gaire, N.P., Dhakal, Y.R., Shah, S.K., Fan, Z.X., Bräuning, A., Thapa, U.K., Bhandari, S., Aryal, S., Bhuju, D.R., Drought (scPDSI) reconstruction of trans-Himalayan region in western Nepal using *Pinus wallichiana* tree-rings, Palaeogeogr. Palaeoclim. Palaeoecol. 514 (2019) 251–264, https://doi.org/10.1016/j.palaeo.2018.10.026.

Gaire, N.P., Fan, Z.X., Shah, S.K., Thapa, U.K., Rokaya, M.B, Tree-ring record of winter temperature from Humla, Karnali in central Himalaya: a 229 years-long perspective for recent warming trend, Geograf. Ann.: Ser. A Phys. Geogr. 102 (3) (2020) 297–316, https://doi.org/10.1080/04353676.2020.1751446.

Gittens, R., Canonical Analysis, Springer-Verlag, Berlin, 1985.

Goswami, B.N., South Asian monsoon, in: Lau, W.K.M., Waliser, D. (Eds.), Intraseasonal Variability in the Atmosphere-Ocean Climate System, Springer-Verlag Berlin, Heidelberg, 2012, pp. 21–72.

Grießinger, J., Bräuning, A., Helle, G., Hochreuther, P., Schleser, G., Late Holocene relative humidity history on the southeastern Tibetan plateau inferred from a tree-ring $\delta^{18}O$ record: recent decrease and conditions during the last 1500 years, Quat. Int. 430 (2017) 52–59, https://doi.org/10.1016/j.quaint.2016.02.011.

Grove, R.H., The great El Niño of 1789–93 and its global consequences: reconstructing an extreme climate event in world environmental history, Mediev. Hist. J. 10 (1–2) (2007) 75–98, https://doi.org/10.1177/097194580701000203.

Harris, I., Osborn, T.J., Jones, P., Lister, D., Version 4 of the CRU TS monthly high-resolution gridded multivariate climate dataset, Sci. Data 7 (2020) 109, https://doi.org/10.1038/s41597-020-0453-3.

Holmes, R., Computer-assisted quality control in tree-ring dating and measurement, Tree-ring Bull. 43 (1983) 69–78.

Kagawa, A., Sano, M., Nakatsuka, T., Ikeda, T., Kubo, S., An optimized method for stable isotope analysis of tree rings by extracting cellulose directly from cross-sectional laths, Chem. Geol. 393–394 (2015) 16–25, https://doi.org/10.1016/j.chemgeo.2014.11.019.

Kumar, K.K., Rajagopalan, B., Cane, M.A., On the weakening relationship between the Indian monsoon and ENSO, Science 284 (5423) (1999) 2156–2159, https://doi.org/10.1126/science.284.5423.2156.

Kumar, K.K., Rajagopalan, B., Hoerling, M., Bates, G., Cane, M., Unraveling the mystery of Indian monsoon failure during El Niño, Science 314 (5796) (2006) 115–119, https://doi.org/10.1126/science.1131152.

Kumar, K.K., Soman, M.K., Kumar, K.R., Seasonal forecasting of Indian summer monsoon rainfall: a review, Weather 50 (12) (1995) 449–467, https://doi.org/10.1002/j.1477-8696.1995.tb06071.x.

Laumer, W., Andreu, L., Helle, G., Schleser, G.H., Wieloch, T., Wissel, H., A novel approach for the homogenization of cellulose to use micro-amounts for stable isotope analyses, Rapid Commun. Mass Spectrom. 23 (13) (2009) 1934–1940, https://doi.org/10.1002/rcm.4105.

Lieberman, V., Strange Parallels: Integration of the Mainland Southeast Asia in Global Context, c.800-1830, 1, Cambridge University Press, Cambridge, 2003.

Loader, N.J, Robertson, I., Barker, A.C., Switsur, V.R., Waterhouse, J.S., An improved technique for the batch processing of small wholewood samples to α-cellulose, Chem. Geol. 136 (3-4) (1997) 313–317, https://doi.org/10.1016/S0009-2541(96)00133-7.

Managave, S., Shimla, P., Yadav, R.R., Ramesh, R., Balakrishnan, S., Contrasting centennial-scale climate variability in High Mountain Asia revealed by a tree-ring oxygen isotope record from Lahaul-Spiti, Geophys. Res. Lett. 47 (4) (2020) e2019GL086170, https://doi.org/10.1029/2019GL086170.

Mann, M.E., Rutherford, S., Climate reconstruction using 'pseudoproxies', Geophys. Res. Lett. 29 (10) (2002) 139(1)-139(4), https://doi.org/10.1029/2001GL014554 .

McCarroll, D., Loader, N.J., Stable Isotopes in Tree Rings, Quat. Sci. Rev. 23 (2004) 771–801. https://doi.org/10.1016/j.quascirev.2003.06.017.

Medina, S., Houze Jr., R., Kumar, A., Niyogi, D., Summer monsoon convection in the Himalayan region: terrain and land cover effects, Q. J. R. Meteorol. Soc. 136 (2010) 593–616.

Meko, D., Dendroclimatic reconstruction with time varying predictor subsets of tree indices, J. Clim. 10 (4) (1997) 687–696, https://doi.org/10.1175/1520-0442(1997)010<0687:DRWTVP>2.0.CO;2.

Mishra, V., Thirumalai, K., Singh, D., Aadhar, S., Future exacerbation of hot and dry summer monsoon extremes in India, NPJ Clim. Atmos. Sci. 3 (10) (2020) , https://doi.org/10.1038/s41612-020-0113-5.

Naidu, C.V., Raju, A.D., Satyanarayana, G.C., Kumar, P.V., Chiranjeevi, G., Suchitra, P., An observational evidence of decrease in Indian summer monsoon rainfall in the recent three decades of global warming era, Glob. Planet. Change 127 (2015) 91–102, https://doi.org/10.1016/j.gloplacha.2015.01.010.

Osborn, T., Barichivich, J., Harris, I., van der Schrier, G., Jones, P., 2017. Monitoring global drought using the self-calibrating Palmer Drought Severity Index, in: Blunden, J., Arndt, D.S., (Eds.), State of the Climate in 2016. Bull. Am. Meteorol. Soc. 98(8), S32–S33, https://doi.org/10.1175/2017BAMSStateoftheClimate.1.

Palmer, J.G., Cook, E.R., Turney, C.S., Allen, K., Fenwick, P., Cook, B.I., O'Donnell, A., Lough, J., Grierson, P., Baker, P., Drought variability in the eastern Australia and New Zealand summer drought atlas (ANZDA, C.E. 1500-2012) modulated by the Interdecadal Pacific Oscillation, Environ. Res. Lett. 10 (12) (2015) 124002, http://dx.doi.org/10.1088/1748-9326/10/12/124002.

Palmer, W.C., Meteorological Drought, U.S. Department of Commerce Weather Bureau, Washington, DC, 1965 Research Paper No. 45.

Pandey, U., Shah, S.K., Mehrotra, N., Tree-ring studies from Kashmir Valley: present status and future perspectives, Geophytology 46 (2) (2016) 207–220.

Parthasarathy, B., Sontakke, N.A., Monot, A.A., Kothawale, D.R., Droughts/floods in the summer monsoon season over different meteorological subdivisions of India for the period 1871–1984, J. Climatol. 7 (1987) 57–70, https://doi.org/10.1002/joc.3370070106.

Pattanayak, S., Puttaiah, E., Analysis of India's quest for ensuring food security, Int. J. Innov. Res. Dev. 3 (6) (2014) 313–319.

Ramesh, R., Bhattacharya, S.K., Gopalan, K., Dendroclimatological implications of isotope coherence in trees from Kashmir Valley, India, Nature 317 (1985) 802–804, https://doi.org/10.1038/317802a0.

Ramesh, R., Bhattacharya, S.K., Gopalan, K., Climatic correlations in the stable isotope records of silver fir (*Abies pindrow*) trees from Kashmir, India, Earth Planet. Sci. Lett. 79 (1-2) (1986) 66–74, https://doi.org/10.1016/0012-821X(86)90041-5.

Rayner, N.A., Parker, D.E., Horton, E.B., Folland, C.K., Alexander, L.V., Rowell, D.P., Kent, E.C., Kaplan, A., Global analyses of sea surface temperature, sea ice, and night marine air temperature since the late nineteenth century, J. Geophys. Res.-Atmos. 108 (D14) (2003) 4407, https://doi.org/10.1029/2002JD002670.

Roden, J.S., Lin, G., Ehleringer, J.R., A mechanistic model for interpretation of hydrogen and oxygen isotope ratios in tree-ring cellulose, Geochim. Cosmochim. Acta 64 (1) (2000) 21–35, https://doi.org/10.1016/S0016-7037(99)00195-7.

Rutherford, S., Mann, M.E., Delworth, T.L., Stouffer, R.J., Climate field reconstruction under stationary and nonstationary forcing, J. Clim. 16 (2003) 462–479, https://doi.org/10.1175/1520-0442(2003)016<0462:CFRUSA>2.0.CO;2.

Rydval, M., Gunnarson, B.E., Loader, N.J., Cook, E.R., Druckenbrod, D.L., Wilson, R., Spatial reconstruction of Scottish summer temperatures from tree rings, Int. J. Climatol. 37 (3) (2017) 1540–1556, https://doi.org/10.1002/joc.4796.

Sano, M., Dimri, A.P., Ramesh, R., Xu, C., Li, Z., Nakatsuka, T., Moisture source signals preserved in a 242-year treering δ^{18}O chronology in the western Himalaya, Glob. Planet. Change 157 (2017) 73–82, https://doi.org/10.1016/j.gloplacha.2017.08.009.

Sano, M., Ramesh, R., Sheshshayee, M., Sukumar, R., Increasing aridity over the past 223 years in the Nepal Himalaya inferred from a tree-ring δ^{18}O chronology, Holocene 22 (7) (2012) 809–817, https://doi.org/10.1177/0959683611430338.

Sano, M., Sheshshayee, M.S., Managave, S., Ramesh, R., Sukumar, R., Sweda, T., Climatic potential of δ^{18}O of *Abies spectabilis* from the Nepal Himalaya, Dendrochronologia 28 (20) (2010) 93–98, https://doi.org/10.1016/j.dendro.2009.05.005.

Sano, M., Tshering, P., Komori, J., Fujita, K., Xu, C., Nakatsuka, T., May–September precipitation in the Bhutan Himalaya since 1743 as reconstructed from tree-ring cellulose δ^{18}O, J. Geophys. Res. Atmos. 118 (15) (2013) 8399–8410, https://doi.org/10.1002/jgrd.50664.

Sano, M., Xu, C., Dimri, A.P., Ramesh, R., Summer monsoon variability in the Himalaya over recent centuries, in: Dimri, A.P., Bookhagen, B., Stoffel, M., Yasunari, T. (Eds.), Himalayan Weather and Climate and Their Impact on the Environment, Springer Nature, Switzerland, 2020, pp. 261–280. AG 2020, https://doi.org/10.1007/978-3-030-29684-1 .

Schneider, T., Analysis of incomplete climate data: estimation of mean values and covariance matrices and imputation of missing values, J. Clim. 14 (5) (2001) 853–871, https://doi.org/10.1175/1520-0442(2001)014<0853:AOICDE>2.0.CO;2.

Seftigen, K., Björklund, J., Cook, E.R., Linderholm, H.W., A tree-ring field reconstruction of Fennoscandian summer hydroclimate variability for the last millennium, Clim. Dyn. 44 (2015a) 3141–3154, https://doi.org/10.1007/s00382-014-2191-8.

Seftigen, K., Cook, E.R., Linderholm, H.W., Fuentes, M., Björklund, J., The potential of deriving tree-ring-based field reconstructions of droughts and pluvials over Fennoscandia, J. Clim. 28 (9) (2015b) 3453–3471, https://doi.org/10.1175/JCLI-D-13-00734.1.

Shah, S.K., Mehrotra, N., Bhattacharyya, A., Tree-ring studies from Eastern Himalaya: prospects and challenges, Himalayan Res. J. 2 (1) (2014) 76–87.

Shah, S.K., Pandey, U., Mehrotra, N., Wiles, G.C., Chandra, R., A winter temperature reconstruction for the Lidder Valley, Kashmir, Northwest Himalaya based on tree-rings of *Pinus wallichiana*, Clim. Dyn. 53 (2019) 4059–4075, https://doi.org/10.1007/s00382-019-04773-6.

Singh, J., Singh, N., Chauhan, P., Yadav, R.R., Bräuning, A., Mayr, C., Rastogi, T., Tree-ring δ^{18}O records of abating June–July monsoon rainfall over the Himalayan region in the last 273 years, Quat. Int. 532 (2019) 48–56, https://doi.org/10.1016/j.quaint.2019.09.030.

Sinha, A., Kathayat, G., Cheng, H., Breitenbach, Sebastian F.M., Berkelhammer, M., Mudelsee, M., Biswas, J., Edwards, R.L., Trends and oscillations in the Indian summer monsoon rainfall over the last two millennia, Nat. Commun. 6 (2015) 6309, https://doi.org/10.1038/ncomms7309.

Sinha, N., Chattopadhyay, R., Chakraborty, S., Bay of Bengal branch of Indian summer monsoon and its association with spatial distribution of rainfall patterns over India, Theor. Appl. Climatol. 137 (2018) 1895–1907 https://doi.org/10.1007/s00704-018-2709-9.

Smerdon, J.E., Cook, B.I., Cook, E.R., Seager, R., Bridging past and future climate across paleoclimatic reconstructions, observations, and models: a hydroclimate case study, J. Clim. 28 (8) (2015) 3212–3231, https://doi.org/10.1175/JCLI-D-14-00417.1.

Stahle, D.W., Cook, E.R., Burnette, D.J., Villanueva, J., Cerano, J., Burns, J.N., Griffin, D., Cook, B.I., Acuna, R., Torbenson, M.C., Szejner, P., The Mexican Drought Atlas: tree-ring reconstructions of the soil moisture balance during the late pre-Hispanic, colonial, and modern eras, Quat. Sci. Rev. 149 (2016) 34–60, https://doi.org/10.1016/j.quascirev.2016.06.018.

Stokes, M.A., Smiley, T.L., An Introduction to Tree-Ring Dating, The University of Chicago Press, Chicago, 1968.

Sun, Q., Shan, Y., Sein, K., Su, Y., Zhu, Q., Wang, L., Sun, J., Gu, Z., Chu, G., A 530 year long record of the Indian Summer Monsoon from carbonate varves in Maar Lake Twintaung, Myanmar, J. Geophys. Res. Atmos. 121 (10) (2016) 5620–5630, https://doi.org/10.1002/2015JD024435.

The Imperial Gazetteer of India vol. III, The Indian Empire, Economic, Chapter X: Famine, Clarendon Press, Oxford, 1907, pp. 475–502.

Thompson, L.G., Yao, T., Mosley-Thompson, E., Davis, M.E., Henderson, K.A., Lin, P.N., A high-resolution millennial record of the South Asian monsoon from Himalayan ice cores, Science 289 (5486) (2000) 1916–1919, https://doi.org/10.1126/science.289.5486.1916.

Touchan, R., Anchukaitis, K.J., Meko, D.M., Sabir, M., Attalah, S., Aloui., A., Spatiotemporal drought variability in northwestern Africa over the last nine centuries, Clim. Dyn. 37 (2011) 237–252, https://doi.org/10.1007/s00382-010-0804-4.

Trouet, V., Oldenborgh, G.J.V., KNMI climate explorer: a web based research tool for high-resolution paleoclimatology, Tree-Ring Res. 69 (1) (2013) 3–13, https://doi.org/10.3959/1536-1098-69.1.3.

van der Schrier, G., Barichivich, J., Briffa, K.R., Jones, P.D., A scPDSI-based global data set of dry and wet spells for 1901–2009, J. Geophys. Res. Atmos. 118 (10) (2013) 4025–4048, https://doi.org/10.1002/jgrd.50355.

Vuille, M., Werner, M., Bradley, R., Keimig, F., Stable isotopes in precipitation in the Asian monsoon region, J. Geophys. Res. 110 (2005) D23108, https://doi.org/10.1029/2005JD006022.

Wang, L., Liu, Y., Li, Q., Song, H., Cai, Q., Sun, C., Fang, C., Liu, R., A 210-year tree-ring $\delta^{18}O$ record in North China and its relationship with large-scale circulations, Tellus B: Chem. Phys. Meteorol. 72 (1) (2020) 1–15, https://doi.org/10.1080/16000889.2020.1770509.

Webster, P.J., Magaña, V.O., Palmer, T.N., Shukla, J., Tomas, R.A., Monsoons: processes, predictability, and the prospects for prediction, J. Geophys. Res. 103 (C7) (1998) 14451–14510, https://doi.org/10.1029/97JC02719.

Wells, N., Goddard, S., Hayes, M.J., A self-calibrating Palmer Drought Severity Index, J. Clim. 17 (12) (2004) 2335–2351, https://doi.org/10.1175/1520-0442(2004)017<2335:ASPDSI>2.0.CO;2.

Wieloch, T., Helle, G., Heinrich, I., Voigt, M., Schyma, P., A novel device for batch-wise isolation of α-cellulose from small-amount wholewood samples, Dendrochronologia 29 (2) (2011) 115–117, https://doi.org/10.1016/j.dendro.2010.08.008.

Xu, C., Sano, M., Dimri, A.P., Ramesh, R., Nakatsuka, T., Shi, F., Guo, Z., Decreasing Indian summer monsoon on the northern Indian sub-continent during the last 180 years: evidence from five tree-ring cellulose oxygen isotope chronologies, Clim. Past. 14 (5) (2018) 653–664, https://doi.org/10.5194/cp-14-653-2018.

Xu, C., Sano, M., Nakatsuka, T., Tree-ring cellulose $\delta^{18}O$ of *Fokienia hodginsii* in northern Laos: a promising proxy to reconstruct ENSO? J. Geophys. Res. 116 (2011) D24109, https://doi.org/10.1029/2011JD016694.

Xu, C., Sano, M., Nakatsuka, T., A 400-year record of hydroclimate variability and local ENSO history in northern Southeast Asia inferred from tree-ring $\delta^{18}O$, Palaeogeogr. Palaeoclim. Palaeoecol. 386 (2013) 588–598, https://doi.org/10.1016/j.palaeo.2013.06.025.

Wilson, R., D'Arrigo, R., Buckley, B., Büntgen, U., Esper, J., Frank, D., Luckman, B., Payette, S., Vose, R., Youngblut, D., A matter of divergence: Tracking recent warming at hemispheric scales using tree-ring data, J. Geophys. Research: Atmos. 112 (D17) (2007) D17103. https://doi.org/10.1029/2006JD008318.
Xu, H., Hong, Y., Hong, B., Decreasing Asian summer monsoon intensity after 1860 AD in the global warming epoch, Clim. Dyn. 39 (2012) 2079–2088, https://doi.org/10.1007/s00382-012-1378-0.
Zhang, Z., Mann, M.E., Cook, E.R., Alternative methods of proxy-based climate field reconstruction: application to summer drought over the conterminous United States back to AD 1700 from tree-ring data, Holocene 14 (4) (2004) 502–516, https://doi.org/10.1191/0959683604hl727rp.
Zhao, H., Moore, G.W.K., Reduction in Himalayan snow accumulation and weakening of the trade winds over the Pacific since the 1840s, Geophys. Res. Lett. 33 (17) (2006) L17709, https://doi.org/10.1029/2006GL027339.
Zhou, T., Zhang, L., Li, H., Changes in global land monsoon area and total rainfall accumulation over the last half century, Geophys. Res. Lett. 35 (16) (2008) L16707, https://doi.org/10.1029/2008GL034881.

CHAPTER

12

Bond events and monsoon variability during Holocene—Evidence from marine and continental archives

Upasana S. Banerji, D. Padmalal

National Centre for Earth Science Studies, Ministry of Earth Sciences, Thiruvananthpuram, Kerala, India

12.1 Introduction

The millennial-scale climate oscillation is a testimony for the coherent interaction among the earth's atmosphere, oceans, and cryosphere (Hemming, 2004). The high frequency of climate instability for the late Pleistocene has been revealed from the recurrence of Dansgaard–Oeschger (D/O) and Heinrich (H) events (Broecker et al., 1992; Johnsen et al., 1992; Dansgaard et al., 1993). The D/O events were associated with the abrupt temperature variations constrained within the glacial part in the Greenland ice core, whereas the H-events were preceded by the ocean surface cooling and thus, H-events prevailed as a result of climate variability rather than causing climate perturbations (Bond et al., 1993, 1999; Broecker, 1994). The H-events were (Bond et al., 1993) deduced from lithic clasts and foraminifera in the sediment cores of west Portugal, suggesting that the percentages of lithics $>180\ \mu m$ in these cores signified a series of Ice Rafted Debris (IRD) events (Heinrich, 1988; Broecker et al., 1992). Thus, a close coupling of D/O and H events was established (Bond et al., 1993). Spanning for hundreds to 1000 yr, the H-events reappear on an average of ~7000-yr intervals (Bond and Lotti, 1995). The Younger Dryas is considered as the last D/O event and its abrupt termination was followed by the onset of Holocene epoch (Broecker, 1994).

By the end of 1990s, Gerard Bond and his research team investigated the Holocene climatic variability of North Atlantic region. The study on sea bed deposits revealed certain unusual layers of IRD [count of quartz and hematite stained grains (HSG)] that were frequently observed at regular intervals (Bond et al., 2001). The transportation of such coarser material in the oceanic basin could reach through the floating icebergs or through the ice-rich Arctic surface water masses that can reach near the Great Britain during cool climate which upon melting unloads the debris (detritus) over the sea floor causing deposition of coarser layers (Lning and Vahrenholt, 2016). Further, the synchronicity between repetitive debris deposit pattern and weak solar activity was invoked which led to the cold phase in North Atlantic region (Bond et al., 2001) and based on which, a ~1500-yr cycle was postulated that prevailed throughout the Holocene Epoch (Wanner and Bütikofer, 2008). These cycles of the Holocene epoch were treated as equivalents of the Pleistocene D/O cycles (Alley, 2005).

Holocene Climate Change and Environment. DOI: https://doi.org/10.1016/B978-0-323-90085-0.00016-4

12.1.1 Holocene epoch

The Holocene epoch is the only geochronologic unit in the earth's geological history whose boundary has been defined based on climatostratigraphy. In contrast, biostratigraphy has been implemented to demarcate other Phanerozoic boundaries (Ramesh et al., 2010). The term 'holocènes' was first coined by Paul Gervais (1867–1869, p. 32), which means "entirely recent," referring to the warm episode which commenced with the end of the last glacial period (Walker et al., 2018). The first formal subdivisions of the Holocene were from Northern Europe and the substage boundaries were defined by radiocarbon dated chronozone and palynologically define biozones viz. Preboreal and Boreal (10,000–8000 ^{14}C yr BP), Atlantic and Sub-boreal (8000–2500 ^{14}C yr BP) and sub-Atlantic (post 2500 ^{14}C yr BP; Mangerud et al., 1974). However, such chronostratigraphic subdivisions based on biological evidence can provide local or regional scale climate and may lag in providing global scenario (Björck et al., 1998; Wanner et al., 2008). Recently, formal ratification of the Holocene epoch into three subdivisions has been proposed by Global Boundary Stratotype Sections and Points based on abrupt shifts in climate evident from ice core of Greenland and speleothem of NE India and thus named as Greenlandian (11.7–8.2 ka), Northgrippian (8.2–4.2 ka), and Meghalayan (4.2 ka to 1950 AD; Walker et al., 2019) stages. Basically, the Holocene subdivisions are made at ~11.8 ka (onset of Holocene), ~8.2 ka and ~4.2 ka and these abrupt climatic events are generally encountered in most of the global records (Walker et al., 2018). The abrupt climatic event of 8.2 ka and 4.2 ka nearly in agreement with the two of the Bond events (BEs) evidenced from the North Atlantic Ocean (Bond et al., 2001). Bond and his team numbered the cold phases of the Holocene epoch from 0 to 8 wherein 0 corresponds to Little Ice Age (LIA: ~0.5 ka) and 1 corresponds to Dark Age Cold Period (DACP: ~1.4 ka), while 2, 3, 4, 5a, 5b, 6, 7, and 8 correspond to ~2.8, ~4.4, ~5.5, ~7.5, ~8.1, ~9.4, ~10.3, and ~11.1 ka (Fig. 12.2A), respectively (Bond et al., 2001; Wanner et al., 2011). Though, their drivers remain uncertain, some of the BEs demonstrate hemisphere scale climatic teleconnections (Wanner et al., 2008; Fan et al., 2016). Further, the nine documented BEs and associated cooling phases have been linked with the reduced thermohaline circulation in the North Atlantic region (Zielhofer et al., 2019). The carbon-isotope composition of benthic foraminifera from the deep sea sediment core of the subpolar north-eastern Atlantic suggested that deep-water production varied on a centennial–millennial timescale and the identified cold events (Oppo et al., 2003) and were mostly in agreement with the BEs (Bond et al., 2001). Based on BEs, a quasiperiodic "1500-yr" cycle has been recognised in the IRD records of Holocene (Bond et al., 2001). This 1500-yr cycle plausibly represents a combination of two of the solar cycles viz. the Eddy cycle (1000 yr) and the Hallstatt cycle (2300 yr; Debret et al., 2009; Obrochta et al., 2012).

A close correlation between the millennial-scale climate events of North Atlantic region and the Indian Summer Monsoon (ISM; Overpeck et al., 1996; Schulz et al., 1998) suggests a mechanistic link during the last ice age (Gupta et al., 2003). Unlike the ice age, a stable climate prevailed during the Holocene epoch wherein the amplitude of North Atlantic climate variability was also trivial compared to the late Pleistocene D/O and H events. Thus, it is necessary to comprehend the link of ISM variability recorded in the continental and marine archives with the BEs, that was exclusively observed as a cold event of the North Atlantic Ocean during Holocene epoch.

12.1.2 Summer and winter monsoons

Indian landmass is surrounded by the northern Indian Ocean consisting of two distinct basins viz. the Arabian sea and the Bay of Bengal. Over the north of 10°S Indian Ocean (north Indian Ocean), the wind reverses its direction twice a year. The wind blows from the southwest during May to September (summer monsoon) while March–April and October are the periods of transition with weak winds. Winter monsoon prevails during November–February wherein wind blows from northeast (Shankar et al., 2002). During summer season, continuous monsoon current flows eastward from the western Arabian Sea to the Bay of Bengal while during winter, the currents flow westward, from the eastern boundary of the Bay of Bengal to the western Arabian Sea. Thus, these currents transfer water masses between the two highly distinct basins of northern Indian Ocean, the Bay of Bengal, and the Arabian Sea (Shankar et al., 2002). Such distinct characteristics of the Arabian Sea and the Bay of Bengal invoked extensive investigation on sediments of northern Indian Ocean to decipher the ISM variability for the Holocene epoch.

The ISM is generally associated with >70% rainfall over India and nearby areas, and plays a pronounced role in the large-scale circulation pattern, thereby having enormous implications on South Asia's agriculture and economy (Lau and Waliser, 2011). Over the past three centuries, the monsoon was considered as a gigantic land–sea breeze associated with the differential heating between the ocean and the land (Halley, 1753). However, the seasonal migration of the Intertropical Convergence Zone (ITCZ) is the accepted mechanism that manifested the ISM rainfall (Sikka and Gadgil, 1988; Gadgil, 2003; Fleitmann et al., 2007). An investigation on daily variation of the maximum cloud zone using environmental satellite imageries suggested two favorable locations of cloud bands or ITCZ, one over the heated Indian subcontinent and other over the warm equatorial Indian Ocean (Sikka and Gadgil, 1988). The intraseasonal northward migration of these cloud bands (ITCZ) from equatorial Indian ocean to Indian subcontinent is the most prominent feature that occurs at an interval of 2–6 weeks (Gadgil, 2003). Generally, the cloud radiation feedbacks or the feedback between vertical stability of the moist tropical atmosphere and mid-tropospheric heating by the deep clouds are the major feedbacks that operate the Continental Tropical Convergence Zone (CTCZ) over the Indian subcontinent and the oceanic ITCZ (Krishnamurti and Bhalme, 1976; Gadgil and Francis, 2012).

The commencement of the ISM onset phase is marked during June over Kerala and thus the gateway of ISM. It takes nearly 40−45 days for the northward progression of ISM over the Indian landmass from its onset over Kerala to its culmination over the western part of the Indian monsoon zone (Ding, 2006). Generally, the CTCZ gets established in the region between latitudes 18.0°N and 28.0°N and longitudes 65.0°E and 88.0°E during the end of the onset phase of the monsoon over the Indian landmass. Within this region, the position of the CTCZ varies from July to August (peak ISM months) and thus witnesses remarkable rainfall fluctuations between active and break spells. This zone has been identified as the Core Monsoon Zone (Rajeevan et al., 2010).

Later, the CTCZ retreats southward and reaches its southernmost extent during January. The reversal of pressure gradient results in Northeast Monsoon (NEM) causing NEM rainfall from October to December, over the SE India especially Tamil Nadu coast and southern Andhra Pradesh (Srinivasan and Ramamurthy, 1973). Unlike ISM, the NEM leads to nearly 11% of the all-India annual rainfall (Rajeevan et al., 2012); nevertheless, it contributes almost 30−60% of the annual mean in the regions of Tamil Nadu, coastal, and southern Andhra Pradesh (Srinivasan and Ramamurthy, 1973; Rajeevan et al., 2012). Further, southern Karnataka and Kerala also receive some additional rainfall during NEM along with major contribution from the ISM (Srinivasan and Ramamurthy, 1973). A negative correlation between ISM and NEM has been observed based on the rainfall series for the past 100 yrs (Dhar and Rakhecha, 1983).

The North-western India also witnesses winter precipitation during December–February which is linked with large-scale interaction between the tropical air masses and the mid-latitude resulting in the formation of synoptic systems known as the western disturbances (WDs; Kar and Rana, 2014). These WDs are usually the low-pressure systems in the lower troposphere in the subtropics moving toward the Himalayan region (Dimri et al., 2015). The WD is a cyclonic storm linked with the mid-latitude subtropical westerly jet, which leads to extreme precipitation over northern India that generally gets improved over the Himalayas due to its orographic land–atmospheric interactions (Quamar, 2019). It causes snowfall in the Himalayan region during winters and replenishes the regional water resources (Dimri et al., 2015). The WD originates from the Mediterranean Sea or Atlantic Ocean (Agnihotri and Singh, 1982) and thus, the moisture source for the WD is generally from the Black Sea, the Mediterranean, and North Atlantic (Dixit and Tandon, 2016).

12.2 Scope

The predictions for the daily weather pattern in the tropics are restricted to 2–3 days, while the seasonal pattern of mean monsoon circulation in the tropical regions is relatively more predictable (Rajeevan, 2001). Observational studies have deciphered teleconnections of ISM with various surface boundary conditions such as land surface temperature (Rajeevan et al., 1998), snow cover of Eurasia and Himalaya (Bamzai and Shukla, 1999), and sea surface temperatures (SSTs) of Indian Ocean (Rajeevan et al., 2002) and east Pacific Ocean (Rasmusson and Carpenter, 1983). However, studies suggested that the boundary conditions may not be the sole forcing factor for the monsoon circulation but it could also be governed by the internal dynamics thereby limiting in the reliable prediction of the ISM rainfall (Webster et al., 1998; Mohan and Goswami, 2000). The present climate scenario and its variability has been deciphered based on the available instrumental data and which was further extended based on natural archives in order to enhance the predictability. The strengthening of varied aspects of ISM variability and its teleconnections could allow better climate prediction initiatives.

12.2.1 Instrumental data

An increase of global mean annual surface temperature by 0.55°C between 1860 and 1990 AD (Parker et al., 1994) with significant climate transformation since the postindustrial era (1750 AD) led to changes in the precipitation pattern (Solomon et al., 2007). Such variability revealed that the last two centuries have been significantly impacted by natural climate perturbations along with the anthropogenic influence. Further, an increase of nearly 0.5°C during last century (1901−2003 AD) in all India mean annual temperature has been invoked based on the detailed analysis of the instrumental data by India Meteorological Department (Kothawale and Rupa Kumar, 2005), while the instrumental rainfall data showed a reduction in ISM since 1960 AD (Sontakke et al., 2008). The limited extension of the instrumental data up to last two centuries bolstered the climatic reconstructions based on natural archives such as tree rings and corals that can generally yield multidecadal to century-scale climate history.

12.2.2 Tree ring and corals: annual to decadal resolution climate reconstruction

The tree growth is affected by rainfall and temperature (Fritts, 1972) and thus, the intensities of these parameters control the diametric growth of the tree during the growing season. The quantification of the annual growth by the tree is a function of ring width that acts as a proxy for the growth limiting conditions (Ramesh et al., 2010). A control of ISM has been elucidated based on the tree ring width of Teak from Madhya Pradesh which demonstrated a major drought episodes between 1835 and 1997 (Shah et al., 2007). The extended rainfall reconstruction based on $\delta^{18}O$ of Teak cellulose from southern India suggested high monsoon in later part of the LIA (Managave et al., 2011). Another dendroclimatological attempt on Teak from Southern India depicted low growth as a result of drought (low ISM) events associated with El Niño during last 526 yr (Borgaonkar et al., 2010). The tree ring from Himalayan region has provided climate and rainfall records for the last millennium. The ring width indices of *Cedrus deodara* from Uttaranchal, western Himalaya revealed a strong relationship with spring precipitation during a time span of 1731–1986 with highest during 1977–1986 and lowest during 1932–1941 (Singh and Yadav, 2005). However, a similar reconstruction for the time span of 1310–2004 from Kinnaur, Himachal Pradesh suggested enhanced precipitation trends during late 20th century thereby underpinning the regional differences (Singh et al., 2009; Ramesh et al., 2010). Most of the tree ring investigations conducted from the Himalayan region are on conifers but a recent study on *Toona ciliata Roem* from Kalimpong, Eastern Himalaya described that during 1824–2003 AD, the mean minimum temperature of winter season as a significant factor in modulating the radial growth of the Toon tree in the region (Shah and Mehrotra, 2017). The dendroclimatic study spanning since 1507 AD for the eastern Himalaya from Arunachal Pradesh and Sikkim indicated that the coolest period was during 1801–1810 AD while 1978–1987 AD was the warmest period (Bhattacharyya and Chaudhary, 2003). The tree ring study of Khasi pine, from Shillong suggested narrower ring during the years subsequent to the major earthquake of 1897 thereby underscoring its significance in determining the paleoseismic signatures (Chaudhary and Bhattacharyya, 2002).

The reconstruction of interannual to decadal scale ISM variability and clues on ocean–atmospheric processes has been provided by scleractinian corals ($CaCO_3$ skeleton forming hard corals). Till date numerous studies on corals from Indian ocean have been attempted to address the impacts of ocean atmospheric processes on Asian monsoon (Cantin et al., 2010; Hua et al., 2005; Marshall and McCulloch, 2001; Tierney et al., 2015; Zinke et al., 2014, and references therein). But limited number of studies have been conducted on the northern Indian ocean especially on the Indian corals (Suresh and Mathew, 1993; Chakraborty and Ramesh, 1997; 1998; Ahmad et al., 2011; Sagar et al., 2016) that are capable of providing prominent signatures of ISM variability along with other climate variables. The oxygen isotopic study on Lakshadweep corals demonstrated a significant control of SST and marked influence of wind induced mixing on oxygen isotopes ($\delta^{18}O$). However, correlation of ISM over isotopic variations in the corals was seldom observed (Chakraborty and Ramesh, 1997; Ahmad et al., 2011). The tree rings and corals respond well in accordance with the global climate system vis-à-vis ISM variability during the last few centuries. Even though, these studies invoked the ISM weakening during BE-0 (LIA), the longer records in multimillennial timescales are required to delineate the ISM response toward the BE. Such multimillennial longer records with signatures of BEs will not only yield the recurrence of abrupt cooling events and its associated impacts on ISM but also help in projecting the future cooling events and its possible global repercussions.

12.3 Methodology

The instrumental data along with tree rings and corals from the ISM influenced region have provided significant information on the ISM variability and its linkage with the other climate variables in multidecadal to multi centennial scales. However, these records seldom extended beyond the last millennium and thus provided information only on last the BE, that is, BE-0 associated with the LIA. In order to investigate the signatures of BEs in Indian Ocean and Indian subcontinent proxies and also to decipher the impressions of BEs, in the present review, the discussion is limited to those records that has provided a climate history of at least a millennium time span during Holocene. This is due to the fact that the last BE was observed during LIA that corresponds to ~500 yr BP and a ~1500 yr cyclicity has been recorded from the ice rafted debris representing BEs during Holocene epoch (Bond et al., 2001). Additionally, the BEs being recorded in the North Atlantic Ocean and with some signatures observed in the Mediterranean region, it is also necessary to first scrutinize the impression and intensity of each of the BEs in the aforementioned records.

In the present review, published studies addressing ISM variability from northern Indian Ocean and Indian subcontinent are collated and investigated. Since the Indian subcontinent and the nearby region witness North Atlantic influence exclusively through WD with moisture source mediated from the Mediterranean region. Thus, the Holocene climate variability recorded from North Atlantic Ocean and Mediterranean region has been discussed to provide a better understanding of the BE engrossed in those locations. Following which, the natural archive-based studies from northern Indian Ocean and Indian subcontinent (broadly divided into marine and continental proxies) are discussed. Nearly 10 paleoclimate studies from North Atlantic Ocean and Mediterranean region (Fig. 12.1A; Table 12.1) and ~60 studies from northern Indian Ocean and Indian Subcontinent (Fig. 12.1B; Table 12.2) have been scrutinized, in order to decipher the link of Indian monsoon with the BEs that is an exclusive signature of North Atlantic and Mediterranean regions. The response of the natural archives toward the BEs are scrutinized to comprehend the paleoclimate signatures vis-à-vis ISM variability during the Holocene epoch.

12.4 Observations

The WD being a part of large-scale subtropical westerly jet that causes winter precipitation in northern and north western India, plays a crucial role in production of Rabi crops, in addition to maintain Himalayan glaciers (Midhuna et al., 2020). Generally, weakening of ISM has occurred in tandem with the migration of midlatitude WDs into the tropics (Dimri et al., 2015) and a negative correlation between wave kinetic energy of WDs and weekly averaged all India summer rainfall has been invoked (Bawiskar et al., 2005). The moisture source for WDs being associated with the North Atlantic and Mediterranean regions raises the necessity to address the impacts of North Atlantic climate on the Indian climate system during the Holocene epoch.

12.4.1 North Atlantic and Mediterranean regions

Even though, the solar variability has been the major forcing during the Holocene epoch (Bond et al., 2001; Rohling et al., 2002; Mayewski et al., 2004), the oscillations in the North Atlantic Deep Water (NADW) production rates and transport of heat to the poles either triggered or amplified the Holocene climate fluctuations (Bond et al., 1997; Schulz and Paul, 2002; Oppo et al., 2003). The transformation of surface water to deep water in the North Atlantic region leads to heat release from the ocean to the atmosphere which plausibly amplified the climate variability in millennial scale during glacial times (Broecker, 1990) and recently during Holocene epoch (Alley et al., 1997; Keigwin and Boyle, 2000;

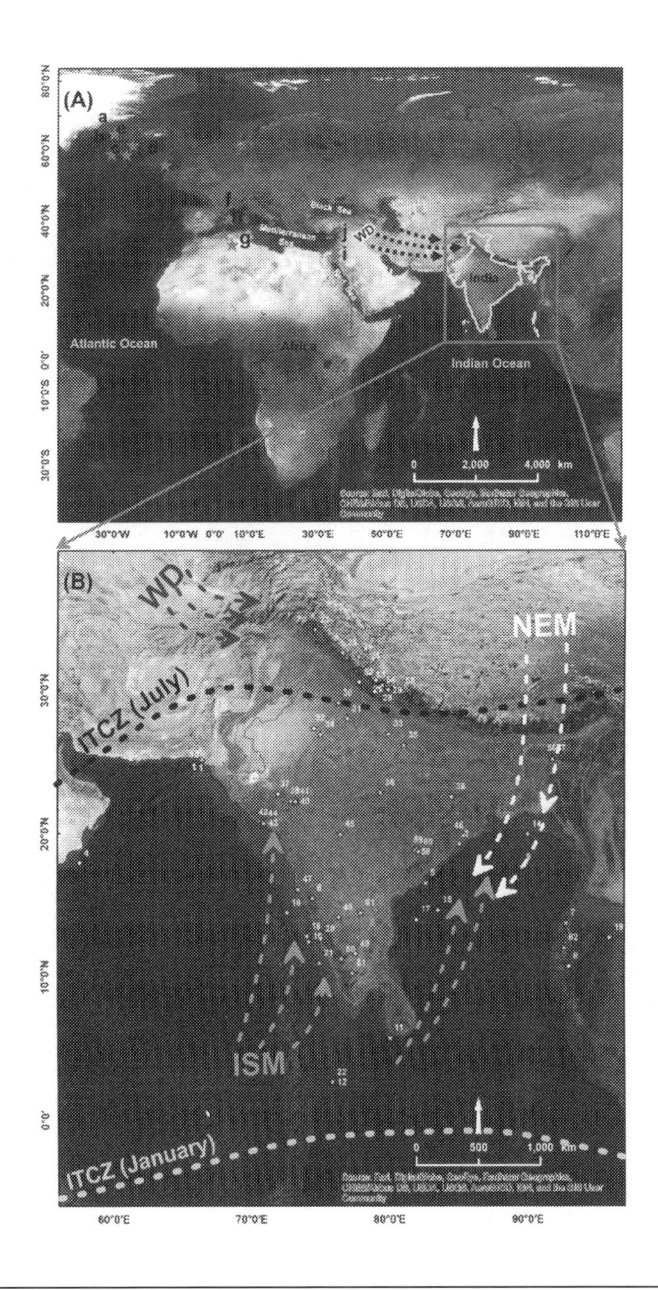

FIGURE 12.1

(A) Locations of the studies from North Atlantic Ocean and Mediterranean Sea discussed in Fig. 12.2. (B) Locations of the studies discussed in the present work are shown in the figure. The figure also depicts the position of ITCZ during July and January. The orange, white, and blue arrows represent Indian summer monsoon (ISM), North East Monsoon (NEM), and the Western Disturbances (WD) respectively. The details of the studies in (A) and (B) are given in Table 12.1 and Table 12.2 respectively wherein the Sl no. refers to the respective locations in the map.

Table 12.1 Details of the studies from north Atlantic and Mediterranean region discussed in the review wherein each of the study is represented by the table's Sl. No. in Fig. 12.1A.

Region	Sl. No.	Archive	Proxy Id	Latitude	Longitude	Location	References
North Atlantic	a.	Marine Sediment	MC52-VM 29-191	64.78°N	29.57°W	E. North Atlantic	(Bond et al., 2001)
	b.		LO09-14	58.94°N	30.41°W	Reykjanes Ridge	(Andersen et al., 2004)
	c.		MD 952015	58.76°N	25.96°W	Gardar contour drift	(Giraudeau et al., 2000)
	d.		ODP 980	55.82°N	14.70°W	Feni Abyssal Drift	(Oppo et al., 2003)
	e.		NEAP4K	61.50°N	24.17°W	Bjorn Drift	(Hall et al., 2004)
Mediterranean	f.	Marine Sediment	MD99-2343	40.50°N	04.03°E	N. Minorica, NW Mediterranean Sea	(Frigola et al., 2007)
	g.	Lake Sediment	SiA	33.05°N	05.00°E	Sidi Ali Lake, Morocco	(Zielhofer et al., 2019)
	h.	Marine Sediment	MD95-2043	36.15°N	02.62°E	W. Mediterranean Sea	(Fletcher et al., 2013)
	i.	Marine Sediment	GeoB5804-4	29.50°N	34.96°E	N. Gulf of Aqaba, Red Sea	(Lamy et al., 2006)
	j.	Speleothem	SoC	31.45°N	35.03°E	Soreq Cave, Nahal Qanah	(Bar-Matthews et al., 1997)

Table 12.2 List of the studies discussed in the chapter wherein each of the study is represented by the table's Sl. No. in Fig. 12.1B. (Note: N: north; W: west; E: east; NW: northwest; NE: northeast; SE: southeast; Eq.: equatorial; C: central)

Proxy	Sl. No.	Proxy Id	Latitude (°N)	Longitude (°E)	Location	References1
Marine Proxy					**Sediment**	
	1.	56KA	24.83	65.92	NW. Arabian Sea	(Von Rad et al., 1999; Doose-rolinski et al., 2001; Luckge et al., 2001)
	2.	SO90-63 KA	24.61	65.98	NE Arabian Sea	(Burdanowitz et al., 2019)
	3.	VC-04	19.31	85.12	W. BOB	(Ankit et al., 2017)
	4.	RC-27/14	17.99	57.59	W. Arabian Sea	(Altabet et al., 1999)
	5.	NGHP-16A	16.59	82.68	W. Bay of Bengal	(Ponton et al., 2012)
	6.	SK27B/8	15.53	74.5	SE. Arabian Sea	(Caratini et al., 1994)
	7.	SN SC-03/2008	13.79	92.9	N. Andaman Sea	(Achyuthan et al., 2014)
	8.	SSK50-GC3A	10.76	93.1	W. Andaman Sea	(Sebastian et al., 2019)
	9.	BoB-88	17.68	89.61	C. Bay of Bengal	(Li et al., 2020)
	10.	3104	12.5	74.2	E. Arabian Sea	(Agnihotri et al., 2003b)
	11.	SK-304A/05	5.76	80.08	Eq. Central IO	(Chandana et al., 2017)
					Foraminifera	
	12.	63KA; 41KL	24.83	65.92	N. Arabian Sea	(Overpeck et al., 1996; Sirocko et al., 1996; Staubwasser et al., 2002)
	13.	KL-126	19.97	90.03	N. Bay of Bengal	(Kudrass et al., 2001)
	14.	VM-29/19	14.71	83.58	N. Bay of Bengal	(Rashid et al., 2011)
	15.	AAS9/21	14.51	72.65	E. Arabian Sea	(Govil and Naidu, 2010)
	16.	SK218/1	14.04	82	W. Bay of Bengal	(Govil and Naidu, 2011)
	17.	SK291/GC11	12.87	74.1	E. Arabian Sea	(Saravanan et al., 2020)
	18.	RC-12/344	12.77	96.07	Andaman Sea	(Rashid et al., 2007)
	19.	SN-6	12.68	75.1	E. Arabian Sea	(Tiwari et al., 2015)
	20.	SK-237/GC04	10.98	75	E. Arabian Sea	(Saraswat et al., 2013)
	21.	SS-3827G	2.7	75.91	Eq. Indian Ocean	(Tiwari et al., 2006b)
Continental Proxy					**Sediment/Peat**	
	22.	BS	34.25	74.67	Manasbal Lake, Kashmir	(Babeesh et al., 2019)
	23.	TMD	32.91	78.32	Tso Moriri, Jammu & Kashmir	(Leipe et al., 2014)
	24.	TN	32.66	76.66	Triloknath, Himachal Pradesh	(Bali et al., 2017)
	25.	CPT	32.4	77.8	Chandra peat, Lahaul Himalaya	(Rawat et al., 2015a)

(continued on next page)

Table 12.2 (*continued*)

Proxy	Sl. No.	Proxy Id	Latitude (°N)	Longitude (°E)	Location	References1
	26.	KT	30.73	79.06	Kedarnath, higher Himalayas, Uttarakhand	(Srivastava et al., 2017)
	27.	BTA	30.16	79.25	Benital lake, Uttarakhand	(Bhushan et al., 2018)
	28.	PD	30.05	79.93	Pinder Valley, Higher Himalaya	(Rühland et al., 2006)
	29.	RW	29.14	76.36	Riwasa, Haryana	(Dixit et al., 2014a)
	30.	KD	28.03	77	Kotla Dahar, Haryana	(Dixit et al., 2014b)
	31.	DW	27.4	74.58	Didwana, Rajasthan	(Wasson et al., 1984)
	32.	SA	27	80	Sanai Tal, Uttar Pradesh	(Sharma et al., 2004)
	33.	BH	26.95	75.09	Sambhar, Rajasthan	(Sinha et al., 2006)
	34.	BrT	26.2	81.07	Baraila Tal, Uttar Pradesh	(Misra et al., 2020)
	35.	NoT	22.9	79.4	Nonia Tal, MP	(Kumar et al., 2019)
	36.	NAL	22.8	72	Nal lake, Gujarat	(Prasad et al., 1997)
	37.	MN	22.61	84.54	Mahanadi, Orissa	(Tripathi et al., 2014)
	38.	KTH	22.32	72.94	Kothiyakhad, Gujarat	(Prasad et al., 2007)
	39.	BK	22.32	73.28	Timbi lake, Gujarat	(Sridhar et al., 2020)
	40.	LN	22.3	73.25	Lower Narmada, Gujarat	(Laskar et al., 2013a)
	41.	RH	20.82	71.23	Rohisa, Gujarat	(Banerji et al., 2019)
	42.	VV	20.75	71	Vasoj, Gujarat	(Banerji et al., 2015)
	43.	DV	20.74	70.98	Diu Island, Gujarat	(Banerji et al., 2017)
	44.	LO	19.98	76.51	Lonar Lake, Maharashtra	(Menzel et al., 2014; Prasad et al., 2014; Sarkar et al., 2015)
	45.	CHI	19.87	85.4	Chilka, Orissa	(Pandey et al., 2014)
	46.	HD	16.13	73.47	Hadi, Malwan	(Limaye and Kumaran, 2012)
	47.	TK	14.2	76.4	Thimmannanayakanakere, Karnataka	(Warrier and Shankar, 2009)
	48.	Enn	11.65	77.59	Ennamangalam Lake, Tamil Nadu	(Mishra et al., 2019)
	49.	NP	11.25	76.58	Sandynallah, Nilgiri Peat	(Sukumar et al., 1993)
	50.	KK	10.27	77.37	Kukkal Lake, Tamilnadu	(Rajmanickam et al., 2017)
					Speleothem	
	51.	SAH	30.6	77.87	Sahiya Cave, Uttarakhand	(Kathayat et al., 2017)
	52.	SA-1	30.38	79.53	Sainji cave, Central Indian Himalaya	(Kotlia et al., 2015)
	53.	TUL	30.02	79.45	Chulerasim cave, Lesser Himalaya	(Kotlia et al., 2012)
	54.	DH2	29.98	80.87	Dharamjali cave, east Uttarakhand	(Sanwal et al., 2013)
	55.	KM-A	25.26	91.82	Mawmluh Cave, Meghalaya	(Berkelhammer et al., 2012)

(*continued on next page*)

Table 12.2 (*continued*)

Proxy	Sl. No.	Proxy Id	Latitude (°N)	Longitude (°E)	Location	References1
	56.	WS-B	25.25	91.87	Wah Shikar Cave, Meghalaya	(Sinha et al., 2011)
	57.	KOT-1	19	82	Kotumsar Cave, Chhattisgarh	(Band et al., 2018)
	58.	JHU-1	18.87	81.87	Jhumar Cave, Chhattisgarh	(Sinha et al., 2011)
	59.	Gt	18.75	82.17	Gupteshwar Cave, Orissa	(Yadava and Ramesh, 2005)
	60.	NK-1305	14.52	77.99	Nakarallu Cave, Andhra Pradesh	(Sinha et al., 2018)
	61.	AN4	12.08	92.75	Baratang cave, Andaman Island	(Laskar et al., 2013b)
	62.	BT_8	30.79	77.78	Bittoo cave, Uttarakhand	(Kathayat et al., 2016)

Bond et al., 2001). However, the mechanism behind the millennial-scale climate fluctuations during the Holocene epoch is still debatable.

A composite record from three sediment cores of sub polar North Atlantic region revealed a highly variable surface conditions during the Holocene epoch with recurring cool events centered at ~10.4, ~9.8, ~8.3, ~7.9, ~6.4, ~4.7, ~4.3, and 2.8 kyr (Fig. 12.2B) with thermal optimum record during 7.5–5 kyr (Andersen et al., 2004). A near similar cooling events during 8.5–8 ka, 6–4.8 ka, and 3.5–2.8 ka but with warm climate between 10 ka and 6 ka (Fig. 12.2C) has been revealed by the investigation of coccoliths from the sediment core of south Iceland over the Gardar contour drift (Giraudeau et al., 2000). Further, a concurrent variation in the coccoliths (*E. huxleyi*) concentration and HSG concentration implied similar surface circulation at Gardar contour drift (Giraudeau et al., 2000) and sediment core site MC52-VM29-191 (Fig. 12.2A) of North Atlantic region (Bond et al., 2001), respectively. The instances of high HSG in MC52-VM29-191 of North Atlantic region also corroborated with the depleted $\delta^{13}C$ in benthic foraminifera of Feni Drift, subpolar north-eastern Atlantic demonstrating several episodes of reduced NADW contribution (Fig. 12.2D) such as ~9.3, ~8, ~5, and ~2.8 kyr and invoking the possible linkage of surface with the deep waters (Oppo et al., 2003). Similarly, the granulometric proxy from the sedimentary records of subpolar North Atlantic demonstrated higher flow intensity of Iceland–Scotland overflow waters during ~7200, ~5300, ~4600, ~4100, ~3000, and 2700–1400 yr BP (Fig. 12.2E), which were in agreement with the peaks of drift ice deposition events (Hall et al., 2004).

The western Mediterranean region has been sensitive toward glacial events thereby indicating a possible teleconnection of western Mediterranean regions with that of North Atlantic climate (Rohling et al., 1998, 2002; Moreno et al., 2004). The geochemical investigation on sediment core near Minorca Island, western Mediterranean Sea demonstrated stratification of water masses leading to slowdown of deep-water circulation during 12–10.5 ka due to increased sea level and humidity and later the deep-water circulation was recovered till 7 ka. Signatures of stable condition with high terrigenous inputs and deep-water circulation observed during 7–4 ka (Fig. 12.2F), following which a drier climate with progressive decrease in the detrital flux prevailed since last 4 ka (Frigola et al., 2007). The $\delta^{18}O$ values for the Lake Sidi Ali, Morocco, western Mediterranean revealed an in-phase variation of winter rain minima with the BEs of ~11.4, ~10.2, ~9.3, ~8.2, and ~6.0 ka (BEs 8–4), while an out-phase variation of winter rain (Fig. 12.2G) maxima during ~4.2, ~2.7, ~1.2 ka, and LIA correspond to BEs 3–0 (Zielhofer et al., 2019). The western Mediterranean marine core elucidated several episodes of declined forest during early-Holocene (~10.1, ~9.2, ~8.3, and ~7.4 cal. ka BP) and mid–late-Holocene (~5.4–4.5 ka, 3.7–2.9 ka, and 1.9–1.3 ka) period (Fig. 12.2H). Further, the study underscored a strong sun–climate relation during the early-Holocene period while the surface ocean circulation dynamics of North Atlantic ocean along with the internal variabilities in the strength of deep-water convection plausibly played a major role during mid to late-Holocene epoch (Fletcher et al., 2013). The investigation of the sediment core from Gulf of Aqaba, NE Red Sea demonstrated low sand accumulation and weak water column stratification due to reduced aridity (Fig. 12.2I) possibly associated with the present day more positive phase of North Atlantic Oscillation (Lamy et al., 2006). The intervals of low sand accumulation and weak water column stratification associated with weak aridity are in agreement with the prevalence of repeated humid intervals during the last 3500 yr BP, as inferred from the $\delta^{18}O$ variations in planktonic foraminifera from southern Israel (Schilman et al., 2001). The $\delta^{18}O$ of speleothem from Soreq cave, Israel demonstrated high rainfall during 12,000–10,000 yr BP interrupted by an arid climate during ~11,500 yr BP. Later, extensive rainfall conditions prevailed during 10,000–7000 yr BP with an intermittent short arid event at ~8000 yr BP (Fig. 12.2J) while warmer and drier conditions, similar to present day climate prevailed since last 7000 yr BP (Bar-Matthews et al., 1997).

The studies from North Atlantic Ocean invariably demonstrated the signatures of BEs during the Holocene epoch, even the studies from Mediterranean region has also acted as a crucial recorder of BE. Further, the oxygen isotopic study of corals from Red sea showed a strong coherence with ISM that led to an understanding of strong control of Asian monsoon on eastern Saharan and the eastern Mediterranean subsidence and aridity (Felis et al., 2000). Thus, the possibility of recording BEs in marine and continental archive from Indian region can seldom be overlooked.

12.4.2 Marine archives from Northern Indian Ocean

Ocean covering >70% of the earth not only sustains the hydrological cycle and life but also provides significant cues on past climate variabilities. Nearly 6–11 billion metric tons of sediments accumulate in the ocean every year. This sedimentary archive yields information on paleoclimate variability of the ocean and also the nearby landmass (Bradley, 2015).

12.4.2.1 *Marine sediment*

Marine sediments consist of both biogenic and terrigenous materials. Biogenic fraction includes planktonic and benthic organisms that yields clues on past climate and past oceanographic conditions. While the terrigenous fraction provides information on humidity–aridity variations over the landmass by delineating the mode of transportation of sediments such as those brought by wind, fluvial activity, and ice rafting (Bradley, 2015). The marine sediments are considered as an important archive as it remains undisturbed and not influenced by anthropogenic interferences thereby providing past information of climate and oceanographic conditions.

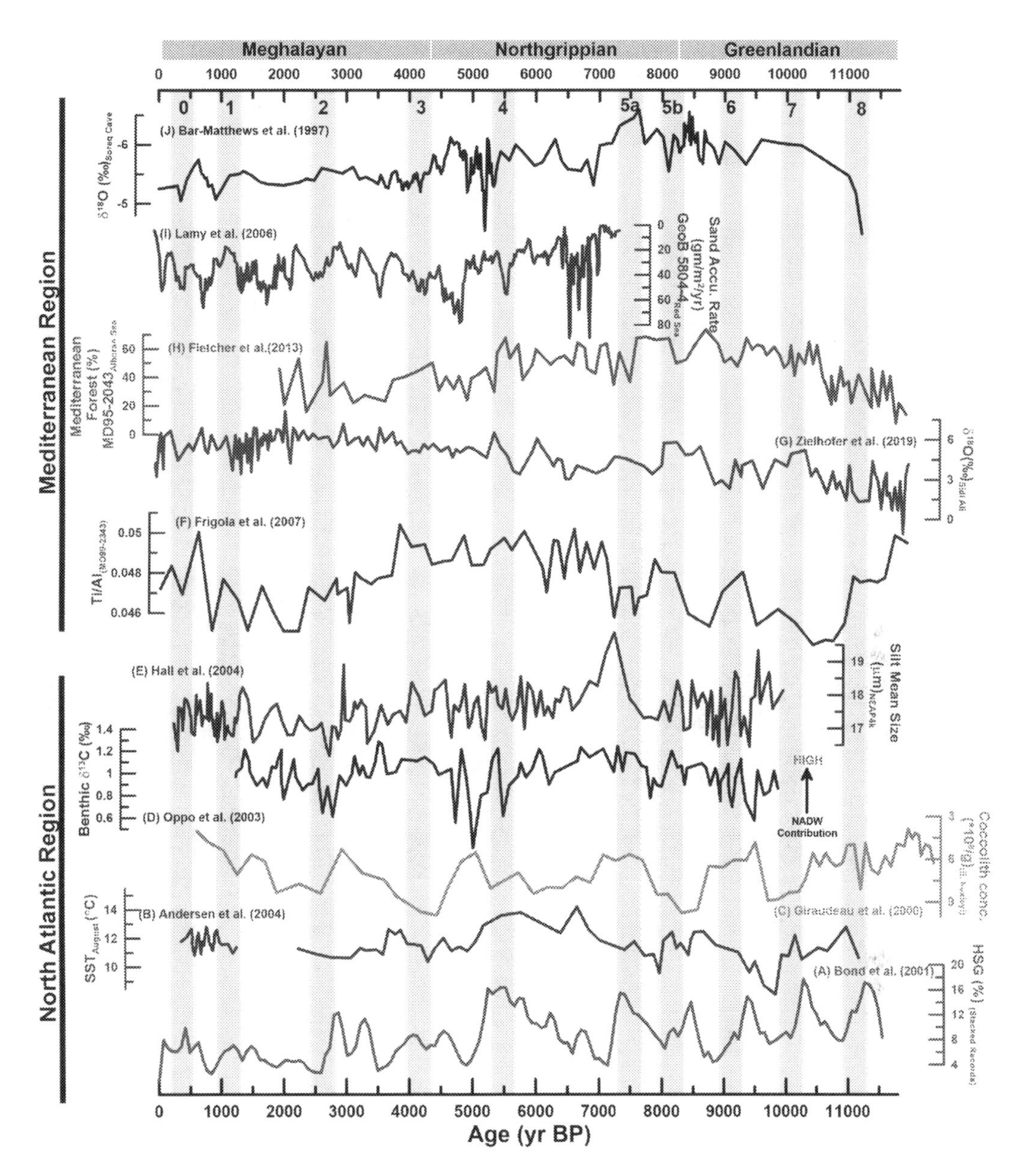

FIGURE 12.2

This figure shows collated comparison of natural archives from the North Atlantic region such as hematite stained grains (HSG) of VM 29-191, Reykjanes Ridge, Gardar contour drift, Feni Abyssal Drift, Bjorn Drift with the Mediterranean records such as NW Mediterranean, Sidi Ali Lake, western Mediterranean sea, Red Sea, and Soreq Cave. The yellow vertical bands represent the climate perturbations attested as the Bond events.

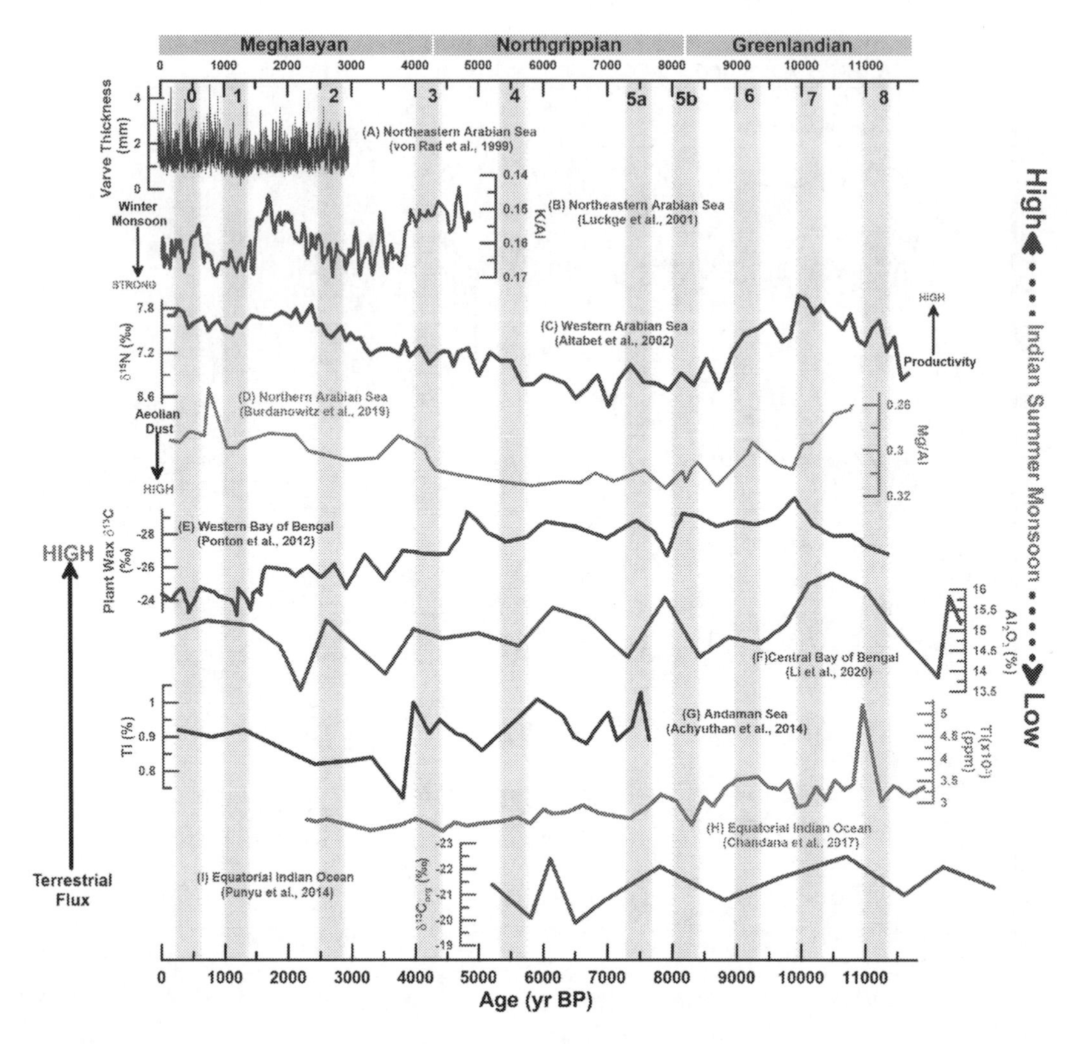

FIGURE 12.3

Comparison of the paleoceanographic reconstruction made from the marine sediment cores of Arabian Sea, Bay of Bengal and equitorial Indian Ocean from the northern Indian Ocean. The yellow vertical bands represent the climate perturbations attested as the Bond events.

The study of varved sediments off Karachi, Pakistan demonstrated reduced precipitation during 4000–3500 yr BP, 200 BC–100 AD, and 300–1600 AD followed by enhanced precipitation during cold periods of LIA (Fig. 12.3A; von Rad et al., 1999). The geochemical approach on the same sediment core indicated cooler SSTs and improved winter precipitation during 3900–3000 yr BP (Fig. 12.3B) following which summer monsoon intensification during 2000–1500 yr BP (Luckge et al., 2001).

The century to millennial scale climate oscillations with enhanced denitrification associated with ISM strengthening during the early-Holocene (11−9 ka) has been inferred based on $\delta^{15}N$ (Fig. 12.3C), total chlorins, and nitrogen from the Oman margin sediment core (Altabet et al., 2002). Based on the geochemical signatures (Mg/Al) on the sediment core from Northeast Arabian Sea the study elucidated climate aridification during ~5 ka, 4.4–3.8 ka, and 1.2–1 ka (Fig. 12.3D) wherein the latter was associated with the southern shift of ITCZ resulting in increased influence of westerlies on the sedimentation process (Burdanowitz et al., 2019). A gradual aridification after the early-Holocene humid climate toward 1700 yr BP has been decoded from the $\delta^{13}C$ plant wax in the sediment core from western Bay of Bengal (Fig. 12.3E) with prominent aridity signatures observed after 4000 yr BP (Ponton et al., 2012).

Warm and wet climate with increased terrigenous flux sourced from Himalayan region during 10–7.5 ka (Fig. 12.3F) and simultaneous reduction of terrigenous flux from the peninsular region to the central Bay of Bengal has been inferred based on multiproxy approach on the sediment core (Li et al., 2020). Early-Holocene wet climate associated with ISM intensification was also observed in the sediment cores collected from offshore Mahanadi basin, Western Bay of Bengal (Phillips et al., 2014), and from the western Andaman Sea (Sebastian et al., 2019), the latter also indicated wet climate during 5–3.5 kyr BP followed by ISM weakening since last 3.5 kyr BP (Sebastian et al., 2019). Similar observations of climate amelioration was observed during 6500–5500 cal yr BP and 4059−2342 cal yr BP interrupted by dry climate during 5500−4100 cal yr BP and 2342 cal yr BP to present (Fig. 12.3G) from the sediment core of northern Andaman Sea (Achyuthan et al., 2014). Similarly, a dry climate after 3100 cal yr BP preceded by wet climate during 6800–3100 cal yr BP has been decoded from the continental shelf sediment core of eastern Indian coast near Rushikulya river (Ankit et al., 2017). The warm and humid climate during mid-Holocene from the Andaman Sea and continental shelf of east Indian peninsula corroborates well with the sediment core from off Karwar, western India (Caratini et al., 1994).

The geochemical investigation on the sediment core collected from off Sri Lanka demonstrated enhanced terrestrial contribution during early-Holocene (Fig. 12.3H) underscoring ISM strengthening triggered by increased solar insolation during the onset of Holocene epoch (~11 ka) preceded by high in situ productivity during LGM (Chandana et al., 2017). Nearly similar observations of ISM strengthening have been inferred during 10–8 ka from Eastern Arabian Sea (Agnihotri et al., 2003). The ISM intensification from 14,500 to 2000 cal yr BP plausibly associated with the strengthening of the Walker Circulation as indicated by the investigation from the equatorial Arabian Sea (Tiwari et al., 2006b).

12.4.2.2 Foraminifera

Ocean surface water hydrography, water mass properties, and upwelling play a significant role in the distribution of the planktonic foraminifera in the ocean and seas (Johannessen et al., 1994; Pak and Kennett, 2002; Yu et al., 2007; Dissard et al., 2021). Generally, the paleoclimate and paleooceanographic reconstruction using foraminiferal shell has been made either by quantifying species assemblage or by measuring geochemical (e.g., Mg/Ca) and isotopic variabilities (e.g., $\delta^{18}O$; Staubwasser et al., 2002; Guiot and de Vernal, 2007; Rashid et al., 2007; Bradley, 2015). The oxygen isotopic investigation of *G. ruber* in sediment core retrieved from the continental margin, off Pakistan indicated ISM weakening events during early-Holocene, that is, ~11,350 cal yr BP and 8150–8400 cal yr BP (Fig. 12.4A) wherein the latter was associated with the abrupt reduction in the Indus river discharge

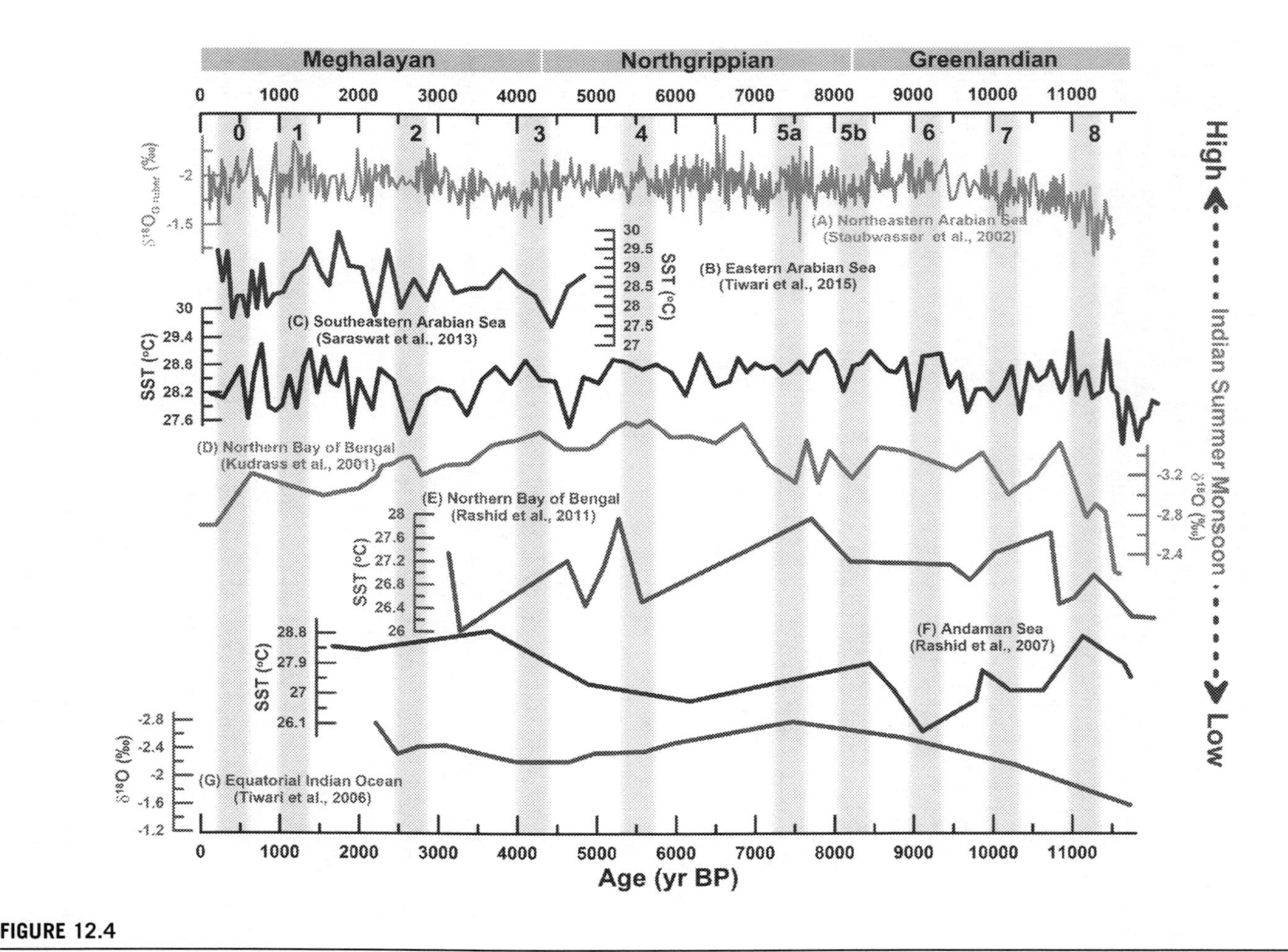

FIGURE 12.4

Comparison of reconstructed paleoceanographic conditions using foraminifera extracted from the marine sediment core of the Northern Indian Ocean. The yellow vertical bands represent the climate perturbations attested as the Bond events.

(Staubwasser et al., 2002). The foraminiferal study the sediment core from off Oman, western Arabian Sea revealed the strengthened monsoon during 9.5 ka followed by a gradual declining trend toward 5.5 ka corroborating with the warm episode of the North Atlantic region, while gradual monsoon weakening after 5.5 ka was in tandem with the reduction of solar insolation (Overpeck et al., 1996). Climate aridity centered at 4600 yr BP followed by marginal improvement of ISM in the subsequent 900 yrs has been inferred from eastern Arabian Sea based on salinity and SST reconstruction. While during ~2000 yr BP, increased SST and sea surface salinity (SSS; Fig. 12.4B) underscored severe arid climate (Tiwari et al., 2015). A reduction in SST by 1°C during ~9.7 ka with subsequent rise during ~9.2 ka and a gradual decrease in SST till 2.2 ka (Fig. 12.4C) has been demonstrated through the paired analysis of Mg/Ca and $\delta^{18}O_{foram}$ from the eastern Arabian Sea sediment core (Saraswat et al., 2013). Increased detrital contribution with warm climate prevailed during 11–10 ka followed by a stable ISM between 7 and 5.6 ka. Further, a dry phase with weak ISM centered at 4.6 ka and during 2.5–2.3 ka has been demonstrated by the SST and SSS reconstruction from sediment core of SE Arabian Sea (Kessarkar et al., 2013).

The depleted $\delta^{18}O$ in foraminifera extracted from sediment core of northern Bay of Bengal demonstrated warm climate and ISM intensification during early to mid-Holocene epoch (Fig. 12.4D; Kudrass et al., 2001). Similar observations of early-Holocene wet climate have been inferred based on reconstructed SST and SSS from paired Mg/Ca and $\delta^{18}O$ analysis on the foraminifera extracted from northern Bay of Bengal sediment core. Further the study also underscored an arid climate during 4.8–3.1 ka due to ISM weakening (Fig. 12.4E) leading to reduction in Ganga, Brahmaputra, and Meghna outflow to the Bay of Bengal (Rashid et al., 2011). Further, the SST reconstruction from Andaman Sea also underscored early-Holocene ISM strengthening (Fig. 12.4F) which subsequently weakened after 5.5 ka (Rashid et al., 2007). Both studies from the Andaman Sea and the Bay of Bengal suggested SST fluctuation within ~2.5°C during the Holocene epoch with SST warming of ~1.4°C decoded from western Bay of Bengal at ~11.3 ka (Govil and Naidu, 2011). Additionally, enhanced monsoon during 12–2 ka documented from the western Bay of Bengal (Govil and Naidu, 2011) was in agreement with the reconstructed $\delta^{18}O_{sw}$ records of eastern Arabian Sea (Govil and Naidu, 2010). Another study on foraminiferal population from the sediment core of eastern Arabian Sea elucidated a short term cooling at ~6000 yr and enhanced wind induced surface mixing during 3200–530 yr BP along with a dominant periodicity of ~1500 yr during the mid to late-Holocene epoch suggesting possible role of precessional forcing (Saravanan et al., 2020). The depleted $\delta^{18}O$ in planktonic foraminifera from the sediment core of the equatorial Indian Ocean invoked ISM intensification during 10–7 ka and 3–2.2 ka (Fig.12.4G) interrupted by weak ISM, as depicted from enriched $\delta^{18}O$ (Tiwari et al., 2006a).

The marine records from the two distinct basins of Northern Indian ocean viz. the Arabian Sea and the Bay of Bengal indicated an ISM intensification during early-Holocene with a gradual declining trend in the ISM. Further, northern Arabian Sea records indicated winter precipitations during 3900–3000 yr BP caused by westerlies associated with the Mediterranean moisture source (von Rad et al., 1999; Lückge et al., 2001) while arid climate during 4.4–3.8 cal yr BP has been invoked from the NE Arabian Sea (Burdanowitz et al., 2019). The possible influence of natural forcings such as solar and precissional cycles is engrossed in the marine records of northern Indian ocean (Agnihotri et al., 2002; Saravanan et al., 2020). Further, the SST during the Holocene epoch fluctuated within ~2.5°C for both Andaman Sea and Bay of Bengal (Govil and Naidu, 2011) suggesting near similar response to climate perturbation.

12.4.3 Continental archives from Indian subcontinent

The records of ISM have been decoded from several continental archives such as tree rings, speleothems, lacustrine sediments, etc. (Singhvi and Kale, 2010; Singhvi et al., 2010). However, high temporal resolution with strong chronological control from the continental records is the major lacuna in reconstructing ISM variability vis-à-vis climate perturbations (Bhushan et al., 2018).

12.4.3.1 Sediments and peats

Layered accumulation of the sediment prevailing at the coastal region, lakes, wetlands, estuaries, and inland (Bradley, 2015; Maya et al., 2017) records the ambient climatic conditions thereby documenting the climate variabilities. Additionally, over the past few decades, the understanding of peatland facilitated the applicability of peats in the paleoclimate and paleoenvironmental reconstruction (Booth et al., 2010). Peatlands are widely distributed between mid and high latitudes, and are sensitive to hydroclimatic variabilities, providing centennial to millennial-scale climate records (Booth et al., 2010). Peat is a salient feature of tropical coast line persisted globally in the Holocene sequence (Sukumar et al., 1993; Mascarenhas, 1997). Numerous studies on sediment sections/cores (Nair et al., 2010; Mohan et al., 2014; Banerji et al., 2017; Misra et al., 2018) and peat sections (Sukumar et al., 1993; Phadtare and Pant, 2006; Rühland et al., 2006; Padmalal et al., 2011; Srivastava et al., 2017) have been attempted to decode the ISM variability.

Recent multiproxy study from Manasbal Lake sediments, Kashmir valley deciphered high lake levels due to wet and cold phases with intense precipitation during 3500−2500 and 1800−1300 yr BP with intermittent dry and cold intervals during ~3345−3300 yrs BP and ~2500−1800 yrs BP (Fig. 12.5A). The fluctuation in the climate was associated with the influence of WD along with the localized katabatic winds rather than the ISM (Babeesh et al., 2019). The palynological investigation from Tso Moriri Lake revealed moisture enhancement with the termination of Younger Dryas which sustained during 11–9.6 ka and later it gradually reduced toward present with nominal increase in humidity during 1–0.4 ka (Fig. 12.5B). The gradual decline in the moisture trend was concurrent with the reducing trend of solar insolation during Holocene epoch (Leipe et al., 2014). Similar observation of increased moisture with warm and wet climate during 10,398–5770 cal yr BP with cold-dry climate and weak ISM during 8810–8117 cal yr BP, ~1839–1260 cal yr BP, and 852–239 cal yr BP amid ISM strengthening periods viz. ~3172–1839 cal yr BP, ~1260–852 cal yr BP, and last 239 cal yr (Fig. 12.5C) has been elucidated from Chandra Peat, Northwest Himalaya (Rawat et al., 2015b). Similarly, the ISM strengthening with warm and wet climate during 6500−5400 cal yr BP, 3800–3000 cal yr and MWP interspersed by weak ISM with cold and dry climate has been demonstrated by the peat section from Kedarnath, Higher Himalaya. Further, the study surmized climate reorganization during last 5500 yr BP associated with the ENSO triggered variability led to widespread aridity during mid-Holocene epoch (Srivastava et al., 2017). Another peat section from Pinder valley, Higher Himalaya suggested that cool and moist climate during LIA while the last two centuries witnessed improved seasonal runoff and associated feedback mechanism instigated by increased snow and ice melt and not due to ISM changes (Rühland et al., 2006).

The geochemical proxies from Benital Lake, Uttarakhand revealed nearly six phases of ISM strengthening episodes during 10,000–9600, 9500–9200, 8600–5800, 5000–4200, 3500–2400, and 1800–1000 cal yr BP (Fig. 12.5D). Further the study underscored association of North Atlantic BEs with weak ISM periods (Bhushan et al., 2018). The palynological investigation on lacustrine sediments from Triloknath glacier valley, Lahaul region divulged warm and less-humid climate during 5379–3167

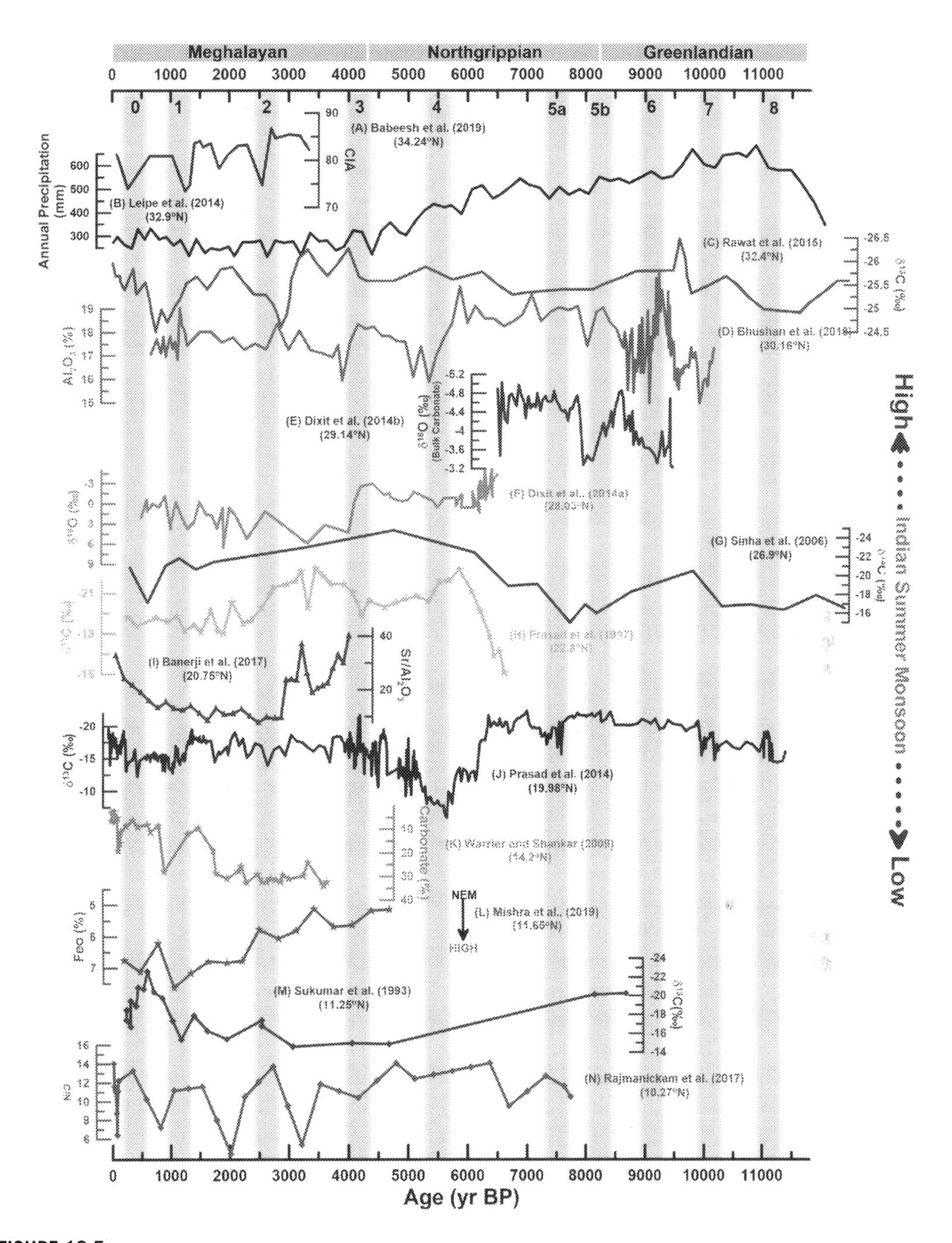

FIGURE 12.5

Past climate reconstruction made from the sediments and peats from the continental region of Indian landmass. The yellow vertical bands represent the climate perturbations attested as the Bond events.

yr BP followed by severe arid conditions till 2228 cal yr BP, whereas from 2228 to 300 cal yr BP, a two-step ISM intensification has been deciphered (Bali et al., 2017). The $\delta^{18}O$ study of Ostracoda from the Lake Riwasa sediments of Haryana, northern India, showed an intermittent dry event after 8.3 ka (Fig. 12.5E) which was associated with the 8.2 ka global event (transition between Greenlandian and Northgrippian stage) resulted due to the weak Atlantic Meridional Overturning Circulation (AMOC) caused by sudden discharge of fresh water flux in the North Atlantic region (Barber et al., 1999; Dixit et al., 2014a). Similarly, another drying event at ~4.1 ka (Fig. 12.5F) has been decoded from the Kotladhar Lake (Dixit et al., 2014b) located south of Riwasa Lake (Dixit et al., 2014a) which was in agreement with global 4.2 ka (transition between Northgrippian and Meghalayan stage) dry event (Wanner et al., 2008; Misra et al., 2018; Banerji et al., 2020). Sanai Tal Lake, central Ganga Plain, Uttar Pradesh demonstrated an improved ISM during 10,000–5800 yr BP preceded and followed by a dry phase during 11,500–10,500 yr BP and 5000–2000 yr BP, respectively, while re-establishment of ISM strengthening prevailed at ~1700 yr BP (Sharma et al., 2004). A recent biotic proxy based study on Baraila Tal, Central Ganga Plains revealed ISM strengthening with lake expansion during 11.5–10.3 ka and 7–5 ka which was interspersed by dry climate during 9–8 ka. The aquatic pollen was absent in the periods ~8.9 ka and ~7.7 ka (Misra et al., 2020). The Sambhar Playa of Rajasthan, western India showed a semihumid condition with fluctuating lake levels during ~16–7.5 kyr BP followed by an arid phase with high salinity during 7.5–6.8 ka. Later, a humid and a semiarid phase has been deciphered during 6.8–2.5 ka and post 2.5 ka (Fig. 12.5G), respectively (Sinha et al., 2006). The climatic variabilities were in agreement with salinity and water level fluctuations of Didwana Lake, Rajasthan (Wasson et al., 1984). A transition from arid to humid climatic conditions during 6.6–6 ka followed by dry climate with short spells and wet climate during 6–4.8 ka and 4.8–3 ka (Fig. 12.5H) have been demonstrated from Nal Lake, central Gujarat, western India (Prasad et al., 1997). The last 2 ka witnessed the possible prevalence of present climate with short spells (Prasad et al., 1997). Similarly, the multiproxy approach on the relict mudflats of southern Saurashtra underscored warm and humid climate during 4710–2825 and 1835–1500 cal yr BP interspersed by an arid climate (Banerji et al., 2015). The observations were in agreement with the geochemical investigation of an active mudflat of Diu island (Fig. 12.5I) with warm and wet climate during 4105–2640 cal yr BP and 1930–355 cal yr BP interspersed by an arid phase (Banerji et al., 2017). Two of the dry phases associated with ISM weakening during late Holocene has been identified at 2.1 ka and 1.3 ka from Narmada valley, Gujarat (a). Recent study from Diu mudflat revealed a similar ISM weakening period during 1800–1300 cal yr BP along with ISM strengthening during 2000–1800 cal yr BP and climate warming during last 200 cal yr BP (Banerji et al., 2021a). Moreover, the coupling of both summer and winter precipitation during 3400–3000 cal yr BP and LIA has been inferred from Mahi Estuary (Prasad et al., 2007) and Rohisa mudflats of Gujarat, respectively (Banerji et al., 2019).

Recently, a multiproxy approach on Timbi Lake sediments, Dhadhar river basin in the mainland Gujarat demonstrated occurrence of three flash flood events between 4830 and 2730 yr BP within the arid phase (Sridhar et al., 2020). Similar observations of episodic flood events during 4600–4200 yr BP with subsequent decline in the flood were demonstrated based on landscape studies of Hakra-Ghaggar and Indus rivers (Giosan et al., 2012). Pulse of higher precipitation has been evidenced during 2730–1730 yr BP from Timbi Lake following which monsoon strengthening and humid climate during 1730–880 yr BP and higher lake levels were deciphered during 880–360 yr BP. A reduced humid condition was observed after 360 yr BP (Sridhar et al., 2020). The Lonar Lake, Maharashtra investigated for the paleoclimate reconstruction demonstrated wet climate during 11–6.2 ka followed by a stepwise monsoon weakening and prevalence of semiarid conditions during 5.6–3.9 ka with a prolonged drought event during 4.6–3.9 ka (Fig. 12.5J). The drought was terminated with the gradual onset of humidification till 2 ka but the last 1.4 ka witnessed anthropogenic interference (Prasad et al., 2014). A similar observation of monsoon strengthening during ~11.4–9.5 ka followed by a stepwise ISM weakening with significantly reduced monsoon observed during 9.5–2 ka (~8.1 and 6.3 ka BP, ~6.3 to 4.7 ka BP, and ~3.0 to 2.0 ka BP) has been deciphered from Nonia Tal (Kumar et al., 2019).

In the peninsular region, the palynological investigation demonstrated warm and humid climate during 7220−4770 yr BP followed by an arid climate during 3500–1500 cal yr BP (Limaye and Kumaran, 2012). Similarly, the palynological study on a sedimentary sequence near Mahanadi, Orissa elucidated a warm and humid climate during 5840–4380 yr BP which was associated with the high monsoonal activity. Following which a marginal increased humidity was suggested during 4380–3230 cal yr BP with subsequent decline in the humidity between 3230 and 1860 cal yr BP (Tripathi et al., 2014).The observations from the Mahanadi river section were in agreement with the studies from Chilka lagoon (Pandey et al., 2014) and Cauvery delta (Srivastava and Farooqui, 2013). A similar observations of ISM strengthening during 6400 cal yr BP followed by weak monsoon during 4390–2600 cal yr BP but with an intermittent monsoon spell during 3800–2600 cal yr BP has been deciphered from the Vellayani lake, Kerala, Southern India (Banerji et al., 2021b).

The gradual onset of humid climate after 2 ka associated with the intensified rainfall (Fig. 12.5K) has been deciphered from Thimmannanayakanakere Lake of Karnataka, Southern India, prior to which a dry climate persisted since 4 ka (Warrier and Shankar, 2009). The arid phase prior to 2 ka is also inferred from the $\delta^{13}C$ on peat deposits from Nilgiris, southern India (Fig. 12.5M) which further elucidated prevalence of wet period during ~3000, 1500, and 900 cal yr BP (Sukumar et al., 1993). A contrary signature of increased monsoon identified as NEM strengthening during last ~ 4800 yr BP (Fig. 12.5L) has been delineated from the multiproxy study of Ennamangalam Lake sediments, Tamil Nadu. Further, they hypothesized that the increased ENSO events and SST of the equatorial Indian Ocean possibly triggered ISM weakening with a subsequent strengthening of NEM (Mishra et al., 2019). These observations can be linked with the multiproxy study from Kukkal Lake, Tamil Nadu which inferred ISM strengthening during 9–8.7 ka (Fig. 12.5N) followed by the monsoon weakening (Rajmanickam et al., 2017).

The lacustrine sediments from the Indian subcontinent have been extensively explored for the Holocene climate reconstruction using several proxies such as geochemical, palynology and isotopes. The studies demonstrated ISM intensification during early-Holocene followed by weak ISM during the globally known 4.2 ka event. The last two millennia experienced prevalence of ISM strengthening during MWP followed by its weakening during LIA. Further, the lake sediments and peats of northern India are found to be more sensitive towards the North Atlantic climatic episodes plausibly due to its direct influence from the western disturbances.

12.4.3.2 Speleothem

Speleothems are the cave deposits existing for millions of years and are mainly referred to as stalagmites, stalactites and flowstone (Ramesh et al., 2010). Unlike aragonite and gypsum speleothem, calcite speleothem have received wide attention for paleoclimatic reconstruction (Ford, 2004). There has been several attempts to reconstruct ISM using speleothems from Chhattisgarh (Yadava and Ramesh, 2006; Sinha et al., 2007), Orissa (Yadava and Ramesh, 2005; 2006), Karnataka (Yadava and Ramesh, 2006), western Himalaya (Yadava et al., 2007), Central Himalaya (Sanwal et al., 2013;

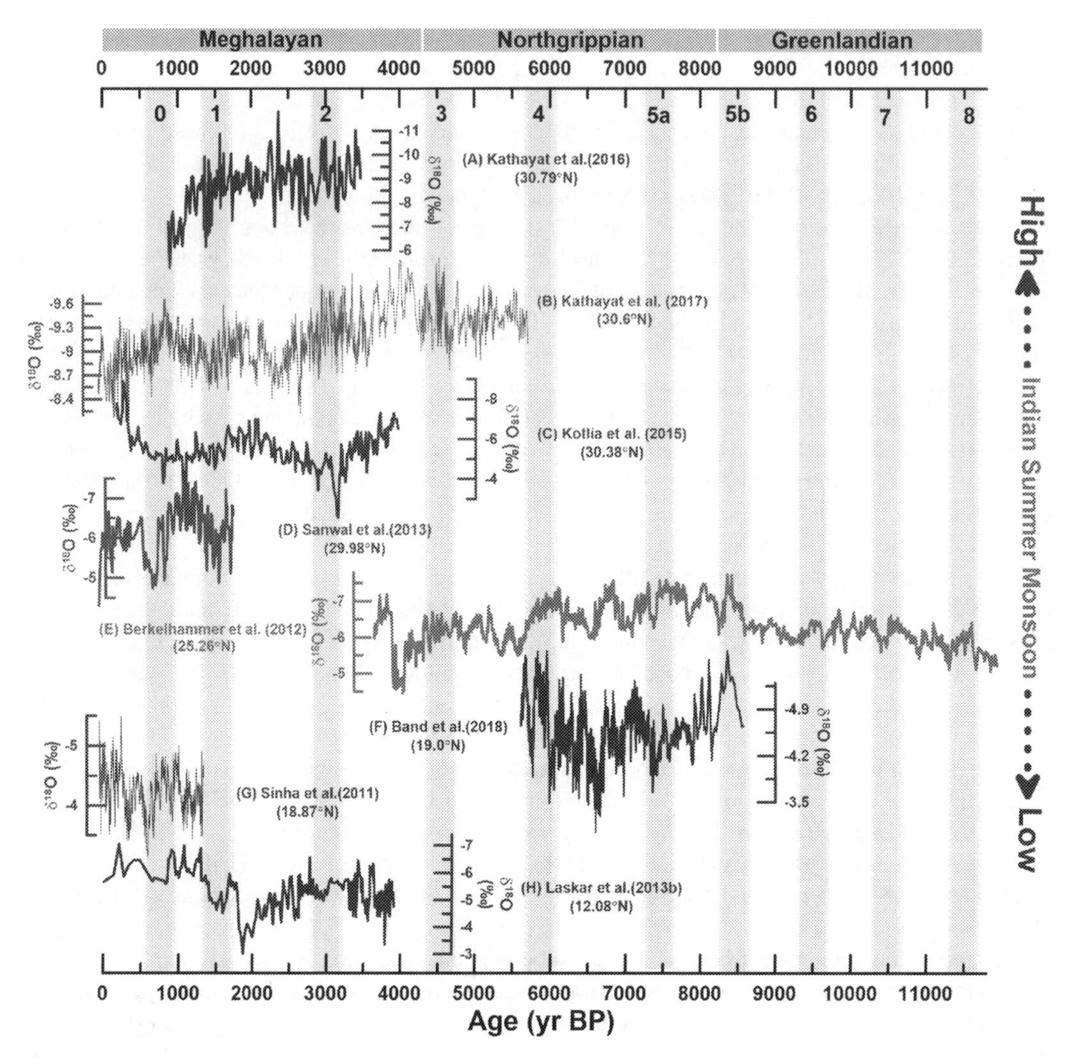

FIGURE 12.6

Comparison of the studies conducted on speleothems from continental region of India. Compilation is made only for those records that archived at least ~1000 yr time span of climate. The yellow bands represent the climate perturbations attested as the Bond events.

Kotlia et al., 2015), Uttar Pradesh (Yadava and Ramesh, 2006), and Madhya Pradesh (Yadava and Ramesh, 2006). The $\delta^{18}O$ of Bittoo cave speleothem in northern India (Fig. 12.6A) showed a climate consistency of with that of the Chinese speleothem which is significantly influenced by the East Asian monsoon (EAM) thereby underscoring the fact that the Asian monsoon (EAM and ISM) responded collectively toward the Northern Hemisphere summer insolation (Kathayat et al., 2016). The $\delta^{18}O$ of speleothem from the Sahiya cave, Uttarakhand revealed strong ISM period during ~800–1200, 1600–2300, 2900–3300, and 3800–4800 yr BP (Fig. 12.6B) which correlated with the Northern Hemisphere temperature reconstruction (Kathayat et al., 2017). Similarly, the study from the Sainji cave, central Himalaya demonstrated reduced precipitation during 4−3 ka followed by an increased precipitation during 3–2 ka. While during 1450−750 AD, an improved precipitation resulted due to (Fig. 12.6C) the increased influence of westerlies (Kotlia et al., 2015). The prevalence of the wet climate due to westerlies was in agreement with the previous work on speleothem from Chulerasim cave (Kotlia et al., 2012) and Dharamjali Cave (Sanwal et al., 2013), Kumaun Himalaya. The drastic enrichment of $\delta^{18}O$ in the speleothem from Mawmluh cave, Meghalaya with initial enrichment at ~4300 yr BP followed by a more pronounced shift at ~4100 yr BP demonstrated ISM weakening which has been associated with the possible prevalence of negative phase of Indian Ocean Dipole (IOD). Further, the study underscored a synchronicity of ISM weakening with the climate reported from other locations such as North Africa, the Middle East, the Tibetan Plateau, southern Europe, and North America (Berkelhammer et al., 2012). Thus, the conspicuous shift in the Meghalayan speleothem during ~4.2 ka has been recognized as a Global Boundary Stratotype Sections and Points which was termed as Meghalayan Stage (Walker et al., 2019). The study from the speleothem of Kotumsar cave, Chhattisgarh revealed monsoon weakening at ~8.2 ka followed by strengthened monsoon with intermittent drought events at ~7.7, ~7.3, ~6.5, and ~5.7 ka (Fig. 12.6F) which were in agreement with the North Atlantic climate variabilities (Band et al., 2018). The progressive drying during the late-Holocene epoch associated with the southward migration of the ITCZ triggered by reduced northern hemispheric summer insolation has been revealed by the Nakarallu cave, Andhra Pradesh (Sinha et al., 2018). Likewise, the late-Holocene arid climate at ~2000 yr BP and 1730 yr BP with strengthened ISM during 3400–3000 yr BP and ~600 yr BP has been demonstrated by Gupteshwar cave, Orissa (Yadava and Ramesh, 2005). The $\delta^{18}O$ study of Jhumar cave of central India revealed enhanced monsoon while Wah Shikar cave of North East India demonstrated reduced monsoon during the late 17th century. Such contrasting monsoon scenarios have been attributed to the planetary-scale ITCZ changes which can directly influence ISM precipitation by controlling the frequency distribution of active and break monsoon periods (Sinha et al., 2011). The speleothem from Andaman island suggested monsoon reduction during 1800–2100 yr BP, ~1500 and 400–800 yr BP while monsoon strengthening was recorded between 800 and 1200 yr BP (Laskar et al., 2013b).

The speleothem studies have provided improved resolution for the climate reconstruction during the Holocene epoch. The speleothem from northern India responded analogous to the Chinese speleothem studies, wherein the latter is significantly impacted by EAM (Kathayat et al., 2016) thereby suggesting a close association of ISM with EAM. A contrasting scenario of ISM variability between Jhumar, central India and Wah Shikar cave, NE India during late 17th century was associated with the ITCZ migration and its position thereby causing contrasting ISM intensities (Sinha et al., 2011).

12.5 Discussion

A climate synchronicity between the North Atlantic and the Mediterranean region has been attested from several studies. The western Mediterranean witnessed episodes of drier atmospheric conditions during the early-Holocene epoch as demonstrated from the evidences of declining forest cover (Fletcher et al., 2013). This in turn is supported by the North Atlantic ice-rafting, high-latitude cooling events associated with the meltwater pulse interrupting AMOC, and regional atmospheric climate perturbations over Europe and Greenland (Bond et al., 2001; Rasmussen et al., 2007; Boch et al., 2009; Lang et al., 2010). Thus, an arid climate prevailed in the western Mediterranean region accompanied by North Atlantic cooling during the early-Holocene epoch which were comparable with the climate of glacial periods (Fletcher et al., 2013).

12.5.1 Imprints of Bond events in marine and continental archives

The strengthened ISM during the early-Holocene epoch with gradual declining trend interrupted by several ISM weakening periods has been inferred from both marine and continental proxies. The most prominent period inferred as weak ISM are the 4.2 ka and 8.2 ka which not only represented the transition of Greenlandian–Northgrippian and Northgrippian–Meghalayan Stages but also is associated with BEs 3 and 5b, respectively. Generally, the 8.2 ka cooling event is associated with the abrupt freshwater flux from the Lake Agassiz through the Hudson Bay into the North Atlantic ocean as a result of glacial outburst (Bauer et al., 2004) which plausibly interrupted the production of NADW and depreciated the AMOC thereby a short cooling event was triggered (Barber et al., 1999). However, no such evidence of freshwater flux to the North Atlantic ocean has been attested for the 4.2 ka event (Mayewski et al., 2004). But the 4.2 ka global aridity event not only observed in North America, China, Africa, South America, and Antarctica (Mayewski et al., 2004; Staubwasser and Weiss, 2006) but has been linked with the decline of several civilizations viz. Greece and Crete (Early Bronze Age civilizations), Egypt (Old Kingdom), and the Mesopotamia (Akkadian Empire; Cullen et al., 2000; Weiss and Bradley, 2001; Marshall et al., 2011). The 8.2 ka cooling event is significantly registered in marine records while the northern India has been found to be sensitive toward recording such climatic events that prevailed for shorter time period (Banerji et al., 2020). On the contrary, the 4.2 ka event possibly of its longer time span, has been registered well in both marine and continental records (Banerji et al., 2020 and references therein).

Globally, the LIA (650–100 yr BP; Lamb, 1965; Moreno et al., 2014) being attested as BE-0 has been associated with glacial extension, cool winters and deteriorating harvests (Nichols, 1972; Grove, 1988; Lamb, 1995). However, the Indian continental records revealed discordant climate, that is, wet climate associated with WDs in the northern and western region while dry climate in most of the peninsular region (Banerji et al., 2020). Such discrepancy in the LIA climate has been associated with the southward migration of the ITCZ that led to increased winter precipitation caused by WDs with reduced ISM strength (Fleitmann et al., 2007; Kotlia et al., 2012; Banerji et al., 2019; 2020). The DACP is frequently identified climate anomaly, demonstrated as wet conditions in Europe and North Africa while dry climates in Mediterranean and China/Tibet regions (Helama et al., 2017). The DACP climate anomaly has been associated with the BE-1 but the ISM response during DACP has remained poorly addressed while limited studies have identified ISM weakening during DACP event (Moreno et al., 2012; Misra et al., 2018; Banerji et al., 2019; 2020; Naidu et al., 2020). Nevertheless, both marine and continental archives (Figs. 12.3−12.6) have responded in tandem with the DACP event but remained unnoticed possibly due to its shorter duration or major climate variability prior to the DACP event.

The weakening of ISM during 2000−4000 yr BP has been inferred from northern Indian ocean (Luckge et al., 2001; Tiwari et al., 2006a; 2015; Kessarkar et al., 2013; Ankit et al., 2017) and the continental records (Yadava and Ramesh, 2005; Limaye and Kumaran, 2012; Laskar et al., 2013b; Pandey et al., 2014; Tripathi et al., 2014; Banerji et al., 2015; 2017; Kathayat et al., 2017; Srivastava et al., 2017; Bhushan et al., 2018; Babeesh et al., 2019). On the contrary, a flash floods during 4830–2730 yr BP in Timbi Lake, mainland Gujarat has been elucidated within the globally established arid event which is in line with the observations made in Hakkra-Ghaggar and Indus rivers (Giosan et al., 2012). The time span of 2000–3000 yr BP corresponds well with the BE-2 (Bond et al., 1997) wherein strengthening of westerlies over Siberia and North Atlantic (Meeker and Mayewski, 2002) along with advancement of the alpine glacier (Denton and Karlén, 1973) has been demonstrated. Moreover, a pulse of wet climate during 3000–4000 yr BP has been inferred from Andaman Sea (Achyuthan et al., 2014), eastern Indian continental shelf (Ankit et al., 2017), equatorial Indian Ocean (Tiwari et al., 2006a), off Karwar Coast (Caratini et al., 1994) and Diu active mudflat (Banerji et al., 2017) and Vasoj relict mudflats (Banerji et al., 2015) of western Gujarat. However, a plausible prevalence of winter precipitation with weak ISM signatures during 3660–3400 cal yr BP has been suggested from Kothiyakhad, mainland Gujarat (Prasad et al., 2007) and during 3900–3000 yr BP from NE Arabian sea (Luckge et al., 2001). While a monsoon spell during 3560–3180 cal yr BP (Banerji et al., 2015) and 3230–2790 cal yr BP (Banerji et al., 2017) from western Gujarat and during 3800–2600 cal yr BP has been invoked from a recent study conducted on Vellayani lake, southern India (Banerji et al., 2021b). The BE-4 represented by the time range between 5000 and 6000 yr BP has been reported as arid climate from NE Arabian sea (Burdanowitz et al., 2019), northern Andaman Sea (Achyuthan et al., 2014), equatorial Indian Ocean (Tiwari et al., 2006a), Kedarnath (Srivastava et al., 2017) and Benital Lake (Bhushan et al., 2018) Uttarakhand, Lonar Lake, Maharashtra (Prasad et al., 2014) and the Kotumsar cave, Chattisgarh (Band et al., 2018). On the contrary, warm and humid climate during 5840–4380 yr BP has been elucidated from Mahanadi, Orissa which possibly represented the continuation of the Holocene climate optimum ranging between 8000 and 5000 yr BP (Tripathi et al., 2014). Moreover, an additional arid phase during 7.5–6.8 ka and 6.6–6.0 ka has been deciphered from the Sambhar Playa (Sinha et al., 2006) and Nal Lake (Prasad et al., 1997), respectively, which represented BE-5a. Even though, the ISM response during this time frame has been unintentionally being overlooked but the records from Northern Atlantic has revealed climate cooling at ~7.9 ka and ~6.4 ka (Andersen et al., 2004) which also demonstrated higher flow intensity of Iceland–Scotland overflow water during ~7200 yr BP (Hall et al., 2004).

The BEs 6–8 represented by Holocene time span of 9000–11,800 yr BP has been reported by a few of the paleoclimate archives (Staubwasser et al., 2002; Sharma et al., 2004; Bhushan et al., 2018). However, the imprint of reduced ISM has been recorded by other studies as well, which could be decoded after comparing it with other climatic records and prevalent of BEs (Kudrass et al., 2001; Altabet et al., 2002; Sinha et al., 2006; Rashid et al., 2007; Berkelhammer et al., 2012; Saraswat et al., 2013; Punyu et al., 2014; Rawat et al., 2015a; Li et al., 2020; Figs. 12.3–12.6).

Marine archives have recorded and responded toward the BE simultaneously, on the contrary, the limited extend of the continental proxy restricted in addressing its response toward the BEs.

12.5.2 Natural forcing and climate variables

The majority of the natural climate forcing operative during the Holocene epoch were orbital, solar and volcanic that led to significant global climate variability (Wanner et al., 2008). Additionally, the anthropogenically induced Green House Gases (GHGs) also might have acted as the climate forcing but based on the simulation, a negative (~-0.7 W/m^2) forcing during early–mid-Holocene compared to the preindustrial time for the GHGs has been inferred (LeGrande and Schmidt, 2009). The orbital parameters such as eccentricity, obliquity, and precession (Milankovitch cycles) with periodicities of 400,000–100,000 yr, 40,000 yr, and 20,000 yr, respectively, is associated with the gravitational force by Jupiter, Saturn and other planets acting on Earth (Wanner et al., 2008). The increased solar irradiance received by northern hemisphere in boreal summer (low solar irradiance during austral summer in the southern hemisphere) during the commencement of Holocene epoch was associated with the orbital parameters that led to warm climate and enhanced monsoon circulation and precipitation (Zeuner, 1945). The millennial-scale low frequency cooling during the Holocene has been associated with the orbitally induced reduction in the solar insolation during the boreal summer (Wanner et al., 2008). The Holocene epoch which is interspersed by ~8–10 cold events on multidecadal to centennial timescales (Wanner et al., 2011), but the mechanism associated with these cold events has remained unaddressed or poorly focused (Wanner et al., 2015). In view of this, the Milankovitch cycles associated with the orbital parameters pertaining cyclicities from 20 kyr to ~100 kyr, belongs to higher time span than the centennial to millennial-scale climate variations of Holocene epoch. Thus, the solar and volcanic forcing has been primarily focused for addressing the abrupt cooling events of Holocene and the climate perturbations associated with the orbital parameters is out of scope in the present review.

The ice accumulated at high mountains and polar regions for several centennial to millennial timescales has extensively been investigated to reconstruct the past climatic conditions through exhaustive physical and chemical analysis (Singhvi and Kale, 2010; Bradley, 2015). Their evidences of general relationship between the rapid temperature fluctuations in the Greenland ice records of northern high latitudes and low-latitude monsoonal climate variability is well established (Schulz et al., 1998). The drilled ice cores have been investigated for addressing trapped gases, windblown dust and various chemical species and isotopes such as sulfates and oxygen which enables to demonstrate the past atmospheric composition, paleovolcanism (Grootes et al., 1993; Zielinski et al., 1994; Zielinski and Mershon, 1997; Alley, 2004) and climatic variabilities in the form of past temperature reconstruction (Alley, 2004). Exhaustive information on the global climate perturbations and associated climate variables has been extracted from the Greenland and Antarctica ice cores which are further compared with tropical climatic archives (Staubwasser et al., 2002; Holmgren et al., 2003; Ji et al., 2005; Herzschuh, 2006; Lal et al., 2006; Tiwari et al., 2006b; Martín-Puertas et al., 2010; Menzel et al., 2014; Sandeep et al., 2017). The applicability of $\delta^{18}O$ and δD as temperature proxies has been well established for the polar ice cores (Kukla et al., 2002; Alley, 2004) however, for lower latitudes, these isotopes behave as a function of precipitation rather than temperature changes (Rozanski et al., 1993; Qin et al., 2000; Shichang et al., 2000; Tian et al., 2001). The ice cores from Tibetan plateau (Dunde, Guliya, and Dasuopu) has been studied to reconstruct the climate and monsoon variabilities (Yao et al., 1997; Liu et al., 1998; Yang et al., 2007; Seong et al., 2009; Grigholm et al., 2015). Broadly the studies have underpinned moist conditions between 9000 and 6000 yr BP due to monsoon intensification while numerous drought periods have been indicated during last millennium, but the twentieth century has been interpreted as warming period (Singhvi and Kale, 2010). The ice cores have been extensively utilized for the reconstruction of past volcanism beyond the historical and political evidences (Cole-dai, 2010). Usually the earth cooling caused by volcanic activities can prevail either due to absorption of solar radiation by volcanic aerosols (Crowley, 2000) or due to reflection of incoming solar radiation by the H_2SO_4 clouds formed in the troposphere due to the volcanic activities (Zielinski, 2000). Mostly to decipher past volcanic activity through ice cores, the sulfate measurements are preferred to over the acidic contents as a substantial proportion of the sulfates can be sourced from volcanism (Delmas et al., 1985).

The sulfate content estimated in the GISP2 ice core retrieved from Greenland revealed several volcanic events out of which 85% occurred during the last two millennia while 30% occurred during 1–7000 BC corresponds to documented volcanic eruptions (Zielinski et al., 1994; Zielinski and Mershon, 1997).

The crustal stresses readjustment caused by the glacial unloading has been linked with the enhanced volcanism which is significantly recorded in the GISP2 ice core (Grove, 1976). The time span during 15,000–8000 yrs BP (Fig. 12.7A) identified as the commencement of deglacial period, Holocene, has greatest number of volcanic signals (Zielinski et al., 1996). The high volcanic sulfate peaks between 11.8 and 8.0 ka was in agreement with the prevalence of deeper Icelandic low (Fig. 12.7E) leading to high HSG (Fig. 12.7C) in North Atlantic sediments corresponding to BEs 8 to 6. Possibly, the volcanic eruption due to crustal readjustments and deglaciation of the Laurentide ice sheet during the commencement of the Holocene epoch (Zielinski et al., 1996) significantly impacted the climate and ocean circulation near the North Atlantic region. The prevalence of cold events during BE-6 and BE-7 is also inferred from sediment cores of sub polar North Atlantic (Andersen et al., 2004). The comparison of marine and terrestrial archives indicated that the BE-6, 7, and 8 (Figs. 12.3–12.6) have been observed as ISM weakening periods. These events remain abstain with limited discussions in the literatures as of now.

On the basis of selected studies providing temperature and humidity/precipitation for a time span of the last 10 kyr, it is revealed that there were nearly six conspicuous events of cold period viz. 8.6–8.0 ka, 6.5–5.9 ka, 4.8–4.5 ka, 3.3–2.5 ka, 1.75–1.35 ka (DACP), and 0.70–0.15 ka (LIA) (Wanner et al., 2011). Among these, the cold event 8.6−8.0 ka, 3.3−2.5 ka and LIA were in agreement with the BEs 5b, 2 and 1 (Wanner et al., 2011). The BE-5b associated with the 8.2 ka has been linked to the reduced production of NADW and declining the AMOC and triggered a short cooling event (Barber et al., 1999). The reduction of NADW was also inferred from depleted values for the foraminiferal $\delta^{13}C$ extracted from the sediment core of Feni drift, subpolar NE Atlantic Ocean (Oppo et al., 2003). The BE-5b varied in its intensity and geographical extent but was seen as cooling at higher latitudes (cool poles) and dry in lower latitudes (dry tropics; Mayewski et al., 2004).

The BE-2 (3.3–2.5 ka) is globally known for glacial advances at several locations such as Swiss Alps (Holzhauser et al., 2005), Norway (Matthews et al., 2005), Western North America (Koch and Clague, 2006), Patagonia (Mercer, 1982). The cooling and glacial advance during this period has been associated with reduced solar activities during 2900–2800 yr BP (van Geel et al., 2000). The solar forcing has played a crucial role in modulating climate (Herschel, 1800) wherein the solar activity is controlled by a series of quasiperiodical cycles ranging from 11 to 2300 yrs and termed as Schwabe (11 yr), Hale (22 yr), Gleissberg (87 yr), Suess (210 yr), Eddy (1000 yr), and Hallstatt cycles (2300 yr; Lning and Vahrenholt, 2016). Further, the BE-2 has been considered to be analogue to the LIA with low solar radiation (Fig. 12.7B) during Maunder Minima (1630–1715 AD) (van Geel et al., 2000; Wanner et al., 2008). Generally, the past solar activity beyond the optical observational period has been reconstructed based on the cosmogenic isotopes in the ice cores (^{10}Be, ^{14}C, and ^{36}Cl) generated by cosmic rays (Abreu et al., 2010; Lning and Vahrenholt, 2016). Increase in the ^{10}Be and marginal increase in $\Delta^{14}C$ during BE-3 ($\sim$4.2 ka) have been linked with the possible reduction in the solar output leading to abrupt climate perturbation (Mayewski et al., 2004). Further, a strengthened Siberian High during BE-3 has been revealed by continental sourced non-sea salt potassium (nss K^+; Fig. 12.7I) in Greenland ice cores (Mayewski et al., 1994; 1997). Thus, the strengthening and expansion of the Siberian High possibly blocked the moisture laden westerlies from reaching western Asia thereby leading to enhanced summer droughts with reduced ISM in India and nearby regions (Persoiu et al., 2019).

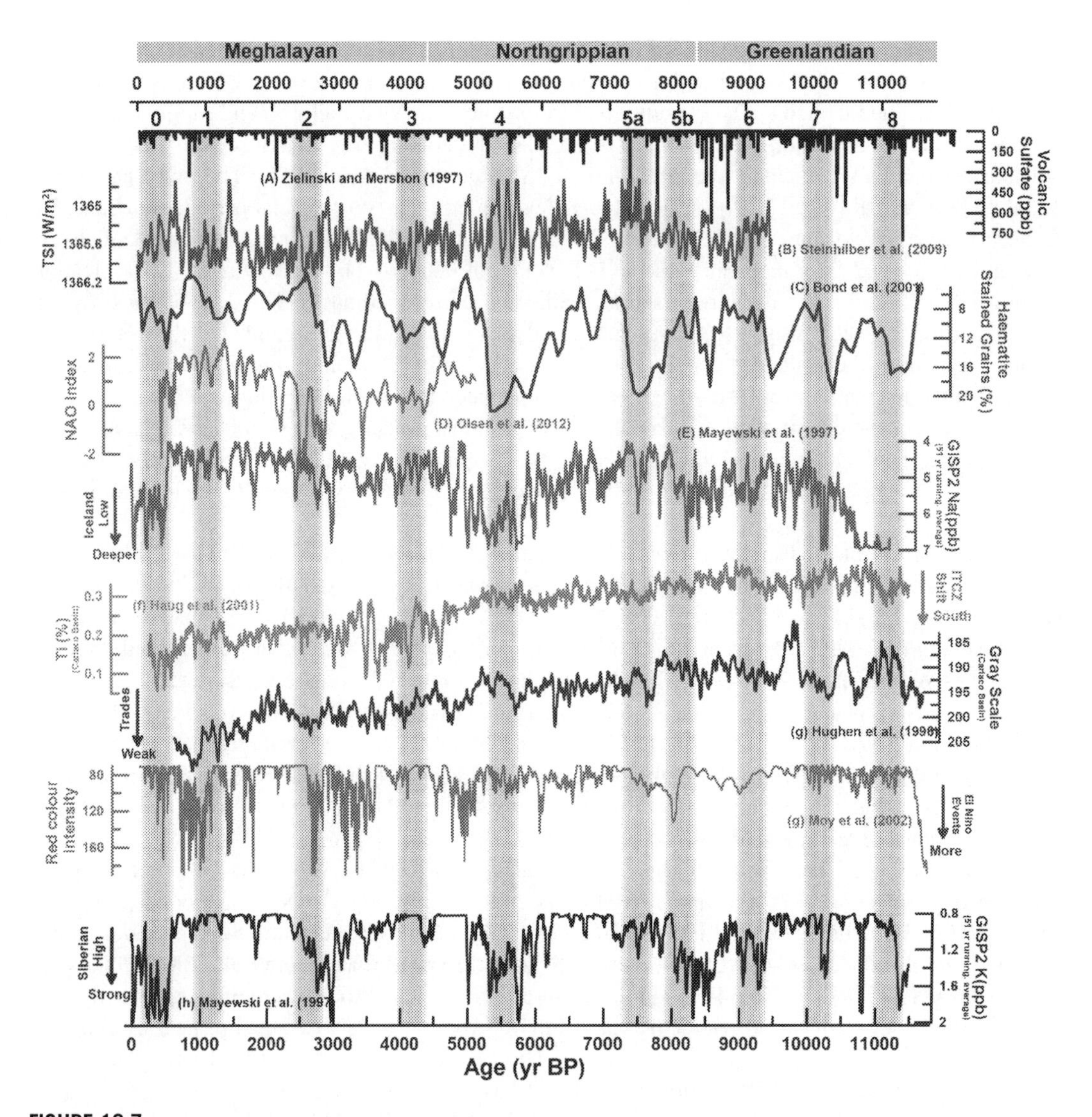

FIGURE 12.7

Comparison of global proxies such as (A) volcanic forcing from GISP2 ice core sulfate records, (B) total solar insolation, (C) hematite stained grains from North Atlantic, (D) reconstructed NAO index, (E and I) Siberian high and Icelandic low represented by K and Na content in GISP2 ice core, (F and G) ITCZ migration from Cariaco basin, and (H) red color intensity representing ENSO events from the Laguna Pallcacocha. The yellow bands represent the climate perturbations attested as the Bond events.

Other high frequency climate variability modes that have substantially impacted the ISM are El Niño Southern Oscillation (ENSO), North Atlantic Oscillation (NAO) and Indian Ocean Dipole (IOD). The oscillation of anomalous warming of surface ocean in the central and eastern Pacific with enhanced convection in the atmosphere has been termed as El Niño while reduced convection in the atmosphere associated with the unusual cooling of ocean surface waters has been defined as La Niña and the cumulative phenomenon of El Niño and La Niña has been described as ENSO (Gadgil and Francis, 2012). An increased El Niño events during late-Holocene compared to middle Holocene (Wanner et al., 2008) has been demonstrated through study of the lithic flux in marine sediment, off Peru (Rein et al., 2004) and coupled ocean–atmospheric model (Clement et al., 2000). The digital scanning of the sediment core raised from Laguna Pallcacocha of southern Ecuador (Fig. 12.7H) revealed increased El Niño events 3500–2500 yr BP (Moy et al., 2002). The period corresponds to BE-2 which further reveals ISM weakening from marine and continental records (Luckge et al., 2001; Kessarkar et al., 2013; Banerji et al., 2015; 2017; Ankit et al., 2017; Srivastava et al., 2017; Babeesh et al., 2019).

The climate variability in North Atlantic region is controlled by the NAO which can be defined as the pressure difference between the Icelandic low and the Azores high (Hurrell and Deser, 2009), that in turn governs the track of westerly wind position across Europe (Mercer, 1982; Vicente-Serrano et al., 2011; Dermody et al., 2012; Olsen et al., 2012; Smith et al., 2016). Generally, warm and wet climate over NW Europe and dry conditions over southern Europe are associated with the positive phase of NAO (NAO+) which forces winter storm tracks (Atlantic westerly jets) to migrate northward (Hurrell and Deser, 2009). While the negative phase of NAO (NAO-) is characterized by weak winter storm track (Smith et al., 2016) but enhanced precipitation over the western Mediterranean (Trigo et al., 2002; Nieto-Moreno et al., 2013). The stable positioning of the Atlantic Westerly Jets during the Holocene epoch reveals that the phenomenon of NAO must have prevailed over longer timescales beyond the instrumental extent (Olsen et al., 2012;). The $\delta^{18}O$ values for the speleothem from Northern Iberia revealed NAO- conditions leading to increased rainfall during 2.1–2.2 ka, ~1.4 ka, and 0.5–0.3 ka while low rainfall associated with the NAO+ during 2.0–1.5 ka (Smith et al., 2016). The time span for NAO- phases corresponds to BE-2, BE-1 (DACP) and BE-0 (LIA) while NAO+ corresponds to the MWP (Smith et al., 2016). Among these, a prominent NAO- feature during BE-2 and BE-0 (Fig. 12.7D) is also reflected from the reconstructed NAO index using lake sediment record of southwestern Greenland (Olsen et al., 2012). Further, a shift from NAO+ to NAO- phase during the transition from MWP to LIA has been suggested by many others (Trouet et al., 2009; Brauer et al., 2011). Further, several studies have revealed that the LIA witnessed southward migration of the ITCZ such as from Atlantic (Arbuszewski et al., 2013), Pacific (Linsley et al., 1994), Oman (Fleitmann et al., 2007; Ivory et al., 2017), Indonesian (Konecky et al., 2013; Wundsch et al., 2014), and Indian (Kotlia et al., 2012; Banerji et al., 2019, 2020). The titanium (Ti) concentration and Gray scale from the Cariaco basin (Fig. 12.7F and G) revealed a gradual migration of ITCZ toward south during the Holocene epoch with significant excursions in Ti at LIA revealing weak rainfall and run off accompanied by increased upwelling (Peterson et al., 2000; Haug et al., 2001). The southward migration of the ITCZ was further supported by the late-Holocene expansion of evergreen forest toward southward at the southern margin of Amazonia (Mayle et al., 2000). The observations were in agreement with the increased precipitation during LIA suggested by the sedimentary records of Titicaca Lake (south of the Cariaco basin) (Baker et al., 2001). Generally, the southward migration of the ITCZ has been linked with the increased El Niño events (Fedorov and Philander, 2000; Haug et al., 2001) which is in agreement with the increased red color intensity of Laguna Pollach (Fig. 12.7H). The volcanic eruptions have been responsible for the southward migration of the ITCZ (Mukherjee et al., 1987) along with other natural forcing (orbital and solar). The conspicuous shift of ITCZ during LIA has been associated with the sudden eruption of the mount Samalas during 1258 AD located in Indonesia which possibly led to southward migration of ITCZ thereby resulting in the onset of BE-0 (LIA) (Lavigne et al., 2013; Banerji et al., 2019). The ITCZ shift to southward led to weak ISM but elevated the influence of winter precipitation and thus, caused increase influence of WD over the Indian subcontinent (Lückge et al., 2001; Kotlia et al., 2012; Babeesh et al., 2019; Banerji et al., 2019). The study on western Mediterranean Sea sediment core revealed that the early-Holocene climate has been mainly controlled by the changing solar irradiance amplified by the meltwater dynamics and change in the ocean circulation while the mid–late-Holocene (~ last 6ka) controlled by oceanic forcing and ocean–atmospheric processes similar to NAO like patterns (Fletcher et al., 2013). Another study from western Mediterranean suggested dry winters during early to mid-Holocene while wet winters during mid to late-Holocene epoch (Zielhofer et al., 2019). It can be elucidated that the impression of North Atlantic climate and influence of WD over Indian subcontinent is more prominently observed during mid to late-Holocene epoch with winter precipitation during LIA and possibly during 2–4 ka.

Along with the ENSO and NAO, the IOD has also played a crucial role in influencing the ISM variability in the Indian subcontinent. The IOD has been identified as the coupled mode of ocean–atmosphere in the tropical Indian Ocean with the reversal of the winds and SST gradient across the basin (Saji et al., 1999). In general, the IOD has been instigating extreme weather events over Indian landmass and nearby regions with positive IOD phase (pIOD) being associated with ISM intensification while vis-à-vis for the negative IOD phase (nIOD) (Ashok et al., 2001; Brown et al., 2009; Ogwang et al., 2015). Based on climate simulations it has been suggested that the El Niño conditions can lead to initiation of IOD events by decreasing the thermocline depth in the Indonesian throughflow region leading to an eastward migration of the Pacific Walker circulation cell (Fischer et al., 2005; Behera et al., 2006).

The future climate projections indicated stronger IOD, however, the amplitude had large intermodal spread (Ihara et al., 2009). Recently, a model-based study conducted to project the IOD characteristics during the mid-Holocene (8200–4200 yr BP) indicated pIOD like pattern (Iwakiri and Watanabe, 2019) which was in agreement with the proxy record (Abram et al., 2007; 2009). While nIOD phase has been invoked during mid-17th century (Li et al., 2017) based on the SST reconstructions from the Arabian Sea and the Indo-Pacific Warm Pool (Oppo et al., 2009; Tierney et al., 2015). Despite several studies addressing ISM and IOD interrelations, yet the association of ISM with the IOD is still ambiguous (Ashok et al., 2001; Jourdain et al., 2013; Behera and Ratnam, 2018). Thus, the limited attempts to delineate the role of IOD in ISM predictions and its implications on ISM variability defy to decode the behavioral pattern and teleconnection of ISM-IOD with the Bond Events.

12.6 Learning and knowledge outcomes

The ISM has been highly variable during the Holocene epoch as a function of natural forcing and ocean–atmospheric processes which led to the migration of the ITCZ. The signatures of BE significantly registered in the North Atlantic and Mediterranean regions, have also been deciphered in the northern Indian Ocean and Indian subcontinent proxies revealing a teleconnection of ISM with North Atlantic climate. Thus, the impacts of BE on the planetary scale position of the ITCZ need to be extensively explored. Moreover, the linkage of ENSO of Pacific Ocean with ISM has been established previously, but the present review also suggests a close connection and interplay of NAO and ENSO in modulating the ISM intensities. Among the BEs, the BE-5, 3, 2, 1, and 0 are well studied as ISM weakening with increased winter precipitation during BE-0 and possibly during BE-2 while BE-4, 6, 7, and 8 need to be explored further even though limited studies have deciphered weak ISM during these periods. Further, the BEs-2, 4, 5b and 6 have been identified as the possible evidence of reduced NADW contribution (Oppo et al., 2003). Thus, its signatures in the marine and continental archives in recording ISM, need to be explored to decipher the changes in the thermohaline circulation during Holocene epoch. Archives such as corals and tree rings that can generate interannual, multidecadal and multicentennial climate perturbations beyond instrumental records need to be explored not only to calibrate the instrumental data but also to quantify the climatic and/or oceanic parameters such SST, SSS, and rainfall.

Here we have made an attempt to pave understanding on the impacts of North Atlantic climate over the ISM influenced region. But ultimately, it would be ideal to quantify the qualitative inferences supported by statistical and modeling approach to individually discriminate the natural forcing and climate variables and its impacts on the ISM by using spatiotemporal datasets. However, it is premature to proceed with such quantitative and modeling approach as it need sufficient spatio-temporal dataset with enhanced resolution and reliable chronologic controls. Thus, several climate variables such as temperature, rainfall, etc. need to be studied from different locations that receive varied ISM and winter monsoon intensities followed by quantification and modeling. Such an approach can provide reliable and longer future climate projections. Additionally, it can also help discriminating the intertwining impacts of natural and anthropogenic forcings on the climate change. Thus, the future of climate research lies in extensive studies on natural archives from marine and continental regions with improved resolution and chronological constrains.

Acknowledgments

We dedicate this review paper to the fond memory of our beloved colleague and friend (Late) Dr. Arulbalaji P. This research was supported by the National Centre for Earth Science Studies (NCESS), Ministry of Earth Sciences, Government of India. The authors are grateful to the Director, NCESS for his support and encouragement for the present study. We thank the editors and anonymous reviewers for their constructive comments and suggestion that enhanced the manuscript. We have attempted to include most of the relevant research contributions and if any of the relevant publication has been missed in the present review, it was unintentional.

References

Abram, N.J., Gagan, M.K., Liu, Z., Hantoro, W.S., Mcculloch, M.T., Suwargadi, B.W., Seasonal characteristics of the Indian Ocean Dipole during the Holocene epoch, Nature 445 (2007) 299–302.

Abram, N.J., McGregor, H.V., Gagan, M.K., Hantoro, W.S., Suwargadi, B.W., Oscillations in the southern extent of the Indo-Pacific Warm Pool during the mid-Holocene, Quat. Sci. Rev. 28 (2009) 2794–2803.

Abreu, J.A., Beer, J., Ferriz-Mas, A., Past and future solar activity from cosmogenic radionuclides, SOHO-23: Understanding a Peculiar Solar Minimum, ASP Conference Series, 428, 2010, p. 287.

Achyuthan, H., Nagasundaram, M., Gourlan, A.T., Eastoe, C., Ahmad, S.M., Padmakumari, V.M., Mid-Holocene Indian Summer Monsoon variability off the Andaman Islands, Bay of Bengal, Quat. Int. 349 (2014) 232–244.

Agnihotri, C.L., Singh, M.S., Satellite study of western disturbances, Mausam 33 (1982) 249–254.

Agnihotri, R., Dutta, K., Bhushan, R., Somayajulu, B.L.K.K., Evidence for solar forcing on the Indian monsoon during the last millennium, Earth Planet. Sci. Lett. 198 (2002) 521–527.

Agnihotri, R., Sarin, M.M., Somayajulu, B.L.K., Jull, A.J.T., Burr, G.S., Late-Quaternary biogenic productivity and organic carbon deposition in the eastern Arabian Sea, Palaeogeogr. Palaeoclimatol. Palaeoecol. 197 (2003) 43–60.

Ahmad, S.M., Padmakumari, V.M., Raza, W., Venkatesham, K., Suseela, G., Sagar, N., Chamoli, A., Rajan, R.S., High-resolution carbon and oxygen isotope records from a scleractinian (Porites) coral of Lakshadweep Archipelago, Quat. Int. 238 (2011) 107–114.

Alley, R.B., 2004. GISP2 ice core temperature and accumulation data. IGBP PAGES/World Data Center for Paleoclimatology Data Contribution Series 13.

Alley, R.B., Abrupt climate changes, oceans, ice, and us, Oceanography 17 (2005) e206.

Alley, R.B., Mayewski, P.A., Sowers, T., Stuiver, M., Taylor, K.C., Clark, P.U.A., Holocene climatic instability: a prominent, widespread event 8200 yr ago, Geology 25 (1997) 483–486.

Altabet, M.A., Higginson, M.J., Murray, D.W., The effect of millennial-scale changes in Arabian Sea denitrication on atmospheric CO_2, Nature 415 (6868) (2002) 159–162.

Altabet, M.A., Pilskaln, C., Thunell, R., Pride, C., Sigman, D., Chavez, F., Francois, R., The nitrogen isotope biogeochemistry of sinking particles from the margin of the eastern North Pacific. Deep-Sea Research Part I, Oceanographic Research Papers 46 (1999) 655–679.

Andersen, C., Koç, N., Moros, M., A highly unstable Holocene climate in the subpolar North Atlantic: evidence from diatoms, Quat. Sci. Rev. 23 (2004) 2155–2166.

Ankit, Y., Kumar, P., Anoop, A., Mishra, P.K., Varghese, S., Mid-late Holocene climate variability in the Indian monsoon: evidence from continental shelf sediments adjacent to Rushikulya river, eastern India, Quat. Int. 443 (2017) 155–163.

Arbuszewski, J.A., Demenocal, P.B., Cléroux, C., Bradtmiller, L., Mix, A., Meridional shifts of the Atlantic intertropical convergence zone since the Last Glacial Maximum, Nat. Geosci. 6 (2013) 959–962.

Ashok, K., Guan, Z., Yamagata, T., Impact of the Indian Ocean dipole on the relationship between the Indian monsoon rainfall and ENSO, Geophys. Res. Lett. 28 (23) (2001) 4499–4502.

Babeesh, C., Achyuthan, H., Resmi, M.R., Nautiyal, C.M., Shah, R.A., Late Holocene paleoenvironmental changes inferred from Manasbal Lake sediments, Kashmir Valley, India, Quat. Int. 507 (2019) 156–171.

Baker, Andy, JohnHellstrom, C., BryceKelly, F.J., Gregoire Mariethoz, Valerie, Trouet., A Composite Annual-Resolution Stalagmite Record of North Atlantic Climate over the Last Three Millennia, Sci. Rep. 510307 (2015) 1–8.

Baker, P.A., Seltzer, G.O., Fritz, S.C., Dunbar, R.B., Grove, M.J., Tapia, P.M., Cross, S.L., Rowe, H.D., Broda, J.P., The history of South American tropical precipitation for the past 25,000 years, Science 291 (5504) (2001) 640–643.

Bali, R., Chauhan, M.S., Mishra, A.K., Ali, Nawaz, S., Tomar, A., Khan, I., Singh, Sen, D., Srivastava, P., Vegetation and climate change in the temperate-subalpine belt of Himachal Pradesh since 6300 cal. yrs. BP, inferred from pollen evidence of Triloknath palaeolake, Quat. Int. 444 (2017) 11–23.

Bamzai, A.S., Shukla, J., Relation between Eurasian snow cover, snow depth, and the Indian summer monsoon: an observational study, J. Clim. 12 (1999) 3117–3132.

Band, S., Yadava, M.G., Lone, M.A., Shen, C.C., Sree, K., Ramesh, R., High-resolution mid-Holocene Indian Summer Monsoon recorded in a stalagmite from the Kotumsar Cave, Central India, Quat. Int. 479 (2018) 19–24.

Banerji, U.S., Arulbalaji, P., Padmalal, D., Holocene climate variability and Indian Summer Monsoon: an overview, Holocene 30 (2020) 744–773.

Banerji, U.S., Bhushan, R., Jull, A.J.T., Signatures of global climatic events and forcing factors for the last two millennia on the active mudflats of Rohisa, southern Saurashtra, Gujarat, western India, Quat. Int. 507 (2019) 172–187.

Banerji, U.S., Bhushan, R., Joshi, K.B., Shaji, J., Jull, A.J.T., Hydroclimate variability during the last two millennia from the mudflats of Diu Island, Western India, Geological Journal (2021a).

Banerji, U.S., Bhushan, R., Jull, A.J.T., Mid–late Holocene monsoonal records from the partially active mudflat of Diu Island, southern Saurashtra, Gujarat, western India, Quat. Int. 443 (2017) 200–210.

Banerji, U.S., Pandey, S., Bhushan, R., Juyal, N., Mid-Holocene climate and land-sea interaction along the southern coast of Saurashtra, western India, J. Asian Earth Sci. 111 (2015) 428–439.

Banerji, U.S., Shaji, J., Arulbalaji, P., Maya, K., Vishnu Mohan, S., Dabhi, A. J., Shivam, A., Bhushan, R., Padmalal, D., Mid-late Holocene evolutionary history and climate reconstruction of Vellayani lake, south India. Quaternary International (2021b) 018. https://doi.org/10.1016/j.quaint.2021.03.

Barber, D., Dyke, A., Hillaire-Marcel, C., Jennings, A.E., Andrews, J.T., Kerwin, M.W., Bilodeau, G., McNeely, R., Southon, J., Morehead, M.D., Gagnon, J.M., Forcing of the cold event of 8,200 years ago by catastrophic drainage of Laurentide lakes, Nature 400 (1999) 344–348.

Bar-Matthews, M., Ayalon, A., Kaufman, A., Late Quaternary paleoclimate in the Eastern Mediterranean region from stable isotope analysis of speleothems at Soreq Cave, Israel, Quat. Res. 47 (1997) 155–168.

Bauer, E., Ganopolski, A., Montoya, M., Simulation of the cold climate event 8200 years ago by meltwater outburst from Lake Agassiz, Paleoceanography 19 (3) (2004) PA3014.

Bawiskar, S.M., Chipade, M.D., Mujumdar, V.R., Puranik, P.V, Bhide, U.V, Kinetic energy of extratropical waves and their effect on the Indian monsoon rainfall, Mausam 56 (2005) 337–342.

Behera, S.K., Luo, J.J., Masson, S., Rao, S.A., Sakuma, H., Yamagata, T., A CGCM study on the interaction between IOD and ENSO, J. Clim. 19 (2006) 1688–1705.

Behera, S.K., Ratnam, J.V., Quasi-asymmetric response of the Indian summer monsoon rainfall to opposite phases of the IOD, Sci. Rep. 8 (2018) 1–8.

Berkelhammer, M., Sinha, A., Stott, L., Cheng, H., Pausata, F.S.R., Yoshimura, K., An abrupt shift in the Indian monsoon 4000 years ago, Geophys. Monogr. Ser. 198 (2012) 75–87.

Bhattacharyya, A., Chaudhary, V., Late-summer temperature reconstruction of the Eastern Himalayan region based on tree-ring data of *Abies densa*, Arctic Antarctic Alpine Res. 35 (2003) 196–202.

Bhushan, R., Sati, S.P., Rana, N., Shukla, A.D., Mazumdar, A.S., Juyal, N., High-resolution millennial and centennial scale Holocene monsoon variability in the Higher Central Himalayas, Palaeogeogr. Palaeoclimatol. Palaeoecol. 489 (2018) 95–104.

Björck, S., Walker, M.J., Cwynar, L.C., Johnsen, S., Knudsen, K.L., Lowe, J.J., Wohlfarth, B., An event stratigraphy for the Last Termination in the North Atlantic region based on the Greenland ice-core record: a proposal by the INTIMATE group, J. Quat. Sci. 13 (1998) 283–292.

Boch, R., Spötl, C., Kramers, J., High-resolution isotope records of early Holocene rapid climate change from two coeval stalagmites of Katerloch Cave, Austria, Quat. Sci. Rev. 28 (2009) 2527–2538.

Bond, G., Broecker, W., Johnsen, S., McManus, J., Labeyrie, L., Jouzel, J., Bonani, G., Correlations between climate records from North Atlantic sediments and Greenland ice, Nature 363 (1993) 210–211.

Bond, G., Kromer, B., Beer, J., Muscheler, R., Evans, M.N., Showers, W., Hoffmann, S., Lotti-Bond, R., Hajdas, I., Bonani, G., Persistent solar influence on North Atlantic climate during the Holocene, Science 294 (2001) 2130–2136.

Bond, G., Showers, W., Cheseby, M., Lotti, R., Almasi, P., Priore, P., Cullen, H., Hajdas, I., Bonani, G., A pervasive millennial-scale cycle in North Atlantic Holocene and glacial climates, Science 278 (1997) 1257–1266.

Bond, G.C., Lotti, R., Iceberg discharges into the North Atlantic on millennial time scales during the last glaciation, Science 267 (1995) 1005–1010.

Bond, G.C., Showers, W., Elliot, M., Evans, M., Lott, R., Hajdas, L., Bonani, G., Johnson, S., The north Atlantic's 1-2 Kyr climate rhythm: Relation to heinrich events, dansgaard/oeschger cycles and the little ice age, Geophys. Monogr. Ser. 112 (1999) 35–58.

Booth, R.K., Jackson, S.T., Notaro, M., Using peatland archives to test paleoclimate hypotheses, PAGES News 18 (2010) 1–5.

Borgaonkar, H.P., Sikder, A.B., Ram, S., Pant, G.B., El Niño and related monsoon drought signals in 523-year-long ring width records of teak (*Tectona grandis* L.F.) trees from south India, Palaeogeogr. Palaeoclimatol. Palaeoecol. 285 (2010) 74–84.

Bradley, R.S., Paleoclimatology: Reconstructing Climates of the Quaternary, Paleoclimatology: Reconstructing Climates of the Quaternary, 3rd, Elsevier Inc., Amsterdam, Netherlands, 2015.

Brauer, A., Zorita, E., Mart, C., Hydrological evidence for a North Atlantic oscillation during the Little Ice Age outside its range observed since 1850, Clim. Past Discuss. 7 (2011) 4149–4171.

Broecker, W., Bond, G., Klas, M., Clark, E., McManus, J., Origin of the northern Atlantic's Heinrich events, Clim. Dyn. 6 (1992) 265–273.

Broecker, W.S., A salt oscillator in the glacial Atlantic? 1. The concept, Paleoceanography 5 (4) (1990) 469–477.

Broecker, W.S., Massive iceberg discharges as triggers for global climate change, Nature 372 (1994) 421.

Brown, J., Lynch, A.H., Marshall, A.G., Variability of the Indian Ocean Dipole in coupled model paleoclimate simulations, J. Geophys. Res. Atmos. 114 (2009) 1–12.

Burdanowitz, N., Gaye, B., Hilbig, L., Lahajnar, N., Lückge, A., Rixen, T., Emeis, K.C., Holocene monsoon and sea level-related changes of sedimentation in the northeastern Arabian Sea, Deep-Sea Res. Part II: Top. Stud. Oceanogr. 166 (2019) 6–18.

Cantin, N.E., Cohen, A.L., Karnauskas, K.B., Tarrant, A.M., McCorkle, D.C., Ocean warming slows coral growth in the central Red Sea, Science 329 (2010) 322–325.

Caratini, C., Bentaleb, I., Fontugne, M., Morzadeckerfourn, M.T., Pascal, J.P., Tissot, C., A less humid climate since Ca.3500-Yr-Bp from marine cores off Karwar, Western India, Palaeogeogr. Palaeoclimatol. Palaeoecol. 109 (1994) 371–384.

Chakraborty, S., Ramesh, R., Environmental significance of carbon and oxygen isotope ratios of banded corals from Lakshadweep, India, Quat. Int. 37 (1997) 55–65.

Chakraborty, S., Ramesh, R., Stable isotope variations in a coral (*Favia speciosa*) from the Gulf of Kutch during 1948–1989 AD: environmental implications, Proc. Indian Acad. Sci.-Earth Planet. Sci. 107 (1998) 331–341.

Chandana, K.R., Bhushan, R., Jull, A.J.T., Evidence of poor bottom water ventilation during LGM in the equatorial Indian Ocean, Front. Earth Sci. 5 (2017) 84.

Chaudhary, V., Bhattacharyya, A., Suitability of *Pinus kesiya* in Shillong, Meghalaya for tree-ring analyses, Curr. Sci. 83 (2002) 1010–1015.

Clement, A.C., Seager, R., Cane, M.A., Suppression of El Niño during the mid-Holocene by changes in the Earth's orbit, Paleoceanography 15 (6) (2000) 731–737.

Cole-dai, J., Volcanoes and climate, WIREs Clim. Change 1 (2010) 824–839.

Crowley, T.J., Causes of climate change over the past 1000 years, Science 289 (2000) 270–277.

Cullen, H.M., Menocal, P.B.De, Hemming, S., Brown, F.H., Guilderson, T., Sirocko, F., Climate change and the collapse of the Akkadian empire: evidence from the deep sea, Geology 28 (2000) 379–382.

Dansgaard, W., Johnsen, S.J., Clausen, H.B., Dahl-Jensen, D., Gundestrup, N.S., Hammer, C.U., Hvidberg, C.S., Steffensen, J.P., Sveinbjörnsdottir, A.E., Jouzel, J., Bond, G., Evidence for general instability of past climate from a 250-kyr ice-core record, Nature 364 (1993) 218–220.

Debret, M., Sebag, D., Crosta, X., Massei, N., Petit, J.R., Chapron, E., Bout-Roumazeilles, V., Evidence from wavelet analysis for a mid-Holocene transition in global climate forcing, Quat. Sci. Rev. 28 (2009) 2675–2688.

Delmas, R.J., Legrand, M., Aristarain, A.J., Zanolini, F., Volcanic deposits in Antarctic snow and ice, J. Geophys. Res. 90 (1985) 901–920.

Denton, G.H., Karlén, W., Holocene climatic variations-their pattern and possible cause, Quat. Res. 3 (1973) 155–205.

Dermody, B.J., Boer, H.J.De, Bierkens, M.F.P., Weber, S.L., Wassen, M.J., Dekker, S.C., A seesaw in Mediterranean precipitation during the Roman Period linked to millennial-scale changes in the North Atlantic, Clim. Past 8 (2012) 637–651.

Dhar, O.N., Rakhecha, P.R., Foreshadowing northeast monsoon rainfall over Tamil Nadu, India, Monthly Weather Rev. 111 (1983) 109–112.

Dimri, A.P., Niyogi, D., Barros, A.P., Ridley, J., Mohanty, U.C., Yasunari, T., Sikka, D.R., Western disturbances: a review, Rev. Geophys. 53 (2015) 225–246.

Ding, Yihui, Synoptic systems and weather. In: *The Asian Monsoon*, Springer, Berlin, Heidelberg, 2006, pp. 131–201.

Dissard, D., Reichart, G.J., Menkes, C., Mangeas, M., Frickenhaus, S., Bijma, J., Mg/Ca, Sr/Ca and stable isotope from planktonic foraminifera *T. sacculifer*: testing a multi-proxy approach for inferring paleo-temperature and paleo-salinity, Biogeosciences 18 (2021) 423–439.

Dixit, Y., Hodell, D.A., Sinha, R., Petrie, C.A., Abrupt weakening of the Indian summer monsoon at 8.2 kyrB.P, Earth Planet. Sci. Lett. 391 (2014a) 16–23.

Dixit, Y., Hodell, D.A., Petrie, C.A., Abrupt weakening of the summer monsoon in northwest India $\sim$ 4100 yr ago, Geology 42 (2014b) 339–342.

Dixit, Y., Tandon, S.K., Hydroclimatic variability on the Indian subcontinent in the past millennium: review and assessment, Earth-Sci. Rev. 161 (2016) 1–15.

Doose-rolinski, H., Rogalla, U., Scheeder, G., Lfickge, A., RadVon, U., High-resolution temperature and evaporation changes during the late Holocene in the northeastern Arabian Sea, Paleoceanography 16 (2001) 358–367.

Fan, J., Xiao, J., Wen, R., Zhang, S., Wang, X., Cui, L., Li, H., Xue, D., Yamagata, H., Droughts in the East Asian summer monsoon margin during the last 6 kyrs: link to the North Atlantic cooling events, Quat. Sci. Rev. 151 (2016) 88–99.

Fedorov, A.V., Philander, S.G., Is El Niño changing? Science 288 (2000) 1997–2002.

Felis, T., Pätzold, J., Loya, Y., Fine, M., Nawar, A.H., Wefer, G., A coral oxygen isotope record from the northern Red Sea documenting NAO, ENSO, and North Pacific teleconnections on Middle East climate variability since the year 1750, Paleoceanography 15 (2000) 679–694.

Fischer, A.S., Terray, P., Guilyardi, E., Gualdi, S., Delecluse, P., Two independent triggers for the Indian Ocean dipole/zonal mode in a coupled GCM, J. Clim. 18 (2005) 3428–3449.

Fleitmann, D., Burns, S.J., Mangini, A., Mudelsee, M., Kramers, J., Villa, I., Neff, U., Al-Subbary, A., Buettner, A., Hippler, D., Matter, A., Holocene ITCZ and Indian monsoon dynamics recorded in stalagmites from Oman and Yemen (Socotra), Quat. Sci. Rev. 26 (2007) 170–188.

Fletcher, W.J., Debret, M., Fernanda, M., Goñi, S., Mid-Holocene emergence of a low-frequency millennial oscillation in western Mediterranean climate: implications for past dynamics of the North Atlantic atmospheric westerlies, Holocene 23 (2) (2013) 153–166.

Ford, D., Paleoenvironments: spleoethems, in: Gunn, J. (Ed.), Caves and Karst Science, Fitzroy, Dearborn, New York, 2004.

Frigola, J., Moreno, A., Cacho, I., Canals, M., Sierro, F.J., Flores, J.A., Grimalt, J.O., Hodell, D.A., Curtis, J.H., Holocene climate variability in the western Mediterranean region from a deepwater sediment record, Paleoceanography 22 (2007) 1–16.

Fritts, H.C., Tree Rings and Climate, Sci. Am. 226 (1972) 92–100.

Gadgil, S., The Indian monsoon and its variability, Annu. Rev. Earth Planet. Sci. 31 (2003) 429–467.

Gadgil, S., Francis, P., Oceans and the Indian monsoon, Monsoon Monogr. II (2012) 1–58.

Giosan, L., Clift, P.D., Macklin, M.G., Fuller, D.Q., Constantinescu, S., Durcan, J.A., Stevens, T., Duller, G.A.T., Tabrez, A.R., Gangal, K., Adhikari, R., Alizai, A., Filip, F., VanLaningham, S., Syvitski, J.P.M., Fluvial landscapes of the Harappan civilization, Proc. Natl. Acad. Sci. U.S.A 109 (26) (2012) E1688–E1694.

Giraudeau, J., Cremer, M., Manthé, S., Labeyrie, L., Bond, G., Coccolith evidence for instabilities in surface circulation south of Iceland during Holocene times, Earth Planet. Sci. Lett. 179 (2000) 257–268.

Govil, P., Naidu, P.D., Evaporation-precipitation changes in the eastern Arabian Sea for the last 68 ka: implications on monsoon variability, Paleoceanography 25 (2010) 1–11.

Govil, P., Naidu, P.D., Variations of Indian monsoon precipitation during the last 32kyr reflected in the surface hydrography of the Western Bay of Bengal, Quat. Sci. Rev. 30 (2011) 3871–3879.

Grigholm, B., Mayewski, P. A., Kang, S., Zhang, Y., Morgenstern, U., Schwikowski, M., Sneed, S., Twentieth century dust lows and the weakening of the westerly winds over the Tibetan Plateau, Geophys. Res. Lett. 42 (7) (2015) 2434–2441.

Grootes, P.M., Stuiver, M., White, J.W.C., Johnsen, S., Jouzel, J., Comparison of oxygen isotope records from the GISP2 and GRIP Greenland ice cores, Nature 366 (1993) 552–554.

Grove, E.W., Deglaciation—A possible triggering mechanism for volcanism. In: Andean and Antarctic Volcanology Problems. International Association of Volcanologists Symposium, Santiago, Chile, 1976, pp.88–97.

Grove, J.M., The 'Little Ice Age', Methuen, 1988.

Guiot, J., de Vernal, A., Transfer Functions: Methods for Quantitative Paleoceanography Based on Microfossils, Developments in Marine Geology, Elsevier B.V, 2007, pp. 523–563.

Gupta, A.K., Anderson, D.M., Overpeck, J.T., Abrupt changes in the Asian southwest monsoon during the Holocene and their links to the North Atlantic Ocean, Nature 421 (2003) 354–357.

Hall, I.R., Bianchi, G.G., Evans, J.R., Centennial to millennial scale Holocene climate-deep water linkage in the North Atlantic, Quat. Sci. Rev. 23 (2004) 1529–1536.

Halley, E., An historical account of the trade winds, and monsoons, observable in the seas between and near the Tropicks, with an attempt to assign the physical cause of the said winds, Philos. Trans. R. Soc. Lond. 16 (1753) 153–168.

Haug, G.H., Hughen, K.A., Sigman, D.M., Peterson, L.C., Rohl, U., Southward migration of the Intertropical Convergence Zone through the Holocene, Science 293 (5533) (2001) 1304–1308.

Heinrich, H., Origin and consequences of cyclic ice rafting in the Northeast Atlantic Ocean during the past 130,000 years, Quat. Res. 29 (1988) 142–152.

Helama, S., Jones, P.D., Briffa, K.R., Dark Ages Cold Period: a literature review and directions for future research, Holocene 27 (2017) 1600–1606.

Hemming, S.R., Heinrich events: massive late Pleistocene detritus layers of the North Atlantic and their global climate imprint, Rev. Geophys. 42 (2004) RG1005.

Herschel, W., Observations tending to investigate the nature of the Sun, in order to find the causes or symptoms of its variable emission of light and heat; with remarks on the use that may possibly be drawn from solar observations., Philos. Trans. R. Soc. Lond. 91 (1800) 265–318.

Herzschuh, U., Palaeo-moisture evolution in monsoonal Central Asia during the last 50,000 years, Quat. Sci. Rev. 25 (2006) 163–178.

Holmgren, K., Lee-Thorp, J.a., Cooper, G.R.J., Lundblad, K., Partridge, T.C., Scott, L., Sithaldeen, R., Talma, a.S., Tyson, P.D., Persistent millennial-scale climatic variability over the past 25,000 years in Southern Africa, Quat. Sci. Rev. 22 (2003) 2311–2326.

Holzhauser, H., Magny, M., Zumbühl, H.J., Glacier and lake-level variations in west-central Europe over the last 3500 years, Holocene 15 (2005) 789–801.

Hua, Q., Woodroffe, C.D., Smithers, S.G., Barbetti, M., Fink, D., Radiocarbon in corals from the Cocos (Keeling) Islands and implications for Indian Ocean circulation, Geophys. Res. Lett. 32 (2005) 1–4.

Hurrell, J.W., Deser, C., North Atlantic climate variability: the role of the North Atlantic Oscillation, J. Mar. Syst. 78 (2009) 28–41.

Ihara, C., Kushnir, Y., Cane, M.A., Pena La De, V. H., Climate change over the equatorial indo-pacific in global warming, J. Clim. 22 (2009) 2678–2693.

Ivory, S.J., Braconnot, P., Marti, O., Anne-marie, L., Timing of the southward retreat of the ITCZ at the end of the Holocene Humid Period in Southern Arabia: Data-model comparison, Quat. Sci. Rev. 164 (2017) 68–76.

Iwakiri, T., Watanabe, M., Strengthening of the Indian Ocean dipole with increasing seasonal cycle in the Mid-Holocene, Geophys. Res. Lett. 46 (2019) 8320–8328.

Ji, J., Shen, J., Balsam, W., Chen, J., Liu, L., Liu, X., Asian monsoon oscillations in the northeastern Qinghai-Tibet Plateau since the late glacial as interpreted from visible reflectance of Qinghai Lake sediments, Earth Planet. Sci. Lett. 233 (2005) 61–70.

Johannessen, T., Jansen, E., Flatøy, A., Ravelo, A.C., The Relationship between Surface Water Masses, Oceanographic Fronts and Paleoclimatic Proxies in Surface Sediments of the Greenland, Iceland, Norwegian Seas, in: Zahn, R., Pedersen, T.F., Kaminski, M.A., Labeyrie, L. (Eds.), Carbon Cycling in the Glacial Ocean: Constraints on the Ocean's Role in Global Change, NATO ASI Series (Series I: Global Environmental Change), 17, Springer, Berlin, Heidelberg, 1994.

Johnsen, S.J., Clausen, H.B., Dansgaard, W., Fuhrer, K., Gundestrup, N., Hammer, C.U., Stauffer, B., Steffensen, J.P., Irregular glacial interstadials recorded in a new Greenland ice core, Nature 259 (1992) 311–313.

Jourdain, N.C., Gupta, A.S., Taschetto, A.S., Ummenhofer, C.C., Moise, A.F., Ashok, K., The Indo-Australian monsoon and its relationship to ENSO and IOD in reanalysis data and the CMIP3/CMIP5 simulations, Clim. Dyn. 41 (2013) 3073–3102.

Kar, S.C., Rana, S., Interannual variability of winter precipitation over northwest India and adjoining region: impact of global forcings, Theor. Appl. Climatol. 116 (2014) 609–623.

Kathayat, G., Cheng, H., Sinha, A., Spötl, C., Edwards, R.L., Zhang, H., Li, X., Yi, L., Ning, Y., Cai, Y., Lui, W.L., Breitenbach, S.F.M., Indian monsoon variability on millennial-orbital timescales, Sci. Rep. 6 (2016) 1–7.

Kathayat, G., Cheng, H., Sinha, A., Yi, L., Li, X., Zhang, H., Li, H., Ning, Y., Edwards, R.L., The Indian monsoon variability and civilization changes in the Indian subcontinent, Sci. Adv. 3 (2017) 1–9.

Keigwin, L.D., Boyle, E.A., Detecting Holocene changes in thermohaline circulation, Proc. Natl. Acad. Sci. U.S.A 97 (2000) 1343–1346.

Kessarkar, P.M., Purnachadra Rao, V., Naqvi, S.W.A., Karapurkar, S.G., Variation in the Indian summer monsoon intensity during the Bølling-Ållerød and Holocene., Paleoceanography 28 (3) (2013) 413–425.

Koch, J., Clague, J.J., Are insolation and sunspot activity the primary drivers of Holocene glacier fluctuations, PAGES Newslett. 14 (2006) 20–21.

Konecky, B.L., Russell, J.M., Rodysill, J.R., Vuille, M., Bijaksana, S., Huang, Y., Intensification of southwestern Indonesian rainfall over the past millennium, Geophys. Res. Lett. 40 (2013) 386–391.

Kothawale, D.R., Rupa Kumar, K., On the recent changes in surface temperature trends over India, Geophys. Res. Lett. 32 (2005) L18714.

Kotlia, B.S., Ahmad, S.M., Zhao, J.-X., Raza, W., Collerson, K.D., Joshi, L.M., Sanwal, J., Climatic fluctuations during the LIA and post-LIA in the Kumaun Lesser Himalaya, India: evidence from a 400 yr old stalagmite record, Quat. Int. 263 (2012) 129–138.

Kotlia, B.S., Singh, A.K., Joshi, L.M., Dhaila, B.S., Precipitation variability in the Indian Central Himalaya during last ca. 4,000 years inferred from a speleothem record: impact of Indian Summer Monsoon (ISM) and Westerlies, Quat. Int. 371 (2015) 244–253.

Krishnamurti, T.N., Bhalme, H.N., Oscillations of a monsoon system. Part I. Observational aspects, J. Atmos. Sci. 33 (1976) 1937–1954.

Kudrass, H.R., Hofmann, A., Doose, H., Emeis, K., Ostseeforschung, I., Modulation and amplification of climatic changes in the Northern Hemisphere by the Indian summer monsoon during the past 80 k.y, Geology 29 (2001) 63–66.

Kukla, G., Bender, M., de Beaulieu, J., Bond, G., Broecker, W., Cleveringa, P., Gavin, J., Herbert, T., Imbrie, J., Jouzel, J., Keigwin, L., Knudsen, K., McManus, J., Merkt, J., Muhs, D., Muller, H., Poore, R., Porter, S., Seret, G., Shackleton, N., Turner, C., Tzedakis, P., Winograd, I., Last interglacial climates, Quaternary Research 58 (2002) 2–13.

Kumar, K., Agrawal, S., Sharma, A., Pandey, S., Indian summer monsoon variability and vegetation changes in the core monsoon zone, India, during the Holocene: a multiproxy study, Holocene 29 (2019) 110–119.

Lal, D., Charles, C., Vacher, L., Goswami, J.N., Jull, A.J.T., McHargue, L., Finkel, R.C., Paleo-ocean chemistry records in marine opal: Implications for fluxes of trace elements, cosmogenic nuclides (^{10}Be and ^{26}Al), and biological productivity, Geochim. Cosmochim. Acta 70 (2006) 3275–3289.

Lamb, H.H., The early medieval warm epoch and its sequel, Palaeogeogr. Palaeoclimatol. Palaeoecol. 1 (1965) 13–37.

Lamb, H.H., Climate, History and the Modern World, second ed., Routledge, London, 1995.

Lamy, F., Arz, H.W., Bond, G.C., Bahr, A., Pätzold, J., Multicentennial-scale hydrological changes in the Black Sea and northern Red Sea during the Holocene and the Arctic/North Atlantic Oscillation, Paleoceanography 21 (2006) 1–11.

Lang, B., Bedford, A., Brooks, S.J., Jones, R.T., Richardson, N., Birks, H.J.B., Marshall, J.D., Early-Holocene temperature variability inferred from chironomid assemblages at Hawes Water, northwest England, Holocene 20 (2010) 943–954.

Laskar, A.H., Yadava, M.G., Sharma, N., Ramesh, R., Late-Holocene climate in the Lower Narmada valley, Gujarat, western India, inferred using sedimentary carbon and oxygen isotope ratios, The Holocene 23 (2013a) 1115–1122.

Laskar, A.H., Yadava, M.G., Ramesh, R., Polyak, V.J., Asmerom, Y., A 4 kyr stalagmite oxygen isotopic record of the past Indian Summer Monsoon in the Andaman Islands, Geochem. Geophys. Geosyst. 14 (2013b) 3555–3566.

Lau, W.K.-M., Waliser, D.E., Intraseasonal Variability in the Atmosphere-Ocean Climate System, Springer Science & Business Media, Berlin, Germany, 2011.

Lavigne, F., Degeai, J.-P., Komorowski, J.-C., Guillet, S., Robert, V., Lahitte, P., Oppenheimer, C., Stoffel, M., Vidal, C.M., Surono, Pratomo, I, Wassmer, P., Hajdas, I., Hadmoko, D.S., Belizal, E.de, Source of the great A.D. 1257 mystery eruption unveiled, Samalas volcano, Rinjani Volcanic Complex, Indonesia, Proc. Natl. Acad. Sci. U.S.A 110 (2013) 16742–16747.

LeGrande, A.N., Schmidt, G.A., Sources of Holocene variability of oxygen isotopes in paleoclimate archives, Clim. Past 5 (2009) 441–455.

Leipe, C., Demske, D., Tarasov, P.E., A Holocene pollen record from the northwestern Himalayan lake Tso Moriri: implications for palaeoclimatic and archaeological research, Quat. Int. 348 (2014) 93–112.

Li, J., Liu, S., Shi, X., Chen, M.Te, Zhang, H., Zhu, A., Cui, J., Khokiattiwong, S., Kornkanitnan, N., Provenance of terrigenous sediments in the central Bay of Bengal and its relationship to climate changes since 25 ka, Prog. Earth Planet. Sci. 7 (2020) 1–16.

Li, K., Liu, X., Wang, Y., Herzschuh, U., Ni, J., Liao, M., Xiao, X., Late Holocene vegetation and climate change on the southeastern Tibetan Plateau: implications for the Indian Summer Monsoon and links to the Indian Ocean Dipole, Quat. Sci. Rev. 177 (2017) 235–245.

Limaye, R.B., Kumaran, K.P.N., Mangrove vegetation responses to Holocene climate change along Konkan coast of south-western India, Quat. Int. 263 (2012) 114–128.

Linsley, B.K., Dunbar, R.B., Wellington, G.M., Mucciarone, D.A., A coral-based reconstruction of Intertropical Convergence Zone variability over Central America since 1707, J. Geophys. Res. 99 (C5) (1994) 9977–9994.

Liu, K., Yao, Z., Thompson, L.G., A pollen record of Holocene climatic changes from the Dunde ice cap, Qinghai-Tibetan Plateau, Geology 26 (1998) 135–138.

Lückge, A., Doose-Rolinski, H., Khan, A.A., Schulz, H., von Rad, U., Monsoonal variability in the northeastern Arabian Sea during the past 5000 years: geochemical evidence from laminated sediments, Palaeogeogr. Palaeoclimatol. Palaeoecol. 167 (2001) 273–286.

Lning, S., Vahrenholt, F., The Sun's role in climate, in: Easterbrook, D. (Ed.), Evidence-Based Climate Science: Data Opposing CO_2 Emissions as the Primary Source of Global Warming, Second Edition, Elsevier Inc, Amsterdam, Netherlands, 2016, pp. 283–305.

Managave, S.R., Sheshshayee, M.S., Ramesh, R., Borgaonkar, H.P., Shah, S.K., Bhattacharyya, A., Response of cellulose oxygen isotope values of teak trees in differing monsoon environments to monsoon rainfall, Dendrochronologia 29 (2011) 89–97.

Mangerud, J.A.N., Andersen, S.T., Berglund, B.E., Donner, J.J., Quaternary stratigraphy of Norden, a proposal for terminology and classification, Boreas 3 (1974) 109–126.

Marshall, J.F., McCulloch, M.T., Evidence of El Niño and the Indian Ocean Dipole from Sr/Ca derived SSTs for modern corals at Christmas Island, eastern Indian Ocean, Geophys. Res. Lett. 28 (2001) 3453–3456.

Marshall, M.H., Lamb, H.F., Huws, D., Davies, S.J., Bates, R., Bloemendal, J., Boyle, J., Leng, M.J., Umer, M., Bryant, C., Late Pleistocene and Holocene drought events at Lake Tana, the source of the Blue Nile, Global Planet. Change 78 (2011) 147–161.

Martín-Puertas, C., Jiménez-Espejo, F., Martínez-Ruiz, F., Nieto-Moreno, V., Rodrigo, M., Mata, M.P., Valero-Garcés, B.L., Late Holocene climate variability in the southwestern Mediterranean region: an integrated marine and terrestrial geochemical approach, Clim. Past 6 (2010) 807–816.

Mascarenhas, A., Significance of peat on the western continental shelf of India, J. Geol. Soc. India 49 (1997) 145–152.

Matthews, J.A., Berrisford, M.S., Quentin Dresser, P., Nesje, A., Olaf Dahl, S., Elisabeth Bjune, A., Bakke, J., John,, H., Birks, B., Lie, Ø., Dumayne-Peaty, L., Barnett, C., Holocene glacier history of Bjørnbreen and climatic reconstruction in central Jotunheimen, Norway, based on proximal glaciofluvial stream-bank mires, Quat. Sci. Rev. 24 (2005) 67–90.

Maya, K., Mohan, S.V., Limaye, R.B., Padmalal, D., Kumaran, K.P.N., Geomorphic response to sea level and climate changes during Late Quaternary in a humid tropical coastline: terrain evolution model from Southwest India, PLoS ONE 12 (5) (2017) e0176775.

Mayewski, P.A., Meeker, L.D., Twickler, M.S., Lyons, W.B., Prentice, M., Whitlow, S., Yang, Q., Lyons, W.B., Prentice, M., Major features and forcing of high-latitude northern hemisphere atmospheric circulation using a 110,000-year-long glaciochemical series, J. Geophys. Res.: Oceans 102 (1997) 26345–26366.

Mayewski, P.A., Meeker, L.D., Whitlow, S.L., Twickler, M.S., Morrison, M.C., Bloomfield, P., Bond, G.C., Alley, R.B., Gow, A.J., Meese, D.A., Grootes, P.M., Changes in Atmospheric circulation and ocean ice cover over the North Atlantic during the last 41,000 years, Science 263 (1994) 1747–1751.

Mayewski, P.A., Rohling, E.E., Stager, J.C., Karlén, W., Maasch, K.A., Meeker, L.D., Meyerson, E.A., Gasse, F., Kreveld, S.van, Holmgren, K., Lee-Thorp, J., Rosqvist, G., Rack, F., Staubwasser, M., Schneider, R.R., Steig, E.J., Holocene climate variability, Quat. Res. 62 (2004) 243–255.

Mayle, F.E., Burbridge, R., Killeen, T.J., Millennial-scale dynamics of southern Amazonian rain forests, Science 290 (2000) 2291–2294.

Meeker, L.D., Mayewski, P.A., A 1400-year high-resolution record of atmospheric circulation over the North Atlantic and Asia, The Holocene 12 (3) (2002) 257–266.

Menzel, P., Gaye, B., Mishra, P.K., Anoop, A., Basavaiah, N., Marwan, N., Plessen, B., Prasad, S., Riedel, N., Stebich, M., Wiesner, M.G., Linking Holocene drying trends from Lonar Lake in monsoonal central India to North Atlantic cooling events, Palaeogeogr. Palaeoclimatol. Palaeoecol. 410 (2014) 164–178.
Nichols, H., 1972. Times of Feast, Times of Famine. A History of Climate since the Year 1000. Emmanuel le Roy Ladurie. Translated from the French by Barbara Bray. Doubleday, Garden City, NY, 1971. xxiv, 426 pp.+ plates. $10.
Mercer, J.H., Holocene Glacier Variations in Southern South America, Striae 18 (1982) 35–40.
Midhuna, T.M., Kumar, P., Dimri, A.P., A new Western Disturbance Index for the Indian winter monsoon, J. Earth Syst. Sci. 129 (1) (2020) 1–14.
Mishra, P.K., Ankit, Y., Gautam, P.K., Lakshmidevi, C.G., Singh, P., Anoop, A., Inverse relationship between south-west and north-east monsoon during the late Holocene: geochemical and sedimentological record from Ennamangalam Lake, southern India, Catena 182 (2019) 104117.
Misra, P., Farooqui, A., Sinha, R., Khanolkar, S., Tandon, S.K., Millennial-scale vegetation and climatic changes from an Early to Mid-Holocene lacustrine archive in Central Ganga Plains using multiple biotic proxies, Quat. Sci. Rev. 243 (2020) 106474.
Misra, P., Tandon, S.K., Sinha, R., Holocene climate records from lake sediments in India: assessment of coherence across climate zones, Earth-Sci. Rev. 190 (2018) 370–397.
Mohan, R.S.A., Goswami, B.N., A common spatial mode for intra-seasonal and inter-annual variation and predictability of the Indian summer monsoon, Curr. Sci. 79 (8) (2000) 1106–1111.
Mohan, S.V., Padmalal, D., Maya, K., Baburaj, B., Sea level oscillations, climate change and landform evolution in the western coastal lowlands of Trivandrum block in Peninsular India, Indian Journal of Geo-Marine Sciences 43 (2014) 1145–1151.
Moreno, A., Cacho, I., Canals, M., Grimalt, J.O., Sanchez-Vidal, A., Millennial-scale variability in the productivity signal from the Alboran Sea record, Western Mediterranean Sea, Palaeogeogr. Palaeoclimatol. Palaeoecol. 211 (2004) 205–219.
Moreno, A., Prez, A., Frigola, J., Nieto-Moreno, V., Rodrigo-Gmiz, M., Martrat, B., Gonzlez-Sampriz, P., Morelln, M., Martn-Puertas, C., Corella, J.P., Belmonte, nchel, Sancho, C., Cacho, I., Herrera, G., Canals, M., Grimalt, J.O., Jimnez-Espejo, F., Martnez-Ruiz, F., Vegas-Vilarrbia, T., Valero-Garcs, B.L., The medieval climate anomaly in the Iberian Peninsula reconstructed from marine and lake records, Quat. Sci. Rev. 43 (2012) 16–32.
Moreno, J., Fatela, F., Leorri, E., de la Rosa, J.M., Pereira, I., Arajo, M.F., Freitas, M.C., Corbett, D.R., Medeiros, A., Marsh benthic foraminifera response to estuarine hydrological balance driven by climate variability over the last 2000yr (Minho estuary, NW Portugal), Quat. Res. (United States) 82 (2014) 318–330.
Moy, C.M., Seltze, G.O., Rodbell, D.T., Anderson, D.M., Variability of El Niño/Oscillation activity at millennial timescales during the Holocene epoch, Nature 420 (2002) 162–165.
Mukherjee, B.K., Indira, K., Dani, K.K., Low-latitude volcanic Eruptions and their effects on Sri Lankan rainfall during the north-east monsoon, J. Climatol. 7 (1987) 145–155.
Naidu, P.D., Ganeshram, R., Bollasina, M.A., Panmei, C., Nürnberg, D., Donges, J.F., Coherent response of the Indian monsoon rainfall to Atlantic multi-decadal variability over the last 2000 years, Sci. Rep. 10 (1) (2020) 1–11.
Nair, K.M., Padmalal, D., Kumaran, K.P.N., Sreeja, R., Limaye, R.B., Srinivas, R., Late quaternary evolution of Ashtamudi-Sasthamkotta lake systems of Kerala, south west India, J. Asian Earth Sci. 37 (2010) 361–372.
Nieto-Moreno, V., Martinez-Ruiz, F., Giralt, S., Gallego-Torres, D., Garcia-Orellana, J., Masque, P., Ortega-Huertas, M., Climate imprints during the "Medieval Climate Anomaly" and the "Little Ice Age" in marine records from the Alboran Sea basin, Holocene 23 (2013) 1227–1237.

Obrochta, S.P., Miyahara, H., Yokoyama, Y., Crowley, T.J., A re-examination of evidence for the North Atlantic "1500-year cycle" at Site 609, Quat. Sci. Rev. 55 (2012) 23–33.

Ogwang, B.A., Ongoma, V., Xing, L., Ogou, F.K., Influence of mascarene high and Indian Ocean dipole on East African extreme weather events, Geogr. Pannon. 19 (2015) 64–72.

Olsen, J., Anderson, N.J., Knudsen, M.F., Variability of the North Atlantic oscillation over the past 5,200 years, Nat. Geosci. 5 (2012) 808–812.

Oppo, D.W., McManus, J.F., Cullen, J.L., Deep water variability in Holocene Epoch, Nature 422 (2003) 277–278.

Oppo, D.W., Rosenthal, Y., Linsley, B.K., 2,000-Year-long temperature and hydrology reconstructions from the Indo-Pacific warm pool, Nature 460 (7259) (2009) 1113–1116.

Overpeck, J., Anderson, D., Trumbore, S., Prell, W., The southwest Indian Monsoon over the last 18 000 years, Clim. Dyn. 12 (1996) 213–225.

Padmalal, D., Kumaran, K.P.N., Nair, K.M., Baijulal, B., Limaye, R.B., Mohan, S.V., Evolution of the coastal wetland systems of SW India during the Holocene: evidence from marine and terrestrial archives of Kollam coast, Kerala, Quat. Int. 237 (2011) 123–139.

Pak, D.K., Kennett, J.P., A foraminiferal isotopic proxy for upper water mass stratification, J. Foraminiferal Res. 32 (2002) 319–327.

Pandey, S., Scharf, B.W., Mohanti, M., Palynological studies on mangrove ecosystem of the Chilka Lagoon, east coast of India during the last 4165 yrs BP, Quat. Int. 325 (2014) 126–135.

Parker, D.E., Jones, P.D., Folland, C.K., Bevan, A., Interdecadal changes of surface temperature since the late nineteenth century, J. Geophys. Res.: Atmos. (1984-2012) 99 (1994) 14373–14399.

Persoiu, A., Ionita, M., Weiss, H., Atmospheric blocking induced by the strengthened Siberian High led to drying in west Asia during the 4.2 ka BP event—a hypothesis, Clim. Past 15 (2019) 781–793.

Peterson, L.C., Haug, G.H., Hughen, K.A., Rohl, U., Rapid changes in the hydrologic cycle of the tropical Atlantic during the last glacial, Science 290 (2000) 1947–1951.

Phadtare, N.R., Pant, R.K., A century-scale pollen record of vegetation and climate history during the past 3500 years in the Pinder Valley, Kumaon Higher Himalaya, India, J. Geol. Soc. India 68 (2006) 495–506.

Phillips, S.C., Johnson, J.E., Giosan, L., Rose, K., Monsoon-influenced variation in productivity and lithogenic sediment flux since 110ka in the offshore Mahanadi Basin, northern Bay of Bengal, Mar. Petrol. Geol. 58 (2014) 502–525.

Ponton, C., Giosan, L., Eglinton, T.I., Fuller, D.Q., Johnson, J.E., Kumar, P., Collett, T.S., Holocene aridification of India, Geophys. Res. Lett. 39 (2012) 1–6.

Prasad, S., Anoop, A., Riedel, N., Sarkar, S., Menzel, P., Basavaiah, N., Krishnan, R., Fuller, D., Plessen, B., Gaye, B., Röhl, U., Wilkes, H., Sachse, D., Sawant, R., Wiesner, M., Stebich, M., Prolonged monsoon droughts and links to Indo-Pacific warm pool: a Holocene record from Lonar Lake, central India, Earth Planet. Sci. Lett. 391 (2014) 171–182.

Prasad, S., Kusumgar, S., Gupta, S.K., A mid to late Holocene record of palaeoclimatic changes from Nal Sarovar: a palaeodesert margin lake in western India, J. Quat. Sci. 12 (1997) 153–159.

Prasad, V., Phartiyal, B., Sharma, A., Evidence of enhanced winter precipitation and the prevalence of a cool and dry climate during the mid to late Holocene in mainland Gujarat, India, Holocene 17 (2007) 889–896.

Punyu, V.R., Banakar, V.K., Garg, A., Equatorial Indian Ocean productivity during the last 33 kyr and possible linkage to Westerly Jet variability, Mar. Geol. 348 (2014) 44–51.

Qin, D.H., Mayewski, P.A., Kang, S.C., Ren, J.W., Hou, S.G., Yao, T.D., Yang, Q.Z., Jin, Z.F., Mi, D.S., Evidence for recent climate change from ice cores in the central Himalaya, Ann. Glaciol. 31 (2000 31) (2000) 153–158.

Quamar, M.F., Vegetation dynamics in response to climate change from the wetlands of Western Himalaya, India: Holocene Indian summer monsoon variability, Holocene 29 (2019) 345–362.

Rajeevan, M., Prediction of Indian summer monsoon: status, problems and prospects, Curr. Sci. 81 (2001) 1451–1458.

Rajeevan, M., Gadgil, S., Bhate, J., Active and break spells of the Indian summer monsoon, J. Earth Syst. Sci. 119 (2010) 229–247.

Rajeevan, M., Pai, D.S., Thapliyal, V., Spatial and temporal relationships between global land surface air temperature anomalies and Indian summer monsoon rainfall, Meteorol. Atmos. Phys. 66 (1998) 157–171.

Rajeevan, M., Pai, D.S., Thapliyal, V., Predictive relationships between Indian Ocean sea surface temperatures and Indian summer monsoon rainfall, Mausam 53 (2002) 337–348.

Rajeevan, M., Unnikrishnan, C.K., Bhate, J., Niranjan Kumar, K., Sreekala, P.P., Northeast monsoon over India: variability and prediction, Meteorol. Applic. 19 (2012) 226–236.

Rajmanickam, V., Achyuthan, H., Eastoe, C., Farooqui, A., Early-Holocene to present palaeoenvironmental shifts and short climate events from the tropical wetland and lake sediments, Kukkal Lake, Southern India: geochemistry and palynology, Holocene 27 (2017) 404–417.

Ramesh, R., Tiwari, M., Chakraborty, S., Managave, S.R., Yadava, M.G., Sinha, D.K., Retrieval of south Asian monsoon variation during the holocene from natural climate archives, Curr. Sci. 99 (2010) 1770–1786.

Rashid, H., England, E., Thompson, L., Polyak, L., Late glacial to Holocene Indian summer monsoon variability based upon sediment, Terres. Atmos. Ocean Sci. 22 (2011) 215–228.

Rashid, H., Flower, B.P., Poore, R.Z., Quinn, T.M., A ~25 ka Indian Ocean monsoon variability record from the Andaman Sea, Quat. Sci. Rev. 26 (2007) 2586–2597.

Rasmusson, E.M., Carpenter, T.H., The relationship between eastern equatorial Pacific sea surface temperatures and rainfall over India and Sri Lanka, Monthly Weather Rev. 111 (1983) 517–528.

Rasmussen, S.O., Vinther, B.M., Clausen, H.B., Andersen, K.K., Early Holocene climate oscillations recorded in three Greenland ice cores, Quat. Sci. Rev. 26 (2007) 1907–1914.

Rawat, S., Gupta, A.K., Srivastava, P., Sangode, S.J., Nainwal, H.C., A 13,000 year record of environmental magnetic variations in the lake and peat deposits from the Chandra valley , Lahaul: implications to Holocene monsoonal variability in the NW Himalaya, Palaeogeogr. Palaeoclimatol. Palaeoecol. 440 (2015a) 116–127.

Rawat, S., Gupta, A.K., Sangode, S.J., Srivastava, P., Nainwal, H.C., Late Pleistocene-Holocene vegetation and Indian summer monsoon record from the Lahaul, Northwest Himalaya, India, Quat. Sci. Rev. 114 (2015b) 167–181.

Rein, B., Lückge, A., Sirocko, F., A major Holocene ENSO anomaly during the Medieval period, Geophys. Res. Lett. 31 (2004) L17211.

Rohling, E., Mayewski, P., Abu-Zied, R., Casford, J., Hayes, A., Holocene atmosphere-ocean interactions: records from Greenland and the Aegean Sea, Clim. Dyn. 18 (2002) 587–593.

Rohling, E.J., Fenton, M., Jorissen, F.J., Bertrand, P., Ganssen, G., Caulet, J.P., Magnitudes of sea-level lowstands of the past 500,000 years, Nature 394 (1998) 162–165.

Rozanski, K., Araguás-Araguás, L., Gonfiantini, R., Isotopic Patterns in Modern Global Precipitation, in: Swart, P.K., Lohmann, K.C., Savin, J.S. (Eds.), Climate Change in Continental Isotopic Records, Geophysical Monograph Series, American Geophysical Union, Washington, D.C., United States, 1993, pp. 1–36.

Rühland, K., Phadtare, N.R., Pant, R.K., Sangode, S.J., Smol, J.P., Accelerated melting of Himalayan snow and ice triggers pronounced changes in a valley peatland from northern India, Geophys. Res. Lett. 33 (2006) L15709.

Sagar, N., Hetzinger, S., Pfeiffer, M., Ahmad, S.M., Dullo, W.-C., Garbe-Schonberg, D., High-resolution Sr/Ca ratios in a Porites lutea coral from Lakshadweep Archipelago, southeast Arabian Sea: an example froma region experiencing steady rise in the reef temperature, J. Geophys. Res.: Oceans 121 (2016) 252–266.

Saji, N.H., Goswami, B.N., Vinayachandran, P.N., Yamagata, T., A dipole mode in the tropical Indian Ocean, Nature 401 (1999) 360–363.

Sandeep, K., Shankar, R., Kumar, A., Yadava, M.G., Ramesh, R., Jani, R.A., Weijian, Z., Xuefeng, L., A multi-proxy lake sediment record of Indian summer monsoon variability during the Holocene in southern India, Palaeogeogr. Palaeoclimatol. Palaeoecol. 476 (2017) 1–14.

Sanwal, J., Kotlia, B.S., Rajendran, C., Ahmad, S.M., Rajendran, K., Sandiford, M., Climatic variability in Central Indian Himalaya during the last ~1800 years: evidence from a high resolution speleothem record, Quat. Int. 304 (2013) 183–192.

Saraswat, R., Lea, D.W., Nigam, R., Mackensen, A., Naik, D.K., Deglaciation in the tropical Indian Ocean driven by interplay between the regional monsoon and global teleconnections, Earth Planet. Sci. Lett. 375 (2013) 166–175.

Saravanan, P., Gupta, A.K., Zheng, H., Majumder, J., Panigrahi, M.K., Kharya, A., A 23000 year old record of paleoclimatic and environmental changes from the eastern Arabian Sea, Mar. Micropaleontol. 160 (2020) 101905.

Sarkar, S., Prasad, S., Wilkes, H., Riedel, N., Stebich, M., Basavaiah, N., Sachse, D., Monsoon source shifts during the drying mid-Holocene: Biomarker isotope based evidence from the core 'monsoon zone' (CMZ) of India, Quat. Sci Rev. 123 (2015) 144–157.

Schilman, Y.B., Bar-Matthews, M., Almogi-Labin, A., Schilman, B., Bar-Matthews, M., Almogi-Labin, A., Luz, B..

Schulz, M., Paul, A., Holocene Climate Variability on Centennial-to-Millennial Time Scales: 1. Climate Records from the North-Atlantic Realm, in: Wefer, G., Berger, W., Behre, K.E., Jansen, E. (Eds.), Climate Development and History of the North Atlantic Realm, Springer-Verlag, Berlin Heidelberg, 2002, pp. 41–54.

Schulz, H., von Rad, U., Erlenkeuser, H., Correlation between Arabian Sea and Greenland climate oscillations of the past 110,000 years, Nature 393 (1998) 54–57.

Sebastian, T., Nagender Nath, B., Venkateshwarlu, M., Miriyala, P., Prakash, A., Linsy, P., Kocherla, M., Kazip, A., Sijinkumar, A.V., Impact of the Indian Summer Monsoon variability on the source area weathering in the Indo-Burman ranges during the last 21 kyr – a sediment record from the Andaman Sea, Palaeogeogr. Palaeoclimatol. Palaeoecol. 516 (2019) 22–34.

Seong, Y.B., Owen, L.A., Yi, C., Finkel, R.C., Quaternary glaciation of Muztag Ata and Kongur Shan: evidence for glacier response to rapid climate changes throughout the late glacial and holocene in westernmost Tibet, Bull. Geol. Soc. Am. 121 (2009) 348–365.

Shah, S.K., Bhattacharyya, A., Chaudhary, V., Reconstruction of June-September precipitation based on tree-ring data of teak (*Tectona grandis* L.) from Hoshangabad, Madhya Pradesh, India, Dendrochronologia 25 (2007) 57–64.

Shah, S.K., Mehrotra, N., Tree–ring studies of Toona ciliata from subtropical wet hill forests of Kalimpong, eastern Himalaya, Dendrochronologia 46 (2017) 46–55.

Shankar, D., Vinayachandran, P.N., Unnikrishnan, A.S., The monsoon currents in the north Indian Ocean, Prog. Oceanogr. 52 (2002) 63–120.

Sharma, S., Joachimski, M., Sharma, M., Tobschall, H.J., Singh, I.B., Sharma, C., Chauhan, M.S., Morgenroth, G., Lateglacial and Holocene environmental changes in Ganga plain, Northern India, Quat. Sci. Rev. 23 (2004) 145–159.

Shichang, K., Wake, C.P., Dahe, Q., Mayewski, P.A., Tandong, Y., Monsoon and dust signals recorded in Dasuopu glacier, Tibetan Plateau, J. Glaciol. 46 (2000) 222–226.

Sikka, D.R., Gadgil, S., On the maximum cloud zone and the ITCZ over Indian, longitudes during the southwest monsoon, Monthly Weather Rev. 108 (1988) 1840–1853.

Singh, J., Yadav, R.R., Spring precipitation variations over the western Himalaya, India, since A.D. 1731 as deduced from tree rings, J. Geophys. Res. D: Atmos. 110 (2005) 1–8.

Singh, J., Yadav, R.R., Wilmking, M., A 694-year tree-ring based rainfall reconstruction from Himachal Pradesh, India, Clim. Dyn. 33 (2009) 1149–1158.

Singhvi, A.K., Kale, V.S., Paleoclimate Studies in India: Last Ice Age to the Present, Indian National Science Academy, New Delhi, India, 2010.

Singhvi, A.K., Rupakumar, K., Thamban, M., Gupta, A.K., Kale, V.S., Yadav, R.R., Bhattacharyya, A., Phadtare, N.R., Roy, P.D., Chauhan, M.S., Chauhan, O., Instrumental, Terrestrial and Marine Records of the Climate of South Asia during the Holocene: Present Status, Unresolved Problems and Societal Aspects, Global Environmental Changes in South Asia, Springer, Dordrecht, 2010.

Sinha, A., Berkelhammer, M., Stott, L., Mudelsee, M., Cheng, H., Biswas, J., The leading mode of Indian Summer Monsoon precipitation variability during the last millennium, Geophys. Res. Lett. 38 (2011) L15703.

Sinha, A., Cannariato, K.G., Stott, L.D., Cheng, H., Edwards, R.L., Yadava, M.G., Ramesh, R., Singh, I.B., A 900-year (600 to 1500 A.D.) record of the Indian summer monsoon precipitation from the core monsoon zone of India, Geophys. Res. Lett. 34 (2007) 1–5.

Sinha, N., Gandhi, N., Chakraborty, S., Krishnan, R., Yadava, M.G., Ramesh, R., Abrupt climate change at ~2800 yr BP evidenced by a stalagmite record from peninsular India, Holocene 28 (2018) 1720–1730.

Sinha, R., Smykatz-Kloss, W., Stüben, D., Harrison, S.P., Berner, Z., Kramar, U., Doris, S.Ben, Harrison, S.P., Berner, Z., Kramar, U., Late Quaternary palaeoclimatic reconstruction from the lacustrine sediments of the Sambhar playa core, Thar Desert margin, India, Palaeogeogr. Palaeoclimatol. Palaeoecol. 233 (2006) 252–270.

Sirocko, F., Garbe-Schonberg, D., Mcintyre, A., Molfino, B., Teleconnections between the subtropical moonsoons and high-latitude climates during the last deglaciation, Sci. 272 (1996) 526–529.

Smith, A.C., Wynn, P.M., Barker, P.A., Leng, M.J., Noble, S.R., Tych, W., North Atlantic forcing of moisture delivery to Europe throughout the Holocene, Sci. Rep. 6 (2016) 1–7.

Solomon, S., Qin, D., Manning, M., Chen, Z., Marquis, M., Averyt, K. B., Miller, H. L., Climate change 2007: Synthesis Report. Contribution of Working Group I, II and III to the Fourth Assessment Report of the Intergovernmental Panel on Climate Change. Summary for Policymakers. *Climate change 2007: Synthesis Report. Contribution of Working Group I, II and III to the Fourth Assessment Report of the Intergovernmental Panel on Climate Change*, Summary for Policymakers.Chicago, 2007.

Sontakke, N.A., Singh, N., Singh, H.N., Instrumental period rainfall series of the Indian region (AD 1813-2005): revised reconstruction, update and analysis, Holocene 18 (2008) 1055–1066.

Sridhar, A., Thakur, B., Basavaiah, N., Seth, P., Tiwari, P., Chamyal, L.S., Lacustrine record of high magnitude flood events and climate variability during mid to late Holocene in the semiarid alluvial plains, western India, Palaeogeogr. Palaeoclimatol. Palaeoecol. 542 (2020) 109581.

Srinivasan, V., Ramamurthy, K., Northeast Monsoon, The Deputy Director General of Observations (Forecasting), Poona, Maharashtra, 1973.

Srivastava, J., Farooqui, A., Late Holocene mangrove dynamics and coastal environmental changes in the North-eastern Cauvery River Delta, India, Quat. Int. 298 (2013) 45–56.

Srivastava, P., Agnihotri, R., Sharma, D., Meena, N., Sundriyal, Y.P.P., Saxena, A., Bhushan, R., Sawlani, R., Banerji, U.S., Sharma, C., Bisht, P., Rana, N., Jayangondaperumal, R., 8000-year monsoonal record from Himalaya revealing reinforcement of tropical and global climate systems since mid-Holocene, Sci. Rep. 7 (2017) 1–11.

Staubwasser, M., Sirocko, F., Grootes, P.M., Erlenkeuser, H., South Asian monsoon climate change and radiocarbon in the Arabian Sea during early and middle Holocene, Paleoceanography 17 (2002) 15-1-15–12.

Staubwasser, M., Weiss, H., Holocene climate and cultural evolution in late prehistoric-early historic West Asia, Quat. Res. 66 (2006) 372–387.

Sukumar, R., Ramesh, R., Pant, R.K., Rajagopalan, G., A $\delta^{13}C$ record of late Quaternary climate change from tropical peats in southern India, Nature 364 (6439) (1993) 703–706.

Suresh, V.R., Mathew, K.J., Skeletal extension of staghorn coral *Acropora formosa* in relation to environment at Kavaratti atoll (Lakshadweep), Indian J. Mar. Sci. 22 (1993) 176–179.

Tian, L., Masson-Delmotte, V., Stievenard, M., Yao, T., Jouzel, J., Tibetan Plateau summer monsoon northward extent revealed by measurements of water stable isotopes, J. Geophys. Res. Atmos. 106 (2001) 28081–28088.

Tierney, J.E., Abram, N.J., Anchukaitis, K.J., Evans, M.N., Giry, C., Kilbourne, K.H., Saenger, C.P., Wu, H.C., Zinke, J., Tropical Corals 400 yrs reconstructed from coral archives, Paleoceanography 30 (2015) 226–252.

Tiwari, M., Nagoji, S.S., Ganeshram, R.S., Multi-centennial scale SST and Indian summer monsoon precipitation variability since the mid-Holocene and its nonlinear response to solar activity, Holocene 25 (2015) 1415–1424.

Tiwari, M., Ramesh, R., Somayajulu, B.L.K., Jull, A.J.T., Burr, G.S., Paleomonsoon precipitation deduced from a sediment core from the equatorial Indian Ocean. Geo-Marine Letters 26 (2006a) 2330.

Tiwari, M., Ramesh, R., Bhushan, R., Simayajulu, B.L.K., Jull, AJ.T., Burr, G.S., Paleoproductivity variations in the equatorial Arabian Sea: implications for East African and Indian Summer Rainfalls and the El Nino Frequency, Radiocarbon 48 (2006b) 17–19.

Trigo, R.M., Osborn, T.J., Corte-real, J.M., The North Atlantic oscillation influence on Europe: climate impacts and associated physical mechanisms, Clim. Res. 20 (1) (2002) 9–17.

Tripathi, S., Basumatary, S.K., Singh, V.K., Bera, S.K., Nautiyal, C.M., Thakur, B., Palaeovegetation and climate oscillation of western Odisha, India: a pollen data-based synthesis for the Mid-Late Holocene, Quat. Int. 325 (2014) 83–92.

Trouet, V., Esper, J., Graham, N.E., Baker, A., Scourse, J.D., Frank, D.C., Medieval Climate Anomaly, Science 78 (2009) 1256–1260.

van Geel, B., Heusser, C.J., Renssen, H., Schuurmans, C.J.E., Climatic change in Chile at around 2700 BP and global evidence for solar forcing: a hypothesis, Holocene 10 (2000) 659–664.

Vicente-Serrano, S.M., López-Moreno, J.I., Drumond, A., Gimeno, L., Nieto, R., Morán-Tejeda, E., Lorenzo-Lacruz, J., Beguería, S., Zabalza, J., Effects of warming processes on droughts and water resources in the NW Iberian Peninsula (1930-2006), Clim. Res. 48 (2011) 203–212.

von Rad, U., Schaaf, M., Michels, K.H., Schulz, H.H., Berger, W.H.W, Sirocko, F., A 5000-yr record of climate change in varved sediments from the oxygen minimum zone off Pakistan, Northeastern Arabian Sea, Quat. Res. 51 (1999) 39–53.

Walker, M., Gibbard, P., Head, M.J., Berkelhammer, M., Björck, S., Cheng, H., Cwynar, L.C., Fisher, D., Gkinis, V., Long, A., Lowe, J., Newnham, R., Rasmussen, S.O., Weiss, H., Formal subdivision of the Holocene series/epoch: a summary, J. Geol. Soc. India 93 (2019) 135–141.

Walker, M., Head, M.J., Berkelhammer, M., Björck, S., Cheng, H., Cwyna, L., Fisher, D., Gkinis, V., Long, A., Lowe, J., Newnham, R., Rasmussen, S.O., Weiss, H., Formal ratification of the subdivision of the Holocene series/epoch (Quaternary System/Period): two new Global Boundary Stratotype Sections and Points (GSSPs) and three new stages/subseries, Episodes 41 (2018) 213–223.

Wanner, H., Bütikofer, J., Holocene bond cycles: real or imaginary? Geograf. Sbornik 113 (2008) 338–350.

Wanner, H., Beer, J., Bütikofer, J., Crowley, T.J., Cubasch, U., Flückiger, J., Goosse, H., Grosjean, M., Joos, F., Kaplan, J.O., Küttel, M., Müller, S.A., Prentice, I.C., Solomina, O., Stocker, T.F., Tarasov, P., Wagner, M., Widmann, M., Mid to Late Holocene climate change: an overview, Quat. Sci. Rev. 27 (2008a) 1791–1828.

Wanner, H., Mercolli, L., Grosjean, M., Ritz, S.P., Holocene climate variability and change; a data-based review, J. Geol. Soc. 172 (2) (2015) 254–263.

Wanner, H., Solomina, O., Grosjean, M., Ritz, S.P., Jetel, M., Structure and origin of Holocene cold events, Quat. Sci. Rev. 30 (2011) 3109–3123.

Warrier, A.K., Shankar, R., Geochemical evidence for the use of magnetic susceptibility as a paleorainfall proxy in the tropics, Chem. Geol. 265 (2009) 553–562.

Wasson, R.J., Smith, G.I., Agrawal, D.P., Late quaternary sediments, minerals, and inferred geochemical history of Didwana Lake, Thar Desert, India, Palaeogeogr. Palaeoclimatol. Palaeoecol. 46 (1984) 345–372.

Webster, P.J., Magaña, V.O., Palmer, T.N., Shukla, J., Tomas, R.A., Yanai, M., Yasunari, T., Monsoons: processes, predictability, and the prospects for prediction, J. Geophys. Res.: Oceans 103 (1998) 14451–14510.

Weiss, H., Bradley, R.S., What drives societal collapse? Science 291 (5504) (2001) 609–610.

Wundsch, M., Biagioni, S., Behling, H., Reinwarth, B., Franz, S., Bierbass, P., Daut, G., Mausbacher, R., Haberzettl, T., ENSO and monsoon variability during the past 1.5 kyr as reflected in sediments from Lake Kalimpaa, Central Sulawesi (Indonesia), Holocene 24 (2014) 1743–1756.

Yadava, M.G., Ramesh, R., Monsoon reconstruction from radiocarbon dated tropical Indian speleothems, Holocene 15 (2005) 48–59.

Yadava, M.G., Ramesh, R., Stable oxygen and carbon isotope variations as monsoon proxies: a comparative study of speleothems from four different locations in India, J. Geol. Soc. India 68 (2006) 461–475.

Yadava, M.G., Ramesh, R., Pandarinath, K., A positive "amount effect" in the Sahayadri (Western Ghats) rainfall, Curr. Sci. 93 (2007) 560–564.

Yang, B., Braeuning, A., Yao, T., Davis, M.E., Correlation between the oxygen isotope record from Dasuopu ice core and the Asian Southwest Monsoon during the last millennium, Quat. Sci. Rev. 26 (2007) 1810–1817.

Yao, T., Thompson, L.G., Shi, Y., Qin, D., Jiao, K., Yang, Z., Tian, L., Thompson, E.M., Climate variation since the Last Interglaciation recorded in the Guliya ice core, Sci. China Ser. D: Earth Sci. 40 (1997) 662–668.

Yu, J., Elderfield, H., Hönisch, B., B/Ca in planktonic foraminifera as a proxy for surface seawater pH, Paleoceanography 22 (2) (2007) PA2202.

Zeuner, F.E., The Pleistocene Period: Its Climate, Chronology, and Faunal Successions, Ray Society, London, UK, 1945, p. 322.

Zielhofer, C., Köhler, A., Mischke, S., Benkaddour, A., Mikdad, A., Fletcher, W.J., Western Mediterranean hydro-climatic consequences of Holocene ice-rafted debris (Bond) events, Clim. Past 15 (2019) 463–475.

Zielinski, G.A., Use of paleo-records in determining variability within the volcanism-climate system, Quat. Sci. Rev. 19 (2000) 417–438.

Zielinski, G.A., Mayewski, P.A., Meeker, L.D., Whitlow, S., Twickler, M.S., A 110,000-yr record of explosive volcanism from the GISP2 (Greenland) ice core, Quat. Res. 45 (1996) 109–118.

Zielinski, G.A., Mayewski, P.A., Meeker, L.D., Whitlow, S., Twickler, M.S., Morrison, M., Meese, D.A., Gow, A.J., Alley, R.B., Record of volcanism since 7000 BC from the GISP2 Greenland ice core and implications for the volcano-climate system, Science 264 (1994) 948–952.

Zielinski, G.A., Mershon, G.R., Paleoenvironmental implications of the insoluble microparticle record in the GISP2 (Greenland) ice core during the rapidly changing climate of the Pleistocene-Holocene transition, Geol. Soc. Am. Bull. 109 (1997) 547–559.

Zinke, J., Loveday, B.R., Reason, C.J.C., Dullo, W.C., Kroon, D., Madagascar corals track sea surface temperature variability in the Agulhas current core region over the past 334 years, Sci. Rep. 4 (2014) 1–8.

CHAPTER

13 Indian Summer Monsoon variability on different timescales and deciphering its oscillations from irregularly spaced paleoclimate data using different spectral techniques

Phanindra Reddy A[a,b], Naveen Gandhi[a]

[a] *Indian Institute of Tropical Meteorology, Pune, India.* [b] *Savitribai Phule Pune University, Pune, India*

13.1 Introduction

Monsoon rainfall dynamics and their implications on the global population are serious concerns to climate change scientists and policymakers (Wang et al., 2014). Indian Summer Monsoon (ISM) is one of the major components of the global monsoon and Earth climate system, which carries moisture and heat from tropical oceans to subtropical regions. Indian Summer Monsoon rainfall (ISMR) contributes more than 70% of the annual precipitation over the Indian landmass. Half of the Indian population depends on agriculture, which accounts for around 18% of the gross domestic product (Madhusudhan, 2015). Monsoon's onset, retreat, strength, precipitation, and variability have a major impact on the Indian Economy, particularly on the Agriculture sector (Gadgil and Gadgil, 2006; Madhusudhan, 2015). Societal collapse and fall of kingdoms over the Indian subcontinent documented in the historical records are closely coinciding with the prolonged weaken ISMR over Indian and surrounding regions. A terrestrial climate-proxy record study has elaborated on the Indian monsoon's intriguing features and related cultural changes for the last ~5700 years (Kathayat et al., 2017). Major Indian devastating droughts and decadal length famines coincide with the multidecadal scale weakened ISMR (Sinha et al., 2007). Several climate-proxy studies documented the existence of decadal to multidecadal length scale droughts (e.g., Prasad et al., 2014; Shi et al., 2014; Sinha et al., 2011a; Tan et al., 2018). The declining trends in ISMR for a century to multicentury timescales have induced several prolonged mega-droughts over ISM region (Band et al., 2018; Prasad et al., 2014; Sinha et al., 2011a). This invokes the significance of decadal, multidecadal, and centennial to multicentennial scale ISMR variability.

Earth's climatic phenomena such as El Niño Southern Oscillation (ENSO), North Atlantic Oscillation (NAO), Pacific Decadal Oscillation (PDO), Northern Hemispheric Temperature (NHTemp), and natural forcing such as solar activity shows significant impacts on ISMR at different timescales. Understanding the forcing mechanisms controlling ISMR on different timescales is vital to predicting the ISMR variability. Several studies (Kumar et al., 1999, 2006; Krishnamurthy and Goswami, 2000; Lau and Nath, 2000) documented the robust negative relation between ISMR and ENSO on annual to decadal timescales. (Joseph et al., 2013) hypothesized that the interannual to interdecadal ISMR

Holocene Climate Change and Environment. DOI: https://doi.org/10.1016/B978-0-323-90085-0.00025-5

variations resulted from monsoon heat source interactions, deep convection, mid-latitude westerlies, low-level Jetstream and Asia Pacific Wave. In another study, the interdecadal ISMR variability is attributed to air–sea interactions in the eastern equatorial Indian Ocean and the Indian subcontinent via equatorial Walker and regional monsoon Hadley circulations (Krishnamurthy and Goswami, 2000). A terrestrial proxy study from the central Indian cave indicated a possible linkage between multidecadal power in ISMR and multidecadal variability in the North Atlantic Sea Surface Temperature (NAsst; Berkelhammer et al., 2010). However, Sankar et al. (2016) shown the nonstationary relationship between ISMR and Atlantic Multi-decadal Variability on multidecadal timescales for the last 500 years. Instrumental records show a weaker connection between Atlantic Multi-decadal Oscillation (AMO) and ISMR after mid-1990s (Luo et al., 2018b). Over the past 100 years, the model runs suggest the AMO-ISMR connection is intrinsic to the climate system and North Pacific is the key bridging player in such connection (Luo et al., 2018a). An intensive study using proxy data suggests ISMR multidecadal mode of variability is likely to be an integral part of global multidecadal variability and might arise from ocean–atmospheric–land interactions (Goswami et al., 2015). Another study from central India suggested that ISM prolonged (decadal to centennial-scale) droughts are a manifestation of Indo-Pacific Warm Pool through changing meridional overturning circulation and position of the anomalous Walker cell (Prasad et al., 2014). Similarly, proxy-based reconstruction (Shi et al., 2017) has shown ISMR exhibited interannual to centennial-scale variations and are connected to ENSO on interannual, PDO on decadal, AMO on multidecadal, total solar irradiance (TSI) on centennial timescales. Simultaneously, the solar influence on ISMR variability through dynamical processes through the atmosphere as a bridge cannot be neglected (Kodera, 2004; Hiremath et al., 2015). Another rigorous statistical analysis study (Malik and Brönnimann, 2017) concludes that TSI influences the monsoon on multidecadal timescales but mainly operated through AMO. On interannual to decadal (Hiremath et al., 2015), centennial (Shi et al., 2017; Gupta et al., 2013; Sinha et al., 2018) timescales ISMR variability is linked to variations in solar activity. A tree-ring reconstruction of the South Asian Summer Monsoon Index (Shi et al., 2014) suggests the solar influence on multidecadal periods by affecting the land-ocean thermal contrast. Sinha et al. (2015) suggested multicentennial scale ISMR variations are closely related to NHTemp changes and could be the important driver on these timescales.

Most of the earlier studies mainly focused on a single location to location basis datasets to study the relation between ISMR and climate phenomena/natural forcing (CPNF). This study reviewed the climate forcing agents controlling ISMR variability on decadal to multicentennial timescales, using multiarchive records, including terrestrial cave records, ocean sediment records, and several CPNF datasets. Widely used spectral techniques for the irregularly spaced paleoclimate datasets are used to determine the common mode of variability in ISMR and CPNF. We also reviewed these spectral techniques for analyzing the irregularly spaced datasets and documented their merits and demerits.

13.2 Scope

High-resolution long-term terrestrial archives are important in understanding the ISMR variability on different temporal and spatial scales. Among the terrestrial archives, the speleothem preserves the rainfall variability on very high temporal resolution and a longer period. Many studies have shown that the oxygen isotopic variations of speleothems mimic the ISM variability (Sinha et al., 2005, 2007, 2015; Yadava and Ramesh, 2005; Berkelhammer et al., 2010; Kathayat et al., 2016; Joshi et al., 2017). Besides, oxygen isotopic variations of speleothem explained the paleoclimate excursions and have shown strong coherence with other proxy records (Rashid et al., 2011; Bird et al., 2014; Evans et al., 2015). Here, we have chosen oxygen isotopic variations of stalagmites as a direct proxy of ISMR to exploit the monsoon variability on different timescales. To assess the forcing mechanisms controlling ISMR variability at different timescales, high temporally resolved and long-term data of climate phenomena are necessary. However, this is not the case. Either we have climate phenomena data for a short period with high temporal resolution or a long period with a low temporal resolution that limits our understanding of climate variables' role on ISMR variability.

Mathematical and statistical techniques are required to find periodic components of the time-varying signal. French mathematician Joseph Fourier documented the modern spectral method in 19th century (Fourier, 2009) Later, he extended his theory to fast Fourier transforms and discrete Fourier transforms, which revolutionized modern applied mathematics. Earlier algorithms were developed only for evenly spaced time-varying signals. One widely used spectral technique for evenly spaced time series is the Blackman–Tukey method (Jenkins and Wats, 1968). Unfortunately, irregular observations of natural processes (e.g., Variable star data, Geophysical systems) make it challenging to employ mathematical tools and techniques, which are developed for evenly spaced data. Interpolation is one of the methods to transform irregularly spaced data to evenly spaced data. However, interpolation significantly alters the spectral content (peak position and power) of the time-varying natural signal (Schulz and Stattegger, 1997; DeLong et al., 2009). We have discussed and reviewed commonly used five spectral methods suitable for irregularly spaced data in the following sections. Subsequently, we also discuss the spectral variations in ISMR and possible forcing mechanisms that drives ISMR variability on different timescales.

13.3 Methodology and processes

13.3.1 Spectral methods used for irregularly spaced time-varying natural signals

Several methods have been developed to date to find the periodic components of irregularly spaced data. The fundamental and common approach for spectral analysis is the "linear least square method." First of its kind, to study the periodic behavior of irregularly spaced variable star data (Lomb, 1976) was introduced in the late 20th century. Later (Scargle, 1982, 1989) enhanced Lomb's least square method, and now it is known as Lomb–Scargle Fourier transform (LSFT). Here we have used five different spectral methods to estimate the power in ISMR, viz., 1. SPECTRUM (Schulz and Stattegger, 1997), 2. REDFIT3.8 (Schulz and Mudelsee, 2002), 3. American Association of Variable Star Observers (AAVSO) Time Series (TS) analysis code (Foster, 1995; Templeton, 2004), 4. NESToolbox (Rehfeld et al., 2011; Rehfeld and Kurths, 2014). 5. WAVEPAL (Lenoir and Crucifix, 2018a, 2018b). A comparison of inbuilt features and techniques of the spectral methods is listed in Table 13.1. A brief discussion on these spectral techniques is provided in the following sections.

13.3.1.1 SPECTRUM

Schulz and Stattegger (1997) designed a DOS-supported menu-driven PC program SPECTRUM, based on LSFT with Welch-Overlapped-Segment-Averaging (WOSA; Welch, 1967) procedure that allows

Table 13.1 Comparison of inbuilt features of widely used Spectral methods for paleoclimate datasets.

In built features	SPECTRUM	REDFIT 3.8	TS AAVSO	NESToolbox	WAVEPAL
Polynomial fitting and detrending	Linear detrending	Linear detrending	Polynomial fitting and calculate residuals	—	Polynomial fitting and calculate residuals
Fourier transform method	LSFT	LSFT	DCDFT	LSFT	Extended LS periodogram
Segmentation	WOSA	WOSA	—	—	WOSA
Statistical significance tests	χ^2 test	χ^2 test, MC false alarm, Rednoise from AR(1) process	—	Rednoise from AR(1) process	Analytical confidence levels, MC false alarm

performing spectral analysis on irregularly spaced time series in the frequency domain. The LSFT uses the least square approach on original data by minimizing the sum of the squares of the difference between the Fourier series and the data (please see Eq. (2) and (3) of Schulz and Stattegger, 1997). WOSA procedure is applied to reduce the computational power while performing Fourier analysis, which involves segmenting the dataset into user-defined segments (n_{50}) of length N_{seg} that overlap each other by 50%, hence $N_{seg} = 2N/(n_{50} + 1)$, where '$N$' being the total length of the dataset. Then it performs the Fourier transform to obtain modified periodograms of each segment, and then average them. Over sampling factor (*ofac*) that determines the number of frequencies in LSFT. Similarly, highest frequency (*hifac*) determines the maximum frequency for LSFT (Schulz and Stattegger, 1997). N_{seg}, *ofac*, and *hifac* together decides the number of frequencies "K" (please see Eq. (2) of Schulz and Stattegger, 1997) in spectral analysis of SPECTRUM software, where $K = \frac{ofac*hifac*N_{seg}}{2}$. SPECTRUM can perform various time-series analyses such as harmonic, univariate, and bivariate spectral, frequency, coherency, phase, and cross-spectrum. It also has options for linear trend removal for each WOSA segment, choice of choosing *ofac*, number of WOSA segmentation, *hifac*, and many spectral windowing techniques (e.g., Welch, Hanning, Rectangular, Triangular, and Blackman–Harris) to suppress the spectral leakage. The number of frequencies of spectrum for an irregularly spaced dataset is computed based on $\langle f_{Nyq} \rangle = 1/(2\langle \Delta t \rangle)$, being average Nyquist frequency; where $\langle \Delta t \rangle$ mean of sampling interval of the irregularly spaced dataset. SPECTRUM estimates the spectral peaks and spectral variance (σ^2) of time series. This program does not have feature for computing rednoise spectra. However, it calculates the confidence levels of the spectrum based on the χ^2 distribution (please see Eq (6) of Schulz and Stattegger, 1997). Also, this software limits the length of the dataset up to 3000 data points. Schulz and Stattegger (1997) document a detailed explanation of SPECTRUM.

13.3.1.2 REDFIT3.8

REDFIT3.8 (here after REDFIT) is Windows-based spectral analysis program, which estimates rednoise spectra and spectrum of irregularly spaced time series. This program is an enhanced version of SPECTRUM (Schulz and Stattegger, 1997). Like SPECTRUM, this program can perform cross-spectral (cross-amplitude-, coherency-, and phase-spectrum) and harmonic analyses based on the same spectral estimate procedure, that is, LSFT. It has the same parameter choices, as in SPECTRUM; like *ofac*, N_{seg} and spectral window types, but it automatically removes the linear trend from each WOSA segment before to take the Fourier transforms. REDFIT uses a first-order autoregressive process (henceforth AR(1)) to estimate rednoise spectra for irregular spaced datasets. This algorithm combines the cross-spectral analysis with a Monte-Carlo (MC) approach for estimating uncertainties of spectral parameters. Based on sampling times of original dataset, using an ensemble of N_{sim} MC simulations AR(1) time series is generated with a fixed persistence time ("tau"; Mudelsee, 2002) and average "tau" is calculated from "tau" estimates of each WOSA segment which are bias-corrected (Mudelsee, 2010). Then, the average ensemble spectrum deviation from the known theoretical spectrum is used for bias correction. In this study, we use σ^2/c, as the REDFIT estimated bias corrected spectral variance. Additional features of REDFIT than SPECTRUM provide theoretical rednoise spectra and MC confidence levels for the yielded spectral variance of irregularly spaced datasets. Unlike the SPECTRUM, REDFIT uses the available amount of machine memory, limiting the length of the dataset. Schulz and Mudelsee (2002) provide a detailed description of REDFIT.

13.3.1.3 TS AAVSO

Foster (1995) has introduced a method for removing the false peaks from a power spectrum using *CLEANest* algorithm. It is shown to be a useful technique for detecting and describing multiperiodic signals. Later, AAVSO incorporated this algorithm into a Fourier analysis code TS, publicly available from the website (https://www.aavso.org/software-directory). *CLEANest* algorithm removes the user-defined unwanted frequencies (false peaks) from the data one at a time by performing n number of iterations. The algorithm needs the CLEANest spectrum, which forms by summing up the "discrete" and "residual" spectra (Foster, 1995). Discrete spectrum is obtained by "model" the observed data with more than one trail frequency simultaneously with the help of "Date-Compensated Discrete Fourier Transform" (DCDFT; Ferraz-Mello, 1981). For a set of n trail frequencies, data projected onto $2n + 1$ dimensional subspace to generate nth-order "model function."

Residual spectrum is obtained by subtracting the "model function" from the original data and Fourier transform of the residuals are obtained using DCDFT. The first iteration of CLEANest algorithm finds the strongest single peak, subtracted first to produce the CLEANest(1) spectrum. The spectrum is then scanned to verify that this most substantial remaining peak is statistically significant. Then the original data are analyzed to find the pair of frequencies that best fits the model data. This "model function" is subtracted, and these residuals are Fourier analyzed to produce the CLEANest (2) spectrum. This procedure (CLEANest (3), (4), etc., spectra) continues until the model function includes all the statistically significant frequencies. This process is called the sequential CLEANest algorithm. Eventually, a higher-order CLEANest spectrum should provide better frequency estimates of the observed data. Other advantageous features installed in TS are the choosing desired polynomial fit, period (frequency) range, resolution, CLEANest (deleting false frequencies) and SLICK spectrum algorithms to determine the range of true frequencies in irregularly spaced datasets. The number of frequencies TS can resolve ranges from 0 to 1/4T, "T" is the total time period of the original dataset. TS program does not have inbuilt options to get the confidence limits of the spectral power.

13.3.1.4 NESToolbox

Gaussian-kernel-based autocorrelation (Rehfeld et al., 2011) is a widely used statistical technique for estimating autocorrelation and cross-correlation of the irregularly spaced datasets. To estimate autocorrelation function (ACF), the original irregularly spaced time series is resampled on a time grid and linearly interpolated with the mean sampling time interval (Δt) of the dataset. After resampling, the absolute squared LSFT is employed to find the Lomb–Scargle (LS) periodogram. The minimum and maximum frequencies are adopted as recommended values from Scargle (1989); these frequencies depend on the original time-series' minimum and maximum observation times. Prior to linear interpolation, while slotting for correlation analysis Gaussian weight (h; width parameter) for observation times is considered. The Fourier transform is applied to ACF, and this statistical method has been incorporated in NESToolbox (Rehfeld et al., 2011; Rehfeld and Kurths, 2014). This method uses a AR(1) process to surrogate rednoise time series for a given irregular spaced datasets, and confidence levels can be estimated. Detailed methodology of this statistical technique has been documented in Rehfeld et al. (2011) and Rehfeld and Kurths (2014) and the toolbox are available at the *url* (https://tocsy.pik-potsdam.de/nest.php).

13.3.1.5 WAVEPAL

Recently a new Python-based toolbox, WAVEPAL has been developed for frequency analysis of irregularly spaced time series (Lenoir and Crucifix, 2018a). This method is based on the LS periodogram, which extended to consider the polynomial trend, periodic component, and background noise. The irregularly spaced time series is subsampled on to regular time grid using the greatest common divisor of all time intervals. The periodic component may oscillate around a more complex trend in the climate and paleoclimate signals. An operator (Ferraz-Mello, 1981) is used to model the signals as a combination of the periodic component plus noise. This operator is designed to account for the correlation between the trend and the periodic component of the signal. However, an extended LS periodogram is not sufficient enough to handle the very noisy datasets and spectral leakage, so, tapered WOSA method is employed to deal with the datasets. Prior to averaging the WOSA segmented periodograms, unlike SPECTRUM or REDFIT, this method gives weight to each periodogram of WOSA segment to get a reliable representation of the squared amplitude of the periodic component. Besides, the WOSA technique reduces the periodogram variance and analytical confidence levels are estimated from the continuous autoregressive-moving-average process. An approximate proportionality is derived from WOSA periodogram. The squared amplitude of the periodic component is estimated from least-squares methods to extend the new weighted WOSA periodogram. It has the option of ten window techniques to control the spectral leakage. The frequencies and resolution are based on the WOSA segmentation (please see Eqs. (71)–(74) Lenoir and Crucifix, 2018a).

13.3.2 Spectral techniques, their stability, bias correction, and peak finding

At first, we tested these spectral techniques' stability and efficacy by changing various external and internal parameters. We randomly removed some data points from the evenly spaced dataset Dandak, using *Python* random number generator. Regular and generated irregular time series were then tested with the LS Fourier spectral technique of SPECTRUM and REDFIT software. We also tested the SPECTRUM software for its different *ofac*, N_{seg} and we noted down the individual results and discussed them in consequent sections. To review these spectral methods, we have used ISM region speleothems data. Speleothem datasets have been detrended before employing any spectral technique. While using SPECTRUM, REDFIT, and WAVEPAL algorithms, Hanning window techniques, with 90% and 95%

confidence levels, have been tested. A set of 2000 MC simulations were used to bias correct the LS spectrum and to test the significant levels of spectrum in REDFIT. A theoretical rednoise has been calculated to find the confident peaks in spectral components of analyzed datasets. For testing the stability and comparison of spectral methods we used Dandak data as it is well-studied (Berkelhammer et al., 2010; Sinha et al., 2015) and an equally spaced paleoclimate dataset.

13.3.3 Description of data used

This study has used various proxy data from well-studied and high temporally resolved terrestrial and marine records. Mainly, speleothem records have been used from the Indian subcontinent and Arabian Peninsula region to find the spectral power in the ISMR (Fig. 13.1). Reconstructed ISM datasets from marine sediment cores around the Indian subcontinent are used to complement the terrestrial records. We have also used several reconstructed Earth's climatic phenomena and TSI known to have an influence on the ISMR variability on different temporal scales. Details of the datasets are given in Table 13.2.

13.4 Results

13.4.1 SPECTRUM results

We tested the SPECTRUM program's stability by randomly removing the data points (383 out of 1383) of evenly spaced time series of Dandak record. Both the regular and irregular datasets follow the normal distribution and the time series show no change in the observable amplitude variations. Then we ran the SPECTRUM program for these datasets and found no change in the position of the spectral peaks of regular and irregular datasets with an insignificant change in the spectral variance (Fig. 13.2). The changes in spectral variance are higher (lower) for high periodicities (low periodicities) in regular (irregular) spaced datasets (Fig. 13.2). Since N_{seg}, *ofac*, and *hifac* decide the number of frequencies, we fixed *hifac* value to 1 for rest of the analysis and changed values for *ofac* and N_{seg} accordingly (Fig. 13.3). This experiment shows that the spectral variance varies differently with changing the *ofac* and N_{seg} parameters. Higher N_{seg} gives the lower amplitude of variance and lower spectral resolution (Fig. 13.3). Variations in *ofac* (2–4) and N_{seg} values (3–12), shift the spectral peaks as high as ~135 to ~166 years on higher periodicity range, respectively, besides there are no such shifts in lower periodicities. For example, if we consider constant *ofac*2, N_{seg} (3 and 16), the spectral variance is as high as ~12.25. And if we choose, *ofac* (2 and 4), and N_{seg} 3, there are no significant changes in the spectral variance. Irrespective of *ofac*, N_{seg} values used for WOSA, SPECTRUM program, alters the spectral variance and spectral peak position (Fig. 13.3). Therefore, we chose *ofac* 2 and N_{seg} 3 for better spectral resolution of the datasets. The rednoise (dashed line) is shown up to 90 years period (Fig. 13.2 inset), whereas peaks after 90 years are not above the rednoise (figure not shown).

13.4.2 REDFIT results

To find the different periodicities in Dandak record, we have set the REDFIT parameters, *hifac* to 1, *ofac* to 2, and N_{seg} as 3. The spectral results obtained from REDFIT are bias-corrected spectral variance (σ^2/c). These results are not different from the SPECTRUM results as the spectral technique used for

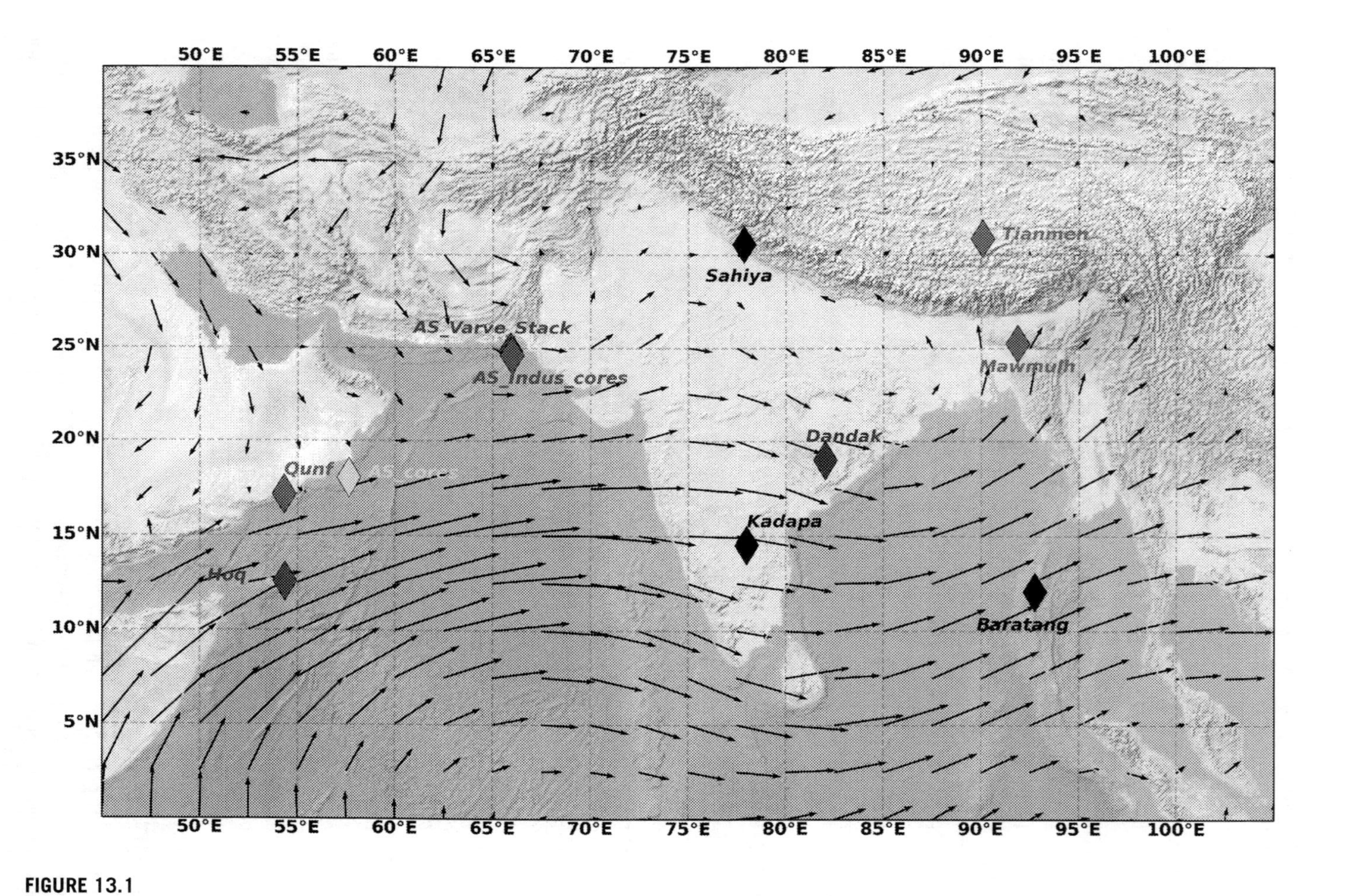

FIGURE 13.1

Indian subcontinent map showing ISM proxy record locations as diamond symbols used for this study. The same color coding of the proxy sites used as spectral analysis of the records in the subsequent figures. The black arrows are June to September climatological winds at 850 hPa for the period of 1979–2020.

Table 13.2 Details of datasets used in the present study.

Record	Site name	Region	Lat (°)	Long (°)	Data availability (period in BP)	Reference
All India rainfall (AIR)	India Meteorological Department (IMD) stations	All India	–	–	–56 to 137	(Sontakke et al., 2008)
DAN-D and JHU-1 composite (henceforth Dandak)	Dandak cave (Jhumar cave)	Central India	19 (18°52′)	82 (81°52′)	–57 to 1325	(Sinha et al., 2011b)
SAH-A and SAH-B composite	Sahiya cave	North India	30°36′	77°52′	–56 to 2090	(Sinha et al., 2015)
NK-1305	Kadapa cave	South India	14.52°	77.99°	172 to 3180	(Sinha et al., 2018)
AN4 and AN8 composite	Baratang cave	Andaman and Nicobar Islands	12°05′	92°45′	0 to 3911	(Laskar et al., 2013)
Q5	Qunf cave	Southern Oman	17°10′	54°18′	2700 to 10,558	(Fleitmann et al., 2003)
STM1	Hoq cave	Socatra Island, Yemen	12°35′	54°21′	1 to 3994	(Rampelbergh et al., 2013)
TM-18	Tianmen cave	Tibetan Plateau	30°55′	90°04′	4147 to 9045	(Cai et al., 2012)
KM-A	Mawmulh cave	North East India	25°15′	91°52′	3653 to 12,395	(Berkelhammer et al., 2012)
Sediment core	Arabian Sea 723 A site hole	Continental Margin, Oman	18°03.079′	57°36.561′	560 to 10,877	(Gupta et al., 2003)
Varve stack count data	SO 90-56KA and SO 90-39KG	North East Arabian Sea	24°50′	65°55′	–42 to 4883	(Rad et al., 1999)
Arabian Sea cores Indus river	63KA, 41KL, 42KG	North East Arabian Sea	24°37′	65°59′	–44 to 12,754	(Staubwasser et al., 2002, 2003)
NAO	Lake SS1220	North western Greenland	67°	–51°	316 to 5196	(Olsen et al., 2012)
SOI	Pacific Ocean	Equatorial Pacific	–	–	–5 to 1900	(Yan et al., 2011)
NHtemp	GSIP2	Greenland	72°36′	–38°30′	–43 to 11,512	(Kobashi et al., 2017)
Solar irradiance	Tree rings and Ice cores	Global	–	–	17 to 9389	(Steinhilber et al., 2012)
PDO	Southern California and Western Canada	North America	34°04′ and 52°	–116°29′ and –116°27′	–46 to 957	(MacDonald and Case, 2005)
NIO SST	SK 157/4	Equatorial Indian Ocean	02°40′	78°	0 to 130,000	(Saraswat et al., 2013)

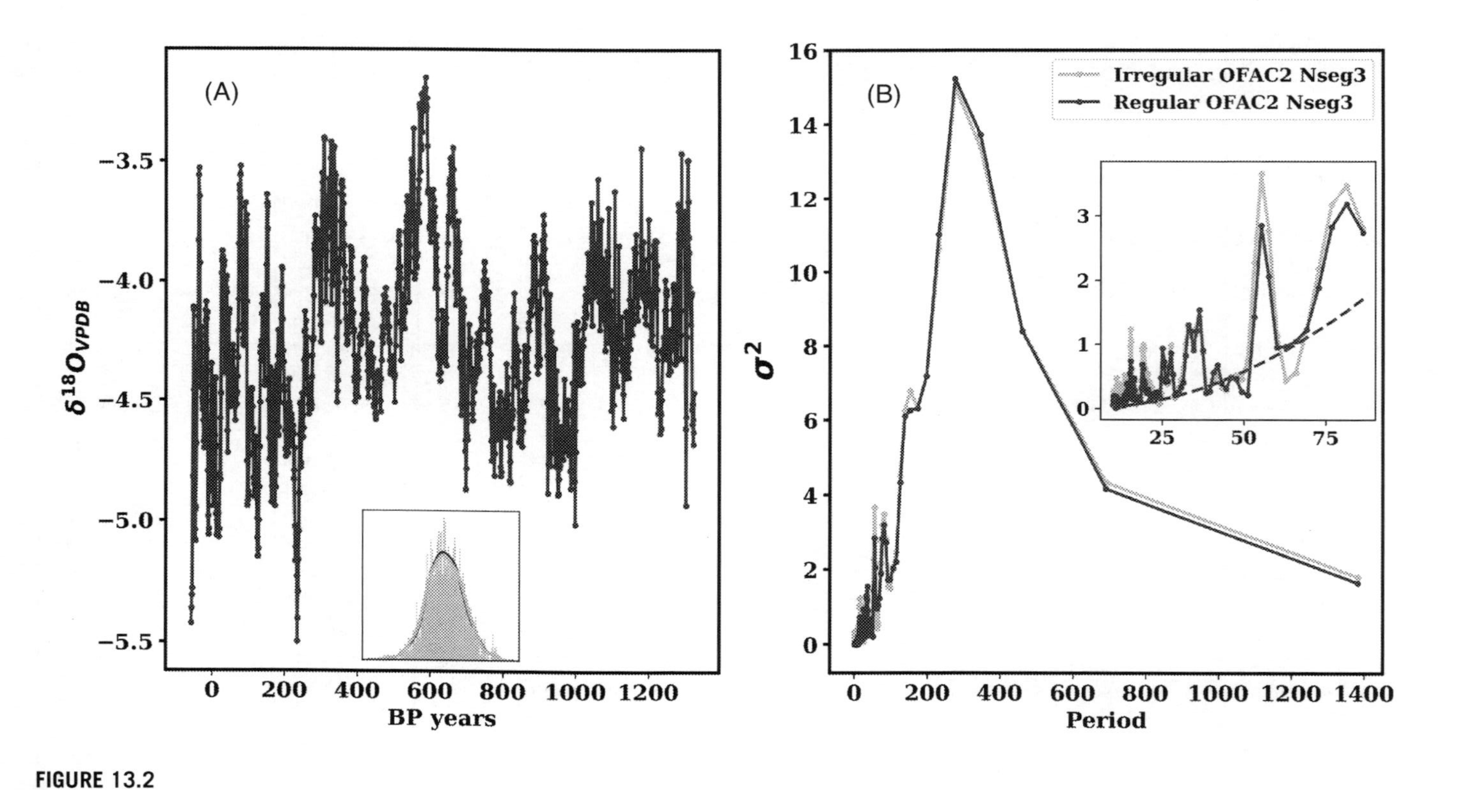

FIGURE 13.2

Time-series analysis of Dandak–Jhumar composite (Sinha et al., 2011b). (A) Time series of Dandak record (−57 to 1325 BP) equally spaced (green) and irregularly (383 out of 1383 data points are removed) spaced (dark green) with their respective frequency distributions (inset figure). (B) Regularly and irregularly spaced Dandak data spectral analysis using SPECTRUM software for given parameters (as *ofac* 2 and N_{seg} 3) and zoomed frequency range (inset figure), the dashed lines are theoretically calculated rednoise. The y-label σ^2 is spectral variance of power spectrum.

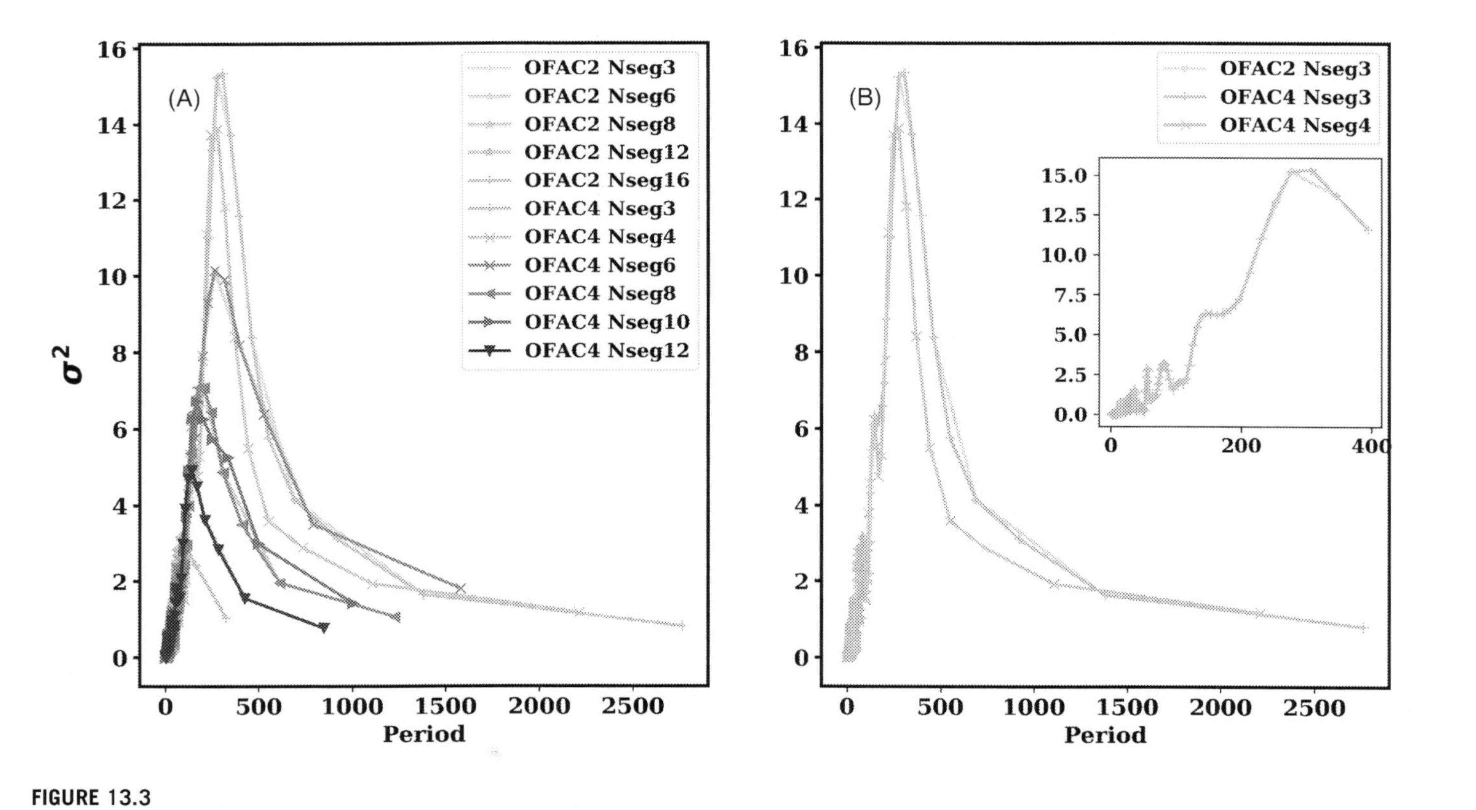

FIGURE 13.3

Spectral analysis of Dandak record using SPECTRUM software. (A) Spectral variations of Dandak record for a variety of *ofac* and N_{seg} combinations. (B) Spectral variations of Dandak record for widely used *ofac* and N_{seg} combinations and zoomed frequency range (inset figure). The y-label σ^2 is spectral variance of power spectrum.

REDFIT is the same as SPECTRUM (Schulz and Mudelsee, 2002). The spectral peak positions did not show any change, and the spectral variance shows no considerable variation (Fig. 13.4). However, spectral variance obtained for regularly spaced time series is lower than that obtained for generated irregularly spaced time series of Dandak record (figure not shown). The choice of inbuilt functions like theoretical rednoise spectra, MC simulations, and significant levels in REDFIT have improved the level of understanding for obtained spectral variance.

13.4.3 TS AAVSO results

The TS program's spectral results for Dandak record data are shown in Fig. 13.4. Before applying the TS spectral analysis technique on Dandak data, a linear trend has been removed, and residual data have been used to find the spectral power, with 1-year spectral resolution and period range 1–1383 years. The TS program derived the frequencies similar to the rest of the programs; however, TS results show additional frequencies in the low-frequency range (Fig. 13.4C, M, and R). In contrast, all the spectral techniques that use the WOSA segmentation couldn't resolve additional frequencies on the lower frequency range since the frequency resolution is dependent on WOSA segmentation. For example, NESToolbox does not use any segmentation technique while calculating the power spectrum, showing additional frequency on the lower frequency range (Fig. 13.4N). Besides, the frequency position shifts in the lower frequency range in TS results (third and last rows Fig. 13.4), the same as in the case for NESToolbox. We suspect these differences in peak width and position on lower frequency range would have occurred due to mixing of multiple frequencies (Kovacs, 1981), WOSA segmentation and windowing techniques. The spectral peaks derived from TS are above the calculated theoretical rednoise (not shown).

13.4.4 NESToolbox results

Python-based NESToolbox software is used to calculate the spectral peak in the irregularly spaced datasets. Before doing spectral analysis, we detrended (linear) the data and calculated the ACF of the respective dataset using the default parameters such as Gaussian Kernel estimator (gXCF) with 0.25 Kernel size, *ofac* 2, and lag vector as the total length of the dataset. The ACF values have been used to find the power spectral density by taking the Fourier transform of ACF. These absolute values of Fourier transform of given ACF are equivalent to power spectral density of the dataset from which the ACF is obtained by the Wiener–Khinchin theorem (Rehfeld et al., 2011; Rehfeld and Kurths, 2014). The computed frequencies for ACF depend on the lag vector and are obtained by minimizing the nearest power of 2 for the given lag vector, limiting the spectral frequencies of the dataset (Fig. 13.4).

13.4.5 WAVEPAL results

As mentioned earlier, WAVEPAL is a Python-based package to find the spectral peaks of irregularly spaced time series without interpolating the data. Before doing spectral analysis, we removed the first-order polynomial trend and calculated the trend vectors. Then we have used the following parameters to find the weighted WOSA periodogram, viz., WOSA segments as 3, minimal coverage of 50% for each WOSA segment, temporal length of WOSA segment (D) as half of the total period of the

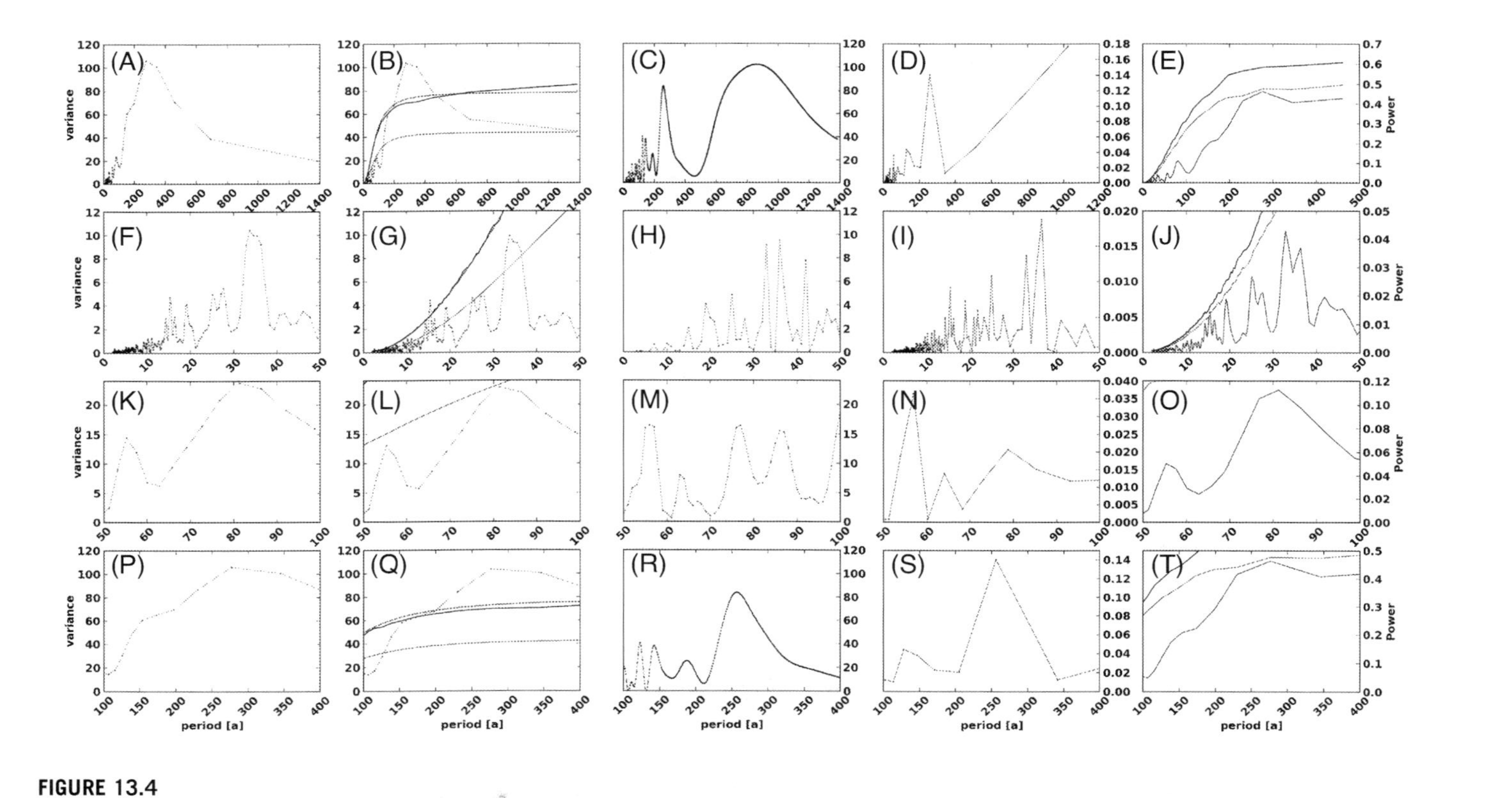

FIGURE 13.4

Comparison of spectral analyses of normalized Dandak record (Sinha et al., 2011b) using various spectral analysis software. Left to right columns show the results from SPECTRUM, REDFIT, TS AAVSO, NESToolbox, and WAVEPAL. Top to bottom rows represent (1) total spectral content, (2) periodicities within 0–50 years, (3) periodicities within 50–100 years, (4) priorities within 100–400 years, respectively. Red, blue, and black lines in the REDFIT results (second column) represent theoretical rednoise, χ^2 test and MC false alarm 90% confidence levels. Gray and black lines in WAVEPAL results (fifth column) are 90% and 95% MC confidence levels, respectively. The y-label in first two columns is spectral variance and bias-corrected spectral variance of power spectrum from SPECTRUM and REDFIT, respectively.

dataset, minimum frequency as 1/(total period of the dataset), maximum frequency as 1/(maximum temporal resolution of the dataset), and Hanning window. Confidence levels (90% and 95%) with 2000 MC simulations, minimum autocorrelation of 0.2 are obtained by continuous autoregressive-moving-average ($p = 1$, $q = 0$) parameters. The frequencies are limited and depend on the temporal length of WOSA segment. None of the WAVEPAL spectral peaks' obtained power is passing 95% MC confidence limits; however, ~16 year periodicity is passing 90% MC confidence levels. We notice that sharp and finite periodicities are observed in low-period ranges, similar to TS and NESToolbox. This similarity occurred as the TS and WAVEPAL use a similar operator, which considers the signal as a periodic component plus a constant trend and noise (see Section 13.1; Ferraz-Mello, 1981; Foster, 1995; Lenoir and Crucifix, 2018a). Whereas, differences in the peak width and position at higher periodic range with respect to TS and NESToolbox, would have occurred due to WOSA segmentation, which is also the case in SPECTRUM and REDFIT.

13.4.6 Periodicities in ISMR

Composite speleothem record from central India (Sinha et al., 2011b) has been used to find the monsoon spectral variability on different timescales using five spectral methods (Fig. 13.4). We used common parameters to compare all five spectral methods' spectral density and data were normalized before applying each technique. For example, we have used the number of WOSA segments as three and Hanning window for SPECTRUM, REDFIT, and WAVEPAL. Although there are variations in the spectral power associated with the spectral peaks generated, a strong coherence in spectral peak position has been found from all selected methods (Fig. 13.4). However, a minor shift in peak position is observed for higher periodicities in TS and NESToolbox (Figs. 13.4M, N, R, and S). Hence it can be stated that ISMR varies on these observed periodicities. The differences in spectral power are attributed to their methodologies, particularly resampling, bias correction, scaling. For example, there is a difference in spectral variance among SPECTRUM and REDFIT results because of the bias correction of spectral variance (first and second columns of Fig. 13.4). However, there is no change in the location of spectral peaks (note the chosen parameters for SPECTRUM and REDFIT are the same). All methods show similar peak positions for low periodicities (0–50 years; Figs. 13.4). Significant differences are observed among the spectral techniques as we move to 50–400 years periods. TS and NESToolbox show similar spectral peak positions, and SPECTRUM, REFIT, and WAVEPAL show similar periodicities. Although the chosen methods show similar spectral peaks, the peak significance needs to be derived separately for TS program. The rest of the programs have inbuilt options for significant levels and/or rednoise spectra for given irregular spaced time series, which has improved understanding for obtained spectral power. Since the REDFIT considers the spectral leakage problem, WOSA segmentation, confidence levels, rednoise from AR(1) process and certain known periods (decadal to multidecadal) of ISMR are passing confidence limits and rednoise, we, therefore, are inclined to use REDFIT program (with $ofac2$, $N_{seg}3$, Welch window) to find spectral power of all the datasets.

Spectral analysis of ISMR from different records shows periodicities on 2–3 years to multicentennial timescales (Figs. 13.5–13.9). Similar to the earlier studies (Munot and Kothawale, 2000; Azad et al., 2010), our analysis from instrumental and cave records show 2–20-year periodicities (Fig. 13.5). High temporal resolution AIR and Kadapa datasets show significant periodicities of 2–3 years in ISMR (Fig. 13.5). A strong statistically significant peak at ~5 years in Kadapa data and a statistically

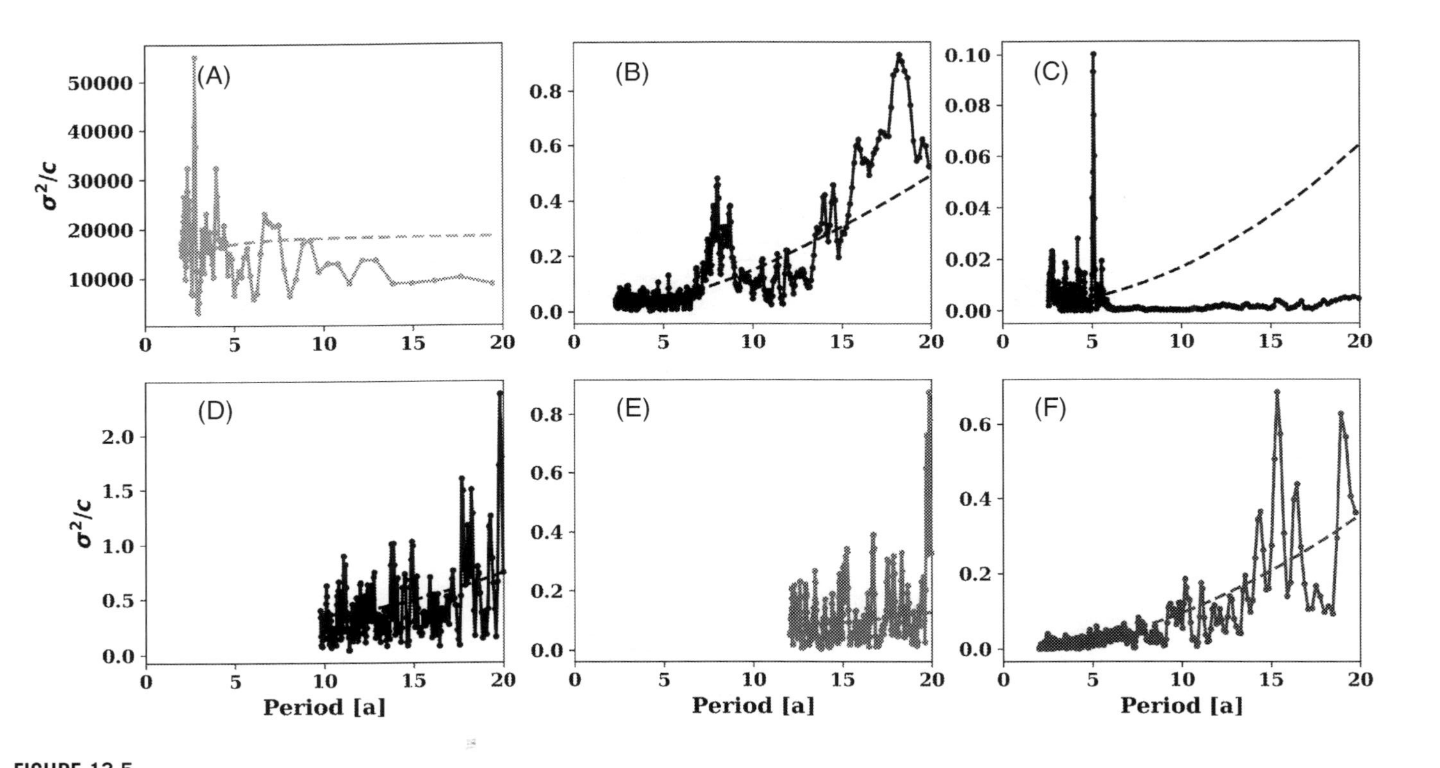

FIGURE 13.5

Comparison of periodicities (2–20 years) in (A) AIR and oxygen isotopic composition obtained from (B) Sahiya, (C) Kadapa, (D) Baratang, (E) Qunf, and (F) Dandak cave records, obtained from REDFIT. Respective dashed lines are rednoise spectra derived from AR(1) process using REDFIT. The y-label σ^2/c is bias corrected spectral variance of power spectrum.

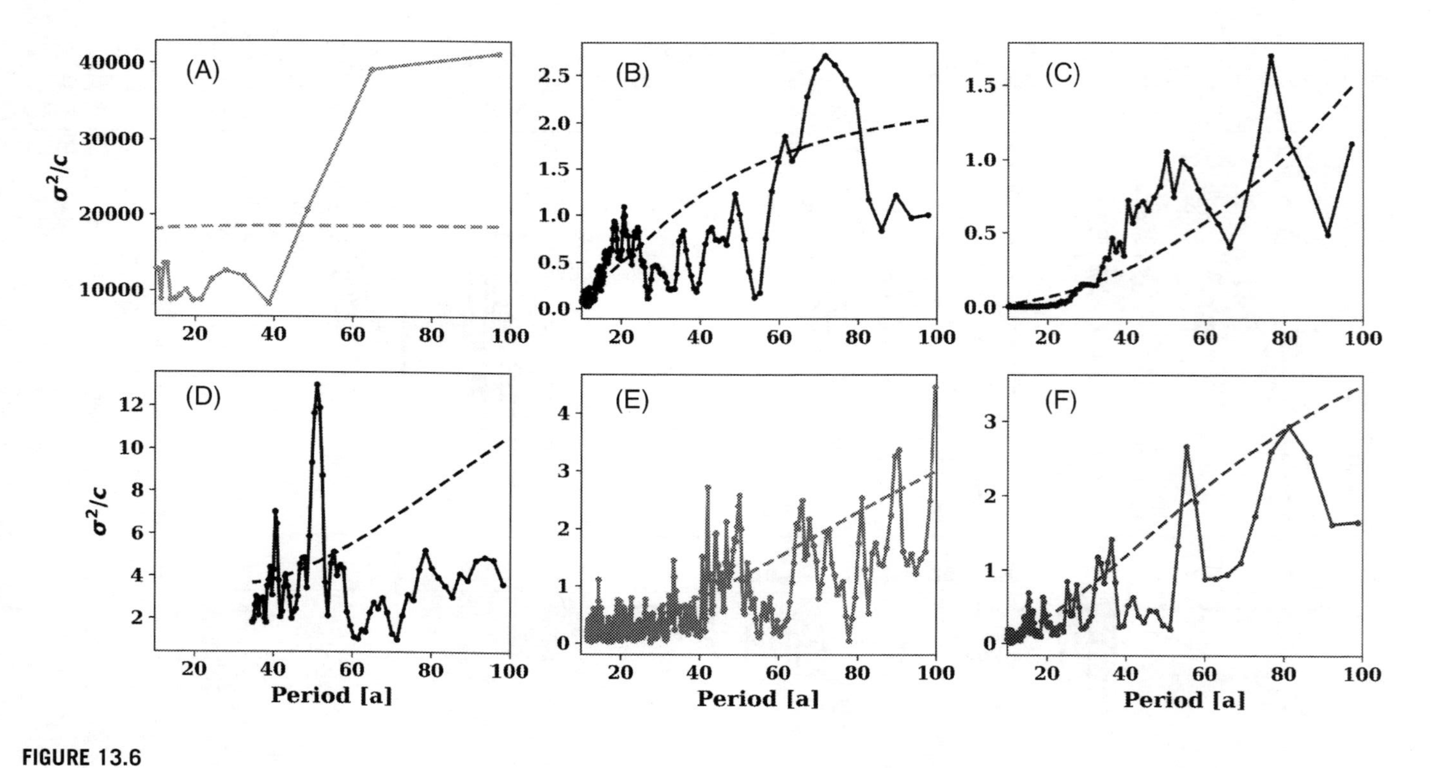

FIGURE 13.6

Comparison of periodicities (10–100 years) in (A) AIR and oxygen isotopic composition obtained from (B) Sahiya, (C) Kadapa, (D) Baratang, (E) Qunf, and (F) Dandak cave records, obtained from REDFIT. Respective dashed lines are rednoise spectra derived from AR(1) process using REDFIT. The y-label σ^2/c is bias corrected spectral variance of power spectrum.

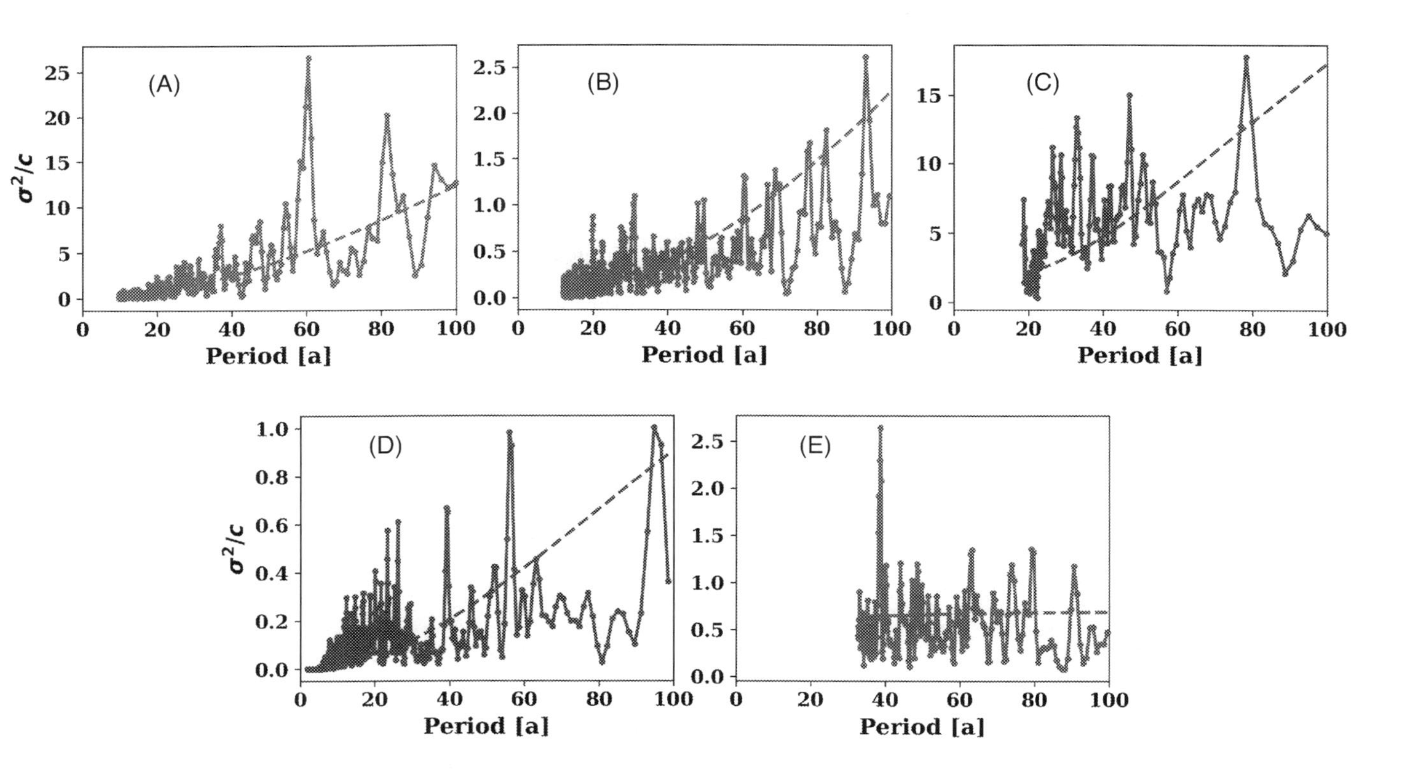

FIGURE 13.7

Comparison of periodicities (0–100 years) in oxygen isotopic composition obtained from (A) Tianmen, (B) Mawmulh, (C) Hoq cave records, (D) Varve stack thickness data from northern Arabian Sea sediment, and (E) Planktonic oxygen isotopic data from Indus river delta, obtained from REDFIT. The y-label σ^2/c is bias corrected spectral variance of power spectrum. Respective dashed lines are rednoise spectra derived from AR(1) process using REDFIT.

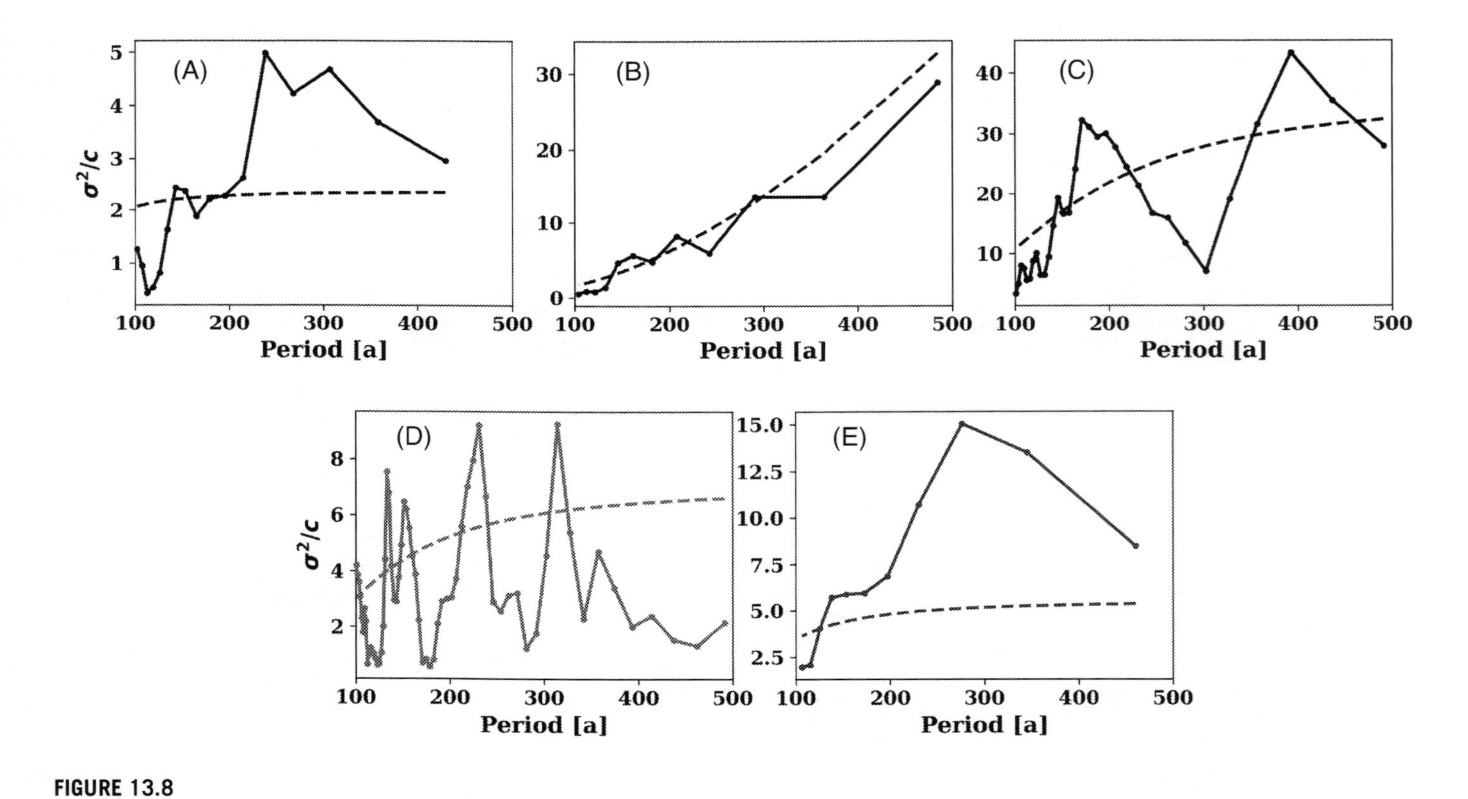

FIGURE 13.8

Comparison of periodicities (100–500 years) in oxygen isotopic composition obtained from (A) Sahiya, (B) Kadapa, (C) Baratang, (D) Qunf, and (E) Dandak cave records, obtained from REDFIT. The y-label σ^2/c is bias corrected spectral variance of power spectrum. Respective dashed lines are rednoise spectra derived from AR(1) process using REDFIT.

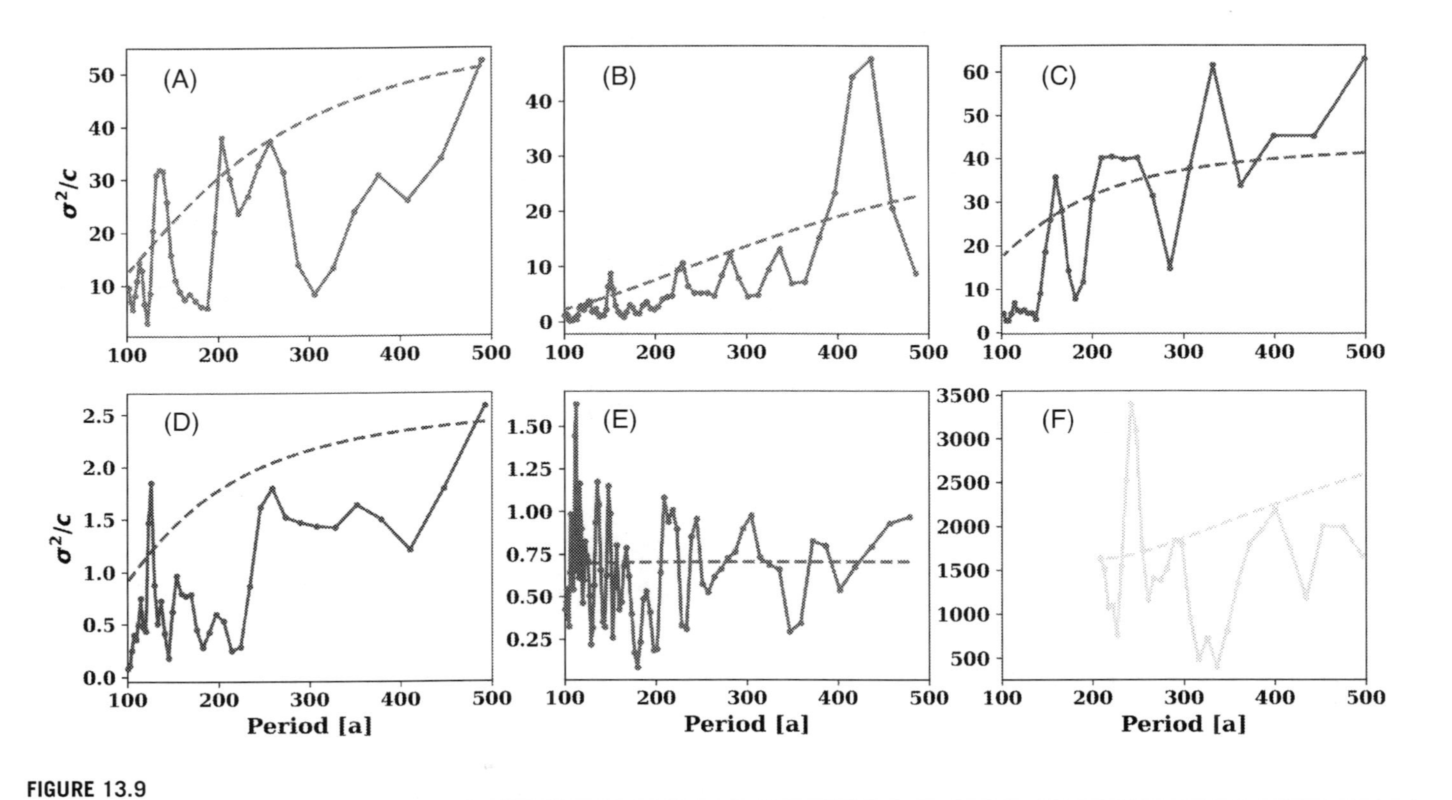

FIGURE 13.9

Comparison of periodicities (100–500 years) in oxygen isotopic composition obtained from (A) Tianmen, (B) Mawmulh, (C) Hoq cave records, (D) Varve stack thickness data from northern Arabian Sea sediment, (E) Planktonic oxygen isotopic data from Indus river delta, and (F) G. bulloides proxy data from Arabian Sea sediment core, obtained from REDFIT. The y-label σ^2/c is bias corrected spectral variance of power spectrum. Respective dashed lines are rednoise spectra derived from AR(1) process using REDFIT.

significant peak at ~7.5 years in AIR, Sahiya and Dandak data are observed. On the other hand, 8–20 years peaks are also observed from Sahiya and Dandak datasets. ISMR exhibited decadal to multidecadal variability and was well captured by all the chosen cave records (Figs. 13.6 and 13.7). Irrespective of the temporal resolution of selected datasets, the synchronous 40–80-year multidecadal signal suggests that this periodicity is one of the ISMR modes of variability. Similar 40–80-year multidecadal signals in ISMR are observed in a proxy study (Goswami et al., 2015). We have also analyzed distinct proxy data from the Arabian peninsula (Rampelbergh et al., 2013), the North-Eastern region of India (Berkelhammer et al., 2012; Cai et al., 2012), and the AS sediments (Rad et al., 1999; Gupta et al., 2003; Staubwasser et al., 2003) for better understanding of the multidecadal mode of ISMR. Tianmen, Mawmulh, Hoq cave records and Varve stack data showed strong statistically significant multidecadal periodicities, supporting the results from Indian mainland records.

The available instrumental rainfall record of ISMR, span over 193 years, is insufficient to resolve centennial to multicentennial scale variability. Spectral analysis of cave records and AS sediment data suggests ISMR varied on 100–500-year periodicity, that is, centennial to multicentennial timescales. The ISMR variability on centennial to multicentennial timescales obtained from the different cave records and AS sediment data presented in Figs. 13.8 and 13.9. Statistically significant peaks at ~150 years periodicity are common in all the records. Several different periodicities ~200, ~ 250, ~300, and ~350 are also observed. The results are in the line with the earlier studies (Gupta et al., 2003; Fleitmann et al., 2007; Sinha et al., 2015).

13.5 Discussion

13.5.1 Forcing mechanisms on ISMR with different timescales

13.5.1.1 2–20 years periodicities

ISMR variability on 2–3 years (Fig. 13.5) may be due to atmospheric oscillation (quasi-biennial oscillation) or solar activity influence (Hiremath and Mandi, 2004). The analysis from 30-year ISMR instrumental data also suggests 2–3 years periodicities in the North-West and West-Central India (Munot and Kothawale, 2000). The Kadapa cave record also shows a strong ~5-year periodicity. Southern Oscillation Index (SOI) shows ~3–8 years (figure not shown) and ENSO type periodicities (3–7 years) also shown to dominate North-East and Peninsular India (Munot and Kothawale, 2000). Thus, the ~5-year periodicity obtained from the Kadapa record might have links with ENSO. Spectrum analysis of SOI, PDO, and NHTemp datasets has shown 5–20 years periodicities (Fig. 13.10). Many parts of the Pacific region, mainly the tropical and North Pacific Ocean, exhibited quasi-decadal oscillation (QDO; Hasegawa and Hanawa, 2006; Jin et al., 2020). Likewise, SOI also displays QDO (Brassington, 1997). The review of Pacific Decadal Variability suggests the QDO from northwest North Pacific and adjacent seas are linked with NAO (Minobe et al., 2004). On the other hand, phases of QDO Niño-3.4 index have a strong influence on ENSO behavior (Hasegawa and Hanawa, 2006). Since the spectral analysis of ISMR also shows the oscillations of 5–20 years (Fig. 13.5), the origin of these periodicities of ISMR might have linked with the tropical and north Pacific and North Atlantic.

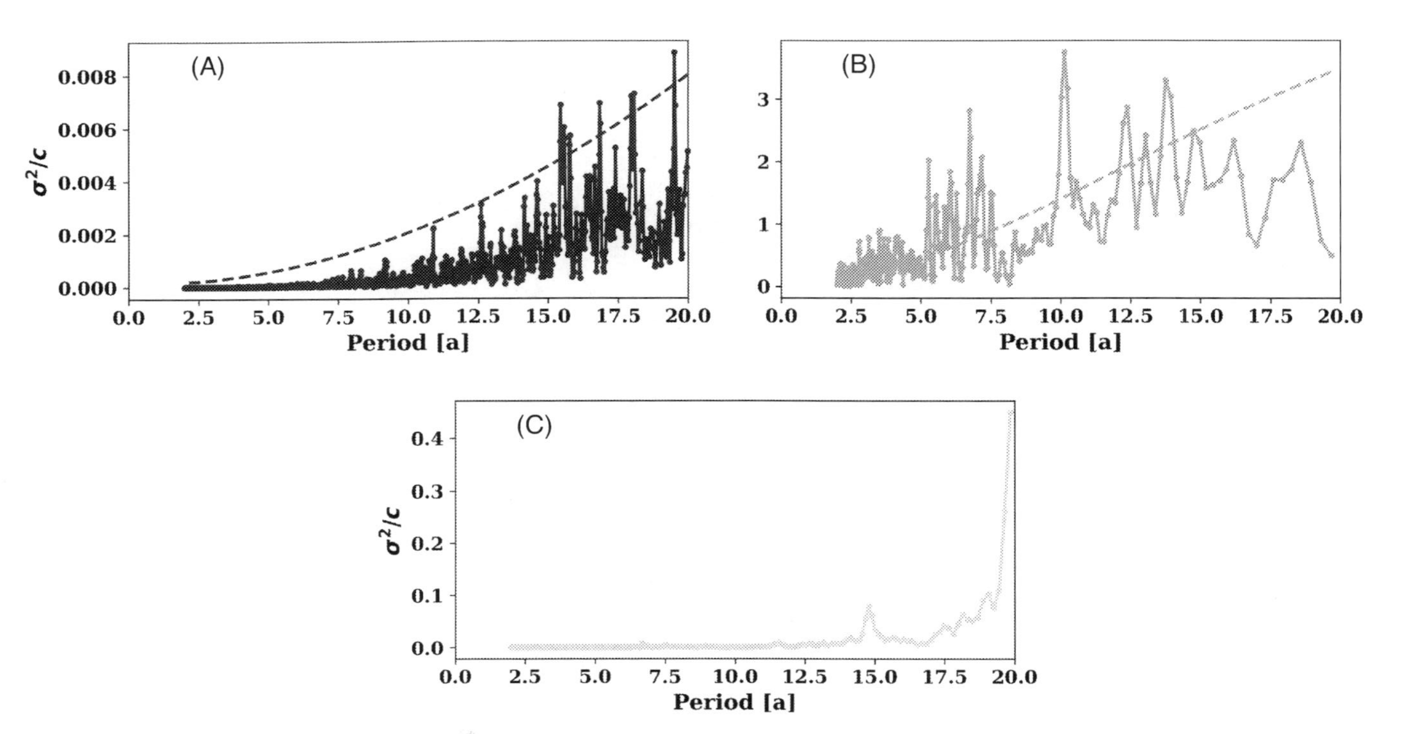

FIGURE 13.10

2–20 years of periodicities of different climate phenomenon. Spectrum derived from REDFIT for (A) NHtemp, (B) PDO, and (C) SOI, obtained from REDFIT. The y-label σ^2/c is bias corrected spectral variance of power spectrum. Respective dashed lines are rednoise spectra derived from AR(1) process using REDFIT.

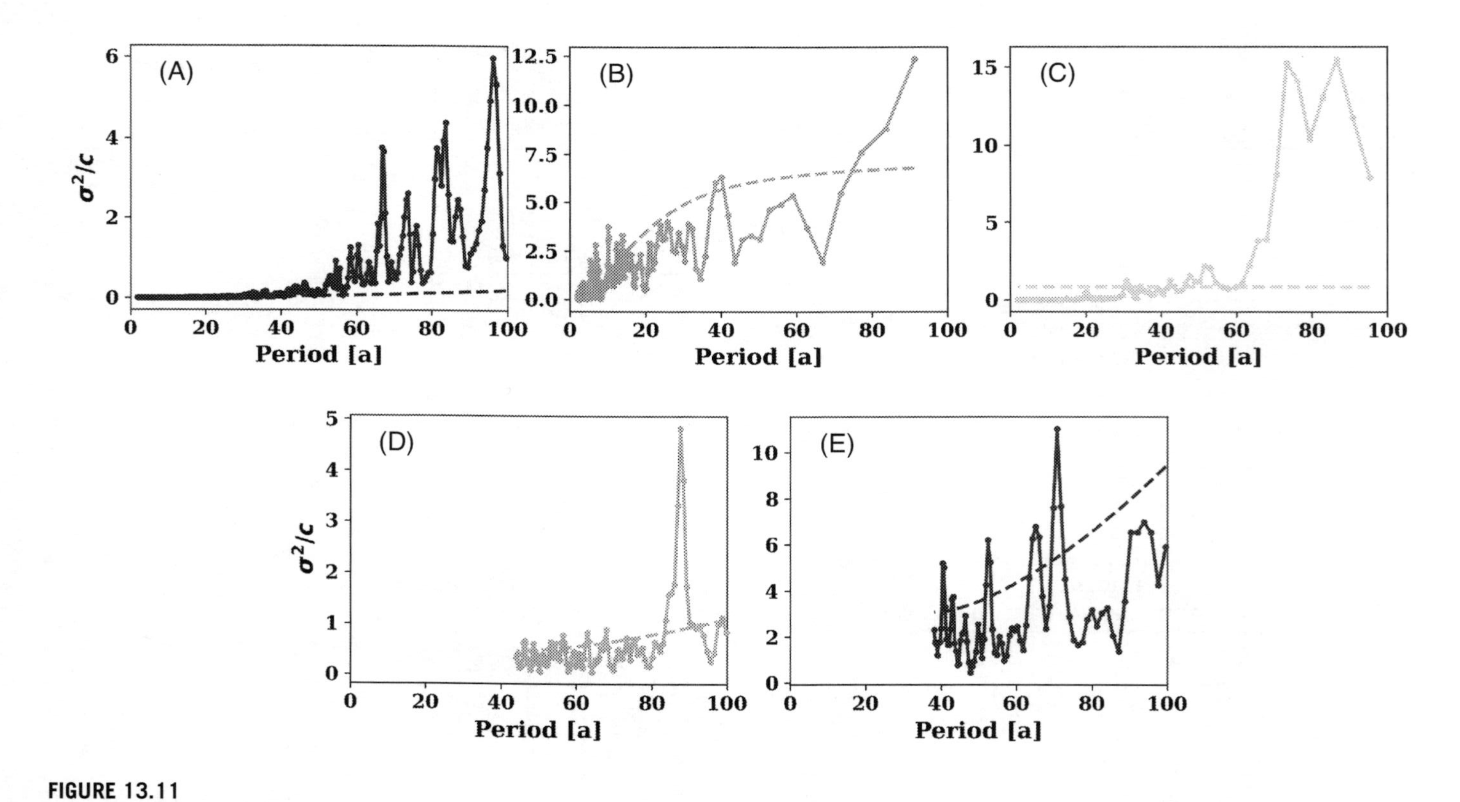

FIGURE 13.11

Decadal to multidecadal variability of climate variables. Spectrum derived from REDFIT for (A) NHTemp, (B) PDO, (C) SOI, (D) solar irradiance, and (E) NAO, obtained from REDFIT. The y-label σ^2/c is bias corrected spectral variance of power spectrum. Respective dashed lines are rednoise spectra derived from AR(1) process using REDFIT.

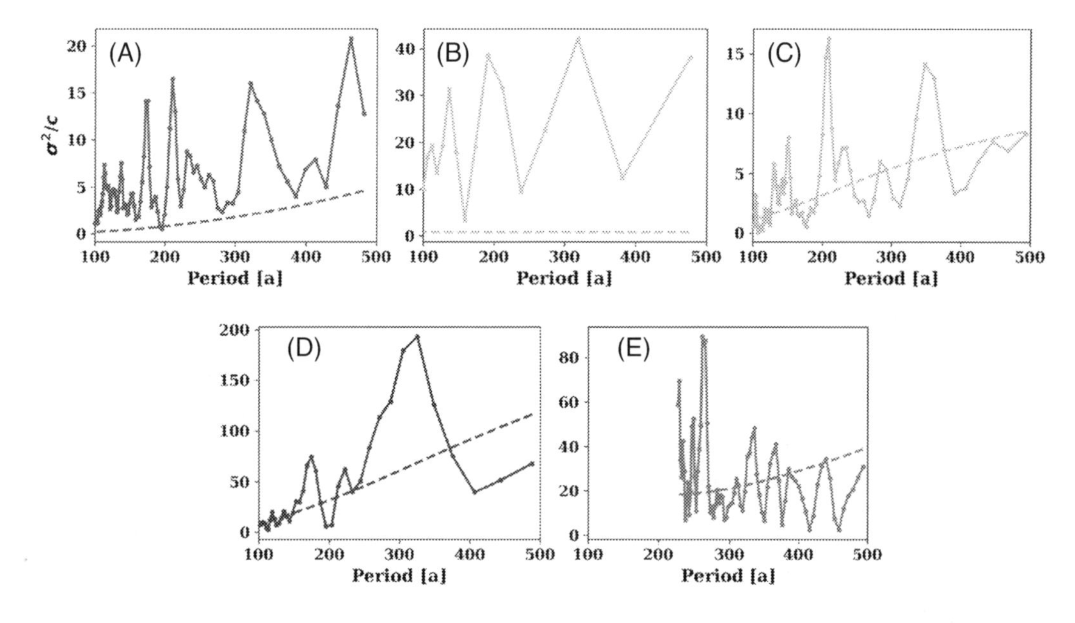

FIGURE 13.12

Centennial to multicentennial variability of climate variables. Spectrum derived from REDFIT for (A) NHTemp, (B) SOI, (C) TSI, (D) NAO, and (E) NIOSST, obtained from REDFIT. The y-label $\sigma^2/c\sigma^2/c$ is bias corrected spectral variance of power spectrum. Respective dashed lines are rednoise spectra derived from AR(1) process using REDFIT.

13.5.2 Decadal to multidecadal

Decadal to multidecadal variations are identified from the instrumental rainfall, distinct proxy data from ISM region and different climatic phenomena and solar activity (Figs. 13.6, 13.7, and 13.11). A common mode (40–80 years cycle) of observed ISMR multidecadal variability is closely related to NAO, NHTemp, PDO, and SOI multidecadal variability (Fig. 13.11). Several mechanisms are put forward to explain the multidecadal variability of ISMR. For example, a recent study using cave records from central India suggests that multidecadal power in ISMR is driven by NAsst multidecadal variability (Berkelhammer et al., 2010). Anomalous changes in NAsst's strongly effects the Tibetan Plateau surface temperatures, in turn the meridional thermal gradient between Tibetan Plateau and tropical Indian Ocean. The modulations in thermal gradient affect the onset and variations of ISMR (Feng and Hu, 2008). Another study revealed that ISMR decadal/multidecadal variability is related to PDO/AMO (Shi et al., 2017). Proxy data shows significant negative correlation between PDO and ISMR on multidecadal timescales over last millennia (Shi et al., 2017). The cold (warm) phases of decadal part of PDO are known to influences the ISMR, by decreasing (increasing) rainfall over India. Warm phase of PDO on decadal timescales, influence the winds that are extending from North Pacific. This affects the moisture flow north of 20°N that leads to reduced ISMR (Krishnamurthy and Krishnamurthy, 2017). ENSO also shows positive relationships with ISM on a multidecadal scale (Berkelhammer et al., 2013). Our analysis and earlier studies suggest 40–80 year ISMR multidecadal oscillation is related to coupled land–ocean–atmosphere system interactions.

13.5.3 Centennial to multicentennial forcings

Distinct proxy data indicate centennial to multicentennial scale ISMR variability (Figs. 13.8 and 13.9). Solar irradiance, SOI, NAO, and NHTemp also signify the forcings on ISMR variability on these timescales (Fig. 13.12). NAO shows robust oscillations on 100–500 years, and SOI also shows strong oscillation on these timescales. Several proxy studies suggested that the centennial to multicentennial scale variations in ISMR are related to solar forcing (e.g., Gupta et al., 2013; Shi et al., 2017; Sinha et al., 2018; Tan et al., 2018). A study from the Arabian Peninsula shows centennial variations in monsoon precipitation are associated with NAsst (Fleitmann et al., 2003). Further, "Active-Break" dominated centennial-scale ISMR variability is related to internal dynamics sensitive to slowly evolving external boundary conditions (Sinha et al., 2011b). Climate modeling simulations and observations (Malik and Brönnimann, 2017) show a strong relation between AMO and ISMR variability on centennial timescales. Another model simulation study suggests the abrupt changes in ISM are closely related to cold events in the North Atlantic at centennial timescales (Zhang et al., 2016). Our analysis suggests that the centennial to multicentennial ISMR variability covary with the North Atlantic circulation and ENSO variations (Fig. 13.12).

13.6 Learning and knowledge outcomes

This work would be an important document to learn about the importance of ISMR variability on different timescales and their significance. Further, this work discusses the forcing mechanisms of ISMR variability on different timescales. The comprehensive review of commonly used spectral methods would undoubtedly help researchers choose a suitable spectral method for irregularly spaced data analysis.

Acknowledgments

We thank the Director, Indian Institute of Tropical Meteorology, for support and encouragement. We thank AAVSO and Michael Schulz for providing TS and SPECTRUM & REDFIT programs and acknowledge WAVEPAL and NESToolbox authors for providing the software for free. The authors would like to thank Asha Pinto for helping in the installation of WAVEPAL and Prof. M.S. Santhanam, IISER, Pune, for his valuable suggestions. The authors are also grateful to the anonymous reviewers for their comments and suggestions. This work was supported by PACMEDY Project under the Ministry of Earth Sciences, India. The National Oceanic and Atmospheric Administration (NOAA)/National Climatic Data Center (NCDC) paleoclimatology program is acknowledged for proxy data.

References

Azad, S., Vignesh, T.S., Narasimha, R., Periodicities in Indian monsoon rainfall over spectrally homogeneous regions, Int. J. Climatol. 30 (2010) 2289–2298, doi:10.1002/joc.2045.

Band, S., Yadava, M.G., Ahmad, M., Shen, C., Sree, K., Ramesh, R., High-resolution mid-Holocene Indian Summer Monsoon recorded in a stalagmite from the Kotumsar Cave, Central India, Quat. Int. 479 (2018) 19–24, doi:10.1016/j.quaint.2018.01.026.

Berkelhammer, M., Sinha, A., Mudelsee, M., Cheng, H., Yoshimura, K., Biswas, J., On the low frequency component of the ENSO-Indian monsoon relationship; a paired proxy perspective, Clim. Past 9 (2013) 3103–3123, doi:10.5194/cpd-9-3103-2013.

Berkelhammer, M., Sinha, A., Mudelsee, M., Cheng, H., Edwards, R.L., Cannariato, K., Persistent multidecadal power of the Indian Summer Monsoon, Earth Planet. Sci. Lett. 290 (2010) 166–172, doi:10.1016/j.epsl.2009.12.017.

Berkelhammer, M., Sinha, A., Stott, L., Cheng, H., Pausata, F.S.R.R., Yoshimura, K., An abrupt shift in the Indian monsoon 4000 years ago, Geophys. Monogr. Ser. 198 (2012) 75–87, doi:10.1029/2012GM001207.

Bird, B.W., Polisar, P.J., Lei, Y., Thompson, L.G., Yao, T., Finney, B.P., Bain, D.J., Pompeani, D.P., Steinman, B.A., A Tibetan lake sediment record of Holocene Indian summer monsoon variability, Earth Planet. Sci. Lett. 399 (2014) 92–102, doi:10.1016/j.epsl.2014.05.017.

Brassington, G.B., The modal evolution of the Southern Oscillation, J. Clim. 10 (1997) 1021–1034, doi:10.1175/1520-0442(1997)010<1021:TMEOTS>2.0.CO;2.

Cai, Y., Zhang, H., Cheng, H., An, Z., Lawrence Edwards, R., Wang, X., Tan, L., Liang, F., Wang, J., Kelly, M., Edwards, R.L., Wang, X., Tan, L., Liang, F., Wang, J., Kelly, M., The Holocene Indian monsoon variability over the southern Tibetan Plateau and its teleconnections, Earth Planet. Sci. Lett. 335–336 (2012) 135–144, doi:10.1016/j.epsl.2012.04.035.

DeLong, K.L., Quinn, T.M., Mitchum, G.T., Poore, R.Z., Evaluating highly resolved paleoclimate records in the frequency domain for multidecadal-scale climate variability, Geophys. Res. Lett. 36 (2009) 1–6, doi:10.1029/2009GL039742.

Evans, D., Bhatia, R., Stoll, H., Müller, W., LA-ICPMS Ba/Ca analyses of planktic foraminifera from the Bay of Bengal: implications for late Pleistocene orbital control on monsoon freshwater flux, Geochem. Geophys. Geosyst. 16 (2015) 2598–2618, doi:10.1002/2015GC005822.

Feng, S., Hu, Q., How the North Atlantic Multidecadal Oscillation may have influenced the Indian summer monsoon during the past two millennia, Geophys. Res. Lett. 35 (2008) 1–5, doi:10.1029/2007GL032484.

Ferraz-Mello, S., Estimation of periods from unequally spaced observations, Astronom. J. 86 (1981) 619, doi:10.1086/112924.

Fleitmann, D., Burns, S.J., Mangini, A., Mudelsee, M., Kramers, J., Villa, I., Neff, U., Al-Subbary, A.A., Buettner, A., Hippler, D., Matter, A., Holocene ITCZ and Indian monsoon dynamics recorded in stalagmites from Oman and Yemen (Socotra), Quat. Sci. Rev. 26 (2007) 170–188, doi:10.1016/j.quascirev.2006.04.012.

Fleitmann, D., Burns, S.J., Mudelsee, M., Neff, U., Kramers, J., Mangini, A., Matter, A., Holocene forcing of the Indian monsson recorded in a stalgmite from Southern Oman, Science 300 (2003) 1737–1739.

Foster, G., The Cleanest Fourier Spectrum, 4th, Astron. J., 109 (1995) 1889–1902.

Fourier, J., *The Analytical Theory of Heat* (Cambridge Library Collection – Mathematics) (A. Freeman, Trans.). Cambridge University Press, Cambridge, 2009. doi:10.1017/CBO9780511693205.

Gadgil, Sulochana, Gadgil, Siddhartha, The Indian monsoon, GDP and agriculture, Econ. Polit. Wkly. 41 (2006) 4887–4895, doi:10.2307/4418949.

Goswami, Bhupendra H., Kriplani, Ramesh H., Boargaonkar, Hemant P., Bhaskar, Preethi, Multi-Decadal Variability in Indian Summer Monsoon Rainfall Using Proxy Data. Climate Change: Multidecadal and Beyond. World Scientific Series on Asia-Pacific Weather and Climate, Singapore, 2015.

Gupta, A.K., Anderson, D.M., Overpeck, J.T., Abrupt changes in the Asian southwest monsoon during the Holocene and their links to the North Atlantic Ocean, Nature 421 (2003) 354–357, doi:10.1038/nature01340.

Gupta, A.K., Mohan, K., Das, M., Singh, R.K., Solar forcing of the Indian summer monsoon variability during the Ållerød period, Sci. Rep. 3 (2013) 1–5, doi:10.1038/srep02753.

Hasegawa, T., Hanawa, K., Impact of quasi-decadal variability in the tropical Pacific on ENSO modulations, J. Oceanogr. 62 (2006) 227–234, doi:10.1007/s10872-006-0047-5.

Hiremath, K.M., Mandi, P.I., Influence of the solar activity on the Indian Monsoon rainfall, New Astron. 9 (2004) 651–662, doi:10.1016/j.newast.2004.04.001.

Hiremath, K.M., Manjunath, H., Soon, W., Indian summer monsoon rainfall: dancing with the tunes of the Sun, New Astron. 35 (2015) 8–19, doi:10.1016/j.newast.2014.08.002.

Jenkins, G.M., Wats, D.G., Spectral Analysis and Its Applications, Holden-Day. San Francisco (1968).

Jin, C., Wang, B., Liu, J., Emerging Pacific Quasi-Decadal Oscillation over the past 70 years, Geophys. Res. Lett. 48 (2) (2020), doi:10.1029/2020GL090851.

Joseph, Porathur Vareed, Gokulapalan, Bindu, Nair, Archana, Wilson, Shinu Sheela, Variability of Summer Monsoon Rainfall in India on Inter-Annual and Decadal Time Scales, Atmos. Ocean. Sci. Lett. 6 (5) (2013) 398–403, doi:10.3878/j.issn.1674-2834.13.0044.

Joshi, L.M., Kotlia, B.S., Ahmad, S.M., Wu, C.C., Sanwal, J., Raza, W., Singh, A.K., Shen, C.C., Long, T., Sharma, A.K., Reconstruction of Indian monsoon precipitation variability between 4.0 and 1.6 ka BP using speleothem δ18O records from the Central Lesser Himalaya, India, Arab. J. Geosci. 10 (2017) 0–16, doi:10.1007/s12517-017-3141-7.

Kathayat, G., Cheng, H., Sinha, A., Spötl, C., Edwards, R.L., Zhang, H., Li, X., Yi, L., Ning, Y., Cai, Y., Lui, W.L., Breitenbach, S.F.M., Indian monsoon variability on millennial-orbital timescales, Sci. Rep. 6 (2016) 4–10, doi:10.1038/srep24374.

Kathayat, G., Cheng, H., Sinha, A., Yi, L., Li, X., Zhang, H., Li, H., Ning, Y., Edwards, R.L., The Indian monsoon variability and civilization changes in the Indian subcontinent, Sci. Adv. 3 (2017) 1–9, doi:10.1126/sciadv.1701296.

Kodera, K., Solar influence on the Indian Ocean Monsoon through dynamical processes, Geophys. Res. Lett. 31 (2004), https://doi.org/10.1029/2004GL020928.

Kobashi, T., Menviel, L., Jeltsch-Thömmes, A., Vinther, B.M., Box, J.E., Muscheler, R., Nakaegawa, T., Pfister, P.L., Döring, M., Leuenberger, M., Wanner, H., Ohmura, A., Volcanic influence on centennial to millennial Holocene Greenland temperature change. Sci. Rep. 7 (2017) 1–10. https://doi.org/10.1038/s41598-017-01451-7.

Kovacs, G., Frequency shift in Fourier analysis, Astrophys. Space Sci. 78 (1981) 175–188.

Krishnamurthy, L., Krishnamurthy, V., Indian monsoon's relation with the decadal part of PDO in observations and NCAR CCSM4, Int. J. Climatol. 37 (2017) 1824–1833, doi:10.1002/joc.4815.

Krishnamurthy, V., Goswami, B.N., Indian monsoon-ENSO relationship on interdecadal timescale, J. Clim. 13 (2000) 579–595, doi:10.1175/1520-0442(2000)013<0579:IMEROI>2.0.CO;2.

Kumar, K.K., Rajagopalan, B., Cane, M.A., On the weakening relationship between the indian monsoon and ENSO, Science 284 (1999) 2156–2159, doi:10.1126/science.284.5423.2156.

Kumar, K.K., Rajagopalan, B., Hoerling, M., Bates, G., Cane, M., Unraveling the mystery of Indian monsoon failure during El Niño, Science 314 (2006) 115–119, doi:10.1126/science.1131152.

Laskar, A.H., Yadava, M.G., Ramesh, R., Polyak, V.J., Asmerom, Y., A 4 kyr stalagmite oxygen isotopic record of the past Indian Summer Monsoon in the Andaman Islands. Geochemistry, Geophys. Geosystems 14 (2013) 3555–3566. https://doi.org/10.1002/ggge.20203.

Lau, N.C., Nath, M.J., Impact of ENSO on the variability of the Asian-Australian Monsoons as simulated in GCM experiments, J. Clim. 13 (2000) 4287–4309, doi:10.1175/1520-0442(2000)013<4287:IOEOTV>2.0.CO;2.

Lenoir, G., Crucifix, M., A general theory on frequency and time–frequency analysis of irregularly sampled time series based on projection methods – part 1: frequency analysis, Nonlin. Processes Geophys. 25 (2018a) 145–173, doi:10.5194/npg-25-145-2018.

Lenoir, G., Crucifix, M., A general theory on frequency and time–frequency analysis of irregularly sampled time series based on projection methods – part 2: extension to time–frequency analysis, Nonlin. Processes Geophys. 25 (2018b) 175–200, doi:10.5194/npg-25-175-2018.

Lomb, N.R., Least-squares frequency analysis of unequally spaced data, Astrophys. Space Sci. 39 (1976) 447–462, doi:10.1007/BF00648343.

Luo, F., Li, S., Gao, Y., Svendsen, L., Furevik, T., Keenlyside, N., The connection between the Atlantic Multidecadal Oscillation and the Indian Summer Monsoon since the Industrial Revolution is intrinsic to the climate system, Environ. Res. Lett. 13 (2018a), doi:10.1088/1748-9326/aade11.

Luo, F.F., Li, S., Furevik, T., Weaker connection between the Atlantic Multidecadal Oscillation and Indian summer rainfall since the mid-1990s, Atmos. Oceanic Sci. Lett. 11 (2018b) 37–43, doi:10.1080/16742834.2018.1394779.

MacDonald, G.M., Case, R.A., Variations in the Pacific Decadal Oscillation over the past millennium. Geophys. Res. Lett. 32 (2005) 1–4. https://doi.org/10.1029/2005GL022478.

Madhusudhan, L., Agriculture role on Indian economy, Bus. Econ. J. 06 (2015) 1000176, doi:10.4172/2151-6219.1000176.

Malik, A., Brönnimann, S., Factors affecting the inter-annual to centennial timescale variability of Indian summer monsoon rainfall, Clim. Dyn. 50 (2017) 1–18, doi:10.1007/s00382-017-3879-3.

Minobe, S., Schneider, N., Deser, C., Liu, Z., Mantua, N., Nakamura, H., Nonaka, M., 2004. Pacific decadal variability: a review. In: Proceedings of the First International CLIVAR Conference, Baltimore, MD, USA.

Mudelsee, M., TAUEST: a computer program for estimating persistence in unevenly spaced weather/climate time series, Comput. Geosci. 28 (2002) 69–72, doi:10.1016/S0098-3004(01)00041-3.

Mudelsee, M., Climate Time Series Analysis, Springer, Netherlands (2010).

Munot, A.A., Kothawale, D.R., Intra-seasonal, inter-annual and decadal scale variability in summer monsoon rainfall over India, Int. J. Climatol. 20 (2000) 1387–1400, doi:10.1002/1097-0088(200009)20:11<1387::AID-JOC540>3.0.CO;2-Z.

Olsen, J., Anderson, N.J., Knudsen, M.F., Variability of the North Atlantic Oscillation over the past 5,200 years. Nat. Geosci. 5 (2012) 808–812. https://doi.org/10.1038/ngeo1589.

Prasad, S., Anoop, A., Riedel, N., Sarkar, S., Menzel, P., Basavaiah, N., Krishnan, R., Fuller, D., Plessen, B., Gaye, B., Röhl, U., Wilkes, H., Sachse, D., Sawant, R., Wiesner, M.G., Stebich, M., Prolonged monsoon droughts and links to Indo-Pacific warm pool: a Holocene record from Lonar Lake, central India, Earth Planet. Sci. Lett. 391 (2014) 171–182, doi:10.1016/j.epsl.2014.01.043.

Rashid, H., England, E., Thompson, L., Polyak, L., Late glacial to Holocene Indian summer monsoon variability based upon sediment, Terres. Atmos. Ocean Sci. 22 (2011) 215–228, doi:10.3319/TAO.2010.09.17.02(TibXS)1.

Rehfeld, K., Kurths, J., Similarity estimators for irregular and age-uncertain time series, Clim. Past 10 (2014) 107–122, doi:10.5194/cp-10-107-2014.

Rehfeld, K., Marwan, N., Heitzig, J., Kurths, J., Comparison of correlation analysis techniques for irregularly sampled time series, Nonlin. Processes Geophys. 18 (2011) 389–404, doi:10.5194/npg-18-389-2011.

Sankar, S., Svendsen, L., Gokulapalan, B., Joseph, P.V., Johannessen, O.M., The relationship between Indian summer monsoon rainfall and Atlantic multidecadal variability over the last 500 years, Tellus A: Dyn. Meteorol. Oceanogr. 68 (2016) 31717, doi:10.3402/tellusa.v68.31717.

Saraswat, R., Lea, D.W., Nigam, R., Mackensen, A., Naik, D.K., Deglaciation in the tropical Indian Ocean driven by interplay between the regional monsoon and global teleconnections. Earth Planet. Sci. Lett. (2013) 1–10. https://doi.org/10.1016/j.epsl.2013.05.022.

Scargle, J.D., Statistical aspects of spectral analysis of unevenly spaced data, Astrophys. J. 263 (1982) 835–853, doi:10.1086/160554.

Scargle, J.D., Studies in Astronomical Time Series Analysis. III. Fourier Transforms, Autocorrelation Functions, and Cross-Correlation Functions of Unevenly Spaced Data, Astrophys. J. 343 (1989) 874–887, doi:10.1086/167757.

Schulz, M., Mudelsee, M., REDFIT: estimating red-noise spectra directly from unevenly spaced paleoclimatic time series, Comput. Geosci. 28 (2002) 421–426, doi:10.1016/S0098-3004(01)00044-9.

Schulz, M., Stattegger, K., SPECTRUM: spectral analysis of unevenly spaced paleoclimatic time series, Comput. Geosci. 23 (1997) 929–945, doi:10.1016/S0098-3004(97)00087-3.

Shi, F., Fang, K., Xu, C., Guo, Z., Borgaonkar, H.P., Interannual to centennial variability of the South Asian summer monsoon over the past millennium, Clim. Dyn. 49 (2017) 2803–2814, doi:10.1007/s00382-016-3493-9.

Shi, F., Li, J., Wilson, R.J.S., A tree-ring reconstruction of the South Asian summer monsoon index over the past millennium, Sci. Rep. 4 (2014) 1–8, doi:10.1038/srep06739.

Sinha, A., Berkelhammer, M., Stott, L., Mudelsee, M., Cheng, H., Biswas, J., The leading mode of Indian Summer Monsoon precipitation variability during the last millennium, Geophys. Res. Lett. 38 (2011b) 2–6, doi:10.1029/2011GL047713.

Sinha, A., Cannariato, K.G., Stott, L.D., Cheng, H., Edwards, R.L., Yadava, M.G., Ramesh, R., Singh, I.B., A 900-year (600 to 1500 A.D.) record of the Indian summer monsoon precipitation from the core monsoon zone of India, Geophys. Res. Lett. 34 (2007) 1–5, doi:10.1029/2007GL030431.

Sinha, A., Cannariato, K.G., Stott, L.D., Li, H.C., You, C.F., Cheng, H., Edwards, R.L., Singh, I.B., Variability of Southwest Indian summer monsoon precipitation during the Bølling-Ållerød, Geology 33 (2005) 813–816, doi:10.1130/G21498.1.

Sinha, A., Kathayat, G., Cheng, H., Breitenbach, S.F.M., Berkelhammer, M., Mudelsee, M., Biswas, J., Edwards, R.L., Trends and oscillations in the Indian summer monsoon rainfall over the last two millennia, Nat. Commun. 6 (2015) 1–8, doi:10.1038/ncomms7309.

Sinha, A., Stott, L., Berkelhammer, M., Cheng, H., Edwards, R.L., Buckley, B., Aldenderfer, M., Mudelsee, M., A global context for megadroughts in monsoon Asia during the past millennium, Quat. Sci. Rev. 30 (2011a) 47–62, doi:10.1016/j.quascirev.2010.10.005.

Sinha, N., Gandhi, N., Chakraborty, S., Krishnan, R., Yadava, M.G., Ramesh, R., Abrupt climate change at ~2800 yr BP evidenced by a stalagmite record from peninsular India, Holocene 28 (2018) 1720–1730, doi:10.1177/0959683618788647.

Tan, L., Cai, Y., Cheng, H., Edwards, L.R., Lan, J., Zhang, H., Li, D., Ma, L., Zhao, P., Gao, Y., High resolution monsoon precipitation changes on southeastern Tibetan Plateau over the past 2300 years. Quat. Sci. Rev. 195 (2018) 122–132. https://doi.org/10.1016/j.quascirev.2018.07.021.

Sontakke, N.A., Singh, N., Singh, H.N., Instrumental period rainfall series of the Indian region (AD 1813-2005): Revised reconstruction, update and analysis. Holocene 18 (2008) 1055–1066. https://doi.org/10.1177/0959683608095576.

Staubwasser, M., Sirocko, F., Grootes, P.M., Erlenkeuser, H., South Asian monsoon climate change and radiocarbon in the Arabian Sea during early and middle Holocene. Paleoceanography 17 (4) (2002) 1063. doi:10.1029/2000PA000608.

Staubwasser, M., Sirocko, F., Grootes, P. M., Segl, M., Climate change at the 4.2 ka BP termination of the Indus valley civilization and Holocene south Asian monsoon variability, Geophys. Res. Lett. 30 (8) (2003) 1425, doi:10.1029/2002GL016822.

Steinhilber, F., Abreu, J.A., Beer, J., Brunner, I., Christl, M., Fischer, H., Heikkila, U., Kubik, P.W., Mann, M., McCracken, K.G., Miller, H., Miyahara, H., Oerter, H., Wilhelms, F., 9,400 Years of Cosmic Radiation and Solar Activity From Ice Cores and Tree Rings. Proc. Natl. Acad. Sci. 109 (2012) 5967–5971. https://doi.org/10.1073/pnas.1118965109.

Templeton, M., Time-series analysis of variable star data, JAAVSO 32 (2004) 41–54.

Van Rampelbergh, M., Fleitmann, D., Verheyden, S., Cheng, H., Edwards, L., De Geest, P., De Vleeschouwer, D., Burns, S.J., Matter, A., Claeys, P., Keppens, E., Mid- to late Holocene Indian Ocean Monsoon variability recorded in four speleothems from Socotra Island, Yemen, Quat. Sci. Rev. 65 (2013) 129–142, doi:10.1016/j.quascirev.2013.01.016.

Von Rad, U., Schaaf, M., Michels, K.H., Schulz, H., Berger, W.H., Sirocko, F., A 5000-yr record of climate change in varved sediments from the oxygen minimum zone off Pakistan, Northeastern Arab. Sea 51 (1999) 39–53 https://doi.org/10.1006/qres.1998.2016.

Wang, P.X., Wang, B., Cheng, H., Fasullo, J., Guo, Z.T., Kiefer, T., Liu, Z.Y., The global monsoon across timescales: coherent variability of regional monsoons, Clim. Past 10 (2014) 2007–2052, doi:10.5194/cp-10-2007-2014.

Welch, P.D., The use of fast Fourier transform for the estimation of power spectra: a method based on time averaging over short, modified periodograms, IEEE Trans. Audio Electroacoust. 15 (1967) 70–73, doi:10.1109/TAU.1967.1161901.

Yadava, M.G., Ramesh, R., Monsoon reconstruction from radiocarbon dated tropical Indian speleothems, Holocene 15 (2005) 48–59, doi:10.1191/0959683605h1783rp.

Yan, H., Sun, L., Wang, Y., Huang, W., Qiu, S., Yang, C., A record of the Southern Oscillation Index for the past 2,000 years from precipitation proxies. Nat. Geosci. 4 (2011) 611–614. https://doi.org/10.1038/ngeo1231.

Zhang, X., Jin, L., Jia, W., Centennial-scale teleconnection between North Atlantic sea surface temperatures and the Indian summer monsoon during the Holocene, Clim. Dyn. 46 (2016) 3323–3336, doi:10.1007/s00382-015-2771-2.

https://www.aavso.org/software-directory.

CHAPTER

14 Monsoon variability in the Indian subcontinent—A review based on proxy and observational datasets

Ankit Yadav[a], Bulbul Mehta[a], Ambili Anoop[a], Praveen K. Mishra[b]

[a] *Indian Institute of Science Education and Research Mohali, Manauli, Punjab, India.* [b] *Wadia Institute of Himalayan Geology, Dehradun, Uttarakhand, India*

14.1 Introduction

Indian monsoon system shapes the livelihood of approximately one-fifth of the world's population (Katzenberger et al., 2020). The variability in monsoonal precipitation plays a crucial role in the socioeconomic stability of the country (Ghosh et al., 2016). The understanding of interannual and interdecadal monsoon variability has been studied using instrumental observations and paleoclimate proxy records (e.g., Gadgil, 2003; Gadgil and Gadgil, 2006; Goswami et al., 2006; Staubwasser and Weiss, 2006; Nageswararao et al., 2019). The observational data indicate a significant increase in the frequency of extreme climatic events (flood and drought) in the recent decades and is expected to increase in future (Goswami et al., 2006; Madhura et al., 2014). However, there remains an uncertainty in the impacts of intrinsic natural factors and anthropogenic influence on the monsoon variability (e.g., Krishnan et al., 2020). To delineate the contributions from natural forcing factors and anthropogenically induced perturbations, long-term high-resolution past climate reconstruction is required.

The long-term paleoclimate data from different archives (e.g., lake/marine sediment cores, stalagmites) using various climate-sensitive proxies provide detailed information regarding the past climatic variability, its forcing factors, and the mechanisms and dynamics of climate changes (Fleitmann et al., 2007; Laskar et al., 2013; Prasad et al., 2014; Mishra et al., 2015a; Sandeep et al., 2017; Ali et al., 2018; Ghosh et al., 2020; Jha et al., 2020). The intricate interaction of various components in the atmosphere, geosphere, and hydrosphere is capable of contributing to the climate variability on varying timescales. The anthropogenic activity also plays a crucial role in nonlinear short-scale climate variability especially during the late-Holocene (Krishnan et al., 2020). The terrestrial paleoclimate archives from the Indian subcontinent are extensively studied to address the hydrological changes associated with the monsoon variability (Ghosh et al., 2014; Dixit and Tandon, 2016; Bhushan et al., 2018; Mishra et al., 2019). The quantitative assessment and understanding of climatic fluctuations in response to (a) external (e.g., solar forcing); and (b) global circulation patterns (e.g., El Nino southern oscillation—ENSO; Indian Ocean dipole—IOD; North Atlantic Oscillation (NAO), Active–Break spells) and anthropogenic elements can be derived based from observational datasets and terrestrial paleoclimate datasets in the Indian subcontinent (Clemens et al., 1991; Schott et al., 2009; Abram et al., 2020; Ha et al., 2020, and reference within). ENSO strongly affects monsoon rainfall through an interaction between equatorial atmospheric circulation (Walker circulation) over tropical Pacific and meridional circulation (Hadley circulation)

Holocene Climate Change and Environment. DOI: https://doi.org/10.1016/B978-0-323-90085-0.00001-2

in south Asia (Shukla and Paolino, 1983; Stachnik and Schumacher, 2011). The El Niño (positive phase) event is associated with warming of the tropical Pacific Ocean along with a destabilization of ocean-atmosphere circulations resulting in the weakening of zonal Walker circulations and reduction in precipitation. In contrast, the increase of sea surface temperature (SST) gradient and strengthened Walker circulations results in increasing monsoon rainfall during La Niña (negative phase; Ummenhofer et al., 2011; Hernández et al., 2020). Similar to the ENSO in the Pacific Ocean, the local linked phenomenon in the Indian Ocean is known as IOD. The IOD in the tropical Indian Ocean corresponds to the anomalous changes in the zonal SST and zonal surface winds (Saji et al., 1999; Abram et al., 2015). The quantitative apportionment of the IOD is classically defined as Indian Ocean Dipole Mode Index calculated as the spatially averaged difference in SST over Western (10°N–10°S, 50°E–70°E) and Eastern (0°–10°S, 90°E–110°E) Indian Ocean. The positive phase of IOD is marked with the warming of the western Indian Ocean as compared to the eastern part of the ocean, whereas reverse condition occurs during the negative phase of IOD. The occurrence of IOD events is closely interlinked to the strengthening and weakening of walker circulations over the tropical oceans (Vinayachandran et al., 2009; Sreekala et al., 2012). For instance, the circulation over the Indian Ocean weakens during positive IOD years and strengthens during negative IOD years (Mohtadi et al., 2017; Hernández et al., 2020).

An integrated approach of understanding the modern dynamics of teleconnections and reliable holistic paleoclimate reconstruction based on proxy records is essential to delineate past boundary conditions in order to develop a comprehensive understanding on different modes of climate variability. The information gained using this approach can help to reduce uncertainty on future climate prediction. Further, this will help policymakers and environmentalist to build a robust framework to understand the societal implication of climate change. In this chapter, we provide a comprehensive assessment of modern climate observations and terrestrial paleoclimate data to understand the forcing factors and teleconnections governing the monsoon precipitation variability in the Indian subcontinent (Fig. 14.1A).

The major objectives of the study are to (1) assess the observational data and paleoclimate proxy behavior (surface run-off and evaporative) in order to disentangle the past boundary conditions; (2) characterize the role of teleconnections in modern day precipitation changes; (3) outline the dynamics of natural forcing factors responsible for hydroclimatic fluctuations during the Holocene; and (4) understand the impact of amplified anthropogenic stressors on monsoonal rainfall over Indian subcontinent.

14.2 Scope

The understanding of interannual to millennial to multi-millennial time scales climate variability of the Indian summer monsoon is fundamental for water resources availability. The changes in large-scale circulation patterns and best-known modes of variability have significant impact on the spatial and temporal ISM precipitation changes over Indian subcontinent. Over the last decades, several studies from terrestrial environments using different proxies have aimed to reconstruct the past hydrological conditions over Holocene to better understand the Spatio-temporal variability of these modes and their impacts. However, the association of climate and a given proxy vary over time because the sensitivity of a proxy to a mean climate state and associated climate drivers are different. Therefore, the calibration and testing of proxies according to change in local meteorological conditions and their link with instrumental/observational data is crucial to establish a better perspective on the temporal evolution of the climate system and its response to various teleconnections. This review aims to provide a better

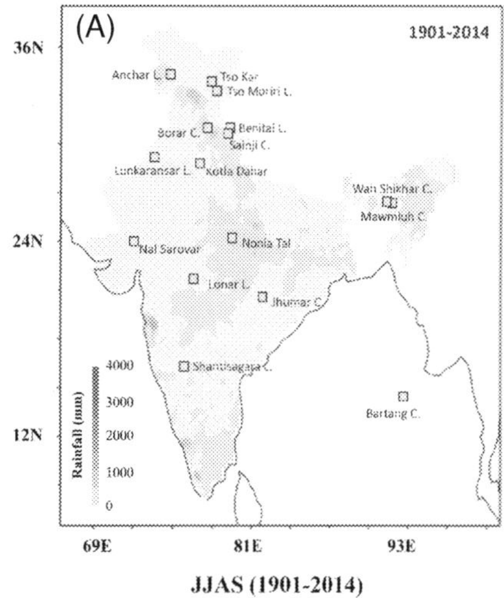

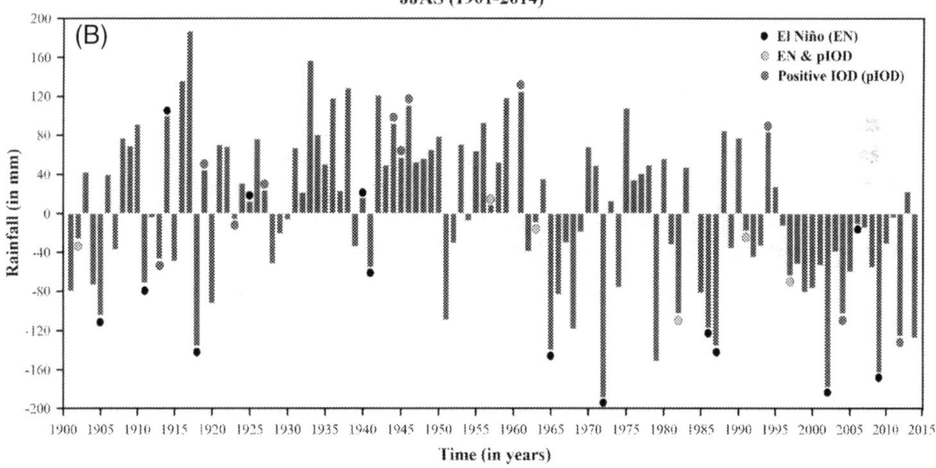

FIGURE 14.1

(A) The spatial distribution of mean rainfall from the Indian subcontinent; and (B) rainfall anomaly based on the datasets from the year 1901 to 2014 (source: IMD). The squares (in A) represent paleoclimate sites discussed in the text. The blue (red) bars in B are positive (negative) anomaly of rainfall.

understanding on Spatio-temporal variability in ISM rainfall, using different proxies from different archives in the Indian subcontinent during the Holocene period.

14.3 Methodology and processes

The daily rainfall dataset covering 1901–2014 with a spatial resolution of $0.25° \times 0.25°$ obtained from India Meteorological Department, Ministry of Earth Sciences (Government of India) is used to compute the modern precipitation dynamics in the Indian subcontinent (Pai et al., 2014; Fig. 14.2). The SST dataset (1854–2018 AD) was obtained from US National Oceanic, and Atmospheric Administration, Extended Reconstructed Sea Surface Temperature (ERSST) having a spatial resolution $2° \times 2°$ (Huang et al., 2017; Fig. 14.3). The Climate Data Operators (v. 1.9.8) and MATLAB (v. 2018a) based scripts were used to analyze and process the observational datasets (Schulzweida, 2019). Furthermore, Grid Analysis and Display System (v. 2.2.0) with ncview (v. 2.1.7) was used to visualize and interpolate the NetCDF data and its metadata to understand the spatial and temporal variability of rainfall pattern.

Further, the long-term climate data derived from wide range of proxy methods differ temporally and spatially, and limitations related to sample resolution, chronological uncertainties, and/or hiatus in sedimentation pertain in these paleoclimate records (Gradie and Tedesco, 1982; Hajdas, 2008; Wang et al., 2010; Mischke et al., 2013, 2017). The terrestrial paleoclimate proxies from diverse climatic regimes in the Indian subcontinent has been used to comprehend the paleohydrological changes (Fig. 14.1A). In order to provide a consistent, quality-controlled dataset, we selected terrestrial records based on certain criteria: (a) the dataset should represent moisture contribution either directly or in terms of the changes in precipitation/evaporation (P/E) in the region; (b) the dataset should have continuous records with the absence of any hiatus; (c) if multiple records are available from the same region, then the records with high-resolution datasets are chosen; and (d) since the study is focused only on Holocene climate records, therefore data points beyond 11,700 cal BP are not considered for further calculations. Additionally, based on the applicability of the proxies in the climate reconstruction, the datasets have been divided into two categories: (i) detrital—as represented by Al_2O_3 (a proxy for detrital input in lake system); magnetic parameters (χfd and χlf denote catchment contribution into the lake basins); and (ii) proxy for evaporative or drier environment (e.g., carbonate content and ^{18}O isotope).

The understanding of the behavior and sensitivity of proxies in various environmental settings is critical for reliable climate reconstructions. However, the synthesis of climate records based on multiple proxies is difficult, as the proxies show various degree of amplitude depending on factors such as type of archives, proxy sensitivity, and noise in the calibrated data (i.e., dependent on instrument sensitivity and lab standards). The biasness associated with the varying amplitude of a proxy and irregularly sampled time series may result in the unprecedented partiality in the results (Amrhein, 2020). Therefore, to compare all paleoclimate records, the datasets have been normalized using "*min–max normalization*" method (Covelli and Fontolan, 1997). The method is based on linear transformation, which converts the original data in the normalized scale between 0 and 1 to reduce the amplitude effect

$$\text{Normalized } value = (X_i - \text{min}_{\text{data}})/(\text{max}_{\text{data}} - \text{min}_{\text{data}}) \tag{14.1}$$

where X_i is the original value; min_{data} is the minimum value in the data; and max_{data} is the maximum value in the data.

14.4 Observations and results

In paleoclimate studies, the careful investigation of proxy data is crucial for climate reconstruction, as the proxy response is often associated with nonlinear interaction between various factors, such as (i) proxy sensitivity, (ii) interaction between biotic and abiotic components, (iii) role of varying environmental conditions, and (iv) anthropogenic activity (Lotter, 2003; Birks and Birks, 2006; Mudelsee et al., 2012). The nonstationarity of the proxy behavior limits its utilization to understand their response

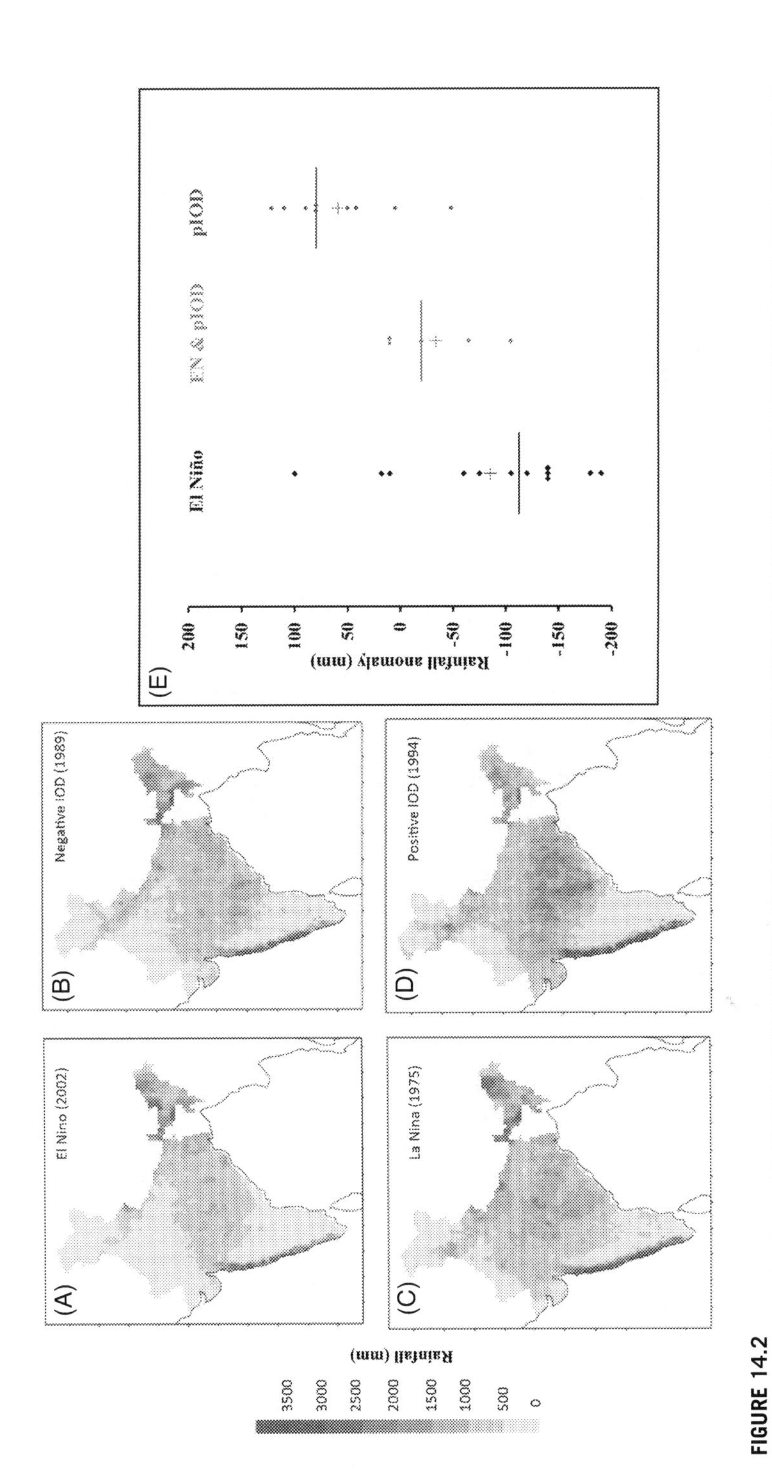

FIGURE 14.2

Total rainfall (JJAS) during (A) El Niño; (B) negative IOD; (C) La Niña; and (D) positive IOD years; (E) rainfall anomalies shown as diamond for El Niño (black), co-occurring El Niño and pIOD (green), and pIOD (red) events.

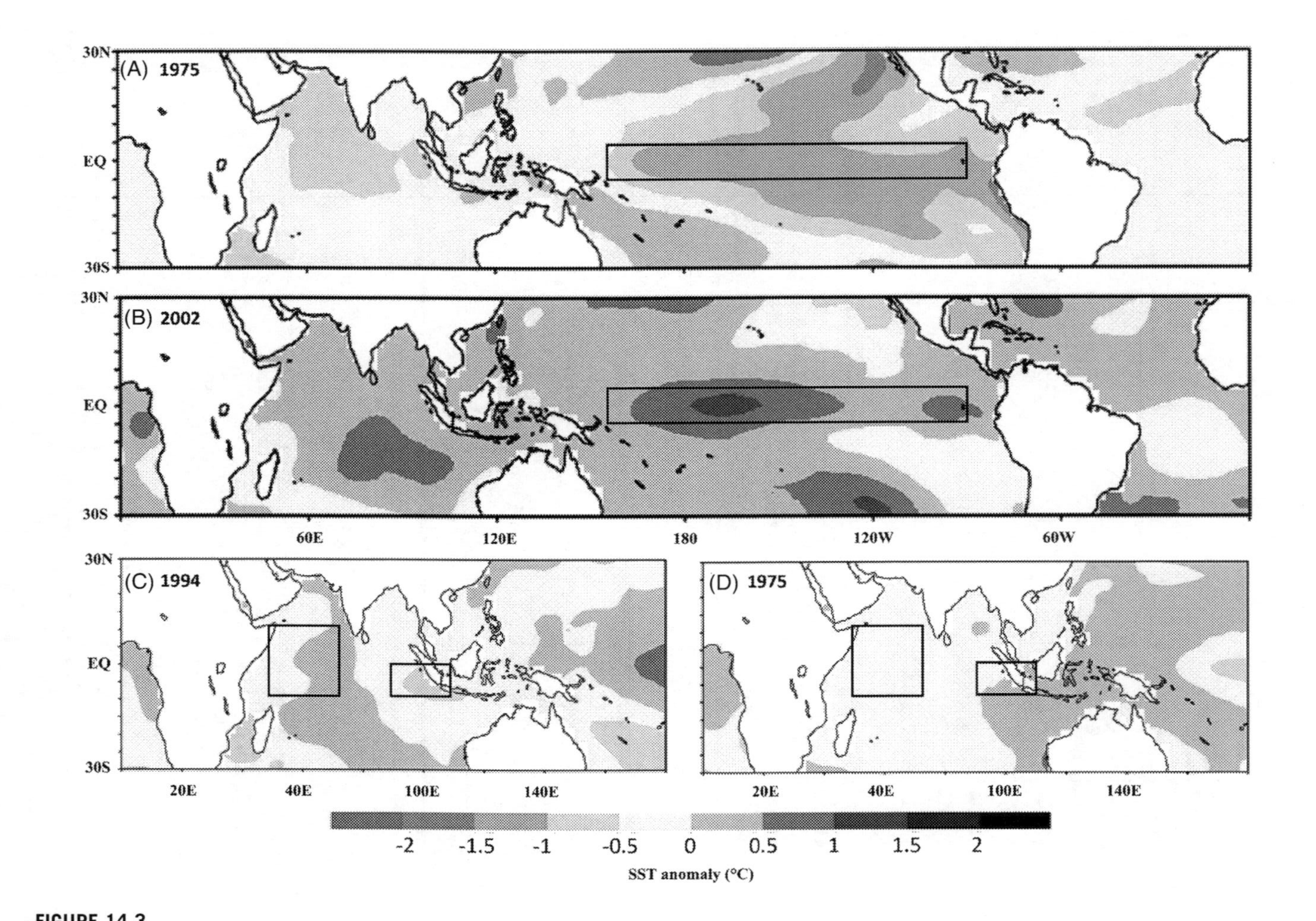

FIGURE 14.3

SST anomaly during (A) La Nina; (B) El Nino; (C) positive IOD; and (D) negative IOD condition.

over the varied climatic conditions (Wu et al., 2006; Chen et al., 2015; Prasad et al., 2020). Therefore, it is crucial to apprehend the proxy response based on the development of the proxies according to change in the climate variables, and thus can be comprehend to understand various external (e.g., sediment transportation, weathering in the catchment), as well as the internal processes (e.g., lake mixing, stratification, aquatic productivity). The behavior of the proxies (e.g., elemental concentrations and ratios, stable isotopes ($\delta^{13}C$ and $\delta^{18}O$) in varying environmental conditions) can be linked to either temperature or precipitation change in the region (Leipe et al., 2013; Basu et al., 2015; Wang et al., 2016; Saini et al., 2017; Sandeep et al., 2017; Misra et al., 2020). However, in this chapter, we have only focused on the proxies representing moisture variability in the Indian monsoon realm. The climate-sensitive proxies directly affected by the rainfall include elemental concentrations (Al_2O_3, TiO_2), elemental ratios (e.g., Mg/Al, Ca/Al), and magnetic parameters (χ_{fd} and χ_{lf}). Conversely, proxies such as carbonate content, stable isotopes of oxygen, and carbon are influenced by local hydrological changes and physicochemical conditions of the lake system (Opitz et al., 2012; Prasad et al., 2014, 2020; Warrier et al., 2017; Bhushan et al., 2018; Kämpf et al., 2020). Therefore, a multilevel approach (i.e., station-based measurements and proxy calibrations) is crucial to understand the proxy response in diverse environmental conditions.

14.4.1 Proxy for surface runoff

The lake sediments are characterized by the cumulative effect of both the catchment (e.g., weathering, sediment transport) and the in situ lake processes imprinting the environmental conditions (e.g., sediment mixing, evaporation, and productivity changes; Sun et al., 2010; Whitlock et al., 2012; Minyuk et al., 2013; Wennrich et al., 2013; Mishra et al., 2014; Lü et al., 2016). The intense precipitation enhances the chemical weathering in the watershed resulting in an increased concentration of dissolved ions in the stream water (Das and Haake, 2003). The elemental concentrations and ratios (e.g., Al, Ti, Fe, etc.) have been previously used as a proxy indicating higher surface runoff and consequently higher moisture contribution in the region (Das and Haake, 2003; Jin et al., 2006; Limmer et al., 2012; Mir and Mir, 2019; Ankit et al., 2017). For instance, variability in weathering proxies (such as Zr/Al, Ti/Al, and Rb/Sr) and the increase in Al_2O_3 concentration in Benital Lake (NW Himalaya) during the early-Holocene has been interpreted as an enhanced surface runoff associated with the intensification of ISM precipitation (Bhushan et al., 2018; Fig. 14.4A). Likewise, the study from Tso Moriri Lake used the relative contribution of Ti% to understand the ISM intensification in the region (Mishra et al., 2015a). In contrast, the variability of Al_2O_3 from Lonar core sediments (central India) represent the shoreline proximity (Prasad et al., 2014; Fig. 14.4C). This inference has been established based on the modern calibration of detrital input (Al_2O_3) into the lake basin (Basavaiah et al., 2014; Prasad et al., 2014).

The paleorecord from central (Nonia Tal) and peninsular India (Shantisagara Lake) have demonstrated the applicability of environmental magnetic parameters as a proxy to reconstruct paleohydrological changes (Sandeep et al., 2017; Kumar et al., 2018; Fig. 14.4B, D). However, the contribution of magnetic minerals from different sources (bacterial magnetite, authigenic greigite, diagenesis of magnetic minerals, and industrial waste; Basavaiah et al., 2014; Warrier et al., 2017) limits the utilization of these proxies as precipitation indicators. Therefore, modern calibrations linking magnetic mineral assemblages to climate variables are warranted for the use of magnetic proxy for paleoprecipitation

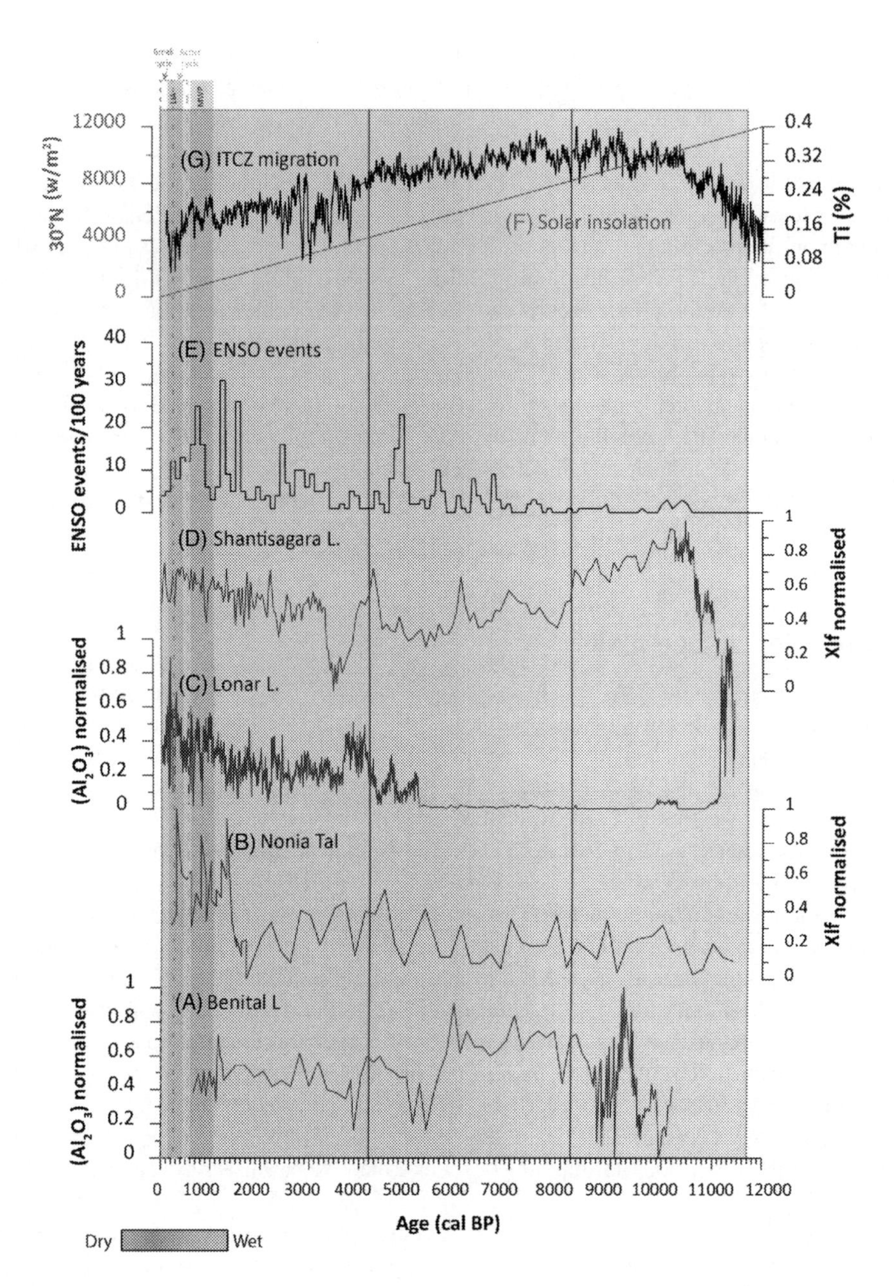

FIGURE 14.4

Temporal variability of detrital proxies and forcing factors representing Holocene climate variability. (A) Benital Lake (Bhushan et al., 2018); (B) Nonia Tal (Kumar et al., 2018); (C) Lonar Lake (Prasad et al., 2014); (D) Shantisagara Lake (Sandeep et al., 2017); (E) ENSO events (Moy et al., 2002); (F) Solar insolation curve (Berger and Loutre, 1991); and (G) Ti (%) from Cariaco basin off the Venezuelan coast (Haug et al., 2014).

reconstruction (Sandeep et al., 2017). Overall, elemental ratios and their variability provide valuable information on the hydrological changes in the Indian subcontinent.

14.4.2 Evaporative proxies

The water-level in hydrological systems is largely controlled by the precipitation (P) and evaporation (E) processes. The relative contribution between these two components (P/E ratio) provides crucial information about past hydrological changes in the lake basin. The evaporative process in a lake system can lead to changes in the physicochemical condition of the lake water, and increased pH condition (alkaline; >7 pH) results in the residual water supersaturated with Ca^{2+} and CO_3^{2-} and precipitation of carbonate mineral (Wetzel, 2001; White, 2013). The carbonate mineral content and their isotope ($\delta^{13}C$ and $\delta^{18}O$) data have been widely used to understand the paleoprecipitation changes (Dixit et al., 2014; Mishra et al., 2014, 2015a; Prasad et al., 2014; Babeesh et al., 2017; Pillai et al., 2017). The high correlation between $\delta^{13}C$ and $\delta^{18}O$ in the lake sediment is characteristic of a hydrologically closed system (Talbot, 1990; Mishra et al., 2015a). However, factors such as lake alkalinity and lake productivity may also influence $\delta^{13}C$ and $\delta^{18}O$ composition of lake sediments (Li and Ku, 1997). Conversely, the use of $\delta^{13}C$ and $\delta^{18}O$ in speleothem as a proxy for paleoclimatic reconstruction is complicated due to the complex interaction between cave microclimate, rainfall condition in the region, infiltration, and drip seasonality (Lachniet, 2009). In addition, changes in the moisture sources, rainfall amount, and temperature-dependent fractionation between water and calcite may also affect the isotopic composition of drip water and thus $\delta^{18}O$ of precipitated carbonate (Yadava and Ramesh, 2005a; Lachniet, 2009; Berkelhammer et al., 2012; Liu et al., 2018).

The variation in $\delta^{18}O$ and/or carbonate content records from lake sediments or speleothems from Indian subcontinent provides evidence for significant spatial and temporal variability in monsoonal precipitation over Indian subcontinent (Enzel et al., 1999; Fleitmann et al., 2003; Berkelhammer et al., 2012; Dixit et al., 2014; Band et al., 2018; Dutt et al., 2020). The majority of results shows an early-Holocene wet phase (depleted $\delta^{18}O$ or decreased carbonate concentration) followed by declining monsoon precipitation during mid–late-Holocene punctuated by two major colder events around 8200 and 4200 cal BP (Berkelhammer et al., 2012; Misra et al., 2018; Banerji et al., 2020; and reference therein).

14.4.3 Observation and reanalysis data

The spatio-temporal changes in Indian Summer Monsoon rainfall to the various modes of teleconnections have been typically identified using observational and modeled data. The temporally and spatial averaged summer monsoon rainfall (Reference year—1972) over Indian subcontinent for a period of 1901–2014 is illustrated in Fig. 14.1A, B. The considerable fluctuations in interannual and intraseasonal monsoonal precipitation over Indian subcontinent are related to the various large-scale convective and circulation patterns in the atmosphere (e.g., ENSO, IOD, NAO, Active and Break spells; Gadgil and Joseph, 2003; Gadgil et al., 2004; Krishnan et al., 2006; Rajeevan et al., 2010). However, ENSO and IOD exert dominant control over the precipitation in the Indian subcontinent (Krishnaswamy et al., 2014; Hrudya et al., 2020). To assess the teleconnection patterns with regional and global circulation features, independent IOD and ENSO years are identified and plotted in Fig. 14.2. Further, the co-occurrence of ENSO events to the pIOD (positive IOD) mode is explored to understand the impact of IOD on the ENSO-Indian monsoon teleconnections (Ummenhofer et al., 2011). The assessment of the influences for the dominant Indo-pacific modes of climate variability, anomalous rainfall during El Niño, pIOD, and co-occurring El Niño with pIOD events is presented in Fig. 14.2E. Additionally, SST anomalies derived from the ERSST datasets for the respective years are plotted in Fig. 14.3. The spatial distribution of monsoonal precipitation during the year 2002 (El Niño) and 1989 (negative IOD) shows subsidence in the ISM rainfall (Fig. 14.2A, B). In contrast, the surplus in the summer monsoon rainfall is observed for the year 1975 (La Niña) and 1994 (positive IOD; Fig. 14.2C, D). The SST anomaly in the tropical pacific for the respective years of El Niño and La Niña clearly shows the bimodal variation in the SST (Fig. 14.3). Although ENSO is considered to be closely tied to ISM, but its negative inter-relationship has weakened in the recent decades probably due to interplay of atmosphere–ocean interactions over the Indian Ocean (Hrudya et al., 2020).

14.5 Discussion

The global circulation model shows that the long-term ISM evolution is largely influenced by orbital parameters (e.g., solar insolation), Tibet-Himalaya landscape evolution, atmospheric CO_2 concentration, and glacial-age surface boundary conditions (Prell and Kutzbach, 1992). The Holocene climate is impacted by various teleconnections (e.g., ENSO, IOD, and NAO) and anthropogenic inputs (Von Rad et al., 1999; Conroy et al., 2008b; Laskar et al., 2013; Menzel et al., 2014; Srivastava et al., 2017; Amir et al., 2020; Hernández et al., 2020). However, the interpolated instrumental data demonstrated the strong influence of ENSO and IOD on regional precipitation in the Indian monsoon realm (Hernández et al., 2020; Hrudya et al., 2020). The majority of the proxy datasets from spatially different climate record from lake sediments (e.g., Tso Moriri, Tso Kar, Shantisagara Lake, Benital Lake; Wünnemann et al., 2010; Mishra et al., 2015b; Sandeep et al., 2017; Bhushan et al., 2018) and speleothems (Borar, Oman, and Mawmluh cave; Fleitmann et al., 2007; Berkelhammer et al., 2010; Singh, 2018) shows increased precipitation during early-Holocene (11.5–8.2 cal ka; Walker et al., 2012) in response to northward migration of Intertropical Convergence Zone and maximum solar insolation (Berger and Loutre, 1991; Haug et al., 2014; Figs. 14.4 and 14.5). However, a few of the climatic records (e.g., Chopta valley in NE India, Tso Kar in NW Himalayas) demonstrate a significant lag in the early-Holocene peak probably due to the age uncertainties, interaction between different moisture sources, and influence of various teleconnections (Wünnemann et al., 2006; Menzel et al., 2014; Ali et al., 2018).

The mid-Holocene (8.2–4.2 cal ka; Walker et al., 2012) climate variability in the Indian monsoon domain is characterized by a declining phase of monsoonal precipitation in response to reduced solar insolation (Fleitmann et al., 2007; Demske et al., 2009; Ponton et al., 2012; Sandeep et al., 2017; Figs. 14.4 and 14.5). In contrast to the drying trend in the ISM dominated region, the second peak of monsoon precipitation (during mid-Holocene) has been observed in several climate records from NW Himalaya (Demske et al., 2009; Anoop et al., 2013; Bhushan et al., 2018). The paleoclimate records from Nal Sarovar and Lunkaransar Lake in NW India and paleolake sediments from Kotla Dahar region (in the west of Indo-Gangetic plain) have also shown a significant higher precipitation condition during the mid-Holocene probably due to the contribution from mid-latitude westerlies (winter precipitation; Dimri et al., 2016; Fig. 14.1A). However, these hydrological changes are not consistent in all paleo-records from the Indian subcontinent. The modern observational data from northern and northwestern India demonstrate that the region received significant precipitation during nonmonsoonal months (i.e., December, January, and February, DJF) associated with western disturbance (Midhuna et al., 2020). Therefore, these regions are characterized by distinct climatic regimes with differing precipitation due

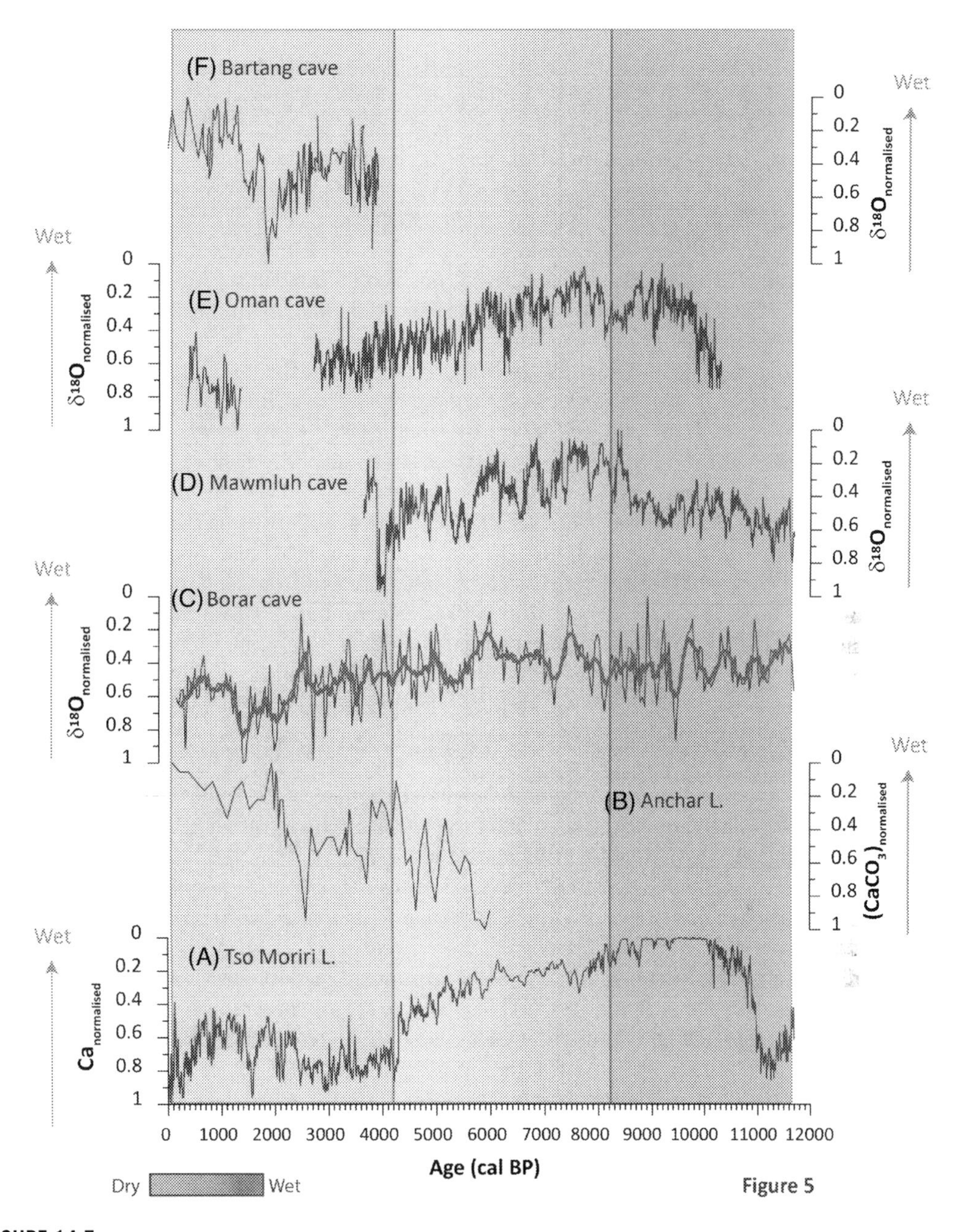

FIGURE 14.5

Temporal variability of evaporative proxies in various lakes and speleothem records. (A) Tso Moriri Lake (Mishra et al., 2015a); (B) Anchar Lake (Lone et al., 2020); (C) Borar cave (Singh, 2018); (D) Mawmluh cave (Berkelhammer et al., 2012); (E) Oman cave (Fleitmann et al., 2007); and (F) Bartang cave (Laskar et al., 2013).

to complex interaction between ISM and mid-latitude westerlies (Benn and Owen, 1998; Dixit et al., 2014; Mishra et al., 2015b; Srivastava et al., 2017; Bhushan et al., 2018; Lone et al., 2020). The result is well corroborated with the paleorecords from the arid central Asia (region influenced by only mid-latitude westerlies) showing highest moisture availability from mid-latitude westerlies during the mid-Holocene (Chen et al., 2008). Furthermore, the $\delta^{18}O$ record from central and peninsular India (Lonar and Shantisagara Lake) shows an overall decline in monsoonal rainfall with short fluctuation of ISM activity during mid-Holocene probably linked to the strengthening of El Niño events (Prasad et al., 2014; Emile-Geay et al., 2016; Crétat et al., 2020).

The late-Holocene witnessed a declining trend in ISM variability with several short-term climatic events resulted due to the interaction between natural (solar insolation, Intertropical Convergence Zone migration coupled with El Niño events) and amplified anthropogenic (e.g., increase in the greenhouse gases, such as CO_2 and CH_4) stressors (Yadava and Ramesh, 2005b; Rehfeld et al., 2013; Sinha et al., 2015; Kathayat et al., 2016). In NW Himalaya, the speleothem records from Sainji cave indicates the weakening of monsoon from 4200 to 3200 cal BP resulting in the disappearance of Harappan and Ghaggar civilization (Kotlia et al., 2015). The drying trend is also supported by the lake sediment records from NW Himalaya (Tso Moriri and Tso Kar Lake; Demske et al., 2009; Mishra et al., 2015b) and southern Tibetan Plateau (Paru Co, Nir'pa Lake; Bird et al., 2014, 2017). In central India, the declining ISM and intensification of regional aridity are well recorded in Lonar Lake (4600–3900 and 1400–600 cal BP); Nonia Tal (~3000–2000 cal BP) and sediment record from Godavari basin (~4000 to ~1700 cal BP; Ponton et al., 2012; Prasad et al., 2014; Kumar et al., 2018) in response to variation in the seasonality of northern hemisphere insolation, and strengthening of El-Niño events (Conroy et al., 2008a). Additionally, the latter phase of the late-Holocene is characterized by several short-term centennial scale climatic events such as MCA or MWP (~1350–650 cal BP) and LIA (450–150 cal BP) as evident from paleoclimate records from Indian subcontinent (Wanner et al., 2008; Polanski et al., 2013; Dixit and Tandon, 2016; Rajmanickam et al., 2017; Singh et al., 2020). The MWP is marked by warmer climate along with higher precipitation, whereas LIA is depicted as colder climate event linked to solar minima, change in thermohaline circulation and volcanic activity (Anderson et al., 2002; Sinha et al., 2011; Dixit and Tandon, 2016; Rajmanickam et al., 2017; Ali et al., 2018; Ghosh et al., 2018a; Gupta et al., 2019). These small events were recorded in several terrestrial as well as marine records from the ISM dominated region (Rawat et al., 2015; Dixit and Tandon, 2016; Rajmanickam et al., 2017; Giosan et al., 2018; Singh et al., 2020). Although their timing and interval are not consistent throughout the region indicates interplay between different moisture regimes, and role of various teleconnections (such as ENSO, IOD, and Active–Break spells; Tiwari et al., 2003; Sinha et al., 2011; Dixit and Tandon, 2016). However, the ISM rainfall variability in the late 20th century is being associated with IOD events, corroborated with the weakening relationship of ENSO and ISM rainfall (Ashok et al., 2001, 2004; Ashok and Saji, 2007). Moreover, the reconstruction of IOD intensity and its variability during the Holocene would provide a quantitative understanding in the context of natural and forced climatic variability (Abram et al., 2015).

14.6 Learning and knowledge outcomes

The variability in the monsoon precipitation and its impact on socioeconomic status has motivated to perform a detailed investigation into understanding of teleconnections, forcing factors, and their validation in terms of observational data. In this study, we have presented an overview of hydrological changes since Holocene using proxy reconstruction methods and analysis of observational datasets to understand the boundary conditions and different mode of climate variability responsible for climatic fluctuations over the Indian subcontinent. The understanding of proxy-based climate reconstruction and the outlined mechanisms is essential to understand the present climatic condition with minimal uncertainty. However, due to multiple factors, such as (i) proxy behavior and their response to modern environmental condition; (ii) uncertainties in chronological data; and (iii) lack of interdisciplinary approach, it is challenging to understand the boundary condition in the past, and their implication for future climatic study. In addition, the calibration and testing of proxies according to change in modern meteorological conditions and their link with instrumental/observational data are also crucial to establish a better perspective on the temporal evolution of climate system and its response to various teleconnections. Further, the study has highlighted the close correspondence between ISM rainfall and teleconnections such as IOD and ENSO. However, along with modern observation data, high-resolution paleoclimate data are required to validate the response of these dipole modes for future climate variability.

Acknowledgments

The authors are thankful to Dr. K.P.N. Kumaran for inviting us to contribute a chapter to this book. A.A. and P.K.M. gratefully acknowledge the financial support provided by INSPIRE Faculty Fellowship from Department of Science and Technology.

References

Abram, N.J., Bronwyn, C., Dixon, M.G., Rosevear, B.P., Michael, K.G., Wahyoe, S.H., Steven, J.P., Optimized coral reconstructions of the Indian Ocean Dipole: An assessment of location and length considerations. Paleoceanogr. Paleoclimatol. 30 (10) (2015) 1391–1405. doi:org/10.1002/2015PA002810.

Abram, N.J., Hargreaves, J.A., Wright, N.M., Thirumalai, K., Ummenhofer, C.C., England, M.H., Palaeoclimate perspectives on the Indian Ocean Dipole, Quat. Sci. Rev. 237 (2020) 106302, doi:10.1016/j.quascirev.2020.106302.

Ali, S.N., Dubey, J., Ghosh, R., Quamar, M.F., Sharma, A., Morthekai, P., Dimri, A.P., Shekhar, M., Arif, M., Agrawal, S., High frequency abrupt shifts in the Indian summer monsoon since Younger Dryas in the Himalaya, Sci. Rep. 8 (2018) 1–8, doi:10.1038/s41598-018-27597-6.

Amir, M., Paul, D., Malik, J.N., Geochemistry of Holocene sediments from Chilika Lagoon, India: inferences on the sources of organic matter and variability of the Indian summer monsoon, Quat. Int. (2020), doi:10.1016/j.quaint.2020.08.050.

Amrhein, D.E., How large are temporal representativeness errors in paleoclimatology? Clim. Past 16 (2020) 325–340, doi:10.5194/cp-16-325-2020.

Anderson, D.M., Overpeck, J.T., Gupta, A.K., Increase in the Asian southwest monsoon during the past four centuries, Sci. 297 (5581) (2002) 596–599.

Ankit, Y., Kumar, P., Anoop, A., Mishra, P.K., Varghese, S., Mid-late Holocene climate variability in the Indian monsoon: evidence from continental shelf sediments adjacent to Rushikulya river, eastern India, Quat. Int. 443 (2017) 155–163, doi:10.1016/j.quaint.2016.12.023.

Anoop, A., Prasad, S., Krishnan, R., Naumann, R., Dulski, P., Intensified monsoon and spatiotemporal changes in precipitation patterns in the NW Himalaya during the early-mid Holocene, Quat. Int. 313–314 (2013) 74–84, doi:10.1016/j.quaint.2013.08.014.

Ashok, K., Guan, Z., Saji, N.H., Yamagata, T., Individual and combined influences of ENSO and the Indian Ocean dipole on the Indian summer monsoon, J. Clim. 17 (16) (2004) 3141–3155.

Ashok, K., Guan, Z., Yamagata, T., Impact of the Indian Ocean dipole on the relationship between the Indian monsoon rainfall and ENSO, Geophys. Res. Lett. 28 (2001) 4499–4502, doi:10.1029/2001GL013294.

Ashok, K., Saji, N.H., On the impacts of ENSO and Indian Ocean dipole events on sub-regional Indian summer monsoon rainfall, Nat. Hazards 42 (2007) 273–285.

Babeesh, C., Lone, A., Achyuthan, H., Geochemistry of Manasbal lake sediments, Kashmir: weathering, provenance and tectonic setting, J. Geol. Soc. India 89 (2017) 563–572, doi:10.1007/s12594-017-0645-4.

Band, S., Yadava, M.G., Lone, M.A., Shen, C.C., Sree, K., Ramesh, R., High-resolution mid-Holocene Indian Summer Monsoon recorded in a stalagmite from the Kotumsar Cave, Central India, Quat. Int. 479 (2018) 19–24, doi:10.1016/j.quaint.2018.01.026.

Banerji, U.S., Arulbalaji, P., Padmalal, D., Holocene climate variability and Indian Summer Monsoon: an overview, Holocene 30 (5) (2020) 744–773.

Basavaiah, N., Wiesner, M.G., Anoop, A., Menzel, P., Nowaczyk, N.R., Deenadayalan, K., Brauer, A., Gaye, B., Naumann, R., Riedel, N., Stebich, M., Prasad, S., Physicochemical analyses of surface sediments from the Lonar Lake, central India—implications for palaeoenvironmental reconstruction, Fundam. Appl. Limnol. 184 (2014) 51–68, doi:10.1127/1863-9135/2014/0515.

Basu, S., Agrawal, S., Sanyal, P., Mahato, P., Kumar, S., Sarkar, A., Carbon isotopic ratios of modern C3–C4 plants from the Gangetic Plain, India and its implications to paleovegetational reconstruction, Palaeogeogr. Palaeoclimatol. Palaeoecol. 440 (2015) 22–32, doi:10.1016/j.palaeo.2015.08.012.

Benn, D.I., Owen, L.A., The role of the Indian summer monsoon and the mid-latitude westerlies in Himalayan glaciation: review and speculative discussion. Journal of the Geolo. Soc. 155 (1998) 353–363.

Berger, A., Loutre, M.F., Insolation values for the climate of the last 10 million years, Quat. Sci. Rev. 10 (1991) 297–317, doi:10.1016/0277-3791(91)90033-Q.

Berkelhammer, M., Sinha, A., Mudelsee, M., Cheng, H., Edwards, R.L., Cannariato, K., Persistent multidecadal power of the Indian Summer Monsoon, Earth Planet. Sci. Lett. 290 (2010) 166–172, doi:10.1016/j.epsl.2009.12.017.

Berkelhammer, M., Sinha, A., Stott, L., Cheng, H., Pausata, F.S.R., Yoshimura, K., An abrupt shift in the Indian monsoon 4000 years ago, Geophys. Monogr. Ser. 198 (2012) 75–87, doi:10.1029/2012GM001207.

Bhushan, R., Sati, S.P., Rana, N., Shukla, A.D., Mazumdar, A.S., Juyal, N., High-resolution millennial and centennial scale Holocene monsoon variability in the Higher Central Himalayas, Palaeogeogr. Palaeoclimatol. Palaeoecol. 489 (2018) 95–104, doi:10.1016/j.palaeo.2017.09.032.

Bird, B.W., Lei, Y., Perello, M., Polissar, P.J., Yao, T., Finney, B., Bain, D., Pompeani, D., Thompson, L.G., Late-Holocene Indian summer monsoon variability revealed from a 3300-year-long lake sediment record from Nir'pa Co, southeastern Tibet, Holocene 27 (2017) 541–552, doi:10.1177/0959683616670220.

Bird, B.W., Polisar, P.J., Lei, Y., Thompson, L.G., Yao, T., Finney, B.P., Bain, D.J., Pompeani, D.P., Steinman, B.A., A Tibetan lake sediment record of Holocene Indian summer monsoon variability, Earth Planet. Sci. Lett. 399 (2014) 92–102, doi:10.1016/j.epsl.2014.05.017.

Birks, H.H., Birks, H.J.B., Multi-proxy studies in palaeolimnology, Veg. Hist. Archaeobot. 15 (2006) 235–251, doi:10.1007/s00334-006-0066-6.

Chen, F., Xu, Q., Chen, J., Birks, H.J.B., Liu, J., Zhang, S., Jin, L., An, C., Telford, R.J., Cao, X., Wang, Z., Zhang, X., Selvaraj, K., Lu, H., Li, Y., Zheng, Z., Wang, H., Zhou, A., Dong, G., Zhang, J., Huang, X., Bloemendal, J., Rao, Z., East Asian summer monsoon precipitation variability since the last deglaciation, Sci. Rep. 5 (2015) 1–11, doi:10.1038/srep11186.

Chen, F., Yu, Z., Yang, M., Ito, E., Wang, S., Madsen, D.B., Huang, X., Zhao, Y., Sato, T., Birks, H.J.B., Boomer, I., Chen, J., An, C., Wu, B., Holocene moisture evolution in arid central Asia and its out-of-phase relationship with Asian monsoon history. Quat. Sci. Rev. 27 (2008) 351–364. doi:10.1016/j.quascirev.2007.10.017.

Clemens, S., Prell, W., Murray, D., Shimmield, G., Weedon, G., Forcing mechanisms of the Indian Ocean monsoon, Nat. 353 (6346) (1991) 720–725.

Conroy, J.L., Overpeck, J.T., Cole, J.E., Shanahan, T.M., Steinitz-Kannan, M., Holocene changes in eastern tropical Pacific climate inferred from a Galápagos lake sediment record, Quat. Sci. Rev. 27 (2008b) 1166–1180, doi:10.1016/j.quascirev.2008.02.015.

Covelli, S., Fontolan, G., Application of a normalization procedure in determining regional geochemical baselines. Environ. Geol. 30 (1–2) (1997) 34–45.

Crétat, J., Braconnot, P., Terray, P., Marti, O., Falasca, F., Mid-Holocene to present-day evolution of the Indian monsoon in transient global simulations, Clim. Dyn. 55 (9) (2020) 2761–2784.

Das, B.K., Haake, B.-G., Geochemistry of Rewalsar Lake sediment, Lesser Himalaya, India: implications for source-area weathering, provenance and tectonic setting, Geosci. J. 7 (2003) 299–312, doi:10.1007/BF02919560.

Demske, D., Tarasov, P.E., Wünnemann, B., Riedel, F., Late glacial and Holocene vegetation, Indian monsoon and westerly circulation in the Trans-Himalaya recorded in the lacustrine pollen sequence from Tso Kar. Palaeogeogr. Palaeoclimatol. Palaeoecol. 279 (2009) 172–185. doi:10.1016/j.palaeo.2009.05.008.

Dimri, A.P., Yasunari, T., Kotlia, B.S., Mohanty, U.C., Sikka, D.R., Indian winter monsoon: present and past, Earth Sci. Rev. 163 (2016) 297–322.

Dixit, Y., Hodell, D.A., Petrie, C.A., Abrupt weakening of the summer monsoon in northwest India ~4100 yr ago, Geology 42 (2014) 339–342, doi:10.1130/G35236.1.

Dixit, Y., Tandon, S.K., Hydroclimatic variability on the Indian subcontinent in the past millennium: review and assessment, Earth Sci. Rev. 161 (2016) 1–15, doi:10.1016/j.earscirev.2016.08.001.

Dutt, S., Gupta, A.K., Cheng, H., Clemens, S.C., Singh, R.K., Tewari, V.C., Indian summer monsoon variability in northeastern India during the last two millennia, Quat. Int. xx (2020) xx–xx.

Emile-Geay, J., Cobb, K.M., Carré, M., Braconnot, P., Leloup, J., Zhou, Y., Harrison, S.P., Corrège, T., McGregor, H.V., Collins, M., Driscoll, R., Links between tropical Pacific seasonal, interannual and orbital variability during the Holocene, Nat. Geosci. 9 (2) (2016) 168–173.

Enzel, Y., Ely, L.L., Mishra, S., Ramesh, R., Amit, R., Lazar, B., Rajaguru, S.N., Baker, V.R., Sandler, A., High-resolution holocene environmental changes in the Thar Desert, northwestern India, Science 284 (1999) 125–128, doi:10.1126/science.284.5411.125.

Fleitmann, D., Burns, S.J., Mangini, A., Mudelsee, M., Kramers, J., Villa, I., Neff, U., Al-Subbary, A.A., Buettner, A., Hippler, D., Matter, A., Holocene ITCZ and Indian monsoon dynamics recorded in stalagmites from Oman and Yemen (Socotra), Quat. Sci. Rev. 26 (2007) 170–188, doi:10.1016/j.quascirev.2006.04.012.

Fleitmann, D., Burns, S.J., Mudelsee, M., Neff, U., Kramers, J., Mangini, A., Matter, A., Holocene forcing of the Indian monsoon recorded in a stalagmite from Southern Oman, Science 300 (2003) 1737–1739, doi:10.1126/science.1083130.

Gadgil, S., The Indian monsoon and its variability, Annu. Rev. Earth Planet. Sci. 31 (2003) 429–467, doi:10.1146/annurev.earth.31.100901.141251.

Gadgil, S., Gadgil, S., The Indian monsoon, GDP and agriculture, Econ. Polit. Wkly. 41 (2006) 4887–4895.

Gadgil, S., Joseph, P.V., On breaks of the Indian monsoon, Proc. Indian Acad. Sci. Earth Planet. Sci. 112 (2003) 529–558, doi:10.1007/BF02709778.

Gadgil, S., Vinayachandran, P.N., Francis, P.A., Gadgil, S., Extremes of the Indian summer monsoon rainfall, ENSO and equatorial Indian Ocean oscillation, Geophys. Res. Lett. 31 (2004) 2–5, doi:10.1029/2004GL019733.

Ghosh, R., Biswas, O., Paruya, D.K., Agrawal, S., Sharma, A., Nautiyal, C.M., Bera, M., Bera, S., Hydroclimatic variability and corresponding vegetation response in the Darjeeling Himalaya, India over the past ~2400 years, Catena 170 (2018) 84–99, doi:10.1016/j.catena.2018.05.043.

Ghosh, R., Paruya, D.K., Khan, M.A., Chakraborty, S., Sarkar, A., Bera, S., Late Quaternary climate variability and vegetation response in Ziro Lake Basin, Eastern Himalaya: a multiproxy approach, Quat. Int. 325 (2014) 13–29, doi:10.1016/j.quaint.2014.01.046.

Ghosh, S., Sanyal, P., Roy, S., Bhushan, R., Sati, S.P., Philippe, A., Juyal, N., Early Holocene Indian summer monsoon and its impact on vegetation in the Central Himalaya: Insight from δD and δ13C values of leaf wax lipid. The Holocene 30 (7) (2020) 1063–1074.

Ghosh, Subimal, Vittal, H, Sharma, Tarul, Karmakar, Shubhankar, Kasiviswanathan, K.S., Dhanesh, Y., Sudheer, K.P., Gunthe, S.S., Indian Summer Monsoon Rainfall: Implications of Contrasting Trends in the Spatial Variability of Means and Extremes. Plos One 7 (11) (2016) e0158670. doi:doi.org/10.1371/journal.pone.0158670.

Giosan, L., Orsi, W.D., Coolen, M., Wuchter, C., Dunlea, A.G., Thirumalai, K., Munoz, S.E., Clift, P.D., Donnelly, J.P., Galy, V., Fuller, D.Q., Neoglacial climate anomalies and the Harappan metamorphosis, Clim. Past 14 (2018) 1669–1686, doi:10.5194/cp-14-1669-2018.

Goswami, B.N., Venugopal, V., Sangupta, D., Madhusoodanan, M.S., Xavier, P.K., Increasing trend of extreme rain events over India in a warming environment, Science 314 (2006) 1442–1445, doi:10.1126/science.1132027.

Gradie, J., Tedesco, E., Compositional structure of the asteroid belt, Science 216 (1982) 1405–1407, doi:10.1126/science.216.4553.1405.

Gupta, A.K., Dutt, S., Cheng, H., Singh, R.K., Abrupt changes in Indian summer monsoon strength during the last ~900 years and their linkages to socio-economic conditions in the Indian subcontinent, Palaeogeogr. Palaeoclimatol. Palaeoecol. 536 (2019) 109347.

Ha, K.J., Kim, B.H., Chung, E.S., Chan, J.C., Chang, C.P., Major factors of global and regional monsoon rainfall changes: natural versus anthropogenic forcing, Environ. Res. Lett. 15 (3) (2020) 034055.

Hajdas, I., Radiocarbon dating and its applications in Quaternary studies, E&G Quat. Sci. J. 57 (2008) 2–24, doi:10.3285/eg.57.1-2.1.

Haug, G.H., Haug, G.H., Hughen, K.A., Sigman, D.M., Peterson, L.C., Ro, U., Southward Migration of the Intertropical Convergence Zone Through the Holocene. Sci. 293 (5533) (2014) 1304–1308. doi:10.1126/science.1059725.

Hernández, A., Martin-Puertas, C., Moffa-Sánchez, P., Moreno-Chamarro, E., Ortega, P., Blockley, S., Cobb, K.M., Comas-Bru, L., Giralt, S., Goosse, H., Luterbacher, J., Martrat, B., Muscheler, R., Parnell, A., Pla-Rabes, S., Sjolte, J., Scaife, A.A., Swingedouw, D., Wise, E., Xu, G., Modes of climate variability: synthesis and review of proxy-based reconstructions through the Holocene, Earth Sci. Rev. 209 (2020) 103286, doi:10.1016/j.earscirev.2020.103286.

Hrudya, P.H., Varikoden, H., Vishnu, R., A review on the Indian summer monsoon rainfall, variability and its association with ENSO and IOD, Meteorol. Atmos. Phys. 133 (2020) 1–14, doi:10.1007/s00703-020-00734-5.

Huang, B., Thorne, P.W., Banzon, V.F., Boyer, T., Chepurin, G., Lawrimore, J.H., Menne, M.J., Smith, T.M., Vose, R.S., Zhang, H.M., Extended reconstructed Sea surface temperature, Version 5 (ERSSTv5): upgrades, validations, and intercomparisons, J. Clim. 30 (2017) 8179–8205, doi:10.1175/JCLI-D-16-0836.1.

Jha, D.K., Sanyal, P., Philippe, A., Multi-proxy evidence of Late Quaternary climate and vegetational history of north-central India: Implication for the Paleolithic to Neolithic phases. Quat. Sci. Rev. 229 (2020) 106–121.

Jin, Z., Li, F., Cao, J., Wang, S., Yu, J., Geochemistry of Daihai Lake sediments, Inner Mongolia, North China: implications for provenance, sedimentary sorting, and catchment weathering. Geomorph. 80 (2006) 147–163. doi:10.1016/j.geomorph.2006.02.006.

Kämpf, L., Plessen, B., Lauterbach, S., Nantke, C., Meyer, H., Chapligin, B., Brauer, A., Stable oxygen and carbon isotopes of carbonates in lake sediments as a paleoflood proxy, Geology 48 (2020) 3–7, doi:10.1130/G46593.1.

Kathayat, G., Cheng, H., Sinha, A., Spötl, C., Edwards, R.L., Zhang, H., Li, X., Yi, L., Ning, Y., Cai, Y., Lui, W.L., Indian monsoon variability on millennial-orbital timescales. Sci. Rep. 6 (1) (2016) 17. doi:10.1038/srep24374.

Katzenberger, A., Schewe, J., Pongratz, J., Levermann, A., Robust increase of Indian monsoon rainfall and its variability under future warming in CMIP-6 models, Earth Syst. Dyn. Discuss. 12 (2) (2020) 367–386.

Kotlia, B.S., Singh, A.K., Joshi, L.M., Dhaila, B.S., Precipitation variability in the Indian Central Himalaya during last ca. 4,000 years inferred from a speleothem record: impact of Indian Summer Monsoon (ISM) and Westerlies, Quat. Int. 371 (2015) 244–253.

Krishnan, R., Ramesh, K.V., Samala, B.K., Meyers, G., Slingo, J.M., Fennessy, M.J., Indian Ocean-monsoon coupled interactions and impending monsoon droughts. Geophysical. Res. Lett. 33 (2006) 2–5. doi:10.1029/2006GL025811.

Krishnan, R., Sanjay, J., Gnanaseelan, C., Mujumdar, M., Kulkarni, A., Chakraborty, S., Assessment of Climate Change Over the Indian Region: A Report of the Ministry of Earth Sciences (MOES), Government of India, Springer, Singapore, 2020, doi:10.1007/978-981-15-4327-2.

Kumar, K., Agrawal, S., Sharma, A., Indian Summer Monsoon Variability and Vegetation Changes in the Core Monsoon Zone, India, During the Holocene: A Multiproxy Study. The Holocene 29 (1) (2018) 110–119. doi:10.1177/0959683618804641.

Krishnaswamy, J., Vaidyanathan, S., Rajagopalan, B., Bonell, M., Sankaran, M., Bhalla, R.S., Badiger, S., Non-Stationary and Non-Linear Influence of Enso and Indian Ocean Dipole on the Variability of Indian Monsoon Rainfall and Extreme Rain Events. Clim. Dyna. 45 (2014) 175–184. doi:10.1007/s00382-014-2288-0.

Lachniet, M.S., Climatic and environmental controls on speleothem oxygen-isotope values, Quat. Sci. Rev. 28 (2009) 412–432, doi:10.1016/j.quascirev.2008.10.021.

Laskar, A.H., Yadava, M.G., Sharma, N., Ramesh, R., Late-Holocene climate in the Lower Narmada valley, Gujarat, western India, inferred using sedimentary carbon and oxygen isotope ratios, Holocene 23 (2013) 1115–1122, doi:10.1177/0959683613483621.

Leipe, C., Demske, D., Tarasov, P.E., Members, H.P., A Holocene pollen record from the northwestern Himalayan lake Tso Moriri: implications for palaeoclimatic and archaeological research. Quat. Int. 348 (2013) 93–112. doi:10.1016/j.quaint.2013.05.005.

Li, H.C., Ku, T.L., δ13C-δ18O covariance as a paleohydrological indicator for closed-basin lakes, Palaeogeogr. Palaeoclimatol. Palaeoecol. 133 (1997) 69–80, doi:10.1016/S0031-0182(96)00153-8.

Limmer, D.R., Köhler, C.M., Cooper, M.J., Tabrez, A.R., Clift, P.D., Geochemical record of Holocene to Recent sedimentation on the Western Indus continental shelf, Arab. Sea 13 (2012) 1–26, doi:10.1029/2011GC003845.

Liu, D., Wang, Y., Cheng, H., Edwards, R.L., Kong, X., Chen, S., Liu, S., Contrasting patterns in abrupt Asian summer monsoon changes in the last glacial period and the Holocene, Paleoceanogr. Paleoclimatol. 33 (2) (2018) 214–226.

Lone, A.M., Achyuthan, H., Shah, R.A., Sangode, S.J., Kumar, P., Chopra, S., Sharma, R., Paleoenvironmental shifts spanning the last ~6000 years and recent anthropogenic controls inferred from a high-altitude temperate lake: Anchar Lake, NW Himalaya, Holocene 30 (2020) 23–36, doi:10.1177/0959683619865599.

Lotter, A.F., Multi-proxy climatic reconstructions. In: Mackay, A.W., Battarbee, R.W., Birks, H.J.B., Old-field(eds)Global, F. (Eds.), Change in the Holocene. E. Arnold, London, (2003) pp. 373–383.

Lü, X., Versteegh, G.J.M., Song, J., Li, X., Yuan, H., Li, N., Geochemistry of Middle Holocene sediments from south Yellow Sea: implications to provenance and climate change, J. Earth Sci. 27 (2016) 751–762, doi:10.1007/s12583-015-0577-0.

Madhura, R.K., Krishnan, R., Revadekar, J.V., Mujumdar, M., Goswami, B.N., Changes in western disturbances over the Western Himalayas in a warming environment, Clim. Dyn. 44 (2014) 1157–1168, doi:10.1007/s00382-014-2166-9.

Menzel, P., Gaye, B., Mishra, P.K., Anoop, A., Basavaiah, N., Marwan, N., Plessen, B., Prasad, S., Riedel, N., Stebich, M., et al., Linking Holocene drying trends from Lonar Lake in monsoonal central India to North Atlantic cooling events, Palaeogeogr. Palaeoclimatol. Palaeoecol. 410 (2014) 164–178.

Midhuna, T.M., Kumar, P., Dimri, A.P., A new Western Disturbance Index for the Indian winter monsoon, J. Earth Syst. Sci. 129 (2020) 1–14, doi:10.1007/s12040-019-1324-1.

Minyuk, P.S., Borkhodoev, V.Y., Wennrich, V., Inorganic data from El' gygytgyn Lake sediments: stages 6-11. Climate of the Past 10 (2) (2013) 393–433. doi:10.5194/cpd-9-393-2013.

Mir, I.A., Mir, R.A., Geochemistry of surface sediments in parts of Bandipora–Ganderbal areas, Kashmir valley, western Himalaya: Implications for provenance and weathering, J. Earth Syst. Sci. 128 (2019) 1–16, doi:10.1007/s12040-019-1248-9.

Mischke, S., Lai, Z., Aichner, B., Heinecke, L., Mahmoudov, Z., Kuessner, M., Herzschuh, U., Radiocarbon and optically stimulated luminescence dating of sediments from Lake Karakul, Tajikistan, Quat. Geochronol. 41 (2017) 51–61, doi:10.1016/j.quageo.2017.05.008.

Mischke, S., Weynell, M., Zhang, C., Wiechert, U., C-reservoir effects in Tibetan Plateau lakes, Quat. Int. 313–314 (2013) 147–155, doi:10.1016/j.quaint.2013.01.030.

Mishra, P.K., Ankit, Y., Gautam, P.K., Lakshmidevi, C.G., Singh, P., Anoop, A., Inverse relationship between south-west and north-east monsoon during the late Holocene: geochemical and sedimentological record from Ennamangalam Lake, southern India, Catena 182 (2019) 104–117, doi:10.1016/j.catena.2019.104117.

Mishra, P.K., Anoop, A., Jehangir, A., Prasad, S., Menzel, P., Schettler, G., Naumann, R., Weise, S., Andersen, N., Yousuf, A.R., Gaye, B., Limnology and modern sedimentation patterns in high altitude Tso Moriri Lake, NW Himalaya—implications for proxy development, Fundam. Appl. Limnol. 185 (2014) 329–348, doi:10.1127/fal/2014/0664.

Mishra, P.K., Anoop, A., Schettler, G., Prasad, S., Jehangir, A., Menzel, P., Naumann, R., Yousuf, A.R., Basavaiah, N., Deenadayalan, K., et al., Reconstructed late Quaternary hydrological changes from Lake Tso Moriri, NW Himalaya, Quat. Int. 371 (2015b) 76–86.

Mishra, P.K., Prasad, S., Anoop, A., Plessen, B., Jehangir, A., Gaye, B., Menzel, P., Weise, S.M., Yousuf, A.R., Carbonate isotopes from high altitude Tso Moriri Lake (NW Himalayas) provide clues to late glacial and Holocene moisture source and atmospheric circulation changes, Palaeogeogr. Palaeoclimatol. Palaeoecol. 425 (2015a) 76–83.

Misra, S., Bhattacharya, S., Mishra, P.K., Misra, K.G., Agrawal, S., Anoop, A., Vegetational responses to monsoon variability during Late Holocene: inferences based on carbon isotope and pollen record from the sedimentary sequence in Dzukou valley, NE India, Catena 194 (2020), doi:10.1016/j.catena.2020.104697.

Misra, V., Bhardwaj, A., Mishra, A., Local onset and demise of the Indian summer monsoon, Clim. Dyn. 51 (5) (2018) 1609–1622.

Mohtadi, M., Prange, M., Schefuß, E., Jennerjahn, T.C., Late Holocene slowdown of the Indian Ocean Walker circulation, Nat. Commun. 8 (2017) 1–7, doi:10.1038/s41467-017-00855-3.

Moy, C.M., Seltzer, G.O., Rodbell, D.T., Anderson, D.M., Variability of El Niño/Southern Oscillation activity at millennial timescales during the Holocene epoch, Nature 420 (2002) 162–165, doi:10.1038/nature01194.

Mudelsee, M., Fohlmeister, J., Scholz, D., Effects of dating errors on nonparametric trend analyses of speleothem time series, Clim. Past 8 (2012) 1637–1648, doi:10.5194/cp-8-1637-2012.

Nageswararao, M.M., Sannan, M.C., Mohanty, U.C., Characteristics of various rainfall events over South Peninsular India during northeast monsoon using high-resolution gridded dataset (1901–2016), Theor. Appl. Climatol. 137 (2019) 2573–2593, doi:10.1007/s00704-018-02755-y.

Opitz, S., Wünnemann, B., Aichner, B., Dietze, E., Hartmann, K., Herzschuh, U., IJmker, J., Lehmkuhl, F., Li, S., Mischke, S., Plotzki, A., Stauch, G., Diekmann, B., Late Glacial and Holocene development of Lake Donggi Cona, north-eastern Tibetan Plateau, inferred from sedimentological analysis, Palaeogeogr. Palaeoclimatol. Palaeoecol. 337–338 (2012) 159–176, doi:10.1016/j.palaeo.2012.04.013.

Pai, D.S., Sridhar, L., Badwaik, M.R., Rajeevan, M., Analysis of the daily rainfall events over India using a new long period (1901–2010) high resolution (0.25° × 0.25°) gridded rainfall data set, Clim. Dyn. 45 (2014) 755–776, doi:10.1007/s00382-014-2307-1.

Pillai, A.A.S., Anoop, A., Sankaran, M., Sanyal, P., Jha, D.K., Ratnam, J., Mid-late Holocene vegetation response to climatic drivers and biotic disturbances in the Banni grasslands of western India, Palaeogeogr. Palaeoclimatol. Palaeoecol. 485 (2017) 869–878, doi:10.1016/j.palaeo.2017.07.036.

Polanski, S., Fallah, B., Prasad, S., Cubasch, U., Simulation of the Indian monsoon and its variability during the last millennium, Clim. Past Discuss. 9 (2013) 703–740, doi:10.5194/cpd-9-703-2013.

Ponton, C., Giosan, L., Eglinton, T.I., Fuller, D.Q., Johnson, J.E., Kumar, P., Collett, T.S., Holocene aridification of India, Geophys. Res. Lett. 39 (2012) 1–6, doi:10.1029/2011GL050722.

Prasad, S., Anoop, A., Riedel, N., Sarkar, S., Menzel, P., Basavaiah, N., Krishnan, R., Fuller, D., Plessen, B., Gaye, B., Röhl, U., Wilkes, H., Sachse, D., Sawant, R., Wiesner, M.G., Stebich, M., Prolonged monsoon droughts and links to Indo-Pacific warm pool: a Holocene record from Lonar Lake, central India, Earth Planet. Sci. Lett. 391 (2014) 171–182, doi:10.1016/j.epsl.2014.01.043.

Prasad, S., Marwan, N., Eroglu, D., Goswami, B., Mishra, P.K., Gaye, B., Anoop, A., Basavaiah, N., Stebich, M., Jehangir, A., Holocene climate forcings and lacustrine regime shifts in the Indian summer monsoon realm, Earth Surf. Processes Landforms 45 (15) (2020) 3842–3853, doi:10.1002/esp.5004.

Prell, W.L., Kutzbach, J.E., Sensitivity of the Indian monsoon to forcing parameters and implications for its evolution, Nature 360 (1992) 647–652.

Rajeevan, M., Gadgil, S., Bhate, J., Active and Break Spells of the Indian Summer Monsoon. NCC research report: March 2008, J. Earth Syst. Sci. 119 (2010) 229–247.

Rajmanickam, V., Achyuthan, H., Eastoe, C., Farooqui, A., Early-Holocene to present palaeoenvironmental shifts and short climate events from the tropical wetland and lake sediments, Kukkal Lake, Southern India: Geochemistry and palynology, Holocene 27 (2017) 404–417, doi:10.1177/0959683616660162.

Rawat, S., Gupta, A.K., Srivastava, P., Sangode, S.J., Nainwal, H.C., A 13,000 year record of environmental magnetic variations in the lake and peat deposits from the Chandra valley, Lahaul: implications to Holocene monsoonal variability in the NW Himalaya, Palaeogeogr. Palaeoclimatol. Palaeoecol. 440 (2015) 116–127, doi:10.1016/j.palaeo.2015.08.044.

Rehfeld, K., Marwan, N., Breitenbach, S.F.M., Kurths, J., Late Holocene Asian summer monsoon dynamics from small but complex networks of paleoclimate data, Clim. Dyn. 41 (2013) 3–19, doi:10.1007/s00382-012-1448-3.

Saini, J., Günther, F., Aichner, B., Mischke, S., Herzschuh, U., Zhang, C., Mäusbacher, R., Gleixner, G., Climate variability in the past 19,000 yr in NE Tibetan Plateau inferred from biomarker and stable isotope records of Lake Donggi Cona, Quat. Sci. Rev. 157 (2017) 129–140, doi:10.1016/j.quascirev.2016.12.023.

Saji, N.H., Goswami, B.N., Vinayachandran, P.N., Yamagata, T., A dipole mode in the tropical Indian Ocean. Nature 401 (6751) (1999) 360–363. doi:doi.org/10.1038/43854.

Sandeep, K., Shankar, R., Warrier, A.K., Yadava, M.G., Ramesh, R., Jani, R.A., Weijian, Z., Xuefeng, L., A multi-proxy lake sediment record of Indian summer monsoon variability during the Holocene in southern India, Palaeogeogr. Palaeoclimatol. Palaeoecol. 476 (2017) 1–14, doi:10.1016/j.palaeo.2017.03.021.

Schott, F.A., Xie, S.P., McCreary Jr., J.P., Indian Ocean circulation and climate variability, Rev. Geophys. 47 (1) (2009) G1002.

Schulzweida, U., CDO User Guide (Version 1.9. 6), Max Planck Institute for Meteorology, Hamburg, Germany, 2019.

Shukla, J., Paolino, D.A., The southern oscillation and long range forecasting of the summer monsoon rainfall over India, Am. Meteorol. Soc. 111 (1983) 1830–1837.

Singh, A.K., High Resolution Palaeoclimatic Changes in Selected Sectors of the Indian Himalaya by Using Speleothems—Past Climatic Changes Using Cave Structures, Springer International Publishing, Cham, Switzerland, 2018.

Singh, S., Gupta, A.K., Dutt, S., Bhaumik, A.K., Anderson, D.M., Abrupt shifts in the Indian summer monsoon during the last three millennia, Quat. Int. 558 (2020) 59–65, doi:10.1016/j.quaint.2020.08.033.

Sinha, A., Berkelhammer, M., Stott, L., Mudelsee, M., Cheng, H., Biswas, J., The leading mode of Indian Summer Monsoon precipitation variability during the last millennium, Geophys. Res. Lett. 38 (2011) 2–6, doi:10.1029/2011GL047713.

Sinha, A., Kathayat, G., Cheng, H., Breitenbach, S.F.M., Berkelhammer, M., Mudelsee, M., Biswas, J., Edwards, R.L., Trends and oscillations in the Indian summer monsoon rainfall over the last two millennia, Nat. Commun. 6 (2015) 1–8, doi:10.1038/ncomms7309.

Sreekala, P.P., Rao, S.V.B., Rajeevan, M., Northeast monsoon rainfall variability over south peninsular India and its teleconnections, Theor. Appl. Climatol. 108 (2012) 73–83, doi:10.1007/s00704-011-0513-x.

Srivastava, P., Agnihotri, R., Sharma, D., Meena, N., Sundriyal, Y.P., Saxena, A., Bhushan, R., Sawlani, R., Banerji, U.S., Sharma, C., Bisht, P., Rana, N., Jayangondaperumal, R., 8000-year monsoonal record from Himalaya revealing reinforcement of tropical and global climate systems since mid-Holocene, Sci. Rep. 7 (2017) 1–11, doi:10.1038/s41598-017-15143-9.

Stachnik, J.P., Schumacher, C., A comparison of the Hadley circulation in modern reanalyses, J. Geophys. Res. Atmos. 116 (2011) 1–14, doi:10.1029/2011JD016677.

Staubwasser, M., Weiss, H., Holocene climate and cultural evolution in late prehistoric-early historic West Asia, Quat. Res. 66 (2006) 372–387, doi:10.1016/j.yqres.2006.09.001.

Sun, Q., Wang, S., Zhou, J., Chen, Z., Shen, J., Xie, X., Wu, F., Chen, P., Sediment geochemistry of Lake Daihai, north-central China: implications for catchment weathering and climate change during the Holocene, J. Paleolimnol. 43 (2010) 75–87, doi:10.1007/s10933-009-9315-x.

Talbot, M.R., A review of the palaeohydrological interpretation of carbon and oxygen isotopic ratios in primary lacustrine carbonates. Chemical Geology: Isotope Geoscience Section 80 (4) (1990) 261–279.

Tiwari, M., Managave, S., Yadava, M.G., Ramesh, R., Spatial and temporal coherence of paleomonsoon records from marine and land proxies in the Indian region during the past 30 ka. Platinum Jubilee publication of the Indian Academy of sciences, Bangalore, India, (2003) pp. 517–535.

Ummenhofer, C.C., Sen Gupta, A., Li, Y., Taschetto, A.S., England, M.H., Multi-decadal modulation of the El Nĩo-Indian monsoon relationship by Indian Ocean variability, Environ. Res. Lett. 6 (2011) 034006, doi:10.1088/1748-9326/6/3/034006.

Vinayachandran, P.N., Francis, P.A., Rao, S.A., Indian Ocean dipole: processes and impacts. In: Current trends in science, platinum jubilee special volume of the Indian Academy of Sciences. Indian Academy of Science, Bangalore, India, (2009) pp. 569–589.

Von Rad, U., Schaaf, M., Michels, K.H., Schulz, H., Berger, W.H., Sirocko, F., A 5000-yr record of climate change in varved sediments from the oxygen minimum zone off Pakistan, Northeastern Arabian Sea, Quat. Res. 51 (1999) 39–53, doi:10.1006/qres.1998.2016.

Walker, M.J.C., Berkelhammer, M., Björck, S., Cwynar, L.C., Fisher, D.A., Long, A.J., Lowe, J.J., Newnham, R.M., Rasmussen, S.O., Weiss, H., Formal subdivision of the Holocene Series/Epoch: a discussion paper by a Working Group of INTIMATE (Integration of ice-core, marine and terrestrial records) and the Subcommission on Quaternary Stratigraphy (International Commission on Stratigraphy), J. Quat. Sci. 27 (2012) 649–659, doi:10.1002/jqs.2565.

Wang, M.D., Liang, J., Hou, J.Z., Hu, L., Distribution of GDGTs in lake surface sediments on the Tibetan Plateau and its influencing factors, Sci. China Earth Sci. 59 (2016) 961–974, doi:10.1007/s11430-015-5214-3.

Wang, Y., Liu, X., Herzschuh, U., Asynchronous evolution of the Indian and East Asian Summer Monsoon indicated by Holocene moisture patterns in monsoonal central Asia, Earth Sci. Rev. 103 (2010) 135–153, doi:10.1016/j.earscirev.2010.09.004.

Wanner, H., Beer, J., Bütikofer, J., Crowley, T.J., Cubasch, U., Flückiger, J., Goosse, H., Grosjean, M., Joos, F., Kaplan, J.O., Küttel, M., Müller, S.A., Prentice, I.C., Solomina, O., Stocker, T.F., Tarasov, P., Wagner, M., Widmann, M., Mid- to Late Holocene climate change: an overview, Quat. Sci. Rev. 27 (2008) 1791–1828, doi:10.1016/j.quascirev.2008.06.013.

Warrier, A.K., Sandeep, K., Shankar, R., Climatic periodicities recorded in lake sediment magnetic susceptibility data: further evidence for solar forcing on Indian summer monsoon, Geosci. Front. 8 (2017) 1349–1355, doi:10.1016/j.gsf.2017.01.004.

Wennrich, V., Francke, A., Dehnert, A., Juschus, O., Leipe, T., Vogt, C., Brigham-Grette, J., Minyuk, P.S., Melles, M., Modern sedimentation patterns in Lake El'gygytgyn, NE Russia, derived from surface sediment and inlet streams samples, Clim. Past 9 (2013) 135–148, doi:10.5194/cp-9-135-2013.

Wetzel, R.G., Limnology: Lake and River Ecosystems (3rd Edn.), Academic Press, Cambridge, 2001.

White, W.M., Geochemistry, Cambridge University Press, Cambridge, 2013.

Whitlock, C., Dean, W.E., Fritz, S.C., Stevens, L.R., Stone, J.R., Power, M.J., Rosenbaum, J.R., Pierce, K.L., Bracht-Flyr, B.B., Holocene seasonal variability inferred from multiple proxy records from Crevice Lake, Yellowstone National Park, USA, Palaeogeogr. Palaeoclimatol. Palaeoecol. 331–332 (2012) 90–103, doi:10.1016/j.palaeo.2012.03.001.

Wu, J., Lin, L., Gagan, M.K., Schleser, G.H., Wang, S., Organic matter stable isotope (δ13C, δ 15N) response to historical eutrophication of Lake Taihu, China, Hydrobiologia 563 (2006) 19–29, doi:10.1007/s10750-005-9133-8.

Wünnemann, B., Demske, D., Tarasov, P., Kotlia, B.S., Reinhardt, C., Bloemendal, J., Diekmann, B., Hartmann, K., Krois, J., Riedel, F., Arya, N., Hydrological evolution during the last 15 kyr in the Tso Kar lake basin (Ladakh, India), derived from geomorphological, sedimentological and palynological records, Quat. Sci. Rev. 29 (2010) 1138–1155, doi:10.1016/j.quascirev.2010.02.017.

Yadava, M.G., Ramesh, R., Monsoon reconstruction from radiocarbon dated tropical Indian speleothems, Holocene 15 (2005) 48–59.

CHAPTER

15 Holocene hydroclimatic shifts across the Indian subcontinent: A review based on interarchival coherences

Pavani Misra[a], Aqib J. Ansari[a], Ambili Anoop[b], Praveen K. Mishra[c]

[a] *Department of Earth Sciences, Indian Institute of Technology Kanpur, Kanpur, India.* [b] *Indian Institute of Science Education and Research Mohali, Manauli, Punjab, India.* [c] *Wadia Institute of Himalayan Geology, Dehradun, Uttarakhand, India*

15.1 Introduction

The climate in the Indian subcontinent is primarily governed by a complex monsoon system and fluctuations in monsoonal intensities results in a widespread impact on the country's socioeconomic structure. However, the modern meteoric dataset is limited in terms of the time frame and may provide climatic information spanning only the last 150 years. The evaluation of monsoon over longer timescales, such as, over the late Quaternary period can be phenomenal in comprehending the natural variations in the monsoonal system.

The Earth experiences cycles of glacial and interglacial periods due to various external forcing functions such as the orbital forcing, solar insolation etc. (Ji et al., 2006; Wanner et al., 2008). The most recent glacial period terminated at around 11.7 kyr BP, which marked the onset of the current interglacial called the Holocene (Rasmussen et al., 2006; Pearce et al., 2013). Although Holocene has witnessed some prominent climatic shifts (e.g., Holocene Climate Optimum [HCO], 8.2 kyr event, 4.2 kyr event, medieval warm period, Little Ice Age [LIA]), but the magnitude of these changes has been small, therefore this epoch is regarded as the most stable interglacial period in the last 400 kyr (Petit et al., 1999). The Holocene epoch records the rise and fall of many important civilizations including the Mesopotamian Akkadian Empire, Egyptian Old Kingdom, the Early Bronze Age civilizations of Greece and Crete, and Liangzhu culture in the lower Yangtze valley (Hodell et al., 1995; Cullen et al., 2000; deMenocal, 2001; Weiss and Bradley, 2001; Drysdale et al., 2006). This brings out the direct impact that climate has over the establishment and growth of the societies. Therefore, a comprehensive analysis of the late Quaternary is of utmost importance for understanding future climatic trends.

The last glacial period reached its maximum (LGM—Last Glacial Maximum) around 20 kyr BP, after which a gradual transition from the glacial to the current interglacial period began with increasing temperatures and retreating glaciers. The transitional period of glacial retreat between the LGM and Holocene was initially punctuated by the Oldest Dryas, a cooling event between ~17.0 and ~14.9 kyr BP (Dortch et al., 2013). Toward the end of the last stadial, the temperature abruptly rose again between ~14.6 and 12.8 kyr BP (Rasmussen et al., 2006). This warm and wet period is called the Bølling–Allerød

Holocene Climate Change and Environment. DOI: https://doi.org/10.1016/B978-0-323-90085-0.00001-2

event, which was shortly punctuated by the Older Dryas between ~13.9 and ~13.7 kyr BP (Dortch et al., 2013). Around 12.8 kyr BP, the Allerød event was disrupted by a brief cold period of the Younger Dryas which terminated at about 11.7 kyr BP, marking the beginning of the present interstadial (Rasmussen et al., 2006; Pearce et al., 2013). The early Holocene is marked by a gradual strengthening of the monsoon and globally a peak in its intensity is recorded during the HCO between ~9 and ~5 kyr BP. However, at 8.2 kyr BP, both the marine and terrestrial paleoclimate dataset record sudden dryness and reduced temperatures in and around the North Atlantic Ocean (Alley et al., 1997). This abrupt drought phase is called the 8.2 kyr event when the excessive melting of Laurentide ice sheet released a huge amount of freshwater into Labrador Sea slowing the Atlantic Meridional Overturning circulation (Klitgaard-Kristensen et al., 1998; Erlenkeuser et al., 1998; von Grafenstein et al., 1998; Dixit et al., 2014a). A gradual weakening in monsoon intensity is recorded during ~5 kyr BP, giving way to a very similar condition comparable to the present-day climate. Another abrupt peak of aridity has been recorded at ~4.2 kyr BP, when strong El-Nino and a negative IOD (Indian Ocean Dipole) have been linked to the weakening of the monsoon (Morrill et al., 2003; Fisher et al., 2008; MacDonald, 2011). The collapse of many civilizations around the world such as the Early Bronze Age in Greece, the Liangzhu culture in the Yangtze valley, the Mesopotamian Akkadian Empire, the Egyptian Old Kingdom and the Indus valley civilization have been closely associated with this event of aridification (Wenxiang and Tungsheng, 2004; Staubwasser and Weiss, 2006; Marshall et al., 2011; Berkelhammer et al., 2012; Ponton et al., 2012). Based on these distinct climatic phases, the International Commission on Stratigraphy (Cohen et al., 2018) recently divided the Holocene epoch into three ages: (i) the Greenlandian age (11.7–8.2 kyr BP), (ii) the Northgrippian age (8.2–4.2 kyr BP), and (iii) the Meghalayan age (4.2 kyr BP to present).

Proxy datasets for the late Holocene period show two alternate phases of warm and cold climate since ~2.5 kyr BP. The warm phase between 2.5 and 1.6 kyr BP is recorded in Spain, North Sea, Greenland, and the Tibetan Plateau (Hass, 1996; Seidenkrantz et al., 2007; Martín-Puertas et al., 2009; Wang et al., 2013) and called the Roman Climate Optimum or the Roman Warm Period (RWP). RWP was followed by a 550 years long cold episode recorded in parts of North America, Arctic, Europe, North Atlantic, North Pacific, and China. This episode is known as the Dark Ages Cold Period, which lasted between 1.5 and 1.05 kyr BP (Helama et al., 2017). Subsequently, the period between 1.05 and 0.6 kyr BP records a period of warming known as the Medieval Climate Anomaly (MCA; Graham et al., 2011). The climatic amelioration during MCA not only aided the colonization of Vikings in Greenland but also records the propagation of vineyards up to England, Germany and France (Fagan, 2000). The last phase of cooling in the Holocene is known as the LIA which is recorded between 0.45 and 0.1 kyr BP (Graham et al., 2011). Apart from fluctuations in the solar insolation, increased episodes of volcanism have been linked to this cold event (Crowley et al., 2008). The major phases of climatic shifts globally observed in the last 20 kyr are summarized here and the timing of these episodes have been used in discussing the later sections of the chapter.

15.2 Scope

The work presented in this chapter is a compilation of the oxygen isotopic records from three different climate archives (speleothems, lake and marine sediments) produced for the Late Pleistocene – Holocene epoch across the Indian subcontinent. The individual isotopic records show fluctuations which may or may not be related to regional or global scale changes and instead may only be associated with some local forcing. Therefore, in order to gain insights into the forcings involved, a comparison of the isotopic records of each archive is discussed, followed by an inter-archival analysis. The intra- and inter-archival comparison of the oxygen isotopic records from across the country provide a reasonable understanding of the trends in climatic fluctuations since Last Glacial Maximum, drivers responsible for spatial scale changes in precipitation patterns, and the timing of the major hydroclimatic events recorded in the Indian subcontinent. The comparative analysis of climate archives not only brings out the synchronicities observed in monsoonal trends through time and space, but also emphasizes the importance of Northeast (NE) monsoon in supporting the agrarian economy of peninsular India. This chapter discusses in detail the relationship between the regional patterns of fluctuation in the strength of Indian summer monsoon (ISM) and its teleconnections to understand the role of external forcings.

15.3 Present-day climatic regimes within the Indian subcontinent

The modern-day climate across the Indian subcontinent is primarily governed by a complex monsoonal system. The country receives its rainfall from three major monsoonal zones: (i) Indian Summer Monsoon (ISM) zone; (ii) Northeast Monsoon zone, and (iii) the westerly winds (Fig. 15.1). The ISM provides about 80% of the annual rainfall received across the Indian subcontinent. Since a major portion of the country's economy is dependent on agricultural produce, it is of utmost importance to understand the modern-day precipitation pattern and its driving factors that may not only have a direct impact on the country's economic growth but also be responsible for the livelihood of a vast population.

Slightest changes in the solar activity not only impacts the Earth's climate but also determines the strength of the monsoonal winds (Rind and Overpeck, 1993; Bond et al., 2001). The periods of intensified ISM are directly linked with increased solar insolation (Solanki et al., 2004; Wanner et al., 2008). The monsoonal trade winds converge from both the hemispheres along a low-pressure belt forming the Intertropical Convergence Zone (ITCZ). The seasonal migration of this zone greatly influences the strength and timing of the ISM. The past climate records from various archives suggest a northward migration of the ITCZ along with a strengthened monsoon in the northern hemisphere during the early Holocene (Sirocko et al., 1993; Overpeck et al., 1996; Haug et al., 2001). During the mid- to late-Holocene transition, a weakened ISM and a decrease in solar insolation in the northern hemisphere was observed marking a steady migration of the ITCZ toward the southern hemisphere (Fleitmann et al., 2003, 2007; Cheng et al., 2009).

15.4 Indian Summer Monsoon and its teleconnections

The ISM is remotely influenced by various external forcing functions. One of the earliest teleconnections in global climate studies exists between the southern oscillation and the seasonal mean summer monsoon precipitation over India (Walker, 1924). The previous work on understanding the correlation between the El Niño Southern Oscillation (ENSO) and ISM show that a weakened ISM is linked to the warm phase of the Pacific event (El Nino), whereas a strengthened ISM is observed during the cold Pacific event (La-Nina) (Sikka, 1980; Angell, 1981; Rasmusson and Carpenter, 1983; Ropelewski and Halpert, 1987). The warming in the central and eastern Pacific Ocean leads to an eastward shift of the trade winds that results in the downwelling over the western Pacific and a consequent reduction in the monsoonal

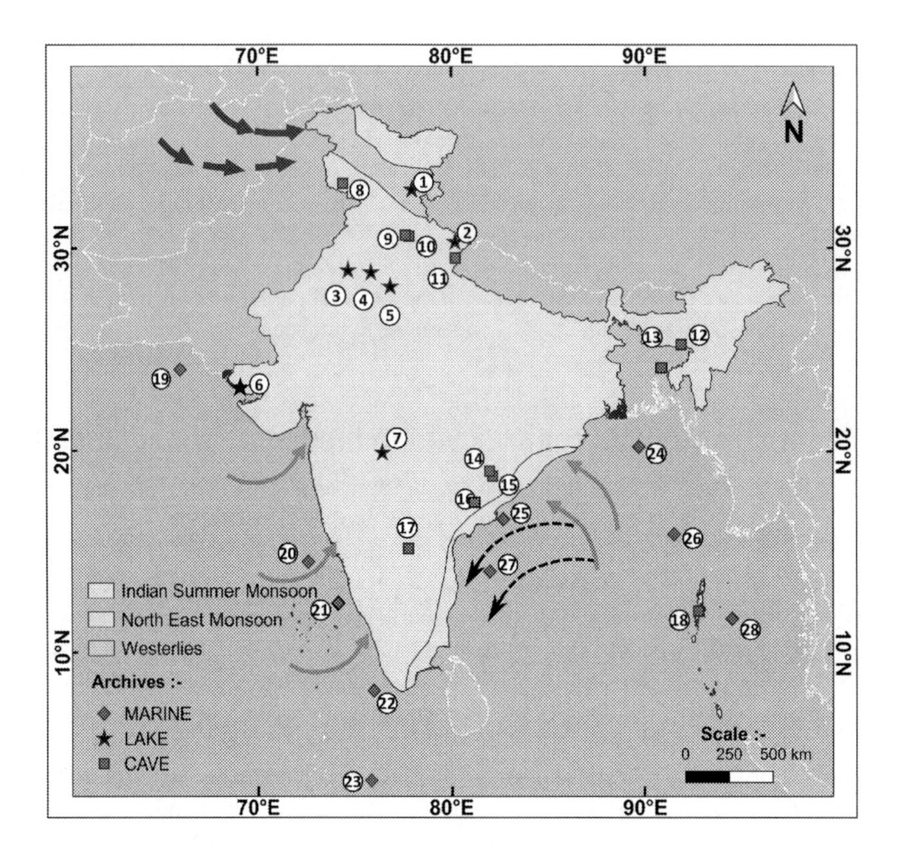

FIGURE 15.1

Map of India showing the present-day climatic regimes of the country as mentioned in the figure legend. The locations of various studies compiled in this chapter are plotted on the map. The star symbol represents the lacustrine records, the square symbol is for the cave deposits, and the diamond symbol represents the marine records compiled in this chapter. The blue arrows show the westerly winds, orange arrows denote the two branches (Arabian Sea and Bay of Bengal) of ISM, and the dashed black arrows indicate the NE monsoon winds (redrawn from Gunnell et al., 2007).

precipitation over India (Lau and Wang, 2006; Achuthavarier and Krishnamurthy, 2011). During the cooling phase of ENSO (La Nina), anomalously stronger trade winds over the western Pacific cause excessive upwelling of cold waters towards the eastern coast. This cold event in the eastern Pacific is observed to be in sync with a strengthened ISM (Rasmusson and Carpenter, 1983; Ropelewski and Halpert, 1987).

Similar irregularities in the sea surface temperatures (SSTs) are observed over the Indian Ocean (Saji et al., 1999; Webster et al., 1999; Murtugudde et al., 2000). A temperature dipole (warm and the cold pole) forms along the equatorial belt of the Indian Ocean between the Arabian Sea in the west and the east Indian Ocean. A warmer western part of the ocean results in a positive IOD and vice-versa. A positive IOD is observed to be directly proportional to a strengthened ISM, whereas its negative state correlates well with a weakened ISM. Apart from the ENSO–ISM and IOD–ISM relations, some studies also suggest the existence of an ENSO–IOD linkage (Behera and Yamagata, 2003). The positive IOD is often linked with the warm phase of ENSO, whereas the negative IOD with its cold phase.

To add to these complex forcing–feedback relationships, the Indian landmass also has a vast geomorphic diversity—with Himalaya in the north, Thar Desert in the west, the Indo-Gangetic Plains, the NE Himalaya and Brahmaputra floodplains in the east, the Central Indian Peninsula, and the coastal belt in southern India. There are some dominant monsoon forcings that determine the strength of the monsoonal winds in each of these geomorphic zones. In the present day, the monsoonal winds that reach the Himalaya travel from the Bay of Bengal (BOB) fan, thus the eastern Himalaya are much humid in comparison to the western part of this range (Hofer and Messerli, 2006; Rees and Collins, 2006). Precipitation in the form of snow feeds the glacial region in the higher altitudes. The western Himalayan glaciers build up only in winters and melt in summers but the eastern Himalayan glaciers continue to simultaneously accumulate and melt during summers. Monsoon weakening, delay, or failure in this region can result in a major drought.

In northwest India toward west Rajasthan, lies the Thar Desert covered with alluvium and eolian sands. This arid region hosts many ephemeral lakes and eolian landforms, some of which have been studied for reconstructing the past climatic history of this region (Roy and Singhvi, 2016). In contrast to the present-day arid climate of western India, the northeast part of the country is very humid and some regions receive rainfall as high as 12,000 mm. The ENSO events are known to modulate the ISM intensity in Central India, but the amount of precipitation in this region is relatively less affected by these events (Myers et al., 2015).

Episodes of extremely weakened monsoon (arid phase) over Ganga Plains have been strongly linked to a simultaneous occurrence of an El-Nino event. However this teleconnection does not hold true in most of the cases for an intensified monsoon period over the Indo-Gangetic Plains (Misra et al., 2020). Previously, the Ganga Plains have experienced many episodes of floods and droughts with no association to an ENSO event (Bhatla et al., 2015). The changes in precipitation over Ganga Plains have also been linked to an intraseasonal ocean–atmosphere phenomenon known as the Madden Julian Oscillation (MJO), which is known to impact the strength and timing of the monsoonal winds over the tropical region (Mishra et al., 2016; Singh and Bhatla, 2020). The amplitude of this oscillation is defined using the Real-time Multivariate MJO index which categorizes MJO into active and suppressed phases (LaFleur et al., 2015). Climate studies done across the Central Indian region show that apart from the solar insolation and ENSO, ISM is also influenced by climatic fluctuations in the North Atlantic region (Menzel et al., 2014). The southern peninsular India receives its rainfall from both the ISM and NE monsoon zone (Chidambaram et al., 2009). The NE monsoon contributes ~30% of the average annual rainfall received in the southern part of the country (Sreekala et al., 2012). Therefore, the agricultural production in this region is not only dependent on ISM but is also greatly influenced by the NE monsoon.

15.5 Methodology

Paleoclimatic studies have been carried out globally for reconstructing natural climate shifts over different timescales using various climate archives such as the tree rings, speleothems, ice cores, marine sediments, fluvial, and lake sediments. These records impart detailed spatial and temporal information regarding climate change which help us better in differentiating between natural and anthropogenic forcings. In order to understand regional fluctuations, it is important to put together and comprehensively analyze the available records. The oxygen ($\delta^{18}O$) isotopes are sensitive to changes in the local hydrological cycle and have been widely used as a proxy to understand the paleoprecipitational changes (Gat and Gonfiantini, 1981; Gat, 1996; Mook and Rozanski, 2000; Anoop et al., 2013). The temporal changes in the $^{18}O/^{16}O$ ratio of meteoric, surface, soil, and groundwaters act as valuable indicators of paleoclimatic shifts in the proxy material of various climate archives such as ice cores, lake/marine sediment cores, speleothems etc. (Gat and Gonfiantini, 1981; Gat, 1996).

This chapter attempts to compile the oxygen isotopic records from both the terrestrial and marine archives spanning over the last 20 kyr BP studied across India, in order to improve our understanding of the local and regional forcing functions (Table 15.1). In this study we (i) summarize the factors governing the spatial variability of precipitation; and (ii) discuss the available ($\delta^{18}O$) data on Holocene climate variability with special focus on the timing of hydroclimatic shifts across the Indian subcontinent.

15.5.1 Sources of uncertainty in climate records

Apart from the local geomorphic setting and climatic regime, there are also additional factors that add to the uncertainty in the paleoclimate dataset. Some of the common factors that propagate uncertainties include—uneven sampling, analytical errors, uncertainty in dating and calibration, and uncertainty related to proxy response in different depositional environments. All these factors add up and induce larger uncertainties while assessing the synchronicities in the available past climatic records. Although reliance on interpolation between two dates provides a better understanding of climatic fluctuations through time but lesser number of dates can also lead to erroneous interpretations. Nevertheless, if the calibration uncertainties are unknown, it is reasonable to depend upon the originally published timescales for comparative analyses (Franke and Donner, 2019).

15.6 Observations based on interproxy coherences

In the past few decades, many paleoclimate researchers have utilized stable isotopes from various climate archives to quantify the fluctuations in precipitation and to comprehend the hydroclimatic variability through time. Particularly, the oxygen isotopic records have been produced to evaluate the past variations in temperature, precipitation, and salinity (Urey, 1947; Scheurle and Hebbeln, 2003; Yang et al., 2016). The analysis of oxygen isotopic records can prove to be a powerful tool in correlating and assessing interarchival synchronicities across various climatic and geomorphic zones of the country. In a summer-monsoon dominated country like India, the enriched values of $\delta^{18}O_{carb}$ (‰) suggest higher evaporation (weakened precipitation) and vice versa (Zachos et al., 1994).

15.6.1 Lake sediments

Oxygen isotopic record for the Holocene epoch from the Himalaya is available from two lakes, namely, Tso Moriri and Burfu (Fig. 15.2). It is important to understand that multiple variables control the isotopic shifts in the Himalayan region. The $\delta^{18}O$ (‰) records from this proglacial part of the country is complex to interpret as fluctuations observed could be associated either with glacial melting or direct precipitation (Misra et al., 2019). Therefore, multiproxy analyses are required from this region to validate the isotopic

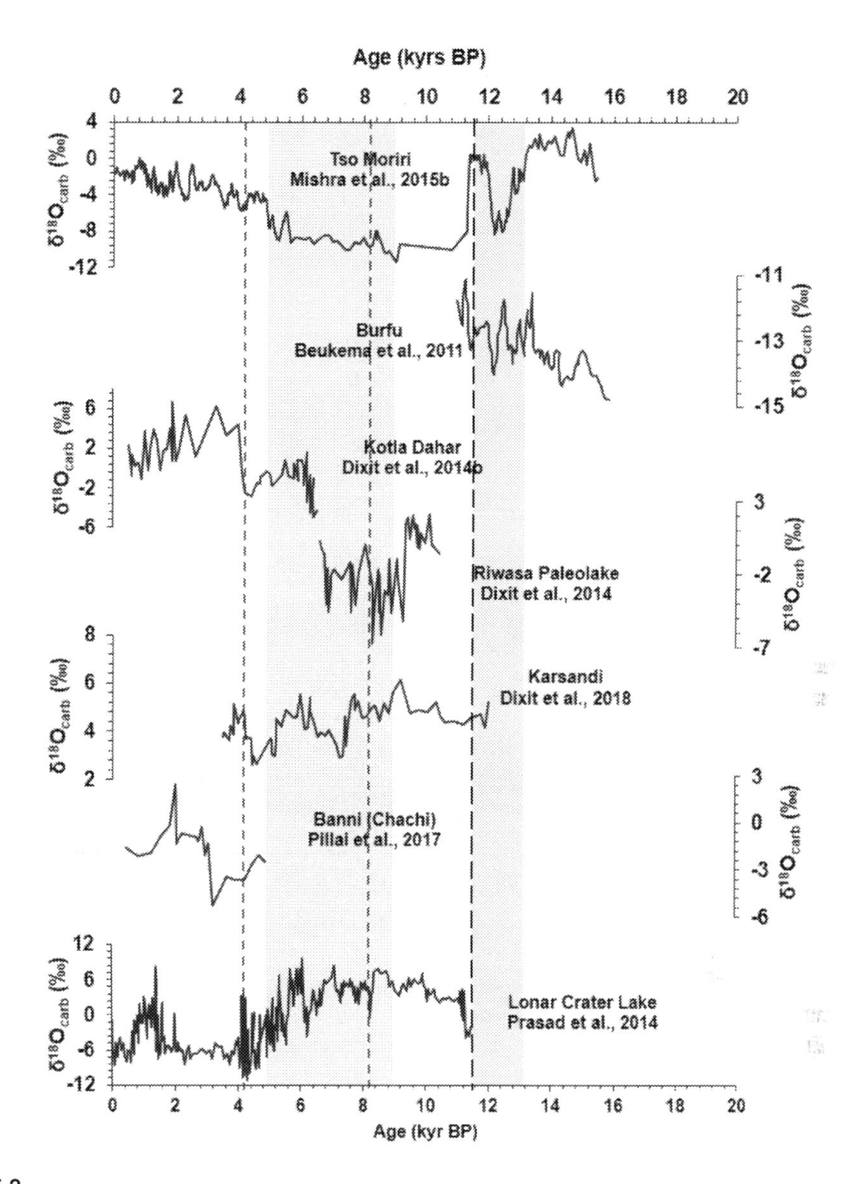

FIGURE 15.2

Intra-archival comparison of oxygen isotopic records from the lacustrine deposits across the Indian subcontinent. The records include Tso Moriri (Mishra et al., 2015b), Burfu (Beukema et al., 2011), Karsandi (Dixit et al., 2018), Riwasa (Dixit et al., 2014a), Kotla Dahar (Dixit et al., 2014b), Banni (Pillai et al., 2017), and Lonar (Prasad et al., 2014). The black dashed line represents the Holocene–Pleistocene boundary, the two red dashed lines represent the abrupt drying events globally recorded at 8.2 and 4.2 kyr BP. The shaded regions represent the Younger Dryas (between 13 and 11.7 kyr BP) and the Holocene Climate Optimum (between 9 and 5 kyr BP).

Table 15.1 List of paleoclimate records compiled in this chapter categorized under different climate archives.

Site no.	Lake	Lat	Long	References
1	Tso Moriri	32.9	78	Mishra et al., 2015b
2	Burfu	30.3	80.2	Beukema et al., 2011
3	Karsandi	28.9	74.7	Dixit et al., 2018
4	Riwasa	28.8	75.9	Dixit et al., 2014a
5	Kotla Dahar	28.1	76.9	Dixit et al., 2014b
6	Banni (Chachi)	23.55	69.85	Pillai et al., 2017
7	Lonar	19.9	76.5	Prasad et al., 2014
Site no.	Speleothem	Lat	Long	References
8	Kalakot	33.22	74.43	Kotlia et al., 2016
9	Tityana	30.64	77.67	Joshi et al., 2017
10	Sahiya	30.6	77.86	Kathayat et al., 2017
11	Dharamjali	29.52	80.21	Sanwal et al., 2013
12	Mawmluh	25.26	91.88	Dutt et al., 2015
13	Kotumsar	19	82	Band et al., 2018
14	Gupteshwar	18.75	82.16	Yadava et al., 1999
15	Valmiki	15.15	77.82	Lone et al., 2014
16	Baratang	12.08	92.75	Laskar et al., 2013
17	Wah Shikhar	25.25	91.86	Sinha et al., 2011
18	Jhumar	18.86	81.86	Sinha et al., 2011
Site no.	Marine	Lat	Long	References
19	W Arabian Sea (63KA/41KL)	24.6	65.9	Staubwasser et al., 2002
20	E Arabian Sea (AAS9/21)	14.5	72.65	Govil and Naidu, 2010
21	E Arabian Sea (3268G5)	12.5	74.16	Sarkar et al., 2000
22	SE Arabian Sea (3904)	8.13	76.03	Chauhan et al., 2009
23	E Arabian Sea (SS3827G)	3.7	75.9	Tiwari et al., 2005
24	N Bay of Bengal (SO126-39KL)	20.2	89.7	Weldeab et al., 2019
25	Mouth of Godavari, BOB (NGHP-16A)	16.6	82.7	Ponton et al., 2012
26	Bay of Bengal (31/11)	15.86	91.5	Chauhan, 2003
27	Bay of Bengal (SK218/1)	14.03	82.0	Govil and Naidu, 2011
28	Bay of Bengal (SK168)	11.4	94.2	Sijinkumar et al., 2016

signals. Instead of strengthened precipitation, the depleted $\delta^{18}O$ (‰) values from these proglacial records at times may suggest increased inflow of glacial meltwater in the lacustrine system.

Initially, around 16 kyr BP, opposite isotopic signals are observed at both lakes (Fig. 15.2). Tso Moriri shows enriched values while depleted values are observed at Burfu. Since Burfu is mainly a glacial-fed lake (Pant et al., 2006), the depletion observed in the $\delta^{18}O$ (‰) values suggests an influx from glacial melting. Between 16 and ~14 kyr BP, the proportion of glacial meltwater input increases at Tso Moriri but decreases at Burfu suggesting a drier climate at Tso Moriri but increased precipitation at Burfu (Fig. 15.2).

A short-wet phase is recorded at Tso Moriri between 13 and 11.7 kyr BP (Mishra et al., 2015a), whereas Burfu records a cold period between 12.7 and 12.3 kyr BP which has been linked to the Younger Dryas event (Beukema et al., 2011). However, the Younger Dryas at Tso Moriri is recorded between 11.7 and 11.2 kyr BP (Mishra et al., 2015a). Apart from the difference in the dominant climatic drivers observed by researchers at a time at Tso Moriri and Burfu, the presence of millennial-scale lags and incoherence observed can also be attributed to differences in the robustness of the chronological framework and their stratigraphic resolution. After 11.3 kyr BP, a synchronicity is observed in the isotopic signals with contribution from direct precipitation at both the lakes suggesting a warmer climate with intensified monsoon (Fig. 15.2; Beukema et al., 2011; Mishra et al., 2015a). The $\delta^{18}O$ (‰) paleolake records from Riwasa, Kotla Dahar, Karsandi, and Banni (Chachi) together give insight into the Holocene climatic fluctuations from the eastern margin of Thar desert region (Fig. 15.2). The variations in the balance of input to evaporative loss (P/E ratio) have a significant effect on the lake's isotope balance in arid regions (Holmes et al., 2007; Liu et al., 2009). In hydrologically closed limnological systems, enriched (high) values generally reflect preferential evaporative loss of the ^{16}O (Pillai et al., 2017). The $\delta^{18}O_{carb}$ (‰) records of Riwasa and Karsandi suggest initiation of lake filling started after 11 kyr BP and a peak in wetness was recorded between 9.4 and 8.3 kyr BP at Riwasa (Dixit et al., 2014a, 2018). A coherence in wetness is observed between Riwasa and Tso Moriri, as the latter also records a wet period between 11 and 8.5 kyr BP (Dixit et al., 2014a; Mishra et al., 2015b). Between 11 and 9 kyr BP, depleted $\delta^{18}O$ (‰) values and lake deepening is also recorded at Lonar lake in Central India (Prasad et al., 2014). Prevalence of dry conditions is observed at Karsandi throughout between 9 and 5.1 kyr BP, however an abrupt enrichment in $\delta^{18}O$ (‰) values is recorded at Riwasa at ~8.2 kyr BP (Dixit et al., 2014a, 2018). A signal of ISM weakening is also recorded at Tso Moriri in the higher Himalaya between 8.5 and 5.5 kyr BP, but wet conditions continued to prevail at Lonar lake until ~6.2 kyr BP after which ISM weakening was reported (Prasad et al., 2014). Nevertheless, unlike Riwasa, other lake records do not show abrupt drying at 8.2 kyr BP and instead record prolonged phases of drying and wetting around that time. During the mid- to late-Holocene transition, the ISM weakened as enrichment in $\delta^{18}O$ (‰) values is coherently recorded at Tso Moriri, Riwasa, Karsandi, and Lonar (Fig. 15.2). Approximately, between ~6 and ~4 kyr BP, Tso Moriri, Kotla Dahar, and Lonar lakes record gradual enrichment in $\delta^{18}O$ (‰) with increased evaporation suggesting a long-term regional reduction in precipitation (Dixit et al., 2014b; Prasad et al., 2014; Mishra et al., 2015a, 2015b). However, there is a slightly contrasting signal observed at Karsandi which records wetter climate at ~5.1 kyr BP (Dixit et al., 2018). Lonar lake in central India records a prolonged drought between 4.6 and 3.9 kyr BP, and Kotla Dahar near the desert boundary records abrupt enrichment in $\delta^{18}O$ (‰) values at 4.1 kyr BP suggesting a sudden aridification in the region (Fig. 15.2). Overall, there is a spatial coherence in aridity recorded during the mid to late Holocene transition. During the late Holocene, all records show a gradual increase in $\delta^{18}O$ (‰) values indicating a weakened monsoon and a consequent increase in aridity across the country (Prasad et al., 2014; Mishra et al., 2015b; Pillai et al., 2017).

15.6.2 Speleothems

The $\delta^{18}O$ (‰) of speleothems shows the fluctuations recorded in the monsoonal strength since the LGM (Fig. 15.3). Multiple stalagmites from Mawmluh cave in northeast India have been studied (Breitenbach et al., 2010; Berkelhammer et al., 2012; Dutt et al., 2015; Lechleitner et al., 2017) and the longest

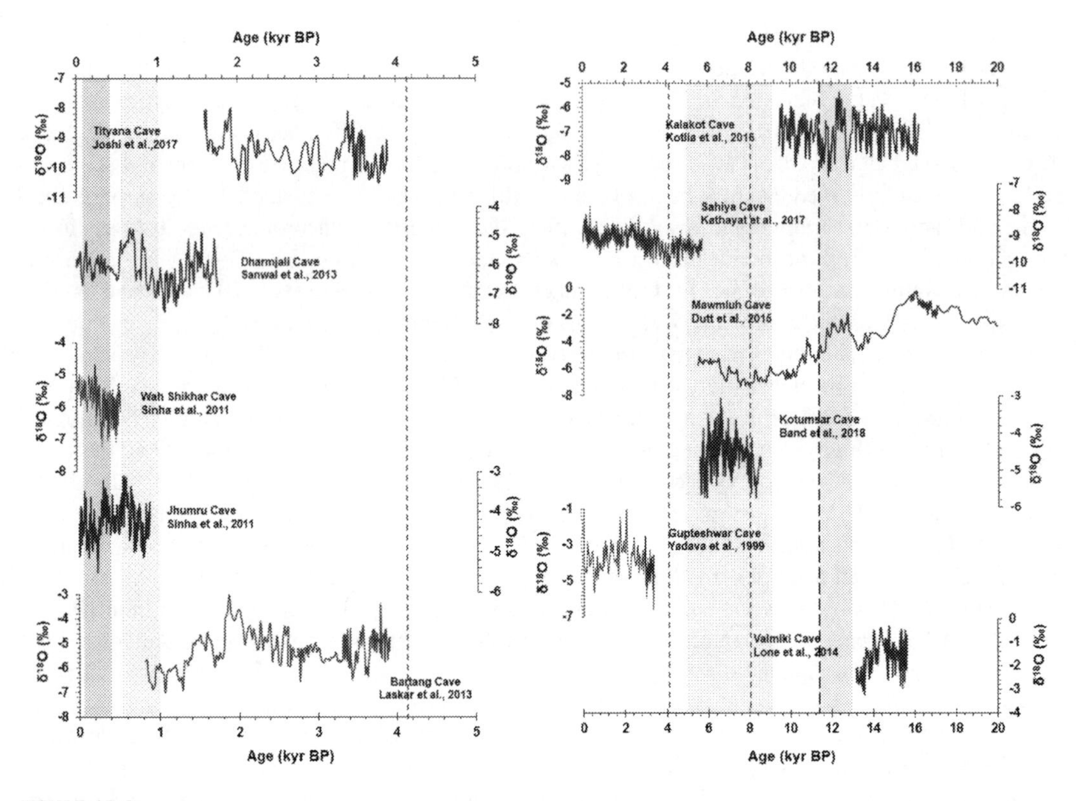

FIGURE 15.3

Intra-archival comparison of the oxygen isotopic records from cave deposits of India. The records in the above plot are from Kalakot (Kotlia et al., 2016), Tityana (Joshi et al., 2017), Sahiya (Kathayat et al., 2017), Dharamjali (Sanwal et al., 2013), Mawmluh (Dutt et al., 2015), Kotumsar (Band et al., 2018), Gupteshwar (Yadava et al., 1999), Valmiki (Lone et al., 2014), Baratang (Laskar et al., 2013), Wah Shikhar, and Jhumar (Sinha et al., 2011). The black dashed line represents the Holocene–Pleistocene boundary, the two red dashed lines represent the abrupt drying events globally recorded at 8.2 and 4.2 kyr BP. The shaded regions represent the Younger Dryas (between 13 and 11.7 kyr BP) and the Holocene Climate Optimum (between 9 and 5 kyr BP).

available record extends up to ~33 kyr BP which has mainly been used in the present analysis (Dutt et al., 2015). At Mawmluh, an initial strengthening in ISM circulation was recorded around LGM followed by an enrichment in $\delta^{18}O$ (‰) values between 17 and 15 kyr BP suggesting weakening of ISM which has been linked to the Heinrich event H1 (Dutt et al., 2015). The Kalakot cave in the NW Himalaya records a slightly wet period from ~16 kyr BP to ~14 kyr BP (Kotlia et al., 2016). An initial weak phase of ISM is recorded at Valmiki cave in southern India between 15.7 and 15.3 kyr BP, followed by an overall depletion in $\delta^{18}O$ (‰) values between 15.3 and ~14.7 kyr BP with short episodic drying reported at ~15.1 and ~14.8 kyr BP (Lone et al., 2014).

A depletion in $\delta^{18}O$ (‰) values at Mawmluh is reported between 15 and 12.9 kyr BP suggesting ISM strengthening which has been linked with the Bølling–Allerød interstadials (Dutt et al., 2015).

After initially fluctuating signals, the Kalakot stalagmite shows enriched $\delta^{18}O$ (‰) values between 14.3 and 13.9 kyr BP recording the Older Dryas event followed by a warm and wet Allerød interstadial up to ~12.7 kyr BP (Kotlia et al., 2016). Mawmluh and Kalakot stalagmites record a positive shift in $\delta^{18}O$ (‰) values between ~12.9 and ~11.5 kyr BP and ~12.7 and ~12.2 kyr BP, respectively, bracketing the cold and dry Younger Dryas (Fig. 15.3). The Kalakot cave in NW Himalaya shows strengthened ISM circulation between 12.2 and 9.5 kyr BP with a negative shift in $\delta^{18}O$ (‰) values. Mawmluh cave also records depleted $\delta^{18}O$ (‰) values between 10 and 6.5 kyr BP (Fig. 15.3). These signals of wetness at both sites can be linked to the warm phase of early Holocene (Dutt et al., 2015). However, Kotumsar Cave in Central India records a gradual decline in ISM intensity between 8.5 and ~7.3 kyr BP due to the southward migration of the ITCZ (Band et al., 2018). Mawmluh shows slight enrichment in $\delta^{18}O$ (‰) values between ~6.5 and ~5.5 kyr BP, when a depletion in $\delta^{18}O$ (‰) values is reported from Kotumsar stalagmite (Dutt et al., 2015; Band et al., 2018). The Sahiya Cave in Lesser Himalaya initially records a weakened ISM between 5.7 and 4.8 kyr BP and a brief phase of drying at 5.7 kyr BP is also reported from Kotumsar (Kathayat et al., 2017; Band et al., 2018). ISM strengthening associated with the late phase of the mid-HCO is recorded at Sahiya Cave between 4.8 and 3.8 kyr BP (Kathayat et al., 2017). Tityana Cave in Central Lesser Himalaya records enriched $\delta^{18}O$ values between 4 and 3.4 kyr BP suggesting a weak phase of ISM (Fig. 15.3). Between ~3.4 and ~2.8 kyr BP, both Tityana and Sahiya Cave records show depleted $\delta^{18}O$ values suggesting a strengthening of ISM (Joshi et al., 2017; Kathayat et al., 2017). A stalactite studied from Gupteshwar cave in Orissa, east India also records intensified ISM between 3.4 and ~3 kyr BP (Yadava and Ramesh, 2005), showing synchronicity with Tityana and Sahiya Cave (Fig. 15.3). In the past two millennia, Sahiya Cave in Himalaya records a warm and wet phase with depleted $\delta^{18}O$ values and increased solar insolation between 2.3 and 1.6 kyr BP that has been linked to the RWP (Kathayat et al., 2017), whereas Baratang Cave in the Andaman Islands reports a sharp decrease in rainfall between 2.1 and 1.8 kyr BP (Laskar et al., 2013). A coherence in depleted $\delta^{18}O$ values and thus wetness is recorded between ~1.2 and ~0.8 kyr BP at Sahiya, Dharamjali, and Baratang cave which corresponds to the MCA (Laskar et al., 2013; Sanwal et al., 2013; Kathayat et al., 2017). Enrichment in $\delta^{18}O$ values suggesting aridity related to the LIA is reported between ~0.45 and ~0.05 kyr BP from Dharamjali, Sahiya, Jhumar, Wah Shikar, and Bartang cave records (Sinha et al., 2011; Laskar et al., 2013; Sanwal et al., 2013; Kathayat et al., 2017).

15.6.3 Marine sediments

Globigerinoides ruber, a planktonic foraminifera is commonly found in marine sediments and the oxygen isotopic composition of this organism is used as a proxy for paleotemperature reconstructions. Some of the $\delta^{18}O$ records of *G. ruber* available from the Arabian Sea and BOB have also been assessed to understand the coherences in climatic signals in the past 20 kyr BP (Fig. 15.4). In the current assessment, six oxygen isotopic records of *G. ruber* extend up to ~20 kyr BP—two from the Eastern Arabian Sea (EAS) and the remaining four from the BOB (Fig. 15.4).

The marine sediment cores studied from the EAS and the BOB show a positive shift in $\delta^{18}O$ values under low temperatures and high salinity between ~20 and ~16 kyr BP suggesting a weakened ISM since LGM (Chauhan, 2003; Tiwari et al., 2005; Govil and Naidu, 2010; Govil and Naidu, 2011). Two distinct episodes of warming with enriched $\delta^{18}O$ values centered at ~19 and ~17 kyr BP are reported from the EAS record corresponding to the Arabian Sea Warming (ASW) events—ASW 2 and ASW

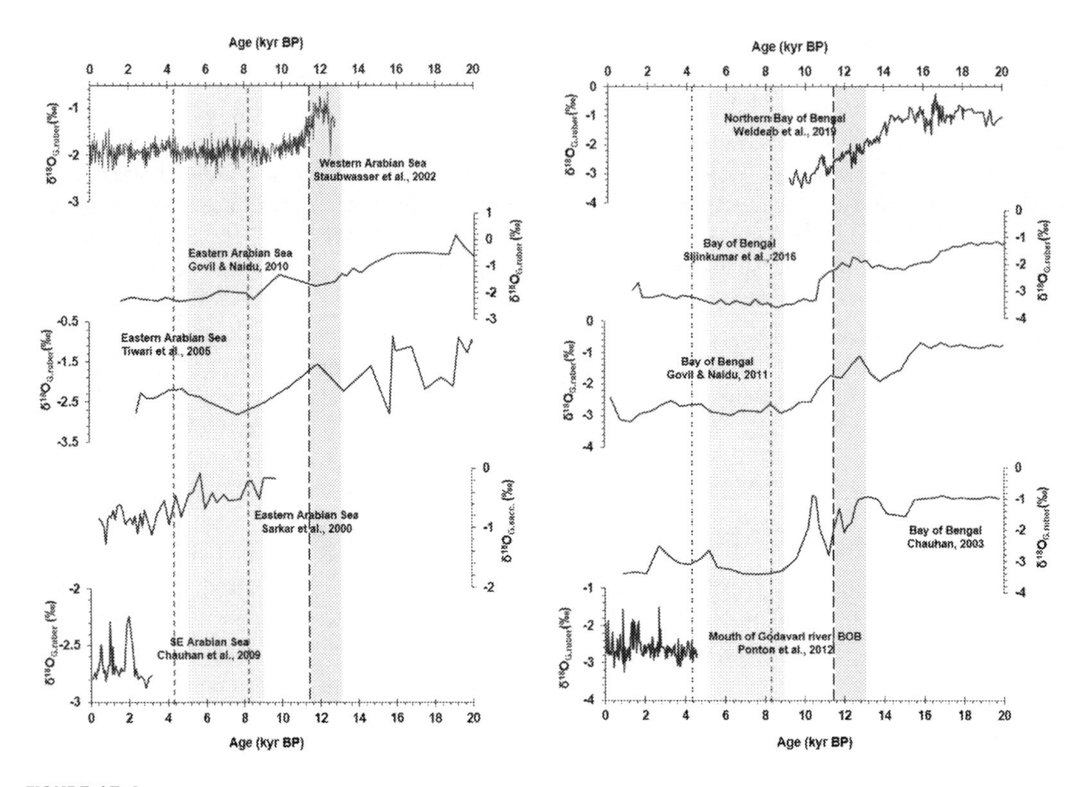

FIGURE 15.4

Intra-archival comparison of the oxygen isotopic records from (A) Arabian Sea and (B) Bay of Bengal. The marine core records from Arabian Sea include Western Arabian Sea (Staubwasser et al., 2002), Eastern Arabian Sea (Sarkar et al., 2000; Tiwari et al., 2005; Govil and Naidu, 2010), and SE Arabian Sea (Chauhan et al., 2009). Studies from Bay of Bengal include the marine cores from northern Bay of Bengal (Weldeab et al., 2019), Mouth of Godavari, BOB (Ponton et al., 2012), Bay of Bengal (Chauhan, 2003; Govil and Naidu, 2011; Sijinkumar et al., 2016). The black dashed line represents the Holocene–Pleistocene boundary, the two red dashed lines represent the abrupt drying events globally recorded at 8.2 and 4.2 kyr BP. The shaded regions represent the Younger Dryas (between 13 and 11.7 kyr BP) and the Holocene Climate Optimum (between 9 and 5 kyr BP).

1, respectively (Govil and Naidu, 2010). The marine record from the EAS shows a negative shift in $\delta^{18}O$ values ~15.5 kyr BP while the western BOB record shows a rise in SSTs with a simultaneous depletion in $\delta^{18}O$ values ~14.7 kyr BP, both related to the Bolling/Allerød event when the conditions became warmer resulting in increased river discharge and initiation of deglaciation (Chauhan, 2003; Tiwari et al., 2005; Govil and Naidu, 2011). A lead–lag relationship can be observed in the records from the Arabian Sea and BOB, possibly due to chronological uncertainties. Both the Arabian Sea and BOB records show enriched $\delta^{18}O$ values between ~12.8 and ~10.5 kyr BP suggesting prevalence of a cold and arid climate with a weakened ISM during the Younger Dryas event (Staubwasser et al., 2002; Chauhan, 2003; Tiwari et al., 2005; Govil and Naidu, 2010, 2011; Weldeab et al., 2019). Early

Holocene is marked by a general trend of gradual depletion in $\delta^{18}O$ values coupled with increased fluvial discharge due to intensification of ISM in both the Arabian Sea and BOB records (Fig. 15.4). An EAS record indicates increased discharge from Indus river to Arabian Sea from about 9.4 kyr BP which abruptly ceases at around 8.4 kyr BP, possibly recording the brief 8.2 kyr cold and arid event (Staubwasser et al., 2002). During the middle Holocene, between 9 and 6 kyr BP, the marine record from the northern BOB region also shows signature of fresher surface water than the modern salinity values further pointing toward a strengthened monsoonal circulation (Sijinkumar et al., 2016). Overall the $\delta^{18}O$ values of marine records from both the Arabian Sea and BOB region point towards an increased precipitation up to about 2 kyr BP after which an enrichment in $\delta^{18}O$ values is observed reflecting slightly arid conditions that are similar to present day (Govil and Naidu, 2010, 2011).

15.7 Discussion

The paleoclimate archives (lake/marine sediment cores and cave deposits) used in our compilation provide past climatic information at different time spans and temporal resolutions. For assessing the climate of the late Pleistocene to Holocene period, the long-term marine archives (millennial- to multimillennial-scale resolution) are relatively weakly resolved in comparison to the lake sediment (centennial- to millennial-scale resolution) and speleothem (decadal-scale resolution) records. Therefore, it must be noted that only general trends in shifts can be observed and the interarchival assessment cannot be done on finer resolutions.

Between 20 and 17 kyr BP, the cave record from NE India suggests ISM intensification, whereas the marine records from both the EAS and BOB reflect weaker circulation of monsoonal winds. However, two warm events namely, ASW 2 and ASW 1 in Arabian Sea centered at ~19 and ~17 kyr BP have been reported that have been linked to the weakening of the NE monsoon. A marine sediment record from the northern BOB (Weldeab et al., 2019), a stalagmite studied from NE India (Dutt et al., 2015) and a lake record from the western Himalaya (Mishra et al., 2015a, 2015b), all report a weakening of ISM around 16 kyr BP. The speleothem and marine record both correlate this shift to the Heinrich event H1. ISM strengthening under warmer conditions between ~15.5 and ~14 kyr BP have been reported from both the continental (Lone et al., 2014; Dutt et al., 2015; Kotlia et al., 2016) and marine records (Chauhan, 2003; Tiwari et al., 2005; Govil and Naidu, 2011). This warm phase has been linked to the Bolling–Allerød event. In the Himalayan region, a striking coherence in the cold and dry signal associated with the Younger Dryas event can be observed from both, a lacustrine deposit (lake Burfu) and a speleothem record (Kalakot cave) between 12.7 and ~12.2 kyr BP. ISM weakening during YD has also been reported from the marine records, however for a longer duration from ~12.8 to ~10.5 kyr BP (Staubwasser et al., 2002; Chauhan, 2003; Tiwari et al., 2005; Govil and Naidu, 2010; 2011; Weldeab et al., 2019). The lack of congruence observed here can be regarded to the different temporal resolution of different archives. The transitional phase from YD to early Holocene (between ~12 and ~9 kyr BP) records strengthened monsoonal winds from both the speleothem and lake record in the NW Himalaya (Dutt et al., 2015; Mishra et al., 2015a, 2015b). Between 10 and ~6.5 kyr BP, Mawmluh cave in NE India (Dutt et al., 2015) and a marine record from the northern BOB (Sijinkumar et al., 2016) show wet conditions, whereas dry conditions are recorded from the Karsandi lake in NW India (Dixit et al., 2018). This contrast can be attributed to the difference in the climatic and geomorphic setting between the eastern and western part of the country. Signal of ISM intensification is observed between 9.4 and ~8.3 kyr BP, when a coherence in peak wetness is reported from lakes near the Thar margin (Dixit et al., 2014a, 2018) and from a marine archive which records a very high fluvial discharge into the Arabian Sea (Staubwasser et al., 2002). The abrupt 8.2 kyr arid event has been recorded in both, a paleolake in west India and a marine core in SE Arabian Sea (Staubwasser et al., 2002; Dixit et al.,2014a). The signature of this short-lived event is however not so distinctly reported from other regions. The lake and cave records in the Himalaya and NE India (Dutt et al., 2015; Kathayat et al., 2017) record a weakened ISM between 6.5 and 5.5 kyr BP, when slightly wet conditions are reported from the archives in Central India (Band et al., 2018; Prasad et al., 2014). During the late Holocene, a lake near Thar Desert (Dixit et al., 2014b) records the abrupt drying at 4.1 kyr BP when prolonged phases of dryness are reported from a lake in Central India (Prasad et al., 2014) and a cave in Central lesser Himalaya (Joshi et al., 2017) between ~4.5 and ~3.5 kyr BP. All climate archives exhibit a decline in monsoonal strength after ~3 kyr BP and the climate became arid, similar to present day.

15.8 Learning and knoledge outcomes

The compilation of paleoclimate data presented in this chapter reflects changes in monsoonal wind intensity and related climatic shifts in both terrestrial and marine settings over the last 20 kyr BP. These changes are externally driven by a variety of forcing functions operating at different temporal and spatial scales. Various paleoclimate records suggest that solar forcing is evidently the major driver of the small-scale changes in the climate (Bond et al., 2001; Gupta et al., 2003). However, apart from the changes in solar insolation, topography, and elevation mainly governs the climate system in large mountain basins such as the Himalaya which act as an orographic blockade for atmospheric circulation for both the ISM and westerly winds (Jianchu et al., 2009). The temperature in the northern hemisphere is affected by the extent of sea–ice in the North Atlantic Ocean and these changes are transported through westerly winds which may not only cause delay in the onset of ISM but may also impact the amount of precipitation over the Indian landmass (Gupta et al., 2003; Pausata et al., 2011; Dutt et al., 2015). The enriched oxygen isotopic signatures with high salinity values recorded in marine archives post LGM suggest a weaker NE monsoon circulation which reduced the flow of depleted water from the BOB region (Saher et al., 2007; Govil and Naidu, 2010). The speleothem record from NE India shows that the ISM circulation in the region is linked to solar insolation and North Atlantic oscillations related to fluctuations in the atmospheric pressure, outgoing longwave radiations (OLRs) and SSTs over the North Atlantic Ocean (Srivastava et al., 2002). A significant correlation also exists between the OLR anomalies over the North Atlantic Ocean and ISM—a positive OLR anomaly is linked to a weakened monsoon over India and vice versa (Srivastava et al., 2002). The cave deposit also shows that ENSO influences the ISM moisture pathways especially during peak monsoon periods (Berkelhammer et al., 2010). Back trajectory analyses have shown that northeast India receives its moisture from the northern BOB during a Central Pacific El-Nino, whereas when the El-Nino is in Eastern Pacific, the ISM moisture to NE India is sourced from the southern BOB and Indian Ocean (Myers et al., 2015).

Around 15 kyr BP, an intensification of ISM observed especially toward the eastern Thar margin is related to a rise in SST of the Indian Ocean, increased northern hemisphere summer insolation and a consequent northward migration of the ITCZ (Bryson and Swain, 1981). The early Holocene ISM strengthening especially in western India is attributed to increased solar insolation associated with the

Earth's precession cycle (Dixit et al., 2018). Although the summer monsoon gradually reduced and fluctuating SST in the Indian Ocean were observed but with increased winter insolation between 10 and 6 kyr BP, water bodies maintained stable water levels (Bryson and Swain, 1981; Singh et al., 1990). In the last 6 kyr BP, an increase in salinity is observed in sediment records suggesting the southward shift of the ITCZ, which increased the frequency and magnitude of ENSO events, thus gradually reducing the areal coverage and intensity of ISM, especially in this region (Roy and Singhvi, 2016). Aerosol absorption (especially desert dust particles) in the north and northwest India is also known to modulate ENSO, in turn affecting the strength of ISM (Kim et al., 2016).

The strengthened phases of ISM observed across Central India are linked with the warming in the North Atlantic and a weakened ISM is correlated with colder periods in the North Atlantic region (Menzel et al., 2014). Periods of weakened ISM in Central India have also been correlated with the warm phase of the Indo-Pacific Warm Pool (Prasad et al., 2014). The records compiled from the southern part of the country suggest that phases of weakened ISM are compensated to some extent by increased moisture influx from the northeast monsoon precipitation (Dhar and Rakhecha, 1983; Basu et al., 2017; Mishra et al., 2019). An intensified NE monsoon precipitation is also positively linked to ENSO and IOD (Dhar and Rakhecha, 1983; Kripalani and Kumar, 2004; George and Ault, 2011; Yadav, 2012). The warming of the equatorial Indian Ocean during ENSO and that of the western Indian Ocean during the IOD, both result in an intensified NE monsoon precipitation (Kripalani and Kumar, 2004; Yadav, 2012; Mishra et al., 2019).

The distinct manifestation of the globally reported abrupt events at Riwasa and Kotla Dahar at 8.2 and 4.1 kyr BP, respectively, while no prominent signature of these short-lived events in other records suggests that these desert lakes are much sensitive to shifts in hydrological balance when compared to records from other parts of the country.

Acknowledgments

The authors are grateful to Dr. K.P.N. Kumaran for providing the opportunity to contribute a chapter to this book. We extend our acknowledgement to Dr. S. Prasad and Prof. R. Sinha for their support and guidance in this work. P.M. thanks Mr. Zeeshan Ahmad for his help in preparing the map.

References

Achuthavarier, D., Krishnamurthy, V., Role of Indian and Pacific SST in Indian Summer Monsoon intraseasonal variability, J. Clim. 24 (2011) 2915–2930, doi:10.1175/2010JCLI3639.1.

Alley, R.B., Mayewski, P.A., Sowers, T., Stuiver, M., Taylor, K.C., Clark, P.U., Holocene climatic instability: a prominent, widespread event 8200 yr ago, Geology 25 (1997) 483–486.

Angell, J.K., Comparison of variations in atmospheric quantities with sea surface temperature variations in the Equatorial Eastern Pacific, Monthly Weather Rev. 109 (1981) 230–243.

Anoop, A., Prasad, S., Plessen, B., Basavaiah, N., Gaye, B., Naumann, R., Menzel, P., Weise, S., Brauer, A., Palaeoenvironmental implications of evaporative gaylussite crystals from Lonar Lake, central India, J. Quat. Sci. 28 (2013) 349–359.

Band, S., Yadava, M.G., Lone, M.A., Shen, C.C., Sree, K., Ramesh, R., High-resolution mid-Holocene Indian Summer Monsoon recorded in a stalagmite from the Kotumsar Cave, Central India, Quat. Int. 479 (2018) 19–24, doi:10.1016/j.quaint.2018.01.026.

Basu, S., Anoop, A., Sanyal, P., Singh, P., Lipid distribution in the lake Ennamangalam, south India: indicators of organic matter sources and paleoclimatic history, Quat. Int. 443 (2017) 238–247, doi:10.1016/j.quaint.2016.08.045.

Behera, S.K., Yamagata, T., Influence of the Indian Ocean dipole on the Southern Oscillation, J. Meteorol. Soc. Jpn. Ser. II 81 (2003) 169–177, doi:10.2151/jmsj.81.169.

Berkelhammer, M., Sinha, A., Mudelsee, M., Cheng, H., Edwards, R.L., Cannariato, K., Persistent multidecadal power of the Indian Summer Monsoon, Earth Planet. Sci. Lett. 290 (2010) 166–172.

Berkelhammer, M., Sinha, A., Stott, L., Cheng, H., Pausata, F.S.R., Yoshimura, K., An abrupt shift in the Indian monsoon 4000 years ago, Geophys. Monogr. Ser. 198 (2012) 75–87, doi:10.1029/2012GM001207.

Beukema, S.P., Krishnamurthy, R.V., Juyal, N., Basavaiah, N., Singhvi, A.K., Monsoon variability and chemical weathering during the late Pleistocene in the Goriganga basin, higher central Himalaya, India, Quat. Res. 75 (2011) 597–604, doi:10.1016/j.yqres.2010.12.016.

Bhatla, R., Singh, M., Mall, R.K., Tripathi, A., Raju, P.V.S., Variability of summer monsoon rainfall over Indo-Gangetic plains in relation to El-Nino/La-Nina. Nat. Hazards 78 (2015) 837–853. doi:10.1007/s11069-015-1746-2.

Bond, G., Kromer, B., Beer, J., Muscheler, R., Evans, M.N., Showers, W., Hoffmann, S., Lotti-Bond, R., Hajdas, I., Bonani, G., Persistent solar influence on North Atlantic climate during the Holocene, Science 294 (5549) (2001) 2130–2136.

Breitenbach, S.F.M., Adkins, J.F., Meyer, H., Marwan, N., Kumar, K.K., Haug, G.H., Strong influence of water vapor source dynamics on stable isotopes in precipitation observed in Southern Meghalaya, NE India, Earth Planet. Sci. Lett. 292 (2010) 212–220, doi:10.1016/j.epsl.2010.01.038.

Bryson, R.A., Swain, A.M., Holocene variations of monsoon rainfall in Rajasthan, Quat. Res. 16 (1981) 135–145.

Chauhan, O.S., Past 20,000-year history of Himalayan aridity: evidence from oxygen isotope records in the Bay of Bengal, Curr. Sci. 84 (2003) 90–93.

Chauhan, O.S., Vogelsang, E., Basavaiah, N., Kader, U.S.A., Reconstruction of the variability of the southwest monsoon during the past 3 ka, from the continental margin of the southeastern Arabian Sea, J. Quat. Sci. 25 (2009) 798–807.

Cheng, H., Fleitmann, D., Edwards, R.L., Wang, X., Cruz, F.W., Auler, A.S., Mangini, A., Wang, Y., Kong, X., Burns, S.J., Matter, A., Timing and structure of the 8.2. kyr B.P. event inferred from $\delta^{18}O$ records of stalagmites from China, Oman, and Brazil, Geology 37 (2009) 1007–1010, doi:10.1130/G30126A.1.

Chidambaram, S., Prasanna, M.V., Ramanathan, A.L., Vasu, K., Hameed, S., Warrier, U.K., Srinivasamoorthy, K., Manivannan, R., Tirumalesh, K., Anandhan, P., Johnsonbabu, G., A study on the factors affecting the stable isotopic composition in precipitation of Tamil Nadu, India, Hydrol. Processes 23 (2009) 1792–1800, doi:10.1002/hyp.7300.

Cohen, K.M., Harper, D.A.T., Gibbard, P.L., Fan, J.X., 2018. ICS International Chronostratigraphic Chart 2018/07.

Crowley, T.J., Zielinski, G., Vinther, B., Udisti, R., Kreutz, K., Cole-Dai, J., Castellano, E., Special section: data-model comparison, PAGES News 16 (2008) 22–23, doi:10.1029/2002GL0166335.

Cullen, H.M., deMenocal, P.B., Hemming, S., Brown, F.H., Guilderson, T., Sirocko, F., Climate change and the collapse of the Akkadian empire: evidence from the deep sea, Geology 28 (2000) 379–382, doi:10.1130/0091-7613(2000)28<379:CCATCO>2.0.CO;2.

deMenocal, P.B., Cultural responses to climate change during the late holocene, Science 292 (2001) 667–673, doi:10.1126/science.1059827.

Dhar, O.N., Rakhecha, P.R., Foreshadowing Northeast monsoon rainfall over Tamil Nadu, India., Monthly Weather Rev. 111 (1983) 109–113.

Dixit, Y., Hodell, D.A., Giesche, A., Tandon, S.K., Gázquez, F., Saini, H.S., Skinner, L.C., Mujtaba, S.A.I., Pawar, V., Singh, R.N., Petrie, C.A., Intensified summer monsoon and the urbanization of Indus Civilization in northwest India, Sci. Rep. 8 (2018) 1–8, doi:10.1038/s41598-018-22504-5.

Dixit, Y., Hodell, D.A., Petrie, C.A., Abrupt weakening of the summer monsoon in northwest India ~4100 yr ago, Geology 42 (2014b) 339–342, doi:10.1130/G35236.1.

Dixit, Y., Hodell, D.A., Sinha, R., Petrie, C.A., Abrupt weakening of the Indian summer monsoon at 8.2 kyr B.P. Earth Planet. Sci. Lett. 391 (2014a) 16–23. doi:10.1016/j.epsl.2014.01.026.

Dortch, J.M., Owen, L.A., Caffee, M.W., Timing and climatic drivers for glaciation across semi-arid western Himalayan-Tibetan orogen, Quat. Sci. Rev. 78 (2013) 188–208, doi:10.1016/j.quascirev.2013.07.025.

Drysdale, R., Zanchetta, G., Hellstrom, J., Maas, R., Fallick, A., Pickett, M., Cartwright, I., Piccini, L., Late Holocene drought responsible for the collapse of Old World civilizations is recorded in an Italian cave flowstone, Geology 34 (2006) 101–104, doi:10.1130/G22103.1.

Dutt, S., Gupta, A.K., Clemens, S.C., Cheng, H., Singh, R.K., Kathayat, G., Lawrence Edwards, R., Abrupt changes in Indian summer monsoon strength during 33,800 to 5500years B.P., Geophys. Res. Lett. 42 (2015) 5526–5532, doi:10.1002/2015GL064015.

Erlenkeuser, H., Muïler, J., Johnsen, S., Grafenstein, U.Von, Erlenkeuser, H, Muïler, J, Jouzel, J., Johnsen, S, The cold event 8200 years ago documented in oxygen isotope records of precipitation in Europe and Greenland, Clim. Dyn. 14 (1998) 73–81.

Fagan, B., The Little Ice Age: How Climate Made History 1300-1850, Nueva, York, 2000.

Fisher, D., Osterberg, E., Dyke, A., Dahl-Jensen, D., Demuth, M., Zdanowicz, C., Bourgeois, J., Koerner, R.M., Mayewski, P., Wake, C., Kreutz, K., Steig, E., Zheng, J., Yalcin, K., Goto-Azuma, K., Luckman, B., Rupper, S., The Mt Logan Holocene-late Wisconsinan isotope record: tropical Pacific-Yukon connections, Holocene 18 (2008) 667–677, doi:10.1177/0959683608092236.

Fleitmann, D., Burns, S.J., Mangini, A., Mudelsee, M., Kramers, J., Villa, I., Neff, U., Al-Subbary, A.A., Buettner, A., Hippler, D., Matter, A., Holocene ITCZ and Indian monsoon dynamics recorded in stalagmites from Oman and Yemen (Socotra), Quat. Sci. Rev. 26 (2007) 170–188, doi:10.1016/j.quascirev.2006.04.012.

Fleitmann, D., Burns, S.J., Mudelsee, M., Neff, U., Kramers, J., Mangini, A., Matter, A., Holocene forcing of the Indian monsoon recorded in a stalagmite from southern Oman, Science 300 (2003) 1737–1739, doi:10.1126/science.1083130.

Franke, J.G., Donner, R.V., Correlating paleoclimate time series: sources of uncertainty and potential pitfalls, Quat. Sci. Rev. 212 (2019) 69–79, doi:10.1016/j.quascirev.2019.03.017.

Gat, J.R., Oxygen and hydrogen isotopes in the hydrologic cycle, Annu. Rev. Earth Planet. Sci. 24 (1996) 225–262, doi:https://10.1146/annurev.earth.24.1.225.

Gat, J.R., Gonfiantini, R., Stable Isotope Hydrology: Deuterium and Oxygen-18 in the Water Cycle 210, IAEA. Tech. Rep. Ser. 1981, pp. 177–201.

George, S.S., Ault, T.R., Is energetic decadal variability a stable feature of the central Pacific Coast's winter climate? J. Geophys. Res. Atmos. 116 (2011) 1–6, doi:10.1029/2010JD015325.

Govil, P., Naidu, P.D., Evaporation-precipitation changes in the eastern Arabian Sea for the last 68 ka: implications on monsoon variability, Paleoceanography 25 (2010) 1–11, doi:10.1029/2008PA001687.

Govil, P., Naidu, P.D., Variations of Indian monsoon precipitation during the last 32kyr reflected in the surface hydrography of the Western Bay of Bengal, Quat. Sci. Rev. 30 (2011) 3871–3879, doi:10.1016/j.quascirev.2011.10.004.

Graham, N.E., Ammann, C.M., Fleitmann, D., Cobb, K.M., Luterbacher, J., Support for global climate reorganization during the "Medieval Climate Anomaly.", Clim. Dyn. 37 (2011) 1217–1245, doi:10.1007/s00382-010-0914-z.

Gunnell, Y., Anupama, K., Sultan, B., Response of the South Indian runoff-harvesting civilization to northeast monsoon rainfall variability during the last 2000 years: instrumental records and indirect evidence, Holocene 17 (2007) 207–215.

Gupta, A.K., Anderson, D.M., Overpeck, J.T., Abrupt changes in the Asian southwest monsoon during the Holocene and their links to the North Alantic Ocean, Nature 421 (2003) 354–357, doi:10.1038/nature01340.

Hass, H.C., Northern Europe climate variations during late Holocene: evidence from marine Skagerrak, Palaeogeogr. Palaeoclimatol. Palaeoecol. 123 (1996) 121–145, doi:10.1016/0031-0182(95)00114-X.

Haug, G.H., Hughen, K.A., Sigman, D.M., Peterson, L.C., Röhl, U., Southward migration of the intertropical convergence zone through the Holocene, Science 293 (2001) 1304–1308.

Helama, S., Jones, P.D., Briffa, K.R., Dark Ages Cold Period: a literature review and directions for future research, Holocene 27 (2017) 1600–1606, doi:10.1177/0959683617693898.

Hodell, D.A., Curtis, J.H., Brenner, M., Possible role of climate in the collapse of Classic Maya civilization, Nature 375 (1995) 391–394.

Hofer, T., Messerli, B., Floods in Bangladesh: History, Dynamics and Rethinking the Role of the Himalayas, Ecology 29 (2006) 254–283.

Holmes, J.A., Zhang, J., Chen, F., Qiang, M., Paleoclimatic implications of an 850-year oxygen-isotope record from the northern Tibetan Plateau, Geophys. Res. Lett. 34 (23) (2007) 1–5.

Ji, J., Balsam, W., Chen, X., Chen, J., Chen, Y., Wang, H., Rate of solar insolation change and the glacial/interglacial transition, Geophys. Res. Lett. 33 (2006) 3–6, doi:10.1029/2005GL025401.

Jianchu, X., Shrestha, A., Eriksson, M., Climate change and its impacts on glaciers and water resource management in the Himalayan Region, International Hydrological Programme of UNESCO and Hydrology and Water Resources Programme of WMO (2009) 44–54.

Joshi, L.M., Kotlia, B.S., Ahmad, S.M., Wu, C.C., Sanwal, J., Raza, W., Singh, A.K., Shen, C.C., Long, T., Sharma, A.K., Reconstruction of Indian monsoon precipitation variability between 4.0 and 1.6 ka BP using speleothem $\delta^{18}O$ records from the Central Lesser Himalaya, India, Arab. J. Geosci. 10 (2017) 1–16, doi:10.1007/s12517-017-3141-7.

Kathayat, G., Cheng, H., Sinha, A., Yi, L., Li, X., Zhang, H., Li, H., Ning, Y., Edwards, R.L., The Indian monsoon variability and civilization changes in the Indian subcontinent, Sci. Adv. 3 (2017) 1–9, doi:10.1126/sciadv.1701296.

Kim, M.K., Lau, W.K.M., Kim, K.M., Sang, J., Kim, Y.H., Lee, W.S., Amplification of ENSO effects on Indian summer monsoon by absorbing aerosols, Clim. Dyn. 46 (2016) 2657–2671, doi:10.1007/s00382-015-2722-y.

Klitgaard-Kristensen, D., Sejrup, H.P., Haflidason, H., Johnsen, S., Spurk, M., A regional 8200 cal. yr BP cooling event in northwest Europe, induced by final stages of the Laurentide ice-sheet deglaciation? J. Quat. Sci. 13 (1998) 165–169 https://doi.org/10.1002/(SICI)1099-1417(199803/04)13:2%3C165::AID-JQS365%3E3.0.CO;2-%23.

Kotlia, B.S., Singh, A.K. Sanwal, J., Raza, W., Ahmad, S.M., Joshi, L.M., Sirohi, M., Sharma, A.K., Sagar, N., Stalagmite inferred high resolution climatic changes through Pleistocene-Holocene transition in Northwest Indian Himalaya, J. Earth Sci. Clim. Change 07 (2016) 1–7, doi:10.4172/2157-7617.1000338.

Kripalani, R.H., Kumar, P., Northeast monsoon rainfall variability over south peninsular India vis-à-vis the Indian Ocean dipole mode, Int. J. Climatol. 24 (2004) 1267–1282, doi:10.1002/joc.1071.

Lafleur, D.M., Barrett, B.S., Henderson, G.R., Some climatological aspects of the Madden-Julian oscillation (MJO), J. Clim. 28 (2015) 6039–6053, doi:10.1175/JCLI-D-14-00744.1.

Laskar, A.H., Yadava, M.G., Ramesh, R., Polyak, V.J., Asmerom, Y., A 4 kyr stalagmite oxygen isotopic record of the past Indian Summer Monsoon in the Andaman Islands, Geochem. Geophys. Geosyst. 14 (2013) 3555–3566, doi:10.1002/ggge.20203.

Lau, N.-C., Wang, B., Interactions between the Asian monsoon and the El Niño/Southern Oscillation, Asian Monsoon 12 (2006) 479–512, doi:10.1007/3-540- 37722-0_12.

Lechleitner, F.A., Breitenbach, S.F.M., Rehfeld, K., Ridley, H.E., Asmerom, Y., Prufer, K.M., Marwan, N., Goswami, B., Kennett, D.J., Aquino, V.V., Polyak, V., Haug, G.H., Eglinton, T.I., Baldini, J.U.L., Tropical rainfall over the last two millennia: evidence for a low-latitude hydrologic seesaw, Sci. Rep. 7 (2017) 1–9, doi:10.1038/srep45809.

Liu, W., Li, X., Zhang, L., An, Z., Xu, L., Evaluation of oxygen isotopes in carbonate as an indicator of lake evolution in arid areas: the modern Qinghai Lake, Qinghai–Tibet Plateau, Chem. Geol. 268 (2009) 126–136, doi:10.1016/j.chemgeo. 2009.08.004.

Lone, M.A., Ahmad, S.M., Dung, N.C., Shen, C.C., Raza, W., Kumar, A., Speleothem based 1000-year high resolution record of Indian monsoon variability during the last deglaciation, Palaeogeogr. Palaeoclimatol. Palaeoecol. 395 (2014) 1–8, doi:10.1016/j.palaeo.2013.12.010.

MacDonald, G., Potential influence of the Pacific Ocean on the Indian summer monsoon and Harappan decline, Quat. Int. 229 (2011) 140–148, doi:10.1016/j.quaint.2009.11.012.

Marshall, M.H., Lamb, H.F., Huws, D., Davies, S.J., Bates, R., Bloemendal, J., Boyle, J., Leng, M.J., Umer, M., Bryant, C., Late Pleistocene and Holocene drought events at Lake Tana, the source of the Blue Nile, Global Planet. Change 78 (2011) 147–161, doi:10.1016/j.gloplacha.2011.06.004.

Martín-Puertas, C., Valero-Garcés, B.L., Brauer, A., Mata, M.P., Delgado-Huertas, A., Dulski, P., The Iberian-Roman Humid Period (2600-1600 cal yr BP) in the Zoñar Lake varve record (Andalucía, Southern Spain), Quat. Res. 71 (2009) 108–120, doi:10.1016/j.yqres.2008.10.004.

Menzel, P., Gaye, B., Mishra, P.K., Anoop, A., Basavaiah, N., Marwan, N., Plessen, B., Prasad, S., Riedel, N., Stebich, M., Wiesner, M.G., Linking Holocene drying trends from Lonar Lake in monsoonal central India to North Atlantic cooling events, Palaeogeogr. Palaeoclimatol. Palaeoecol. 410 (2014) 164–178, doi:10.1016/j.palaeo.2014.05.044.

Mishra, P.K., Ankit, Y., Gautam, P.K., Lakshmidevi, C.G., Singh, P., Anoop, A., Inverse relationship between south-west and north-east monsoon during the late Holocene: geochemical and sedimentological record from Ennamangalam Lake, southern India, Catena 182 (2019) 104117.

Mishra, P.K., Anoop, A., Schettler, G., Prasad, S., Jehangir, A., Menzel, P., Naumann, R., Yousuf, A.R., Basavaiah, N., Deenadayalan, K., Wiesner, M.G., Gaye, B., Reconstructed late Quaternary hydrological changes from Lake Tso Moriri, NW Himalaya, Quat. Int. 371 (2015a) 76–86, doi:10.1016/j.quaint.2014.11.040.

Mishra, P.K., Prasad, S., Anoop, A., Plessen, B., Jehangir, A., Gaye, B., Menzel, P., Weise, S.M., Yousuf, A.R., Carbonate isotopes from high altitude Tso Moriri Lake (NW Himalayas) provide clues to late glacial and Holocene moisture source and atmospheric circulation changes, Palaeogeogr. Palaeoclimatol. Palaeoecol. 425 (2015b) 76–83, doi:10.1016/j.palaeo.2015.02.031.

Mishra, S.K., Sahany, S., Salunke, P., Linkages between MJO and summer monsoon rainfall over India and surrounding region, Meteorol. Atmos. Phys. 129 (2016) 283–296, doi:10.1007/s00703-016-0470-0.

Misra, P., Farooqui, A., Sinha, R., Khanolkar, S., Tandon, S.K., Millennial-scale vegetation and climatic changes from an Early to Mid-Holocene lacustrine archive in Central Ganga Plains using multiple biotic proxies, Quat. Sci. Rev. 243 (2020) 106474.

Misra, P., Tandon, S.K., Sinha, R., Holocene climate records from lake sediments in India: assessment of coherence across climate zones, Earth Sci. Rev. 190 (2019) 370–397, doi:10.1016/j.earscirev.2018.12.017.

Mook, W., Rozanski, K., Environmental Isotopes in the Hydrological Cycle III, IAEA, Technical documents in Hydrology, 2000, pp. 1–10.

Morrill, C., Overpeck, J.T., Cole, J.E., A synthesis of abrupt changes in the Asian summer monsoon since the last deglaciation, Holocene 13 (2003) 465–476, doi:10.1191/0959683603hl639ft.

Murtugudde, R., McCreary, J.P., Busalacchi, A.J., Oceanic processes associated with anomalous events in the Indian Ocean with relevance to 1997-1998, J. Geophys. Res.: Oceans 105 (2000) 3295–3306, doi:10.1029/1999jc900294.

Myers, C.G., Oster, J.L., Sharp, W.D., Bennartz, R., Kelley, N.P., Covey, A.K., Breitenbach, S.F.M., Northeast Indian stalagmite records Pacific decadal climate change: Implications for moisture transport and drought in India, Geophys. Res. Lett. 42 (2015) 4124–4132, doi:10.1002/2015GL063826.

Overpeck, J., Anderson, D., Trumbore, S., Prell, W., The southwest Indian Monsoon over the last 18000 years, Clim. Dyn. 12 (1996) 213–225, doi:10.1007/BF00211619.

Pant, R.K., Juyal, N., Basavaiah, N., Singhvi, A.K., Late Quaternary glaciation and seismicity in the Higher Central Himalaya: evidence from Shalang basin (Goriganga), Uttaranchal, Curr. Sci. 90 (2006) 1500–1505.

Pausata, F.S.R., Li, C., Wettstein, J.J., Kageyama, M., Nisancioglu, K.H., The key role of topography in altering North Atlantic atmospheric circulation during the last glacial period, Clim. Past 7 (2011) 1089–1101, doi:10.5194/cp-7-1089-2011.

Pearce, C., Seidenkrantz, M.S., Kuijpers, A., Massé, G., Reynisson, N.F., Kristiansen, S.M., Ocean lead at the termination of the Younger Dryas cold spell, Nat. Commun. 4 (2013) 1–6, doi:10.1038/ncomms2686.

Petit, J.R., Jouzel, J., Raynaud, D., Barkov, N.I., Barnola, J.M., Basile, I., Bender, M., Chappellaz, J., Davisk, M., Delaygue, G., Delmotte, M., Kotlyakov, V.M., Legrand, M., Lipenkov, V.Y., Lorius, C., Pé, L., Ritz, C., Saltzmank, E., Stievenard, M., Climate and atmospheric history of the past 420,000 years from the Vostok ice core, Antarctica. The recent completion of drilling at Vostok station in East, Nature 399 (1999) 429–436.

Pillai, A.A.S., Anoop, A., Sankaran, M., Sanyal, P., Jha, D.K., Ratnam, J., Mid-late Holocene vegetation response to climatic drivers and biotic disturbances in the Banni grasslands of western India, Palaeogeogr. Palaeoclimatol. Palaeoecol. 485 (2017) 869–878, doi:10.1016/j.palaeo.2017.07.036.

Ponton, C., Giosan, L., Eglinton, T.I., Fuller, D.Q., Johnson, J.E., Kumar, P., Collett, T.S., Holocene aridification of India, Geophys. Res. Lett. 39 (2012) 1–6, doi:10.1029/2011GL050722.

Prasad, S., Anoop, A., Riedel, N., Sarkar, S., Menzel, P., Basavaiah, N., Krishnan, R., Fuller, D., Plessen, B., Gaye, B., Röhl, U., Wilkes, H., Sachse, D., Sawant, R., Wiesner, M.G., Stebich, M., Prolonged monsoon droughts and links to Indo-Pacific warm pool: a Holocene record from Lonar Lake, central India, Earth Planet. Sci. Lett. 391 (2014) 171–182, doi:10.1016/j.epsl.2014.01.043.

Rasmusson, E.M., Carpenter, T.H. The relationship between Eastern Equatorial Pacific sea surface temperatures and rainfall over India and Sri Lanka. Monthly Weather Review 111 (3), (1983) 517–528.

Rasmussen, S.O., Andersen, K.K., Svensson, A.M., Steffensen, J.P., Vinther, B.M., Clausen, H.B., Siggaard-Andersen, M.L., Johnsen, S.J., Larsen, L.B., Dahl-Jensen, D., Bigler, M., Röthlisberger, R., Fischer, H., Goto-Azuma, K., Hansson, M.E., Ruth, U., A new Greenland ice core chronology for the last glacial termination, J. Geophys. Res. Atmos. 111 (2006) 1–16, doi:10.1029/2005JD006079.

Rees, H.G., Collins, D.N., Regional differences in response of flow in glacier-fed Himalayan rivers to climatic warming, Hydrol. Processes 20 (2006) 2157–2169, doi:10.1002/hyp.6209.

Rind, D., Overpeck, J., Hypothesized causes of decade-to-century-scale climate variability: climate model results, Quat. Sci. Rev. 12 (1993) 357–374, doi:10.1016/S0277-3791(05)80002-2.

Ropelewski, C.F., Halpert, M.S., Global and regional precipitation patterns associated with the El Niño Southern Oscillation, Monthly Weather Rev. 115 (1987) 1606–1626.

Roy, P.D., Singhvi, A.K., Climate variation in the Thar desert since the last glacial maximum and evaluation of the Indian monsoon, TIP 19 (2016) 32–44, doi:10.1016/j.recqb.2016.02.004.

Saher, M.H., Jung, S.J.A., Elderfield, H., Greaves, M.J., Kroon, D., Sea surface temperatures of the western Arabian Sea during the last deglaciation, Paleoceanography 22 (2007) 1–12, doi:10.1029/2006PA001292.

Saji, N.H., Goswami, B.N., Vinayachandran, P.N., Yamagata, T., A dipole mode in the tropical Indian Ocean, Nature 401 (1999) 360–363.

Sanwal, J., Kotlia, B.S., Rajendran, C., Ahmad, S.M., Rajendran, K., Sandiford, M., Climatic variability in Central Indian Himalaya during the last ~1800 years: evidence from a high resolution speleothem record, Quat. Int. 304 (2013) 183–192, doi:10.1016/j.quaint.2013.03.029.

Sarkar, A., Ramesh, R., Somayajulu, B.L.K., Agnihotri, R., Jull, A.J.T., Burr, G.S., High resolution Holocene monsoon record from the eastern Arabian Sea, Earth Planet. Sci. Lett. 177 (3-4) (2000) 209–218.

Scheurle, C., Hebbeln, D., Stable oxygen isotopes as recorders of salinity and river discharge in the German Bight, North Sea, Geo-Mar. Lett. 23 (2003) 130–136.

Seidenkrantz, M.S., Aagaard-Sørensen, S., Sulsbrück, H., Kuijpers, A., Jensen, K.G., Kunzendorf, H., Hydrography and climate of the last 4400 years in a SW Greenland fjord: Implications for Labrador Sea palaeoceanography, Holocene 17 (2007) 387–401, doi:10.1177/0959683607075840.

Sijinkumar, A.V., Clemens, S., Nath, B.N., Prell, W., Benshila, R., Lengaigne, M., δ^{18}O and salinity variability from the Last Glacial Maximum to Recent in the Bay of Bengal and Andaman Sea, Quat. Sci. Rev. 135 (2016) 79–91, doi:10.1016/j.quascirev.2016.01.022.

Sikka, D.R., Some aspects of the large scale fluctuations of summer monsoon rainfall over India in relation to fluctuations in the planetary and regional scale circulation parameters, Proc. Indian Acad. Sci. (Earth Planet. Sci.) 89 (2) (1980) 179–195.

Singh, G., Wasson, R.J., Agrawal, D.P., Vegetational and seasonal climatic changes since the last full glacial in the Thar Desert, northwestern India. Review of Palaeobotany and Palynology 64 (1–4) (1990) 351–358.

Singh, M., Bhatla, R., Intense rainfall conditions over Indo-Gangetic Plains under the influence of Madden–Julian oscillation, Meteorol. Atmos. Phys. 132 (2020) 441–449, doi:10.1007/s00703-019-00703-7.

Sinha, A., Berkelhammer, M., Stott, L., Mudelsee, M., Cheng, H., Biswas, J., The leading mode of Indian Summer Monsoon precipitation variability during the last millennium, Geophys. Res. Lett. 38 (2011) 1–5, doi:10.1029/2011GL047713.

Sirocko, F., Sarntheln, M., Erlenkeuser, H., Lange, H., Duplessyll, J.C., Century-scale events in monsoonal climate over the past 24,000 years, Nature 364 (1993) 322–324.

Solanki, S.K., Usoskin, I.G., Kromer, B., Schüssler, M., Beer, J., Unusual activity of the Sun during recent decades compared to the previous 11,000 years, Nature 43 (2004) 1084–1087.

Sreekala, P.P., Rao, S.V.B., Rajeevan, M., Northeast monsoon rainfall variability over south peninsular India and its teleconnections, Theor. Appl. Climatol. 108 (2012) 73–83, doi:10.1007/s00704-011-0513-x.

Srivastava, A.K., Rajeevan, M., Kulkarni, R., Teleconnection of OLR and SST anomalies over Atlantic Ocean with Indian summer monsoon, Geophys. Res. Lett. 29 (2002) 29–38, doi:10.1029/2001GL013837.

Staubwasser, M., Sirocko, F., Grootes, P.M., Erlenkeuser, H., South Asian monsoon climate change and radiocarbon in the Arabian Sea during early and middle Holocene, Paleoceanography 17 (2002) 15-1–15-12, doi:10.1029/2000pa000608.

Staubwasser, M., Weiss, H., Holocene climate and cultural evolution in late prehistoric-early historic West Asia, Quat. Res. 66 (2006) 372–387, doi:10.1016/j.yqres.2006.09.001.

Tiwari, M., Ramesh, R., Somayajulu, B.L.K., Jull, A.J.T., Burr, G.S., Solar control of southwest monsoon on centennial timescales, Curr. Sci. 89 (2005) 1583.

Urey, H.C., The thermodynamic properties of isotopic substances, J. Chem. Soc. (1947) 562–581.

Von Grafenstein, U., Erlenkeuser, H., Müller, J., Jouzel, J., Johnsen, S., The cold event 8200 years ago documented in oxygen isotope records of precipitation in Europe and Greenland, Clim. Dyn. 14 (1998) 73–81, doi:10.1007/s003820050210.

Walker, G.T., Correlations in seasonal variations of weather: a further study of world weather, Mem. Indian Meteorol. Dep. 24 (1924) 275–332.

Wang, S., Ge, Q., Wang, F., Wen, X., Huang, J., Abrupt climate changes of Holocene, Chin. Geogr. Sci. 23 (2013) 1–12, doi:10.1007/s11769-013-0591-z.

Wanner, H., Beer, J., Bütikofer, J., Crowley, T.J., Cubasch, U., Flückiger, J., Goosse, H., Grosjean, M., Joos, F., Kaplan, J.O., Küttel, M., Müller, S.A., Prentice, I.C., Solomina, O., Stocker, T.F., Tarasov, P., Wagner, M., Widmann, M., Mid- to Late Holocene climate change: an overview, Quat. Sci. Rev. 27 (2008) 1791–1828, doi:10.1016/j.quascirev.2008.06.013.

Webster, Peter J., Moore, Andrew M., Loschnigg, Johannes P., Leben, R.R., Coupled ocean-atmosphere dynamics in the Indian Ocean during 1997-98, Nature 401 (1999) 356–360.

Weiss, H., Bradley, R.S., What drives societal collapse? Science 291 (2001) 609–610, doi:10.1126/science.1058775.

Weldeab, S., Rühlemann, C., Bookhagen, B., Pausata, F.S.R., Perez-Lua, F.M., Enhanced Himalayan glacial melting during YD and H1 recorded in the Northern Bay of Bengal, Geochem. Geophys. Geosyst. 20 (2019) 2449–2461, doi:10.1029/2018GC008065.

Wenxiang, W., Tungsheng, L., Possible role of the "Holocene Event 3" on the collapse of Neolithic Cultures around the Central Plain of China, Quat. Int. 117 (2004) 153–166, doi:10.1016/S1040-6182(03)00125-3.

Yadav, R.K., Why is ENSO influencing Indian northeast monsoon in the recent decades? Int. J. Climatol. 32 (2012) 2163–2180, doi:10.1002/joc.2430.

Yadava, M.G., Ramesh, R., Monsoon reconstruction from radiocarbon dated tropical Indian speleothems. Holocene 15, (2005) 48–59. doi:10.1191/0959683605h1783rp.

Yadava, M.G., Ramesh, R., Speleothems – useful proxies for past monsoon rainfall, J. Sci. Ind. Res. 58 (1999) 339–348.

Yang, H., Johnson, K.R., Griffiths, M.L., Yoshimura, K., Interannual controls on oxygen isotope variability in Asian monsoon precipitation and implications for paleoclimate reconstructions, J. Geophys. Res.: Atmos. 121 (2016) 8410–8428.

Zachos, J.C., Stott, L.D., Lohmann, K.C., Evolution of Early Cenozoic marine temperatures, Paleoceanography 9 (1994) 353–387, doi:10.1029/93PA03266.

CHAPTER

16

Application of precipitation isotopes in pursuit of paleomonsoon reconstruction: An Indian perspective

Supriyo Chakraborty[a], Amey Datye[a], Charuta Murkute[a], Subrota Halder[a], Anant Parekh[a], Nitesh Sinha[b,c], P.M. Mohan[d]

[a] *Indian Institute of Tropical Meteorology, Ministry of Earth Sciences, Pune, India.* [b] *Center for Climate Physics, Institute for Basic Science, Busan, Republic of Korea.* [c] *Pusan National University, Busan, Republic of Korea.* [d] *Pondicherry University, Port Blair, India*

16.1 Introduction

Indian monsoon rainfall is characterized by different modes of variability, starting from the diurnal scale to tens of thousands of years. Understanding the nature of the variability is necessary for a better prediction of the monsoon rainfall, especially on a longer timescale. Our understanding of the future scenario of the monsoon system and its effect on socioeconomic factors very much depends on how well we know its past patterns of changes both in temporal and spatial domains. Therefore an integrated approach of available instrumental data and proxy derived records is a key to understand better the interactions between various components of the monsoon system and the underlying mechanism.

Long-term rainfall information going beyond the instrumental observation and historical account is usually sourced from the paleoclimatic investigation. An alternative approach simulates the past climate using the general circulation models that provide further insight, which complements the proxy observations. Such kind of simulation is available that investigated monsoon variability specifically for the last millennium, mid-Holocene, and the Last Glacial Maximum (Tejavath et al., 2020; Jalihal et al., 2020). However, the model-derived climate variables, such as the Indian monsoon rainfall, often show large discrepancies (Tejavath et al., 2019), and hence they need to be validated against the observational study. Oxygen isotopic records of speleothem are believed to offer a reliable means as far as monsoon rainfall is concerned (Kaushal et al., 2018). Hence, one of the fundamental requirement is to produce well-constrained speleothem isotopic records for a considerable period in the past. But, the speleothem isotopic data only provide a relative measure of monsoon variability (Sinha et al., 2007). To get quantitative estimates of rainfall, a calibration equation between rainfall and its isotopic variability (Yadva and Ramesh, 2015), that is, the *amount effect* (Dansgaard, 1964), needs to be established.

The amount effect is defined as the observed decrease in rainfall isotopic values (denoted as δ^{18}Op) with increased rainfall amount (dδ^{18}Op/dP; P stands for precipitation; Dansgaard, 1964; Rozanski et al., 1993). The traditional method of examining the amount effect is done by measuring the precipitation isotopes and establishing a linear relation with the rainfall amount (Johnson and Ingram, 2004). This

Holocene Climate Change and Environment. DOI: https://doi.org/10.1016/B978-0-323-90085-0.00002-4

is usually defined as the depletion of precipitation–δ^{18}O for every 100 mm of rainfall on a monthly timescale (Lachinet, 2009). Several researchers investigated the P versus $\delta^{18}O_p$ relation and estimated the amount effect for the Indian subcontinent; Fig. 16.1A shows the P–$\delta^{18}O_p$ correlation values for selected sites in India and the neighboring countries. A notable feature is that the north and central Indian sites (except Kanpur) are characterized by high values (i.e., precipitation isotopes suffer large depletions, about -4 to -8‰ per 100 mm of rainfall), while the sites in Bangladesh show moderate effect (-2 to -3‰ per 100 mm rainfall). The southern Indian sites, including those in Sri Lanka, are characterized by a low amount effect (typically <-2‰ per 100 mm rainfall). Two factors that may significantly affect the rainfall–isotope relation are (i) the spatial scale of the rain used in establishing their connections and (ii) the sample size. In most cases, the isotopic values of precipitation were correlated with the rainfall data on a point scale. Though the correlation coefficient (r) in some cases shows reasonably high values, the low sample size consisting of one or two observational years may not yield an expected value for a given region. For example, the $|r|$ values for the South Indian and Sri Lankan sites are typically >0.5. But, the Port Blair site in the Bay of Bengal (henceforth denoted as BoB) has a value of 0.3. One likely reason for having a low value in the case of Port Blair data is a large number of samples ($n = 550$) for Port Blair against 15–39 in the case of the sites in Sri Lanka was used. Obviously, large sample size in a given location would be a better representative than a zone characterized by a low sample size.

It has been reported by some investigators that the amount effect though weak on a point scale, but gets improved on a larger scale (Moerman et al., 2013). The basic premise for such a behavior is that the observed isotopic values in precipitation and water vapor integrate convective activities over time and space (Galewsky et al., 2016). Hence according to Galewsky et al. (2016), the amount effect is not a purely local phenomenon. This is especially relevant in monsoon regions where deep convection is associated with more isotopically depleted water vapor and precipitation in downstream air mass trajectories (op cit.). Hence, it is an essential task that the amount effect for a given region is examined over a larger spatial and longer timescale, which, by and large, remained unaddressed for the Indian subcontinent. This is required for better quantification of past rainfall variability.

16.2 Scope of the work

To achieve this task for the Andaman Islands region, we examine the linear relation pattern between the rainfall and its isotopic values, how the correlation changes if the precipitation is integrated over broader spatial scales. The reason for choosing the Andaman Island site is that this island harbors a large number of caves consisting of stalactites and stalagmites, together called the speleothems. If a well-constrained amount effect is established for this region, the isotopic records of speleothem could be used to quantify the past rainfall variability with better precision. Second, the isotopic values of precipitation in this region are mostly controlled by the oceanic processes, considering the area of the island is small relative to the size of the BoB basin. Moistures produced by terrestrial processes have different isotopic signature than those made in the marine environment. Hence, the isotopic values of vapor and precipitation are expected to be influenced by less complexity in the marine environment than they would experience in the terrestrial region (Sinha and Chakraborty, 2020).

Further, it is believed that the isotopic values derived from rainwater essentially carry the isotopic information of the moisture (Kendall and McDonnell, 2012) rather than the rainfall event (Dar and Ghosh, 2017). Rainwater collected at a single site represents a point scale event, but the overlying

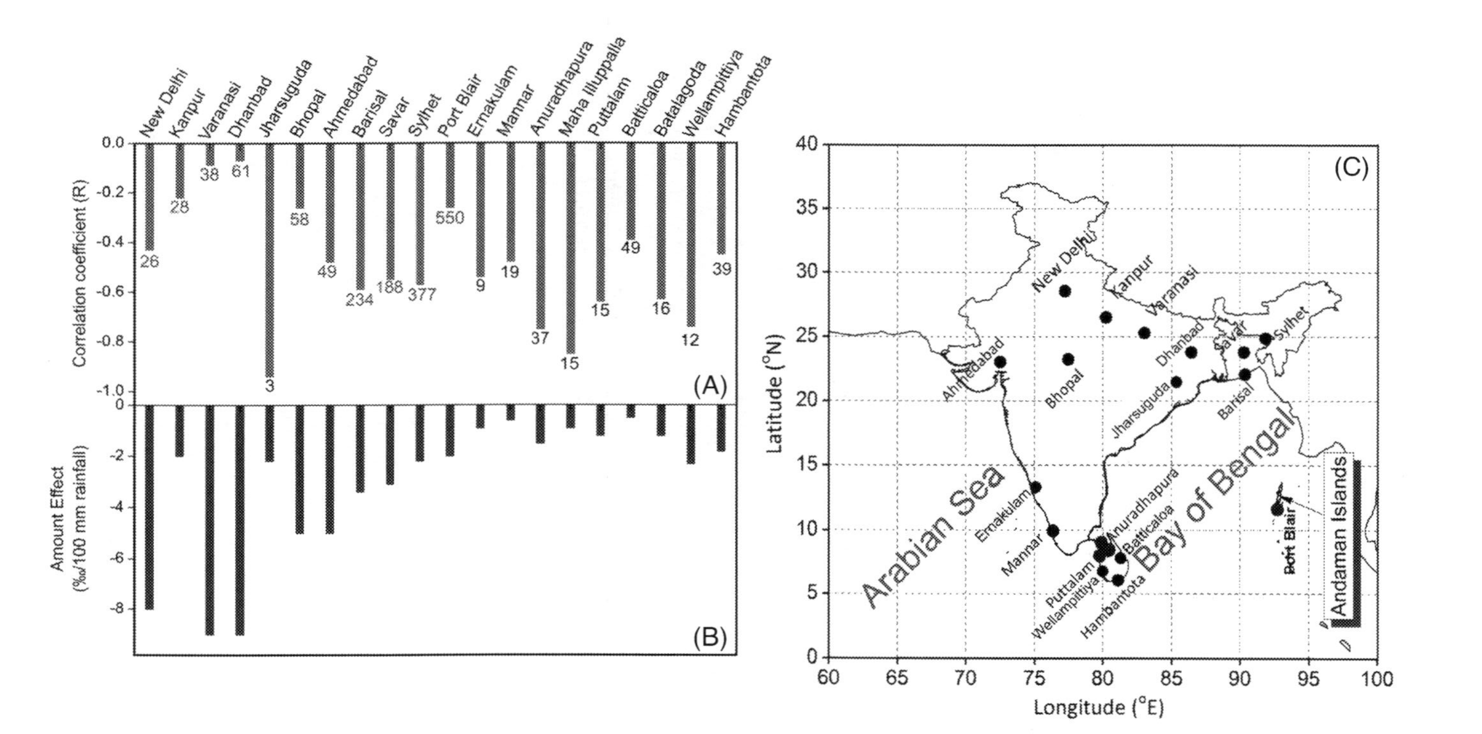

FIGURE 16.1

(A) The correlation coefficient (*r*) between the rainfall and its isotopic values for various sites across India and neighboring countries. The numbers below the bars represent the number of samples used to establish relationships. Daily (monthly) scale samples are shown in red (black). (B) The amount effect, that is, the slope of the regression lines of the parameters as mentioned earlier, representing the depletion in isotope values (in permil; ‰) for each 100 mm of rainfall. The right panel (C) shows the location of the sites. The data were obtained from Lekshmy et al. (2015), Midhun et al. (2018), Yadava and Ramesh (2015), and (Sinha et al., 2019b).

moistures integrate its dynamical process over a large area. Hence the isotopic composition of rainfall is considered better to represent the moisture dynamics than the rainfall event. The chapter also discusses this aspect using moisture dynamical characteristics.

16.3 Methodology and processes

16.3.1 Precipitation sampling and precipitation data

Daily rainwater samples accumulated during the last 24 h were collected every day at 8:30 am at the Pondicherry University campus, Port Blair, the Andaman Islands. An ordinary rain gauge with a provision of radiation shielding was used for this purpose; the rainfall value was obtained by measuring the collected water sample using a calibrated cylinder. Rainfall data based on an automated rain gauge was also obtained from the India Meteorological Department. The areal distance between the sample collection site and the Port Blair India Meteorological Department site is approximately 4 km. The period of rainfall collection was May to October (the summer monsoon season), spanning from 2012 to 2018.

In addition to the gauge measurement, the TRMM/GPM (Tropical Rain Measurement Mission/Global Precipitation Measurement) satellite rainfall data were also used. TRMM/GPM dataset is available since 1998 and provides rainfall information on a high spatial ($0.25° \times 0.25°$) and temporal (hourly/daily) resolution (Huffman et al., 2007; Adler et al., 2003). Since the data are available on a gridded form, it is possible to integrate the rainfall information on the desired scale, which is not feasible for rain gauge data. The TRMM/GPM data are available both on the land and marine environment. Hence, these data have been used to examine the amount effect on all scales. We have used one more dataset, the Global Precipitation Climatology Project monthly precipitation version 2.3, independent of TRMM/GPM satellite data product (Pendergrass et al., 2020). This dataset combines observations and satellite precipitation data into $2.5° \times 2.5°$ global grids. The precipitation unit is mm/day. This data have been used to determine the evaporation (E) by precipitation (P) ratio.

16.3.2 Evaporation flux

Evaporation from the ocean surface was calculated from the latent heat flux (LHF, W/m^2) from Tropflux (Praveen Kumar et al., 2012). The following formula was used to determine evaporation (mm/day). Evaporation (E) = LHF/L; where L is the latent heat of vaporization ($=2.260 \times 10^6$ J/kg). According to sign convention, precipitation flux is positive downward, and evaporation flux is negative upward. Hence, -1 was multiplied to E to make it positive for calculation purposes.

16.3.3 Moisture flux divergence

Moisture flux divergence (kg/kg/s) at 850 hPa is calculated from u, v, q obtained from ERA5 (Hersbach et al., 2020) and scaled to 10^9. The following formula was used (Wallace and Hobbs, 2006):

$$\text{Moisture flux divergence } \left(s^{-1}\right) = \frac{\partial(qu)}{\partial x} + \frac{\partial(qv)}{\partial y} \tag{16.1}$$

where u and v (m/s) are the zonal and meridional wind components, q (kg/kg) is the specific humidity.

16.3.4 Isotopic analysis of rainwater

The isotopic abundance is reported in "δ" notation. It is defined as the relative difference between the heavy to the light isotopic value of the sample and reference material. The difference is normalized with the isotopic ratio of the reference material and then multiplied by a factor of 1000, expressed in ‰ (permil) notation. As an example, the oxygen isotopic value of precipitation is defined as follows:

$$\delta^{18}\mathrm{Op}\ (‰)=\left[\frac{(^{18}\mathrm{O}/^{16}\mathrm{O})_{\mathrm{precipitation}}-(^{18}\mathrm{O}/^{16}\mathrm{O})_{\mathrm{reference}}}{(^{18}\mathrm{O}/^{18}\mathrm{O})_{\mathrm{reference}}}\right]\times 1000$$

The reference material is the Vienna Standard Mean Ocean Water, whose $^{18}O/^{16}O$ ratio (0.0020052; Criss, 1999) is precisely known.

It is to be noted that both the oxygen and hydrogen isotopic values of precipitation are controlled in a similar manner; hence δD_p and $\delta^{18}O_p$ are usually always strongly correlated. The linear equation involving these two parameters for a given region is called the Local Meteoric Water Line (LMWL). The slope and intercept of an LMWL are generally determined by the equilibrium process and the environmental condition of the moisture source, respectively (Criss, 1999). A secondary parameter, d-excess (Dansgaard 1964) is commonly used to identify the source moisture, which is defined as d-excess $= \delta D - 8 * \delta^{18}O$.

The isotopic analysis of the rainwater was carried out at the Stable Isotope Laboratory of the Indian Institute of Tropical Meteorology, Pune. Initially, a Delta-V Plus Isotope Ratio Mass Spectrometer (Thermo Fisher Instrument, Bremen, Germany) was used for this purpose. Later, the measurements were carried out in a laser isotope analyzer of the Los Gatos Research (Model: TIWA-45-EP). The precision of oxygen isotopes ($\delta^{18}O_p$) was <0.1, and that for hydrogen isotopes was <1.0‰, respectively. The preliminary analysis of the precipitation isotopic data of Port Blair was presented in Chakraborty et al. (2016), Sinha et al. (2019a, 2019b), Munksgaard et al. (2019), Sinha and Chakraborty (2020). Statistical analysis of this dataset is provided in Chakraborty et al. (2020).

16.3.5 Analysis of rainfall data

To examine the rainfall–isotope relation, the oxygen isotopic records (δ^{18}Op) were correlated with the rainfall record first with the rain gauge data and then with the TRMM-derived rainfall time series. The TRMM rainfall data were integrated for the following spatial scales: (i) point scale; that is, the gridded data extracted at the sampling site (11.6°N, 92.73°E); (ii) 2.5° × 2.5° (small square box in Fig. 16.2); (iii) 5° × 5° (the mid-sized square box in Fig. 16.2, and then with a large spatial scale covering an area of approx. 1500 km × 3000 km (the large box in Fig. 16.2). Since the monsoon circulation is predominantly south-westerly, the boxes were chosen in the south-west direction of the sampling site.

To obtain the statistical correlation between the isotopic and rainfall records, the TRMM-derived rainfall data were treated as follows. First, a rainfall climatology was created on a daily scale for a specific box. Then the rainfall anomaly was calculated by subtracting the climatological value separately for each year. The use of the anomaly, rather than the absolute value of rainfall, gives an opportunity to compare the year-to-year variability on a common scale.

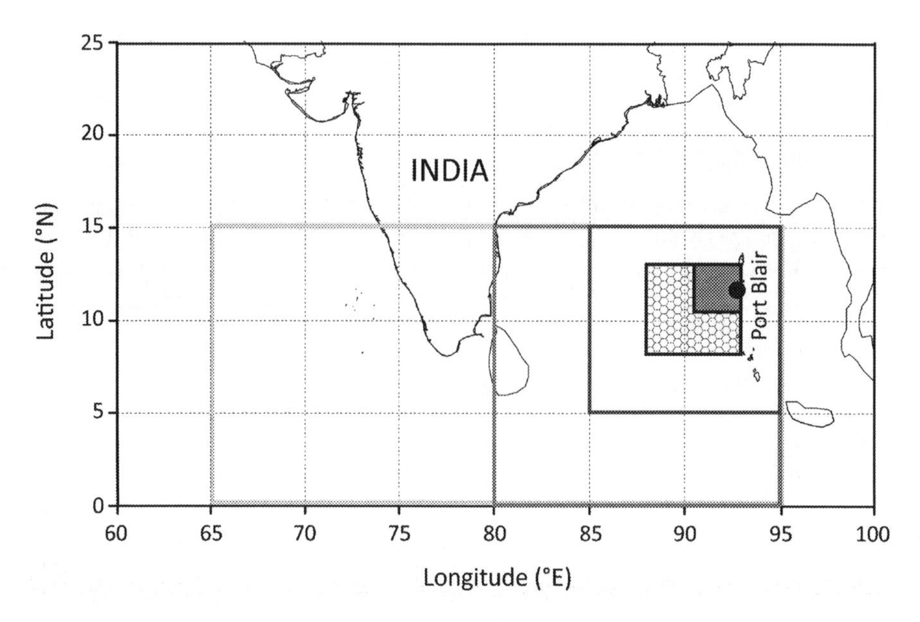

FIGURE 16.2

Examination of the amount effect for a varying spatial domain. The area has been varied from a small scale (gray box) to a very large scale (box with green border). The TRMM-derived rainfall over these boxes were regressed with the precipitation isotope values at Port Blair to get the equations for different areas.

16.4 Results

Fig. 16.3 presents the correlation coefficient (r) calculated between the rainfall anomaly and $\delta^{18}O_p$. The r values, when calculated with the rain gauge data, are shown separately as gray diamonds, varying from -0.08 to -0.28. The same for the TRMM-derived rainfall data has been shown as line plots. As shown in this figure, the correlation is low when the rainfall anomaly is calculated over the sampling site (area $\approx$ zero). Afterward, the r-value steadily increases when rainfall anomaly is calculated over higher regions. When the area equals $2.5° \times 2.5°$, the r-value systematically increases for all the years. When the area is taken as $5° \times 5°$, the r-value in general increases, but for the years 2012 and 2016, it shows a slight decreasing trend. If the area is further increased, the r-value shows inconsistent behavior; it stays nearly the same for 2012, increases for 2016 and 2017, and decreases for the remaining years. When the area reaches a continental scale ($30° \times 15°$), the r-value decreases at a higher rate; but the year 2015 and 2017 still show an increasing trend.

The above analysis has been done on a daily scale, which has little relevance for paleomonsoon study as no proxy effectively provides rainfall estimation with such high resolution. However, the above study provides a guideline of spatial scale where the correlation is expected to be maximum. Accordingly, spatial scales of $5° \times 5°$ and beyond are chosen for doing monthly scale analysis. Toward this, the monthly composite of $\delta^{18}O_p$ was calculated for each month, May to October, and for all the years (2012–2018). Then a single time series of $\delta^{18}O_p$ (also δD_p) was constructed representing the monsoon

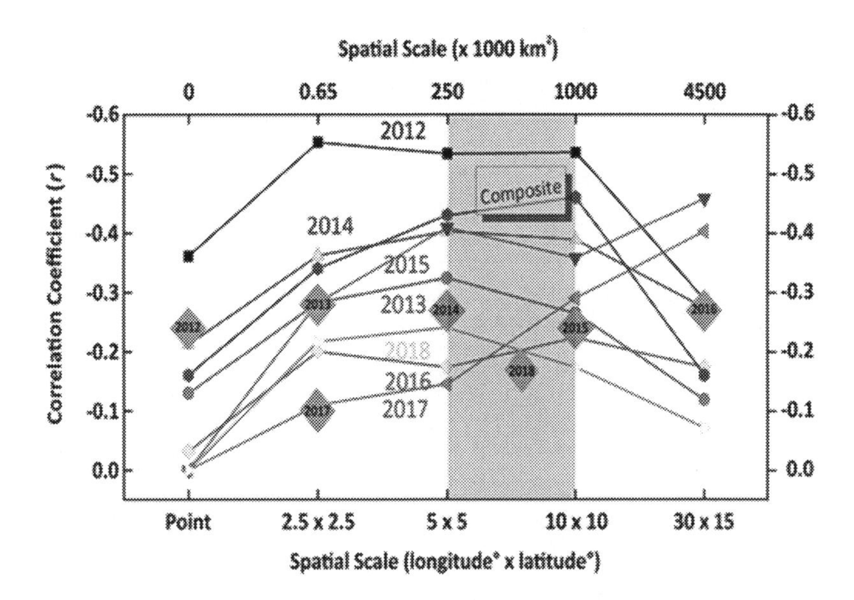

FIGURE 16.3

Spatial dependence of the amount effect. The *x*-axis is the spatial domain in which the rainfall anomaly has been calculated. The corresponding boxes are shown in Fig. 16.2. The correlation coefficient (*r*) between the rainfall anomaly and $\delta^{18}O_p$ is shown on the *y*-axis. The *r* values for the gauge rainfall vs. $\delta^{18}O_p$ (point scale) are shown as a gray diamond for the corresponding year (*x*-coordinates for all these cases are point scale only). The *r* values for the rainfall–$\delta^{18}O_p$ relationship on a monthly scale are dark yellow, labeled as "composite."

season. A total of 42 data points were obtained; however, May-2016 and Oct-2018 were found to be outliers; hence they were discarded. Similarly, rainfall time series on a monthly scale was prepared for the following gird boxes using the Global Precipitation Climatology Project data: (i) 88–93°E, 8–13°N (box with honeycomb shading in Fig. 16.2); (ii) 85–95°E, 5–15°N (box with blue border); (iii) 80–95°E, 0–15°N (box with red border); (iii) 65–95°E, 0–15°N (box with green border). Similarly, moisture dynamical parameters, such as moisture flux divergence and evaporation (*E*) were also computed for the above grid boxes. The correlation for the monthly values has been discussed and shown in a separate figure later; however, it (Fig. 16.3; labeled as "composite"; dark yellow line) has been shown here for comparison with the daily scale data.

16.5 Discussion

Dansgaard, in his seminal paper in 1964, noted that the amount effect is dominant over the ocean and can be well observed on tropical islands. However, contrary to Dansgaard's hypothesis, the amount effect is rather poor in the Andaman Islands, which belong to the tropical region of BoB. The Andaman Island region is known to be characterized by intense convective activity (Uma et al., 2016). Munksgaard et al. (2019) synthesized the precipitation isotope data from across the tropics and examined their association

Table 16.1 Interannual variability of the slope and intercept of the local meteoric water lines (LMWL) and the d-excess–$\delta^{18}O_p$ relationships at Port Blair, Andaman Islands. Correlation coefficient for LMWL are shown in the fourth column. The same for d-excess–$\delta^{18}O$ are shown in 6th column.

Year	LMWL		*r*	*R*	*n*	d-excess vs. $\delta^{18}O$
	Slope	Intercept				Slope
2012	7.10 ± 0.20	9.42 ± 0.80	0.97	-0.46	82	-0.90 ± 0.19
2013	6.65 ± 0.17	4.86 ± 0.63	0.97	-0.64	90	-1.35 ± 0.17
2014	6.58 ± 0.19	3.81 ± 0.71	0.96	-0.65	111	-0.17 ± 0.19
2015	7.49 ± 0.22	8.22 ± 0.78	0.95	-0.20	120	-0.51 ± 0.22
2016	6.36 ± 0.18	1.86 ± 0.51	0.95	-0.64	116	-1.64 ± 0.18
2017	7.57 ± 0.12	5.90 ± 0.43	0.98	-0.33	101	-0.43 ± 0.12
2018	7.29 ± 0.09	6.56 ± 0.37	0.99	-0.56	116	-0.71 ± 0.09

The r *values in all cases are significant at 99.99% level.*

with the stratiform fraction of rainfall. A strong inverse correlation was found between the stratiform rainfall area fraction and the precipitation isotopes of the corresponding regions at all (total seven) sites except the Andaman Islands (op cit.). The $\delta^{18}O_p$ of the Andaman Islands showed no correlation with the stratiform rainfall area fraction. These studies indicate that the strong convective activities are the probable cause of yielding a low $\delta^{18}O_p$–P correlation in this part of the BoB.

The pattern of *r*-values, as illustrated in Fig. 16.3 in general, shows consistent behavior with that of the corresponding LMWL. Table 16.1 summarizes the characteristics of the LMWL and the regression coefficient of d-excess versus $\delta^{18}O_p$. The value of the slope of an LMWL when close to 8 (that of the Global Meteoric Water Line; Craig, 1961) is believed to have undergone an equilibrium process and suffered little raindrop evaporation. As seen in this table, the year 2017 had the maximum value of the slope (7.57). On the other hand, when raindrops undergo significant evaporation, the d-excess–$\delta^{18}O_p$ line would be characterized by a higher correlation and a steeper slope (Gat, 2010). This is indeed observed in the case of the year 2017, showing the highest value (7.57) in LMWL but lowest slope (-0.43) in case of d-excess–$\delta^{18}O_p$ line indicating minimal raindrop evaporation. Similar behavior was also observed in the case of 2015 (7.49 and -0.51, respectively) and at the same time, other years had LMWL slope $\leq$7.1 and relatively higher slopes for d-excess–$\delta^{18}O_p$ line. This may be the reason for having different characteristics in *r*-value for these years (2015 and 2017; Fig. 16.3), which started with the low values but progressively increased with the increase in the spatial domain.

16.5.1 Examination of the amount effect on a monthly scale

In this section, we discuss the monthly scale precipitation isotopes and their dependency on the precipitation amount and the moisture dynamical parameters. As shown in Fig. 16.3, the *r*-value on a monthly scale ($\delta^{18}O_p$ was averaged monthly) is almost always higher than that observed on a daily scale except for the year 2012. This means that the amount effect is better manifested when $\delta^{18}O_p$ is integrated on a higher temporal scales. Such kind of behavior could be explained in terms of moisture dynamics.

On a global scale, evaporation (E) equals precipitation (P), but on a regional scale, the $E = P$ equation may not be balanced. In the case of the BoB, P often exceeds E, especially during the monsoon season (Parekh et al., 2016), yielding a negative $E - P$. The deficit is balanced by the horizontal flow of moisture from the neighboring areas. This means that moistures are generated far apart and transported to the sampling station (i.e., the Andaman Island), but presumably within the geographical boundary of the BoB. Such kind of dynamical nature of moisture transport is studied by means of moisture divergence, defined in Eq. (16.1). We have taken negative of this quantity (termed as Moisture Flux Convergence or MFC) to have the same sign slopes when precipitation isotopic values are regressed against precipitation or MFC.

As mentioned earlier, for the BoB, horizontally transported moisture is significant during the monsoon season, making the quantity E/P usually less than one. This is in sharp contrast to the Arabian Sea, where E/P is typically more than one throughout the year (e.g., Parekh et al., 2016). Moistures that travel a long distance progressively lose their heavier isotopes due to rainouts. This process, known as the Rayleigh distillation (Criss, 1999) makes the moistures arriving over the Andaman Islands significantly depleted in heavier isotopes. So the precipitation over the Andaman Islands and, in general, over the entire BoB makes the surface water isotopically depleted compared to that of the Arabian Sea (Achyutan et al., 2013; Sengupta et al., 2013). This is, however, also controlled by the freshwater influx, and it is known that the Bay receives a massive freshwater discharge during the summer monsoon season (Sreelekha et al., 2018). Hence, the study of the isotopic behavior of moisture and the ensuing rainfall must take into account the role of the transported moisture. A theoretical framework describing the effect of the moisture convergence on the precipitation isotopes has been developed by Moore et al. (2014) based on an isotope-enabled circulation model. These authors proposed that convergence of vapor predominantly determines the isotopic composition of precipitation, especially in the areas where rainfall exceeds evaporation.

We have tested the Moore et al. (2014) hypothesis by investigating the effect of moisture transport processes for the three spatial regions ($5° \times 5°$; $10° \times 10°$; $15° \times 15°$) within the domain of the BoB. Additionally, another region ($30° \times 10°$) belonging to both BoB and the Arabian Sea (Fig. 16.2) has been considered. Monthly values of precipitation $\delta^{18}O_p$ of the observational years were regressed with the area-averaged precipitation and the MFC for the corresponding areas. One of the consequences of the Moore et al. (2014) hypothesis is a linear relation between E/P and the precipitation isotopes, in which Moore et al. used hydrogen isotopes (i.e., δD_p) in their analysis. Accordingly, precipitation δD_p was regressed with the E/P parameter. The result is shown in Fig. 16.4.

The first column in Fig. 16.4 shows the P–$\delta^{18}O_p$ relation for a varying spatial scale. The correlation coefficient (˜-0.30) is lower in the case of $5° \times 5°$ and $10° \times 10°$ grid boxes but relatively higher (-0.47) on a $15° \times 15°$ scale. On the other hand, when the scale is increased to $30° \times 15°$, covering a part of the Arabian Sea, the correlation turns positive and becomes insignificant. Similar behavior was also observed in the case of MFC–$\delta^{18}O_p$ correlation. The r-value, in this case (MFC–$\delta^{18}O_p$), is significantly higher than the previous case (P–$\delta^{18}O_p$), though the maximum value was obtained for a spatial scale of $10° \times 10°$ (-0.72). The r-value became poor and insignificant for the $30° \times 15°$ scale. On the other hand, the relation between the δD_p and E/P parameter also shows very similar behavior. High correlations were found consistently until $15° \times 15°$ grid box; afterward, the r-value decreased and turned negative for the largest spatial scale of $30° \times 15°$ box and lost its significance. The reason for the breakdown of the linear correlation between the precipitation isotopes and the atmospheric variables in the case of the bigger grid size is presumably the multiple moisture source arising from the Arabian

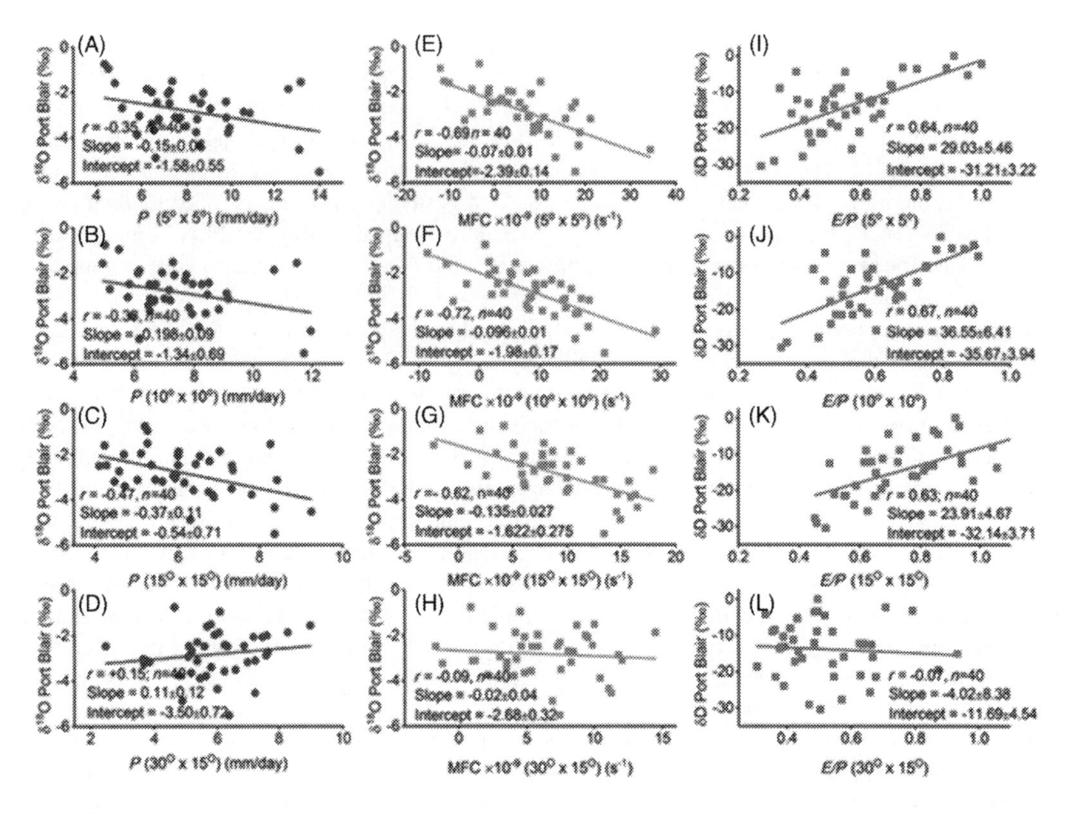

FIGURE 16.4

Linear correlation coefficients are estimated between the precipitation isotopes and the atmospheric variables. The left column shows $\delta^{18}O_p$ vs. P averaged over an area of 5° longitude × 5° latitude (A), 10° longitude × 10° latitude (B), 10° longitude × 15° latitude) (C), 30° longitude × 15° latitude (D). The middle column and third column represent the same for $\delta^{18}O_p$ vs. MFC and δD vs. E/P, respectively.

Sea and also from southern peninsular India. As mentioned earlier, the Arabian Sea has a different E/P (>1) characteristics than the BoB ($E/P < 1$). On the other hand, southern peninsular India generates a significant amount of vapors through the process of evapotranspiration. When these moistures produce rainfall over these regions, the precipitation isotopes observed at the Andaman Islands would no longer maintain a simple, functional relation. We have also estimated the evaporation parameter separately for the Arabian Sea, BoB, and the southern peninsular Indian region and confirm the above speculation (result not shown).

The variation of r-values has been shown in the form of line plots in Fig. 16.5. An important observation of this exercise is that the precipitation isotopes maintain a reasonably strong relation for a large spatial scale extending to an area up to 15° × 15° equivalent to the entire southern portion of the BoB. The relation is even stronger with the moisture dynamical parameters, which may have paleoclimatic implications. Most of the proxy-based climate reconstructions endeavor to reconstruct

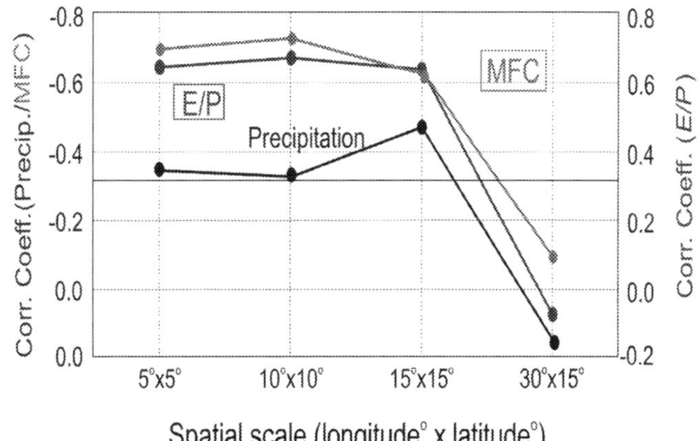

FIGURE 16.5

The correlation coefficient between the precipitation isotopes (point scale) and the atmospheric variables with increasing spatial scale. On a temporal scale, all the parameters represent the monthly values. The correlation coefficient between $\delta^{18}O_p$ and precipitation and MFC is shown on the left *y*-axis; the scale has been inverted to illustrate a higher correlation pointing upward. The correlation between δD_p and E/P ratio has been shown on the right *y*-axis. The gray horizontal line represents the correlation value significant at 95%.

past monsoon rainfall, but this study shows that isotopic proxies could be used to better understand the moisture dynamical characteristics to have a better perspective of the past monsoon system (Fig. 16.5).

A careful observation of the scatter plots between the MFC versus $\delta^{18}O_p$ and *E*/*P* versus δD_p reveals that the distributions are not similar. In the case of MFC versus $\delta^{18}O_p$, the points are reasonably uniform, but in the case of *E*/*P* versus δD_p, they show "mid-level bulging," which indicates nonuniform surface evaporation. A strong convergence leads to enhanced precipitation, leading to moisture saturation; as a result, evaporation is constrained. However, moisture convergence is mostly controlled by the wind field; evaporation plays a minor role in this regard. As a result, its distribution is less constrained, and hence it is more or less uniformly distributed.

16.5.2 Estimation of amount effect for the Andaman Islands region

The linear correlations calculated between $\delta^{18}O_p$ and precipitation (Fig. 16.4A, B, C) can be used to determine the amount effect for the Andaman Island region. The values are obtained for all the spatial domains and are shown graphically (except for the grid box: 30° × 15°, comprising both the Arabian Sea and the BoB) in Fig. 16.6. The amount effect obtained using a model (Iso-GSM; Yoshimura et al., 2008) derived precipitation isotopes and used by Laskar et al. (2013) for interpreting their Andaman Island's speleothem data, matches well with our estimate, shown in this figure as an open circle. Though these two estimates (Laskar et al. considered an area of 2° × 2° and our estimate is for 5° × 5°) agree well, the effect does not remain the same when the spatial scale is increased steadily. As shown in Fig. 16.6, the trend experiences a declining slope. This means that precipitation isotopes would suffer more depletions for distant areas, even if the area-averaged rainfall remains the same. In other words, a given

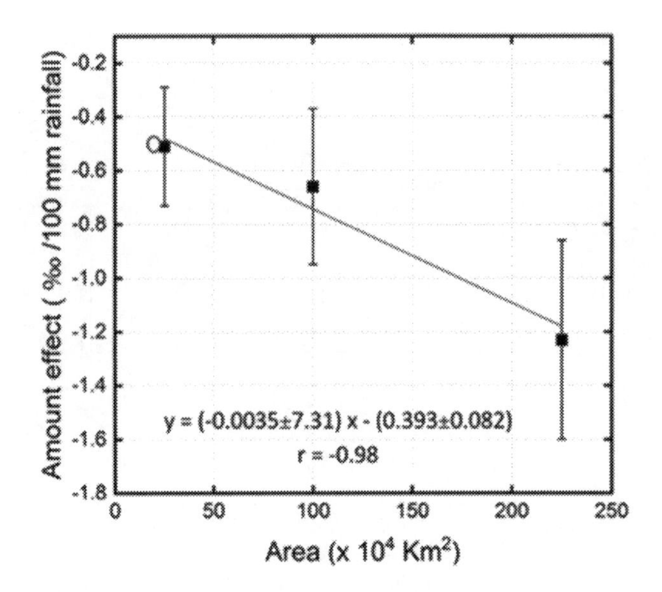

FIGURE 16.6

Spatial dependency of the amount effect as obtained from this study. On a shorter spatial scale (up to about 5° × 5°), the value is ca. -0.5‰/100 mm rainfall, which matches well with that (shown as an open circle; Laskar et al., 2013) estimated based on an isotope enabled circulation model. However, the value changes with the increasing area in which precipitation is being considered.

isotope signal, say a depletion of 1‰ observed at the Andaman Island, would indicate an increase in rainfall of 200 mm in its neighboring areas. But the same signal, if applied for a large area of 15° × 15°, the rainfall would decrease by about 100 mm. Also, the error associated with these estimates could be significant (see Fig. 16.6). These two characteristics, viz., the uncertainty in amount effect and the spatial nonlinearity, may have important implications for paleomonsoon reconstruction. The study shows that the point scale observations do respond to large-scale processes; however, rainfall reconstructions on an absolute term is scale-dependent. In this context, however, the following point is to be noted. The calculation has been done on a seasonal scale, that is, for the summer monsoon season, during which the wind is predominantly south-westerly. During the winter, the easterly wind delivers a significant amount of rainfall in the Andaman Islands region. So if the same exercise is done on an annual scale, the correlation values obtained (Fig. 16.4) are likely to be different. But we did not consider the winter season first because the contribution of winter rainfall to the annual rainfall is only 22%. Second, on an annual scale, the number of observations was only seven (2012–2018); this is too small to make statistically significant calculations. Long-term data are required to refine the calculations, especially for annual timescales.

The climatic processes acting on short spatial scales are not necessarily the same operating on a large scale. Rainfall reconstruction over a relatively short spatial scale (up to about 500 km × 500 km) is expected to provide a realistic estimate. But, over a larger spatial scale, the rainfall estimation could be significantly underestimated. Hence the comparison of rainfall reconstructions using proxy records separated by a few thousand kilometers may not identify the *same* processes. To circumvent this problem, it is essential to have a robust network of speleothem records across the monsoon domain. Each record can produce realistic monsoon information on a local to regional scale. Then integration of these results on a paleoclimatic data assimilation framework (Steiger et al., 2017) could produce a reliable reconstruction of large-scale processes. Climate model simulation of proxy records also supports this proposition. Midhun et al. (2021) analyzed a large number of speleothem isotopic records to examine how the climate modes, such as ENSO, impact the local isotopic content of precipitation. Their analysis demonstrates that the magnitude of the climate signal captured by an individual site is relatively small. On the other hand, if a network of $\delta^{18}O$ speleothem records are analyzed, it would show enhanced skill in reconstructing the ENSO variance.

16.6 Learning and knowledge outcomes

Analysis of precipitation isotopes over an island site in the BoB was carried out from 2012 to 2018. Linear correlation analysis was carried out between the precipitation isotopes and three atmospheric variables, viz. precipitation, MFC, and the evaporation/precipitation ratio, to examine the sensitivity of the precipitation isotopes on these variables. The results show that the precipitation isotopes respond better to MFC and *E*/*P* ratio than precipitation. Since most of the isotope-based paleoclimatic investigations endeavor to reconstruct monsoon rainfall alone, reconstruction of moisture fluxes and evaporation process may also be possible which will help gain a better understanding of the monsoon systems. Precipitation isotopes also respond to large-scale system, extending up to the geographical boundary of the BoB. Beyond this boundary, the sensitivity weakens presumably due to the admixture of moistures from multiple sources. Analysis of precipitation isotopes, and precipitation over a varying spatial domain helped estimate the amount effect for the respective domain. It was shown that the amount effect possesses a nonlinear behavior; as a result, the rainfall estimated for a large area could be significantly underestimated.

Acknowledgments

The Indian Institute of Tropical Meteorology is fully supported by the Earth System Science Organization, Ministry of Earth Sciences, Government of India. ERA5: Fifth generation of ECMWF atmospheric reanalyses of the global climate. Comments received from K. Ashok, Jasper A. Wassenburg, and two anonymous reviewers helped to improve the quality of the presentation. Copernicus Climate Change Service Climate Data Store (CDS), date of access: Oct 10, 2020. This work is part of a project work of the International Atomic Energy Agency, CRP F31006.

References

Achyuthan, H., Deshpande, R.D., Rao, M.S., Kumar, B., Nallathambi, T., Sashi Kumar, K., Ramesh, R., Ramachandran, P., Maurya, A.S., Gupta, S.K., Stable isotopes and salinity in the surface waters of the Bay of Bengal: implications for water dynamics and paleoclimate, Mar. Chem. 149 (2013) 51–62, https://doi.org/10.1016/j.marchem.2012.12.006.

Adler, R.F., Huffman, G.J., Chang, A., Ferraro, R., Xie, P., Janowiak, J., Rudolf, B., Schneider, U., Curtis, S., Bolvin, D., Gruber, A., Susskind, J., Arkin, P., The version 2 Global Precipitation Climatology Project (GPCP) monthly precipitation analysis (1979–Present), J. Hydrometeor. 4 (2003) 1147–1167.

Chakraborty, S., Birmal, S., Burman, D., Pramit, K., Datye, A., Fousiya, A.A., Aravind, G.H., Mohan, P.M., Trivedi, N., Trivedi, R.K, Statistical analysis of the precipitation isotope data with reference to the Indian Subcontinent, IntechOpen "Hydrology" (2020) 172–190, https://doi.org/10.5772/intechopen.93831.

Chakraborty, S., Sinha, N., Chattopadhyay, R., et al., Atmospheric controls on the precipitation isotopes over the Andaman Islands, Bay of Bengal, Sci Rep. 6 (2016) 19555, doi:10.1038/srep19555.

Craig, H., Isotopic variations in meteoric water, Science 133 (3465) (1961) 1702–1703, doi:10.1126/science.133.3465.1702.

Criss, R.E., Principals of Stable Isotope Distributions, Oxford University Press, New York, 1999.

Dansgaard, W., Stable isotopes in precipitation, Tellus 16 (1964) 436–468.

Dar, S.S., Ghosh, P., Estimates of land and sea moisture contributions to the monsoon rain over Kolkata, deduced based on isotopic analysis of rainwater, Earth Syst. Dyn. 8 (2017) 313–321, doi:10.5194/esd-8-313-2017.

Galewsky, J., Steen-Larsen, H.C., Field, R.D., Worden, J., Risi, C., Schneider, M., Stable isotopes in atmospheric water vapor and applications to the hydrologic cycle, Rev. Geophys. 54 (2016) 809–865, doi:10.1002/2015RG000512.

Gat, J.R., Isotope Hydrology: A Study of the Water Cycle, Imperial College Press, London, 2010, p. 189.

Hersbach, H., Bell, B., Berrisford, P., et al., The ERA5 global reanalysis, Q. J. R. Meteorol. Soc. 146 (2020) 1999–2049, https://doi.org/10.1002/qj.380.

Huffman, G.J., et al., The TRMM Multisatellite Precipitation Analysis (TMPA): quasi-global, multiyear, combined-sensor precipitation estimates at fine scales, J. Hydrometeor. 8 (2007) 38–55, https://doi.org/10.1175/JHM560.1.

Jalihal, C., Srinivasan, J, Chakraborty, A., Different precipitation response over land and ocean to orbital and greenhouse gas forcing, Sci. Rep. 10 (2020) 11891, https://doi.org/10.1038/s41598-020-68346-y.

Johnson, Kathleen, Ingram, B., Spatial and Temporal Variability in the Stable Isotope Systematics of Modern Precipitation in China: Implications for Paleoclimate Reconstructions, Earth and Plan. Sci. Lett. 220 (2004) 365–377, doi:10.1016/S0012-821X(04)00036-6.

Kaushal, N., Breitenbach, S.F.M., Lechleitner, F.A., Sinha, A., Tewari, V.C., Ahmad, S.M., Berkelhammer, M., Band, S., Yadava, M., Ramesh, R., Henderson, G.M., The Indian Summer Monsoon from a Speleothem $\delta^{18}O$ Perspective – a review, Quaternary 1 (3) (2018) 29, https://doi.org/10.3390/quat1030029.

Kendall, C., McDonnell, J.J., Isotope Tracers in Catchment Hydrology. Elsevier Science, Amsterdam, The Netherlands, 2012, p. 1585.

Lachniet, M.S., Climatic and environmental controls on speleothem oxygen-isotope values, Quat. Sci. Rev. 28 (2009) (2009) 412–432.

Laskar, A.H., Yadava, M.G., Ramesh, R., Polyak, V.J., Asmerom, Y, A 4 kyr stalagmite oxygen isotopic record of the past Indian Summer Monsoon in the Andaman Islands, Geochem. Geophys. Geosyst. 14 (2013) 3555–3566, doi:10.1002/ggge.20203.

Lekshmy, P.R., Midhun, M., Ramesh, R., Spatial variation of amount effect over peninsular India and Sri Lanka: role of seasonality, Geophys. Res. Lett. 42 (2015) 5500–5507, doi:10.1002/2015GL064517.

Midhun, M., Lekshmy, P.R., Ramesh, R., Yoshimura, K., Sandeep, K.K., Kumar, S., et al., The effect of monsoon circulation on the stable isotopic composition of rainfall, J. Geophys. Res.: Atmos. 123 (2018) 5205–5221 https://doi.org/10.1029/2017JD027427.

Midhun, M., Stevenson, S., Cole, J.E., Oxygen isotopic signatures of major climate modes and implications for detectability in speleothems, Geophys. Res. Lett. 48 (2021) e2020GL089515, https://doi.org/10.1029/2020GL089515.

Moerman, J.W., Cobb, K.M., Adkins, J.F., Sodemann, H., Clark, B., Tuen, A.A., Diurnal to interannual rainfall $\delta 18O$ variations in northern Borneo driven by regional hydrology, Earth Planet. Sci. Lett. 369–370 (2013) 108–119.

Moore, M., Kuang, A., Blossey, P.N., A moisture budget perspective of the amount effect, Geophys. Res. Lett. 41 (2014) 1329–1335, doi:10.1002/2013GL058302.

Munksgaard, N.C., Kurita, N., Sánchez-Murillo, R., et al., Data descriptor: daily observations of stable isotope ratios of rainfall in the tropics, Sci. Rep. 9 (2019) 14419 https://doi.org/10.1038/s41598-019-50973-9.

Parekh, A., Chowdary Jasti, S., Ojha, S., Fousiya, T.S., Gnanaseelan, C., Tropical Indian Ocean surface salinity bias in climate forecasting system coupled models and the role of upper ocean processes, Clim. Dyn. 46 (2016) 2403–2422, doi:10.1007/s00382-015-2709-8.

Pendergrass, A., Wang, J.-J.National Center for Atmospheric Research Staff, The Climate Data Guide: GPCP (Monthly): Global Precipitation Climatology Project, 2020 Retrieved from https://climatedataguide.ucar.edu/climate-data/gpcp-monthly-global-precipitation-climatology-project (Accessed on 15 Oct 2020).

Praveen Kumar, B., Vialard, J., Lengaigne, M., et al., TropFlux: air-sea fluxes for the global tropical oceans—description and evaluation, Clim. Dyn. 38 (2012) 1521–1543, https://doi.org/10.1007/s00382-011-1115-0.

Rozanski, K., Araguás-Araguás, L., Gonfiantini, R., Isotopic patterns in modern global precipitation, in: Swart, P.K., Lohmann, K.L., McKenzie, J., Savin, S. (Eds.), Climate Change in Continental Isotopic Records, American Geophysical Union, Washington, DC, 1993, pp. 1–37.

Sengupta, S., Parekh, A., Chakraborty, S., Ravi Kumar, K., Bose, T., Vertical variation of oxygen isotope in Bay of Bengal and its relationships with water masses, J. Geophys. Res. Oceans 118 (2013) 6411–6424, doi:10.1002/2013JC008973.

Sinha, A., Cannariato, K.G., Stott, L.D., Cheng, H., Edwards, R.L., Yadava, M.G., Ramesh, R., Singh, I.B., A 900-year (600 to 1500 A.D.) record of the Indian summer monsoon precipitation from the core monsoon zone of India, Geophys. Res. Lett. 34 (2007) L16707, doi:10.1029/2007GL030431.

Sinha, N., Chakraborty, S., Isotopic interaction and source moisture control on the isotopic composition of rainfall over the Bay of Bengal, Atmos. Res. V. 35 (2020) 104760, https://doi.org/10.1016/j.atmosres.2019.104760.

Sinha, N., Chakraborty, S., Chattopadhyay, R., et al., Isotopic investigation of the moisture transport processes over the Bay of Bengal, J. Hydrol. XV (2019a) 2, doi:10.1016/j.hydroa.2019.100021.

Sinha, N., Chakraborty, S., Mohan, P.M., Modern rain-isotope data from Indian island and the mainland on the daily scale forthe summer monsoon season, Data Brief Volume 23 (2019b) 103793, PII:S2352-3409(19)30144-1 https://doi.org/10.1016/j.dib.2019.103793.

Sree Lekha, J., Buckley, J.M., Tandon, A., Sengupta, D., Subseasonal dispersal of freshwater in the northern Bay of Bengal in the 2013 summer monsoon season, J. Geophys. Res. Oceans 123 (2018) 6330–6348 https://doi.org/10.1029//.

Steiger, N.J., Steig, E.J., Dee, S.G., Roe, G.H., Hakim, G.J., Climate reconstruction usingdata assimilation of waterisotope ratios from ice cores, J. Geophys. Res. Atmos. 122 (2017) 1545–1568, doi:10.1002/2016JD026011.

Tejavath, C.T., Ashok, K., Chakraborty, S. et al. A PMIP3 narrative of modulation of ENSO teleconnections to the Indian summer monsoon by background changes in the Last Millennium. Clim. Dyn. 53 (2019) 3445–3461, https://doi.org/10.1007/s00382-019-04718-z.

Tejavath, C., Upadhyay, P., Karumuri, A., The past climate of the Indian region as seenfrom the modelling world, Curr. Sci. 119 (2) (2020) 316–327, doi:10.18520/cs/v119/i2/316-327.

Uma, R., Lakshmi Kumar, T.V., Narayanan, M.S., Understanding convection features over Bay of Bengal using sea surface temperature and atmospheric variables, Theor. Appl. Climatol. 125 (2016) 469–478, https://doi.org/10.1007/s00704-015-1518-7.

Wallace, J.W., Hobbs, P.V., Atmospheric Science: An Introductory Survey, Academic Press, Burlington, USA, 2nd Edition, 2006 ISBN 0-12-732951-X.

Yadva, M.G., Ramesh, R., Monsoon reconstruction from radiocarbondated tropical Indian speleothems, Holocene 15 (1) (2015) 48–59.

Yoshimura, K., Kanamitsu, M., Noone, D., Oki, T., Historical isotope simulation using reanalysis atmospheric data, J. Geophys. Res. 113 (2008) D19108, doi:10.1029/2008JD010074.

CHAPTER

17 Mangrove Ostracoda species fluctuations, habitual adaptation, and its environmental implications—A review

S.M. Hussain, Mohammed Noohu Nazeer, K. Radhakrishnan, A. Rajkumar, V. Sivapriya
Department of Geology, School of Earth and Atmospheric Sciences, University of Madras, Chennai, India

17.1 Introduction

Mangroves are intertidal forests distributed across the globe in tropical and subtropical regions. They thrive in protected tidal lagoons, embayment areas, and estuaries (Kennish, 1994). Carbon dioxide (CO_2) aids the growth of mangrove vegetation and nearly 70 mangrove species or hybrids of mangroves have been globally reported (Duke et al., 1998). Twenty-four species of true mangroves were observed from Indian subcontinent; of these species belong to family Rhizophoracea showed maximum richness (Jyotiskona and Soumyajit, 2014). The mangrove root systems and their related biota act to capture, accumulate, and stabilize sediments suspended in the intertidal water. They are tolerant toward increased salinity, high temperature, extreme tides, faster sedimentation, and muddy anaerobic soils (Giri et al., 2010).

Crab species in mangrove sediments aids in the production of oxygen, reduce the levels of sulfites and initiate the growth of new roots of the young mangroves (Kristensen and Alongi, 2006). The bioturbation process by small benthic living forms enhances the nitrogen production necessary for the growth of temperate macrophytes (Mermillod and Lemoine, 2010). Mangroves can influence on the soil accretion processes through the accumulation of organic matter. Distinctive root system of mangroves is prime characteristics of different mangroves genera and the functional root types include prop roots as in *Rhizophora*, pneumatophores as in *Avicennia*, knee roots as in *Bruguiera*, and plant roots observed in *Xylocarpus*, which develop along various hydrological and geomorphological gradients (Ken et al., 2014).

Mangroves act as barrier offer against catastrophic events, such as tsunami, tropical cyclones, and tidal bores and can enhance shoreline erosion (Alongi, 2014). Not only mangrove forest is helping in balancing the ecosystem, they also play an important role in supporting the sustainability and livelihoods of human being which are used for food, timber, fuel, and medicine (Alongi, 2002).

17.2 Scope of the study

The micro size and sensitiveness of Ostracoda species to feeble environmental fluctuations mark Ostracoda a very useful indictor in deciphering the paleoenvironment. Since the habitat of most of the

Holocene Climate Change and Environment. DOI: https://doi.org/10.1016/B978-0-323-90085-0.00008-5

Ostracoda species is benthic, they explain in detail the bottom bed and bathymetrical characteristics. Despite the monographs, research and review papers on Ostracoda from different environments, the studies pertaining Ostracoda from the Mangrove environment are sparse. In order to identify the paucity in Ostracoda studies and to fill the research gap of Ostracoda from Mangrove environment the chapter will be highly advantageous.

17.3 Results and Discussion

17.3.1 Holocene Ostracoda

Ostracods are tiny bivalve Crustaceans; inhabit all types of aquatic environments. They are retrieved from brackish water environment such as coastal lagoons, coastal lakes, mangrove environments, and open estuaries. Mostly they are benthic. The chief factors governing the Ostracoda distribution are water temperature, salinity, and substrate (Yassini and Jones, 1995). Marginal marine environments like marsh, rivers, coastal lagoons, deltas, estuaries, mangrove islands, salt marshes, and fluvial marine assemblages are characterized by several species which are peculiar to these environments.

Ostracoda shell trace elements and their concentrations are fluctuating with respect to seasonal environmental changes (Xia et al., 1997). Calcium partitioning of Mg, Sr, Fe, and U in Ostracoda shells is an indicator for temperature, salinity, oxygenation, and organic matter decay. The relationships between the molar ratios of Mg/Ca occur in positive correlation with the temperature of the host-water (Zhongning et al., 1995). Mg and Sr content in Ostracoda genus *Krithe* can unfold the paleotemperature and paleosalinity (Oscar et al., 2018). Mg/Ca partitioning in Ostracoda shell is an important tool for determining the paleotemperature of bottom waters (Chivas et al., 1983; Holmes, 1992; De Deckker et al., 1999; Rosenthal and Linsley, 2013). Climate-driven body size studies display that Ostracoda shell size varies with temperature (Gene et al., 2010). Thalassocypridinian Ostracod tribe live mostly in brackish environments and have well-developed setae and a relatively small adult body size (0.5–0.7 mm). The tribe is diverse and species are abundant in brackish-water environments at low latitudes, the group is relatively poorly known. So far, 49 species belonging to seven genera have been recorded worldwide, mainly from the tropical regions (Maddocks, 2005; Smith et al., 2006; Hu and Tao, 2008).

Morphologically, Ostracoda possesses a segmented body with 5–7 pairs of appendages, aids in their locomotion, food intake, and as a sensory organ of the organism (Pokorny, 1978). Mostly, Ostracoda are bottom dwellers. They are oval to round shaped, with a bivalved carapace enclosing their soft body, hinged in the dorsal region. Both the valves are attached together by adductor muscles, which leave an imprint in the shells when the organism dies. Valves are divided into outer and inner lamella, which are separated by the line of concrescence. The exoskeleton of Ostracoda is known as carapace and made up of rhomb-shaped calcite called crystallites (Oertli, 1972).

Unlike other Arthropods, head and thorax of Ostracoda are undistinguishable and pygidium is absent. Morphotypes can be identified and distinguished by the shell ornamentations along with the pore patterns. The pores along with the appendages in Ostracoda help the organism to interact with the environment. Bradley and Benson (1971) identified that the shells of Ostracoda are made of three distinct layers. These layers are differentiated from one another. Ostracoda body is covered by cuticle, an epidermis secretion.

Ostracoda possesses a well-developed digestive system and reproductive system. However, Ostracoda has no special circulatory system and heart is absent in most of the groups (Pokorny, 1978). Sexual dimorphism is present in Ostracoda and the reproduction takes place both sexually and asexually. The adult Ostracoda can grow up to 1 mm in size. The microcrustacean in their life, shed their exoskeleton eight times, during their growth stages until, they become adults. This process of removal of their exoskeleton is known as moulting or ecdysis. Ostracoda moulting results in shedding of the complete exoskeleton of carapace and appendages (Smith et al., 2015). Most Ostracoda are omnivorous and most often scavengers (Schmit et al., 2007). The scavenger forms will consume the accumulation of the organic matter. Stratigraphically, Ostracoda ranges from the Lower Paleozoic to the Recent times. During the ages they multiplied, many taxa got extinct and some species survived over major extinction phases and ice ages.

Being so small and fragile in nature Ostracoda needs very special methods of collection, separation, illustration, and identification. This chapter discusses about the Ostracoda species of various mangrove environments around the world and its biogeographical distribution along with the relationship to the species fluctuations and habitual adaptations.

17.3.2 Ostracoda species fluctuation in mangrove environment

From Indian subcontinent, the Ostracoda species belonging to suborder Podocopa are exhibiting maximum diversity. *Cytherelloidea leroyi* belongs to suborder Platycopa is also a mangrove Ostracoda species from Indian terrain. Pitchavaram mangroves in India exhibit diverse biological entities which have 13 species of mangrove trees, along with 73 taxa of other plants, 52 species of bacteria, 23 species of fungi, 82 species phytoplankton, 22 species of seaweeds, 3 species of seagrass, 117 fish species, and 200 bird species (Kathiresan, 2000). This massive distribution of diverse living forms highlights the importance of mangroves in ecobalancing. The destruction of mangrove forest already made the species thriving on the environment endangered. Besides Pitchavaram, Pulicat, and Muthupet, mangrove forests are found in Mullapallam region and Punnakayal region in Tamil Nadu district, India. The dominant Ostracoda species from the stretch includes *Hemicytheridea bhatiai, Hemicytheridea paiki, Cytherelloidea leroyi, Neomonoceratina iniqua, Neomonoceratina jaini, Jankeijcythere mckenziei, Kalingella mckenzie, Callistocythere flavidofuscaintricatoides, Tanella gracilis, Keijella reticulata, Phlyctenophora orientalis, and different species of Ostracoda. The Ostracoda genera Loxoconcha* is mainly found occurring in the shallow marine environments from tropical to subarctic regions.

Ostracoda species observed from Wadi Al Gemal and Abu Ghoson from the Red Sea coast of Egypt elucidate that dense vegetation and muddy substrate yield diverse assemblages (Sobhi and Mohammed, 2012). Their study shows that growth of green algae, sea grass, etc. influences the occurrence and growth of Ostracoda. The dominant Ostracoda genus from the terrain are *Xestoleberis, Ghardaglaia, Loxoconcha, Quadracythere, Hiltermannicythere, Loxocorniculum, Paranesidea*, and *Neonesidea. Xestoleberis, Loxoconcha, Paranesidea*, and *Neonesidea* are common in Indian subcontinent also. Ostracoda species reported with respect to mangrove species from different locations across the globe are tabled (Table 17.1).

Three *Thalassocypridine* species were reported from *Mangalocypriaryukyuensis* sp. nov., *Paracyprialongiseta* sp. nov., and *Paracypriaplumosa* sp. nov. from Japan (Hirutha and Kakui, 2016). They have discussed the anatomical and morphological characters of Ostracoda in detail. Ostracoda from mangroves are understudied, and a very literature is available describing the distribution of the microfauna.

Table 17.1 Ostracoda species reported with respect to mangrove species from different locations across the globe.

S. no.	Ostracoda species	Location	Predominant mangrove species	Important findings	References
1.	Multiple species	Pulicat, Andhra Pradesh, India	1. *Avicennia marina* 2. *Excoecaria agallocha* 3. *Rhizopora mucronata*	Higher organic matter percentage favors maximum Ostracoda population	Hussain et al., 2019 Rajyalakshmi and Basha, 2016
2.	1. *Neomonoceratina iniqua* 2. *Hemicytheridea paiki*	Muthupett, Tamil Nadu India	1. *Avicennia marina* 2. *Acanthus ilicfolius* 3. *Egicera scorniculatun* 4. *Ercoecaria agallocha* 5. *Rhizopora mucronata*	35 ostracod taxa of the order Podocopida are identified	Hussain et al., 2012
3.	Podocopid Ostracoda	Wadi Al Gemal and Abu Ghoson, Red Sea coast, Egypt	1. *Avicennia marina* 2. *Rhizopora mucronata*	Dense vegetation and muddy substrate yield diverse Ostracod assemblages	Ahmed, 2015 Sobhi and Mohammed, 2012
4.	Thalassocypridine species	Ryuku Islands, Japan	1. *Rhizopora stylosa* 2. *Kandelia cancel* 3. *Bruguiera gymnorrhiza* 4. *Avicennia marina*	Six mangrove species were recorded dominant from the area	Miyawaki, 1980 Hirutha and Kakui, 2016
5.	Multiple species	Pitchavaram, Tamil Nadu	1. *Avicennia marina* 2. *Rhizopora* sp.	Twenty-nine Ostracoda species along with their ecological conditions were discussed	Arul et al., 2003 Kathiresan, 2000

17.3.3 Habitual adaptation of Ostracoda from mangrove environment

Interestingly about 80% of the sediments carried by the tides are retained in mangrove forests (Furukawa et al., 1997). Tides may carry microfossil groups along with the sediments in many cases. The tidal turbulence and root advancement will be impacted in the Ostracoda distribution and occurrence in mangrove environments. The biodiffusion in mangrove environment differed significantly between mangrove stages, co-varying mainly with density and functional richness (Aschenbroich et al., 2017). They observed the reworking rates of the sediments by the infaunal organisms along the French Guiana coast. The mangrove roots as well as different species thriving on the mangrove environment impact on the bioturbation at various rates.

Ostracoda biodiversity and species composition varies concerning the climate and addresses the changes in deep-sea ecosystems (Cronin et al., 1999). Valve chemistry, ornamentation, and preservation on Ostracoda shell imply on the physicochemical conditions of the environment in which the organism grow (De Deckker, 2002).Long-term cooling trends in the deep ocean will have an impact on Ostracoda carapace (Gene et al., 2010).

Carapace (Exoskeleton of Ostracoda) shape marks the mode of life of Ostracoda species. Swimming Ostracoda is characteristically spherical in shape, whereas oblong and elongated forms are bottom dwellers (De Deckker, 2002). Mangrove environments are attributed with benthic Ostracoda species in general. Recent study of mangrove environment exhibits a loss of 20% of benthic biodiversity with extinction of four Phyla (Cladocera, Kynorincha, Priapulida, Tanaidacea), and a loss of 80% of decomposition rates of the benthic biomass and of the trophic resources mediated by microbes (Laura et al., 2018).

Actinocythereis scutigera, Pterigocythereis chennaiensis, Bradleya andamanae, Cytherella semitalis, Ruggieria indopacifica, and *Bairdoppilata alcyonicola* are intertidal Ostracoda genus which are found in mangrove environments. True mangrove Ostracoda possesses well developed pore patterns for the environmental interactions. Morphometrical analysis reveals that mangrove Ostracoda possess different ornamentation pattern. Smooth type of pore patterns are a few in true mangrove Ostracoda species. However, *Bairdoppilata alcyonicola* the intertidal species are sometimes found associated with mangroves. The dominant ornamentation pattern exhibited by Ostracoda in the mangrove environment is reticulate and punctuate forms. Ostracoda recovered from mangroves are dominantly benthic forms shallow water thriving forms. The cold and deep water loving genera *Krithe* are usually absent in the mangrove region. *Actinocythereis scutigera* and *Bairdoppilata alcyonicola* are coral loving Ostracoda species, but are reported from mangrove environment. This can be attributed mainly due to tidal influence. During high tide, the sediments along with Ostracoda specimens are carried into the mangrove swamps. Mangroves retain a good percent of sediments received during high tides.

It is a fact that not all mangrove forests are rich in Ostracoda species. About 24 of the 50 well grown mangrove species across the globe are found occurring in Sundarbans (Jyotiskona and Soumyajit, 2014). However, this diversity is not yielding higher number of Ostracoda. Mangrove distribution in Munroe islands s patchy and *Avicennia marina* is the dominant species from the area (Fig. 17.1). We worked on Munroe island mangroves, Kerala (Fig.17.1A) and Sundarbans, West Bengal (Fig.17.1B) in India. *Hemicytheridea paiki* is the only Ostracoda species recovered from the Sundarbans.

The intertidal form *Bairdoppilata alcyonicola* (Fig. 17.2) occur in adjacent mangrove sediments due to tidal activity. The ornamentation structures of different species of Ostracoda are illustrated in Fig. 17.3. *Actinocythereis scutigera* is an intertidal Ostracoda species, exhibits blunt spines which help in their locomotion and buoyancy (Fig. 17.3A). The alar forming *Bythoceratina* (Fig. 17.3E) and ridge form *Cytheropteron* (Fig. 17.3D) are common in shelf edge of major oceans. Sieve pored forms *Loxoconcha* is common from mangrove environment (Fig. 17.3K). Smoothly ornamented forms *Bairdoppilata alcyonicola* (Fig. 17.3L) and *Krithe* (Fig. 17.3M) are found occurring from different marine settings; from intertidal and deep waters, respectively.

17.4 Learning and knowledge outcomes

Well-developed roots enhance the disturbance of sediment strata as they emerge from the basal nodes of stem. As a result of the distinctive and functional root growth, the sediments are ploughed by the mangrove plants. This in turn will affect the chronological control of Ostracoda assemblages in the sediment strata. Entire mixing up of the sediments caused as a result of bioturbation will question the principle of superposition in the sediment strata. Ostracoda distribution and taxonomy from

FIGURE 17.1

(A) *Avicennia marina* the most common mangrove species across the globe. (B) Sundarban mangrove forest, West Bengal, India (during low tide).

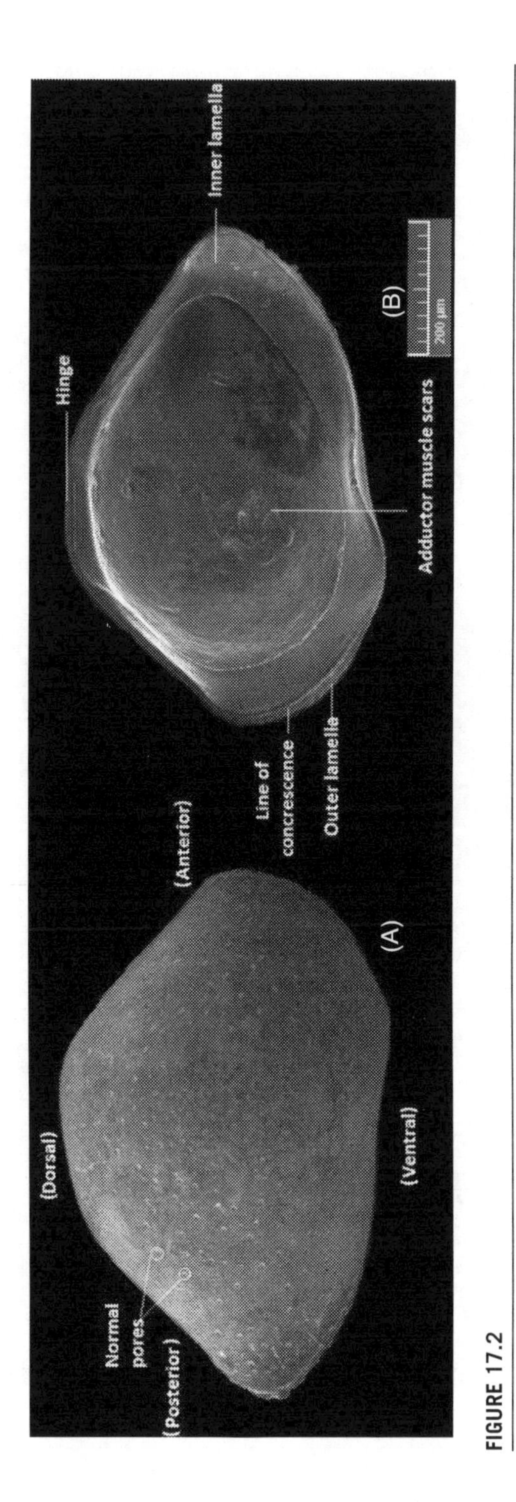

FIGURE 17.2
Morphology of left valve of Ostracoda species *Bairdoppilata alcyonicola* (intertidal Ostracoda species).

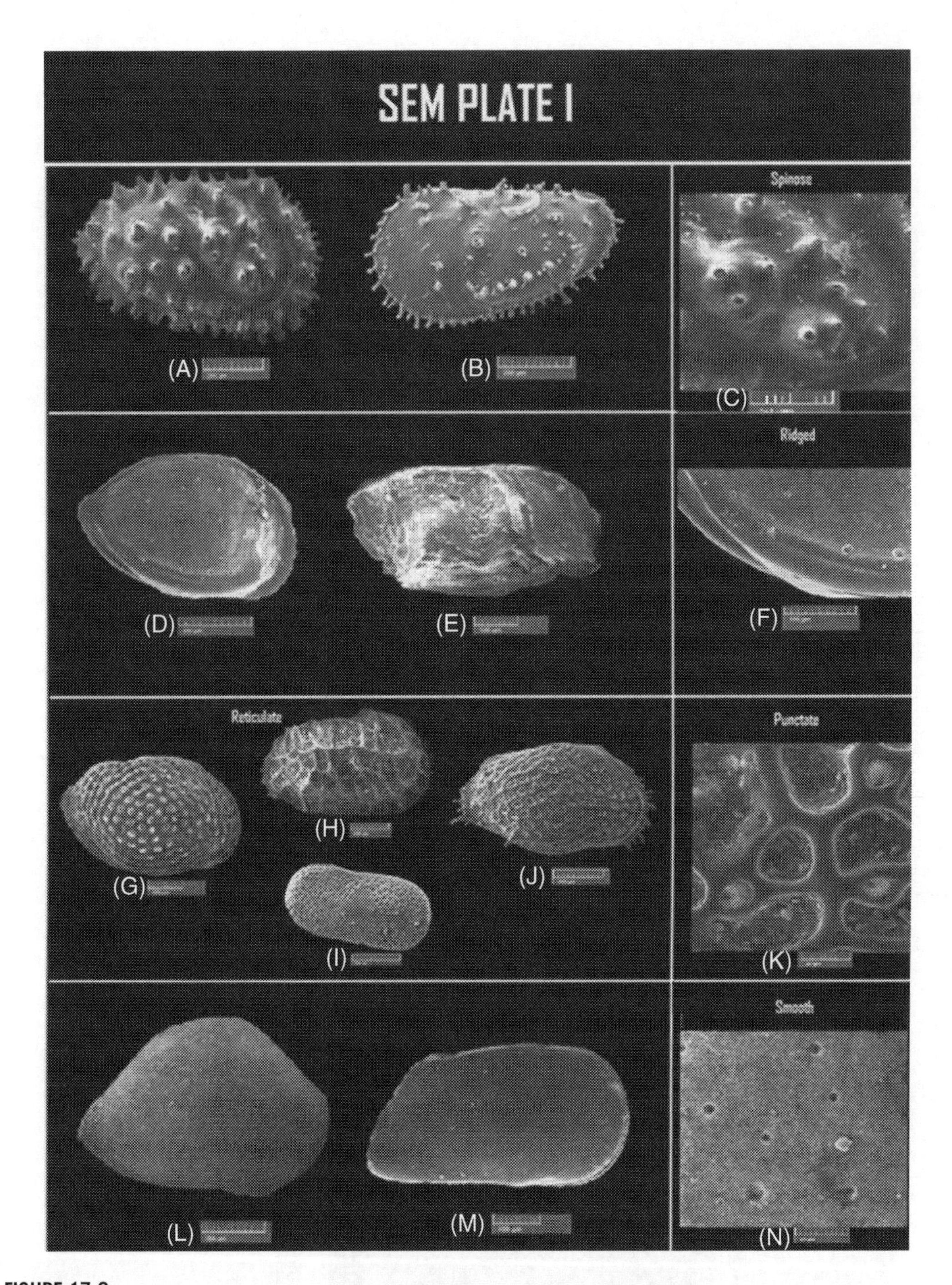

FIGURE 17.3

SEM images and ornamentation pattern of different Ostracoda species. SEM Plate I: Description: A: *Actinocythereis scutigera*, B: *Pterigocythereis chennaiensis*, C: spinose structures in *Actinocythereis scutigera*, D: *Cytheropteron* sp., E: *Bythoceratina* sp., F: ridged structure in *Cytheropteron*, G: *Loxoconcha* sp., H: *Bradleya andamanae*, I: *Cytherella semitalis*, J: *Ruggieria indopacifica*, K: sieve pores in *Loxoconcha* sp., L:

the mangrove environment is understudied from different mangrove environment around the world. However, the deterioration of mangrove environment also impacting on the diversity and distribution of Ostracoda.

Mangroves are deteriorating with an alarming rate, so as the species dependent and thriving on them. Human activities in coastal region along with the sea level changes are the prime causes which enables the shift in mangrove environments in the current scenario. Ever since the last glaciation the sea level rise is happening and influencing the mangrove ecosystem and biota. The study of the Ostracoda species from mangrove environment is sparse compared with the other geological settings. It is a fact that the actual role of Ostracoda from the mangrove environment is not understood completely. However, the species diversity of Ostracoda varies with respect to the Mangrove forest. *Avicennia marina* is a more or less common mangrove species across the globe. However, the diversity and distribution of Ostracoda assemblages in mangrove species vary concerning to the regional factors and not dependent on the mangrove species.

References

Ahmed, S.M.K., Mangroves of the Red Sea, Springer Earth System Sciences, Verlag, Berlin Heidelberg, 2015, pp. 585–597.

Alongi, D.M., Carbon cycling and storage in mangrove forests, Annu. Rev. Mar. Sci. 6 (2014) 195–219.

Alongi, D.M., Present state and future of the world's mangrove forests, Environ. Conserv. 29 (2002) 331–349.

Arul, B., Sridhar, S.G.D., Hussain, S.M., Darwin, F.A., Periakali, P., Distribution of recent benthic Ostracoda from the sediments of Pitchavaram mangroves, Tamil Nadu, Southeast coast of India, Bull. Pure Appl. Sci. 22 (2003) 55–73.

Aschenbroich, A., Michaud, E., Gilbert, F., Fromard, F., Alt, A., Le Garrec, V., Thouzeau, G., Bioturbation functional roles associated with mangrove development in French Guiana, South America, Hydrobiologia 794 (1) (2017) 179–202.

Bradley, S.P.C., Benson, R.H., Terminology for surface features in ornate Ostracods, Lethaia 4 (3) (1971) 249–286.

Chivas, A.R., De Deckker, P., Shelley, J.M.G., Magnesium, strontium and barium partitioning in nonmarine Ostracode shells and their use in paleoenvironmental reconstructions—a preliminary study, in: Maddocks, R.F. (Ed.), Applications of Ostracoda, University of Houston, Geosciences, Dordrecht, Netherlands, 1983, pp. 238–249.

Cronin, T.M., DeMartino, D.M., Dwyer, G.S., Rodriguez, L., Deep sea Ostracoda species, response to late Quaternary climate change, Mar. Micropaleontol. 37 (1999) 231–249.

De Deckker, P., The Ostracoda: applications in Quaternary research, Geophys. Monogr. 131 (2002) 121–134.

De Deckker, P., Chivas, A.R., Shelley, J.M.G., Uptake of Mg and Sr in the euhyaline ostracod Cyprideis determined from in vitro experiments, Paleogeogr. Paleoclimatol. Paleoecol. 148 (1999) 105–116.

Duke, N.C., Ball, M.C., Ellison, J.C., Factors influencing biodiversity and distributional gradients in mangroves, Global Ecol. Biogeogr. Lett. **7** (1998) 27–47.

Furukawa, K., Wolanski, E., Mueller, H., Currents and sediment transport in mangrove forests, Estuar. Coast. Shelf Sci. 44 (1997) 301–310.

Gene, H., Wicaksono, S.A., Julia, B.E., Kenneth, M.G., Climate-driven body-size trends in the ostracod fauna of the deep Indian ocean, Palaeontology 53 (6) (2010) 1255–1268.

Giri, C., Ochieng, E., Tieszen, L.L., Zhu, Z., Singh, A., Loveland, T., Masek, J., Duke, N., Status and distribution of mangrove forests of the world using earth observation satellite data, Global Ecol. Biogeogr. 20 (2010) 154–159.

Hiruta Shimpei, F., Kakui, K., Three new brackish-water thalassocypridine species (Crustacea: Ostracoda: Paracyprididae) from the Ryukyu Islands, southwestern Japan, Zootaxa 4169 (3) (2016) 515–539.
Holmes, J.A., Non marine Ostracods as quaternary paleoenvironmental indicators, Prog. Phys. Geogr. 16 (4) (1992) 405–431.
Hu, C.H., Tao, H.J., Studies of the ostracod fauna of Taiwan and its adjacent seas, J. Natl. Taiwan Museum Special Public. Ser. 13 (2008) 1–910.
Hussain, S.M., Kalaiyarasi, A., Madha Suresh, V., 2012. Distribution of Ostracoda in the Mullipallam creek, Southeast coast of India: implications on microenvironment. In: Second International Symposium, pp. 340–341.
Hussain, S.M., Mahalakshmi, P., Selvasundaram, S., Distribution of Ostracods in the mangrove location of Pulicat Lagoon, Tamil Nadu, Southeast coast of India, Asian Acad. Res. J. Multidiscip. 5 (2019) 234–250.
Jyotiskona, B., Soumyajit, C., True mangrove species of Sundarban Delta, West Bengal, Eastern India, J. Species List Distrib. 10 (2) (2014) 329–334.
Kathiresan, K., A review of studies on Pitchavaram mangrove, Southeast India, Hydrobiiology 430 (1) (2000) 185–205.
Ken Krauss, W., Karen McKee, L., Catherine Lovelock, E., Donald Cahoon, R., Neil Saintilan, R.R., Luzhen, C., How mangrove forests adjust to rising sea level, New Phytol. 202 (2014) 19–34.
Kennish, M.J., Practical Handbook of Marine Science, second ed., CRC press, Boca Raton, Florida, 1994, pp. 1–566.
Kristensen, E., Alongi, D.M., Control by fiddler crabs (*Ucavocans*) and plant roots (*Avicennia marina*) on carbon, iron, and sulfur biogeochemistry in mangrove sediment, Limnol. Oceanogr. 51 (2006) 1557–1571.
Laura, C., Beatrice, G., Eugenio, R., Marco Lo, M., Caterina, C., Silvestro, G., Roberto, D., Impact of mangrove forests degradation on biodiversity and ecosystem functioning, Sci. Rep. 8 (2018) 1–11.
Maddocks, R.F., New thalassocypridine Ostracoda from anchialine caves of the Loyalty Islands, New Caledonia (Podocopida, Paracypridinae), Micropaleontology 51 (3) (2005) 169–216.
Mermillod-Blondin, F., Lemoine, D.G., Ecosystem engineering by tubificid worms stimulates macrophyte growth in poorly oxygenated wetland sediments, Funct. Ecol. 24 (2010) 444–453.
Miyawaki, Vegitation of Japan (in Japanese), vol. 1, *Shibundo Publishers*, Tokyo, 1980, pp. 118–120.
Oertli, H.J., The conservation of ostracode tests-observations made under the scanning electron microscope, Bull. Am. Paleontol. 65 (5) (1972) 549–575.
Oscar, B., Simon, R.A.T., Aurora, E.C., Elizabeth, R., Sergio, V., Henry, E., The distribution and coordination of trace elements in Krithe ostracods and their implications for paleothermometry, Geochim. Cosmochim. Acta 236 (1) (2018) 230–239.
Pokorny, V., Ostracodes, in: Haq, B.U., Boersma, A. (Eds.), Introduction to Marine Micropaleontology, *Elsevier*, New York, 1978, pp. 109–149.
Rajyalakshmi, E., Basha, S.K.M., Floral diversity of mangrove ecosystem of Pulicat Lake, Andhra Pradesh, Imperial J. Interdiscip. Res. 2 (5) (2016) 164–169.
Rosenthal, Y., Linsley, B.Scott Elias (Chief editor), Carry Mock (Associate Editor), Mg/Ca and Sr/Ca paleothermometry from calcareous marine fossils, Encyclopedia of Quaternary Sciences, 2nd ed., Elsevier, USA, 2013, pp. 871–883.
Schmit, O., Rossetti, G., Vandekerkhove, J., Mezquita, F., Food selection in Eucyprisvirens (Crustacea: Ostracoda) under experimental conditions, Hydrobiologia 585 (2007) 135–140.
Smith, A.J., Horne, D.J., Martens, K., Schön, I., Class Ostracoda, in: Thorp, J., Rogers, D.C. (Eds.), Ecology and General Biology: Thorp and Covich's Freshwater Invertebrates, Academic Press, London, 2015, pp. 757–780.
Smith, R.J., Kamiya, T., Horne, D.J., Living males of the 'ancient asexual' Darwinulidae (Ostracoda: Crustacea), Proc. R. Soc. B: Biol. Sci. 273 (2006) 1569–1578.

SobhiHelal, A., El-Wahab, M.A., Distribution of Podocopid Ostracods in mangrove ecosystems along the Egyptian Red sea coast, Crustaceana 85 (14) (2012) 1669–1696.

Xia, J., Ito, E., Engstrom, D.R., Geochemistry of ostracode calcite: part 1. The experimental determination of oxygen isotope fractionation, Geochem. Cosmochem. Acta 61 (2) (1997) 377–382.

Yassini, I., Jones, B.G., Foraminiferida and Ostracaoda from estuarine and shelf environments on the south eastern coast of India, University of Wollongong, Wollongong, N.S.W., 1995.

Zhongning, D., Chigang, R., Quanhong, P., Fujia, Y., Quantitative micro-PIXE and micro-probe analysis of ostracode shells to reconstruct the paleoenvironment, Nucl. Instrum. Methods Phys. Res. 104 (1995) 619–624.

CHAPTER

18 Late Quaternary landscape evolution of Peninsular India: A review based on fluvial archives

M.R. Resmi[a], Hema Achyuthan[b], Gaurav Chauhan[c], Hritika Deopa[a]

[a] *School of Earth Sciences, Banasthali Vidyapith, 304022, Rajasthan, India.* [b] *Institute for Ocean Management (IOM), Anna University, Chennai, 600025, Tamil Nadu, India.* [c] *Department of Earth and Environmental Science, K.S.K.V. Kachchh University, Bhuj, Kachchh, 370001, India*

18.1 Introduction

Evolution of any fluvial system depends primarily on exogenic (e.g., climatic perturbations and base-level changes) and endogenic (e.g., tectonic rock uplift) process operating on the Earth's surface (Bishop, 1995; Gregory et al., 2006; Bridgland and Westaway, 2008). A critical question arises how landscapes respond to these exogenic and endogenic processes over a millennial timescale. Understanding the controlling factors on the evolution of the fluvial system is essential for surmising the landform development (Schumm, 1977; Lavé and Avouac, 2000; Bookhagen et al., 2005; Bookhagen and Burbank, 2006; Behr et al., 2010; Zielke et al., 2010; Kirby and Whipple, 2012). However, the fluvial channel network gives a critical feedbacks on climatic and tectonic forcing across the landscape (e.g., Whipple and Tucker, 1999; Attal et al., 2011; Whipple et al., 2013; Nennewitz et al., 2018; Joshi et al., 2013).

During the recent years, Peninsular India was the subject of neotectonic (Kale et al., 1986, 1994, 1996; Kale, 1990, 1999a, 2003, 2005; Kale and Subbarao, 2004; Kale and Shejwalkar, 2007, 2008; Resmi et al., 2017a, 2017b) and paleoclimatic investigations (Gunnell and Radhakrishna, 2001; Juyal et al., 2010; Ray and Srivastava, 2010; Resmi and Achyuthan, 2018a, 2018b). Fluvial systems are very sensitive to tectonic processes such as folding and faulting (Burbank and Anderson, 2011; Pérez-Peña et al., 2010). These processes are accountable for river incision, asymmetries of the catchments, river diversions, formation of the fluvial terraces, development of knickpoints among other effects (e.g., Cox, 1994; Burbank and Anderson, 2011; Clark et al., 2004; Salvany, 2004; Schoenbohm et al., 2004; Whipple et al., 2013; Gailleton et al., 2019; Joshi et al., 2021). Also, river incision in such regions is related to other processes like climatic changes, stream piracy, base-level lowering, etc. (e.g., Starkel, 2003). The incision of bedrock coupled with the development of strath terraces indicates the rock uplift related to tectonic movements (Pazzaglia et al., 1998; Hancock and Anderson, 2002; Dortch et al., 2011; Srivastava et al., 2013). At the valley scale, river aggradation and incision of valley fill sediments are controlled by climatic variations (Pratt-Sitaula et al., 2004; Scherler et al., 2015). The channel avulsion/migration is the most significant river process operating on the Himalayan rivers (Gole and Chitale, 1966; Galgali, 1986; Valdiya and Kotlia, 2001; Chaudhary et al., 2015) and Peninsular

Holocene Climate Change and Environment. DOI: https://doi.org/10.1016/B978-0-323-90085-0.00009-7

rivers (Resmi et al., 2017a, 2020). Several studies have attributed that neotectonic movements, aggradation, flood, and base-level changes as plausible causes for avulsion and channel migration (Jorgensen et al., 1994; Wells and Dorr, 1987; Sinha, 1996). Understanding the riverine processes with their spatial and temporal variations is the key to delineate the geomorphic evolution of a river basin.

Studies along the rivers draining Peninsular India—namely, the Narmada, Tapi, Mahanadi, Godavari, Krishna, Cauvery, and its adjoining rivers, which receive water from the Indian Summer Monsoon and North East Monsoon—have given an understanding of aggradation and incision phases along with their relationship to climate and tectonics. In this chapter, we attempt to understand how geomorphic markers can be used as a quantitative proxy for understanding the long-term landscape evolution of Peninsular Indian rivers.

18.2 Scope

Peninsular Indian Rivers are distinctive in terms of their hydrology and sediment transport characteristics. Many of these rivers have received international attention since last several decades and the evolutionary trajectories of Peninsular rivers are ranging from millions of years to millennia. Hence, this chapter highlights the major research contributions on various aspects of fluvial landform evolution since the Quaternary Period. This work also evaluates the significance of geomorphic markers to understand evolution trajectories of the major river basins of Peninsular India.

18.3 Methodology and data sources

Here, we identified the previous work from Peninsular Indian Rivers, particularly on fluvial Geomorphological aspect. From earlier work, we are concerned primarily with river catchment processes, identification of geomorphic markers such as terrace formation, gorges, longitudinal profiles, morphotectonic indices, paleochannels, etc. We did not include papers that concerned only tectonic activity and climatic fluctuations without reference to more classical fluvial processes and landforms. Only a few papers were directly concerned with the origin of geomorphic landforms and fluvial evolution. Hence in the present study, we discussed the evolution of the Peninsular Rivers on temporal and spatial scales. Besides, under the spatial scale, we evaluated the landforms, formed due to tectonic or climatic forcing? The tectonic controls on these cratonic rivers were evaluated based on the investigations of the longitudinal profiles, morphotectonic indices, terrace formation, paleochannels, etc. Climatic control is evaluated by the increase in river discharge also confirmed by the presence of large flood plains, stream diversion, etc.

18.4 Observations and results

The Indian Peninsula (Fig. 18.1), a tectonically stable landmass, constitute an amalgamation of the Precambrian crustal blocks (Subrahmanya, 1996). It was an integral part of Gondwanaland, sandwiched between Africa and Antarctica (Gunnell and Radhakrishna, 2001). The intraplate region consists of relict mountains (Aravalli and Mahendragiri Hills), block mountains (Satpura, Nilgiri, Shevaroy, and

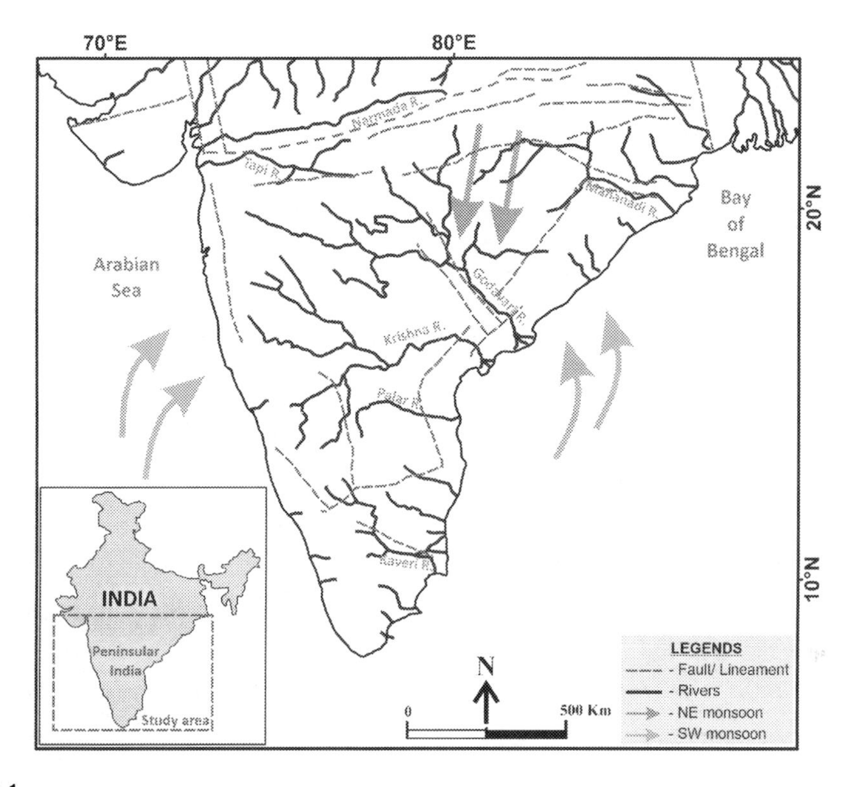

FIGURE 18.1

Major rivers of Peninsular India (modified after (Rehfeld et al., 2011; Kale and Vaidyanadhan, 2014; Kayal, 2016; Resmi and Achyuthan, 2018a)).

Biligirirangan Hill Ranges), cratons and basins (Dharwar, Bastar, the southern extremity of Singhbhum, Cuddapahs), Rift valleys of Narmada, Tapi, Godavari, Mahanadi, and Damodar river grabens. The central part of Peninsular India consists of structural ridges and intervening valleys of Archean to Proterozoic igneous and metasedimentary rocks. Whereas, the eastern and western parts of Peninsular India are characterized by Gondwana and Tertiary rocks exposed along the uplands, patchy outcrops, and coastal cliffs which are further encircled by Quaternary sediments (Ramasamy, 1987).

The tectonic antiquity of Peninsular India is not yet well documented. Valdiya (2001) observed that Peninsular Indian tectonism initiated during the breakup of India from the Gondwanaland. In the Peninsular region, tectonic activity is related with the Himalayan Orogeny, the stress builds up and it leads to the reactivation of fault and shear system during Quaternary Period (Radhakrishna, 1993; Singh and Rajamani, 2001). Geologically, Peninsular India consists of several Precambrian cratonic blocks and is surrounded by rifts and Proterozoic fold belts (Kale and Vaidyanadhan, 2014). It is composed of Precambrian rocks, Cretaceous extrusive volcanic rocks (Deccan basalt), lower Triassic to Upper Carboniferous sedimentary rocks, and Quaternary alluvium. The major geomorphic landforms in Peninsular India are block mountains, plateaus, mesas, cuestas, steep escarpments, and valleys (Kale and Vaidyanadhan, 2014).

Peninsular India displays a diverse climatic regime; a large part of the Peninsula is characterized by dry and subhumid conditions. The central and south-central part show semiarid conditions, and the western section is represented by the arid climate (Gadgil and Joshi, 1983). The areas near to the Western Ghats receive more than 3000 mm rainfall, whereas the annual rainfall in the central part is less than 500 mm (Gadgil and Joshi, 1983). Peninsular India is entirely characterized by monsoon-fed rivers with moisture sources derived from South West and North East monsoon. The South West monsoon is the period of high discharge especially for Peninsular Rivers, In contrast rivers in the northern side of Tamil Nadu and Coastal side of Andhra Pradesh are fed by North East monsoon. Despite the extremely variable discharges, the Peninsular Rivers are steadier and less capricious, due to very low channel slope and more resistant channel boundaries (Kale, 1998; Gupta et al., 1999). Another important characteristic is most of the rivers have ample channel capacity to occupy rare, large flows and even large floods are inadequate to fill the whole channel (Gupta et al., 1999). The river channel variations are not common and of much smaller magnitude than the Himalayan rivers (Deodhar and Kale, 1999).

The entire landscape of Peninsular India reveals a regional complexity of various components such as the lithology, geological history, climate and are distinctly different in terms of hydrological and basin-scale physical characteristics. The physiography of Peninsular India is characterized predominantly by eastward flowing drainages and its associated landforms. However, in the present study, we have grouped Peninsular River basins based on their geological evolution, the diverse tectonic and climatic settings into three parts, namely (a) West flowing Peninsular Rivers, (b) East flowing Peninsular Rivers, (c) Southern Peninsular Rivers.

18.5 Discussion

18.5.1 West flowing Peninsular rivers

The northern boundary of Peninsular India is demarcated by Proterozoic Satpura mountain belt, which is ENE–WSW trending tectonically active zone (Mohanty, 2010; Roy et al., 2006). This zone was reactivated during the Cretaceous Period and resulted in the formation of ENE–WSW trending rift zones (Waghmare and Carlo, 2008). Narmada and Tapi rivers are the two largest westward flowing rivers of Peninsular India (Fig. 18.1), which have a typical elongated shape and controlled by rift valleys (Tandon and Sinha, 2007). These major rivers are controlled by Son–Narmada faults/lineaments and Tapi north fault, which is having an ENE–WSW trend (Kothyari and Rasatogi, 2013). The westerly flowing Narmada and Tapi rivers flow within the fault-bounded rift blocks (Choubey, 1974; Radhakrishna and Ramakrishnan, 1988; Jain et al., 2020; Acharyya and Roy, 2000; Chamyal et al., 2002).

The interaction between sedimentation processes due to climatic and tectonic forcing, have played a major role in the geomorphic evolution of west-flowing Peninsular Rivers. Tectonic uplifts occurred in the south of the Narmada River leads to the northward shifting of its course (Maurya et al., 2000). The structurally controlled nature of the Narmada River is also revealed by the presence of straight channels in the upper to middle reaches, incised meanders, gorges, unpaired terraces, knickpoints, deflection of river channels, alluvial cliffs, deeply entrenched meanders, gullies, and abandoned cliffs (Chamyal et al., 1997, 2002; Kothyari and Rastogi, 2013, Jain et al., 2020; Fig. 18.2A–C). Similarly, the lower reaches of the Narmada River basin is also experienced a tectonic subsidence during the Late Pleistocene (Chamyal et al., 2002). Development of rectangular and trellis

FIGURE 18.2

(A) Photograph showing knickpoint in a tributary stream of Narmada river. (B) Entrenched meander in the alluvial zone of NSF near Tejpur. (C) Alluvial plains of Narmada.

drainage patterns along the river course is also corroborating the influence of neotectonic activity in the river basin (Tricart et al., 1974). Besides, a thick aggradation of point bar, overbank deposits, and floodplain deposits of the Quaternary period is suggesting that monsoon precipitation also influenced the evolution of landforms in Narmada river (Patnaik, 2009; Chamyal et al., 2003; Fig.18.3).

Tapi river is the second-largest westward flowing river of Peninsular India (Sridhar and Chamyal, 2018). The neotectonic activity within the Tapi basin is manifested by the occurrence of freequent earthquakes within the basin and the presence of several geomorphic features such as escarpments, piedmont-plains, colluvial plains, bedrock gorges, tilting and warping of Holocene and Pleistocene fluvial terraces, incision of valleys and development of two to three levels of strath terraces (Sridhar and Chamyal, 2018; Kothyari and Rastogi, 2013) (Fig. 18.3). Ghatak and Ghatak (2005) have reported the red paleosol horizon interbedded with ash beds near Palasur as the oldest Quaternary unit which directly overlies the Deccan basalt. Further, the paleoseismic features like liquefaction features with sand dykes, convolute lamination, small-scale thrust faults reported Tapi river course are the manifestation of the

FIGURE 18.3

(A) Quaternary terraces in Lower Narmada basin. (B) Tapi fault scarp parallel to the Tapi river and two levels of terrace.

neotectonism in the region (Ghatak and Ghatak, 2005). Among other Peninsular rivers, the Tapi River is distinguished for its large and extra ordinary floods during the monsoon season and it occurred in its recent past also. Thus, Tapi River displays all the hydrologic and geomorphic characteristics of a flood dominated river. The major signal of paleoflood evidences from the basin is slack water deposit in the bed rock gorges. The morphological characteristics of the bedrock and the alluvial channels of the Tapi River are also maintained by extreme floods (Kale, 2003).

18.5.2 East flowing Peninsular rivers

Eastern Peninsular Rivers such as Mahanadi, Godavari, and Krishna Rivers originate from the Western Ghats, and takes a South-Easterly drainage direction. The Mahanadi basin is formed due to the rifting and breaking up of Gondwanaland, situated on the East Coast of India (Lisker and Stefan, 2001).

The river respond to large-magnitude monsoon floods by increasing their width and width–depth ratio (Deodhar and Kale, 1999). Fluvial aggradation (>30–10 ka BP) and incision phases in the basin is contrast with major climatic events such as LGM, Holocene monsoon variability, and subsequent sea-level changes (Kale and Rajaguru, 1987). The dynamic nature of alluvial channels of the Godavari River is reflected by the presence of large flood plains, frequent channel migrations, sharp knee-bend like deflections in the river channel, channel bar deposits consists predominantly of conglomerates, pebbles, and sandy deposits along with alluvial terraces. The occurrence of paleochannels and the sharp rectilinear bends in the Godavari River indicates its response to neotectonic activity during the Quaternary period (Sangode et al., 2013). Development of various gorges and water-induced seismic activity along the fault zones also corroborates the ongoing tectonic activity in the region (Hickman et al., 1995; Noir et al., 1997; Floyd et al., 2001; Janssen et al., 2005; Kale and Vaidyanathan, 2014). Krishna River shows various geomorphic markers of active tectonics such as gullies (filled by the silt and clay deposits), gorges (Jaldurg and the Srisailam gorges), knickpoints (near Wai and Ozarde area), several potholes along the river course, waterfalls (in the head region), and entrenched meanders. The incised alluvial channels of the Krishna river display channel-in-channel physiography. Some of the gorges are incised across structural and topographic highs of the Eastern Ghat by the Godavari, and Krishna rivers suggesting that either these rivers are antecedent or superimposed drainages (Kale and Vaidyanathan, 2014). Another significant geomorphic features developed due to tectonic forcing are terraces along the banks but are discontinuous because of the tributaries, hills, and rock outcrops.

18.5.3 Southern Peninsular Rivers

The rivers of southern Peninsular India are flowing over a stable craton. The major characteristic features of these rivers are broad and shallow valleys with low gradients. Only the coastal rivers, heading in the Western or Eastern Ghats, are short, steep, and swift-flowing. All the rivers draining Southern Peninsular India are monsoon-fed and, therefore, active only during the monsoon season.

The Southern Peninsular rivers, such as Chaliyar, Bharathapuzha, Cauvery, and Palar River, exhibit a periodic uplift during the Quaternary Period. Generally, in the tectonically active region, the bedrock channel network indicates the interaction between landforms, elevation, and erosion rates (Ambili and Narayana, 2014). The convex longitudinal profiles of Chaliyar river and its major tributaries indicate a rapid uplift in the region. Likewise, the structurally controlled knickpoints and high steepness index also suggest a regional uplift and high rate of the incision. However, the high rate of incision and formation of knickpoints in this region are also associated with intensification of South-West monsoon and change in sea level during the Pleistocene time. Therefore, the regional uplift followed by intensification of monsoon along with base-level changes during late Quaternary period triggered river incision and lateral migration/avulsion of stream channels at many places especially within the Nilambur valley (Ambili and Narayana, 2014). Presence of the topographic features such as active strike-slip fault, fault scarp, river ponding, presence of V-shaped valleys, frequent recurring of landslides, active lineaments in the Chaliyar river also corroborate the ongoing tectonic activity in the river basin. Channel avulsion has been noticed as a typical, and one of the key fluvial processes operating on Peninsular India. The avulsion of channels and the associated shift in the Bharathapuzha river also indicate that a shift in slope is induced by the reverse fault. This reverse movement of the fault might also results in the formation of the waterfall near the fault zone (John and Rajendran, 2008).

FIGURE 18.4

(A) Scenic view of the Cauvery flowing at Mekedatu. (B) Cauvery flowing at Mekadatu through a narrow gorge. (C) and (D) Cauvery waterfalls at Hogenakkal.

Cauvery River, the major Southern Peninsular river system, exhibits differential tectonism (Kale et al., 2014). The river and its tributaries have major knickpoints along the topographic transition zones (middle reaches) from the plateau to plains, which are marked as waterfalls (Kale et al., 2014). Incised valleys in the middle reach of the Cauvery River and formation of hanging valleys and high stream gradient index among its tributaries in response to incision of the main channel indicates active fluvial erosion associated with tectonic forcing. Moreover, the two prominent breaks in the longitudinal profile of the Cauvery River at Sivasamudram and Hogenakal, and several knickpoints on the tributaries in this domain provide compelling evidence for tectonically triggered disequilibrium conditions (Fig. 18.4A–D). Due to the orographic effect, the middle reaches of the river experience an increased monsoonal precipitation as compared to higher and lower reaches (Kale, 2012). Therefore, the rapid erosion in the middle reaches is the combined effect of flow competency and rapid uplift. Various lines of evidence indicate that the Cauvery catchment and the adjoining areas have been subjected to tectonic activities from the time of separation of India from the Gondwanaland (Valdiya, 2001). However, the deposition at

FIGURE 18.5

(A) View of the Palar river major paleochannel. (B) and (C) show major aggradational phases in paleochannel and Palar river during Quaternary Period.

Biligundlu (upper reaches of Cauvery) was induced by enhanced sediment supply from upstream during the humid phase in Quaternary Period. The Cauvery River was aggrading as a means of increasing its gradient back in response to partial damming by movements along the Hogenakal fault is difficult to ascertain (Valdiya, 1998).

Resmi et al. (2017a) investigated the morphologic characteristics of the Palar River. The study suggests that the drainage network of the Palar River basin is tectonically controlled and the knickpoints along with the longitudinal stream profiles of uniform lithology shows neotectonic activity along the active lineaments. The active strike-slip and reverse fault in Palar River drainage basin have produced geomorphic markers of active tectonism such as paleochannels (Fig. 18.5A–C), aggradation, river ponding, etc. The occurrence of paleochannels also indicates that tectonically and monsoon triggered clockwise migration of the river has occurred in the late mid–late-Holocene period. Since the region receives dominant rainfall from NE monsoon, the landscape evolution in the Palar basin has been attributed to the interplay between tectonic activity and NE monsoons variability.

18.6 Learning and knowledge outcomes

Peninsular Indian River systems were developed through different evolutionary processes, which ultimately controlled their dynamics and morphology. The present chapter emphasizes the major processes for the evolution of fluvial landforms. The major observations include:

West flowing Peninsular Rivers: The isostatic response to erosional process represents the complex form of tectonic–topographic feedback, which ultimately results in the evolution of Narmada and Tapi Rivers. Neotectonic uplifts/subsidence likely imply the influence of isostatic and tectonic forces in fluvial environments. Here, the tectonic forces lead to the development of major faults, rift valleys, gorges, knickpoints, and shifting of the river course. In addition, the evolution of these rivers due to enhanced monsoon conditions during the Quaternary period is also evident.

East flowing Peninsular Rivers: Climatic variations predominantly controlled the channel morphology and sediment supply of these rivers. Especially, the extreme flood events favored in shaping the stable alluvial channels. Tectonics, in combination with lithology, also governs the topography of East flowing Peninsular river system.

Southern Peninsular Rivers: The geomorphic features of Southern Peninsular rivers are pointing that the evolution is primarily due to tectonic and climate forcing. Major faults, active lineaments, channel avulsion, rapid incision, presence of series of knickpoints, deep gorges, hanging valleys, and other drainage anomalies are given ample evidences for tectonic activity during the Quaternary Period. However, our studies also indicate that climatically induced landforms are dominated during the Holocene period. This is further corroborated by the presence of Holocene flood plains of Cauvery River and paleochannels of the Palar River, etc.

The recognition and depiction of landscape evolution processes are both timely and critical. Conclusively, here we illustrated the diversity in fluvial activity and subsequent evolution of Peninsular Indian rivers. Hence, the significant remarks that come out from our studies are as follows:

- Understanding the tectono-climatic changes is key to surmise the Peninsular River's evolution and its implications on the modern-day river dynamics.
- Isostatic and tectonic forcing are the major plausible causes for Uplift/subsidence and subsequent erosion. Tectonic forcing includes the development of structural controls, and topographic stress fields, which ultimately leads to the evolution of fluvial landforms in Peninsular India.
- The monsoon variation also played an important role in shaping the alluvial channels predominately since the Holocene. The major geomorphic markers of Peninsular India that illustrate monsoonal variations includes channel meanders, incised valleys, channel migration, alluvial terraces, and avulsion.
- Finally, uplift/subsidence and reactivation of pre-existing faults, climate variability, and the sea-level changes have played a significant role in the evolution of Peninsular Indian rivers.

Acknowledgments

We would like to acknowledge Dr. Subhash Bhandari (K.S.K.V. Kachchh University, Bhuj) and Dr. Girish Ch. Kothyari (Institute of Seismological Research, Gandhinagar) for providing the field photos of Narmada and Tapi river basin.

References

Acharya, S.K., Roy, A., Tectonothermal history of the Central Indian Tectonic Zone and reactivation of major fault/shear zones, J. Geol. Soc. India 55 (2000) 239–256.

Ambili, V., Narayana, A.C., Tectonic effects on the longitudinal profiles of the Chaliyar River and its tributaries, southwest India, Geomorphology 217 (37) (2014) 47, doi:10.1016/j.geomorph.2014.04.013.

Attal, M., Cowie, P.A., Whittaker, A.C., Hobley, D., Tucker, G.E., Roberts, G.P., Testing fluvial erosion models using the transient response of bedrock rivers to tectonic forcing in the Apennines, Italy, J. Geophys. Res. 116 (F2) (2011) 1–17, doi:10.1029/2010JF001875.

Behr, W.M., Rood, D.H., Fletcher, K.E., Guzman, N., Finkel, R., Hanks, T.C., Hudnut, K.W., Kendrick, K.J., Platt, J.P., Sharp, W.D., Weldon, R.J., Uncertainties in slip-rate estimates for the Mission Creek strand of the southern San Andreas fault at Biskra Palms Oasis, southern California, Bulletin 122 (9-10) (2010) 1360–1377.

Bishop, P., Drainage rearrangement by river capture, beheading and diversion, Prog. Phys. Geogr. 19 (1995) 449–473, doi:10.1177/030913339501900402.

Bookhagen, B., Burbank, D.W., Topography, relief, and TRMM-derived rainfall variations along the Himalaya, Geophys. Res. Lett. 33 (8) (2006).

Bookhagen, B., Thiede, R.C., Strecker, M.R., Late Quaternary intensified monsoon phases control landscape evolution in the northwest Himalaya, Geology 33 (2) (2005) 149–152.

Bridgland, D., Westaway, R., Climatically controlled river terrace staircases: a worldwide Quaternary phenomenon, Geomorphology 98 (3-4) (2008) 285–315.

Burbank, D.W., Anderson, R.S., Tectonic Geomorphology, John Wiley and Sons, Oxford, U.K., 2011.

Chamyal, L.S., Khadkikar, A.S., Malik, J.N., Maurya, D.M., Sedimentology of the Narmada alluvial fan, western India, Sediment. Geol. 107 (3-4) (1997) 263–279.

Chamyal, L.S., Maurya, D.M., Bhandari, S., Raj, R., Late Quaternary geomorphic evolution of the lower Narmada valley, Western India: implications for neotectonic activity along the Narmada–Son Fault, Geomorphology 46 (3-4) (2002) 177–202.

Chamyal, L.S., Maurya, D.M., Raj, R., Fluvial systems of the drylands of western India: a synthesis of Late Quaternary environmental and tectonic changes, Quat. Int. 104 (1) (2003) 69–86.

Chaudhary, S., Shukla, U.K., Sundriyal, Y.P., Srivastava, P., Jalal, P., Formation of paleovalleys in the Central Himalaya during valley aggradation, Quat. Int. 371 (2015) 254–267.

Choubey, V.D., Long-distance correlation of Deccan basalt flows, Central India: reply, Geol. Soc. Am. Bull. 85 (6) (1974) 1008–1010.

Clark, M.K., Schoenbohm, L.M., Royden, L.H., Whipple, K.X., Burchfiel, B.C., Zhang, X., Tang, W., Wang, E., Chen, L., Surface uplift, tectonics, and erosion of eastern Tibet from large-scale drainage patterns, Tectonics 23 (1) (2004), doi:10.1029/2002TC001402.

Cox, R.T., Analysis of drainage-basin symmetry as a rapid technique to identify areas of possible Quaternary tilt-block tectonics: an example from the Mississippi Embayment, Geol. Soc. Am. Bull. 106 (5) (1994) 571–581.

Deodhar, L.A., Kale, V.S., Downstream adjustments in allochthonous rivers: Western Deccan Trap upland region, India, in: Miller, A.J., Gupta, A. (Eds.), Varieties Fluvial Form, Wiley, New York, 1999, pp. 295–315.

Dortch, J.M., Owen, L.A., Dietsch, C., Caffee, M.W., Bovard, K., Episodic Fluvial Incision of Rivers and Rock Uplift in the Himalaya and Transhimalaya, Geological Society of London U.K., 168 (3) (2011) 783–804, doi:10.1144/0016-76492009-158.

Dortch, J.M., Owen, L.A., Dietsch, C., Caffee, M.W., Bovard, K., Episodic Fluvial Incision of Rivers and Rock Uplift in the Himalaya and Transhimalaya, Geological Society of London, U.K., 2011, pp. 783–804, doi:10.1144/0016-76492009-158.

Floyd, J.S., Mutter, J.C., Goodliffe, A.M., Taylor, B., Evidence for fault weakness and fluid flow within an active low-angle normal fault, Nature 411 (2001) 779–783.

Gadgil, S., Joshi, N.V., Climatic clusters of the Indian region, J. Climatol. 3 (1) (1983) 47–63, doi:10.1002/joc.3370030105.

Gailleton, B., Mudd, S.M., Clubb, F.J., Peifer, D., Hurst, M.D., A segmentation approach for the reproducible extraction and quantification of knickpoints from river long profiles, Earth Surf. Dyn. 7 (1) (2019) 211–230.

Galgali, V.G., 1986. River training and flood regulation on the Kosi River. In: Proceedings of the 53rd Research and Development Session, Bhubaneshwar, Orissa, October, New Delhi. Central Board of Irrigation and Power, pp. 181–197.

Ghatak and Ghatak, Morphometric evolution of active tectonic domain from Central India: a case study from Gavilgarh Fault Zone, Geol. Surv. Ind. Spl. Pub. 85 (2005) 237–248.

Gole, C.V., Chitale, S.V., Inland delta building activity of the Kosi River, J. Hydraulic Div. Am. Soc. Civil Eng. HY 2 (1966) 111–126.

Gregory, K.J., Macklin, M.G., Walling, D.E., Past hydrological events related to understanding global change, Catena 66 (May, special issue) (2006c) 187.

Gunnell, Y., Radhakrishna, B.P., Sahyadri, the Great Escarpment of the Indian Subcontinent. Patterns of landscape development in the Western Ghats, Geol. Soc. India Mem. 47 (1, 2) (2001) 1054.

Gupta, A., Kale, V.S., Rajaguru, S.N., The Narmada river, India, through space and time, in: Miller, A.J., Gupta, A. (Eds.), Varieties of Fluvial Form, Wiley, New York, 1999, pp. 113–143.

Hancock, G.S., Anderson, R.S., Numerical modeling of fluvial strath-terrace formation in response to oscillating climate, Geol. Soc. Am. Bull. 114 (2002) 1131–1142.

Hickman, S., Sibson, R., Bruhn, R., Introduction to Special Section: Mechanical Involvement of Fluids in Faulting, Journal of Geophysical Research: Solid Earth, US Geological Survey, 1995, pp. 12,831–12,840.

Jain, V., Sonam, A.S., Singh, A., Sinha, R., Tandon, S., Evolution of modern river systems: an assessment of 'landscape memory'in Indian river systems, Episodes J. Int. Geosci. 43 (1) (2020) 535–551.

Janssen, C., Romer, R.L., Hoffmann-Rothe, A., Mingram, B., Dulski, P., Moller, P., Al-Zubi, H., The role of fluids in faulting deformation: a case study from the Dead Sea Transform (Jordan), Int. J. Earth Sci. (Geol Rundsch) 94 (2005) 243–255.

John, B., Rajendran, C.P., Geomorphic indicators of neotectonism from the Precambrian terrain of Peninsular India: a case study from the Bharathapuzha Basin, Kerala, J. Geol. Soc. India 71 (2008) 827–840.

Jorgensen, D.W., Harvey, M.D., Flamm, L., Morphology and dynamics of the Indus river: implications for the Mohen jo Daro site, in: Shroder, J., Kazmi, A. (Eds.), Himalayas to the Sea: Geology, Geomorphology and Quaternary, Routledge, London, 1994, pp. 288–326.

Joshi, N., Kothyari, G.C., Pant, C.C., Drainage conformation and transient response of river system in thrust segmentation of Northwest Himachal Himalaya, India, Quat. Int. 575 (2021) 37–50.

Joshi, P.N., Maurya, D.M., Chamyal, L.S., Morphotectonic segmentation and spatial variability of neotectonic activity along the Narmada–Son Fault, Western India: Remote sensing and GIS analysis, Geomorphology 180 (2013) 292–306.

Juyal, N., Sundriyal, Y., Rana, N., Chaudhary, S., Singhvi, A.K., Late Quaternary fluvial aggradation and incision in the monsoon-dominated Alaknanda valley, Central Himalaya, Uttrakhand, India, J. Quat. Sci. 25 (8) (2010) 1293–1304.

Kale, V., On the link between extreme floods and excess monsoon epochs in South Asia, Clim. Dyn. 39 (5) (2012) 1107–1122.

Kale, V.S., Morphological and hydrological characteristics of some allochthonous river channels, western Deccan Trap upland region, India. Geomorphology 3 (1) (1990) 31–43.

Kale, V.S., Monsoon floods in India: a hydro geomorphic perspective, in: Kale, V.S. (Ed.), Flood Studies in India, 41, Geological Society of India, Memoir, Bangalore, 1998, pp. 229–256.

Kale, V.S., Long-period fluctuations in monsoon floods in the Deccan Peninsula, India, J. Geol. Soc. India 53 (1999) 5–16.

Kale, V.S., Geomorphic effects of monsoon floods on Indian rivers. in: Flood Problem and Management in South Asia, Mirza M.M.Q., Dixit A., Nishat A. (Eds.), Springer, Dordrecht, 2003, pp. 65–84. https://doi.org/10.1007/978-94-017-0137-2_3.

Kale, V.S., Mishra, S., Baker V.R., et al., Sedimentary records of palaeofloods in the bedrock gorges of the Tapi and Narmada rivers, central India, Current Science 84 (8) (2003).

Kale, V.S., The sinuous bedrock channel of the Tapi River, Central India: its form and processes, Geomorphology 70 (3-4) (2005) 296–310.

Kale and Shejwalkar, Western Ghat escarpment evolution in the Deccan Basalt Province: Geomorphic observations based on DEM analysis, J. Geological Soc. of India 70 (2007) 459–473.

Kale and Shejwalkar, Uplift along the western margin of the Deccan Basalt Province: Is there any geomorphometric evidence? J. Earth System Sci. 117(6) (2008) 959–971, doi:10.1007/s12040-008-0081-3.

Kale, V.S., Ely, L.L., Enzel, Y., Baker, V.R., Geomorphic and hydrologic aspects of monsoon floods on the Narmada and Tapi Rivers in Central India. Geomorphology and Natural Hazards, in: Geomorphology and Natural Hazards, Morisawa M. (Ed.), Elsevier, USA, 1994, pp. 157–168.

Kale, V.S., Ely, L.L., Enzel, Y., Baker, V.R., Palaeo and historical flood hydrology, Indian Peninsula, 115, Geological Society, London, Special Publications, 1996, pp. 155–163.

Kale, V.S., Rajaguru, S.N., Late Quaternary alluvial history of the northwestern Deccan upland region, Nature 325 (6105) (1987) 612.

Kale, V.S., Rajaguru, S.N., Rajagopalan, G., Late Holocene evidence of neotectonics in the Upper Vashishthi Valley (Western Maharashtra), Curr. Sci. (Bangalore) 55 (24) (1986) 1240–1241.

Kale, V.S., Subbarao, K.V., Some observations on the recession of the Western Ghat escarpment in the Deccan trap region, India: based on geomorphological evidence, Trans. Jpn. Geomorphol. Union 25 (3) (2004) 231–245.

Kale, V.S., Survase, V. and Upasani, D. 2014. Geological mapping in the Koyna–Warna region. ACWADAM Technical Report no: 2014-C1; DOI: 10.13140/2.1.3098.0809, 163p.

Kale, V.S., Vaidyanadhan, R., The Indian Peninsula: Geomorphic Landscapes in: Landscapes and Landforms of India. Springer, Dordrecht, World Geomorphological Landscapes, Kale V. (Eds.), Springer Science, Dordrecht, 2014, doi:https://doi.org/10.1007/978-94-017-8029-2_6.

Kayal, J.R, IGCP Project 559 - Seismic Images, Geological Survey of India, Kolkata, India (2016).

Kirby, E, Whipple, K.X., Expression of active tectonics in erosional landscapes, J. Struct. Geol. 44 (2012) 54–75.

Kothyari, G.C., Rastogi, B.K., Tectonic control on drainage network evolution in the Upper Narmada Valley: implication to neotectonics, Geogr. J. 2013 (2013).

Lavé, J., Avouac, J.P., Active folding of fluvial terraces across the Siwaliks Hills, Himalayas of central Nepal, J. Geophys. Res.: Solid Earth 105 (B3) (2000) 5735–5770.

Lisker, F., Fachmann, S., Phanerozoic history of the Mahanadi region, India, J. Geophys. Res.: Solid Earth 106 (B10) (2001) 22027–22050.

Maurya, D.M., History of tectonic evolution of Gujarat alluvial plains, western India during Quaternary: a review, J. Geol. Soc. India 55 (2000) 343–366.

Mohanty, S., Tectonic evolution of the Satpura Mountain Belt: a critical evaluation and implication on supercontinent assembly, J. Asian Earth Sci. 39 (6) (2010) 516–526.

Nennewitz, M., Thiede, R.C., Bookhagen, B., Fault activity, tectonic segmentation, and deformation pattern of the western Himalaya on Ma timescales inferred from landscape morphology, Lithosphere 10 (5) (2018) 632–640.

Noir, J., Jacques, E., Be'Kri, S., Adler, P.M., King, G.C.P, Fluid flow triggered migration of events in the 1989 Dobi earthquake sequence of Central Afar, Geophys. Res. Lett. 24 (1997) 2335–2338.

Patnaik, R., Chauhan, P.R., Rao, M.R., Blackwell, B.A.B., Skinner, A.R., Sahni, A., Chauhan, M.S., Khan, H.S., New geochronological, paleoclimatological, and archaeological data from the Narmada Valley hominin locality, Central India, J. Human Evol. 56 (2009) 114–133, doi:10.1016/j.jhevol.2008.08.023.

Pazzaglia, F.J., Gardner, T.W., Merritts, D., Bedrock fluvial incision and longitudinal profile development over geologic time scales determined by fluvial terraces, in: Geophysical Monograph-American Geophysical Union, Wohl, E., Tinkler, K. (Eds.), vol. 107, 1998, pp. 207–235.

Pérez-Peña, J.V., Azor, A, Azañón, J.M., Keller, E.A., Active tectonics in the Sierra Nevada (Betic Cordillera, SE Spain): insights from geomorphic indexes and drainage pattern analysis, Geomorphology 119 (2010) 74–87.

Pratt-Sitaula, B., Burbank, D.W., Heimsath, A., Ojha, T., Landscape disequilibrium on 1000–10,000-year scales Marsyandi River, Nepal, central Himalaya, Geomorphology 58 (1-4) (2004) 223–241.

Radhakrishna, B.P., Neogene uplift and geomorphic rejuvenation of the Indian Peninsula, Curr. Sci. 64 (1993) 787–793.

Radhakrishna, B.P., Ramakrishnan, M., Archaean-Proterozoic boundary in India, J. Geol. Soc. India 32 (4) (1988) 263–278.

Ramasamy, S.M., Panchanathan, S., Palanivelu, R., 1987. Pleistocene earth movements in Peninsular India—evidences from Landsat MSS and thematic mapper data. In: IGARSS'87-International Geoscience and Remote Sensing Symposium, 2, pp. 1157–1161.

Ray, Y., Srivastava, P., Widespread aggradation in the mountainous catchment of the Alaknanda–Ganga river system: timescales and implications to Hinterland–foreland relationships, Quat. Sci. Rev. 29 (17-18) (2010) 2238–2260.

Rehfeld, K., Marwan, N., Heitzig, J., Kurths, J., Comparison of correlation analysis techniques for irregularly sampled time series, Nonlinear Processes in Geophysics 18 (3) (2011) 389–404.

Resmi, M.R., Achyuthan, H., The north-east monsoon variations since the Holocene: inferred from the paleo and active channels of the Palar River, southern peninsular India, Holocene 28 (6) (2018a) 895–913.

Resmi, M.R., Achyuthan, H., Lower Palar river sediments, southern peninsular, India: geochemistry, source-area weathering, provenance and tectonic setting, J. Geol. Soc. India 92 (1) (2018b) 83–91.

Resmi, M.R., Achyuthan, H., Babeesh, C., Holocene evolution of the Palar river, southern Indian Peninsular India: tracking history of migration, Provenance, weathering and tectonics, Quaternary İnternational 575 (2020) 358–374, doi:10.1016/j. quaint.2020.09.010.

Resmi, M.R., Achyuthan, H., Jaiswal, M.K., Middle to late Holocene paleochannels and migration of the Palar river, Tamil nadu: implications of neotectonic activity, Quat. Int. 443 (2017a) 211–222.

Resmi, M.R., Achyuthan, H., Jaiswal, M.K., Holocene tectonic uplift using Geomorphometric parameters and GIS: Palar river basin, southern Peninsular India, Zeitsch. Geomorphol. 61 (3) (2017b) 243–265.

Roy, A., Kagami, H., Yoshida, M., Roy, A., Bandyopadhyay, B.K., Chattopadhyay, A., Khan, A.S., Huin, A.K., Pal, T., Rb–Sr and Sm–Nd dating of different metamorphic events from the Sausar Mobile Belt, central India: implications for Proterozoic crustal evolution, J. Asian Earth Sci. 26 (1) (2006) 61–76.

Salvany, J.M., Tilting neotectonics of the Guadiamar drainage basin, SW Spain, Earth Surf. Processes Landforms 29 (2) (2004) 145–160.

Sangode, S.J., Meshram, D.C., Kulkarni, Y.R., Gudadhe, S.S., Malpe, D.B, Herlekar, M.A., Neotectonic response of the Godavari and Kaddam rivers in Andhra Pradesh, India: implications to Quaternary reactivation of old fracture system, J. Geol. Soc. India 81 (2013) 459–471.

Scherler, D., Bookhagen, B., Wulf, H., Preusser, F., Strecker, M.R., Increased late Pleistocene erosion rates during fluvial aggradation in the Garhwal Himalaya, northern India, Earth Planet. Sci. Lett. 428 (2015) 255–266.

Schoenbohm, L.M., Whipple, K.X., Burchfiel, B.C., Chen, L., Geomorphic constraints on surface uplift, exhumation, and plateau growth in the Red River region, Yunnan Province, China, Geol. Soc. Am. Bull. 116 (7-8) (2004) 895–909.

Schumm, The Fluvial System, John Willey & Sons, London (1977) 338.

Singh, P., Rajamani, V., REE geochemistry of recent clastic sediments from the Kaveri floodplains, Southern India: implication to source area weathering and sedimentary processes, Geochim. Cosmochim. Acta 65 (2001) 3093–3108.

Sinha, R., Channel avulsion and floodplain structure in the Gandak-Kosi interfan, north Bihar plains, India, Zeitsch. Geomorphol. SB 103 (1996) 249–268.

Sridhar, A., Chamyal, L.S., Implications of palaeohydrological proxies on the late Holocene Indian Summer Monsoon variability, western India, Quat. Int. 479 (2018) 25–33.

Srivastava, P., Ray, Y., Phartiyal, B., Sharma, A., Late Pleistocene-Holocene morphosedimentary architecture, Spiti River, arid higher Himalaya, Int. J. Earth Sci. 102 (7) (2013) 1967–1984.

Starkel, L., Climatically controlled terraces in uplifting mountain areas, Quat. Sci. Rev. 22 (20) (2003) 2189–2198.

Subramanya, K.R., Active intraplate deformation in south India, Tectonophysics 262 (1996) 231–241.

Tandon, S.K., Sinha, R., Large rivers: geomorphology and management. in: Large Rivers: Geomorphology and Management, Gupta A (Ed). John Wiley and Sons, Chichester, England, 2007, pp. 7–28.

Tricart, J., Beaver, S.H., Derbyshire, E., Structural Geomorphology, Longman, London, 1974, p. 303.

Valdiya, K.S., Late Quaternary movements and landscape rejuvenation in southeastern Karnataka and adjoining Tamilnadu in southern Indian Shield, J. Geol. Soc. India 51 (1998) 139–166.

Valdiya, K.S., River response to continuing movements and the scarp development in central Sahyadri and adjoining coastal belt, J. Geol. Soc. India 57 (2001) 13–30.

Valdiya, K.S., Kotlia, B.S., Fluvial geomorphic evidence for Late Quaternary reactivation of a synclinally folded nappe in Kumaun Lesser Himalaya, J. Geol. Soc. India 58 (4) (2001) 303–317.

Waghmare, S.Y., Carlo, L., Geomagnetic investivation in the seismoactive area of Narmada-Son lineament, Central India, J. Indian Geophys. Union 12 (1) (2008) 1–10.

Wells, N.A., Dorr, J.A., Shifting of the Kosi River, northern India, Geology 15 (1987) 204–207.

Whipple, K., DiBiase, R., Crosby, B., Bedrock rivers, in: Shroder, J., Wohl, E. (Eds.), Treatise on Geomorphology, vol. 9, Academic Press, San Diego, CA, 2013, pp. 550–573.

Whipple, K.X., Tucker, G.E., Dynamics of the stream-power river incision model: implications for height limits of mountain ranges, landscape response timescales, and research needs, J. Geophys. Res. 104 (1999) 17661–17674, doi:10.1029/1999JB900120.

Zielke, O., Arrowsmith, J.R., Ludwig, L.G., Akçiz, S.O., Slip in the 1857 and earlier large earthquakes along the Carrizo Plain, San Andreas fault, Science 327 (5969) (2010) 1119–1122.

CHAPTER

19 The imprints of Holocene climate and environmental changes in the South Mahanadi Delta and the Chilika lagoon, Odisha, India—An overview

Siba Prasad Mishra, Kumar Chandra Sethi

Department of Civil Engineering, Centurion University of Technology and Management, Jatni, Bhubaneswar, Odisha, India

19.1 Introduction

The 11,800-years-old Holocene epoch (between Pleistocene and the present Anthropocene) comprises a vast terrestrial and marine sedimentary cover of the earth. The geologic and stratigraphic records divulge the history of human domination and the recent interglacial phase of the Quaternary (Lyell, 1833; Davis, 2011; Barnosky, 2013; Head, 2017; Fairbridge et al., 2018). Pleistocene geochronology and stratigraphy are often beyond the record. But the Holocene records and pieces of evidence are available at sedimentary terrains, archeological sites and also registered well in geomorphic patterns, regional sea-level changes (RSLR), anastomosis of paleochannels, lithologic and fossil records, flora/vegetation expansions, and faunal/avifaunal migrations.

On a geologic perspective, lagoons are fugacious hydrologic land-forms, and the ephemeral water body sustains for a few thousand years (Kjerfve et al., 1989; Borzenkova et al., 2015). Like many other tropical coasts, the coastal lands of the Indian subcontinent are also endowed with many lagoons and backwater bodies. A review of the existing literature reveals that although many studies are available on the lagoons of the west coast of India (Nair et al., 2010; Padmalal et al., 2014, and the references therein), systematic studies on the lagoons in the east coast of India are scarce. Therefore, an attempt has been made here to examine the Holocene climate and environmental changes of the south Mahanadi delta (SMD) and the Chilika lagoon in the east coast of India as a case study site. The SMD extends from the Kuakhai river's near Bhubaneswar up to the Chilika lagoon. It lies between the bifurcation point of the river at Uttara and the Eastern Ghats Belt Hills. The SMD rivers (the Daya and the Bhargovi) contribute on average 61% of the total inland flow through the lagoon, whereas 70–80% of the total sediment flow through the inland drainage channels. The lagoon and the Mahanadi delta area enjoy a tropical climate, and burgeoning demography.

Holocene Climate Change and Environment. DOI: https://doi.org/10.1016/B978-0-323-90085-0.00015-2

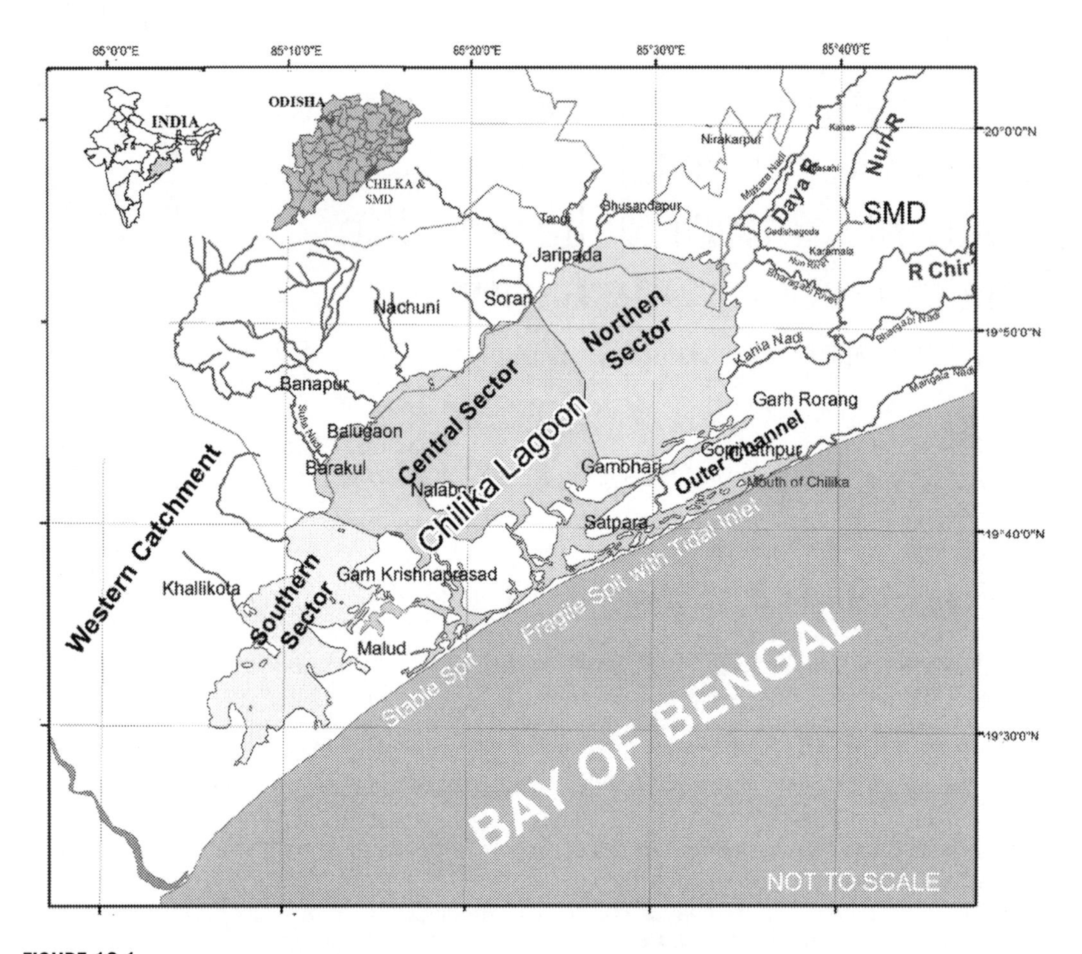

FIGURE 19.1

The map of the Chilika lagoon and the adjoining South Mahanadi Delta (SMD).

19.1.1 The geographical and geomorphological setting of the study area

The area of focus of the present study is the Chilika lagoon (19.84°N, 85.47°E) which spreads adjacent to the SMD. The shore line of the area is curved and extends for a length of 64.3 km. The area is bounded on the south east by the Bay of Bengal (BoB) and west by the northern discontinuous hillocks of the Eastern Ghats belts (EGB). The hillocks intermittently occur/emerge the lake body or river course and act like spurs to accrue sediment at their toe. The Daya and Bhargovi, streams join the western swamps of the Chilika lagoon (Fig. 19.1).

The total areal spread of the Chilika lagoon was of 1500 km^2 (including water body, sandy barrier spits, NW swamps) during the early Holocene, which was later reduced to about 1000 km^2. The Mangalajodi area, which is a swamp and a bird sanctuary at present was within the water spread of the Lagoon once (Fig. 19.1). The geomorphological features along with probable physiography of the study area are depicted in Table 19.1.

The present Kuakhai drainage system in the SMD and its paleochannel anastomosis indicate many abandoned meandering of channels, intermediary transitions, dendritic channel form and lateral channel formation at river mouths and sand bars due to regional sea level rise (RSLR) in BoB. The configurational shifts and the morphological changes have been reconstructed along paleocoast lines since mid-Holocene. The rivers Sunamunhi, the Ratnachira are gradually becoming defunct with evolution of the Rajua and the Makara rivers along with the Gangua systems.

19.1.2 Previous work

Proxy information on the Holocene climate and environmental changes can be deduced from the sedimentary archives of the Chilika lagoon and the SMD (Fairbridge et al., 1977; Van-Heteren et al., 2000; Dougherty, 2014; Costas et al., 2016; https://www.ncdc.noaa.gov/global-warming/mid-holocene-warm-period). The loss of mangroves signifies the weakening of the summer monsoon during 3.5–1.5 KYBP and later there was an increase in the trend due to strengthened summer monsoon in India (Limaye et al., 2012; Monalisha, 2018). Post-Holocene temperature reconstruction was made by different authors in Northern India and reported that warming from 900 YBP to 500 YBP, cooling from 500 to 100 YBP and later the present warm period attributed due to rapid deglaciation in Himalayan region (Borgaonkar et al., 2009; Kothawale et al., 2016; Sanjay et al., 2020). The carbon dating records divulge a humid period during LIA (650–100 YBP) with centennial-scale wet events (two) that could be matched with the medieval warm period (MWP; 1050–650 YBP; Yang et al., 2009).

Paleoclimatic studies reveal that there was an abrupt shift of intertropical convergence zone (ITCZ) and change in Indian summer monsoon during the Holocene epoch (Hahn et al., 1976; Gupta et al., 2003; Ponton et al., 2012; Srivastav et al., 2020). There was geospatial and temporal close relationship existed between the westerly Jet stream and Indian summer monsoon (ISM; Schiemann et al., 2009; Cheng et al., 2016; Kathayat et al., 2017; Banerji, 2020). Many uncertainties in climate that varied during the Holocene periods in different zones of India, and also in the mid-latitudes and Central Asia were reported (An et al., 2000; Liet al., 2012; Dimiri, et al., 2016; Dixit et al., 2016).

The North East Arabian Sea is the epicenter of the Indian winter monsoon (IWM), and when ISM overlaps, it is considered to be in antiphased (Boll et al., 2014; Nicole et al., 2019; Stefan et al., 2020). The ISM (strengthening or weakening) is dependent on its activity based on the positioning of ITCZ (Northerly shift active ISM and vice versa) in Northern Hemisphere in periods of strong insolation, which was encountered during the mid-Holocene in India. In addition to the overriding and long-term orbital influences, forcing from solar variability, El Niño Southern Oscillation (ENSO), and North Atlantic Oscillation (NAO) have been suggested to be responsible for centennial to millennial monsoon strength variability (Wang et al., 2005; Boll et al., 2014).

19.2 Scope of the study

The present study deals with the Holocene climate changes that have transpired with the Chilika lagoon and its northwesterly neighbor, the SMD. The study includes the paleoimprints, temperature variations, spatial positioning of the ITCZ, shift of Indian summer monsoon, alterations in coastal hydrodynamics, regional sea-level fluctuations, ecological vicissitudes, and geomorphologic changes in the study area.

Table 19.1 The geomorphological features of the Chilika Lagoon and its chronologic bearing.

Morphological feature	Landforms	Name of places/stretch of the area	Process of formation	Probable chronology of the event (YBP)
North east corner				
Fragile barrier spit (outer channel) OC-1	Sandy with tidal inlets, Isles in OC	Sipakuda to Motto (≈32–34 km)	Waves, tides, littoral sediment drift	500 YBP to present
Landward side barrier spit West bank OC-1	Settlements in sands, mangroves, and parallel dunes near Arakhakuda	12.1 km, Satapada to Arakhakuda	Waves, tides, littoral drift, human activities	≈1850–1300 YBP
OC-2,OC-1, marshes sand dunes, Dahikhia river	Mangrove, marshes, flats, settlement on sand dunes and d/c	≈5 km, Arakhakuda to Panash Pada	Waves, tides, littoral drift, anthropogenic activities	≈2200–2100 YBP
Marine shelf and terraces near spit	shelf-breaks, shallow old/new mud flats	≈0.1–3 km width; Many patches along beach	Marine/fluvial activity	≈8000–6000 YBP
Southern side				
Beach and stable barrier spit	Frontal beach, lateral channels, back dunes, settlements, vegetation	≈32 km; Prayagi to Sipakuda width 150 m to 15 km (Palur canal)	Aeolian activity, waves and tides	≈2200 YBP
Lagoon				
Hillocks, islands and stretched plains, the triple point sand dunes	Islands with hillocks, habitations, gherri bundh, marshy islands	≈223 km^2, ≈203 Nos Northern Southern, Central, and OC	Pre-Cambrian, Aeolian activity, waves, fluvial, tidal	Hillocks – Pre-Cambrian; sandy isles: 4200 to recent
Brackish water body	Fresh, brackish and saline	Lake body varying ≈750 to 1050 km^2; floor of lake muddy sandy, rocky; volume ≈4 km^3	Inland fresh water, waves, tides, storm surges	OC: 4–5 days; N-sector -132 days; lake floor: 4300 to present
Swamps of the Daya and the Bhargovi	Substratum sand, and silt, mud and organic matter, channels and water bodies	Badagotha–Satapada–Chupuringi (JaggnathSadak)–Mangalajodi	Flood sediment recent from flood with organic matter; anthropogenic stress	≈1700 YBP to recent
Lake periphery	Mud flats and delta, marshes, channels, hillocks	Mangala Jodi–Rambha–Prayagi	Inland flow, anthropogenic stress	550 YBP to present

Source: Arya and Lakhotia, 2006; Khandelwal et al., 2008; Mohanty et al., 2016.

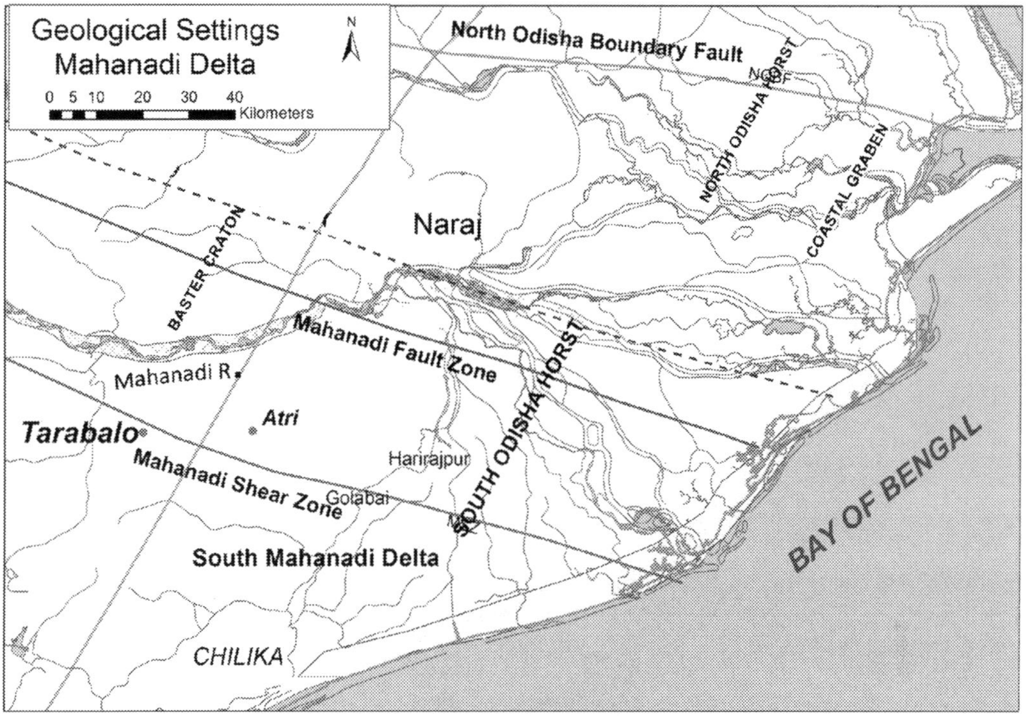

FIGURE 19.2

Geological setting of the Mahanadi Delta showing major fault zones.

The study also comprises the physiognomy, water resources, landform changes of the area, and the meteorological extremes that both the lagoon and the delta have encountered. Since the literature and history cover for a period of about 1000–1500 YBP in details, only a vague picture has been constructed based on the proxies during the Anthropocene period. The present study is a compromise between the available segregated pieces of literature, anthropogenic excavations and historical findings to reconstruct the climate and environmental changes of the study area.

19.3 Methodology

The aim of the present study is to construct a data base on climate changes based on proxies (environmental reconstruction, chronology, monitoring, and construction of models) to elucidate the ecological changes in the Chilika lagoon and the SMD catchments during the Holocene. The Holocene climate and its effect on areal changes in the lagoon need to be searched for its origin, mechanism, and causes of health threat. The panoramic brackish water lagoon and its Holocene climatic vicissitudes during different stages from its formation, LGM period, medieval climatic maximum, LIA, and present age require detailed study. However, for evaluating the present changes, the available survey of India topo sheets (1930) have been georeferenced and compared with the latest LU/LC with the latest available map of 2015. The geological map of the study area is depicted in Fig. 19.2.

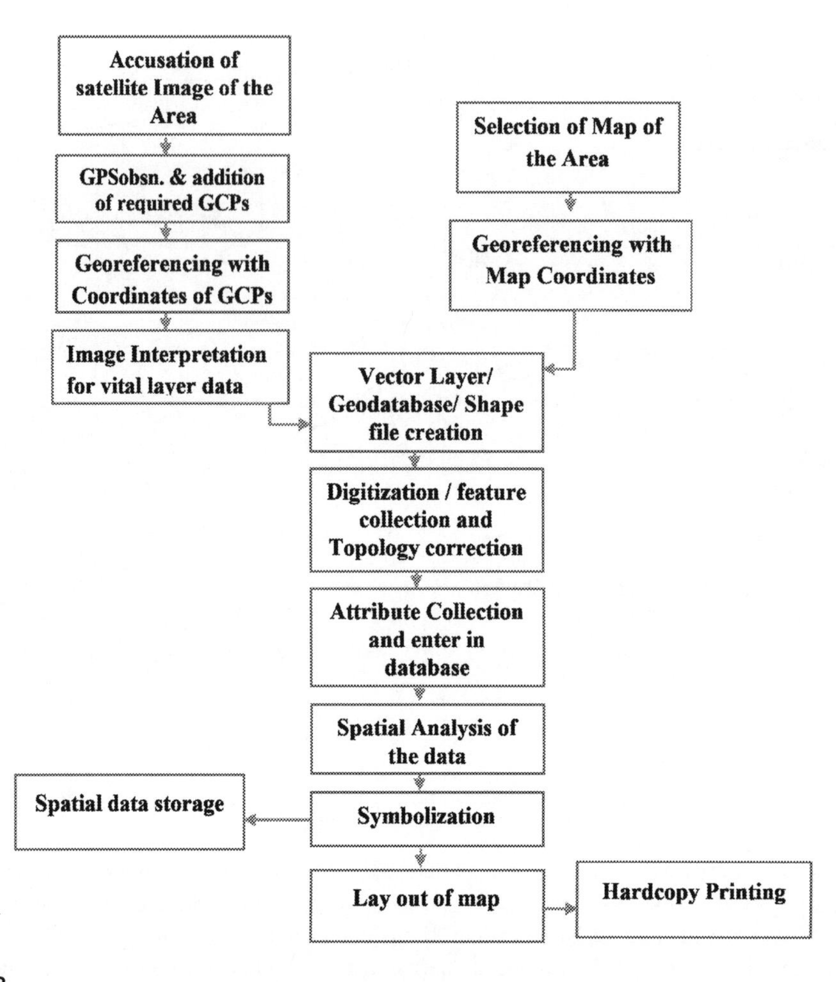

FIGURE 19.3

Flow chart showing the methodology of map processing with attributes.

Some of the paleochannels in the north western corner have been identified by systematic field works using base maps prepared from topographic maps and imageries. Radio carbon dates and lithological characteristics collected from published works by different researchers have been used extensively in the study. Those defunct coastal paleochannels are the Sunamunhi river, the Malini river (Puri), and the Ratnachira river (left loop channel), conversion of the Sar and the Samang inland lakes to human utilities, the East Kania, Bali Nai, and the Musha Nai in the SMD. The present study uses topo map 1930 and 2018–2019. The digitization of the survey of India topo maps were done by the software Micro-station. Later the digitized map was imported to ARC GIS software and figures were prepared. The figures were achieved to proper quality by using 3D paint and Photoshop.

Except few topo maps, no other documentary records are available for the study during Holocene periods. To exhibit the geomorphologic changes between Holocene and the Anthropocene period the

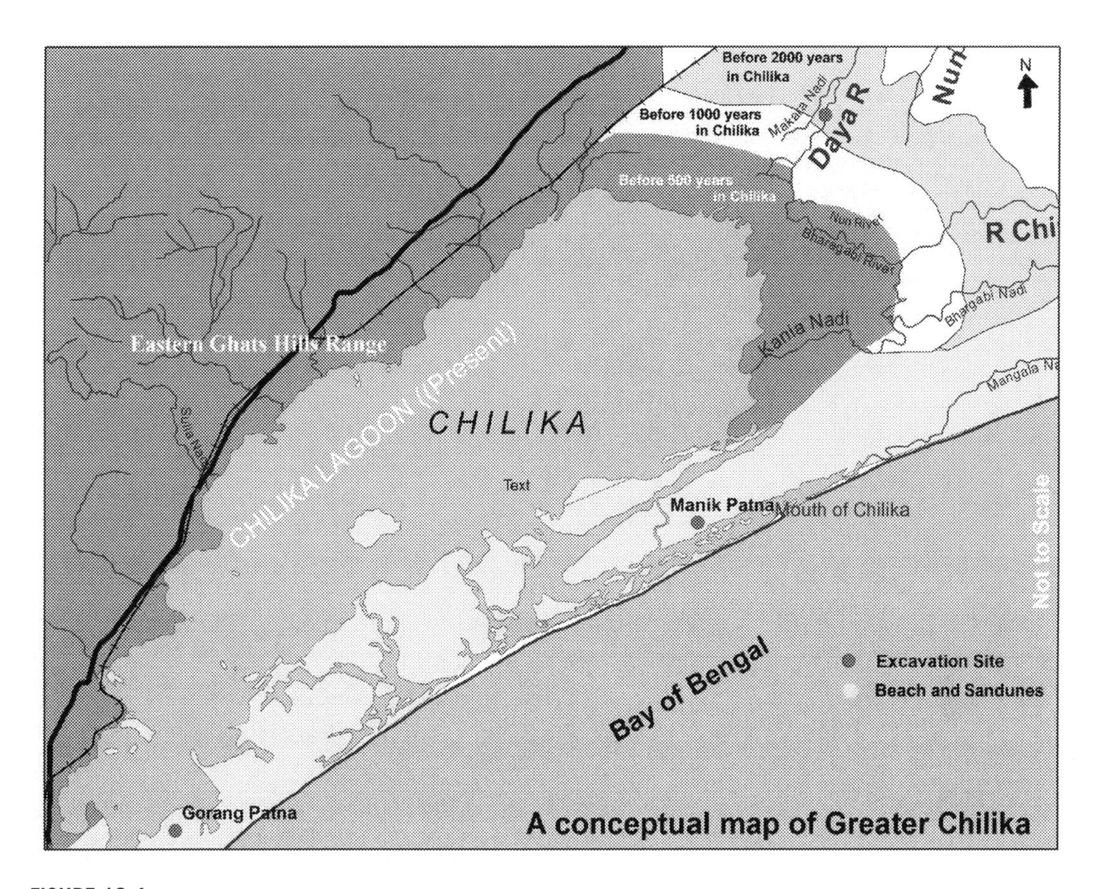

FIGURE 19.4

Late Holocene changes of the Greater Chilika with exploration sites.

proxy records can be found from old topographic maps. For present records, the attributes available from the satellite imageries can be easily found. The two procedures applied in the present study are either from hard copy map to feature class or from satellite imagery maps. The methods are given in Fig. 19.3.

For current analysis of satellite imagery to attribute maps, the processes involved are download /collect satellite Imageries, selection of ground control points (GCPs), geographic positioning system, observation and computation of GCPs, geo-referencing of imagery with GCPs and creation of layers and further steps like previous map generation.

The Chilika lagoon has reduced its lacustrine area, depth, and salinity due to variation in flow exchanges. The greater Chilika of 1500 km^2 has reduced in area to 700 km^2 during summer (Sadakata et al., 1997). The geomorphology changes in the late-Holocene and the present have been compared and depicted in Fig. 19.4.

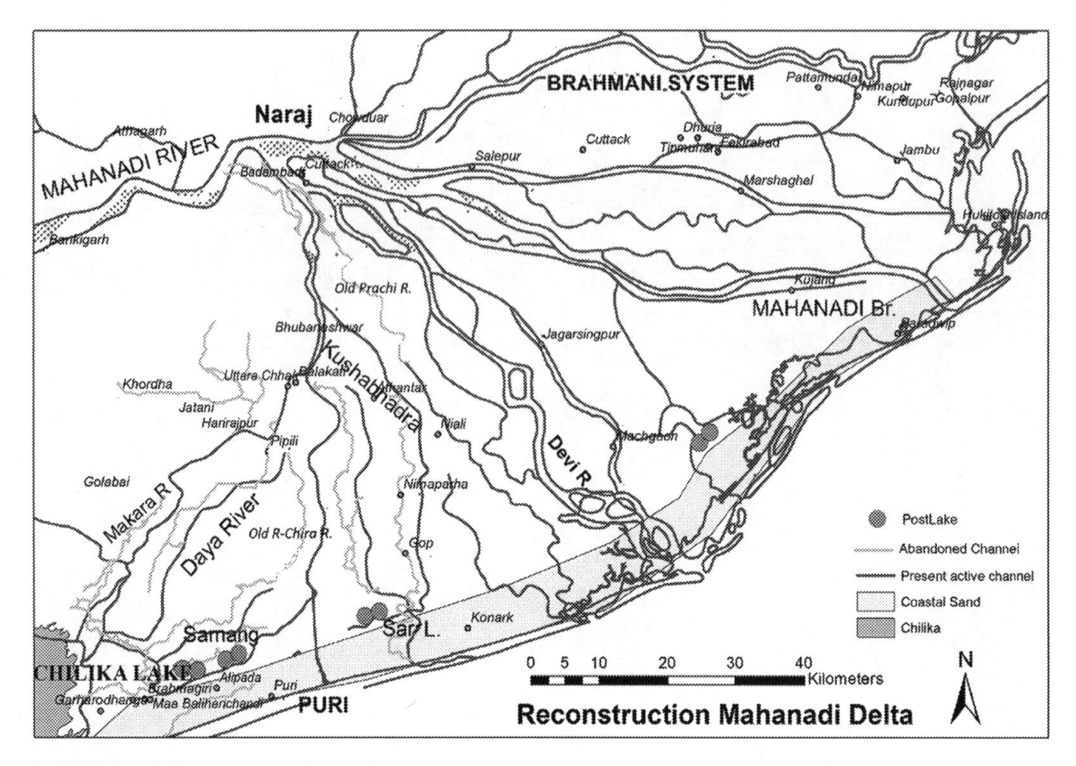

FIGURE 19.5

Reconstruction of Mahanadi delta during Holocene epoch (Source: Modified after Mahalik et al., 1996).

19.4 Observation and results

19.4.1 Geoenvironmental setting of the study area

19.4.1.1 Imprints of Holocene paleodrainage in SMD

The SMD is characterized by distributary channels of varied age categories which are evolved during various phases of Holocene epoch. They are: (1) buried channels (Old Kathajodi phase), (2) dead channels (Sukhabhadra and Burdha phase), (3) Defunct channels (Prachi and Alaka phase), and (4) present active channels (Kuakhai–Devi–Birupa). The present Kuakhai drainage system in the SMD and their paleochannel anastomosis indicate many inactive meandered channels, intermediary transitions, dendritic channel form, and lateral channel formation at river mouths, sand bars formed due to RSLR in BoB. The configurational shifts and the morphological changes have been reconstructed along paleocoastlines from mid Holocene. The old systems were Old Kathajodi–Burdha–Alaka–Prachi and Ratnachira system. Present system has been reconstructed as Birupa–Mahanadi–Kathajodi–Devi–Kuakhai system, where the SMD comprises (Kuakhai–Daya–Bhargovi–Kushabhadra) system (Mahalik et al., 1996; Mishra and Dwibedy, 2015; Fig. 19.5).

19.4.2 Holocene temperature changes

Three main temperature anomalies were transpired during pre-, mid-, and late-Holocene periods. The periods during Holocene epoch were from 8.2 KYBP cooling events, 4.2 KYBP drying events, and 2.5 KYBP warm and dry events (Achyuthan et al., 2014; Banerjee et al., 2020). From Pollen studies of the offshore region of Mahanadi basin, it was inferred that there was high insolation with minimum precession, maximum obliquity, and wettest period during 11.3–4.2 KYBP and low insolation with maximum/minimum precessions during post-Holocene period from 4.2 KYBP to present with relatively dry period. Climatic anomaly was recorded during little ice age (LIA) with more humidity and lowering of temperature, https://darchive.mblwhoilibrary.org/bitstream/handle/1912/7606/ (Fig. 19.6). There was sluggish rainfall in the Mahanadi Basin and arid climate existed during 3.5 KYBP (Naidu, 1999). Probably this is the period when there was rapid growth of barrier spits from southern fringe of the Chilika lagoon. During this fag end of the Holocene period the Mahanadi delta has prograded by maximum growth rate (Hazara et al., 2020), in its northwest and southern fringes (Mishra et al., 2020).

19.4.2.1 ITCZ effect during Holocene

The ITCZ is the band of troughs of low-pressure area (LOPAR) that exists parallel to equator where the NE and the SE trade winds converge, but oscillate northerly or southerly with respect to the position of sun. In India, during summer, the ITCZ positions over Gangetic plains forming cumulonimbus clouds and have thunderstorms and spells of showers.

The little Ice age was from $\approx$1850 to 1350 YBP covering the Moghul period in India as per the paleo proxy, carbon dating, lithology, and limnology records. The climate change was due to shift of the manifestation of ITCZ from west to east in the Northern Hemisphere and dry period in the Saharan desert and the African east coast. The position of the ITCZ, had changed the frequency and amplitude of the Indian Ocean Dipole (IOD), Madden-Julian Oscillation (MJO) and ENSO activities in the north Indian Ocean. SWM (SW monsoon) had rapid changes with declining flood flow having frequencies 650–450 YBP, 1100–1000 YBP, and 2200–1800 YBP during the late-Holocene (Gupta, 2003; Wanner et al., 2008; Chauhan et al., 2010; Mishra, et al., 2020). The annual shift of Indian Summer Monsoon in the Indian subcontinent during Holocene is depicted in Fig. 19.7.

19.4.2.2 The shift of Indian summer monsoon

Three phases of wet climate followed by three segments of dry conditions prevailed over India during Holocene epoch. The period from 11.7 to 9.1 KYBP, 8.1 to 6.5 KYBP, and 6.35 to 5.00 KYBP were wet periods, and the dry periods were noticed between 9.00 and 8.10 KYBP, 6.650 and 6.350 KYBP, and ~5.00 and 4.00 KYBP. The rest of the periods were of normal years. During mid-Holocene, it was observed that there were migrations of monsoon activities from north Indian Ocean as there was shift in ITCZ from west to east. The civilizations those were built up in west (Mohenjo-Daro, Harappa, etc.) and very large water bodies like Aral Sea have become dried up and gradually desertification started in such areas.

Ali et al. (2018) reported about five positive and three negative shifts of ITCZ. The positive shifts were noticed since early Holocene. The first was observed toward end of Younger Dryas (~11.7–11.4 KYBP), stable ISM between ~11 and 6 KYBP, and declining ISM between 6 and 3 KYBP. During MWP and LIA periods, there was strong monsoon in Indian subcontinent. Still there were regional

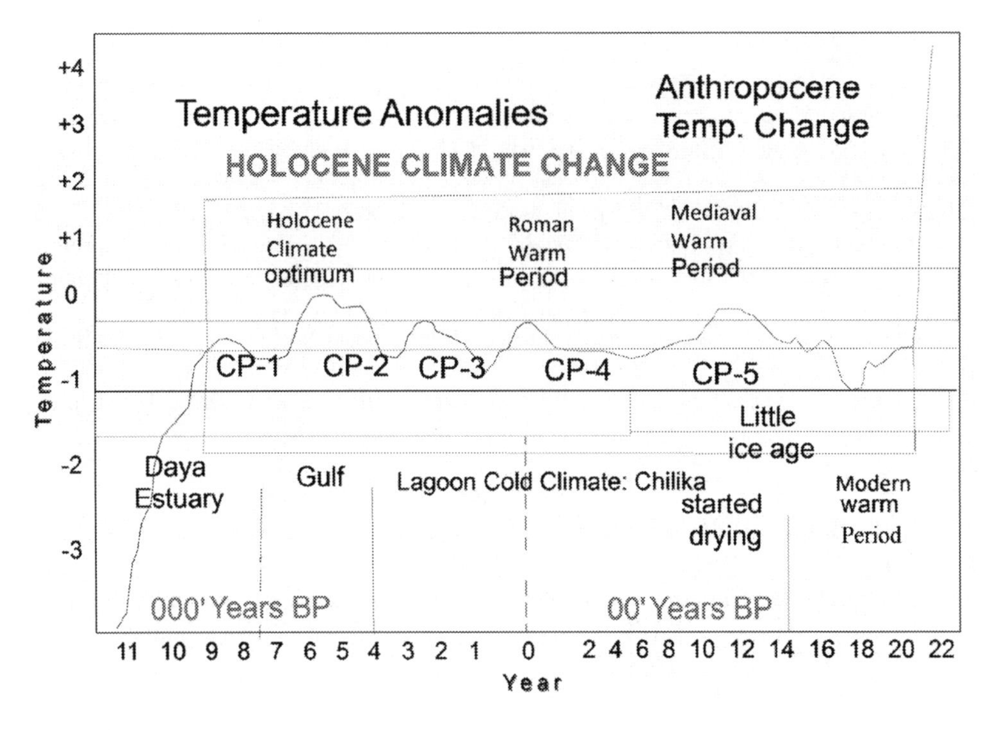

FIGURE 19.6

The Holocene temperature changes in Chilika converting the gulf to a lagoon.

vicissitudes in ISM which were coupled with southward fluctuation in mean position of the ITCZ. During the LIA there was two low solar irradiance period observed (90 years each). During the coldest years Spörer Minimum (1450–1540) and similarly the Maunder Minimum (1645–1715), the Chilika barrier spits were naturally constructed as there was sea level retrogression. The live glorious ports in coasts of the lagoon became dead or dormant.

The Holocene monsoon intensification transpired at 9.5 and 9.1 KYBP. Then monsoon debilitated gradually up to 7.0 KYBP in India. Later the weak monsoon during post- and mid-Holocene periods was ~8.0, 7.0, 5.5 and 3.5 KYBP. There were significant periodicities of summer monsoon during the latter part of Holocene (2.2, 1.6, 1.35, 0.95, 0.75, 0.47, 0.32, 0.22, 0.156, 0.126, 0.113, 0.104, and 0.092 KYBP (Chauhan et al., 2010; Thamban et al., 2007).

19.4.2.3 ***Holocene coastal dynamics***

The coasts are dynamic landforms with special fragile ecosystem that chances geospatially corresponding to change in oceanography, river mouths, littoral sands, and natural extremes, (Kumaran et al., 2012). The regional sea level changes were observed above ≈31.0 m during ≈10,965 YBP (early Holocene) and depleted fast between 8000 and 6000 YBP by ≈−8.7 m. It was due to fast melting of continental ice sheet from late glacial to pre-Holocene (Sparrenbom et al., 2006; Gawali, 2019).

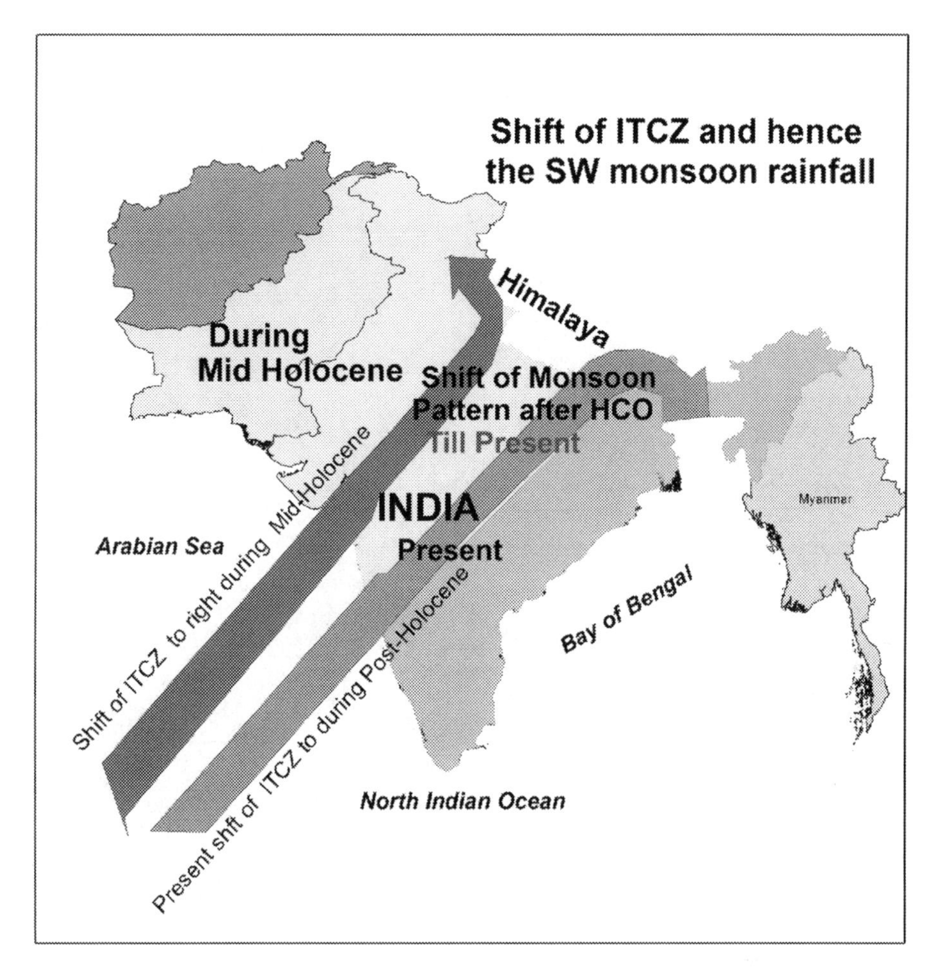

FIGURE 19.7

Shift of ITCZ from west to east (Source: https://scied.ucar.edu/docs/why-monsoons.happen).

Proxies from the Greenland Ice-Sheet Project (GISP2) ice core had exhibited four fast transitions in climate that emerged from 50 KYBP till the beginning of 11.8 KYBP. During this period there were slower rise of methane 200–300 ppb for 200–300 years. This slow rise would have brought swift climate changes in the terrestrial biome by decomposition of methane hydrate in sediments (Brook et al., 2000).

There was continuous change of shoreline along east coast of India being governed by the RSLR, inland sediment flow, littoral drift, climatic extremes and sun-earth geometry. The coastal shore line of the Chilika lagoon (≈64–65 km), the SMD, and Puri coast up to Rama Chandi (≈55 km) are on a constant change either inland or seaward depending upon changes in sea level. The prominent changes in shoreline have been noticed in the entire Mahanadi delta. These changes are well preserved as strand lines in the area (Fig. 19.8).

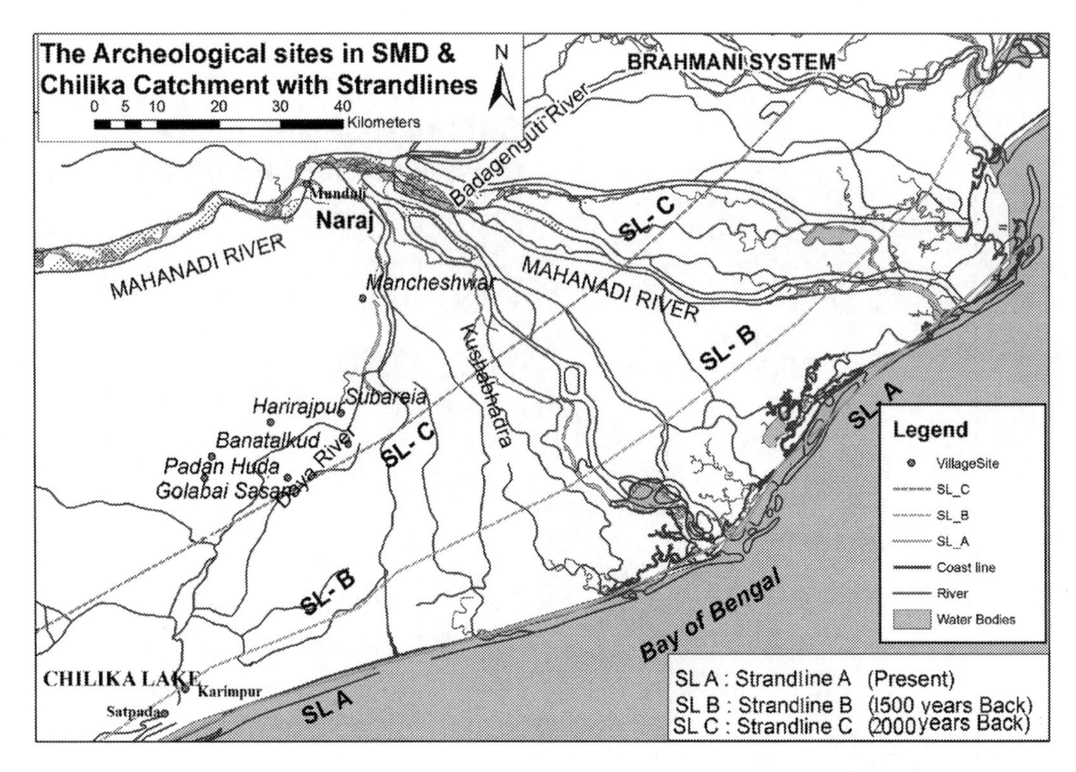

FIGURE 19.8

The archeological sites and strand line based on sand ridges in SMD and Chilika.

The RSLR in the SMD and the Chilika basin are represented in five strand lines (A to E) with five regression phases leading over four transgression phases of amplitude (up to 4 m) from mid Holocene to present. The present coastal landform was submerged under BoB as evidenced by the presence of many inland imprints of paleosignatures in the inland of SMD. The fluctuations in the MSL during 7300–2500 YBP have made conducive environment for the formation of the Chilika lagoon. Subsequently, the scenario of the Chilika lagoon and the associated SMD has been drastically changed in the recent past. The hydrological changes in in1930 and 2018 have been demonstrated in the form of a cartoon for a comparative assessment (Fig. 19.9).

19.4.2.4 Regional sea level changes

Loveson and Nigam (2019) have reported that the regional sea level fluctuations were with moderate rate of sea level rising @ $\approx$ 0.88 m/century during early Holocene. There was sharp upwelling of RSLR during 10.8–9.2 KYBP. Later a steep surge in RSLR, was observed @ 2.06 m/100 years accompanied by an intermittent sluggish RSLR phase ($\approx$1100 years) of $\approx$0.82 m/100 years from 9.2 KYBP to 8.1 KYBP. Then with progress of a warm period there was abrupt escalation of RSLR @ $\approx$ 2.22 m/100 years during 8.1K to 7.2, KYBP. From 7.2 KYBP to present five dominant RSLR regression phases are identified over transgression phases in the Mahanadi delta. The climate change history tells about three

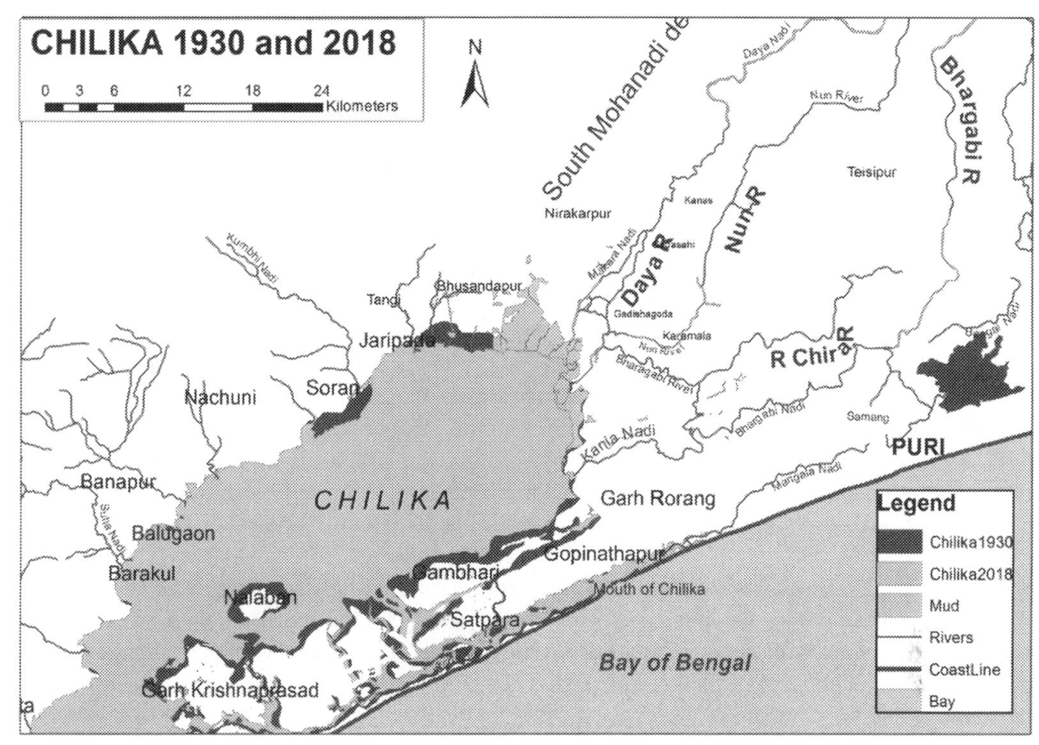

Fig 9: Post Holocene changes in the hydrology of Chilika Lagoon and south Mahanadi delta

FIGURE 19.9

Late Holocene changes in the hydrology of Chilika lagoon and south Mahanadi delta.

segments of climatic fluxes in the Mahanadi delta, that is, during LIA (temporal), during temperature drifts, that is, 11.0–8.2 KYBP (warm, dry and stable climate) with 1.0–3.5°C above mid Holocene levels (8.0–4.2 KYBP). Later during post-Holocene period the temperature was of decreasing trend but with climatic uncertainty, which is continuing till date. Two mid-Holocene high stands of height 3–4 m (regional sea level) succeeded by 2 m fall and further rise of 3 m MSL during early stages of late Holocene have been marked by Banerjee et al. (2000) and Woodroffe et al. (2005).

Parallel to the coastal front along Puri coast, three lagoons have identified in the estuary of the rivers, the Daya, the Bhargovi and the Kushabhadra during late Holocene. They are the Chilika lagoon, the Samang Lake (Now dead) and the Sar Lake (Now a swamp) in the south Mahanadi delta (Fig. 19.5). During pre-Holocene they were estuaries. The inland sand ridges are identified in the region. The connecting lines are drawn as strand lines (Fig. 19.8). The penultimate strandline (SL-B) passes through Magarmunha lagoon outlet, the inland outer channel, the apex of estuarine lakes of the river Bhargovi and Kushabhadra, respectively. The inland lakes which were actively controlling the river systems gradually got silted up during last 100 years and they have become agricultural land, settlements, and swamps during the latter stages.

19.4.3 The Chilika lagoon and the south Mahanadi delta

19.4.3.1 Holocene ecology changes in SMD and the lagoon

The terrestrial flora and fauna, aqua fauna, and avifaunal fauna did not exhibit any transformations during the relatively short-lived Holocene. The diversification, extinction and distribution of flora, fauna and many invasive alien species have been observed in the study area during Holocene epoch, http://www.geologypage.com/2014/05/holocene-poch.html#ixzz6bHw7iPkx. With Holocene regional sea level fluctuations, there were abrupt changes in estuarine ecology. Erosion and accretion along coastal beach of Chilika lagoon has been a regular phenomenon since its formative stages. The span from the apex of the delta to coast can be fluvial, fluviomarine, and marine, with its own ecosystem different from the other. Studies of proxies disclose the wetland ecological variations with time and climate along with increasing human accomplishments from mid-Holocene onward. Distraction of ecosystem during Holocene started with demographic growth in the last few tens of centauries; however the system recouped back when the population growth became mild (Cui et al., 2017; Dixit et al., 2018; Laug et al., 2020). The SMD system comprises of 90–50 km as fluvial, 50–20 km as fluviomarine, and 20 km to coast dominated by marine ecosystem. Climatic disparities during the Holocene reveals in the variation of the lake levels and vegetation. There was formation of multifaceted hydrologic anastomosed drainage network belated to the MWP and LIA in the late Holocene. The lagoon, lakes, swamps, and the islands have their own ecological suites. Vegetation variations has been reported with fluctuation in monsoon rainfall during the entire Holocene in the SMD and the transformation processes from estuarine environment to marine lacustrine lagoon had physiognomic changes in flora, fauna, aquafauna, and avifauna availability (Borzenkova et al., 2015).

19.4.3.2 The archeological standings in SMD and Chilika

The Neolithic and Chalcolithic evidences (Bronze age) are found from Daya valley and Chilika catchment area. The areas are Golabai Sasan, Manikpatna (west coast of Chilika), Sisupalgarh (Bhubaneswar), and Liakhia (Kushabhadra river mouth), Harirajpur (near Jatni) and in total 44 sites have been excavated, out of which 13 sites are in the western bank of Chilika in Khordha district, nine in the Puri district in the NW sector of Chilika lagoon, one site in the Cuttack district and the rest in the southern coast of Chilika. The findings from the excavation sites are colored wares, Terracotta objects, boat building materials, glass wares and marine remnants of Chalcolithic period, indicating urban development and maritime activities along the coasts of Chilika (Tripathi, 2000; Behera, 2013; Pattanaik et al., 2014). The archeological excavations in SMD link to Sua Barei, Pipili (20°09′20.6″N lat, 85°49′42.2″E long), Banga-Harirajapur-Jatni (20°08′33.82″N lat; 85°43′37.42″E long), Banatala Kuda, Khurdha (200 05′43″N lat; 850 38′25″E long), Padan Huda, Delang (20°03′34.37″N lat; 85°33′56.80"E long.), Golbai Sasan, Khurdha (20°01′45″N lat, 85 33.22"E long.), and Bardhya Kuda, Chilika (19°26′13.58″ lat; 85°07′09.53" E long). These are the prominent places where human imprints have been traced out (Rao et al., 1995; Sinha, 2000; Mishra and Bihari, 2019). The archeological places are shown in Fig. 19.8.

19.4.3.3 The physiognomy of the lagoon

The Chilika area comprises of the EGB Hills range and parts of the Daya-Bhargovi delta (Doab VII and Doab VIII). The area comprises of extensive fresh water swamps, fresh water lacustrine area, Tidal flats, brackish water marshes, islands, temporal isles, huge brackish water body, mangroves, transitional vegetation, barrier islands, weed infested areas, spits, and dunes of fragile fresh-brackish-marine water geo-bio system. The lagoon has five biophysical zones/regimes viz: brackish water area, lagoon fringe shallow waters, swamps in the river outlets, tidal flats, sand dunes, islands, and marine areas between lagoon and BoB (Paul et al., 2014).

Reconstruction of the east flowing rivers in SMD has been done based on the ancient morphological features like drainage channels, strandlines, beach ridges and abandoned channels, monuments, and antiquities (Babu, 1978; Bharali, 1991; Mahalik et al., 1996; Mishra et al., 2019). The rebuilding geomorphology of the SMD along with Chilika lagoon has been studied by many researchers (Mishra and Jena, 2015; Hazara, 2020). The study shows inactive and active drainage channels in the area (Figs. 19.4 and 19.8).

19.4.3.4 Morphological changes of Chilika lagoon

The Chilika lagoon (the largest brackish water body in Asia) has gone through multiple manifestations from estuary to gulf then to a lagoon. Later the characteristics of the leaky lagoon has transformed to either restricted or choked after the development of barrier spit. The lagoon was an estuary during last part of Pleistocene and later became a part of BoB during early Holocene. In the Holocene climatic optimum (from 7000 to 5000 YBP), the estuary might have transformed to a gulf due to prominent sea level changes. During mid-Holocene warm periods, development of barrier spits from south might have occurred, which gradually separated the sea and the gulf. Later during late Holocene (3750 $\pm$ 200 years) the bay might have transformed to a lagoon. The navigation history tells that the lagoon was young and active till 500 YBP. Later like the other lagoons along the east coast, a greater part of the Chilika had faced the problem of shrinking, sinking, and reduced its size from early 15th century onward.

19.4.3.5 The water spread of Chilika lagoon

The Greater Chilika had a water spread of 1500 km^2. Its western coast was close to the EGB Hills. So, there was less shrinkage from that direction. The diminution of the lagoon was attributed to extension of barrier spit, retreating beach, sediment accretion, and extension of SMD encroaching from northwestern direction. The deterioration started because of the development of a huge swamp/marsh land extending up to 20–25 km inland. The shrinkage could be linked with climate change like recurrence cyclones, erratic spells of rain, depleting salinity, rising temperature, decreasing mangroves, and many more (Sadakata et al., 1997; Nayak, 2014; Rao et al., 2020). The average water spread area of the Chilika lagoon was 1165 km^2 during floods and 906 km^2 during summer (Andrews, 1846).

The Chilika system has 52 numbers major inflowing rivers, rivulets, and drainage channels to the lagoon. The SMD has five major drains (the Bhargovi, the Ratnachira, the Nuna, the Daya, and the Makara) debouching in to the lagoon. About 61% of its total flood flow to the lagoon is contributed from the Mahanadi system. The flushing flow for the lagoon required from the Mahanadi system is about 2830 cumec in Kuakhai river (main distributary in SMD) to restrict sedimentation in the lagoon. The necessity of flushing flow for the Chilika lagoon can flush the sediment and the aquatic plants to the BoB via tidal inlets when the flow was above 2830 cumec in the Kuakhai system.

19.4.4 Hydrology and geomorphology of SMD

19.4.4.1 Hydrologic system

Anastomosis of the Mahanadi drainage can be divided into two systems—the Mahanadi and the Kathajodi. The interfluves of the Kathajodi have two subsystems the Kuakhai and the Devi. The southernmost, the Kuakhai river could be divided in to three sub–sub systems, the Kushabhadra, the Bhargovi, and the Daya. The SMD is the doab VII. The deltaic land is located essentially between the two rivers the Bhargovi and the Daya with two major drainage systems/rivulets—the Ratnachira and the Nuna. The extension part of Daya is the doab VIII and is the land between the Daya and a combination of Rajua and the Makara river systems. During summer, the rivers are almost dry out; however, during 1970 onward scanty seepage from irrigation water enters the lagoon from SMD.

19.4.4.2 Droughts in Odisha during Anthropocene

History reports that there was abrupt variability in the SW monsoon rainfall in SMD leading to droughts, floods causing social delirium. It is also observed that high flood years are also drought years in the lower Mahanadi basin. Mythology also depicts about severe droughts in the last few centuries (Cook et al., 2010; Sinha et al., 2011). Toward 18th century, considering the poor collection of revenue from Odisha, the British engineers inferred that the open flood plains without embankments had caused havoc and responsible for both drought and flood. The other cause was regarding the intensity (reached its minimum) of Northern Hemisphere summer insolation in the late Holocene.

El Niño frequency/intensity also increased in late Holocene (Moy et al., 2002). Between 1846 to1932, there was marked land use changes made by the people in India. Due to continuous floods (90 spells from low to very high) and droughts (24 spells from mild to very severe) followed by severe attack of famines (three spells in 1864–1866, 1869, and 1874) in the SMD during 1855–1945 (Mahalanobis, 1940). But during the period from 1803 to 1864 there were 12 famine/crop loss years, that is, 1803, 1806, 1807, 1809, 1817, 1828, 1834, 1836 and 1837, 1839, 1840, 1863, with climatic imposition that a famine follows a flood. During the period 1751–1803 there were three major famines in 1770 (great famine), 1780 and 1792 taking lives of one million people, mostly from Odisha (Griffiths, 1952; https://shodhganga.inflibnet.ac.in/bitstream/10603/189922/10/10_chapter).

The rivers are not embanked prior to 1850 except natural levees and all the flood plains are deposited with nutrient-rich silts during a flood; and as a result, the area was under fertile regime all over the Mahanadi delta. The Chilika was in full brim throughout the monsoon period and getting flushing floods during monsoon. Large opening of spits allows a thorough fare of saline and inland flow. As a result, the lagoon had maintained its character for a long period.

19.4.4.3 Floods in SMD during Anthropocene

The river Mahanadi has a catchment area of 141,600 km^2. The river flows through five states such as Maharashtra (250 km^2), Madhya Pradesh (130 km^2), Chhattisgarh (74,970 km^2), Jharkhand (650 km^2), and Odisha (65,600 km^2) for a length of 851 km up to apex of the delta (Naraj; Nayak et al., 2006). The Hirakud dam constructed for providing flushing flood of 2830 cumec in Kuakhai branch (Mohapatra, 2015; Mishra and Jena, 2017). The peak flood statistics of 166 years (1855–2020) reveals that the south Mahanadi basin had undergone dry spells toward second half of both 19th and 20th century, but strong wet spells were prevailed during 1911–1947 and 2001–2014, with active monsoon conditions (Table 19.2).

Table 19.2 Various types floods in the SMD since observation history (from 1855).

Type	Qty (Th. Cumec)	Year of occurrence	Frequency
Low	17–20	1867, 1869, 1886, 1887, 1906, 1956, 1964, 1979, 1981, 1987, 1988, 1989, 1999, 2000, 2002, 2010, 2012, 2013, 2017	18
Medium	20–34	1856, 1862, 1863, 1871, 1875, 1876, 1877, 1880, 1881, 1884, 1885, 1891, 1893, 1894, 1900, 1904, 1907, 1908, 1909, 1910, 1912, 1913, 1915, 1917, 1918, 1921, 1928, 1930, 1931, 1932, 1935, 1936, 1942, 1943, 1945, 1948, 1950, 1952, 1953, 1958–1963, 1967, 1968, 1969, 1971–1978, 1981, 1983–1986, 1990, 1991, 1993–1995, 1997, 1998, 2002, 2004, 2005, 2007, 2009, 2011, 2013, 2014, 2020	79
High	34–40	1866, 1874, 1879, 1895, 1911, 1914, 1919, 1925, 1934, 1939, 1944, 1946, 1947, 1955, 1967, 1980, 1992, 2001, 2003, 2006, 2011	20
Very high	>40	1831, 1855, 1872, 1892, 1896, 1920, 1926, 1929, 1933, 1937, 1940, 1982, 2008	12
Very low	<17	Gap years not mentioned above	35

In the last 30 years, the monsoon was normal in SMD. There were less changes in the flow scenario and tidal inlet activities in the connecting lagoon because intermittent flushing flow was there in the years 1920, 1926, 1929, 1933, 1937, and 1940. Very high floods >40,000 cumec was absent from 1940 to 1950. The above peak flows show that during last part of 18th century to 1950 (end of Holocene), high and very high floods were frequent whereas after 1950 onward the frequency of high floods are less.

19.4.4.4 *Geomorphology of south Mahanadi delta*

The river Kuakhai, the bifurcate branch of river Kathajodi, that emerges from right hand limb at 5 km, is a spilled one and evolve at a height of about 2 m higher level at origin. The rise in bed allows only flood flow of the parent Mahanadi river when reaches 2500–3000 cumec flow. Kuakhai flows for about 16 km and branches out before debouching into BoB at Ramachandi as the river Kushabhdra. The Daya and the Bhargovi are the bifurcated branches at Utara. The Bhargovi is flowing east for 87 km and the Daya is running 60 km to SE maintaining the land contour and both are joining with the lagoon. The Bhargovi run through a flat gradient terrain and meandered a 90° at Gobakund. The river Daya shows meandering at its head and tail reaches but in the middle reach it flows through a steep gradient, while flowing through mountainous ranges between Dhauli and Kanti (Mahalik et al., 2006; Singh et al., 2011). Geomorphic highs, beach ridges (Ghoradia, Puri, Konark, and present coast line) with lineaments and meandering lobes are identified within SMD, along with other morphological features. The rate of progradation of the Chilika coast is estimated @9.1 km/millennium (Mahalik et al., 2006; Somanna et al., 2016).

New alluvial plains have developed toward south of Daya and proliferating toward the Makara and the Malaguni river which indicates that the delta is growing toward south. Geotechnical studies infer that the soil characteristics of the right embankment of the Daya river are of lateritic nature. But due to progradation of the delta the soil started blanketing with alluvium since 18th century. Agricultural fields

Table 19.3 The drilling sights, the hydrologic system, the C14 age of core samples indicating SMD and Chilika age.

No.	Sampling location	Origin to fall	Length of the component	Radio carbon date in BP	Probable formation yrs in BP	Remarks (about formation age
1	KushaBhadra river	Balianta-Ramachandi	56	1220 ± 180	≈750	Young, active, 8–2.5 m high
2	Bhargovi river	Uttara-Badagoth	85.5	590 ± 150–1220 ± 180	2300–750	Old, inactive for last 40 km, 8–2.5 m high
3	Daya river	Uttara-Badagoth	60 km	2300–750	2300–750	Young, active, avulsive, 8–2.5 m high
4	Southern spit	Rambha-sipakuda	34 km	3750 ± 200	3750	Stable, post-Holocene origin
5	Raghunathpur	Chilika west bank	Nachuni	3750 ± 200	3750	Polen-grain got; archeology finding
6	19°43^{1}N, 85.37°E	Satpada	Northern sec (satapada)	538 ± 5	<550	Low lying swampy area in N-sector
7	Centre of Northern sec.	≈230° long and 850 lat	Lake central Northern sec	11,245 ± 180	13,160	Light gray silt/clay a part of landform
8	Malaguni R 10 km west bank	Golabai	Lake periphery	4300	4300	Boat building Arch. finding at Golabai

Sources: Dobemeir et al., 2003; Khandelwal et al., 2008; Mishra and Jena, 2015.

with regular cultivation have developed towards west of Daya which were of lateritic soil 100–250 years back. The gully drainage channels and formation of the Makara, Rajua, and Weikhia systems during early 18th century became active and then started discharging more water to south during Anthropocene.

19.4.4.5 SMD—the anastomosed delta

The Daya and the Bhargovi rivers are multichannel drainage systems in deltaic alluvial plains. They are the results of the processes of avulsion and anastomosis drainages like Ratnchira, Nuna, Gangua, etc., by bypassing or splitting. The extension of Daya river system could be possible due to formation of Rajua, Makara, and Weikhia channels beyond southern territory of the delta. Development of Bhubaneswar city, new drainage system named Gangua drain; now adding substantial urban flood waters debouching to the river Daya directly. Toward the tail end (39 km) of the southern coastal deltaic zone the rivers has been anastomosed forming multiple parallel channels submerging in the northern corner of the lagoon (Samantray et al., 2014; Mishra and Jena, 2017).

19.4.4.6 Geochronology

The radio carbon dating reported by different authors are compiled and presented in the Table 19.3. The radio carbon studies have been done along the tail reaches (Kushabhadra and Bhargovi rivers), middle reach (Daya river), the stable southern spit, along west coast near Nachuni, Satapada, Golabai, and old

Table 19.4 Lithostratigraphy of sediment and 14C data in YBP Formation age estimation.

No.	Bore hole core/sampling level	Sampling depth (m)	Radio carbon date in YBP	Calculated age in YBP	Mean age in YBP	Inference
1	Upper layer (Anthropocene soil organic deposit)	0.0–0.36	-	Recent	Recent	Gray to black sheet type silt and flood Sediment recent
2	Below surface layer (pre little ice age)	0.61–0.66	214C data in YBP 100 ± 35	1900–1580	1740	Gray to black laminated silt and flood sediment and organic deposit
3	After formation from gulf to lagoon	1.17–1.23	2810 ± 35	2770–2420	2595	Gray silt, during formation of lagoon as a leaky one
4	Pre-Holocene estuarine Status	3.3–3.4	7150 ± 105	8105–7825	7965	Dark gray silt of estuary and riverine soil stratum
5	Pre-Holocene estuarine Status	6.53–6.65	7795 ± 85	8695–8425	8560	Dark gray silt of estuary and riverine soil stratum
6	Pre-Holocene estuarine Status	7.58–7.74	11,245 ± 180	13,350–1970	13,160	Upper late-glacial chrono zone Pre-Daya as river

Sources: Khandelwal et al., 2008; Kjerfve and Magill, 1989; Chilika Development Authority, 2012.

river Malaguni. The probable formation years are estimated corroborating with the results obtained from carbon dating (Table 19.3).

The results of the radiocarbon date indicate that Daya is the oldest river in the SMD and the river Kushabhadra is the youngest. The presence of a port at the mouth of the river Kushabhadra (at Khalakata Patana and Ramachandi temple) during 11th century shows that the river was active river in historical past (Sinha, 1992; Tripathi, 2000; Patnaik et al., 2014).

Studies of researchers like Khandelwal (2008) and Amir et al. (2020) reveal that the Chilika lagoon was a land mass till mid-Holocene and became a lagoon during Late Holocene.

19.5 Discussion

19.5.1 The Holocene changes in Chilika lagoon

The Chilika lagoon is a unique biotope of ecological biodiversity, shallow, pear shaped, wind driven, and micro tidal lagoon having water holding present capacity of about 4 km^3. The average water spread area of the lake is about 1045 km^2. The barrier spit was half stable and rest fragile running with conjoint parallel channels, one on shore and the other about 5 km inland representing the oldest manifestation of the lagoon, probably dates back to about 5000–4000 YBP (Wanner et al., 2008).

The landform changes and sediment core data were collected and studied in detail. The oldest date obtained was 13,500 YBP which indicates a pre-Holocene age. But during mid-Holocene (5000–2500 YBP), there was major change in the monsoon precipitation and there was rise in bay level due to RSLR (Khandelwal et al., 2008). During fag end of the Holocene epoch, the Chilika Lake dried up faster (Pandey, 2014). Pande et al. (2014) have reported that the mangrove vegetation along Chilika coast was formed between 4165 and 2549 YBP as the climate was warm and humid with plenty of monsoon rainfall. Later the paleoclimate changed for 200–300 years. After 2549 YBP, the climate change revealed that the mangroves started developing in the area with stable climate and constant RSLR in the Chilika domain. The study may be concluded with the Chilika as a large marine water body formed during 4165–2549 YBP. But the interruption in monsoon rainfall started developing barrier spit from the southern sector from Rambha side as the hill range along Balugaon to Rambha acted as a series of spurs and started depositing sediment from Prayagi sector when the large water body was converted to a leaky lagoon.

During LGM period to present the pollen records indicate that there was a change from ground herbaceous plant to fresh water taxa along the Lake Koleru coast (Andhra Pradesh), which infer a change from dry to wet climate along the east coast of India during 8000 YBP. But later from 8.0 to 4.9 KYBP there was Holocene RSLR and change from deciduous dry to mangrove wet forest. Since Chilika Lake is bounded by EGB Hills in the western bank and adjacent to the Lake Koleru, the west coast must have challenged by identical transformations from LGM period to present and formed the resulting climate changes and relative RSLR of the BoB during mid-Holocene period (Rao Kakani et al., 2020). During LIA there was retreat of the coast which favored formation of the barrier spit that separated the lagoon from the BoB (Wanner et al., 2008; Cheng et al., 2018).

Multiproxy study the southern coasts of India deciphers three phases of climate change during the period 3100 YBP to present. There was a wet climate during 3100–2500 YBP, which is succeeded by dry climate between 2500 and 1000 YBP. From 1000 YBP onward the area enjoyed a pleasant climate. Radio carbon dating of the fragile outer channel gave an age range of 540–550 YBP whereas southern sector gave an age of 3750 YBP. From this it can be revealed that the Chilika lagoon has undergone three phases (Bhattacharjee et al., 2015). The first phase was the phase of formation, the second phase was the phase of conversion to a lagoon (i.e., gulf to leaky by formation of spit from southern end) and the third phase was cyclic phases of opening and closing of the tidal inlets.

19.5.2 Holocene changes in SMD

The Mahanadi basin has formed as a result of the direct rifting and disruptions Gondwana lands and successive drifting of the Indian plate during pre-Cambrian age (Nayak et al., 2012). In the deltaic region, the southern distributaries are incapable to accommodate the peak flood beyond 40,000 Cumec in the Mahanadi system. About 15,000 Ha remained inundated for months together in the NW periphery of the lagoon. Channels became defunct with evolution of new channels. The imprints of the Ratnachira (left) channel of about 16.2 km length exists which has become defunct and dead from early 19th century. The southern delta had undergone geomorphic changes in past when the rivulet Ratnachira was a prominent river prior to the pre-Holocene epoch (Mahalik et al., 1996; Sinha et al., 2020) which became unimportant later when the Daya emerged as a main river in the southern fringe of the delta. The formation of Makara river during fag end of Holocene has tried to create its own deltaic deposits and helping in the northerly extension of the Mahanadi delta. The south Mahanadi delta is under expansion in its right fringe and the adjacent drainage channel the Makara started gaining prominence (Mishra et al., 2015).

19.6 Learning and knowledge outcomes

The geology, RSLR, SWM, and the geomorphology of the Chilika lagoon and SMD were studied. The fluctuations in south west monsoon and dry years in the delta could not provide adequate flushing flow in to the lagoon during later part of the Holocene epoch, hence started sedimentation. The shift in the channels and rivers to south indicated continuous trend of erosion and accretion of the shoreline along with the RSLR due to global warming, littoral drift along with anthropogenic activities within the delta. The death of the Bhargovi and the shore parallel river was prominent along with extension of Mahanadi delta to south were concluded. The evolution of the Makara, the Rajua river and diminution of lagoon dimensions during post-Holocene period. The lagoon has changed the geomorphology whose health is vulnerable at present.

Acknowledgments

Our sincere thanks to the Water Resources Department Odisha, the CDA Authority toward their data source, and Dr. KPN Kumaran whose persistent persuasion has made this project to stand.

References

Achyuthan, H., Nagasundaram, M., Gourlan, A.T., Eastoe, C., Ahmad, S.M., Padmakumari, V.M., Mid-holocene Indian summer monsoon variability off the andaman islands, Bay of Bengal, Quat. Int. 349 (2014) 232–244.

Ali, H., Mishra, V., Increase in subdaily precipitation extremes in india under 1.5 and 2.0 C warming worlds, Geophys. Res. Lett. 45 (14) (2018) 6972–6982.

Amir, Mohd., Paul, D., Malik, J.N., Geochemistry of Holocene sediments from Chilika Lagoon, India: inferences on the sources of organic matter and variability of the Indian summer monsoon, Quat. Int. In press, doi:10.1016/j.quaint.2020.08.050. (Available online 5 September 2020).

An, Z., Porter, S.C., Kutzbach, J.E., Xihao, W., Suming, W., Xiaodong, L., Xiaoqiang, L., Weijian, Z., Asynchronous Holocene optimum of the East Asian monsoon, Quat. Sci. Rev. 19 (8) (2000) 743–762.

Andrew, S., Orissa, Its Geography, Statistics, History, Religion and Antiquities, John Snow, 35 Paternoster Row, sold by Brooks, Leicester, Noble, Boston. Source: British Library, 1846.

Arya, R., Lakhotia, S.C., Imprints of Chilika lake in the offshore region – a geomorphic evidence, Curr. Sci. 90 (9) (2006) 1180–1182.

Babu, P.V.L.P., 1978. Geomorphic evolution of the Mahanadi delta. Symp. Morphology and Evolution of Landforms. Dept. Geol., Delhi Univ., Delhi, 217-221.

Banerjee, P.K., Holocene and Late Pleistocene relative sea level fluctuations along the east coast of India, Mar. Geol. 167 (2000) 243–260.

Banerji, U.S., Arulbalaji, P, Padmalal, D, Holocene climate variability & Indian summer monsoon: an overview, Holocene 30 (5) (2020) 744–773, doi:10.1177/0959683619895577.

Barnosky, A.D., Paleontological Evidence for Defining the Anthropocene, Geological Society, London, 2013, p. 395. Special Publications, http://dx.doi.org/10.1144/SP395.6.

Behera, A., Newly Discovered Archaeological Sites in Coastal Odisha; Odisha Review; May 2013, Govt. of Odisha, 2013, http://magazines.odisha.gov.in/Orissareview/2013/may/engpdf/62-66.pdf.

Bharali, B., A brief review of Mahanadi delta and the deltaic sediments in Mahanadi basin, Mem. Geol. Soc. Ind. 22 (1991) 31–49.

Bhattacharyya, A, Sandeep, K, Misra, S, et al., Vegetational and climatic variations during the past 3100 years in southern India: evidence from pollen, magnetic susceptibility and particle size data, Environ. Earth Sci. 74 (4) (2015) 3559–3572.

Boll, A., Lückge, A., Munz, P., Forke, S., Schulz, H., Ramaswamy, V., et al., Late Holocene primary productivity and sea surface temperature variations in the northeastern Arabian Sea: implications for winter monsoon variability, Paleoceanography 29 (2014) 778–794, doi:10.1002/2013PA002579.

Borgaonkar, H.P., Somaru, R., Sikder, A.B., Assessment of tree-ring analysis of high elevation Cedrusdeodara D. Don from Western Himalaya (India) in relation to climate and glacier fluctuations, Dendrochronologia 27 (1) (2009) 59–69.

Borzenkova, I., Eduardo, Z., Olga, B., Laimdota, K., Dalia, K., Tiiu, K., Denis, K., et al., Climate change during the Holocene (Past 12,000 years), in: The BACC II Author Team (Eds.), 2nd Assessment of C.C. for the Baltic Sea Basin, Reg. Climate Studies, Springer, Cham, (2015) 24–49, https://doi.org/10.1007/978-3-319-16006-1_2.

Brook, E.J., Harder, S, Sevennghaus, J, Steig, E.J., Suche, C.M., On the origin and timing of rapid changes in atmospheric methane during the last glacial period, Global Biogeochem Cycles 14 (2) (2000) 559–572.

Chauhan, O.S., Dayal, A.M., Basavaiah, N., et al., Indian summer monsoon and winter hydrographic variations over past millennia resolved by clay sedimentation, Geochem. Geophys. Geosyst. 11 (9) (2010) 003067.

Cheng, H., Spötl, C., Breitenbach, S.F.M., Sinha, A., Wassenburg, J.A., et al., Climate variations of Central Asia on orbital to millennial timescales, Sci. Rep. 5 (2016) 36975.

Cheng, X., McCreary, J.P., Qiu, B., Qi, Y., Du, Y., Chen, X., Dynamics of eddy generation in the central Bay of Bengal, J. Geophys. Res. Oceans 123 (2018) 6861–6875, https://doi.org/.

Chilika Development Authority, External group meeting to develop indicators to assess coastal eco-system health, Organizer CDA and National Center for Sustainable Dev., MOEF, (2012) 25–27, Chandraput, Balugaon, Odisha.

Cook, E.R.K.J., Anchukaitis, B.M., Buckley, R.D., D'Arrigo, G.C.J., Wright, W.E., Asian monsoon failure and mega-drought during the last millennium, Science 328 (5977) (2010) 486–489, doi:10.1126/science.1185188.

Costas, S., Ferreira, Ó., Plomaritis, T., Leorri, E., Coastal barrier stratigraphy for Holocenehigh-resolution sea-level reconstruction, Sci. Rep. 6 (2016) 38726, https://doi.org/10.1038/srep38726.

Cui, M., Wang, Z., Nageswara, R.K., et al., A mid- to late-Holocene record of vegetation decline and erosion triggered by monsoon weakening and human adaptations in the south-east Indian Peninsula, Holocene 27 (12) (2017) 1976–1987, doi:10.1177/0959683617715694.

Davis, R.V., Inventing the present: historical roots of the Anthropocene; Earth sciences history, J. Hist. Earth Sci. Soc. 30 (1) (2011) 63–84, doi:10.17704/eshi.30.1.p8327x7042g3q989.

Dimri, A.P., Yasunari, T., Kotlia, B.S., Mohanty, U.C., Sikka, D.R., Indian winter monsoon: present and past, Earth Sci. Rev. 163 (2) (2016) 298–305, https://doi.org/10.1016/j.earscirev.2016.10.008.

Dixit, Y., Hodell, D.A., Giesche, A., et al., Intensified summer monsoon and the urbanization of Indus Civilization in northwest India, Sci Rep 8 (2018) 4225 https://doi.org/10.1038/s41598-018-22504-5.

Dixit, Y., Tandon, S.K., Hydroclimatic variability on the Indian subcontinent in the past millennium: review and assessment, Earth Sci. Rev. 161 (2016) 1–15.

Dobemeir, C., Raith, M.M., Yosida, M., Windely, B.F., Dasgupta, S., Crustal architecture and evolution of eastern Ghats belt and adjacent regions of India, J. Geol. Soc. India, 73 (2003) 145–168.

Dougherty, A.J, Extracting a record of Holocene storm erosion and deposition preserved in the morpho-stratigraphy of a prograde coastal barrier, Continent. Shelf Res. 86 (2014) 116–131, doi:10.1016/j.csr.2013.10.014.

Fairbridge, R.W., Agenbroad, L.D., Holocene epoch. Geochronology, Encyclopædia Britannica (2018) https://www.britannica.com/science/Holocene-Epoch.

Fairbridge, R.W., Hillaire-Marcel, C., An 8000-yr paleoclimatic record of the/'Double-Hale/' 45-yr solar cycle, Nature 268 (1977) 413–416.

Gawali, P.B., Lakshmi, B.V., Deenadayalan, K., Climate change and monsoon: looking into its antecedents, SAGE Open 9 (1) (2019) 1–13 https://doi.org/10.1177/2158244018822246.

Griffiths, P., The British Impact on India, first ed., Routledge, Taylor & Francis eBooks, 1952, https://doi.org/10.4324/9780429057656.

Gupta, A.K., Anderson, D.M., Overpeck, J.T., Abrupt changes in the Asian southwest monsoon during the Holocene and their links to the North Atlantic Ocean, Nature 421 (6921) (2003) 354–357, https://doi.org/10.1038/nature01340.

Hahn, D.G., Shukla, J., An apparent relationship between Eurasian snow cover and Indian monsoon rainfall, J. Atmos. Sci. 33 (12) (1976) 2461–2462.

Hazra, S., Das, S., Ghosh, A., Raju, P.V., Patel, A., The Mahanadi delta: a rapidly developing delta in India, in: Nicholls, R., Adger, W., Hutton, C., Hanson, S. (Eds.), Deltas in the Anthropocene, Palgrave Macmillan, Cham, 2020, https://doi.org/10.1007/978-3-030-23517-8_3.

Head, M.J., Aubry, M.P., Walker, S.M., Brian, R.P., A case for formalizing subseries (subepochs) of the Cenozoic Era, Episodes 40 (1) (2017) 22–27, doi:10.18814/epiiugs/2017/v40i1/017004.

Kathayat, G, Cheng, H, Sinha, A, et al., The Indian monsoon variability and civilization changes in the Indian subcontinent, Sci. Adv. 3 (12) (2017) e1701296, doi:10.1126/sciadv.1701296.

Khandelwal, A., Mohanti, M., Garcı á-Rodrı ́guez, F., Scharf, B.W., Vegetation history and sea level variations during the last13,500 years inferred from a pollen record at Chilika Lake, Orissa, India, Veget. Hist Archaeobot. 17 (4) (2008) 335–344, doi:10.1007/s00334-007-0127-5.

Kjerfve, B., Magill, K., Geographic and hydrographic characteristics of shallow coastal lagoons, Mar. Geol. 88 (1989) 187–199.

Kothawale, D.R., Deshpande, N.R., Kolli, R.K., Long term temperature trends at maj., med., small cities and hill stations in India during the period 1901–2013, Am J Clim Change 5 (3) (2016) 383–398, doi:10.17632/4gt974j755.1.

Kumaran, K.P.N., Limaye, R.B., Padmalal, D., India's fragile coast with special reference to late quaternary environmental dynamics, Proc. Indian Nat Sci. Acad. 78 (3) (2012) 343–352.

Laug, A., Schwarz, A., Lauterbach, S., Engels, S., Schwalb, A., Ecosystem shifts at two mid-Holocene tipping points in the alpine Lake Son Kol (Kyrgyzstan, Central Asia), Holocene 30 (10) (2020) 1410–1419, doi:10.1177/0959683620932973.

Li, Y., Wang, N., Chen, H., Li, Z., Zhou, X., et al., Tracking millennial–scale climate change by analysis of the modern summer precipitation in the marginal regions of the Asian monsoon, J Asian Earth Sci 58 (2012) 78–87.

Limaye, R.B., Kumaran, K.P.N., Mangrove vegetation responses to Holocene climate change along Konkan coast of south-western India, Quat. Int. 263 (2012) 114–128, doi:10.1016/j.quaint.2012.01.034.

Loveson, V.J., Nigam, R., Reconstruction of Late Pleistocene and Holocene Sea Level Curve for the East Coast of India. J. Geol. Soc. India 93 (2019) 507–514, https://doi.org/10.1007/s12594-019-1211-z.

Lyell, C., Principles of Geology, 3, John Murray, London, 1833, pp. 398–558.

Mahalanobis, P.C., Rain storms and river floods in Orissa, Sankhya 5 (1) (1940) 1–20 http://hdl.handle.net/10263/1914.

Mahalik, N.K., A study of the morphological features and bore hole cuttings in understanding the evolution and geological processes in Mahanadi Delta, East Coast of India, J. Geol. Soc. India 67 (2006) 595–603.

Mahalik, N.K., Das, C., Wataru, M., Geomorphology and evolution of Mahanadi Delta, India, Geosci. Osaka City Univ. 39 (6) (1996) 111–122 Article.

Mishra, S.P., Dwibedy, S., Geo-hydrology of south Mahanadi delta and Chilika Lake, Odisha, India, Int. J. Adv. Res. 3 (11) (2015) 430–445.

Mishra, S.P., Jena, J.G., Morphological reconstruction of southern Mahanadi delta and Chilika lagoon, India – a critical study, Int. J. Adv. Res. 3 (5) (2015) 691–702.

Mishra S.P., Jena, J.G., Geophysical Changes of Chilika Lagoon in Post Naraj Barrage Period: Synopsis of the Thesis Submitted in Partial Fulfilment of the Requirements for the Doctoral Degree; SOAA Univ., 2015, https://pdfs.semanticscholar.org/46eb/e5382.

Mishra, S.P., Jena, J.G., 2017. Management of probabilistic peak flood: Mahanadi branches draining into Chilika Lagoon, Odisha, India. In: Proceedings of Inter. Conf. on: Global Civil Eng. Challenges in Sustainable Dev. and Climate Change" (ICGCSC - 17th -18th March 2017).

Mishra, S.P., Ojha, A.C., Fani, an outlier among pre-monsoon intra-seasonal cyclones over Bay of Bengal, Int. J. Emerg. Tech. 11 (2) (2020) 271–282 ISSN No. (Online): 2249-3255; 11(2); 271-282.

Mishra, S.P., Sethi, B.K., Barik, K.K., Delta partitioning, geospatial changes, anastomosis of Mahanadi tri-delta, India, Int. J. Earth Sci. Eng. 12 (01) (2019) 21–40 Caffetinnova org., doi:10.21276/ijee.2019.12.0103.

Mishra, U.K., Behari, S., Understanding nature and attributes of early historical Orissa: an archaeological perspective; from the Odisha historical research journal, Odisha State Museum, BBSR LVIII (1 & 2) (2019) 1–32 2019.

Mohanty, M.M., Mohanty, P., Patnaik, A.K., Samal, R.N., et al., Hydrodynamics, temperature/salinity variability and residence time in the Chilika lagoon during dry and wet period: measurement and modeling, Cont. Shelf Res. 125 (2016) 28–43, doi:10.1016/j.csr.2016.06.017.

Mohapatra, R., Why Hirakud dam failed to check flood? Down to Earth (2015) https://www.downtoearth.org.in/news/why-hirakud-dam-failed-to-check-flood-33982.

Monalisha, M., Panda, G.K., Coastal erosion and shoreline change in Ganjam coast along East Coast of India, J. Earth Sci. Cli. Ch. 9 (2018) 467, https://doi.org/10.4172/2157-7617.1000467.

Moy, C.M., Seltze, G.O., Rodbell, D.T., Anderson, D.M., Variability of El Nino/oscillation activity at millennial timescales during the Holocene epoch, Nature 420 (2002) 62–165.

Naidu, P.D., A review on Holocene climate changes in Indian subcontinent, Mem. Geol. Soc. India 42 (1999) 303–314.

Nair, K.M., Padmalal, D., Kumaran, K.P.N., Sreeja, R., Limaye, R.B., Srinivas, R., Late quaternary evolution of Ashtamudi-Sasthamkotta lake systems of Kerala, south west India, J. Asian Earth Sci. 37 (2010) 361–372, doi:10.1016/j.jseaes.2009.09.004.

Nayak, G.K., Rao, C.R., Rambabu, H.V., Aeromagnetic evidence for the arcuate shape of Mahanadi Delta, India, Earth Planet Special 58 (2006) 1093–1098, https://doi.org/10.1186/BF03352615.

Nayak, P.K., The Chilika Lagoon social-ecological system: an historical analysis, Ecol. Soc. 19 (1) (2014) 1–13, 1 http://dx.doi.org/10.5751/ES-05978-190101 .

Nayak, S., Das, S., Bastia, R., Kar, B., 2012. Lava delta below 85^0 ridge, Mahanadi offshore basin, identification characterization and implication of hydrocarbon prospectivity. In: 9th Biennial Int. Conf. and Exposition on Petroleum Geophysics. Hyd-2012, 212.

Nicole, B., Gayea, B., Hilbig, L., Lahanjar, N., et al., Holocene monsoon and sea level-related changes of sedimentation in the northeastern Arabian Sea, Deep Sea Res. Part II: Top. Stud. Oceanogr. 166 (2019) 6–18, https://doi.org/10.1016/J.DSR2.2019.03.003.

Padmalal, D., Kumaran, K.P.N., Nair, K.M., Limaye, R.B., Vishnu Mohan, S., Baijulal, B., Anooja, S., Consequences of sea level and climate changes on the morphodynamics of a tropical coastal lagoon during Holocene: An evolutionary model, Quat. Int. 333 (2014) 156–172, doi:10.1016/j.quaint.2013.12.018.

Pandey, S., Burkhard, W.S., Mohanti, M.M., Palynological studies on mangrove ecosystem of the Chilka Lagoon, east coast of India during the last 4165 yrs BP, Geol. Quat. Int. 325 (2014) 126–135, doi:10.1016/J.QUAINT.2013.09.001.

Patnaik, S.K., Tripathy, B., Pradhan, G.C., Archaeological excavations conducted by OIMSEAS, in: Sunil, K.P. (Ed.), Buddhsm & Maritime Heritage of South East Asia, Odisha an Perspective, Saujanya Books, New Delhi, 2014, pp. 174–189.

Paul A.K., Islam S.M., Jana S., 2014. An Assessment of Physiographic Habitats, Geomorphology and Evolution of Chilika Lagoon (Odisha, India) Using Geospatial Technology; Ch-6 of book; Remote Sensing and Modeling Advances in Coastal and Marine Resources, Springer, Cham.

Ponton, C., Giosan, L., Eglinton, T.I., Fuller, D.Q., Johnson, J.E., Kumar, P., Collett, T.S., Holocene aridification of India, Geophys. Res. Lett. 39 (2012) L03704.

Rao Kakani, N.M.M., Sinha, P., Mohanty, U.C., Mishra, S., Occurrence of more heat waves over the central east coast of India in the recent warming era, Pure Appl. Geophys. 177 (2020) 1143–1155, https://doi.org/10.1007/s00024-019-02304-2.

Rao Nageswar, K., Pandey, S., Kubo, S., Saito, Y., Naga, V.K., Demudu, G., Hema Malini, B., et al., Paleoclimate and Holocene relative sea-level history of the east coast of India, J. Paleolimnol. 64 (2) (2020) 71–89, doi:10.1007/s10933-020-00124-2.

Rao, N.K., Rao, K.S., Gurreddy, M., Geomorphic evolution and changing environment of Chilika lake, East. Geogr. 1995 (5&6) (1995) 1–19.

Sadakata, N., Rao Kakani, N., Radiocarbon ages of the sub-surface sediments in the Krishna Delta, India, and their geomorphological implications, Quart. J. Geogr. 49 (3) (1997) 163–170, https://doi.org/10.5190/tga.49.163.

Samantaray, D., Chatterjee, C., Singh, R., Gupta, P.K., Panigrahi, S., Flood risk modeling for optimal rice planning for delta region of Mahanadi river basin in India, Nat. Hazards 76 (1) (2014) 347–372, doi:10.1007/s11069-014-1493-9.

Sanjay, J., Ravedkar, J.V., Temperature changes in India, in: Krishnan, R., Sanjay, J., Gnanaseelan, C., Mujumdar, M., Kulkarni, A., Chakraborty, S. (Eds.), Assessment of Climate. Change Over the Indian Region, Springer, Singapore, 2020, https://doi.org/10.1007/978-981-15-4327-2_2.

Schiemann, R., Lüthi, D., Schär, C., Seasonality and interannual variability of the westerly jet in the Tibetan Plateau region, J. Clim. 22 (2009) 2940–2957.

Singh, S.B., Veeraiah, B., Dhar, R., Prakash, L., Deep resistivity sounding studies for probing deep fresh aquifers in the coastal areas of Orissa, India, J. Hydrol. 19 (2011) 355–366.

Sinha, A.M., Berkelhammer, L., Stott, L., Mudelsee, M., Cheng, H., Biswas, J., The leading mode of Indian summer monsoon precipitation variability during the last millennium, Geophys. Res. Lett. 38 (15) (2011) 1–5 L15703, doi:10.1029/2011GL047713.

Sinha, B.K., Archaeology of Orissa, in Basa, K.K., Mohanty, P. (Eds.), Pratibha Prakashan, New Delhi, 2000, pp. 322–355.

Sinha, B.K., Khalkatapatna—a small port on the coast of Orissa, in B. Nayak & N.C. Ghosh, editors, New Trends in Indian Art and Archaeology, Aditya Prakashan, New Delhi, vol. 12, 1992, pp. 423–428.

Sinha, R., Chandrasekaran, R., Awasthi, N., Geomorphology, land use/land cover and sedimentary environments of the Chilika basin, in: Finlayson, C., Rastogi, G., Mishra, D., Pattnaik, A. (Eds.), Ecology, Conservation, and Restoration of Chilika Lagoon, India. Wetlands: Ecology, Conservation and Management, 6, Springer, Cham, (2020) 231–250, https://doi.org/10.1007/978-3-030-33424-6_10.

Somanna, K., Reddy, T.S., Rao, M.S., Geomorphology and evolution of the modern Mahanadi delta using remote sensing data, Int. J. Sci. Res. (IJSR) 5 (2) (2016) 1329–1335.

Sparrenbom, C.J., Bennike, O., Björck, S., Lambeck, K., Relative sea-level changes since 15 000 cal. yr BP in the Nanortalik area, southern Greenland, J. Quat. Sci. 21 (2006) 29–48.

Srivastava, P., Jovane, L., Misinterpreting proxy data for paleoclimate signals: a comment on Shukla et al., Holocene 30 (12) (2020) 1866–1873, doi:10.1177/0959683620941165.

Stefan, L., Nils, A., Yiming, V.W., Thomas, B., Larsen, T., Schneider, R.R., An $\sim$130 kyr record of surface water temp. and δ18O from the northern Bay of Bengal: investigating the linkage between heinrich events and weak monsoon intervals in Asia, Paleoceanogr. Paleoclimatol. 35 (2020) 2, doi:10.1029/2019PA003646.

Thamban, M., Kawahata, H., Rao, V.P., Indian summer monsoon variability during the Holocene as recorded in sediments of the Arabian Sea: timing and implications, J. Oceanogr. 63 (6) (2007) 1009–1020.

Tripathi, S., Maritime Archaeology:, Historical Description of Seafaring of the Kalingans, Kaveri Books, New Delhi, 2000.

Van-Heteren, S., Huntley, D.J., van de Plassche, O., Lubberts, R.K., Optical dating of dune sand for the study of sea-level change, Geology 28 (2000) 411–414.

Wang, P., Clemens, S., Beaufort, L., Braconnot, P., Ganssen, G, Jian, Z, et al., Evolution and variability of the Asian monsoon system: state of the art and outstanding issues, Quat. Sci. Rev. 24 (2005) 595–629.

Wanner, H., Beer, J., Bütikofera, J., Thomas, J., Ulrich, C., Jacqueline, C., Flückigere, J., et al., Mid- to late Holocene climate change: an overview, Quat. Sci. Rev. 27 (19-20) (2008) 1791–1828, https://doi.org/10.1016/j.quascirev.2008.06.013.

Woodroffe, S.A., Horton, B.P., 2005. Holocene sea-level changes in the Indo-Pacific. Retrieved from; http://repository.upenn.edu/ees_papers/21

Yang, B., Wang, J., Brauning, A., Dong, Z., Esper, J., Late Holocene climatic and environmental changes in arid central Asia, Quat. Int. 194 (2009) 68–78.

https://shodhganga.inflibnet.ac.in/bitstream/10603/189922/10/10_chapter

https://www.ncdc.noaa.gov/global-warming/mid-holocene-warm-period

CHAPTER

20

Holocene changes in fluvial geomorphology, depositional environments, and evolution of coastal wetlands—A multiproxy study from Southwest India

K. Maya[a], D. Padmalal[a], M. Vandana[a], S. Vishnu Mohan[b], V.R. Vivek[a], Ruta B. Limaye[c], K.P.N. Kumaran[c]

[a]*National Centre for Earth Science Studies, Ministry of Earth Sciences, Thiruvananthapuram, India.* [b]*Department of Geology, University of Kerala, Thiruvananthapuram, India.* [c]*PalynoVision, Mon Amour, Erandaane, Pune, Maharashtra, India*

20.1 Introduction

The southwest coast of India is known for many unique features–the occurrence of economically viable beach placer deposits of strategic minerals, silica sand, and many minor mineral resources like brick/tile clays and fine aggregates. Furthermore, many stretches of the coast, its expansive backwater bodies, cliffs, and pocket beaches make the coast an area of outstanding natural beauty. The coast had been subjected to several spells of sea-level oscillations in the late Quaternary period and that perhaps shaped the coast to the present level (Merh, 1992; Banerjee, 2000; Kumaran et al., 2005; Nair et al., 2006; Padmalal et al., 2013, Arul et al., 2020). Lagoons and backwaters are considered as an important ecosystem in many low-lying coasts of the world and India is not an exception (Bird, 1994; Kjerfve, 1994; Joseph et al., 2009; Padmalal et al., 2014). Further, lagoons and their adjoining wetlands offer a wide range of services to mankind, including inland fishery and navigation (Dill, 1990; Gönenc and Wolflin, 2005). Of the lagoons, backwaters and coastal wetlands of India, nearly half of them are reported from the southwest coast, especially along the Kerala coast. Most of them have been developed and modified as a result of the climate and sea-level changes to which the coast has been subjected to during the Holocene epoch. A review of the literature reveals that although many studies exist in the literature on the various aspects of coastal wetlands/lagoons (Kurien, 1971; Kurup, 1982; Mallik and Suchindan, 1984; Sankaranarayanan et al., 1986; Seralathan and Padmalal, 1995; Nair et al., 1998; Narayana and Priju, 2006; Nair and Kumaran, 2006; Padmalal et al., 2011; Limaye et al., 2016), practically not many studies have been made to address the origin and evolution of the coastal wetlands that support life and economic prosperity of the region. Inadequate base line data on these life-sustaining systems make the life of the people miserable, especially during the extreme climatic events. A multiparametric study has been carried out in one of the significant wetlands on the southwest coast—the Kuttanad Kole Wetland (KKW), a Ramsar wetland of international importance—using

Holocene Climate Change and Environment. DOI: https://doi.org/10.1016/B978-0-323-90085-0.00026-7

25 drilled core data together with spatial distribution and chronology of lime shell (subfossil shell) deposits interbedded within the Holocene sedimentary sequences of the area.

20.2 Scope

India has a long coast line and the width of the coastal zone vary considerably from place to place. Sea-level oscillations in the late Quaternary period, especially in the Holocene, had a significant imprint on the present coastal geomorphology/landforms. Many stretches on the southwestern coast of India receive the status of "Area of Outstanding Natural Beauty" because of the presence of the interlacing network of rivers, tidal channels, and expansive backwaters. At the same time, mining of the strategic beach placer minerals, silica sands, lime shells, and fine aggregates from the coastal areas not only degrades its natural beauty but impose marked changes in the coastal landform features. Many of the coastal wetlands, which act as room for floodwaters of the extreme rainfall events in the last few decades, are shrinking at alarming rates due to urban sprawl and consequent land reclamation for building constructions and other developmental activities. Lack of adequate knowledge on the evolution of these wetland systems and their subsurface stratigraphic architecture is a major lacuna challenging wise decision making on the developmental processes in the current ever-changing climate scenarios. Therefore, the present multiparametric study of the unique KKW in southwest India will be equally relevant both for the academic community and also for the environmental managers and decision makers at various levels.

20.3 Study area

The KKW is located in the southern expansive reach of the Vembanad Lake (also known as Vembanad lagoon or Vembanad Kayal)—the largest, coast parallel wetland (Ramsar) in India. The KKW spreads over an area of about 854 km^2 (Nath et al., 2016) between 9°17′–9°38′N latitudes and 76°17′–76°38′E longitudes (Fig. 20.1). It is a unique wetland system in India where paddy cultivation is being carried out 1–2 m below mean sea level. In an earlier study, Chattopadhyay and Sidharthan (1985) classified the KKW and its Western strand plains/barrier beaches into six microphysiographic units such as (1) Upper Kuttanad, (2) Lower Kuttanad, (3) Eastern transitional zone, (4) Western coastal sand dune, (5) Coastal marshes with sand dune, and (6) Kari Land. While the Upper Kuttanad is generally fluvial dominated, the lower Kuttanad is clay dominated (lagoonal), and the western coastal sand dunes is sand (beach/strand plains) dominated. The other zones generally have complex textural and environmental characteristics. However, for this study which depends generally on borehole cores retrieved from specific locations, a fivefold classification is followed: (1) Upper Kuttanad and floodplains, (2) Lower Kuttanad, (3) North Kuttanad, (4) Pathiramanal Island, (5) Western coastal strand plains/barrier beaches. This is for better presentation and discussion of the results generated from the studied cores and evolution of the geomorphic features of KKW.

The Upper Kuttanad and nearby areas influenced by the Achankovil, Pamba, and Manimala rivers fall under the Upper Kuttanad and flood plains sector. The lowland reaches of the southern rivers and parts of the Pachcha–Edathua areas influenced mainly by the rivers are included under this category for addressing the fluvial processes and their influences. The lower Kuttanad zone comprises the southern,

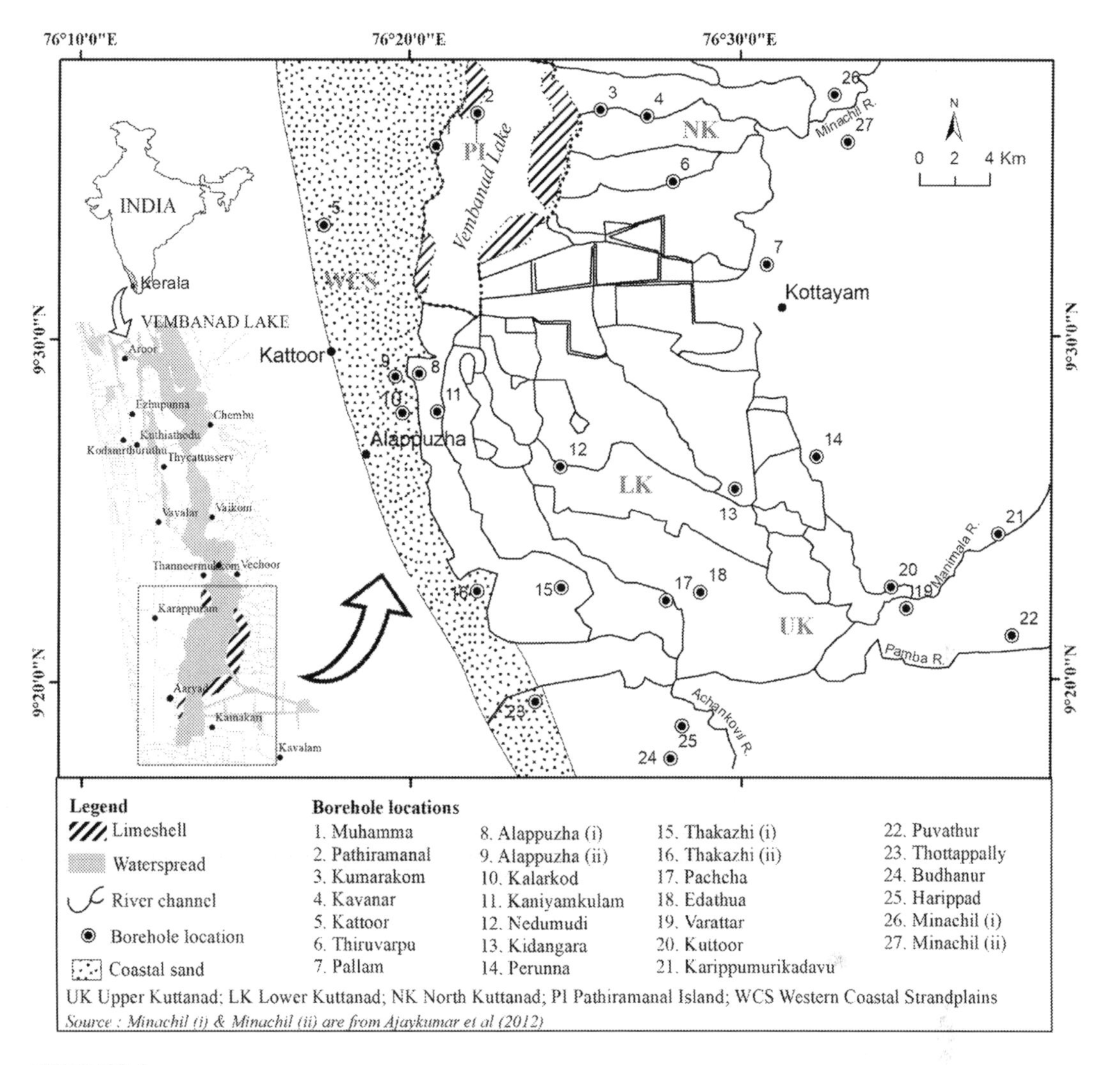

FIGURE 20.1

Study area showing borehole locations and distribution of limeshell (subfossil shells) in Kuttanad Kole (Ramsar) Wetland (KKW) in SW India.

generally rice cultivated lands and the kayal lands of the KKW. The Minachil river and its distributary influenced areas around Kumarakom–Thiruvarpu are categorized under the North Kuttanad zone. The southern river influenced areas of the Upper Kuttanad and northern Minachil river influenced areas of the Lower Kuttanad and Eastern transitional zones referred to by Chattopadhyay and Sidharthan (1985) have many similarities in their subsurface sediments. The Western coastal sand dune zone and coastal marshes with sand dunes of Chattopadhyay and Sidharthan (1985) is essentially made up of older strand plains and the barrier beaches adjoining to it. But certain southern reaches in the southern side are transformed into marshes. We have not covered too much in the marsh dominated areas and the present study is generally confined to the Western strand plains and barrier beaches. Apart from this, a core is also taken from the Vembanad Lake, that is, in the Pathiramanal island. Geologically the study area is formed of Neogene and late Quaternary sediments, flanked by Precambrian crystalline along the eastern border. The present climate of the region is humid tropical with a variation of temperature between 21°C and 36°. The study area receives an average annual rainfall of about 3000 mm. Out of the total rainfall contribution, a substantial proportion is from the southwest monsoon (60%) and the remaining from northeast monsoon and also from summer showers (Sreejith, 2013).

20.4 Materials and methods

Extensive fieldwork was carried out in the entire Kuttanad and nearby areas for the collection of data on the geomorphic aspects of the region for choosing borehole drilling sites. A total of 25 borehole cores with a depth range of 2.0–40.0 m were collected using rotary drilling or using a specially fabricated coring device (Fig. 20.1). The cores used in the present study were collected through many state and centrally sponsored projects for addressing different aspects of the coastal lowlands of SW India. Uncontaminated sediment samples were collected from the area and sectioned at 10 cm sampling intervals and subsamples representing the various litho units were packed in neatly labeled polyethylene bags and deep frozen till the time of analysis. The samples have been subjected to sedimentological, mineralogical, and palynological studies following standard procedures. The method suggested by Lewis (1984) was followed for textural analysis of the samples. Heavy mineral sample preparations from fine sand fractions were done following Carver (1971) and the heavy mineral species identification was performed following Mange and Maurer (1992). The organic carbon and $CaCO_3$ contents were estimated following the method suggested by El Wakeel and Riley (1957) and Jackson (1967), respectively. The sediment samples were processed by conventional methods as suggested by Traverse (2007) and were examined for palynomorphs and nonpollen palynomorphs (NPPs). The identification of spores and pollen, and NPPs were carried out following published records (Thanikaimoni et al., 1984; Williams and Bujak, 1985; Nayar, 1990; Tissot et al., 1994; Kumaran, 2001; Limaye et al., 2007). Radiocarbon (^{14}C) dates of organic-rich sediments, wood and shells at specific levels of the borehole cores were determined at Birbal Sahni Institute of Paleobotany, Lucknow (India) and Physical Research Laboratory, Ahmedabad (India).

20.5 Results

The following sections deal with the results of the multiparametric study carried out using the borehole cores retrieved from the KKW and the adjoining areas. The locations of the core and the microphysiographic units are depicted in Fig. 20.1. A few data were collected from the published work of Ajaykumar et al. (2012).

20.5.1 Borehole lithology and textural characteristics

The cores retrieved from the KKW exhibit distinct lithological assemblages in different microphysiographic zones. Fig. 20.2 shows the lithological characteristics of selected cores from the area. The sand, silt, and clay contents in the borehole cores are summarized in Table 20.1.

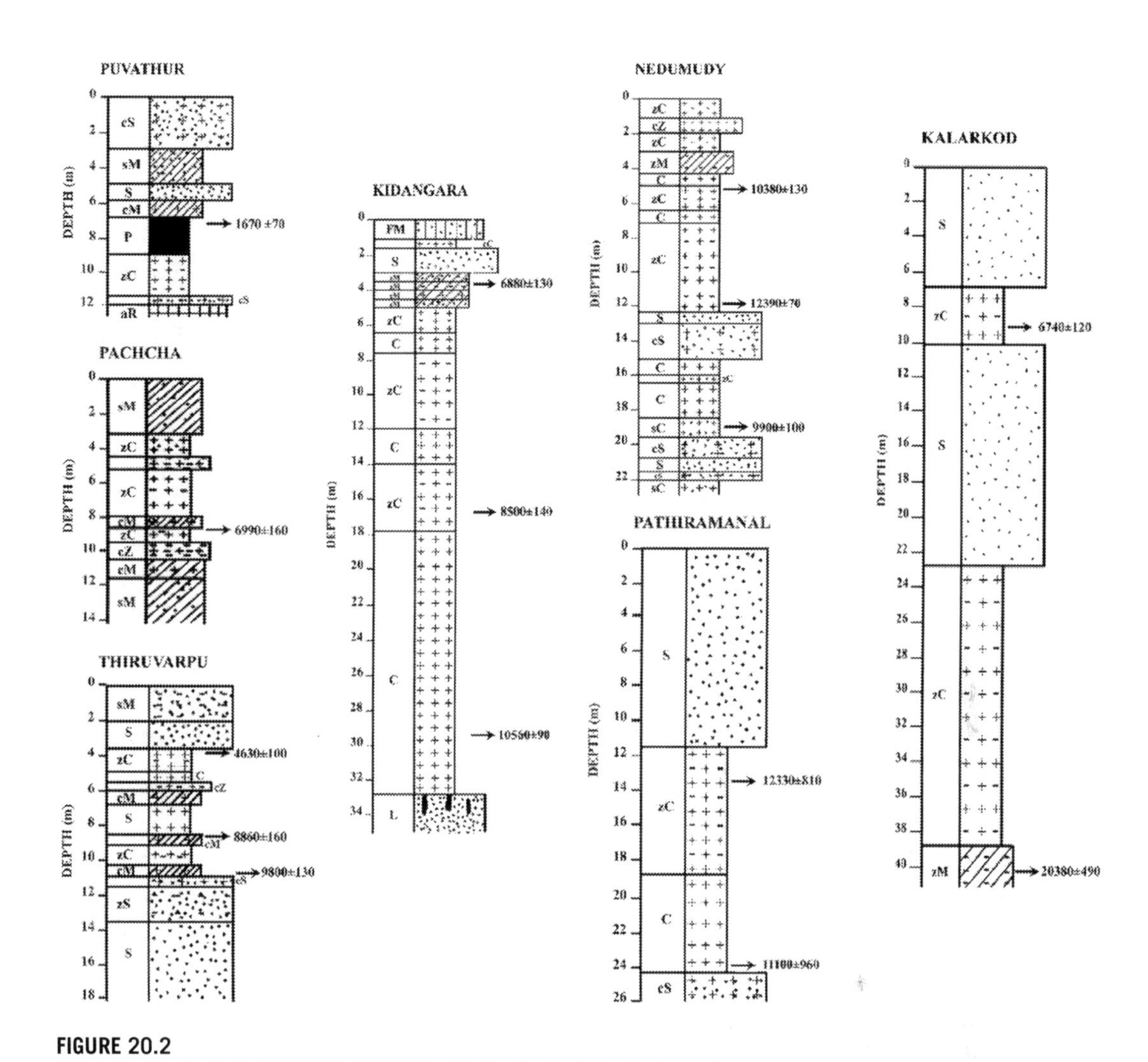

FIGURE 20.2

Lithologs of selected borehole cores collected from the Kuttanad Kole Wetland. S, sand; cS, clayey sand; zS, silty sand; sM, sandy mud; zM, silty mud; cM, clayey mud; cZ, clayey silt; zC, silty clay; C, clay; P, peat; L, laterite; aR, altered rock; FM, Filled material.

20.5.1.1 Upper Kuttanad and flood plains

In this section, a total of seven cores collected from the Upper Kuttanad and nearby areas were used for detailed analysis. Three cores are from the floodplains of the Manimala, Pamba, and Achankovil rivers (one core each), two from the upper reaches of the Upper Kuttanad (Varattar and Kuttoor), and two (Pachcha and Edathua) cores from areas close to the downstream end of the Upper Kuttanad. The core collected from the Achankovil is a short core and it is generally sand dominated. The other cores are 12.00 m long in Puvathur and 9.00 m long in Karippumuri. These cores are generally made up of yellowish-brown silty sand or silty clay at the top followed by medium to coarse sand. The

Table 20.1 Sand, silt, and clay contents of the borehole cores samples of Kuttanad Kole Wetlands.

		Sand and sand dominated			Silt and silt dominated			Clay and clay dominated			Mud and mud dominated		
Physiographic zone	Core location	Sand (%)	Silt (%)	Clay (%)	Sand (%)	Silt (%)	Clay (%)	Sand (%)	Silt (%)	Clay (%)	Sand (%)	Silt (%)	Clay (%)
Upper Kuttanad	Varattar	66.49–83.31 (78.11)	1.26–11.85 (4.02)	1.05–18.26 (5.41)							27.68–30.32 (29)	29.55–31.12 (30.33)	40.13–41.2 (40.66)
	Karippumri	80.21–90.94 (86.33)	1.83–7.12 (4.17)	6.9–12.67 (9.49)							19.21–38.4 (28.6)	31.4–44.56 (37.40)	20.11–48.37 (33.99)
	Puvathur	50.14–87.94 (73.54)	3.27–18.06 (10.17)	8.79–32.38 (16.28)				0.56–16.35 (6.79)	27.33–37.53 (31.42)	56.32–70.03 (61.79)	5.56–40.47 (23.01)	22.66–46.29 (34.47)	36.87–48.15 (42.51)
	Kuttoor	74.24–79.14 (79.69)	13.82–17.2 (15.51)	7.04–8.56 (7.8)	2.25–5.74 (3.90)	52.57–61.65 (57.38)	32.61–45.18 (38.71)	0.93–3.79 (1.99)	38.1–43.87 (40.28)	54.68–60.97 (57.73)	2.23–49.24 (28.37)	26.43–49.91 (40.14)	16.05–48.66 (31.43)
	Pachcha				0.73–6.47 (3.6)	50.52–56.78 (53.65)	36.75–48.75 (42.75)	0.47–5.60 (3.95)	36.32–47.32 (43.97)	47.41–58.08 (52.08)	8.73–44.2 (33.15)	28.39–43.86 (34.49)	24.41–47.41 (32.36)
	Edathua	90.82	0.69	8.49				9.95–12.37 (11.16)	32.55–35.84 (34.19)	54.21–55.08 (54.64)	5.71–14.33 (8.93)	48.48–49.61 (48.90)	37.04–44.78 (42.17)
Lower Kuttanad	Alappuzha (i)	77.7–95.9 (87.08)	2.85–5.22 (3.63)	0.9–19.35 (9.28)				0.48–36.5 (9.81)	10.3–42.28 (17.94)	50.4–86.6 (72.26)	21.1	31.55	47.35
	Kaniyamkulam	64.85–72.03 (68.44)	3.47–9.45 (6.46)	24.5–25.7 (25.1)				0.26–37.78 (8.97)	7.87–49.52 (34.67)	47.79–63.39 (56.35)	24.28–43.33 (34.17)	21.32–45.98 (30.7)	20.25–47.85 (35.13)
	Nedumudi	58.54–79.89 (71.56)	0.53–8.63 (5.16)	7.79–40.93 (23.27)	5.44–5.82 (5.63)	53.03–56.61 (54.82)	38.17–41.53 (39.85)	0.52–39.26 (7.12)	4.53–46.7 (11.395)	50.13–94.52 (71.47)	13.34	46.29	40.37
	Kidangara	76.64–94.4 (86.02)	1.17–14.26 (5.06)	1.5–20.26 (8.91)				0.27–22.69 (4.56)	5.31–44.7 (23.37)	50.49–89.93 (72.08)	14.48–44.86 (32.26)	34.28–48.1 (39.41)	20.76–45.49 (28.32)
	Thakazhi (i)	59.27–86.7 (74.4)	2.74–14.99 (7.95)	10.56–29.41 (17.65)				6.13	38.46	55.41	1.6–39.33 (17.04)	26.59–50.39 (40.88)	34.08–52.75 (42.08)

(*continued on next page*)

Table 20.1 (*continued*)

Physiographic zone	Core location	Sand and sand dominated			Silt and silt dominated			Clay and clay dominated			Mud and mud dominated		
		Sand (%)	Silt (%)	Clay (%)	Sand (%)	Silt (%)	Clay (%)	Sand (%)	Silt (%)	Clay (%)	Sand (%)	Silt (%)	Clay (%)
North Kuttanad	Kumarakom							0.46–5.25 (1.99)	21.9–47.28 (31.73)	78.88–76.11 (66.28)	13.52	44.05	42.43
	Thiruvarpu	71.3–74.52 (72.38)	8.95–13.39 (11.56)	15.31–16.53 (16.05)				0.69–18.83 (6.26)	28.81–46.81 (37.71)	51.83–69.19 (59.02)			
	Pallam							0.19–2.1 (0.845)	7.16–39.72 (23.55)	58.18–92.65 (73.60)			
Vembanad	Pathiramanal	93.34–94.99 (94.29)	1.35–3.69 (2.11)	2.35–4.57 (3.59)				0.9–17.41 (4.66)	5.27–44.52 (22.04)	51.35–92.25 (75.97)			
Western coastal sand dunes	Muhamma	55.03–98.17 (73.24)	0.98–10.1 (5.54)	0.7–37.08 (20.84)				0.31–25.01 (9.16)	1.13–26.87 (13.87)	54.51–97.9 (75.97)	3.39–44.55 (26.25)	23.74–48.71 (32.37)	30.8–47.9 (41.39)
	Kattoor	65.7–99.27 (80.62)	0.43–24.33 (9.32)	0.25–24.47 (10.01)				1.2–33.29 (9.83)	13.85–27.34 (18.88)	51.93–77.79 (71.28)			
	Alappuzha (ii)	86.92–97.93 (93.69)	0.55–7.82 (2.87)	0.78–10.05 (3.43)				4.23–19.94 (12.07)	20.96–30.15 (26.06)	51.43–71.69 (61.32)			
	Kalarkod	82.67–90 (85.87)	3.5–7.33 (5.13)	6.0–10.0 (9.00)				1.48–4.89 (2.55)	20.1–47.59 (28.8)	50.6–77.5 (68.8)	4.86	48.86	46.45
	Thakazhi (ii)	51.18–97.71 (77.53)	0.63–21.62 (9.53)	1.66–27.2 (12.54)				1.60–13.93 (6.15)	22.98–33.01 (28.05)	55.42–74.76 (65.79)	29.47–34.59 (32.4)	25.26–32.23 (29.74)	33.68–45.27 (37.86)
	Thottappally	77.5–94.97 (90.08)	0.58–10.97 (4.43)	1.75–13.7 (5.48)	2.97–23.99 (14.45)	50.6–52.46 (51.27)	25.3-46.46 (34.29	0.59–9.27 (3.31)	25.98–44.55 (32.35)	51.85–72.43 (64.35)	20.23–46.72 (34.49)	10.14–43.77 (32.87)	10.5–43.14 (32.64)

sand, silt, and clay contents of this layer are given in Table 20.1. No detailed textural analysis has been carried out in the Budhanur core (Achankovil river). In Puvathur and Karippumuri cores, the top oxidized sequence of floodplain sediments rests on an organic-rich/peaty layer with high loading of organic carbon. Decomposed vegetative debris often with huge tree trunks is also common in the subsurface formations. For example, many dug out wells in the floodplains and exposed sections at its base are embedded with buried wood/tree trunks and, branches and vegetative debris. In an earlier study, Kumaran et al. (2016) reported occurrence of a huge buried tree (*Cynometra* sp.) in uprooted position in a well near the Puvattur borehole core site. Similar such occurrences of tree trunks of *Dipterocarpus*, and many other unidentified plants have been reported from the exposed sequences of the Manimala and Achankovil rivers (Kumaran et al., 2014). The cores from Varattar and Kuttoor are sited close to the southwestern margins of the upper Kuttanad. The Varattar core is a shallow one and is very similar to the top layers of the floodplain cores described earlier. The core exhibits a fining upward sequence with channel sand (sand 79.84%) at the base, and silt and clay (mud) dominated sequence at the top (sand 30.32%; silt + clay 69.98%). The bottom most channel sand is slightly gravelly and often mixed with pottery pieces presumably derived from the upstream inhabited areas of river basin. Recently, many such artifacts indicating early historical human habitations have been reported from many parts of the Pamba–Manimala rivers basins. The Kuttoor core is collected from the upstream reaches of the upper Kuttanad. The occasional presence of shell fragments is noticed at certain levels in the lagoonal clay dominated part of the core. The Edathua and Pachcha borehole cores, *stricto sensu*, fall in the lower Kuttanad area of Chattopadhyay and Sidharthan (1985), but close to the lower boundary of the Upper Kuttanad. However, in the present study, these cores are also considered along with Upper Kuttanad and flood plain cores. Further, the stratigraphic analysis clearly indicates that these cores exhibit a similarity to the Kuttoor core, although the cores contain more marine shells in the lagoonal clays. The upper layer of these two cores is yellowish brown and with fluvial mud dominated sediments. This is followed by lagoonal clays with broken and intact shells of marine pelecypods and gastropods. The averages and ranges of silt and clay dominated sediments of Pachcha and Edathua cores are summarized in Table 20.1.

20.5.1.2 Lower Kuttanad

Five borehole cores (Nedumudi, Alappuzha (i), Kaniyamkulam, Thakazhi (i), and Kidangara) were examined for finding out the stratigraphic architecture of the lower Kuttanad region. The cores are generally composed of clay dominated sediments with interlayers of sand dominated sediments at certain levels. The Kaniyamkulam core is interlayered with a shell bed of the pelecypod *Villoritta* sp. Occurrence of *Villoritta* sp. is noticed at many locations in the paddy lands and kayal lands of KKW. The core collected from Kidangara is the longest among the cores from the Lower Kuttanad. It is having a core length of 35 m and is composed of 1–2 m thick yellowish-brown sediment at the top with average sand, silt, and clay content of 60.85%, 15.46%, and 23.67%, respectively. The organic carbon and $CaCO_3$ contents are substantially low in the layer at the top which is generally in oxidized state. It is then followed by lagoonal clay dominated sediments with decayed wood and broken/unbroken shells of pelecypods and gastropods. Subfossil shells of the genus *Nerita* sp. is noticed at 9.9 m level of the core. The average sand, silt, and clay contents of the lagoonal sediments are 27.29%, 21.63%, and 51.07%, respectively. The entire sequence rests over lateritic basement. The Nedumudi core is collected from almost in the central part of the KKW and close to the Kayal lands. The core is 23 m long and is composed essentially of 19.75 m thick silt and clay dominated sediments intervened at 13–15 m level by a sand layer. Like the case of Kidangara core, the clay dominant sequence of the Nedumudi core also rests over a lateritic basement.

The Thakazhi (i) core was collected from the southern part, close to the Western coastal sand dune. The core is 27.5 m long and composed of a yellowish brown, fluvial sediment of 2 m thick at the top (sand 10.2%, silt 50.39%, and clay 39.41%) followed by littoral sands with broken and unbroken shells. The layer accounts for 1.6–86.7% (38.3%) of sand, 2.74–50.39% (27.17%) silt, and 10.56–55.41% (34.52%) of clay. The entire sedimentary sequence rests over a hard lateritized basement. The Kaniyamkulam core is 6 m long and was collected essentially from the eastern boundary of Kayal lands. The core is composed of silt and clay dominated sediments which are intervened by 1–1.5 m thick shell bed containing entirely of matured valves of the pelecypod *Villoritta* sp. Mining of subfossil shells for manufacture of lime-based products is common in the area. The Alappuzha core (i) is collected from an area located between the eastern margin of the Western strand plains and the southern Vembanad Lake. The core begins with ˜1 m thick sand dominated layer (sand 94.99%, silt 1.59%, and clay 3.42%) followed by silt and clay dominated bed with broken and unbroken shells. Shells of *Villoritta* sp. unidentified gastropods and other pelecypods are recorded from the upper part of the silt and clay dominated layer. The layer was intervened at the middle by a sand dominated layer of 1.3 m thickness. The presence of unbroken shells of the pelecypod *Venus* sp. was recorded from the layer.

20.5.1.3 North Kuttanad

The North Kuttanad is essentially influenced by the Minachil river and its tributaries in the eastern side and tidal processes in the west. Out of the four borehole cores retrieved from the zone, the one collected from Kumarakom is the deepest and composed essentially of organically rich, clay dominated sediments. Occasional presence of broken shell/shell dust is noticed in the upper part of the core. The entire sequence rests over lateritized Neogene sandstones. The Thiruvarpu core was collected from a location close to one of the distributary channels of the Minachil river. The core was composed of a clay dominated layer (sand 6.26%, silt 37.71%, and clay 59.02%) sandwiched between two comparatively thin sand layers. The clay dominated middle layer embeds many broken and unbroken molluscan shells. Among the shells, gastropods like *Cantharus, Fusinus, Nerita*, and *Drupa*, and pelecypods like *Venus, Lucina*, and *Ostrea* the predominant ones. The entire sequence lies above lateritized Neogene sediments. The sediment characteristics of two dated cores with length of ˜6 m collected from the floodplains of the Minachil river by Ajaykumar et al. (2012) were also used in this study to address the riverine input into the receiving waterbody of the Holocene epoch. The core locations are depicted in Fig. 20.1. The core is composed essentially of alluvial clay–sand sequence. The channel sand is intervened by organic rich, 0.6 m thick, clay with decayed wood and other vegetative debris. The clay dominated top layer is yellowish brown and composed of silts and clays with appreciable amount of fine to very fine sand. The Pallam core (14 m) is collected from the eastern boundary in the widest part of the southern Vembanad Lake. The core is composed essentially of clay dominated sediment with broken and unbroken shells of pelecypods and gastropods. The entire sequence rests over a laterite basement.

20.5.1.4 Pathiramanal Island in Vembanad Lake

The Pathiramanal borehole core collected from the Pathiramanal Island in the Vembanad Lake is 23.0 m long and sited within the lake close to the eastern border of the Western coastal strand plains. The island is separated from the Western coastal strand plains by ˜500 m wide channel. The core is composed of 11.5 m thick, light gray, medium to fine grained sand with occasional presence of shells of the pelecypod *Villoritta* sp. The layer accounts for 0.9–94.99% of sand, 1.35–44.52% of silt, and 2.35–92.5 (40.01)% of clay. The sand dominated upper layer is followed by 12.75 m thick, silt and clay dominated lagoonal sediments which rest over a lateritic clayey sand. On average, the layer contains 42.79% of sand, 17.2% silt, and 40.01% of clay.

20.5.1.5 Western coastal strand plains and barrier beaches

A total of six cores are examined from the Western coastal strand plains to understand its subsurface sediment architecture. The cores are composed generally of a top sandy layer, often interlayered at certain levels by comparatively finer clastics like silt and clay or both. In the Muhamma core, there are two major litho units–19 m thick upper sand dominated layer, intervened at 3.5–5.5 m level by mud layer, and 12.5–13.0 m level by clayey sand layer. The sediments often contain broken and unbroken shells of marine affinity. The average sand, silt, and clay contents of this layer are 55.16%, 9.09%, and 35.75%, respectively. The sand dominated layer is followed by silt and clay (mud) dominated sediments with 26.25% of sand, 32.37% of silt, and 41.49% of clay. The Kattoor core is collected from almost the middle part of the Western coastal strand plains. The entire sequence is composed of light gray, sand dominated sediments intervened at 7–9 m level by a composite sequence with alteration of mud and sand dominated layers. Broken shells/shell dust are occasionally seen in the borehole core sediments. The sand, silt, and clay contents of the sand dominated layers are 65.7–99.27% (80.62), 0.43–24.33 (9.32%), and 0.25–24.47 (10.01%), respectively. The Alappuzha (ii) core, collected from the Western margins of the coastal strand plains are composed of 14.75 m thick, coarse to medium grained sand at the top followed by grayish black, clay dominated sediments (>17.25 m) with broken and unbroken shells. The top marine/littoral sand layer contains 95.59% of sand, 3.63% of silt, and 0.78% of clay whereas the clay dominated sediments at the bottom content 12.07% sand, 26.06% silt and 61.32% clay.

The Kalarkod core (40 m), collected from the Western coastal strand plains, is located about 5 km east of the present coastline. The core is composed of a 23 m thick sand layer intervened at 7–10 m level by a clay dominated layer with shell (broken and unbroken) contents. The sand, silt, and clay contents of this layer are 2.63%, 26.87%, and 70.5%, respectively. The sand, silt, and clay contents of the intervening clay dominated layer are 2.4%, 20.1%, and 77.5%, respectively. The sand dominated layer is followed downward by a 17 m thick silt and clay dominated layer with occasional presence of shells (broken and unbroken). This layer accounts for 4.86% sand, 48.86% silt, and 46.45% clay (Table 20.1).

The Thottapally (43 m) borehole core is drilled on the strand plains south of the Thottapally spillway. The core begins with 2.5 m thick, organic-rich clay dominated sediments, followed by 18 m thick coarse to medium grained, quartzose sand with substantially high content of heavy mineral placers and shell pockets at different levels. A *Turritella* bed is noticed at 10 m below ground level (bgl). The sand, silt, and clay dominated upper layer are 20.23%, 40.95%, and 38.82%, respectively. The entire sequence rests over a 22.5 m thick clay dominated layer. The sand, silt, and clay contents of the clay dominated layer are 0.59–25.27% (9.46%), 25.98–52.46% (39.12%), and 25.3–72.43 (51.41%), respectively.

20.5.2 Organic carbon (C-org) and $CaCO_3$

Organic carbon and $CaCO_3$ are integral components of coastal sediments. They generally reach depositional environments from allogenic and authigenic sources. Table 20.2 summarizes the ranges and averages of C-org and CaCO3 in selected borehole cores of KKW. In the Upper Kuttanad and nearby floodplains, three cores are studied for C-org and $CaCO_3$. They are Karippumuri, Puvathur and Pachcha cores. In the Karippumuri and Puvathur cores, the C-org content exhibits a wide variation of 0.32 – 34.94%. The highest values of C-org are noticed for the peat samples in both the cores (Karippumuri: 19.99%, Puvathur: 34.94%). The sand dominated sediments generally record the lowest C-org values. In the case of Pachcha core C-org values vary from 0.74% to 3.37% (av. 2.31%). The lowest value of 0.74% is noticed for the yellowish brown, fluvial sediment at the top. The lagoonal sediments register a C-org content of 2.37–3.37%. The content of $CaCO_3$ varies from 1.5% to 26.0%. The highest $CaCO_3$ value is noticed at 8.5 m bgl where the sample is rich in calcareous shells. Out of the four cores examined for C-org and $CaCO_3$ contents in the Lower Kuttanad, except Kidangara, the other cores exhibit almost the same C-org contents in the range of 0.26–4.16% Alappuzha (i) 1.94%; Kaniyamkulam 1.52%; Nedumudi 0.46%. The Kidangara core, which is located close to the eastern periphery of the KKW, records wide variation of C-org (0.04–11.57%; 4.10%). The sand dominated sediments record the lowest content of C-org whereas the silt and clay dominant ones the highest. The content of $CaCO_3$ shows wide variation in the Alappuzha (ii) and Kidangara cores where shell pockets and broken shell contents are plenty at certain levels. The shell bed in the Kaniyamkulam core is exempted from $CaCO_3$ measurements as the layer is almost entirely composed of shells of *Villoritta* sp. In the North Kuttanad, the cores retrieved from Kumarakom and Pallam are studied for the C-org and $CaCO_3$ contents. The C-org content varies from 1.44% to 4.69% in Kumarakom and from 2.54% to 6.06% in Pallam. $CaCO_3$ data are available only for the former (Kumarakom core) and varies from 0.25% to 2.63%. The Pathiramanal island core accounts for an average C-org content of 3.02% (1.00–6.20%) and $CaCO_3$ content of 10.40% (3.00–10.40%). Out of the four cores studied for C-org and $CaCO_3$ contents in the Western coastal strand planes, the lagoonal clay of the Alappuzha core, lying below the coastal sands, accounts for the highest content of C-org with average value of 5.20%. However, the other cores recorded substantially low values of C-org, as majority of them are sand dominated or with poor accumulation of organic matter in the depositional site. The $CaCO_3$ content was higher (16.25%) in the Kalarkod core compared to other cores.

20.5.3 Heavy mineralogy

The heavy minerals in sediments are one of the most effective tools to discriminate the provenance/source rock characteristics of sediments (Blatt et al., 1972; Pettijohn, 1985). Taking this into account, an attempt has been made here to examine the heavy mineral contents in the sediments of KKW which receives sediments from multiple provenances. Table 20.3 summarizes the ranges and averages of heavy minerals in selected cores of KKW. Two cores are examined for heavy mineral contents in the Upper Kuttanad and floodplains. The fine sediment fraction of the Puvathur core in the floodplain of Pamba river is made up essentially of inosilicates (except at the lowermost level where opaques dominate in the heavy mineral crop), followed by opaques. Garnet and zircon are noticed in minor amounts. Traces of sillimanite and biotite are noticed at certain levels of the cores. The second borehole examined is from Pachcha which is made up of a three tier association of heavy mineral suite. The upper,

Table 20.2 Ranges and averages of organic carbon ($C_{-org.}$) and $CaCO_3$ in selected borehole cores of Kuttanad Kole Wetland.

Physiographic zone	Core location	Sand and sand dominated		Silt and silt dominated		Clay and clay dominated		Mud and mud dominated		Peat
		$C_{-Org.}$ %	$CaCO_3$ %	$C_{-Org.}$ %	$CaCO_3$ %	$C_{-Org.}$ %	$CaCO_3$ %	$C_{-Org.}$ %	$CaCO_3$ %	$C_{-Org.}$ %
Upper Kuttanad	Karippumri *Kadavu*	0.35–1.12 (0.58)						0.32–10.1 (1.96)		2.2–18.99 (11.06)
	Pachcha			2.67–3.37 (3.02)	1.5–2.5 (2)	2.87–3.31 (3.09)	1.5–3.5 (2.5)	0.74–3.04 (2.05)	4–50 (26.66)	
Lower Kuttanad	Alappuzha (i)	0.49–2.75 (1.62)	3.5–4.5 (4)			0.52–2.78 (2.04)	0.5–10 (4.04)			
	Kaniyamkulam	0.26–0.92 (0.59)	1.5	2.46	0.5			0.26–2.63 (0.93)	1–4.25 (1.87)	
	Nedumudi	0.19–1.58 (0.64)	0.51–2.12 (1.39)	3.66–4.16 (3.91)	0.15–0.29 (0.22)	0.1–3.93 (2.78)	0.15–4.55 (1.41)	2.95	2.4	
	Kidangara	0.04–5.66 (2.17)	0.19–3.04 (1.92)		0.27–11.57 (4.64)	0.36–7.84 (4)	7.23–10.77 (8.94)	0.46–4.47 (2.26)		
North Kuttanad	Kumarakom					1.44–4.08 (2.9)	0.25–2.63 (1.62)	3.63–4.69 (4.16)	1.22–2.25 (1.73)	
	Pallam					2.54–6.06 (4.03)				
Vembanad	Pathiramanal	1.2–1.5 (1.35)	3–3.2 (3.1)			1–6.2 (3.85)	5.9- 10.4 (8.15)			
Western coastal sand dunes	Muhamma	0.9–1.4 (1.1)	1.5–4 (3.13)			2.1–4 (3.05)	8–10.4 (9.2)	1–4.2 (2.6)	6–8.2 (7.1)	
	Kattoor	0.08–3.45 (0.51)				3.12–3.95 (3.59)				
	Alappuzha (ii)	0.2–0.48 (0.34)				7.98–19.95 (12.47)				
	Kalarkod	1.46	10–16 (13)			0.33	20–24 (22)	2.66		
	Thottappally	0.09–0.35 (0.2)	0.25–2.5 (1.35)	1.72	7.13	2.18–2.43 (2.31)	0.5–1.75 (1.12)	0.79–9.09 (5.8)	0.39–8.5 (3.13)	

Table 20.3 Ranges and averages of heavy mineral contents in selected borehole cores of Kuttanad Kole Wetland.

Sl. no.	Bore hole core	Depth (m)	THM (wt.%)	TLM (wt.%)	Number %						
					Opaques	Sillimanite	Zircon	Monazite	Ionosilicate	Garnet	Others
1	Puvathur	0–3	5.21–9.11 (7.39)	90.89–94.79 (92.59)	18.89–38.28 (27.6)	0.56	0.73–1.67 (1.06)		57.81–77.78 (68.39)	1.11–3.3 (2.51)	
		3–10	4.91	95.09	15.8		0.29		81.9	1.72	
		11.5–12	2.88	97.12	78.95		2.63		15.79	2.63	
2	Pachcha	2–3	8.5–9.6 (9.05)	90.4–91.5 (90.95)	28–42 (35)	17–22 (19.5)	3–4 (3.5)		12–42 (27)	2–4 (3)	
		3–8.5	2.1	97.9	40	49	7			4	
		12–14	5.5–13.4 (8.33)	86.6–94.5 (91.6)	69–78 (73.33)	16–25 (21.33)	2.5–5 (4.1)			0.5	
3	Kidangara	1.3–4.5	4.17–8.67 (6.34)	91.33–95.83 (93.65)	10.06–23.4 (16.48)	1.16–6.38 (2.78)	0.58–1.19 (0.78)		58.33–79.88 (69.29)	0.58–1.18 (0.88)	1.12–2.13 (1.35)
		34–34.5	61.38–70.31 (65.84)	29.69–38.62 (34.15)	100						
4	Thakazhi	0–6.5	3.44–8.14 (5.74)	91.86–95.56 (93.92)	10.48–64.29 (39.73)	4.19- 25.9 (16.99)	0.89–1.32 (1.11)		2–24 (9.56)	1.4–3.3 (2.35)	
		10–20	1.87–2.84 (2.35)	97.16–98.13 (97.64)	42.4–60.42 (51.41)	30–41.6 (35.8)	8.02–9.6 (8.81)				
		21–27	11.56–26.96 (19.26)	73.04–88.44 (80.74)	86.56–92.83 (89.69)	3.87–6.45 (5.16)	1.15–5.37 (3.51)				
5	Kumarakom	0–10	4.64–9.12 (6.88)	90.88–95.36 (93.12)	3.95–6.56 (5.25)	3.98–4.92 (4.5)			4.37–5.72 (5.04)		

(continued on next page)

Table 20.3 (*continued*)

Sl. no.	Bore hole core	Depth (m)	THM (wt.%)	TLM (wt.%)	Number %						
					Opaques	Sillimanite	Zircon	Monazite	Ionosilicate	Garnet	Others
		39–45	6.93–14.41 (9.63)	85.59–93.07 (90.33)	61.75–91.44 (77)	2.56–27.32 (14.12)	1.71–4.52 (3.29)	1.28–2.0 (1.71)			
6	Pathiramanal	1.5–11.3	2.56–5.43 (3.50)	94.57–97.44 (96.49)	36.27–43.96 (39.23)	45.93–53.29 (49.87)	2.05–5.66 (3.67)	0.69	0.64–2.67 (1.41)	0.89	
		24–25	6.43	93.57	89.73	4.40	1.47				
7	Muhamma	3–18	0.06–3.36 (1.36)	96.64–99.64 (98.63)	4.15–43.77 (25.1)	47.06–88.89 (66.01)	2.02–5.65 (3.73)		0.94–1.01 (0.96)	0.84–1.09 (0.96)	
		18–30	0.94–1.16 (1.05)	98.84–99.06 (98.95)	7.28–14.87 (11.07)	53.13–59.21 (56.17)	3.94–4.68 (4.31)				
8	Kalarkod	1–22.5	0.69–12.33 (3.25)		9.49–65.42 (43.65)	25.27–54.6 (40.47)	0–3.01 (1.35)				
9	Thottappally	2–3	11.07	88.93	43.41	46.57	2.87			2.15	
		3–12	0.92–45.29 (15.83)	54.71–99.08 (84.14)	26.14–70.63 (47.73)	21.59–63.83 (45.64)	0.48–3.31 (1.8)	0.41–0.48 (0.45)	1.21–1.29 (1.25)	1.64–1.8 (1.73)	
		22–40	2.67–9.67 (4.07)	93.53–97.33 (95.99)	25.61–38.57 (32.49)	10.26–52.25 (29.87)					

river influenced yellowish brown, silt and clay dominated layer is composed of opaques, inosilicates, and sillimanite as the major minerals. At the same time, middle lagoonal clay dominated layer is with opaques and sillimanite as the major heavies. Minor amounts of zircon and garnet are also noticed in the layer. Interestingly the bottom part, represented by lateritized Neogene sediments, is dominated by opaques with appreciable quantities of sillimanite.

Of the four cores examined in the Lower Kuttanad, the Kidangara core is located in the eastern part, Nedumudi in the central and Thakazhi (i) and Alappuzha (i) in the western side. The Kidangara core is composed dominantly of inosilicates with appreciable quantity of opaques. Minor amount of zircon and sillimanite are also noticed in the core. In Nedumudi, except the top layer, the rest of the core is composed of opaques and sillimanite with minor amounts of zircon. The top layer, on the other hand, is composed essentially of inosilicates. The heavy mineral crop in the Alappuzha (i) core is made up essentially of opaques and sillimanite which together constitute more than 85% of the heavy mineral residue. Zircon is present in small amounts whereas inosilicate is confined only to the top layer. Like the Alappuzha (i), the Thakazhi (i) core is also composed of mainly opaques and sillimanite with minor quantities of zircon. The top layer of the core is dominated by opaques, sillimanite, and inosilicates with minor quantities of zircon. Only one core—Kumarakom core—is studied for heavy mineral content for representing North Kuttanad. The clay dominated upper part is with minor amounts of sillimanite and inosilicates. The lower, lateritized Neogene sediments are made up essentially of opaques and sillimanites with minor amount of zircon and trace amounts of monazite. The lone core collected within the Vembanad Lake is from the Pathiramanal island and is composed of opaques and sillimanite as the major minerals and zircon and inosilicates as minor minerals. Traces of monazite and garnet are noticed at certain levels. Three representative cores collected from Muhamma, Kalarkode, and Thottappalli are examined in the case of Western coastal strand plains and barrier beaches. The heavies in these cores are made up of sillimanite and opaques with minor amount of zircon. Traces of inosilicates are noticed in the top of the Muhamma core. Occurrences of garnet, monazite, and rutile are noticed in traces in the Thottappalli core at certain levels.

20.5.4 Palynology

The recovery of organic contents in the sedimentary archives of all the studied boreholes of KKW provided signatures of biological remains indicating preferential niches and environmental changes. These biological entities include pollen, spores, NPPs, and organic walled parts (palynodebris) that could be easily retrieved during the palynological processing. Though the overall organic recovery has been fairly consistent throughout the Holocene, their paucity toward the late-Holocene reflected the changing scenario of depositional environment due to sea-level oscillations and the subsequent modifications of the landforms and vegetation cover. Significant features of palynological and NPP remain indicated changes in ecological facies and depositional environments and these included hinterland, freshwater swamp, floodplains, estuarine, and shallow marine conditions. The fairly abundant occurrence of *Cullenia exarillata* (Bombacaceae) pollen at certain intervals of various boreholes is significant particularly while inferring climatic conditions such as heavy rainfall and wet period. Being a member of wet evergreen forest and enjoying a high precipitation range (>3000–5000 mm), the occurrence and relative abundance of *Cullenia* pollen in the marine, marine–brackish–lagoonal facies indicates heavy rainfall (Fig. 20.3). As this plant grows very far from the sea coast and is presently

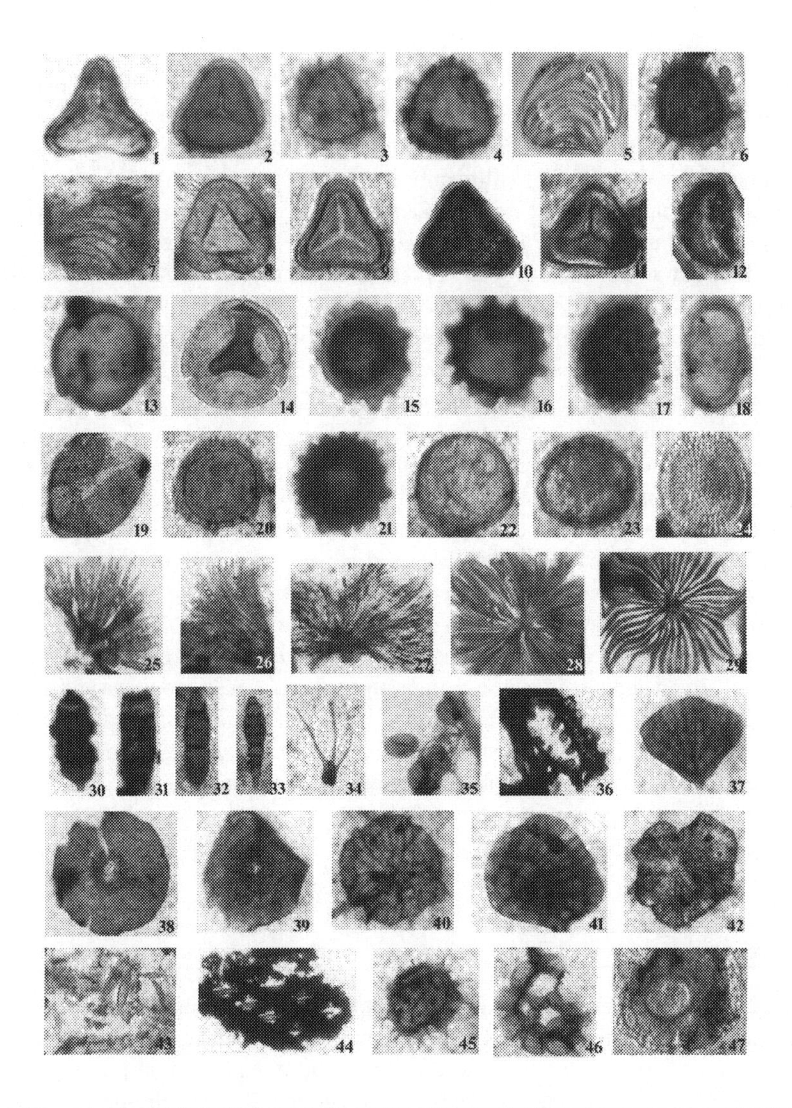

FIGURE 20.3

Selected pollen, spores, and other nonpollen palynomorphs (NPPs) of specific ecological/environmenta1 conditions of the study area. (All photomicrographs are enlarged ca. 500 × except fungal remains, which are, enlarged ca. 1000 × unless otherwise specified). 1, 9. *Cyathea* sp. (Cyatheaceae), 2–4. *Pteris* sp. (Pteridaceae), 5, *Ceratopteris thalictroides*—fresh water flood plain indicator, 6. *Anemia phyllitidis* (Anemiaceae), 7. *Schizaeisporites multistriatus* (Schizaeceae), 8. *Glechenia* sp. 10. *Pteridacidites* sp. (Pteridaceae), 11. *Cyadhidites* sp. 12. *Polypodium vulgare* (Polypodiaceae), 13, 14, 22. *Cullenia exarillata* heavy rainfall indicator, 15, 21. Euphorbiaceae pollen, 16. Malvaceae pollen, 17. *Compositoipollenites* sp. (Compositae), 18. *Sonneratia* sp. Mangrove facies, 19. *Liliacidites* sp. (Liliaceae), 20, 23. Meliaceae pollen, 24. *Croton* sp. (Euphorbiaceae), 25, 27, 29. *Gleotrichia* sp. morphotypes (Cyanobacteria), 26, 28. *Rivularia* sp. morphotypes (Cyanobacteria), 30, 33. *Multicellaesporites* sp. (fungal spore), 31. *Pluricellaesporites* sp. (fungal restricted to relatively higher altitudes, the pollen grains must have been transported and eventually deposited into the wetlands. The intervals of 10.0–16.0 m in Kumarakom, 2.40–2.50 m in Pallam of North Kuttanad, and 6.40–6.50 m in Puvathur of Upper Kuttanad are considerably important in this aspect. In fact, the radiocarbon dates of a few samples of the above interval at other locations fall within the period (10–4 ky BP) of increased monsoon over the Indian subcontinent.

The contribution of mangrove elements helped distinguish estuarine facies in some of the borehole samples. Heavy accumulation of organic matter, probably contributed by the mangrove vegetation. This observation is significant in Thakazhi (10.0–11.0 m) and a major portion of the Kumarakam borehole. As the mangroves are the most common contributors of organic matter in estuaries and situated generally in the intertidal region, they are directly related to the sea level at the time of deposition. The spores of *Ceratopteris thalictroides* at certain intervals along with a few NPP indicated the transition from estuarine facies to freshwater flood plains as a result of freshwater influx toward the wetland in the KKW.

The NPP constitutes organic walled microfossils that are frequently encountered in the palynological preparations during pollen analysis. Like pollen and spores they too are resistant to corrosion and get preserved but not destroyed during maceration. They provided an alternative source of paleoinformation where there was a scarcity of pollen and spores in the sediments. Freshwater algae dominated the assemblage while fungi and invertebrates were sparsely represented depending on the facies. The occurrence of *Botryococcus, Pediastrum*, and colonies of *Rivularia* (cyanobacteria) at certain intervals in the subsurface sediments has considerable significance to ascertain the hydrological changes associated with the evolution of KKW. Cyanobacteria (*Rivularia* type) infer phosphate eutrophication of water bodies of the coastal plains and their relative abundance may be attributed to nutrient-rich habitats around the coastal plains resulting into local bloom associated with rainfall conditions/freshwater influx. *Pediastrum* being allochthonous, indicates freshwater stream activity, gets transported as freshwater influxes into the lagoons during the pluvial periods of high rainfall. We have observed that their implications are significant in pollen pauper sediments deposited in arid and fewer rainfall periods of the late-Holocene ~3500–2500 ky BP. The intervals of the boreholes belonging to Nedumudi (4.90–5.00 m) and Kidangara (3.40–3.50 m) in the lower Kuttanad are significant while drawing inference to this aspect. *Glomus* cf. *fasciculatum* and thecamoebians are found to be useful in the interpretation of soil conditions associated with aridity/stressed environment particularly toward late-Holocene (<3500 yr BP). Further, their occurrence along with palynomorphs of aquatic and terrestrial vascular plants in the KKW is of immense significance in the reconstruction and the Holocene evolution of the region associated with the climatic changes and sea-level oscillations.

spore), 32. *Phragmospore* sp. (fungal spore), 34. *Frasnacritetrus* sp. (fungal spore), 35. *Glomus* sp. (fungal spore) erosion indicator, 36. *Meliolinites* sp. (fungal spore), 37, 41. *Kutchiathyrites eccentricus* (fungal fruiting body), 38. *Notothyrites* sp. (fungal fruiting body), 39. *Phragmothyrites* sp. (fungal fruiting body), 40. *Callimothallus assamicus* (fungal fruiting body), 42. *Callimothallus* sp. (fungal fruiting body), 43. Sponge spicules (indicator of dry period), 44. Poaceae cuticle (charcoal) (suggestive of dry period), 45. *Spiniferites* sp. (Dinoflagellate) (indicator of marine environment), 46. Foraminiferal lining (indicator of marine environment), 47. *Thecamoeba* sp. (indicator of pollution, acidity, and stress).

FIGURE 20.4

Drilling borehole cores from Pallam site of Kuttanad Kole Wetland (A); *Villoritta* sp., excavated during construction of an open well at Kaniyamkulam (B); White shell (*Villoritta* sp.) embedded in the matrix of yellowish-brown clays (C); White shell deposits in the paddy lands of Kuttanad Kole Wetlands (D).

20.5.5 Lime shell occurrences

The KKW contain economically viable deposits of subfossil shells of the Pelecypod *Villoritta* sp. along with the peripheral areas of the southern Vembanad Lake (Fig. 20.4A and B). Patches of shell deposits also occur in the lower part of the lower Kuttanad (Fig. 20.4C and D). Although the dominant species is *Villoritta* sp. trace quantities of gastropod and other bivalve shells are also noticed at certain places. The deposits occur as isolated patches/lenticles/lenticels in the late-Holocene sediments. Fig. 20.1 shows the spatial distribution of subfossil shells in the Vembanad Lake. The thickness of the shell deposits varies from a few centimeters to a few meters. Major shell occurring areas are Thannirmukkam South, Kumarakom, Kainakari, Punnamada, and adjoining paddy land areas like Chithira, Marthandam and Rani Kayals, Mathi Kayal, Kakka Kayal, Arupanku, C-block, and Kuppappuram.

There are many attempts made in the past to quantify the lime shell occurrences in the KKW. In 1941, the Travancore Cements Limited in collaboration with the Department of Research of the erstwhile University of Travancore conducted a detailed survey of the lime shell deposit of the KKW. They estimated a total deposit of 4.5 million tons in an area of about 93 km^2 (Anonymous, 1984). Geological Survey of India, during 1962–1963 made a reinvestigation and reported that the area contains a shell reserved of 2–2.5 million tons. In 1976 the Department of Mining and Geology has proved the existence of fairly thick deposits of lime shell in Kulasekharamangalam (0.182 million tons), Pallippuram (0.104 million tons), Vechoorpadam (0.257 million tons), and Thanneermukkam (0.249 million tons) area of Alappuzha and Kottayam districts. Borehole drilling investigation carried out as a part of this study revealed that limeshell deposits occur generally 0.5–2 m below the ground surface/lake bed and deposited over lagoonal clays. The deposit occurs as planar or lenticular beds. But in the paddy lands in the peripheral areas of the Vembanad Lake, lime shell often occurs just a few centimeters below ground surface.

20.5.6 Geochronology

Selected samples of shells, wood fragments, peat, and sediment samples were subjected to radiocarbon dating for establishing the chronological bearing of the depositional environments. The radiocarbon dates of the borehole cores collected from the Upper Kuttanad and the floodplains gave an age of Holocene (Table 20.4). The organic carbon-rich fluvial sediments/peats within the sequence gave a late-Holocene age (Karippumuri 3100 ± 150 yrs BP; 9 m bgl/Puvathur 1670 ± 70 yrs BP; 7 m bgl). In the Upper Kuttanad and nearby deltaic areas, the borehole core yielded a late-Holocene age for the organic carbon-rich basal part of the fluvial sediments (Kuttoor 3687 ± 64 m yrs BP; 9 m bgl) whereas middle–early-Holocene ages for the underlying lagoonal sediments (Kuttoor: 7249 ± 37 yrs BP; 11 m bgl/Pachcha 6990 ± 160; 8.5 m bgl; Table 20.4).

Out of the four borehole cores collected from the Lower Kuttanad, the cores collected from its eastern and central part gave a late-Pleistocene–early-Holocene age (Kidangara 6880 ± 130–10,560 ± 90 yrs BP; Nedumudi 9900 ± 100–12,390 ±70 yrs BP) whereas the cores collected from the Lower Kuttanad, close to the Western coastal strand plains gave a middle–late-Holocene age (Kaniyamkulam 3360 ± 110 yrs BP; Alappuzha (i) 4840 ± 170–6780 ± 160 yrs BP). Further, the cores at certain levels exhibited inverted ages. This might be attributed either due to the reworking of older marine sediments from the coastal/nearshore areas into the eastern or central part of the Lower Kuttanad areas where the width of the lagoon attains its maximum or due to marginal contributions of organic matter-rich sediments from the Neogene older formations in the eastern hinterlands. Discrimination of these two sources requires detailed investigation which is beyond the scope of the present study. Radiocarbon dates of only two borehole cores are available for the North Kuttanad region of the present study. This region is also influenced by fluvial processes essentially by the Minachil river. To address the depositional processes of the physiographic unit, the published radiocarbon dates of wood embedded within the organic matter-rich layers intervened in channel sands by Ajaykumar et al. (2012) are also used. The wood sample collected from a floodplain sequence exposed on the left bank of the river yielded an age of 2888 ± 78 yrs BP (depth 6 m bgl); whereas, the section exposed on the right bank of the river yielded an age of 5570 ± 30 yrs BP (depth 6 m bgl; Ajaykumar et al., 2012). The Thiruvarpu borehole core, considered for the present study gave an age range of 4630 ± 100–9800 ± 130 yrs BP. The organic carbon-rich sediments used for dating yielded lagoonal characteristics. The shells collected from the

Table 20.4 Radiocarbon dates of samples from selected depths of the bore hole cores of Kuttanad Kole Wetland.

Sl. no.	Bore hole location	Bore hole depth	Material	^{14}C age	Calibrated age; one sigma (Cal yr BP)	Mid-point (Cal yr BP)
			Upper Kuttanad and flood plain			
1	Puvathur	7	Peat	1670 ± 70	1397–1731	1564
2	Pachcha	8.5	Shell	6990 ± 160	7571–8070	7820
3	Karippumuri	9	Peat	3100 ± 150	2922–3635	3278
			Lower Kuttanad			
4	Alappuzha (ii)	2.35	Sediment	3990 ± 150	4082–4843	4462
5	Alappuzha (ii)	5.25	Sediment	4840 ± 170	5258–5932	5595
6	Alappuzha (ii)	8.25	Sediment	5670 ± 140	6195–6785	6490
7	Alappuzha (ii)	10.95	Sediment	6780 ± 160	7417–7951	7684
8	Alappuzha (ii)	12.95	Sediment	6630 ± 230	7149–7939	7574
9	Alappuzha (ii)	15.45	Sediment	6620 ± 200	7156–7876	7516
10	Kaniyamkulam		Shell	3360 ± 110	3371–3872	3621
11	Nedumudi	4.95	Sediment	10380 ± 130	11767–12617	12192
12	Nedumudi	11.95	Sediment	12390 ± 70	14123–14872	14497
13	Nedumudi	18.95	Sediment	9900 ± 100	11162–11759	11460
14	Kidangara	3.45	Sediment	6880 ± 130	7555–7958	7756
15	Kidangara	15.95	Sediment	8500 ± 140	9126–9891	9508
16	Kidangara	29.45	Sediment	10560 ± 90	12371–12707	12539
			North Kuttanad			
17	Thiruvarpu	3.5	Sediment	4630 ± 100	5040–5586	5313
18	Thiruvarpu	8.5	Sediment	8860 ± 160	9538–10251	9894
19	Thiruvarpu	11.4	Sediment	9800 ± 130	10752–11641	11196
20	Pallam	11.5	Sediment	8520 ± 250	8978–10225	9601
21	Kavanar	6.5	Sediment	2180 ± 70	2036–2335	2185
22	Kavanar	9.5	Shell	7090 ± 100	7693–8058	7875
23	Minachil (i)*	6	Wood	2888 ± 78	2842–3232	3037
24	Minachil (ii)*	6	Wood	5570 ± 30	6302–6404	6353
			Vembanad			
25	Pathiramanal	13.75	Sediment	12330 ± 810	12716–16766	14714
26	Pathiramanal	23.95	Sediment	11100 ± 960	10414–15518	12966
			Western coastal sand dune zones			
27	Muhamma	3.95	Sediment	6800 ± 180	7413–7976	7694
28	Muhamma	21.45	Sediment	8320 ± 110	9032–9499	9265
29	Muhamma	25.45	Sediment	9700 ± 130	10660–11391	11025
30	Kattoor	9	Sediment	2400 ± 110	2300–2744	2522
31	Kalarkod	8.45	Sediment	6740 ± 120	7426–7832	7629
32	Kalarkod	40	Sediment	20380 ± 490	23451–25664	24557
33	Thottappalli	26	Sediment	>40000		
34	Thakazhi (ii)	10	Sediment	35180 ± 2850	34126–43823	38974
35	Thakazhi (ii)	15.5	Sediment	>40000		

*Source: Ajaykumar et al. (2012).

lagoonal sediments of the Kavanar core also gave middle–early-Holocene date (7090 ± 100 yrs BP; 9.5 m bgl). The Pallam borehole core retrieved from the southeastern margin of the North Kuttanad recorded an early-Holocene age (8520 ± 250 yrs BP; 11.5 m bgl) for the organic carbon-rich lagoonal sediments. The most notable observation is that in the river influenced regions of the North Kuttanad, fluvial sedimentation in the lagoonal environment was noticed since the onset of Holocene Climatic Optimum; whereas in the southern side, especially in the Lower Kuttanad, fluvial sedimentation is noticed over older lagoonal sediments only during late-Holocene. The lone core collected within the Pathiramanal island in Vembanad Lake gave a late-Pleistocene–early-Holocene age. The lithological sequence of the core gives a similarity with respect to the nearby Muhamma core retrieved from the Western coastal strand plains, indicating the fact that this borehole site was a part of the Western coastal strand plains and was later separated from it during the evolutionary phases of lake system.

The radiocarbon dates of the Muhamma and the Kattoor borehole cores that are collected from the northern part of the western coastal strand plains gave an early–late-Holocene age (9700 ± 130–2400 ± 110 yrs BP). These cores show a lagoonal clay with marine shells/shell fragments over which deposition of littoral sands occurred. The littoral sands, in turn, embed interlayers of clay beds at different levels yielding an age of middle–late-Holocene. For instance, the clayey interlayer in the Muhamma core gave an age of 6800 ± 180 yrs BP at 3.95 m bgl; whereas the one at Kattoor gave an age of 2400 ± 110 yrs BP at 9 m bgl. This presumably indicates the bar breaching at different geological ages (middle–late-Holocene) during the developmental stages of the coastal geomorphic units. Interestingly, the only date representing the Last Glacial Maxima is recorded at Kalarkod (20,380 ± 490; 40 m bgl). At the same time, similar to the Muhamma core, the organic carbon-rich interlayer within the thick pile of littoral sand at Kalarkod yielded an age of 6740 ± 120 yrs BP; 8.45 m bgl. The Thakazhi and Thottapalli borehole cores retrieved from the southern side of the Western coastal strand plains yielded Pleistocene ages indicating the fact that this part of the coast, especially representing the dated levels of the core, may be either older or received significant level of contribution of sediments from the recorded Neogene sediments in the hinterlands.

20.6 Discussion

The KKW witnessed marked changes in depositional environments in the late Quaternary period. The region was an embayment during pre-Holocene which received sediments both from marine–littoral and terrestrial (especially in the North Kuttanad) sources. The spread of the sand–silt–clay plots in the ternary diagram of Pejrup (1988) is indicative of the existence of varied hydrodynamic conditions prevailed in the different subenvironments of the KKW (Fig. 20.5). As seen from Fig. 20.5, moderate level turbulent condition prevailed in the marginal areas of the KKW which is influenced by riverine processes. The segregation of sample plots of the Upper Kuttanad and flood plains cores in the II and III sectors of the ternary diagram supports this view. Contrary to this, the sample plots of the Lower Kuttanad, North Kuttanad, and Pathiramanal Island cores spread in the sectors I and II of the ternary diagrams indicating the prevalence of a calm, sheltered environment of deposition during most part of the depositional period (Fig. 20.5). The sample plots of the cores retrieved from the Western coastal strand plains spreads in all four sectors (there is only one sample point in the sector IV). This indicates that the region enjoys calm and quiet environment during the deposition of the underlying clays, but the

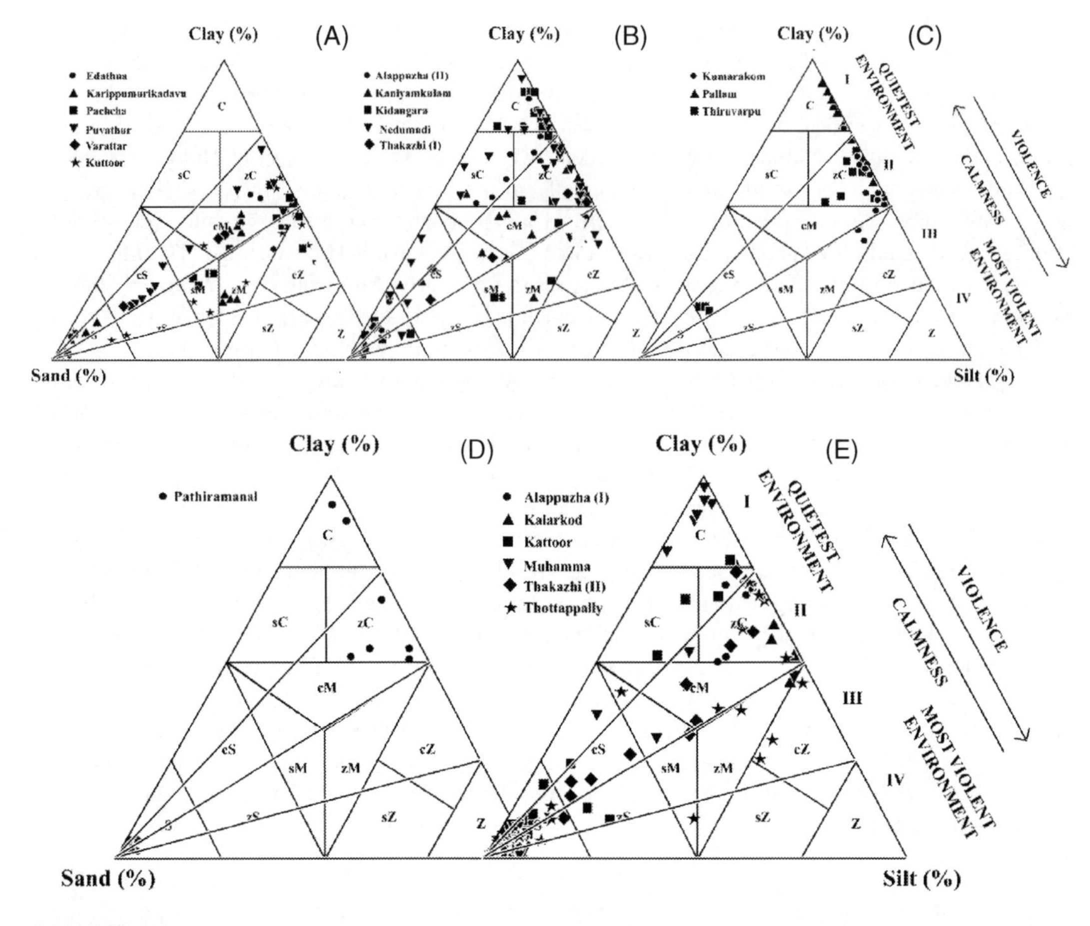

FIGURE 20.5

Sediment types (after Picard, 1971) of the borehole cores retrieved from the Kuttanad Kole Wetland together with area the energy regime (Pejrup, 1988) prevailed during the deposition of sediments in the different sub-environment of Kuttanad Kole Wetland (A) Upper Kuttanad, (B) Lower Kuttanad, (C) North Kuttanad, (D) Pathiramanal Island (Vembanad Lake), and (E) Western coastal strand plains and barrier beaches.

energy regime changed from moderately to most violent during the sea-level rise and fall events which was responsible for the build-up of barrier beach in the region.

The KKW is one of the unique geomorphic features in the coastal lowlands in the northern side of the Achankovil Shear Zone (ASZ). The southern side of the ASZ, represented by the Trivandrum Block, is made up essentially of the khondalite suite of rocks with the indicator mineral sillimanite (specifically in khondalites); whereas the northern side of it is made up of the Madurai block composed of the charnockite suite of rocks with indicator mineral inosilicates (e.g., hypersthene in charnockite). These minerals along with the other light and heavy minerals will be of much use in the decoding of the source rock characteristics of sediments in the KKW and adjoining regions. For example, occurrence

Table 20.5 Radiocarbon ages of subfossil lime shells of Kuttanad Kole Wetland.

Sl. no.	Bore hole location	Material	^{14}C age	References
1	Vechur	Shell	3710 ± 90	Rajendran et al. (1989)
2	Muhamma	Shell	3130 ± 100	Rajendran et al. (1989)
3	Kaniyamkulam	Shell	3360 ± 110	Maya et al. (2007)
4	Vembanad Lake (Near Pathiramanal)	Shell	4780 ± 120	Maya et al. (2007)

of sillimanite in the heavy mineral population in the northern side of the ASZ is resulted from the northward drift of sediments originated from the khondalite suite of rocks in the Trivandrum block and brought later by the longshore currents. At the same time, the occurrence of inosilicates (pyriboles) in the top layers (alluvial) of the borehole cores retrieved from the northern side of the ASZ, especially in the Upper and Lower Kuttanad regions might have derived from denudation of the charnockite group of rocks in the hinterlands.

The nature and characteristics of heavy mineral suite could give insights into the climatic conditions prevailing in the hinterland river catchments in addition to address the littoral processes that are in operation in the region. Interestingly, the sand and mud dominated sediments in the KKW and nearby areas of the Western coastal strand plains/barrier beaches contain sillimanites in appreciable quantities which are not found in the hinterland rocks north of the ASZ. This is a clear indication of the prevalence of longshore drift of sediments and barrier build up that are active during the Holocene epoch. But the presence of inosilicates in the lagoonal sediments is an indication of change in provenance in the Upper Kuttanad region. Further, the west flowing Achankovil, Pamba, and Manimala rivers took a marked northerly shift before debouching the southern expansive reach of the Vembanad Lake. These rivers while entering the KKW branch out and flow north westerly direction for about 20 km before joining the Vembanad Lake. Instead of covering this long distance, the rivers could have been directly debouched into the Arabian Sea near Thottappally flowing a distance of just 10 km. The river influenced part (Upper Kuttanad) of KKW is blanketed with 2–5 m thick alluvial sands and clays over a fairly thick deposit of lagoonal sediments. The entire sequence lies above lateritized crystallines or Neogene sediments. Radiocarbon dates of the lagoonal sediments in KKW and southern Vembanad lagoon (1670 ± 70–10,560 ± 90 yrs BP) show Holocene age. The dating of lime shell deposit (essentially composed of *Villoritta* sp.) varies from 3130 ± 100 yrs BP to 4780 ± 80 yrs BP (Table 20.5). The bivalve *Villoritta* sp. thrives usually in the fresh water end of an estuary (Kurian et al., 1985) with a salinity range of 3–16‰ and change in salinity is one of the most important factors that promote spawning and proliferation of the clam species (Laxmilatha et al., 2005). In KKW, limeshell deposits are generally found as stratified deposits often attaining a thickness of a few meters (Maya et al., 2007). But at certain parts of the lake, the deposit occurs in the sediment column as sparsely disseminated or even found as pockets and/or lenticular bodies. Maya et al. (2007) mapped many such occurrences of lenticular limeshell bodies in the Kaniyamkulam and nearby areas of the KKW (Fig. 20.6). The uneven nature of the subsurface limeshell bed may be attributed to the changing salinity conditions as the distributary channel migration during the evolutionary phase of the KKW during the middle–late-Holocene.

The content of palynological and NPP elements in the sedimentary archive shows records of heavy rainfall and sea-level rise during the early–middle-Holocene period. Fig. 20.7 shows a fourfold evolutionary model prepared from the observations and multiparametric studies carried out for this

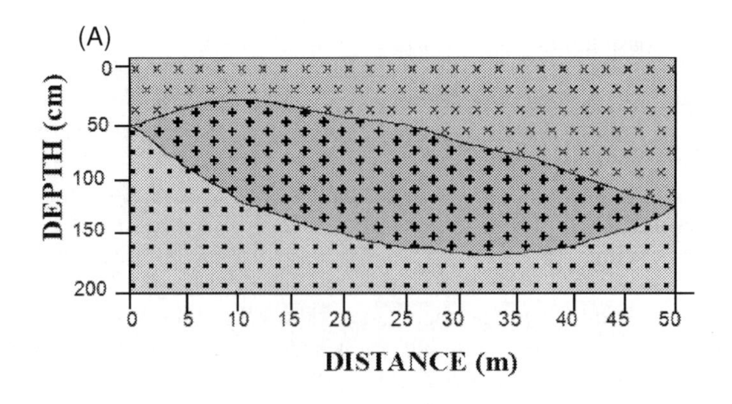

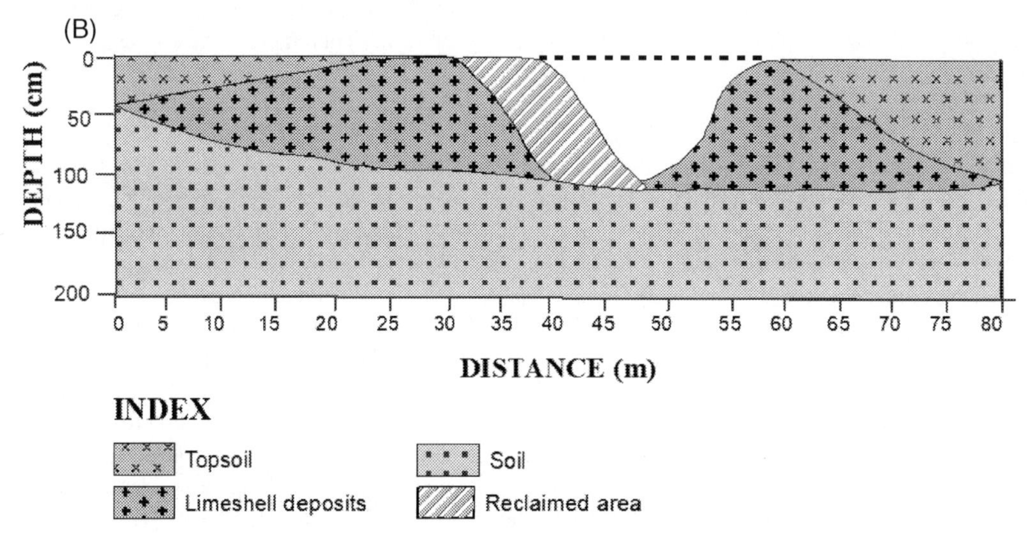

FIGURE 20.6

Sections showing sub-surface occurrence of limeshell deposit in Kaniyamkulam area in Lower Kuttanad, SW India. (A) Subsurface occurrence of lenticular limeshell bed; (B) A limeshell bed after partial mining.

work. During pre-Holocene, the KKW was an embayment boarded on the east by lateritized hillocks. The rivers were flowing through the southern side of the topographic high in Pachcha–Edathua areas that acted as a divide between the river course and the embayment. During early–middle-Holocene, the topographic high might have eroded due to the continuous wave action of the transgressive waters. Also, the sea-level rise and subsequent build-up of a barrier spit by the longshore drift might have separated the landward part of the embayment into a lagoon. The erosion of the topographic high that acted as drainage divide and subsidence of the Central Block of the South Kerala Sedimentary Basin on which the Kuttanad wetland locates were responsible for the north–north westerly diversion of the Achankovil, Pamba, and Manimala rivers. It is revealed that some of the faults, especially the one near the deeper part of the basin dissected the late Quaternary sediments (Fig. 20.8). This was an indication of the fault

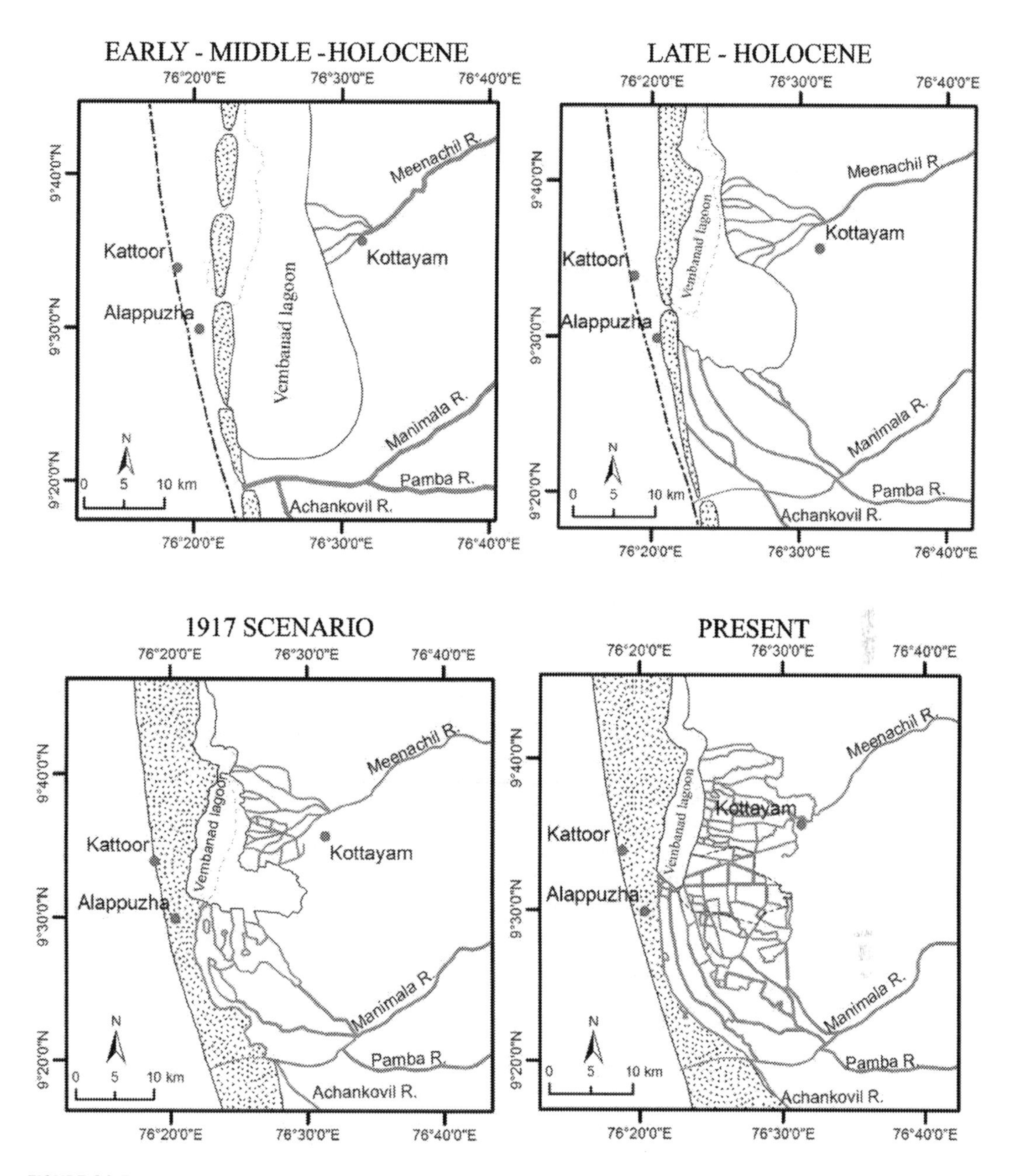

FIGURE 20.7

Fourfold Holocene evolutionary model of the Kuttanad Kole Wetland. Note the northward diversion of the fluvial channels that occurred during late-Holocene.

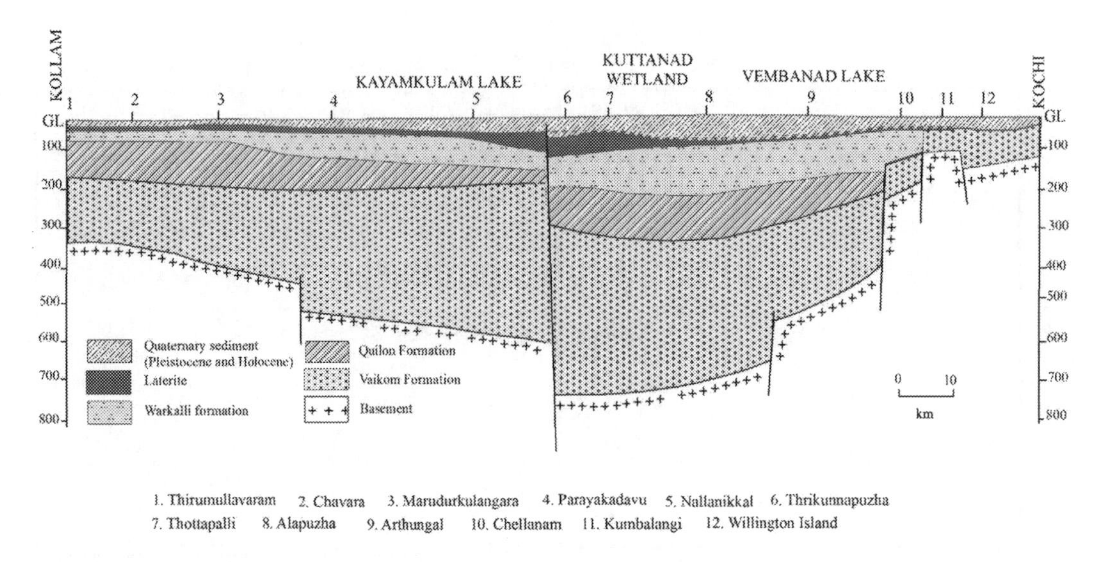

FIGURE 20.8

Geological cross section between Kollam and Kochi showing the structural pattern of Quaternary, Neogene, basement (modified after (Nair et al., 2007)).

activation and subsidence. Thus, the southern expanded arm of the then existing Vembanad Lake was turned into the major depocenter for the sediments brought by the Achankovil, Pamba, and Manimala rivers. The occurrence lime shell deposits of *Villoritta* sp. with an age range of 3130 ± 100 yrs BP to 4780 ± 80 yrs BP in the southern side of Vembanad lagoon and adjoining area of KKW is an indication of the timing of river course changes and estuarine mixing that favored the proliferation *Villoritta* sp. that thrives in the freshwater end of an estuary, where freshwater mixes with marine waters (Kurian et al., 1985; Laxmilatha et al., 2005; Sheeba and Padmalal, 2007). The sedimentary record of the KKW and the adjoining coastal lands as well as the frequent episodic flood events to which the region is subjected during the recent decades reveal that the southern part of KKW is not yet turned into a full-fledged deltaic system, but a delta is in the making in the Kuttanad region. At the same time, the northern Kuttanad received alluvial build up in the river confluence of Minachil since the early–middle-Holocene.

20.7 Learning and knowledge outcomes

- The KKW and its adjoining areas of the coastal lands and waterbodies are characterized by many important geoenvironmental features.
- The Achankovil, Pamba, and Manimala rivers when enter the KKW, branches out and takes a north westerly direction and flows a long distance, before joining with the Vembanad Lake, instead of flowing straight to the Arabian Sea near Thottappally just covering a few kilometers.

- The river influenced, south-eastern part (Upper Kuttanad) of KKW is blanketed 2–5 m thick alluvial sands and clays over a fairly thick deposit of lagoonal clays often with shells. The entire sequence lies either above lateritized crystallines or above Neogene sediments.
- Heavy mineralogical study of the borehole cores retrieved from the Upper Kuttanad areas reveals that the riverine sediments are dominated by inosilicates (pyroxenes and amphiboles) which is indicative of sediment inputs from charnockitic source rocks in the hinterlands. On the contrary the sand fraction separated out from the lagoonal sediments is dominated by opaques and sillimanites indicating sediment source south of ASZ and brought up by littoral currents.
- Radio-carbon dates of the lagoonal sediments in KKW and southern Vembanad lagoon (1670 $\pm$ 70 yrs BP to 10,560 $\pm$ 90 yrs BP) give Holocene age. The dating of lime shell deposit (essentially composed of *Villoritta* sp.) varies from 3130 $\pm$ 100 yrs BP to 4780 $\pm$ 80 yrs BP.
- The bivalve *Villoritta* sp. thrives usually in the freshwater end of an estuary. This clearly indicates that freshwater from rivers reached the southern Vembanad lagoon and the KKW during the beginning of the late-Holocene.
- The content of palynological and NPP elements in the sediment archive shows records of heavy rainfall and sea-level rise during the early–middle-Holocene period.
- During pre-Holocene, the KKW was an embayment boarded on the east by lateritized hillocks. Further, the Achankovil, Pamba and Manimala rivers were flowing through the southern side of a topographic high that acted as a divide between the river course and the embayment.
- During the early–middle-Holocene, the topographic high might have eroded due to the continuous wave impinging of the transgressive waters. Also, the sea-level rise and subsequent build-up of a barrier spit by the long shore drift have separated out the landward part of the employment into a lagoon.
- The erosion of the topographic high that acted as drainage divide and subsidence of the Central Block of the South Kerala Sedimentary Basin on which the Kuttanad wetland spreads were responsible for the north–north westerly diversion of the distributaries of Achankovil, Pamba, and Manimala rivers. The diversions were from two points—one from the eastern end (near Pachcha) of the buried topographic high and the other from its western end, near Kuttoor.
- The expansive, southern arm of the then existed Vembanad lagoon was turned into the major depocenter for the sediments brought by the Achankovil, Pamba, and Manimala rivers.
- The sedimentary record of the KKW and the adjoining coastal lands as well as the frequent flood events to which the region is subjected during every heavy monsoon period reveal that the KKW is not yet turned into a full-fledged deltaic system, but a delta is in the making in the Kuttanad region.

Acknowledgments

The authors thank the Director, National Centre for Earth Science Studies (NCESS), Thiruvananthapuram for extending facilities and support. Heartfelt gratitude is extended to Dr. K.M. Nair (Late), Former Director, CESS and collaborating investigator for many projects with Dr. K.P.N. and D.P. for his guidance, encouragements, and support in the studies of the Kerala basin. D.P. thanks DST, New Delhi for financial support in the form of sponsored Research Project No. ESS/23/VES/006/98, along with Dr. K.M. Nair; K.P.N. thanks KSCSTE, GoK for the project No. 56/2004/KSCSTE dated 05.03.2004, along with Dr. K.M. Nair; K.P.N. and D.P. thanks CSIR, New Delhi for the Project No. 24 (0275)/05/EMR-11; K.P.N. thanks CSIR, New Delhi for the Emeritus Scientist Scheme No.

21 (0828)/10/EMR-II dated 23.12.2010. R.B.L. acknowledges Department of Science and Technology, New Delhi (Govt. of India) for providing financial support in the form of DST Women Scientist Scheme (WOS-A, SR/WOS-A/ES-08/2013. D.P. and K.M. thank CESS and KSCSTE for various internally funded projects through which a few borehole cores and field-level data are collected for the present study.

References

Ajaykumar, B., Sreedharan, K., Mohan, M., Paul, J., Thomas, A.P., Nair, P.K.K., Evaluation of the Holocene environmental changes of the southwest coast, India: a palaeopalynological approach, J. Earth Syst. Sci. 121 (2012) 1093–1103.

Anonymous, Resource Atlas of Kerala, Centre for Earth Science Studies, Trivandrum, 1984.

Arulbalaji, P., Banerji, U.S., Maya, K., Padmalal, D, Signatures of late Quaternary land-sea interactions and landform dynamics along southern Kerala coast, SW India, Quat. Int. 575–576 (2020) 270–279 Doi/org/10.1016/j.quaint.2020.05.001.

Banerjee, P.K., Holocene and Late Pleistocene relative sea level fluctuations along the east coast of India, Mar. Geol. 167 (2000) 243–260.

Bird, E.C.F., Physical setting and geomorphology of coastal lagoons, in: Kjerfve, B.J. (Ed.), Coastal Lagoon Processes, Elsevier, Amsterdam, 1994, pp. 9–39.

Blatt, H., Middleton, G.V., Murray, R.C., Origin of Sedimentary Rocks, Prentice Hall Pvt. Ltd, New Jersey, 1972, p. 634.

Carver, R.E., Procedures in Sedimentary Petrology, Wiley – Inter Science, New York, 1971, p. 478.

Chattopadhyay, S., Sidharthan, S., Regional Analysis of the Greater Kuttanad, Kerala, Centre for Earth Science Studies, Thiruvananthapuram, India, 1985.

Dill, W.A., Inland Fisheries of Europe, FAO, Rome, 1990 EIFAC Technical Paper No. 52.

El Wakeel, S.K., Riley, J.P., The determination of organic carbon in marine muds, J. conseil int. pour l'explor. mer 22 (1957) 180–183.

Gönenc, I.E., Wolflin, J.P., Coastal Lagoons: Ecosystem Processes and Modelling for Sustainable Use and Development, CRC Press, Boca Raton, Florida, USA, 2005.

Jackson, M.L., Soil Chemical Analysis, Prentice-Hall of India Private Limited, New Delhi, 1967.

Joseph, A., Balachandran, K.K., Mehra, P., Prabhudesai, R.G., Kumar, V., Agarvadekar, Y., Dabholkar, N., Amplified Msf tides at Kochi backwaters on the southwest coast of India, Curr. Sci. 97 (6) (2009) 776–784.

Kjerfve, B., Coastal Lagoon Processes, Elsevier, Amsterdam, The Netherlands, 1994.

Kumaran, K.P.N., 2001. Report on Palynological Analysis of Subsurface Samples of Quaternary Sea Level Oscillations, Geological and Geomorphological Evolution of South Kerala Sedimentary Basin (DST Project ESS/23/VES/006/98), p. 22.

Kumaran, K.P.N., Nair, K.M., Shindikar, M., Limaye, R.B., Padmalal, D., Stratigraphical and palynological appraisal of the Late Quaternary mangrove deposits of the west coast of India, Quat. Res. 64 (3) (2005) 418–431.

Kumaran, K.P.N., Padmalal, D., Limaye, R.B., Jennerjahn, T., Gamre, P.G., Tropical peat and Peatland development in the floodplains of the greater Pamba basin, South-Western India during the Holocene, PLoS One 11 (5) (2016) e0154297.

Kumaran, K.P.N., Padmalal, D., Nair, M.K., Limaye, R.B., Guleria, J.S., Srivastava, R., Shukla, A., Vegetation response and landscape dynamics of Indian Summer Monsoon variations during Holocene: an eco-geomorphological appraisal of tropical evergreen forest subfossil logs, PLoS One 9 (4) (2014) e93596.

Kurian, N.P., Hameed, T.S.S., Baba, M., A study of monsoonal beach processes around Alleppey, Proc. Indian Acad. Sci. 94 (1985) 323–332.

Kurien, C.V., Ecology of benthos in a tropical estuary, Proc. Indian Natl. Sci. Acad. 38 (1971) 158–163.
Kurup, M.B., Studies in the Systematics and Biology of the Fishes of Vembanad Lake (Ph.D. thesis). Cochin University of Sciences and Technology, Kochi, 1982.
Laxmilatha, P., Velayudhan, T.S., Kripa, V., Jenni, B., Alloycious, P.S., Biology of the black clam, *Villoritta cyprinoides* (Gray) in the backwaters of Vembanad Lake, Indian J. Fish. 52 (3) (2005) 361–366.
Lewis, D.W., Practical Sedimentology, Hutchinson Ross Publishing Company, Pennsylvania, 1984.
Limaye, R.B., Kumaran, K.P.N., Nair, K.M., Padmalal, D., Non-pollen palynomorphs (NPP) as potential palaeoenvironmental indicators in the late quaternary sediments of west coast of India, Curr. Sci. 92 (10) (2007) 1370–1382.
Limaye, R.B., Padmalal, D., Kumaran, K.P.N, Late Pleistocene Holocene monsoon variations on climate, landforms and vegetation cover in southwestern India: an overview, Quat. Int. 443 (2016) 143–154.
Mallik, T.K., Suchindan, G.K., Some sedimentological aspects of Vembanad Lake, Kerala, West coast of India, Indian J. Mar. Sci. 13 (1984) 159–163.
Mange, M.A., Maurer, H.F.W., Heavy Minerals in Colour, Chapman and Hall, New York, 1992.
Maya, K., Babu, K.N., Padmalal, D., Sreebha, S., Occurrence and Mining of Limeshell Resources of Vembanad Lake, Kerala, Centre for Earth Science Studies, Thiruvananthapuram, India, 2007, p. 66.
Merh, S.S., Quaternary sea level changes along Indian coasts, Proc.-Indian Natl. Sci. Acad. Part A 58 (5) (1992) 461–472.
Nair, K.M., Kumaran, K.P.N., Effects of Holocene Climate and Sea Level Changes on Ecology, Vegetation and Landforms in Coastal Kerala, Kerala State Council for Science Technology and Environment, Thiruvananthapuram, 2006, p. 41 p. Project Completion Report.
Nair, K.M., Padmalal, D., Kumaran, K.P.N., Quaternary geology of South Kerala sedimentary basin—an outline, J. Geol. Soc. India 67 (2006) 165–179.
Nair, K.M., Padmalal, D., Kumaran, K.P.N., 2007. Response of post middle Miocene tectonics on stratigraphy and geomorphology of onshore-offshore regions of SW coast of India. In: Proceedings: IGCP Symposium 514, Dec 2007. CESS, p. 17e20.
Nair, K.M., Sajikumar, S., Padmalal, D., Silting up of a Holocene mega lagoon along Kerala coast. In: National Seminar on Coastal Evolution, Process and Products, vol. 12, 1998.
Narayana, A.C., Priju, C.P., Landform and shoreline changes inferred from satellite images along the Central Kerala coast, J. Geol. Soc. India 68 (2006) 35–49.
Nath, V.A., Aparna, B., Thampatti, M., Chemical and biological characterization of acid sulphate Kuttanad soils, Asian J. Soil Sci. 11 (2) (2016) 269–276.
Nayar, T.S., Pollen flora of Maharashtra State, India. In: International Bioscience Series 14. Today & Tomorrow's Printers and Publishers, New Delhi, (1990) p. 157.
Padmalal, D., Kumaran, K.P.N., Nair, K.M., Baijulal, B., Limaye, R.B., Vishnu Mohan, S., Evolution of the coastal wetland systems of SW India during the Holocene: evidence from marine and terrestrial archives of Kollam coast, Kerala, Quat. Int. 237 (2011) 123–139.
Padmalal, D., Kumaran, K.P.N., Nair, K.M., Limaye, R.B., Mohan, S.V., Baijulal, B., Anooja, S., Consequences of sea level and climate changes on the morphodynamics of a tropical coastal lagoon during Holocene: an evolutionary model, Quat. Int. 333 (2014) 156–172.
Padmalal, D., Nair, K.M., Kumaran, K.P.N., Sajan, K., Mohan, S.V., Maya, K., Limaye, R.B., Climate and sea level changes in a Holocene Bay Head Delta, Kerala, Southwest coast of India, Climate Change and Island and Coastal Vulnerability, Springer, Dordrecht, 2013, pp. 191–208.
Pejrup, M., The triangular diagram used for classification of estuarine sediments: a new approach, in: de Boer, P.L., Van Gelder, A., Nio, S.D. (Eds.), Tide Influenced Sedimentary Environments and Facies, Reidel, Dordrecht, 1988, pp. 289–300.
Pettijohn, F.J., Sedimentary Rocks, CBS Publishers and Distributers, DelhiIndia, 1985, p. 625.
Picard, M.D., Classification of fine-grained sedimentary rocks, J. Sediment. Res. 41 (1) (1971) 179–195.

Rajendran, C.P., Rajagopalan, G., Narayanswamy, Quaternary geology of Kerala: evidence from radiocarbon dates, J. Geol. Soc. India 33 (1989) 218–225.

Sankaranarayanan, V.N., Udayavarma, P., Balachandran, K.K., Pylee, A., Joseph, T., Estuarine characteristics of the lower reaches of the river Periyar (Cochin backwaters), Indian J. Mar. Sci. 15 (1986) 166–170.

Seralathan, P., Padmalal, D., Geochemistry of Fe and Mn in the surficial sediments of tropical river and estuary, Central Kerala, India – a granulometric approach, Environ. Geol. 254 (1995) 270–276.

Sheeba, S., Padmalal, D., Impact of environmental degradation on the clam resources in the Paravur lake (Kerala), southwest coast of India. In: Proceedings of Kerala Environment Congress, 2007, pp. 288–294.

Sreejith, K.A., Human impact on Kuttanad wetland ecosystem—an overview, Int. J. Sci. Technol. 2 (4) (2013) 679–690.

Thanikaimoni, G., Caratini, C., Venkatachala, B.S., Ramanujam, C.G.K., Kar, R.K., Selected tertiary angiosperm pollen from India and their relationship with African tertiary pollen, Travaux sect. sci. Tech. 19 (1984) 1–93.

Tissot, C., Chikhi, H., Nayar, T.S., Pollen of Wet Evergreen Forests of the Western Ghats of India, Publications du départment d'écologie, Institut Français de Pondicherry, Pondichéry, 1994.

Traverse, A., Palaeopalynology, second ed., Springer, Dordrecht, The Netherlands, 2007.

Williams, G.L., Bujak, J.P., 1985. Bolli, H.M., Saunders, J.B., Perch-Nielsen, K. (Eds.), Mesozoic and Cenozoic dinoflagellates. Plankton Stratigraphy,Cambridge university press, Australlia, 847–964.

CHAPTER

21 Holocene evolution of coastal wetlands—A case study from Southern Kerala, India

S. Vishnu Mohan[a], M.S. Aneesh[b], K. Maya[b], K.P.N. Kumaran[c], D. Padmalal[b]

[a] *Department of Geology, Sree Narayana College, Chempazhanthy, Kerala, India.* [b] *National Centre for Earth Science Studies, Ministry of Earth Sciences, Thiruvananthapuram, Kerala, India.* [c] *PalynoVision, Mon Amour, Erandaane, Pune, Maharashtra, India*

21.1 Introduction

Earth has been subjected to dramatic changes in its environment during Quaternary period—the time span representing the last 2 million years of earth history. The climatic changes of the Quaternary period were influenced greatly by geomorphic evolution, sedimentation rate, sea-level changes, and also the developmental history of the biological world including that of human beings (Savin, 1977; Padmalal et al., 2013; Mohan, 2015). In recent years, apart from the scientific and academic communities, planners and policy makers are also becoming increasingly aware of the relevance of Quaternary studies because of its immense application potential in the fields of climate studies, and agricultural and natural resources management. The increased concern about the adverse effects of global climate change demands in-depth studies on the past geological and environmental conditions. Though the effects of global warming and related climatic changes affect many countries like India, the study of past climate, especially that of Quaternary period, has received much importance in India in the recent past only. It is now widely realized that a better understanding of the Quaternary geological events is essential for the accurate reconstruction of past climate and sea-level changes that affect human societies directly and indirectly. Although the nature of interaction between climate, sea-level oscillations, and human society is mediated by a wide range of cultural processes and varied greatly for different biophysical settings and social organizations, the socioeconomic impact of climate and sea-level changes needs special attention for framing appropriate climate change adaptation strategies.

Although the south west coast of India has fairly thick deposit of late Quaternary sediments of 70–80 m in the South Kerala Sedimentary Basin and its adjoining coastal lowlands, not many studies have been made to unfold its paleoclimatic and paleoenvironmental aspects till the beginning of the present century (Jayalakshmi et al., 2004). But later, many researchers who have worked on the different aspects of Quaternary sediments in the coastal lands of south west India (Joseph and Thrivikramji, 2002; Nair and Padmalal, 2003; Kumaran et al., 2005; Nair et al., 2006; Limaye et al., 2010) revealed that like other parts of the tropics, this coastal segment was also affected significantly by climate and sea-level changes. These changes had a strong bearing on the human inhabitation and/or migration in the area. Recent advances in archeological investigations in the Pattanam–Kodungallur stretch of Central Kerala (Shajan, 1998; Shajan et. al., 2004; Abraham, 2006) also gave indications of shifts in human settlements

Holocene Climate Change and Environment. DOI: https://doi.org/10.1016/B978-0-323-90085-0.00024-3

in accordance with the changing climates and/or sea-level fluctuations in the middle–late-Holocene. Therefore, investigations of the late Quaternary deposits that are profusely developed in the coastal lowlands of South Kerala Sedimentary Basin are not only helpful in strengthening our understanding on paleoclimate and sea-level changes, but also throw light on how ancient human civilization responded to these millennia-scale geological events. Further, high-resolution data on climate and sea-level variability could also be used for fine tuning the climate and sea-level prediction models developed essentially from instrumental measurements.

Kerala forms the southwestern segment of the complex western coastline. The state has a coastline of 590 km (Soman, 2002). Of this a cumulative 360 km length of coastline is very dynamic and fluctuates seasonally causing considerable damage to life and property and loss arable land. The coastal features identified include beaches, beach cliffs, islands, shore platforms, estuaries, lagoons, mud flats, and tidal flats. Of the various lineaments identified over the Kerala region, the fault lineaments parallel to the coast as well as those at right angles to it are found to be neotectonically active and influence the configuration of the shoreline. Apart from the known factors of slopes, morphometry, lithology, and storm-wave generation, the nonparametric factors responsible for coastal erosion include presence or absence of mud banks and instability of off-shore regions (Mohan et al., 2016; Shynu et al., 2017).

Although considerable studies have been carried out on the geology of the pre-Cambrian crystallines and Tertiary sediments of Kerala, not enough work has been carried out on the late Quaternary deposits of the region. But it is a fact that most of the economically valuable deposits like beach placers, glass sands, lime shells, etc. are associated with the late Quaternary deposits of the state (Nair et al., 2010; Padmalal et al., 2011, 2014a). Therefore, a better understanding of the origin and evolution on the Quaternary deposits in the coastal lands of Kerala is very essential for the sustainable extraction and judicious management of the resources in addition to planning developmental initiatives. Further, an understanding on the Quaternary deposits is also essential for the assessment of groundwater potential of the coastal lowlands of the state. Considering these, an attempt has been made here to investigate the sedimentological and geochemical characteristics of subsurface sediments of a few borehole cores, red sand sediments, beach and strand line sediments collected from the coastal lands of Kazhakuttom–Kaniyapuram belt (Southern Kerala) to decode the sea-level changes and coastal evolution. Also this area is influenced by sediment discharges from the hinterlands and by coastal–nearshore processes.

21.2 Scope

The present study was carried out on the late Quaternary sediments in the coastal lowlands of southern Kerala. The outcome of the study will bridge at least a part of the knowledge gap in the late Quaternary evolution of coastal wetlands in the cliffed coasts of southwest India. The high-resolution paleoclimatic data derived from the study of late Quaternary sediments will be useful for all climatic related studies of Kerala in particular and the Indian ocean rim countries in general.

21.3 Study area

Thiruvananthapuram district, which hosts the study area, is located between North latitudes 8°17′–8°51′ and East longitudes 76°41″–77°17′. The district boundary is shared in the north by Kollam district, east by Thirunelveli district, south by Kanyakumari district, and west by the Lakshadweep Sea. The coastal parts of the district are with well-developed beaches and pocket beaches at many places. The width of the Thiruvananthapuram coast varies from 25 m to 100 m. The area selected for the present study falls within the coastal lands of southern Kerala in the central part of Thiruvananthapuram district around the Kazhakuttom–Kaniyapuram areas (Fig. 21.1).

Contrasting geological features are characteristics of the study area. The study area is particularly made up of pre-Cambrian crystallines, Neogene, and Quaternary sediments (Fig. 21.2). The pre-Cambrian crystalline rocks are composed mainly of garnet–biotite gneisses, khondalites, and charnockites. Generally these rock types are exposed in the eastern and southern parts of Kollam and Thiruvananthapuram districts. As mentioned earlier, the Neogene sediments are represented by Quilon and Warkalli Formations of lower Miocene age. The Warkalli Formation is composed of sandstones and clays. The Quilon Formation lying below the Warkalli is represented by fossiliferous limestones and sandy carbonaceous clays. The pre-Cambrian crystallines and Neogene sediments are lateritized at the top. The Quaternary deposits are represented by alluvial clays, sandy clays, peat, and sand.

The Cenozoic sequence is broadly grouped into four, (1) Vaikom Formation, (2) Quilon Formation, (3) Warkalli Formation, and (4) Vembanad Formation (Najeeb, 1999). The lower most sequence is Vaikom Formation and consists of sandstones with pebbles and gravel beds, clays and lignite, and carbonaceous clay. It is overlain by Quilon Formation constituting limestone, marl, clays/calcareous clays with marine and lagoonal fossils. The Warkalli Formation has laterite capping over the crystallines and Neogene sediments. The Vembanad formation consists mainly of sands, clays, molluscan shell beds, riverine alluvium, and floodplain deposit.

21.4 Materials and methods

The surface samples collected from the study area were washed, dried, and subjected to coning and quartering for obtaining a representative portion. The subsamples were sieved for 15 minutes on a mechanical Ro-tap sieve shaker using a standard set of ASTM Endecott sieves arranged in the descending order of mesh size at half phi (1/2 ϕ) intervals. It is then subjected to textural (Folk and Ward, 1957) and mineralogical studies (Mange and Maurer, 1992). CM pattern of the sediment is worked out following Pessaga (1964). After documenting the lithological characteristics, the borehole cores were sectioned at 10.0 cm interval and the subsamples were packed in neatly labeled polyethylene bags for further analysis. Utmost care was taken to avoid contamination during subsampling, packing, and processing for various analytical procedures. Subsamples from selected depths were subjected to textural analysis following Lewis (1984). The ternary diagram of Picard (1971) was used for the classification of sediments. A few representative samples from Core III (Kaniyapuram) were subjected to XRF analysis for the determination of major (in oxide forms) and minor/trace elements in National Centre for Earth Science Studies, Thiruvananthapuram, Kerala (India). The XRF facility consists of a Bruker S4 Pioneer Sequential Wavelength-Dispersive X-Ray Diffractometer equipped with goniometer and Spectra Plus software.

21.5 Field work and sampling techniques

Systematic fieldwork was carried out in the study area for the collection of primary and secondary data. A total of three borehole cores were used in the present study of which Kalpananagar (C I) and

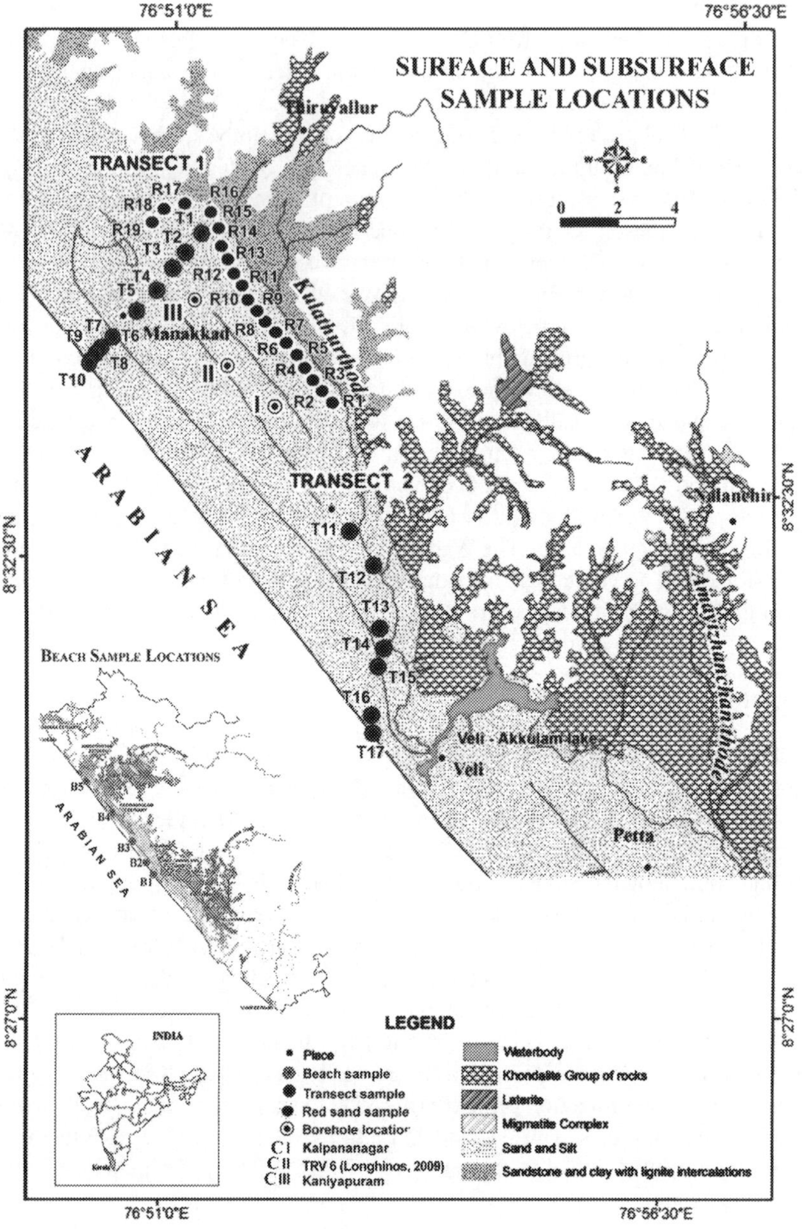

Figure 21.1

Location of the study area and sampling stations.

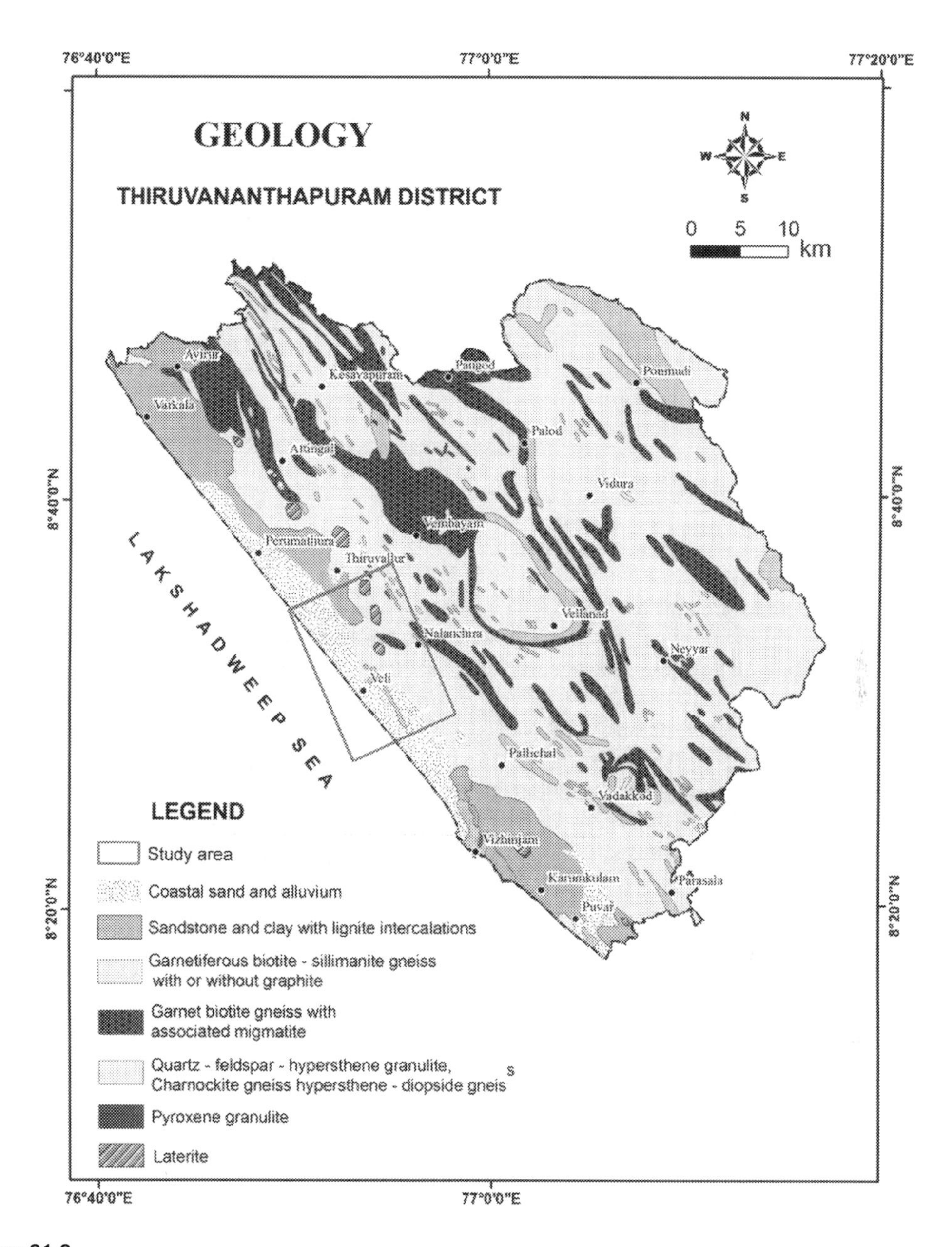

Figure 21.2

Geological map of the Thiruvananthapuram district. Note the coastal sands and alluvium where the present study has undertaken.

Kaniyapuram (C III) borehole core were collected using rotary drilling from the coastal plain of the study area (Fig. 21.1 and 21.3). Information on the TRV 6 borehole core is obtained from Longhinos (2009). In addition to this a total of 41 surface samples were also collected from Kazhakuttom–Kaniyapuram belt for detailed textural, mineralogical, and geochemical study which includes 19 red sand samples, 5 beach samples, and 17 samples. The samples were collected from the two coast perpendicular transects at Kaniyapuram (Transect 1) and Kulathur (Transect 2) of Kazhakuttom–Kaniyapuram belt in Thiruvananthapuram district (Fig. 21.1). The samples taken in perpendicular transects to the beach represents all the features of microenvironments of the beach. Out of the beach samples, one sample each was collected from Kulathur (B1), Pallithura (B2), Puthenthoppu (B3), Perumathura (B4), and Onnampalam (B5) beaches which are coming under Kazhakuttom–Kaniyapuram belt (Fig. 21.1).

21.6 Results

21.6.1 Borehole lithology and sediment texture

The borehole core retrieved from Kalpananagar (Fig. 21.1; CI) is 13.5 m long and is sited on ridge between two swales through which two coast parallel streams are flowing. The borehole core is composed mainly of sand-dominated sediments intervened at 8–13 m below ground level (bgl) by a grayish black, organic matter rich, sandy mud (sM). This lies over a reddish brown, clayey sand (cS) layer (Fig. 21.4). The variations of sand, silt, and clay reveal marked changes in its percent contents in the downcore (Fig. 21.5). The content of sand, silt, and clay ranges from 44.35% to 87.25%, 5.43–24.39%, and 3.52–31.26%, respectively (Table 21.1). The sand in the upper layer is generally white to off white, medium to fine sand resembling beach origin. Fig. 21.6 shows the Ternary diagram showing granulometric properties. The borehole core retrieved from Kaniyapuram is 15 m long and is also sited at the crest of the same ridge. The borehole is composed dominantly of sandy sediments (Fig. 21.4) which rest over lateritized clayey sediments. The sand content of core sample ranges from 79.87% to 93.62% and the lowest value is recorded at a depth of 15 m and highest value is at 1.5 m and 4.5 m (bgl; Table 21.1). The maximum concentration of sand is noticed at the top most part of the core and the content exhibited a general decreasing downward trend. The silt content of the core sample ranges from 4.78% to 12.63% and the lowest value is recorded at a depth of 4.5 m while the highest is noticed at 15 m bgl. The content of clay was substantially low throughout the core. Fig. 21.6 shows the ternary diagram indicating the variations noticed in the Kaniyapuram borehole core.

21.6.2 Surface sediments—granulometric characteristics

In the Kaniyapuram borehole core, particle size varies from granule to very fine sand. The variation of textural grades in Kaniyapuram borehole shows that majority of the sediments are medium to fine sand. The mean size of subsurface sediments ranges from 1.09Φ to 2.27Φ. The coarser sediments are noticed at a depth of 15 m bgl (1.09Φ), while comparatively finer sediments are seen at 12 m depth (2.27Φ). The analysis of mean size reveals that majority of the samples are composed of medium sand. Standard deviation (sediment sorting) shows a minimum value of 0.83Φ at a depth of 4.5 m and maximum value

Figure 21.3

Some selected scanning electron photomicrographs of Kaniyapuram borehole (A) At a depth of 1.5m. (B) High magnification image at depth 1.5m, (C & F) At a depth of 9m. (D & E) High magnification image at depth 9m.

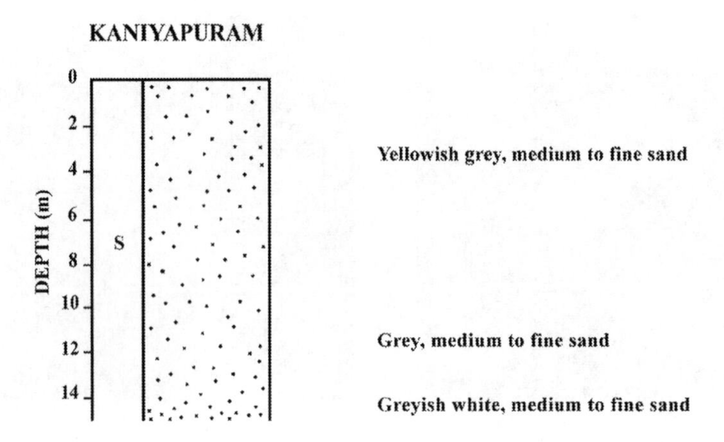

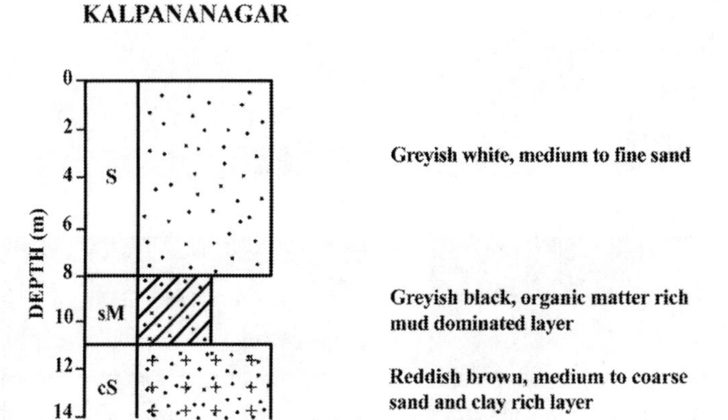

Figure 21.4

Lithological characteristics of Kaniyapuram and Kalpananagar borehole cores retrieved from the study area. S Sand, cS Clayey sand, sM Sandy mud.

of 1.42Φ at a depth of 15 m. Majority of the samples in the Kaniyapuram borehole cores are moderately sorted and a few are poorly sorted. Skewness varies from -0.545 to 0.112 with an average of -0.31. An analysis of skewness reveals that more than half of the samples are nearly symmetrical, and remaining are fine skewed and very coarsely skewed. The kurtosis values of the core vary from 0.91 to 1.08 with an

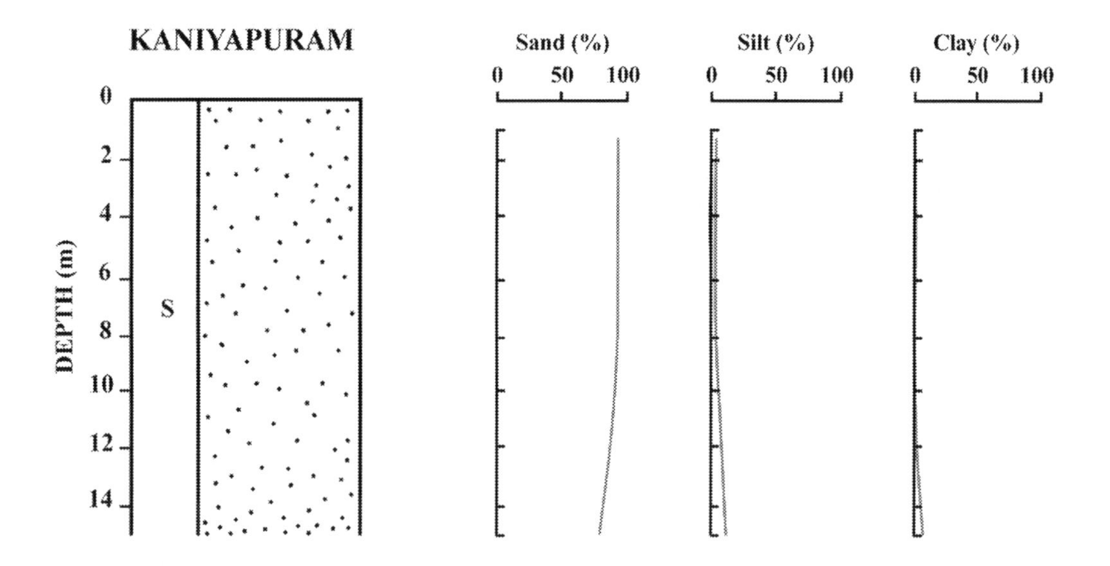

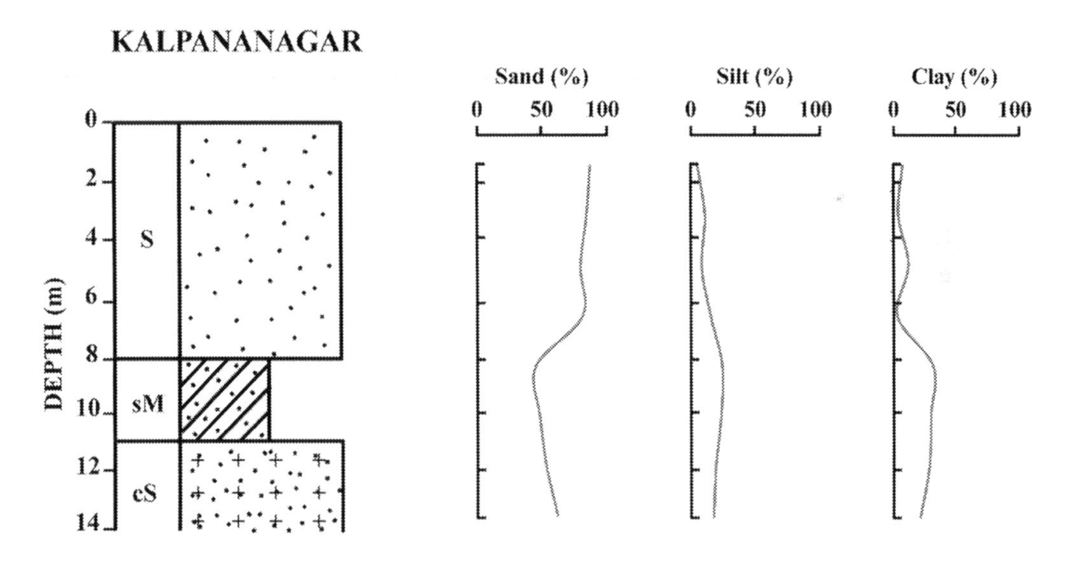

Figure 21.5

Downcore variation of sand, silt and clay in Kaniyapuram and Kalpananagar borehole core.

Table 21.1 Sand, silt, and clay contents in the subsurface sediments of borehole cores retrieved from study area.

	Grain size				Statistical parameters[b]			
Borehole core/ sample depth (m)	**Sand (%)**	**Silt (%)**	**Clay (%)**	**Sediment type[a]**	**Mean (x)**	**Sorting (x)**	**Skewness**	**Kurtosis**
Kalpananagar								
1.5	87.25	5.43	7.32	S	NA	NA	NA	NA
3.0	84.25	11.44	4.31	S	NA	NA	NA	NA
4.5	79.35	8.73	11.92	S	NA	NA	NA	NA
6.0	80.97	15.51	3.52	S	NA	NA	NA	NA
9.0	44.35	24.39	31.26	sM	NA	NA	NA	NA
10.5	47.35	23.39	29.26	sM	NA	NA	NA	NA
12.0	52.33	19.41	28.26	cS	NA	NA	NA	NA
13.5	61.25	17.39	21.36	cS	NA	NA	NA	NA
Kaniyapuram								
1.5	93.62	5.46	0.92	S	1.403	0.853	0.026	1.069
4.5	93.62	4.78	1.60	S	1.184	0.833	0.053	0.948
9.0	93.42	5.45	1.13	S	1.531	1.077	0.112	1.084
12.0	88.82	9.06	2.12	S	2.268	1.595	–0.545	0.956
15.0	79.87	12.63	7.50	S	1.096	1.416	0.044	0.914

NA, not analyzed; S, sand; sM, sandy mud; cS, clayey sand.
[a]Picard (1971).
[b]Folk and Ward (1957).

average of 0.99 (Table 21.2). The minimum kurtosis value is noticed at a depth of 15 m and maximum at 9 m depth.

In the case of red sands, particle size also varies from granule to very fine sand, however, the presence of granule is noticed in few samples (R3 and R7). In general, the spatial variation of grain size fraction exhibits marked variation and majority of the sample shows medium sand. Mean size ranges from 0.99x to 1.60x and most of the samples are in the category of medium sand and only one sample shows coarse grained texture (R18). Standard deviation varies from 0.92x to 0.68x and it is seen that 94.74% of samples exhibit moderately well sorted particle dispersal pattern and remaining are generally poorly sorted. Skewness varies from -0.049 to 0.045, the finely skewed category forms the majority of samples. Kurtosis values vary from 0.807 to 1.227 (Table 21.2) and the samples generally fall in the platykurtic to leptokurtic category.

Like the case of red sands, the beach sands are also composed of very coarse to fine sand. The spatial variation of grain size fraction shows that majority of the samples fall in medium sand fraction and the highest concentration is noticed in the Puthenthoppu beach. Mean size of the beach sediments range from 1.16Φ to 1.98Φ (Table 21.2). Coarser sediments are noticed in Onnampalam beach (1.66Φ) whereas comparatively finer sediments in the Perumathura beach (1.98Φ). The phi mean exhibits a general decreasing trend toward north showing that there is a coarsening of sediments toward northern stations as wave energy is higher in that direction. The variation of standard deviation in the beach sands

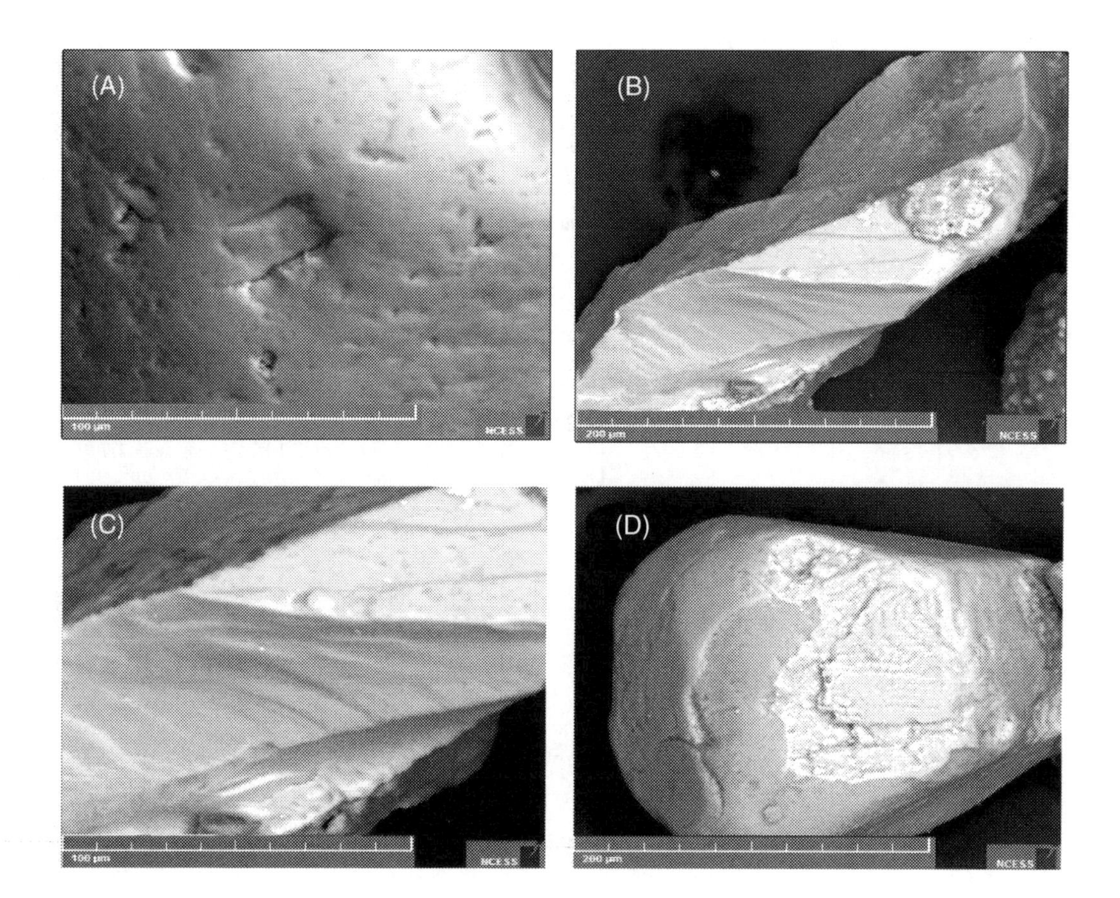

Figure 21.6

Some selected scanning electron photomicrographs of Kaniyapuram borehole core (A) High magnification image at depth 9m, (B, C, & D) High magnification image at a depth of 15m.

from Kulathur beach to Onnampalam beach are summarized in Table 21.2. The standard deviation shows a minimum value of 0.39Φ in Puthenthopu beach and maximum value of 0.55Φ in Perumathura beach. Majority of the samples are moderately well sorted. The skewness values of the beach samples range from -0.10 to 0.15, that is, the sediments are finely skewed to very coarsely skewed. Kulathur beach shows a maximum skewness value while the Perumathura beach exhibits the minimum. The kurtosis value of the beach samples varies from 1.04 to 1.41 (Table 21.2). The samples fall in the very mesokurtic to leptokurtic category with minimum kurtosis noticed for the Pallithura samples and maximum to the Puthenthopu beach.

Along the Kaniyapuram and Kulathur transects, the particle size varies from fine pebble to very fine sand. Mean size of the sediment samples from the Kaniyapuram transect ranges from -0.4Φ to 1.68Φ (Table 21.2). In this transect, Sample No. 1 shows the maximum phi value (1.68Φ) and Sample No. 2 shows the minimum value (-0.4Φ). An analysis of mean size in the Kaniyapuram transect reveals that

Table 21.2 Statistical parameters and granulometric characteristics of the surface sediments collected from study area.

Sample name	Statistical parameters				Granulometric characteristics
	Mean (ϕ)	Sorting (ϕ)	Skewness	Kurtosis	
Red sand samples					
R1	1.598	0.83	0.045	1.133	0.47% very coarse sand, 19.92% coarse sand, 53.08% medium sand, 22.99% fine sand, 3.53% very fine sand
R2	1.509	0.733	0.026	1.254	0.25% very coarse sand, 20.61% coarse sand, 58.26% medium sand, 18.54% fine sand, 2.34% very fine sand
R3	1.488	0.785	0.009	1.224	0.59% very coarse sand, 21.58% coarse sand, 56.82% medium sand, 19.24% fine sand, 1.77% very fine sand
R4	1.479	0.771	0.012	1.243	0.21% very coarse sand, 21.72% coarse sand, 57.99% medium sand, 18.04% fine sand, 2.05% very fine sand
R5	1.499	0.766	0.01	1.277	1.02% very coarse sand, 19.57% coarse sand, 59.16% medium sand, 18.22% fine sand, 2.03% very fine sand
R6	1.363	0.749	–0.045	1.192	0.45% very coarse sand, 25.25% coarse sand, 57.52% medium sand, 15.61% fine sand, 1.17% very fine sand
R7	1.301	0.763	–0.044	1.105	0.78% very coarse sand, 28.43% coarse sand, 55.26% medium sand, 13.47% fine sand, 2.06% very fine sand
R8	1.376	0.917	0.026	0.973	1.07% granule, 3.1% very coarse sand, 28.69% coarse sand, 44.79% medium sand, 18.98% fine sand, 3.36% very fine sand
R9	1.216	0.747	–0.028	0.994	0.66% very coarse sand, 34.16% coarse sand, 53.73% medium sand, 10.09% fine sand, 1.36% very fine sand
R10	1.154	0.679	–0.08	0.868	0.54% very coarse sand, 37.34% coarse sand, 55.55% medium sand, 5.95% fine sand, 0.63% very fine sand
R11	1.408	0.774	–0.02	1.240	0.91% very coarse sand, 23.55% coarse sand, 57.32% medium sand, 16.26% fine sand, 1.97% very fine sand
R12	1.227	0.752	–0.035	1.01	0.92% very coarse sand, 32.98% coarse sand, 54.1% medium sand, 10.59% fine sand, 1.40% very fine sand
R13	1.218	0.729	–0.049	0.992	0.55% very coarse sand, 33.37% coarse sand, 55.53% medium sand, 9.71% fine sand, 0.83% very fine sand
R14	1.354	0.794	–0.022	1.127	0.16% granule, 1.12% very coarse sand, 26.67% coarse sand, 54.47% medium sand, 15.37% fine sand, 2.22% very fine sand

(continued on next page)

Table 21.2 (*continued*)

Sample name	Statistical parameters				Granulometric characteristics
	Mean (ϕ)	Sorting (ϕ)	Skewness	Kurtosis	
R15	1.457	0.836	0.025	1.18	0.87% very coarse sand, 24.67% coarse sand, 53.15% medium sand, 17.94% fine sand, 3.38% very fine sand
R16	1.247	0.785	−0.013	0.992	1.14% very coarse sand, 33.29% coarse sand, 50.27% medium sand, 13.61% fine sand, 1.68% very fine sand
R17	1.418	1.003	0.176	1.205	1.16% very coarse sand, 31.94% coarse sand, 46.65% medium sand, 9.99% fine sand, 10.27% very fine sand
R18	0.989	0.721	0.055	0.807	5.55% very coarse sand, 45.66% coarse sand, 42.89% medium sand, 5.19% fine sand, 0.71% very fine sand
R19	1.233	0.788	−0.011	1.021	1.54% very coarse sand, 33.08% coarse sand, 51.34% medium sand, 11.46% fine sand, 2.59% very fine sand
Beach sand samples					
Kulathur (B1)	1.41	0.54	0.15	1.27	1.36% very coarse sand, 17.73% coarse sand, 69.24% medium sand, 10.29% fine sand, 1.34% very fine sand
Puthenthoppu (B2)	1.43	0.39	−0.22	1.41	0.09% very coarse sand, 13.33% coarse sand, 80.14% medium sand, 6.29% fine sand, 0.13% very fine sand
Pallithura (B3)	1.4	0.42	−0.12	1.04	0.12% very coarse sand, 15.66% coarse sand, 77.27% medium sand, 6.78% fine sand, 0.13% very fine sand
Perumathura (B4)	1.98	0.55	−0.10	1.11	0.16% very coarse sand, 6.34% coarse sand, 41.61% medium sand, 49.48% fine sand, 2.32% very fine sand
Onnampalam (B5)	1.16	0.54	0.06	0.98	1.88% very coarse sand, 38.87% coarse sand, 51.43% medium sand, 7.66% fine sand, 0.08% very fine sand
Kaniyapuram transect samples (Transect 1)					
T1	1.68	1.3	0.17	1.24	0.47% fine pebble ,1.95% granule, 6.53% very coarse sand, 18.31% coarse sand , 34.94% medium sand, 24.72% fine sand, 13.04% very fine sand
T2	−0.4	1.87	−0.31	1.26	1.97% fine pebble, 5.86% granule, 12.79% very coarse sand, 27.03% coarse sand, 27.92% medium sand, 15.99% fine sand, 8.41% very fine sand
T3	1.21	0.68	0.11	1.09	0.45% granule, 1.79% very coarse sand, 36.62% coarse sand, 50.63% medium sand, 8.50% fine sand, 1.99% very fine sand

(*continued on next page*)

Table 21.2 (*continued*)

Sample name	Statistical parameters				Granulometric characteristics
	Mean (ϕ)	Sorting (ϕ)	Skewness	Kurtosis	
T4	1.14	0.66	0.1	1.01	0.80% fine pebble, 0.70% granule, 3.26% very coarse sand, 37.21% coarse sand, 48.22% medium sand, 9.49% fine sand, 0.48% very fine sand
T5	1.21	0.74	0.1	1.04	0.34% granule, 3.26% very coarse sand, 36.82% coarse sand, 45.02% medium sand, 12.07% fine sand, 2.47% very fine sand
T6	1.06	0.96	–0.37	1.73	1.68% granule, 5.65% very coarse sand, 38.55% coarse sand, 44.85% medium sand, 8.12% fine sand, 1.13% very fine sand
T7	1.38	0.74	–0.12	0.9	0.04% granule, 2.84% very coarse sand, 30.80% coarse sand, 44.62% medium sand, 20.50% fine sand, 1.17% very fine sand
T8	0.64	1.59	–0.3	1.38	0.89% fine pebble, 14.55% granule, 6.21% very coarse sand, 20.20% coarse sand, 41.19% medium sand, 14.98% fine sand, 1.93% very fine sand
T9	1.62	0.59	–0.11	1.12	0.2% fine pebble, 0.16% granule, 1.07% very coarse sand, 12.13% coarse sand, 62.96% medium sand, 22.14% fine sand, 1.31% very fine sand
T10	1.43	0.39	–0.22	1.41	0.09% very coarse sand, 13.33% coarse sand, 80.14% medium sand, 6.29% fine sand, 0.13% very fine sand
Kulathur transect samples (Transect 2)					
T11	1.8	0.93	0.25	1.02	0.27% fine pebble, 1.57% granule, 2.9% very coarse sand, 13.6% coarse sand, 44.53% medium sand, 24.72% fine sand, 12.34% very fine sand
T12	1.30	0.63	0.27	1.10	0.33% fine pebble, 0.35% granule, 1.08% very coarse sand, 27.23% coarse sand, 57.98% medium sand, 11.56% fine sand, 1.43% very fine sand
T13	2.01	1.06	0.29	1.91	1.12% fine pebble, 2.46% granule, 2.35% very coarse sand, 5.44% coarse sand, 30.30% medium sand, 51.76% fine sand, 6.53% very fine sand
T14	1.44	0.62	0.3	1.11	0.36% fine pebble, 0.08% granule, 0.19% very coarse sand, 21.76% coarse sand, 59.65% medium sand, 15.92% fine sand, 1.70% very fine sand
T15	1.14	0.50	0.12	1.00	0.66% very coarse sand, 43.26% coarse sand, 60.80% medium sand, 51.30% fine sand, 0.13% very fine sand

(*continued on next page*)

Table 21.2 (*continued*)

Sample name	Statistical parameters				Granulometric characteristics
	Mean (ϕ)	Sorting (ϕ)	Skewness	Kurtosis	
T16	1.92	0.56	0.21	1.17	2.85% coarse sand, 57.42% medium sand, 34.32% fine sand, 5.39% very fine sand
T17	1.41	0.54	0.15	1.27	0.35% granule, 1.01% very coarse sand, 17.73% coarse sand, 69.24% medium sand, 10.29% fine sand, 1.34% very fine sand

Phi mean: -1 to 0 (very coarse sand), 0 to 1 (coarse sand), 1 to 2 (medium sand), 2 to 3 (fine sand), 3 to 4 (very fine sand); sorting: <0.35 (very well sorted), 0.35 to 0.50 (well sorted), 0.50 to 0.71 (moderately well sorted), 0.71 to 1.00 (moderately sorted), 1.00 to 2.00 (poorly sorted), 2.00 to 4.00 (very poorly sorted), >4.00 (extremely poorly sorted); Skewness: 0.30 (very finely skewed), 0.30 to 0.10 (finely skewed), 0.10 to -0.10 (nearly symmetrical), -0.10 to -0.30 (coarse skewed), <-0.30 (very coarse skewed); Kurtosis: <0.67 (very platykurtic), 0.67 to 0.90 (platykurtic), 0.90 to 1.11 (mesokurtic), 1.11 to 1.50 (leptokurtic), 1.50 to 3.00 (very leptokurtic), >3.00 (extremely leptokurtic).

80% of the samples are of medium sand, 10% very coarse sand and 10% coarse sand. In the Kulathur transect, the mean size of the sediments range from 1.14Φ to 2.01Φ. Here also, medium sand is the dominant size class which is followed by fine sand. The standard deviation in the Kaniyapuram transect ranges from 0.39Φ to 1.87Φ (Table 21.2) with about 30% samples each are of poorly sorted, moderately sorted, and moderately well sorted, and the remaining 10% is well sorted. The well sorted samples are noticed toward the beach end of transect, whereas the poorly sorted ones are closer toward the landward end. In the Kulathur transect, the sorting ranges from 0.5Φ to 1.06Φ with 71.42% of samples fall in the moderately well sorted and the remaining samples in the poorly to moderately sorted category. Among the transect samples, samples in the Kaniyapuram transect are very coarsely skewed to finely skewed with skewness ranges from 0.17 to -0.37 (Table 21.2). The skewness values of Kulathur transect ranges from 0.12 to 0.3 and all the samples of this transect are finely skewed. In the Kaniyapuram transect, the kurtosis values ranges from 0.9 to 1.73 with 50% of the samples fall in the leptokurtic category, 40% in the mesokurtic category, and 10% in the very leptokurtic category. The ranges of kurtosis values in the Kulathur transect vary from 1 to 1.91 with an average of 1.23. In this transect, 57.14% samples are mesokurtic, 28.58% samples leptokurtic, and 14.28% samples very leptokurtic (Table 21.2).

21.6.3 Heavy mineralogy

The total heavy mineral (THM) content in the fine sand fraction of the borehole core retrieved from Kalpananagar site shows an increasing trend toward the bottom of the core. THM content varies from 49.52% to 96.18% (Table 21.3). Among the eight samples analyzed, lowest THM content is noticed at 1.5 m depth (49.52%), and the highest at 9 m depth (96.18%). Opaques and sillimanite together constitute about 90% of total heavies followed by zircon. Monazite, rutile, garnet, and inosilicates are recorded only in traces. In the case of Kaniyapuram borehole core, THM content shows an opposite trend, the values are increasing toward the top. THM content ranges from 5.83% to 27.50% and the maximum concentration is noticed at a depth of 1.5 m and minimum at 12 m (Table 21.3). Opaques are dominant minor category in the heavy mineral population, followed by sillimanite and garnet. Garnet

Table 21.3 Concentration of heavy minerals in the borehole cores retrieved from study area (THM and TLM are in weight percentage and others are in number percentage).

Borehole depth	THM (wt.%)	TLM (wt.%)	Heavy minerals (number %)								
			Opaque	Sillimanite	Zircon	Monazite	Rutile	Garnet	Inosilicate	Biotite	Others
Kalpananagar											
1.5	49.52	50.47	78.41	15.56	3.49	0.32	0.95	0.63	–	–	0.63
3.0	69.72	30.27	53.54	41.41	4.04	–	–	–	–	–	1.01
4.5	92.54	7.45	41.42	55.03	1.78	–	0.59	0.30	–	–	0.59
6.0	92.85	7.14	52.79	43.65	2.03	–	–	–	–	–	1.02
9.0	96.18	3.81	30.10	66.56	1.67	0.33	–	–	–	–	0.67
10.0	95.66	4.33	34.34	63.02	1.51	–	0.38	–	–	–	0.75
12.0	92.91	7.08	51.53	43.25	2.76	0.61	0.31	0.31	–	–	0.92
13.5	88.77	11.22	58.33	36.86	2.88	0.64	–	–	0.32	–	0.96
Kaniyapuram											
1.5	27.50	72.50	82.71	13.55	0.93	0.47	1.87	0.47	–	–	–
4.5	14.41	85.59	74.77	7.48	4.21	0.93	0.47	11.68	–	–	–
9.0	15.24	84.76	76.04	11.98	0.92	–	0.92	9.68	–	0.46	–
12.0	5.83	94.17	76.07	19.15	1.06	–	0.53	3.19	–	–	–
15.0	7.35	92.65	86.18	9.21	2.63	0.66	0.66	0.66	–	–	–

content is high at the middle layers of the borehole core. Zircon, monazite, rutile, and inosilicates are recorded only in traces.

The coastal sediment comprises two categories of sands—beach sand and strand plain/old coastal sand. The THM content in the fine sand fraction of beach sand coming under the Kazhakuttom–Kaniayapuram belt varies from 0.22% to 33.67% (Table 21.4). The sample collected from Kulathur exhibits the highest content of THM (33.67%), whereas Perumathura sample exhibits the lowest THM content (0.22%). The heavy mineral assemblage in the beach sands consists of opaques, sillimanite, and zircon as the major minerals in which the first two members dominate over the third (i.e., zircon). Garnet, rutile, and monazite are present in minor concentration and inosilicates occur as trace in the heavy mineral crop (Table 21.4). In the case of red sands, THM content ranges from 16.93% to 51.03%. The heavy mineral assemblage in the red sands consists of opaque and sillimanite as the major minerals and zircon, rutile, and monazite as the minor minerals.

Apart from the beach sands and red sands, a total of 15 sediment samples from two different transects (Kaniyapuram and Kulathur transects) have also been collected and studied. The THM occurrences in the sediments collected from Kaniyapuram transect vary from 0.5% to 41.94%. Among the total heavies, the major constituents are opaques and sillimanite and the minor one are rutile, monazite, garnet, and inosilicates (Table 21.4). The THM occurrences recorded in the Kulathur transect vary from 10.18% to 74.24% (Table 21.4). Like the case of other samples in Kulathur transect, opaques dominate the heavy mineral suite, whereas sillimanite is comparatively lower than in the Kaniyapuram transect (Table 21.4). Another interesting observation is that the garnet content is on the higher side in Kulathur transect especially in sample number T13. In general, THM content increases toward the beach in all two transects. The mineral percentage varies from ~1% to ~90% with an average of 22%. Among the 15 samples collected from the two transects, the highest THM content is noticed in the Kulathur transect in the sample T14 (74.24%), whereas the lowest concentration is in Kaniyapuram transect samples T2 (0.5%; Table 21.4).

21.6.4 SEM analysis

The fine sand grains from Kaniyapuram borehole were scanned under different magnifications for the identification of surface textural features. The scanning electron photomicrographs of sand grains (fine sand fraction) of Core C III are depicted in Fig. 21.6. The magnified images of the grains at 400×, 600×, and 800× reveal a spectrum of surface textural features. Fig. 21.6A and B shows scanning electron photomicrographs of Kaniyapuram borehole at a depth of 1.5 m. Majority of grains are rounded to subrounded and demonstrate typical microrelief features of beach origin/marine activity. The different beach activities by means of wave and current action are responsible for its high roundness and sphericity. Fig. 21.6C–F shows images of the grains retrieved from the Core C III at a depth of 9 m. It has comparatively low sphericity and degree of roundness. This indicates that the grains underwent comparatively low amount of wear and tear as they are originated from the hinterlands through the branches of the Kulathur *thodu*. The high magnification study shows triangular depressions, bulbous edges, linear grooves, pitted surfaces, and V-shaped impact pits. The Plate 3 (B, C, and D) shows photomicrographs of Kaniyapuram borehole at a depth of 15 m. They are highly angular to subangular with very low degree of roundness due to low amount of winnowing activity to which the grains are subjected. The sediments at this level might have been derived from the hinterland drainages. The sands

Table 21.4 Concentration of heavy minerals in the surface samples collected from study area (THM and TLM are in weight percentage and others are in number percentage).

Sample name	THM (wt.%)	TLM (wt.%)	Heavy minerals (number %)								
			Opaque	Sillimanite	Zircon	Monazite	Rutile	Garnet	Inosilicate	Biotite	Others
Red sand samples											
R2	51.03	48.97	77.20	11.40	7.82	2.93	0.65	–	–	–	–
R4	50.15	49.85	74.91	21.31	2.41	1.03	0.34	–	–	–	–
R7	32.20	67.80	69.31	26.73	1.98	0.99	0.99	–	–	–	–
R9	41.10	58.90	75.78	21.74	1.24	0.62	0.62	–	–	–	–
R10	38.86	61.14	76.36	18.18	3.64	–	1.82	–	–	–	–
R11	26.17	73.83	67.43	29.72	1.14	0.57	1.14	–	–	–	–
R13	25.81	74.19	66.84	29.47	2.63	0.53	–	0.53	–	–	–
R17	25.37	74.63	45.45	49.20	3.74	1.07	0.53	–	–	–	–
R18	16.93	83.07	53.53	41.18	3.82	1.18	0.29	–	–	–	–
Beach sand samples											
Kulathur (B1)	33.67	66.32	60.56	31.92	4.23	0.47	0.47	1.41	–	–	0.94
Puthenthoppu (B2)	5.89	94.1	32.16	64.91	1.17	–	–	–	1.17	–	–
Pallithura (B3)	5.61	94.39	25.88	68.24	2.35	–	–	–	–	–	4.27
Perumathura (B4)	0.22	99.78	20.00	70.83	2.50	–	2.50	–	–	–	4.17
Onnampalam (B5)	1.80	98.20	40.45	49.44	2.25	–	–	2.25	–	–	5.61

(continued on next page)

Table 21.4 (*continued*)

Sample name	THM (wt.%)	TLM (wt.%)	Heavy minerals (number %)								
			Opaque	Sillimanite	Zircon	Monazite	Rutile	Garnet	Inosilicate	Biotite	Others
Kaniyapuram transect samples (Transect 1)											
T1	6.67	93.32	76.12	20.6	2.09	0.3	–	0.3	–	–	0.6
T2	0.50	99.49	80.00	7.56	9.33	0.44	1.33	0.44	–	–	0.81
T3	25.6	74.39	82.38	12.64	2.68	0.77	0.38	0.38	–	–	0.77
T4	41.94	58.05	39.76	54.22	2.01	–	–	1.61	1.61	–	0.8
T5	5.29	94.7	68.09	24.82	4.61	1.06	0.35	0.35	–	–	0.71
T6	19.52	80.47	69.55	19.92	6.02	0.75	0.38	1.88	0.75	–	0.75
T8	6.04	93.95	57.66	34.27	4.84	0.4	1.21	0.81	–	–	0.81
T9	5.89	94.1	32.16	64.91	1.17	–	–	–	1.17	–	0.58
Kulathur transect samples (Transect 2)											
T11	10.18	89.81	61.33	33.6	3.52	0.78	–	–	–	–	0.78
T12	11.53	88.46	66.11	28.51	3.31	0.41	0.83	–	–	–	0.83
T13	74.24	25.75	85.02	4.41	4.85	1.32	–	3.52	–	–	0.88
T14	68.25	31.74	85.45	1.82	10.00	1.82	–	–	–	–	0.91
T15	65.56	34.43	84.55	4.83	8.21	0.97	–	0.48	–	–	0.97
T16	58.95	41.05	83.17	12.50	1.44	–	–	1.92	–	–	0.96
T17	33.67	66.32	60.56	31.92	4.23	0.47	0.47	1.41	–	–	0.94

Table 21.5 Concentration of major elements (expressed as oxides) and trace elements in the Kaniyapuram borehole core.

	Sample depth (m)					The standard values			
Elements	**1.5**	**4.5**	**9**	**12**	**15**	**SINGO**	**PAAS**	**NASC**	**UC**
				Major elements (%)					
SiO_2	93.12	90.09	88.56	90.4	78.85	**72.65**	**62.80**	**64.80**	**66.00**
TiO_2	1.39	1.23	1.44	1.99	0.82	**0.42**	**1.00**	**0.70**	**0.50**
Al_2O_3	2.43	4.43	5.56	4.85	14.23	**13.33**	**18.90**	**16.90**	**15.2**
MnO	0	0.01	0.01	0	0	**0.08**	**0.11**	**0.06**	**0.08**
Fe_2O_3	1.47	1.88	1.64	0.63	1.69	**2.6**	**7.22**	**5.65**	**5.00**
CaO	0.33	0.35	0.35	0.05	0.31	**1.09**	**1.30**	**3.63**	**4.20**
MgO	0.11	0.13	0.13	0	0.09	**0.6**	**2.20**	**2.86**	**2.20**
Na_2O	0.28	0.27	0.31	0.01	0.24	**3.54**	**1.20**	**1.14**	**3.90**
K_2O	0.04	0.15	0.13	0.06	0	**4.87**	**3.70**	**3.97**	**3.40**
P_2O_5	0.06	0.06	0.06	0.03	0.05	**0.13**	**0.16**	**0.13**	-
				Trace elements (ppm)					
V	128	125	139	158	111	**28**	**150**	**130**	**60**
Cr	36	54	53	84	52	**27.7**	**110**	**125**	**35**
Ni	3	4	6	7	3	**11**	**55**	**58**	**20**
Cu	3	6	11	0	3	**BDL**	**50**	**BDL**	**25**
Zn	75	0	0	3	0	**40.2**	**85**	**BDL**	**71**
Ga	0	0	2	7	0				
Rb	16	18	17	16	11	**348**	**160**	**125**	**112**
Sr	34	39	43	34	37	**103**	**200**	**142**	**350**
Y	3	4	3	2	4	**57**	**27**	**35**	**22**
Zr	420	632	946	0.22	326	**159**	**210**	**200**	**190**
Ba	0	3	24	67	84	**552**	**650**	**636**	**550**
La	26	14	37	57	31	**56.5**	**38**	**31.1**	**30**

The standard values of SINGO, average Singo granite (data from Nagudi et al., 2000); PAAS, average post-Archean Australian shale; UC, upper crust (data from Taylor and McLennan, 1985); and NASC, average North American shale (data from Gromet et al., 1984) are also quoted in the table.

are of first cycle sediments derived from the weathering front. The high magnification study of grains shows conchoidal fractures and fresh grain breakage surfaces. These grains appear to be subjected to high degree of diagenetic changes.

21.6.5 Geochemistry

The major and trace element composition of selected samples of Kaniyapuram borehole and red sands is given in Tables 21.5 and 21.6. X-ray fluorescence studies of the subsurface sediments of the Core C III reveal that the major elements (in oxide forms) such as SiO_2, Al_2O_3, and Fe_2O_3 together constitute more than 70% of the sediments. The other elements are present only in substantially lower percentages.

Table 21.6 Concentration of major elements (expressed as oxides) and trace elements in selected red sand samples.

	Sample name			The standard values			
Elements	**R12**	**R14**	**R16**	**SINGO**	**PAAS**	**NASC**	**UC**
			Major elements (%)				
SiO_2	92.23	76.82	80.9	**72.65**	**62.8**	**64.8**	**66**
TiO_2	0.68	1.78	1.83	**0.42**	**1.00**	**0.70**	**0.50**
Al_2O_3	4.28	11.88	9.09	**13.33**	**18.9**	**16.9**	**15.2**
MnO	0.00	0.03	0.05	**0.08**	**0.11**	**0.06**	**0.08**
Fe_2O_3	0.68	4.07	4.28	**2.6**	**7.22**	**5.65**	**5.00**
CaO	0.06	0.2	0.11	**1.09**	**1.30**	**3.63**	**4.20**
MgO	0.12	0.36	0.25	**0.6**	**2.20**	**2.86**	**2.20**
Na_2O	0.01	0.12	0.06	**3.54**	**1.20**	**1.14**	**3.90**
K_2O	0.03	0.26	0.17	**4.87**	**3.70**	**3.97**	**3.40**
P_2O_5	0.09	0.15	0.15	**0.13**	**0.16**	**0.13**	-
			Trace elements (ppm)				
V	94	162	168	**28**	**150**	**130**	**60**
Cr	42	99	82	**27.7**	**110**	**125**	**35**
Ni	21	52	42	**11**	**55**	**58**	**20**
Cu	0	5	9	**BDL**	**50**	**BDL**	**25**
Zn	0	18	1	**40.2**	**85**	**BDL**	**71**
Ga	0	10	5				
Rb	20	37	34	**348**	**160**	**125**	**112**
Sr	34	37	35	**103**	**200**	**142**	**350**
Y	2	6	7	**57**	**27**	**35**	**22**
Zr	354	467	451	**159**	**210**	**200**	**190**
Ba	0	76	47	**552**	**650**	**636**	**550**
La	20	25	37	**56.5**	**38**	**31.1**	**30**

The standard values of SINGO, average Singo granite (data from Nagudi et al., 2000); PAAS, average post-Archean Australian shale; UC, upper crust (data from Taylor and McLennan, 1985); and NASC, average North American shale (data from Gromet et al., 1984) are also quoted in the table.

The maximum concentration of SiO_2 (93.12%) is noticed at a depth of 1.5 m bgl. The concentration of Al_2O_3 and Fe_2O_3 shows a gradual increase toward the bottom of the core. The concentration of SiO_2, Al_2O_3, and Fe_2O_3 ranges from 78.85% to 93.12%, 2.43–14.23%, and 0.63–1.88%, respectively. The abundance of major elements in the core of the order $SiO_2 > Al_2O_3 > Fe_2O_3 > TiO_2 > CaO > Na_2O > MgO > K_2O > P_2O5 > MnO$ (Table 21.5). The downcore variation as shown in Table 21.6 reveals that the oxides of SiO_2 and TiO_2 show almost similar trend.

The concentrations of trace elements in the borehole core are of the order Zr > V > Cr > Ba > Zn > Sr > La > Rb > Cu > Ni > Ga > Y (Table 21.5). The elements like V, Cr, and Ni show comparatively less concentration because of the abundance of sand and sand-dominated sediments. This is due to the greater surface area of finer particles which favors metal absorptivity. The other constituents do not show any specific trend in the sediments. Generally all trace elements show enhanced concentrations in the wetland core in comparison to the estuarine cores. The sequence of the Core C III is dominated with sand sediments. The sand-rich sediments in the near shore cores exhibit enhanced concentration of TiO_2, Fe_2O_3, and CaO that are possibly derived from the littoral or near shore environments.

Apart from Kaniyapuram borehole sediments, selected samples of red sand sediments are also subjected to X-ray fluorescence studies. The major elements (in oxide forms) such as SiO_2, Al_2O_3, and Fe_2O_3 together constitute more than 70% of the total geochemical. The other elements are present only in substantially lower percentages. The abundance of major elements in the borehole core is of the order $SiO_2 > Al_2O_3 > Fe_2O_3 > TiO_2 > MgO > K_2O > P_2O_5 > CaO > Na_2O > MnO$ (Table 21.6). In trace elements, Zr shows maximum concentration (354–467 ppm) and the elements like V, Cr, and Ni show comparatively less concentration. The concentrations of trace elements in the borehole core are of the order Zr > V > Cr > Ba > Ni > Sr > Rb > La > Zn > Ga > Y > Cu.

21.7 Discussion

21.7.1 Sedimentary processes and depositional environments

The subsurface sediments of Kazhakuttom–Kaniyapuram coast and its adjoining areas show a wide spectrum of textural classes, viz., sand, clayey sand, sandy mud, etc. As seen from the ternary plots of sand, silt, and clay in the hydrodynamic facies of Pejrup's (1998) model, the sample plots are segregated mainly in one sector—close to the sand-dominated sector (Fig. 21.7). There are four different sectors in the ternary diagram indicating different energy fields from I to IV (quietest to most violent). In the present study, the ternary plots of the borehole sediments retrieved from the Kazhakuttom–Kaniyapuram coast are generally segregated in the Sector II and III indicating a comparatively moderate energy regime existed during the deposition of sediments. However, a few samples in the Core (C III) deviate marginally from this general trend because of occasional turbulence/high energy regime in the depositional environments.

The downcore variations of heavy minerals revealed marked changes in the quantity and quality of minerals in the borehole sediments of the C I and C III resemble the beach environment. The most abundant heavy mineral groups in the borehole sediments are opaques and sillimanite. One of the striking observations in the examination of heavy mineral species is the presence of comparatively large amounts of garnet in the bottom of Kaniyapuram borehole. The garnet-rich minerals are derived mainly from the khondalite suite of rocks in the hinterlands and may be eroded, transported, and deposited by Kulathoor *thodu*. The surface sediments collected from the beach (Kulathur to Onnampalam) also show a wide spectrum of heavy minerals. Among the heavy mineral assemblage in the beach sands, opaque, sillimanite, and zircon are major minerals (minerals with number percentage >5%), garnet, rutile, and monazite are minor minerals (minerals with number percentage 1–5%), and inosilicates as the trace minerals. Spatial distribution of the heavy–light mineral concentration reveals that the content of THMs is higher in the southern sides of the study area around the place called Karimanal, whereas silica-rich white sands are deposited in the Veli and further north. The observed sediment dispersal pattern might have attributed to the peculiar energy conditions that prevailed in the region. This peculiar energy regime is favorable for the segregation of high dense heavy minerals in situ and transportation of low dense lighter components toward the direction of the long shore current. The comparatively low

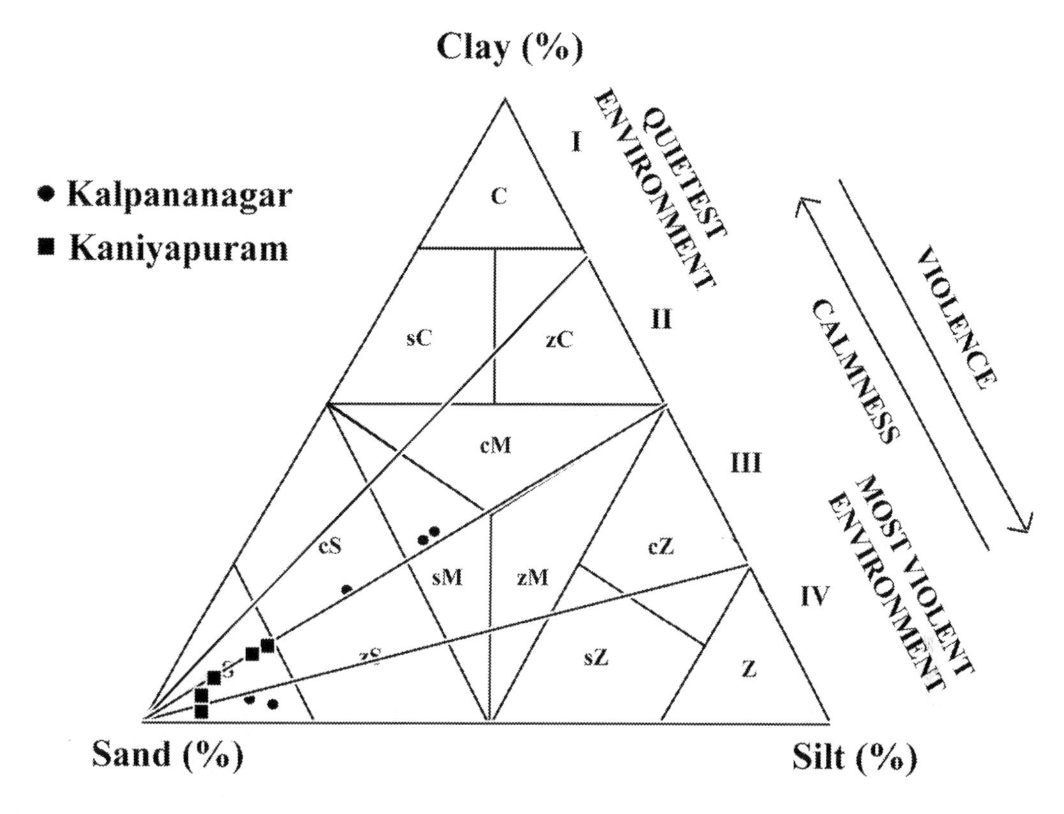

Figure 21.7

Sediment types (after Picard, 1971) of the borehole cores retrieved from the study area, along with depositional environments as per Pejrup (1988).

content of heavy mineral residue in the older coastal plains may be attributed to the role of eolian activity in segregating lighter minerals over the heavier/denser minerals (Anooja et al., 2013).

The scanning electron microscopic studies (SEM) of the sand grains from the subsurface sediments retrieved from Kaniyapuram borehole gives clear indication of beach environment at the top and riverine input of sediments at the bottom. Two types of ternary diagrams of Nesbitt et al. (1996), Nesbitt and Young (1982, 1984, 1989, 1996) are used here to deduce weathering trends: (1) A–CN–K [Al_2O_3–($CaO + Na_2O$)–K_2O] and (2) A–CNK–FM [Al_2O_3–($CaO + Na_2O + K_2O$)–$Fe_2O_3 + MgO$] diagrams. In both the diagrams, the samples fall close to the Al_2O_3 sector, indicating intense weathering in the source area before being deposited in the present site (Fig. 21.8A and B). The ternary plots of the present study are seen clustered more close toward the Al_2O_3 end. The interrelationship between WIP and CIA is depicted in Fig. 21.9. The CIA and WIP values for the borehole sediments of Kaniyapuram borehole falls in the field of WIP 0 and 10, and CIA 80 and 100 (Fig. 21.9; Table 21.7) indicating the maturity of the sediments, whereas the plots of the red sand sample spread in the field 0 and 10 for WIP and 95 and 100 for CIA (Fig. 21.9; Table 21.7). This reiterates that the bottom sediments of the borehole core and

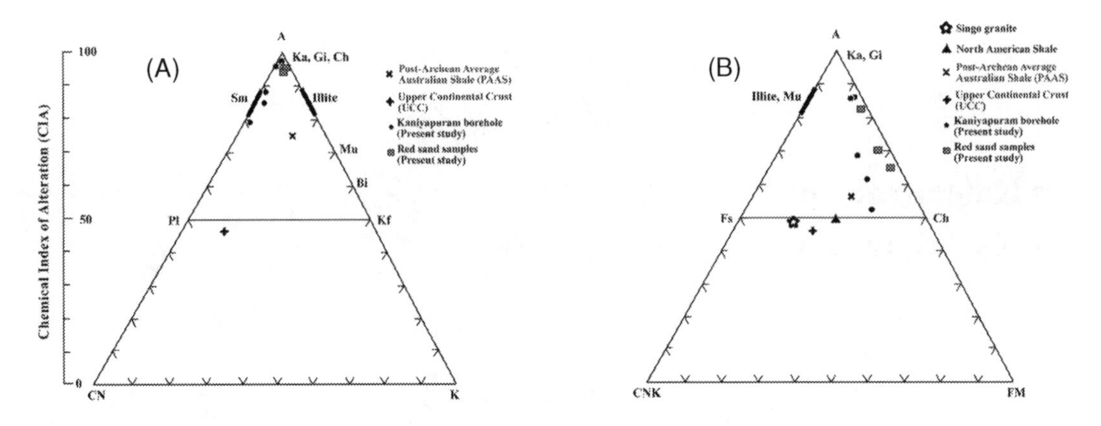

Figure 21.8

A (Al_2O_3) – CN ($CaO+Na_2O$) – K (K_2O) diagram (A) and A (Al_2O_3) – CNK ($CaO+Na_2O+K_2O$) – FM (Fe_2O_3+MgO) diagram (B) of the Kaniyapuram borehole sediments and red sand samples collected from the study area. Ka Kaolinite, Gi Gibbsite, Ch Chlorite, Il Illite, Mu Muscovite, Kf Potash feldspar, Pl Plagioclase, Sm Smectite, Fs Feldspar, Ch Charnockite (after Nesbitt and Young, 1982, 1984, 1989, 1996; Nesbitt et al., 1996).

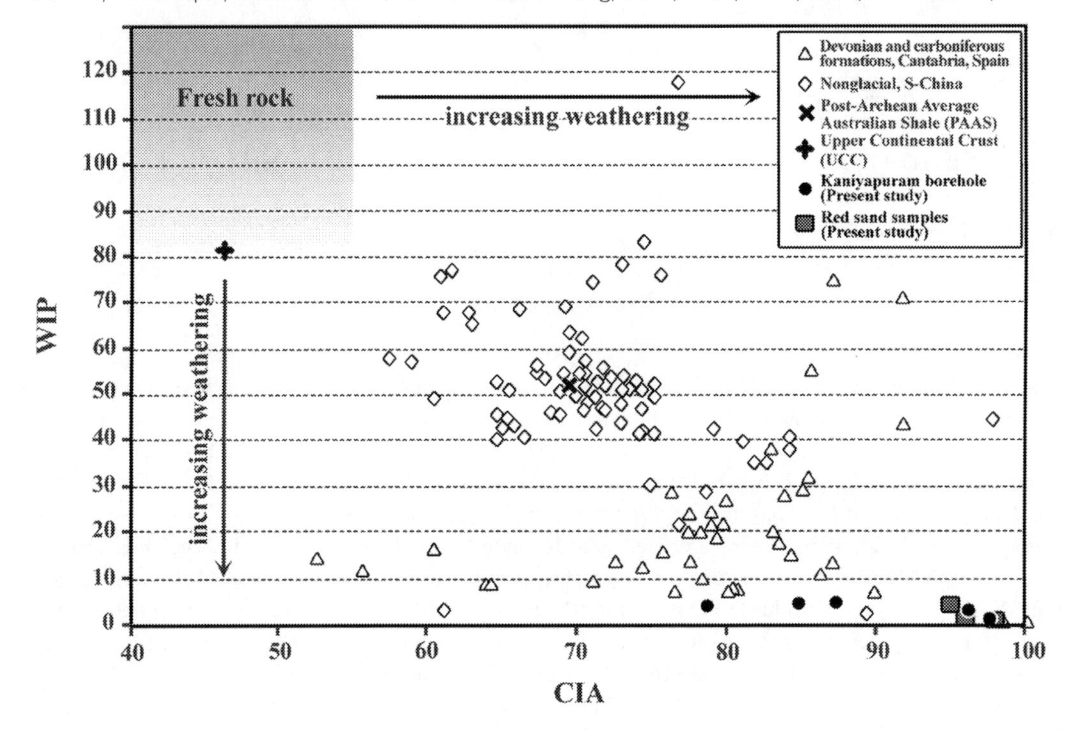

Figure 21.9

Interrelationship between two weathering proxies – Weathering Index of Parker (WIP) and Chemical Index of Alteration (CIA) worked out for the Kaniyapuram borehole sediments and red sand samples.

Table 21.7 Sediment type, Weathering Index of Parker (WIP), Chemical Index of Alteration (CIA), and Chemical Index of Weathering (CIW) of the borehole sediments and red sands.

Depth (m)/sample location	Sediment type	WIP[a]	CIA[b]	CIW[c]
		Kaniyapuram borehole		
1.5	Sand	4.07	78.90	79.93
4.5	Sand	5.01	85.19	87.72
9	Sand	5.21	87.56	89.39
12	Sand	0.73	97.59	98.78
15	Sand	3.25	96.28	96.28
		Red sands		
R12	NA	0.77	97.72	98.39
R14	NA	4.63	95.35	97.38
R16	NA	2.84	96.39	98.16

[a] *After Parker (1970).*
[b] *After Nesbitt and Young (1982).*
[†] *After Nyakariu and Koeberl (2001).*
NA, not applicable.

red sand samples are more altered/weathered. The SEM images of the Kaniyapuram borehole at a depth of 1.5 m show that majority of the grains are rounded to subrounded showing high amount of wear and tear in a high energy environment. The degree of roundness is high at the top and decreases toward the bottom. It demonstrates typical beach-born characteristics with curved grooves and pits. At a depth of 9 m, the sand grains show comparatively less sphericity and low degree of roundness. The mineralogical studies revealed occurrence of garnet from hinterlands. All these point to the hinterland influence in contributing sediments in the core at this level. The sand grains at a depth of 15 m are highly angular to subangular with conchoidal fractures. Like the samples at 9 m depth, the nature and characteristics of the sand grains points toward its hinterland source.

21.7.2 Sea-level changes and coastal evolution

The Indian coastline, stretching over 6500 km, is the sixth longest in the world and has been subjected to several episodes of sea-level variations that followed the Last Glacial Maximum (LGM). Since the LGM, it has been undergoing physical changes, and the present coastal geomorphology has evolved largely in the background of the postglacial transgression over the pre-existing topography of the shore, coast, and offshore zones (Merh, 1992; Hashimi et al., 1995; Banerjee, 1993; Somayajulu, 2002; Nair et al., 2006). The Holocene sea fluctuated in the course of the last 10,000 years, and its oscillations between 3000 and 6000 BP have left significant imprints in the sedimentary archives, shedding light on the evolution of the coast. The sea-level and climate changes have strongly influenced the coastal marine processes. Much of the coastal land and associated landforms around the world have been carved out during the late Quaternary and the geomorphic evolution has greatly influenced the history of humanity (Woodroffe, 2002).

Table 21.8 Geochronological bearing of the subsurface samples from the study area.

Sl. no.	Reference	Location	Depth (m)	Material	Age (yrs BP)
1	Longhinos (2009)	TRV 6	2.0	Sediment	1444 ± 269^{a}
2	Longhinos (2009)	TRV 6	5.0	Sediment	5169 ± 849^{a}
3	Longhinos (2009)	TRV 6	10.0	Sediment	7500 ± 350^{b}
4	Longhinos (2009)	TRV 6	11.0	Sediment	81738 ± 1916^{a}
5	Longhinos (2009)	TRV 6	12.0	Sediment	9560 ± 250^{b}
6	Longhinos (2009)	TRV 6	20.0	Sediment	16626 ± 2980^{a}

[a] *OSL date.*
[b] *Radiocarbon date.*

The sea-level changes and its after effects play a vital role in the geological evolution of the study area. The coastal wetlands falling within the Kazhakuttom–Kaniyapuram belt have been evolved and modified due to Holocene sea-level fluctuations. The lithological sequences of the boreholes retrieved from the study area and the SEM studies of the subsurface sediments show indications of sea-level changes to which the sand-dominated sediments in the upper part of the sequences were subjected. The dated core of Longhinos (2009) which is sited 3 km southeast of the Kaniyapuram borehole site shows a late Quaternary age (Table 21.8). Generally ^{14}C dating method was chosen specifically to those litho units rich in organic carbon and OSL dating technique was adopted to decipher the age of those litho units without organic carbon. On comparison and correlation of the dates from the two methods of dating employed reveals, reasonable correspondence among the datasets. The borehole taken as reference has a total depth of 30 m, in which up to 10 m constitutes 80–95% of sand having quartz as 96% of the mineral population. It is very much similar to the borehole cores taken for present study which is also having sand-dominated upper layers. The timing of sediment deposition in the two borehole cores is fit mutually and can be laterally linked. The age–depth relations in the samples dated support the vertical accretion of sediments in the central belt of Thiruvananthapuram coastal plain. In summary it can be stated that the 20–30 m thick sand-dominated sediment pile along the central belt of the Thiruvananthapuram coastal plain was deposited during late-Pleistocene to late-Holocene. The present study area typically belongs to central belt (Kazhakuttom–Kaniyapuram belt) of the Thiruvananthapuram coastal plain and the sand-dominated subsurface sediments in this region are of Holocene age. It indicates the role of Holocene geological events which gave rise to the landform features of the study area. The studies reveal that the sea level in the west coast was about 120 m below the present level during the LGM, that is, approximately 18,000 yrs BP (Subrahmanya, 1987; Merh, 1992; Wagle, 1990, and several others). This might have lowered the base level of erosion of the west flowing rivers of Kerala, which were debouching sediments far into the sea at that time (Nair et al., 2010). The heavy rainfall and river discharges during the early–middle-Holocene together with sea-level rise of ~6000 yrs BP might have aggravated deposition of alluvial sediments at the river confluence (Shi et al., 2001; Padmalal et al., 2011; Mohan et al., 2015, 2017). Subsequent lowering of sea level and the arid climate at the beginning of the Meghalayan time might have exposed many elevated areas in the uplands to eolian activity and the formation of beach ridges in the study area (Banerji, 1993; Padmalal et al., 2011, 2014a, 2014b; Veena et al., 2014; Mohan et al., 2015, 2017).

The most important coastal wetland in the Kazhakuttom–Kaniyapuram belt is Kadinamkulam lagoon or locally known as Kadinamkulam *kayal*. The Kadinamkulam *kayal*, a temporary lagoon in the southern part of Kerala is largest of its kind in Thiruvananthapuram district. A major portion of the basin lies southwardly and it shares boundary with the Anchuthengu *kayal* on the north and the Veli *kayal* on the south and connected by Parvathiputhanar canal. The mean depth ranges from 2.11 to 0.27 m and the nature of the mouth depends on the extent of sand bar formation and it normally gets closed completely for varying periods during the premonsoon with the diminishing freshwater discharge. The Kadinamkulam lagoon remains connected with the Arabian Sea for varying periods depending on rainfall and river discharge. It is the one of the important back water system in the south-western coast of India and the coastal area encompassing the Kadinamkulam lagoon comprises a spectrum of geomorphic units such as barrier beaches, ridges and runnels, lagoons, and flood plains. The lagoon and its adjoining coastal lands are known for coconut husk retting and production of various coir products. The sedimentological, geochemical, and geochronological studies reveal that the sea-level and climate changes have brought significant modifications in the geomorphic settings of the system. The depositional regimes and the evolutionary phases showed marked differences in the northern and southern reaches of the lagoon. The northern half evolved during the transgressive–regressive events of middle–late-Holocene from a sheltered coastal water body surrounded by thick coastal vegetation. The proposed three stage evolutionary model of the Kadinamkulam lagoon indicates that it evolved from an embayment of the Arabian Sea by the progradation and development of ridges and runnels during transgressive–regressive cycles under heavy rainfall events of the Holocene.

Geological history of the area has been interpreted by collating the information acquired from various sources, such as, physiography of the coastal plain, geologic correlation and chronology with adjacent dated borehole, sea-level changes, and evidences from textural, mineralogical, and geochemical studies. The overall evolution of Kadinamkulam lagoon can be divided into three scenarios, (1) Pre-LGM scenario, (2) Middle–late-Holocene scenario, (3) Present scenario (Fig. 21.10). The coast in southwest India is endowed with an interlacing network of backwater bodies (locally known as *kayals*) and drainage channels, many of which evolved during the Holocene (Joseph and Thrivikramaji, 2002). The backwater bodies are either estuaries or lagoons which are open to the sea seasonally or perennially, separated by spits and bars. These wetlands play a pivotal role in the socioeconomic and environmental scenarios of the region and therefore, their conservation is essential for the sustenance of the people in the area. Among these water bodies, the Kadinamkulam lagoon is one of the important back water systems in the entire south-western coast of India. Previously, there was no conceptual model or hypothesis to explain how the Kadinamkulam lagoon has evolved in the recent past.

The evolution of coastal lands with valleys and wetlands were influenced by many local and regional factors like changes in climate, sea level, local, and regional tectonics (Narayana and Priju, 2006). Many evidences exist in the Kerala coast to reveal that the region was subjected to several spells of sea-level rises and falls during Holocene. However, the coastal lagoons along west coast of India have not received much attention with respect to their antiquity and morphodynamics of the recent past. Recently, attempts have been initiated to address the evolution of the wetland systems of southwest India in relation to sea-level and climate changes (Padmalal et al., 2011). There has been little integration of physical evolution with biological changes in lagoon development along the west coast of India. Although the process of formation of the Kadinamkulam lagoon dates to middle–late-Holocene, the present geomorphic configuration is the outcome of the interplay of late Quaternary climatic and sea-level changes, especially during the later phases of the LGM. The subsurface data of

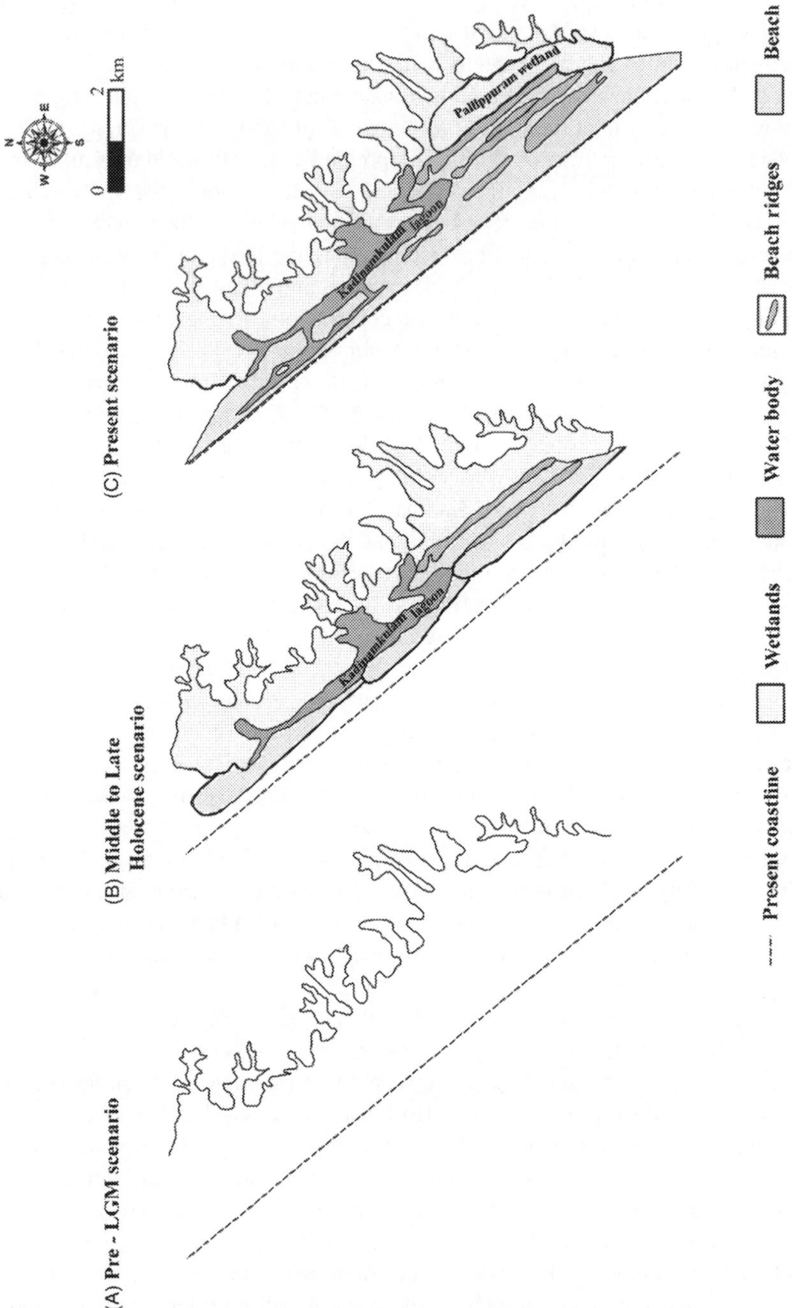

Figure 21.10
Evolutionary stages of Kadinamkulam Lagoon.

the boreholes cores provided sufficient lithological and mineralogical information to distinguish facies changes within a geochronological framework while addressing the evolution of the Kadinamkulam lagoon.

The marine transgression in the beginning of Holocene has had a profound impact on the antiquity and evolution of the Kadinamkulam lagoon. The presence of a marine embayment and altitude of the location almost equal to the ambient sea level facilitated the earliest marine influence during the early-Holocene (9.5–8.0 ka BP). Although sedimentation appears to have kept pace with rise in sea level, it did not result in open marine conditions in this water body, due to the mixing of marine waters and the heavy influx of fresh water. High rainfall from 10.0 to 5.5 ka BP is responsible for the high rate of sediment and organic matter accumulation in the coastal lands of southwest India (Nair et al., 2010; Padmalal et al., 2011). The Holocene epoch started with an episode of heavy precipitation, continuing almost until ~5.5 ka BP (Prell and Kutzbach, 1987; Kale, 2005). This period is often referred to as the Holocene Climatic Optimum, and the final phase of it coincided with sea-level rise of 3–4 m with respect to the present level (Mathur et al., 2004; Nair, 2007). As seen from the previous discussions, the early-Holocene witnessed heavy rainfall which had not only broadened the valleys developed over the Neogene and Archean crystallines but also brought in huge volumes of sediments into the broad estuarine basins that are entrenched over the uplifted southern blocks located south of the Achankovil shear zone.

During early–middle-Holocene, the present study area was fully covered by the sea due to maximum marine transgression. The interactions of sea-level changes and the paleoclimatic events had significant impact on the hydrological regimes and sediment accumulation pattern and the subsequent evolution of the Kadinamkulam lagoon. The high rainfall, rising sea level, and the northerly currents of the early–middle-Holocene have had considerable roles in the accretion and development of ridges and runnels in the nearshore region. The large southern half of the pre-Kadinamkulam lagoon evolved from an embayment of the Arabian Sea due to prolongation of a series of barrier spits, ridges, and runnel systems. This is not only evident from the sandy apron over the late-Pleistocene–Holocene silt- and clay-dominated sediments, but also from the peculiar suite of heavy minerals in the subsurface sediments.

The rising sea level coupled with northward drifting littoral currents in the early–middle-Holocene were responsible for the progradation and development of sand barriers across the embayment and further enhanced the formation of Kadinamkulam lagoon. The sea-level rise continued even in the beginning of the middle-Holocene. The rise in sea level during middle-Holocene at about 6000 yrs BP had aggravated deposition of river-borne materials/alluvium in the south eastern side of the coast. Toward the end of middle–late-Holocene beach ridges formed at the south eastern portion of the coast and a part of Kadinamkulam lagoon began to form in the north eastern portion of the coast due to the lowering of sea level.

The Kadinamkulam lagoon has evolved to its present form during Holocene. Prior to the beginning of Holocene, there existed an expansive embayment of pre-Kadinamkulam basin. An abundant supply of sediments in the early-Holocene resulted in the formation of the barrier system, whereas reduced supply caused breaching of the barrier spit and opening of the lagoon (Carter et al., 1989; Vaz et al., 2002). The open lagoon setting of the middle-Holocene Kadinamkulam lagoon, similar to the one illustrated by Nichols and Allen (1981), transformed the system into a partly closed lagoon with limited inlet–outlet systems in the regressive phase of the late-Holocene. In the present scenario, Kadinamkulam lagoon extends further near to the present coastal line due to continuous long shore drift toward north. The subsequent development of barrier beaches and constant supply of sediments brought in by the littoral drift and also the anthropogenic activities in the last few decades played a major role in shaping the Kadinamkulam lagoon in its present form. The coastal areas of southern Kerala selected for the present study registers many geological events of the post-LGM period that are intact within its sedimentary archive. The most noteworthy development is entrenchment of estuarine/wetland basins over the Neogene sediments. Sedimentation in these basins initiated from early-Holocene onward. In short, the present configuration of the wetlands and associated landform features of the study area were the result of the combined effects of sea level, climate changes, and neotectonism. It is true that the exact role of the last mentioned natural driving force (neotectonism) has to be discriminated through more focused research in the Trivandrum block located south of the Achankovil Shear Zone.

21.8 Learning and knowledge outcomes

The geological evolution of the study area has been interpreted by collating the information acquired from various sources, such as, physiography of the coastal plain, geologic correlation, and chronology with adjacent dated borehole, sea-level changes, and evidences from textural, mineralogical and geochemical studies. The study explains the origin and evolution of coastal wetlands in the Kazhakuttom–Kaniyapuram belt and provides an evolution model for the coast. The most important coastal wetland coming in the area is Kadinamkulam lagoon or locally known as Kadinamkulam *kayal*. The coastal wetlands coming under Kazhakuttom–Kaniyapuram belt has been evolved and modified due to Holocene sea-level fluctuations. The heavy rainfall and river discharges during the early–middle-Holocene time together with sea-level rise of ~6000 yrs BP might have aggravated deposition of alluvial sediments at the river confluence of Kulathur Thodu. Later, the lowering of sea level in the middle- and late-Holocene might have exposed many elevated areas in the form of ridges and runnels of the present strandlines.

The process of formation of the Kadinamkulam lagoon dates back to middle–late-Holocene. The present geomorphic configuration is the outcome of the interplay of late Quaternary climatic and sea-level changes, especially during the later phases of the LGM. The subsurface sedimentary architecture, geomorphology of the Kadinamkulam coast, and the chronology of samples from southern to the northern part indicate that the present day lagoon has evolved from an embayment of the Arabian Sea that existed prior to the Holocene. Although the process of lagoon development was initiated by natural events of pre-Holocene, sea-level and climate changes have transformed it considerably into its present form during the Holocene. Sediment supply and sea-level fluctuations have been the dominant factors responsible for the development of various geomorphic features along the coastal plains of southern Kerala. The different units detected by means of sedimentological and mineralogical approaches and by SEM surface texture analysis and X-ray fluorescence studies have been found to be useful for developing a three-phase model for the evolution of the wetland systems of the study area during the Holocene. The presence of a marine embayment and altitude of the location almost near to the ambient sea level facilitated the earliest marine influence during the early-Holocene. The subsequent formation of sand barrier spits and beach ridges across the pre-Holocene embayment paved the way for development of the Pallipuram wetland and then the Kadinamkulam lagoon. The Akathumuri lagoon (kayal) and the beach barrier systems might be the youngest wetland developed in the region. The northward shift of the Pallipuram wetland, the Kadinamkulam lagoon, and the Akathumuri lagoon indicate the northward drift of coastal sediments since early-Holocene.

Acknowledgments

The D.P., K.M., and M.S.A. thank the Director, National Centre for Earth Science Studies (NCESS), Thiruvananthapuram for providing facilities. K.P.N.K. thanks the CSIR, HRDG for its support in the form of ES project [21(0828)/10/EMR-II]. V.M.S. acknowledges The Principal, Sree Narayana College, Chempazhanthy, Thiruvananthapuram and Head, Department of Geology, Sree Narayana College, Chempazhanthy, Thiruvananthapuram for support and facilities.

References

Abraham, S.A., Structuring an intensive surface survey: strategies for investigating the early historic port site of Pattanam, Kerala, Adharam 1 (2006) 8–32.

Anooja, S., Padmalal, D., Maya, K., Mohan, V.S., Baburaj, B., Heavy mineral contents and provenance of Late Quaternary sediments of southern Kerala, Southwest India, J. GeoMar. Sci. 42 (2013) 749–757.

Banerjee, P.K., Imprints of Late Quaternary climate and sea level changes on east and south Indian coast, Geo-Mar. Lett. 13 (1) (1993) 56–60.

Carter, R.W.G., Forbes, D.L., Jennings, S.C., Orford, J.D., Show, J., Taylor, R.E., Barrier and lagoon coast evolution under differing relative sea-level regimes: examples from Ireland and Nova Scotia, Mar. Geol. 88 (1989) 221–242.

Folk, R.L., Ward, W.C., Brazos river bar: a study in the significance of grain size parameters, J. Sed. Pet. 27 (1957) 3–26.

Gromet, L.P., Dymek, R.F., Haskin, L.A., Korotev, R.L., The 'North American Shale Composite': Its compilation, major and trace element characteristics. Geochi. et Cosmochi. 48 (1984) 2469–2482.

Hashimi, N.H., Nigam, R., Nair, R.R., Rajagopalan, G., Holocene sea level fluctuations on western Indian continental margin: an update, J. Geol. Soc. India 46 (1995) 157–162.

Jayalakshmi, K., Nair, K.M., Kumai, H., Santosh, M., Late Pleistocene-Holocene paleoclimatic history of the Southern Kerala Basin, Southwest India, Gond. Res. 7 (2) (2004) 585–594, doi:10.1016/s1342-937x(05)70808-x.

Joseph, S., Thrivikramji, K.P.Narayana, A.C. (Ed.), Kayals of Kerala coastal land and implication to Quaternary sea level changes, Mem. Geol. Soc. India 49 (2002) 51–64.

Kale, V.S., 2005. Late Quaternary climate changes: the Indian scene. Indian Institute of Geomorphologist. In: Proceedings of the 18th Convention and National Seminar on Quaternary Climate Changes and Landforms. Manonmaniam Sundaranar University, Abstract Volume, 2005, pp. 15–28.

Kumaran, K.P.N., Nair, K.M., Shindikar, M., Limaye, R.B., Padmalal, D., Stratigraphical and palynological appraisal of Late Quaternary mangrove deposits of the west coast of India, Quat. Res. 64 (3) (2005) 418–431.

Lewis, D.W., Practical Sedimentology, Hutchinson Ross Publishing Company, Pennsylvania, 1984.

Limaye, R.B., Kumaran, K.P.N., Nair, K.M., Padmalal, D., Cyanobacteria as potential biomarker of hydrological changes in the Late Quaternary sediments of South Kerala Sedimentary Basin, India, Quat. Int. 213 (2010) 79–90.

Longhinos, B., Geology and geochronology of silica sands of coastal plain of Thiruvananthapuram district, Kerala, India, with special reference to late Quaternary environment, PhD Thesis, Cochin University of Science Technology, 2009.

Mange, M.A., Maurer, H.F.W., Heavy Minerals in Colour, Chapman & Hall, London, 1992.
Mathur, U.B., Pandey, D.K., Bahadur, T., Falling late Holocene sea level along Indian coast, Curr. Sci. 87 (2004) 439–440.
Merh, S.S., Quaternary sea level changes along Indian coast, Proc. Indian Acad. Sci. 58 (1992) 461–472.
Mohan, V.S., 2015. Late Quaternary geology of the coastal lands of southern Kerala, India with special reference to palaeoclimate and coastal evolution, PhD Thesis, Cochin University of Science and Technology (unpublished).
Mohan, V.S., Limaye, R.B., Maya, K., Padmalal, D., Kumaran, K.P.N., Tracing signatures of mud bank formation in the Late Holocene sediments along Kerala coast, SW India, 33rd Convention of Indian Association of Sedimentologist, Banaras Hindu University, Varanasi, 2016.
Mohan, V.S., Limaye, R.B., Padmalal, D., Ahmad, S.M., Kumaran, K.P.N., Holocene climatic vicissitudes and sea level changes in the south western coast of India: appraisal of stable isotopes and palynology, Quat. Int. 443 (2017) 164–176.
Nagudi, B., Koeberl, C., Kurat, G., Petrography and geochemistry of the Singo granite, Uganda: Interpretations and implications for the origin. J. African Earth Sci. 36 (2000) 73–87.
Nair, K.K., Quaternary geology and geomorphology of coastal plains of Kerala, Geol. Surv. India Special Publ. 88 (2007) 1–73.
Nair, K.M., Padmalal, D., Quaternary Sea Level Oscillations Geological and Geomorphological Evolution of South Kerala Sedimentary 4 Basin, Department of Science and Technology, Government of India, 2003 Project Completion Report.
Nair, K.M., Padmalal, D., Kumaran, K.P.N., Quaternary geology of South Kerala Sedimentary Basin—an outline, J. Geol. Soc. India 67 (2006) 165–179.
Nair, K.M., Padmalal, D., Kumaran, K.P.N., Sreeja, R., Limaye, R.B., Srinivas, R., Late Quaternary evolution of Ashtamudi, Sasthamkotta lake systems of Kerala, south west India, J. Asian Earth Sci. 37 (2010) 361–372.
Najeeb, K.M., Ground Water Exploration in Kerala on 31.3.1999, Technical Report, Central Ground Water Board, Kerala Regions, Trivandrum, 1999.
Narayana, A.C., Priju, C.P., Evolution of the coastal landforms and sedimentary environments of the late Quaternary period along central Kerala, southwest coast of India, J. Coastal Res. 39 (2006) 1898–1902.
Nesbitt, H.W., Young, G.M., Early Proterozoic climates and plate motions inferred from major element chemistry of lutites, Nature 299 (1982) 715–717.
Nesbitt, H.W., Young, G.M., Prediction of some weathering trends of plutonic and volcanic rocks based on thermodynamic and kinetic considerations, Geochem. Cosmochem. Acta 48 (1984) 1523–1534.
Nesbitt, H.W., Young, G.M., Formation and diagenesis of weathering profiles, J. Geol. 97 (2) (1989) 129–147.
Nesbitt, H.W., Young, G.M., Petrogenesis of sediments in the absence of chemical weathering: effects of abrasion and sorting on bulk composition and mineralogy, Sededimentology 43 (2) (1996) 341–358.
Nesbitt, H.W., Young, G.M., McLennan, S.M., Keays, R.R., Effects of chemical weathering and sorting on the petrogenesis of siliciclastic sediments, with implications for provenance studies, J. Geol. 104 (1996) 525–542.
Nichols, M., Allen, G., Sedimentary process in coastal lagoons, UNESCO, Technical Paper in Marine Science 33, Coastal Lagoon Research, 1981, pp. 27–80.
Nyakairu, G.W.A., Koeberl, C., Mineralogical and chemical composition and distribution of rare earth elements in clay-rich sediments from central Uganda. Geoche. J. 35 (2001) 13–28.
Padmalal, D., Kumaran, K.P.N., Limaye, R.B., Baburaj, B., Mohan, V.S., Maya, K., Effect of Holocene climate and sea level changes on landform evolution and human habitation: central Kerala, India, Quat. Int. 325 (2014a) 162–178.
Padmalal, D., Kumaran, K.P.N., Nair, K.M., Baijulal, B., Limaye, R.B., Mohan, V.S., Evolution of the coastal wetland systems of SW India during Holocene: evidence from marine and terrestrial archives of Kollam coast, Kerala, Quat. Int. 237 (2011) 123–139.

Padmalal, D., Kumaran, K.P.N., Nair, K.M., Limaye, R.B., Mohan, V.S., Baijulal, B., Anooja, S., Consequences of sea level and climate changes on the morphodynamics of a tropical coastal lagoon during Holocene: an evolutionary model, Quat. Int. 333 (2014b) 156–172.
Padmalal, D., Maya, K., Mohan, V.S., Late Quaternary Climate, Sea Level Changes and Coastal Evolution—A Case Study From SW India, Monograph, Centre for Earth Science Studies, Thiruvananthapuram, India, 2013.
Parker, A., An index of weathering for silicate rocks. Geo. Mag. 107 (1970) 501–504.
Passega, R., Grain size representation by CM patterns as a geological tool, J. Sed. Pet. 34 (1964) 830–837.
Pejrup, M., The triangular diagram used for classification of estuarine sediments: a new approach, in: de Boer, P.L., Van Gelder, A., Nio, S.D. (Eds.), Tide-Influenced Sedimentary Environments and Facies, Reidel, Dordrecht, 1988, pp. 289–300.
Picard, M.D., Classification of fine grained sedimentary rocks, J. Sed. Pet. 41 (1971) 179–195.
Prell, W.L., Kutzbach, J.E., Monsoon variability over the past 150,000 years, J. Geophys. Res. 92 (1987) 8411–8425.
Savin, S.M., The history of the Earth's surface temperature during the past 100 million years, Anl. Rev. Earth Planet. Sci. 5 (1) (1977) 319–355.
Shajan, K.P., 1998. Studies on the Late Quaternary sediments and sea level changes of the central Kerala coast, India. Ph.D. thesis, Cochin University of Science and Technology, Kochi.
Shajan, K.P., Selvakumar, V., Cheriyan, P.J., Locating the ancient Port of Muziris: fresh findings from Pattanam, J. Roman Arch. 17 (2004) 312–320.
Shi, P., Du, Y., Wang, D., Gan, Z., Annual cycle of mixed layer in South China Sea, J. Trop. Oceanogr. 20 (1) (2001) 10–17.
Shynu, R., Rao, V.P., Samiksha, S.V., Vethamony, P., Naqvi, S.W.A., Kessarkar, P.M., Dineshkumar, P.K., Suspended matter and fluid mud off Alleppey, southwest coast of India, Estuar. Coast. Shelf Sci. 185 (2017) 31–43.
Soman, K., Geology of Kerala, Geological Society of India Publications, Bangluru, 2002.
Somayanjalu, B.L.K., Present and past (late Quaternary) sea level—available new information from Indian coastal regions, Mem. Geol. Soc. India (49) (2002) 1–16.
Subrahmanya, K.R., Evolution of the Western Ghats, India—a simple model, J. Geol. Soc. India 29 (1987) 446–449.
Taylor, S.R., McLennan, S.H., The continental crust: Its composition and evolution. Blackwell, Oxford, 1985.
Vaz, G.G., Mohapatra, G.P., Hariprasad, M., Geomorphology and evolution of Barrier–Lagoon coast in a part of north Andhra Pradesh, Geol. Soc. India 49 (2002) 31–40.
Veena, M.P., Achyuthan, H., Eastoe, C., Farooqui, A., A multi-proxy reconstruction of monsoon variability in the late Holocene, South India, Quat. Int. 325 (2014) 63–73.
Wagle, B.G., Beach rocks of central west coast of India, Geo-Mar. Lett. 10 (1990) 111–115.
Woodroffe, C.D., Coasts Form, Process and Evolution, Cambridge University Press, UK, 2002.

CHAPTER

22 Late Quaternary geoarcheology and palynological studies of some saline lakes of the Thar Desert, Rajasthan, India

B.C. Deotare, Sheila Mishra, S.N. Rajaguru

Formerly of Deccan College Post-Graduate and Research Institute (now Deemed to be University), Pune, India

22.1 Introduction

The Thar Desert, an easternmost extent of the mid latitude Saharan desert belt, is locally known as "Maru Bhumi" (land of death). Geographically the Thar Desert (24°30′N and 30°N, 69°30′E and 78°E) covers approximately an area of 0.32 million sq km and is bounded on the north by the alluvial plain of the Satluj river and its principal tributary the river Ghaggar, on the west by sand dunes and the Indus delta of Pakistan and on the south by the Luni river originating in the eastern Aravalli hill ranges and debouching into the Great Rann of Kutchch in Gujarat. In our communication we have covered the Central Southern Zone of the Thar Desert having only the river Luni in the south (Fig. 22.1).

Climatically the major part of the Thar Desert is arid and the mean annual rainfall varies from about 500 mm in the extreme eastern margin to around 150 mm in the western border with Pakistan. About 90% of the rainfall is from the Indian Summer Monsoon (ISM) between the months of July and September and has a high annual variability with a high rate of evapotranspiration (PET) (~1500 mm/yr to >2000 mm/yr). The summer being the months of March to June, is very hot (40°C–50°C) and windy (25–30 km/h) while the winter months of October to February are moderately cold (3°C–10°C) with some westerly winter rains (<5–8%). Droughts are more common than good rainy years.

In a sense the Thar is not a true desert but rather a semiarid desert which enjoyed semiarid climate during the early Quaternary and arid to semiarid climate during the late Quaternary. It experienced humid to subhumid climate (>1200 mm annual rainfall) during the late Tertiary. In this chapter, we have used terms like semiarid (with rainfall 800 mm), wet semiarid (with rainfall around 800–1000 mm), and arid (with rainfall <400 mm). We are aware of these simplified climatic terms which need to be made precise after getting high-resolution proxy data on pollen, paleosols, fluvial, eolian, and lacustral sediments, and on varieties of calcrete which are part and parcel of Quaternary sediments preserved in the Thar.

22.1.1 Vegetation

Although the Thar Desert is continuous with the mid latitudinal desert belt, it belongs to the Oriental biogeographic realm while Arabia and the Sahara are part of the Saharo-Arabian biogeographic realm

Holocene Climate Change and Environment. DOI: https://doi.org/10.1016/B978-0-323-90085-0.00006-1

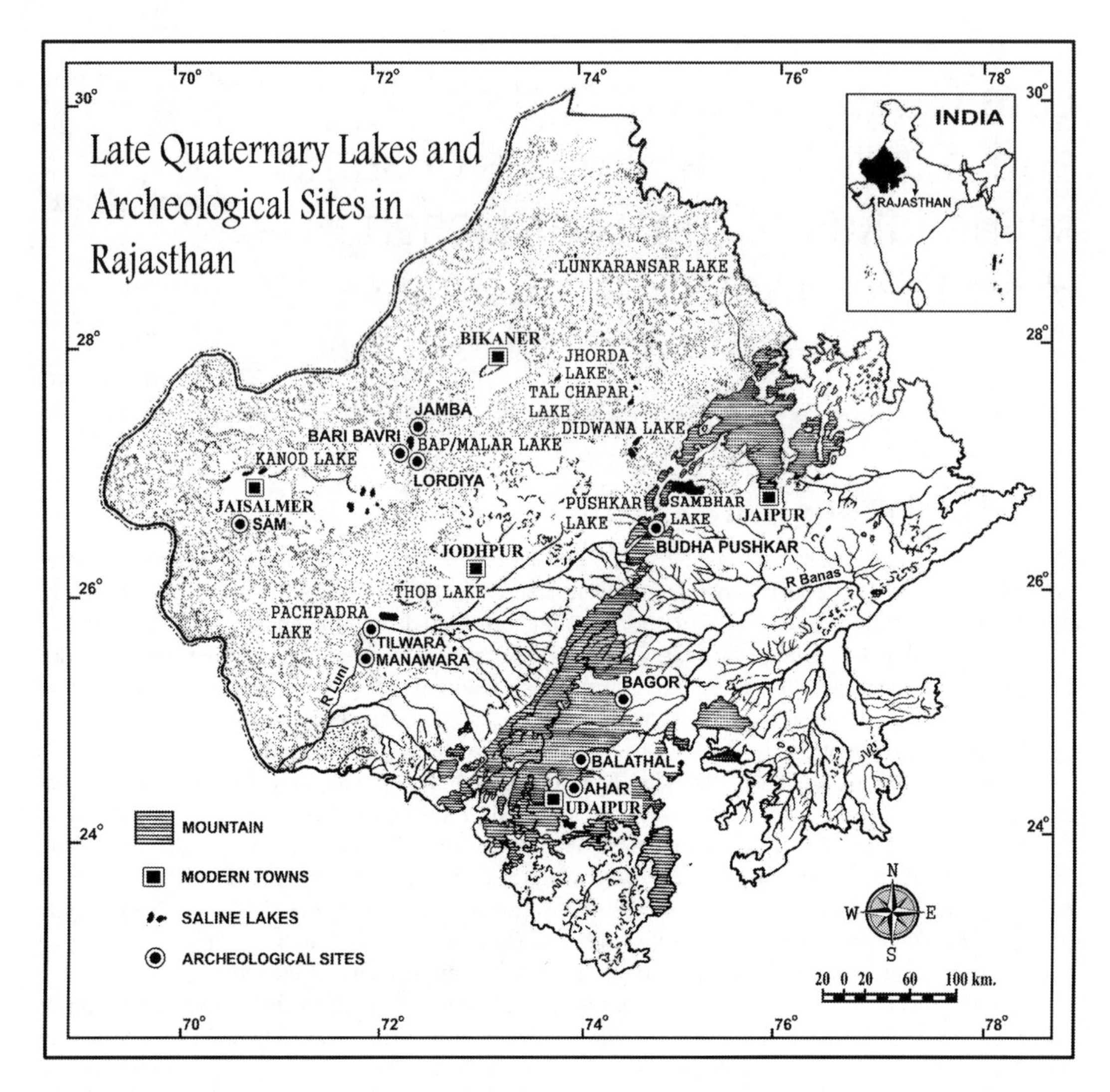

FIGURE 22.1

Distribution of some late Quaternary saline lakes and Archeological sites in Rajasthan.

(Holt et al., 2013). In spite of the common climatic factors, fauna and flora of the Thar Desert are distinct. The Thar Desert also has more endemic species than the Saharo-Arabian Desert (Khan and Frost, 2001). The parts of the Thar Desert which have annual rainfall up to 500 mm normally would not be considered "desertic," however the extreme seasonality of the rainfall means that in spite of relatively abundant annual rainfall totals, the overall climate remains arid (Dhir et al., 2018). In spite of the arid/semiarid climate and irregular monsoon rains, the Thar Desert has a good vegetation cover. The typical desert landscape transforms into grassland with a typical desert flora. The sandy soil cover seems to be good percolator of rainwater and thus provides well aerated soil and water for seasonal as well as perennial plants to survive in the typical arid climate. The majority of the vegetation comprising both monocot (15%) and dicotyledonous (76–84%) families have dominance with families like Compositae, Gramineae, Fabaceae, etc. (Bhandari, 1978). Most notable are *Acacia* sp. *Aerva tomentosa, Calligonum polygonoides, Ephedra* sp., etc. The biota of particular shrubby trees like *Khejari* and *Rohida* and innumerable varieties of grasses (including medicinal types) are well adapted to arid conditions. Tropical thorn forests are found in arid and semiarid regions of western Rajasthan. The major tree species are *Acacia nilotica, Prosopis cineraria, Capparis aphylla, Zizyphus* spp. The monsoonal climate of the Indian desert both increases the reliability and amount of rainfall compared to the Saharo-Arabian desert zone, but also, with extreme seasonality of climate assures that in spite of moderate annual rainfall amounts, aridity is extreme for most of the months of the year. The well adapted vegetation of the region is able to provide a cover of scrubs, grasses and ephemerals even in the driest areas of the Thar (Dhir et al., 2018).

22.1.2 Landforms of the Thar Desert

The Aravalli mountain ranges in the eastern margin form the catchment area of the Luni river and its ephemeral tributaries. The bedrock includes Precambrian gneiss, schists, quartzite, etc.; Proterozoic sandstones, limestones, quartzite sandstone; Mesozoic sandstones, shales, limestone, and Tertiary shale, sandstone, etc. The major landforms of the Thar Desert are rocky pediments (4.82%) with inselbergs, tors, groups of hills composed of hard bedrock which cover 2.02% of the total area of the Thar. Desert pavements with anomalous pebbly cobbly gravel beds rich in well-rounded quartz, chert, jasper, and fossil wood occupy about 3.31% of the area. Sand dunes cover 70% and aggraded alluvial plains (both active as well as older) have spatial extent of about 15 %. Saline lakes form a small component (~1%) of the desert landforms (Dhir et al., 2018; Kar, 2014).

During the late Tertiary a few structural basins were created in rocky pediplains or erosional surfaces of the Thar Desert. These structural basins preserve thick (~350 in the eastern margin to 80–120 m in the western margin) fluvioeolian sediments of Neogene and Quaternary periods (Wadhavan and Sharma, 1997). The Neogene tectonic movement formed series of parallel, NE–SW trending horst and graben structures.

Ferricrete suggests humid climate >1000 mm/yr and evapotranspiration less than precipitation. It has developed over shale and sandstone of late Tertiary age and over boulder cobbly beds representing paleochannels of a powerful drainage from the eastern Aravalli hill ranges of Plio–Pleistocene age (Rajaguru et al., 2015). These features are certainly older than hardpan calcrete of early Pleistocene age in the Thar Desert (Singhvi and Kar, 2004).

The Luni river flows through a structurally weak zone and has been flowing since the Pliocene (Jain et al., 2005). The Luni was perennial, braided with near channel flood plains during the middle-Pleistocene and shifted to an aggradational mode and deposited fluvioeolian sediments during the late-Pleistocene. The Luni has also preserved stone artifacts—Acheulian to Mesolithic of mid-Pleistocene to mid-Holocene age, respectively (Misra, 1962; Mishra et al., 1999).

22.2 Scope

In this chapter, we have summarized and discussed investigations on various lakes in the Thar Desert, both that done by others, as well as carried out by us, some of which was in collaboration with others (see acknowledgements). While exploring the surrounding area around lakes, a number of Paleolithic and microlithic sites were discovered and documented in the field and stone tools studied in the Department of Archaeology of Deccan College, Post-Graduate and Research Institute, Pune. The excavations were conducted at sites like Bari Bavri, Jamba, etc. and pits/trenches were taken in all lakes for stratigraphic succession and collected samples for laboratory investigations.

Paleoenvironmental evidence comes primarily from the lake records which are poor in archeological evidence. On the other hand archeological sites provide limited paleoenvironmental evidence. This chapter primarily reviews the lake records. The archeological evidence is not reviewed here but is discussed in relation to the lake records.

22.3 Observations

22.3.1 Saline lakes

We have first summarized findings of Sambhar, Didwana, Tal Chaper, Lunkaransar, Jhorda and then fully described Bap-Malar, Kanod, Pachpadra, and Thob where our team has carried out multidisciplinary studies including archeology in late 1994 (Fig. 22.1).

22.3.1.1 Sambhar Lake

Sambhar is the largest inland lake in western India and covers approximately an area of about 225 sq km. It is located 80 km southwest of Jaipur and 64 km northeast of Ajmer. The lake is elliptically shaped, about 35 km long and width varying from 3 to 10 km. The circumference of the lake is 96 km and it is surrounded by the Aravalli hills on all sides. Today it receives about 500 mm average rainfall from ISM with PET of about 1400 mm. Yet it has fluctuating hydrology–freshwater–saline water throughout the Holocene. It is primarily of structural origin “occurs along major curvilinear strike-slip faults and may have originated as a pull apart depression” (Achyuthan et al., 2007, p. 495).

It has preserved five fluvial, two eolian, and four lacustral units with an approximate thickness of >10 m and these sediments rest unconformably on mica schist of Precambrian age (Sundaran et al., 1996). The basal unit is represented by strongly calcretised rubble with schist clasts in a matrix of yellowish silt. The overlying lacustral unit with gypsum crystals is capped by alternating fluviolacustral and eolian facies.

Sinha et al. (2004, 2006) recovered a 10 m core and dated the beginning of lake formation by C^{14} dating method. Geochemical studies including stable isotope data, ratios of Na/Al, Na/Ti, Na/K, and evaporate mineralogy with the support of conventional C^{14} dates on organic matter suggested that the lake experienced a relatively dry phase with increased salinity around 18 ka (LGM) and between 7.5 and 6.8 ka. The lake was almost full of freshwater around 6 ka and even at present never gets completely dry.

In his pioneer studies Singh (1971), Singh et al. (1974) suggested that pollen like *Syzigium cumini, Oldenlandia, Mimosa rubicaulis* indicated a higher amount of rainfall during the mid-Holocene. They also correlated flourishing Mature Harappan culture (ca. 2500–1700 BC) with the lake full stage (Singh 1971).

Apart from Singh’s pioneering palynological studies on the Sambhar Lake, Mukherjee and Rakshit (2012) also carried out pollen studies on lake sediments collected through shallow drilling from the surface of the lake. The basal most sediments consisting of well laminated grayish, brownish silty clays show high percentage of pollen, particularly of *Artemisia, Oldelandia, Asteraceae, Poaceae, Cyperaceae,* etc. These pollen species indicate wet semiarid climatic condition, not subhumid or humid as commented by other investigators and also by Singh et al. (1974).

The overlying sediments are grayish to greenish silty sand and are poor in pollen thereby indicating semiarid climate not very different from the present with an annual rainfall of around 500 mm. Briefly, Sambhar Lake in the easternmost border of the Thar Desert has provided very good record of climatic change from arid to wet semiarid to semiarid during late-Pleistocene, mid-Holocene, and late-Holocene, respectively.

22.3.1.2 Didwana Lake

This lake is located 100 km east of Nagaur and nearly 122 km northwest of Sambhar Lake and is about 250 km northeast of Jodhpur. It receives about 400 mm of annual rainfall mainly from ISM with high rate of evapotranspiration (PET around 1600 mm), winter rainfall is less than 10%. The lake is elongated with NE–SW orientation, and 7 km long and 2.5 km wide. It is bordered by stabilized obstruction/linear dunes and low lying hillocks composed of carbonaceous phyllites/shales with intrusions of quartzite and vein quartz. Morphologically the lake seems to have formed in a structurally controlled curvilinear fault depression in a pediment surface developed on Precambrian rocks (Achyuthan et al., 2007). At present the lake is ephemeral during the ISM and becomes totally dry for 6–8 months when it is mined for sodium salts used in the tanning industry.

Singh et al. (1974) made pioneering palynological studies and later on Wasson et al. (1984) worked on geochemistry and evaporate mineralogy of lake sediments and a revised paper (Singh et al., 1990) based on palynology and evaporate mineralogy was published.

Wasson et al. (1984) describe the stratigraphy obtained in 6 m deep pit in the lake and divided stratigraphy into 15 members. The entire sequence rests on what is locally called "murud" which is a mixture of sand and calcrete, interpreted as an alluvial deposit by Wasson et al. and eolian deposit by Singh et al. (1990). The lowermost unit, unit 15 rests on this and is sand mixed with mineral evaporates. A ~1 m thick halite layer is found from 460 to 570 cm and is labeled member 14. The upper 10 members from 0 to 340 cm are well-laminated silty clays without saline minerals. Color grades from olive gray (10Y 5/2, 10Y5/4), at the top to black below (10YR 7/1). In the lower part, Members 11–13 from 340 to 470 cm are gray (N6/0) and whitish gray (N 7/0), lack laminations and are rich in evaporite minerals.

In the revised study the lowest strata, below 4.5 m was dated to 9.3–12.8 ka BP by radiocarbon method (Singh et al., 1990). During this time the lake was saline and the surrounding vegetation was more or less devoid of trees. Overlying lacustral sediments dated between 9.7 to 7.5 ka BP exhibit a greater representation of fresh water plants and a weakening of evaporites. These two parameters indicate less aridity with improved ISM. The period between 7.5 and 6.2 ka BP experienced significant changes in lake hydrology. The pollen of *Calligonum, Ephedra,* and *Avera* which are typical of an arid phase and totally disappeared and were replaced by *Gramineae* and *Artemesia* species thereby indicating the existence of fresh water conditions in the lake due to improved ISM and possibly increased winter rains from the westerlies winds from western mid latitude regions (Wasson et al., 1984; Dhir et al., 2018).

After this peak period of fresh water, the lake level started declining, as indicated by increased saline minerals and pollen like *Calligonum, Ephedra*, and *Avera.* The lake turned totally dry by 4 ka BP due to weakening of ISM and also of winter rains. A saline water phase occurred between 6.2 ka and 4 ka.

After 4 ka BP, during the late-Holocene, the lake intermittently held freshwater during good monsoons but otherwise remained a salty whitish flat during the hot summer. *Artemesia* which was prominent in mid-Holocene, totally disappeared after 4 ka thereby indicating insignificance of winter rains at Didwana.

Changes in lake hydrology of Didwana are also manifest in a well-excavated linear/obstruction dune (locally known as 16R point) by Misra and Rajaguru (1989), Singhvi et al. (2010), Blinkhorn (2013). The 20 m deep trench with a number of OSL dates has preserved evidence of human activity in the form of stone artifacts (lower, middle, and late Paleolithic) during relatively wet climatic phases. The post-LGM phase shows decreased human activity when the Didwana Lake was hyper saline and surrounding landscape was almost without any trees. Microliths are scattered on presently stabilized dune surface when lake was holding fresh water during the mid-Holocene. Most of the detailed record from 16 R section predates the Didwana Lake record.

22.3.1.3 Tal Chapar Lake

Tal Chapar Lake is 60 km north of Didwana Lake and is an elliptical saline lake, 5–6 km long and 2–3 km wide and it is fed by ground water and ephemeral streams originating in nearby low hillocks fed by summer rains. It is bordered by stabilized dunes of late-Pleistocene age and microliths occur on the stable surface of dunes. The area receives a mean annual rainfall of 360 mm and evapotranspiration of around 1600 mm. The water table is 10 m below the dry lake bed and the lake basin has formed on Precambrian rocks due to geo-structural control (Achyuthan et al., 2007).

The Quaternary sediments (~8 m thickness) consist of laminated silt layers (0.5 m thick) rich in organic matter and interlayering cross bedded sands (1.5–2 m thick). Sand and silt layers are separated by laminated silt–clay layers with gypsum/calcite. This is common in upper 4 m of deeper section of dug well for extracting salt (Achyuthan et al., 2007).

Conventional radiocarbon dates at Tal Chapar were obtained from organic clay beds dated to 7190 ± 155–150 and 9903 ± 360–350 at a depth of 1.35 m and 1.80 m, respectively. A date of 13,090 ± 310–300 at a depth of 2.37 m from the surface of the well was also obtained. Interestingly Tal Chapar continued to be ephemeral lake till 1000 yrs BP (Achyuthan et al., 2007).

All dates from 1 ka to 13 ka are within a depth of 2.5 m from the surface. Lacustral or fluviolacustral sediments (>14 ka) continue to a depth of 10 m below the surface. These sediments, as in the case of Didwana and Bap-Malar, if dated in future will throw additional light on the evolutionary history of these lakes at least since the LGM (18–21 ka) or even beyond.

Both Tal Chapar and Didwana more or less depict the same paleoenvironmental history during terminal Pleistocene and the Holocene. Both lakes are of tectonic origin, developed on erosional depressions in pediments developed on Precambrian rocks. The lacustral phase, probably because of blocking of ephemeral (?) or disorganized low order fluvial system of linear/obstruction dunes of later Pleistocene age, commenced sometime after the end of the LGM (after 18 ka) when ISM started strengthening between 16 and 12 ka. Tal Chapar already existed around 14 ka and Didwana also may show the same story if lower layers are dated in the future.

In short, Eastern Margin saline lakes, occurring in the present semiarid climatic zone, have preserved an interesting paleomonsoon history: very weak around 18 ka, moderate during 16–9 ka, getting stronger between 9 ka and 7 k and strong between 7 ka and 6 ka. The relatively wet phase can be labeled as wet semiarid and not as subhumid (Achyuthan et al., 2007) or humid (Singh et al., 1974, 1990). The role of westerly winter rains during wet semiarid phase is recorded also in other saline lakes in arid western margin of the Thar Desert.

22.3.1.4 Jhorda Lake

Jhorda (27°28′40″N, 73°49′35″E), around 28 km to the north of Nagaur, is a small interdunal depression in between Chau and Jhorda villages. It is approximately 1500 m in length and 300 m in width. A pit was taken in the eastern part of the depression where mining for calcrete was going on. In the middle of September it was already dry. A visit to this lake was made by Lisa Ely, Yehuda Enzel, S.N. Rajaguru, and Sheila Mishra in September 1995. As per personal communication from Lisa Ely, the details of the section and age estimation from radiocarbon dating are as follows:

102–54 cm: Carbonate layer. This is sandier and well consolidated from top to bottom with maximum oxidized root traces in the middle. Radiocarbon date from the base of the layer is 9.3 ka and 5.5 ka BP from the top of this layer. There is a sharp boundary with the underlying dune sand at 102 cm.
54–27 cm: Dark brown massive clay with blocky structure. The lower part has more carbonate and there is a sharp boundary with the underlying carbonate. A sample collected from 39 to 42 cm gave a radiocarbon date of about 4 ka BP.
27–0 cm: Silty sand. A bulk sample from 18 to 21 cm depth gave a radiocarbon date of around 1.2 ka BP.

This interdunal depression is a minor feature in the present landscape. Yet it has preserved paleoclimatic history in tune with other eastern margin saline lakes. The basal sandy carbonate seems to have formed during the early-Holocene semiarid climate and the brackish water. The dark brown clay represents mid-Holocene wet semiarid climate with fresh water in the lake which started drying around 4 ka BP.

22.3.1.5 Lunkaransar Lake

This lake is located about 100 km north of Bikaner and exploited for selenite mining. Singh et al. (1974) studied Lunkaransar and carried out detailed palynological studies. Enzel et al. (1999) carried out detailed mineralogical, stable isotope and sedimentological analysis on the sediments exposed to depth of 6 m below the surface of which the top 3 m were assessable to study. Prasad and Enzel (2006) reviewed and compared different paleoenvironmental records and summarized the Lunkaransar study. The sedimentary deposits were divided into four lithological units. Basal lacustral layers consist of finely laminated clay/silt separated by fine laminae of gypsum, authegenic evaporate as well as partly allogenic (wind blown). This unit was C^{14} dated between ~11.5 and 9.4 cal. ka BP. The overlying lithounit consists of laminated silt/clay layers interspersed with gypsum—both authegenic and allogenic origin as in unit I. This unit was dated between ~9.4–7.2 cal. ka BP by radiocarbon method. Unit II is relatively richer in gypsum than unit I. Lithounit III is predominantly of silt–clay laminae and almost devoid of gypsum lamina. This unit is dated between 7.2 and 5.3 cal. ka BP. An increased inflow of eolian sand is observed slightly before 6 cal. ka BP. The fourth unit (~5.3–1.5 cal. ka BP) is relatively rich in eolian gypsum dust and nonlaminated sandy silt.

Briefly the lake held fresh water between 7.2 and 5.3 ka BP and was saline before 7.2 and after 5.3 ka BP. The lake turned ephemeral, at times getting totally dry after 5.3 cal ka BP. "The $\delta 13$ C of organic matter shows a decrease from -14% to -20% around 7.2 ka BP and abrupt increase around 6 cal. ka BP" (Prasad and Enzel, 2006, p. 5).

22.3.1.6 *Bap-Malar Lake*

Bap-Malar Lake (27°15′ to 27°22′N and 72°22′ to 72°27′E) is a closed dry basin with an area of about 78 sq km located about 20 km north of Phalodi and about 150 km northwest of district town of Jodhpur in western Rajasthan. It has formed as a structural shallow depression in a rolling pediment surface developed on sandstone of Proterozoic age. It receives ephemeral streams from Lordiya and Bari Bavri which have preserved middle Paleolithic and early historic sites, respectively, which have been discussed later.

The lake is surrounded by low stabilized dunes on eastern and southwestern sides and the source of water to the lake is monsoonal rains (200–250 mm/yr) and the subsurface groundwater. At present, the water table is 8–10 m below the dry surface of the lake and the PET is around 1800 mm/yr.

22.3.1.6.1 Composite stratigraphy

Total thickness as exposed in W10 well of the lake sediments is about 6 m below the surface and these sediments of fluviolacustral origin are divided into four lithounits (Fig. 22.2).

Lithounit I (6–3.30 m below the surface) rests unconformably on weathered pinkish sandstone. The unit consists of alternating layers of silt and reworked medium to fine grained sand and coarsely laminated silt and sandy layers (20–40 cm) are thicker than silt layers (10–20 cm). Authigenic gypsum sand occurs between silt and overlying sand. Sand is dominated by quartz, feldspar, and muscovite while silt is dominated by clay minerals like kaolinite and illite.

Lithounit II (3.30–1.80 m below the surface) consists of well laminated silt (20 cm thick) capped by weakly laminated sandy silt (60 cm thick) which is overlain by well laminated silt (30 cm). The uppermost layer (40 cm thick) is represented by finely laminated clayey silt with excellent development of selenite (crystalline gypsum) along bedding plane.

Lithounit II grades to lithounit III (180–60 cm below the surface) and consists of broadly laminated silt (light gray) with presence of selenite along bedding planes.

Lithounit IV is a pedogenized phase of lithounit III. The pedogenesis is indicated by well-developed blocky peds.

These fluviolacustral sediments are distinctly alkaline (pH ranging from 7.8 to 10), low in organic carbon (0.1–0.6 %), and low to moderate in calcium carbonate (from 2% to 38%) content. Over all chemical and textural studies show that these sediments are deficient in organic carbon and low in clay content.

The XRD studies show predominance of detrital minerals like quartz, feldspar, muscovite, and chlorite suggesting a proximal source, that is, from surrounding dune and rocky pediments. Authigenic evaporites (gypsum, halite, calcite, and dolomite) were useful for interpreting depositional environment.

An inverse relationship exists in proportions of gypsum to halite and gypsum to calcite. Gypsum from 0 to 120 cm depth is in traces, while halite is predominant and calcite occurs in moderate amount. Between 120 and 225 cm depth, gypsum is dominant and halite and calcite are in minor amounts. More or less, similar relationship of gypsum, halite, and calcite is noticeable in samples from Kanod.

Ninety-five samples were analyzed from 10 stratigraphic profiles for the extraction of pollen, only a few yielded microfossils at a depth of 235–250 cm from W10 profile (Fig. 22.2). Most of the facies

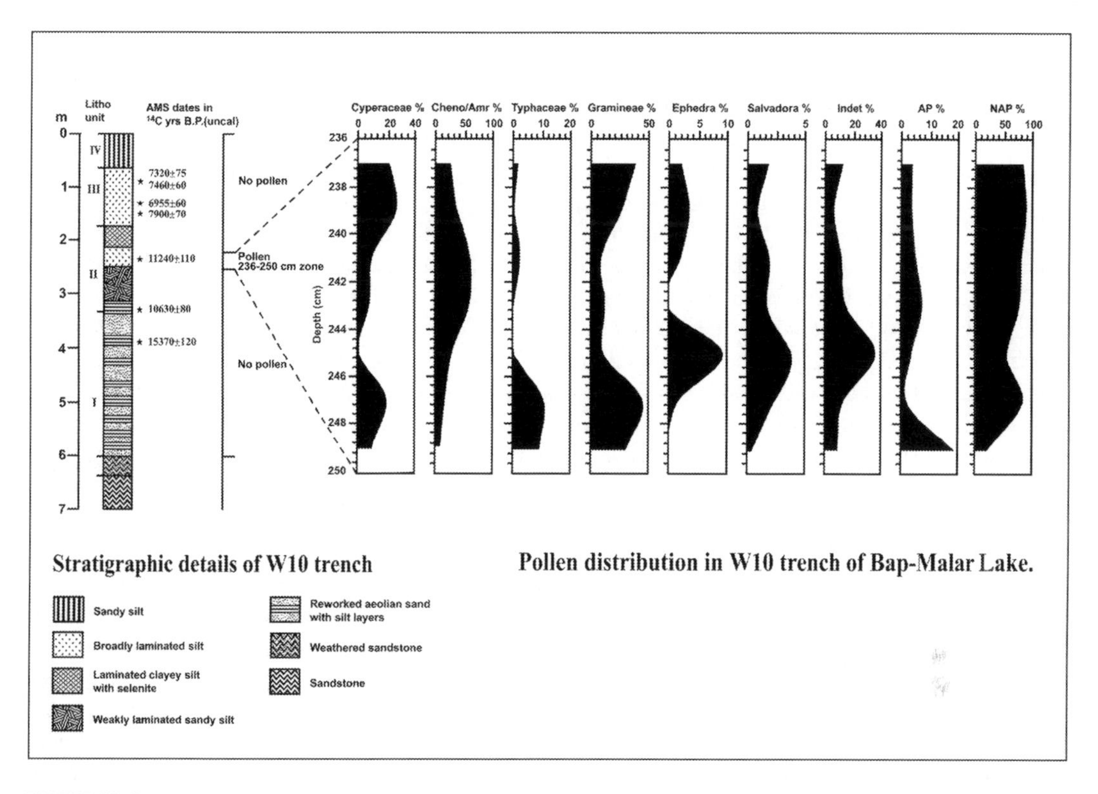

FIGURE 22.2

Pollen distribution with stratigraphic details in W10 trench of Bap-Malar Lake (after Deotare et al., 2004).

did not yield adequate pollen and spores excepting zone III. Pollen are represented by morphotypes belonging to *Chenopodiaceae, Gramineae, Cyperaceae, Compositae, Typhaceae, Mimosae, Rhamnaceae*, etc. and spores are mostly of fungal origin. These microfossils by and large indicate local ponds and stagnant water bodies in the lake proper and also in the surrounding regions (Deotare et al., 2004).

A total of six AMS dates (uncalibrated) show that the lake started drying around 7.3 ka BP and it held deep water (yet shallow) with retention of calcium and sodium minerals but less gypsum between 7.9 and 7.3 ka BP. On the other hand, the lake turned shallower and more saline (as indicated by increased proportion of authigenic selenite) between 10/11 ka BP and around 8 ka BP. The lower part of unit I is about 2 m thick and is undated but certainly much older than 15 ka BP.

The stable carbon isotopic values on seven samples from W10 profile(w.r.t. P.D.B.) range from -15 to -18‰, thereby suggesting significantly increased proportion of C4 plants in Bap-Malar region during the Pleistocene-Holocene transitional phase. This suggests predominance of C4 types indicating arid climate.

The depositional environment of lake sediments and the relative proportion of evaporites like gypsum, halite, and calcite suggest significant changes in paleohydrological conditions since LGM. Drying episodes of the lake are indicated by moderate pedogenesis of sediments of upper part of unit

IV (0–30 cm), by the development of gypsum crystals (selenite) across the bedding plane of silt fine-sand laminae in upper part of unit II (180–220 cm) and by weak pedogenesis of sandy laminae of unit I. Weak pedogenesis is indicated by coating of powdery carbonate over sand grains. Drying of lake of unit IV is of longer duration as indicated by moderate pedogenesis after 6 ka. This phase of drying needs to be precisely dated in future work. On the other hand, other two drying phases are of shorter duration and indicate temporary fluctuations in hydrology of lake.

A Perennial stage of the lake is represented by finely laminated silt and fine sand with reduced concentration of gypsum but enhanced proportion of halite and calcite. Silt and fine sand laminae of unit II were deposited between 8 ka and 7 ka, around 11 ka and sometimes after the LGM at a depth of 5.2–5.3 m. Though the lake was perennial with ground water above the surface of lake, it was still saline as indicated by the presence of calcite, halite, and traces of gypsum crystallites. The perennial water of early-Holocene (8 ka and 7 ka) was the most conspicuous hydrological change in Bap-Malar Lake which by and large remained primarily saline and shallow like a puddle with fluctuating ground water particularly between 15 ka and 11–10 ka when it was distinctly ephemeral with weak occasional flooding (Deotare et al., 1998, 2004; Kajale and Deotare, 1995).

22.3.1.7 ***Kanod Lake***

Kanod Lake (27°08′14″N, 71°14′E), with an area of about 32 sq km, is a crescentic elongated depression, located 35 km northeast of Jaisalmer (Dist. Jaisalmer). The mean annual rainfall in the area is 160 mm and with potential evapotranspiration in excess of 2000 mm/yr. Ephemeral channels with small catchments cuts through the pediments surrounding the lake and drain into it. The lake surface remains practically dry, excepting the monsoon months of August–September when occasional storm rain, fill the lake with water up to about 0.5 m. The mean annual winter rainfall is ~30 mm and does not contribute effectively to surface runoff. The Kanod Lake is thus dependent more on groundwater discharge (brine with a salinity $>$ 25 g/L) than the surface runoff during summer monsoons. Thus groundwater constitutes the major component of the hydrology of this lake.

Kanod Lake is a tectonically evolved shallow linear depression formed in fluvially eroded pediment surface formed on sandstone sometime during the Pleistocene. This lake has received sediments from the low order drainage, mostly ephemeral, on the south-western and western part, and marginally from stabilized dunes of the late-Pleistocene age.

The composite stratigraphy is based on detailed description of sediments of fluviolacustral origin with total thickness of about 2.5 m which is divided into five (I–V) lithounits. Basement rock could not be observed due to the presence of saline ground water at a depth >2.5 m. We examined five test pits, measuring 2 m × 2 m for studying lake stratigraphy and for collecting samples for sedimentological, palynological and C^{14} (AMS) dating.

The lowermost lithounit I (50 cm thick) consists of brownish gray reworked fluvioeolian sand from local ephemeral streams. Lithounit I is disconformably capped by lithounit II (~50 cm thick) consisting of finely laminated grayish clayey silt interlayering with whitish gypsum sand. The overlying lithounit III (20 cm thick) is dominated by finely laminated light grayish clayey silt without any gypsum laminae. Lithounit IV (110 cm thick) is disconformably caps lithounit III and consists of broadly laminated grayish clayey silt with alternating laminae (<0.5 cm thick) of gypsum sand throughout. The uppermost part of lithounit IV is capped by grayish well-laminated silt (~15 cm thick) with development of selenite crystals across laminations. Overlying lithounit V (about 35 cm thick) consists of laminated clayey silt with excellent development of gypsum crystals across laminations. Uppermost

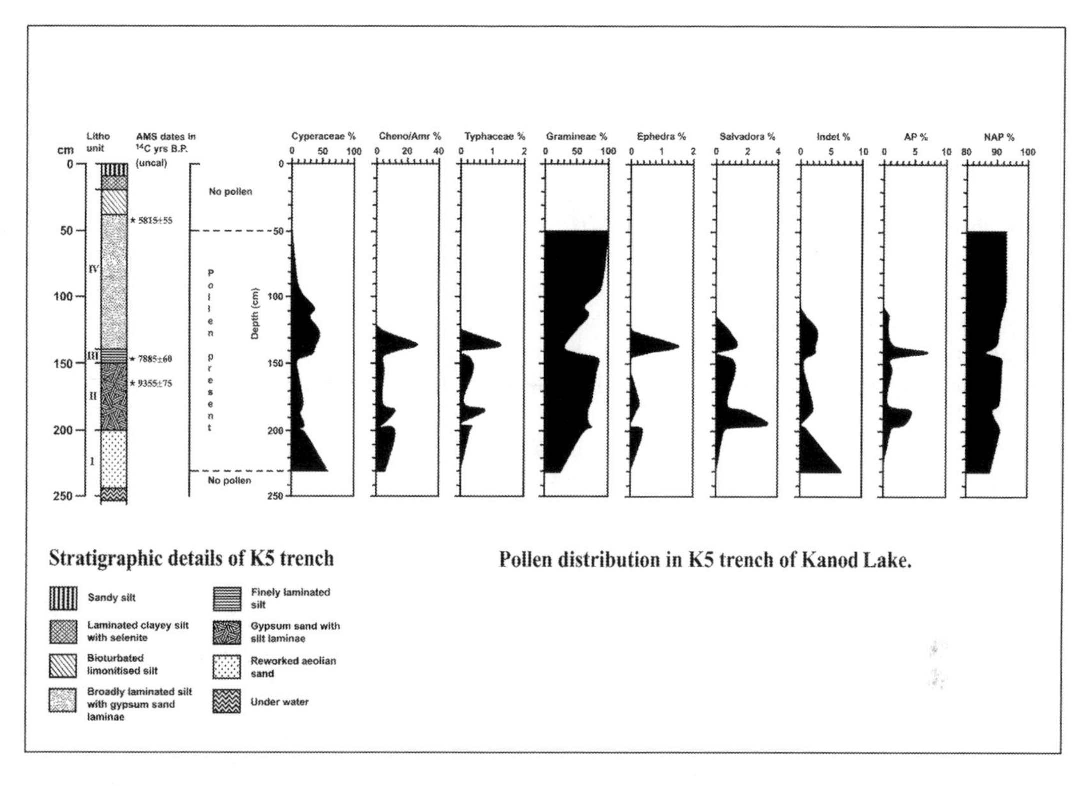

FIGURE 22.3

Pollen distribution with stratigraphic details in K5 trench of Kanod Lake (after Deotare et al., 2004).

10 cm of unit V is brownish gray sand silt with signs of pedogenesis as indicated by crumb peds (Deotare et al., 2004).

Overall, the lake sediments were strongly alkaline (pH varying from 7.8 to 8.3) and medium to high in organic carbon (0.3–1.3%) and calcium carbonate varies from 3% to 25%.

The XRD studies of lacustral sediments show that quartz is the dominant mineral present, followed by aragonite, calcite, gypsum, feldspar, muscovite, and halite. Quartz, muscovite, and feldspars are present in all units, particularly 5–10 cm and 205–250 cm depth from the surface of dried lake. These minerals, therefore, represent derivation from local dunes while other minerals are authegenic origin particularly halite, calcite, gypsum, and aragonite. Presence of powdery gypsum laminae suggests development in a shallow water body, while well-developed selenite crystals of gypsum across laminations indicate near drying or shallowing of water depth in the lake. It appears that halite and calcite have an inverse relationship with gypsum. This aspect of inverse relationship of gypsum with halite and calcite is discussed in summary part of this section.

As the organic carbon was moderate to high, we attempted AMS radio carbon dates on three samples from two trenches (Fig. 22.3). Kanod Lake originated at the beginning of the early-Holocene

(~9–10 ka), dried completely around 6 ka and was saline throughout its existence during early-Holocene. In order to understand bioenvironment we carried out pollen analysis from potentially organic rich samples collected from five trenches and also stable isotope studies on four samples.

The $\delta^{13}C$ values range from –14.7‰ to –16.2‰ thereby indicating dominance of grass cover in the surrounding area of Kanod Lake between 10 ka and 6 ka. Pollen and spores in lithounits II, III, and IV are represented mainly by *Cyperaceae, Chenopodiaceae Amaranthaceae, Gramineae, Pinaceae, Typhaceae,* etc. (Fig. 22.3; Deotare et al., 2004).

It shows that this shallow water saline lake existed between 9.5 ka BP and around 5.8 ka BP, or little later as indicated by uppermost 40 cm thick undated clayey silty sediments. On the whole, the rate of sedimentation is low (with an average of 2 cm in 40 years, 1 cm in 20 years) indicating very shallow water depth over the lake surface. The lake appears to have preserved a fresh water phase around 7.8 ka BP and may indicate short phase of strong rains which might have flooded the entire lake (Deotare et al., 2004).

We also studied two more salt lakes Pachpadra and Thob in order to get paleoenvironmental information for occupational or primary context Mesolithic site of Tilwara situated close to the Luni river in Barmer district. The site of Tilwara was excavated by Late Prof. V.N. Misra (1971). We have briefly described the site in the succeeding section on man–land relationship to Holocene environmental changes.

22.3.1.8 Pachpadra salt basin

Pachpadra salt basin is situated about 8 km west of Pachpadra village (25°55′N, 72°12′E) and 20 km northwest of Balotra town in Barmer district of Rajasthan. This is a vast basin spread over an area of about 82 sq km and is one of the major salt producing centers in western Rajasthan. This basin is a part of the old Luni drainage system and present course of Luni river is about 3 km from the western boundary of the basin. The basin is surrounded by high dunes and has an opening to Luni river near Soda ki Dhani village. Basically it is an abandoned channel and not a typical lake, and the profile developed essentially because of channel migration which is a characteristic feature of arid/semiarid desert where discharge of river is highly fluctuating.

The general slope of the basin is from northeast to southwest. We examined a number of pit sections taken for extracting salt and selected one section (about 5 m deep) in southwest part of the basin (Deotare and Kajale, 1996), the stratigraphy of this fluviolacustral deposit from top to bottom is as follows:

0–80 cm dried sandy silty lake bed
80–110 cm yellowish well sorted eolian sand Disconformity
110–400 cm alternating layers of sandy silt (about 20 cm thick) and silty clay (about 15 cm thick) with traces of salt Disconformity
400–480 cm yellowish unconsolidated well sorted sand (probably reworked eolian sand by flood water grades to pebbly sandy gravel deposited as a paleochannel of river Luni).

Unfortunately these sediments were found to be poor in organic carbon (<0.3%) and hence found deficient in pollen (Deotare and Kajale, 1996). We also could not date these organically poor, moderately oxidized fluvial sediments deposited by the abandoned channel of river Luni. In the absence of pollen and absolute dates we have avoided any interpretation of the Pachpadra salt basin (Deotare and Kajale, 1996; Kajale and Deotare, 1995).

22.3.1.9 *Thob Lake*

Thob Lake (26°05′N, 72°20′E) covering 8 sq km area is located 20 km north of Pachpadra basin and about 35 km north of Balotra town in Barmer district. It receives <300 mm of annual rainfall with PET around 1800 mm/yr. It probably originated in a shallow erosional depression on a rolling pediment surface developed on sandstone. Thob Lake is commercially mined for selenite gypsum. We examined two mining pits to a depth of about 3.5 m which is the water level and the composite stratigraphy from the surface is as follows:

0–80 cm weakly pedogenized yellowish silty sand, probably reworked fluvioeolian sand. This unit represents dried lake bed.

80–250 cm silty sandy clay with interlayering of dark brown silty clay without gypsum and whitish/grayish clay with gypsum crystal developed over subhorizontal bedding planes.

250–320 cm well-laminated clay with variation in color from white, gray, dark brown, and yellow. The clay laminae varies in thickness from 5 to 20 cm and has preserved powdery authogenic gypsum along bedding planes in whitish layer while the lower clay facies are gypsum free (Kajale and Deotare 1995, 1997).

Unfortunately we could not get any radiocarbon date for the clay beds but we succeeded in extracting a significant number of pollen grains, particularly from dark brown nongypseous clay facies.

The variation in pollen from the profile is an indication of varied plant assemblages which in turn reflect the local hydrological environment. Pollen grain of *Ephedra* sp. a typical desert plant, at the lowermost level is suggestive of relatively dry conditions. The relatively high proportion of Cyperaceae (36%), Gramineae (23.5%), and Cheno-Amaranthaceae (10%) type pollen at 2.9–3.1 m level indicate wet hydrological local conditions (Deotare and Kajale, 1996).

22.3.2 Archeology around saline lakes

As stated earlier, the saline lakes started holding water at least since 15 ka BP with a fluctuating hydrology, at times saline, brackish and at times with fresh water almost until 4 ka BP. We now describe some Stone Age sites which occurred either on the banks of ephemeral bedrock streams feeding lakes or stable dune surrounding lakes. The sites discovered and described below are Lordiya, Manawara, Sam, Jamba, Bagor, Tilwara, Bari Bavri, and Budha Pushkar (see Table 22.1).

22.3.2.1 *Lordiya*

The paleolithic site of Lordiya (27°05′21″N, 72°22′52″E) is located 6 km southeast of Phalodi on the Phalodi–Jodhpur highway. It is situated on the right bank of a dry channel, 150–200 m wide and 3–5 m deep cut into siliceous cherty limestone of Paleozoic age. The channel originates 8 km upstream of the site on a limestone plateau and it has initially a north-easterly trend but after flowing about 3–4 km, it takes westerly turn and gives an impression of a southerly feeder of Bap-Malar Lake.

This ephemeral stream drains into the Bap-Malar Lake which is about 20 km northwest. The earlier record of archeological material from this region was essentially confined to the sporadic occurrence of middle Paleolithic artifacts. The first stone age site was found at Baridhani in Indian desert proper discovered in 1962 by Bhatiya and Sahu, geologists from Punjab University which was later published by Mohapatra et al. (1963) and noticed again by Mishra et al. (1993). The presence of Mesolithic sites in

Table 22.1 Archeological sites discussed in the text.

Site name	Lat. long	Approx. age	Archeological material	Geomorphic context	References
Lordiya	27°05′21″N, 72°22′52″E	Middle- and late-Pleistocene	Miniature handaxe and flakes	Dry channel on Rocky pediment	Deotare et al., 1998
Manawara	25°47′31″N, 71°56′35″E	Early–middle-Holocene	11–3 ka	Older Luni gravel buried by a dune	Mishra et al., 1999
Sam	26°48′37″N 70°25′34″E	Terminal Pleistocene	14 ka	Stabilized dune surface	Dhir et al., 2012, p. 72
Jamba	27°18′N, 72°31′E	Early-Holocene	~7 ka	Stabilized dune surface	Deotare et al., 1998, 2004
Bagor	25°23′N, 74°23′E	Early–middle-Holocene	7–4, 5 ka	Stabilized dune surface	Misra and Kanungo, 2019
Tilwara	25°52′N, 72°50′E	Late-Holocene	Undated	Stabilized dune surface	Misra and Kanungo, 2019
Bari Bavri	27°16′N 72°13′E	Late-Holocene	~2 ka	Ephemeral channel	Deotare et al., 1999
Budha Pushkar	26°30′N 74°36E	Late-Holocene	700 yrs	Reworked dune	Singhvi et al., 1994, Allchin et al., 1972

FIGURE 22.4

Late Acheulian handaxe and other flake tools from Lordiya (after Deotare et al., 2004).

desert may indicate as yet undiscovered Chalcolithic settlements. Most probably the sites are the result of occasional sorties of hunters or pastoralist into the desert following a good monsoon.

The association of middle Paleolithic sites with depressions which today only hold saline water such as Hokra, Sambhar in dry zone and Thob, Baridhani in the desert and associated with streams which never flow at all in the present conditions, supports the idea of a more humid climate in middle Paleolithic times. Allchin et al. (1978), Misra (1962, 1965), Mohapatra et al. (1963), and Mishra et al. (1993) all consider their middle Paleolithic finds to be associated with a more humid climate than the present.

The most noteworthy artifact from Lordiya is an ovate-shaped handaxe. It is fairly small ($11.5 \times 7.5 \times 3.0$ cm) with asymmetrical outline, straight edges and acute edge angle ($\sim 30°$). This means that a strong and sharp edge was available for use. The handaxe is made on quartzitic sandstone with a desert varnish (5YR3/4 dark reddish brown) on both sides, although on one side the varnish is uneven (5YR4/4 reddish brown). The flake scars are sharp without any sign of wind abrasion, indicating burial before the. LGM phase of dune mobility and absence of significant transport from its findspot. This belongs to the late Acheulian tradition which should predate the Last Interglacial as middle Paleolithic artifacts from 16R fossil dune site near Didwana (Nagaur Dist.) have been dated to Last Interglacial period (Singhvi et al., 2010). Along with this handaxe, a number of flakes, scrapers, prepared cores, blades made on quartzitic sandstone were also recovered (Fig. 22.4). On typological grounds the artifacts belong to the middle Paleolithic cultural tradition (Deotare et al., 1998). It was interesting to observe that when the flake collection was sorted into those showing wind abrasion versus those without wind abrasion a difference in technology could be observed with radial flaking found on the wind abraded flakes and absence of wind abrasion on the other flakes which showed single direction flaking. The interesting point is that the artifacts of differing technology, although found together, experienced different weathering regimes after their discard.

Geomorphologically, the artifact bearing gravel represents a paleochannel flowing during exceptionally heavy rains, a characteristic feature of the arid landscapes. We tentatively date the middle Paleolithic site of Lordiya to early–late-Pleistocene (pre-LGM) when the Thar Desert was relatively enjoying wet yet arid climate (Misra, 1962; Allchin et al., 1978; Deotare et al., 1998, 2004; Dhir et al., 2010; Blinkhorn et al., 2014, 2017).

22.3.2.2 Manawara

The site of Manawara (25°47′31″N, 71°56′35″E) is a locality on the Bhuka to Karna stretch of the Luni river on the left bank. In this stretch the river is entrenched in the older cemented gravel and this gravel is exposed between the dirt road and the river channel. The gravel is made up primarily subrounded of calcrete clasts along with some angular local rhyolites. The surface of the gravel is strewn with calcrete coated Chalcolithic pottery and animal bones. Dating of different components gave an interesting story. The gravel itself was dated to between 10 and 11 ka BP by OSL while the gastropods from the gravel dated from 7.5–8.0 ka by radiocarbon method. The pottery was characteristic of the Chalcolithic period and, confirming this, a TL date of 3.4 ± 3 ka was obtained on one of the sherds. The animal bones were dominated by domesticated species and had fluorine phosphate values between 1 and 1.5 which is also consistent with a Chalcolithic age for the gravel occupation. In the light of these absolute dates, the site can be interpreted as follows:

1. The gravel dates to the beginning of the Holocene humid phase.
2. The gastropods are not coeval with the phase of stream activity but colonized the gravel when it retained water during the Holocene optimum.
3. The dry exposed channel was occupied by people during the Chalcolithic period.
4. This gravel is overlain by eolian sand dating to around 2.8 ka.
5. Local villagers informed us that the gravel was exposed by the stripping away of the overlying sand layer during the flood of 1979 (Mishra et al., 1999).

Around Manawara similar evidence as Lordiya was found for the earlier period. On a hill near Bajawa village, a small weathered handaxe was found on the surface. A gravel dated to around 80 ka BP, exposed around Karna village, yielded a few undiagnostic flakes similar to those from the surface as at Lordiya.

22.3.2.3 Sam

During our exploration in 1980 we discovered a rich microlithic site at Sam, a popular spot for viewing the sunset in the western margin of the Thar Desert. The site of Sam is located 44 km west–southwest of Jaisalmer. It receives annual rainfall of about 150 mm with high rate of evaporation (PET > 2000 mm/yr). Recently Blinkhorn et al. (2017) found Microlithic site at Katoti in the eastern margin of the Thar Desert and dated it to ~21 ka BP by OSL method. This is the only well dated and earliest (the LGM) Microlithic site in the Thar Desert.

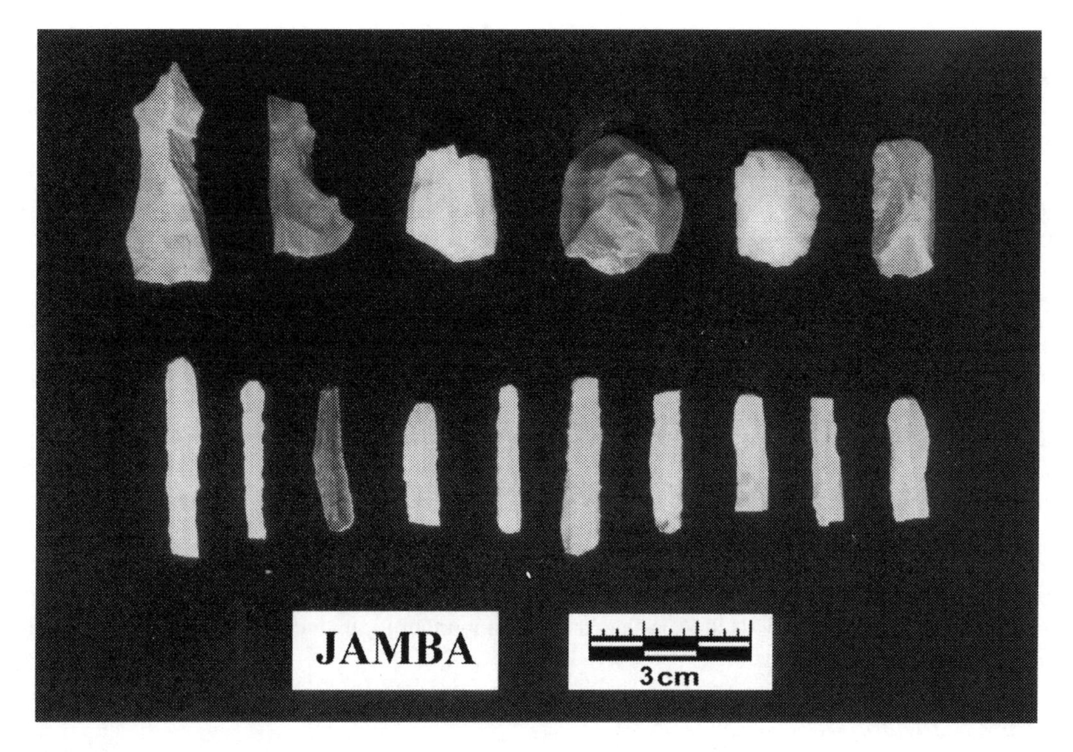

FIGURE 22.5

Microlithic cores and blades from Jamba (after Deotare et al., 2004).

Dunes around Sam ranges in height from 10 to 25 m and length up to 10 km and have been classified as hair pin parabolics by Wasson et al. (1983) while Kar (1987) placed them in complex longitudes, transitional to parabolics. Whatever may be the exact origin of the classical dunes, we could locate a rich scatter of microliths on the stable dune surface with sign of weak paleosol (calcisol or biocrusted soil). Microliths consist of microblades, flutted cores, retouched blades (3–5 cm in length) etc. and are made on quartz, chert, and chalcedony. We carried out a small excavation (1 × 1 × 1 m) and got a preliminary TL date of 15 ka BP of dune sample at a depth of 0.8 m below the microlithic scatter. Tentatively we ascribe a date of about 14 ka to this scatter as it has been observed that time range between sand accumulation (due to strong winds) generated during the terminal Pleistocene and stabilization due to rains as a result of ISM is not long (Shrivastav et al., 2020; Chawla et al., 1992).

22.3.2.4 *Jamba*

It is located on the old dune surface surrounded by the dune. A rich scatter of microliths with potsherds and bone fragments were discovered on a stabilized dune surface located about 6 km east of Bap-Malar Lake. Microliths made on chalcedony, chert, and quartz, consist of flakes with or without retouch, fluted cores on nodules and flakes, and a variety of points and blades (Fig. 22.5). The pottery found along with microliths is nondiagnostic for any specific cultural tradition. The animal bones are fragmentary and partially mineralized and the fluorine/phosphate ratio is within 1–2, suggestive of early–mid-Holocene age of the microliths (Deotare et al., 1998). A TL date of ~7 ka was obtained just 1 m below the microlith. Thus the fluorine/phosphate ratio and luminescence age indicate an early-Holocene (~7 ka BP) age of the microlithic site at Jamba (Deotare et al., 2004).

The interdunal depression might be the area in which the surface runoff water accumulates during the rainy season. This interdunal area might have served as the source of water to the prehistoric man/microlithic people.

22.3.2.5 Bagor

A recently published book—*Mesolithic Age in South Asia* by late Misra and Kanungo (2019) provides an overview of various primary, semiprimary microlithic sites (Mesolithic cultural stage) in India in general and the Rajasthan in particular. Bagor (25°23′N, 74°23′E) is located on stable dune surface bordering river Kothari, a tributary of river Banas in Bhilwara district of Eastern Rajasthan. The site is located on the eastern slope of the Aravalli hills and receives about 500 mm rainfall and is in the present semiarid parts of the Eastern Rajasthan. The site has yielded large (thousands) numbers of geometric microliths along with animal bones of birds, sheep, buffalo, pig, chital, blackbuck, etc. The site has been dated by C^{14} method and lies between 7 ka BP and 4.5 ka BP (covering early- and mid-Holocene periods). We have included this site in our communication as it is the only well-dated Mesolithic site of the early–mid-Holocene periods in Rajasthan and it is important for understanding microlithic sites in the Thar Desert.

22.3.2.6 Tilwara

In the Thar Desert, Misra excavated a microlithic site of Tilwara (25°52′N, 72°50′E) situated on the left bank of river Luni, 16 km west of Balotra in Barmer district. The area receives annual precipitation of about 200 mm and evapotranspiration of about 2000 mm/yr. The site is located on low dune capping the old floodplain of the Luni river. In the excavation Misra and Kanungo (2019) found microliths without pottery between 15 and 50 cm depth from the surface and microliths with pottery between 0 and 15 cm. The stone tools consist of blades and bladelets and are made on quartz, chert, and rhyolite. The blades are produced from the quartz cobbles by pressure technique. Animal bones include those of cattle, goat, pig, spotted deer, jackal, or dog. The primary site of Tilwara could not be dated by C^{14} method. On the basis of close association with pottery, particularly in upper part of the habitational deposit (0–15 cm), the site is dated to around 4 ka BP and microlith bearing deposit (15–50 cm) is likely to be considerably older (?) (p. 75). According to us it could be early–mid-Holocene (5–6 ka).

22.3.2.7 Bari Bavri

The site of Bari Bavri (27°16′N, 72°13′E) is located about 10 km northwest of Phalodi, taluka town of Jodhpur district. A scatter of potsherds (Fig. 22.6), bones and stone artifacts were observed on the surface of the 1.5 km wide cobbly pebbly sandy channel with insignificant noncohesive banks. A small pit (2 × 2 m) was taken at the center of the site which revealed a 1.2 m thick habitational deposit with an intact hearth with large amount of charcoal from 90 to 70 cm from the surface. Just outside the above trial pit, a well baked brick of 37 × 25 × 7 cm dimension was also recovered from the surface which belongs to the early historic period. Some of the archeological material at the site therefore might be slightly earlier in age than the dated archeological component (Deotare et al., 1999).

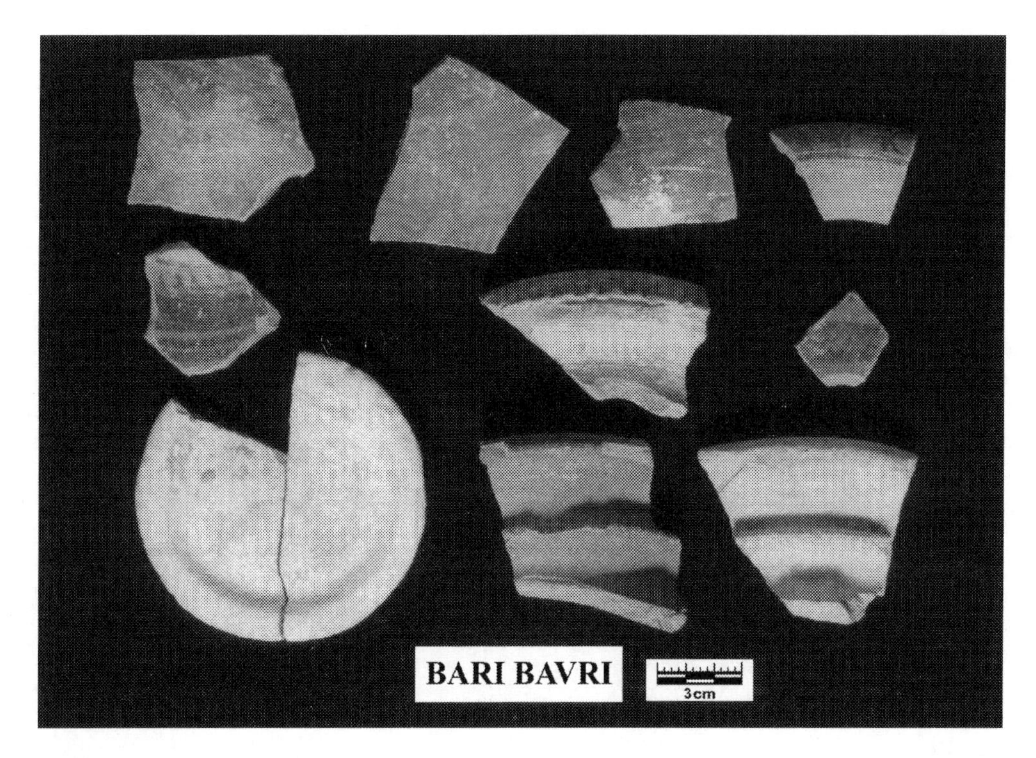

FIGURE 22.6

Potsherds from early historic site of Bari Bavri (after Deotare et al., 2004).

Our discovery of the site of Bari Bavri is interesting from the point of view of understanding man–environment relationship during the early historic period (~500 CE) in the arid core of the Thar Desert (Deotare et al., 2004). This site is located on the right bank of an ephemeral channel draining into the Bap-Malar Lake. This area receives mean annual rainfall less than 200 mm and has evapotranspiration rate of about 2000 mm per annum. The rainfall is highly erratic, stormy, and the variation is also very high.

The three C^{14} dates of the charcoal from the hearth as well as from test pit, confirmed the archeological dating of the site by giving value of 1490 ± 95, 1400 ± 95, and 1490 ± 85 BP at 70–75, 20–30, and 10–15 cm, depth, respectively (Deotare et al., 1999).

The feeder streams of Bari Bavri are mostly bedrock channels with a high width/depth ratio. The bedrock lithology has controlled the width/depth ratio to some extent. Bari Bavri channel is wide (1.5 km) near the archeological site as it has incised over relatively softer sandstone. On the other hand, the Bari Bavri channel increases its depth due to the presence of relatively hard cherty limestone, occurring 2 km downstream of the archeological site. This wide and shallow channel has preserved alluvial fills with archeological materials. The archeological contents of fluvial sediments suggest that the stream has been predominantly aggrading during the late-Pleistocene and late-Holocene (around 1.5 kyr or little later). The same stream has been in an erosional phase during the late-Holocene.

In spite of these complex hydrological factors involved in ephemeral stream behavior in arid region, we have tried to interpret available geoarcheological evidence at Bari Bavri against the known information on paleoenvironment during the early-Holocene in the Thar Desert.

22.3.2.8 *Budha Pushkar*

In support of Misra's hypothesis of symbiotic relationship between farming-based societies and pastoral Stone Age nomads, an interesting microlithic site of Budha Pushkar is located on stable dunes bordering fresh water lake of Pushkar, a famous pilgrimage lake in Ajmer district. The site was well investigated by Allchin et al. (1978) but the team did not carry out any excavations of microlithic sites around Budha Pushkar. We (Singhvi et al., 1994) carried out excavation (1 × 1 × 12.5 m depth) on a stable dune with rich scatter of microlithic artifacts covered by reworked modern sand in which 10 meter (10 m) thick red brown moderately pedogenized eolian sand below reworked modern sand was found. The sand continues up to a depth of 12.5 m with well-developed calcrete bands, rhizo calcreteous of ground water, and pedogenic origin.

The sand rests unconfirmably on bedrock of Precambrian age. We dated the sand sample by TL method at a depth of 30 cm (with microliths), 6.50 m, 7.50 m, 9.50 m, 11.70 m, and 12.30 m below the reactivated sand samples. Respective dates are 0.7 ka (30 cm depth), 16 ka, 13.1 ka, 15.2 ka, 22 ka, 27 ka, and 34 ka (at 12.30 m). Though two dates show reversals, overall dune accumulation has taken place during the late-Pleistocene (34–13.1 ka; Singhvi et al., 1994).

22.4 Discussion

22.4.1 Lakes

Saline lakes of the Thar Desert have provided one of the best paleoenvironmental records for the late Quaternary (~30–4 ka BP) in western India where geomorphic proxies like cut and fill colluvioalluvial terraces, eolian dune and loess like silts, paleosols such as vertisols, calcisols, and duricrusts like calcretes and ferricrets have been used for reconstruction of past environments, particularly climate. Even though local factors such as hardrock geology including structure, groundwater conditions, landscape, and vegetation cover have played important role in the origin of these shallow puddle like lakes, the majority of the lakes have responded sensitively to rainfall (monsoon as well as westerly winter rains) in the last 15,000 yrs or little more. The lake of Sambhar, the oldest (30 ka), one has not only remained a fresh water lake for a longer (>3 ka) duration during the early–mid-Holocene, but has never got totally dried up in its long history. Didwana Lake has preserved excellent evidence of hyper salinity (in the form of halite bed) during arid terminal Pleistocene. Pollen and evaporate mineral data of Didwana establish fresh water conditions for about 2000 yrs between 6.7 and 4.7 ka BP. The Tal Chapar Lake, on the other hand, shows existence of more or less fresh water conditions for about 3 ka from 9 ka to 6 ka during early- and mid-Holocene. Lunkaransar Lake remained fluctuating saline/freshwater throughout its existence from 9 ka to ~5 ka. A freshwater phase in Lunkaransar Lake prior to its drying up at around 5 ka, is in fair agreement with the Didwana Lake record (Engel et al., 1999).

Bap-Malar certainly existed around 15 ka as a shallow lake. Silt and sand layers were contributed by ephemeral streams originating in the surrounding pediments with a regolith rich in weathered sandstone. On the other hand, sand was brought by ephemeral streams draining stable dunes of the late-Pleistocene age. Calibrated dates probably support our hypothesis that the lake started holding water much earlier than the LGM probably during relatively wet climatic phase of the later late-Pleistocene age.

The shallow saline lake of Bap-Malar is likely to be older than 15 ka as suggested by presence of alternating layers of silt and sand continuing up to 2 m below the layer dated to 15 ka. We tentatively propose that the Bap-Malar Lake came into existence sometime after 18 ka (LGM) when ISM was getting active around 16–17 ka BP. Our hypothesis needs further testing by getting additional AMS dates on organic material from silt layers occurring between 5 m and 6 m (Unit I) from the surface.

The saline lake of Bap-Malar was more or less contemporary with eastern margin lakes and remained saline throughout its existence for about 8 ka years. On the other hand, Kanod Lake in the heart of arid western margin of the Thar also remained saline throughout its existence for about 4 ka years and preserved relatively good pollen. Pollen from Kanod and $\delta^{13}C$ data from Bap-Malar and Kanod shows that western arid margin was dominated by grass and scrub vegetation even during relatively wet climate of 7–6 ka BP.

22.4.2 Human–environment relationship

The Thar Desert witnessed strong eolian activity and sand accumulation between 14 ka and 11 ka due to the intensification of summer monsoons. The rainfall was relatively intense during early-Holocene (i.e., between 8 and 6 ka) and became weak after 5 ka. These climatic fluctuations have affected hydrology of the Bap-Malar Lake during the Holocene.

There is some interrelationship between hydrological changes of the Bap-Malar Lake and the behavior of the feeder stream like Bari Bavri. The other important aspect of our study is an understanding of the preservation of an intact hearth, a well baked brick and unabraded pot sherds of late Gupta period (~500 CE.) in the sandy/gravelly channel of the Bari Bavri. It appears that high width/depth ratio of the channel reduced erosive capacity during high intensity floods. Thus the preservation of archeological materials in a wide sandy gravelly channel during the fifth century CE in an undisturbed geomorphic context probably indicates that channel aggradation due to intense and short floods took place with a long time interval in between subsequent flooding events.

The alluvial fill about 2 km downstream of the above-mentioned archeological site occurs in a slightly narrow and incised bedrock valley. As shown earlier, this fill contains abraded potsherds derived from the site in the upstream area. This fill is later than the channel aggradation described above. The aggradation took place probably due to blocking of the channel by eolian sand further downstream toward the Bap-Malar Lake. The last phase of major dune development has occurred around 600–700 BP in the Thar Desert (Thomas et al., 1999). It is therefore, quite likely that fill containing late Gupta potsherds might have been formed as a result of blocking of the stream during the major eolian activity assignable to around 600-700 BP.

In brief, microlithic activity goes back to peak of Last Glacial Maximum at Katoti in eastern margin of the Thar Desert and surprisingly occurs at the beginning of the strengthening of ISM around 14 ka at Sam in the western margin of the Thar. On the whole, compared to the large number of middle Paleolithic sites of pre-LGM period (covering 75–25 ka BP), the microlithic sites are rare during the terminal Pleistocene. Human response to one of the driest arid period of the Thar was markedly poor. The site of Sam therefore, acquires importance, showing that the late-Paleolithic hunter gatherers with microlithic technology could face a harsh arid environment with unpredictable sources of perennial fresh water. The nearest saline lake of Kanod is about 50–55 km northeast of Sam and our investigation shows that the Kanod had started holding shallow and saline water during the terminal Pleistocene–early-Holocene transitional phase. Kanod Lake continues to remain saline with fluctuating water depth till mid-Holocene. At present it appears that the site of Sam can be categorized as a seasonal camping site when interdunal depressions might be holding fresh water after good (yet rare) spells of summer rains. Future multidisciplinary studies of sites like Sam will throw good light on human adaptation in a dunal country without obvious perennial source of fresh water in nonrainy period, a familiar environmental factor in arid core of the Thar with strong winds and very high potential evapotranspiration (>2000 mm/yr).

One of the possible source of fresh water is the abandoned village Kuldhara where a 300 years old settlement of the Paliwal community exists on the ephemeral bedrock stream with 2.5 m high knick point. The abandoned village is 6 km off the Jaisalmer–Sam road. Within the deserted village, there is interesting evidence for storing rainwater in a natural depression in bedrock, locally known as *Khadin*. Such natural depressions in limestone are commonly seen on way to Sam dunes. There is a strong possibility that such depressions might have held rainwater even during the terminal Pleistocene when Indian Monsoon was gradually getting active.

The Paliwal community had constructed a small dam like structure for storing rainwater some 300 years back. It shows that the present dry stream also was active during the late Medieval period. We therefore suggest that the Kuldhara stream could also have served as one of the sources of fresh water for microlithic people at Sam which is in the heart of well-developed dune field about 10 km of Kuldhara.

Compared to the site of Sam, other Microlithic sites like Jamba, Didwana, Tilwara of early–mid-Holocene age (7–4 ka) provide very good examples of human adaptation to excellent environmental conditions such as good and reliable monsoonal and also winter rainfall. Increased depth of fresh water in saline lakes and probably good vegetation cover on the rocky as well as dunal landscape played as an important environmental factor. Though Kanod, Bap-Malar, and Lunkaransar Lakes got dried up earlier than the lakes of Didwana and Tal Chapar, the hunter gatherers (or pastoral nomads) continued to adapt even after 4 ka in spite of reduction in monsoonal rains and vegetation cover. The site of Tilwara is a very good example in the western margin of the Thar. It had enough water sources in the ephemeral Luni river and also in nearby lake of Pachpadra.

The majority of Microlithic sites are located on stabilized sand dunes all along the lake boundaries such as Bap-Malar, Sambhar, Pachpadra, and some on rocky pediments around Kanod. The higher density of sites during this periods may also reflects a greater mobility necessitated during dry season due to restricted availability of water in the area. The microlithic site of Jamba dated to 7 ka matches reasonably well with the lake full stage of Bap-Malar Lake (Deotare et al., 1998, 2004). Similarly, the occurrence of microliths with pottery is also a very common feature of the sites around these lakes. So it can be surmized that Microlithic occupants were primarily foragers who adopted the pastoral way of life.

Similar types of evidence have been reported from the surrounding of Thari region in the Thar Desert of Sindh, Pakistan. Biagi and Veesar, 1998 have surveyed the above area and discovered many sites on the surface of the sand dunes surrounding the salt water basin of Ganero and Jamal Shah Sim. The majority of these sites consist of microliths of different shapes and sizes belonging to Microlithic period and some of the flint scatters are older, may be attributed to end of Upper Paleolithic. The Microlithic sites with ceramic tradition in the Thar Desert of India and Pakistan open a new avenue for further research. An early historic site dated to around 500 CE around Bap-Malar also focuses on man–land relationship during the late-Holocene (Deotare et al., 1999).

Even in 19th century, Tod (1829-1832) quoted Elphinstone to describe northern Thar Desert of 1808 "thus moving from Shekhawati (eastern part of the Thar Desert) to Pugal (250 km to the west) was all a terrain of hills and valley of loose and heavy sand. The sand hills are said to shift their position and alter shape as these are affected by wind: and in summer the passage is rendered dangerous by the clouds of moving sand, but when I saw hills in winter they seem to have great degree of permanence for they bore grass, besides shrubs and trees, which together gave them an appearance that sometime amounted to vendur" (Dhir et al., 2018, p. 40–41). This description of the Thar in 1808 clearly demonstrates that the Thar Desert could have remained relatively hospitable for nomadic pastorals even during the late-Holocene which was significantly drier than the mid-Holocene. In spite of lakes getting saline or totally dry, the interdunal depressions, the structural depressions in rocky troughs must have stored fresh water after strong short lived monsoonal rains probably even in hot windy summers. The biota of shrubby trees like *Khejari* and *Rohida* and innumerable varieties of grasses (including medicinal types) is well adapted to hardy condition naturally. Tropical thorn forest is found in arid and semiarid regions of western Rajasthan. The tree species are *Acacia nilotica, Prosopis cineraria, Capparis aphylla, Zizyphus* spp.

Despite edaphic and climatic constraints, the Thar has an impressive cover of vegetation. Even the driest areas with less than 100 mm of annual rainfall have a respectable cover of scrubs, grasses, and ephemerals (Dhir et al., 2018, p. 81).

The Thar with good vegetation cover was also rich in mega as well as microvertebrates. "Amongst the large–medium mammals, most prominent are the Indian gazelle, black buck, Nil Gai (Blue Bull), Jackel, fox, desert cat etc." (Dhir et al., 2018, p. 90). We have already shown that similar type of fauna has been exploited by microlithic pastorals at Bagor and Tilwara (Misra and Kunango, 2019). One can therefore, argue that bioenvironment during the late Quaternary was favorable for hunter gatherers for the last 15 ka years or even more.

As far as raw material used for making stone artifacts (microliths) is concerned, it is observed that anomalous boulder beds provided cobbles of siliceous rocks like chert, chalcedony, rhyolite, quartzite, etc. were locally available at most of the microlithic sites in the Thar Desert. The site of Sam however provides an example of long distance (several kms) movement of microlithic people for obtaining proper siliceous rocks and minerals.

In summary, Microlithic people of the Thar appear to have adapted to harsh arid climate of the LGM (as at Katoti), to strong monsoon winds without much moisture of the terminal Pleistocene (as at Sam) and to good monsoonal rains of early–mid-Holocene (as at Jamba and Didwana). It appears that Stone Age culture (Acheulian to Microlithic) survived in the arid Thar Desert at least for the last 600 ka years. Fluvioeolian landforms with saline lakes, rocky pediments, and low hills of Aravalli range with fairly good bioenvironment were ideal environmental factors for hunter gatherers even in harsh arid climate. The ISM has never failed totally and therefore the Thar Desert today is the most densely populated arid region in the world.

We would like to quote Misra "excavations at Bagor and Tilwara have provided excellent evidence of the simultaneous evidence of technologically, economically and socially different modes of life which is so characteristic in India today, and was equally, if not more true in the past. These societies with highly contrasting modes of life were not living in isolation but were interacting with each other and had a symbiotic relationship" (Misra and Kanungo 2019, p. 77).

Our studies thus support above-mentioned hypothesis of Misra, that is, continuation of symbiotic relationship between advanced settled communities and nomadic pastorals even during Medieval period in the eastern margin of the Thar Desert.

22.5 Learning and knowledge outcomes

The arid Thar Desert remained in the shadow of semiarid/arid climatic zone almost throughout the Quaternary and yet hunter gatherers continued to carry out their activity at least for the last 600 ka years, with a reasonable decline during the LGM. Easy availability of suitable rocks and minerals for making stone artifacts, availability of surface fresh water, at least seasonally even in the arid western margin during the LGM and favorable geomorphic features like eolian sand sheets and dunes for nourishing growth of deep rooted vegetation were key environmental parameters useful for human activity even in harsh arid climate. The saline lakes played a very important role during terminal Pleistocene when the fluvial activity was almost dormant in the Thar. Saline lakes favored a perched water table of precipitation during ISM and westerly winter. The salt encrusted surface of the dried lake might have served as a source of salt for animals. Though chronological input as saline lake is as yet of low resolution, they have preserved paleoclimatic signatures of terminal Pleistocene, mid-Holocene (particularly 7–6 ka technically known as climatic optimum), and late-Holocene 4 ka in response to global climatic changes.

In the concluding part of this small communication, we have attempted man–environment relationship that existed between hunter gatherers and shallow lakes surviving for more than 15,000 years. Probably this part of arid/semiarid Thar Desert is unique in the context of Indian Prehistory which is dominated by interrelationship between hunter gatherers and riverine landscape blessed by ISM.

Acknowledgments

We dedicate this chapter to Late Professor V.N. Misra (Formerly of the Deccan College Post-Graduate and Research Institute, Pune) who initiated S.N. Rajaguru in the Thar Desert in early 80s of the 20th century. We express our deep gratitude for academic interaction during two decades of group research in the Thar to Drs. D.P. Agrawal and A.K. Singhvi (Formerly of Physical Research Laboratory, Ahmedabad), R.P. Dhir and Amal Kar (Formerly of Central Arid Zone Research Institute, Jodhpur), S.K. Tandon (Formerly of Delhi University), R.K. Wadhavan (Formerly of Geological Survey of India, Jaipur), and Drs. R.J. Wasson, Gurdip Singh, Martin Williams of Australia. We profusely thank Drs. M.D. Kajale (Formerly of Deccan College Post-Graduate and Research Institute, Pune), Hema Achyuthan (Formerly of Anna University, Chennai), and Savita Ghate (Formerly of Ambedkar College, Pune) for their active participation in field and laboratory studies. We thank the University Grants Commission, Department of Science and Technology, New Delhi and the Deccan College Post-Graduate and Research Institute, Pune for financial and logistic support.

We are grateful to Dr. Liza Ely of Washington State University, Seattle, USA and Dr. Yehuda Enzel of Jerusalem University, Israel for allowing us to use unpublished field and lab investigations of Lake Jhorda. We are also thankful to Mr. Deodatta Phule for timely map and line drawings. We are grateful to the authorities of the Indian Academy of Sciences for permitting us to reproduce figures and photos from published article of Deotare et al. (2004) for this chapter. We are grateful to Editors and Reviewers for their valuable suggestions.

This chapter is a product of selective review—author's observations mainly carried out in late 20th century. View expressed in this chapter reflects our understanding and perspective on the saline lakes and Stone Age culture of the Thar Desert.

References

Achyuthan, H., Kar, A., Eastoe, C., Late Quaternary-Holocene lake-level changes in the eastern margin of the Thar Desert, India, J. Paleolimnol. 38 (2007) 493–507.

Allchin, B., Goudie, A.S., Hedge, K.T.M., The Prehistory and Palaeogeography of the Great Indian Desert, Academic Press, London, 1978.

Allchin, B., Hegde, K.T.M., Goudie, Andrew, et al., Prehistory and Environmental Change in Western India: a Note on the Budha Pushkar Basin, Rajasthan. Man 7 (4) (1972) 541–564.

Bhandari, M., Flora of the Indian Desert, Scientific Publishers, Jaipur, 1978.

Biagi, P., Veesar, G.M., An archaeological survey in the neighbourhood of Thari in the Thar Desert (Sindh, Pakistan). Ancient Sindh 5 (1998–99) 93–118.

Blinkhorn, J., A new synthesis of evidence for the upper Pleistocene occupation of 16R dune and its southern Asian context, Quat. Int. 300 (2013) 282–291.

Blinkhorn, J., Late Middle Palaeolithic surface sites occurring on dated sediment formations in the Thar Desert, Quat. Int. 350 (2014) 94–104.

Blinkhorn, J., Achyuthan, H., Ditchfield, P., Petraglia, M., Palaeoenvironmental dynamics and Palaeolithic occupation at Katoati, Thar Desert, India, Quat. Res. 87 (2017) 298–313.

Chawla, S., Dhir, R.P., Singhvi, A.K., Luminescence chronology of sand profiles in the Thar Desert and their implications, Quat. Sci. Rev. 11 (1992) 25–32.

Deotare, B.C., Kajale, M.D., Quaternary pollen analysis and palaeoenvironmental studies on the salt basins at Pachpadra and Thob, western Rajasthan, India, preliminary observations, Man Environ. 21 (1996) 24–31.

Deotare, B.C., Kajale, M.D., Kshirsagar, A., Rajaguru, S.N., Geoarchaeological and palaeoenvironmental studies around Bap-Malar playa, district Jodhpur, Rajasthan, Curr. Sci. 75 (1998) 316–320.

Deotare, B.C., Kajale, M.D., Kusumgar, Sheela, Rajaguru, S.N., Late Holocene environment and culture at Bari Bavri, western Rajasthan, India, Man Environ. 24 (1) (1999) 27–38.

Deotare, B.C., Kajale, M.D., Rajaguru, S.N., Kusumgar, S., Jull, A.J.T., Donahue, J.D., Palaeoenvironmental history of Bap-Malar and Kanod playas of western Rajasthan, Thar desert, Proc. Indian Acad. Sci. (Earth Planet. Sci.) 113 (3) (2004) 403–425.

Dhir, R., Alappat, L., Mukarjee, P., Raghav, R., Srivastava, P., Wadhawan, S., Singhvi, A., 2012. Field guide—Thar Desert. Desert dune systems, past dynamics and chronology. In: International Workshop on Desert Dune Systems. Ahmedabad, India.

Dhir, R.P., Joshi, D.C., Kathju, S., Thar Desert in Retrospect and Prospect, Scientific Publishers, Jodhpur, 2018.

Dhir, R.P., Singhvi, A.K., Andrews, J.E., Kar, A., Sareen, B.K., Tandon, S.K., Kailath, A., Thomas, J.V., Multiple episodes of aggradation and calcrete formation in Late Quaternary aeolian sands, Central Thar Desert, Rajasthan, India, J. Asian Earth Sci. 37 (2010) 10–16.

Enzel, Y., Ely, L.L., Mishra, S., Ramesh, R., Amit, R., Lazar, B., Rajaguru, S.N., Baker, V.R., Sandler, A., High resolution Holocene environmental changes in the Thar Desert, Northwestern India, Science 284 (1999) 125–128.

Holt, B.G., Lessard, J.-P., Borregaard, M.K., Fritz, S.A., Araújo, M.B., Dimitrov, D., Fabre, P.H., Graham, C.H., Graves, G.R., Jønsson, K.A., Nogués-Bravo, D., Wang, Z., Whittaker, R.J., Fjeldså, J., Rahbek, C., An update of Wallace's zoogeographic regions of the World, Science 339 (2013) 74–78.

Jain, M., Tandon, S.K., Singhvi, A.K., Mishra, S., Bhatt, S.C., Quaternary alluvial stratigraphical development in a desert setting: a case study from the Luni River basin, Thar Desert of western India, Special Public. Int. Assoc. Sedimentol. 35 (2005) 349–371.
Kajale, M.D., Deotare, B.C., Field observations and lithostratigraphy of three salt lake deposits in Indian desert of western Rajasthan, Bull. Deccan Coll. Post-Grad. Res. Inst. 53 (1995) 117–134.
Kajale, M.D., Deotare, B.C., Late Quaternary environmental studies on salt lakes in western Rajasthan, India: a summarised view, J. Quat. Sci. 12 (1997) 405–412.
Kar, A., Origin and transformation of longitudinal sand dunes in the Indian desert, Zeits. Geomorphol. 31 (1987) 311–337 .
Kar, A., The Thar or the Great Indian Sand Desert, Landscapes and Landforms of India, Springer, Dordrecst, 2014, pp. 79–90.
Khan, T.I., Frost, S., Floral biodiversity: a question of survival in the Indian Thar Desert, Environmentalist 21 (2001) 231–236.
Mishra, S., Jain, M., Tandon, S.K., Singhvi, A.K., Joglekar, P.P., Bhatt, S.C., Kshirsagar, A., Naik, S., Deshpande-Muhkerjee, A., Prehistoric cultures and late Quaternary environments in the Luni basin around Balotra, Man Environ. 24 (1999) 39–50.
Mishra, S., Rajaguru, S., Ghate, S., Stone age Jaisalmer: implications for human adaptation to deserts, Bull. Deccan Coll. Res. Inst. 53 (1993) 259–266.
Misra, V.N., Palaeolithic culture of Western Rajputana, Bull. Deccan Coll. Post-Grad. Res. Inst. 21 (1962) 86–156.
Misra, V.N., Govindgarh, a palaeolithic site in Western Rajasthan, J. Asiatic Soc. Bombay 38 (1965) 205–208.
Misra, V.N., Kanungo, A.K., The Mesolithic Age in South Asia: Tradition and Transition, Aryan Books International, Delhi, 2019.
Misra, V.N., Rajaguru, S.N., Palaeoenvironment and Prehistory of the Thar Desert, Rajasthan, India. In: Frifelt, Karen, Sorensen, Per (Eds.), South Asian Archaeology 1985. Scandinavian Institute of Asian Studies Occasional Papers (4), Curzon Press London, 1989, pp. 296–320.
Mohapatra, G., Bhatia, S., Sahu, B., The discovery of a stone age site in the Indian Desert, Res. Bull. Panjab Univ. 14 (1963) 205–223.
Mukherjee, P., Rakshit, P., Late Quaternary climatic vicissitudes as deciphered from study of lacustrine sediments of Sambhar Lake, Rajasthan, India, Indian J. Geosci. 66 (2012) 213–224.
Prasad, S., Enzel, Y., Holocene palaeoclimates of India, Quat. Res. 66 (2006) 442–453.
Rajaguru, S.N., Deo, S.G., Gaillard, C., Pleistocene geoarchaeology of Thar Desert, Ann. Arid Zone 53 (2015) 63–76.
Singh, G., The Indus Valley culture (paleobotanical study of climatic changes), Puratattva 4 (1971) 68–76.
Singh, G., Joshi, R.D., Chopra, S.K., Singh, A.B., Late Quaternary history of vegetation and climate of the Rajasthan Desert, India, Philos. Trans. R. Soc. Lond. Ser. B 267 (1974) 443–501.
Singh, G., Wassan, R.J., Agrawal, D.P., Vegetation and seasonal climatic changes since the last full glacial in the Thar Desert, Northwestern India, Rev. Palaeobot. Palynol. 64 (1990) 351–358.
Singhvi, A.K., Banerjee, D., Rajaguru, S.N., Kishan Kumar, V.S., Luminescence chronology of a fossil dune at Budha Pushkar, Thar Desert: palaeoenvironmental and archaeological implications, Curr. Sci. 66 (1994) 770–773.
Singhvi, A.K., Kar, A., The aeolian sedimentation record of the Thar desert, Proc. Indian Acad. Sci. (Earth Planet. Sci.) 113 (2004) 371–401.
Singhvi, A.K., Williams, M.A.J., Rajaguru, S.N., Misra, V.N., Chawla, S., Stokes, S., Chauhan, N., Francis, T., Ganjoo, R.K., Humphreys, G.S., A ~200 ka record of climatic change and dune activity in the Thar Desert, India, Quat. Sci. Rev. 29 (2010) 3095–3105.
Sinha, R., Smykatz-Kloss, W., Stuben, D., Harrison, S.P., Berner, Z., U., K., Late Quaternary palaeoclimatic reconstruction from the lacustrine sediments of the Sambhar Playa Core, Thar Desert margin, India, Palaeogeogr. Palaeoclimatol. Palaeoecol. 233 (2006) 252–270.

Sinha, R., Stueben, D., Berner, Z., Palaeohydrology of the Sambhar Playa, Thar Desert, India, using geomorphological and sedimentological evidences, Geol. Soc. India 64 (2004) 419–430.

Srivastava, A., Thomas, D.S.G., Durcan, J.A., Bailey, R.M., Holocene palaeoenvironmental changes in the Thar Desert: an integrated assessment incorporating new insights from aeolian systems, Quat. Sci. Rev. 233 (2020) 106214.

Sundaram, R., Rakshit, P., Pareek, S., Regional stratigraphy of Quaternary deposits in parts of Thar desert, Rajasthan, J. Geol. Soc. India 48 (1996) 203–210.

Thomas, J., Kar, A., Kailath, A., Juyal, N., Rajaguru, S., Late Pleistocene-Holocene history of aeolian accumulation in the Thar Desert, India, Zeits. Geomorphol. Supplementband 116 (1999) 181–194.

Tod, Col. J., Annals and Antiquities of Rajasthan, 1 and 2. Routledge and Kegan Paul, London, 1829-1832 reprinted 1957, pp. 1–1862.

Wadhavan, S.K., Sharma, H.S., Quaternary stratigraphy and morphology of desert ranns and evaporite pans in central Rajasthan India, Man Environ. 22 (1997) 1–10.

Wasson, R.J., Rajaguru, S.N., Misra, V.N., Agrawal, D.P., Dhir, R.P., Singhvi, A.K., Kameswara Rao, K., Geomorphology, late Quaternary stratigraphy and paleoclimatology of the Thar dunefield, Zeits. Geomorphol. Supplementband 45 (1983) 117–151.

Wasson, R.J., Smith, G.I., Agrawal, D.P., Late Quaternary sediments, minerals, and inferred geochemical history of Didwana Lake, Thar Desert, India, Palaeogeogr. Palaeoclimatol. Palaeoecol. 46 (1984) 345–372.

CHAPTER

Late Quaternary geoarchaeological and palaeoenvironmental aspects of Kurnool Basin in the Indian Peninsula

23

Ravi Korisettar

ICHR Senior Academic Fellow, Sri Saranakripa, Sivagiri, Dharwad, Karnataka , India

23.1 Introduction

The Kurnool Basin in southern Indian Peninsula preserves evidence for a continuous hominin occupation from the Lower Palaeolithic to modern times. The Billa Surgam (Billasurgam) cave complex in the basin is a key site for systematic geological, palaeontological, and archaeological research in response to paradigm shifts in human biocultural studies, since the 1840s. The nineteenth century search for hominin remains from Billa Surgam caves had also led to the discovery of Neolithic and Megalithic sites at Patapadu and a limestone cave at Errajari Gabi in the Jurreru Valley (Haslam et al., 2010a). During the first decade of the twenty-first century a series of Palaeolithic sites interstratified with a rhyolite tuff, painted rock shelters with microlithic occupation and Neolithic-Megalithic sites, were discovered in the Jurreru Valley ((Fig. 23.1; Petraglia et al., 2009b). Fieldwork was carried out in the east-west trending two subvalleys along a 50-km stretch of the Erramalai Hills in the Kurnool District of Andhra Pradesh: one lying to the north of the range around Billa Surgam and the other to the south around the villages of Patapadu and Jwalapuram (Fig. 23.2).

The rhyolite tuff was ejected from Toba caldera in Sumatra, Indonesia 74 ka ago, designated Youngest Toba Tuff (YTT). The YTT is found dispersed as air-fall tephra over vast areas covering Lake Malawi in East Africa, Indian Ocean, and South China Sea. It is also found deposited with varying thickness on the continental ancient shallow water bodies on the Indian Peninsula (Fig. 23.3). Glass chemistry, stratigraphy, and grain-size data, its distinctive biotite composition in particular, have conclusively resolved its age and source (Lane et al., 2013; Mathews et al., 2012; Westgate and Pearce 2017).

The ongoing debate on out of Africa expansion of modern humans which suggested that Indian subcontinent was at the geographical crossroads of hominin dispersal between Africa and Australia necessitated renewed investigations at Billa Surgam caves, considering the potential of desiccating calcic deposits for preservation of fossil hominin/faunal remains. Further, report of the presence of YTT by Rao and Rao (1992) in the contiguous Jurreru Valley provided an impetus to launch a collaborative initiative between Karnatak University (India) and the University of Cambridge (UK)—"The Kurnool District Archaeological Project" (KAP).

Holocene Climate Change and Environment. DOI: https://doi.org/10.1016/B978-0-323-90085-0.00007-3

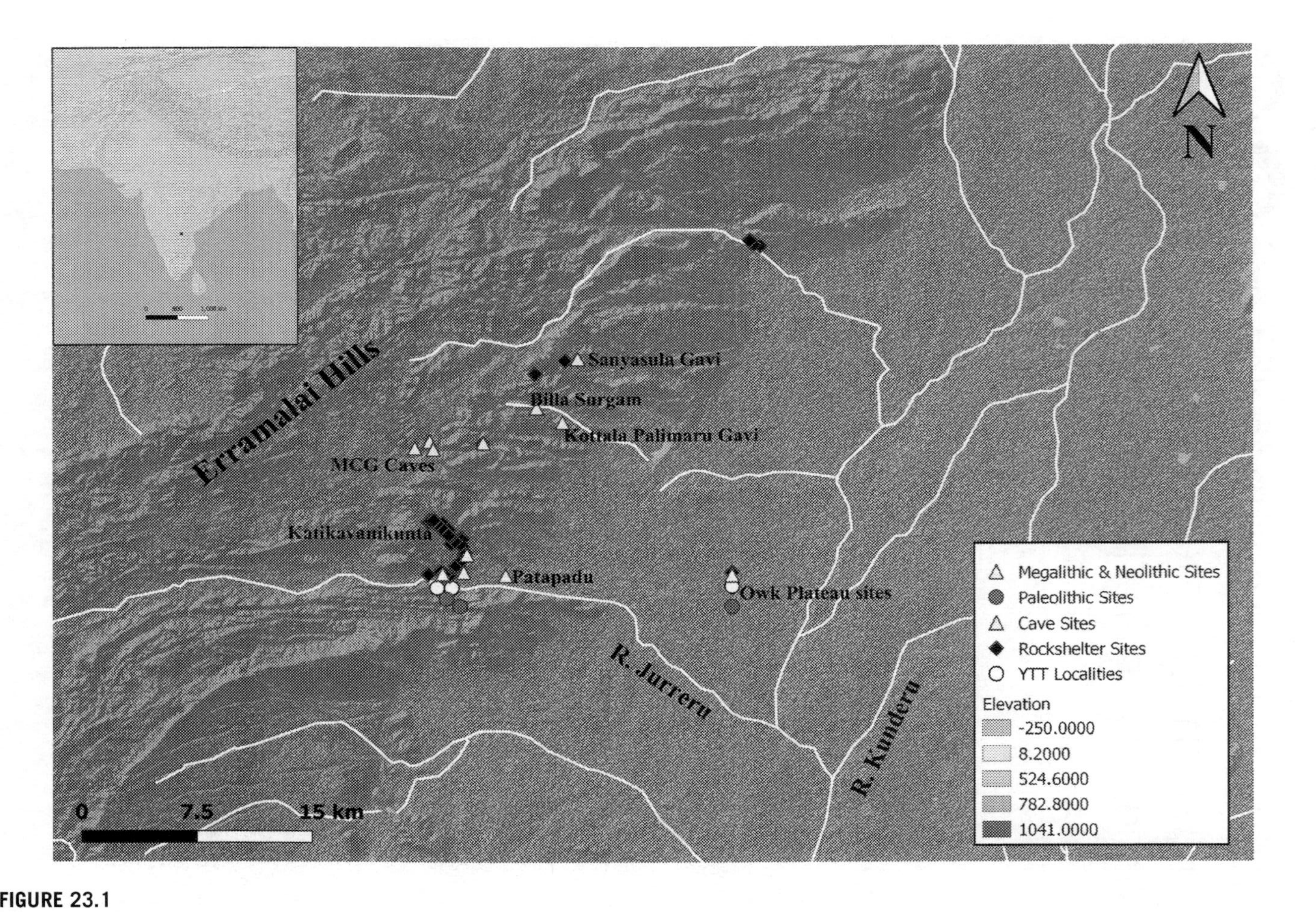

FIGURE 23.1

A dense cluster of archaeological sites is located at the eastern end of the antiformal valley. The Jurreru river enters the open plain at Yaganti village. Hundreds of rock shelters occur along the northern plateau with evidence for hominin occupation ranging in time from the Late Pleistocene to Late Holocene. Source: Prepared by the author and Anil Devara, The M.S. University of Baroda, Vadodara.

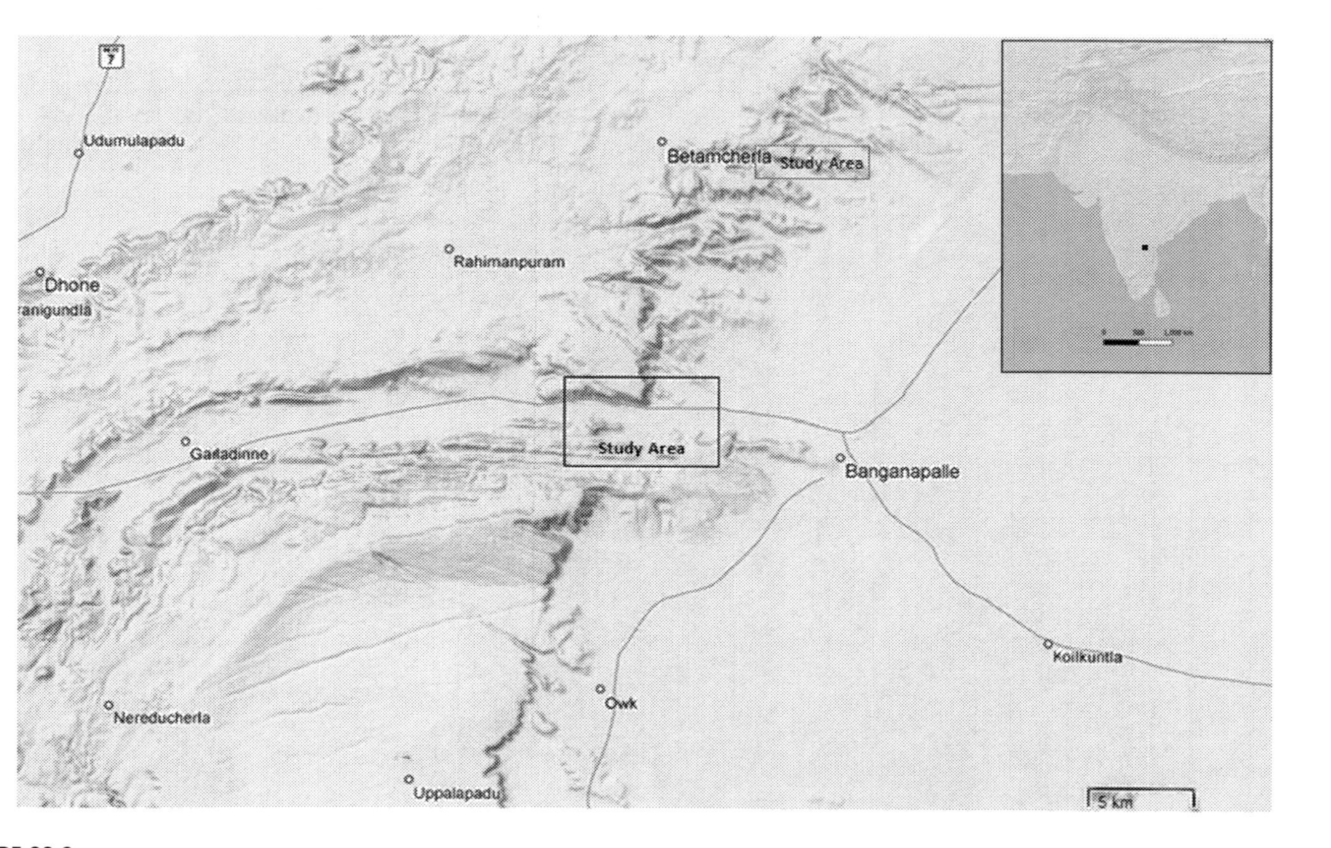

FIGURE 23.2

Map of the study area in the Erramalai range of the Kurnool Basin, Andhra Pradesh. Source: The Kurnool District Archaeological Project.

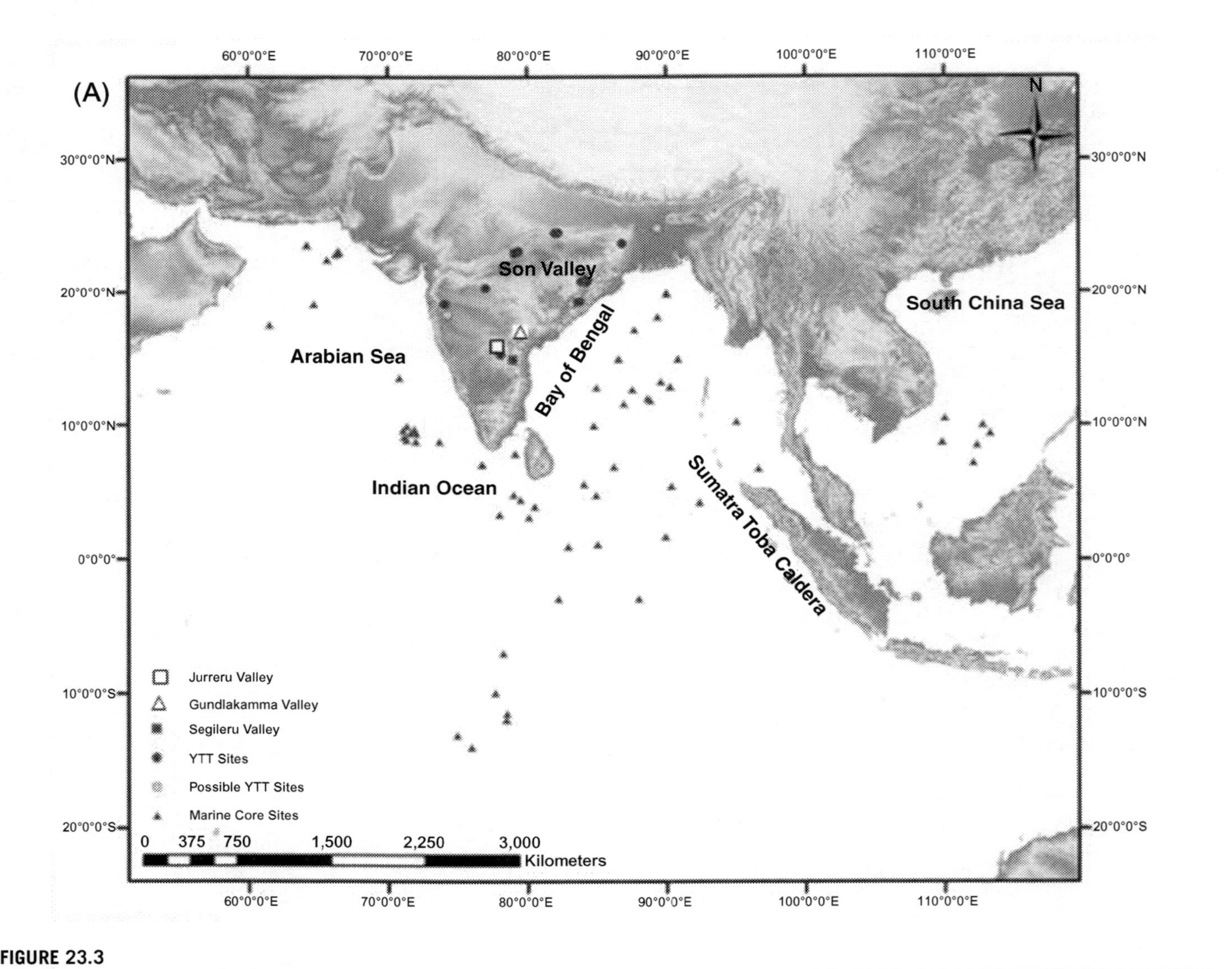

FIGURE 23.3

YTT occurs as cryptotephra between Lake Malawi in East Africa, and beyond Sumatra in South China Sea. It is preserved as cryptotephra in the distal areas and as ignimbrite in the proximal area. The black dots are sites where YTT occurs and open circle in Sumatra is a source region at Toba Caldera (based on original map by Westgate and Pearce 2017).

The KAP team's initial reconnaissance surveys led by Ravi Korisettar and Michael Petraglia in the Billa Surgam and Jurreru Valleys resulted in the discovery of: (1) Middle Palaeolithic assemblages interstratified with the YTT in the Jurreru Valley; (2) hundreds of painted rock shelters with microlithic occupation; (3) Late Holocene, Neolithic, and Iron Age settlements in the caves and rock shelters; and (iv) identification of potential spots for excavation in the Billa Surgam cave and rock shelter floor deposits. Therefore the follow-up research focused on: (1) determining the age of the Middle Palaeolithic in the Jurreru Valley through OSL dating and tephrochronology, (2) assessing the impact of YTT air-fall on contemporary hominin populations and regional environmental change, (3) attempt OSL dating of the Billa Surgam cave and rock shelter floor deposits, (4) examine the geographical importance of the Indian subcontinent as crossroads in the out of Africa dispersal of modern humans, and (5) critically assess suggested routes of dispersal out of Africa. While YTT's presence as an isochron was well recognized, understanding the environment of deposition, reconstruction of pre-YTT landscape, delineation of intersite variation within the study area, development of lithic technology, hominin behavioral change, rise of behavioral modernity, and understanding of Palaeolithic transitions became the prime objectives (Petraglia et al., 2012a; Korisettar 2015, 2016; Korisettar and Janardhana 2014).

23.2 Scope

The Indian subcontinent was considered a region of crossroads between Africa and Australia and began to attract the world attention to trace the dispersal of modern humans out of Africa. It was in this context the Billa Surgam complex of caves in the Kurnool Basin once again became the focus of multidisciplinary geoarchaeological investigations in search of hominin remains. The first phase of investigations suggested the presence of Upper Palaeolithic remains in the cave floor deposits (Haslam et al., 2010a). The second phase of research involved independent investigations by paleontologists into the faunal record of the limestone caves on the one hand (Prasad 1996; Prasad and Verma 1969; Prasad and Yadagiri 1986) and the search for definitive Upper Palaeolithic, Mesolithic, and Neolithic assemblages on the other (Soundararajan and Joshi, (1956-57); Murty, (1974, 1975, 1985); Murty and Reddy, (1975); Reddy, (1977). The KAP's investigations represent the third phase of interdisciplinary research that has ushered in a paradigm shift in Indian prehistoric archaeology (Korisettar, 2012).

The last 200,000 years span of Quaternary prehistory of the Kurnool Basin in particular and the Indian subcontinent in general is presented in the chapter. Geoarchaeology of the Palaeolithic to the end of Iron Age archaeological sites has been carried out to delineate man-land relationships. Tephrochronology, OSL, and Accelarator Mass Spectrometric (AMS) radiocarbon dating methods have been profitably applied. Archaeological record from the Late Acheulian to the Iron Age has been placed against the environmental background. Assessing the impact of YTT air-fall tephra on the hominin populations and understanding depositional environments are some of the highlights of this research. Population stability and faunal continuity during Mid- to Late Quaternary does not support the drastic consequences of the YTT impact. Interdisciplinary research into the cave floor deposits of Billa Surgam has facilitated in recognizing the Kurnool Basin as one of the major prehistoric refugia in the Indian subcontinent with potential for probing the Early Pleistocene contexts in the cave floor deposits.

23.3 Study area

23.3.1 Regional geography and geology

The Proterozoic Kurnool Basin is a subbasin lying on the western portion of the larger Cuddapah Basin. The rock strata of the Kurnool Supergroup and Cuddapah Supergroup have undergone a varying degree of deformation. The litho sequence of the Kurnool Basin includes a succession of quartzite, sandstone, shale, and limestone. They overlie unconformably on the older Cuddapah sedimentaries. The Erramalai and Nallamalai Hills mark the western and eastern boundaries of the Kurnool Basin respectively. The Erramalai range marks the orographic and geologic boundary between the Achaean granite-gneiss formations to the west and the Proterozoic sedimentary and igneous rock formations of the Kurnool Basin to the east. The basin topography including the Kurnool and Cuddapah Basins is characterized by a series of parallel hill ranges such as Erramalai, Nallamalai, and Velikonda Hills, and are part of the Eastern Ghats. A network of rivers, some of which having a source in the headwater springs (the Kundu river or Kunderu), drain these valleys.

The Kurnool Super Group comprised six formations: Banganapalli Quartzites, Narji Limestone, Owk Shale, Paniam Quartzite, Koilakuntla Limestone, and Nandyal Shale from the base upwards. The dominant lithology in the study area is represented by Paniam Quartzites, Owk Shale, Narji and Koilkuntla Limestones including a variety of cryptocrystalline silica rich minerals and dykes, and sills of dolerite. In the Kurnool Basin, the degree of deformation of rock strata is lower than in the eastern portion of the Cuddapah Basin, conjugate fractures are observed. The Jurreru River flows through an antiformal axis. The Paniam Quartzite capping the plateau has developed dip-slope valleys, forming major passes in the plateau (Ramam and Murty, 1997).

The siliceous Narji Limestone was the most preferred rock for making Palaeolithic artifacts. A variety of cryptocrystalline minerals and dyke rocks were the most preferred lithic raw materials of the Holocene Mesolithic hunter–gatherers and Neolithic-Megalithic agro-pastoral communities. The plateau quartzites overlying the softer limestone and shale have moved down owing to dissolution, undercutting of softer shale, and scarp retreat processes. They occupy various positions along the slope of the escarpments and preserve evidence of hominin occupation dating from the Late Pleistocene, ~40, 000 years ago (Clarkson et al., 2009). Systematic transect surveys were useful in documenting the rock art sites on the plateau around Katikavanikunta-Yaganti valley (Blinkhorn et al., 2010).

23.4 Billa Surgam investigations

23.4.1 Late Middle Pleistocene to Late Holocene faunal and archaeological record

A series of karstic caves have formed in the Narji Limestone formation as a result of dissolution via subsurface and surface water flow. The well-known caves are (1) Yerrajari Gabi in the Yaganti-Jurreru Valley, (2) a series of small and large caves including the Bill Surgam complex near Betamcherla, and (3) Belum caves in the Chitravati Valley. The lithological junctions in the rock strata governed the formation of groundwater seepage and the consequent perennial spring activity gave rise to the formation of perennial water bodies in the valley bottoms that sustained high floral and faunal biomass.

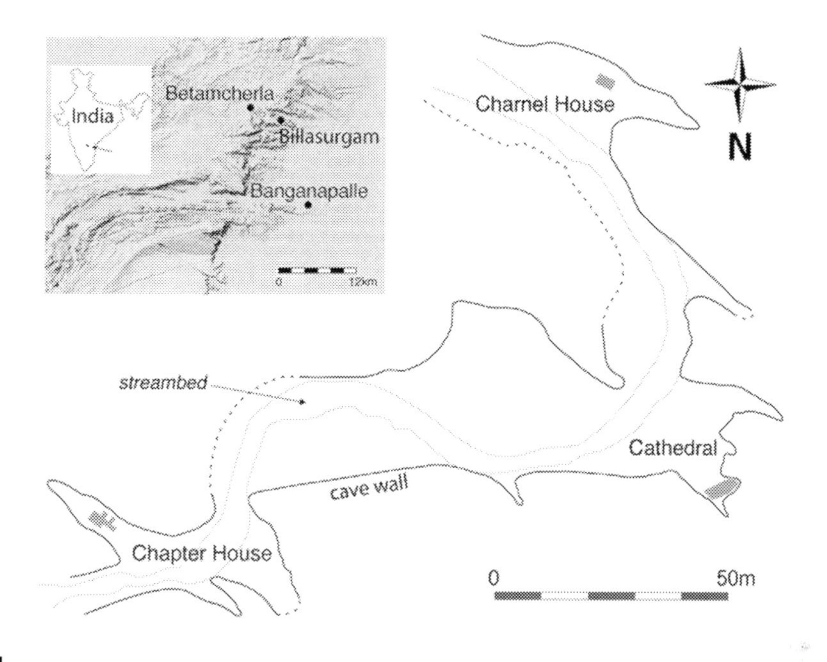

FIGURE 23.4

The complex of caves at Bill Surgam in the northern part of the study area. Source: The Kurnool District Archaeological Project.

23.4.2 Field investigations in the Billa Surgam Valley

For the purpose of this study, the Billa Surgam Valley is demarcated by the village of Kanumakinda Kottala to the west and the Billa Surgam cave complex to the east in the Erramalai Hills. The valley is flanked by steep-sided escarpments. The cave complex is situated along the foot of the southern escarpment, locally called Yerrakonda. The lithological succession includes sandstone, shale, and quartzite. There is no through-flowing river in the valley but for the monsoon-fed seasonal streams and local water bodies in the valley bottom fed by groundwater seepage.

The Billa Surgam (N15°26.153′ E 7°11.122) is a complex of caves: Cathedral, Charnel House, Purgatory, Chapter House North, etc. (Fig. 23.4). Of these the Charnel House Cave and Cathedral Cave floors were excavated to greater depths to reach the bedrock. The objectives included reconfirming the significance of previous researchers' findings, including faunal and artifact assemblages, systematic taxonomic classification of the fauna, and fixing their chronology through OSL and other geochronological methods. Excavations and analyses of cave floor deposits did not, however, yield diagnostic and definite lithic artifacts of the Palaeolithic period, as opposed to Murty and Reddy (1975).

23.4.3 Laboratory studies and dating

The faunal studies have shed new light on the nature of habitat, faunal continuity, and extinctions, with implications for habitat fragmentation during the last 200 ka (Roberts et al., 2014). The composite

Table 23.1 Count of macrofaunal species recovered from Charnel House Cave according to MIS approximated from OSL dating (after Roberts et al., 2014 S1 Appendix: 24; after Smith et al., 2017). Source: The Kurnool District Archaeological Project.

Taxon	MIS 1–2	MIS 3	MIS 5
Axis axis	1	0	0
Bosephalus sp.	0	0	1
Equus sp.	0	1	0
Erinaceus sp.	0	0	1
Semnopithecus entellus	0	0	18
Herpestes griseus	0	0	1
Hystrix crassidens	1	1	1
Lepus sp.	6	1	1
Rhinoceros sp.	0	1	1
Tetracerus quadricornis	0	1	0
Total	8	5	24

stratigraphy of the two caves—Cathedral and Charnel House caves—has revealed a 11-m succession of deposits characterized by silt-dominated cave earths (silt-grade siliciclastic deposits), with intercalations of angular blocky limestone breccias and lenses of well-rounded waterworn pebble conglomerate (details of sediment characteristics are given in Smith et al., 2017). OSL dating of the strata and the identification of YTT cryptotephra at a depth of 3.5 m from the top in the Charnel House floor deposits have enabled the chronometry of the sedimentary strata, ranging in time from >200 ka to the present, from the base to the top (Lane et al., 2011).

The composite section (Fig. 23.5) shows eight levels of OSL dating, four from Charnel House and another four from Cathedral. The top four (from Charnel House) OSL dates are supported by the YTT date of 74 ka, and define a time bracket of 74–57 ka. The upper levels of the Charnel House are dated to the Holocene by the presence of Neolithic pottery. The Neolithic occupation of these caves is also supported by the presence of ceramic sherds and a dated nested diamond-shaped petroglyph (5000 BP) on the wall of Cathedral cave (Tacon et al., 2013). The deposits of the Cathedral cave have greater time-depth, extending up to 210 ka. Thus, for the first time, these dates have provided an absolute timeframe to the faunal succession that has been documented since the last quarter of the nineteenth century. The composite stratigraphy and OSL chronology were correlated with the oxygen isotopic stages (MIS 7 to MIS1). Gaps or interruptions in the sedimentary sequence are attributed to the removal of upper strata of sediments by previous excavations.

The absence of lithic artifacts in the excavations was compensated by abundant micro- and macropalaeontological remains. The macrofaunal remains, including those from published sources and KAP's excavations have now been studied in detail (Tables 23.1 and 23.2). Taphonomical considerations suggest that these faunal remains are preserved in a secondary context, and that only about 181 of 5281 faunal remains were identifiable to species or genera level (Roberts et al., 2014). This is considerably low when compared with the large volume of faunal remains recovered from both the caves. It is observed

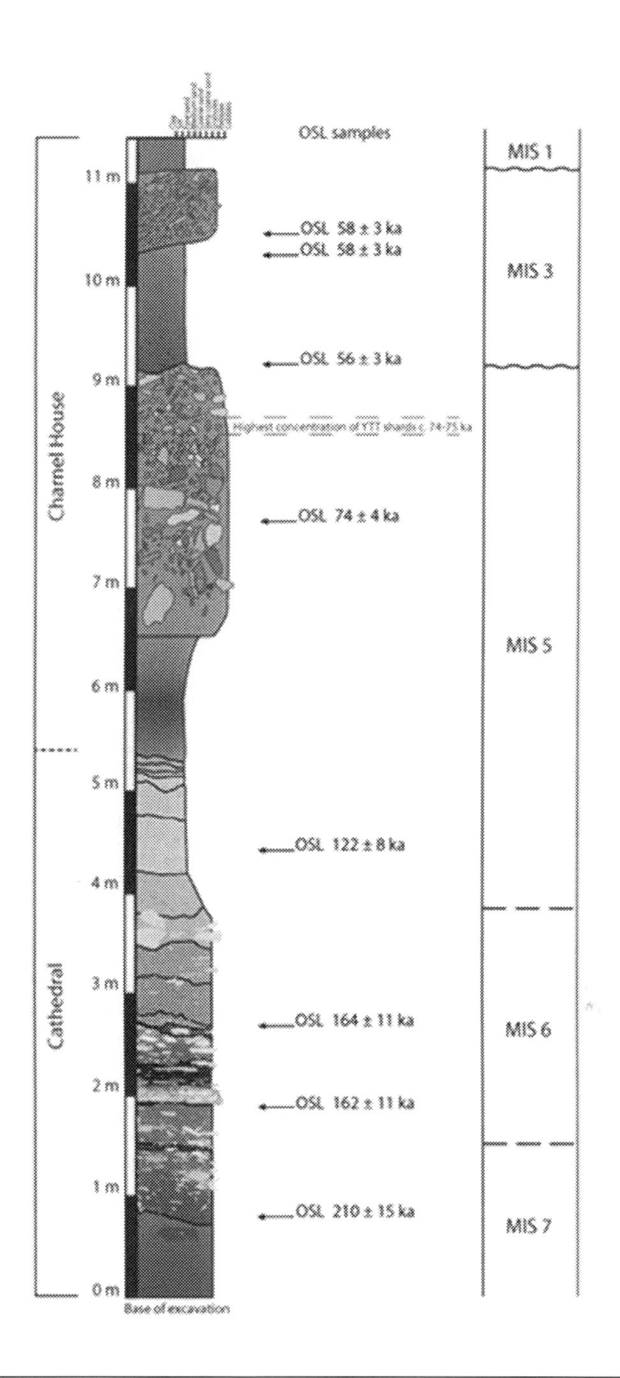

FIGURE 23.5

Composite lithology of Cathedral and Charnel House cave floor deposits correlated with MIS from 7-1, interruption in the sequence was caused by the removal of upper sediments from the Cathedral by previous excavations. The profile for the first time is dated by tephrochronology of the YTT and Optically Stimulated Luminescence (OSL) dating of sediments below and above the lithounit containing cryptotephra. Source: The Kurnool District Archaeological Project.

Table 23.2 Count of macrofaunal species recovered from Cathedral Cave according to MIS approximated from OSL dating (after Roberts et al., 2014, S1 Appendix: 23; after Smith et al., 2017). Source: The Kurnool District Archaeological Project.

Taxon	MIS 5	MIS 6	MIS 7
Antilope cervicapra	2	9	0
Axis axis	3	0	0
Bos sp.	2	2	0
Bosephalus sp.	2	4	1
Canis sp.	1	1	0
Equus sp.	4	5	2
Erinaceus sp.	1	0	0
Felis chaus	0	2	0
Gazella sp.	3	3	1
Herpestes griseus	1	1	0
Hystrix crassidens	18	8	0
Lepus sp.	1	4	0
Muntiacus sp.	0	2	0
Panthera pardus	1	0	0
Rhinoceros sp.	18	2	0
Semnopithecus entellus	11	4	1
Sus sp.	6	15	0
Ursid	0	1	0
Vulpes sp.	1	1	0
Total	75	64	5

that the majority of the fauna lived on the plateau above the caves and was incorporated into the cave floor through sink holes and ephemeral stream activity (Smith et al., 2017).

The presence of YTT shards in a stratum of the Charnel House deposits show a compositional match with the YTT samples recovered from Jwalapuram in the Jurreru Valley. The limestone rubble deposit that yielded cryptotephra of YTT also yielded an OSL age of 74 ka, confirming the dating of the cryptotephra to the YTT source. The sequence of OSL dates for the composite litholog enabled Roberts et al., (2014) to approximately correlate the strata from the base upwards to MIS7-MIS1. Sedimentological changes observed in the sequence are attributed to corresponding global cooling and warming climatic phases (Fig. 23.5).

Tables 23.1 and 23.2 present a revised faunal succession based on reexamination of earlier faunal identifications: reidentification of *Cenocephalus* sp., as *Theropithecus gelada* (now extinct in India). Roberts et al., (2014) also note the presence of *Rhinoceros* and *Equus* spp., (during MIS5) and *Semnopithecus entellus* (during MIS2/1). The faunal composition of cave floor deposits has implications for landscape changes. It is inferred that during the last 125,000 years landscapes around Billa Surgam supported a variety of grazing animals and their predators and it suggests that there was a mix of grassland and forest vegetation, with sufficient wetlands in the area.

FIGURE 23.6

A large topographic depression existed at the valley bottom prior to the deposition of pre-YTT sediments. Some part of the valley floor was much deeper (Locality 3) than other in the upstream sector. Despite the Late Holocene dry climatic conditions, the groundwater table is high (4 m below the surface). Source: The Kurnool District Archaeological Project.

23.5 Investigations in the Jurreru Valley

The study area was demarcated between the Jurreru Dam to the west and Patapadu village to the east (Fig. 23.6). Between the Jurreru Dam and Jwalapuram village, the valley is bounded by Peddakonda to the north and Eddulakonda to the south. At this point the valley floor lies approximately 250 m above sea level. The Eddulakonda hills comprised limestone and shale beds intruded by dolerite dykes, reaching a maximum height of 350 m Above Mean Sea Level (AMSL). The Peddakonda comprises limestone, shale, and quartzite. It rises to a height of 480 m AMSL. Quaternary fluvio-lacustrine sediments are preserved in the valley bottom. Systematic transect survey in the 4.5-km stretch of the valley around the village of Jwalapuram, roughly covering an area of 25,000 m^2 (Shipton et al., 2010) and excavations at a number of designated localities, has revealed a continuous succession of hominin occupation from Lower Palaeolithic to Iron Age through the Neolithic. In addition, hundreds of painted rock shelters dating from the Late Pleistocene have been subject to systematic documentation and comparative study (Tacon et al., 2010).

Terminal Pleistocene channel incision has revealed a succession of low-energy, suspended load sediments interstratified with the YTT. The coarser sediment accumulation is observed toward the southern foothills, the Jwalapuram Tank breccia (JWP Tank, Fig. 23.7). The geochemistry and phytolith analysis of pre- and post-YTT sediments, and the reconstruction of pre-YTT topography of the valley bottom have indicated a paludal environment characterized by marsh land with not so deep standing water body at the time of air-fall YTT (Blinkhorn et al., 2012).

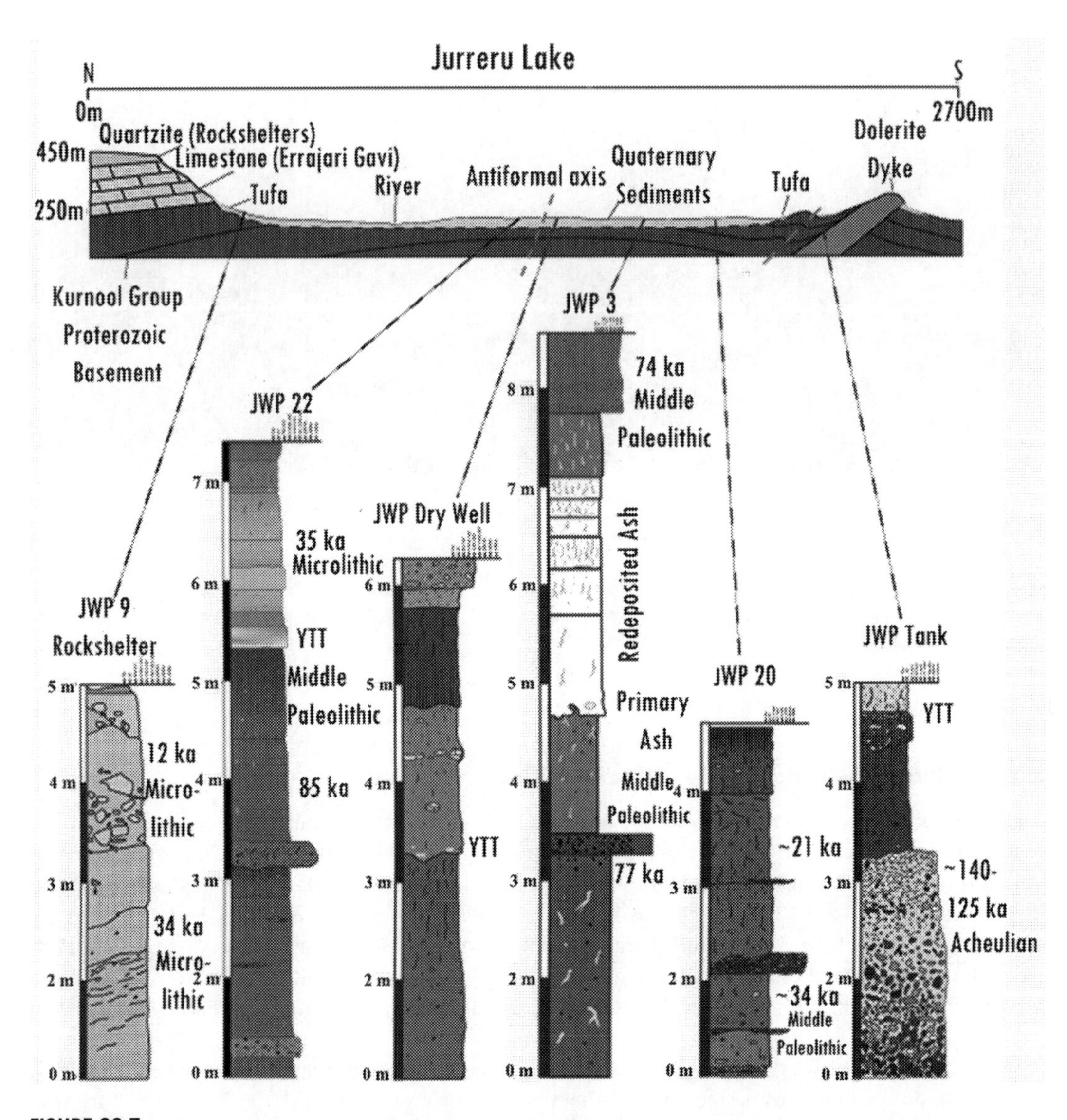

FIGURE 23.7

Series of valley bottom sedimentary profiles from localities designated JWP9, JWP22, Dry Well, JWP3, JWP20, and JWP Tank from the footslopes of Peddakonda to the footslopes of Eddulakonda. The Middle Paleolithic assemblages occur interstratified with the YTT, also constrained by OSL chronology. JWP9 is a rock shelter microlithic site. JWP Tank is an Acheulian site associated with colluvial deposit. Source: The Kurnool District Archaeological Project.

23.5.1 Stratified YTT and depositional environment

In the Jurreru Valley, YTT occurs as a primary ash fall deposit (3–5 cm). Emplacement and reworking has led to an artificial thickness of 2.5 m (JWP3, Fig. 23.7). It immediately overlies the primary ash deposit. Quarry pits dug into the valley bottom sediments revealed buried Middle Palaeolithic assemblages in the pre-YTT low-energy sediments with interfingering of sandy pebbly gravel, at a depth of 1 m below the base of the YTT (Petraglia et al., 2007; Petraglia et al., 2009b).

The primary ash layer comprising 90% glass is characterized by the absence of sedimentary structures. It occurs as a continuous layer with sharp lower and upper contacts. The redeposited tuff shows sedimentary structures such as hardpans, cross-bedding, fine laminations, mud cracks, and fine-scale load structures indicating influence of fluvial, lacustrine, and Aeolian modes of deposition. Rhizoliths and leaf impressions are well preserved in the reworked deposit. Bioturbation is observed in the reworked ash deposit. The absence of these features in the basal 3–5 cm is considered good evidence for recognizing the primary ash deposit, which maintains a sharp contact with preexisting topography (JWP3, Fig. 23.8). The presence of hardpans above the primary layer is interpreted as intervals of deposition during drier seasons. However, uniformity in grain size throughout the 2.5 m deposit, including the basal primary ash is suggestive of reworking having taken place immediately after it blanketed the valley slopes and the lake bottom. In the Jurreru Valley, the YTT is devoid of detrital material incorporated into it.

The present topography of the valley and uneven thickness of the ash deposits from west to east suggest the presence of topographic depressions and attest the existence of a palaeolake or deeper depressions in the valley bottom prior to the ash fall (Fig. 23.6). The sedimentological analysis also revealed fluctuations in water depth in response to fluctuations in the monsoon regimes in conformity with stadial and interstadial periods. Response to ash fall event is indicated by the fluctuating shore line and the presence of lithic assemblages on the ancient beach. At locality JWP3, there is substantial evidence of retreated lakeshore and human occupation around 77 ka. The Jwalapuram Lake was not only fed by the seasonal monsoon circulation but also springs emanating from hill slopes flanking the valley. These springs continued to provide freshwater resources consequent upon the ash fall that caused chemical pollution of the standing water bodies. The fault springs along the southern escarpments and tufaceous beds at both the southern and northern flanks attest to uninterrupted groundwater seepage into the valley bottom both during pre-YTT and post-YTT time periods.

The presence of a fairly low-energy and shallow hydrological environment prevailed during the Late Quaternary. It is attested by the sedimentary features of the pre-YTT sediments and fossil plant impressions preserved in it. This is also indicated by the abundant plant fossils preserved in the ash deposit. Plant material got incorporated into the ash deposit following redeposition of the initial air-fall that was accompanied by torrential precipitation from the upslope to the north and south of the valley. The absence of artifact horizons within the redeposited ash indicated that it was a rapid event that took place soon after the initial event. Obviously the steep slopes flanking the valley bottom and the increased spring discharge triggered immediate redeposition of the ash which blanketed the valley slopes. The absence of clastic deposits between the lower primary ash fall and the reworked ash indicated little time gap between primary ash fall and subsequent reworked ash. The upslope spring activity and access to high-quality lithic raw materials were the reckoning factors that governed the survival of hominin populations despite the hazardous consequence of air-fall tephra.

The lake bottom lithounits in the Jurreru Valley have been designated, Strata A-D, from top to bottom at the key site of JWP3 (Fig. 23.8). Blinkhorn et al., (2012) carried out topographic analysis of

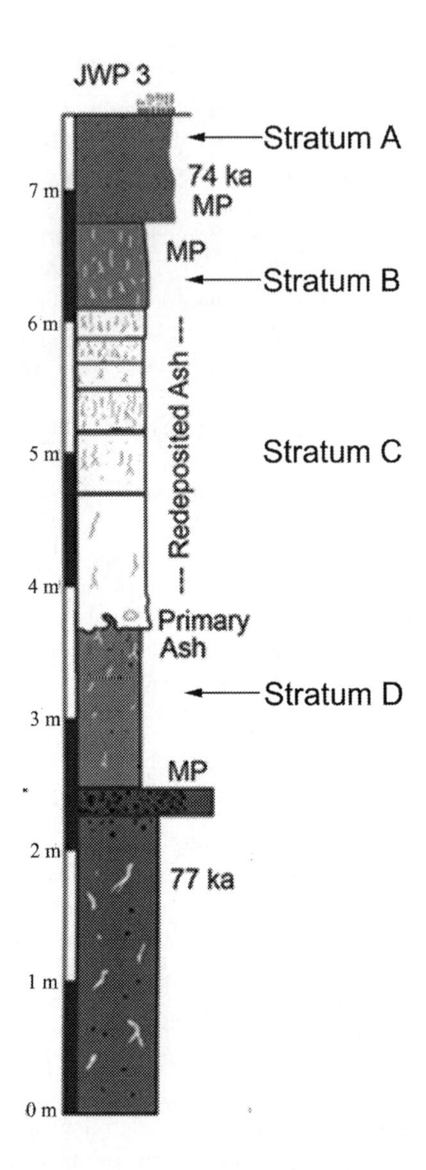

FIGURE 23.8

The succession of strata at key site of Locality 3 (JWP3). Primary YTT (3-4 cm) layer is at the base of Stratum C, showing a sharp contact (disconformity) with Stratum D. The Middle Palaeolithic assemblage occurs associated with a sandy-pebbly gravel 1 m below the base of the YTT in Stratum D. Source: The Kurnool District Archaeological Project.

pre-ash-fall surface through total station survey at 60 sites. They observed a maximum of 4.91 m variation from west to east between sites JWP22 (87.88 m) and JWP1 (82.97 m). This pronounced depression toward the southeast is not observed in the modern topography. The existence of deeper standing water bodies toward the southeast, for example JWP3 –JWP1, favored accumulation of both initial and emplaced air-fall tephra and continuation of post-YTT accumulation of silty sand deposits. The OSL date of 74 ka for the post-YTT deposit at JWP3 attests to continuity of depositional environments with only minor changes in the depositional facies at some point on the lake bottom. Stable carbon isotopic analysis of the pre-Toba sediment, which contains a Middle Palaeolithic assemblage, at locality JWP3, indicated a trend toward C_4 woodland vegetation environment. In stratum D (Fig. 23.8), representing lacustral clay revealed prevalence of mosaic environments with increasing level of C_4 vegetation and just prior to the ash fall grasslands constituted 50% of the landscape. Similarly phytolith record from infra YTT deposit has indicated a mosaic of vegetation. Sediment samples were subject to a variety of analytical techniques including particle size, magnetic susceptibility, loss on ignition, percentage carbonate, stable carbon and oxygen isotopes of carbonate nodules, rhizoliths, phytoliths as well as mega and micro biological remains (Jones, 2010; Haslam et al., 2010b, 2010c). Added to these investigations was the OSL dating of soil/sediment samples sandwiching the YTT.

23.6 Palaeolithic sites and dated lithic and symbolic artifact assemblages

In the Jurreru Valley 2 Acheulian sites (JWP Tank and PTP2), seven Middle Palaeolithic sites, three rock shelter sites, and five Neolithic sites were included in the detailed documentation and study through survey and excavation. Of these excavations at localities JWP 3, JWP17, 20–23, and JWP9 have added significant new knowledge to our understanding of the Middle Palaeolithic hominin adaptation to the fluvio-lacustrine landscapes in the region (Fig. 23.7).

23.6.1 Jwalapuram tank Acheulian occurrence

Large cutting tools of the Acheulian technocomplex occur embedded in the piedmont colluvium and on the surface along the southern hill flanks of the Jurreru River. These artifacts constitute the earliest evidence of human habitation in the valley. The artifacts are found dispersed along 1-km stretch of the foothill, which is capped by reworked residual volcanic ash. Acheulian occupations appear to have occurred on the foothill zone as these localities in piedmont situations are known to be areas of perennial spring activity in the past. The artifacts do not show battered edges indicating point provenance. The assemblage consists of handaxes, knives, and choppers of different sizes. The tools have been made on locally occurring quartzite and siliceous limestone. However some were made on cobble and a few others were on large flakes and tabular limestone. An OSL date indicates that Acheulian hominins survived late in the valley and were present during MIS 6 (Roberts et al., 2010).

23.6.2 Stratified Middle Palaeolithic occupation floors

Surface occurrence of Middle Palaeolithic artifacts was documented from the eroded floodplain surfaces, formerly buried under the sub-recent alluvium. Excavations were carried out at JWP 3, 20, 21, 22, 23, etc., (Fig. 23.7), which confirmed the existence of assemblages buried under YTT, as well as above. Lithounits in the excavated trenches were designated Strata A-D, from top to bottom (Fig. 23.8). At Locality 3, five trenches (3 and 3A–D) were dugout in an area of 600 m^2. The Middle Palaeolithic artifacts were recovered from the first three trenches from below the YTT. At Locality 3, the depth of the trench reached 3.6 m below the tephra, where water table was encountered. Roughly about 1 m below the YTT a thin lens of sandy pebbly gravel, OSL dated to ~77 ka, contained 215 Middle Palaeolithic artifacts and a piece of red ochre (with striations). The post-Toba layer contained 108 stone artifacts of similar character (Haslam et al., 2010b, 2010c, 2012).

At Locality 22, a trench measuring 10 × 10 m was dug down to a depth of ~ 6.5 m. The YTT layer lies between 1.8 and 1.9 m below the surface. Below this lies the undulating paleosol surface with abundant scatter of artifacts. Further below the sediment is darker and contained coarser materials. What is important to note here is the association of the occupation surface with the paleosol. The OSL ages for the samples range from 78 to 85 ka, consistent with the ages for the pre-YTT Middle Palaeolithic layer at JWP3. A total number of 1628 artifacts were collected from the occupation surface, the largest so far at Jwalapuram open air stations. The large quantity of lithic debitage and evidence for repeated use of the surface clearly indicate the in situ nature of the workshop activity at the locality.

23.6.3 Microlithic sequence at Jwalapuram 9 (JWP9)

Excavation of a painted rock shelter, JWP 9, has revealed a proliferation of microlithic artifacts beginning from about 35 ka. A 4 × 4 m trench to a depth of 3.3 m was excavated at the base of the rock shelter. The section was divisible into five strata (A–E; Fig. 23.9) on the basis of associated cultural material, and stratum F at the base was identified as sterile calcretised bedrock. A suite of AMS radiocarbon dates has provided the most secure timeframe to trace the development of a distinctive "Late Palaeolithic" industry, behavioral modernity, and the first example of well-dated Pleistocene human fossils in India. The proliferation of the microlithic industry, beads of stone and shell, worked bone (harpoon and awl) is unique to this site.

Neolithic to Iron Age cultural materials were documented from the upper two strata, A and B, belonging to the Late Holocene period. Similar evidence is documented from a series of rock shelters in the region. Stratum C comprised calcium carbonate encrusted sediment, with a stone feature indicating structural activity. AMS radiocarbon dating of land snails and bivalve shelves (*Unio* sp.) has provided ages of 20–12 ka to stratum C and around 34 ka to stratum D. There is abundance of microlithic artifacts and mussel shells. Bone fragments are found throughout the stratum, including pieces of worked and cut bone and antler. Lower strata (D–E) are more compact, but show a decrease in the number of lithics, bone and shell.

Fragments of modern human cranium and a tooth reported from here come from stratum C, bracketed by ages of 20 and 12 ka cal. BP. These fragments show calcined marks, indicating cremation practices. A total of 2732 vertebrate animal remains and 1644 mollusc shells have been recovered from excavations. Wild animals are represented by *Gazella gazelle, Antilope cervicapra, Boselaphus tragocamelus, Sus* sp., *Tetracerus quadricornis* and *Muntiacus muntjak*. Small- and medium-sized ungulates dominate the assemblage. Only a few bones are derived from carnivore and nonhuman primates. Change in the macrovertebrate assemblage through time reveals shift in habitat preference from Strata D to C, suggesting broad environmental changes in the region during the later Pleistocene shift from MIS3-2. A solitary example of uniserial harpoon was found around the same depth as the initial proliferation

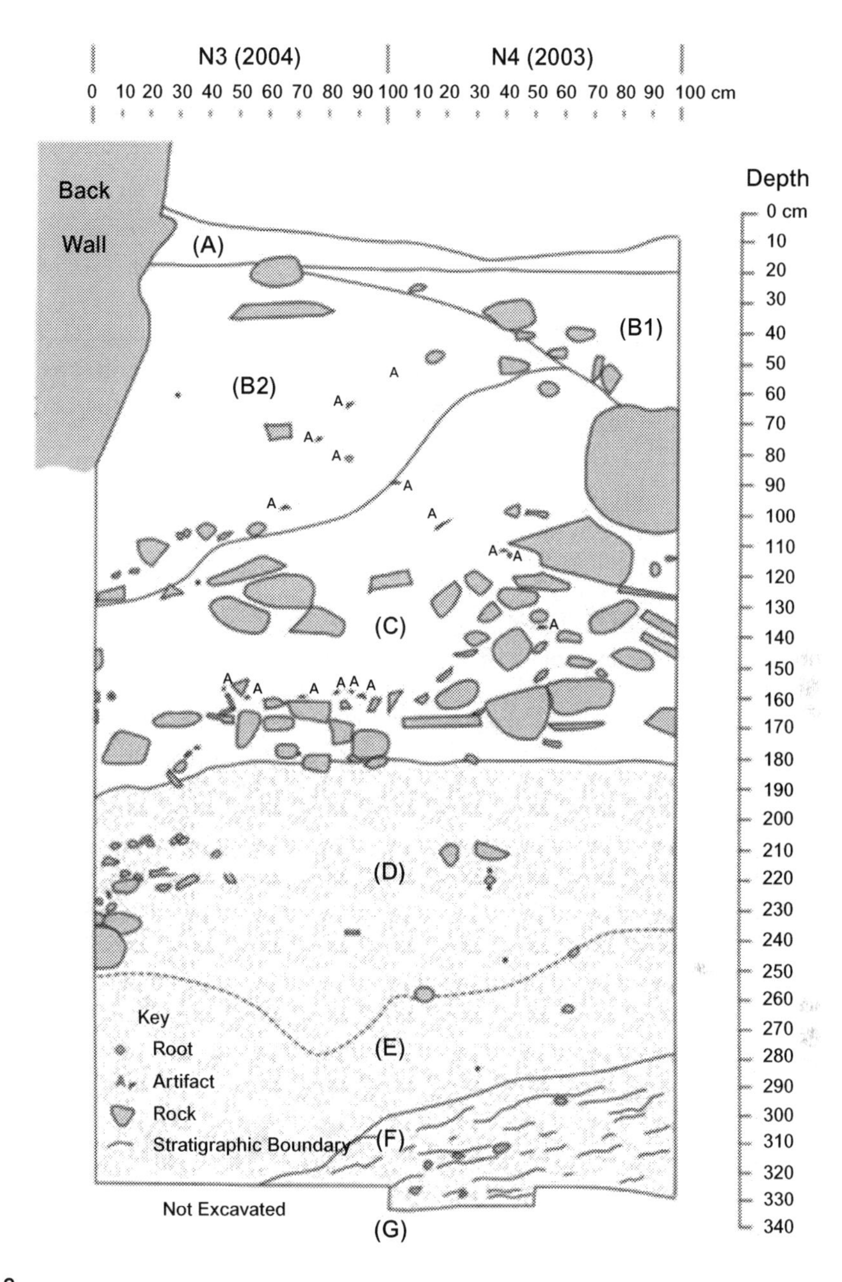

FIGURE 23.9

A 1-m-wide trench was dug to a depth of nearly 4 m, where a calcretised sterile bedrock was encountered at the base of the rock shelter JWP9. Excavation was carried out in spits and based on material cultural remains the strata were subdivide into A–G, which represent cultural levels of hominin occupation dating from 34 ka (microlithic assemblage) to the Late Holocene Neolithic –Megalithic periods (2000–500 BC). Source: The Kurnool District Archaeological Project.

of backed artifacts. The mollusc assemblage record from the site reveals the intensity of site use, with emphasis on freshwater bivalves collection through the period of occupation at the rock shelter. Large lands nails constituted an important component of diet during the much of periods covered by strata D and C. The oldest known stone beads are documented from Stratum C—25 limestone and bone beads, along with stone drills and a suspected bone awl. A large number of bead blanks are also recovered. A striated red ochre crayon has also been found at the interface between strata C and D (Clarkson et al., 2009).

23.7 The Middle and Late Palaeolithic typology and technology

23.7.1 Middle Palaeolithic assemblages

More than 2000 Middle Palaeolithic artifacts were collected from pre-and post-YTT lithounits from excavated sites in the valley bottom. The attribute analysis and typological variation between pre- and post-YTT contexts have revealed little or no change in the lithic technology. Change is documented from assemblages dating from about 38 ka. A gradual decrease in the size of the flakes and preference to crypto-crystalline silica rocks and tendency toward microlithization occurs, as revealed by the assemblage from rock shelter site of JWP9 rock shelter (Fig. 23.10).

The assemblage is devoid of diminutive handaxes that are known to occur in other parts of southern India. If present they constitute less than 0.5%. The assemblage is dominated by prepared cores, expedient cores, retouched flakes, scrapers, and points. Limestone is the dominant raw material (95%), followed by dolerite (hammer stones), chert, quartzite, chalcedony, quartz, etc. The core types include Levallois, multi- and single-platform cores, bidirectional, discoidal and semi-discoidal cores, facetted unidirectional core, and micro-blade core (post-YTT). Flakes have the mean length of 23 mm, possess dorsal ridge, and retain cortex. Platforms are typically large revealing the use of hard-hammer percussion. Retouched flakes can be broadly classified as scrapers and notches. Points include examples of deliberately tanged pieces. These are common in the Indian Middle Palaeolithic assemblages (James and Petraglia, 2005). It has been observed that the core reduction technologies are diverse in India. Both Levallois and discoidal techniques were most common.

The post-YTT context artifact assemblage at Locality 3, for instance, comprises largely multiplatform cores, faceted single platform cores, and core fragments. The dominance of limestone as a chief raw material is greatly reduced and one observes increasing emphasis on chert, and chalcedony makes its appearance. The metrical characteristics of the flakes are similar to those found below the YTT. At locality 21, chalcedony dominates the assemblage followed by chert and limestone. Flake retouching has resulted in a high proportion of side and end scrapers (Petraglia et al., 2009b; Haslam et al., 2010b, 2010c; Jones, 2010).

The pre-YTT assemblages revealed preference for larger flakes with lateral secondary working on the ventral surface. The post-YTT assemblages revealed retouching on the distal ends. However there is overall strong evidence of little technological change between pre- and post-YTT assemblages. There is no observable change in the shape of the flakes and reduction or retouch intensity and attest to cultural continuity with same population continuity despite the impact of the air-fall tephra. Obviously the severity of the super-eruption appears to have little impact on the continuity of human population equipped with the Middle Palaeolithic technology. Several scholars have suggested that

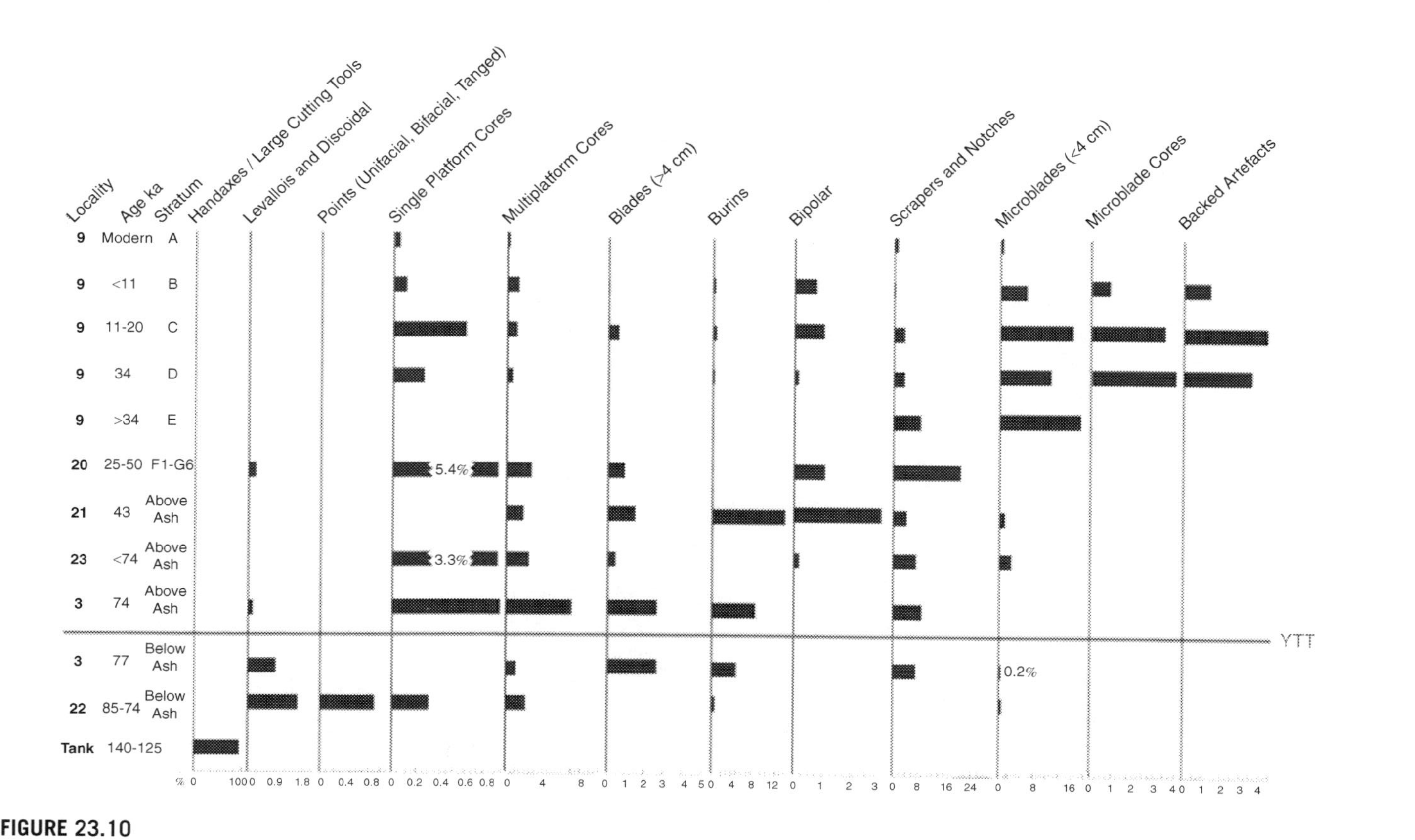

FIGURE 23.10
Summary of long-term technological change in the Jurreru Valley from 140 ka. Source: The Kurnool District Archaeological Project.

post-YTT volcanic winter, coinciding with the onset of MIS4 cooling phase, led to changes in terrestrial environments, hominin populations, and changes in social and ritual behaviors in them (Shea, 2008). Both the pre- and post-YTT assemblages are characterized by nonstandardized typologies as is common with the Middle Palaeolithic assemblages. Prepared core technology for flake removal is predominant. Radial flaking is common and flakes show different forms of preparation. Burins, microliths, and blades though present constitute a minority. Utilized pieces of hematite are also present in the pre-YTT levels along with the artifacts of the Middle Palaeolithic.

23.7.2 Late Palaeolithic Microlithic technology

More than 53,000 microlithic artifacts have been recovered from JWP9. The artifact deposition is near unimodal over the depth of the site, with a pronounced peak of deposition between 1.9 and 1.3 m, bracketed by AMS dates of 20 and 12 ka. Microblades are dominant from the base to the top and therefore form a continuous record from prior to 34 ka to until the Holocene. Backed artifacts and burins appear after 34 ka and remain at peak until 20 ka. Preliminary use-wear and residue analysis of the mircoblades have shown that backed artifacts were hafted using a resinous mastic.

Single-platform cores and bidirectional and heavily rotated cores are present. In the lower levels, retouched scrapers and notches are common and decrease upwards in the stratigraphy. Clarkson et al., (2009) observe clear technological change in the production of backed artifacts; asymmetric backed artifacts including trapezes and triangles are predominant between 2.2.0 and 1.80 m depth; symmetric artifacts including lunates begin to increase between 1.80 and 1.50 m. Points produced by backing the lateral margins are only present in stratum C (at depth of 1.82–1.10 m). Microblade production also shows change through time, suggesting a regional variation in their production. It is observed that ridges on microblades increased through time, "from an average 1 ridge in Stratum E to around 1.5 in Stratum B. Microblades become narrower through time. As mircoblades also become narrower over time (median width of microblades reduces from 8.6 to 7 mm over the sequence), increasing numbers of dorsal ridges suggest blade removals became over closely spaced over time. This is supported by a rise in the proportion of elongated parallel flake scars on microblade cores from around 40% to over 60% of scars through time" (Clarkson et al., 2009, p. 340). The use of chert, chalcedony, and crystal quartz is greatest in Strata C and D.

23.8 Neolithic Kunderu Tradition during the Late Holocene

During the Holocene, the South Deccan in general and the Kurnool Basin in particular are predominantly hot and semi-arid, and lie in the rain shadow of both summer and winter monsoon circulation. It is largely a plain open country until the western margins of the Kurnool Basin, represented by the Erramalai Hills. Both topography and hydrological conditions dramatically change from lower rainfall in the Erramalai range to relatively higher rainfall toward the Nallamalai range. The undulating topography, rainfall, and groundwater movement have rendered the Kurnool Basin a relatively well-watered region, characterized by higher vegetation diversity and animal biomass. The vegetation association is represented by *Anogeissus latifolia* and *Hardwickia bipinnata*, and grades to *Tectona grandis* and *Terminalia tomentosa* toward the Nallamalais. This was perhaps the reason why hunter–gatherer way of life survived much longer than the region to the west of the Erramalai range, that is, the region of Ashmound Tradition Neolithic.

More than 1200 Neolithic sites in the southern states of Telangana, Andhra Pradesh, Karnataka, and Tamil Nadu have been documented so far (Korisettar et al., 2002). Recent archaeological research on the Southern Neolithic has identified three distinctive spatially contiguous traditions in the developmental stages of agro-pastoral economies between 3000 and 1200 BC, in the larger Southern Neolithic tradition. They are (1) the Ashmound Tradition, the earliest among them, (2) the Kunderu Tradition, and (3) the Hallur Tradition. These three traditions are chronologically well constrained by a suite of AMS radiocarbon dates (Fuller et al., 2007; Korisettar, 2021). The Ashmound Tradition began earlier than the other two and survived for almost two millennia. These three traditions reveal successful adaptation to the overall semiarid environments of the Late Mid- and Early Late Holocene fluctuating monsoon regimes. A systematic archaeobotanical sampling and laboratory identification of a suite of food crops, both native and introduced, have been identified from a series of Neolithic sites along an east–west transect between the Western and Eastern Ghats. This has considerably enhanced our understanding of the origins of agriculture as well as documenting evidence for identifying the Ashmound Tradition as an independent center of agricultural evolution during the period between the early third and mid-second millennium BC (Fuller et al., 2007).

Neolithic sites of the Kunderu Tradition were first identified by a distinctive ceramic "Patpadu Ware," and later it was identified as Variant 5 among regional ceramic traditions of the Southern Neolithic (Fuller et al., 2001; Fuller et al., 2004; Allchin, 1962). The Neolithic–Megalithic sites are located in the valleys drained by the river Kunderu and its tributaries, caves, and rock shelters. A series of Neolithic sites along the Kunderu and its tributary streams, including Chintalapalli, Hanmantharaopeta, Injedu, Mandlem, Pandipadu, Peddamudiyam, Tangatur, Rupanagudi, and Singanapalle, were documented by Venkatasubbaiah (2007). A series of painted rock shelters and caves revealed evidence of human occupation dating from the Late Pleistocene to Late Holocene time periods. The latter included a number of Neolithic–Megalithic sites. Along the southern flanks of the Jurreru Valley, on the foot hill of the Eddulakonda, two Neolithic sites were designated as Patapadu I and II, Mogasarayanigondi II and Deyyapu Gundu. Rock shelters along the northern flanks of the Valley, including JWP9 and JWP12, revealed Neolithic remains (Fig. 23.9). Evidence for Neolithic occupation of the caves was represented by potsherds of "Patpadu Ware," microliths and rock art at Billa Surgam and Sanyasula Gavi 1 and 2, in the northern part of the study area. Direct AMS dates were obtained for mung bean from Sanyasula Gavi (1900–1800 BC) and rock art at Billa Surgam (2200 BC) (Fuller et al., 2007; Tacon et al., 2013).

Archaeobotanical data obtained from a number of these sites indicated the cultivation of two pulses: *Macrotyloma uniflorum* (horse gram) and *Vigna radiata* (mung bean); and two millets: *Brachiaria ramosa* (browntop millet) and *Setaria verticillata* (foxtail millet). Evidence for sawa millet (*Echinochloa colona*) and black gram (*Vigna mungo*) was also reported. A number of fragments probably of gathered fruits are also known, including ber (*ziziphus*), *jamun* (*Syzygium cumini*), cucumbers (*Cucumis* cf., *prophetarum*), etc. Faunal remains included cattle bones in predominance, sheep and goat in minority, and occasionally chicken bones were also identified, suggestive of adaptation to the semiarid grassland ecosystem of the southern Deccan plain open country. The majority of the cultivated crops suggest their cultivation during the summer monsoon and with the addition of wheat and barley winter cultivation ensued. Although there is no conclusive statement about the processes leading to the emergence of the Kunderu Tradition, the available radiocarbon chronology shows nearly 400 year time gap between this and Ashmound Tradition during which period trade relations may have been operating gradually leading to the emergence of agricultural way of life, but with a different settlement system (Fuller et al., 2001).

23.9 Challenges and solutions

The first report of the occurrence of stratified YTT in the alluvial sediments of the Son Valley, in the Vindhya Basin was made by Williams and Clarke (1984) . Following this Basu et al., (1987) extended their survey to the Narmada Valley. Its discovery in the Kukdi Valley in the upland western Deccan Volcanic Province by Korisettar et al., (1989a, 1989b) led to documenting its presence in a number of stratified alluvial deposits across the Peninsula (Acharyya and Basu, 1993). It is recognized as an isochron and marker bed in the Late Quaternary deposits and associated Palaeolithic assemblages in India (Westgate et al., 1998; Petraglia et al., 2007; Blinkhorn et al., 2014; Westgate and Pearce, 2017; Anil et al., 2019). It is best preserved in low-energy lacustral sediments or low-energy fluvial deposits of the river basins as well as on topographic depressions of the plateau (Owk Plateau, Kurnool Basin) of hill ranges in the Kurnool Basin (Blinkhorn et al., 2014).

The presence of the YTT in diverse geomorphic contexts across the Indian Peninsula and its association with Palaeolithic assemblages, for the first time, provided scope for establishing the absolute timeframe for the Palaeolithic succession in India. Although different dating laboratories produced divergent dates for the same deposit (Korisettar et al., 1989a, 1989b; Horn et al., 1993; Shane et al., 1995; Westgate et al., 1998; Westaway et al., 2011; Gatti, 2013; Mark et al., 2014) it has now been resolved that all the tephra deposits in India belong to the 74 ka super-eruption of Toba caldera (Westgate and Pearce 2017).

The YTT occurs interstratified with Palaeolithic artifacts, according to some with the Acheulian (Gaillard et al., 2010; Jangra et al. 2018). Does this imply the young age of the Acheulian (74 ka) or the stratigraphic relationship between the Acheulian conglomerate and the superimposed tephra needs to be carefully assessed taking into consideration the high-energy fluvial environments in the Kukdi (Kukadai) Valley and set in an dominantly erosional basin not only at Bori but also at other tephra-associated Acheulian sites in upland western Deccan Volcanic Province. Westgate and Pearce (2017) have presented specific data to establish the identity and age of the Bori tephra to be YTT, which is missing in the statements made by Jangra et al., (2018). Information on the depositional environments, pretephra landscape, grain size data of the sediments, biotite composition, and the sedimentary features of the tephra would have certainly strengthened their argument. The Jurreru Valley studies have clearly established the difference between the primary and reworked tephra and its stratigraphic association with the Middle Palaeolithic industry preserved in a low-energy depositional environment (Petraglia et al., 2012a). Similarly Vijaya Kumari et al., (2017) mention the occurrence of 5- to 6-m-thick YTT in the Sagileru Valley, without describing the sedimentary features of the YTT and associated Palaeolithic assemblages. Their work leaves much to be desired in the light of the studies of Blinkhorn et al., (2014) and Geethanjali et al., (2019). Evidently the vast majority of documented stratified Middle Palaeolithic sites in the Kurnool Basin are associated with the YTT (Blinkhorn et al., 2014; Anil et al., 2019; Clarkson et al., 2020; Anil et al., in press).

23.9.1 Dating the Middle and Late Palaeolithic in the Jurreru Valley

Armed with multiple dating methods, an absolute time-frame for the stratified lithic assemblages and Pleistocene fauna was successfully accomplished. Sediments below and above the YTT were sampled for OSL dating from all the excavated localities. Dating was crucial to ascertain the age of the YTT as well as the interstratified Middle Palaeolithic assemblages. At Locality JVP3A-200, the Middle Palaeolithic assemblage occurs at a depth of 1.8 m beneath the YTT. The OSL date of 77± 6 ka was determined for the lower time limit for the assemblage. The post-YTT Middle Palaeolithic assemblage at the same locality occurring at 1.1 m above the YTT horizon has been OSL dated to 74 ± 7 ka, thus confirming both the age of the YTT and the interstratified Middle Palaeolithic assemblages. Additional post-Toba assemblages have also been dated to 34 ± 3 ka suggesting its upper time limit in the Jurreru Valley. Rock shelter excavations at JWP9 have revealed the beginning of microlithization around this time (Fig. 23.10). At locality JWP3, the pre-YTT Middle Palaeolithic assemblages is dated to 85± 6 ka. This is the earliest dated occupation floor in the Jurreru Valley.

The uninterrupted Palaeolithic succession in the Jurreru Valley clearly reveals the continuity of hominin occupation right from the Acheulian to the Iron Age through the Late Palaeolithic and Neolithic. Faunal continuity in the Billa Surgam environs supports the stability of the environment during the last 200 ka (Roberts et al., 2014). The composite lithostratigraphy of the Quaternary comprises the piedmont colluvium containing Large Cutting Tools, being the basal lithounit, upon which the litho-sequence seen at Locality 3 completes the Pleistocene stratigraphy. Holocene or sub-recent alluvium is not well preserved, but wherever preserved it is represented by a dark-brown clayey sediment containing microlithic artifacts of the Holocene age. The later Pleistocene calcareous alluvia and the cemented rock shelter lithounits (Strata E, D, and C) are of a comparable age, with more or less similar lithic assemblages.

The Jwalapuram Tank large cutting tools are similar to other Acheulian assemblages found across south India. The artifacts still retain sharp edges indicating a minimal displacement from the original place of discard. It is probable that this cultural material lies deeply buried underneath the alluvium. Recent dating of the Acheulian site at Attirampakkam bearing similar artifact assemblages has placed the earliest colonization of the subcontinent at >1.5 Ma (Pappu, et al., 2011). Its late survival has been documented at Jwalapuram (140 ka) and Son Valley (130 ka) (Haslam, et al., 2011).

The presence of well-stratified YTT in the alluvial sequence containing Middle Palaeolithic artifact assemblages is unique to the Indian subcontinent. While the YTT constitutes an important isochron, environmental change and its impact on hominin populations has also been debated (Ambrose, 1998; Balter, 2010; Jones 2007a, 2007b, 2010, 2012). However, the continuity of human populations after the Toba eruption, as reflected in the continuity of Middle Palaeolithic technological traditions, indicates that hominin populations in the Jurreru Valley, lying at the distal end of the ash fall, survived the so-called volcanic winter. This vital evidence also suggests modern humans were already in the subcontinent before the ash fall event (Petraglia et al., 2007). However, changes in drainage pattern and local geomorphic changes are documented in the sedimentary record. The lacustral environments that existed prior to ash fall were filled up with ash, which led to drainage pattern changes and local landform changes, are documented in the sedimentary record (Haslam et al., 2011). The Middle Palaeolithic hominins response to the changed geomorphic environments is reflected in the alteration in the tool-type frequencies and introduction of new lithic raw materials. OSL dates have established a time span between 85 ka and 38 ka at Jwalapuram.

23.9.2 Are the Pre-YTT Middle Palaeolithic assemblages in secondary context?

At Locality 21, excavation revealed a 3.2-m-deep section with six distinctive strata (A–F) above the YTT stratum. An OSL date of 38 ka has been obtained for this deposit containing a total of 131 Middle

Palaeolithic artifacts. This discrepancy in the age of post-YTT deposit between Localities 3 (74 ka) and 21 (38 ka) led to questioning the primary context of Middle Palaeolithic assemblages, as the artifacts came from sands and gravels and that they are in fluvial context (Williams et al., 2009, 2010; Petrglia, et al., 2012b). Their objections have been well addressed by Haslam and Petraglia, (2010) and Petraglia et al., (2012b). Unlike the alluvial context YTT sites in central India the Jurreru Valley sites are part of low-energy deposits in a stable and lacustral topographic setting, with a gentle gradient toward the southern foot hills.

The pre-YTT land surface documented by Blinkhorn et al., (2012) clearly indicated the unevenness of the valley bottom topography along the antiformal axis of the valley, the depth of such topographic depressions could not be reached owing to high water tables. Obviously such deeper depressions in the past were fed by springs along the fault line and geological contacts between quartzites and limestones along sides of the steep escarpments flanking the valley. Bed rock was not reachable at none of the excavated localities. Existence of deep water lacustral topographic features in the past and that the lake shores were areas of human activity is a distinct possibility. The Middle Palaeolithic assemblage from Locality 3 was associated with a thin sandy-pebbly subangular gravel, locally derived, under short episodes of lake-level fluctuations. The valley bottom dramatically widens out to the east of Jwalapuram, where the river becomes much more intermittent. It was largely a closed lake basin, surrounded by topographic highs, during most of the Pleistocene and opened up, following filling up of these depressions by the lacustrine silts and sands. Drainage reorganization and channel incision during the Terminal Pleistocene wet phase have partially exposed the lacustrine sediments on the valley bottom. The present Jurreru channel pattern is not older than the Terminal Pleistocene.

Further evidence of the primary context of Middle Palaeolithic sites associated with YTT comes from the contiguous river valleys: (1) Sagileru Valley between Nallamalai and Velikonda ranges and (2) the Gundlakamma, Maneru, and Paleru to the east, northeast of the Nallamalai Hills on the East Coast of the Peninsula. A succession of Palaeolithic cultures has been well documented from these river valleys early on (Reddy and Sudarsen 1978; Rao 1979, 1987). The post-YTT Middle Palaeolithic assemblages in the Sagileru Valley are found to be blade dominated (Blinkhorn et al., 2014). Current research in the Gundlakamma Valley (Fig. 23.11) has also revealed similar continuity of Middle Palaeolithic assemblages in the pre- and post-YTT contexts. According to Anil Devara (personal communication on 12 January, 2021; Anil et al., 2019) the post-YTT assemblages from the Gundlakamma Valley are similar to the Jurreru Middle Palaeolithic assemblages and are characterized by unidirectional and bidirectional Levallois core reduction technology. He has also observed disconformity between the YTT and the overlying fine sediment, indicating only a little depositional time gap existing among these layers.

23.9.3 When did the Middle Palaeolithic technology originate?

Absolute dates for the Middle Palaeolithic are now available from YTT contexts from the Jurreru, Gundlakamma, and Sagileru Valleys and from non-YTT contexts at Attirampakkam and Paleru Valley by OSL, Thermoluminescence (TL), Tephrochronology, and pIR-Infrared Stimulated Luminescence (IRSL) methods. While the Jurreru, Sagileru, and Gundlakamma contexts are constrained by the YTT, Attirampakkam assemblages are constrained by TL (395–172 ka) and the Paleru Valley assemblages are constrained by pIR-IRSL (200–250 ka), projecting much deeper antiquity to the beginning of the Middle Palaeolithic. These results are posing challenges to the existing theories of origins of anatomically

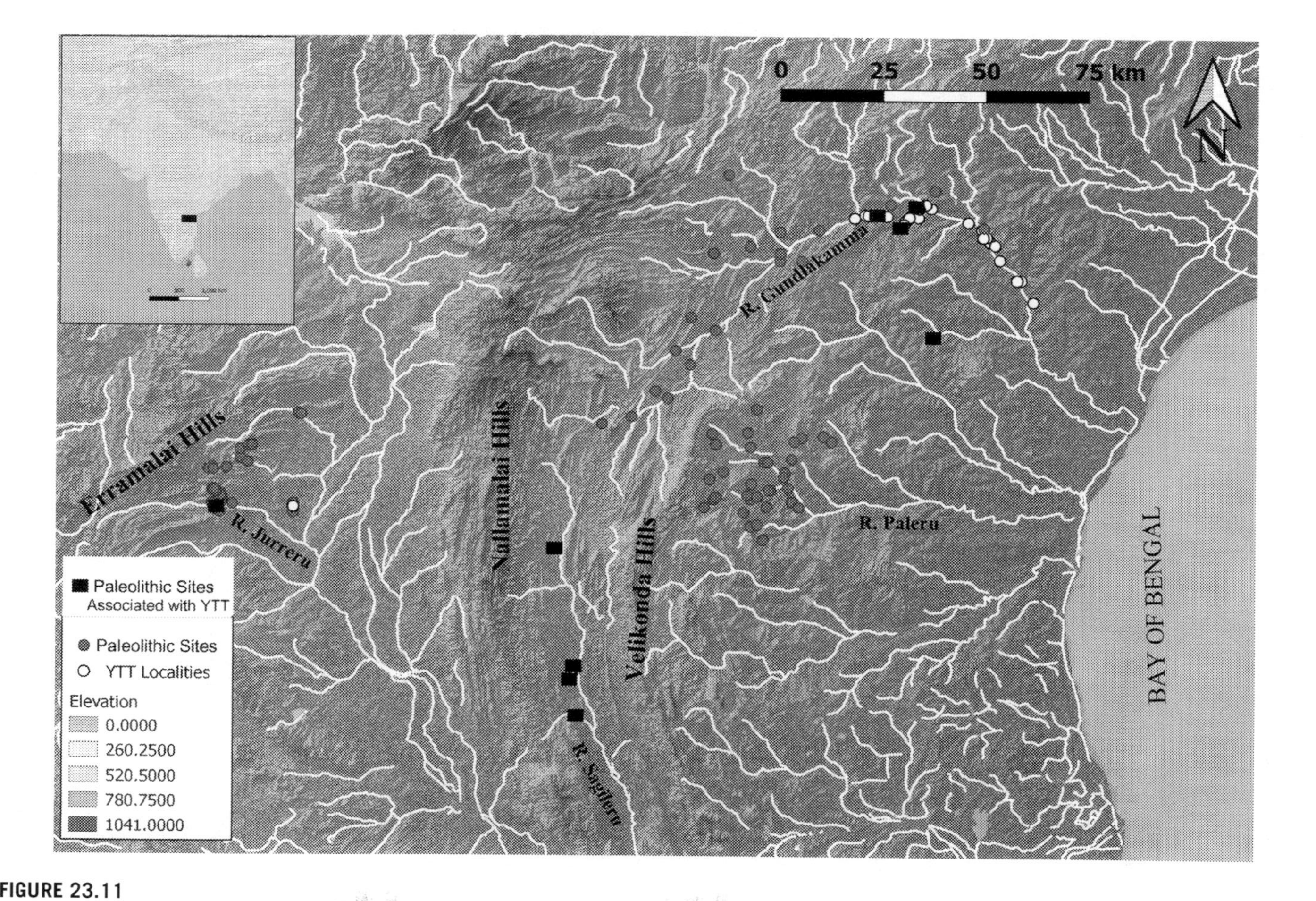

FIGURE 23.11

The three contiguous regions, two in the Cuddapah Basin (A) between Erramalai and Nallamalai ranges, (B) between Nallamalai and Velikonda ranges, and (C) the East Coast plains were previously known for the Paleolithic settlements. The fallout of the Jurreru Valley investigations has led to identifying YTT interstratified sites in the previously investigated areas. Prepared by the author and Anil Devara, The M.S. University of Baroda, Vadodara.

modern humans and their exit from Africa. Future research needs to reckon these dates from India and assess our current understandings. The big question about the makers of the Middle Palaeolithic is as yet difficult to answer.

23.10 Post KAP developments in Indian prehistory

The OSL, TL, and IRSL dating of Acheulian bearing alluvial sediments at Bamburi 1, Nakjhar Khurd, and Patpara in the Middle Son Valley (Vindhya Purana Basin) has placed the upper time limit of the Late Acheulian around 140 ka (Petraglia et al., 2012a). This corresponds with the similar age for the late Acheulian at Jwalapuram (Kurnool Purana Basin) and several other sites in India (Haslam et al., 2011). On the other hand the earliest date for the Acheulian at Attirampakkam (Upper Gondwana formations) and Isampur (Bhima Purana Basin) is placed at 1.5–1.7 Ma and 1.2 Ma, respectively (Pappu et al., 2011; Paddayya et al., 2002). These time brackets indicate the long survival of the archaic hominins till the end of MIS6 and clearly attest that the refugia, such as the Purana-Gondwana Basins, were governed by favorable paleoenvironments throughout the Lower and Middle Pleistocene and that both archaic and modern hominins coexisted till the latter expanded beyond the refugia during and after MIS5 interglacial.

Although the Middle Son Valley has received much attention of Quaternarists and archaeologists since the early work of Martin Williams and M.F. Clarke (1984) and the joint archaeological research by G.R. Sharma and J.D. Clark (1983) resulting in the discovery of a range Palaeolithic to Neolithic sites, it became the focus of renewed interdisciplinary research stimulated by the findings from the Jurreru Valley in the Kurnool Basin (Petraglia et al., 2012a). The OSL and luminescence chronology of the Quaternary formations and excavations of associated lithic assemblages helped refine the relationship among Quaternary formations, archaeological sites, and the YTT. Unlike in the Jurreru Valley the YTT deposits in the Son Valley occur as isolated and discontinuous beds not directly associated with the Middle Palaeolithic assemblages.

In a significant development the beginning of the emergence of Middle Palaeolithic technology at Attirampakkam in Tamil Nadu has been dated to 395,000 years ago. It has considerably pushed back the beginning of the Middle Palaeolithic (Akhilesh et al., 2018). They suggest a time bracket from 395,000 to 172,000 years ago for Middle Palaeolithic technology and recognize substantial behavioral changes as compared with the preceding Acheulian large-flake technology. Furthermore the Gundlakamma Valley pre-YTT Middle Palaeolithic assemblages are showing a dating range from 200 to 246 ka (personal communication: Anil Devara 12 January, 2021). Implications of these dates are not clear, yet they clearly suggest the possibility of older Middle Palaeolithic assemblages deeply buried in the pre-YTT lithounits at Jwalapuram. These developments have led to a rethinking of the current Out of Africa models as well as the direct association of modern humans with the rise of Levallois technology outside of Africa.

The publication of 35 ka date for the beginning of microlithization at Jwalapuram by Clarkson et al., (2009) came in as a flash in the pan and accorded worldwide recognition to the early dating of microliths from Patne (Sali, 1989) in the Tapi Valley of Deccan Volcanic Province, that had remained a sleeping beauty till then. This stimulated the OSL dating of microlithic contexts at some of the previously known sites in India. Dates earlier than JWP9 were determined for Mehtakheri (48 ka) (Mishra et al., 2013), Mahadebbera (34–25 ka), and Kana (42 ka) (Basak et al., 2014). These dates are comparable to the microlithic sites at Patne (28 ka) and are in conformity with the dates for the oldest microliths at

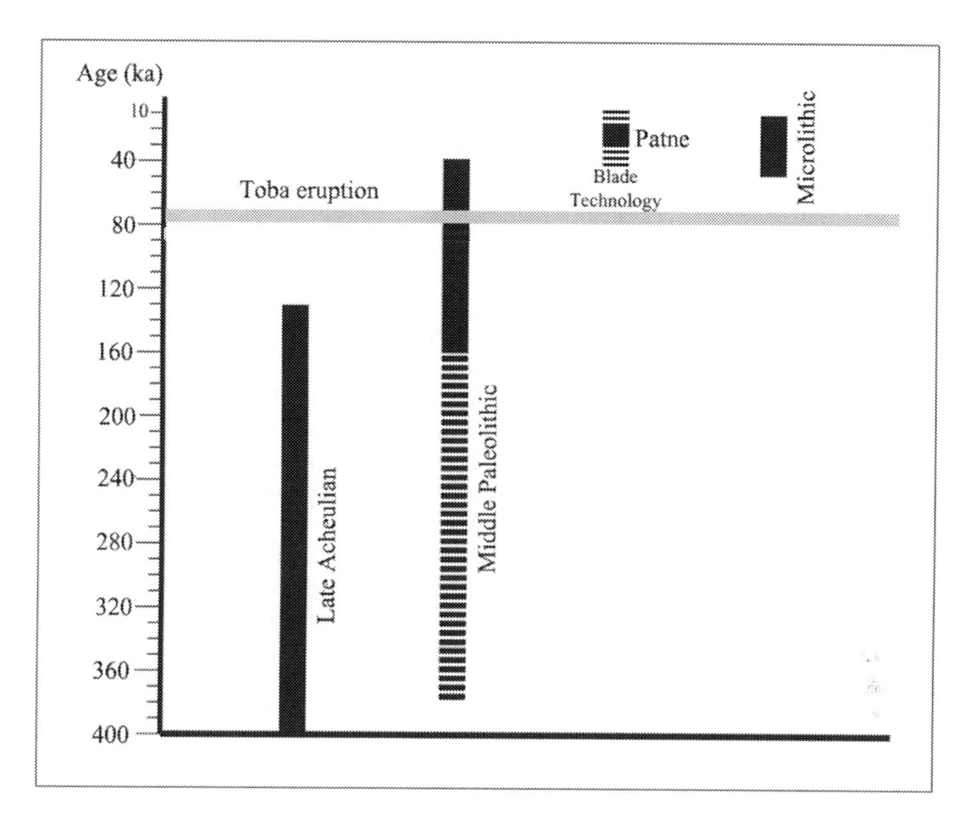

FIGURE 23.12

Chart showing temporal overlap between Late Acheulian and Early Middle Palaeolithic, pre- and post-YTT continuity of the Middle Palaeolithic and early beginning of microblade technologies as documented from Palaeolithic sites on the Indian Peninsula. Prepared by the author and Anil Devara, The M.S. University of Baroda, Vadodara.

Fa-Hien Lena cave (38 ka; Perera et al. 2011) in Sri Lanka (now revised to 48 ka; Wedage et al., 2019) (Fig. 23.12). These dates have implications for early microlithization in the subcontinent and represent convergence of technologies in the Old World context as an adaptive response to both demographic and climate deterioration.

23.10.1 Out of Africa into the Indian Subcontinent: the organized search for hominin remains

Post 1987 paradigm shift (Cann et al., 1987) focused on molecular analysis of hominin fossils from African and Eurasian sites. Reinvestigations at a number of paleoanthropological sites and aDNA studies (on Neanderthals and modern human fossils) focused on human ancestry and population histories in different geographical environments of the world. This necessitated multipronged investigations into the archaeological, paleoanthropological, palaeontological, palaeoenvironmental, paleoecological, geochronological, and paleoanthropological context of hominins preserved in a variety of geomorphic contexts. The Kurnool Basin geoarchaeology was one of these renewed investigations. The impact of these investigations led to discovering Palaeolithic contexts associated with the YTT in the contiguous valleys (Fig. 23.11; Blinkhorn et al., 2014; Anil et al., 2018, 2019).

Jwalapuram rock shelter (JWP9) is one of the deeply stratified rock shelter sites in India. AMS dating of microlithic levels indicated that microlithic technology developed from the previous Middle Palaeolithic tradition between 38 and 35 ka (Clarkson et al., 2009; Petraglia et al., 2009a). The proliferation of microlithic technology during the time period 35–12 ka also coincides with significant demographic increase in the subcontinent (Petraglia et al., 2009a). The early emergence (70 ka) of microlithic technologies and symbolic artifacts in Africa led to postulating coastal dispersal of anatomically modern humans around 45 ka, coincident with mtDNA coalescent ages (Endicott et al., 2007). On the other hand the continuous development of Middle Palaeolithic to early microlithic technologies revealing in-situ development of microlithic technology and symbolic behavior in India has generated an alternative perspective on the timing of the exit of anatomically modern humans out of Africa as well as the emergence of modern human behavior. These datasets opened up an argument for the Middle Palaeolithic expansion of modern humans out of Africa around MIS5 as against post-Toba microlithic expansion (Mellars, 2006a, 2006b; Clarkson et al., 2017). Haslam et al., (2017) discuss the issue of modern human dispersal of modern humans out of Africa in terms of failed and successful dispersal events and agree with the genetic and microlithic model of post Toba expansion and suggest an Indian staged dispersal model, as an alternative to a two-stage dispersal model (Mishra et al., 2013).

Alongside the debate on timing of dispersal out of Africa, the routes of dispersal out of Africa into the Indian subcontinent have also received considerable attention. Korisettar (2007, 2015, 2016, 2017) has argued for inland routes of dispersal taking into consideration the geoenvironmental constraints, limited access to high-quality lithic resources, along the western coastal landforms and spatial and temporal variation in the distribution of Palaeolithic and microlithic industries in India and Sri Lanka. The issue of emergence of microlithization, the routes of dispersal, and the geographical importance of the Indian subcontinent based on the cultural and environmental datasets from the region have been addressed by Clarkson et al. (2017). While they disagree with the two-stage dispersal model (Mishra et al., 2013), they are noncommittal on the suggested Indian staged dispersal model (Haslam et al., 2017). They reiterate the pre-Toba entry of modern humans into the subcontinent and present the evidence of gradual evolution of microlithic technologies as a response to demographic increase and decreasing monsoon regimes during the Late Pleistocene, beginning MIS3, in terms of an adaptive model. Homogeneity among the members of the KAP is reflected in generating the critical datasets and at the same time heterogeneity in the way these datasets are interpreted by them is the hallmark of an eclectic approach to place the subcontinent at the center of global debates.

23.11 Learning and knowledge outcomes

A decade-long and ongoing investigations carried out in the Kurnool Basin have made significant contribution to Indian prehistory. The first successful application of the OSL dating of the Indian Palaeolithic succession is the hallmark of these investigations (Petraglia et al., 2007). A series of stratified Lower, Middle and Late Palaeolithic, Mesolithic, Neolithic, and Iron Age-Early Historic sites were excavated (Petraglia et al., 2009b). The highlight of these investigations is the implications of core reduction technological similarities between African and Indian dated Middle Palaeolithic industries suggesting a much earlier expansion of modern humans into the Indian subcontinent (MIS5 interglacial). The technological analysis of the long-enduring microlithic assemblages (dated to >38 ka) from rock shelter excavations indicated in situ evolution from the Middle to Late Paleolithic as an adaptive response to the onset of MIS4 drier conditions (Petraglia et al., 2009a). Further, the rock shelter excavations have revealed evidence for the rise of behavioral modernity represented by personal ornaments, projectiles, production of rock art, and evidence of intentional burial in the Late Pleistocene context, hitherto not reported from any site in India (Korisettar, 2015). The first documentation of hundreds of painted rock shelters with occupational deposits ranging from Late Palaeolithic to Iron Age has shed fresh light on the field study of prehistoric rock art in India (Tacon, et al., 2010; Korisettar, 2018; Koshi et al., 2018). Identification of extinct and living animal fauna from the Billa Surgam cave floor deposits spanning the last 200,000 years (Roberts et al., 2014). The Late Holocene transition to an agricultural way of life was characterized by the domestication of local minor millets and ubiquitous pulses (Fuller et al., 2007). More importantly, the evidence drawn from the Jurreru, Sagileru, Gundlakamma, and Son Valley supports the argument that distal areas of YTT deposition made little impact on local hominin populations and their cultural development during the Late Quaternary. Furthermore there existed a network of refugia across the Indian Peninsula, the Proterozoic Purana, and Gondwana Basins, where hominin populations were not affected by the impact of air-fall tephra as well as long- and short-term climatic changes during the Quaternary (Korisettar, 2007).

In summary it can be stated that the KAP achieved significant success in the areas of: (1) establishing a firm absolute chronology of faunal succession in the Billa Surgam cave record, in terms of extinction and continuity; (2) the application of optically stimulated luminescence dating (OSL) and tephrochronology of both the faunal and archaeological contexts in the Billa Surgam and Jurreru Valleys; (3) resolving the issues relating to the Upper/Late Palaeolithic transition; (4) establishing the usefulness of the YTT as a chronometer; (5) extend the time depth of microlithic technology, through OSL and AMS radiocarbon dating; (6) debating the impact of air-fall tephra and the onset of the so-called volcanic winter on hominin populations; (7) suggesting much earlier expansion of *Homo sapiens* into the Indian subcontinent with Middle Palaeolithic technology; (8) the new dates for the early evolution of Levallois technology from Attirampakkam have added a new dimension to the out of Africa debate; (9) documenting hundreds of painted rock shelters through total station survey; (10) uninterrupted continuation of hominin occupation of the Kurnool basin both during pre- and post-YTT event and (11) identifying a regional Neolithic Kunderu Tradition and its chronology. The region has proved its potential for further research and has opened up opportunities for penetrating deep in time.

Acknowledgments

I thank Navnith Kurmaran for his forbearance and for inviting me to contribute to this prestigious volume. I appreciate the editorial advice and suggestions from Dr. D. Padmalal. I place on record my appreciation of Anil Devara for helping me in the preparation of maps and sharing information on his ongoing research in eastern Andhra Pradesh. This chapter is, perhaps, a cumulative summary of their outstanding contribution. I place on record my appreciation of constructive comments from the reviewers and editors. The members of the Kurnool District Archaeological Project are remembered for their productive contribution and cordial association.

References

Acharyya, S.K., Basu, P.K., Toba ash on the Indian subcontinent and its implications for correlation of Late Pleistocene alluvium, Quat. Res. 40 (1993) 10–19.

Akhilesh, K., Pappu, S., Rajapara, H.M., Gunnell, Y., Shukla, A.D., Singhvi, A.K., Early Middle Palaeolithic culture in India around 385-172 ka reframes out of Africa models, Nature 554 (7690) (2018) 97–101.

Allchin, F.R., Painted pot from Patpad, Andhra Pradesh. Antiquity 36 (1962) 221–224.

Ambrose, S.H., Late Pleistocene human population bottlenecks, volcanic winter and differentiation in modern humans, J. Hum. Evol. 34 (1998) 623–651.

Anil, D., Ajithprasad, P., Vrushab, M., Ravindra, D., Stratified Palaeolithic sites from the upper Paleru river basin, Prakasam district, Andhra Pradesh: a preliminary report, Herit. J. Multidiscip. Stud. Archaeol. 6 (2018) 93–108.

Anil, D., Ajithprasad, P., Vrushab, M., Gopesh, J., Middle Palaeolithic sites associated with Youngest Toba Tuff deposits from the middle Gundlakamma valley Andhra Pradesh, India, Herit. J. Multidiscip. Stud. Archaeol. 7 (2019) 1–14.

Anil, D., Ajithprasad, P., Vrushab, M., Palaeolithic assemblages associated with Youngest Toba Tuff deposits from the Upper Gundlakamma river basin, Andhra Pradesh, India. In: Nupur Tiwari, Vivek Singh and Shashi Mehra (Eds.), Quaternary Geoarchaeology of India. The Geological Society of London Special Publication. London 2021.

Balter, M., Of two minds about Toba's impact, Science 327 (2010) 1187–1188.

Basak, B., Srivastava, P., Dasgupta, S., Kumar, A., Rajaguru, S.N., Earliest dates and implications of microlithic industries of late Pleistocene from Mahadebbera and Kana, Purulia district, West Bengal, Curr. Sci. 107 (7) (2014) 1167–1171.

Basu, P.K., Biswas, S., Acharyya, S.K., Late Quaternary ash beds from Son and Narmada basins, Madhya Pradesh. Indian Minerals 41 (1987) 66–72.

Blinkhorn, J., Bora, J., Koshy, J., Korisettar, Ravi, Boivin, N., Petraglia, M., Systematic transect survey enhances the investigation of rock art in its landscape: an example from the Katavani Kunta valley, Kurnool District. Man Environ. 34 (2) (2010) 1–12.

Blinkhorn, J., Parker, A,G., Ditchfield, P., Haslam, M., Petraglia, M., Uncovering a landscape buried by the super-eruption of Toba, 74,000 years ago: a multi-proxy environmental reconstruction of landscape heterogeneity in the Jurreru valley, south India, Quat. Int. 258 (2012) 135–147.

Blinkhorn, J., Smith, V.C., Achyuthan, H., Shipton, C., Jones, S.C., Ditchfield, P., Petraglia, M., Discovery of Youngest Toba Tuff localities in the Sagileru valley, south India, in association with Palaeolithic industries, Quat. Sci. Rev. 105 (2014) 239–243.

Cann, R.L., Stoneking, M., Wilson, A.C., Mitochondrial DNA and human evolution, Nature 325 (1987) 31–36.

Clarkson, C., Harris, C., Shipton, C., Lithics at the crossroads: a review of technological transitions in South Asia, in: Korisettar, Ravi (Ed.), Beyond Stones and More Stones, The Mythic Society, Bengaluru, 2017, pp. 323–359.

Clarkson, C., Petraglia, M., Harris, C., Shipton, C., Norman, K., The South Asian microliths: Homo sapiens dispersal or adaptive response?, in: Robinson, E., Sellet, F. (Eds.), Lithic Technological Organisation and Palaeoenvironmental Change, Studies in Human Ecology and Adaptation 9 (2017) doi:10.100/978-3-319-64407-3_3.

Clarkson, C., Petraglia, M., Korisettar, R., Haslam, M., Boivin, N., Crowther, A., Ditchfield, P., Fuller, D., Miracle, P., Harris, C., Connell, K., James, H., Koshy, J., The oldest and longest enduring microlithic sequence in India: 35000 years of modern human occupation and change at the Jwalapuram Locality 9 rock shelter, Antiquity 83 (2009) 326–348.

Clarkson, Chris, Harris, Clair, Li, Bo, Neudorf, C.M., Roberts, R.G., Lane, C., Norman, K., Pal, J., Jones, S., Shipton, C., Koshy, J., Gupta, M.C., Mishra, D.P., Dubey, A.K., Boivin, N., Petraglia, M., Human occupation of northern India spans the Toba super-eruption ~74,000 years ago, Nat. Commun. 11 (2020) 961, https://doi.org/10.1038/s41467-020-14668-4.

Endicott, P., Metspalu, M., Kivisild, T., Genetic evidence on modern human dispersals in South Asia: Y chromosome and mitochondrial DNA perspectives: the world through the eyes of two haploid genomes, in: Petraglia, M.D., Allchin, B. (Eds.), The Evolution and History of Human Populations in South Asia, Springer, Amsterdam, 2007, pp. 229–244.

Fuller, D.Q., Boivin, N., Korisettar, R., Dating the Neolithic of south India: new radiometric evidence for key economic, social and ritual transformations, Antiquity 81 (2007) 755–778.

Fuller, D.Q., Venkatasubbaiah, P.C., Korisettar, Ravi, The beginning of agriculture in the Kunderu basin: evidence from archaeological survey and archaeobotany, Puratattva 31 (2001) 1–8.

Fuller, D., Korisettar, R., Vankatasubbaiah, P.C., Jones, M.K., Early plant domestications in southern India: some preliminary archaeobotanical results, Veg. Hist. Archaeobot. 13 (2004) 115–129.

Gaillard, C., Mishra, S., Singh, M., Deo, S., Abbas, R., Lower and early Middle Pleistocene Acheulian in the Indian sub-continent, Quat. Res. 223-224 (2010) 234–241.

Gatti, E., 2013. Geochemical and Sedimentological Investigations of Youngest Toba Tuff ash fall deposits (Ph.D. thesis). https://doi.org/10.17863/CAM.16436.

Geethanjali, K., Achyuthan, H., Jaiswal, M.K., The Toba tephra as a late Quaternary stratigraphic marker: investigations in the Sagileru river basin, Andhra Pradesh, India, Quat. Res. 513 (2019) 107–123.

Haslam, M., Petraglia, M.D., Comment on "Environmental impact of the 73 ka Toba super-eruption in South Asia", by M. Williams et al. [Palaeogeography, Palaeoclimatology, Palaeoecology 284 (2009) 295-314], Palaeogeogr. Palaeoclimatol. Palaeoecol. 296 (2010) 199–203.

Haslam, M., Oppenheimer, S., Korisettar, R., Out of Africa, into South Asia: a review of archaeological ad genetic evidence for the dispersal of *Homo sapiens* into the Indian subcontinent, in: Korisettar, R. (Ed.), Beyond Stones and More Stones, The Mythic Society, Bengaluru, 2017, pp. 117–149. vol. 1.

Haslam, M., Clarkson, C., Petraglia, M., Korisettar, R., Jones, S., Shipton, C., Ditchfield, P., Ambrose, S.H., The 74 ka Toba super-eruption and southern Indian hominins: archaeology, lithic technology and environments at Jwalapuram Locality 3, J. Archaeol. Sci. 37 (2010a) 3370–3384.

Haslam, M., Clarkson, C., Petraglia, M., Korisettar, R., Bora, J., Boivin, N., Ditchfield, P., Jones, S., Mackay, A., Indian lithic technology prior to the 74,000 BP Toba super-eruption: searching for an early modern human signature, in: Boyle, K., Gamble, C., Bar-Yosef, O. (Eds.), The Upper Palaeolithic Revolution in Global Perspective: Essays in honour of Paul Mellars, McDonald Institute for Archaeological Research, Cambridge, 2010b, pp. 73–84.

Haslam, M., Korisettar, R., Petraglia, M., Smith, T., Shipton, C., Ditchfield, P., In Foote's steps: the history, significance and recent archaeological investigation of the Billa Surgam caves in southern India, South Asian Stud. 26 (1) (2010c) 1–10.

Haslam, M., Roberts, R.G., Shipton, C., Pal, J.N., Fenwick, J., Ditchfield, P., Boivin, N., Dubey, A.K., Gupta, M.C., Petraglia, M., Late Acheulean hominins at the marine isotope stage 6/5e transition in north-central India, Quat. Res. 75 (2011) 670–682.

Haslam, M., Clarkson, C., Roberts, R.G., Bora, J., Korisettar, R., Ditchfield, P., Chivas, A.R., Harris, C., Smith, V., Oh, A., Eksambekar, S., Boivin, N., Petraglia, M., A southern Indian Middle Palaeolithic occupation surface sealed by the 74 ka Toba eruption: further evidence from Jwalapuram locality 22, Quat. Int. 258 (2012) 148–164.

Horn, P., Muller-Sohnius, D., Storzer, D., Zoller, L., K-Ar., Fission track- and thermoluminesence ages of Quaternary volcanic tuffs and their bearing on Acheulian artefacts from Bori, Kukdi valley, Pune district, India, Z. Dtsh. Geol. Ges. 144 (1993) 326–329.

James, H.V.A., Petraglia, M.D., Modern human origins and the evolution of behavior in the later Pleistocene record of South Asia, Curr. Anthropol. 46 (2005) S3–S27.

Jangra, B., Deo, S.G., Rajaguru, S.N., Joglekar, J., Sengupta, S., Raphael, J.T., Analyzing and contextualizing the lithic assemblage from the Acheulian site at Bori, Pune District, Maharashtra, Herit. J. Multidiscip. Stud. Archaeol. 6 (2018) 109–128.

Jones, S., The Toba supervolcanic eruption: Tephra fall deposits in India and paleoanthropological implications, in: Petraglia, M.D., Allchin, B. (Eds.), The Evolution and History of Human Populations in South Asia, Springer, Amsterdam, 2007a, pp. 173–200.

Jones, S. A Human Catastrophe? The Impact of the 74,000 Year-old Super-Volcanic Eruption of Toba on Hominin Populations in India. (Ph.D. thesis), University of Cambridge, 2007b.

Jones, S., Palaeoenvironmental response to the 74 ka Toba ash-fall in the Jurreru and Middle Son valleys in southern and north-central India, Quat. Res. 73 (2010) 336–350.

Jones, S., Local and regional scale impacts of the ~74 ka Toba supervolcanic eruption on hominin populations and habitats in India, Quat. Int. 258 (2012) 100–118.

Korisettar, R., Towards developing a basin model for Palaeolithic settlement of the Indian subcontinent: geodynamics, monsoon dynamics, habitat diversity and dispersal routes, in: Petraglia, M.D., Allchin, B. (Eds.), The Evolution and History of Human Populations in South Asia: Interdisciplinary Studies in Archaeology, Biological Anthropology, Linguistics and Genetics, Springer, Dordrecht, the Netherlands, 2007, pp. 69–96.

Korisettar, R., Paradigm shifts in the geoarchaeology of the Palaeolithic of the Indian subcontinent: building a grand Pleistocene cultural sequence, Puratattva 42 (2012) 12–43.

Korisettar, R., Antiquity of modern humans and behavioural modernity in the Indian subcontinent: implications of the Jwalapuram evidence, in: Kaifu, Y., Izuho, M., Goebel, T., Sato, H., Ono, A. (Eds.), Emergence and Diversity of Human Behaviour in Palaeolithic Asia, Garland Publishing Inc., New York & London, 2015, pp. 80–93.

Korisettar, R., Out of Africa into South Asia, in: Schug, G.W., Walimbe, S.R. (Eds.), A Companion to South Asia in the Past, Wiley Blackwell, Chichester, 2016, pp. 60–71.

Korisettar, R., The genus *Homo* and the African exodus, in: Korisettar, R. (Ed.), Beyond Stones and More Stones, The Mythic Society, Bengaluru, 2017, pp. 63 116. vol. 1.

Korisettar, R., Prehisotric rock art studies in India: V.S. Wakankar ad After, in: Korisettar, R. (Ed.), Beyond Stones and More Stones, The Mythic Society, Bengaluru, 2018, pp. 257–297. vol. 2.

Korisettar, R., Ancient agriculture in the Indian subcontinent: the archaeobotanical evidence, in: Hollander, David, Howe, Timothy (Eds.), Companion to Ancient Agriculture, John Wiley & Sons, Inc., Chichester, 2021, pp. 577–610 .

Korisettar, R., Janardhana, B., Jwalapuram, in: Chakrabarti, D.K., Lal, M. (Eds.), History of India: Prehistoric Roots, Vivekananda International Foundation and Aryan Books International, New Delhi, 2014, pp. 424–437.

Korisettar, R., Venkatasubbaiah, Fuller, D.Q., Brahmagiri and beyond: the archaeology of the southern Neolithic, in: Settar, S., Korisettar, Ravi (Eds.), Indian Archaeology in Retrospect, Indian Council of Historical Research and Manohar, New Delhi, 2002, pp. 151–238. Vol. 1.

Korisettar, R., Venkatesan, T.R., Mishra, S., Rajaguru, S.N., Somayajulu, B.L.K., Tandon, S.K., Gogte, V.D., Ganjoo, R.K., Kale, V.S., Discovery of a Tephra bed in the Quaternary alluvial sediments of Pune district (Maharashtra), Peninsular India, Curr. Sci. 58 (10) (1989a) 564–567.

Korisettar, R., Mishra, S., Rajaguru, S.N., Gogte, V.D., Ganjoo, R.K., Venkatesan, T.R., Tandon, S.K., Somayajulu, B.L.K., Kale, V.S., Age of the Bori volcanic ash and Acheulian culture of the Kukdi valley, Bulletin of the Deccan College Postgraduate and Research Institute 49 (1989b) 135–138.

Koshi, J., Koshi, M., korisettar, R., Kumar, A., Masethung, R., Painted rock shelters in the Erramalai Tablelands (Mesas), Proterozoic Kurnool Basin, Andhra Pradesh, in: Korisettar, R. (Ed.), Beyond Stones and More Stones, The Mythic Society, Bengaluru, 2018, pp. 316–343. vol. 2.

Lane, C., Haslam, M., Petraglia, M., Ditchfield, P., Smith, V., Korisettar, Ravi, Cryptotephra from the 74 ka BP Toba super-eruption in the Billa Surgam caves, southern India, Quat. Sci. Rev. 30 (2011) 1819–1824.

Lane, C.S., Chorn, B.T., Johnson, T.C., Ash from the Toba super-eruption in Lake Malawi shows no volcanic winter in East Africa at 75 ka, Proc. Natl. Acad. Sci. 110 (2013) 8025–8029.

Mark, D.F., Petraglia, M., Smith, V.C., Morgan, L.E., Barfod, D.N., Ellis, B.S., Pearce, N.J., Pal, J.N., Korisettar, R., A high-precision 40Ar/39Ar age for the Young Toba Tuff and dating of ultra-distal tephra: forcing of

Quaternary climate and implications for hominin occupation of India, Quat. Geochronol. . 21 (2014) 90–103, doi:10.1016/j.quageo.2012.12.004.

Mathews, N.E., Smith, V.C., Costa, A., Durant, A.J., Pyle, D.M., Pearce, J.G., Ultra-distal tephra deposits from super-eruption: examples from Toba, Indonesia and Taupo Volcanic Zone, New Zealand, Quat. Int. 258 (2012) 54–79.

Mellars, P., Going east: new genetic and archaeological perspectives on the modern human colonization of Eurasia, Science 313 (2006a) 796–800.

Mellars, P., Why did modern humans populations disperse from Africa ca. 60,000 year ago? A new model, Proc. Natl. Acad. Sci. 103 (2006b) 9381–9386.

Mishra, S., Chauhan.N., Singhvi, A.K., Continuity of microblade technology in the Indian Subcontinent since at least 45 ka: implications for the dispersal of modern humans, PLoS One 8 (2013) e69280.

Murty, M.L.K., A late Pleistocene cave site in southern India, Proc. Am. Philos. Soc. 118 (1974) 196–230.

Murty, M.L.K., Late Pleistocene fauna of Kurnool caves, South India, in: Clason, A.T. (Ed.), Archaeozoological Studies: Papers of the Archaeozoological Conference 1974, held at the Biologisch-Archaeologisch Instituut of the State University of GroningenNorth Holland Publishing Company, Amsterdam, 1975, pp. 132–138.

Murty, M.L.K., Ethnoarchaeology of Kurnool cave areas, south India, World Archaeol. 17 (2) (1985) 192–205.

Murty, M.L.K., Reddy, K.T., The significance of lithic finds in the cave areas of Kurnool, India. Asian Perspect. 18 (1975) 214–226.

Paddayya, K., Blackwell, B.A.B., Jhaldiyal, R., Petraglia, M.D., Fevrier, S, Skinner, A.R., Recent findings on the Acheulain of the Hunsgi and Baichbal valleys, Karnataka, with special reference to the Isampur excavation and its dating, Curr. Sci. 83 (2002) 641–647.

Pappu, S., Gunnell, Y., Akhilesh, K., Braucher, R., Taieb, M., Demory, F., Thouveny, N., Early Pleistocene presence of Acheulian hominins in south India, Science 331 (2011) 1596–1599.

Perera, N., Kourampas, N., Simpson, I.A., Deraniyagala, S.U., Bulbeck, D., Kamminga, J., Perera, J., Fuller, D.Q., Szabo, K., Oliveira, N.V., People of the ancient rainforest: late Pleistocene foragers at the Batadomba-lena rockshelter, Sri Lanka, J. Hum. Evol. 61 (3) (2011) 254–269.

Petraglia, M., Korisettar, R., Pal, J.N., The Toba super-eruption of 74, 000 years ago: climate change, environments and evolving humans, Quat. Int. 258 (2012a) 1–5.

Petraglia, M., Ditchfield, P., Jones, S., Korisettar, R., Pal, J.N., The Toba volcanic super-eruption, environmental change, and hominin occupation history in India over the last 1,40,000 years, Quat. Int. 258 (2012b) 135–147.

Petraglia, M., Korisettar, R., Boivin, N., Clarkson, C., Ditchfield, P., Jones, S., Koshy, J., Lahr, M.M., Oppenheimer, C., Pyle, D., Roberts, R., Schwenninger, J.-L., Arnold, L., White, K., Middle Palaeolithic assemblages from the Indian subcontinent before and after the Toba super-eruption, Science 317 (2007) 114–116.

Petraglia, M., Clarkson, C., Boivin, N., Haslam, M., Korisettar, R., Chaubey, G., Ditchfield, P., Fuller, D., James, H., Jones, S., Kivisild, T., Koshy, J., Lahr, M.M., Metspalu, M., Roberts, R.G., Arnold, L., Population increase and environmental deterioration correspond with microlithic innovations in South Asia ca. 35,000 years ago, Proc. Natl. Acad. Sci. 106 (2009a) 12261–12266.

Petraglia, M., Korisettar, R., Kasturi Bai, M., Boivin, N., Bora, J., Clarkson, C., Cunningham, K., Ditchfield, P., Fuller, D., Hampson, J., Haslam, M., Jones, S., Koshy, J., Miracle, P., Oppenheimer, C., Roberts, R., White, K., Human occupation, adaptation and behavioral change in the Pleistocene and Holocene of South India: recent investigations in the Kurnool District, Andhra Pradesh, Eurasian Prehist. 6 (2009b) 119–166.

Prasad, K.N., Pleistocene cave fauna from peninsular India, J. Caves Karst Stud. 58 (1996) 30–34.

Prasad, K.N., Verma, K.K., 1969. The Kurnool Cave Fauna. Unpublished report.

Prasad, K.N., Yadagiri, P., Pleistocene cave fauna, Kurnool district, Records of the Geological Survey of India 115 (1986) 71–77.

Reddy, K.T., Sudarsen, V., Prehistoric investigations in Sgileru Basin, Man Environ. 2 (1978) 32–40.

Ramam, P.K., Murty, V.N., Geology of Andhra Pradesh, Bangalore, Geological Society of India, 1997.

Rao, V.V.M., Stone Age Cultures of Prakasam District, Andhra Pradesh, Andhra University, Visakhapatnam, 1979 Unpublished Ph.D. thesis),.

Rao, V.V.M., Middle Palaeolithic sites in Paleur valley, Prakasam District, Andhra Pradesh, Man Environ. IX (1987) 41–47.

Rao, V., Rao, C.V.N.K., Palaeontological Studies of the cave fauna of Kurnool district, Andhra Pradesh, Records of the Geological Survey of India 135 (240) (1992) 240–241.

Reddy, K.T., Billasurgam: an Upper Palaeolithic cave site in south India, Asian Perspect 20 (1977) 206–277.

Roberts, R., Fenwick, J., Arnold, L., Jacobs, Z., Jafari, Y., Numerical dating of sediments associated with volcanic ash and stone artefacts in southern and northeastern India, Paper presented at the Toba super-eruption conference, University of Oxford, 2010 20 February 2010.

Roberts, P., Delson, E., Miraclef, P., Ditchfield, P., Roberts, R.G., Jacobs, Z., Blinkhorn, J., Ciochon, R.L., Fleagle, J.G., Frost, S.R., Gilbert, C.C., Gunnell, G.F.Harrison T., Korisettar, R., Petraglia, M.D., Continuity of mammalian fauna over the last 200, 00 y in the Indian subcontinent, , Proc. Natl. Acad. Sci. 111 (16) (2014) 5848–5853.

Sali, S.A., The Upper Palaeolithic and Mesolithic Cultures of Maharashtra, Deccan College, Pune, 1989.

Shane, P., Westgate, J., Williams, M., Korisettar, Ravi, New geochemical evidence for the identity of the Youngest Toba Tuff at archaeological sites on the Indian subcontinent, Quat. Res. 44 (1995) 200–204.

Sharma, G.R., Clark, J.D., Palaeoenvironments and Prehistory of the Middle Son Valley, Abinash Prakashan, Allahabad, 1983.

Shea, J., Transitions or turnovers? Climatically forced extinctions of *Homo sapiens* and Neanderthals, Quat. Sci. Rev. 27 (2008) 2253–2270.

Shipton, C., Bora, J., Koshy, J., Petraglia, M., Haslam, M., Korisettar, R., Systematic transect survey of the Jurreru Valley, Kurnool District, Andhra Pradesh, Man Environ. 35 (2010) 24–36.

Smith, T., Koshy, J., Aravazhi, P., Excavating limestone caves in India: the Billasurgam experience, in: Korisettar, Ravi (Ed.), Beyond Stones and More Stones, The Mythic Society, Bengaluru, 2017, pp. 310–322.

Soundararajan, K.V., Joshi, R.V., Indian Archaeology A Review 1956-57, Archaeological Survey of India, New Delhi, 1956.

Tacon, P.S.C., Boivin, N., Hampson, J., Blinkhorn, J., Korisettar, Ravi, Petraglia, M., New rock art discoveries in the Kurnool District, Andhra Pradesh, India, Antiquity 84 (2010) 335–350.

Taçon, Paul S.C., Boivin, N., Petraglia, M., Blinkhorn, J., Chivas, A., Fink, D., Higham, T., Ditchfield, P., Korisettar, Ravi, Mid-Holocene age obtained for nested diamond pattern petroglyphs in the Billasurgam cave complex, Kurnool District, southern India, J. Archaeol. Sci. 40 (4) (2013) 1787–1796.

Venkatasubbaiah, P.C., South Indian Neolithic Culture: Pennar Basin, Andhra Pradesh. Bharatiya Kala Prakashan, Delhi (2007).

Vijaya, Kumari, Madhu, Teliki, Brahmaiah, Tallapalli, Occurrence of volcanic ash in Sagileru river valley, Cuddapah district, southern India, Int. J. Sci. Res. 6 (2017) 462–466 .

Wedage, O., Picin, A., Blinkhorn, J., Douka, K., Deraniyagala, S., Kourampas, N., Perera, N., Simpson, I., Boivin., Petraglia, M., Roberts, P., Microliths in the South Asian rainforest ~45-4 ka: new insights from Fa-Hien Lena Cave, Sri Lanka, PLoS one 14 (10) (2019) e0222606 https://doi.org/10.1371/journal.pone.0222606.

Westaway, R., Mishra, S., Deo, S., Bridgland, D.R., Methods for determination of the age of Pleistocene tephra, derived from eruption of Toba, in central India, J. Earth System Sci. 120 (3) (2011) 503–530.

Westgate, J.A., Pearce, J.G., Quaternary tephrochronology of the Toba tuffs and its significance with respect to archaeological studies in Peninsular India, in: Korisettar, Ravi (Ed.), Beyond Stones and More Stones, The Mythic Society, Bengaluru, 2017, pp. 199–233 .

Westgate, J.A., Shane, P.A.R., Pearce, N.J.G., Perkins, W.T., Korisettar, R., Chesner, C.A., Williams, M.A.J., Acharyya, S.K., All Toba tephra occurrences across Peninsular India belong to the 75,000 yr B.P. eruption, Quat. Res. 50 (1998) 107–112.

Williams, M., Clarke, M.F., Late Quaternary environments in north-central India, Nature 308 (1984) 633–635.

Williams, M., Ambrose, S.H., van der Kaars, S., Ruehlemann, C., Chattopadhyaya, U.C., Pal, J.N., Chauhan, P., Environmental impact of the 73 ka Toba super-eruption in South Asia, Palaeogeogr. Palaeoclimatol. Palaeoecol. 284 (2009) 295–314.

Williams, M., Ambrose, S.H., van der Kaars, S., Ruehlemann, C., Chattopadhyaya, U.C., Pal, J.N., Chauhan, P., Reply to the comment on 'Environmental impact of the 73 ka Toba super-eruption in South Asia', Palaeogeogr. Palaeoclimatol. Palaeoecol. 296 (2010) 204–211.

CHAPTER

Holocene vegetation, climate, and culture in Northeast India: a pollen data–based review

24

Anjali Trivedi
Birbal Sahni Institute of Palaeosciences, Lucknow , Uttar Pradesh, India

24.1 Introduction

The Holocene, covering about 11,500 years onwards, began to recede ice sheets and ended with the last glaciation. This period has continuous and fast climatic fluctuations that impacted the natural archives, play a vital role in human evolution/civilization and crop economy. It is well recognized that the empirical records on climate are only available for the last 150 years that is since 1863 CE for India. The densely inhabited Indian subcontinent's socioeconomy is primarily influenced by the Indian Summer Monsoon (ISM). It brings a significant amount of summer rainfall that accounts for ~70%–80% of the total annual rainfall (Webster et al., 1998; Gadgil, 2006). The ISM significantly influenced the agrarian-based Indian economy and is considered the most potent climate system nourished governing complete forest vegetation cover (Behre, 1981; Champion & Seth, 1968; Gadgil, 2006; Lamb, 1965; Yang et al., 2008; Xie et al. 2019; Trivedi et al., 2020). Proxy data from geographically diverse regions of the Indian subcontinent have been used to understand the climate variability during the late Quaternary (Chauhan, 2000; Prasad and Enzel, 2006; Juyal et al., 2009; Bhattacharayya et al., 2011, Srivastava et al., 2018). However, regional high-resolution paleoclimatic records are limited; yet, it suggested that abrupt climate variability has been catastrophic and associated with the rise and fall of the ancient human civilizations (William, 1999; Farmar, 2004; Prasad et al., 2014; Sarkar et al., 2016).

The northeast (NE) India, known for its diverse and most extensive lush forest cover, comprises 64% of the total geographical area (Jain et al., 2013). It is a global biodiversity hotspot, meeting region of temperate east Himalayan flora, paleoArctic flora of Tibetan highland, and wet evergreen flora of south-east Asia and Yunnan forming a bowl of biodiversity (Hara, 1965; Hooker, 1906). NE India has mostly relied on pollen-based reconstructions from sedimentary sequences (e.g., Basumatary et al., 2015; Bhattacharyya et al., 2007; Chauhan and Sharma, 1996; Dixit and Bera, 2011, 2012a, 2012b; Ghosh et al., 2014; Mehrotra et al., 2014; Mishra et al., 2020 and references therein). The problem gets further enhanced in the NE India region due to limited climate reconstruction records with poorly constrained chronologies throughout the region, which experiences maximum rainfall during the monsoon. Henceforth, to develop our understanding of the Holocene ISM variability and enhance our predictive capabilities for its future variability, there is a prerequisite to reconstruct the paleovegetation-climate vis-á-vis climate-culture relationship before the instrumental period. For this chapter (Fig. 24.1), a list of the pollen study (Table 24.1) pursued from diversified geographical regions of NE India and their allocation based on altitudinal categorization and vegetation types is shown in Fig. 24.2

Holocene Climate Change and Environment. DOI: https://doi.org/10.1016/B978-0-323-90085-0.00019-X

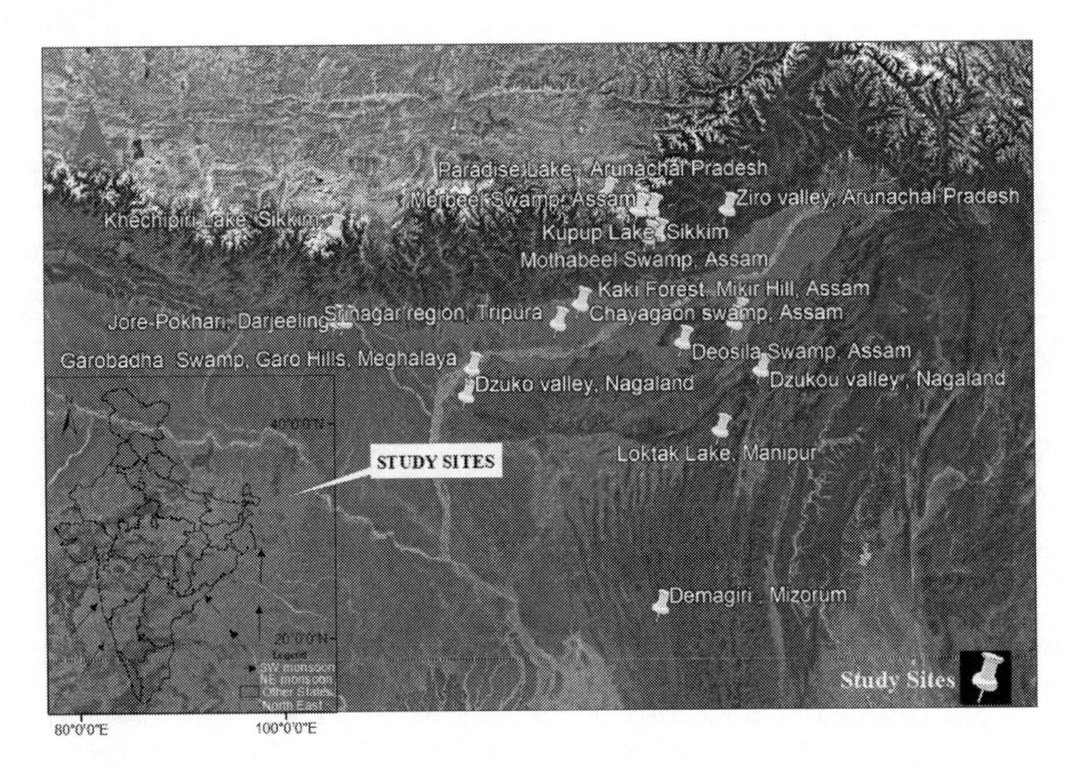

FIGURE 24.1

Map of northeast India with pins indicating the pollen study sites.

24.2 Scope

This chapter addresses the Holocene vegetation, climate history, and cultural aspects of Northeast (NE) India based on updated pollen proxy records. Northeast India palaeovegetation reconstructions primarily play vital roles to decode past oscillations on different ecotones ranging from tropical rain forest to subtropical to alpine flourishing of different altitudes with varied precipitation and temperature gradient. These data can be worked out the temporal and spatial phytogeographical distribution of some of the prominent forest constituents in the region and the impact of anthropogenic activities on the natural resources, and the inception and pace of agricultural practices in the area, vastly influenced by monsoon fluctuations. The present pollen datasets based review of Northeast India provides a baseline for the direction of climate changes.

24.3 General climatic conditions

NE India is experiencing three major seasons namely, winter, summer, and rainy season. There is a climatic contrast between the valleys and the mountainous region. As NE India is lying near the

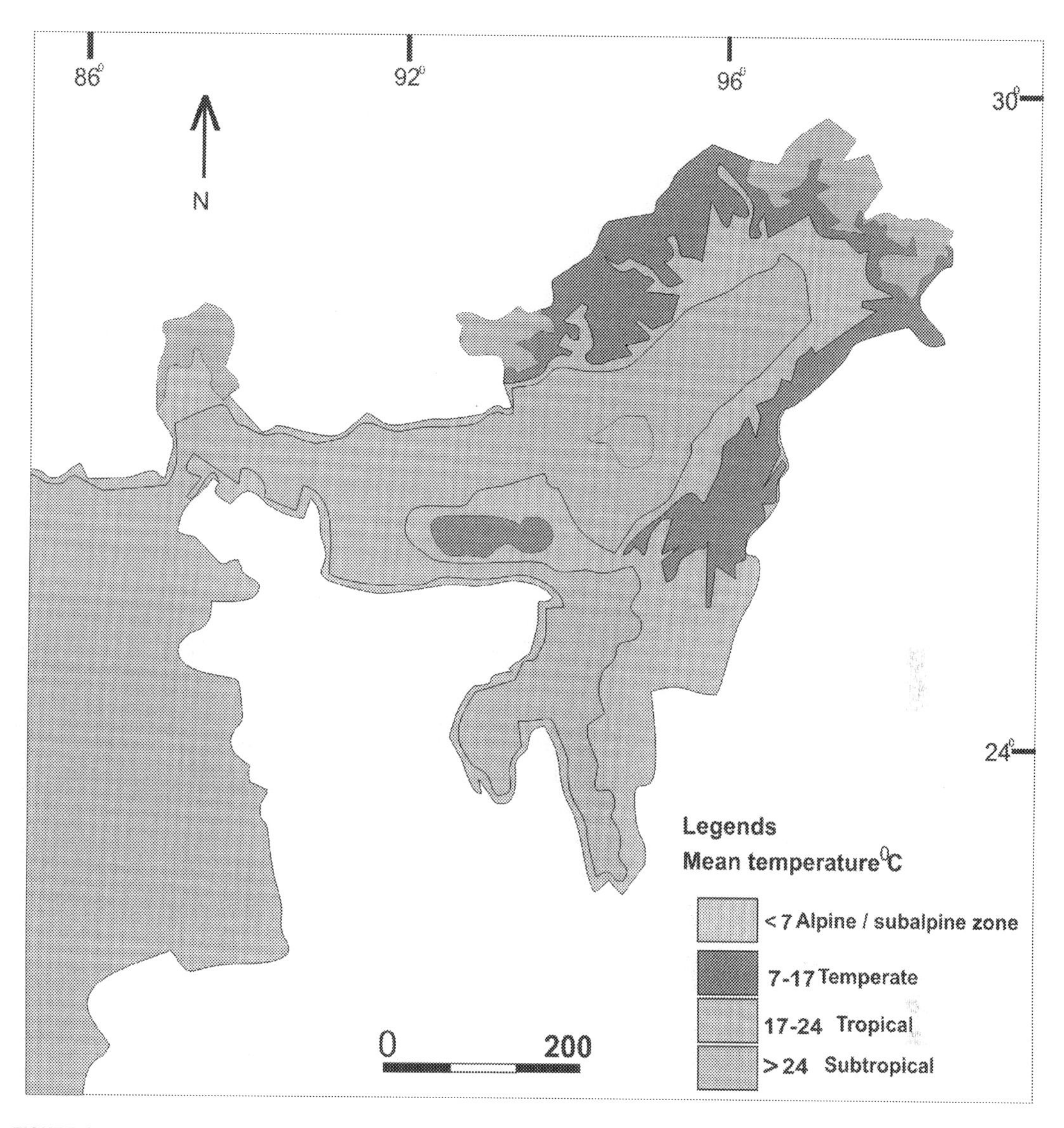

FIGURE 24.2

Mean annual temperature of northeast India.

tropics, it displays tropical weather, especially in Tripura, Mizoram, Manipur, and southern Assam (Dikshit and Dikshit, 2014). The high altitude locale of the Arunachal Pradesh, Sikkim, and West Bengal encounters cold to temperate climate (Fig. 24.2) because of the eastern Himalayas augmentation. NE India experiences a temperature as low as 9 °C in the coldest months and can reach 32 °C in the summer months (Rao, 2006). The January temperature in Assam valley is around 16 °C and the summer temperatures vary between 30 and 33 °C. The moderate temperature and relatively high humidity occur in the plains of the Brahmaputra and Barak rivers.

Arunachal Pradesh and Nagaland experience a mean summer temperature of around 20 °C with a mean minimum of 15 °C. The mountains above 2000 m ASL experience snowfall. The heavy fog above the mountain regions all through the year is common (Rao, 2006).

24.4 Monsoon

NE India is the country's rainiest region that receives over 90% of its annual rain from the summer monsoon (Dikshit & Dikshit, 2014; Mani, 1981). There is an average rainfall of 3000–4000 mm, with more than 75% in the monsoon months (June–September). The region has high variability in summer rainfall (Fig. 24.3) attributed to orographic influence and variation in location, timing, and intensity (Mani, 1981).

The Southwest monsoon arrives at India and bifurcates—-one Arabian Sea branch and the second Bay of Bengal branch. The Bay of Bengal branch flows over the bay, heading toward western and southern parts of the region such as Tripura, Meghalaya plateau, the west of Brahmaputra, Sikkim, West Bengal, and Bangladesh. The water-laded winds arrive with large amounts of rain at the Eastern Himalayas to the state of Arunachal Pradesh, which receives mean annual rainfall varying from 1400 mm to 6000 mm. Cherrapunji, in Meghalaya, receives a mean annual rainfall of 11,445 mm, of which over 95% is received during summer from the Indian monsoon. (Dikshit & Dikshit, 2014; Rao, 2006). Based on an average rainfall received, different parts of the NE India are divided into the following categories: (1) Wet areas, the areas that receive over 3000 mm rainfall covering the southern part of Meghalaya, the North-Eastern part of the Brahmaputra corridor, Lakhimpur, and Central Mizoram. (2) Humid areas receive rainfall between 2000 and 3000 mm, which include the western part of Brahmaputra valley and the northern part of Meghalaya plateau, Nagaland, Manipur, and Tripura. (3) Subhumid areas receiving rainfall between 1500 and 2000 mm, which include the central part of Brahmaputra valley. (4) Moderately rainy areas receive ~1500 mm of rain caused by the rain shadow effect. It consists of two places: One in the lee of Barail range and Mikir, and Kopili valley (Dikshit & Dikshit, 2014).

24.5 Methodology

Of 30 sedimentary profiles investigated so far from different forest composition and geological settings in the NE India, only 19 (Table 24.1) have rendered the seminal database affecting the climatic variability and significant vegetation developments covering different time intervals from the Holocene. The remaining profiles have culled because of lack of radiocarbon dates or too incompetent to elucidate Holocene climate variabilities and their impact on the vegetation.

The inferences drawn have also been concerning cultural shifts because of anthropogeny and crop economy, which have significantly affected the deviating climatic conditions during the past.

This review comprises the following states of this region: Arunachal Pradesh, Assam, Manipur, Meghalaya, Mizoram, Nagaland, and Tripura. Furthermore, the current study also included the Sikkim and West Bengal to enhance spatial coverage. It will deliver thoughtful enduring connections between

Table 24.1 Details of the pollen study sites with present vegetation types occurring in the surroundings, used for the present review.

S. No.	Pollen study site	Latitude (N)	Longitude (E)	Altitude amsl (m)	Present vegetation type
1.	Merbeel Swamp, Assam	27°36′	95°30′	140	Moist tropical deciduous Sal forest
2.	Mothabeel Swamp, Assam	27°17′	95°41′	100	Moist and dry tropical deciduous Sal forest
3.	Dangrithan Swamp, Assam	27° 18′	95°49′	200	Moist and dry tropical deciduous Sal forest
4.	Deosila Swamp, Assam	26°00′	90°80′	100	Moist and dry tropical deciduous Sal forest
5.	Dabaka Swamp, Lower Brahmaputra flood plain, Assam	26°11′	92°80′	130	Dense tropical mixed deciduous forest
6.	Chayagaon swamp, Assam	26°33′	91°42′	46	Moist and dry tropical deciduous Sal forest
7.	Kaki Forest, Mikir Hill, Assam	26°17′	93°50′		Moist semievergreen and moist evergreen forest
8.	Loktak Lake, Manipur	24°55′	93°83′	768	Dense tropical mixed deciduous forest
9.	Srinagar region, Tripura	22°98′	91°69′	939	Tropical evergreen forest
10.	Khecheopalri Lake, Sikkim	27°37′	88°20′	1700	Broad-leaved mixed temperate forest
11.	Kupup Lake, Sikkim	27°37′	88° 76′	4000	Alpine scrubby vegetation
12.	Mirik Lake, Darjeeling	26°89′	88°18′	1494	Mixed broad-leaved forests with *Cryptomeria japonica* forest
13.	Jore-Pokhari, Darjeeling	26°75′	88°25′	2260	Mixed broad-leaved forests with *Cryptomeria japonica* forest
14.	Paradise Lake, Arunachal Pradesh	27°50′	92°10′	4176	Subalpine and Rhododendron
15.	Ziro valley, Arunachal Pradesh	27°32′	93°49′	1557	Mixed subtropical broad-leaved and pine forest
16.	Demagiri , Mizoram	22°52′	92°28′	900	Tropical evergreen forest
17.	Garobadha Swamp, Garo Hills, Meghalaya	25°52′	90°10′	1700	Mixed deciduous
18.	Dzuko valley, Nagaland	25°33′	90°04′	2547	Temperate to subalpine plant species
19.	Dzukou valley, Nagaland	25°35′	94°05′	2438	Temperate to subalpine plant species

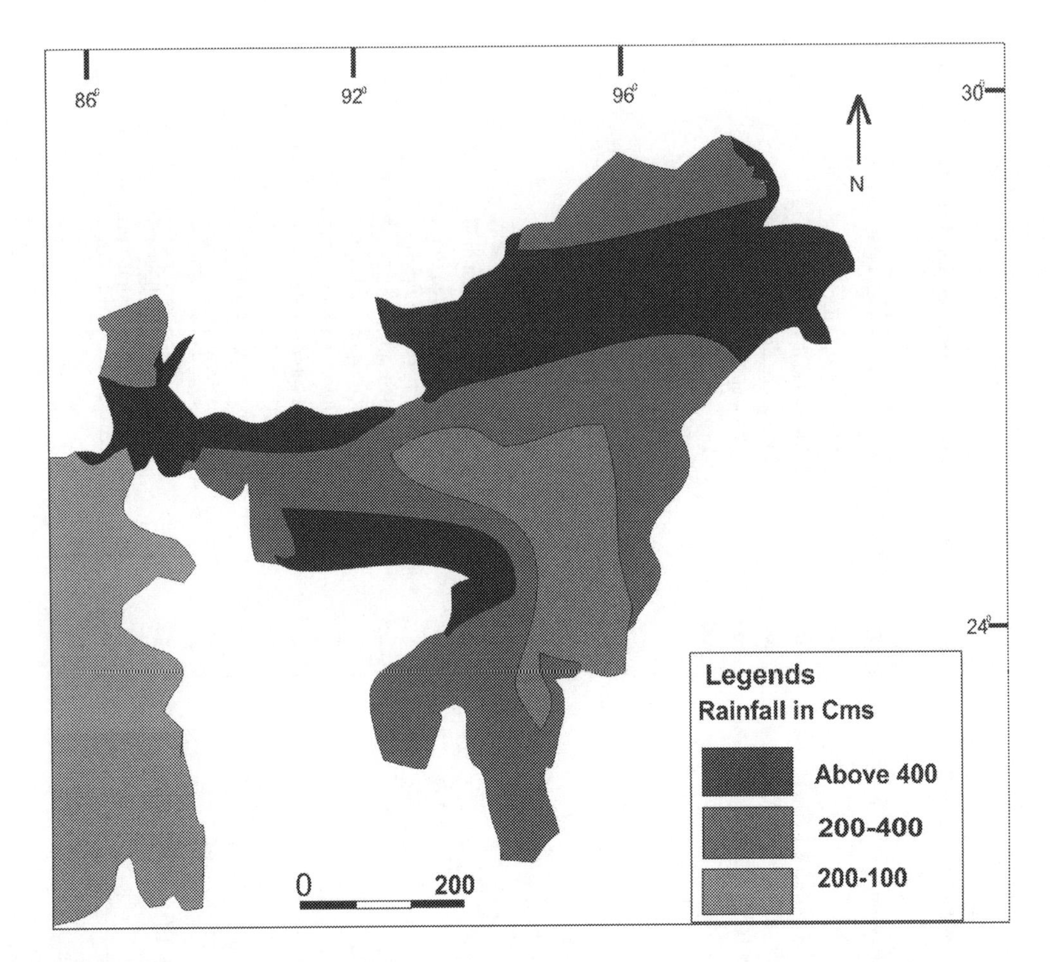

FIGURE 24.3

Mean annual precipitation of northeast India.

the distribution of temperature and precipitation dynamics in different altitudinal measures. A standard scale was created (Fig. 24.4) and plotted in color codes, these states were marked accordingly to bring out a detailed understanding of the Holocene pollen proxy-based climate archives.

24.6 Holocene vegetation and climatic records

24.6.1 Assam

24.6.1.1 Merbeel Swamp, Jeypore Reserve

Pollen analysis of lacustrine sedimentary profile was carries out by Bera and Dixit (2011). The vegetation history inferred that Tropical mixed deciduous to a semievergreen forest occupied the region

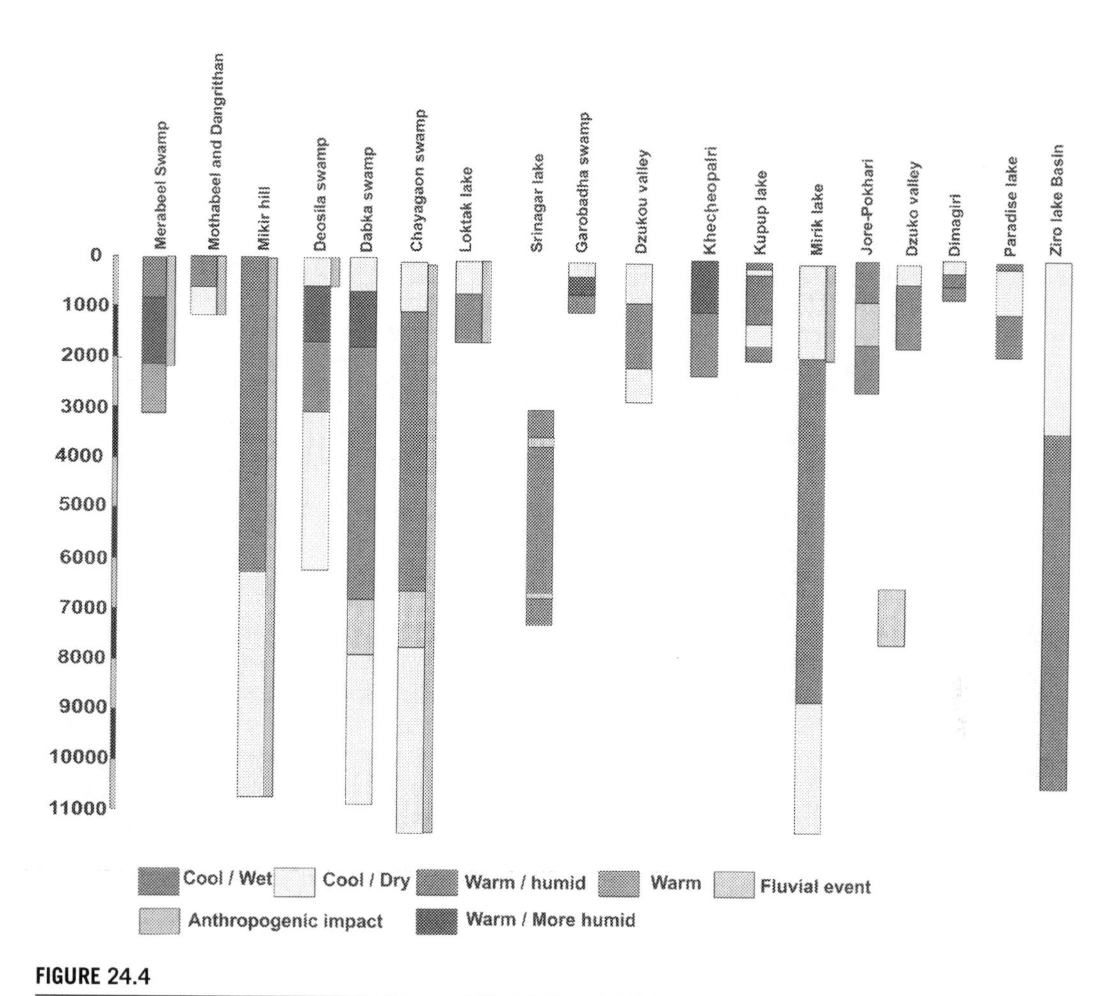

FIGURE 24.4

Vertical comparative chart showing climate change and human impact of the lakes stand in Northeast India.

since 3000 years BP, under three climatic phases, that is, relatively cool and dry, the onset of warm, and increased warm and humid. Around 3000–2200 years BP, Lauraceae, *Ilex, Mesua, Symplocos,* Ericaceae, *Carya alba* and limited *Gleichenia*, and *Pteris pentaphylla* are cool-preference taxa, which showed relatively cooler and drier climatic conditions. *Cerealia* (cultigens), along with other typical cultural pollen taxa, supports the initiation of agricultural practice. Around 2200 and 859 years BP, *Elaeocarpus,* Meliaceae, *Emblica, Dillenia*, and Oleaceae appeared that represent tropical mixed plant assemblage that portrays a warmer phase in the region. More tropical tree taxa such as *Syzygium, Terminalia, Madhuca, Adina*, and Dipterocarpus, etc. invaded the territory after 850 years BP with increased warm and humid climate conditions. The increased frequency of grasses, *Tubuliflorae* and *Xanthium*, indicated the extensive rise in pastoral activity during this phase.

24.6.1.2 Mothabeel and Dangrithan reserve forest

Pollen analysis since 1760 years BP conducted by Bera et al. (2011a), from two sedimentary profiles from each area, shows that around 1760 years BP, the area was under high fluvial activity as evidenced by the scarcity of pollen because of occurrences in coarse to fine sandy layers. Between 1100 to 550 years BP, the area under the influence of semiarid and warm climate followed by a mixed assemblage of ferns, conifers, and angiosperms was recovered in two sedimentary profiles. During 1100–550 years BP in the Dangrithan sediments, *Areca catechu* (betel nut) and *Persea bombycina* (silk plant) pollen were present, which are absent from the present-day vegetation in the region. *Carya alba* and *Tsuga*, with a typical subtropical assemblage found in these sediments, might be evidence of an existing migratory route between India and part of Northwest Asia during that period. Around 550 years BP onwards, the semievergreen forest turned to mixed deciduous, as evidenced by the invasion of more deciduous elements, namely, Dipterocarpaceae, *Lagerstroemia, Acacia*, etc., display the beginning of warm and humid climate. The presence of *Cerealia* and other cultural pollen supports the anthropogenic influence in and around the study area since 1100 years BP.

24.6.1.3 Deosila Swamp, Rangjuli reserve forest

The 2.5-m-deep lake soil profile around 6340 years (mid-Holocene) has been pollen analyzed by Dixit and Bera (2011) and it was deduced that the four-fold vegetational and climatic history of *Shorea robusta* dominated forest. Around 6340–2970 years BP, tropical tree savannah-type vegetation constituted grassland with scattered trees such as *Salmalia, Dellenia, Embelica*, and Meliaceae, indicating cool and dry climatic conditions. The infrequent existence of *Artocarpus chaplasha, Symplocos, Ilex, Schima*, and *Shorea robusta portrayed* that the climate was relatively less cool and dry in the upper part of the phase. During 2970–1500 years BP, there was the increament and diversification of tropical mixed deciduous forest taxa such as *S. robusta, Lagerstroemia, Lannea, Semecarpus*, and *Acacia* suggested that the southwest monsoon became stronger, and the climate was more warm and humid. Between 1510 and 540 BP, the establishment of tropical deciduous Sal forest succeeded *Lannea, Lagerstroemia, Terminalia*, Sapotaceae, *Albizia*, and *Adina*, suggesting that the region was under warm and more humid conditions. The high amount of marshy and aquatic pollen and ferns and fungal spores supported good climatic conditions during that period. The climate turned warmer and dry during 540 years BP, possibly due to the weak ISM. The high interference by human throughout this phase is supported by *Cerealia* and *Melastoma, Ziziphus*, and *A. catechu*.

24.6.1.4 Dabaka Swamp, Dabaka reserve forest

The pollen diagram of 3.2-m core constructed by Dixit and Bera (2012a) deduced five-fold development of tropical mixed deciduous forest since 14,120 years BP. The scarcity of palynomorphs and fluvial marker trees such as *Ludwigia octovalvis* and *Botryococcus* supports fluvial activity during 14,120–12,700 years BP. The expansion of tropical tree savanna-type forest supports the region was under a cool and dry climate between 12,700 and 11,600 years BP, corresponding to that of global Younger Dryas. Around 11,600 and 8310 years BP, the inception of tropical mixed deciduous taxa including *S. robusta* and *Lagerstroemia parviflora* suggests less cool and dry climate prevailed in the region. The reappearance of fluvial activity between 8310 and 7100 years BP was witnessed by low pollen and spores in the pollen assemblages. The tropical mixed deciduous forest became flourished after Pluvial activity that is between 7100 and 1550 years BP, corresponded with the global episode Holocene climatic optimum (HCO). The expansion of tropical mixed deciduous forests was sustained beneath an increased warm and humid climate till 768 years BP. Hereafter, the deterioration of tropical mixed deciduous forest started under a warm and relatively dry climatic regime

24.6.1.5 Chayagaon Swamp, Kamrup District

A 3.4-m-deep sedimentary core studied by Dixit and Bera (2012b) from the Western Brahmaputra flood plain revealed climate and vegetation successions since the Late Quaternary. The lack of pollen and spores, with fluvial marker taxa such as *Ludwigia octovalvis, Mimosa pudica, Pleospora, Glomus,* and *Botryococcus* along with frequent *Pseudoschizia*, marked the fluvial activity during 14,895–12,450 years BP. Between 12,450–10,810 cal BP, tropical tree savanna-type vegetation persisted under cool and dry climate, supporting weakened ISM corresponding to the Younger Dryas. Between 10,810–7680 cal BP, comparatively less cool and dry climate succeeded with the initiation of tropical mixed deciduous taxa comprising *Syzygium cumini, Dillenia pentagyna,* and *L. parviflora*. Incipient cereal-based pastoral practices began during this phase. This phase again showed fluvial activity between 7680 and 6780 years BP, as indicated by lack of palynomorphs along with frequent fluvial marker pollen taxa. After that, enhancement in tropical mixed deciduous forest taxa including *Sizygium cumini, Terminalia bellirica* and *Mimusops elengi*, and a relative increase of marshy and aquatic taxa represent the warm and moderately humid climatic conditions between 6780 and 1950 years BP. This phase is well matched with HCO (7000 years BP), expressing high monsoon precipitation. An enhancement in warm and humid climate was conquered between 1950 and 989 years BP. Simultaneously, the tropical mixed deciduous forests swapped the proliferating deciduous tree taxa, concurrently matching with the Medieval Warm Period (MWP). Since 989 cal BP to present the deterioration in tropical mixed deciduous forests shifted climate to warmer and relatively drier conditions. The sudden increment in Cerealia along with ruderal pollen taxa indicates urbanization impact. Forest clearance was evident from the relative increases in *Melastoma, Ziziphus,* and *A. catechu* combined with reduced numbers of arboreal *Shorea, Terminalia, Lagerstroemia, Dillenia,* and *Emblica*.

24.6.1.6 Kaki Forest Division, Mikir Hills

A pollen study of 1.5-m-deep sediment profile (Bera 2003) inferred two-fold climatic oscillations such as arid, semiarid, and warm and humid since 12210 years BP. Mixed tropical and subtropical trees and shrubs comprised the arboreal vegetation. Ferns, including monolete and trilete, are frequently found in the assemblage. During this phase covering ~3000–6000 years BP, the invasion of few tree taxa and the preponderance of grassland taxa reflect the open savanna-type vegetation representing specifically semi-arid climate. From 6000 years BP to the Present, the establishment of tropical/subtropical forest components in the vegetation pattern could result from improved rainfall indicating warm and humid climatic conditions. Cerealia and other cultural pollen are indicative of continuous agricultural practice during the 13000 years BP.

24.6.2 Manipur

24.6.2.1 Loktak Lake

Nautiyal and Chauhan conducted pollen analyses from a 0.73-m in-depth sediment profile (2009) and have recorded the vegetation and climatic fluctuations in the region during the Late Holocene. Between 1650 and 600 years BP, open vegetation comprising chiefly Poaceae and heathland taxa Asteraceae, Chenopodiaceae/Amaranthaceae, etc., along with few trees such as *Holoptelea, Symplocos, Acacia,* etc. occurred in the region adjoining to the lake. The abundance of ferns, fungal, algal remains, and aquatics indicates a humid climate conquered in the region. Recovery of Cerealia and other cultural pollen taxa revealed some sort of anthropogenic activity in the nearby lake. Since 600 years BP, the extension of open vegetation and decline of trees infer the beginning of a less humid climate probably due to a reduction in the summer monsoon. The sharp decline in ferns, fungal, and aquatics represented probable lesser humid conditions, however; the culture pollen taxa do not display any unprecedented change showing agricultural practice as similar as before.

24.6.3 Tripura

24.6.3.1 Srinagar region

The palynological investigation of 1.2-m-deep subsurface sediment from southwest Tripura was conducted by Bhattacharyya et al. (2011). During 7000–3000 years BP, the climate was warm and humid and the vegetation was mainly composed of moist deciduous forests. In the 6800 and 3700–3800 years BP, there were some intermittent rainfall reduction events, which may be due to the weakening of the Indian Summer Monsoon (ISM). The encounter of large-sized Poaceae pollen (above 50 μ) suggests rice cultivation initiation from 5700 BP in this region.

24.6.4 Sikkim Himalaya

24.6.4.1 Khecheopalri Lake

The pollen diagram of a temperate lake sediment profile (Sharma and Chauhan, 1999) suggested that the vegetation mostly contained mixed broad-leaved taxa suggesting that a warm and moist climate continued around 2500 years ago. About 1000 years ago, a high number of *Alnus* and increased frequencies of *Quercus* and *Rhododendron* suggested that the climate had become more humid during that period.

24.6.4.2 Kupup Lake

Sharma and Chauhan (2001) conducted pollen analysis on a 1.20 m alpine deep sedimentary profile and deciphered that since 2000 years BP, *Pinus, Abies, Tsuga, Quercus, Betula, Alnus, Rhododendron*, and *Viburnum*, along with sedges, grasses, Apiaceae, and Asteraceae were the dominant pollen taxa in the assemblage. The paleovegetation reflects a cold and moist climate in the region between 2000 and 1800 years BP, which was similar to the present-day alpine vegetation in the area. During 1800 and 1450 years BP, an improvement in sedges, grasses, and other herbaceous elements and a relative reduction in the number of broad-leaved taxa such as *Betula, Alnus*, and *Rhododendron* showed slightly drier conditions. Perhaps for a short spell, and there might have been a decline in the precipitation. The climate returned to a cold and moist climate during 1450–450 BP, suggested by the rise in the sum of broad-leaved taxa. The amelioration in climate considered to have occurred during AD 700–AD 1200 coincides with the global MWP. The sharp improvement in Cyperaceae, Poaceae, Chenopodiaceae/Amaranthaceae, and Ranunculaceae, and a corresponding decrease of arboreal pollen around 450 and 200 BP, that is, AD 1500 and AD 1750 reflected the impact of the Little Ice Age (LIA) that was recorded for the period AD 1450–AD 850. However, the last phase for 200 years was demonstrated by improvement in *Quercus, Betula, Alnus*, and *Rosaceae* under the cold and moist conditions distinctive of the current alpine zone in the Eastern Himalayan region.

24.6.5 Darjeeling

24.6.5.1 *Mirik Lake*

The vegetation history of a temperate region profile was constructed by Sharma and Chauhan (1994) since the Last Glacial Maximum. During 18,000 and 12,000 years BP, the oak-pine forests were replaced by grasslands reflecting the climate change. Around 11,000 years BP, there was an abrupt descent of broad-leaved taxa and a rise in herbaceous elements because of a short cold spell in the region. The existence of grassland vegetation type suggests that the cold and dry climatic conditions prevailed during the Last Glacial Maximum. The dominance of mixed broad-leaved forest taxa such as *Quercus, Alnus, Betula*, and *Carpinus* during the mid-Holocene supports a transferal from a cold and dry to a warm and moist climate. Around 2000 years BP, the reduction in broad-leaved taxa with a contemporary proliferation in Poaceae, Rosaceae, Cyperaceae, Caryophyllaceae, and Asteraceae suggested minor deterioration in the climate. The anthropogenic activities in the region might have altered this recent vegetation scenario.

24.6.5.2 *Jore-Pokhari*

Chauhan and Sharma (1996) conducted pollen analysis of late Holocene lacustrine sediments in the temperate region of Darjeeling. They opined that around 2500 years BP, the vegetation is dominated by mixed broad-leaved oak forests reflecting that the area was under a warm and humid climate. A short-term cold oscillation is observed between 1600 and 1000 years BP when the broad-leaved tree declined in taxa and a concurrent escalation in the conifers. The retrieval of Cerealia and concomitant culture pollen taxa namely, Chenopodiaceae/Amaranthaceae, Caryophyllaceae, *Artemisia*, and Asteraceae indicated anthropogenic activities at that period. Between 1000 and 300 years BP, amelioration of climate started when the climate turned warm-temperate and humid, and the broad-leaved *Quercus, Betula, Alnus*, and *Rhododendron* increased, whereas *Pinus* decreased.

24.6.6 Arunachal Pradesh

24.6.6.1 *Paradise Lake*

Bhattacharyya et al. (2007) constructed pollen vegetation history during Late Holocene from 1-m-deep sediment profile at the alpine region near Sela Pass. Around 1800 years BP (around AD 240), the conifer-broadleaved forest used to grow in the vicinity of the study site under warm and moist climate, similar to the prevailing present-day conditions, which turned out to be comparatively warmer at 1100 years BP (around AD 985) corresponding to the MWP. The glaciers seem to have receded and tree line might have been closer to the site. Around 550 years BP (around AD 1400) a decrease in *Tsuga, Juniperus*, and *Quercus* suggests a comparatively cooler and less moist climate corresponding to the LIA. It is followed by the amelioration of climate more or less equivalent to the present day.

24.6.6.2 *Ziro Lake Basin*

The paleoclimatic reconstruction was done by Ghosh et al., 2014. The pollen and nonpollen palynomorphs data suggest the prevalence of a moist semi-evergreen forest in the area until the Last Glacial Maximun (LGM). During 10,000–3800 years BP, forest cover expansion indicates the climatic amelioration with the summer monsoon's strengthening. After 3800 years BP, decline in forest cover indicates an increasing trend of dryness.

24.6.7 Mizoram

24.6.7.1 Demagiri Swamp

Palynological analysis of a 2-m-thick sediment profile from tropical southern Mizoram was investigated by Chauhan and Mandaokar (2006). The studies showed that the mixed tropical forests composed of Sapotaceae, *Symplocos, Holoptelea, Lagerstroemia, Dalbergia*, and *thickets of Aspidopterys,* Fabaceae, etc. occurred between 850 and 450 years BP, signifying a warm and humid climate. Between 450 and 250 BP, the woodlands became scarce with the lesser tree taxa possibly, which could be ascribed due to the prevalence of warm and moderately humid climate in response to the weakening of the ISM. From 250 years BP onwards, the additional deterioration in forest ingredients and a simultaneous extension of grasses suggest a warm and less humid climate in the region due to more weakening of the ISM.

24.6.8 Meghalaya

24.6.8.1 Garobadha Swamp, West Garo Hills

The pollen studies conducted by Basumatary and Bera (2010) of 1.3-m sediment core of tropical forest region deciphered three-fold climatic changes that is, the onset of warm and humid, increasing warm and moist, and warm and dry till 1300 years BP. The pollen of *Syzygium cumini, Schima wallichii, Terminalia bellirica* and *Dillenia pentagyna* made up the open vegetation during 1300 years BP, when the first phase occurred, signifying the beginning of warm and humid climatic conditions. Around 752 years BP, there was increased warming, thereby enhancing monsoon in the region. The vegetation had more mixed tropical deciduous arboreals than did phase I. Then, there was a decrease in arboreals, exotics, and marshy/aquatic taxa around 322 years BP, along with a few nonarboreals such as *Artemisia* in the Garobadha Swamp third phase (GA III), which appeared to be relatively drier than GA II.

24.6.9 Nagaland

24.6.9.1 Dzuko valley

Bera et al. (2011b) have studied 1.3-m-deep sedimentary profile from a temperate lake, and deciphered three climatic phases around 1600 years BP. Around 1600 years BP, cool and humid conditions prevailed similar to the present-day temperate climate of Mizoram. Open land vegetation composed of tree taxa such as *Magnolia, Symplocos, Ilex, Carya,* Lauraceae, *Rhododendron*, and dwarf bamboo (*Sinarundinaria rolloana*), along with *Primula, Anemone*, and *Rubus*. The results show that the study area was under the cool and temperate climate around 1600 years BP. During 980 years BP the vegetation turned out to a mixed tropical-subtropical plant assemblage, namely *Emblica,* Sapotaceae, *Elaeocarpus,* Meliaceae, *Lagerstroemia,* Ericaceae, and Oleaceae, which signified changed warmer conditions that prevailed in the region. The pollen of Poaceae, Bambusoideae as well as broad-leaved taxa such as *Quercus, Elaeocarpus*, and Combretaceae together with highland conifers appeared from 450 years BP onwards. These supported an improved warm and humid climate. Consequently, less humid environments in the upper part of the core are designated by higher amounts of *Ephedra* and *Artemisia*.

The second core was analyzed by Mishra et al. (2020), who depicted the recovered pollen data from ~1.4 m sedimentary profile spanning the last ~3150 years BP. According to them, the pollen proxy is suitable to reconstruct climatic and vegetation shifts during the late Holocene. They also postulated that between ~3150 years BP to ~2300 years BP, the region was occupied by pine-oak forest, indicating moderately dry climatic conditions. Subsequently, from ~2300 years BP to ~1000 years BP rise in arboreal tree elements and a gradual decrease in pine taxa suggest moist climatic conditions. From ~1000 years BP onwards, the precipitation declined till date, as indicated by the good recovery of *Pinus*-Oak forest pollen.

24.7 Discussion

The present study provides information about the long-term response to vegetation by the inherent effect of changing climate with the changing latitudes. This representation of pollen studies based totally on altitudinal categorization showed the variation in the records terminology and their temporal and spatial significance.

Additionally, in this review paper, recorded climatic changes in the altitudinal scale based on the pollen records from several sites have been depicted. These records clearly indicate that NE India has been under the monsoon's dynamic influence, inducing vegetation and climate variation throughout the Holocene. Chronologically these archives cover a different time and space within their limited spatial distribution. Fig. 24.4 characterizes the temporal restrictions of the pollen-based records with detailed climate inferences.

The environmental changes recorded from the higher altitudinal reaches of Arunachal and Sikkim were depicted as moderately cold and dry till 2200 years BP (Fig. 24.4). Then, the amelioration started, and it became warm and moist from 2500 years BP to today in most records from this latitude. Few records (Sharma and Chauhan, 2001; Bhattacharyya et al., 2007) also have signals of global events such as the MWP and LIA, displaying how vegetation and climate responded to global climate event. Though, the record from the high-altitude state of Sikkim (Sharma and Chauhan, 2001) exhibited a usually cold dry to the cold, moist fluctuation of climate, along with the signals of LIA and MWP. The climate became cold and wet in recent times.

The study sites in these elevations are significantly influenced by their topography concerning vegetation distribution. Hence, climate history based on vegetation changes is firmly influenced by the undulating topography of these areas in NE India.

Various studies from the Lower area of Assam, Brahmaputra flood plains, and Sikkim related to climate and vegetation fluctuations present a significant shift (Fig. 24.4). The Lower Brahmaputra flood plain was cool and dry from early Holocene till 7680 years BP (Dixit and Bera, 2012a,b). Enhanced fluvial activity was controlled by warm and humid climates in most records. However, the records from the North Bengal, Darjeeling Himalaya site (Sharma and Chauhan, 1994) revealed the beginning of warm temperate and humid climate at 10,000 years BP (Fig. 24.4). Most of the archives exhibited continued warm and humid conditions, which gradually began to shift to dry from 900 years BP. Dixit and Bera (2012b) noted a warm and dry spell in lower Assam since 900 years BP, which they endorsed to the weakening of the Southwest Monsoon, sustained till modern times and began later in other records (Dixit and Bera, 2011, 2012a). The amplified human activities were also stated as a reason for the vegetation shift and alteration in precipitation patterns.

The records from the subtropical belt are not as long and many but show identical shifts toward humid and warm environments, about 1200 years BP, lasting to 350 years BP

(Basumatary and Bera, 2010). The record from Meghalaya also showed a shift from warm and humid to warm and dry toward the present.

The other climate records from the Nagaland–Manipur border concluded that warm and moister conditions in recent times might be attributed to high altitude and lesser anthropogenic activities in the region. After that, Mishra et al. (2020), starting from that 3150 years BP to 2300 years BP moderately dry climate persisted. Subsequently, the moist phase arrived and continued till 1000 cal. From 1000 years BP onwards, the precipitation decline contrasting to Bera et al. (2011b).

The only two records come from Manipur and show a weakening of monsoon and less humid conditions in recent times (Nautiyal and Chauhan, 2009). The palynological studies from the tropical and low altitudes from Tripura have limited records that display an overall warm and humid environment. There are some phases of less humid conditions around 3800 to 3000 years BP (Bhattacharyya et al., 2011). However, in another study relatively less humid climate was observed during recent times from the same region (Chauhan and Mandaokar, 2006). The influences of anthropogenic activities are significant, as agriculture practices in these regions were extensive and may have commenced early. Thus, each pollen-based study gives an appropriate conclusion of deviations in vegetation and climate, effectively but in an unacquainted manner. Although the complete understanding of the climatic fluctuations is concrete, as one compares the different studies on different altitudinal, it becomes tremendously complex to relate the overall climate scenarios in this vast region

24.8 Learning and knowledge outcomes

Pollen analytical data on the floristic changes at various latitudinal positions in NE India clearly indicates that the region has been under the dynamic influence of the summer monsoon, inducing vegetation and climate variation during the Holocene. The floristic turnover is clearly visible at the taxonomic level of plant families during the Holocene. The abundance and diversity of angiosperm families vary strongly with climate and latitudinal changes in various states of the northeastern region of India. The available records in the monsoon-dominated region exhibit human-caused landscape changes. The global signatures of the HCO, MWP, and LIA from the different regions of NE India are now visible in studies being undertaken, making it possible to correlate and compare with other parts of India and the globe. Beforehand, the vast majority of the outcomes were published in in-house journals; however, present investigations are sufficient enough to discover a spot in all the signs with a worldwide outreach.

The outcomes of the contribution indicate that paleo-reconstructions play two key roles. First, they are good paleo-analogs for vegetation changes and landscape dynamics. These can be used for future climate projections under similar temperature conditions, with/without the human impact. Second, they provide a baseline for the direction of changes in the present and thereby provide a means for climate state parameters to model the future. Some recommendations for future efforts are as follows:

1. To develop multiproxy models using various biotic and abiotic to substantiate further the signals of climate-influenced vegetation changes in the region.
2. To study ancient DNA conserved in the sediments to identify the biogeographical pattern of different plant taxa and their specific adaptations in the past.
3. To develop a climate variability model and the corresponding increase in extreme events in an anthropogenically influenced warming environment, which is still in its infancy in the Indian context

Acknowledgments

The author thanks Dr. Vandana Prasad, Director, Birbal Sahni Institute of Palaeosciences, Lucknow, for encouragement and permission to publish this work. I am incredibly grateful to the numerous workers who devoted their lives to palynological research in NE India. The author is indebted to anonymous reviewers for their critical remarks and valuable suggestions to improve this manuscript.

References

Basumatary, S.K., Bera, S.K., Development of vegetation and climatic change in West Garo hills since Late Holocene: pollen sequence and anthropogenic impact, J. Indian Bot. Soc. 89 (2010) 143–148.

Behre, K.-E., The interpretation of anthropogenic indicators in pollen diagrams, Pollen Spores 23 (1981) 225–245.

Bera, S.K., Basumatary, S.K., Gogai, B., Pollen analysis its implications in climate change and vegetation succession in recent past: evidence from surface and subsurface fluvial sediment of tropical forests of eastern Assam, Northeast India, . J. Front. Res. 1 (2011a) 62–74.

Bera, S.K., Basumatary, S.K., Nautiyal, C.M., Dixit, S., Mao, A.A., Gogoi, R., Late Holocene climate and vegetation change in the Dzuko valley, Northeast India, J. Palaeontol. Soc. India 56 (2) (2011b) 143–148.

Bera, S.K., Dixit, S., Pollen analysis of late Holocene lacustrine sediment from Joypore reserve forest, Dibrugarh, Assam, in: Singh, D.S., Chhabra, N.L. (Eds.), Geological Processes and Climate Change, Macmillan Publisher India Ltd, 2011, pp. 85–94.

Bera, S.K., Early Holocene pollen data from Mikir Hills, Assam, India, Palaeobotanist 52 (2003) 121–126.

Bhattacharyya, A., Mehrotra, N., Shah, S.K., Holocene vegetation and climate of South Tripura based on palynological analysis, J. Geol. Soc. India 77 (2011) 521–526.

Bhattacharyya, A., Sharma, J., Shah, S.K., Chaudhury, V., Climatic changes last 1800 years BP from Paradise lake, Sela pass, Arunachal Pradesh, Northeast Himalaya, Curr. Sci. 93 (7) (2007) 983–987.

Champion, H.G., Seth, S.K., A Revised Survey of the Forest Types of India, Manager of Publications Delhi, 1968.

Chauhan, M.S., Sharma, C., Late-Holocene vegetation of Darjeeling (Joree Pokhari), Eastern Himalaya, Paleobotanist 45 (1996) 125–145.

Chauhan, M.S., Pollen evidence of late–quaternary vegetation and climate change in Northeastern Madhya Pradesh, India., Paleobotanist 49 (2000) 491–500.

Chauhan, M.S., Mandaokar, B.D., Pollen proxy records of vegetation and climate change during recent past in southern Mizoram, India. Gondwana Geol. Mag. 21 (2) (2006) 115–119.

Dikshit, K..R., Dikshit, J.K., 2014. North-East India: Land, People and Economy, Advances in Asian Human-Environmental Research. Springer Science and Business Media Dordrecht, pp. 149–173. doi:10.1007/978-94-007-7055-3_6.

Dixit, S., Bera, S.K., Mid-Holocene vegetation and climatic variability in tropical deciduous Sal (*Shorea robusta*) forest of lower Brahmaputra Valley, Assam., J. Geol. Soc. India 77 (2011) 419–432.

Dixit, S., Bera, S.K., Holocene climatic fluctuations from Lower Brahmaputra flood plain of Assam, northeast India, J. Earth Syst. Sci. 121 (1) (2012a) 135–147.

Dixit, S., Bera, S.K., Pollen-inferred vegetation vis-á-vis climate dynamics since Late Quaternary from western Assam, Northeast India: signal of global climatic events, Quat. Int. 286 (2012b) 56–68.

Farmer, S., Spiroat, R., Witzel, M., The collapse of Indian Script thesis: the myth of Harappan literate population, Electron. J. Vedic Stud. 11-2 (2004) 19–57.

Gadgil, S., The Indian Monsoon, GDP and agriculture, Econ. Polit. Wkly. 41 (2006) 4887–4895.

Ghosh, R., Paruya, D.K., Khan, M.A., Chakraborty, S., Sarkar, A., Bera,, S., Late Quaternary climate variability and vegetation response in Ziro Lake Basin, Eastern Himalaya: A multiproxy approach. Quat. International. 325 (2014) 13–29.

Hara, H., Spring Flora of Sikkim Himalaya. Hoikusha Publishing Co. Ltd., Osaka, (1965) ISBN 13: 9784586300662.
Hooker, J.D., A Sketch of the Flora of. British India, London (1906).
Jain, S.K., Kumar, V., Saharia, M., Analysis of rainfall and temperature trends in northeast India, Int. J. Climatol. 33 (2013) 968–978.
Juyal, N., Pant, R.K., Basavaiah, N., Bhushan, R., Jain, M., Saini, N.K., Yadava, M.G., Singhvi, A.K., Reconstruction of Last Glacial to early Holocene monsoon variability from relict lake sediments of the Higher Central Himalaya, Uttrakhand, India, Asian Earth Sci. 34 (2009) 437–449.
Lamb, H.H., The early medieval warm epoch and its sequel, Palaeogeogr. Palaeoclimatol. Palaeoecol. 1 (1965) 13–37.
Mani, A., The climate of the Himalayas, in: Lall, J.S., Moddie, A.D. (Eds.), The Himalayas Aspects of Change, Oxford University Press, Oxford, 1981, pp. 3–15.
Mehrotra, N., Shah, S., K., Bhattacharyya, A., Review of palaeoclimate records from Northeast India based on pollen proxy data of Late Pleistocene-Holocene. Quat. International. 325 (2014) 41–54. doi:10.1016/j.quaint.2013.10.061.
Misra, S., Bhattacharya, S., Mishra, P.K., Misra, K.G., Agrawal, S., Anoop, A., Vegetational responses to monsoon variability during Late Holocene: inferences based on carbon isotope and pollen record from the sedimentary sequence in Dzukou valley, NE India. Catena 194 (2020) 104697.
Nautiyal, C.M., Chauhan, M.S., Late Holocene vegetation and climate change in Loktak Lake region, Manipur, based on pollen and chemical evidence. Palaeobotanist 58 (2009) 2128.
Prasad, S., Anoop, A., Riedel, N., Sarkar, S., Menzel, P., Basavaiah, N., Krishnan, R., Fuller, D., Plessen, B., Gaye, B., Röhl, U., WilkesSachse, H.D., Sawant, R., Wiesner, M.G., Stebich, M., Prolonged monsoon droughts and links to Indo-Pacific warm pool: A Holocene record from Lonar Lake, central India. Earth Sci. Rev. 391 (2014) 171–182.
Prasad, S., Enzel, Y., Holocene paleoclimates of India, Quat. Res. 66 (2006) 442–453.
Rao, V.V.K., Hydropower in the Northeast: Potential and Harnessing Analysis. Background Paper 6 (2006) 60.
Sarkar, A., Mukherjee, A.D., Bera, M.K., Das, B., Juyal, N., Morthekai, P., Deshpande, R.D., Shinde, V.S., Rao, L.S., Oxygen isotope in archeological bioapatites from India: Implications to climate change and decline of Bronze Age Harappan civilization. Sci. Rep. 6 (2016) 26555.
Sharma, C., Chauhan, M.S., Vegetation and climate since Last Glacial Maxima in Darjeeling (Mirik Lake), Eastern Himalaya. In: Proceedings of 29th International Geological Congress Part B (1994) 27–288.
Sharma, C., Chauhan, M.S., Palaeoclimatic inferences from Quaternary paly-nostratigraphy of the Himalayas, in: Dash, S.K., Bahadur, J. (Eds.), The Himalayan Environment, New Age International, New Delhi, 1999, pp. 193–207.
Sharma, C., Chauhan, M.S., Late Holocene vegetation climate Kupup Sikkim, Himalaya, India, J. Palaeontol. Soc. India 46 (2001) 51–58.
Srivastava, P., Agnihotri, R., Sharma, D., Meena, N., Sundriyal, Y.P., Saxena, A., Bhushan, R., Swlani, R., Banerji, U.S., Sharma, C., Bhist, P., Rana, N., Jayangondaperumal, R., 8000 year monsoonal record from Himalaya revealing reinforcement of tropical and global climate systems since mid-Holocene, Sci. Rep. 7 (2018) 14515.
Trivedi, A., Tang, Ye-Na, Qin, Feng, Farooqui, A., Wortley, H.A., Wang, Yu-Fei., Blackmore, S., Li, Cheng-Sen, Yao, Yi-Feng, Holocene vegetation dynamics and climatic fluctuations from Shuanghaizi Lake in the Hengduan Mountains, southwestern China, Palaeogeogr. Palaeoclimatol. Palaeoecol. 507 (2020) 110–135.
Yang, B., Bräuning, A., Dong, Z., Zhang, Z., Keqing, J., Late–Holocene monsoonal temperate glacier fluctuations on the Tibetan Plateau, Glob. Planet. Change 60 (2008) 126–140.

Xie., G., Yao, Y.F., Li, J.F., Yang, J., Bai, J.D., Ferguson, D.K., Trivedi, A., Li, Cheg-Sen., Wang, Yu-Fei., Holocene climate, dynamic landscapes and environmentally driven changes in human living conditions in Beijing, Earth Sci. Rev. 191 (2019) 57–65.

Webster, P.J., Magana, V.O., Palmer, T.N., Shukla, J., Tomas, R.A., Yanai, M.U., Yasunari, T., Monsoons: processes, predictability, and the prospects for prediction, J. Geophys. Res. Oceans 103 (1998) 14451–14510.

William, R.N., The rise and fall of civilizations globalization, in: Nester, W. (Ed.), A Short History of the Modern World, Springer publication, Cham, 1999, pp. 15–18.

CHAPTER

25 Late Holocene vegetation, climate dynamics, and human-environment interaction along Konkan coast, India

Ruta B. Limaye[a,b], K.P.N. Kumaran[a,b], Sharad N. Rajaguru[c,d]

[a] *PalynoVision, Mon Amour, Erandaane, Pune, Maharashtra, India.* [b] *Formerly at Biodiversity & Palaeobiology Group, Agharkar Research Institute, India.* [c] *KamalSudha Apartment, Narayan Peth, Pune, Maharashtra, India.* [d] *Formerly of Deccan College Post-graduate Research Institute (now) Deemed to be University, Pune, India.*

25.1 Introduction

The Late Holocene (< 4000 years) was relatively dry but dynamic because of the anthropogenic influence, particularly along the coastline. The coastal dynamics in terms of vegetation and landscape changes were sparsely understood. There was hardly any such contribution toward this effort although there have been records of dry events during Late Holocene on the basis of marine archives and archeological evidence. The sedimentary archives of the Konkan coastal stretch were selected with an emphasis to address the Late Holocene vegetation and environmental dynamics. The organic matter including pollen grains, spores, and other palynomorphs was identified and compared with their modern analogues for building up of the vegetation history of the terrestrial scenario and also to distinguish any ecological shifts (facies changes) due to sea-level changes and freshwater influx. Non-pollen palynomorph (NPP) signatures provide an alternate option to pollen and spores as most of them indicate the aquatic environment and their response to the hydrological changes as a result of monsoon variability. The ^{14}C dates were used to date the organic samples for chronological control. Further, sedimentological aspects were dealt with to ascertain their utility for correlation of the biological proxies to concentrate and look into the Late Holocene vegetation dynamics and build up the palaeoenvironment and anthropogenic signatures. The monsoon has been found to be variable to a considerable extent since Mid-Holocene. This has been brought to light in the recent literature particularly on the basis of data derived from marine and lacustrine archives (Gupta et al., 2005; Singhvi and Kale, 2009; Kumaran et al., 2013). Signatures of climate change in off-shore sediments have been brought to our notice with the help of foraminifera and carbon and oxygen isotopes (Bhalla et al., 2007; Sujata et al., 2011; Asteman et al., 2013; Osterman and Smith, 2012). In fact, Late Holocene vegetation dynamics of Peninsular India in response to monsoon variations are seldom addressed, due to lack of availability of exposures and potential signatures in the sedimentary archives. Terrestrial archives are rarely tested from the south and western India and complemented with the marine archives and as such the present communication has a wider scope to understand the vegetation response to sea-level oscillations, climate, and anthropogenic factors during the Late Holocene along the Konkan coast. Further, the database generated would be

Holocene Climate Change and Environment. DOI: https://doi.org/10.1016/B978-0-323-90085-0.00012-7

useful for the appraisal of paleobiome in relation to climate change of the Indian Summer Monsoon (ISM) system during the Late Holocene.

Coastal deposits from Koparkhairne and Vasai Creek, Kelshi-Birwadi, Dapoli area (Ratnagiri district), Wadgaon Darya near Pimpalgaon in Sangamner area (Ahmednagar district—Central part) and Adari Lake (Sindhudurg district) were studied. These sites were selected after doing a reconnaissance survey and preliminary analysis of some of the subsurface samples. The following are the details of the study area from North to South Konkan (Fig. 25.1):

- **Vasai creek** [lat. 19°2′N; long. 72°97′E] north-south to the east of the Thane creek, parallel to the sea, keeping a distance of about 6.0–10.0 km from the shores.
- **Koparkhairne** [lat.19°4′28″ N; long. 72°57′59″ E] Thane creek.
- **Wadgaon Darya** [lat. 19°6′N; long. 74°20′E] near Pimpalgaon in Sangamner area, District Ahmednagar.
- **Birwadi** [lat. 17°51′20″N; long. 73°05′55″E] is a coastal village in Dapoli taluka of Ratnagiri district, on a low-order ephemeral stream situated on the northern side 1.0 km from mouth of the Jog River.
- **Kelshi** [lat. 17°90′N; long. 73°11′E] situated around 34.0 km from Dapoli and is toward north of Anjarle. One of the main attractions of Kelshi is a rare coastal dune with past anthropogenic activity.
- **Adari** [lat. 15°50′22″N; long. 73°39′08″E] is 2.0 km from Vengurla, Sindhudurg district.

25.2 Scope

Considering the limited knowledge, the scope of the study has been focused on Late Holocene climate dynamics in response to climate and sea-level changes along the Konkan coast and adjacent hinterlands. The main focus of the theme is the Late Holocene (Meghalayan) dynamics in relation to the sea-level oscillations and their impact on landforms, vegetation, and human intervention. Aspects of Late Holocene climate and paleoenvironment of Konkan coast are very much limited except for a few contributions based on geoarchaeological data.

How far the sea levels and monsoon variability have affected the coastal stretch while shaping the present-day landscape and vegetation during the Meghalayan? How far the coastal vegetation especially the mangrove ecosystem and the freshwater ecosystem responded to the climatic vicissitudes? The role of different components of the vegetation including the nonvascular plants retrieved from sedimentary archives with the help of palynology and geochronology has been looked into while addressing the contribution. How far the anthropogenic factors affected the environment dynamics of the Konkan coast during the recent past? Signatures of any major natural disasters or other episodes reflected in the sedimentary archives too have been looked into while addressing the present theme. These are some of the objectives that have been set up while addressing issues related to Late Holocene study of Konkan.

25.3 Methodology and processes

The subsurface profile from coastal deposits was collected from Koparkhairne (Thane Creek), Vasai Creek (north-south to the east of the Thane Creek), and Adari (Sindhudurg district) by improvising

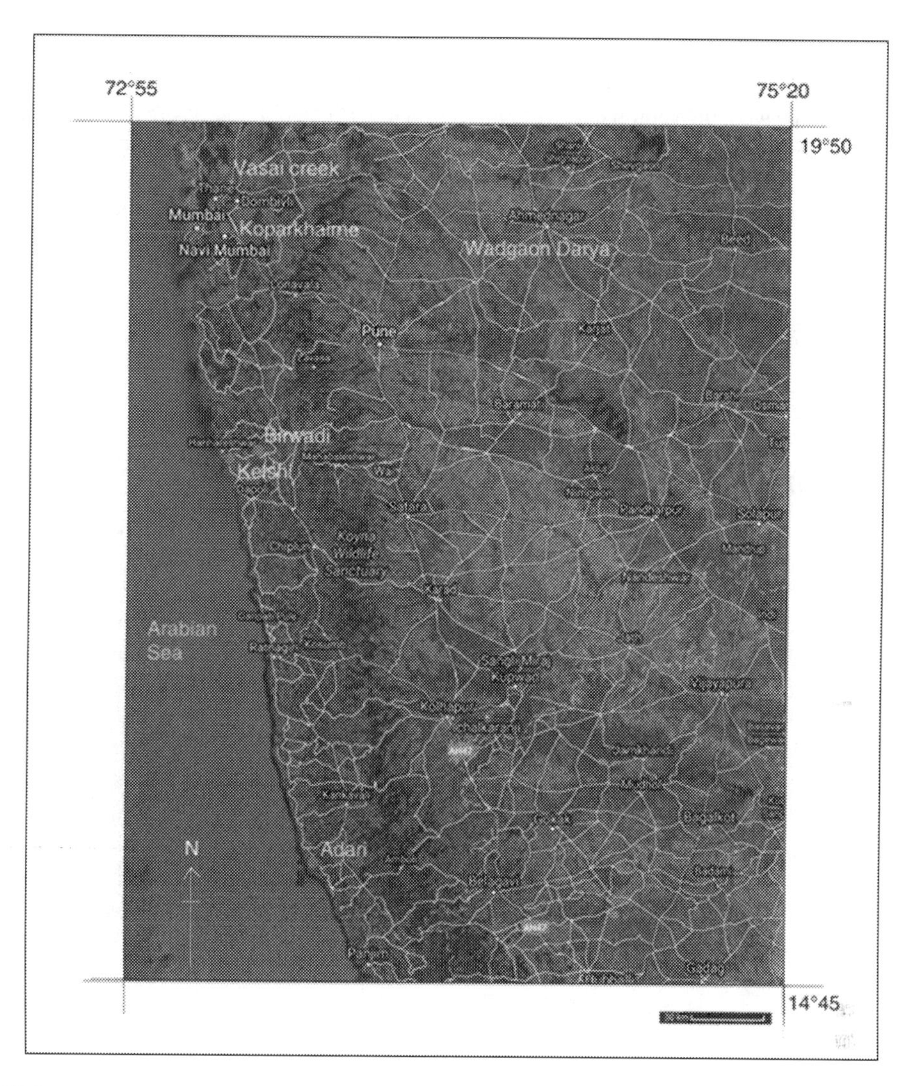

FIGURE 25.1

Location map showing selected sites for the study along the Konkan coast and one area of the hinterland.

special techniques to retrieve uncontaminated and undisturbed soft sediments. Subsurface sampling was carried out penetrating using polyvinyl chloride (PVC) pipe into the sediment bed. Excavated pits and trenches were the main sources for sample collection. Lithosections were prepared with the help of field observations and laboratory analysis. Samples for organic recovery were processed depending upon the lithology using conventional palynological processing techniques adopted by various workers (Gray, 1965; Herngreen, 1983; Brasier, 1980; Traverse, 1988, 2007; Limaye, 2004). These techniques involved mechanical separation, chemical digestion, the concentration of organic materials, and permanent preparation of slides for microscopic examination under light and phase-contrast microscope. Pollen grains, spores, phytoliths, charcoal, fungal spores, fruiting bodies, etc., were identified and photomicrographs were taken. Accordingly, paleopalynological data were gathered. These data were then compared with modern analogs (Thanikaimoni, 1987; Tissot et al., 1994) for a finer resolution of environmental change. Quantitative analysis was carried out for Koparkhairne. Emphasis was on addressing the landscape dynamics of the Dapoli block as studied sites revealed aspects of geomorphological changes associated with the climate dynamics. With the help of available software of SIGMAPLOT, pollen and nonpollen profiles were prepared. To have the age control of these deposits, radiocarbon dating was done at Birbal Sahni Institute of Palaeosciences (BSIP), Lucknow. All the microslides of organic material in the form of palynological preparations are available in the repository of Biodiversity and Palaeobiology of MACS-Agharkar Research Institute, Pune.

25.4 Fieldwork and sampling techniques

Fieldwork was carried out in the North Konkan, Central Konkan, and South Konkan regions to locate the probable sites for collecting subsurface sediments for Late Holocene climate change study and human dimension of coastal ecosystem dynamics using geomorphology of the area and also the present-day vegetation study surrounding the area. Six profiles have been studied two from North Konkan, Vasai creek and Koparkhairne; Wadgaon Darya from central region; and Birwadi, Kelshi, and Adari Lake from South Konkan. Case studies for paleoclimate analysis have been carried out using pollen and non-pollen, sedimentology, geomorphology, and chronology as proxies (Figs. 25.2 to 25.12).

A fieldwork was undertaken between Thane and Vasai, particularly along Ghodbunder River channel and samples have been collected from a location 0.5 km adjacent to Vasai Creek (19°25′N and 72°97′E). It is an estuarine creek, one of the two main distributaries of Ulhas River in Maharashtra state of western India. Samples from a foundation pit near Thane Creek site on the left bank of Thane—Vasai Creek, close to railway station, were collected at a depth of 12.0 m level from the surface (Fig. 25.4). Exposed section shows remarkable compaction of clay. Lithology of the section was studied and lithosection was prepared (Fig. 25.5). ^{14}C dating of two samples was carried out at Birbal Sahni Institute of Palaeosciences, Lucknow, Uttar Pradesh, India.

The Koparkhairne profile (19°4′28″N; 72°57′59″E) of 3.0 m carbonaceous clay was retrieved, closer to New Vashi, Mumbai, North Konkan region. Subsurface samples were collected from a pit at the interval of 0.60 m for palynological and quantitative analysis (Figs. 25.6 and 25.7).

Wadgaon Darya is a village situated in a hilly and plateau area, in Parner taluka, Ahmednagar district, Maharashtra. It has a deep valley near the village. Tufa deposits in Wadgaon Darya (19°6′N and 74°20′E) and near Pimpalgaon in Sangamner area were analyzed for fossil contents (Fig. 25.8).

Birwadi [17°51′20″N; 73°05′55″E], a coastal village in Dapoli taluka of Ratnagiri district, is located on a seasonally flowing low-order stream and is hardly 1.0 km north of the mouth of Jog River. The samples were unlocked from a dug out well constructed by local farmers in 2008. A trench at Birwadi for knowing stratigraphy below the log of wood found at a depth of about 1.5 m below the ground level on the right bank of a runnel ephemeral and originating on low hills near Birwadi and directly flowing into the Arabian Sea. Wood samples located in the well from the bottom part of Savitri Creek at Ambet Village were procured for a geochronological and vegetation analysis. The site was visited by us in May 2015 to understand local geomorphology and to find out a suitable pit for palynological studies (Fig. 25.9).

FIGURE 25.2 Konkan landscape—I.

(A,C,D) Generalized view of Konkan landscape. (B) Exposed section near a stream, Vanand, Dapoli area. (E) Terrace landscape in the valleys, Dapoli area. (F,H) Exposed section, Dapoli area. (G) View of Harnai port.

Huge sand deposits around + 20.0 m were found at Kelshi. This sections and lithosections in the field were studied in Dapoli area at Kelshi (17°55′63″N; 73°03′46″E) for correlating with available records along with the cultural aspects (Figs. 25.10 and 25.11).

Carbonaceous samples from Adari lake (15°50′22″N; 73°39′08″E), 2.0 km away from Vengurla, South of Konkan were studied for a palynological analysis. Subsurface samples were retrieved from the excavated pond with the help of PVC pipes of 6.0 cm diameter to obtain continuous and uncontaminated

FIGURE 25.3 Konkan landscape—II.

(A,B) Exposed section showing Karal formation. (C) Sand dune section, Kelshi. (D) Present vegetation near Birwadi well. (E) The cashew tree (*Anacardium occidentale*) , a tropical evergreen tree. (F) Exposed section near Birwadi. (G) Birwadi well. (H,I) Sea caves as indicators of sea level near Harnai.

cores. Only shallow cores up to 3.0 m length has been obtained by this manual method by penetrating pipes down through the mudflats and lake bottoms and applying physical force on the surface. These shallow cores have been sealed with appropriate directions and labeling. Cores have been cut vertically in the laboratory and appropriate lithosections were prepared.

FIGURE 25.4 Location near by Vasai Creek from where the samples were retrieved for study.

(A,B). Exposed section at Vasai Creek. (C) View of vegetation in and around Vasai Creek. (D) Vasai Creek showing mangrove vegetation.

25.5 Observations and results

At Vasai Creek, the lithology of the section was studied and lithosection was prepared (Figs. 25.4 and 25.5). The palynological assemblage was studied at three levels along the 8.5 m section. The first level (0.0–2.0 m) shows dominance of pteridophytic spores and *Staurastrum* sp. with few foram tests. The second level (2.0–3.5 m) comprises abundance of foram tests with few pteridophytic spores and *Staurastrum* sp. The third level (3.5–8.5 m) shows the dominance of foram tests, few fungal spores, and fungal hyphae. Two samples were radiocarbon dated at BSIP, Lucknow. A 2.0–3.5 m level was dated as 8980 ± 110 years BP and at level 3.5–8.5 m dated as 5980 ± 110 years BP, respectively (Table 25.1).

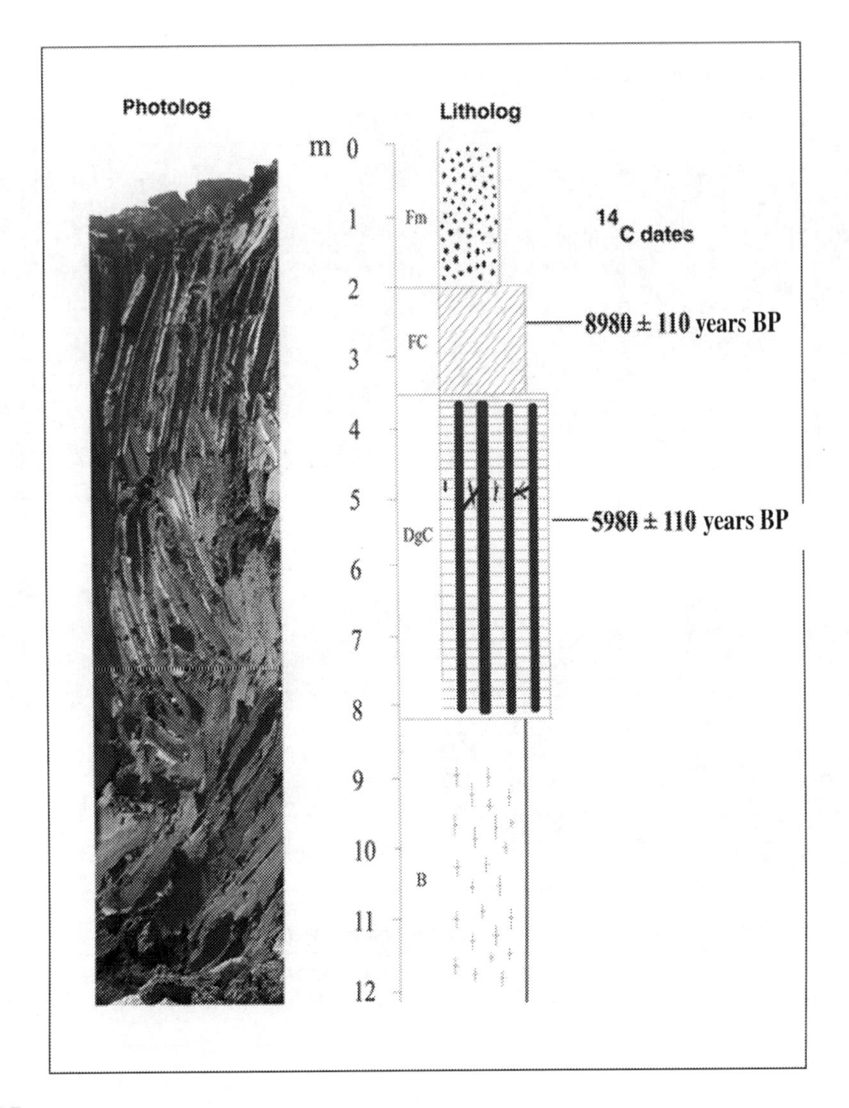

FIGURE 25.5

Photolog and litholog of Vasai Creek section, Thane.

The Koparkhairne profile with dark brown mud, plastic in nature having a total thickness of 3.0 m, was analyzed for palynological studies. The recovery of pollen spore and other organic material from a 3.0 m profile was reasonable as most of the categories are represented. Pollen spectrum represents Late Early to Middle and Late Holocene. The bottom (2.40 m) carbonaceous clays that yielded pollen assemblage gave a ^{14}C date of 6980 ± 370 years BP. The 3.0 m level spore/pollen and dinoflagellate were rather scarce. However, there was a mixed representation of marine and freshwater at 1.80 m level. Foraminiferal linings, *Selenopemphix nephroides, Multispinula quanta, Cymbella* sp. (diatom),

Table 25.1 Palynological assemblage of Vasai Creek.

S. No.	Depth (m)	Palynological observation	^{14}C Date in years BP	Interpretation
1.	0.0–2.0	Dominance of pteridophytic spores and *Staurastrum* sp.; Few foram tests	-	Freshwater environment with slight marine incursion.
2.	2.0–3.5	Abundance of foram tests; few pteridophytic spores and *Staurastrum* sp.	8980 ± 110	Marine environment with slight freshwater incursion.
3.	3.5–8.5	Dominance of foram tests; few fungal spores and fungal hyphae.	5980 ± 110	Marine environment with humid conditions.

and few other dinoflagellates represent the marine forms. This level has been radiocarbon dated as 1710 ± 90 years BP. The spores of *Lygodium flexuosus* were frequent. A marine facies was more pronounced despite the occurrence of certain freshwater forms. At the top level of this sequence, the freshwater source was very well represented due to the presence of spores of pteridophytes and pollen of Angiosperms. The recognizable taxa of pteridophytes belong to Pteridaceae and Parkeriaceae. The pollen taxa of Liliaceae, Pandanaceae, Arecaceae, and Euphorbiaceae were common. The assemblage was also rich in fungal complex and at the same time this section showed signatures of some marine elements. The top-level of 0.60 m had an abundance of structured terrestrial material, amorphous and fungal complex. The middle level of 1.80 m too showed similar assemblage, whereas the bottom level of 3.0 m had abundance of amorphous material. Thus, the palynodebris analysis of Koparkhairne displayed the prevalence of intertidal coastal plain facies (Figs. 25.6 and 25.7).

In the valley at Wadgaon Darya, there is a large cave with stalactite and stalagmite in the temple of Dariyabai and Velhabai. The stalactites formed due to the deposition of calcium salts hanging from the roof of the cave and stalagmite of deposition of calcium salts rising from the floor of the cave were found. The growth of both stalactite and stalagmite was found to be increasing downward and upward direction very slowly. Some of them were connected to each other and took shape of a column. This is a unique feature of limestone morphology termed as tufa deposits in this region. These calcareous tufa deposits were finely laminated with porous calcium carbonate, consisting casts and impressions of twigs and leaves. Though the calcareous tufa deposits were formed due to waterfalls and stream activity in the past, currently such hydrodynamic scenarios are not seen. Database for mega plant fossils was attempted from the collections made during the fieldwork (Fig. 25.8).

In the Birwadi well section, at a depth of 1.2 m, clay-rich uncarbonized plant material (35 cm thick) was encountered. We were informed that this layer is a part of the roof material of a house. This three-layer section was found to be capping 10 wooden rafters with 20–30 cm diameter of good quality Teak (*Tectona grandis*) (Fig. 25.9). It was inferred that the house belonging to a rich man collapsed due to earthquake in tenth Century, as the wood was dated to 960 ± 63 AD by the conventional ^{14}C dating method (Marathe and Rajaguru, 2011). Unfortunately the well where Marathe and Rajaguru discovered remains of the collapsed house constructed and we could not reexamine well stratigraphy, instead in Birwadi itself we found the unlined well and local people have used above-mentioned wooden rafts for the upper part for protecting unlined well. About 1.5-m-thick reddish sandy silt was found to be capping ferricritised breccia, rich in subrounded laterite boulders and pebbles and subangular pebbles of basalt in a matrix of clay and laterite pellets. The ferricritised breccia is about 80–90 cm thick and

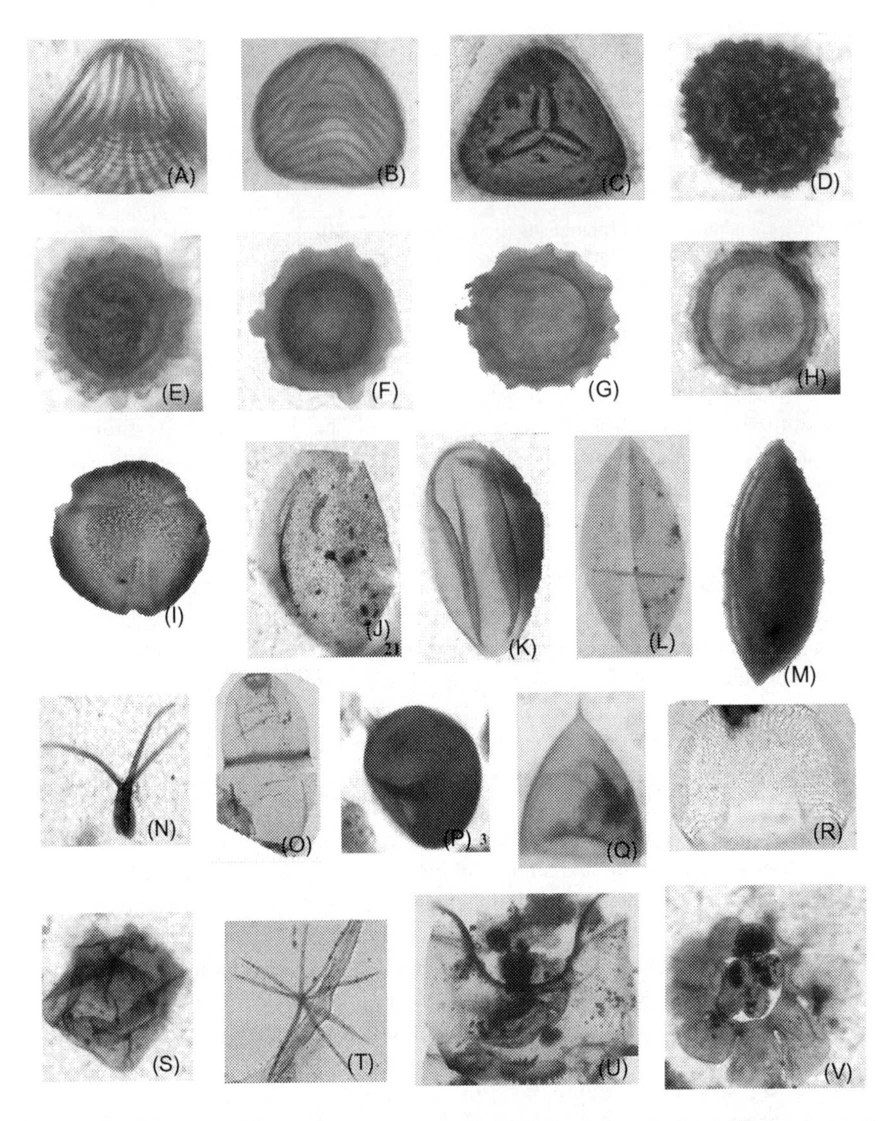

FIGURE 25.6 Palynological assemblage of selected ecological/environmental facies indicator from subsurface sediments of Koparkhairne.

(All photomicrographs are enlarged ca. 500× except fungal remains, which are, enlarged ca. 1000×, unless otherwise specified).

(A,B) *Ceratopteris thalictroides* (Pteridophytic spore); (C) *Pteris* sp. (Pteridophytic spore); (D) *Lygodium* sp. (Pteridophytic spore); (E,F) Euphorbiaceae pollen, a wet evergreen element, high rainfall indicator; (G) (H) Compositae pollen; (I) *Avicennia* sp. (Avicenniaceae); (J) Liliaceae pollen; (K) Arecaceae pollen; (L) Pandanaceae pollen; (M) Palm pollen; (N) *Frasnacritetrus* sp. (Fungal spore); (O) *Dyadosporonites* sp. (Fungal spore); (P) *Glomus* sp. (Fungal spore) erosion indicator; (Q) *Veryhachium* sp. (Phytoplankton); (R) *Cymbella* sp. (Diatom); (S) Dinoflagellate; (T) Trichome; (U) Chironomid; (V) Foraminiferal lining, shallow marine facies.

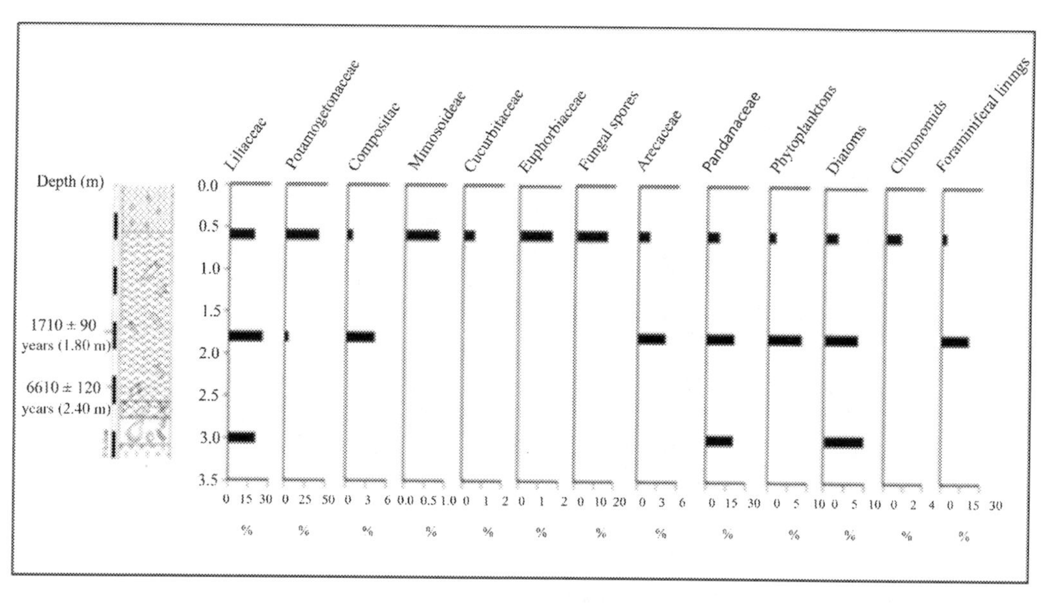

FIGURE 25.7

Pollen and nonpollen palynomorphs (NPP) profile of Koparkhairne.

was unconformably capping compact basalt. Our field studies suggest that the breccia served as a strong foundation for the construction of tenth century collapsed house. Incidentally we also collected a piece of raft wood for dendrochronological and stable isotope studies for knowing the paleoenvironment of the tenth century in the coastal Konkan. The ferricritised breccia was observed on the beach at Birwadi by a local stream when the sea level was low or the sea regressed further west from its present position, sometime during the Early Holocene (before 6 ka BP). It was probably deposited as a hill slope colluvium by a low-order stream originating in surrounding hills around Birwadi sometimes during Late Pleistocene when the sea level was lower by scores of meters than the present one. Field observations around Anjarle, Birwadi helped to tentatively build up a stratigraphy as given Vanand well section (Fig. 25.12).

Huge sand deposits near Kelshi represent nothing but evidence of a rare coastal dune with anthropogenic activity on the west coast of India. The sand deposit is a mixture of fine sand, silt, and sea shells. The sand was deposited on laterite formation or on the basalt (bedrock). It is certain that upper sand is most probably not generated by the tsunami. Detailed mineralogical studies of dune section at depth 0.25 m, 6.2 m, 12.6 m from the exposed section of a dune at Kelshi establish the presence of magnetic minerals such as ilmenite, and their derivatives and nonmagnetic minerals such as quartz, pyroxene (mostly augite), and hypersthene (hornblende), the weight percentage of heavy minerals such as magnetite and ilmenite together changes from 49.6% at 0.25 m level to 10.3% at 12.6 m level. Similarly, nonmagnetic mineral such as quartz varies from 13.5% at 0.25 m level to 2.3% at 12.6 m level of the sand dune. According to these quantitative studies, the bottom zone demonstrates supratidal oceanographic processes, while the top zone represents high-energy littoral processes probably Aeolian based on personal communication from Dr. A.R. Gujar, NIO, Goa) (Fig. 25.10).

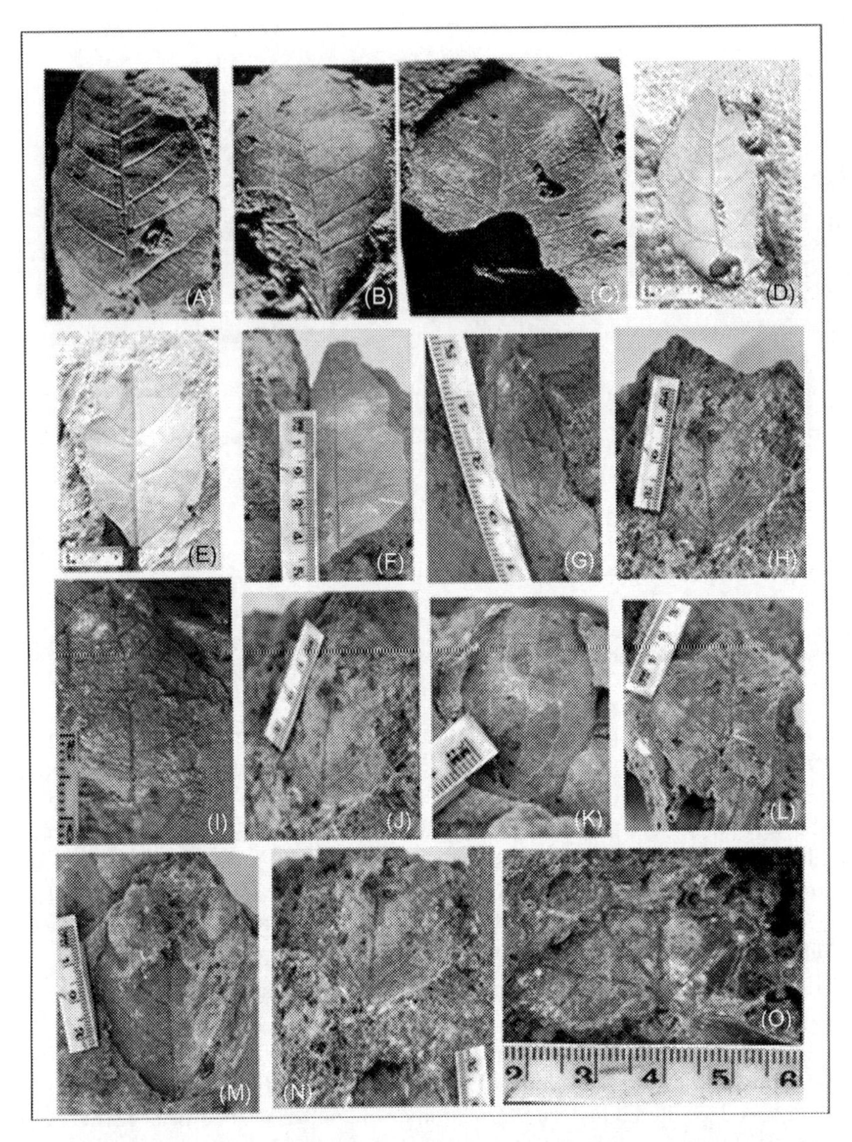

FIGURE 25.8 Plant fossil assemblage from calcareous tufa deposits of the Deccan basalt province.

(A,B) *Ficus glomerata*, (C) *Ficus arnottiana*, (D,E,F) *Ficus racemosa*, (G,I,K,L,M,O) Unidentified leaf impressions, (H, J, N) *Mallotus* sp.

The signatures preserved in the form of pottery at Kelshi may provide evidence of early human settlement along the Konkan coast and its collapse due to geological and/or climate episodes probably of tsunami toward the Medieval period (Fig. 25.11). The charcoal collected from the sand deposits at 4.10 m above mean sea level (MSL) was dated as 1170–990 BP (Deo et al., 2004–2006). The mollusc

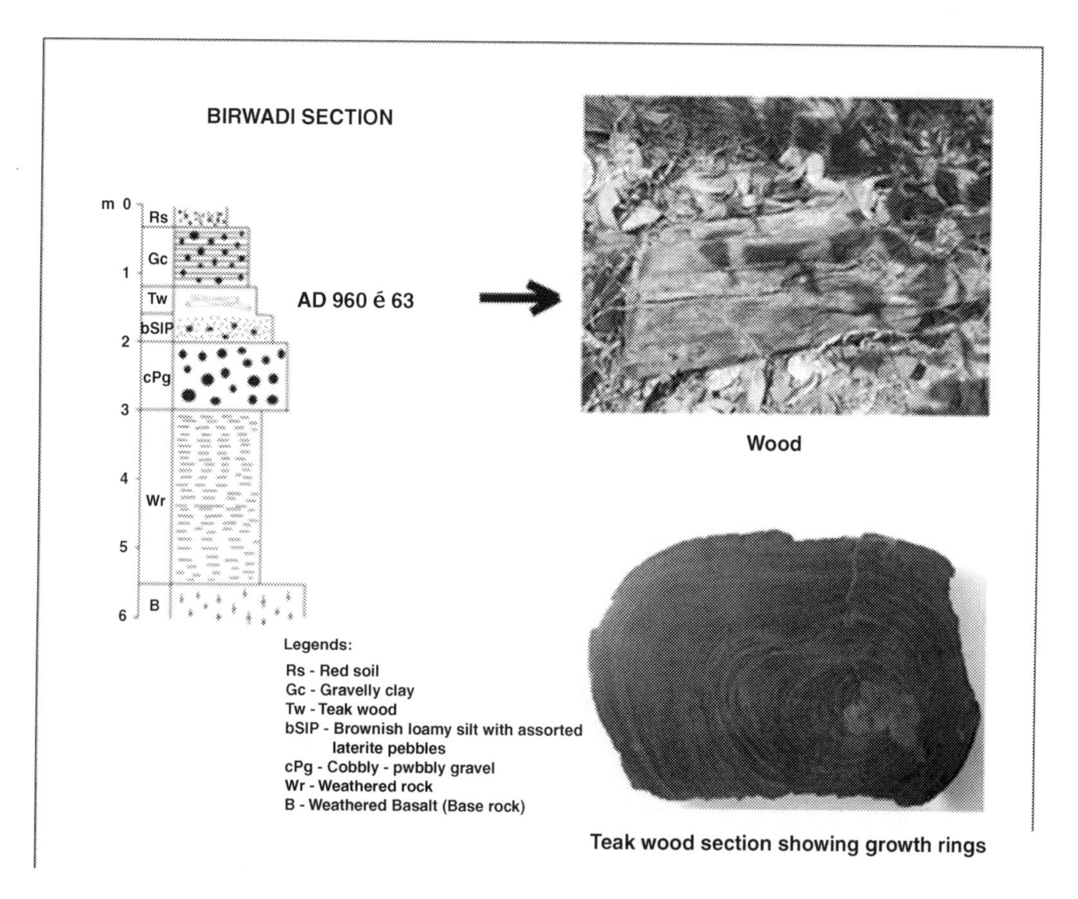

FIGURE 25.9

Lithosection of Birwadi, Dapoli.

shells collected from the top sand deposits were dated 1490 ± 110 years BP and 1690 ± 110 years BP, respectively (Marathe and Rajaguru, 2011; Deo et al., 2011, 2017–2018).

Dark brown lacustrine clay section of Adari with depth 6.09 m contains specks of coaly matter showing dominance of spores and pollen grains. Pteridophytic spores were plenty and they consisted of *Ceratopteris thalictroides*, Polypodiaceae, Schizeaceae, Pteridaceae, Lycopodiaceae, and Selaginellaceae. Bryophytic spores of mosses were too common. Pollen grains identified were of Asteraceae, Euphorbiaceae, Liliaceae, Malvaceae, Portulacaceae, Acanthaceae, and Moraceae types. Dinoflagellates and other planktons *Bacteriastrum* showed marine affinity. A few salt glands and diatoms of the pinnate type also represented in the palynoflorule. Palynological analysis of Adari Lake reflected the dominance of a structured terrestrial material. The 6.0 m level of Adari clay section has been dated as 1700 ± 120 years BP. Adari Lake shows freshwater swamp or floodplain type of environment.

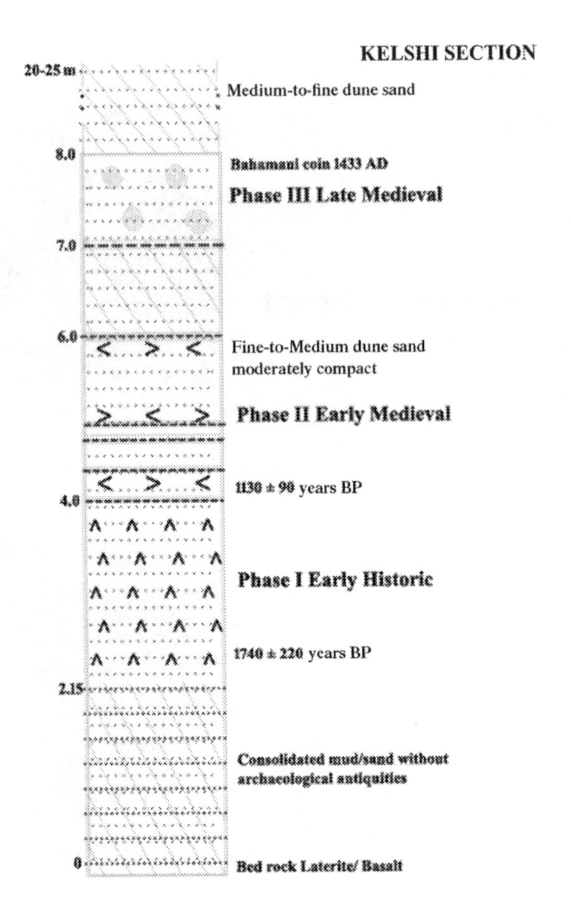

FIGURE 25.10

Lithosection of Kelshi sand dune (modified after Deo et al., 2011).

25.6 Discussion

The present study is focused on vegetation, climate dynamics, and human interaction during the Late Holocene (Meghalayan) along the Konkan coast. The Holocene is the most recent stratigraphic unit within the geological record and covers the time interval from 11.7 ka BP until the present day (Walker et al., 2009) and it refers to the warm episode that began with the end of the last glacial period. Along with the preceding Pleistocene, the Holocene is now formally defined as a Series/Epoch within the Quaternary System/Period (Gibbard et al., 2005).The logical way is to accept and differentiate the Holocene Series/Epochs into three Subseries/Subepochs roughly based on age boundaries for the Early [11,000–8000 years BP]—Middle [7000–4000 years BP]—Late Holocene [3000–present years BP] (Walker et al., 2012). As the newly accepted Meghalayan covering the last 4.2 kyears, the present work falls into the last phase of the Holocene. Late Holocene (< 3 ka years) was relatively dry, but dynamic, because of the anthropogenic influence and the coastal dynamics in terms of vegetation and landscape changes. In fact, the Late Quaternary climate and its impact on landscape and vegetation

FIGURE 25.11 Cultural aspects of Kelshi, evidence for human settlement along Konkan coast.

(A,B,C,F,G) Pieces of pottery; (D) Oyster shells; (E) Vertebral remain.

response in Konkan have been little understood due to the paucity of biological proxies and sediment archives. Palynological data from the coastal deposits of Maharashtra are scarce. Except preliminary account of Vishnu-Mittre and Guzder (1975) and Guzder (1980), no serious effort and study has either been made until 2000. However, subsequent contributions based on subsurface sediments retrieved by new and improvised methods provided a fair amount of palynological data for the paleoecological and paleoenvironmental analysis of Konkan (Limaye, 2004; Kumaran et al., 2001, 2004, 2005, 2013). Despite the above contributions, there exists a gap in our knowledge as far as the Late Holocene

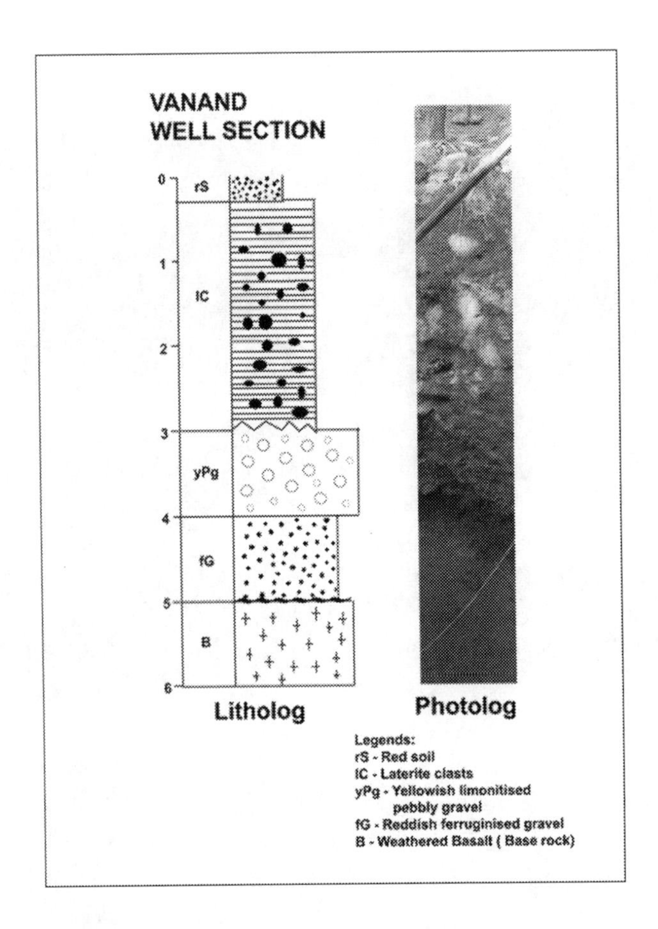

FIGURE 25.12

Vanand well section litholog and photolog, Dapoli area.

of the region is concerned. Although there have been records of dry events during Late Holocene on the basis of marine archives and archeological evidence, Late Holocene climate and vegetation dynamics is yet to be decoded using biological proxies. Nevertheless, pollen data of extant plants are available for identification and comparison purposes, but that too is limited due to incomplete record (Vasanthy, 1976; Nayar, 1990; Pathak, 1994). Considering this limitation, attempts are being made to recover maximum pollen-spore data from the sediments and also neopalynological data from modern analogs. The organic matter including pollen grains, spores, and others can be identified and compared with their modern analogs for building up the vegetation history of the region on one hand to distinguish the ecological shifts (facies changes) due to sea-level changes and freshwater influx, on the other. NPPs signatures are utilized as sediment sequences indicate the aquatic environment and their response to the hydrological changes as a result of monsoon variability. Considering the value of NPP as paleoenvironmental indicators (Van Geel, 2004; Limaye et al., 2007) these microscopic entities offer important proxy data for paleoecological and environmental changes and equally complement palynological data for higher resolution ecological changes and climate of the past. It is observed that their implications are significant in the pollen pauper sediments deposited in arid and low rainfall period of the Late Holocene. ^{14}C and accelerator mass spectrometry (AMS) dating were used to date the organic samples for chronological control. Further, sedimentological aspects were dealt to ascertain their utility for correlation of the biological proxies. The retrieved data are found to be of immense importance for the paleoecological and paleoclimatic interpretation of the Late Holocene deposits of Konkan.

To understand the Late Holocene climate, vegetation and environment studied locations falling in two different geomorphological settings were selected. The studied locations namely Koparkhairne, Vasai Creek, Kelshi, Birwadi, and Adari represent the coastal plains; whereas Wadgaon Darya belongs to the hinterland plateaus on the Deccan Basalt. As the coastal plains in Konkan and the adjacent hinterland regions in western Maharashtra have contrasting landscape scenarios, signatures of vegetation response to climate dynamics and anthropogenic impact during the Holocene could be easily decoded. The coastline is characterized by the presence of wetlands of creeks, bays, and estuaries. The flat coastal plains ranging from 45.0 to 76.0 km in width constitute the Konkan coastal belt and above which stand out residual hills and ranges of the Deccan Volcanics. The coastal tract of Maharashtra, Konkan, exhibits a variety of sediments representing an admixture of different environmental conditions. The older deposits are confined to 8.0–10.0 km inland from the present coast and occur as raised deposits up to 120.0 m above the mean sea-level. The age of these deposits ranges from Neogene to Pleistocene on the basis of palynofloral contents of the contemporaneous deposits that are mainly in the form of peats, carbonaceous clays, and black plastic clays (Limaye, 2004; Kumaran et al., 2005, 2012, 2013, 2016). However, the coastal deposits of the Quaternary are seldom exposed inland except the "Karal"/littoral concrete and beach rocks and they range in age from Early to Late Holocene. Besides the above, soft sediments in the form of bluish-grey clay, peat, and carbonaceous clay do occur in the subsurface at a few pockets along the coast. These sediments have preserved a good record of past life and thus their study has helped to gather a good amount of data to address the effects of sea-level fluctuations, accompanying change of environmental conditions and palaeoclimate. As Koparkhairne, Vasai Creek, Birwadi, and Kelshi locations not far away from the present sea coast, signatures of sea level as well as the mangrove response could be ascertained. Despite the paucity of sediment archives it has been possible to decode aspects of the Late Holocene reasonably. Of these, Kelshi sand dune provided evidence of human occupation and maritime trade (1170 ± 990 years BP, 1490 ± 110 years BP; 1690 ± 110 years BP), while the Birwadi well contained ancient remains of wood used for house construction in the recent past (960 ± 63 years BP) along the Konkan coast in the Late Holocene. The hinterland geomorphic settings were essentially derived from that of the Neogene landscape, and Quaternary sedimentation record is limited to that of calcareous tufa, palaeolake, and few river sedimentary sections (Pawar and Kale, 2006; Rajaguru and Kale, 1985). The hinterland locations of Kaas Plateau and Wadgaon Darya have preserved biological proxies in the form of pollen, spores, NPP, and leaf assemblage and provided vegetation response to Holocene climate variability. Sedimentary archives of Kaas showed that the plateau had sustained seasonal wetland bodies in the Early to Middle Holocene as compared to Late Holocene as the plateau crust topography has been changed since then. Field observations as well as the present-day landscape stand testimony to it. Further the monsoon dynamics too had an impact due to reduced rainfall toward the Late Holocene in the hinterlands. Early Holocene to Late Holocene vegetation and climate along North to South, Konkan coast is shown in Table 25.2.

Table 25.2 Early Holocene to Late Holocene vegetation and climate along North to South, Konkan coast.

S. No.	Location	Geomorphic context	Age (years BP)	Climate/vegetation
1	Vasai Creek	Estuarine flat	5980 ± 110 (Gray clay) to 8980 ± 110 (Gray clay) Mid Holocene	Configurational changes in strong tidal environment. Humid, Higher, and stabilized sea level and higher rainfall (more than 2500 mm related to Holocene Climate Optimum (HCO)
2	Koparkhairne	Estuarine flat	1710 ± 90 (Carbonaceous clay) Late Holocene To 6980 ± 370 (Carbonaceous clay) Mid Holocene	Intertidal coastal plain facies. Lower sea level/ reduced rainfall (2000 – 2500 mm)/dry climate
3	Wadgaon Darya	Fossil Tufa in upland Deccan Plateau	9200 ± 900 (Tufa deposit) Early Holocene	Higher sea level/Higher rainfall related to HCO, Myristica swamps
4	Birwadi	Estuarine flat	960 ± 63 (Teak wood) Late Holocene	Sinking of house due to earthquake 960 BP. Lower sea level/reduced rainfall (2000–2500 mm) dry climate/deciduous forests along with patches of riparian flora
5	Kelshi	Coastal Dune	1170 ± 990 (Charcoal) to 1490 ± 110 1690 ± 110 (Mollusc shells) Late Holocene < 2 ka to < 200 years	Unusual deposition of 12–15 sand over coastal dune. Lower sea level/reduced rainfall (2000–2500 mm) dry climate/deciduous forest
6	Adari	Coastal plain	1700 ± 120 (Gray clay) Late Holocene	Freshwater swamp or flood plain type of environment

25.6.1 Landscape dynamics

The present Konkan has been essentially derived from the combined effect of sea-level changes and the post-Neogene tectonic activities (Pawar, 1993). Evidence of sea caves at a number of places along the coast and raised marine/estuarine terraces in around Velas and Parchuri stand testimony to the above (Rajshekhar and Kumaran, 1998, Kumaran et al., 2004). The record of off-shore peat too showed the impact of sea-level changes on Konkan coast (Deshpande, 1998; Mascarenhas, 1997). Further, there has been a remarkable change in geomorphologic features all along the Konkan stretch. A chain of wetland system of backwaters of limited numbers raised marine terraces, sea caves, and beach rocks/"karal" (littoral concretes) are some of the characteristic features seen all along the Konkan coast. As of now the configuration and features of Konkan coast essentially represent a Mid-to-Late Holocene high stand characterized by barrier spits, open inlets, and estuaries. Some degree of reorientation and landscape dynamics during the late Holocene too has modified the estuarine system. The progressive sedimentation and seasonal inlet closure lead to the development of mangrove swamps in certain pockets toward the Late Holocene (Karlekar and Rajaguru, 2012). The Quaternary sediments that occur all along the Konkan coast are in the form of consolidated and unconsolidated sediments. However their distribution is sparse and the sediment archives for biological proxies are difficult to be retrieved. Essentially one has to improvise special techniques in the form of short coring devices and making trenches for procuring uncontaminated samples (Limaye, 2004).

25.6.2 Vegetation response

The data accrued from the locations investigated and available published data from Konkan provided a fair understanding of vegetation response to climate variability especially during the Late Holocene. Of these the Kaas plateau lake sediment archives were found to be very significant as it the first report from the seasonal plateau lakes based on the sediment records, which has been dealt with elsewhere (Limaye et al., 2021). The pollen and non-pollen spectra show interesting results ranging from Middle to Late Holocene [7865–2655 years BP]. Despite the lack of enough organic carbon in the sediments, the AMS dates could help in distinguishing the facies changes with the help of ecological markers. The Late Holocene sequence at 0.36 m level dated as 2585 years BP was found to be relatively dry. Sedimentary archives and preserved pollen, spores, and NPP showed that the plateau had sustained seasonal wetland bodies in the Early to Middle Holocene as compared to Late Holocene because the plateau crust topography has been changed since then. Field observations, as well as the present-day landscape stand testimony to it as the prevailing landscape no longer provide settings of holding wetland bodies. Further the monsoon dynamics too had an impact due to reduced rainfall toward the Late Holocene in the hinterlands.

Vasai Creek has been probably creek distributary of Ulhas River, since the Late Pleistocene. This section represents 8.0–10.0 m thick Mid to Late Holocene estuarine mudflat above the present Vasai Creek Surface. The fill has been deposited by Vasai Creek, a tributary of River Ulhas, after the sea-level reached its present level around 6.0 ka. ^{14}C dating provided good deal of new information on sea-level changes and also the environment around Thane, Mumbai. The ^{14}C dates obtained at 2.0–3.5 m level is 8980 $\pm$ 110 years BP and 3.5–8.5 m level is 5980 $\pm$ 110 years BP, respectively. ^{14}C date 8980 $\pm$ 110 years BP of a sample (Organic-rich dark brown clay) with a depth of 2.0–3.5 m from the surface shows reversal due to contamination by old vegetational transported material by the Vasai stream from its upper reaches. The entire estuarine fill therefore appears to have formed in fluctuating tidal environment, leading to churning or agitating depositional environment during Mid Late Holocene. The creek fill is affected by configurational changes rather than tectonic factors such as a strong earthquake (as at Birwadi) or complex factors such as Tsunamis, subsidence of offshore zone, and or irregular ISM with increased frequency of cyclonic storms in Arabian seas (As hypothesized for Kelshi).

In ^{14}C dates Kelshi is found to be a unique archaeological site that has helped to recognize and give chronological support in understanding the possibility of Tsunami in this region. Waves exceeding 20.0 m in height attacked Kelshi, and a huge sand deposit of shoreface sand, reaching more than 20.0 m in height above sea-level, was developed after the Tsunami (Sugauwara et al., 2005). A sand dune profile of about 20.0 m at Kelshi has been found to be interesting in terms of a probable geological extreme event as it contains signatures of cultural activities such as pieces of pottery, charcoal, abundant invertebrate, and a few vertebrate faunas. ^{14}C dates obtained from project report (Marathe and Rajaguru, 2011) show that these locations fall within the Late Holocene. The signatures preserved in this profile may provide evidence of early human settlement and its collapse due to geological and or climate episodes. Analyses of samples have not yielded any organic matter for an ecological shift of facies change. However, there is scope for detailed micropaleontological study to decode any major geological event or sea level that might have a major impact on this isolated sand dune formation along the Konkan coast.

The Late Holocene (< 3000 years BP) along the Konkan coast reveals a comparatively poor picture of vegetation as the organic recovery from the sediments is scarce. In fact, arborescent (C_3 plants) are scarce and aquatic plants were developed on the seasonal freshwater bodies that are frequent. Mangrove representation is observed to be poor too as the saltwater balance is badly affected due to the paucity of freshwater inflow. This suggests that the rainfall was much less during the period and perhaps a dry (arid) condition was prevailing at that time. A shift in reduced rainfall resulted in relatively poor vegetation during the Late Holocene, which may be due to threats to the environment due to anthropogenic pressures. In fact, the utility of thecamoebians has been considerably exploited while ascertaining the degradation of environments due to anthropogenic pressures (Patterson et al., 1996; Charman et al., 1998; Woodland et al., 1998; Beyens and Meisterfeld, 2001; Lorencova, 2009; Payne, 2011; Limaye et al., 2017b). The thecamoebians are considered to be potential palaeoecological indicators because of their ability to encyst and survive in adverse environmental conditions including salinity, desiccation, and reduced pH (Medioli et al., 1990). The thecamoebians develop distinct morphotypes in response to environmental stress. As the Arcellaceans live at the sediment-water interface, they are highly responsive to environmental stimuli (Reinhardt et al., 1998). These traits have made them excellent indicators of various natural, chemically polluted, and rehabilitated subenvironments within lakes affected by industrial and mining pollution (Asioli et al., 1996; Patterson et al., 1996; Reinhardt et al., 1998; Kumar and Patterson, 2000). The thecamoebians can also be used as excellent indicators of pollution levels (Collins et al., 1990; Asioli et al., 1996; Patterson et al., 1996). *Centropyxids, Arcella,* and *Cucurbitella tricuspis* types are the most common types of the thecamoebians found in the Holocene archives (Limaye et al., 2017b). The factors that control the geographic distribution of modern Arcellaceans include the vegetation in and around lakes and climatic conditions, which ultimately control water levels, chemistry, trophic levels, and the nature of a thermocline (Collins et al., 1990).The low diversity and dominance by *Centropyxis aculeata* and *Arcella vulgaris* indicate a stressed environment (Dalby et al., 2000). *Cucurbitella tricuspis* is seasonally planktic and readily transported (Kumar and Patterson, 2000).

The calcareous tufa deposits of the Deccan basaltic province provided a different dimension of vegetation response to climate dynamics. Plant fossil assemblage in the form of leaf impressions belonging to *Ficus glomerata, Ficus arnottiana, Ficus racemosa,* and *Mallotus* sp. from calcareous tufa deposits of the Deccan Basalt Province were retrieved from Wadgaon Darya (Fig. 25.8). The leaf assemblage spectrum depicts that of tropical rain forest vegetation that had prevailed until the Early Holocene as these deposits are dated as 9200 $\pm$ 900 years BP (Pawar and Kale, 2006). These leaf impressions suggest of high rainfall region where springs, rivers, and watercourses in the plain might be nearby. The leaves of the trees must have fallen and drifted to the present site and got preserved. Currently, the vegetation in and around this area comprises riparian forest as the rainfall is much reduced almost a dry period can be envisaged from this area. It is very clear that the region had enough rainfall during the early Holocene and has been reduced considerably since then.

The Late Holocene geomorphic history is based on field observations around Anjarle, Birwadi. Continental colluvial—alluvial cobbly and pebbly ferruginous gravel bed resting disconformably on basalt (at places)—appears to be one of the oldest continental deposit exposed in runnel section at Birwadi well from where wood section was collected from a log lying outside the well since 2008. ^{14}C date of wood Teak *T. grandis* is around 900 years BP, which falls within the Late Holocene period (Marathe and Rajaguru, 2011). We think that provenance of minerals such as ilmenite, magnetite, and augite is from locally present rock such as basalt, while quartz and hornblende are originally derived from nonlocal rocks such as gneiss, sandstone, etc. and redeposited in inner-shelf zone close to the present beach sometime during the Quaternary. The presence of freshwater well (now sealed) in the mud/sand flat of estuarine origin suggests that the sea level was slightly lower than the present one and the beach inner-shelf shallow zone was exposed for littoral processes including wind, at least for the last 1700 years. The inverse relationship between weight percentage of heavy minerals with the abnormally increased strength of wind after sixteenth century indicates that upper fine sand (between 8.0 and 20–25 m) was deposited as a catastrophic event in response to either cyclonic storm or a sudden earthquake leading to subsidence of beach offshore interface or a tsunami. In the absence of detailed subsurface geological data, it is difficult to provide any convincing hypothesis for the deposition of upper sand of Late Medieval Age (approximately between 170–150 years BP as suggested by luminescence dates by Dr. Navin Juyal).

The Konkan coastal plains provide a contrasting pattern of vegetation response to climate dynamics as a result of monsoon variability since the Early Holocene. As of now, the Holocene vegetation dynamics is limited to that of Middle and Late Holocene due to paucity of sedimentary archives of Early Holocene. This may be attributed to the erosion of sediments due to heavy rainfall of Holocene climate optimum as reported further south in Kerala Basin (Limaye et al., 2007; Kumaran et al., 2008). Except for Hadi no other location along Konkan provided signatures of vegetation response to sea level and climate variability since Middle Holocene (Limaye and Kumaran, 2012; Limaye et al., 2014). The Koparkhairne pollen and NPP spectrum represents Late Early to Middle and Late Holocene as the bottom (2.40 m) carbonaceous clays that yielded pollen assemblage gave a ^{14}C date of 6980 ± 370 years BP (Calibrated age 7789 years BP), while that of 1.80 m level gave 1710 ± 90 years BP (Calibrated age 1685 years BP). At 3.0 m level, only some freshwater pollen representatives (Liliaceae and Pandanaceae) and diatoms have been observed. However, at 1.80 m level marine influence (foraminiferal linings and phytoplankton) was observed despite freshwater elements were represented by Liliaceae, Potamogetonaceae, Arecaceae, and Pandanaceae. In fact, a mixed representation of marine elements indicates tidal influence to the depositional site is inferred. The presence of rich fungal complex is suggestive of good organic source, high humidity, and fair precipitation. Although this level has not been dated but is definitely younger than 1710 ± 90 years BP and the marine influence at this level has been reduced considerably. Accordingly, it seems that the area has been slowly cut off from the marine influence and exposed subsequently to terrestrial habitat probably due to anthropogenic influence in the recent past. Despite the area being surrounded by mangrove cover, there is hardly any mangrove pollen recovered from this region. This may be attributed to low pollen production and other biotic factors. However, large number of Pteridophytic spores (*C. thalictroides, Lygodium* sp.) produced by ferns that grow below the tall trees in waterlogged environment is found to be present. These ferns are not much affected due to the environmental stress as compared to the mangroves.

The Adari profile represents a Late Holocene age 1700 ± 120 years BP at 6.0 m level. An abundance of pteridophytic spores along with a large number of pollen grains indicates that the depositional environment had been under good vegetation cover contrary to the present-day environment. Hardly there had been any marine influence in the assemblage except a few dinoflagellates and salt glands of mangroves that might have been recruited through streams connecting to the water body. Adari Lake shows freshwater swamp or flood plain type of environment.

25.6.3 Culture and human impact

The biological proxies so far analyzed have indications of land cover change probably due to anthropogenic influence in the coastal plains of Konkan. Of these, the agriculture practice and developmental activities after clearing of mangroves in Adari, Hadi, and Koparkhairne seems to be significant. The gradual shift from estuarine facies to brackish indicated by less saltwater tolerant fern *Acrostichum* and its eventual clearing as decoded from the frequency spores and occurrence of pollen of Poaceae suggested prevalence of agriculture practice along the Konkan. A stressed environment in the form of pollution signatures as a result of industrial activity has also been decoded with the help of testate amoeba at a few locations. From the investigation carried out in this project, the signatures preserved in the form of pottery at Kelshi may provide evidence of early human settlement along the Konkan coast and its collapse due to geological and or climate episodes probably of tsunami toward the Medieval period (Fig. 25.11). Nearby Birwadi not far from Kelshi too indicated the signature of a geological event that resulted in the burial of wood used for house construction. Isotope study and dating of these samples may provide high-resolution climate dynamics of the Late Holocene in Konkan region.

Compared to North Konkan, Southern counterpart (Ratnagiri and Sindhudurg) is good for coastal trade but not for agriculture due to extensive development of laterites particularly below 100 m ASL and also due to gorge-like valleys of east-west flowing major streams. Hunting, gathering, and pastoral culture (microlithic) existed profusely during Mid Holocene and probably even in the Early Late Holocene. So far there has not been a single early farming site (Chalcolithic) reported from South Konkan. Probably only during the Medieval period (< 1000 years) farming has started on a relatively thin pocket of alluvium and in narrow coastal plain. It is hypothesized that the South Konkan responded more to natural environmental factors than anthropogenic ones, particularly agriculture or farming. Overall increase in ISM variability, silting of ports, and catastrophic factors such as earthquakes, cyclonic storms has been responsible for changes in the landscape of South Konkan during the Late Holocene.

The sudden decline in human activity after 500 BP may probably be due to unusually thick (˜15 m at least) medium-to-fine well-sorted quartz dune sand capping lower dune sands interlaying with anthropogenic layers covering a time span of about 1200 years. Reasons for such a thick Aeolian sands are not yet understood properly, in spite of geophysical investigations (Marathe and Rajaguru, 2011), and preliminary mineralogical studies, geophysical studies, involving electrical resistivity and seismic refraction methods showed that the sand is homogenous in texture throughout its thickness (of 20–25 m) and unconformably rests on hard rock probably, laterite or basalt. Geophysical methods could not detect intervening anthropogenic layers. Mineralogical studies show that quartz is a dominant mineral and heavy minerals such as ilmenite, magnetite, and hornblende are present even in the top layers of dune sands. Optically stimulated luminescence (OSL) dating suggests that sand between 8.0 m and 20.0 m has been deposited as a single event.

The source of sand appears to be extensive beach, and probably interface of the beach and offshore zone tentatively; it is suggested that a strong earthquake generating Tsunamis-like situation or a sudden subsidence of beach—offshore marine surface or catastrophic cyclonic storm—may be responsible for the deposition of sand above 8.0 m. At present the problem of 12- to 15-m-thick deposition sometime after 500 BP and most probably between 150 and 170 years BP remains unsolved.

25.7 Learning and knowledge outcomes

The response of plants to climate and environmental changes can be easily ascertained from pollen analysis as the pollen, spores and organic-walled parts are well preserved mainly in subsurface sediments. Besides, NPP provided evidence for changes in ecological facies at few intervals, where pollen signatures are scarce, in most of the studied locations. Comparison with modern analogs revealed that the present ecology is unsuitable for supporting evergreen forests and it can be concluded that coastal plains and associated landforms were covered by thick tropical evergreen forests and were displaced/destroyed in response to monsoon variability (Kumaran et al., 2014; Srivastava et al., 2016). The most important environmental change is relative degree of dryness in the last 2000 years or particularly the last 1500 years with moderate wet spells around 1000 years BP. By and large, 1400 AD to 1800 AD (Little Ice Age) was again dry as evident from the increased frequency of droughts in upland Maharashtra (Maratha period).The mangrove cover too has been drastically reduced as a result of changes in coastal landforms and the hydrodynamic scenarios brought in by the monsoon variations (Limaye et al., 2014). In fact, the Airoli and Vasai Creek profiles did indicate the response of mangroves to sea level and monsoon vicissitudes since last glacial maximum (LGM) (Limaye and Kumaran, 2012; Limaye et al., 2017a). Evidence of anthropogenic impact has been deciphered at Adari, Koparkhairne, and Hadi in the form of changing forest cover into agriculture practice and human occupation and maritime trade at Kelshi. Second signatures of earthquakes at Birwadi, Kelshi, cyclonic storm (Guhagar) are considered to be short events in the Late Holocene. But coastal configurational changes (Deo et al., 2011; Karlekar and Rajaguru, 2012) are equally important in changing the landscape of Konkan during the Late Holocene. Konkan, by and large, was not heavily cultivated in last 2000 years compared to Goa, Kerala, and the upland peninsula. Isotopic study of biological proxies and AMS dates may provide high-resolution climate dynamics of Konkan and adjacent hinterlands.

Acknowledgments

RBL acknowledges Department of Science and Technology, New Delhi (Govt. of India) for providing financial support in the form of DST Women Scientist Scheme (WOS-A, SR/WOS-A/ES-08/2013. KPNK thanks Council of Scientific and Industrial Research, New Delhi for providing financial support in the form of sponsored Research project [24(0275)/05/EMR-II] and Emeritus Scientistship 21(0828)/10/EMR-II. We thank Director, Agharkar Research Institute, Pune for facilities and encouragement. SNR expresses his gratitude to authorities of Deccan College, Pune for providing necessary library and laboratory facilities. We also thank Dr. A.R. Gujar (Formerly of NIO, Goa), Dr. Navin Juyal (Formerly of PRL, Ahmedabad) for academic help and Dr. Karlekar for general discussion on Kelshi. The authors thank the reviewers for their meticulous appraisal of the manuscript and useful suggestions.

References

Asioli, A., Medioli, F.S., Patterson, R.T., Thecamoebians as a tool for reconstruction of palaeoenvironments in some Italian lakes in the foothills of the southern Alps (Orta, Varese and Candia), J. Foramin. Res. 26 (3) (1996) 248–263.

Asteman, I.P., Nordberg, K., Filipsson, H.L., The Little Ice Age: Evidence from a sediment record in Gullmar Fjord, Swedish west coast, . Biogeosciences 10 (2013) 1275–1290.
Beyens, L., Meisterfeld, R., Protozoa: testate amoebae, in: Smol, J.P., Birks, H.J.B., Last, W.M. (Eds.), Tracking Environmental Change Using Lake Sediments, Terrestrial, Algal, and Siliceous Indicators, Kluwer Academic Publishers, Dordrecht, The Netherlands, vol. 3, 2001, pp. 121–153.
Bhalla, S.N., Khare, Shanmukha, D.H., Henriques, P.J., Foraminiferal studies in near shore regions of western coast of India and Laccadives Islands: a review, Indian J. Mar. Sci. 36 (4) (2007) 272–287.
Brasier, M.D., Microfossils, George Allen & Unwin Ltd., London, 1980, p. 193.
Charman, D., Roe, H., Gehrels, W.R., The use of testate amoebae in studies of sea level change: a case study from the Taf Estuary, South Wales, UK, Holocene 8 (1998) 209–218.
Collins, E.S., McCarthy, F.M., Medioli, Scott, D.B., Honig, C.A., Biogeographic distribution of modern thecamoebians in a transect along the eastern North American coast, in: Hemleben, C., Kaminski, M.A., Kuhnt, W., Scott, D.B. (Eds.), Paleoecology, Biostratigraphy, Paleoceanography and Taxonomy of Agglutinated Foraminifera, NATO Advanced Study Institute Series, Springer, Cham, Series C, Mathematical and Physical Sciences, vol. 27, 1990, pp. 783–791.
Dalby, A.P., Kumar, A., Moore, J.M., Patterson, R.T., Preliminary survey of Arcellaceans (Thecamoebians) as limnological indicators in tropical Lake Sentani, Irian Jaya, Indonesia, J. Foramin. Res. 30 (2) (2000) 135–142.
Deo, S.G., Ghate, S., Rajaguru, S.N., Holocene environmental changes and cultural patterns in coastal western India: a geoarchaeological perspective, Quat. Int. 229 (2011) 132–139.
Deo, S.G., Joglekar, P.P., Rajaguru, S.N., Late Holocene human occupation on coastal dune at Kelshi. Konkan coast, Maharashtra, India, J. Indian Ocean Archaeol. 13–14 (2017–2018) 13–28.
Deo, S.G., Joglekar, P.P., Deshpande-Mukherjee, A, Ghate, S., Investigations at Kelshi, Konkan coast 2000 to 2005: a report, Bulletin Deccan College Research Institute 64–65 (2004) 157–179.
Deshpande, G.G., Geology of Maharashtra, Geological Society of India, Bangalore, 1998, p. 223.
Gibbard, P.L., Smith, A.G., Zalasiewicz, J.A., Barry, T.L., Cantrill, D., Coe, A.L., Cope, J.CW., Gale, A.S., Gregory, F.J., Powell, J.H., Rawson, P.F., Stone, P., Waters, C.N., What status for the Quaternary? Boreas 34 (2005) 1–6.
Gray, J., Palynological techniques, in: Kummel, B., Raup, D. (Eds.), Handbook of Palaeontological Techniques, W.H. Freeman and Company, San Francisco, USA, 1965, pp. 471–481.
Gupta, A.K., Das, M., Anderson, D.M., Solar influence on the Indian summer monsoon during the Holocene, Geophys. Res. Lett. 32 (2005) 1–4.
Guzder, S., Quaternary Environment and Stone Age Cultures of the Konkan coastal Maharashtra, India, Deccan College Post-Graduate and Research Institute, Pune, 1980, p. 101.
Herngreen, G.F.W., Palynological preparation techniques, NPD Bull. 2 (1983) 13–34.
Karlekar, S., Rajaguru, S.N., Late Holocene geomorphology of Konkan coast of Maharashtra, Trans. Inst. Indian Geogr. 34 (1) (2012) 21–34.
Kumar, A., Patterson, R.T., Arcellaceans (Thecamoebians): new tools for monitoring long and short term changes in Lake Bottom acidity, Environ. Geol. 39 (2000) 689–697.
Kumaran, K.P.N., Shindikar, M., Limaye, Ruta B., Mangrove associated lignite beds of Malvan, Konkan: evidence for higher sea level during the Late Tertiary (Neogene) along the west coast of India, Curr. Sci. 86 (2) (2004) 335–340.
Kumaran, K.P.N., Limaye, R.B., Padmalal, D., India's fragile coast with special reference to Late Quaternary environmental dynamics, Proc. Indian Natl. Sci. Acad. A 78 (2012) 343–352.
Kumaran, K.P.N., Limaye, R.B., Rajshekhar, C., Rajagopalan, G., Palynoflora and radiocarbon dates of Holocene deposits of Dhamapur, Sindhudurg district, Maharashtra, Curr. Sci. 80 (10) (2001) 1331–1336.

Kumaran, K.P.N., Limaye, R.B., Nair, K.M., Padmalal, D, Palaeoecological and Palaeoclimate potential of subsurface palynological data from the Late Quaternary sediments of South Kerala Sedimentary Basin, southwest India, Curr. Sci. 95 (4) (2008) 515–526.

Kumaran, K.P.N., Nair, K.M., Shindikar, M.R., Limaye, R.B., Padmalal, D., Stratigraphical and palynological appraisal of the Late Quaternary mangrove deposits of west coast of India, Quat. Res. 64 (3) (2005) 418–431.

Kumaran, K.P.N., Limaye, R.B., Punekar, S.A., Rajaguru, S.N., Joshi, S.V., Karlekar, S.N., Vegetation response to South Asian monsoon variations in Konkan, western India during the Late Quaternary: evidence from fluvio-lacustrine archives, Quat. Int. 286 (2013) 3–18.

Kumaran, N.K.P, Padmalal, D., Limaye, R.B., Vishnu Mohan, S., Jennerjahn, T., Gamre, P.G., Tropical peat and peatland development in the floodplains of the greater Pamba Basin, south-western India during the Holocene, PLoS One 11 (5) (2016) e0154297, doi:10.1371/journal.pone.0154297.

Kumaran, N.K.P., Padmalal, D., Nair, M.K., Limaye, R.B., Guleria, J.S., Srivastava, R., Shukla, A., Vegetation response and landscape dynamics of Indian summer monsoon variations during Holocene: an eco-geomorphological appraisal of tropical evergreen forest subfossil logs, PLoS One 9 (4) (2014) e93596, doi:10.1371/journal.pone.0093596.

Limaye, R.B., Contribution to Palaeopalynology and Palaeoecology of Coastal Deposits of Maharashtra, India, University of Pune, Pune, 2004 (Ph.D. thesis unpublished).

Limaye, Ruta.B., Kumaran, K.P.N., Mangrove vegetation responses to Holocene climate change along Konkan coast of South-western India, Quat. Int 263 (2012) 114–128.

Limaye, Ruta.B., Kumaran, K.P.N., Padmalal, D., Mangrove habitat dynamics in response to Holocene sea level and climate changes along southwest coast of India, Quat. Int. 325 (2014) 116–125.

Limaye, Ruta.B., Padmalal, D., Kumaran, K.P.N., Late Pleistocene Holocene monsoon variations on climate, landforms and vegetation cover in southwestern India: an overview, Quat. Int. 443 (2017a) 143–154.

Limaye, Ruta.B., Padmalal, D., Kumaran, K.P.N., Cyanobacteria and testate amoeba as potential proxies for Holocene hydrological changes and climate variability: evidence from tropical coastal lowlands of SW India, Quat. Int. 443 (2017b) 99–114.

Limaye, Ruta.B., Kumaran, K.P.N., Nair, K.M., Padmalal, D., Non-pollen palynomorphs (NPP) as potential palaeoenvironmental indicators in the Late Quaternary sediments of west coast of India, Curr. Sci. 92 (2007) 1370–1382.

Limaye, R.B., Thacker, M., Balasubramanian, K, Padmalal, D., Rajaguru, S.N., Kumaran, K.P.N., Punekar, S.A., Holocene climate dynamics and ecological responses in Kaas Plateau, a UNESCO World Heritage site in the North Western Ghats of India: Evidence from Lacustrine deposits, J. Asian Earth Sci. (2021) (In press).

Lorencova, M., Thecamoebians from recent lake sediments from the Sumava Mts, Czech Republic, Bull. Geosci. 84 (2) (2009) 359–376.

Marathe, A., Rajaguru, S.N., Holocene geoarchaeology of coastal Ratnagiri district, Maharashtra. Project completion report (2011) 88.

Mascarenhas, A., Significance of peat on western continental shelf of India, J. Geol. Soc. India 49 (1997) 145–152.

Medioli, F.S., Scott, D.B., Collins, E.S., McCarthy, F.M.G., Fossil thecamoebians: present status and prospects for the future, in: Hemleben, C., Kaminski, M.A., Kuhnt, W., Scott, D.B. (Eds.), Paleoecology, Biostratigraphy, Paleoceanography and Taxonomy of Agglutinated Foraminifera, North Atlantic Treaty Organization Advanced Study Institute Series, Series C, Springer, Cham, vol. 327, 1990, pp. 813–840. Mathematical and Physical Sciences.

Mittre, V., Guzder, S.J., The stratigraphy and palynology of the coastal mangrove swamps of Bombay and Salsette Islands, Palaeobotanist 22 (2) (1975) 111–117.

Nayar, T.S., Pollen flora of Maharashtra State, India, International Bioscience Series, Today & Tomorrow's Printers and Publishers, vol. XIV (Plates 1-67), 1990, p. 157.

Osterman, L.E., Smith, C.G., Over 100 years of environmental change recorded by foraminifers and sediments in Mobile Bay, Alabama, Gulf of Mexico, USA, Estuar. Coast. Shelf Sci. 115 (2012) 345–358.

Pathak, M.S., Studies on Pollen Flora of Pune district, Maharashtra (Ph.D. Thesis unpublished), University of Pune, 1994.

Patterson, R.T., Baker, T., Burbridge, S.M., Arcellaceans (Thecamoebians) as proxies of arsenic and mercury contamination in northeastern Ontario lakes, J. Foramin. Res. 26 (2) (1996) 172–183.

Pawar, N.J., Geochemistry of carbonate precipitation from the groundwater in basaltic aquifers: an equilibrium thermodynamic approach, J. Geol. Soc. India 41 (1993) 119–131.

Pawar, N.J., Kale, V.S., Waterfall Tufa deposits from the Deccan basalt Province, India: implications for weathering of basalts in the semi-arid tropics, Z. Geomorphol. 145 (2006) 17–36.

Payne, R.J., Can testate amoeba-based palaeohydrology be extended to fens? J. Quat. Sci. 26 (2011) 15–27.

Rajaguru, S.N., Kale, V.S., Changes in the fluvial regime of western Maharashtra Upland rivers during Late Quaternary, J. Geol. Soc. India 26 (1985) 16–27.

Rajshekhar, C., Kumaran, K.P.N, Micropaleontological evidence for tectonic uplift of near-shore deposits around Bankot-Velas, Ratnagiri District, Maharashtra, Curr. Sci. 74 (1998) 705–707.

Reinhardt, E.G., Dalby, A.P., Kumar, A., Patterson, R.T., Arcellaceans as pollution indicators in mine tailing contaminated lakes near Cobalt, Ontario, Canada., Micropalaeontology 44 (1998) 131–148.

Singhvi, A.K., Kale, V.S., Palaeoclimate Studies in India: Last Ice Age to the Present, Indian National Science Academy, New Delhi, 2009, p. 34.

Srivastava, G., Trivedi, A., Mehrotra, R.C., Paudayal, K.N., Limaye, R.B., Kumaran, K.P.N., Yadav, S.K., Monsoon variability over Peninsular India during Late Pleistocene: signatures of vegetation shift recorded in terrestrial archive from the corridors of Western ghats, Palaeogeogr. Palaeoclimatol. Palaeoecol. 443 (2016) 57–65.

Sugauwara, D., Minoura, K., Imamura, F., Takahashi, T., Shuto, N., A huge sand deposit formed by the 1854 Earthquake tsunami in Suruga Bay, Central Japan, ISET J. Earthq. Technol. 42 (4) (2005) 147–158.

Sujata, K.R., Nigam, R., Saraswat, R., Linshy, V.N., Regeneration and abnormality in benthic foraminifera Rosalina leei: implications in reconstructing past salinity changes, Riv. Ital. di Paleontologia Stratigr. 117 (1) (2011) 189–196.

Thanikaimoni, G., 1987. Mangrove Palynology.UNDP/UNESCO Regional Project on Training and Research on Mangrove Ecosystems, RAS/79/002 (Pondicherry), 100 pp.

Tissot, C., Chikhi, H, Nayar, T.S., Pollen of wet evergreen forests of the Western Ghats of India, Publications du dépàrtment décologie, Institut Francais De Pondicherry 35 (1994) 1–133.

Traverse, A., Palaeopalynology, First edn, Unwin Hyman, London, 1988, p. 600.

Traverse, A., Palaeopalynology, Second edn, Springer, Dordrecht, The Netherlands, 2007, p. 813.

Van Geel, B., The palaeoenvironmental indicator value of non pollen Palynomorphs in lake sediments, peat deposits, and archaeological sites, Pollen 14 (2004) 276–277.

Vasanthy, G., Pollen of the South Indian hills, Institut Francais de Pondichery, Travaux de la section scientifique et technique 15 (1976) 1–74.

Walker, M.J.C., Berkelhammer, M., Björck, S., Cwynar, L.C., Fisher, D.A., Long, A.J., Lowe, J.J., Newnham, R.M., Rasmussen, S.O., Weiss, H., Formal subdivision of the Holocene series/epoch: a discussion paper by a working group of INTIMATE (Integration of ice-core, marine and terrestrial records) and the subcommission on quaternary stratigraphy (International Commission on Stratigraphy), J. Quat. Sci. 27 (7) (2012) 649–659.

Walker, M., Johnsen, S., Rasmussen, S.O., Popp, T., Steffensen, J.-P., Gibbard, P., Hoek, W., Lowe, J., Andrews, J., Björck, S., Cwynar, L.C., Hughen, K., Kershaw, P., Kromer, B., Litt, T., Lowe, D.J., Nakagawa, T., Newnham, R., Schwander, J., Formal definition and dating of the GSSP (Global stratotype section and point) for the base of the Holocene using the Greenland NGRIP ice core, and selected auxiliary records, J. Quat. Sci. 24 (2009) 3–17.

Woodland, W., Charman, D., Sims, P., Quantitative estimates of water tables and soil moisture in Holocene peatlands from testate amoebae, Holocene 8 (1998) 261–273.

Appendix

List of reviewers

1. A.P. Pradeepkumar
2. Allu C. Narayana
3. Amzad Laskar
4. Anish K. Warrier
5. Anjali Trivedi
6. Anoop K. Singh
7. Cheng Sen Li
8. Dananjay Sant
9. Deepak Maurya
10. Falugini Bhattacharya
11. H.S. Sharma
12. Harish Chandra Nainwal
13. J.S. Guleria
14. Jagannath Pal
15. Jayendra Singh
16. K. Krishnakumar
17. K. Sandeep
18. Kamal Kumar Barik
19. Kausthab Thirumalai
20. Kodura S. Rao
21. L.S. Chamyal
22. M.R. Rao
23. Mahendra Bhutiyani
24. Manish Tiwari
25. Milap C. Sharma
26. Mohd Akhter Ali
27. Nitesh Kumar Khonde
28. Parth Chauhan
29. Parween Chchetri
30. Pradeep Sreevasthava
31. Prosenjit Ghosh
32. R. Shynu
33. R.K. Saxena
34. R.K. Singh
35. Raghavendra Murthy
36. Rajani Panchang

37. Rakesh Mehrohtra
38. Ratan Kar
39. Ratheesh Kumar
40. Ruta B. Limaye
41. S.M. Hussain
42. Santosh K. Shah
43. Satish Sangode
44. Shiv Mizoram
45. Sonal Khanolkar
46. Su-Ping Li
47. Sumedh Humane
48. Suresh Pillai
49. Upasana Banerji
50. Veena Nair
51. Vikrant Jain
52. Viswapal Santhosh
53. Yi-Feng Yao

Index

Page numbers followed by "*f*" and "*t*" indicate, figures and tables respectively.

J

K

L

M

N

Printed in the United States
by Baker & Taylor Publisher Services